Student Solutions Manual

Kevin M. Bodden
Lewis and Clark Community College

Christopher J. Rigdon
Southern Illinois University, Edwardsville

NINTH EDITION

Calculus

Varberg Purcell Rigdon

Harlow, England • London • New York • Boston • San Francisco • Toronto • Sydney • Dubai • Singapore • Hong Kong
Tokyo • Seoul • Taipei • New Delhi • Cape Town • São Paulo • Mexico City • Madrid • Amsterdam • Munich • Paris • Milan

Editor-in-Chief: Sally Yagan
Acquisitions Editor: Adam Jaworski
Project Manager: Dawn Murrin
Editorial Assistant: Christine Whitlock
Executive Managing Editor: Kathleen Schiaparelli
Senior Managing Editor: Nicole M. Jackson
Assistant Managing Editor: Karen Bosch Petrov
Production Editor: Jessica Barna
Supplement Cover Manager: Paul Gourhan
Supplement Cover Designer: Christopher Kossa
Manufacturing Buyer: Ilene Kahn
Manufacturing Manager: Alexis Heydt-Long

Printed in the United States of America

5 2021

ISBN 0-13-146966-5

Pearson Education Ltd., *London*
Pearson Education Australia Pty. Ltd., *Sydney*
Pearson Education Singapore, Pte. Ltd.
Pearson Education North Asia Ltd., *Hong Kong*
Pearson Education Canada, Inc., *Toronto*
Pearson Educación de Mexico, S.A. de C.V.
Pearson Education—Japan, *Tokyo*
Pearson Education Malaysia, Pte. Ltd.

Table of Contents – Student Solutions Manual

Chapter 9 Infinite Series

Chapter 10 Conics and Polar Coordinates

Chapter 11 Geometry in Space and Vectors

Chapter 12 Derivatives for Functions of Two or More Variables

Chapter 13 Multiple Integrals

Chapter 14 Vector Calculus

Chapter 15 Differential Equations

CHAPTER 0 Preliminaries

0.1 Concepts Review

1. rational numbers

3. If not Q then not P.

Problem Set 0.1

1. $4-2(8-11)+6=4-2(-3)+6$
$=4+6+6=16$

3. $-4[5(-3+12-4)+2(13-7)]$
$=-4[5(5)+2(6)]=-4[25+12]$
$=-4(37)=-148$

5. $\frac{5}{7}-\frac{1}{13}=\frac{65}{91}-\frac{7}{91}=\frac{58}{91}$

7. $\frac{1}{3}\left[\frac{1}{2}\left(\frac{1}{4}-\frac{1}{3}\right)+\frac{1}{6}\right]=\frac{1}{3}\left[\frac{1}{2}\left(\frac{3-4}{12}\right)+\frac{1}{6}\right]$
$=\frac{1}{3}\left[\frac{1}{2}\left(-\frac{1}{12}\right)+\frac{1}{6}\right]$
$=\frac{1}{3}\left[-\frac{1}{24}+\frac{4}{24}\right]$
$=\frac{1}{3}\left(\frac{3}{24}\right)=\frac{1}{24}$

9. $\frac{14}{21}\left(\frac{2}{5-\frac{1}{3}}\right)^2=\frac{14}{21}\left(\frac{2}{\frac{14}{3}}\right)^2=\frac{14}{21}\left(\frac{6}{14}\right)^2$
$=\frac{14}{21}\left(\frac{3}{7}\right)^2=\frac{2}{3}\left(\frac{9}{49}\right)=\frac{6}{49}$

11. $\frac{\frac{11}{7}-\frac{12}{21}}{\frac{11}{7}+\frac{12}{21}}=\frac{\frac{11}{7}-\frac{4}{7}}{\frac{11}{7}+\frac{4}{7}}=\frac{\frac{7}{7}}{\frac{15}{7}}=\frac{7}{15}$

13. $1-\frac{1}{1+\frac{1}{2}}=1-\frac{1}{\frac{3}{2}}=1-\frac{2}{3}=\frac{3}{3}-\frac{2}{3}=\frac{1}{3}$

15. $\left(\sqrt{5}+\sqrt{3}\right)\left(\sqrt{5}-\sqrt{3}\right)=\left(\sqrt{5}\right)^2-\left(\sqrt{3}\right)^2$
$=5-3=2$

17. $(3x-4)(x+1)=3x^2+3x-4x-4$
$=3x^2-x-4$

19. $(3x-9)(2x+1)=6x^2+3x-18x-9$
$=6x^2-15x-9$

21. $(3t^2-t+1)^2=(3t^2-t+1)(3t^2-t+1)$
$=9t^4-3t^3+3t^2-3t^3+t^2-t+3t^2-t+1$
$=9t^4-6t^3+7t^2-2t+1$

23. $\frac{x^2-4}{x-2}=\frac{(x-2)(x+2)}{x-2}=x+2,\ x\neq 2$

25. $\frac{t^2-4t-21}{t+3}=\frac{(t+3)(t-7)}{t+3}=t-7,\ t\neq -3$

27. $\frac{12}{x^2+2x}+\frac{4}{x}+\frac{2}{x+2}$
$=\frac{12}{x(x+2)}+\frac{4(x+2)}{x(x+2)}+\frac{2x}{x(x+2)}$
$=\frac{12+4x+8+2x}{x(x+2)}=\frac{6x+20}{x(x+2)}$
$=\frac{2(3x+10)}{x(x+2)}$

29. **a.** $0\cdot 0=0$ **b.** $\frac{0}{0}$ is undefined.

c. $\frac{0}{17}=0$ **d.** $\frac{3}{0}$ is undefined.

e. $0^5=0$ **f.** $17^0=1$

31.
$$\begin{array}{r} .08\overline{3} \\ 12\overline{)1.000} \\ \underline{96} \\ 40 \\ \underline{36} \\ 4 \end{array}$$

33.

$$21\overline{)3.000000}\quad .\overline{142857}$$

$$\begin{array}{r} 21 \\ \hline 90 \\ \underline{84} \\ 60 \\ \underline{42} \\ 180 \\ \underline{168} \\ 120 \\ \underline{105} \\ 150 \\ \underline{147} \\ 3 \end{array}$$

35.

$$3\overline{)11.0}\quad 3.\overline{6}$$

$$\begin{array}{r} \underline{9} \\ 20 \\ \underline{18} \\ 2 \end{array}$$

37. $x = 0.123123123...$

$$\begin{aligned} 1000x &= 123.123123... \\ x &= 0.123123... \\ \hline 999x &= 123 \\ x &= \frac{123}{999} = \frac{41}{333} \end{aligned}$$

39. $x = 2.56565656...$

$$\begin{aligned} 100x &= 256.565656... \\ x &= 2.565656... \\ \hline 99x &= 254 \\ x &= \frac{254}{99} \end{aligned}$$

41. $x = 0.199999...$

$$\begin{aligned} 100x &= 19.99999... \\ 10x &= 1.99999... \\ \hline 90x &= 18 \\ x &= \frac{18}{90} = \frac{1}{5} \end{aligned}$$

43. Those rational numbers that can be expressed by a terminating decimal followed by zeros.

45. Answers will vary. Possible answer: 0.000001, $\frac{1}{\pi^{12}} \approx 0.0000010819...$

47. Answers will vary. Possible answer: 3.14159101001...

49. Irrational

51. $(\sqrt{3}+1)^3 \approx 20.39230485$

53. $\sqrt[4]{1.123} - \sqrt[3]{1.09} \approx 0.00028307388$

55. $\sqrt{8.9\pi^2+1} - 3\pi \approx 0.000691744752$

57. Let a and b be real numbers with $a < b$. Let n be a natural number that satisfies $1/n < b - a$. Let $S = \{k : k/n > b\}$. Since a nonempty set of integers that is bounded below contains a least element, there is a $k_0 \in S$ such that $k_0/n > b$ but $(k_0 - 1)/n \le b$. Then

$$\frac{k_0 - 1}{n} = \frac{k_0}{n} - \frac{1}{n} > b - \frac{1}{n} > a$$

Thus, $a < \frac{k_0-1}{n} \le b$. If $\frac{k_0-1}{n} < b$, then choose $r = \frac{k_0-1}{n}$. Otherwise, choose $r = \frac{k_0-2}{n}$.

Note that $a < b - \frac{1}{n} < r$.

Given $a < b$, choose r so that $a < r_1 < b$. Then choose r_2, r_3 so that $a < r_2 < r_1 < r_3 < b$, and so on.

59. $r = 4000 \text{ mi} \times 5280 \frac{\text{ft}}{\text{mi}} = 21{,}120{,}000 \text{ ft}$

$$\begin{aligned} \text{equator} &= 2\pi r = 2\pi(21{,}120{,}000) \\ &\approx 132{,}700{,}874 \text{ ft} \end{aligned}$$

61. $V = \pi r^2 h = \pi\left(\frac{16}{2}\cdot 12\right)^2 (270\cdot 12)$

$\approx 93{,}807{,}453.98 \text{ in.}^3$

volume of one board foot (in inches):

$1 \times 12 \times 12 = 144 \text{ in.}^3$

number of board feet:

$$\frac{93{,}807{,}453.98}{144} \approx 651{,}441 \text{ board ft}$$

63. **a.** If I stay home from work today then it rains. If I do not stay home from work, then it does not rain.

b. If the candidate will be hired then she meets all the qualifications. If the candidate will not be hired then she does not meet all the qualifications.

65. a. If a triangle is a right triangle, then $a^2 + b^2 = c^2$. If a triangle is not a right triangle, then $a^2 + b^2 \neq c^2$.

b. If the measure of angle ABC is greater than 0° and less than 90°, it is acute. If the measure of angle ABC is less than 0° or greater than 90°, then it is not acute.

67. a. The statement, converse, and contrapositive are all true.

b. The statement, converse, and contrapositive are all true.

69. a. Some isosceles triangles are not equilateral. The negation is true.

b. All real numbers are integers. The original statement is true.

c. Some natural number is larger than its square. The original statement is true.

71. a. True; If x is positive, then x^2 is positive.

b. False; Take $x = -2$. Then $x^2 > 0$ but $x < 0$.

c. False; Take $x = \frac{1}{2}$. Then $x^2 = \frac{1}{4} < x$

d. True; Let x be any number. Take $y = x^2 + 1$. Then $y > x^2$.

e. True; Let y be any positive number. Take $x = \frac{y}{2}$. Then $0 < x < y$.

73. a. If n is odd, then there is an integer k such that $n = 2k + 1$. Then

$$n^2 = (2k+1)^2 = 4k^2 + 4k + 1$$
$$= 2(2k^2 + 2k) + 1$$

b. Prove the contrapositive. Suppose n is even. Then there is an integer k such that $n = 2k$. Then $n^2 = (2k)^2 = 4k^2 = 2(2k^2)$. Thus n^2 is even.

75. a. $243 = 3 \cdot 3 \cdot 3 \cdot 3 \cdot 3$

b. $124 = 4 \cdot 31 = 2 \cdot 2 \cdot 31$ or $2^2 \cdot 31$

c. $5100 = 2 \cdot 2550 = 2 \cdot 2 \cdot 1275$
$= 2 \cdot 2 \cdot 3 \cdot 425 = 2 \cdot 2 \cdot 3 \cdot 5 \cdot 85$
$= 2 \cdot 2 \cdot 3 \cdot 5 \cdot 5 \cdot 17$ or $2^2 \cdot 3 \cdot 5^2 \cdot 17$

77. $\sqrt{2} = \frac{p}{q}; 2 = \frac{p^2}{q^2}; 2q^2 = p^2$; Since the prime factors of p^2 must occur an even number of times, $2q^2$ would not be valid and $\frac{p}{q} = \sqrt{2}$ must be irrational.

79. Let a, b, p, and q be natural numbers, so $\frac{a}{b}$ and $\frac{p}{q}$ are rational. $\frac{a}{b} + \frac{p}{q} = \frac{aq + bp}{bq}$ This sum is the quotient of natural numbers, so it is also rational.

81. a. $-\sqrt{9} = -3$; rational

b. $0.375 = \frac{3}{8}$; rational

c. $(3\sqrt{2})(5\sqrt{2}) = 15\sqrt{4} = 30$; rational

d. $(1+\sqrt{3})^2 = 1 + 2\sqrt{3} + 3 = 4 + 2\sqrt{3}$; irrational

83. a. Answers will vary. Possible answer: An example
is $S = \{x : x^2 < 5, x \text{ a rational number}\}$.
Here the least upper bound is $\sqrt{5}$, which is real but irrational.

b. True

0.2 Concepts Review

1. $[-1,5); (-\infty,-2]$

3. (b) and (c)

Problem Set 0.2

1. a.

b.

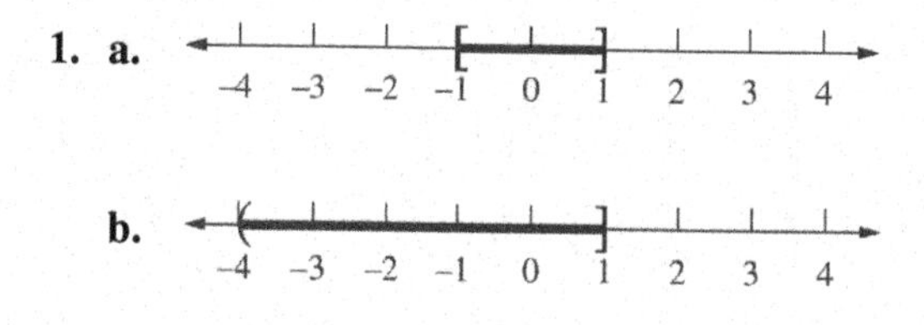

c. 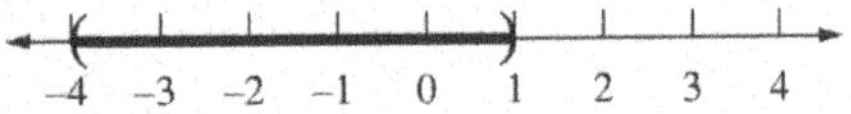

d. 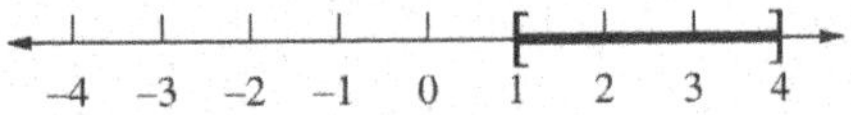

e. 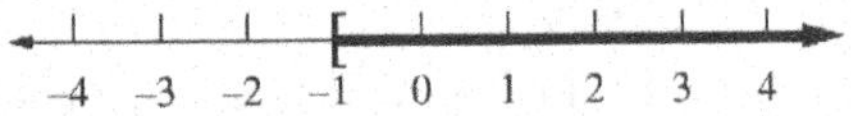

f. 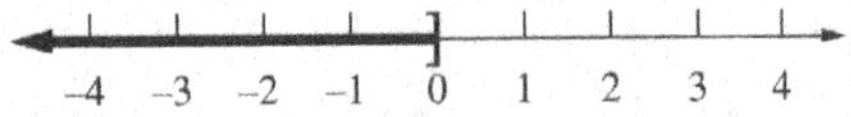

3. $x - 7 < 2x - 5$

$-2 < x; (-2, \infty)$

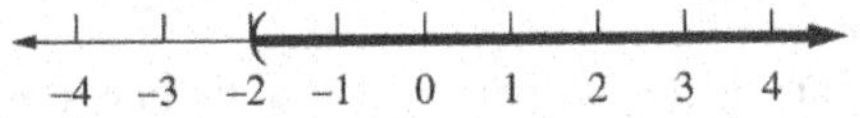

5. $7x - 2 \le 9x + 3$

$-5 \le 2x$

$x \ge -\frac{5}{2}; \left[-\frac{5}{2}, \infty\right)$

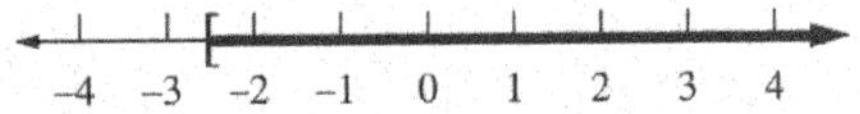

7. $-4 < 3x + 2 < 5$

$-6 < 3x < 3$

$-2 < x < 1; (-2, -1)$

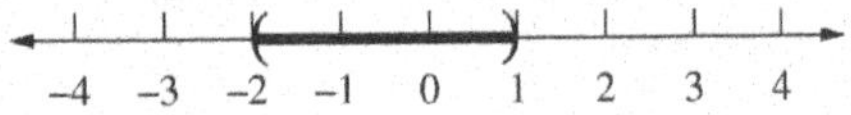

9. $-3 < 1 - 6x \le 4$

$-4 < -6x \le 3$

$\frac{2}{3} > x \ge -\frac{1}{2}; \left[-\frac{1}{2}, \frac{2}{3}\right)$

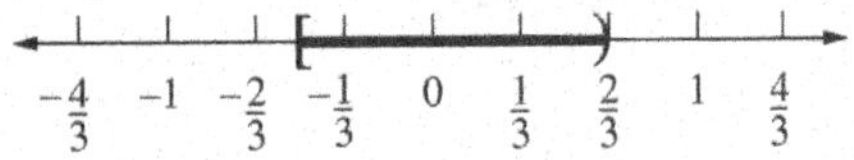

11. $x^2 + 2x - 12 < 0;$

$x = \frac{-2 \pm \sqrt{(2)^2 - 4(1)(-12)}}{2(1)} = \frac{-2 \pm \sqrt{52}}{2}$

$= -1 \pm \sqrt{13}$

$\left[x - \left(-1 + \sqrt{13}\right)\right]\left[x - \left(-1 - \sqrt{13}\right)\right] < 0;$

$\left(-1 - \sqrt{13}, -1 + \sqrt{13}\right)$

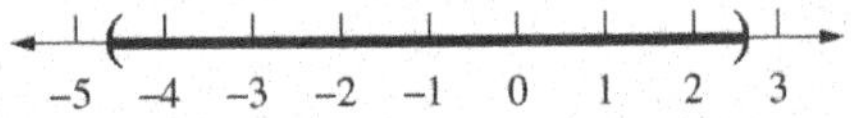

13. $2x^2 + 5x - 3 > 0; (2x - 1)(x + 3) > 0;$

$(-\infty, -3) \cup \left(\frac{1}{2}, \infty\right)$

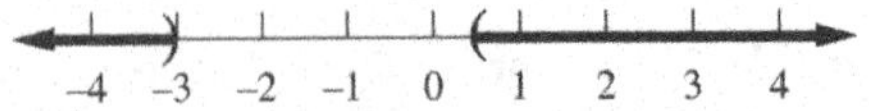

15. $\frac{x+4}{x-3} \le 0; \; [-4, 3)$

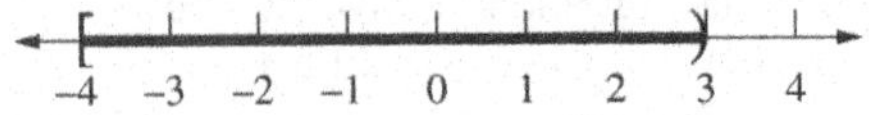

17. $\frac{2}{x} < 5$

$\frac{2}{x} - 5 < 0$

$\frac{2 - 5x}{x} < 0;$

$(-\infty, 0) \cup \left(\frac{2}{5}, \infty\right)$

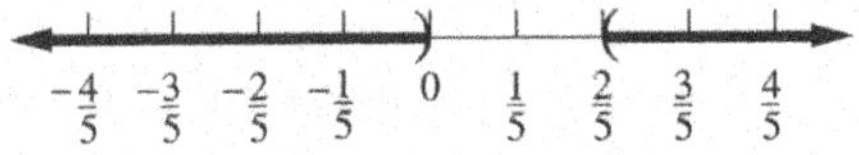

19. $\frac{1}{3x-2} \le 4$

$\frac{1}{3x-2} - 4 \le 0$

$\frac{1 - 4(3x-2)}{3x-2} \le 0$

$\frac{9 - 12x}{3x-2} \le 0; \left(-\infty, \frac{2}{3}\right) \cup \left[\frac{3}{4}, \infty\right)$

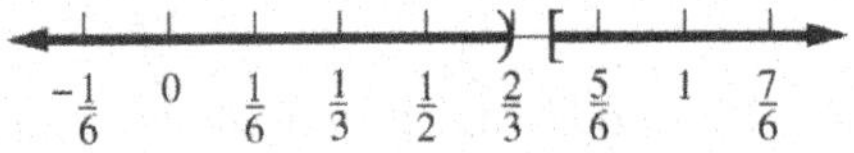

21. $(x+2)(x-1)(x-3) > 0; (-2, 1) \cup (3, 8)$

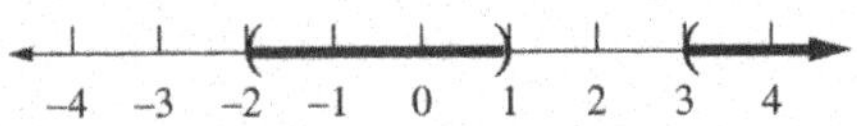

23. $(2x-3)(x-1)^2(x-3) \ge 0; \left(-\infty, \frac{3}{2}\right] \cup [3, \infty)$

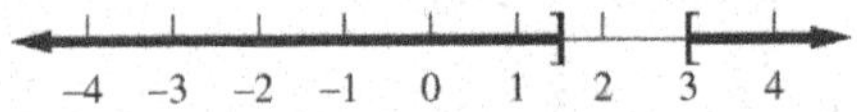

25. $x^3 - 5x^2 - 6x < 0$

$x(x^2 - 5x - 6) < 0$

$x(x+1)(x-6) < 0;$

$(-\infty, -1) \cup (0, 6)$

−2 −1 0 1 2 3 4 5 6

27. a. False. **b.** True.

c. False.

29. a. $\Rightarrow$ Let $a < b$, so $ab < b^2$. Also, $a^2 < ab$. Thus, $a^2 < ab < b^2$ and $a^2 < b^2$. $\Leftarrow$ Let $a^2 < b^2$, so $a \neq b$ Then

$0 < (a-b)^2 = a^2 - 2ab + b^2$

$< b^2 - 2ab + b^2 = 2b(b-a)$

Since $b > 0$, we can divide by $2b$ to get $b - a > 0$.

b. We can divide or multiply an inequality by any positive number.

$a < b \Leftrightarrow \frac{a}{b} < 1 \Leftrightarrow \frac{1}{b} < \frac{1}{a}$.

31. a. $3x + 7 > 1$ and $2x + 1 < 3$
$3x > -6$ and $2x < 2$
$x > -2$ and $x < 1$; $(-2, 1)$

b. $3x + 7 > 1$ and $2x + 1 > -4$
$3x > -6$ and $2x > -5$
$x > -2$ and $x > -\frac{5}{2}$; $(-2, \infty)$

c. $3x + 7 > 1$ and $2x + 1 < -4$
$x > -2$ and $x < -\frac{5}{2}$; $\varnothing$

33. a. $(x+1)(x^2 + 2x - 7) \geq x^2 - 1$

$x^3 + 3x^2 - 5x - 7 \geq x^2 - 1$

$x^3 + 2x^2 - 5x - 6 \geq 0$

$(x+3)(x+1)(x-2) \geq 0$

$[-3, -1] \cup [2, \infty)$

b. $x^4 - 2x^2 \geq 8$

$x^4 - 2x^2 - 8 \geq 0$

$(x^2 - 4)(x^2 + 2) \geq 0$

$(x^2 + 2)(x+2)(x-2) \geq 0$

$(-\infty, -2] \cup [2, \infty)$

c. $(x^2 + 1)^2 - 7(x^2 + 1) + 10 < 0$

$[(x^2 + 1) - 5][(x^2 + 1) - 2] < 0$

$(x^2 - 4)(x^2 - 1) < 0$

$(x+2)(x+1)(x-1)(x-2) < 0$

$(-2, -1) \cup (1, 2)$

35. $|x - 2| \geq 5;$

$x - 2 \leq -5$ or $x - 2 \geq 5$

$x \leq -3$ or $x \geq 7$

$(-\infty, -3] \cup [7, \infty)$

37. $|4x + 5| \leq 10;$

$-10 \leq 4x + 5 \leq 10$

$-15 \leq 4x \leq 5$

$-\frac{15}{4} \leq x \leq \frac{5}{4}; \left[-\frac{15}{4}, \frac{5}{4}\right]$

39. $\left|\frac{2x}{7} - 5\right| \geq 7$

$\frac{2x}{7} - 5 \leq -7$ or $\frac{2x}{7} - 5 \geq 7$

$\frac{2x}{7} \leq -2$ or $\frac{2x}{7} \geq 12$

$x \leq -7$ or $x \geq 42;$

$(-\infty, -7] \cup [42, \infty)$

41. $|5x - 6| > 1;$

$5x - 6 < -1$ or $5x - 6 > 1$

$5x < 5$ or $5x > 7$

$x < 1$ or $x > \frac{7}{5}; (-\infty, 1) \cup \left(\frac{7}{5}, \infty\right)$

43. $\left|\frac{1}{x} - 3\right| > 6;$

$\frac{1}{x} - 3 < -6$ or $\frac{1}{x} - 3 > 6$

$\frac{1}{x} + 3 < 0$ or $\frac{1}{x} - 9 > 0$

$\frac{1+3x}{x} < 0$ or $\frac{1-9x}{x} > 0;$

$\left(-\frac{1}{3}, 0\right) \cup \left(0, \frac{1}{9}\right)$

45. $x^2 - 3x - 4 \geq 0;$

$x = \frac{3 \pm \sqrt{(-3)^2 - 4(1)(-4)}}{2(1)} = \frac{3 \pm 5}{2} = -1, 4$

$(x+1)(x-4) = 0; (-\infty, -1] \cup [4, \infty)$

47. $3x^2 + 17x - 6 > 0;$

$$x = \frac{-17 \pm \sqrt{(17)^2 - 4(3)(-6)}}{2(3)} = \frac{-17 \pm 19}{6} = -6, \frac{1}{3}$$

$(3x - 1)(x + 6) > 0;\ (-\infty, -6) \cup \left(\frac{1}{3}, \infty\right)$

49. $|x-3| < 0.5 \Rightarrow 5|x-3| < 5(0.5) \Rightarrow |5x-15| < 2.5$

51. $|x-2| < \frac{\varepsilon}{6} \Rightarrow 6|x-2| < \varepsilon \Rightarrow |6x-12| < \varepsilon$

53. $|3x-15| < \varepsilon \Rightarrow |3(x-5)| < \varepsilon$

$\Rightarrow 3|x-5| < \varepsilon$

$\Rightarrow |x-5| < \frac{\varepsilon}{3}; \delta = \frac{\varepsilon}{3}$

55. $|6x+36| < \varepsilon \Rightarrow |6(x+6)| < \varepsilon$

$\Rightarrow 6|x+6| < \varepsilon$

$\Rightarrow |x+6| < \frac{\varepsilon}{6}; \delta = \frac{\varepsilon}{6}$

57. $C = \pi d$

$|C - 10| \le 0.02$

$|\pi d - 10| \le 0.02$

$\left|\pi\left(d - \frac{10}{\pi}\right)\right| \le 0.02$

$\left|d - \frac{10}{\pi}\right| \le \frac{0.02}{\pi} \approx 0.0064$

We must measure the diameter to an accuracy of 0.0064 in.

59.

$|x-1| < 2|x-3|$

$|x-1| < |2x-6|$

$(x-1)^2 < (2x-6)^2$

$x^2 - 2x + 1 < 4x^2 - 24x + 36$

$3x^2 - 22x + 35 > 0$

$(3x-7)(x-5) > 0;$

$\left(-\infty, \frac{7}{3}\right) \cup (5, \infty)$

61.

$2|2x-3| < |x+10|$

$|4x-6| < |x+10|$

$(4x-6)^2 < (x+10)^2$

$16x^2 - 48x + 36 < x^2 + 20x + 100$

$15x^2 - 68x - 64 < 0$

$(5x+4)(3x-16) < 0;$

$\left(-\frac{4}{5}, \frac{16}{3}\right)$

63. $|x| < |y| \Rightarrow |x||x| \le |x||y|$ and $|x||y| < |y||y|$ Order property: $x < y \Leftrightarrow xz < yz$ when z is positive.

$\Rightarrow |x|^2 < |y|^2$ Transitivity

$\Rightarrow x^2 < y^2$ $\left(|x|^2 = x^2\right)$

Conversely,

$x^2 < y^2 \Rightarrow |x|^2 < |y|^2$ $\left(x^2 = |x|^2\right)$

$\Rightarrow |x|^2 - |y|^2 < 0$ Subtract $|y|^2$ from each side.

$\Rightarrow (|x| - |y|)(|x| + |y|) < 0$ Factor the difference of two squares.

$\Rightarrow |x| - |y| < 0$ This is the only factor that can be negative.

$\Rightarrow |x| < |y|$ Add $|y|$ to each side.

65. **a.** $|a-b| = |a+(-b)| \le |a| + |-b| = |a| + |b|$

b. $|a-b| \ge ||a|-|b|| \ge |a|-|b|$ Use Property 4 of absolute values.

c. $|a+b+c| = |(a+b)+c| \le |a+b|+|c|$

$\le |a|+|b|+|c|$

67. $\left|\dfrac{x-2}{x^2+9}\right| = \left|\dfrac{x+(-2)}{x^2+9}\right|$

$\left|\dfrac{x-2}{x^2+9}\right| \le \left|\dfrac{x}{x^2+9}\right| + \left|\dfrac{-2}{x^2+9}\right|$

$\left|\dfrac{x-2}{x^2+9}\right| \le \dfrac{|x|}{x^2+9} + \dfrac{2}{x^2+9} = \dfrac{|x|+2}{x^2+9}$

Since $x^2+9 \ge 9, \dfrac{1}{x^2+9} \le \dfrac{1}{9}$

$\dfrac{|x|+2}{x^2+9} \le \dfrac{|x|+2}{9}$

$\left|\dfrac{x-2}{x^2+9}\right| \le \dfrac{|x|+2}{9}$

69. $\left|x^4+\frac{1}{2}x^3+\frac{1}{4}x^2+\frac{1}{8}x+\frac{1}{16}\right|$

$\le |x^4|+\frac{1}{2}|x^3|+\frac{1}{4}|x^2|+\frac{1}{8}|x|+\frac{1}{16}$

$\le 1+\frac{1}{2}+\frac{1}{4}+\frac{1}{8}+\frac{1}{16}$ since $|x| \le 1$.

So $\left|x^4+\frac{1}{2}x^3+\frac{1}{4}x^2+\frac{1}{8}x+\frac{1}{16}\right| \le 1.9375 < 2$.

71. $a \ne 0 \Rightarrow$

$0 \le \left(a-\dfrac{1}{a}\right)^2 = a^2-2+\dfrac{1}{a^2}$

so, $2 \le a^2+\dfrac{1}{a^2}$ or $a^2+\dfrac{1}{a^2} \ge 2$.

73. $0 < a < b$

$a^2 < ab$ and $ab < b^2$

$a^2 < ab < b^2$

$a < \sqrt{ab} < b$

75. For a rectangle the area is ab, while for a square the area is $a^2 = \left(\dfrac{a+b}{2}\right)^2$. From Problem 74, $\sqrt{ab} \le \dfrac{1}{2}(a+b) \Leftrightarrow ab \le \left(\dfrac{a+b}{2}\right)^2$ so the square has the largest area.

77. $\dfrac{1}{R} \le \dfrac{1}{10}+\dfrac{1}{20}+\dfrac{1}{30}$

$\dfrac{1}{R} \le \dfrac{6+3+2}{60}$

$\dfrac{1}{R} \le \dfrac{11}{60}$

$R \ge \dfrac{60}{11}$

$\dfrac{1}{R} \ge \dfrac{1}{20}+\dfrac{1}{30}+\dfrac{1}{40}$

$\dfrac{1}{R} \ge \dfrac{6+4+3}{120}$

$R \le \dfrac{120}{13}$

Thus, $\dfrac{60}{11} \le R \le \dfrac{120}{13}$

0.3 Concepts Review

1. $\sqrt{(x+2)^2+(y-3)^2}$

3. $\left(\dfrac{-2+5}{2}, \dfrac{3+7}{2}\right) = (1.5, 5)$

Problem Set 0.3

1.

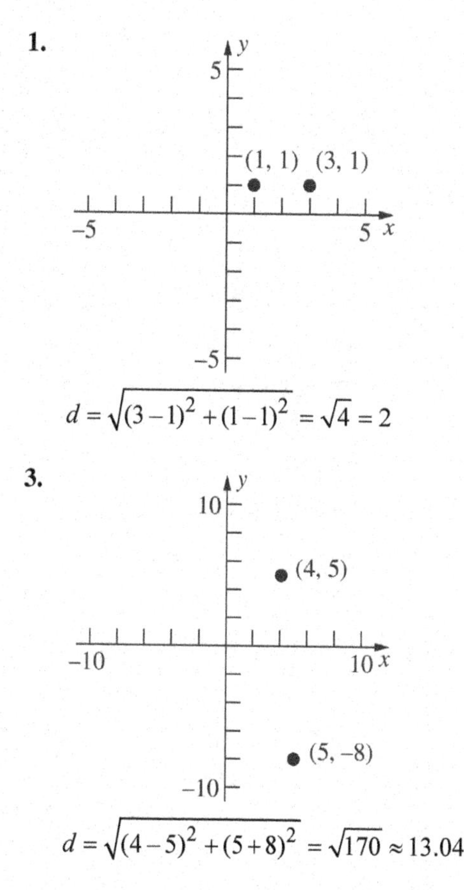

$d = \sqrt{(3-1)^2+(1-1)^2} = \sqrt{4} = 2$

3.

$d = \sqrt{(4-5)^2+(5+8)^2} = \sqrt{170} \approx 13.04$

5. $d_1 = \sqrt{(5+2)^2 + (3-4)^2} = \sqrt{49+1} = \sqrt{50}$

$d_2 = \sqrt{(5-10)^2 + (3-8)^2} = \sqrt{25+25} = \sqrt{50}$

$d_3 = \sqrt{(-2-10)^2 + (4-8)^2}$

$= \sqrt{144+16} = \sqrt{160}$

$d_1 = d_2$ so the triangle is isosceles.

7. (–1, –1), (–1, 3); (7, –1), (7, 3); (1, 1), (5, 1)

9. $\left(\frac{-2+4}{2}, \frac{-2+3}{2}\right) = \left(1, \frac{1}{2}\right);$

$d = \sqrt{(1+2)^2 + \left(\frac{1}{2} - 3\right)^2} = \sqrt{9 + \frac{25}{4}} \approx 3.91$

11 $(x-1)^2 + (y-1)^2 = 1$

13. $(x-2)^2 + (y+1)^2 = r^2$

$(5-2)^2 + (3+1)^2 = r^2$

$r^2 = 9 + 16 = 25$

$(x-2)^2 + (y+1)^2 = 25$

15. center $= \left(\frac{1+3}{2}, \frac{3+7}{2}\right) = (2, 5)$

radius $= \frac{1}{2}\sqrt{(1-3)^2 + (3-7)^2} = \frac{1}{2}\sqrt{4+16}$

$= \frac{1}{2}\sqrt{20} = \sqrt{5}$

$(x-2)^2 + (y-5)^2 = 5$

17. $x^2 + 2x + 10 + y^2 - 6y - 10 = 0$

$x^2 + 2x + y^2 - 6y = 0$

$(x^2 + 2x + 1) + (y^2 - 6y + 9) = 1 + 9$

$(x+1)^2 + (y-3)^2 = 10$

center $= (-1, 3)$; radius $= \sqrt{10}$

19. $x^2 + y^2 - 12x + 35 = 0$

$x^2 - 12x + y^2 = -35$

$(x^2 - 12x + 36) + y^2 = -35 + 36$

$(x-6)^2 + y^2 = 1$

center $= (6, 0)$; radius $= 1$

21. $4x^2 + 16x + 15 + 4y^2 + 6y = 0$

$4(x^2 + 4x + 4) + 4\left(y^2 + \frac{3}{2}y + \frac{9}{16}\right) = -15 + 16 + \frac{9}{4}$

$4(x+2)^2 + 4\left(y + \frac{3}{4}\right)^2 = \frac{13}{4}$

$(x+2)^2 + \left(y + \frac{3}{4}\right)^2 = \frac{13}{16}$

center $= \left(-2, -\frac{3}{4}\right)$; radius $= \frac{\sqrt{13}}{4}$

23. $\frac{2-1}{2-1} = 1$

25. $\frac{-6-3}{-5-2} = \frac{9}{7}$

27. $\frac{5-0}{0-3} = -\frac{5}{3}$

29. $y - 2 = -1(x-2)$

$y - 2 = -x + 2$

$x + y - 4 = 0$

31. $y = 2x + 3$

$2x - y + 3 = 0$

33. $m = \frac{8-3}{4-2} = \frac{5}{2};$

$y - 3 = \frac{5}{2}(x-2)$

$2y - 6 = 5x - 10$

$5x - 2y - 4 = 0$

35. $3y = -2x + 1$; $y = -\frac{2}{3}x + \frac{1}{3}$; slope $= -\frac{2}{3}$;

y-intercept $= \frac{1}{3}$

37. $6 - 2y = 10x - 2$

$-2y = 10x - 8$

$y = -5x + 4;$

slope $= -5$; y-intercept $= 4$

39. a. $m = 2;$

$y + 3 = 2(x-3)$

$y = 2x - 9$

b. $m = -\frac{1}{2}$;

$$y+3 = -\frac{1}{2}(x-3)$$

$$y = -\frac{1}{2}x - \frac{3}{2}$$

c. $2x+3y=6$

$$3y = -2x+6$$

$$y = -\frac{2}{3}x+2;$$

$$m = -\frac{2}{3};$$

$$y+3 = -\frac{2}{3}(x-3)$$

$$y = -\frac{2}{3}x-1$$

d. $m = \frac{3}{2}$;

$$y+3 = \frac{3}{2}(x-3)$$

$$y = \frac{3}{2}x - \frac{15}{2}$$

e. $m = \frac{-1-2}{3+1} = -\frac{3}{4}$;

$$y+3 = -\frac{3}{4}(x-3)$$

$$y = -\frac{3}{4}x - \frac{3}{4}$$

f. $x=3$ **g.** $y=-3$

41. $m = \frac{3}{2}$;

$$y+1 = \frac{3}{2}(x+2)$$

$$y = \frac{3}{2}x+2$$

43. $y = 3(3) - 1 = 8$; (3, 9) is above the line.

45. $2x+3y=4$

$-3x+y=5$

$$\begin{array}{l} 2x+3y=4 \\ \underline{9x-3y=-15} \\ 11x \quad\; = -11 \end{array}$$

$x=-1$

$-3(-1)+y=5$

$y=2$

Point of intersection: (–1, 2)

$3y = -2x+4$

$$y = -\frac{2}{3}x + \frac{4}{3}$$

$$m = \frac{3}{2}$$

$$y-2 = \frac{3}{2}(x+1)$$

$$y = \frac{3}{2}x + \frac{7}{2}$$

47. $3x-4y=5$

$2x+3y=9$

$$\begin{array}{l} 9x-12y=15 \\ \underline{8x+12y=36} \\ 17x \quad\;\; = 51 \end{array}$$

$x=3$

$3(3)-4y=5$

$-4y=-4$

$y=1$

Point of intersection: (3, 1); $3x-4y=5$;

$-4y = -3x+5$

$$y = \frac{3}{4}x - \frac{5}{4}$$

$$m = -\frac{4}{3}$$

$$y-1 = -\frac{4}{3}(x-3)$$

$$y = -\frac{4}{3}x+5$$

49. center: $\left(\frac{2+6}{2}, \frac{-1+3}{2}\right) = (4, 1)$

midpoint $= \left(\frac{2+6}{2}, \frac{3+3}{2}\right) = (4, 3)$

inscribed circle: radius $= \sqrt{(4-4)^2 + (1-3)^2}$

$= \sqrt{4} = 2$

$(x-4)^2 + (y-1)^2 = 4$

circumscribed circle:

radius $= \sqrt{(4-2)^2 + (1-3)^2} = \sqrt{8}$

$(x-4)^2 + (y-1)^2 = 8$

51. Put the vertex of the right angle at the origin with the other vertices at $(a, 0)$ and $(0, b)$. The midpoint of the hypotenuse is $\left(\frac{a}{2}, \frac{b}{2}\right)$. The distances from the vertices are

$$\sqrt{\left(a-\frac{a}{2}\right)^2+\left(0-\frac{b}{2}\right)^2}=\sqrt{\frac{a^2}{4}+\frac{b^2}{4}}=\frac{1}{2}\sqrt{a^2+b^2},$$

$$\sqrt{\left(0-\frac{a}{2}\right)^2+\left(b-\frac{b}{2}\right)^2}=\sqrt{\frac{a^2}{4}+\frac{b^2}{4}}=\frac{1}{2}\sqrt{a^2+b^2}, \text{ and}$$

$$\sqrt{\left(0-\frac{a}{2}\right)^2+\left(0-\frac{b}{2}\right)^2}=\sqrt{\frac{a^2}{4}+\frac{b^2}{4}}=\frac{1}{2}\sqrt{a^2+b^2},$$

which are all the same.

53. $x^2+y^2-4x-2y-11=0$

$$(x^2-4x+4)+(y^2-2y+1)=11+4+1$$

$$(x-2)^2+(y-1)^2=16$$

$$x^2+y^2+20x-12y+72=0$$

$$(x^2+20x+100)+(y^2-12y+36)=-72+100+36$$

$$(x+10)^2+(y-6)^2=64$$

center of first circle: (2, 1)
center of second circle: (–10, 6)

$$d=\sqrt{(2+10)^2+(1-6)^2}=\sqrt{144+25}=\sqrt{169}=13$$

However, the radii only sum to 4 + 8 = 12, so the circles must not intersect if the distance between their centers is 13.

55. Label the points C, P, Q, and R as shown in the figure below. Let $d=|OP|$, $h=|OR|$, and $a=|PR|$. Triangles ΔOPR and ΔCQR are similar because each contains a right angle and they share angle $\angle QRC$. For an angle of $30°$, $\frac{d}{h}=\frac{\sqrt{3}}{2}$ and $\frac{a}{h}=\frac{1}{2}\Rightarrow h=2a$. Using a property of similar triangles, $|QC|/|RC|=\sqrt{3}/2$,

$$\frac{2}{a-2}=\frac{\sqrt{3}}{2}\ \rightarrow\ a=2+\frac{4}{\sqrt{3}}$$

By the Pythagorean Theorem, we have

$$d=\sqrt{h^2-a^2}=\sqrt{3}a=2\sqrt{3}+4\approx 7.464$$

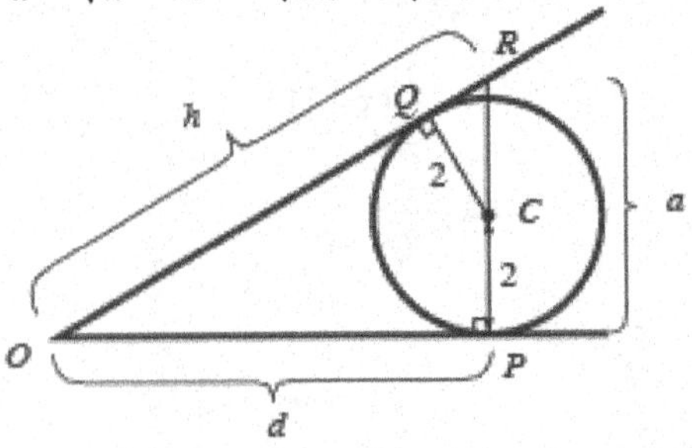

57. Refer to figure 15 in the text. Given ine l_1 with slope m, draw $\triangle ABC$ with vertical and horizontal sides m, 1.
Line l_2 is obtained from l_1 by rotating it around the point A by 90° counter-clockwise. Triangle ABC is rotated into triangle AED. We read off

$$\text{slope of } l_2=\frac{1}{-m}=-\frac{1}{m}.$$

59. Let a, b, and c be the lengths of the sides of the right triangle, with c the length of the hypotenuse. Then the Pythagorean Theorem says that $a^2+b^2=c^2$

Thus, $\frac{\pi a^2}{8}+\frac{\pi b^2}{8}=\frac{\pi c^2}{8}$ or

$$\frac{1}{2}\pi\left(\frac{a}{2}\right)^2+\frac{1}{2}\pi\left(\frac{b}{2}\right)^2=\frac{1}{2}\pi\left(\frac{c}{2}\right)^2$$

$\frac{1}{2}\pi\left(\frac{x}{2}\right)^2$ is the area of a semicircle with diameter x, so the circles on the legs of the triangle have total area equal to the area of the semicircle on the hypotenuse.

From $a^2+b^2=c^2$,

$$\frac{\sqrt{3}}{4}a^2+\frac{\sqrt{3}}{4}b^2=\frac{\sqrt{3}}{4}c^2$$

$\frac{\sqrt{3}}{4}x^2$ is the area of an equilateral triangle with sides of length x, so the equilateral triangles on the legs of the right triangle have total area equal to the area of the equilateral triangle on the hypotenuse of the right triangle.

61. The lengths A, B, and C are the same as the corresponding distances between the centers of the circles:

$A = \sqrt{(-2)^2 + (8)^2} = \sqrt{68} \approx 8.2$

$B = \sqrt{(6)^2 + (8)^2} = \sqrt{100} = 10$

$C = \sqrt{(8)^2 + (0)^2} = \sqrt{64} = 8$

Each circle has radius 2, so the part of the belt around the wheels is

$2(2\pi - a - \pi) + 2(2\pi - b - \pi) + 2(2\pi - c - \pi)$

$= 2[3\pi - (a + b + c)] = 2(2\pi) = 4\pi$

Since $a + b + c = \pi$, the sum of the angles of a triangle.

The length of the belt is $\approx 8.2 + 10 + 8 + 4\pi$

≈ 38.8 units.

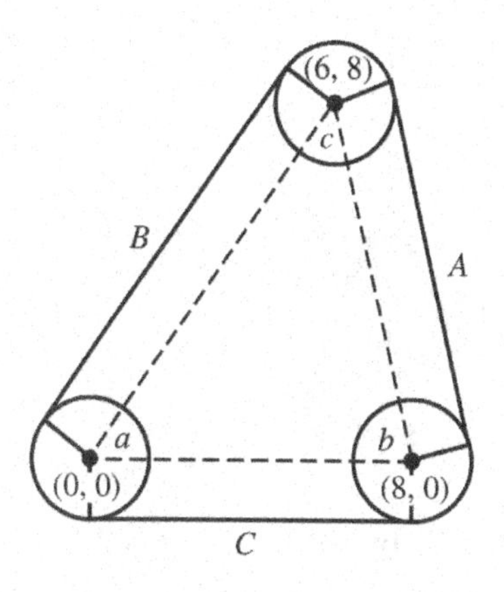

63. $A = 3, B = 4, C = -6$

$$d = \frac{|3(-3) + 4(2) + (-6)|}{\sqrt{(3)^2 + (4)^2}} = \frac{7}{5}$$

65. $A = 12, B = -5, C = 1$

$$d = \frac{|12(-2) - 5(-1) + 1|}{\sqrt{(12)^2 + (-5)^2}} = \frac{18}{13}$$

67. $2x + 4(0) = 5$

$$x = \frac{5}{2}$$

$$d = \frac{\left|2\left(\frac{5}{2}\right) + 4(0) - 7\right|}{\sqrt{(2)^2 + (4)^2}} = \frac{2}{\sqrt{20}} = \frac{\sqrt{5}}{5}$$

69. $m = \dfrac{-2-3}{1+2} = -\dfrac{5}{3}$; $m = \dfrac{3}{5}$; passes through

$$\left(\frac{-2+1}{2}, \frac{3-2}{2}\right) = \left(-\frac{1}{2}, \frac{1}{2}\right)$$

$$y - \frac{1}{2} = \frac{3}{5}\left(x + \frac{1}{2}\right)$$

$$y = \frac{3}{5}x + \frac{4}{5}$$

71. Let the origin be at the vertex as shown in the figure below. The center of the circle is then $(4 - r, r)$, so it has equation $(x - (4 - r))^2 + (y - r)^2 = r^2$. Along the side of length 5, the y-coordinate is always $\frac{3}{4}$ times the x-coordinate. Thus, we need to find the value of r for which there is exactly one x-solution to $(x - 4 + r)^2 + \left(\frac{3}{4}x - r\right)^2 = r^2$.

Solving for x in this equation gives $x = \frac{16}{25}\left(16 - r \pm \sqrt{24\left(-r^2 + 7r - 6\right)}\right)$. There is exactly one solution when $-r^2 + 7r - 6 = 0$, that is, when $r = 1$ or $r = 6$. The root $r = 6$ is extraneous. Thus, the largest circle that can be inscribed in this triangle has radius $r = 1$.

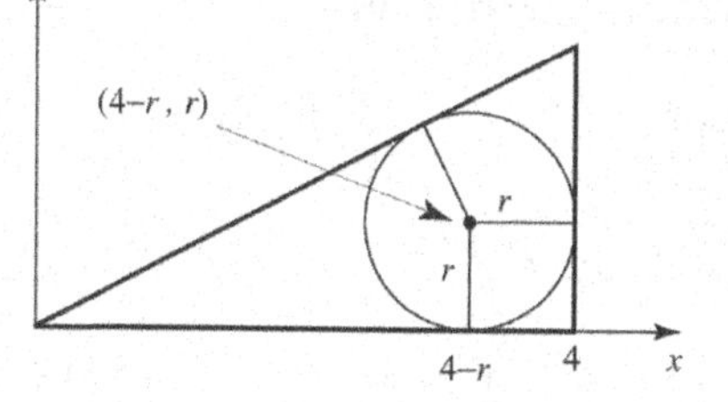

73. $12a + 0b = 36$
$a = 3$
$3^2 + b^2 = 36$
$b = \pm 3\sqrt{3}$
$3x - 3\sqrt{3}y = 36$
$x - \sqrt{3}y = 12$
$3x + 3\sqrt{3}y = 36$
$x + \sqrt{3}y = 12$

75. The midpoint of the side from (0, 0) to $(a, 0)$ is
$\left(\frac{0+a}{2}, \frac{0+0}{2}\right) = \left(\frac{a}{2}, 0\right)$
The midpoint of the side from (0, 0) to (b, c) is
$\left(\frac{0+b}{2}, \frac{0+c}{2}\right) = \left(\frac{b}{2}, \frac{c}{2}\right)$
$m_1 = \frac{c-0}{b-a} = \frac{c}{b-a}$
$m_2 = \frac{\frac{c}{2}-0}{\frac{b}{2}-\frac{a}{2}} = \frac{c}{b-a}$; $m_1 = m_2$

77. $x^2 + (y-6)^2 = 25$; passes through (3, 2)
tangent line: $3x - 4y = 1$
The dirt hits the wall at $y = 8$.

0.4 Concepts Review

1. y-axis

3. 8; –2, 1, 4

Problem Set 0.4

1. $y = -x^2 + 1$; y-intercept = 1; $y = (1 + x)(1 - x)$;
x-intercepts = –1, 1
Symmetric with respect to the y-axis

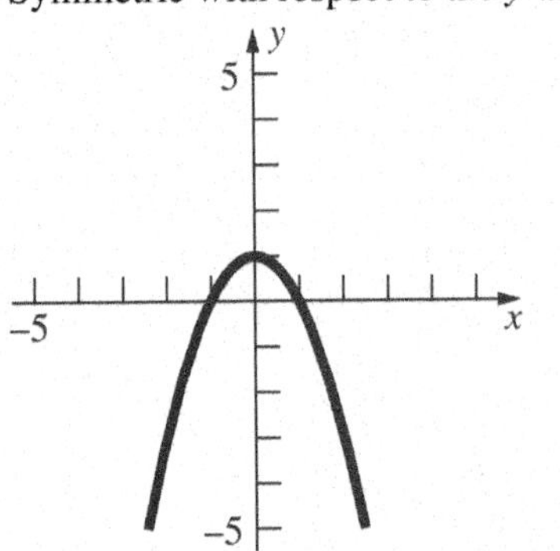

3. $x = -4y^2 - 1$; x-intercept = –1
Symmetric with respect to the x-axis

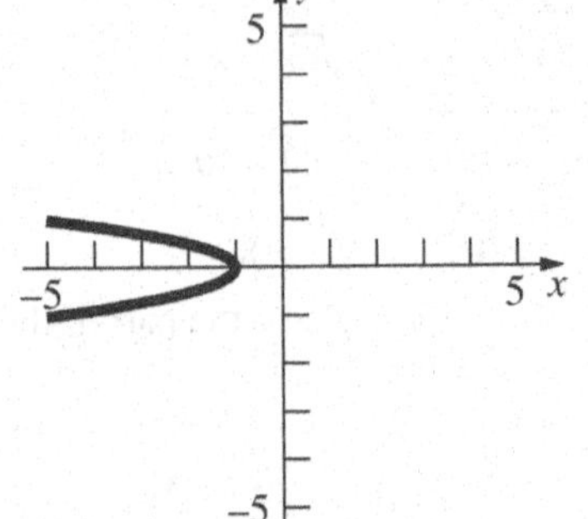

5. $x^2 + y = 0$; $y = -x^2$
x-intercept = 0, y-intercept = 0
Symmetric with respect to the y-axis

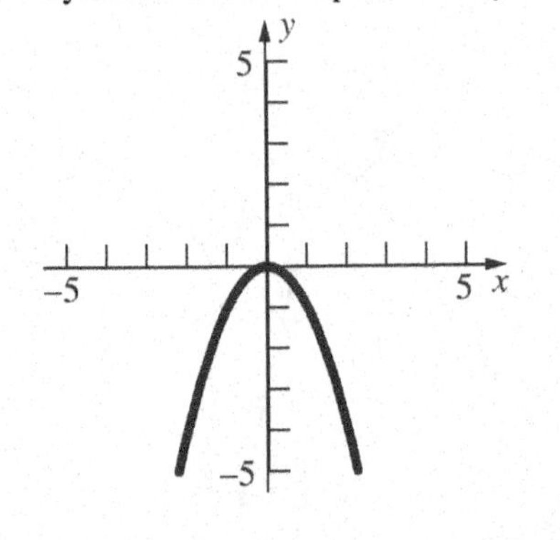

7. $7x^2 + 3y = 0$; $3y = -7x^2$; $y = -\frac{7}{3}x^2$
x-intercept = 0, y-intercept = 0
Symmetric with respect to the y-axis

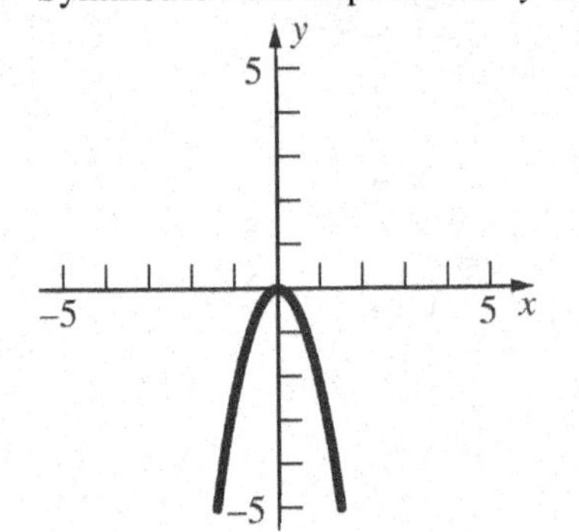

9. $x^2 + y^2 = 4$
x-intercepts = -2, 2; y-intercepts = -2, 2
Symmetric with respect to the x-axis, y-axis, and origin

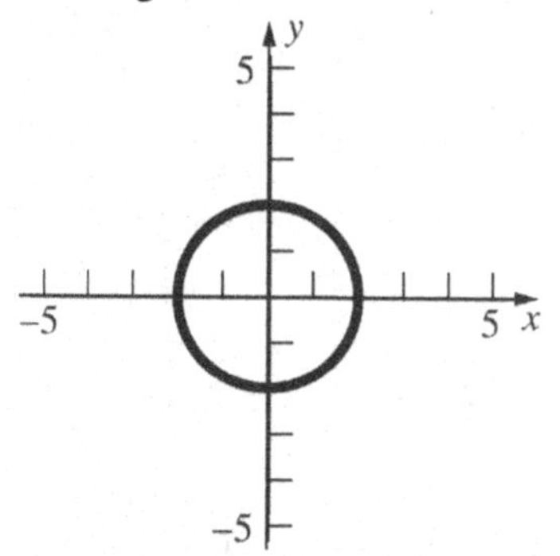

11. $y = -x^2 - 2x + 2$: y-intercept = 2

$$x\text{-intercepts} = \frac{2 \pm \sqrt{4+8}}{-2} = \frac{2 \pm 2\sqrt{3}}{-2} = -1 \pm \sqrt{3}$$

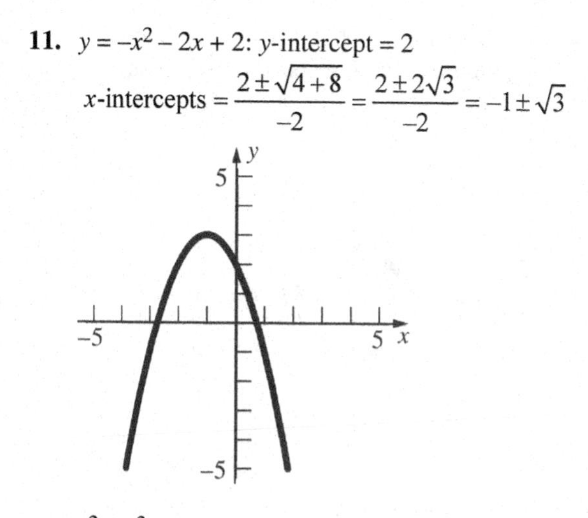

13. $x^2 - y^2 = 4$
x-intercept = -2, 2
Symmetric with respect to the x-axis, y-axis, and origin

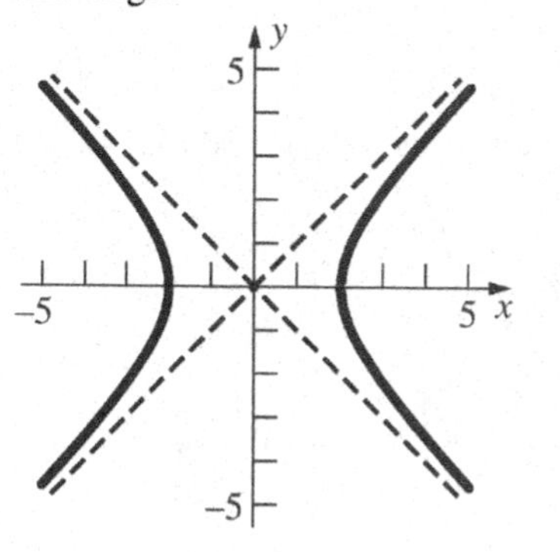

15. $4(x-1)^2 + y^2 = 36$;
y-intercepts $= \pm\sqrt{32} = \pm 4\sqrt{2}$
x-intercepts = –2, 4
Symmetric with respect to the x-axis

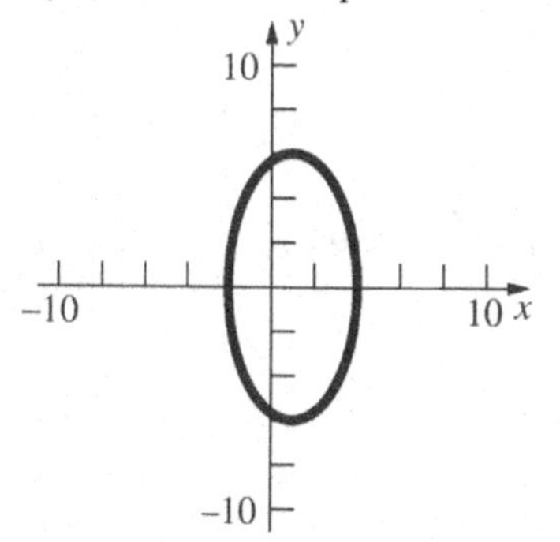

17. $x^2 + 9(y+2)^2 = 36$; y-intercepts = –4, 0
x-intercept = 0
Symmetric with respect to the y-axis

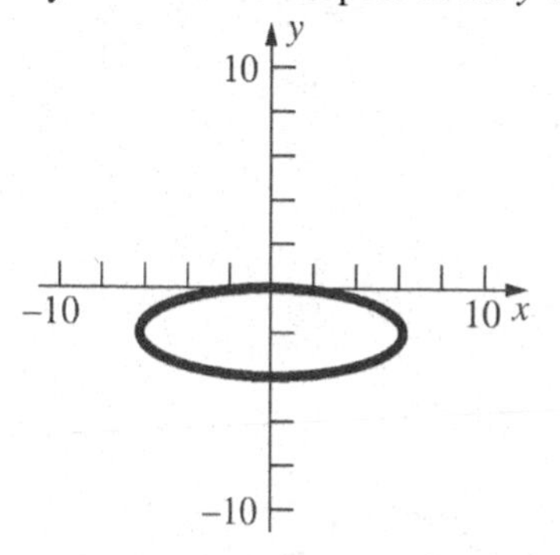

19. $x^4 + y^4 = 16$; y-intercepts $= -2, 2$
x-intercepts $= -2, 2$
Symmetric with respect to the y-axis, x-axis and origin

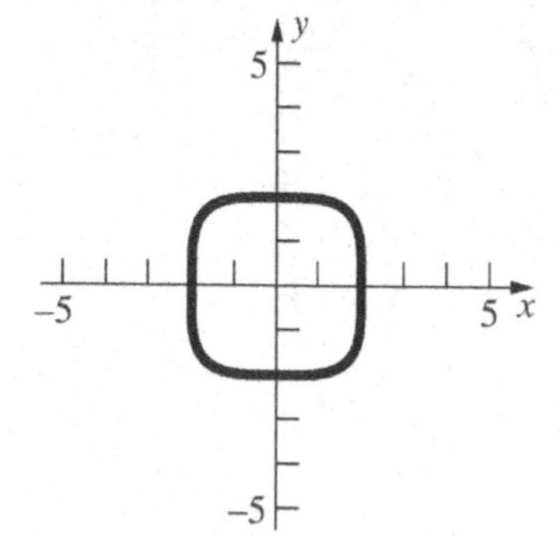

21. $y = \frac{1}{x^2+1}$; y-intercept = 1

Symmetric with respect to the y-axis

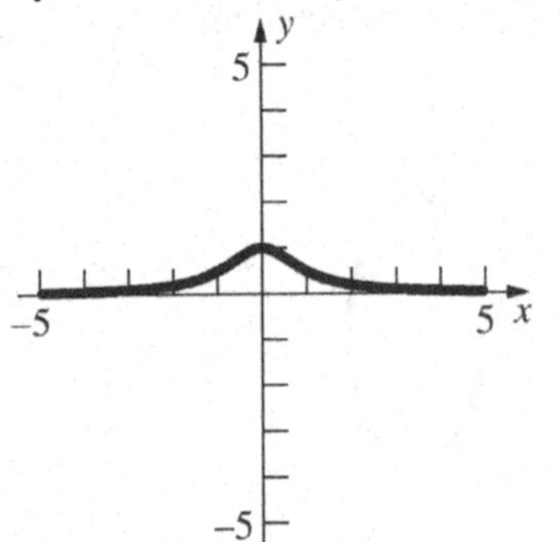

23. $2x^2 - 4x + 3y^2 + 12y = -2$

$2(x^2 - 2x + 1) + 3(y^2 + 4y + 4) = -2 + 2 + 12$

$2(x-1)^2 + 3(y+2)^2 = 12$

y-intercepts = $-2 \pm \frac{\sqrt{30}}{3}$

x-intercept = 1

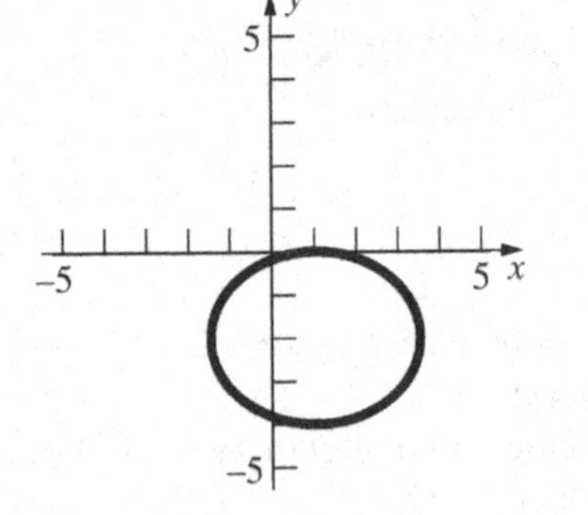

25. $y = (x - 1)(x - 2)(x - 3)$; y-intercept = –6

x-intercepts = 1, 2, 3

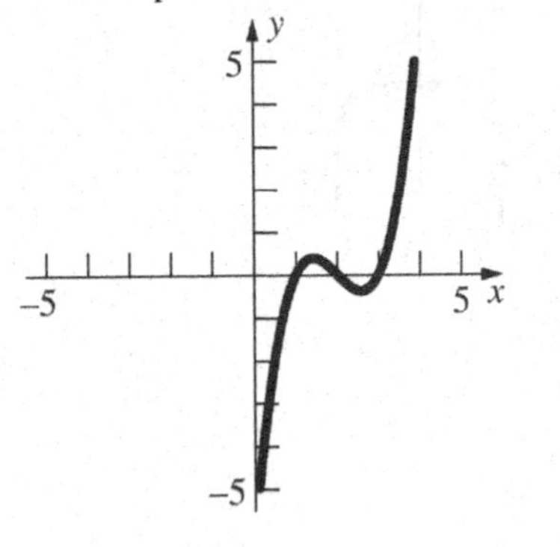

27. $y = x^2(x-1)^2$; y-intercept = 0

x-intercepts = 0, 1

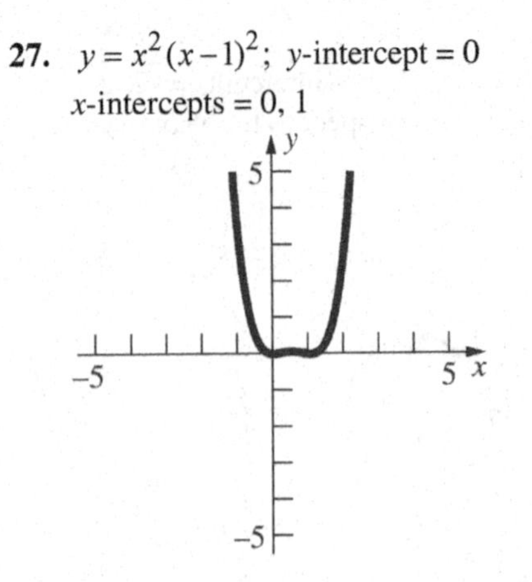

29. $|x| + |y| = 1$; y-intercepts = –1, 1;

x-intercepts = –1, 1

Symmetric with respect to the x-axis, y-axis and origin

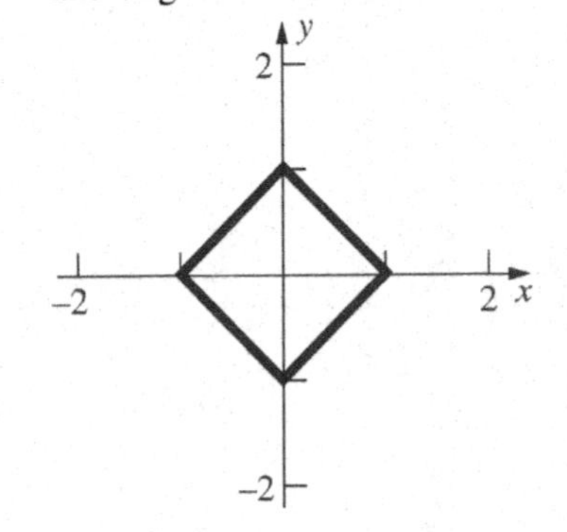

31. $-x + 1 = (x+1)^2$

$-x + 1 = x^2 + 2x + 1$

$x^2 + 3x = 0$

$x(x+3) = 0$

$x = 0, -3$

Intersection points: (0, 1) and (–3, 4)

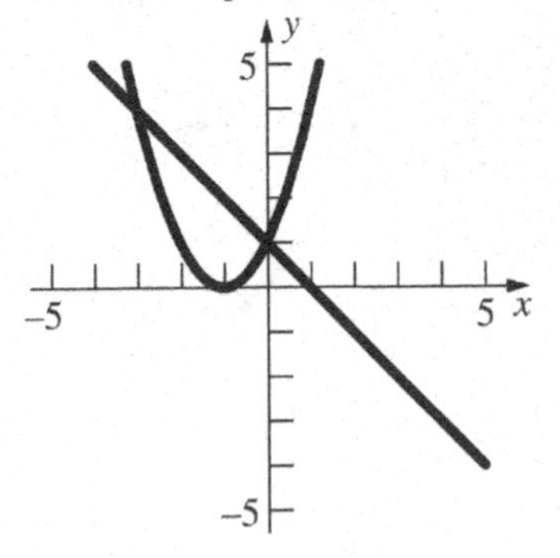

33. $-2x+3=-2(x-4)^2$

$-2x+3=-2x^2+16x-32$

$2x^2-18x+35=0$

$x=\frac{18\pm\sqrt{324-280}}{4}=\frac{18\pm2\sqrt{11}}{4}=\frac{9\pm\sqrt{11}}{2};$

Intersection points: $\left(\frac{9-\sqrt{11}}{2},-6+\sqrt{11}\right)$, $\left(\frac{9+\sqrt{11}}{2},-6-\sqrt{11}\right)$

35. $x^2+x^2=4$

$x^2=2$

$x=\pm\sqrt{2}$

Intersection points: $\left(-\sqrt{2},-\sqrt{2}\right),\left(\sqrt{2},\sqrt{2}\right)$

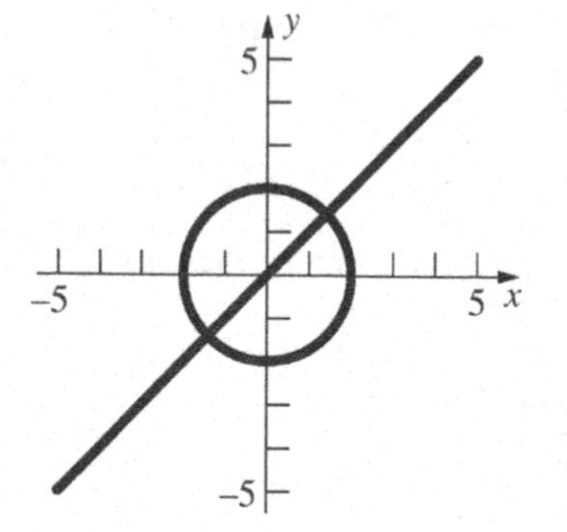

37. $y=3x+1$

$x^2+2x+(3x+1)^2=15$

$x^2+2x+9x^2+6x+1=15$

$10x^2+8x-14=0$

$2(5x^2+4x-7)=0$

$x=\frac{-2\pm\sqrt{39}}{5}\approx-1.65,0.85$

Intersection points:

$\left(\frac{-2-\sqrt{39}}{5},\frac{-1-3\sqrt{39}}{5}\right)$ and

$\left(\frac{-2+\sqrt{39}}{5},\frac{-1+3\sqrt{39}}{5}\right)$

[or roughly (–1.65, –3.95) and (0.85, 3.55)]

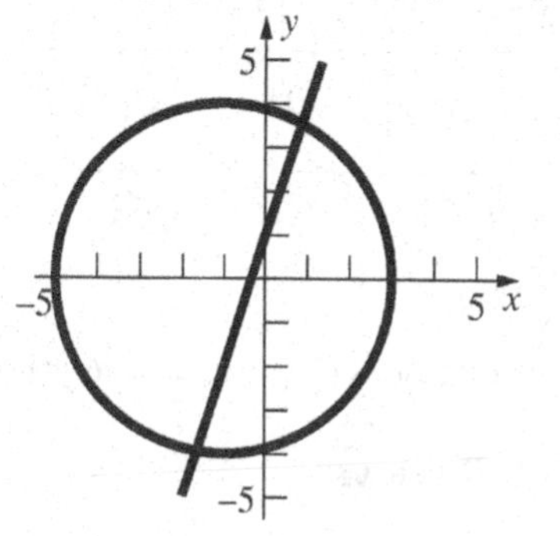

39. a. $y=x^2$; (2)

b. ax^3+bx^2+cx+d, with $a>0$: (1)

c. ax^3+bx^2+cx+d, with $a<0$: (3)

d. $y=ax^3$, with $a>0$: (4)

41. $x^2+2x+y^2-2y=20$; $\left(-2,1+\sqrt{21}\right)$, $\left(-2,1-\sqrt{21}\right),\left(2,1+\sqrt{13}\right),\left(2,1-\sqrt{13}\right)$

$d_1=\sqrt{(-2-2)^2+\left[1+\sqrt{21}-\left(1+\sqrt{13}\right)\right]^2}$

$=\sqrt{16+\left(\sqrt{21}-\sqrt{13}\right)^2}$

$=\sqrt{50-2\sqrt{273}}\approx4.12$

$d_2=\sqrt{(-2-2)^2+\left[1+\sqrt{21}-\left(1-\sqrt{13}\right)\right]^2}$

$=\sqrt{16+\left(\sqrt{21}+\sqrt{13}\right)^2}$

$=\sqrt{50+2\sqrt{273}}\approx9.11$

$$d_3 = \sqrt{(-2+2)^2 + \left[1+\sqrt{21}-\left(1-\sqrt{21}\right)\right]^2}$$
$$= \sqrt{0+\left(\sqrt{21}+\sqrt{21}\right)^2} = \sqrt{\left(2\sqrt{21}\right)^2}$$
$$= 2\sqrt{21} \approx 9.17$$
$$d_4 = \sqrt{(-2-2)^2 + \left[1-\sqrt{21}-(1+\sqrt{13})\right]^2}$$
$$= \sqrt{16+\left(-\sqrt{21}-\sqrt{13}\right)^2}$$
$$= \sqrt{50+2\sqrt{273}} \approx 9.11$$
$$d_5 = \sqrt{(-2-2)^2 + \left[1-\sqrt{21}-\left(1-\sqrt{13}\right)\right]^2}$$
$$= \sqrt{16+\left(\sqrt{13}-\sqrt{21}\right)^2}$$
$$= \sqrt{50-2\sqrt{273}} \approx 4.12$$
$$d_6 = \sqrt{(2-2)^2 + \left[1+\sqrt{13}-\left(1-\sqrt{13}\right)\right]^2}$$
$$= \sqrt{0+\left(\sqrt{13}+\sqrt{13}\right)^2} = \sqrt{\left(2\sqrt{13}\right)^2}$$
$$= 2\sqrt{13} \approx 7.21$$

Four such distances ($d_2 = d_4$ and $d_1 = d_5$).

0.5 Concepts Review

1. domain; range

3. asymptote

Problem Set 0.5

1. a. $f(1) = 1-1^2 = 0$

   b. $f(-2) = 1-(-2)^2 = -3$

   c. $f(0) = 1-0^2 = 1$

   d. $f(k) = 1-k^2$

   e. $f(-5) = 1-(-5)^2 = -24$

   f. $f\left(\frac{1}{4}\right) = 1-\left(\frac{1}{4}\right)^2 = 1-\frac{1}{16} = \frac{15}{16}$

   g. $f(1+h) = 1-(1+h)^2 = -2h-h^2$

   h. $f(1+h)-f(1) = -2h-h^2-0 = -2h-h^2$

   i. $f(2+h)-f(2) = 1-(2+h)^2+3$
   $= -4h-h^2$

3. a. $G(0) = \frac{1}{0-1} = -1$

   b. $G(0.999) = \frac{1}{0.999-1} = -1000$

   c. $G(1.01) = \frac{1}{1.01-1} = 100$

   d. $G(y^2) = \frac{1}{y^2-1}$

   e. $G(-x) = \frac{1}{-x-1} = -\frac{1}{x+1}$

   f. $G\left(\frac{1}{x^2}\right) = \frac{1}{\frac{1}{x^2}-1} = \frac{x^2}{1-x^2}$

5. a. $f(0.25) = \frac{1}{\sqrt{0.25-3}} = \frac{1}{\sqrt{-2.75}}$ is not defined

   b. $f(x) = \frac{1}{\sqrt{\pi-3}} \approx 2.658$

   c. $f(3+\sqrt{2}) = \frac{1}{\sqrt{3+\sqrt{2}-3}} = \frac{1}{\sqrt{\sqrt{2}}}$
   $= 2^{-0.25} \approx 0.841$

7. a. $x^2+y^2 = 1$
   $y^2 = 1-x^2$
   $y = \pm\sqrt{1-x^2}$; not a function

   b. $xy+y+x = 1$
   $y(x+1) = 1-x$
   $y = \frac{1-x}{x+1}$; $f(x) = \frac{1-x}{x+1}$

   c. $x = \sqrt{2y+1}$
   $x^2 = 2y+1$
   $y = \frac{x^2-1}{2}$; $f(x) = \frac{x^2-1}{2}$

   d. $x = \frac{y}{y+1}$
   $xy+x = y$
   $x = y-xy$
   $x = y(1-x)$
   $y = \frac{x}{1-x}$; $f(x) = \frac{x}{1-x}$

9. $\dfrac{f(a+h)-f(a)}{h}=\dfrac{[2(a+h)^2-1]-(2a^2-1)}{h}$

$=\dfrac{4ah+2h^2}{h}=4a+2h$

11. $\dfrac{g(x+h)-g(x)}{h}=\dfrac{\frac{3}{x+h-2}-\frac{3}{x-2}}{h}$

$=\dfrac{\frac{3x-6-3x-3h+6}{x^2-4x+hx-2h+4}}{h}$

$=\dfrac{-3h}{h(x^2-4x+hx-2h+4)}$

$=-\dfrac{3}{x^2-4x+hx-2h+4}$

13. a. $F(z)=\sqrt{2z+3}$

$2z+3\ge 0;\ z\ge -\dfrac{3}{2}$

Domain: $\left\{z\in\mathbb{R}: z\ge -\dfrac{3}{2}\right\}$

b. $g(v)=\dfrac{1}{4v-1}$

$4v-1=0;\ v=\dfrac{1}{4}$

Domain: $\left\{v\in\mathbb{R}: v\ne \dfrac{1}{4}\right\}$

c. $\psi(x)=\sqrt{x^2-9}$

$x^2-9\ge 0;\ x^2\ge 9;\ |x|\ge 3$

Domain: $\{x\in\mathbb{R}: |x|\ge 3\}$

d. $H(y)=-\sqrt{625-y^4}$

$625-y^4\ge 0; 625\ge y^4; |y|\le 5$

Domain: $\{y\in\mathbb{R}: |y|\le 5\}$

15. $f(x)=-4; f(-x)=-4$; even function

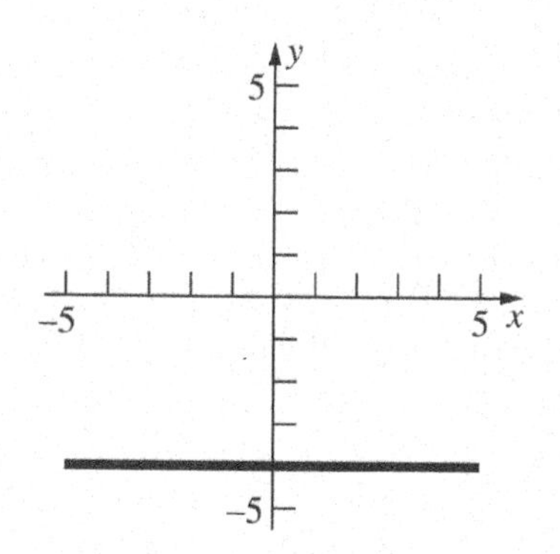

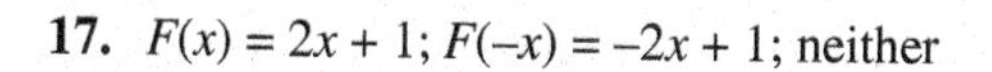

17. $F(x)=2x+1; F(-x)=-2x+1$; neither

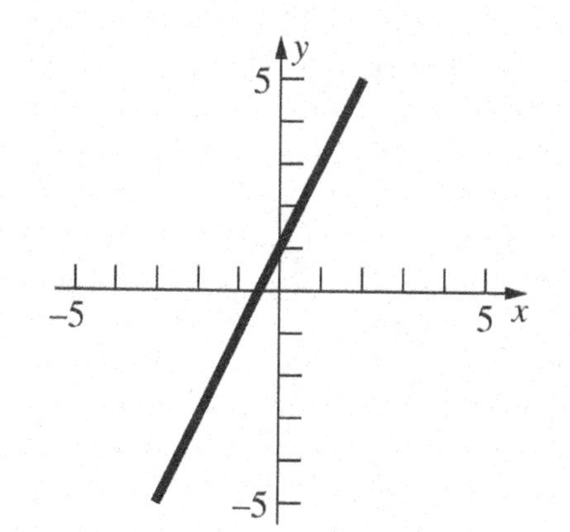

19. $g(x)=3x^2+2x-1; g(-x)=3x^2-2x-1$; neither

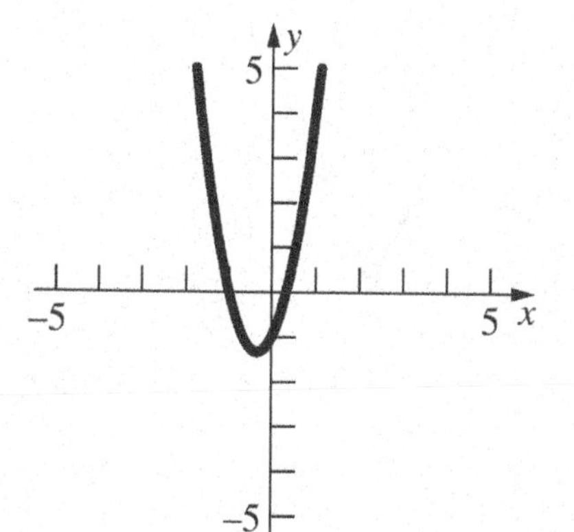

21. $g(x)=\dfrac{x}{x^2-1}; g(-x)=\dfrac{-x}{x^2-1}$; odd

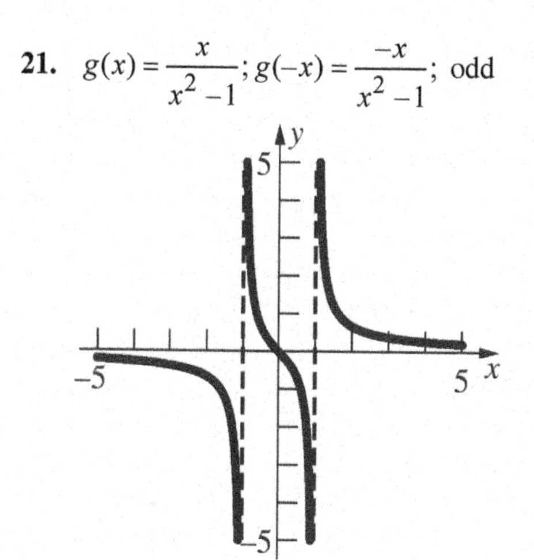

23. $f(w) = \sqrt{w-1}$; $f(-w) = \sqrt{-w-1}$; neither

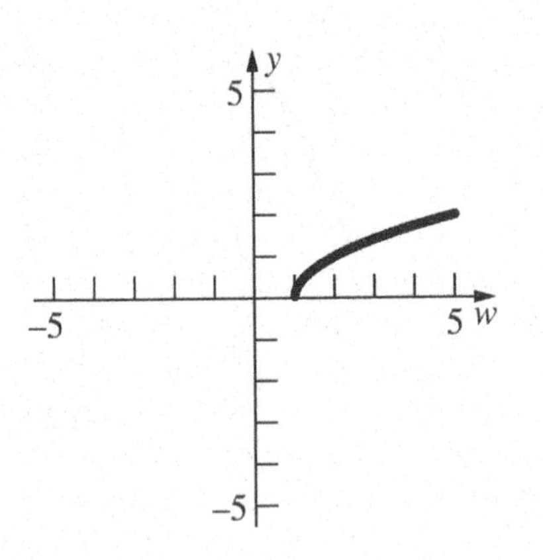

25. $f(x) = |2x|$; $f(-x) = |-2x| = |2x|$; even function

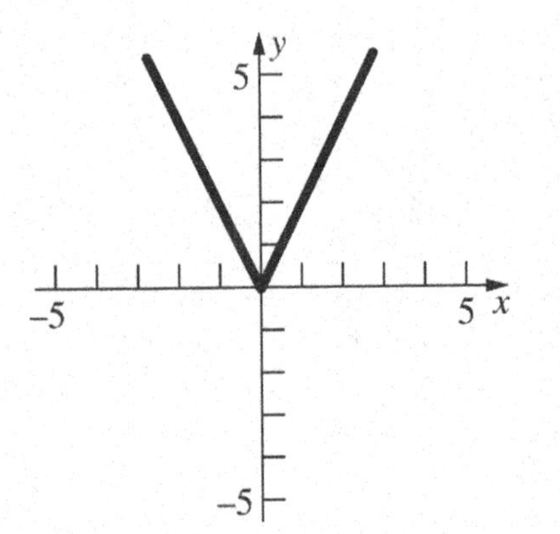

27. $g(x) = \left[\!\left[\frac{x}{2}\right]\!\right]$; $g(-x) = \left[\!\left[-\frac{x}{2}\right]\!\right]$; neither

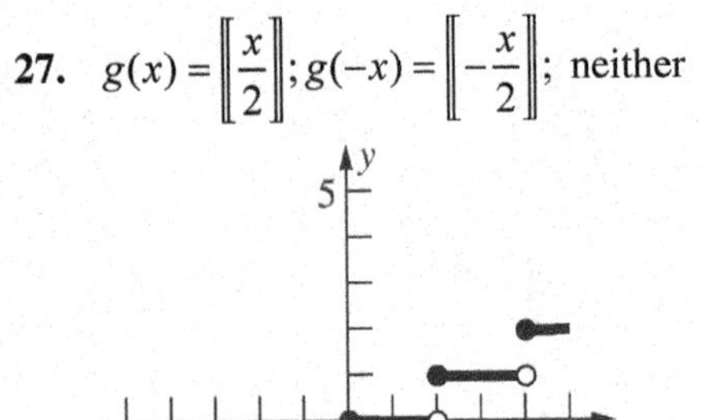
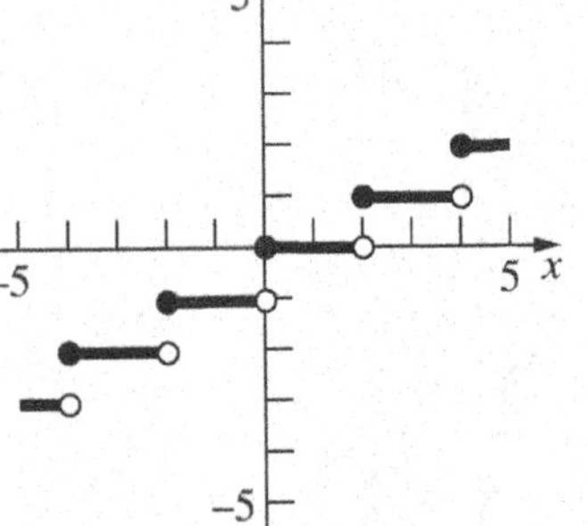

29. $g(t) = \begin{cases} 1 & \text{if } t \le 0 \\ t+1 & \text{if } 0 < t < 2 \\ t^2 - 1 & \text{if } t \ge 2 \end{cases}$ neither

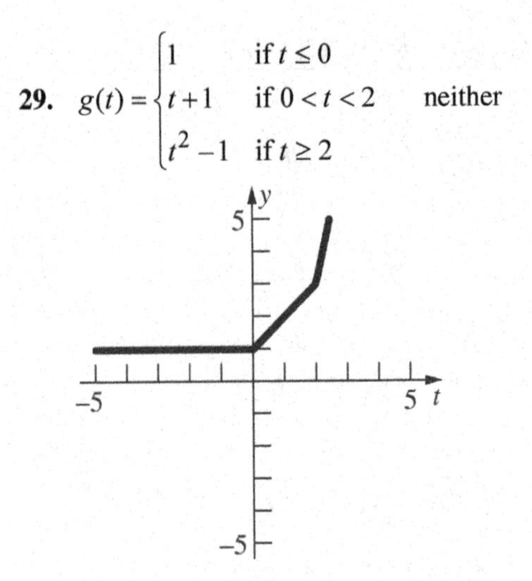

31. $T(x) = 5000 + 805x$
Domain: $\{x \in \text{integers}: 0 \le x \le 100\}$

$$u(x) = \frac{T(x)}{x} = \frac{5000}{x} + 805$$

Domain: $\{x \in \text{integers}: 0 < x \le 100\}$

33. $E(x) = x - x^2$

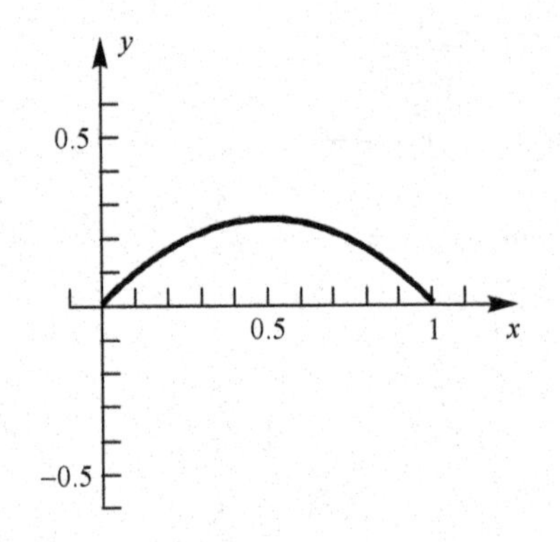

$\frac{1}{2}$ exceeds its square by the maximum amount.

35. Let y denote the length of the other leg. Then

$$x^2 + y^2 = h^2$$
$$y^2 = h^2 - x^2$$
$$y = \sqrt{h^2 - x^2}$$
$$L(x) = \sqrt{h^2 - x^2}$$

37. **a.** $E(x) = 24 + 0.40x$

b. $120 = 24 + 0.40x$
$0.40x = 96$; $x = 240$ mi

39. The area of the two semicircular ends is $\frac{\pi d^2}{4}$.

The length of each parallel side is $\frac{1-\pi d}{2}$.

$$A(d) = \frac{\pi d^2}{4} + d\left(\frac{1-\pi d}{2}\right) = \frac{\pi d^2}{4} + \frac{d - \pi d^2}{2}$$

$$= \frac{2d - \pi d^2}{4}$$

Since the track is one mile long, $\pi d < 1$, so $d < \frac{1}{\pi}$. Domain: $\left\{d \in \mathbb{R} : 0 < d < \frac{1}{\pi}\right\}$

41. a. $B(0) = 0$

b. $B\left(\frac{1}{2}\right) = \frac{1}{2}B(1) = \frac{1}{2}\cdot\frac{1}{6} = \frac{1}{12}$

c.

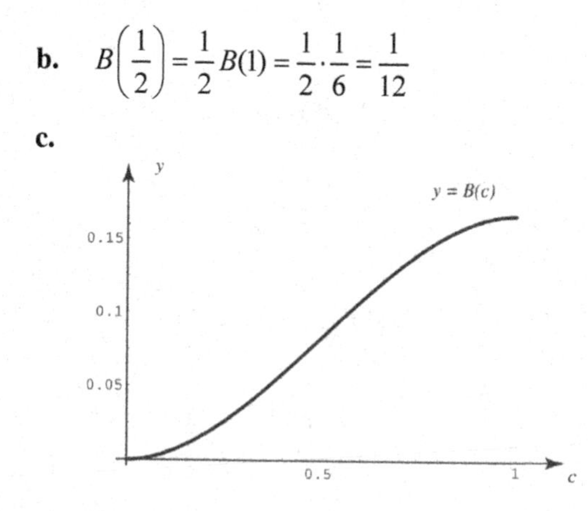

43. For any x, $x + 0 = x$, so $f(x) = f(x + 0) = f(x) + f(0)$, hence $f(0) = 0$.
Let m be the value of $f(1)$. For p in $\mathbb{N}$,
$p = p\cdot 1 = 1 + 1 + \ldots + 1$, so
$f(p) = f(1 + 1 + \ldots + 1) = f(1) + f(1) + \ldots + f(1)$
$= pf(1) = pm$.

$1 = p\left(\frac{1}{p}\right) = \frac{1}{p} + \frac{1}{p} + \ldots + \frac{1}{p}$, so

$$m = f(1) = f\left(\frac{1}{p} + \frac{1}{p} + \ldots + \frac{1}{p}\right)$$

$$= f\left(\frac{1}{p}\right) + f\left(\frac{1}{p}\right) + \ldots + f\left(\frac{1}{p}\right) = pf\left(\frac{1}{p}\right),$$

hence $f\left(\frac{1}{p}\right) = \frac{m}{p}$. Any rational number can be written as $\frac{p}{q}$ with p, q in $\mathbb{N}$.

$$\frac{p}{q} = p\left(\frac{1}{q}\right) = \frac{1}{q} + \frac{1}{q} + \ldots + \frac{1}{q},$$

so $f\left(\frac{p}{q}\right) = f\left(\frac{1}{q} + \frac{1}{q} + \ldots + \frac{1}{q}\right)$

$$= f\left(\frac{1}{q}\right) + f\left(\frac{1}{q}\right) + \ldots + f\left(\frac{1}{q}\right)$$

$$= pf\left(\frac{1}{q}\right) = p\left(\frac{m}{q}\right) = m\left(\frac{p}{q}\right)$$

45. a. $f(1.38) \approx 0.2994$
$f(4.12) \approx 3.6852$

b.

x	$f(x)$
–4	–4.05
–3	–3.1538
–2	–2.375
–1	–1.8
0	–1.25
1	–0.2
2	1.125
3	2.3846
4	3.55

47.

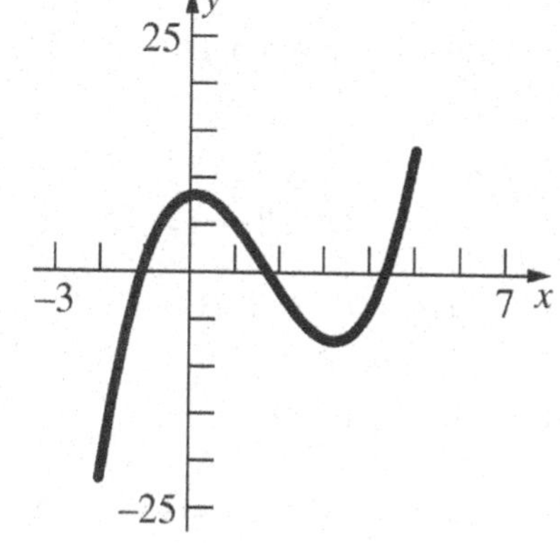

a. Range: $\{y \in \mathbb{R}: -22 \le y \le 13\}$

b. $f(x) = 0$ when $x \approx -1.1, 1.7, 4.3$
$f(x) \ge 0$ on $[-1.1, 1.7] \cup [4.3, 5]$

49.

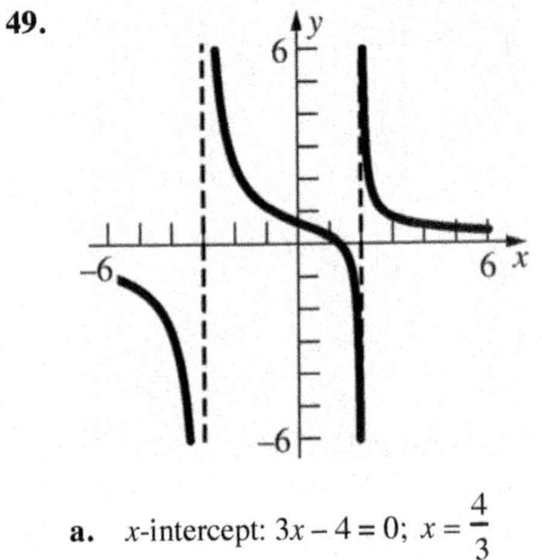

a. x-intercept: $3x-4=0$; $x=\frac{4}{3}$

y-intercept: $\frac{3\cdot 0-4}{0^2+0-6}=\frac{2}{3}$

b. $\mathbb{R}$

c. $x^2+x-6=0$; $(x+3)(x-2)=0$
Vertical asymptotes at $x=-3, x=2$

d. Horizontal asymptote at $y=0$

0.6 Concepts Review

1. $(x^2+1)^3$

3. 2; left

Problem Set 0.6

1. a. $(f+g)(2)=(2+3)+2^2=9$

b. $(f\cdot g)(0)=(0+3)(0^2)=0$

c. $(g/f)(3)=\frac{3^2}{3+3}=\frac{9}{6}=\frac{3}{2}$

d. $(f\circ g)(1)=f(1^2)=1+3=4$

e. $(g\circ f)(1)=g(1+3)=4^2=16$

f. $(g\circ f)(-8)=g(-8+3)=(-5)^2=25$

3. a. $(\Phi+\Psi)(t)=t^3+1+\frac{1}{t}$

b. $(\Phi\circ\Psi)(r)=\Phi\left(\frac{1}{r}\right)=\left(\frac{1}{r}\right)^3+1=\frac{1}{r^3}+1$

c. $(\Psi\circ\Phi)(r)=\Psi(r^3+1)=\frac{1}{r^3+1}$

d. $\Phi^3(z)=(z^3+1)^3$

e. $(\Phi-\Psi)(5t)=[(5t)^3+1]-\frac{1}{5t}$

$=125t^3+1-\frac{1}{5t}$

f. $((\Phi-\Psi)\circ\Psi)(t)=(\Phi-\Psi)\left(\frac{1}{t}\right)$

$=\left(\frac{1}{t}\right)^3+1-\frac{1}{\frac{1}{t}}=\frac{1}{t^3}+1-t$

5. $(f\circ g)(x)=f(|1+x|)=\sqrt{|1+x|^2-4}$

$=\sqrt{x^2+2x-3}$

$(g\circ f)(x)=g\left(\sqrt{x^2-4}\right)=\left|1+\sqrt{x^2-4}\right|$

$=1+\sqrt{x^2-4}$

7. $g(3.141)\approx 1.188$

9. $\left[g^2(\pi)-g(\pi)\right]^{1/3}=\left[(11-7\pi)^2-|11-7\pi|\right]^{1/3}$

≈ 4.789

11. a. $g(x)=\sqrt{x}, f(x)=x+7$

b. $g(x)=x^{15}$, $f(x)=x^2+x$

13. $p=f\circ g\circ h$ if $f(x)=1/x$, $g(x)=\sqrt{x}$,
$h(x)=x^2+1$

$p=f\circ g\circ h$ if $f(x)=1/\sqrt{x}$, $g(x)=x+1$,
$h(x)=x^2$

15. Translate the graph of $g(x)=\sqrt{x}$ to the right 2 units and down 3 units.

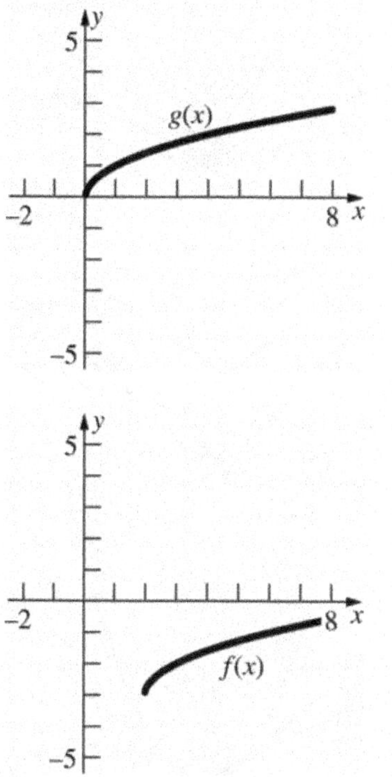

17. Translate the graph of $y=x^2$ to the right 2 units and down 4 units.

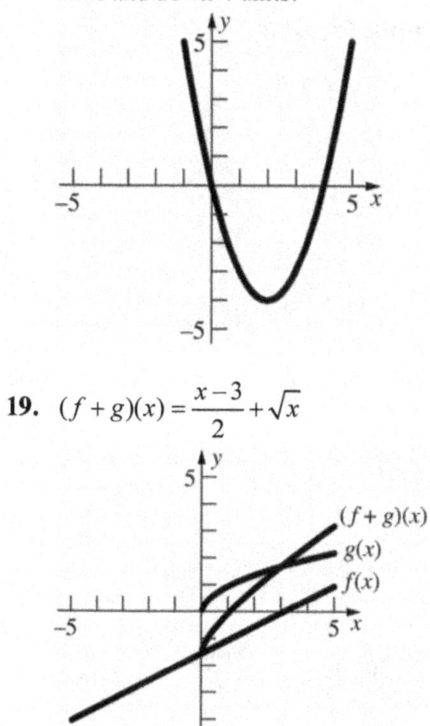

19. $(f+g)(x)=\dfrac{x-3}{2}+\sqrt{x}$

21. $F(t)=\dfrac{|t|-t}{t}$

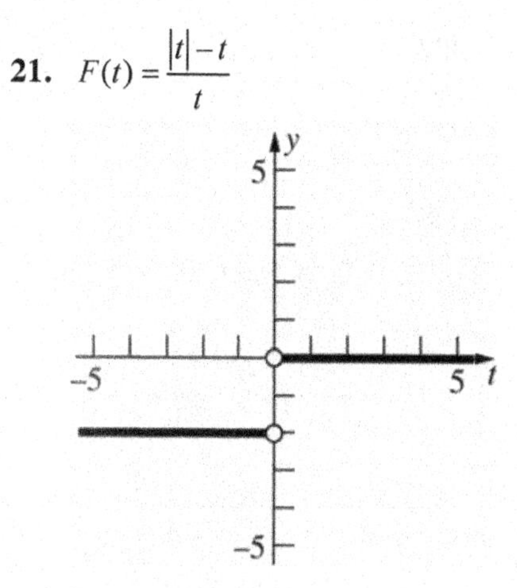

23. a. Even;
$(f+g)(-x)=f(-x)+g(-x)=f(x)+g(x)$
$=(f+g)(x)$ if f and g are both even functions.

b. Odd;
$(f+g)(-x)=f(-x)+g(-x)=-f(x)-g(x)$
$=-(f+g)(x)$ if f and g are both odd functions.

c. Even;
$(f\cdot g)(-x)=[f(-x)][g(-x)]$
$=[f(x)][g(x)]=(f\cdot g)(x)$
if f and g are both even functions.

d. Even;
$(f\cdot g)(-x)=[f(-x)][g(-x)]$
$=[-f(x)][-g(x)]=[f(x)][g(x)]$
$=(f\cdot g)(x)$
if f and g are both odd functions.

e. Odd;
$(f\cdot g)(-x)=[f(-x)][g(-x)]$
$=[f(x)][-g(x)]=-[f(x)][g(x)]$
$=-(f\cdot g)(x)$
if f is an even function and g is an odd function.

25. Not every polynomial of even degree is an even function. For example $f(x)=x^2+x$ is neither even nor odd. Not every polynomial of odd degree is an odd function. For example $g(x)=x^3+x^2$ is neither even nor odd.

27. a. $P=\sqrt{29-3(2+\sqrt{t})+(2+\sqrt{t})^2}$
$=\sqrt{t+\sqrt{t}+27}$

b. When $t=15$, $P=\sqrt{15+\sqrt{15}+27}\approx 6.773$

29. $D(t)=\begin{cases}400t & \text{if } 0<t<1\\ \sqrt{(400t)^2+[300(t-1)]^2} & \text{if } t\ge 1\end{cases}$

$D(t)=\begin{cases}400t & \text{if } 0<t<1\\ \sqrt{250{,}000t^2-180{,}000t+90{,}000} & \text{if } t\ge 1\end{cases}$

31. $f(f(x))=f\left(\frac{ax+b}{cx-a}\right)=\frac{a\left(\frac{ax+b}{cx-a}\right)+b}{c\left(\frac{ax+b}{cx-a}\right)-a}$

$=\frac{a^2x+ab+bcx-ab}{acx+bc-acx+a^2}=\frac{x(a^2+bc)}{a^2+bc}=x$

If $a^2+bc=0$, $f(f(x))$ is undefined, while if $x=\frac{a}{c}$, $f(x)$ is undefined.

33. a. $f\left(\frac{1}{x}\right)=\frac{\frac{1}{x}}{\frac{1}{x}-1}=\frac{1}{1-x}$

b. $f(f(x))=f\left(\frac{x}{x-1}\right)=\frac{\frac{x}{x-1}}{\frac{x}{x-1}-1}$

$=\frac{x}{x-x+1}=x$

c. $f\left(\frac{1}{f(x)}\right)=f\left(\frac{x-1}{x}\right)=\frac{\frac{x-1}{x}}{\frac{x-1}{x}-1}=\frac{x-1}{x-1-x}$

$=1-x$

35. $(f_1\circ(f_2\circ f_3))(x)=f_1((f_2\circ f_3)(x))$
$=f_1(f_2(f_3(x)))$

$((f_1\circ f_2)\circ f_3)(x)=(f_1\circ f_2)(f_3(x))$
$=f_1(f_2(f_3(x)))$
$=(f_1\circ(f_2\circ f_3))(x)$

37.

39.

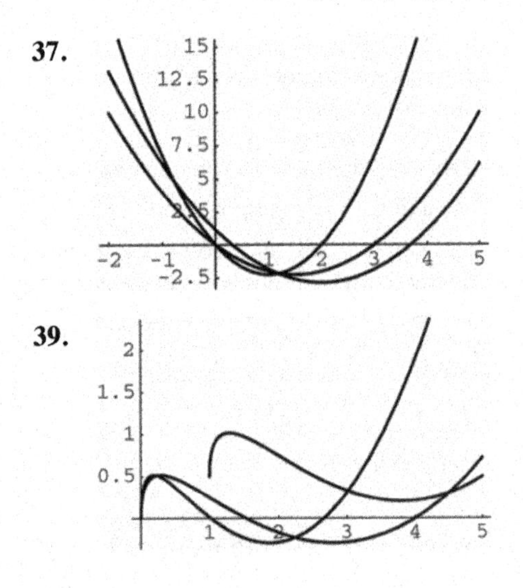

41. a.

b.

c.

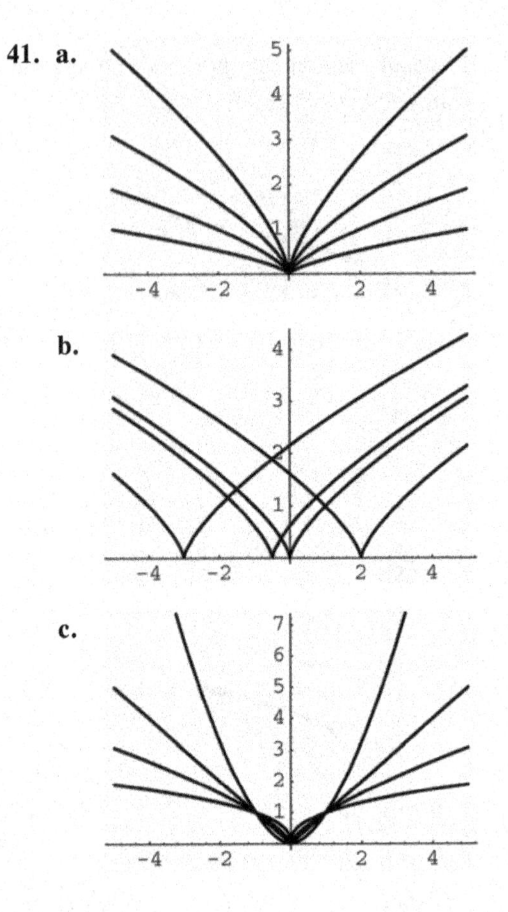

0.7 Concepts Review

1. $(-\infty, \infty)$; $[-1, 1]$

3. odd; even

Problem Set 0.7

1. a. $30\left(\frac{\pi}{180}\right)=\frac{\pi}{6}$

b. $45\left(\frac{\pi}{180}\right)=\frac{\pi}{4}$

c. $-60\left(\frac{\pi}{180}\right)=-\frac{\pi}{3}$

d. $240\left(\frac{\pi}{180}\right)=\frac{4\pi}{3}$

e. $-370\left(\frac{\pi}{180}\right)=-\frac{37\pi}{18}$

f. $10\left(\frac{\pi}{180}\right)=\frac{\pi}{18}$

3. a. $33.3\left(\frac{\pi}{180}\right) \approx 0.5812$

b. $46\left(\frac{\pi}{180}\right) \approx 0.8029$

c. $-66.6\left(\frac{\pi}{180}\right) \approx -1.1624$

d. $240.11\left(\frac{\pi}{180}\right) \approx 4.1907$

e. $-369\left(\frac{\pi}{180}\right) \approx -6.4403$

f. $11\left(\frac{\pi}{180}\right) \approx 0.1920$

5. a. $\frac{56.4 \tan 34.2°}{\sin 34.1°} \approx 68.37$

b. $\frac{5.34 \tan 21.3°}{\sin 3.1° + \cot 23.5°} \approx 0.8845$

c. $\tan(0.452) \approx 0.4855$

d. $\sin(-0.361) \approx -0.3532$

7. a. $\frac{56.3 \tan 34.2°}{\sin 56.1°} \approx 46.097$

b. $\left(\frac{\sin 35°}{\sin 26° + \cos 26°}\right)^3 \approx 0.0789$

9. a. $\tan\left(\frac{\pi}{6}\right) = \frac{\sin\left(\frac{\pi}{6}\right)}{\cos\left(\frac{\pi}{6}\right)} = \frac{\sqrt{3}}{3}$

b. $\sec(\pi) = \frac{1}{\cos(\pi)} = -1$

c. $\sec\left(\frac{3\pi}{4}\right) = \frac{1}{\cos\left(\frac{3\pi}{4}\right)} = -\sqrt{2}$

d. $\csc\left(\frac{\pi}{2}\right) = \frac{1}{\sin\left(\frac{\pi}{2}\right)} = 1$

e. $\cot\left(\frac{\pi}{4}\right) = \frac{\cos\left(\frac{\pi}{4}\right)}{\sin\left(\frac{\pi}{4}\right)} = 1$

f. $\tan\left(-\frac{\pi}{4}\right) = \frac{\sin\left(-\frac{\pi}{4}\right)}{\cos\left(-\frac{\pi}{4}\right)} = -1$

11. a. $(1 + \sin z)(1 - \sin z) = 1 - \sin^2 z$

$= \cos^2 z = \frac{1}{\sec^2 z}$

b. $(\sec t - 1)(\sec t + 1) = \sec^2 t - 1 = \tan^2 t$

c. $\sec t - \sin t \tan t = \frac{1}{\cos t} - \frac{\sin^2 t}{\cos t}$

$= \frac{1 - \sin^2 t}{\cos t} = \frac{\cos^2 t}{\cos t} = \cos t$

d. $\frac{\sec^2 t - 1}{\sec^2 t} = \frac{\tan^2 t}{\sec^2 t} = \frac{\frac{\sin^2 t}{\cos^2 t}}{\frac{1}{\cos^2 t}} = \sin^2 t$

13. a. $\frac{\sin u}{\csc u} + \frac{\cos u}{\sec u} = \sin^2 u + \cos^2 u = 1$

b. $(1 - \cos^2 x)(1 + \cot^2 x) = (\sin^2 x)(\csc^2 x)$

$= \sin^2 x\left(\frac{1}{\sin^2 x}\right) = 1$

c. $\sin t(\csc t - \sin t) = \sin t\left(\frac{1}{\sin t} - \sin t\right)$

$= 1 - \sin^2 t = \cos^2 t$

d. $\frac{1 - \csc^2 t}{\csc^2 t} = -\frac{\cot^2 t}{\csc^2 t} = -\frac{\frac{\cos^2 t}{\sin^2 t}}{\frac{1}{\sin^2 t}}$

$= -\cos^2 t = -\frac{1}{\sec^2 t}$

15. a. $y = \csc t$

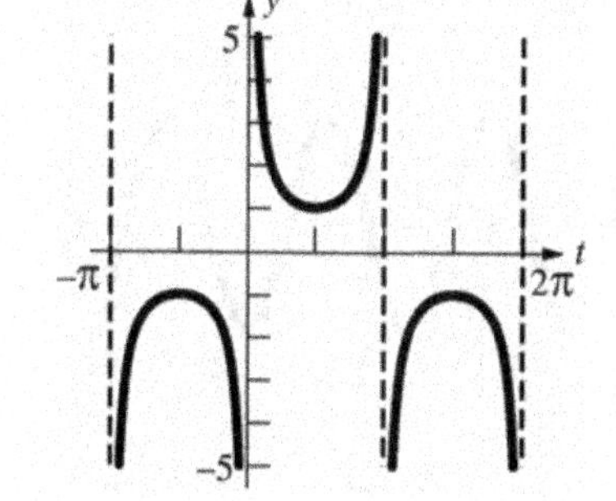

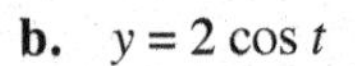

b. $y = 2\cos t$

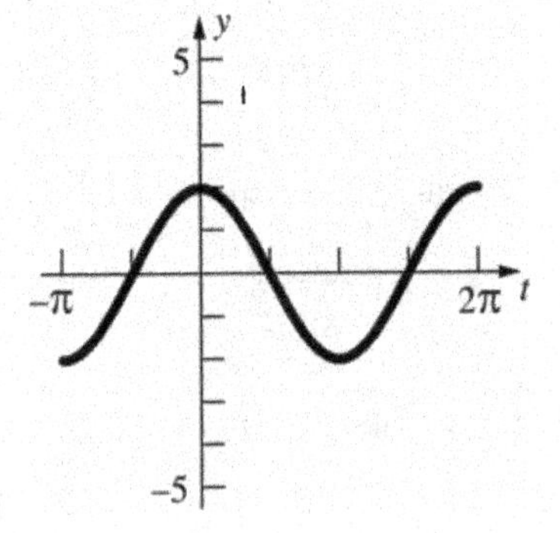

c. $y = \cos 3t$

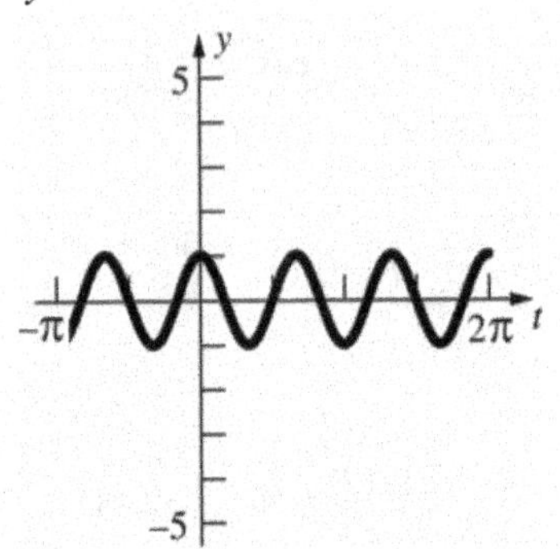

d. $y = \cos\left(t + \frac{\pi}{3}\right)$

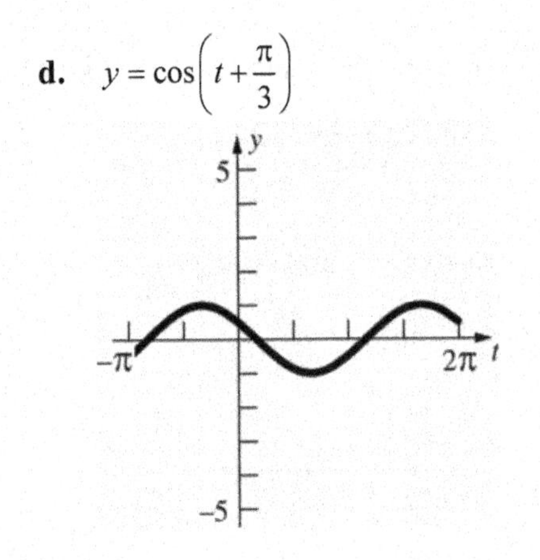

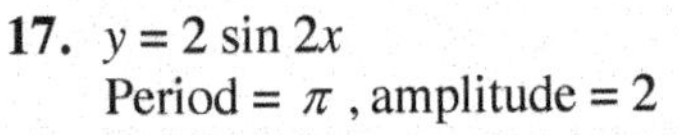

17. $y = 2\sin 2x$
Period = π, amplitude = 2

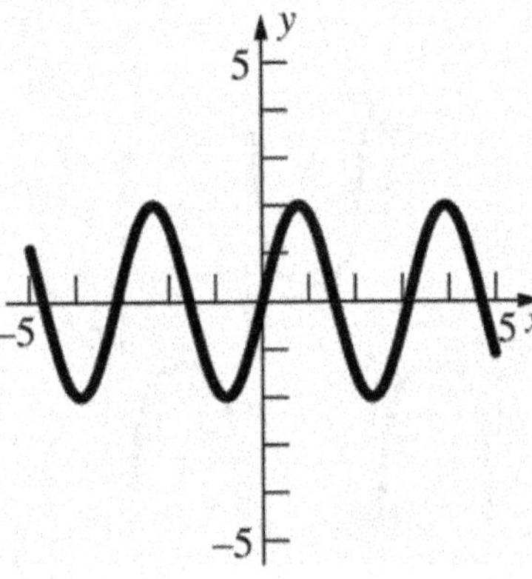

19. $y = 2 + \frac{1}{6}\cot(2x)$

Period = $\frac{\pi}{2}$, shift: 2 units up

21. $y = 21 + 7\sin(2x + 3)$
Period = π, amplitude = 7, shift: 21 units up, $\frac{3}{2}$ units left

23. $y = \tan\left(2x - \frac{\pi}{3}\right)$

Period = $\frac{\pi}{2}$, shift: $\frac{\pi}{6}$ units right

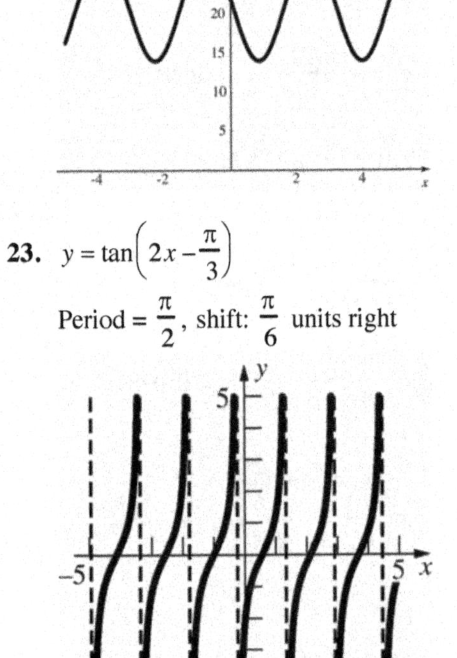

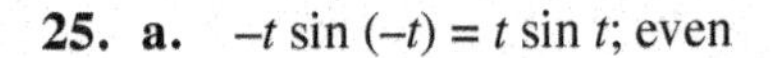

25. a. $-t\sin(-t) = t\sin t$; even

b. $\sin^2(-t) = \sin^2 t$; even

c. $\csc(-t) = \frac{1}{\sin(-t)} = -\csc t$; odd

d. $|\sin(-t)| = |-\sin t| = |\sin t|$; even

e. $\sin(\cos(-t)) = \sin(\cos t)$; even

f. $-x + \sin(-x) = -x - \sin x = -(x + \sin x)$; odd

27. $\cos^2 \frac{\pi}{3} = \left(\cos\frac{\pi}{3}\right)^2 = \left(\frac{1}{2}\right)^2 = \frac{1}{4}$

29. $\sin^3 \frac{\pi}{6} = \left(\sin\frac{\pi}{6}\right)^3 = \left(\frac{1}{2}\right)^3 = \frac{1}{8}$

31. $\sin^2 \frac{\pi}{8} = \frac{1-\cos 2\left(\frac{\pi}{8}\right)}{2} = \frac{1-\cos\frac{\pi}{4}}{2} = \frac{1-\frac{\sqrt{2}}{2}}{2}$

$= \frac{2-\sqrt{2}}{4}$

33. $\tan(t+\pi) = \frac{\tan t + \tan\pi}{1-\tan t\tan\pi} = \frac{\tan t + 0}{1-(\tan t)(0)}$

$= \tan t$

35. $s = rt = (2.5 \text{ ft})(2\pi \text{ rad}) = 5\pi$ ft, so the tire goes 5π feet per revolution, or $\frac{1}{5\pi}$ revolutions per foot.

$\left(\frac{1 \text{ rev}}{5\pi \text{ ft}}\right)\left(60\frac{\text{mi}}{\text{hr}}\right)\left(\frac{1 \text{ hr}}{60 \text{ min}}\right)\left(5280\frac{\text{ft}}{\text{mi}}\right)$

≈ 336 rev/min

37. $r_1t_1 = r_2t_2;\ 6(2\pi)t_1 = 8(2\pi)(21)$

$t_1 = 28$ rev/sec

39. **a.** $\tan\alpha = \sqrt{3}$

$\alpha = \frac{\pi}{3}$

b. $\sqrt{3}x + 3y = 6$

$3y = -\sqrt{3}x + 6$

$y = -\frac{\sqrt{3}}{3}x + 2;\ m = -\frac{\sqrt{3}}{3}$

$\tan\alpha = -\frac{\sqrt{3}}{3}$

$\alpha = \frac{5\pi}{6}$

41. **a.** $\tan\theta = \frac{3-2}{1+3(2)} = \frac{1}{7}$

$\theta \approx 0.1419$

b. $\tan\theta = \frac{-1-\frac{1}{2}}{1+\left(\frac{1}{2}\right)(-1)} = -3$

$\theta \approx 1.8925$

c. $2x - 6y = 12 \qquad 2x + y = 0$

$-6y = -2x + 12 y = -2x$

$y = \frac{1}{3}x - 2$

$m_1 = \frac{1}{3},\ m_2 = -2$

$\tan\theta = \frac{-2-\frac{1}{3}}{1+\left(\frac{1}{3}\right)(-2)} = -7;\ \theta \approx 1.7127$

43. $A = \frac{1}{2}(2)(5)^2 = 25\text{cm}^2$

45. The base of the triangle is the side opposite the angle t. Then the base has length $2r\sin\frac{t}{2}$ (similar to Problem 44). The radius of the semicircle is $r\sin\frac{t}{2}$ and the height of the triangle is $r\cos\frac{t}{2}$.

$A = \frac{1}{2}\left(2r\sin\frac{t}{2}\right)\left(r\cos\frac{t}{2}\right) + \frac{\pi}{2}\left(r\sin\frac{t}{2}\right)^2$

$= r^2\sin\frac{t}{2}\cos\frac{t}{2} + \frac{\pi r^2}{2}\sin^2\frac{t}{2}$

47. The temperature function is

$T(t) = 80 + 25\sin\left(\frac{2\pi}{12}\left(t - \frac{7}{2}\right)\right)$.

The normal high temperature for November 15th is then $T(10.5) = 67.5$ °F.

49. As t increases, the point on the rim of the wheel will move around the circle of radius 2.

a. $x(2) \approx 1.902$

$y(2) \approx 0.618$

$x(6) \approx -1.176$

$y(6) \approx -1.618$

$x(10) = 0$

$y(10) = 2$

$x(0) = 0$

$y(0) = 2$

b. $x(t) = -2\sin\left(\frac{\pi}{5}t\right),\ y(t) = 2\cos\left(\frac{\pi}{5}t\right)$

c. The point is at (2, 0) when $\frac{\pi}{5}t = \frac{\pi}{2}$; that is , when $t = \frac{5}{2}$.

51. **a.** $C\sin(\omega t+\phi)=(C\cos\phi)\sin\omega t+(C\sin\phi)\cos\omega t$. Thus $A=C\cdot\cos\phi$ and $B=C\cdot\sin\phi$.

b. $A^2+B^2=(C\cos\phi)^2+(C\sin\phi)^2=C^2(\cos^2\phi)+C^2(\sin^2\phi)=C^2$

Also, $\dfrac{B}{A}=\dfrac{C\cdot\sin\phi}{C\cdot\cos\phi}=\tan\phi$

c. $A_1\sin(\omega t+\phi_1)+A_2\sin(\omega t+\phi_2)+A_3(\sin\omega t+\phi_3)$

$=A_1(\sin\omega t\cos\phi_1+\cos\omega t\sin\phi_1)$

$+A_2(\sin\omega t\cos\phi_2+\cos\omega t\sin\phi_2)$

$+A_3(\sin\omega t\cos\phi_3+\cos\omega t\sin\phi_3)$

$=(A_1\cos\phi_1+A_2\cos\phi_2+A_3\cos\phi_3)\sin\omega t$

$+(A_1\sin\phi_1+A_2\sin\phi_2+A_3\sin\phi_3)\cos\omega t$

$=C\sin(\omega t+\phi)$

where C and ϕ can be computed from

$A=A_1\cos\phi_1+A_2\cos\phi_2+A_3\cos\phi_3$

$B=A_1\sin\phi_1+A_2\sin\phi_2+A_3\sin\phi_3$

as in part (b).

d. Written response. Answers will vary.

53. 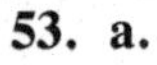

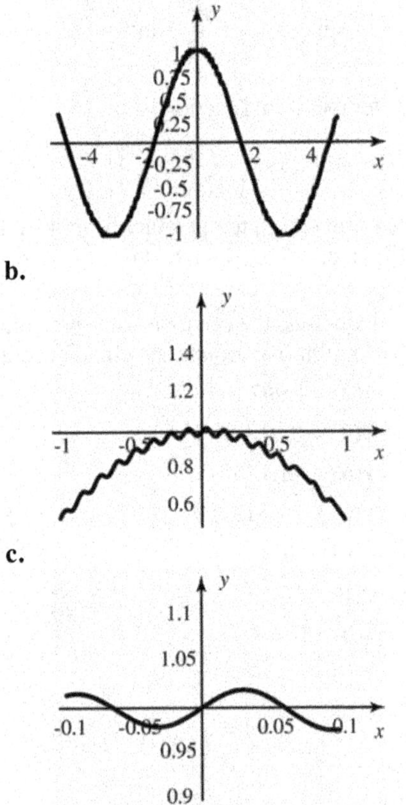

The plot in **(a)** shows the long term behavior of the function, but not the short term behavior, whereas the plot in **(c)** shows the short term behavior, but not the long term behavior. The plot in **(b)** shows a little of each.

55. $f(x)=\begin{cases}4(x-[\![x]\!])+1 & : x\in\left[n,n+\frac{1}{4}\right)\\ -\frac{4}{3}(x-[\![x]\!])+\frac{7}{3} & : x\in\left[n+\frac{1}{4},n+1\right)\end{cases}$

where n is an integer.

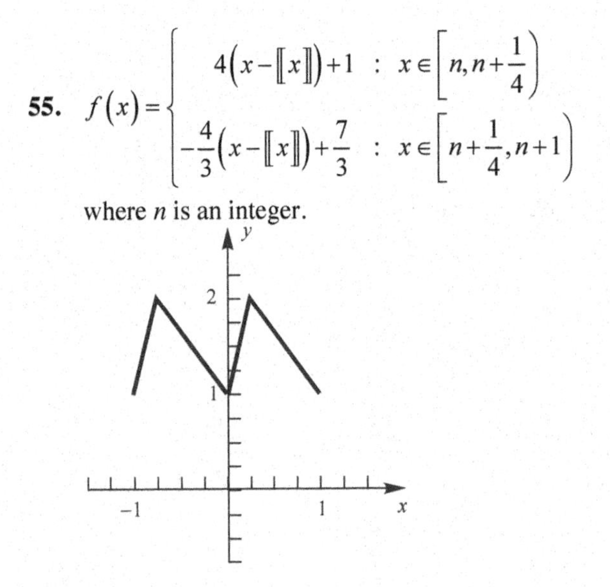

0.8 Chapter Review

Concepts Test

1. False: p and q must be integers.

3. False: If the numbers are opposites ($-\pi$ and π) then the sum is 0, which is rational.

5. False: 0.999... is equal to 1.

7. False: $(a*b)*c = a^{bc}; a*(b*c) = a^{b^c}$

9. True: If x was not 0, then $\varepsilon = \frac{|x|}{2}$ would be a positive number less than $|x|$.

11. True: $a < b < 0; a < b; \frac{a}{b} > 1; \frac{1}{b} < \frac{1}{a}$

13. True: If (a, b) and (c, d) share a point then $c < b$ so they share the infinitely many points between b and c.

15. False: For example, if $x = -3$, then $|-x| = |-(-3)| = |3| = 3$ which does not equal x.

17. True: $|x| < |y| \Leftrightarrow |x|^4 < |y|^4$
$|x|^4 = x^4$ and $|y|^4 = y^4$, so $x^4 < y^4$

19. True: If $r = 0$, then
$$\frac{1}{1+|r|} = \frac{1}{1-r} = \frac{1}{1-|r|} = 1.$$
For any r, $1+|r| \ge 1-|r|$. Since $|r| < 1, 1-|r| > 0$ so $\frac{1}{1+|r|} \le \frac{1}{1-|r|}$;
also, $-1 < r < 1$.

If $-1 < r < 0$, then $|r| = -r$ and $1-r = 1+|r|$, so
$$\frac{1}{1+|r|} = \frac{1}{1-r} \le \frac{1}{1-|r|}.$$
If $0 < r < 1$, then $|r| = r$ and $1-r = 1-|r|$, so
$$\frac{1}{1+|r|} \le \frac{1}{1-r} = \frac{1}{1-|r|}.$$

21. True: If x and y are the same sign, then $||x|-|y|| = |x-y|$. $|x-y| \le |x+y|$ when x and y are the same sign, so $||x|-|y|| \le |x+y|$. If x and y have opposite signs then either $||x|-|y|| = |x-(-y)| = |x+y|$ $(x > 0, y < 0)$ or $||x|-|y|| = |-x-y| = |x+y|$ $(x < 0, y > 0)$. In either case $||x|-|y|| = |x+y|$.
If either $x = 0$ or $y = 0$, the inequality is easily seen to be true.

23. True: For every real number y, whether it is positive, zero, or negative, the cube root $x = \sqrt[3]{y}$ satisfies
$$x^3 = \left(\sqrt[3]{y}\right)^3 = y$$

25. True: $x^2 + ax + y^2 + y = 0$
$$x^2 + ax + \frac{a^2}{4} + y^2 + y + \frac{1}{4} = \frac{a^2}{4} + \frac{1}{4}$$
$$\left(x + \frac{a}{2}\right)^2 + \left(y + \frac{1}{2}\right)^2 = \frac{a^2+1}{4}$$
is a circle for all values of a.

27. True; $y - b = \frac{3}{4}(x-a)$
$$y = \frac{3}{4}x - \frac{3a}{4} + b;$$
If $x = a + 4$:
$$y = \frac{3}{4}(a+4) - \frac{3a}{4} + b$$
$$= \frac{3a}{4} + 3 - \frac{3a}{4} + b = b + 3$$

29. True: If $ab > 0$, a and b have the same sign, so (a, b) is in either the first or third quadrant.

31. True: If $ab = 0$, a or b is 0, so (a, b) lies on the x-axis or the y-axis. If $a = b = 0$, (a, b) is the origin.

33. True: $d = \sqrt{[(a+b)-(a-b)]^2 + (a-a)^2}$
$$= \sqrt{(2b)^2} = |2b|$$

35. True: This is the general linear equation.

37. False: The slopes of perpendicular lines are negative reciprocals.

39. False: $ax + y = c \Rightarrow y = -ax + c$

$ax - y = c \Rightarrow y = ax - c$

$(a)(-a) \neq -1.$

(unless $a = \pm 1$)

41. True: $f(x) = \sqrt{-(x^2 + 4x + 3)}$

$= \sqrt{-(x+3)(x+1)}$

$-(x^2 + 4x + 3) \geq 0$ on $-3 \leq x \leq -1$.

43. True: The domain is $(-\infty, \infty)$ and the range is $[-6, \infty)$.

45. False: The range $(-\infty, \infty)$.

47. True: If $f(x)$ and $g(x)$ are odd functions,
$f(-x) + g(-x) = -f(x) - g(x)$
$= -[f(x) + g(x)]$, so $f(x) + g(x)$ is odd

49. True: If $f(x)$ is even and $g(x)$ is odd,
$f(-x)g(-x) = f(x)[-g(x)]$
$= -f(x)g(x)$, so $f(x)g(x)$ is odd.

51. False: If $f(x)$ and $g(x)$ are odd functions,
$f(g(-x)) = f(-g(x)) = -f(g(x))$, so
$f(g(x))$ is odd.

53. True: $f(-t) = \dfrac{(\sin(-t))^2 + \cos(-t)}{\tan(-t)\csc(-t)}$

$= \dfrac{(-\sin t)^2 + \cos t}{-\tan t(-\csc t)} = \dfrac{(\sin t)^2 + \cos t}{\tan t \csc t}$

55. False: $f(x) = c$ has domain $(-\infty, \infty)$, yet the range has only one value, c.

57. True: $(f \circ g)(x) = (x^3)^2 = x^6$

$(g \circ f)(x) = (x^2)^3 = x^6$

59. False: The domain of $\dfrac{f}{g}$ excludes any values where $g = 0$.

61. True: $\cot x = \dfrac{\cos x}{\sin x}$

$\cot(-x) = \dfrac{\cos(-x)}{\sin(-x)}$

$= \dfrac{\cos x}{-\sin x} = -\cot x$

63. False: The cosine function is periodic, so $\cos s = \cos t$ does not necessarily imply $s = t$; e.g., $\cos 0 = \cos 2\pi = 1$, but $0 \neq 2\pi$.

Sample Test Problems

1. a. $\left(n + \dfrac{1}{n}\right)^n; \left(1 + \dfrac{1}{1}\right)^1 = 2; \left(2 + \dfrac{1}{2}\right)^2 = \dfrac{25}{4};$

$\left(-2 + \dfrac{1}{-2}\right)^{-2} = \dfrac{4}{25}$

b. $(n^2 - n + 1)^2; \left[(1)^2 - (1) + 1\right]^2 = 1;$

$\left[(2)^2 - (2) + 1\right]^2 = 9;$

$\left[(-2)^2 - (-2) + 1\right]^2 = 49$

c. $4^{3/n}; 4^{3/1} = 64; 4^{3/2} = 8; 4^{-3/2} = \dfrac{1}{8}$

d. $\sqrt[n]{\left|\dfrac{1}{n}\right|}; \sqrt[1]{\left|\dfrac{1}{1}\right|} = 1; \sqrt{\left|\dfrac{1}{2}\right|} = \dfrac{1}{\sqrt{2}} = \dfrac{\sqrt{2}}{2};$

$\sqrt[-2]{\left|\dfrac{1}{-2}\right|} = \sqrt{2}$

3. Let $a, b, c,$ and d be integers.

$\dfrac{\frac{a}{b} + \frac{c}{d}}{2} = \dfrac{a}{2b} + \dfrac{c}{2d} = \dfrac{ad + bc}{2bd}$ which is rational.

5. Answers will vary. Possible answer:

$\sqrt{\dfrac{13}{50}} \approx 0.50990...$

7. $\left(\pi - \sqrt{2.0}\right)^{2.5} - \sqrt[3]{2.0} \approx 2.66$

9. $1 - 3x > 0$

$3x < 1$

$x < \dfrac{1}{3}$

$\left(-\infty, \dfrac{1}{3}\right)$

$-\frac{4}{3}$ -1 $-\frac{2}{3}$ $-\frac{1}{3}$ 0 $\frac{1}{3}$ $\frac{2}{3}$ 1 $\frac{4}{3}$

11. $3 - 2x \leq 4x + 1 \leq 2x + 7$

$3 - 2x \leq 4x + 1$ and $4x + 1 \leq 2x + 7$

$6x \geq 2$ and $2x \geq 6$

$x \geq \dfrac{1}{3}$ and $x \leq 3$; $\left[\dfrac{1}{3}, 3\right]$

-4 -3 -2 -1 0 1 2 3 4

13. $21t^2 - 44t + 12 \le -3; 21t^2 - 44t + 15 \le 0;$

$$t = \frac{44 \pm \sqrt{44^2 - 4(21)(15)}}{2(21)} = \frac{44 \pm 26}{42} = \frac{3}{7}, \frac{5}{3}$$

$$\left(t - \frac{3}{7}\right)\left(t - \frac{5}{3}\right) \le 0; \left[\frac{3}{7}, \frac{5}{3}\right]$$

15. $(x+4)(2x-1)^2(x-3) \le 0; [-4, 3]$

17.

$$\frac{3}{1-x} \le 2$$

$$\frac{3}{1-x} - 2 \le 0$$

$$\frac{3 - 2(1-x)}{1-x} \le 0$$

$$\frac{2x+1}{1-x} \le 0;$$

$$\left(-\infty, -\frac{1}{2}\right] \cup (1, \infty)$$

19. For example, if $x = -2$, $|-(-2)| = 2 \ne -2$

$|-x| \ne x$ for any $x < 0$

21. $|t - 5| = |-(5 - t)| = |5 - t|$

If $|5 - t| = 5 - t$, then $5 - t \ge 0$.

$t \le 5$

23. If $|x| \le 2$, then

$$0 \le \left|2x^2 + 3x + 2\right| \le \left|2x^2\right| + |3x| + 2 \le 8 + 6 + 2 = 16$$

also $\left|x^2 + 2\right| \ge 2$ so $\frac{1}{\left|x^2 + 2\right|} \le \frac{1}{2}$. Thus

$$\left|\frac{2x^2 + 3x + 2}{x^2 + 2}\right| = \left|2x^2 + 3x + 2\right| \left|\frac{1}{x^2 + 2}\right| \le 16\left(\frac{1}{2}\right)$$

$$= 8$$

25.

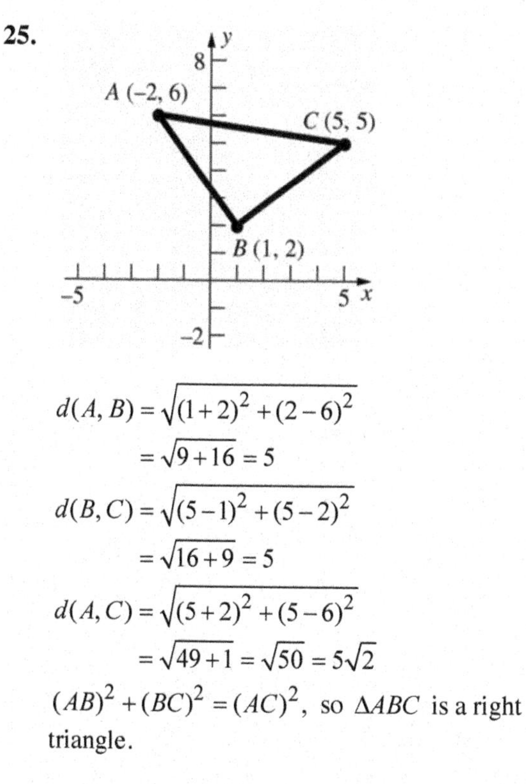

$$d(A, B) = \sqrt{(1+2)^2 + (2-6)^2}$$
$$= \sqrt{9 + 16} = 5$$

$$d(B, C) = \sqrt{(5-1)^2 + (5-2)^2}$$
$$= \sqrt{16 + 9} = 5$$

$$d(A, C) = \sqrt{(5+2)^2 + (5-6)^2}$$
$$= \sqrt{49 + 1} = \sqrt{50} = 5\sqrt{2}$$

$(AB)^2 + (BC)^2 = (AC)^2$, so ΔABC is a right triangle.

27. center $= \left(\frac{2+10}{2}, \frac{0+4}{2}\right) = (6, 2)$

radius $= \frac{1}{2}\sqrt{(10-2)^2 + (4-0)^2} = \frac{1}{2}\sqrt{64 + 16}$

$= 2\sqrt{5}$

circle: $(x-6)^2 + (y-2)^2 = 20$

29.

$$x^2 - 2x + y^2 + 2y = 2$$

$$x^2 - 2x + 1 + y^2 + 2y + 1 = 2 + 1 + 1$$

$$(x-1)^2 + (y+1)^2 = 4$$

center $= (1, -1)$

$$x^2 + 6x + y^2 - 4y = -7$$

$$x^2 + 6x + 9 + y^2 - 4y + 4 = -7 + 9 + 4$$

$$(x+3)^2 + (y-2)^2 = 6$$

center $= (-3, 2)$

$$d = \sqrt{(-3-1)^2 + (2+1)^2} = \sqrt{16 + 9} = 5$$

31. a. $m = \frac{3-1}{7+2} = \frac{2}{9};$

$$y - 1 = \frac{2}{9}(x+2)$$

$$y = \frac{2}{9}x + \frac{13}{9}$$

b. $3x - 2y = 5$

$$-2y = -3x + 5$$

$$y = \frac{3}{2}x - \frac{5}{2};$$

$$m = \frac{3}{2}$$

$$y - 1 = \frac{3}{2}(x + 2)$$

$$y = \frac{3}{2}x + 4$$

c. $3x + 4y = 9$

$$4y = -3x + 9;$$

$$y = -\frac{3}{4}x + \frac{9}{4};\ m = \frac{4}{3}$$

$$y - 1 = \frac{4}{3}(x + 2)$$

$$y = \frac{4}{3}x + \frac{11}{3}$$

d. $x = -2$

e. contains (–2, 1) and (0, 3); $m = \dfrac{3-1}{0+2};$

$$y = x + 3$$

33. The figure is a cubic with respect to y. The equation is **(b)** $x = y^3$.

35.

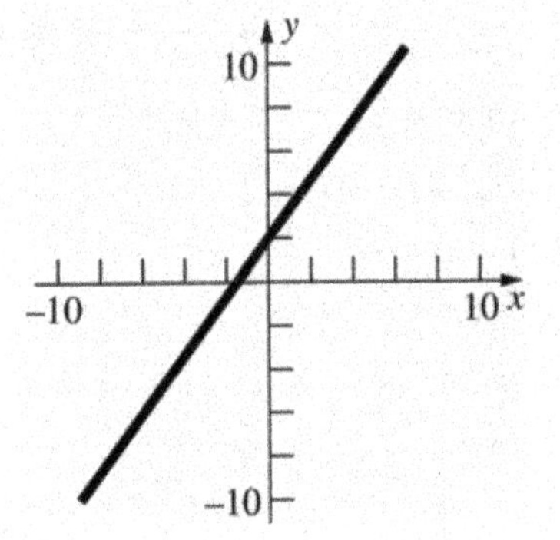

37.

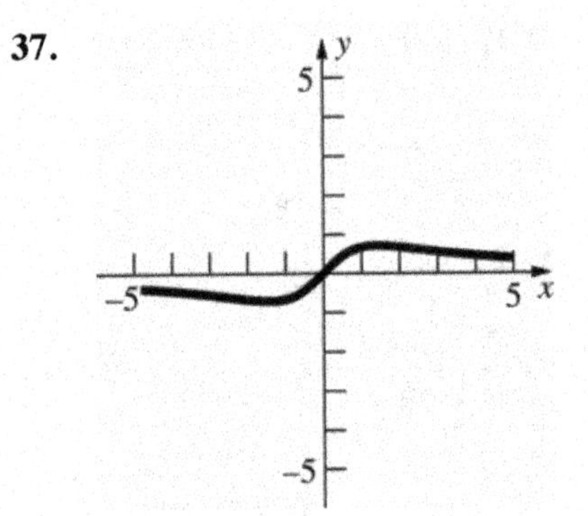

39. $y = x^2 - 2x + 4$ and $y - x = 4$;

$$x + 4 = x^2 - 2x + 4$$

$$x^2 - 3x = 0$$

$$x(x - 3) = 0$$

points of intersection: $(0, 4)$ and $(3, 7)$

41. a. $f(1) = \dfrac{1}{1+1} - \dfrac{1}{1} = -\dfrac{1}{2}$

b. $f\left(-\dfrac{1}{2}\right) = \dfrac{1}{-\frac{1}{2}+1} - \dfrac{1}{-\frac{1}{2}} = 4$

c. $f(-1)$ does not exist.

d. $f(t-1) = \dfrac{1}{t-1+1} - \dfrac{1}{t-1} = \dfrac{1}{t} - \dfrac{1}{t-1}$

e. $f\left(\dfrac{1}{t}\right) = \dfrac{1}{\frac{1}{t}+1} - \dfrac{1}{\frac{1}{t}} = \dfrac{t}{1+t} - t$

43. a. $\{x \in \mathbb{R} : x \neq -1, 1\}$

b. $\{x \in \mathbb{R} : |x| \leq 2\}$

45. a. $f(x) = x^2 - 1$

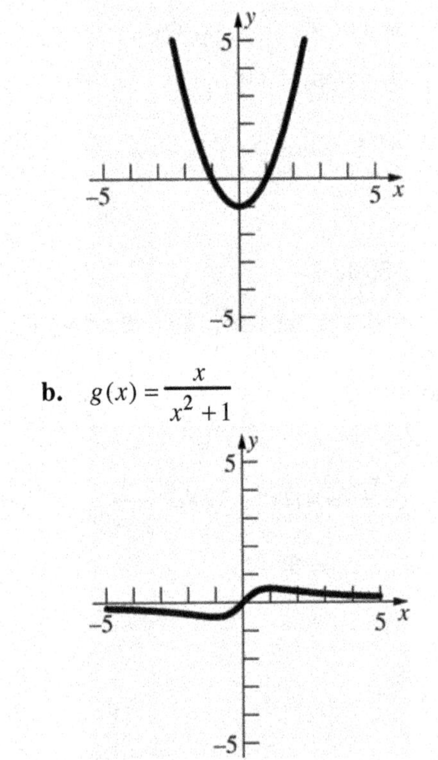

b. $g(x) = \dfrac{x}{x^2 + 1}$

c. $h(x) = \begin{cases} x^2 & \text{if } 0 \le x \le 2 \\ 6 - x & \text{if } x > 2 \end{cases}$

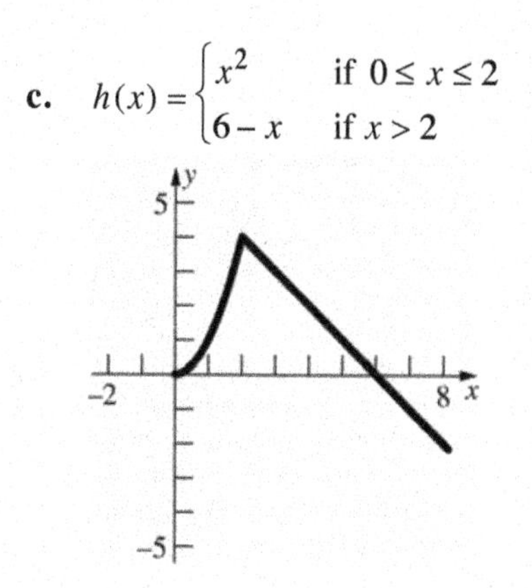

47. $V(x) = x(32 - 2x)(24 - 2x)$
Domain $[0, 12]$

49. a. $y = \frac{1}{4}x^2$

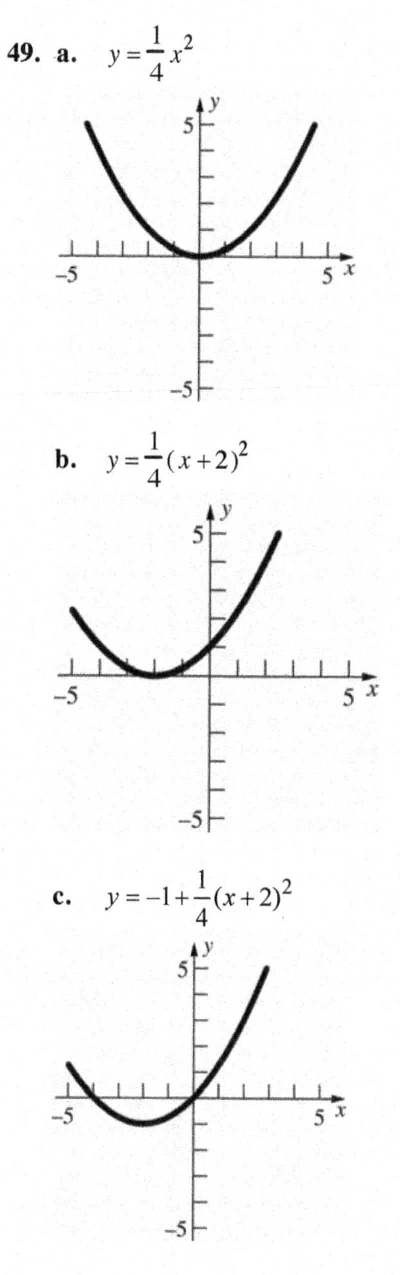

b. $y = \frac{1}{4}(x+2)^2$

c. $y = -1 + \frac{1}{4}(x+2)^2$

51. $f(x) = \sqrt{x}, g(x) = 1 + x, h(x) = x^2, k(x) = \sin x,$
$F(x) = \sqrt{1 + \sin^2 x} = f \circ g \circ h \circ k$

53. a. $\sin(-t) = -\sin t = -0.8$

b. $\sin^2 t + \cos^2 t = 1$
$\cos^2 t = 1 - (0.8)^2 = 0.36$
$\cos t = -0.6$

c. $\sin 2t = 2 \sin t \cos t = 2(0.8)(-0.6) = -0.96$

d. $\tan t = \frac{\sin t}{\cos t} = \frac{0.8}{-0.6} = -\frac{4}{3} \approx -1.333$

e. $\cos\left(\frac{\pi}{2} - t\right) = \sin t = 0.8$

f. $\sin(\pi + t) = -\sin t = -0.8$

55. $s = rt$
$= 9\left(20\frac{\text{rev}}{\text{min}}\right)\left(2\pi\frac{\text{rad}}{\text{rev}}\right)\left(\frac{1 \text{ min}}{60 \text{ sec}}\right)(1 \text{ sec}) = 6\pi$
≈ 18.85 in.

Review and Preview Problems

1. a) $0 < 2x < 4;\ 0 < x < 2$

b) $-6 < x < 16$

3. $x - 7 = 3$ or $x - 7 = -3$
$x = 10$ or $x = 4$

5. $x - 7 = 3$ or $x - 7 = -3$
$x = 10$ or $x = 4$

7. a) $x - 7 < 3$ and $x - 7 > -3$
$x < 10$ and $x > 4$
$4 < x < 10$

b) $x - 7 \le 3$ and $x - 7 \ge -3$
$x \le 10$ and $x \ge 4$
$4 \le x \le 10$

c) $x - 7 \le 1$ and $x - 7 \ge -1$
$x \le 8$ and $x \ge 6$
$6 \le x \le 8$

d) $x - 7 < 0.1$ and $x - 7 > -0.1$
$x < 7.1$ and $x > 6.9$
$6.9 < x < 7.1$

9. **a)** $x-1\neq 0;\ x\neq 1$

b) $2x^2-x-1\neq 0;\ x\neq 1,-0.5$

11. **a)**

$$f(0)=\frac{0-1}{0-1}=1$$
$$f(0.9)=\frac{0.81-1}{0.9-1}=1.9$$
$$f(0.99)=\frac{0.9801-1}{0.99-1}=1.99$$
$$f(0.999)=\frac{0.998001-1}{.999-1}=1.999$$
$$f(1.001)=\frac{1.002001-1}{1.001-1}=2.001$$
$$f(1.01)=\frac{1.0201-1}{1.01-1}=2.01$$
$$f(1.1)=\frac{1.21-1}{1.1-1}=2.1$$
$$f(2)=\frac{4-1}{2-1}=3$$

b)

$$g(0)=-1$$
$$g(0.9)=-0.0357143$$
$$g(0.99)=-0.0033557$$
$$g(0.999)=-0.000333556$$
$$g(1.001)=0.000333111$$
$$g(1.01)=0.00331126$$
$$g(1.1)=0.03125$$
$$g(2)=\frac{1}{5}$$

13. $x-5<0.1$ and $x-5>-0.1$

$x<5.1$ and $x>4.9$

$4.9<x<5.1$

15. **a.** True. **b.** False: Choose $a=0$.

c. True. **d.** True

CHAPTER 1

Limits

1.1 Concepts Review

1. L; c

3. L; right

Problem Set 1.1

1. $\lim_{x\to 3}(x-5) = -2$

3. $\lim_{x\to -2}(x^2+2x-1) = (-2)^2+2(-2)-1 = -1$

5. $\lim_{t\to -1}\left(t^2-1\right) = \left((-1)^2-1\right) = 0$

7. $\lim_{x\to 2}\frac{x^2-4}{x-2} = \lim_{x\to 2}\frac{(x-2)(x+2)}{x-2} = \lim_{x\to 2}(x+2)$
$= 2+2 = 4$

9. $\lim_{x\to -1}\frac{x^3-4x^2+x+6}{x+1} = \lim_{x\to -1}\frac{(x+1)(x^2-5x+6)}{x+1}$
$= \lim_{x\to -1}(x^2-5x+6) = (-1)^2-5(-1)+6 = 12$

11. $\lim_{x\to -t}\frac{x^2-t^2}{x+t} = \lim_{x\to -t}\frac{(x+t)(x-t)}{x+t} = \lim_{x\to -t}(x-t)$
$= -t-t = -2t$

13. $\lim_{t\to 2}\frac{\sqrt{(t+4)(t-2)^4}}{(3t-6)^2} = \lim_{t\to 2}\frac{(t-2)^2\sqrt{t+4}}{9(t-2)^2}$
$= \lim_{t\to 2}\frac{\sqrt{t+4}}{9} = \frac{\sqrt{2+4}}{9} = \frac{\sqrt{6}}{9}$

15. $\lim_{x\to 3}\frac{x^4-18x^2+81}{(x-3)^2} = \lim_{x\to 3}\frac{(x^2-9)^2}{(x-3)^2}$
$= \lim_{x\to 3}\frac{(x-3)^2(x+3)^2}{(x-3)^2} = \lim_{x\to 3}(x+3)^2 = (3+3)^2$
$= 36$

17. $\lim_{h\to 0}\frac{(2+h)^2-4}{h} = \lim_{h\to 0}\frac{4+4h+h^2-4}{h}$
$= \lim_{h\to 0}\frac{h^2+4h}{h} = \lim_{h\to 0}(h+4) = 4$

19.

x	$\frac{\sin x}{2x}$
1.	0.420735
0.1	0.499167
0.01	0.499992
0.001	0.49999992
–1.	0.420735
–0.1	0.499167
–0.01	0.499992
–0.001	0.49999992

$\lim_{x\to 0}\frac{\sin x}{2x} = 0.5$

21.

x	$(x-\sin x)^2/x^2$
1.	0.0251314
0.1	2.775×10^{-6}
0.01	2.77775×10^{-10}
0.001	2.77778×10^{-14}
–1.	0.0251314
–0.1	2.775×10^{-6}
–0.01	2.77775×10^{-10}
–0.001	2.77778×10^{-14}

$\lim_{x\to 0}\frac{(x-\sin x)^2}{x^2} = 0$

23.

t	$(t^2-1)/(\sin(t-1))$
2.	3.56519
1.1	2.1035
1.01	2.01003
1.001	2.001
0	1.1884
0.9	1.90317
0.99	1.99003
0.999	1.999

$\lim_{t\to 1}\frac{t^2-1}{\sin(t-1)} = 2$

25.

x	$(1+\sin(x-3\pi/2))/(x-\pi)$
$1.+\pi$	0.4597
$0.1+\pi$	0.0500
$0.01+\pi$	0.0050
$0.001+\pi$	0.0005
$-1.+\pi$	–0.4597
$-0.1+\pi$	–0.0500
$-0.01+\pi$	–0.0050
$-0.001+\pi$	–0.0005

$$\lim_{x\to\pi}\frac{1+\sin\left(x-\frac{3\pi}{2}\right)}{x-\pi}=0$$

27.

x	$(x-\pi/4)^2/(\tan x-1)^2$
$1.+\frac{\pi}{4}$	0.0320244
$0.1+\frac{\pi}{4}$	0.201002
$0.01+\frac{\pi}{4}$	0.245009
$0.001+\frac{\pi}{4}$	0.2495
$-1.+\frac{\pi}{4}$	0.674117
$-0.1+\frac{\pi}{4}$	0.300668
$-0.01+\frac{\pi}{4}$	0.255008
$-0.001+\frac{\pi}{4}$	0.2505

$$\lim_{x\to\frac{\pi}{4}}\frac{\left(x-\frac{\pi}{4}\right)^2}{(\tan x-1)^2}=0.25$$

29. a. $\lim_{x\to-3} f(x)=2$

b. $f(-3)=1$

c. $f(-1)$ does not exist.

d. $\lim_{x\to-1} f(x)=\frac{5}{2}$

e. $f(1)=2$

f. $\lim_{x\to1} f(x)$ does not exist.

g. $\lim_{x\to1^-} f(x)=2$

h. $\lim_{x\to1^+} f(x)=1$

i. $\lim_{x\to-1^+} f(x)=\frac{5}{2}$

31. a. $f(-3)=2$

b. $f(3)$ is undefined.

c. $\lim_{x\to-3^-} f(x)=2$

d. $\lim_{x\to-3^+} f(x)=4$

e. $\lim_{x\to-3} f(x)$ does not exist.

f. $\lim_{x\to3^+} f(x)$ does not exist.

33.

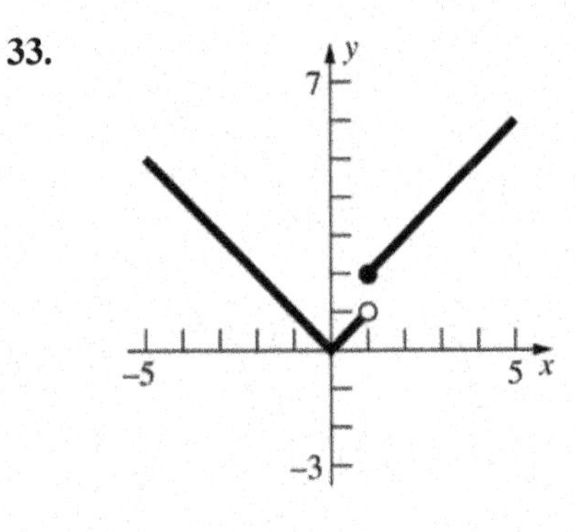

a. $\lim_{x\to0} f(x)=0$

b. $\lim_{x\to1} f(x)$ does not exist.

c. $f(1)=2$

d. $\lim_{x\to1^+} f(x)=2$

35. $f(x)=x-[[x]]$

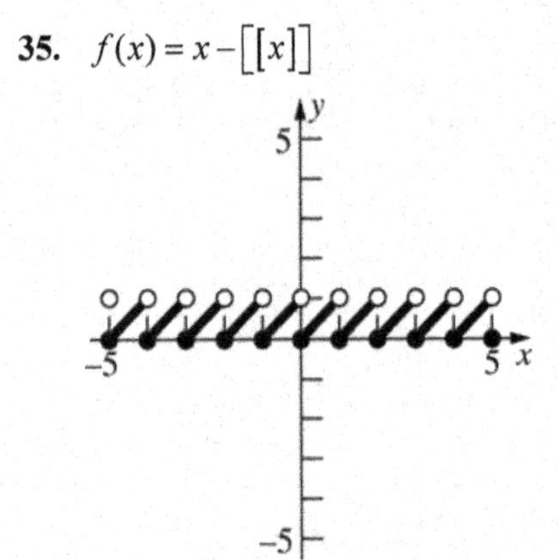

a. $f(0)=0$

b. $\lim_{x\to0} f(x)$ does not exist.

c. $\lim_{x\to0^-} f(x)=1$

d. $\lim_{x\to\frac{1}{2}} f(x)=\frac{1}{2}$

37. $\lim_{x\to 1}\frac{x^2-1}{|x-1|}$ does not exist.

$\lim_{x\to 1^-}\frac{x^2-1}{|x-1|}=-2$ and $\lim_{x\to 1^+}\frac{x^2-1}{|x-1|}=2$

39. a. $\lim_{x\to 1} f(x)$ does not exist.

b. $\lim_{x\to 0} f(x)=0$

41. $\lim_{x\to a} f(x)$ exists for $a=-1, 0, 1$.

43. a. $\lim_{x\to 1}\frac{|x-1|}{x-1}$ does not exist.

$\lim_{x\to 1^-}\frac{|x-1|}{x-1}=-1$ and $\lim_{x\to 1^+}\frac{|x-1|}{x-1}=1$

b. $\lim_{x\to 1^-}\frac{|x-1|}{x-1}=-1$

c. $\lim_{x\to 1^-}\frac{x^2-|x-1|-1}{|x-1|}=-3$

d. $\lim_{x\to 1^-}\left[\frac{1}{x-1}-\frac{1}{|x-1|}\right]$ does not exist.

45. a) 1 b) 0

c) −1 d) −1

47. $\lim_{x\to 0}\sqrt{x}$ does not exist since $\sqrt{x}$ is not defined for $x<0$.

49. $\lim_{x\to 0}\sqrt{|x|}=0$

51. $\lim_{x\to 0}\frac{\sin 2x}{4x}=\frac{1}{2}$

53. $\lim_{x\to 0}\cos\left(\frac{1}{x}\right)$ does not exist.

55. $\lim_{x\to 1}\frac{x^3-1}{\sqrt{2x+2}-2}=6$

57. $\lim_{x\to 2^-}\frac{x^2-x-2}{|x-2|}=-3$

59. $\lim_{x\to 0}\sqrt{x}$; The computer gives a value of 0, but $\lim_{x\to 0^-}\sqrt{x}$ does not exist.

1.2 Concepts Review

1. $L-\varepsilon;\ L+\varepsilon$

3. $\frac{\varepsilon}{3}$

Problem Set 1.2

1. $0<|t-a|<\delta \Rightarrow |f(t)-M|<\varepsilon$

3. $0<|z-d|<\delta \Rightarrow |h(z)-P|<\varepsilon$

5. $0<c-x<\delta \Rightarrow |f(x)-L|<\varepsilon$

7. If x is within 0.001 of 2, then $2x$ *is* within 0.002 of 4.

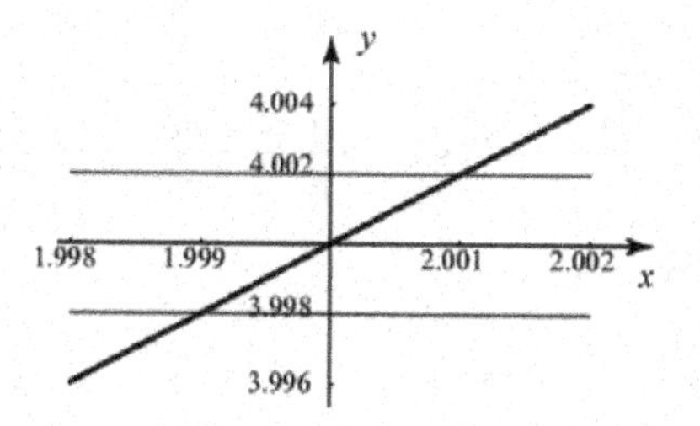

9. If x is within 0.0019 of 2, then $\sqrt{8x}$ is within 0.002 of 4.

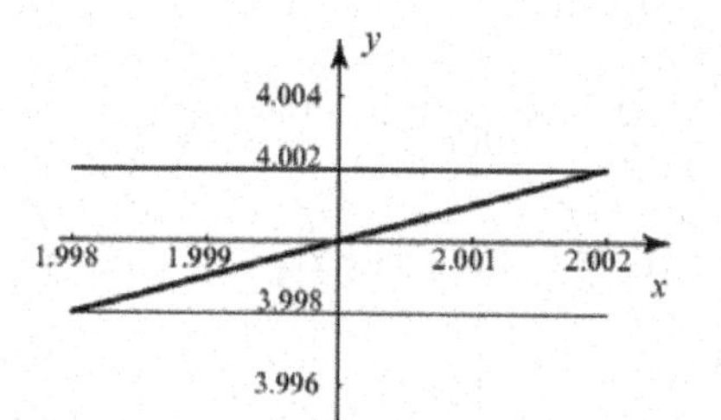

11. $0<|x-0|<\delta \Rightarrow |(2x-1)-(-1)|<\varepsilon$

$|2x-1+1|<\varepsilon \Leftrightarrow |2x|<\varepsilon$

$\Leftrightarrow 2|x|<\varepsilon$

$\Leftrightarrow |x|<\frac{\varepsilon}{2}$

$\delta=\frac{\varepsilon}{2};\ 0<|x-0|<\delta$

$|(2x-1)-(-1)|=|2x|=2|x|<2\delta=\varepsilon$

13. $0<|x-5|<\delta \Rightarrow \left|\frac{x^2-25}{x-5}-10\right|<\varepsilon$

$$\left|\frac{x^2-25}{x-5}-10\right|<\varepsilon \Leftrightarrow \left|\frac{(x-5)(x+5)}{x-5}-10\right|<\varepsilon$$

$$\Leftrightarrow |x+5-10|<\varepsilon$$

$$\Leftrightarrow |x-5|<\varepsilon$$

$\delta=\varepsilon; 0<|x-5|<\delta$

$$\left|\frac{x^2-25}{x-5}-10\right|=\left|\frac{(x-5)(x+5)}{x-5}-10\right|=|x+5-10|$$

$$=|x-5|<\delta=\varepsilon$$

15. $0<|x-5|<\delta \Rightarrow \left|\frac{2x^2-11x+5}{x-5}-9\right|<\varepsilon$

$$\left|\frac{2x^2-11x+5}{x-5}-9\right|<\varepsilon \Leftrightarrow \left|\frac{(2x-1)(x-5)}{x-5}-9\right|<\varepsilon$$

$$\Leftrightarrow |2x-1-9|<\varepsilon$$

$$\Leftrightarrow |2(x-5)|<\varepsilon$$

$$\Leftrightarrow |x-5|<\frac{\varepsilon}{2}$$

$\delta=\frac{\varepsilon}{2}; 0<|x-5|<\delta$

$$\left|\frac{2x^2-11x+5}{x-5}-9\right|=\left|\frac{(2x-1)(x-5)}{x-5}-9\right|$$

$$=|2x-1-9|=|2(x-5)|=2|x-5|<2\delta=\varepsilon$$

17. $0<|x-4|<\delta \Rightarrow \left|\frac{\sqrt{2x-1}}{\sqrt{x-3}}-\sqrt{7}\right|<\varepsilon$

$$\left|\frac{\sqrt{2x-1}}{\sqrt{x-3}}-\sqrt{7}\right|<\varepsilon \Leftrightarrow \left|\frac{\sqrt{2x-1}-\sqrt{7(x-3)}}{\sqrt{x-3}}\right|<\varepsilon$$

$$\Leftrightarrow \left|\frac{(\sqrt{2x-1}-\sqrt{7(x-3)})(\sqrt{2x-1}+\sqrt{7(x-3)})}{\sqrt{x-3}(\sqrt{2x-1}+\sqrt{7(x-3)})}\right|<\varepsilon$$

$$\Leftrightarrow \left|\frac{2x-1-(7x-21)}{\sqrt{x-3}(\sqrt{2x-1}+\sqrt{7(x-3)})}\right|<\varepsilon$$

$$\Leftrightarrow \left|\frac{-5(x-4)}{\sqrt{x-3}(\sqrt{2x-1}+\sqrt{7(x-3)})}\right|<\varepsilon$$

$$\Leftrightarrow |x-4|\cdot\frac{5}{\sqrt{x-3}(\sqrt{2x-1}+\sqrt{7(x-3)})}<\varepsilon$$

To bound $\frac{5}{\sqrt{x-3}(\sqrt{2x-1}+\sqrt{7(x-3)})}$, agree that $\delta \le \frac{1}{2}$. If $\delta \le \frac{1}{2}$, then $\frac{7}{2}<x<\frac{9}{2}$, so

$$0.65<\frac{5}{\sqrt{x-3}(\sqrt{2x-1}+\sqrt{7(x-3)})}<1.65 \text{ and}$$

hence $|x-4|\cdot\frac{5}{\sqrt{x-3}(\sqrt{2x-1}+\sqrt{7(x-3)})}<\varepsilon$

$$\Leftrightarrow |x-4|<\frac{\varepsilon}{1.65}$$

For whatever ε is chosen, let δ be the smaller of $\frac{1}{2}$ and $\frac{\varepsilon}{1.65}$.

$\delta=\min\left\{\frac{1}{2}, \frac{\varepsilon}{1.65}\right\},\ 0<|x-4|<\delta$

$$\left|\frac{\sqrt{2x-1}}{\sqrt{x-3}}-\sqrt{7}\right|=|x-4|\cdot\frac{5}{\sqrt{x-3}(\sqrt{2x-1}+\sqrt{7(x-3)})}$$

$$<|x-4|(1.65)<1.65\delta\le\varepsilon$$

since $\delta=\frac{1}{2}$ only when $\frac{1}{2}\le\frac{\varepsilon}{1.65}$ so $1.65\delta\le\varepsilon$.

19. $0<|x-1|<\delta \Rightarrow \left|\frac{10x^3-26x^2+22x-6}{(x-1)^2}-4\right|<\varepsilon$

$$\left|\frac{10x^3-26x^2+22x-6}{(x-1)^2}-4\right|<\varepsilon$$

$$\Leftrightarrow \left|\frac{(10x-6)(x-1)^2}{(x-1)^2}-4\right|<\varepsilon$$

$$\Leftrightarrow |10x-6-4|<\varepsilon$$

$$\Leftrightarrow |10(x-1)|<\varepsilon$$

$$\Leftrightarrow 10|x-1|<\varepsilon$$

$$\Leftrightarrow |x-1|<\frac{\varepsilon}{10}$$

$\delta=\frac{\varepsilon}{10}; 0<|x-1|<\delta$

$$\left|\frac{10x^3-26x^2+22x-6}{(x-1)^2}-4\right|=\left|\frac{(10x-6)(x-1)^2}{(x-1)^2}-4\right|$$

$$=|10x-6-4|=|10(x-1)|$$

$$=10|x-1|<10\delta=\varepsilon$$

21. $0 < |x+1| < \delta \Rightarrow |(x^2 - 2x - 1) - 2| < \varepsilon$

$|x^2 - 2x - 1 - 2| = |x^2 - 2x - 3| = |x+1||x-3|$

To bound $|x-3|$, agree that $\delta \le 1$.

$|x+1| < \delta$ implies

$|x-3| = |x+1-4| \le |x+1| + |-4| < 1 + 4 = 5$

$\delta \le \frac{\varepsilon}{5}; \delta = \min\left\{1, \frac{\varepsilon}{5}\right\}; 0 < |x+1| < \delta$

$|(x^2 - 2x - 1) - 2| = |x^2 - 2x - 3|$

$= |x+1||x-3| < 5 \cdot \frac{\varepsilon}{5} = \varepsilon$

23. Choose $\varepsilon > 0$. Then since $\lim_{x \to c} f(x) = L$, there is some $\delta_1 > 0$ such that

$0 < |x - c| < \delta_1 \Rightarrow |f(x) - L| < \varepsilon$.

Since $\lim_{x \to c} f(x) = M$, there is some $\delta_2 > 0$ such that $0 < |x - c| < \delta_2 \Rightarrow |f(x) - M| < \varepsilon$.

Let $\delta = \min\{\delta_1, \delta_2\}$ and choose x_0 such that $0 < |x_0 - c| < \delta$.

Thus, $|f(x_0) - L| < \varepsilon \Rightarrow -\varepsilon < f(x_0) - L < \varepsilon$

$\Rightarrow -f(x_0) - \varepsilon < -L < -f(x_0) + \varepsilon$

$\Rightarrow f(x_0) - \varepsilon < L < f(x_0) + \varepsilon$.

Similarly,

$f(x_0) - \varepsilon < M < f(x_0) + \varepsilon$.

Thus,

$-2\varepsilon < L - M < 2\varepsilon$. As $\varepsilon \Rightarrow 0$, $L - M \to 0$, so $L = M$.

25. For all $x \ne 0$, $0 \le \sin^2\left(\frac{1}{x}\right) \le 1$ so

$x^4 \sin^2\left(\frac{1}{x}\right) \le x^4$ for all $x \ne 0$. By Problem 18,

$\lim_{x \to 0} x^4 = 0$, so, by Problem 20,

$\lim_{x \to 0} x^4 \sin^2\left(\frac{1}{x}\right) = 0$.

27. $\lim_{x \to 0^+} |x| : 0 < x < \delta \Rightarrow ||x| - 0| < \varepsilon$

For $x \ge 0$, $|x| = x$.

$\delta = \varepsilon; 0 < x < \delta \Rightarrow ||x| - 0| = |x| = x < \delta = \varepsilon$

Thus, $\lim_{x \to 0^+} |x| = 0$.

$\lim_{x \to 0^-} |x| : 0 < 0 - x < \delta \Rightarrow ||x| - 0| < \varepsilon$

For $x < 0$, $|x| = -x$; note also that $||x|| = |x|$ since $|x| \ge 0$.

$\delta = \varepsilon; 0 < -x < \delta \Rightarrow ||x|| = |x| = -x < \delta = \varepsilon$

Thus, $\lim_{x \to 0^-} |x| = 0$,

since $\lim_{x \to 0^+} |x| = \lim_{x \to 0^-} |x| = 0$, $\lim_{x \to 0} |x| = 0$.

29. Choose $\varepsilon > 0$. Since $\lim_{x \to a} f(x) = L$, there is a $\delta > 0$ such that for $0 < |x - a| < \delta$, $|f(x) - L| < \varepsilon$.

That is, for

$a - \delta < x < a$ or $a < x < a + \delta$,

$L - \varepsilon < f(x) < L + \varepsilon$.

Let $f(a) = A$,

$M = \max\{|L - \varepsilon|, |L + \varepsilon|, |A|\}$, $c = a - \delta$,

$d = a + \delta$. Then for x in (c, d), $|f(x)| \le M$, since either $x = a$, in which case

$|f(x)| = |f(a)| = |A| \le M$ or $0 < |x - a| < \delta$ so

$L - \varepsilon < f(x) < L + \varepsilon$ and $|f(x)| < M$.

31. (b) and (c) are equivalent to the definition of limit.

33. **a.** $g(x) = \dfrac{x^3 - x^2 - 2x - 4}{x^4 - 4x^3 + x^2 + x + 6}$

b. No, because $\dfrac{x+6}{x^4 - 4x^3 + x^2 + x + 6} + 1$ has an asymptote at $x \approx 3.49$.

c. If $\delta \le \frac{1}{4}$, then $2.75 < x < 3$ or $3 < x < 3.25$ and by graphing

$y = |g(x)| = \left|\dfrac{x^3 - x^2 - 2x - 4}{x^4 - 4x^3 + x^2 + x + 6}\right|$

on the interval $[2.75, 3.25]$, we see that

$0 < \left|\dfrac{x^3 - x^2 - 2x - 4}{x^4 - 4x^3 + x^2 + x + 6}\right| < 3$

so m must be at least three.

1.3 Concepts Review

1. 48

3. – 8; – 4 + 5c

Problem Set 1.3

1. $\lim_{x\to 1}(2x+1)$ 4

$= \lim_{x\to 1} 2x + \lim_{x\to 1} 1$ 3

$= 2\lim_{x\to 1} x + \lim_{x\to 1} 1$ 2,1

$= 2(1) + 1 = 3$

3. $\lim_{x\to 0}[(2x+1)(x-3)]$ 6

$= \lim_{x\to 0}(2x+1)\cdot\lim_{x\to 0}(x-3)$ 4, 5

$= \left(\lim_{x\to 0} 2x + \lim_{x\to 0} 1\right)\cdot\left(\lim_{x\to 0} x - \lim_{x\to 0} 3\right)$ 3

$= \left(2\lim_{x\to 0} x + \lim_{x\to 0} 1\right)\cdot\left(\lim_{x\to 0} x - \lim_{x\to 0} 3\right)$ 2, 1

$= [2(0)+1](0-3) = -3$

5. $\lim_{x\to 2}\frac{2x+1}{5-3x}$ 7

$= \frac{\lim_{x\to 2}(2x+1)}{\lim_{x\to 2}(5-3x)}$ 4, 5

$= \frac{\lim_{x\to 2} 2x + \lim_{x\to 2} 1}{\lim_{x\to 2} 5 - \lim_{x\to 2} 3x}$ 3, 1

$= \frac{2\lim_{x\to 2} x + 1}{5 - 3\lim_{x\to 2} x}$ 2

$= \frac{2(2)+1}{5-3(2)} = -5$

7. $\lim_{x\to 3}\sqrt{3x-5}$ 9

$= \sqrt{\lim_{x\to 3}(3x-5)}$ 5, 3

$= \sqrt{3\lim_{x\to 3} x - \lim_{x\to 3} 5}$ 2, 1

$= \sqrt{3(3)-5} = 2$

9. $\lim_{t\to -2}(2t^3+15)^{13}$ 8

$= \left[\lim_{t\to -2}(2t^3+15)\right]^{13}$ 4, 3

$= \left[2\lim_{t\to -2} t^3 + \lim_{t\to -2} 15\right]^{13}$ 8

$= \left[2\left(\lim_{t\to -2} t\right)^3 + \lim_{t\to -2} 15\right]^{13}$ 2, 1

$= [2(-2)^3 + 15]^{13} = -1$

11. $\lim_{y\to 2}\left(\frac{4y^3+8y}{y+4}\right)^{1/3}$ 9

$= \left(\lim_{y\to 2}\frac{4y^3+8y}{y+4}\right)^{1/3}$ 7

$= \left[\frac{\lim_{y\to 2}(4y^3+8y)}{\lim_{y\to 2}(y+4)}\right]^{1/3}$ 4, 3

$= \left(\frac{4\lim_{y\to 2} y^3 + 8\lim_{y\to 2} y}{\lim_{y\to 2} y + \lim_{y\to 2} 4}\right)^{1/3}$ 8, 1

$= \left[\frac{4\left(\lim_{y\to 2} y\right)^3 + 8\lim_{y\to 2} y}{\lim_{y\to 2} y + 4}\right]^{1/3}$ 2

$= \left[\frac{4(2)^3 + 8(2)}{2+4}\right]^{1/3} = 2$

13. $\lim_{x\to 2}\frac{x^2-4}{x^2+4} = \frac{\lim_{x\to 2}(x^2-4)}{\lim_{x\to 2}(x^2+4)} = \frac{4-4}{4+4} = 0$

15. $\lim_{x\to -1}\frac{x^2-2x-3}{x+1} = \lim_{x\to -1}\frac{(x-3)(x+1)}{(x+1)}$

$= \lim_{x\to -1}(x-3) = -4$

17. $\lim_{x\to -1}\frac{(x-1)(x-2)(x-3)}{(x-1)(x-2)(x+7)} = \lim_{x\to -1}\frac{x-3}{x+7}$

$= \frac{-1-3}{-1+7} = -\frac{2}{3}$

19. $\lim_{x\to1}\frac{x^2+x-2}{x^2-1}=\lim_{x\to1}\frac{(x+2)(x-1)}{(x+1)(x-1)}$

$=\lim_{x\to1}\frac{x+2}{x+1}=\frac{1+2}{1+1}=\frac{3}{2}$

21. $\lim_{u\to-2}\frac{u^2-ux+2u-2x}{u^2-u-6}=\lim_{u\to-2}\frac{(u+2)(u-x)}{(u+2)(u-3)}$

$=\lim_{u\to-2}\frac{u-x}{u-3}=\frac{x+2}{5}$

23. $\lim_{x\to\pi}\frac{2x^2-6x\pi+4\pi^2}{x^2-\pi^2}=\lim_{x\to\pi}\frac{2(x-\pi)(x-2\pi)}{(x-\pi)(x+\pi)}$

$=\lim_{x\to\pi}\frac{2(x-2\pi)}{x+\pi}=\frac{2(\pi-2\pi)}{\pi+\pi}=-1$

25. $\lim_{x\to a}\sqrt{f^2(x)+g^2(x)}$

$=\sqrt{\lim_{x\to a}f^2(x)+\lim_{x\to a}g^2(x)}$

$=\sqrt{\left(\lim_{x\to a}f(x)\right)^2+\left(\lim_{x\to a}g(x)\right)^2}$

$=\sqrt{(3)^2+(-1)^2}=\sqrt{10}$

27. $\lim_{x\to a}\sqrt[3]{g(x)}[f(x)+3]=\lim_{x\to a}\sqrt[3]{g(x)}\cdot\lim_{x\to a}[f(x)+3]$

$=\sqrt[3]{\lim_{x\to a}g(x)}\cdot\left[\lim_{x\to a}f(x)+\lim_{x\to a}3\right]=\sqrt[3]{-1}\cdot(3+3)$

$=-6$

29. $\lim_{t\to a}\left[|f(t)|+|3g(t)|\right]=\lim_{t\to a}|f(t)|+3\lim_{t\to a}|g(t)|$

$=\left|\lim_{t\to a}f(t)\right|+3\left|\lim_{t\to a}g(t)\right|$

$=|3|+3|-1|=6$

31. $\lim_{x\to2}\frac{3x^2-12}{x-2}=\lim_{x\to2}\frac{3(x-2)(x+2)}{x-2}$

$=3\lim_{x\to2}(x+2)=3(2+2)=12$

33. $\lim_{x\to2}\frac{\frac{1}{x}-\frac{1}{2}}{x-2}=\lim_{x\to2}\frac{\frac{2-x}{2x}}{x-2}=\lim_{x\to2}\frac{-\frac{x-2}{2x}}{x-2}$

$=\lim_{x\to2}-\frac{1}{2x}=\frac{-1}{2\lim_{x\to2}x}=\frac{-1}{2(2)}=-\frac{1}{4}$

35. Suppose $\lim_{x\to c}f(x)=L$ and $\lim_{x\to c}g(x)=M$.

$|f(x)g(x)-LM|\le|g(x)||f(x)-L|+|L||g(x)-M|$ as shown in the text. Choose $\varepsilon_1=1$. Since $\lim_{x\to c}g(x)=M$, there is some $\delta_1>0$ such that if $0<|x-c|<\delta_1$, $|g(x)-M|<\varepsilon_1=1$ or $M-1<g(x)<M+1$

$|M-1|\le|M|+1$ and $|M+1|\le|M|+1$ so for $0<|x-c|<\delta_1, |g(x)|<|M|+1$. Choose $\varepsilon>0$. Since $\lim_{x\to c}f(x)=L$ and $\lim_{x\to c}g(x)=M$, there exist δ_2 and δ_3 such that $0<|x-c|<\delta_2\Rightarrow$ $|f(x)-L|<\frac{\varepsilon}{|L|+|M|+1}$ and $0<|x-c|<\delta_3\Rightarrow$ $|g(x)-M|<\frac{\varepsilon}{|L|+|M|+1}$. Let $\delta=\min\{\delta_1,\delta_2,\delta_3\}$, then $0<|x-c|<\delta\Rightarrow$

$|f(x)g(x)-LM|\le|g(x)||f(x)-L|+|L||g(x)-M|$

$<(|M|+1)\frac{\varepsilon}{|L|+|M|+1}+|L|\frac{\varepsilon}{|L|+|M|+1}=\varepsilon$

Hence,

$\lim_{x\to c}f(x)g(x)=LM=\left(\lim_{x\to c}f(x)\right)\left(\lim_{x\to c}g(x)\right)$

37. $\lim_{x\to c}f(x)=L\Leftrightarrow\lim_{x\to c}f(x)=\lim_{x\to c}L$

$\Leftrightarrow\lim_{x\to c}f(x)-\lim_{x\to c}L=0$

$\Leftrightarrow\lim_{x\to c}[f(x)-L]=0$

39. $\lim_{x\to c}|x|=\sqrt{\left(\lim_{x\to c}|x|\right)^2}=\sqrt{\lim_{x\to c}|x|^2}=\sqrt{\lim_{x\to c}x^2}$

$=\sqrt{\left(\lim_{x\to c}x\right)^2}=\sqrt{c^2}=|c|$

41. $\lim_{x\to-3^+}\frac{\sqrt{3+x}}{x}=\frac{\sqrt{3-3}}{-3}=0$

43. $\lim_{x\to3^+}\frac{x-3}{\sqrt{x^2-9}}=\lim_{x\to3^+}\frac{(x-3)\sqrt{x^2-9}}{x^2-9}$

$=\lim_{x\to3^+}\frac{(x-3)\sqrt{x^2-9}}{(x-3)(x+3)}=\lim_{x\to3^+}\frac{\sqrt{x^2-9}}{x+3}$

$=\frac{\sqrt{3^2-9}}{3+3}=0$

45. $\lim_{x\to 2^+} \frac{(x^2+1)[\![x]\!]}{(3x-1)^2} = \frac{(2^2+1)[\![2]\!]}{(3\cdot 2-1)^2} = \frac{5\cdot 2}{5^2} = \frac{2}{5}$

47. $\lim_{x\to 0^-} \frac{x}{|x|} = -1$

49. $f(x)g(x) = 1;\ g(x) = \frac{1}{f(x)}$

$\lim_{x\to a} g(x) = 0 \Leftrightarrow \lim_{x\to a} \frac{1}{f(x)} = 0$

$\Leftrightarrow \frac{1}{\lim_{x\to a} f(x)} = 0$

No value satisfies this equation, so $\lim_{x\to a} f(x)$ must not exist.

51. a. $NO = \sqrt{(0-0)^2+(1-0)^2} = 1$

$OP = \sqrt{(x-0)^2+(y-0)^2} = \sqrt{x^2+y^2}$

$= \sqrt{x^2+x}$

$NP = \sqrt{(x-0)^2+(y-1)^2} = \sqrt{x^2+y^2-2y+1}$

$= \sqrt{x^2+x-2\sqrt{x}+1}$

$MO = \sqrt{(1-0)^2+(0-0)^2} = 1$

$MP = \sqrt{(x-1)^2+(y-0)^2} = \sqrt{y^2+x^2-2x+1}$

$= \sqrt{x^2-x+1}$

$\lim_{x\to 0^+} \frac{\text{perimeter of } \Delta NOP}{\text{perimeter of } \Delta MOP}$

$= \lim_{x\to 0^+} \frac{1+\sqrt{x^2+x}+\sqrt{x^2+x-2\sqrt{x}+1}}{1+\sqrt{x^2+x}+\sqrt{x^2-x+1}}$

$= \frac{1+\sqrt{1}}{1+\sqrt{1}} = 1$

b. Area of $\Delta NOP = \frac{1}{2}(1)(x) = \frac{x}{2}$

Area of $\Delta MOP = \frac{1}{2}(1)(y) = \frac{\sqrt{x}}{2}$

$\lim_{x\to 0^+} \frac{\text{area of } \Delta NOP}{\text{area of } \Delta MOP} = \lim_{x\to 0^+} \frac{\frac{x}{2}}{\frac{\sqrt{x}}{2}} = \lim_{x\to 0^+} \frac{x}{\sqrt{x}}$

$= \lim_{x\to 0^+} \sqrt{x} = 0$

1.4 Concepts Review

1. 0

3. the denominator is 0 when $t = 0$.

Problem Set 1.4

1. $\lim_{x\to 0} \frac{\cos x}{x+1} = \frac{1}{1} = 1$

3. $\lim_{t\to 0} \frac{\cos^2 t}{1+\sin t} = \frac{\cos^2 0}{1+\sin 0} = \frac{1}{1+0} = 1$

5. $\lim_{x\to 0} \frac{\sin x}{2x} = \frac{1}{2}\lim_{x\to 0} \frac{\sin x}{x} = \frac{1}{2}\cdot 1 = \frac{1}{2}$

7. $\lim_{\theta\to 0} \frac{\sin 3\theta}{\tan\theta} = \lim_{\theta\to 0} \frac{\sin 3\theta}{\frac{\sin\theta}{\cos\theta}} = \lim_{\theta\to 0} \frac{\cos\theta\sin 3\theta}{\sin\theta}$

$= \lim_{\theta\to 0}\left[\cos\theta\cdot 3\cdot\frac{\sin 3\theta}{3\theta}\cdot\frac{1}{\frac{\sin\theta}{\theta}}\right]$

$= 3\lim_{\theta\to 0}\left[\cos\theta\cdot\frac{\sin 3\theta}{3\theta}\cdot\frac{1}{\frac{\sin\theta}{\theta}}\right] = 3\cdot 1\cdot 1\cdot 1 = 3$

9. $\lim_{\theta\to 0} \frac{\cot\pi\theta\sin\theta}{2\sec\theta} = \lim_{\theta\to 0} \frac{\frac{\cos\pi\theta}{\sin\pi\theta}\sin\theta}{\frac{2}{\cos\theta}}$

$= \lim_{\theta\to 0} \frac{\cos\pi\theta\sin\theta\cos\theta}{2\sin\pi\theta}$

$= \lim_{\theta\to 0}\left[\frac{\cos\pi\theta\cos\theta}{2}\cdot\frac{\sin\theta}{\theta}\cdot\frac{1}{\pi}\cdot\frac{\pi\theta}{\sin\pi\theta}\right]$

$= \frac{1}{2\pi}\lim_{\theta\to 0}\left[\cos\pi\theta\cos\theta\cdot\frac{\sin\theta}{\theta}\cdot\frac{\pi\theta}{\sin\pi\theta}\right]$

$= \frac{1}{2\pi}\cdot 1\cdot 1\cdot 1\cdot 1 = \frac{1}{2\pi}$

11. $\lim_{t\to 0} \frac{\tan^2 3t}{2t} = \lim_{t\to 0} \frac{\sin^2 3t}{(2t)(\cos^2 3t)}$

$= \lim_{t\to 0} \frac{3(\sin 3t)}{2\cos^2 3t}\cdot\frac{\sin 3t}{3t} = 0\cdot 1 = 0$

13. $\lim_{t\to 0} \frac{\sin(3t)+4t}{t\sec t} = \lim_{t\to 0}\left(\frac{\sin 3t}{t\sec t}+\frac{4t}{t\sec t}\right)$

$= \lim_{t\to 0} \frac{\sin 3t}{t\sec t} + \lim_{t\to 0} \frac{4t}{t\sec t}$

$= \lim_{t\to 0} 3\cos t\cdot\frac{\sin 3t}{3t} + \lim_{t\to 0} 4\cos t$

$= 3\cdot 1 + 4 = 7$

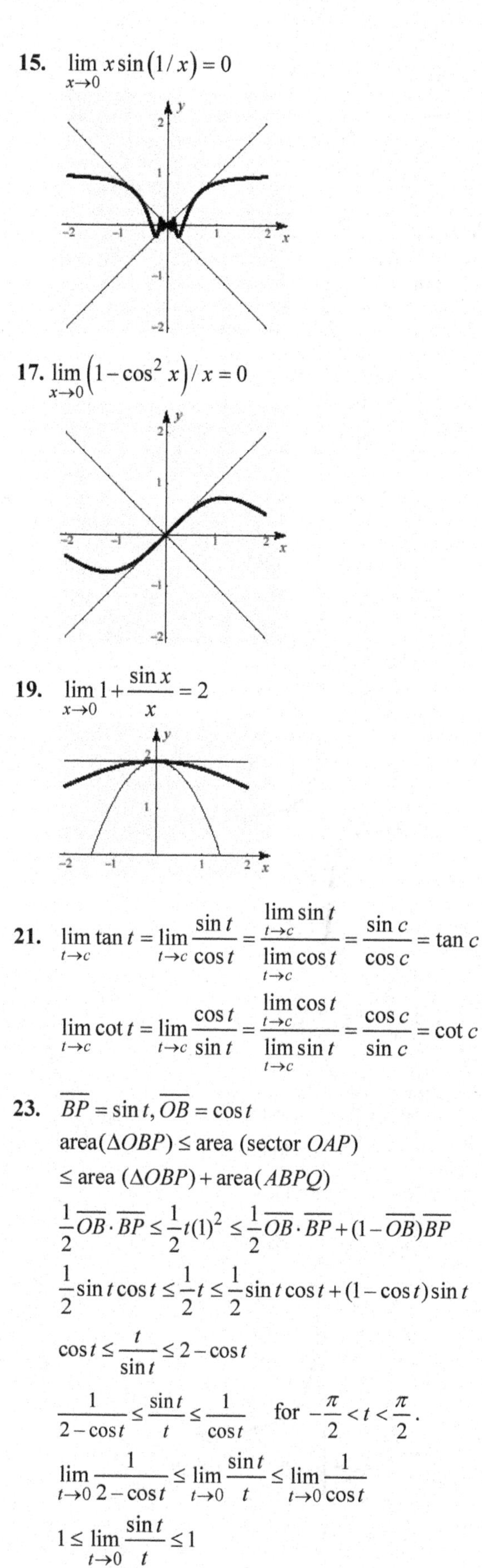

15. $\lim_{x\to 0} x\sin(1/x) = 0$

17. $\lim_{x\to 0}\left(1-\cos^2 x\right)/x = 0$

19. $\lim_{x\to 0} 1+\dfrac{\sin x}{x} = 2$

21. $\lim_{t\to c}\tan t = \lim_{t\to c}\dfrac{\sin t}{\cos t} = \dfrac{\lim_{t\to c}\sin t}{\lim_{t\to c}\cos t} = \dfrac{\sin c}{\cos c} = \tan c$

$\lim_{t\to c}\cot t = \lim_{t\to c}\dfrac{\cos t}{\sin t} = \dfrac{\lim_{t\to c}\cos t}{\lim_{t\to c}\sin t} = \dfrac{\cos c}{\sin c} = \cot c$

23. $\overline{BP} = \sin t, \overline{OB} = \cos t$

area$(\Delta OBP) \le$ area (sector OAP)

$\le$ area $(\Delta OBP) +$ area$(ABPQ)$

$\dfrac{1}{2}\overline{OB}\cdot\overline{BP} \le \dfrac{1}{2}t(1)^2 \le \dfrac{1}{2}\overline{OB}\cdot\overline{BP} + (1-\overline{OB})\overline{BP}$

$\dfrac{1}{2}\sin t\cos t \le \dfrac{1}{2}t \le \dfrac{1}{2}\sin t\cos t + (1-\cos t)\sin t$

$\cos t \le \dfrac{t}{\sin t} \le 2-\cos t$

$\dfrac{1}{2-\cos t} \le \dfrac{\sin t}{t} \le \dfrac{1}{\cos t}$ for $-\dfrac{\pi}{2} < t < \dfrac{\pi}{2}$.

$\lim_{t\to 0}\dfrac{1}{2-\cos t} \le \lim_{t\to 0}\dfrac{\sin t}{t} \le \lim_{t\to 0}\dfrac{1}{\cos t}$

$1 \le \lim_{t\to 0}\dfrac{\sin t}{t} \le 1$

Thus, $\lim_{t\to 0}\dfrac{\sin t}{t} = 1$.

1.5 Concepts Review

1. x increases without bound; $f(x)$ gets close to L as x increases without bound

3. $y = 6$; horizontal

Problem Set 1.5

1. $\lim_{x\to\infty}\dfrac{x}{x-5} = \lim_{x\to\infty}\dfrac{1}{1-\frac{5}{x}} = 1$

3. $\lim_{t\to-\infty}\dfrac{t^2}{7-t^2} = \lim_{t\to-\infty}\dfrac{1}{\frac{7}{t^2}-1} = -1$

5. $\lim_{x\to\infty}\dfrac{x^2}{(x-5)(3-x)} = \lim_{x\to\infty}\dfrac{x^2}{-x^2+8x-15}$

$= \lim_{x\to\infty}\dfrac{1}{-1+\frac{8}{x}-\frac{15}{x^2}} = -1$

7. $\lim_{x\to\infty}\dfrac{x^3}{2x^3-100x^2} = \lim_{x\to\infty}\dfrac{1}{2-\frac{100}{x}} = \dfrac{1}{2}$

9. $\lim_{x\to\infty}\dfrac{3x^3-x^2}{\pi x^3-5x^2} = \lim_{x\to\infty}\dfrac{3-\frac{1}{x}}{\pi-\frac{5}{x}} = \dfrac{3}{\pi}$

11. $\lim_{x\to\infty}\dfrac{3\sqrt{x^3}+3x}{\sqrt{2x^3}} = \lim_{x\to\infty}\dfrac{3x^{3/2}+3x}{\sqrt{2}x^{3/2}}$

$= \lim_{x\to\infty}\dfrac{3+\frac{3}{\sqrt{x}}}{\sqrt{2}} = \dfrac{3}{\sqrt{2}}$

13. $\lim_{x\to\infty}\sqrt[3]{\dfrac{1+8x^2}{x^2+4}} = \sqrt[3]{\lim_{x\to\infty}\dfrac{1+8x^2}{x^2+4}}$

$= \sqrt[3]{\lim_{x\to\infty}\dfrac{\frac{1}{x^2}+8}{1+\frac{4}{x^2}}} = \sqrt[3]{8} = 2$

15. $\lim_{n\to\infty}\dfrac{n}{2n+1} = \lim_{n\to\infty}\dfrac{1}{2+\frac{1}{n}} = \dfrac{1}{2}$

17. $\lim_{n\to\infty}\dfrac{n^2}{n+1} = \lim_{n\to\infty}\dfrac{n}{1+\frac{1}{n}} = \dfrac{\lim_{n\to\infty} n}{\lim_{n\to\infty}\left(1+\frac{1}{n}\right)} = \dfrac{\infty}{1+0} = \infty$

19. For $x > 0$, $x = \sqrt{x^2}$.

$$\lim_{x\to\infty} \frac{2x+1}{\sqrt{x^2+3}} = \lim_{x\to\infty} \frac{2+\frac{1}{x}}{\frac{\sqrt{x^2+3}}{\sqrt{x^2}}} = \lim_{x\to\infty} \frac{2+\frac{1}{x}}{\sqrt{1+\frac{3}{x^2}}}$$

$$= \frac{2}{\sqrt{1}} = 2$$

21. $$\lim_{x\to\infty}\left(\sqrt{2x^2+3}-\sqrt{2x^2-5}\right)$$

$$= \lim_{x\to\infty} \frac{\left(\sqrt{2x^2+3}-\sqrt{2x^2-5}\right)\left(\sqrt{2x^2+3}+\sqrt{2x^2-5}\right)}{\sqrt{2x^2+3}+\sqrt{2x^2-5}}$$

$$= \lim_{x\to\infty} \frac{2x^2+3-(2x^2-5)}{\sqrt{2x^2+3}+\sqrt{2x^2-5}} = \lim_{x\to\infty} \frac{8}{\sqrt{2x^2+3}+\sqrt{2x^2-5}}$$

$$= \lim_{x\to\infty} \frac{\frac{8}{x}}{\frac{\sqrt{2x^2+3}+\sqrt{2x^2-5}}{\sqrt{x^2}}} = \lim_{x\to\infty} \frac{\frac{8}{x}}{\sqrt{2+\frac{3}{x^2}}+\sqrt{2-\frac{5}{x^2}}} = 0$$

23. $$\lim_{y\to-\infty} \frac{9y^3+1}{y^2-2y+2} = \lim_{y\to-\infty} \frac{9y+\frac{1}{y^2}}{1-\frac{2}{y}+\frac{2}{y^2}} = -\infty$$

25. $$\lim_{n\to\infty} \frac{n}{\sqrt{n^2+1}} = \lim_{n\to\infty} \frac{1}{\sqrt{1+\frac{1}{n^2}}} = \frac{1}{\sqrt{1+0}} = 1$$

27. As $x \to 4^+, x \to 4$ while $x-4 \to 0^+$.

$$\lim_{x\to 4^+} \frac{x}{x-4} = \infty$$

29. As $t \to 3^-, t^2 \to 9$ while $9-t^2 \to 0^+$.

$$\lim_{t\to 3^-} \frac{t^2}{9-t^2} = \infty$$

31. As $x \to 5^-, x^2 \to 25, x-5 \to 0^-$, and $3-x \to -2$.

$$\lim_{x\to 5^-} \frac{x^2}{(x-5)(3-x)} = \infty$$

33. As $x \to 3^-, x^3 \to 27$, while $x-3 \to 0^-$.

$$\lim_{x\to 3^-} \frac{x^3}{x-3} = -\infty$$

35. $$\lim_{x\to 3^-} \frac{x^2-x-6}{x-3} = \lim_{x\to 3^-} \frac{(x+2)(x-3)}{x-3}$$

$$= \lim_{x\to 3^-}(x+2) = 5$$

37. For $0 \le x < 1$, $[\![x]\!] = 0$, so for $0 < x < 1$, $\frac{[\![x]\!]}{x} = 0$

thus $\lim_{x\to 0^+} \frac{[\![x]\!]}{x} = 0$

39. For $x < 0$, $|x| = -x$, thus

$$\lim_{x\to 0^-} \frac{|x|}{x} = \lim_{x\to 0^-} \frac{-x}{x} = -1$$

41. As $x \to 0^-, 1+\cos x \to 2$ while $\sin x \to 0^-$.

$$\lim_{x\to 0^-} \frac{1+\cos x}{\sin x} = -\infty$$

43. $\lim_{x\to\infty} \frac{3}{x+1} = 0$, $\lim_{x\to-\infty} \frac{3}{x+1} = 0$;

Horizontal asymptote $y = 0$.

$\lim_{x\to -1^+} \frac{3}{x+1} = \infty$, $\lim_{x\to -1^-} \frac{3}{x+1} = -\infty$;

Vertical asymptote $x = -1$

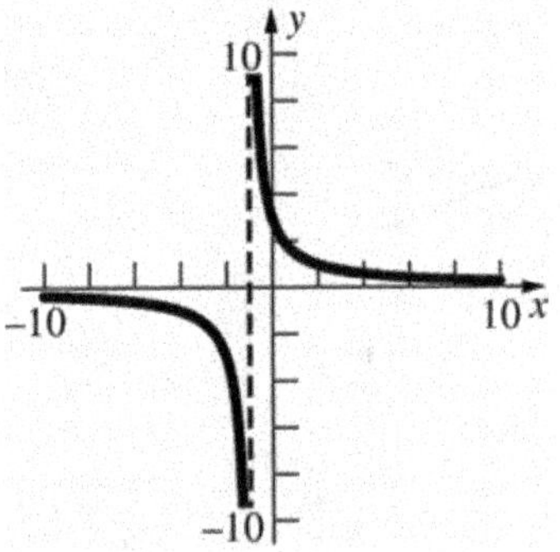

45. $\lim_{x\to\infty} \frac{2x}{x-3} = \lim_{x\to\infty} \frac{2}{1-\frac{3}{x}} = 2$,

$\lim_{x\to-\infty} \frac{2x}{x-3} = \lim_{x\to-\infty} \frac{2}{1-\frac{3}{x}} = 2$,

Horizontal asymptote $y = 2$

$\lim_{x\to 3^+} \frac{2x}{x-3} = \infty$, $\lim_{x\to 3^-} \frac{2x}{x-3} = -\infty$;

Vertical asymptote $x = 3$

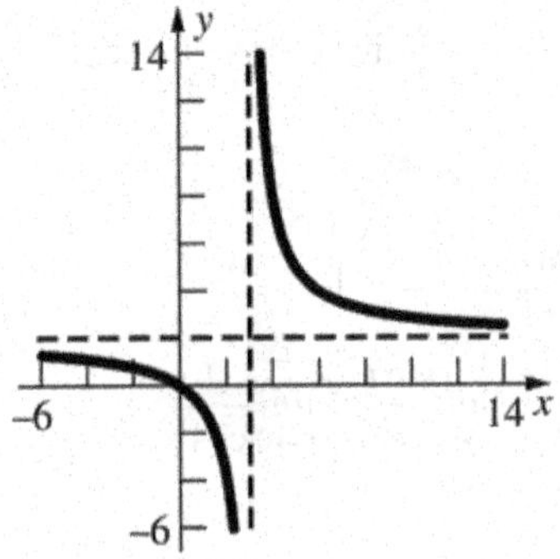

47. $\lim_{x\to\infty} \frac{14}{2x^2+7} = 0,\ \lim_{x\to-\infty} \frac{14}{2x^2+7} = 0;$

Horizontal asymptote $y = 0$

Since $2x^2+7 > 0$ for all x, $g(x)$ has no vertical asymptotes.

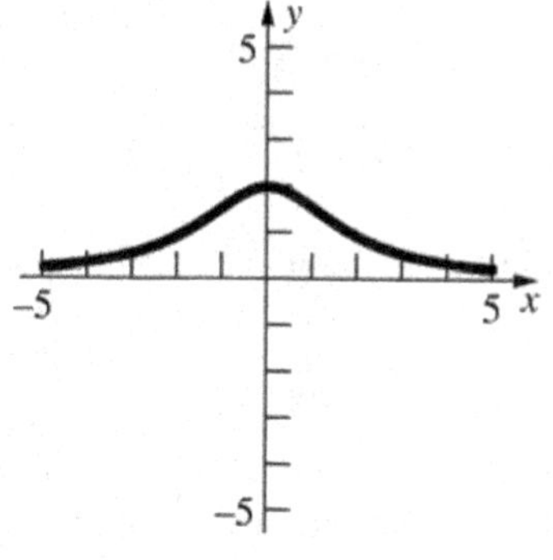

49. $f(x) = 2x + 3 - \frac{1}{x^3 - 1}$, thus

$$\lim_{x\to\infty} [f(x) - (2x+3)] = \lim_{x\to\infty}\left[-\frac{1}{x^3-1}\right] = 0$$

The oblique asymptote is $y = 2x + 3$.

51. a. We say that $\lim_{x\to c^+} f(x) = -\infty$ if to each negative number M there corresponds a $\delta > 0$ such that $0 < x - c < \delta \Rightarrow f(x) < M$.

b. We say that $\lim_{x\to c^-} f(x) = \infty$ if to each positive number M there corresponds a $\delta > 0$ such that $0 < c - x < \delta \Rightarrow f(x) > M$.

53. Let $\varepsilon > 0$ be given. Since $\lim_{x\to\infty} f(x) = A$, there is a corresponding number M_1 such that $x > M_1 \Rightarrow |f(x) - A| < \frac{\varepsilon}{2}$. Similarly, there is a number M_2 such that $x > M_2 \Rightarrow |g(x) - B| < \frac{\varepsilon}{2}$.

Let $M = \max\{M_1, M_2\}$, then

$x > M \Rightarrow |f(x) + g(x) - (A + B)|$

$= |f(x) - A + g(x) - B| \le |f(x) - A| + |g(x) - B|$

$< \frac{\varepsilon}{2} + \frac{\varepsilon}{2} = \varepsilon$

Thus, $\lim_{x\to\infty} [f(x) + g(x)] = A + B$

55. a. $\lim_{x\to\infty} \sin x$ does not exist as $\sin x$ oscillates between –1 and 1 as x increases.

b. Let $u = \frac{1}{x}$, then as $x \to \infty, u \to 0^+$.

$$\lim_{x\to\infty} \sin\frac{1}{x} = \lim_{u\to 0^+} \sin u = 0$$

c. Let $u = \frac{1}{x}$, then as $x \to \infty, u \to 0^+$.

$$\lim_{x\to\infty} x\sin\frac{1}{x} = \lim_{u\to 0^+} \frac{1}{u}\sin u = \lim_{u\to 0^+} \frac{\sin u}{u} = 1$$

d. Let $u = \frac{1}{x}$, then

$$\lim_{x\to\infty} x^{3/2}\sin\frac{1}{x} = \lim_{u\to 0^+}\left(\frac{1}{u}\right)^{3/2} \sin u$$

$$= \lim_{u\to 0^+}\left[\left(\frac{1}{\sqrt{u}}\right)\left(\frac{\sin u}{u}\right)\right] = \infty$$

e. As $x \to \infty$, $\sin x$ oscillates between –1 and 1, while $x^{-1/2} = \frac{1}{\sqrt{x}} \to 0$.

$$\lim_{x\to\infty} x^{-1/2}\sin x = 0$$

f. Let $u = \frac{1}{x}$, then

$$\lim_{x\to\infty} \sin\left(\frac{\pi}{6} + \frac{1}{x}\right) = \lim_{u\to 0^+} \sin\left(\frac{\pi}{6} + u\right)$$

$$= \sin\frac{\pi}{6} = \frac{1}{2}$$

g. As $x \to \infty, x + \frac{1}{x} \to \infty$, so $\lim_{x\to\infty} \sin\left(x + \frac{1}{x}\right)$ does not exist. (See part a.)

h. $\sin\left(x + \frac{1}{x}\right) = \sin x\cos\frac{1}{x} + \cos x\sin\frac{1}{x}$

$$\lim_{x\to\infty}\left[\sin\left(x + \frac{1}{x}\right) - \sin x\right]$$

$$= \lim_{x\to\infty}\left[\sin x\left(\cos\frac{1}{x} - 1\right) + \cos x\sin\frac{1}{x}\right]$$

As $x \to \infty, \cos\frac{1}{x} \to 1$ so $\cos\frac{1}{x} - 1 \to 0$.

From part **b.**, $\lim_{x\to\infty} \sin\frac{1}{x} = 0$.

As $x \to \infty$ both $\sin x$ and $\cos x$ oscillate between –1 and 1.

$$\lim_{x\to\infty}\left[\sin\left(x + \frac{1}{x}\right) - \sin x\right] = 0.$$

57. $\lim_{x\to\infty} \frac{3x^2+x+1}{2x^2-1} = \frac{3}{2}$

59. $\lim_{x\to-\infty} \left(\sqrt{2x^2+3x} - \sqrt{2x^2-5}\right) = -\frac{3}{2\sqrt{2}}$

61. $\lim_{x\to\infty} \left(1+\frac{1}{x}\right)^{10} = 1$

63. $\lim_{x\to\infty} \left(1+\frac{1}{x}\right)^{x^2} = \infty$

65. $\lim_{x\to3^-} \frac{\sin|x-3|}{x-3} = -1$

67. $\lim_{x\to3^-} \frac{\cos(x-3)}{x-3} = -\infty$

69. $\lim_{x\to0^+} (1+\sqrt{x})^{\frac{1}{\sqrt{x}}} = e \approx 2.718$

71. $\lim_{x\to0^+} (1+\sqrt{x})^x = 1$

1.6 Concepts Review

1. $\lim_{x\to c} f(x)$

3. $\lim_{x\to a^+} f(x) = f(a);\ \lim_{x\to b^-} f(x) = f(b)$

Problem Set 1.6

1. $\lim_{x\to3}[(x-3)(x-4)] = 0 = f(3)$; continuous

3. $\lim_{x\to3} \frac{3}{x-3}$ and $h(3)$ do not exist, so $h(x)$ is not continuous at 3.

5. $\lim_{t\to3} \frac{|t-3|}{t-3}$ and $h(3)$ do not exist, so $h(t)$ is not continuous at 3.

7. $\lim_{t\to3} |t| = 3 = f(3)$; continuous

9. $h(3)$ does not exist, so $h(t)$ is not continuous at 3.

11. $\lim_{t\to3} \frac{t^3-27}{t-3} = \lim_{t\to3} \frac{(t-3)(t^2+3t+9)}{t-3}$

$= \lim_{t\to3}(t^2+3t+9) = (3)^2+3(3)+9 = 27 = r(3)$

continuous

13. $\lim_{t\to3^+} f(t) = \lim_{t\to3^+} (3-t) = 0$

$\lim_{t\to3^-} f(t) = \lim_{t\to3^-} (t-3) = 0$

$\lim_{t\to3} f(t) = f(3)$; continuous

15. $\lim_{t\to3} f(x) = -2 = f(3)$; continuous

17. h is continuous on the intervals $(-\infty,-5),\ [-5,4],\ (4,6),\ [6,8],\ (8,\infty)$

19. $\lim_{x\to3} \frac{2x^2-18}{3-x} = \lim_{x\to3} \frac{2(x+3)(x-3)}{3-x}$

$= \lim_{x\to3}[-2(x+3)] = -2(3+3) = -12$

Define $f(3) = -12$.

21. $\lim_{t\to1} \frac{\sqrt{t}-1}{t-1} = \lim_{t\to1} \frac{(\sqrt{t}-1)(\sqrt{t}+1)}{(t-1)(\sqrt{t}+1)}$

$= \lim_{t\to1} \frac{t-1}{(t-1)(\sqrt{t}+1)} = \lim_{t\to1} \frac{1}{\sqrt{t}+1} = \frac{1}{2}$

Define $H(1) = \frac{1}{2}$.

23. $\lim_{x\to-1} \sin\left(\frac{x^2-1}{x+1}\right) = \lim_{x\to-1} \sin\left(\frac{(x-1)(x+1)}{x+1}\right)$

$= \lim_{x\to-1} \sin(x-1) = \sin(-1-1) = \sin(-2) = -\sin 2$

Define $F(-1) = -\sin 2$.

25. $f(x) = \frac{33-x^2}{(\pi-x)(x-3)}$

Discontinuous at $x = 3, \pi$

27. Discontinuous at all $\theta = n\pi + \frac{\pi}{2}$ where n is any integer.

29. Discontinuous at $u = -1$

31. $G(x) = \frac{1}{\sqrt{(2-x)(2+x)}}$

Discontinuous on $(-\infty,-2]\cup[2,\infty)$

33. $\lim_{x\to 0} g(x) = 0 = g(0)$

$\lim_{x\to 1^+} g(x) = 1$, $\lim_{x\to 1^-} g(x) = -1$

$\lim_{x\to 1} g(x)$ does not exist, so $g(x)$ is discontinuous at $x = 1$.

35. Discontinuous at $t = n + \frac{1}{2}$ where n is any integer

37.

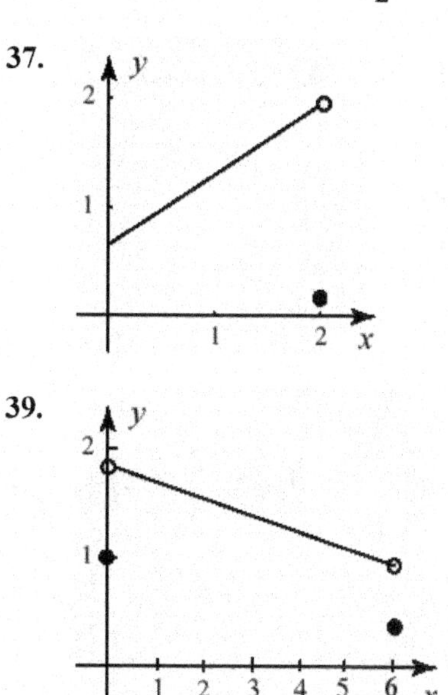

39.

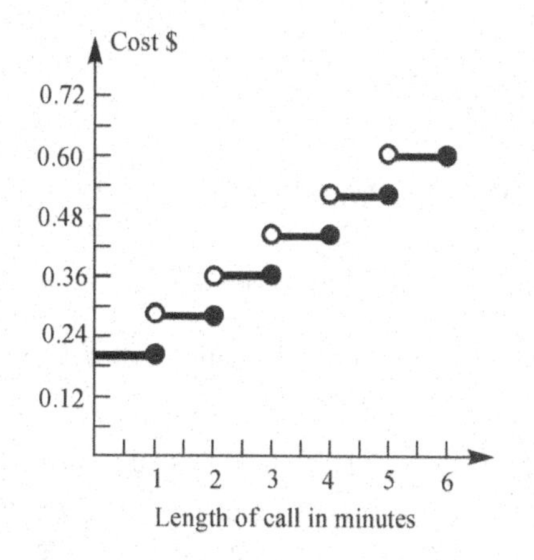

41. Continuous.

43. Discontinuous: removable, define $f(0) = 1$

45. Discontinuous, removable, redefine $g(0) = 1$

47. Discontinuous: nonremovable.

49. The function is continuous on the intervals $(0,1], (1,2], (2,3], \ldots$

51. The function is continuous on the intervals $(0, 0.25], (0.25, 0.375], (0.375, 0.5], \ldots$

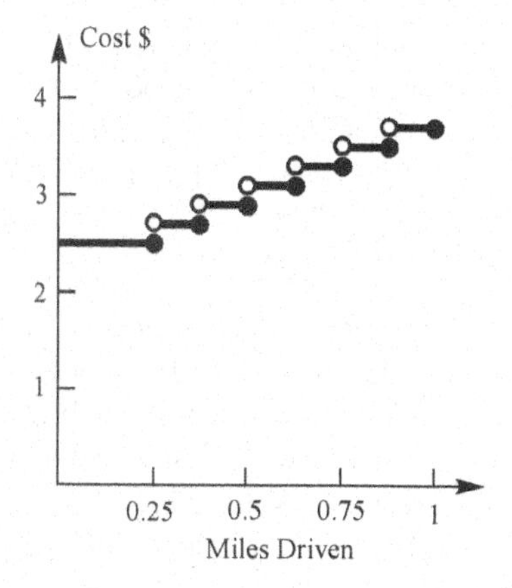

53. Because the function is continuous on $[0, 2\pi]$ and

$(\cos 0)0^3 + 6\sin^5 0 - 3 = -3 < 0$,

$(\cos 2\pi)(2\pi)^3 + 6\sin^5(2\pi) - 3 = 8\pi^3 - 3 > 0$, there is at least one number c between 0 and 2π such that $(\cos t)t^3 + 6\sin^5 t - 3 = 0$.

55. Let $f(x) = \sqrt{x} - \cos x$.. $f(x)$ is continuous at all values of $x \geq 0$. $f(0) = -1, f(\pi/2) = \sqrt{\pi/2}$

Because 0 is between -1 and $\sqrt{\pi/2}$, there is at least one number c between 0 and $\pi/2$ such that $f(x) = \sqrt{x} - \cos x = 0$.

The interval [0.6,0.7] contains the solution.

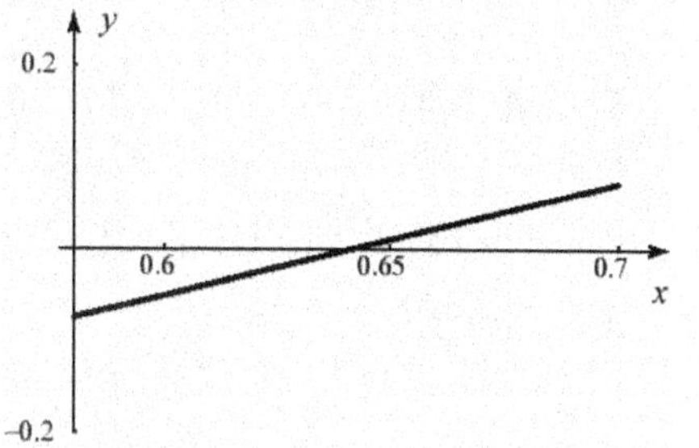

57. Suppose that f is continuous at c, so $\lim_{x\to c} f(x) = f(c)$. Let $x = c + t$, so $t = x - c$, then as $x \to c$, $t \to 0$ and the statement $\lim_{x\to c} f(x) = f(c)$ becomes $\lim_{t\to 0} f(t + c) = f(c)$.

Suppose that $\lim_{t\to 0} f(t + c) = f(c)$ and let $x = t + c$, so $t = x - c$. Since c is fixed, $t \to 0$ means that $x \to c$ and the statement $\lim_{t\to 0} f(t + c) = f(c)$ becomes $\lim_{x\to c} f(x) = f(c)$, so f is continuous at c.

59. Let $g(x) = x - f(x)$. Then, $g(0) = 0 - f(0) = -f(0) \leq 0$ and $g(1) = 1 - f(1) \geq 0$ since $0 \leq f(x) \leq 1$ on $[0, 1]$. If $g(0) = 0$, then $f(0) = 0$ and $c = 0$ is a fixed point of f. If $g(1) = 0$, then $f(1) = 1$ and $c = 1$ is a fixed point of f. If neither $g(0) = 0$ nor $g(1) = 0$, then $g(0) < 0$ and $g(1) > 0$ so there is some c in $[0, 1]$ such that $g(c) = 0$. If $g(c) = 0$ then $c - f(c) = 0$ or $f(c) = c$ and c is a fixed point of f.

61. For x in $[0, 1]$, let $f(x)$ indicate where the string originally at x ends up. Thus $f(0) = a$, $f(1) = b$. $f(x)$ is continuous since the string is unbroken. Since $0 \leq a,\ b \leq 1$, $f(x)$ satisfies the conditions of Problem 59, so there is some c in $[0, 1]$ with $f(c) = c$, i.e., the point of string originally at c ends up at c.

63. Let $f(x)$ be the difference in times on the hiker's watch where x is a point on the path, and suppose $x = 0$ at the bottom and $x = 1$ at the top of the mountain.
So $f(x)$ = (time on watch on the way up) – (time on watch on the way down).
$f(0) = 4 - 11 = -7$, $f(1) = 12 - 5 = 7$. Since time is continuous, $f(x)$ is continuous, hence there is some c between 0 and 1 where $f(c) = 0$. This c is the point where the hiker's watch showed the same time on both days.

65. Yes, g is continuous at R.

$$\lim_{r \to R^-} g(r) = \frac{GMm}{R^2} = \lim_{r \to R^+} g(r)$$

67. a. $f(x) = f(x + 0) = f(x) + f(0)$, so $f(0) = 0$. We want to prove that $\lim_{x \to c} f(x) = f(c)$, or, equivalently, $\lim_{x \to c} [f(x) - f(c)] = 0$. But $f(x) - f(c) = f(x - c)$, so $\lim_{x \to c} [f(x) - f(c)] = \lim_{x \to c} f(x - c)$. Let $h = x - c$ then as $x \to c$, $h \to 0$ and $\lim_{x \to c} f(x - c) = \lim_{h \to 0} f(h) = f(0) = 0$. Hence $\lim_{x \to c} f(x) = f(c)$ and f is continuous at c.
Thus, f is continuous everywhere, since c was arbitrary.

b. By Problem 43 of Section 0.5, $f(t) = mt$ for all t in Q. Since $g(t) = mt$ is a polynomial function, it is continuous for all real numbers. $f(t) = g(t)$ for all t in Q, thus $f(t) = g(t)$ for all t in R, i.e. $f(t) = mt$.

69. Suppose $f(x) = \begin{cases} 1 \text{ if } x \geq 0 \\ -1 \text{ if } x < 0 \end{cases}$. $f(x)$ is discontinuous at $x = 0$, but $g(x) = |f(x)| = 1$ is continuous everywhere.

71. a. Suppose the block rotates to the left. Using geometry, $f(x) = -\frac{3}{4}$. Suppose the block rotates to the right. Using geometry, $f(x) = \frac{3}{4}$. If $x = 0$, the block does not rotate, so $f(x) = 0$.
Domain: $\left[-\frac{3}{4}, \frac{3}{4}\right]$;
Range: $\left\{-\frac{3}{4}, 0, \frac{3}{4}\right\}$

b. At $x = 0$

c. If $x = 0, f(x) = 0$, if $x = -\frac{3}{4}, f(x) = -\frac{3}{4}$ and if $x = \frac{3}{4}, f(x) = \frac{3}{4}$, so $x = -\frac{3}{4}, 0, \frac{3}{4}$ are fixed points of f.

1.7 Chapter Review

Concepts Test

1. False. Consider $f(x) = [\![x]\!]$ at $x = 2$.

3. False: c may not be in the domain of $f(x)$, or it may be defined separately.

5. False: If $f(c)$ is not defined, $\lim_{x \to c} f(x)$ might exist; e.g., $f(x) = \frac{x^2 - 4}{x + 2}$.
$f(-2)$ does not exist, but $\lim_{x \to -2} \frac{x^2 - 4}{x + 2} = -4$.

7. True: Substitution Theorem

9. False: The tangent function is not defined for all values of c.

11. True: Since both $\sin x$ and $\cos x$ are continuous for all real numbers, by Theorem C we can conclude that $f(x) = 2\sin^2 x - \cos x$ is also continuous for all real numbers.

13. True. $2 \in [1,3]$

15. False: Consider $f(x) = \sin x$.

17. False: Since $-1 \le \sin x \le 1$ for all x and $\lim\limits_{x\to\infty} \frac{1}{x} = 0$, we get $\lim\limits_{x\to\infty} \frac{\sin x}{x} = 0$.

19. False: The graph has many vertical asymptotes; e.g., $x = \pm \pi/2, \pm 3\pi/2, \pm 5\pi/2, \ldots$

21. True: As $x \to 1^+$ both the numerator and denominator are positive. Since the numerator approaches a constant and the denominator approaches zero, the limit goes to $+\infty$.

23. True: $\lim\limits_{x\to c} f(x) = f\left(\lim\limits_{x\to c} x\right) = f(c)$, so f is continuous at $x = c$.

25. True: Choose $\varepsilon = 0.001f(2)$ then since $\lim\limits_{x\to 2} f(x) = f(2)$, there is some δ such that $0 < |x-2| < \delta \Rightarrow |f(x) - f(2)| < 0.001f(2)$, or $-0.001f(2) < f(x) - f(2) < 0.001f(2)$
Thus, $0.999f(2) < f(x) < 1.001f(2)$ and $f(x) < 1.001f(2)$ for $0 < |x-2| < \delta$. Since $f(2) < 1.001f(2)$, as $f(2) > 0$, $f(x) < 1.001f(2)$ on $(2-\delta, 2+\delta)$.

27. True: Squeeze Theorem

29. False: That $f(x) \ne g(x)$ for all x does not imply that $\lim\limits_{x\to c} f(x) \ne \lim\limits_{x\to c} g(x)$. For example, if $f(x) = \frac{x^2 + x - 6}{x-2}$ and $g(x) = \frac{5}{2}x$, then $f(x) \ne g(x)$ for all x, but $\lim\limits_{x\to 2} f(x) = \lim\limits_{x\to 2} g(x) = 5$.

31. True: $\lim\limits_{x\to a} |f(x)| = \lim\limits_{x\to a} \sqrt{f^2(x)}$

$$= \sqrt{\left[\lim_{x\to a} f(x)\right]^2} = \sqrt{(b)^2} = |b|$$

Sample Test Problems

1. $\lim\limits_{x\to 2} \frac{x-2}{x+2} = \frac{2-2}{2+2} = \frac{0}{4} = 0$

3. $\lim\limits_{u\to 1} \frac{u^2-1}{u-1} = \lim\limits_{u\to 1} \frac{(u-1)(u+1)}{u-1} = \lim\limits_{u\to 1} (u+1)$
$= 1 + 1 = 2$

5. $\lim\limits_{x\to 2} \frac{1-\frac{2}{x}}{x^2-4} = \lim\limits_{x\to 2} \frac{\frac{x-2}{x}}{(x-2)(x+2)} = \lim\limits_{x\to 2} \frac{1}{x(x+2)}$
$= \frac{1}{2(2+2)} = \frac{1}{8}$

7. $\lim\limits_{x\to 0} \frac{\tan x}{\sin 2x} = \lim\limits_{x\to 0} \frac{\frac{\sin x}{\cos x}}{2\sin x\cos x} = \lim\limits_{x\to 0} \frac{1}{2\cos^2 x}$
$= \frac{1}{2\cos^2 0} = \frac{1}{2}$

9. $\lim\limits_{x\to 4} \frac{x-4}{\sqrt{x}-2} = \lim\limits_{x\to 4} \frac{(\sqrt{x}-2)(\sqrt{x}+2)}{\sqrt{x}-2}$
$= \lim\limits_{x\to 4} (\sqrt{x}+2) = \sqrt{4}+2 = 4$

11. $\lim\limits_{x\to 0^-} \frac{|x|}{x} = \lim\limits_{x\to 0^-} \frac{-x}{x} = \lim\limits_{x\to 0^-} (-1) = -1$

13. $\lim\limits_{t\to 2^-} ([\![t]\!] - t) = \lim\limits_{t\to 2^-} [\![t]\!] - \lim\limits_{t\to 2^-} t = 1 - 2 = -1$

15. $\lim\limits_{x\to 0} \frac{\sin 5x}{3x} = \lim\limits_{x\to 0} \frac{5}{3} \frac{\sin 5x}{5x}$
$= \frac{5}{3} \lim\limits_{x\to 0} \frac{\sin 5x}{5x} = \frac{5}{3} \times 1 = \frac{5}{3}$

17. $\lim\limits_{x\to\infty} \frac{x-1}{x+2} = \lim\limits_{x\to\infty} \frac{1-\frac{1}{x}}{1+\frac{2}{x}} = \frac{1+0}{1+0} = 1$

19. $\lim\limits_{t\to 2} \frac{t+2}{(t-2)^2} = \infty$ because as $t \to 0$, $t+2 \to 4$ while the denominator goes to 0 from the right.

21. $\lim\limits_{x\to \pi/4^-} \tan 2x = \infty$ because as $x \to (\pi/4)^-$, $2x \to (\pi/2)^-$, so $\tan 2x \to \infty$.

23. Preliminary analysis: Let $\varepsilon > 0$. We need to find a $\delta > 0$ such that
$0 < |x-3| < \delta \Rightarrow |(2x+1)-7| < \varepsilon$.
$|2x-6| < \varepsilon \Leftrightarrow 2|x-3| < \varepsilon$
$\Leftrightarrow |x-3| < \frac{\varepsilon}{2}$. Choose $\delta = \frac{\varepsilon}{2}$.

Let $\varepsilon > 0$. Choose $\delta = \varepsilon/2$. Thus,

$$|(2x+1)-7| = |2x-6| = 2|x-3| < 2(\varepsilon/2) = \varepsilon.$$

25. **a.** f is discontinuous at $x = 1$ because $f(1) = 0$, but $\lim_{x\to 1} f(x)$ does not exist. f is discontinuous at $x = -1$ because $f(-1)$ does not exist.

b. Define $f(-1) = -1$

27. **a.** $\lim_{x\to 3}[2f(x) - 4g(x)]$
$= 2\lim_{x\to 3} f(x) - 4\lim_{x\to 3} g(x)$
$= 2(3) - 4(-2) = 14$

b. $\lim_{x\to 3} g(x)\frac{x^2-9}{x-3} = \lim_{x\to 3} g(x)(x+3)$
$= \lim_{x\to 3} g(x)\cdot \lim_{x\to 3}(x+3) = -2\cdot(3+3) = -12$

c. $g(3) = -2$

d. $\lim_{x\to 3} g(f(x)) = g\left(\lim_{x\to 3} f(x)\right) = g(3) = -2$

e. $\lim_{x\to 3}\sqrt{f^2(x) - 8g(x)}$
$= \sqrt{\left[\lim_{x\to 3} f(x)\right]^2 - 8\lim_{x\to 3} g(x)}$
$= \sqrt{(3)^2 - 8(-2)} = 5$

f. $\lim_{x\to 3}\frac{|g(x)-g(3)|}{f(x)} = \frac{|-2-g(3)|}{3} = \frac{|-2-(-2)|}{3}$
$= 0$

29. $a(0) + b = -1$ and $a(1) + b = 1$
$b = -1;\ a + b = 1$
$a - 1 = 1$
$a = 2$

31. Vertical: None, denominator is never 0.

Horizontal: $\lim_{x\to\infty}\frac{x}{x^2+1} = \lim_{x\to -\infty}\frac{x}{x^2+1} = 0$, so $y = 0$ is a horizontal asymptote.

33. Vertical: $x = 1, x = -1$ because $\lim_{x\to 1^+}\frac{x^2}{x^2-1} = \infty$

and $\lim_{x\to -1^-}\frac{x^2}{x^2-1} = \infty$

Horizontal: $\lim_{x\to\infty}\frac{x^2}{x^2-1} = \lim_{x\to -\infty}\frac{x^2}{x^2-1} = 1$, so $y = 1$ is a horizontal asymptote.

35. Vertical: $x = \pm\pi/4, \pm 3\pi/4, \pm 5\pi/4, \ldots$ because $\lim_{x\to \pi/4^-}\tan 2x = \infty$ and similarly for other odd multiples of $\pi/4$.

Horizontal: None, because $\lim_{x\to\infty}\tan 2x$ and $\lim_{x\to -\infty}\tan 2x$ do not exist.

Review and Preview Problems

1. **a.** $f(2) = 2^2 = 4$

b. $f(2.1) = 2.1^2 = 4.41$

c. $f(2.1) - f(2) = 4.41 - 4 = 0.41$

d. $\frac{f(2.1)-f(2)}{2.1-2} = \frac{0.41}{0.1} = 4.1$

e. $f(a+h) = (a+h)^2 = a^2 + 2ah + h^2$

f. $f(a+h) - f(a) = a^2 + 2ah + h^2 - a^2$
$= 2ah + h^2$

g. $\frac{f(a+h)-f(a)}{(a+h)-a} = \frac{2ah+h^2}{h} = 2a + h$

h. $\lim_{h\to 0}\frac{f(a+h)-f(a)}{(a+h)-a} = \lim_{h\to 0}(2a+h) = 2a$

3. **a.** $F(2) = \sqrt{2} \approx 1.414$

b. $F(2.1) = \sqrt{2.1} \approx 1.449$

c. $F(2.1) - F(2) = 1.449 - 1.414 = 0.035$

d. $\frac{F(2.1)-F(2)}{2.1-2} = \frac{0.035}{0.1} = 0.35$

e. $F(a+h)=\sqrt{a+h}$

f. $F(a+h)-F(a)=\sqrt{a+h}-\sqrt{a}$

g. $\dfrac{F(a+h)-F(a)}{(a+h)-a}=\dfrac{\sqrt{a+h}-\sqrt{a}}{h}$

h.
$$\lim_{h\to 0}\frac{F(a+h)-F(a)}{(a+h)-a}=\lim_{h\to 0}\frac{\sqrt{a+h}-\sqrt{a}}{h}$$
$$=\lim_{h\to 0}\frac{\left(\sqrt{a+h}-\sqrt{a}\right)\left(\sqrt{a+h}+\sqrt{a}\right)}{h\left(\sqrt{a+h}+\sqrt{a}\right)}$$
$$=\lim_{h\to 0}\frac{a+h-a}{h\left(\sqrt{a+h}+\sqrt{a}\right)}$$
$$=\lim_{h\to 0}\frac{h}{h\left(\sqrt{a+h}+\sqrt{a}\right)}$$
$$=\lim_{h\to 0}\frac{1}{\sqrt{a+h}+\sqrt{a}}=\frac{1}{2\sqrt{a}}=\frac{\sqrt{a}}{2a}$$

5. a. $(a+b)^3=a^3+3a^2b+\cdots$

b. $(a+b)^4=a^4+4a^3b+\cdots$

c. $(a+b)^5=a^5+5a^4b+\cdots$

7. $\sin(x+h)=\sin x\cos h+\cos x\sin h$

9. a. The point will be at position $(10,0)$ in all three cases ($t=1,2,3$) because it will have made 4, 8, and 12 revolutions respectively.

b. Since the point is rotating at a rate of 4 revolutions per second, it will complete 1 revolution after $\frac{1}{4}$ second. Therefore, the point will first return to its starting position at time $t=\frac{1}{4}$.

11. a. North plane has traveled 600miles. East plane has traveled 400 miles.

b.
$$d=\sqrt{600^2+400^2}$$
$$=721 \text{ miles}$$

c.
$$d=\sqrt{675^2+500^2}$$
$$=840 \text{ miles}$$

CHAPTER 2 The Derivative

2.1 Concepts Review

1. tangent line

3. $\dfrac{f(c+h)-f(c)}{h}$

Problem Set 2.1

1. Slope $= \dfrac{5-3}{2-\frac{3}{2}} = 4$

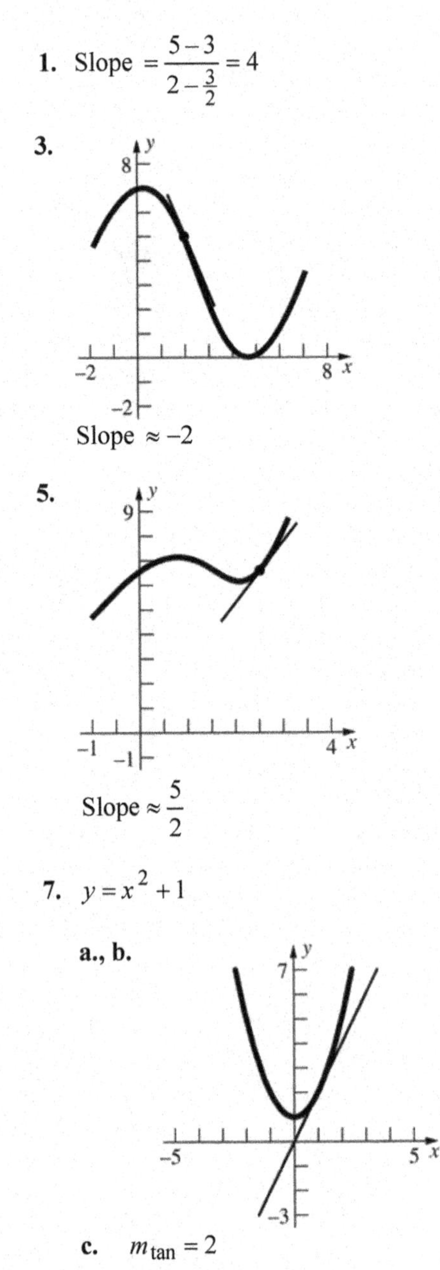

3. Slope ≈ -2

5. Slope $\approx \dfrac{5}{2}$

7. $y = x^2 + 1$

a., b.

c. $m_{\tan} = 2$

d. $m_{\sec} = \dfrac{(1.01)^2 + 1.0 - 2}{1.01 - 1} = \dfrac{0.0201}{.01} = 2.01$

e. $m_{\tan} = \lim\limits_{h \to 0} \dfrac{f(1+h)-f(1)}{h}$

$= \lim\limits_{h \to 0} \dfrac{[(1+h)^2+1]-(1^2+1)}{h}$

$= \lim\limits_{h \to 0} \dfrac{2+2h+h^2-2}{h} = \lim\limits_{h \to 0} \dfrac{h(2+h)}{h}$

$= \lim\limits_{h \to 0} (2+h) = 2$

9. $f(x) = x^2 - 1$

$m_{\tan} = \lim\limits_{h \to 0} \dfrac{f(c+h)-f(c)}{h}$

$= \lim\limits_{h \to 0} \dfrac{[(c+h)^2-1]-(c^2-1)}{h}$

$= \lim\limits_{h \to 0} \dfrac{c^2+2ch+h^2-1-c^2+1}{h}$

$= \lim\limits_{h \to 0} \dfrac{h(2c+h)}{h} = 2c$

At $x = -2$, $m_{\tan} = -4$

$x = -1$, $m_{\tan} = -2$

$x = 1$, $m_{\tan} = 2$

$x = 2$, $m_{\tan} = 4$

11.

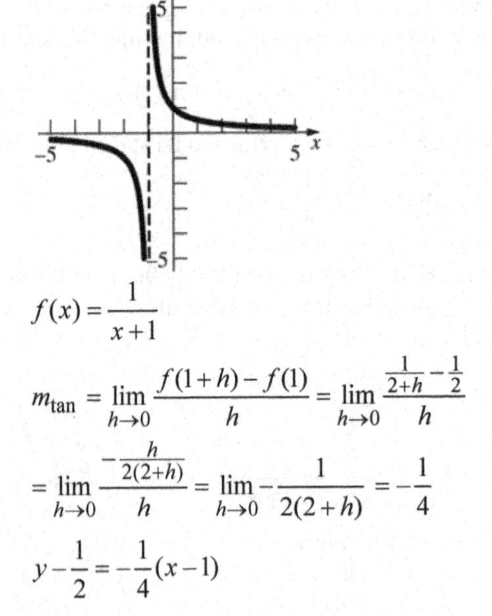

$f(x) = \dfrac{1}{x+1}$

$m_{\tan} = \lim\limits_{h \to 0} \dfrac{f(1+h)-f(1)}{h} = \lim\limits_{h \to 0} \dfrac{\frac{1}{2+h}-\frac{1}{2}}{h}$

$= \lim\limits_{h \to 0} \dfrac{-\frac{h}{2(2+h)}}{h} = \lim\limits_{h \to 0} -\dfrac{1}{2(2+h)} = -\dfrac{1}{4}$

$y - \dfrac{1}{2} = -\dfrac{1}{4}(x-1)$

13. a. $16(1^2)-16(0^2)=16$ ft

b. $16(2^2)-16(1^2)=48$ ft

c. $V_{ave}=\frac{144-64}{3-2}=80$ ft/sec

d. $V_{ave}=\frac{16(3.01)^2-16(3)^2}{3.01-3}=\frac{0.9616}{0.01}=96.16$ ft/s

e. $f(t)=16t^2; v=32c$; $v=32(3)=96$ ft/s

15. a. $v=\lim_{h\to 0}\frac{f(\alpha+h)-f(\alpha)}{h}=\lim_{h\to 0}\frac{\sqrt{2(\alpha+h)+1}-\sqrt{2\alpha+1}}{h}$

$=\lim_{h\to 0}\frac{\sqrt{2\alpha+2h+1}-\sqrt{2\alpha+1}}{h}=\lim_{h\to 0}\frac{(\sqrt{2\alpha+2h+1}-\sqrt{2\alpha+1})(\sqrt{2\alpha+2h+1}+\sqrt{2\alpha+1})}{h(\sqrt{2\alpha+2h+1}+\sqrt{2\alpha+1})}$

$=\lim_{h\to 0}\frac{2h}{h(\sqrt{2\alpha+2h+1}+\sqrt{2\alpha+1})}=\frac{2}{\sqrt{2\alpha+1}+\sqrt{2\alpha+1}}=\frac{1}{\sqrt{2\alpha+1}}$ ft/s

b. $\frac{1}{\sqrt{2\alpha+1}}=\frac{1}{2}$

$\sqrt{2\alpha+1}=2$

$2\alpha+1=4;\ \alpha=\frac{3}{2}$

The object reaches a velocity of $\frac{1}{2}$ ft/s when $t=\frac{3}{2}$.

17. a. $\left[\frac{1}{2}(2.01)^2+1\right]-\left[\frac{1}{2}(2)^2+1\right]=0.02005$ g

b. $r_{ave}=\frac{0.02005}{2.01-2}=2.005$ g/hr

c. $f(t)=\frac{1}{2}t^2+1$

$r=\lim_{h\to 0}\frac{\left[\frac{1}{2}(2+h)^2+1\right]-\left[\frac{1}{2}2^2+1\right]}{h}$

$=\lim_{h\to 0}\frac{2+2h+\frac{1}{2}h^2+1-2-1}{h}$

$=\lim_{h\to 0}\frac{h\left(2+\frac{1}{2}h\right)}{h}=2$

At $t=2$, $r=2$

19. a. $d_{ave}=\frac{5^3-3^3}{5-3}=\frac{98}{2}=49$ g/cm

b. $f(x)=x^3$

$d=\lim_{h\to 0}\frac{(3+h)^3-3^3}{h}$

$=\lim_{h\to 0}\frac{27+27h+9h^2+h^3-27}{h}$

$=\lim_{h\to 0}\frac{h(27+9h+h^2)}{h}=27$ g/cm

21. $a=\lim_{h\to 0}\frac{2(1+h)^2-2(1)^2}{h}$

$=\lim_{h\to 0}\frac{2+4h+2h^2-2}{h}$

$=\lim_{h\to 0}\frac{h(4+2h)}{h}=4$

23. $r_{ave}=\frac{100-800}{24-0}=-\frac{175}{6}\approx -29.167$

29,167 gal/hr

At 8 o'clock, $r\approx\frac{700-400}{6-10}\approx -75$

75,000 gal/hr

25. a. A tangent line at $t=91$ has slope approximately $(63-48)/(91-61)=0.5$. The normal high temperature increases at the rate of 0.5 degree F per day.

b. A tangent line at $t=191$ has approximate slope $(90-88)/30\approx 0.067$. The normal high temperature increases at the rate of 0.067 degree per day.

c. There is a time in January, about January 15, when the rate of change is zero. There is also a time in July, about July 15, when the rate of change is zero.

d. The greatest rate of increase occurs around day 61, that is, some time in March. The greatest rate of decrease occurs between day 301 and 331, that is, sometime in November.

27. In both (a) and (b), the tangent line is always positive. In (a) the tangent line becomes steeper and steeper as t increases; thus, the velocity is increasing. In (b) the tangent line becomes flatter and flatter as t increases; thus, the velocity is decreasing.

29. $A = \pi r^2,\ r = 2t$

$A = 4\pi t^2$

$$\text{rate} = \lim_{h\to 0} \frac{4\pi(3+h)^2 - 4\pi(3)^2}{h}$$

$$= \lim_{h\to 0} \frac{h(24\pi + 4\pi h)}{h} = 24\pi \text{ km}^2/\text{day}$$

31. $y = f(x) = x^3 - 2x^2 + 1$

a. $m_{\tan} = 7$ b. $m_{\tan} = 0$

c. $m_{\tan} = -1$ d. $m_{\tan} = 17.92$

33. $s = f(t) = t + t\cos^2 t$

At $t = 3$, $v \approx 2.818$

2.2 Concepts Review

1. $\dfrac{f(c+h) - f(c)}{h}; \dfrac{f(t) - f(c)}{t - c}$

3. continuous; $f(x) = |x|$

Problem Set 2.2

1. $$f'(1) = \lim_{h\to 0} \frac{f(1+h) - f(1)}{h}$$

$$= \lim_{h\to 0} \frac{(1+h)^2 - 1^2}{h} = \lim_{h\to 0} \frac{2h + h^2}{h}$$

$$= \lim_{h\to 0} (2+h) = 2$$

3. $$f'(3) = \lim_{h\to 0} \frac{f(3+h) - f(3)}{h}$$

$$= \lim_{h\to 0} \frac{[(3+h)^2 - (3+h)] - (3^2 - 3)}{h}$$

$$= \lim_{h\to 0} \frac{5h + h^2}{h} = \lim_{h\to 0} (5+h) = 5$$

5. $$s'(x) = \lim_{h\to 0} \frac{s(x+h) - s(x)}{h}$$

$$= \lim_{h\to 0} \frac{[2(x+h)+1] - (2x+1)}{h} = \lim_{h\to 0} \frac{2h}{h} = 2$$

7. $$r'(x) = \lim_{h\to 0} \frac{r(x+h) - r(x)}{h}$$

$$= \lim_{h\to 0} \frac{[3(x+h)^2 + 4] - (3x^2 + 4)}{h}$$

$$= \lim_{h\to 0} \frac{6xh + 3h^2}{h} = \lim_{h\to 0} (6x + 3h) = 6x$$

9. $$f'(x) = \lim_{h\to 0} \frac{f(x+h) - f(x)}{h}$$

$$= \lim_{h\to 0} \frac{[a(x+h)^2 + b(x+h) + c] - (ax^2 + bx + c)}{h}$$

$$= \lim_{h\to 0} \frac{2axh + ah^2 + bh}{h} = \lim_{h\to 0} (2ax + ah + b)$$

$$= 2ax + b$$

11. $$f'(x) = \lim_{h\to 0} \frac{f(x+h) - f(x)}{h}$$

$$= \lim_{h\to 0} \frac{[(x+h)^3 + 2(x+h)^2 + 1] - (x^3 + 2x^2 + 1)}{h}$$

$$= \lim_{h\to 0} \frac{3hx^2 + 3h^2x + h^3 + 4hx + 2h^2}{h}$$

$$= \lim_{h\to 0} (3x^2 + 3hx + h^2 + 4x + 2h) = 3x^2 + 4x$$

13. $$h'(x) = \lim_{h\to 0} \frac{h(x+h) - h(x)}{h}$$

$$= \lim_{h\to 0} \left[\left(\frac{2}{x+h} - \frac{2}{x}\right) \cdot \frac{1}{h}\right]$$

$$= \lim_{h\to 0} \left[\frac{-2h}{x(x+h)} \cdot \frac{1}{h}\right] = \lim_{h\to 0} \frac{-2}{x(x+h)} = -\frac{2}{x^2}$$

15. $$F'(x) = \lim_{h\to 0} \frac{F(x+h) - F(x)}{h}$$

$$= \lim_{h\to 0} \left[\left(\frac{6}{(x+h)^2 + 1} - \frac{6}{x^2 + 1}\right) \cdot \frac{1}{h}\right]$$

$$= \lim_{h\to 0} \left[\frac{6(x^2+1) - 6(x^2 + 2hx + h^2 + 1)}{(x^2+1)(x^2 + 2hx + h^2 + 1)} \cdot \frac{1}{h}\right]$$

$$= \lim_{h\to 0} \left[\frac{-12hx - 6h^2}{(x^2+1)(x^2 + 2hx + h^2 + 1)} \cdot \frac{1}{h}\right]$$

$$= \lim_{h\to 0} \frac{-12x - 6h}{(x^2+1)(x^2 + 2hx + h^2 + 1)} = -\frac{12x}{(x^2+1)^2}$$

17. $G'(x) = \lim_{h\to 0} \frac{G(x+h) - G(x)}{h}$

$= \lim_{h\to 0}\left[\left(\frac{2(x+h)-1}{x+h-4} - \frac{2x-1}{x-4}\right)\cdot\frac{1}{h}\right]$

$= \lim_{h\to 0}\left[\frac{2x^2+2hx-9x-8h+4-(2x^2+2hx-9x-h+4)}{(x+h-4)(x-4)}\cdot\frac{1}{h}\right] = \lim_{h\to 0}\left[\frac{-7h}{(x+h-4)(x-4)}\cdot\frac{1}{h}\right]$

$= \lim_{h\to 0}\frac{-7}{(x+h-4)(x-4)} = -\frac{7}{(x-4)^2}$

19. $g'(x) = \lim_{h\to 0}\frac{g(x+h)-g(x)}{h}$

$= \lim_{h\to 0}\frac{\sqrt{3(x+h)}-\sqrt{3x}}{h}$

$= \lim_{h\to 0}\frac{(\sqrt{3x+3h}-\sqrt{3x})(\sqrt{3x+3h}+\sqrt{3x})}{h(\sqrt{3x+3h}+\sqrt{3x})}$

$= \lim_{h\to 0}\frac{3h}{h(\sqrt{3x+3h}+\sqrt{3x})} = \lim_{h\to 0}\frac{3}{\sqrt{3x+3h}+\sqrt{3x}} = \frac{3}{2\sqrt{3x}}$

21. $H'(x) = \lim_{h\to 0}\frac{H(x+h)-H(x)}{h}$

$= \lim_{h\to 0}\left[\left(\frac{3}{\sqrt{x+h-2}} - \frac{3}{\sqrt{x-2}}\right)\cdot\frac{1}{h}\right]$

$= \lim_{h\to 0}\left[\frac{3\sqrt{x-2}-3\sqrt{x+h-2}}{\sqrt{(x+h-2)(x-2)}}\cdot\frac{1}{h}\right]$

$= \lim_{h\to 0}\frac{3(\sqrt{x-2}-\sqrt{x+h-2})(\sqrt{x-2}+\sqrt{x+h-2})}{h\sqrt{(x+h-2)(x-2)}(\sqrt{x-2}+\sqrt{x+h-2})}$

$= \lim_{h\to 0}\frac{-3h}{h[(x-2)\sqrt{x+h-2}+(x+h-2)\sqrt{x-2}]}$

$= \lim_{h\to 0}\frac{-3}{(x-2)\sqrt{x+h-2}+(x+h-2)\sqrt{x-2}}$

$= -\frac{3}{2(x-2)\sqrt{x-2}} = -\frac{3}{2(x-2)^{3/2}}$

23. $f'(x) = \lim_{t\to x}\frac{f(t)-f(x)}{t-x}$

$= \lim_{t\to x}\frac{(t^2-3t)-(x^2-3x)}{t-x}$

$= \lim_{t\to x}\frac{t^2-x^2-(3t-3x)}{t-x}$

$= \lim_{t\to x}\frac{(t-x)(t+x)-3(t-x)}{t-x}$

$= \lim_{t\to x}\frac{(t-x)(t+x-3)}{t-x} = \lim_{t\to x}(t+x-3)$

$= 2x-3$

25. $f'(x) = \lim_{t\to x} \frac{f(t)-f(x)}{t-x}$

$= \lim_{t\to x}\left[\left(\frac{t}{t-5}-\frac{x}{x-5}\right)\left(\frac{1}{t-x}\right)\right]$

$= \lim_{t\to x} \frac{tx-5t-tx+5x}{(t-5)(x-5)(t-x)}$

$= \lim_{t\to x} \frac{-5(t-x)}{(t-5)(x-5)(t-x)} = \lim_{t\to x} \frac{-5}{(t-5)(x-5)}$

$= -\frac{5}{(x-5)^2}$

27. $f(x) = 2x^3$ at $x = 5$

29. $f(x) = x^2$ at $x = 2$

31. $f(x) = x^2$ at x

33. $f(t) = \frac{2}{t}$ at t

35. $f(x) = \cos x$ at x

37. The slope of the tangent line is always 2.

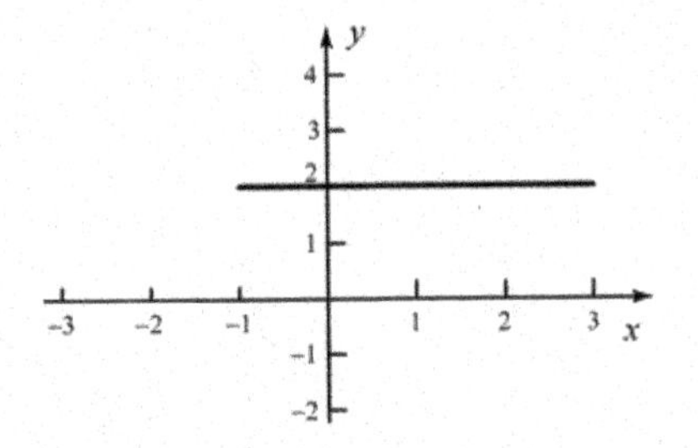

39. The derivative is positive until $x = 0$, then becomes negative.

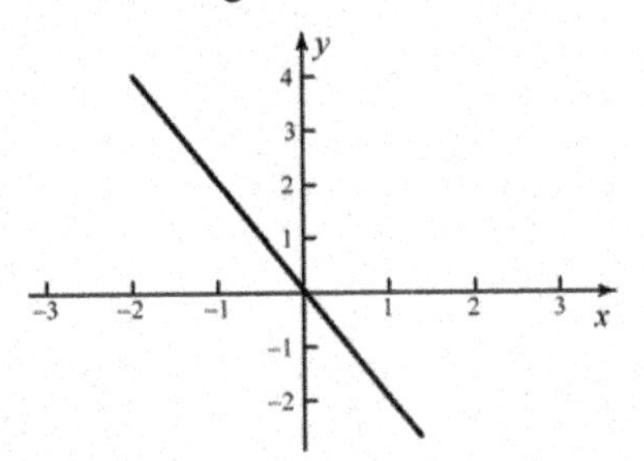

41. The derivative is -1 until $x = 1$. To the right of $x = 1$, the derivative is 1. The derivative is undefined at $x = 1$.

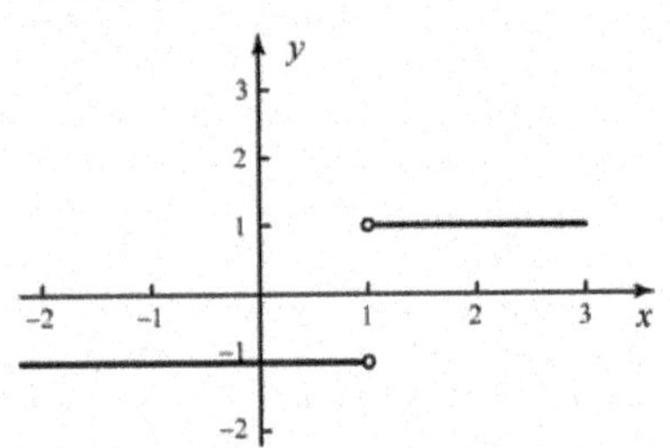

43. The derivative is 0 on $(-3,-2)$, 2 on $(-2,-1)$, 0 on $(-1,0)$, -2 on $(0,1)$, 0 on $(1,2)$, 2 on $(2,3)$ and 0 on $(3,4)$. The derivative is undefined at $x = -2, -1, 0, 1, 2, 3$.

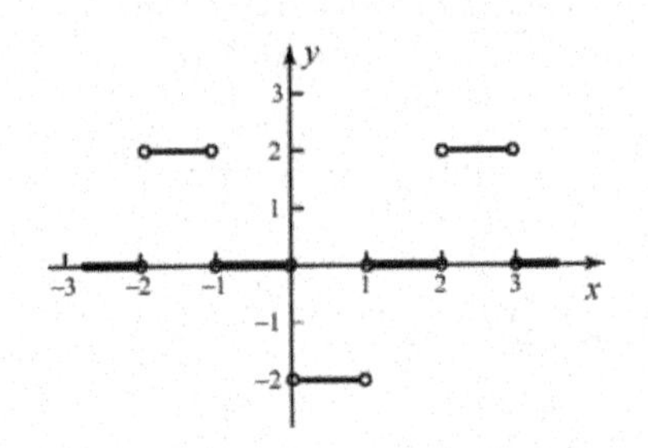

45. $\Delta y = [3(1.5)+2]-[3(1)+2] = 1.5$

47. $\Delta y = 1/1.2 - 1/1 = -0.1667$

49. $\Delta y = \frac{3}{2.31+1} - \frac{3}{2.34+1} \approx 0.0081$

51. $\frac{\Delta y}{\Delta x} = \frac{(x+\Delta x)^2 - x^2}{\Delta x} = \frac{2x\Delta x + (\Delta x)^2}{\Delta x} = 2x + \Delta x$

$\frac{dy}{dx} = \lim_{\Delta x\to 0}(2x+\Delta x) = 2x$

53. $\frac{\Delta y}{\Delta x} = \frac{\frac{1}{x+\Delta x+1} - \frac{1}{x+1}}{\Delta x}$

$= \left(\frac{x+1-(x+\Delta x+1)}{(x+\Delta x+1)(x+1)}\right)\left(\frac{1}{\Delta x}\right)$

$= \frac{-\Delta x}{(x+\Delta x+1)(x+1)\Delta x}$

$= -\frac{1}{(x+\Delta x+1)(x+1)}$

$\frac{dy}{dx} = \lim_{\Delta x\to 0}\left[-\frac{1}{(x+\Delta x+1)(x+1)}\right] = -\frac{1}{(x+1)^2}$

55.

$\frac{\Delta y}{\Delta x} = \frac{\frac{x+\Delta x-1}{x+\Delta x+1} - \frac{x-1}{x+1}}{\Delta x}$

$= \frac{(x+1)(x+\Delta x-1)-(x-1)(x+\Delta x+1)}{(x+\Delta x+1)(x+1)} \times \frac{1}{\Delta x}$

$= \frac{x^2+x\Delta x-x+x+\Delta x-1-\left[x^2+x\Delta x-x+x-\Delta x-1\right]}{x^2+x\Delta x+x+x+\Delta x+1} \times \frac{1}{\Delta x}$

$= \frac{2\Delta x}{x^2+x\Delta x+x+x+\Delta x+1} \times \frac{1}{\Delta x} = \frac{2}{x^2+x\Delta x+x+x+\Delta x+1}$

$\frac{dy}{dx} = \lim_{\Delta x\to 0} \frac{2}{x^2+x\Delta x+x+x+\Delta x+1} = \frac{2}{x^2+2x+1} = \frac{2}{(x+1)^2}$

57. $f'(0) \approx -\frac{1}{2}; f'(2) \approx 1$

$f'(5) \approx \frac{2}{3}; f'(7) \approx -3$

59.

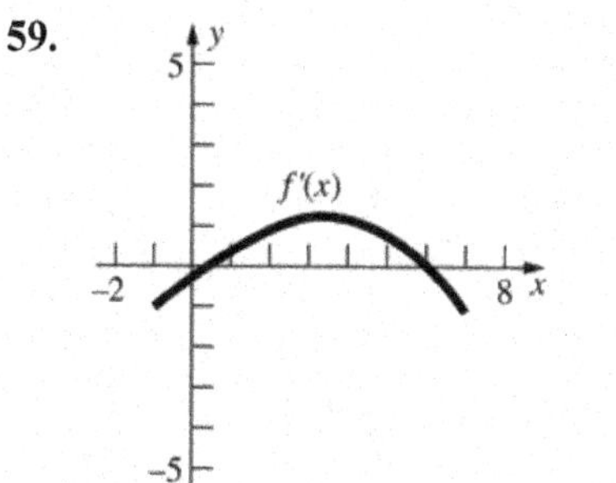

61. a. $f(2) \approx \frac{5}{2}; f'(2) \approx \frac{3}{2}$

$f(0.5) \approx 1.8; f'(0.5) \approx -0.6$

b. $\frac{2.9-1.9}{2.5-0.5} = 0.5$

c. $x = 5$

d. $x = 3, 5$

e. $x = 1, 3, 5$

f. $x = 0$

g. $x \approx -0.7, \frac{3}{2}$ and $5 < x < 7$

63. The derivative is 0 at approximately $t = 15$ and $t = 201$. The greatest rate of increase occurs at about $t = 61$ and it is about 0.5 degree F per day. The greatest rate of decrease occurs at about $t = 320$ and it is about 0.5 degree F per day. The derivative is positive on (15,201) and negative on (0,15) and (201,365).

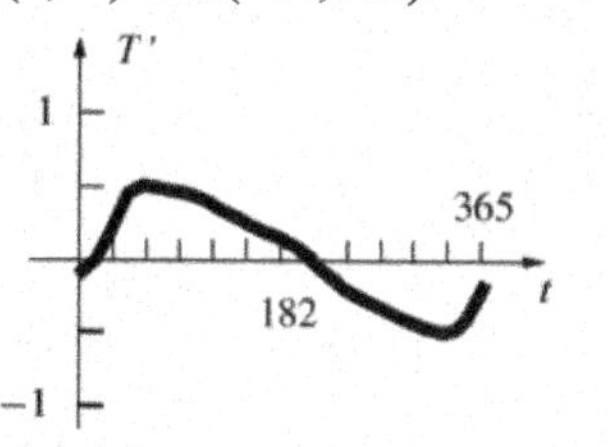

65. The short-dash function has a tangent line with zero slope at about $x = 2.1$, where the solid function is zero. The solid function has a tangent line with zero slope at about $x = 0.4$, 1.2 and 3.5. The long-dash function is zero at these points. The graph shows that the solid function is positive (negative) when the slope of the tangent line of the short-dash function is positive (negative). Also, the long-dash function is positive (negative) when the slope of the tangent line of the solid function is positive (negative). Thus, the short-dash function is *f*, the solid function is $f' = g$, and the dash function is g'.

67. If *f* is differentiable everywhere, then it is continuous everywhere, so
$\lim_{x\to 2^-} f(x) = \lim_{x\to 2^-} (mx+b) = 2m+b = f(2) = 4$
and $b = 4 - 2m$.
For *f* to be differentiable everywhere,

$f'(2) = \lim_{x\to 2} \frac{f(x)-f(2)}{x-2}$ must exist.

$$\lim_{x\to 2^+} \frac{f(x)-f(2)}{x-2} = \lim_{x\to 2^+} \frac{x^2-4}{x-2} = \lim_{x\to 2^+} (x+2) = 4$$

$$\lim_{x\to 2^-} \frac{f(x)-f(2)}{x-2} = \lim_{x\to 2^-} \frac{mx+b-4}{x-2}$$

$$= \lim_{x\to 2^-} \frac{mx+4-2m-4}{x-2} = \lim_{x\to 2^-} \frac{m(x-2)}{x-2} = m$$

Thus $m = 4$ and $b = 4 - 2(4) = -4$

69. $f'(x_0) = \lim_{t\to x_0} \frac{f(t)-f(x_0)}{t-x_0}$, so

$$f'(-x_0) = \lim_{t\to -x_0} \frac{f(t)-f(-x_0)}{t-(-x_0)}$$

$$= \lim_{t\to -x_0} \frac{f(t)-f(-x_0)}{t+x_0}$$

a. If *f* is an odd function,

$$f'(-x_0) = \lim_{t\to -x_0} \frac{f(t)-[-f(-x_0)]}{t+x_0}$$

$$= \lim_{t\to -x_0} \frac{f(t)+f(-x_0)}{t+x_0}.$$

Let $u = -t$. As $t \to -x_0$, $u \to x_0$ and so

$$f'(-x_0) = \lim_{u\to x_0} \frac{f(-u)+f(x_0)}{-u+x_0}$$

$$= \lim_{u\to x_0} \frac{-f(u)+f(x_0)}{-(u-x_0)} = \lim_{u\to x_0} \frac{-[f(u)-f(x_0)]}{-(u-x_0)}$$

$$= \lim_{u\to x_0} \frac{f(u)-f(x_0)}{u-x_0} = f'(x_0) = m.$$

b. If *f* is an even function,

$f'(-x_0) = \lim_{t\to -x_0} \frac{f(t)-f(x_0)}{t+x_0}$. Let $u = -t$, as above, then $f'(-x_0) = \lim_{u\to x_0} \frac{f(-u)-f(x_0)}{-u+x_0}$

$$= \lim_{u\to x_0} \frac{f(u)-f(x_0)}{-(u-x_0)} = -\lim_{u\to x_0} \frac{f(u)-f(x_0)}{u-x_0}$$

$= -f'(x_0) = -m.$

71. 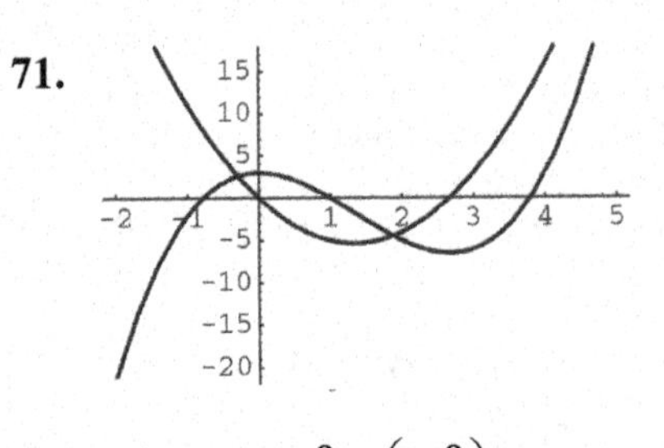

a. $0 < x < \frac{8}{3}$; $\left(0, \frac{8}{3}\right)$

b. $0 \le x \le \frac{8}{3}$; $\left[0, \frac{8}{3}\right]$

c. A function $f(x)$ decreases as x increases when $f'(x) < 0$.

2.3 Concepts Review

1. the derivative of the second; second; $f(x)g'(x) + g(x)f'(x)$

3. $nx^{n-1}h$; nx^{n-1}

Problem Set 2.3

1. $D_x(2x^2) = 2D_x(x^2) = 2 \cdot 2x = 4x$

3. $D_x(\pi x) = \pi D_x(x) = \pi \cdot 1 = \pi$

5. $D_x(2x^{-2}) = 2D_x(x^{-2}) = 2(-2x^{-3}) = -4x^{-3}$

7. $D_x\left(\frac{\pi}{x}\right) = \pi D_x(x^{-1}) = \pi(-1x^{-2}) = -\pi x^{-2}$
$= -\frac{\pi}{x^2}$

9. $D_x\left(\frac{100}{x^5}\right) = 100D_x(x^{-5}) = 100(-5x^{-6})$
$= -500x^{-6} = -\frac{500}{x^6}$

11. $D_x(x^2 + 2x) = D_x(x^2) + 2D_x(x) = 2x + 2$

13. $D_x(x^4 + x^3 + x^2 + x + 1)$
$= D_x(x^4) + D_x(x^3) + D_x(x^2) + D_x(x) + D_x(1)$
$= 4x^3 + 3x^2 + 2x + 1$

15. $D_x(\pi x^7 - 2x^5 - 5x^{-2})$
$= \pi D_x(x^7) - 2D_x(x^5) - 5D_x(x^{-2})$
$= \pi(7x^6) - 2(5x^4) - 5(-2x^{-3})$
$= 7\pi x^6 - 10x^4 + 10x^{-3}$

17. $D_x\left(\frac{3}{x^3} + x^{-4}\right) = 3D_x(x^{-3}) + D_x(x^{-4})$
$= 3(-3x^{-4}) + (-4x^{-5}) = -\frac{9}{x^4} - 4x^{-5}$

19. $D_x\left(\frac{2}{x} - \frac{1}{x^2}\right) = 2D_x(x^{-1}) - D_x(x^{-2})$
$= 2(-1x^{-2}) - (-2x^{-3}) = -\frac{2}{x^2} + \frac{2}{x^3}$

21. $D_x\left(\frac{1}{2x} + 2x\right) = \frac{1}{2}D_x(x^{-1}) + 2D_x(x)$
$= \frac{1}{2}(-1x^{-2}) + 2(1) = -\frac{1}{2x^2} + 2$

23. $D_x[x(x^2 + 1)] = x\,D_x(x^2 + 1) + (x^2 + 1)D_x(x)$
$= x(2x) + (x^2 + 1)(1) = 3x^2 + 1$

25. $D_x[(2x + 1)^2]$
$= (2x + 1)D_x(2x + 1) + (2x + 1)D_x(2x + 1)$
$= (2x + 1)(2) + (2x + 1)(2) = 8x + 4$

27. $D_x[(x^2 + 2)(x^3 + 1)]$
$= (x^2 + 2)D_x(x^3 + 1) + (x^3 + 1)D_x(x^2 + 2)$
$= (x^2 + 2)(3x^2) + (x^3 + 1)(2x)$
$= 3x^4 + 6x^2 + 2x^4 + 2x$
$= 5x^4 + 6x^2 + 2x$

29. $D_x[(x^2+17)(x^3-3x+1)]$

$=(x^2+17)D_x(x^3-3x+1)+(x^3-3x+1)D_x(x^2+17)$

$=(x^2+17)(3x^2-3)+(x^3-3x+1)(2x)$

$=3x^4+48x^2-51+2x^4-6x^2+2x$

$=5x^4+42x^2+2x-51$

31. $D_x[(5x^2-7)(3x^2-2x+1)]=(5x^2-7)D_x(3x^2-2x+1)+(3x^2-2x+1)D_x(5x^2-7)$

$=(5x^2-7)(6x-2)+(3x^2-2x+1)(10x)$

$=60x^3-30x^2-32x+14$

33. $D_x\left(\frac{1}{3x^2+1}\right)=\frac{(3x^2+1)D_x(1)-(1)D_x(3x^2+1)}{(3x^2+1)^2}$

$=\frac{(3x^2+1)(0)-(6x)}{(3x^2+1)^2}=-\frac{6x}{(3x^2+1)^2}$

35. $D_x\left(\frac{1}{4x^2-3x+9}\right)=\frac{(4x^2-3x+9)D_x(1)-(1)D_x(4x^2-3x+9)}{(4x^2-3x+9)^2}$

$=\frac{(4x^2-3x+9)(0)-(8x-3)}{(4x^2-3x+9)^2}=-\frac{8x-3}{(4x^2-3x+9)^2}$

$=\frac{-8x+3}{(4x^2-3x+9)^2}$

37. $D_x\left(\frac{x-1}{x+1}\right)=\frac{(x+1)D_x(x-1)-(x-1)D_x(x+1)}{(x+1)^2}$

$=\frac{(x+1)(1)-(x-1)(1)}{(x+1)^2}=\frac{2}{(x+1)^2}$

39. $D_x\left(\frac{2x^2-1}{3x+5}\right)=\frac{(3x+5)D_x(2x^2-1)-(2x^2-1)D_x(3x+5)}{(3x+5)^2}$

$=\frac{(3x+5)(4x)-(2x^2-1)(3)}{(3x+5)^2}$

$=\frac{6x^2+20x+3}{(3x+5)^2}$

41. $D_x\left(\frac{2x^2-3x+1}{2x+1}\right)=\frac{(2x+1)D_x(2x^2-3x+1)-(2x^2-3x+1)D_x(2x+1)}{(2x+1)^2}$

$=\frac{(2x+1)(4x-3)-(2x^2-3x+1)(2)}{(2x+1)^2}$

$=\frac{4x^2+4x-5}{(2x+1)^2}$

43. $D_x\left(\frac{x^2-x+1}{x^2+1}\right)=\frac{(x^2+1)D_x(x^2-x+1)-(x^2-x+1)D_x(x^2+1)}{(x^2+1)^2}$

$=\frac{(x^2+1)(2x-1)-(x^2-x+1)(2x)}{(x^2+1)^2}$

$=\frac{x^2-1}{(x^2+1)^2}$

45. a. $(f\cdot g)'(0)=f(0)g'(0)+g(0)f'(0)$
$= 4(5) + (–3)(–1) = 23$

b. $(f+g)'(0)=f'(0)+g'(0)=-1+5=4$

c. $(f/g)'(0)=\frac{g(0)f'(0)-f(0)g'(0)}{g^2(0)}$

$=\frac{-3(-1)-4(5)}{(-3)^2}=-\frac{17}{9}$

47. $D_x[f(x)]^2=D_x[f(x)f(x)]$
$=f(x)D_x[f(x)]+f(x)D_x[f(x)]$
$=2\cdot f(x)\cdot D_x f(x)$

49. $D_x(x^2-2x+2)=2x-2$
At $x = 1$: $m_{tan}=2(1)-2=0$
Tangent line: $y = 1$

51. $D_x(x^3-x^2)=3x^2-2x$
The tangent line is horizontal when $m_{tan} = 0$:

$m_{tan}=3x^2-2x=0$

$x(3x-2)=0$

$x=0$ and $x=\frac{2}{3}$

$(0, 0)$ and $\left(\frac{2}{3},-\frac{4}{27}\right)$

53. $y=100/x^5=100x^{-5}$

$y'=-500x^{-6}$

Set y' equal to -1, the negative reciprocal of the slope of the line $y=x$. Solving for x gives

$x=\pm 500^{1/6}\approx \pm 2.817$

$y=\pm 100(500)^{-5/6}\approx \pm 0.563$

The points are $(2.817, 0.563)$ and $(-2.817, -0.563)$.

55. a. $D_t(-16t^2+40t+100)=-32t+40$
$v = –32(2) + 40 = –24$ ft/s

b. $v = –32t + 40 = 0$
$t=\frac{5}{4}$ s

57. $m_{tan}=D_x(4x-x^2)=4-2x$
The line through (2,5) and (x_0, y_0) has slope $\frac{y_0-5}{x_0-2}$.

$4-2x_0=\frac{4x_0-x_0^2-5}{x_0-2}$

$-2x_0^2+8x_0-8=-x_0^2+4x_0-5$

$x_0^2-4x_0+3=0$

$(x_0-3)(x_0-1)=0$

$x_0=1,\ x_0=3$

At $x_0=1$: $y_0=4(1)-(1)^2=3$

$m_{tan}=4-2(1)=2$

Tangent line: $y – 3 = 2(x – 1)$; $y = 2x + 1$

At $x_0=3$: $y_0=4(3)-(3)^2=3$

$m_{tan}=4-2(3)=-2$

Tangent line: $y – 3 = –2(x – 3)$; $y = –2x + 9$

59. $D_x(7-x^2) = -2x$
The line through (4, 0) and (x_0, y_0) has slope $\frac{y_0-0}{x_0-4}$. If the fly is at (x_0, y_0) when the spider sees it, then $m_{\tan} = -2x_0 = \frac{7-x_0^2-0}{x_0-4}$.

$-2x_0^2 + 8x_0 = 7 - x_0^2$
$x_0^2 - 8x_0 + 7 = 0$
$(x_0 - 7)(x_0 - 1) = 0$
At $x_0 = 1: y_0 = 6$
$d = \sqrt{(4-1)^2 + (0-6)^2} = \sqrt{9+36} = \sqrt{45} = 3\sqrt{5}$
≈ 6.7
They are 6.7 units apart when they see each other.

61. The watermelon has volume $\frac{4}{3}\pi r^3$; the volume of the rind is

$$V = \frac{4}{3}\pi r^3 - \frac{4}{3}\pi\left(r - \frac{r}{10}\right)^3 = \frac{271}{750}\pi r^3.$$

At the end of the fifth week $r = 10$, so

$$D_r V = \frac{271}{250}\pi r^2 = \frac{271}{250}\pi(10)^2 = \frac{542\pi}{5} \approx 340 \text{ cm}^3$$

per cm of radius growth. Since the radius is growing 2 cm per week, the volume of the rind is growing at the rate of $\frac{542\pi}{5}(2) \approx 681 \text{ cm}^3$ per week.

2.4 Concepts Review

1. $\frac{\sin(x+h) - \sin(x)}{h}$

3. $\cos x$; $-\sin x$

Problem Set 2.4

1. $D_x(2 \sin x + 3 \cos x) = 2 D_x(\sin x) + 3 D_x(\cos x)$
$= 2 \cos x - 3 \sin x$

3. $D_x(\sin^2 x + \cos^2 x) = D_x(1) = 0$

5. $D_x(\sec x) = D_x\left(\frac{1}{\cos x}\right)$

$$= \frac{\cos x\, D_x(1) - (1) D_x(\cos x)}{\cos^2 x}$$

$$= \frac{\sin x}{\cos^2 x} = \frac{1}{\cos x} \cdot \frac{\sin x}{\cos x} = \sec x \tan x$$

7. $D_x(\tan x) = D_x\left(\frac{\sin x}{\cos x}\right)$

$$= \frac{\cos x\, D_x(\sin x) - \sin x\, D_x(\cos x)}{\cos^2 x}$$

$$= \frac{\cos^2 x + \sin^2 x}{\cos^2 x} = \frac{1}{\cos^2 x} = \sec^2 x$$

9. $D_x\left(\frac{\sin x + \cos x}{\cos x}\right)$

$$= \frac{\cos x\, D_x(\sin x + \cos x) - (\sin x + \cos x) D_x(\cos x)}{\cos^2 x}$$

$$= \frac{\cos x(\cos x - \sin x) - (-\sin^2 x - \sin x \cos x)}{\cos^2 x}$$

$$= \frac{\cos^2 x + \sin^2 x}{\cos^2 x} = \frac{1}{\cos^2 x} = \sec^2 x$$

11. $D_x(\sin x \cos x) = \sin x D_x[\cos x] + \cos x D_x[\sin x]$
$= \sin x(-\sin x) + \cos x(\cos x) = \cos^2 x - \sin^2 x$

13. $D_x\left(\frac{\sin x}{x}\right) = \frac{x D_x(\sin x) - \sin x D_x(x)}{x^2}$

$$= \frac{x \cos x - \sin x}{x^2}$$

15. $D_x(x^2 \cos x) = x^2 D_x(\cos x) + \cos x\, D_x(x^2)$
$= -x^2 \sin x + 2x \cos x$

17. $y = \tan^2 x = (\tan x)(\tan x)$
$D_x y = (\tan x)(\sec^2 x) + (\tan x)(\sec^2 x)$
$= 2 \tan x \sec^2 x$

19. $D_x(\cos x) = -\sin x$
At $x = 1$: $m_{\tan} = -\sin 1 \approx -0.8415$
$y = \cos 1 \approx 0.5403$
Tangent line: $y - 0.5403 = -0.8415(x - 1)$

21. $D_x \sin 2x = D_x(2 \sin x \cos x)$
$= 2[\sin x\, D_x \cos x + \cos x\, D_x \sin x]$
$= -2\sin^2 x + 2\cos^2 x$

23. $D_t(30\sin 2t) = 30D_t(2\sin t\cos t)$

$$= 30\left(-2\sin^2 t + 2\cos^2 t\right)$$

$$= 60\cos 2t$$

$30\sin 2t = 15$

$\sin 2t = \frac{1}{2}$

$2t = \frac{\pi}{6} \rightarrow t = \frac{\pi}{12}$

At $t = \frac{\pi}{12}$; $60\cos\left(2\cdot\frac{\pi}{12}\right) = 30\sqrt{3}$ ft/sec

The seat is moving to the left at the rate of $30\sqrt{3}$ ft/s.

25. $y = \tan x$

$y' = \sec^2 x$

When $y = 0$, $y = \tan 0 = 0$ and $y' = \sec^2 0 = 1$.
The tangent line at $x = 0$ is $y = x$.

27. $y = 9\sin x\cos x$

$y' = 9[\sin x(-\sin x) + \cos x(\cos x)]$

$$= 9\left[\sin^2 x - \cos^2 x\right]$$

$$= 9[-\cos 2x]$$

The tangent line is horizontal when $y' = 0$ or, in this case, where $\cos 2x = 0$. This occurs when $x = \frac{\pi}{4} + k\frac{\pi}{2}$ where k is an integer.

29. The curves intersect when $\sqrt{2}\sin x = \sqrt{2}\cos x$, $\sin x = \cos x$ at $x = \frac{\pi}{4}$ for $0 < x < \frac{\pi}{2}$.

$D_x(\sqrt{2}\sin x) = \sqrt{2}\cos x$; $\sqrt{2}\cos\frac{\pi}{4} = 1$

$D_x(\sqrt{2}\cos x) = -\sqrt{2}\sin x$; $-\sqrt{2}\sin\frac{\pi}{4} = -1$

$1(-1) = -1$ so the curves intersect at right angles.

31. $D_x(\sin x^2) = \lim_{h\to 0}\frac{\sin(x+h)^2 - \sin x^2}{h}$

$$= \lim_{h\to 0}\frac{\sin(x^2 + 2xh + h^2) - \sin x^2}{h}$$

$$= \lim_{h\to 0}\frac{\sin x^2\cos(2xh+h^2) + \cos x^2\sin(2xh+h^2) - \sin x^2}{h} = \lim_{h\to 0}\frac{\sin x^2[\cos(2xh+h^2) - 1] + \cos x^2\sin(2xh+h^2)}{h}$$

$$= \lim_{h\to 0}(2x+h)\left[\sin x^2\frac{\cos(2xh+h^2)-1}{2xh+h^2} + \cos x^2\frac{\sin(2xh+h^2)}{2xh+h^2}\right] = 2x(\sin x^2\cdot 0 + \cos x^2\cdot 1) = 2x\cos x^2$$

33. $f(x) = x\sin x$

a.

b. $f(x) = 0$ has 6 solutions on $[\pi, 6\pi]$
$f'(x) = 0$ has 5 solutions on $[\pi, 6\pi]$

c. $f(x) = x\sin x$ is a counterexample.
Consider the interval $[0, \pi]$.
$f(-\pi) = f(\pi) = 0$ and $f(x) = 0$ has exactly two solutions in the interval (at 0 and π). However, $f'(x) = 0$ has two solutions in the interval, not 1 as the conjecture indicates it should have.

d. The maximum value of $|f(x) - f'(x)|$ on $[\pi, 6\pi]$ is about 24.93.

2.5 Concepts Review

1. $D_t u$; $f'(g(t))g'(t)$

3. $(f(x))^2$; $(f(x))^2$

Problem Set 2.5

1. $y = u^{15}$ and $u = 1 + x$
$D_x y = D_u y \cdot D_x u$
$= (15u^{14})(1)$
$= 15(1+x)^{14}$

3. $y = u^5$ and $u = 3 - 2x$
$D_x y = D_u y \cdot D_x u$
$= (5u^4)(-2) = -10(3-2x)^4$

5. $y = u^{11}$ and $u = x^3 - 2x^2 + 3x + 1$
$D_x y = D_u y \cdot D_x u$
$= (11u^{10})(3x^2 - 4x + 3)$
$= 11(3x^2 - 4x + 3)(x^3 - 2x^2 + 3x + 1)^{10}$

7. $y = u^{-5}$ and $u = x + 3$
$D_x y = D_u y \cdot D_x u$
$= (-5u^{-6})(1) = -5(x+3)^{-6} = -\dfrac{5}{(x+3)^6}$

9. $y = \sin u$ and $u = x^2 + x$
$D_x y = D_u y \cdot D_x u$
$= (\cos u)(2x + 1)$
$= (2x+1)\cos(x^2 + x)$

11. $y = u^3$ and $u = \cos x$
$D_x y = D_u y \cdot D_x u$
$= (3u^2)(-\sin x)$
$= -3\sin x \cos^2 x$

13. $y = u^3$ and $u = \dfrac{x+1}{x-1}$
$D_x y = D_u y \cdot D_x u$
$= (3u^2)\dfrac{(x-1)D_x(x+1) - (x+1)D_x(x-1)}{(x-1)^2}$
$= 3\left(\dfrac{x+1}{x-1}\right)^2\left(\dfrac{-2}{(x-1)^2}\right) = -\dfrac{6(x+1)^2}{(x-1)^4}$

15. $y = \cos u$ and $u = \dfrac{3x^2}{x+2}$

$$D_x y = D_u y \cdot D_x u = (-\sin u)\frac{(x+2)D_x(3x^2) - (3x^2)D_x(x+2)}{(x+2)^2}$$
$$= -\sin\left(\frac{3x^2}{x+2}\right)\frac{(x+2)(6x) - (3x^2)(1)}{(x+2)^2} = -\frac{3x^2 + 12x}{(x+2)^2}\sin\left(\frac{3x^2}{x+2}\right)$$

17. $D_x[(3x-2)^2(3-x^2)^2] = (3x-2)^2 D_x(3-x^2)^2 + (3-x^2)^2 D_x(3x-2)^2$
$= (3x-2)^2(2)(3-x^2)(-2x) + (3-x^2)^2(2)(3x-2)(3)$
$= 2(3x-2)(3-x^2)[(3x-2)(-2x) + (3-x^2)(3)] = 2(3x-2)(3-x^2)(9 + 4x - 9x^2)$

19. $$D_x\left[\frac{(x+1)^2}{3x-4}\right] = \frac{(3x-4)D_x(x+1)^2 - (x+1)^2 D_x(3x-4)}{(3x-4)^2} = \frac{(3x-4)(2)(x+1)(1) - (x+1)^2(3)}{(3x-4)^2} = \frac{3x^2 - 8x - 11}{(3x-4)^2}$$
$$= \frac{(x+1)(3x-11)}{(3x-4)^2}$$

21. $y' = 2(x^2+4)(x^2+4)' = 2(x^2+4)(2x) = 4x(x^2+4)$

23. $D_t\left(\frac{3t-2}{t+5}\right)^3=3\left(\frac{3t-2}{t+5}\right)^2\frac{(t+5)D_t(3t-2)-(3t-2)D_t(t+5)}{(t+5)^2}$

$=3\left(\frac{3t-2}{t+5}\right)^2\frac{(t+5)(3)-(3t-2)(1)}{(t+5)^2}=\frac{51(3t-2)^2}{(t+5)^4}$

25. $\frac{d}{dt}\left(\frac{(3t-2)^3}{t+5}\right)=\frac{(t+5)\frac{d}{dt}(3t-2)^3-(3t-2)^3\frac{d}{dt}(t+5)}{(t+5)^2}=\frac{(t+5)(3)(3t-2)^2(3)-(3t-2)^3(1)}{(t+5)^2}$

$=\frac{(6t+47)(3t-2)^2}{(t+5)^2}$

27. $\frac{dy}{dx}=\frac{d}{dx}\left(\frac{\sin x}{\cos 2x}\right)^3=3\left(\frac{\sin x}{\cos 2x}\right)^2\cdot\frac{d}{dx}\frac{\sin x}{\cos 2x}=3\left(\frac{\sin x}{\cos 2x}\right)^2\cdot\frac{(\cos 2x)\frac{d}{dx}(\sin x)-(\sin x)\frac{d}{dx}(\cos 2x)}{\cos^2 2x}$

$=3\left(\frac{\sin x}{\cos 2x}\right)^2\frac{\cos x\cos 2x+2\sin x\sin 2x}{\cos^2 2x}=\frac{3\sin^2 x\cos x\cos 2x+6\sin^3 x\sin 2x}{\cos^4 2x}$

$=\frac{3(\sin^2 x)(\cos x\cos 2x+2\sin x\sin 2x)}{\cos^4 2x}$

29. $f'(x)=3\left(\frac{x^2+1}{x+2}\right)^2\frac{(x+2)D_x(x^2+1)-(x^2+1)D_x(x+2)}{(x+2)^2}=3\left(\frac{x^2+1}{x+2}\right)^2\frac{2x^2+4x-x^2-1}{(x+2)^2}=\frac{3(x^2+1)^2(x^2+4x-1)}{(x+2)^4}$

$f'(3)=9.6$

31. $F'(t)=[\cos(t^2+3t+1)](2t+3)=(2t+3)\cos(t^2+3t+1)$; $F'(1)=5\cos 5\approx 1.4183$

33. $D_x[\sin^4(x^2+3x)]=4\sin^3(x^2+3x)D_x\sin(x^2+3x)=4\sin^3(x^2+3x)\cos(x^2+3x)D_x(x^2+3x)$

$=4\sin^3(x^2+3x)\cos(x^2+3x)(2x+3)=4(2x+3)\sin^3(x^2+3x)\cos(x^2+3x)$

35. $D_t[\sin^3(\cos t)]=3\sin^2(\cos t)D_t\sin(\cos t)=3\sin^2(\cos t)\cos(\cos t)D_t(\cos t)$

$=3\sin^2(\cos t)\cos(\cos t)(-\sin t)=-3\sin t\sin^2(\cos t)\cos(\cos t)$

37. $D_\theta[\cos^4(\sin\theta^2)]=4\cos^3(\sin\theta^2)D_\theta\cos(\sin\theta^2)=4\cos^3(\sin\theta^2)[-\sin(\sin\theta^2)]D_\theta(\sin\theta^2)$

$=-4\cos^3(\sin\theta^2)\sin(\sin\theta^2)(\cos\theta^2)D_\theta(\theta^2)=-8\theta\cos^3(\sin\theta^2)\sin(\sin\theta^2)(\cos\theta^2)$

39. $D_x\{\sin[\cos(\sin 2x)]\}=\cos[\cos(\sin 2x)]D_x\cos(\sin 2x)=\cos[\cos(\sin 2x)][-\sin(\sin 2x)]D_x(\sin 2x)$

$=-\cos[\cos(\sin 2x)]\sin(\sin 2x)(\cos 2x)D_x(2x)=-2\cos[\cos(\sin 2x)]\sin(\sin 2x)(\cos 2x)$

41. $(f+g)'(4)=f'(4)+g'(4)$

$\approx\frac{1}{2}+\frac{3}{2}\approx 2$

43. $(fg)'(2)=(fg'+gf')(2)=2(0)+1(1)=1$

45. $(f\circ g)'(6)=f'(g(6))g'(6)$

$=f'(2)g'(6)\approx(1)(-1)=-1$

47. $D_xF(2x)=F'(2x)D_x(2x)=2F'(2x)$

49. $D_t\left[(F(t))^{-2}\right]=-2(F(t))^{-3}F'(t)$

51. $\frac{d}{dz}\left[(1+F(2z))^2\right]=2(1+F(2z))\frac{d}{dz}(1+F(2z))$

$=2(1+F(2z))(2F'(2z))=4(1+F(2z))F'(2z)$

53. $\frac{d}{dx}F(\cos x) = F'(\cos x)\frac{d}{dx}(\cos x)$
$= -\sin x F'(\cos x)$

55. $D_x\left[\tan(F(2x))\right] = \sec^2(F(2x))D_x\left[F(2x)\right]$
$= \sec^2(F(2x)) \times F'(2x) \times D_x[2x]$
$= 2F'(2x)\sec^2(F(2x))$

57. $D_x\left[F(x)\sin^2 F(x)\right]$
$= F(x) \times D_x\left[\sin^2 F(x)\right] + \sin^2 F(x) \times D_x F(x)$
$= F(x) \times 2\sin F(x) \times D_x\left[\sin F(x)\right]$
$+ F'(x)\sin^2 F(x)$
$= F(x) \times 2\sin F(x) \times \cos(F(x)) \times D_x\left[F(x)\right]$
$+ F'(x)\sin^2 F(x)$
$= 2F(x)F'(x)\sin F(x)\cos F(x)$
$+ F'(x)\sin^2 F(x)$

59. $g'(x) = -\sin f(x) D_x f(x) = -f'(x)\sin f(x)$
$g'(0) = -f'(0)\sin f(0) = -2\sin 1 \approx -1.683$

61. $F'(x) = -f(x)g'(x)\sin g(x) + f'(x)\cos g(x)$
$F'(1) = -f(1)g'(1)\sin g(1) + f'(1)\cos g(1)$
$= -2(1)\sin 0 + -1\cos 0 = -1$

63. $y = \sin^2 x; \quad y' = 2\sin x \cos x = \sin 2x = 1$
$x = \pi/4 + k\pi,\ k = 0, \pm 1, \pm 2, \ldots$

65. $y' = -2\left(x^2+1\right)^{-3}(2x) = -4x\left(x^2+1\right)^{-3}$
$y'(1) = -4(1)(1+1)^{-3} = -1/2$
$y - \frac{1}{4} = -\frac{1}{2}x + \frac{1}{2}, \quad y = -\frac{1}{2}x + \frac{3}{4}$

67. $y' = -2\left(x^2+1\right)^{-3}(2x) = -4x\left(x^2+1\right)^{-3}$
$y'(1) = -4(2)^{-3} = -1/2$
$y - \frac{1}{4} = -\frac{1}{2}x + \frac{1}{2}, \quad y = -\frac{1}{2}x + \frac{3}{4}$
Set $y = 0$ and solve for x. The line crosses the x-axis at $x = 3/2$.

69. **a.** $(10\cos 8\pi t, 10\sin 8\pi t)$

b. $D_t(10\sin 8\pi t) = 10\cos(8\pi t)D_t(8\pi t)$
$= 80\pi\cos(8\pi t)$
At $t = 1$: rate $= 80\pi \approx 251$ cm/s
P is rising at the rate of 251 cm/s.

71. 60 revolutions per minute is 120π radians per minute or 2π radians per second.

a. $(\cos 2\pi t, \sin 2\pi t)$

b. $(0 - \cos 2\pi t)^2 + (y - \sin 2\pi t)^2 = 5^2$, so
$y = \sin 2\pi t + \sqrt{25 - \cos^2 2\pi t}$

c. $D_t\left(\sin 2\pi t + \sqrt{25 - \cos^2 2\pi t}\right)$
$= 2\pi\cos 2\pi t$
$+ \frac{1}{2\sqrt{25-\cos^2 2\pi t}} \cdot 4\pi\cos 2\pi t \sin 2\pi t$
$= 2\pi\cos 2\pi t\left(1 + \frac{\sin 2\pi t}{\sqrt{25 - \cos^2 2\pi t}}\right)$

73. The minute hand makes 1 revolution every hour, so at t minutes after noon it makes an angle of $\frac{\pi t}{30}$ radians with the vertical. Similarly, at t minutes after noon the hour hand makes an angle of $\frac{\pi t}{360}$ with the vertical. Thus, by the Law of Cosines, the distance between the tips of the hands is

$s = \sqrt{6^2 + 8^2 - 2 \cdot 6 \cdot 8\cos\left(\frac{\pi t}{30} - \frac{\pi t}{360}\right)}$
$= \sqrt{100 - 96\cos\frac{11\pi t}{360}}$
$\frac{ds}{dt} = \frac{1}{2\sqrt{100 - 96\cos\frac{11\pi t}{360}}} \cdot \frac{44\pi}{15}\sin\frac{11\pi t}{360}$
$= \frac{22\pi\sin\frac{11\pi t}{360}}{15\sqrt{100 - 96\cos\frac{11\pi t}{360}}}$

At 12:20,
$\frac{ds}{dt} = \frac{22\pi\sin\frac{11\pi}{18}}{15\sqrt{100 - 96\cos\frac{11\pi}{18}}} \approx 0.38$ in./min

75. $\sin x_0 = \sin 2x_0$

$\sin x_0 = 2\sin x_0 \cos x_0$

$\cos x_0 = \frac{1}{2}$ [if $\sin x_0 \neq 0$]

$x_0 = \frac{\pi}{3}$

$D_x(\sin x) = \cos x$, $D_x(\sin 2x) = 2\cos 2x$, so at x_0, the tangent lines to $y = \sin x$ and $y = \sin 2x$ have slopes of $m_1 = \frac{1}{2}$ and $m_2 = 2\left(-\frac{1}{2}\right) = -1$, respectively. From Problem 40 of Section 0.7, $\tan\theta = \frac{m_2 - m_1}{1 + m_1 m_2}$ where θ is the angle between the tangent lines. $\tan\theta = \frac{-1-\frac{1}{2}}{1+\left(\frac{1}{2}\right)(-1)} = \frac{-\frac{3}{2}}{\frac{1}{2}} = -3$, so $\theta \approx -1.25$. The curves intersect at an angle of 1.25 radians.

77. $y = \sqrt{u}$ and $u = x^2$

$D_x y = D_u y \cdot D_x u$

$= \frac{1}{2\sqrt{u}} \cdot 2x = \frac{2x}{2\sqrt{x^2}} = \frac{x}{|x|} = \frac{|x|}{x}$

79. $D_x|\sin x| = \frac{|\sin x|}{\sin x} D_x(\sin x)$

$= \frac{|\sin x|}{\sin x}\cos x = \cot x|\sin x|$

81. $[f(f(f(f(0))))]'$

$= f'(f(f(f(0)))) \cdot f'(f(f(0))) \cdot f'(f(0)) \cdot f'(0)$

$= 2 \cdot 2 \cdot 2 \cdot 2 = 16$

83. $D_x\left(\frac{f(x)}{g(x)}\right) = D_x\left(f(x)\cdot\frac{1}{g(x)}\right) = D_x\left(f(x)\cdot(g(x))^{-1}\right) = f(x)D_x\left((g(x))^{-1}\right) + (g(x))^{-1}D_x f(x)$

$= f(x)\cdot(-1)(g(x))^{-2}D_x g(x) + (g(x))^{-1}D_x f(x) = -f(x)(g(x))^{-2}D_x g(x) + (g(x))^{-1}D_x f(x)$

$= \frac{-f(x)D_x g(x)}{g^2(x)} + \frac{D_x f(x)}{g(x)} = \frac{-f(x)D_x g(x)}{g^2(x)} + \frac{g(x)}{g(x)}\cdot\frac{D_x f(x)}{g(x)} = \frac{-f(x)D_x g(x)}{g^2(x)} + \frac{g(x)D_x f(x)}{g^2(x)}$

$= \frac{g(x)D_x f(x) - f(x)D_x g(x)}{g^2(x)}$

2.6 Concepts Review

1. $f'''(x), D_x^3 y, \frac{d^3 y}{dx^3}, y'''$

3. $f'(t) > 0$

Problem Set 2.6

1. $\frac{dy}{dx} = 3x^2 + 6x + 6$

$\frac{d^2 y}{dx^2} = 6x + 6$

$\frac{d^3 y}{dx^3} = 6$

3. $\frac{dy}{dx} = 3(3x+5)^2(3) = 9(3x+5)^2$

$\frac{d^2 y}{dx^2} = 18(3x+5)(3) = 162x + 270$

$\frac{d^3 y}{dx^3} = 162$

5. $\frac{dy}{dx} = 7\cos(7x)$

$\frac{d^2 y}{dx^2} = -7^2 \sin(7x)$

$\frac{d^3 y}{dx^3} = -7^3 \cos(7x) = -343\cos(7x)$

7. $\frac{dy}{dx} = \frac{(x-1)(0)-(1)(1)}{(x-1)^2} = -\frac{1}{(x-1)^2}$

$\frac{d^2 y}{dx^2} = -\frac{(x-1)^2(0)-2(x-1)}{(x-1)^4} = \frac{2}{(x-1)^3}$

$\frac{d^3 y}{dx^3} = \frac{(x-1)^3(0)-2[3(x-1)^2]}{(x-1)^6}$

$= -\frac{6}{(x-1)^4}$

9. $f'(x) = 2x; f''(x) = 2; f''(2) = 2$

11. $f'(t) = -\frac{2}{t^2}$

$f''(t) = \frac{4}{t^3}$

$f''(2) = \frac{4}{8} = \frac{1}{2}$

13. $f'(\theta) = -2(\cos\theta\pi)^{-3}(-\sin\theta\pi)\pi = 2\pi(\cos\theta\pi)^{-3}(\sin\theta\pi)$

$f''(\theta) = 2\pi[(\cos\theta\pi)^{-3}(\pi)(\cos\theta\pi) + (\sin\theta\pi)(-3)(\cos\theta\pi)^{-4}(-\sin\theta\pi)(\pi)] = 2\pi^2[(\cos\theta\pi)^{-2} + 3\sin^2\theta\pi(\cos\theta\pi)^{-4}]$

$f''(2) = 2\pi^2[1+3(0)(1)] = 2\pi^2$

15. $f'(s) = s(3)(1-s^2)^2(-2s) + (1-s^2)^3 = -6s^2(1-s^2)^2 + (1-s^2)^3 = -7s^6 + 15s^4 - 9s^2 + 1$

$f''(s) = -42s^5 + 60s^3 - 18s$

$f''(2) = -900$

17. $D_x(x^n) = nx^{n-1}$

$D_x^2(x^n) = n(n-1)x^{n-2}$

$D_x^3(x^n) = n(n-1)(n-2)x^{n-3}$

$D_x^4(x^n) = n(n-1)(n-2)(n-3)x^{n-4}$

$\vdots$

$D_x^{n-1}(x^n) = n(n-1)(n-2)(n-3)...(2)x$

$D_x^n(x^n) = n(n-1)(n-2)(n-3)...2(1)x^0 = n!$

19. a. $D_x^4(3x^3 + 2x - 19) = 0$

b. $D_x^{12}(100x^{11} - 79x^{10}) = 0$

c. $D_x^{11}(x^2-3)^5 = 0$

21. $f'(x) = 3x^2 + 6x - 45 = 3(x+5)(x-3)$

$3(x+5)(x-3) = 0$

$x = -5, x = 3$

$f''(x) = 6x + 6$

$f''(-5) = -24$

$f''(3) = 24$

23. a. $v(t) = \frac{ds}{dt} = 12 - 4t$

$a(t) = \frac{d^2s}{dt^2} = -4$

b. $12 - 4t > 0$
$4t < 12$
$t < 3;\ (-\infty, 3)$

c. $12 - 4t < 0$
$t > 3;\ (3, \infty)$

d. $a(t) = -4 < 0$ for all t

e.

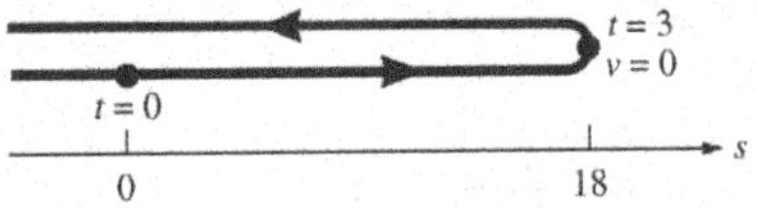

$t = 3$, $v = 0$
$t = 0$
0   18 s

25. a. $v(t) = \frac{ds}{dt} = 3t^2 - 18t + 24$

$a(t) = \frac{d^2s}{dt^2} = 6t - 18$

b. $3t^2 - 18t + 24 > 0$
$3(t-2)(t-4) > 0$
$(-\infty, 2) \cup (4, \infty)$

c. $3t^2 - 18t + 24 < 0$
$(2, 4)$

d. $6t - 18 < 0$
$6t < 18$
$t < 3;\ (-\infty, 3)$

e.

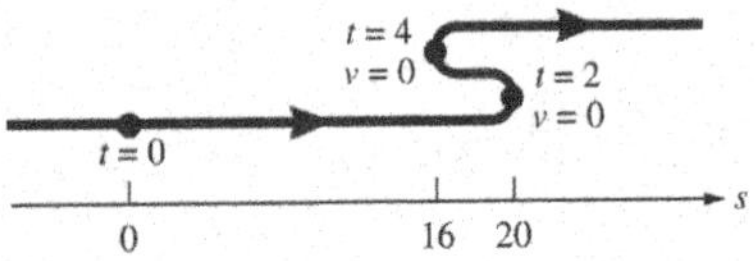

$t = 4$, $v = 0$
$t = 2$, $v = 0$
$t = 0$
0   16   20 s

27. a. $v(t) = \frac{ds}{dt} = 2t - \frac{16}{t^2}$

$a(t) = \frac{d^2s}{dt^2} = 2 + \frac{32}{t^3}$

b. $2t - \frac{16}{t^2} > 0$

$\frac{2t^3 - 16}{t^2} > 0;\ (2, \infty)$

c. $2t - \frac{16}{t^2} < 0;\ (0, 2)$

d. $2 + \frac{32}{t^3} < 0$

$\frac{2t^3 + 32}{t^3} < 0$; The acceleration is not negative for any positive t.

e.

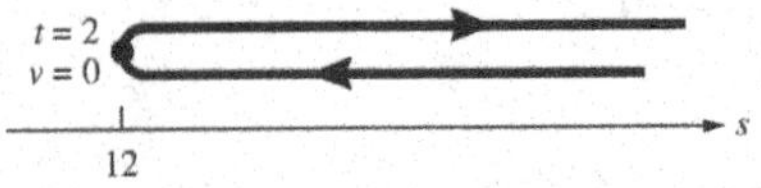

$t = 2$, $v = 0$
12   s

29. $v(t) = \frac{ds}{dt} = 2t^3 - 15t^2 + 24t$

$a(t) = \frac{d^2s}{dt^2} = 6t^2 - 30t + 24$

$6t^2 - 30t + 24 = 0$
$6(t-4)(t-1) = 0$
$t = 4, 1$
$v(4) = -16,\ v(1) = 11$

31. $v_1(t) = \frac{ds_1}{dt} = 4 - 6t$

$v_2(t) = \frac{ds_2}{dt} = 2t - 2$

a. $4 - 6t = 2t - 2$
$8t = 6$
$t = \frac{3}{4}$ sec

b. $|4 - 6t| = |2t - 2|;\ 4 - 6t = -2t + 2$

$t = \frac{1}{2}$ sec and $t = \frac{3}{4}$ sec

c. $4t - 3t^2 = t^2 - 2t$
$4t^2 - 6t = 0$
$2t(2t - 3) = 0$
$t = 0$ sec and $t = \frac{3}{2}$ sec

33. a. $v(t) = -32t + 48$
initial velocity $= v_0 = 48$ ft/sec

b. $-32t + 48 = 0$
$t = \frac{3}{2}$ sec

c. $s = -16(1.5)^2 + 48(1.5) + 256 = 292$ ft

d. $-16t^2 + 48t + 256 = 0$

$t = \frac{-48 \pm \sqrt{48^2 - 4(-16)(256)}}{-32} \approx -2.77, 5.77$

The object hits the ground at $t = 5.77$ sec.

e. $v(5.77) \approx -137$ ft/sec;
speed $= |-137| = 137$ ft/sec.

35. $v(t) = v_0 - 32t$

$v_0 - 32t = 0$

$t = \frac{v_0}{32}$

$v_0\left(\frac{v_0}{32}\right) - 16\left(\frac{v_0}{32}\right)^2 = 5280$

$\frac{v_0^2}{32} - \frac{v_0^2}{64} = 5280$

$\frac{v_0^2}{64} = 5280$

$v_0 = \sqrt{337,920} \approx 581$ ft/sec

37. $v(t) = 3t^2 - 6t - 24$

$\frac{d}{dt}\left|3t^2 - 6t - 24\right| = \frac{\left|3t^2 - 6t - 24\right|}{3t^2 - 6t - 24}(6t - 6)$

$= \frac{|(t-4)(t+2)|}{(t-4)(t+2)}(6t - 6)$

$\frac{|(t-4)(t+2)|(6t-6)}{(t-4)(t+2)} < 0$

$t < -2,\ 1 < t < 4;\ (-\infty, -2) \cup (1, 4)$

39. $D_x(uv) = uv' + u'v$

$D_x^2(uv) = uv'' + u'v' + u'v' + u''v$

$= uv'' + 2u'v' + u''v$

$D_x^3(uv) = uv''' + u'v'' + 2(u'v'' + u''v') + u''v' + u'''v$

$= uv''' + 3u'v'' + 3u''v' + u'''v$

$D_x^n(uv) = \sum_{k=0}^{n}\binom{n}{k} D_x^{n-k}(u) D_x^k(v)$

where $\binom{n}{k}$ is the binomial coefficient

$\frac{n!}{(n-k)!k!}$.

41. a.

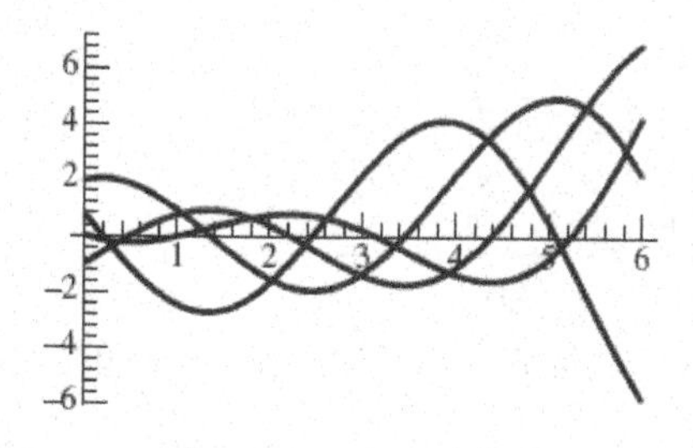

b. $f'''(2.13) \approx -1.2826$

2.7 Concepts Review

1. $\frac{9}{x^3 - 3}$

3. $x(2y)\frac{dy}{dx} + y^2 + 3y^2\frac{dy}{dx} - \frac{dy}{dx} = 3x^2$

Problem Set 2.7

1. $2y\,D_x y - 2x = 0$

$D_x y = \frac{2x}{2y} = \frac{x}{y}$

3. $x\,D_x y + y = 0$

$D_x y = -\frac{y}{x}$

5. $x(2y)D_x y + y^2 = 1$

$D_x y = \frac{1 - y^2}{2xy}$

7. $12x^2 + 7x(2y)D_x y + 7y^2 = 6y^2 D_x y$

$12x^2 + 7y^2 = 6y^2 D_x y - 14xyD_x y$

$D_x y = \frac{12x^2 + 7y^2}{6y^2 - 14xy}$

9. $\frac{1}{2\sqrt{5xy}} \cdot (5x\,D_x y + 5y) + 2D_x y$

$= 2y\,D_x y + x(3y^2)D_x y + y^3$

$\frac{5x}{2\sqrt{5xy}} D_x y + 2D_x y - 2y\,D_x y - 3xy^2 D_x y$

$= y^3 - \frac{5y}{2\sqrt{5xy}}$

$D_x y = \frac{y^3 - \frac{5y}{2\sqrt{5xy}}}{\frac{5x}{2\sqrt{5xy}} + 2 - 2y - 3xy^2}$

11. $x\,D_x y + y + \cos(xy)(x\,D_x y + y) = 0$

$x\,D_x y + x\cos(xy)D_x y = -y - y\cos(xy)$

$D_x y = \frac{-y - y\cos(xy)}{x + x\cos(xy)} = -\frac{y}{x}$

13. $x^3y' + 3x^2y + y^3 + 3xy^2y' = 0$

$y'(x^3 + 3xy^2) = -3x^2y - y^3$

$y' = \dfrac{-3x^2y - y^3}{x^3 + 3xy^2}$

At (1, 3), $y' = -\dfrac{36}{28} = -\dfrac{9}{7}$

Tangent line: $y - 3 = -\dfrac{9}{7}(x - 1)$

15. $\cos(xy)(xy' + y) = y'$

$y'[x\cos(xy) - 1] = -y\cos(xy)$

$y' = \dfrac{-y\cos(xy)}{x\cos(xy) - 1} = \dfrac{y\cos(xy)}{1 - x\cos(xy)}$

At $\left(\dfrac{\pi}{2}, 1\right)$, $y' = 0$

Tangent line: $y - 1 = 0\left(x - \dfrac{\pi}{2}\right)$

$y = 1$

17. $\dfrac{2}{3}x^{-1/3} - \dfrac{2}{3}y^{-1/3}y' - 2y' = 0$

$\dfrac{2}{3}x^{-1/3} = y'\left(\dfrac{2}{3}y^{-1/3} + 2\right)$

$y' = \dfrac{\frac{2}{3}x^{-1/3}}{\frac{2}{3}y^{-1/3} + 2}$

At (1, –1), $y' = \dfrac{\frac{2}{3}}{\frac{4}{3}} = \dfrac{1}{2}$

Tangent line: $y + 1 = \dfrac{1}{2}(x - 1)$

19. $\dfrac{dy}{dx} = 5x^{2/3} + \dfrac{1}{2\sqrt{x}}$

21. $\dfrac{dy}{dx} = \dfrac{1}{3}x^{-2/3} - \dfrac{1}{3}x^{-4/3} = \dfrac{1}{3\sqrt[3]{x^2}} - \dfrac{1}{3\sqrt[3]{x^4}}$

23. $\dfrac{dy}{dx} = \dfrac{1}{4}(3x^2 - 4x)^{-3/4}(6x - 4)$

$= \dfrac{6x - 4}{4\sqrt[4]{(3x^2 - 4x)^3}} = \dfrac{3x - 2}{2\sqrt[4]{(3x^2 - 4x)^3}}$

25. $\dfrac{dy}{dx} = \dfrac{d}{dx}[(x^3 + 2x)^{-2/3}]$

$= -\dfrac{2}{3}(x^3 + 2x)^{-5/3}(3x^2 + 2) = -\dfrac{6x^2 + 4}{3\sqrt[3]{(x^3 + 2x)^5}}$

27. $\dfrac{dy}{dx} = \dfrac{1}{2\sqrt{x^2 + \sin x}}(2x + \cos x)$

$= \dfrac{2x + \cos x}{2\sqrt{x^2 + \sin x}}$

29. $\dfrac{dy}{dx} = \dfrac{d}{dx}[(x^2 \sin x)^{-1/3}]$

$= -\dfrac{1}{3}(x^2 \sin x)^{-4/3}(x^2 \cos x + 2x \sin x)$

$= -\dfrac{x^2 \cos x + 2x \sin x}{3\sqrt[3]{(x^2 \sin x)^4}}$

31. $\dfrac{dy}{dx} = \dfrac{[1 + \cos(x^2 + 2x)]^{-3/4}[-\sin(x^2 + 2x)(2x + 2)]}{4}$

$= -\dfrac{(x + 1)\sin(x^2 + 2x)}{2\sqrt[4]{[1 + \cos(x^2 + 2x)]^3}}$

33. $s^2 + 2st\dfrac{ds}{dt} + 3t^2 = 0$

$\dfrac{ds}{dt} = \dfrac{-s^2 - 3t^2}{2st} = -\dfrac{s^2 + 3t^2}{2st}$

$s^2\dfrac{dt}{ds} + 2st + 3t^2\dfrac{dt}{ds} = 0$

$\dfrac{dt}{ds}(s^2 + 3t^2) = -2st$

$\dfrac{dt}{ds} = -\dfrac{2st}{s^2 + 3t^2}$

35.

$(x + 2)^2 + y^2 = 1$

$2x + 4 + 2y\dfrac{dy}{dx} = 0$

$\dfrac{dy}{dx} = -\dfrac{2x + 4}{2y} = -\dfrac{x + 2}{y}$

The tangent line at (x_0, y_0) has equation

$y - y_0 = -\dfrac{x_0 + 2}{y_0}(x - x_0)$ which simplifies to

$2x_0 - yy_0 - 2x - xx_0 + y_0^{\,2} + x_0^{\,2} = 0$. Since

(x_0, y_0) is on the circle, $x_0^{\,2} + y_0^{\,2} = -3 - 4x_0$,

so the equation of the tangent line is

$-yy_0 - 2x_0 - 2x - xx_0 = 3$.

If (0, 0) is on the tangent line, then $x_0 = -\frac{3}{2}$.
Solve for y_0 in the equation of the circle to get
$y_0 = \pm\frac{\sqrt{3}}{2}$. Put these values into the equation of the tangent line to get that the tangent lines are $\sqrt{3}y + x = 0$ and $\sqrt{3}y - x = 0$.

37. a. $xy' + y + 3y^2y' = 0$

$y'(x + 3y^2) = -y$

$y' = -\frac{y}{x + 3y^2}$

b. $xy'' + \left(\frac{-y}{x+3y^2}\right) + \left(\frac{-y}{x+3y^2}\right) + 3y^2y''$

$+6y\left(\frac{-y}{x+3y^2}\right)^2 = 0$

$xy'' + 3y^2y'' - \frac{2y}{x+3y^2} + \frac{6y^3}{(x+3y^2)^2} = 0$

$y''(x+3y^2) = \frac{2y}{x+3y^2} - \frac{6y^3}{(x+3y^2)^2}$

$y''(x+3y^2) = \frac{2xy}{(x+3y^2)^2}$

$y'' = \frac{2xy}{(x+3y^2)^3}$

39. $2(x^2y' + 2xy) - 12y^2y' = 0$

$2x^2y' - 12y^2y' = -4xy$

$y' = \frac{2xy}{6y^2 - x^2}$

$2(x^2y'' + 2xy' + 2xy' + 2y) - 12[y^2y'' + 2y(y')^2] = 0$

$2x^2y'' - 12y^2y'' = -8xy' - 4y + 24y(y')^2$

$y''(2x^2 - 12y^2) = -\frac{16x^2y}{6y^2 - x^2} - 4y + \frac{96x^2y^3}{(6y^2 - x^2)^2}$

$y''(2x^2 - 12y^2) = \frac{12x^4y + 48x^2y^3 - 144y^5}{(6y^2 - x^2)^2}$

$y''(6y^2 - x^2) = \frac{72y^5 - 6x^4y - 24x^2y^3}{(6y^2 - x^2)^2}$

$y'' = \frac{72y^5 - 6x^4y - 24x^2y^3}{(6y^2 - x^2)^3}$

At (2, 1), $y'' = \frac{-120}{8} = -15$

41. $3x^2 + 3y^2y' = 3(xy' + y)$

$y'(3y^2 - 3x) = 3y - 3x^2$

$y' = \frac{y - x^2}{y^2 - x}$

At $\left(\frac{3}{2}, \frac{3}{2}\right)$, $y' = -1$

Slope of the normal line is 1.

Normal line: $y - \frac{3}{2} = 1\left(x - \frac{3}{2}\right)$; $y = x$

This line includes the point (0, 0).

43. Implicitly differentiate the first equation.
$4x + 2yy' = 0$

$y' = -\frac{2x}{y}$

Implicitly differentiate the second equation.
$2yy' = 4$

$y' = \frac{2}{y}$

Solve for the points of intersection.
$2x^2 + 4x = 6$

$2(x^2 + 2x - 3) = 0$

$(x + 3)(x - 1) = 0$

$x = -3, x = 1$

$x = -3$ is extraneous, and $y = -2, 2$ when $x = 1$.
The graphs intersect at (1, –2) and (1, 2).
At (1, –2): $m_1 = 1, m_2 = -1$
At (1, 2): $m_1 = -1, m_2 = 1$

45. $x^2 - x(2x) + 2(2x)^2 = 28$

$7x^2 = 28$

$x^2 = 4$

$x = -2, 2$

Intersection point in first quadrant: (2, 4)

$y_1' = 2$

$2x - xy_2' - y + 4yy_2' = 0$

$y_2'(4y - x) = y - 2x$

$y_2' = \frac{y - 2x}{4y - x}$

At (2, 4): $m_1 = 2, m_2 = 0$

$\tan\theta = \frac{0-2}{1+(0)(2)} = -2; \theta = \pi + \tan^{-1}(-2) \approx 2.034$

47. $x^2 - xy + y^2 = 16$, when y = 0,

$x^2 = 16$

$x = -4, 4$

The ellipse intersects the x-axis at (–4, 0) and (4, 0).

$2x - xy' - y + 2yy' = 0$

$y'(2y - x) = y - 2x$

$$y' = \frac{y - 2x}{2y - x}$$

At (–4, 0), $y' = 2$

At (4, 0), $y' = 2$

Tangent lines: $y = 2(x + 4)$ and $y = 2(x - 4)$

49. $2x + 2y\frac{dy}{dx} = 0; \frac{dy}{dx} = -\frac{x}{y}$

The tangent line at (x_0, y_0) has slope $-\frac{x_0}{y_0}$, hence the equation of the tangent line is

$y - y_0 = -\frac{x_0}{y_0}(x - x_0)$ which simplifies to

$yy_0 + xx_0 - (x_0^2 + y_0^2) = 0$ or $yy_0 + xx_0 = 1$

since (x_0, y_0) is on $x^2 + y^2 = 1$. If (1.25, 0) is on the tangent line through (x_0, y_0), $x_0 = 0.8$. Put this into $x^2 + y^2 = 1$ to get $y_0 = 0.6$, since $y_0 > 0$. The line is $6y + 8x = 10$. When $x = -2$, $y = \frac{13}{3}$, so the light bulb must be $\frac{13}{3}$ units high.

2.8 Concepts Review

1. $\frac{du}{dt}; t = 2$

3. negative

Problem Set 2.8

1. $V = x^3; \frac{dx}{dt} = 3$

$$\frac{dV}{dt} = 3x^2\frac{dx}{dt}$$

When $x = 12$, $\frac{dV}{dt} = 3(12)^2(3) = 1296$ in.³/s.

3. $y^2 = x^2 + 1^2; \frac{dx}{dt} = 400$

$$2y\frac{dy}{dt} = 2x\frac{dx}{dt}$$

$\frac{dy}{dt} = \frac{x}{y}\frac{dx}{dt}$ mi/hr

When $x = 5$, $y = \sqrt{26}, \frac{dy}{dt} = \frac{5}{\sqrt{26}}(400)$

≈ 392 mi/h.

5. $s^2 = (x + 300)^2 + y^2; \frac{dx}{dt} = 300, \frac{dy}{dt} = 400,$

$$2s\frac{ds}{dt} = 2(x + 300)\frac{dx}{dt} + 2y\frac{dy}{dt}$$

$$s\frac{ds}{dt} = (x + 300)\frac{dx}{dt} + y\frac{dy}{dt}$$

When $x = 300$, $y = 400$, $s = 200\sqrt{13}$, so

$$200\sqrt{13}\frac{ds}{dt} = (300 + 300)(300) + 400(400)$$

$\frac{ds}{dt} \approx 471$ mi/h

7. $20^2 = x^2 + y^2; \frac{dx}{dt} = 1$

$$0 = 2x\frac{dx}{dt} + 2y\frac{dy}{dt}$$

When $x = 5$, $y = \sqrt{375} = 5\sqrt{15}$, so

$\frac{dy}{dt} = -\frac{x}{y}\frac{dx}{dt} = -\frac{5}{5\sqrt{15}}(1) \approx -0.258$ ft/s

The top of the ladder is moving down at 0.258 ft/s.

9. $V = \frac{1}{3}\pi r^2 h; h = \frac{d}{4} = \frac{r}{2}, r = 2h$

$V = \frac{1}{3}\pi(2h)^2 h = \frac{4}{3}\pi h^3; \frac{dV}{dt} = 16$

$$\frac{dV}{dt} = 4\pi h^2\frac{dh}{dt}$$

When $h = 4$, $16 = 4\pi(4)^2\frac{dh}{dt}$

$\frac{dh}{dt} = \frac{1}{4\pi} \approx 0.0796$ ft/s

11. $V = \frac{hx}{2}(20); \frac{40}{5} = \frac{x}{h}, x = 8h$

$V = 10h(8h) = 80h^2; \frac{dV}{dt} = 40$

$\frac{dV}{dt} = 160h\frac{dh}{dt}$

When $h = 3,\ 40 = 160(3)\frac{dh}{dt}$

$\frac{dh}{dt} = \frac{1}{12}$ ft/min

13. $A = \pi r^2; \frac{dr}{dt} = 0.02$

$\frac{dA}{dt} = 2\pi r\frac{dr}{dt}$

When $r = 8.1,\ \frac{dA}{dt} = 2\pi(0.02)(8.1) = 0.324\pi$

≈ 1.018 in.^{2}/s

15. Let x be the distance from the beam to the point opposite the lighthouse and θ be the angle between the beam and the line from the lighthouse to the point opposite.

$\tan\theta = \frac{x}{1}; \frac{d\theta}{dt} = 2(2\pi) = 4\pi$ rad/min,

$\sec^2\theta\frac{d\theta}{dt} = \frac{dx}{dt}$

At $x = \frac{1}{2}, \theta = \tan^{-1}\frac{1}{2}$ and $\sec^2\theta = \frac{5}{4}$.

$\frac{dx}{dt} = \frac{5}{4}(4\pi) \approx 15.71$ km/min

17. a. Let x be the distance along the ground from the light pole to Chris, and let s be the distance from Chris to the tip of his shadow.

By similar triangles, $\frac{6}{s} = \frac{30}{x+s}$, so $s = \frac{x}{4}$

and $\frac{ds}{dt} = \frac{1}{4}\frac{dx}{dt}$. $\frac{dx}{dt} = 2$ ft/s, hence

$\frac{ds}{dt} = \frac{1}{2}$ ft/s no matter how far from the light pole Chris is.

b. Let $l = x + s$, then

$\frac{dl}{dt} = \frac{dx}{dt} + \frac{ds}{dt} = 2 + \frac{1}{2} = \frac{5}{2}$ ft/s.

c. The angular rate at which Chris must lift his head to follow his shadow is the same as the rate at which the angle that the light makes with the ground is decreasing. Let θ be the angle that the light makes with the ground at the tip of Chris' shadow.

$\tan\theta = \frac{6}{s}$ so $\sec^2\theta\frac{d\theta}{dt} = -\frac{6}{s^2}\frac{ds}{dt}$ and

$\frac{d\theta}{dt} = -\frac{6\cos^2\theta}{s^2}\frac{ds}{dt}$. $\frac{ds}{dt} = \frac{1}{2}$ ft/s

When $s = 6,\ \theta = \frac{\pi}{4}$, so

$\frac{d\theta}{dt} = -\frac{6\left(\frac{1}{\sqrt{2}}\right)^2}{6^2}\left(\frac{1}{2}\right) = -\frac{1}{24}$.

Chris must lift his head at the rate of $\frac{1}{24}$ rad/s.

19. Let p be the point on the bridge directly above the railroad tracks. If a is the distance between p and the automobile, then $\frac{da}{dt} = 66$ ft/s. If l is the distance between the train and the point directly below p, then $\frac{dl}{dt} = 88$ ft/s. The distance from the train to p is $\sqrt{100^2 + l^2}$, while the distance from p to the automobile is a. The distance between the train and automobile is

$D = \sqrt{a^2 + \left(\sqrt{100^2 + l^2}\right)^2} = \sqrt{a^2 + l^2 + 100^2}$.

$\frac{dD}{dt} = \frac{1}{2\sqrt{a^2 + l^2 + 100^2}} \cdot \left(2a\frac{da}{dt} + 2l\frac{dl}{dt}\right)$

$= \frac{a\frac{da}{dt} + l\frac{dl}{dt}}{\sqrt{a^2 + l^2 + 100^2}}$. After 10 seconds, $a = 660$ and $l = 880$, so

$\frac{dD}{dt} = \frac{660(66) + 880(88)}{\sqrt{660^2 + 880^2 + 100^2}} \approx 110$ ft/s.

21. $V = \pi h^2\left[r - \frac{h}{3}\right]; \frac{dV}{dt} = -2, r = 8$

$V = \pi rh^2 - \frac{\pi h^3}{3} = 8\pi h^2 - \frac{\pi h^3}{3}$

$\frac{dV}{dt} = 16\pi h\frac{dh}{dt} - \pi h^2\frac{dh}{dt}$

When $h = 3,\ -2 = \frac{dh}{dt}[16\pi(3) - \pi(3)^2]$

$\frac{dh}{dt} = \frac{-2}{39\pi} \approx -0.016$ ft/hr

23. Let P be the point on the ground where the ball hits. Then the distance from P to the bottom of the light pole is 10 ft. Let s be the distance between P and the shadow of the ball. The height of the ball t seconds after it is dropped is $64-16t^2$.

By similar triangles, $\dfrac{48}{64-16t^2}=\dfrac{10+s}{s}$

(for $t>1$), so $s=\dfrac{10t^2-40}{1-t^2}$.

$$\frac{ds}{dt}=\frac{20t(1-t^2)-(10t^2-40)(-2t)}{(1-t^2)^2}=-\frac{60t}{(1-t^2)^2}$$

The ball hits the ground when $t=2$, $\dfrac{ds}{dt}=-\dfrac{120}{9}$.

The shadow is moving $\dfrac{120}{9}\approx 13.33$ ft/s.

25. Assuming that the tank is now in the shape of an upper hemisphere with radius r, we again let t be the number of hours past midnight and h be the height of the water at time t. The volume, V, of water in the tank at that time is given by

$$V=\frac{2}{3}\pi r^3-\frac{\pi}{3}(r-h)^2(2r+h)$$

and so $V=\dfrac{16000}{3}\pi-\dfrac{\pi}{3}(20-h)^2(40+h)$

from which

$$\frac{dV}{dt}=-\frac{\pi}{3}(20-h)^2\frac{dh}{dt}+\frac{2\pi}{3}(20-h)(40+h)\frac{dh}{dt}$$

At $t=7$, $\dfrac{dV}{dt}\approx -525\pi\approx -1649$

Thus Webster City residents were using water at the rate of $2400+1649=4049$ cubic feet per hour at 7:00 A.M.

27. a. Let x be the distance from the bottom of the wall to the end of the ladder on the ground, so $\dfrac{dx}{dt}=2$ ft/s. Let y be the height of the opposite end of the ladder. By similar triangles, $\dfrac{y}{12}=\dfrac{18}{\sqrt{144+x^2}}$, so $y=\dfrac{216}{\sqrt{144+x^2}}$.

$$\frac{dy}{dt}=-\frac{216}{2(144+x^2)^{3/2}}2x\frac{dx}{dt}=-\frac{216x}{(144+x^2)^{3/2}}\frac{dx}{dt}$$

When the ladder makes an angle of 60° with the ground, $x=4\sqrt{3}$ and $\dfrac{dy}{dt}=-\dfrac{216(4\sqrt{3})}{(144+48)^{3/2}}\cdot 2=-1.125$ ft/s.

b. $$\frac{d^2y}{dt^2}=\frac{d}{dt}\left(-\frac{216x}{(144+x^2)^{3/2}}\frac{dx}{dt}\right)=\frac{d}{dt}\left(-\frac{216x}{(144+x^2)^{3/2}}\right)\frac{dx}{dt}-\frac{216x}{(144+x^2)^{3/2}}\cdot\frac{d^2x}{dt^2}$$

Since $\dfrac{dx}{dt}=2$, $\dfrac{d^2x}{dt^2}=0$, thus

$$\frac{d^2y}{dt^2}=\left[\frac{-216(144+x^2)^{3/2}\frac{dx}{dt}+216x\left(\frac{3}{2}\right)\sqrt{144+x^2}(2x)\frac{dx}{dt}}{(144+x^2)^3}\right]\frac{dx}{dt}$$

$$=\frac{-216(144+x^2)+648x^2}{(144+x^2)^{5/2}}\left(\frac{dx}{dt}\right)^2=\frac{432x^2-31,104}{(144+x^2)^{5/2}}\left(\frac{dx}{dt}\right)^2$$

When the ladder makes an angle of 60° with the ground,

$$\frac{d^2y}{dt^2}=\frac{432\cdot 48-31,104}{(144+48)^{5/2}}(2)^2\approx -0.08\text{ ft/s}^2$$

29. $\frac{dV}{dt} = k(4\pi r^2)$

a. $V = \frac{4}{3}\pi r^3$

$\frac{dV}{dt} = 4\pi r^2 \frac{dr}{dt}$

$k(4\pi r^2) = 4\pi r^2 \frac{dr}{dt}$

$\frac{dr}{dt} = k$

b. If the original volume was V_0, the volume after 1 hour is $\frac{8}{27}V_0$. The original radius was $r_0 = \sqrt[3]{\frac{3}{4\pi}V_0}$ while the radius after 1 hour is $r_1 = \sqrt[3]{\frac{8}{27}V_0 \cdot \frac{3}{4\pi}} = \frac{2}{3}r_0$. Since $\frac{dr}{dt}$ is constant, $\frac{dr}{dt} = -\frac{1}{3}r_0$ unit/hr. The snowball will take 3 hours to melt completely.

31. Let l be the distance along the ground from the brother to the tip of the shadow. The shadow is controlled by both siblings when $\frac{3}{l} = \frac{5}{l+4}$ or $l = 6$. Again using similar triangles, this occurs when $\frac{y}{20} = \frac{6}{3}$, so $y = 40$. Thus, the girl controls the tip of the shadow when $y \geq 40$ and the boy controls it when $y < 40$.
Let x be the distance along the ground from the light pole to the girl. $\frac{dx}{dt} = -4$

When $y \geq 40$, $\frac{20}{y} = \frac{5}{y-x}$ or $y = \frac{4}{3}x$.

When $y < 40$, $\frac{20}{y} = \frac{3}{y-(x+4)}$ or $y = \frac{20}{17}(x+4)$.

$x = 30$ when $y = 40$. Thus,

$$y = \begin{cases} \frac{4}{3}x & \text{if } x \geq 30 \\ \frac{20}{17}(x+4) & \text{if } x < 30 \end{cases}$$

and

$$\frac{dy}{dt} = \begin{cases} \frac{4}{3}\frac{dx}{dt} & \text{if } x \geq 30 \\ \frac{20}{17}\frac{dx}{dt} & \text{if } x < 30 \end{cases}$$

Hence, the tip of the shadow is moving at the rate of $\frac{4}{3}(4) = \frac{16}{3}$ ft/s when the girl is at least 30 feet from the light pole, and it is moving $\frac{20}{17}(4) = \frac{80}{17}$ ft/s when the girl is less than 30 ft from the light pole.

2.9 Concepts Review

1. $f'(x)dx$

3. Δx is small.

Problem Set 2.9

1. $dy = (2x + 1)dx$

3. $dy = -4(2x+3)^{-5}(2)dx = -8(2x+3)^{-5}dx$

5. $dy = 3(\sin x + \cos x)^2(\cos x - \sin x)dx$

7. $dy = -\frac{3}{2}(7x^2 + 3x - 1)^{-5/2}(14x + 3)dx$

$= -\frac{3}{2}(14x + 3)(7x^2 + 3x - 1)^{-5/2}dx$

9. $ds = \frac{3}{2}(t^2 - \cot t + 2)^{1/2}(2t + \csc^2 t)dt$

$= \frac{3}{2}(2t + \csc^2 t)\sqrt{t^2 - \cot t + 2}dt$

11.

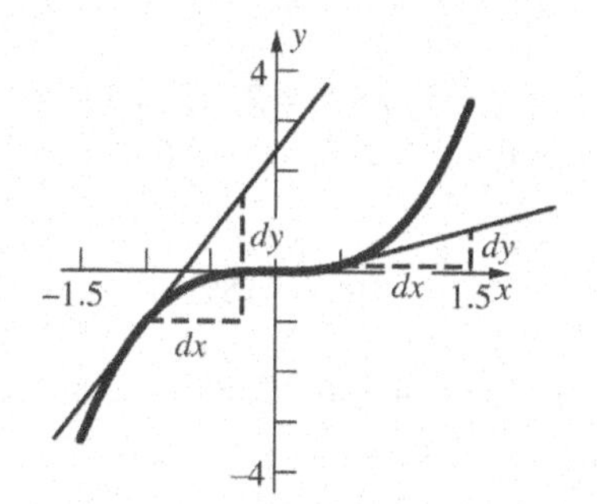

13.

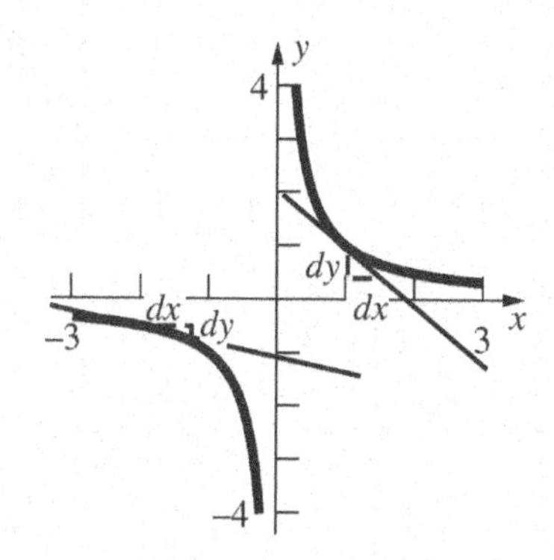

15. a. $\Delta y = \frac{1}{1.5} - \frac{1}{1} = -\frac{1}{3}$

b. $\Delta y = \frac{1}{-1.25} + \frac{1}{2} = -0.3$

17. a. $\Delta y = [(3)^4 + 2(3)] - [(2)^4 + 2(2)] = 67$

$dy = (4x^3 + 2)dx = [4(2)^3 + 2](1) = 34$

b. $\Delta y = [(2.005)^4 + 2(2.005)] - [(2)^4 + 2(2)]$
≈ 0.1706

$dy = (4x^3 + 2)dx = [4(2)^3 + 2](0.005) = 0.17$

19. $y = \sqrt{x};\ dy = \frac{1}{2\sqrt{x}}dx;\ x = 36,\ dx = -0.1$

$dy = \frac{1}{2\sqrt{36}}(-0.1) \approx -0.0083$

$\sqrt{35.9} \approx \sqrt{36} + dy = 6 - 0.0083 = 5.9917$

21. $V = \frac{4}{3}\pi r^3;\ r = 5,\ dr = 0.125$

$dV = 4\pi r^2 dr = 4\pi(5)^2(0.125) \approx 39.27 \text{ cm}^3$

23. $V = \frac{4}{3}\pi r^3;\ r = 6 \text{ ft} = 72 \text{ in.},\ dr = -0.3$

$dV = 4\pi r^2 dr = 4\pi(72)^2(-0.3) \approx -19{,}543$

$V \approx \frac{4}{3}\pi(72)^3 - 19{,}543$

$\approx 1{,}543{,}915 \text{ in}^3 \approx 893 \text{ ft}^3$

25. $C = 2\pi r$; $r = 4000$ mi = 21,120,000 ft, $dr = 2$

$dC = 2\pi dr = 2\pi(2) = 4\pi \approx 12.6\ ft$

27. $V = \frac{4}{3}\pi r^3 = \frac{4}{3}\pi(10)^3 \approx 4189$

$dV = 4\pi r^2 dr = 4\pi(10)^2(0.05) \approx 62.8$ The volume is $4189 \pm 62.8 \text{ cm}^3$.

The absolute error is ≈ 62.8 while the relative error is $62.8/4189 \approx 0.015$ or 1.5%.

29. $s = \sqrt{a^2 + b^2 - 2ab\cos\theta}$

$= \sqrt{151^2 + 151^2 - 2(151)(151)\cos 0.53} \approx 79.097$

$s = \sqrt{45{,}602 - 45{,}602\cos\theta}$

$ds = \frac{1}{2\sqrt{45{,}602 - 45{,}602\cos\theta}} \cdot 45{,}602\sin\theta\, d\theta$

$= \frac{22{,}801\sin\theta}{\sqrt{45{,}602 - 45{,}602\cos\theta}}d\theta$

$= \frac{22{,}801\sin 0.53}{\sqrt{45{,}602 - 45{,}602\cos 0.53}}(0.005) \approx 0.729$

$s \approx 79.097 \pm 0.729$ cm

The absolute error is ≈ 0.729 while the relative error is $0.729/79.097 \approx 0.0092$ or 0.92%.

31. $y = 3x^2 - 2x + 11;\ x = 2,\ dx = 0.001$

$dy = (6x - 2)dx = [6(2) - 2](0.001) = 0.01$

$\frac{d^2y}{dx^2} = 6$, so with $\Delta x = 0.001$,

$|\Delta y - dy| \le \frac{1}{2}(6)(0.001)^2 = 0.000003$

33. Using the approximation

$f(x + \Delta x) \approx f(x) + f'(x)\Delta x$

we let $x = 3.05$ and $\Delta x = -0.05$. We can rewrite the above form as

$f(x) \approx f(x + \Delta x) - f'(x)\Delta x$

which gives

$f(3.05) \approx f(3) - f'(3.05)(-0.05)$

$= 8 + \frac{1}{4}(0.05) = 8.0125$

35. $V = \pi r^2 h + \frac{4}{3}\pi r^3$

$V = 100\pi r^2 + \frac{4}{3}\pi r^3;\ r = 10,\ dr = 0.1$

$dV = (200\pi r + 4\pi r^2)dr$

$= (2000\pi + 400\pi)(0.1) = 240\pi \approx 754 \text{ cm}^3$

37. $f(x) = x^2; f'(x) = 2x;\ a = 2$
The linear approximation is then
$L(x) = f(2) + f'(2)(x-2)$
$= 4 + 4(x-2) = 4x - 4$

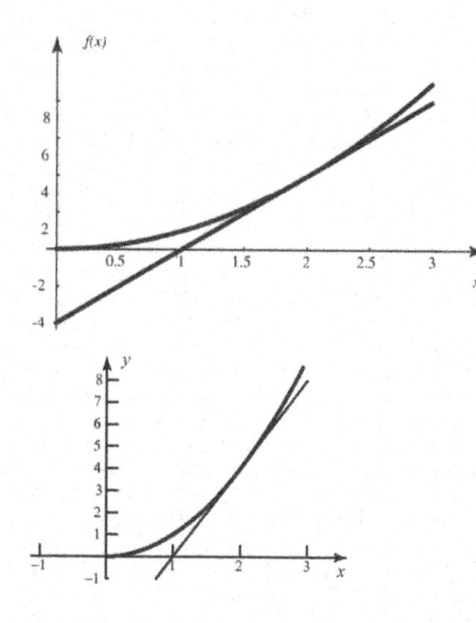

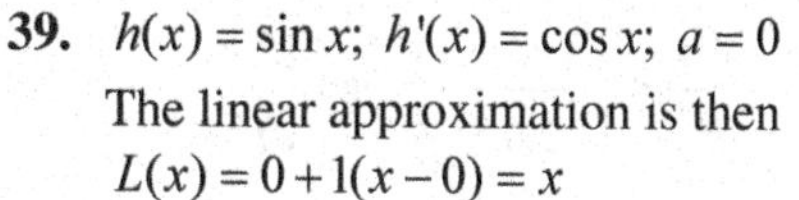

39. $h(x) = \sin x;\ h'(x) = \cos x;\ a = 0$
The linear approximation is then
$L(x) = 0 + 1(x-0) = x$

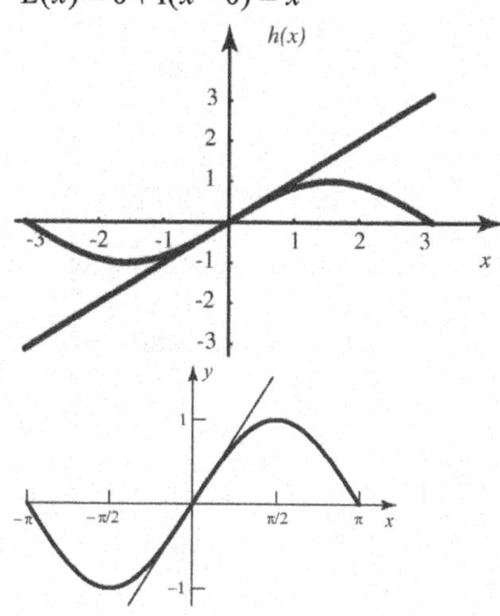

41. $f(x) = \sqrt{1-x^2};$

$f'(x) = \frac{1}{2}\left(1-x^2\right)^{-1/2}(-2x)$

$= \frac{-x}{\sqrt{1-x^2}},\quad a = 0$

The linear approximation is then
$L(x) = 1 + 0(x-0) = 1$

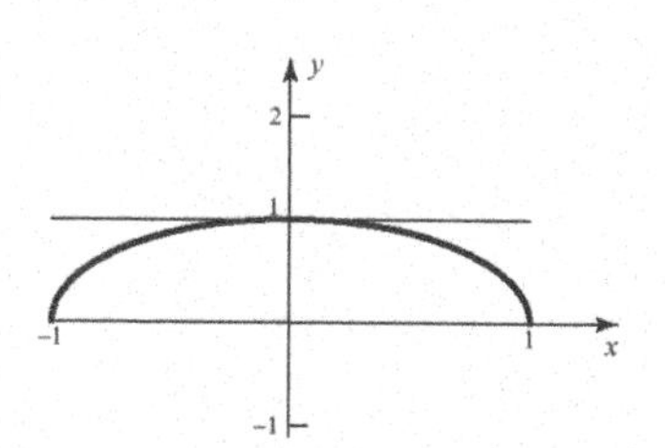

43. $h(x) = x\sec x; h'(x) = \sec x + x\sec x\tan x, a = 0$
The linear approximation is then
$L(x) = 0 + 1(x-0) = x$

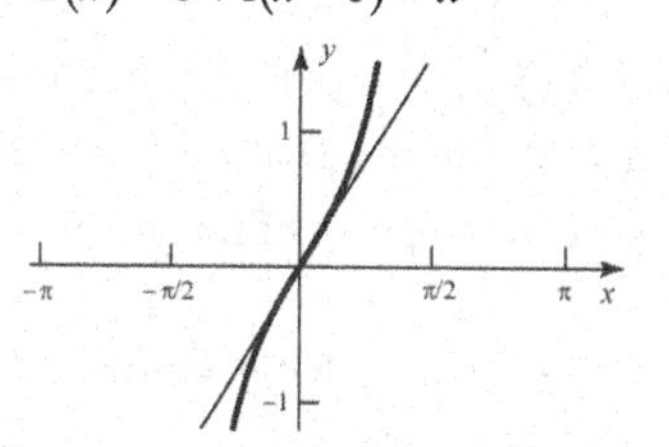

45. $f(x) = mx + b;\ f'(x) = m$
The linear approximation is then
$L(x) = ma + b + m(x-a) = am + b + mx - ma$
$= mx + b \qquad f(x) = L(x)$

47. The linear approximation to $f(x)$ at a is
$L(x) = f(a) + f'(a)(x-a)$
$= a^2 + 2a(x-a)$
$= 2ax - a^2$
Thus,
$f(x) - L(x) = x^2 - \left(2ax - a^2\right)$
$= x^2 - 2ax + a^2$
$= (x-a)^2$
≥ 0

49. a. $\lim_{h\to 0} \varepsilon(h) = \lim_{h\to 0}\left(f(x+h) - f(x) - f'(x)h\right)$
$= f(x) - f(x) - f'(x)0 = 0$

b. $\lim_{h\to 0} \frac{\varepsilon(h)}{h} = \lim\left[\frac{f(x+h) - f(x)}{h} - f'(x)\right]$
$= f'(x) - f'(x) = 0$

2.10 Chapter Review

Concepts Test

1. False: If $f(x) = x^3$, $f'(x) = 3x^2$ and the tangent line $y = 0$ at $x = 0$ crosses the curve at the point of tangency.

3. True: $m_{\tan} = 4x^3$, which is unique for each value of x.

5. True: If the velocity is negative and increasing, the speed is decreasing.

7. True: If the tangent line is horizontal, the slope must be 0.

9. True: $D_x f(g(x)) = f'(g(x))g'(x)$; since $g(x) = x$, $g'(x) = 1$, so $D_x f(g(x)) = f'(g(x))$.

11. True: Theorem 3.2.A

13. False: $(f \cdot g)'(x) = f(x)g'(x) + g(x)f'(x)$

15. True: If $f(x) = x^3 g(x)$, then
$D_x f(x) = x^3 g'(x) + 3x^2 g(x)$
$= x^2[xg'(x) + 3g(x)]$.

17. False:
$D_x y = f(x)g'(x) + g(x)f'(x)$
$D_x^2 y = f(x)g''(x) + g'(x)f'(x)$
$+g(x)f''(x) + f'(x)g'(x)$
$= f(x)g''(x) + 2f'(x)g'(x) + f''(x)g(x)$

19. True: $f(x) = ax^n$; $f'(x) = anx^{n-1}$

21. True:
$h'(x) = f(x)g'(x) + g(x)f'(x)$
$h'(c) = f(c)g'(c) + g(c)f'(c)$
$= f(c)(0) + g(c)(0) = 0$

23. True: $D^2(kf) = kD^2 f$ and
$D^2(f + g) = D^2 f + D^2 g$

25. True:
$(f \circ g)'(2) = f'(g(2)) \cdot g'(2)$
$= f'(2) \cdot g'(2) = 2 \cdot 2 = 4$

27. False: The rate of volume change depends on the radius of the sphere.

29. True:
$D_x(\sin x) = \cos x$;
$D_x^2(\sin x) = -\sin x$;
$D_x^3(\sin x) = -\cos x$;
$D_x^4(\sin x) = \sin x$;
$D_x^5(\sin x) = \cos x$

31. True:
$\lim_{x \to 0} \frac{\tan x}{3x} = \frac{1}{3} \lim_{x \to 0} \frac{\sin x}{x \cos x}$
$= \frac{1}{3} \cdot 1 = \frac{1}{3}$

33. True:
$V = \frac{4}{3}\pi r^3$
$\frac{dV}{dt} = 4\pi r^2 \frac{dr}{dt}$
If $\frac{dV}{dt} = 3$, then $\frac{dr}{dt} = \frac{3}{4\pi r^2}$ so
$\frac{dr}{dt} > 0$.
$\frac{d^2 r}{dt^2} = -\frac{3}{2\pi r^3} \frac{dr}{dt}$ so $\frac{d^2 r}{dt^2} < 0$

35. True:
$V = \frac{4}{3}\pi r^3$, $S = 4\pi r^2$
$dV = 4\pi r^2 dr = S \cdot dr$
If $\Delta r = dr$, then $dV = S \cdot \Delta r$

37. False: The slope of the linear approximation is equal to
$f'(a) = f'(0) = -\sin(0) = 0$.

Sample Test Problems

1. a. $f'(x) = \lim_{h\to 0} \frac{3(x+h)^3 - 3x^3}{h} = \lim_{h\to 0} \frac{9x^2h + 9xh^2 + 3h^3}{h} = \lim_{h\to 0}(9x^2 + 9xh + 3h^2) = 9x^2$

b. $f'(x) = \lim_{h\to 0} \frac{[2(x+h)^5 + 3(x+h)] - (2x^5 + 3x)}{h} = \lim_{h\to 0} \frac{10x^4h + 20x^3h^2 + 20x^2h^3 + 10xh^4 + 2h^5 + 3h}{h}$

$= \lim_{h\to 0}(10x^4 + 20x^3h + 20x^2h^2 + 10xh^3 + 2h^4 + 3) = 10x^4 + 3$

c. $f'(x) = \lim_{h\to 0} \frac{\frac{1}{3(x+h)} - \frac{1}{3x}}{h} = \lim_{h\to 0}\left[-\frac{h}{3(x+h)x}\right]\frac{1}{h} = \lim_{h\to 0} -\left(\frac{1}{3x(x+h)}\right) = -\frac{1}{3x^2}$

d. $f'(x) = \lim_{h\to 0}\left[\left(\frac{1}{3(x+h)^2+2} - \frac{1}{3x^2+2}\right)\frac{1}{h}\right] = \lim_{h\to 0}\left[\frac{3x^2+2-3(x+h)^2-2}{(3(x+h)^2+2)(3x^2+2)}\cdot\frac{1}{h}\right]$

$= \lim_{h\to 0}\left[\frac{-6xh-3h^2}{(3(x+h)^2+2)(3x^2+2)}\cdot\frac{1}{h}\right] = \lim_{h\to 0}\frac{-6x-3h}{(3(x+h)^2+2)(3x^2+2)} = -\frac{6x}{(3x^2+2)^2}$

e. $f'(x) = \lim_{h\to 0}\frac{\sqrt{3(x+h)} - \sqrt{3x}}{h} = \lim_{h\to 0}\frac{(\sqrt{3x+3h}-\sqrt{3x})(\sqrt{3x+3h}+\sqrt{3x})}{h(\sqrt{3x+3h}+\sqrt{3x})}$

$= \lim_{h\to 0}\frac{3h}{h(\sqrt{3x+3h}+\sqrt{3x})} = \lim_{h\to 0}\frac{3}{\sqrt{3x+3h}+\sqrt{3x}} = \frac{3}{2\sqrt{3x}}$

f. $f'(x) = \lim_{h\to 0}\frac{\sin[3(x+h)] - \sin 3x}{h} = \lim_{h\to 0}\frac{\sin(3x+3h)-\sin 3x}{h}$

$= \lim_{h\to 0}\frac{\sin 3x\cos 3h + \sin 3h\cos 3x - \sin 3x}{h} = \lim_{h\to 0}\frac{\sin 3x(\cos 3h - 1)}{h} + \lim_{h\to 0}\frac{\sin 3h\cos 3x}{h}$

$= 3\sin 3x\lim_{h\to 0}\frac{\cos 3h - 1}{3h} + \cos 3x\lim_{h\to 0}\frac{\sin 3h}{h} = (3\sin 3x)(0) + (\cos 3x)3\lim_{h\to 0}\frac{\sin 3h}{3h} = (\cos 3x)(3)(1) = 3\cos 3x$

g. $f'(x) = \lim_{h\to 0}\frac{\sqrt{(x+h)^2+5} - \sqrt{x^2+5}}{h} = \lim_{h\to 0}\frac{\left(\sqrt{(x+h)^2+5}-\sqrt{x^2+5}\right)\left(\sqrt{(x+h)^2+5}+\sqrt{x^2+5}\right)}{h\left(\sqrt{(x+h)^2+5}+\sqrt{x^2+5}\right)}$

$= \lim_{h\to 0}\frac{2xh+h^2}{h\left(\sqrt{(x+h)^2+5}+\sqrt{x^2+5}\right)} = \lim_{h\to 0}\frac{2x+h}{\sqrt{(x+h)^2+5}+\sqrt{x^2+5}} = \frac{2x}{2\sqrt{x^2+5}} = \frac{x}{\sqrt{x^2+5}}$

h. $f'(x) = \lim_{h\to 0}\frac{\cos[\pi(x+h)] - \cos\pi x}{h} = \lim_{h\to 0}\frac{\cos(\pi x + \pi h) - \cos\pi x}{h} = \lim_{h\to 0}\frac{\cos\pi x\cos\pi h - \sin\pi x\sin\pi h - \cos\pi x}{h}$

$= \lim_{h\to 0}\left(-\pi\cos\pi x\frac{1-\cos\pi h}{\pi h}\right) - \lim_{h\to 0}\left(\pi\sin\pi x\frac{\sin\pi h}{\pi h}\right) = (-\pi\cos\pi x)(0) - (\pi\sin\pi x) = -\pi\sin\pi x$

3. a. $f(x) = 3x$ at $x = 1$

b. $f(x) = 4x^3$ at $x = 2$

c. $f(x) = \sqrt{x^3}$ at $x = 1$

d. $f(x) = \sin x$ at $x = \pi$

e. $f(x) = \frac{4}{x}$ at x

f. $f(x) = -\sin 3x$ at x

g. $f(x) = \tan x$ at $x = \frac{\pi}{4}$

h. $f(x) = \frac{1}{\sqrt{x}}$ at $x = 5$

5. $D_x(3x^5) = 15x^4$

7. $D_z(z^3 + 4z^2 + 2z) = 3z^2 + 8z + 2$

9. $D_t\left(\frac{4t-5}{6t^2+2t}\right) = \frac{(6t^2+2t)(4)-(4t-5)(12t+2)}{(6t^2+2t)^2}$

$= \frac{-24t^2+60t+10}{(6t^2+2t)^2}$

11. $\frac{d}{dx}\left(\frac{4x^2-2}{x^3+x}\right) = \frac{(x^3+x)(8x)-(4x^2-2)(3x^2+1)}{(x^3+x)^2}$

$= \frac{-4x^4+10x^2+2}{(x^3+x)^2}$

13. $\frac{d}{dx}\left(\frac{1}{\sqrt{x^2+4}}\right) = \frac{d}{dx}(x^2+4)^{-1/2}$

$= -\frac{1}{2}(x^2+4)^{-3/2}(2x) = -\frac{x}{\sqrt{(x^2+4)^3}}$

15. $D_\theta(\sin\theta + \cos^3\theta) = \cos\theta + 3\cos^2\theta(-\sin\theta)$

$= \cos\theta - 3\sin\theta\cos^2\theta$

$D_\theta^2(\sin\theta + \cos^3\theta)$

$= -\sin\theta - 3[\sin\theta(2)(\cos\theta)(-\sin\theta) + \cos^3\theta]$

$= -\sin\theta + 6\sin^2\theta\cos\theta - 3\cos^3\theta$

17. $D_\theta[\sin(\theta^2)] = \cos(\theta^2)(2\theta) = 2\theta\cos(\theta^2)$

19. $\frac{d}{d\theta}[\sin^2(\sin(\pi\theta))] = 2\sin(\sin(\pi\theta))\cos(\sin(\pi\theta))(\cos(\pi\theta))(\pi) = 2\pi\sin(\sin(\pi\theta))\cos(\sin(\pi\theta))\cos(\pi\theta)$

21. $D_\theta \tan 3\theta = (\sec^2 3\theta)(3) = 3\sec^2 3\theta$

23. $f'(x) = (x^2-1)^2(9x^2-4) + (3x^3-4x)(2)(x^2-1)(2x) = (x^2-1)^2(9x^2-4) + 4x(x^2-1)(3x^3-4x)$

$f'(2) = 672$

25. $\frac{d}{dx}\left(\frac{\cot x}{\sec x^2}\right) = \frac{(\sec x^2)(-\csc^2 x) - (\cot x)(\sec x^2)(\tan x^2)(2x)}{\sec^2 x^2} = \frac{-\csc^2 x - 2x\cot x\tan x^2}{\sec x^2}$

27. $f'(x) = (x-1)^3 2(\sin\pi x - x)(\pi\cos\pi x - 1) + (\sin\pi x - x)^2 3(x-1)^2$

$= 2(x-1)^3(\sin\pi x - x)(\pi\cos\pi x - 1) + 3(\sin\pi x - x)^2(x-1)^2$

$f'(2) = 16 - 4\pi \approx 3.43$

29. $g'(r) = 3(\cos^2 5r)(-\sin 5r)(5) = -15\cos^2 5r\sin 5r$

$g''(r) = -15[(\cos^2 5r)(\cos 5r)(5) + (\sin 5r)2(\cos 5r)(-\sin 5r)(5)] = -15[5\cos^3 5r - 10(\sin^2 5r)(\cos 5r)]$

$g'''(r) = -15[5(3)(\cos^2 5r)(-\sin 5r)(5) - (10\sin^2 5r)(-\sin 5r)(5) - (\cos 5r)(20\sin 5r)(\cos 5r)(5)]$

$= -15[-175(\cos^2 5r)(\sin 5r) + 50\sin^3 5r]$

$g'''(1) \approx 458.8$

31. $G'(x) = F'(r(x)+s(x))(r'(x)+s'(x)) + s'(x)$

$G''(x) = F'(r(x)+s(x))(r''(x)+s''(x)) + (r'(x)+s'(x))F''(r(x)+s(x))(r'(x)+s'(x)) + s''(x)$

$= F'(r(x)+s(x))(r''(x)+s''(x)) + (r'(x)+s'(x))^2 F''(r(x)+s(x)) + s''(x)$

33. $F'(z) = r'(s(z))s'(z) = [3\cos(3s(z))](9z^2)$

$= 27z^2\cos(9z^3)$

35. $V = \frac{4}{3}\pi r^3$

$\frac{dV}{dr} = 4\pi r^2$

When $r = 5$, $\frac{dV}{dr} = 4\pi(5)^2 = 100\pi \approx 314$ m^3 per meter of increase in the radius.

37. $V=\frac{1}{2}bh(12); \frac{6}{4}=\frac{b}{h}; b=\frac{3h}{2}$

$V=6\left(\frac{3h}{2}\right)h=9h^2; \frac{dV}{dt}=9$

$\frac{dV}{dt}=18h\frac{dh}{dt}$

When $h=3,\ 9=18(3)\frac{dh}{dt}$

$\frac{dh}{dt}=\frac{1}{6}\approx 0.167$ ft/min

39. $s=t^3-6t^2+9t$

$v(t)=\frac{ds}{dt}=3t^2-12t+9$

$a(t)=\frac{d^2s}{dt^2}=6t-12$

a. $3t^2-12t+9<0$
$3(t-3)(t-1)<0$
$1<t<3;\ (1,3)$

b. $3t^2-12t+9=0$
$3(t-3)(t-1)=0$
$t=1, 3$
$a(1)=-6, a(3)=6$

c. $6t-12>0$
$t>2;\ (2,\infty)$

41. a. $2(x-1)+2y\frac{dy}{dx}=0$

$\frac{dy}{dx}=\frac{-(x-1)}{y}=\frac{1-x}{y}$

b. $x(2y)\frac{dy}{dx}+y^2+y(2x)+x^2\frac{dy}{dx}=0$

$\frac{dy}{dx}(2xy+x^2)=-(y^2+2xy)$

$\frac{dy}{dx}=-\frac{y^2+2xy}{x^2+2xy}$

c. $3x^2+3y^2\frac{dy}{dx}=x^3(3y^2)\frac{dy}{dx}+3x^2y^3$

$\frac{dy}{dx}(3y^2-3x^3y^2)=3x^2y^3-3x^2$

$\frac{dy}{dx}=\frac{3x^2y^3-3x^2}{3y^2-3x^3y^2}=\frac{x^2y^3-x^2}{y^2-x^3y^2}$

d. $x\cos(xy)\left[x\frac{dy}{dx}+y\right]+\sin(xy)=2x$

$x^2\cos(xy)\frac{dy}{dx}=2x-\sin(xy)-xy\cos(xy)$

$\frac{dy}{dx}=\frac{2x-\sin(xy)-xy\cos(xy)}{x^2\cos(xy)}$

e. $x\sec^2(xy)\left(x\frac{dy}{dx}+y\right)+\tan(xy)=0$

$x^2\sec^2(xy)\frac{dy}{dx}=-[\tan(xy)+xy\sec^2(xy)]$

$\frac{dy}{dx}=-\frac{\tan(xy)+xy\sec^2(xy)}{x^2\sec^2(xy)}$

43. $dy=[\pi\cos(\pi x)+2x]dx$; $x=2$, $dx=0.01$

$dy=[\pi\cos(2\pi)+2(2)](0.01)=(4+\pi)(0.01)$
≈ 0.0714

45. a. $\frac{d}{dx}[f^2(x)+g^3(x)]$

$=2f(x)f'(x)+3g^2(x)g'(x)$

$2f(2)f'(2)+3g^2(2)g'(2)$

$=2(3)(4)+3(2)^2(5)=84$

b. $\frac{d}{dx}[f(x)g(x)]=f(x)g'(x)+g(x)f'(x)$

$f(2)g'(2)+g(2)f'(2)=(3)(5)+(2)(4)=23$

c. $\frac{d}{dx}[f(g(x))]=f'(g(x))g'(x)$

$f'(g(2))g'(2)=f'(2)g'(2)=(4)(5)=20$

d. $D_x[f^2(x)]=2f(x)f'(x)$

$D_x^2[f^2(x)]=2[f(x)f''(x)+f'(x)f'(x)]$

$=2f(2)f''(2)+2[f'(2)]^2$

$=2(3)(-1)+2(4)^2=26$

47. $\sin 15°=\frac{y}{x}, \frac{dx}{dt}=400$

$y=x\sin 15°$

$\frac{dy}{dt}=\sin 15°\frac{dx}{dt}$

$\frac{dy}{dt}=400\sin 15° \approx 104$ mi/hr

49. a. $D_\theta|\sin\theta|=\frac{|\sin\theta|}{\sin\theta}\cos\theta=\cot\theta|\sin\theta|$

b. $D_\theta|\cos\theta|=\frac{|\cos\theta|}{\cos\theta}(-\sin\theta)=-\tan\theta|\cos\theta|$

Review and Preview

1. $(x-2)(x-3)<0$
$(x-2)(x-3)=0$
$x=2$ or $x=3$
The split points are 2 and 3. The expression on the left can only change signs at the split points. Check a point in the intervals $(-\infty,2)$, $(2,3)$, and $(3,\infty)$.
The solution set is $\{x \mid 2<x<3\}$ or $(2,3)$.

–2 –1 0 1 2 3 4 5 6 7 8

3. $x(x-1)(x-2)\le 0$
$x(x-1)(x-2)=0$
$x=0, x=1$ or $x=2$
The split points are 0, 1, and 2. The expression on the left can only change signs at the split points. Check a point in the intervals $(-\infty,0)$, $(0,1)$, $(1,2)$, and $(2,\infty)$. The solution set is $\{x \mid x\le 0 \text{ or } 1\le x\le 2\}$, or $(-\infty,0]\cup[1,2]$.

–5 –4 –3 –2 –1 0 1 2 3 4 5

5. $\dfrac{x(x-2)}{x^2-4}\ge 0$
$\dfrac{x(x-2)}{(x-2)(x+2)}\ge 0$
The expression on the left is equal to 0 or undefined at $x=0$, $x=2$, and $x=-2$. These are the split points. The expression on the left can only change signs at the split points. Check a point in the intervals: $(-\infty,-2)$, $(-2,0)$, $(0,2)$, and $(2,\infty)$. The solution set is $\{x \mid x<-2 \text{ or } 0\le x<2 \text{ or } x>2\}$, or $(-\infty,-2)\cup[0,2)\cup(2,\infty)$.

–5 –4 –3 –2 –1 0 1 2 3 4 5

7. $f'(x)=4(2x+1)^3(2)=8(2x+1)^3$

9. $f'(x)=(x^2-1)\cdot -\sin(2x)\cdot 2+\cos(2x)\cdot(2x)$
$=-2(x^2-1)\sin(2x)+2x\cos(2x)$

11. $f'(x)=2(\tan 3x)\cdot\sec^2 3x\cdot 3=6(\sec^2 3x)(\tan 3x)$

13. $f'(x)=\cos(\sqrt{x})\cdot\frac{1}{2}x^{-1/2}=\dfrac{\cos\sqrt{x}}{2\sqrt{x}}$
(note: you cannot cancel the $\sqrt{x}$ here because it is not a factor of both the numerator and denominator. It is the argument for the cosine in the numerator.)

15. The tangent line is horizontal when the derivative is 0.
$y'=2\tan x\cdot\sec^2 x$
$2\tan x\sec x=0$
$\dfrac{2\sin x}{\cos^2 x}=0$
The tangent line is horizontal whenever $\sin x=0$. That is, for $x=k\pi$ where k is an integer.

17. The line $y=2+x$ has slope 1, so any line parallel to this line will also have a slope of 1. For the tangent line to $y=x+\sin x$ to be parallel to the given line, we need its derivative to equal 1.
$y'=1+\cos x=1$ or $\cos x=0$
The tangent line will be parallel to $y=2+x$ whenever $x=(2k+1)\frac{\pi}{2}$.

19. Consider the diagram:

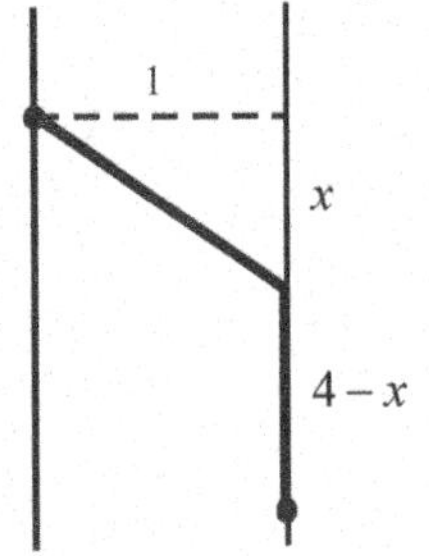

His distance swimming will be $\sqrt{1^2+x^2}=\sqrt{x^2+1}$ kilometers. His distance running will be $4-x$ kilometers. Using the distance traveled formula, $d=r\cdot t$, we solve for t to get $t=\frac{d}{r}$. Andy can swim at 4 kilometers per hour and run 10 kilometers per hour. Therefore, the time to get from A to D will be $\dfrac{\sqrt{x^2+1}}{4}+\dfrac{4-x}{10}$ hours.

21. a. The derivative of x^2 is $2x$ and the derivative of a constant is 0. Therefore, one possible function is $f(x)=x^2+3$.

b. The derivative of $-\cos x$ is $\sin x$ and the derivative of a constant is 0. Therefore, one possible function is $f(x)=-(\cos x)+8$.

c. The derivative of x^3 is $3x^2$, so the derivative of $\frac{1}{3}x^3$ is x^2. The derivative of x^2 is $2x$, so the derivative of $\frac{1}{2}x^2$ is x. The derivative of x is 1, and the derivative of a constant is 0. Therefore, one possible function is $\frac{1}{3}x^3+\frac{1}{2}x^2+x+2$.

CHAPTER 3 Applications of the Derivative

3.1 Concepts Review

1. continuous; closed and bounded

3. endpoints; stationary points; singular points

Problem Set 3.1

1. Endpoints: $-2, 4$
Singular points: none
Stationary points: 0, 2
Critical points: $-2, 0, 2, 4$

3. Endpoints: $-2, 4$
Singular points: none
Stationary points: $-1, 0, 1, 2, 3$
Critical points: $-2, -1, 0, 1, 2, 3, 4$

5. $f'(x) = 2x + 4; 2x + 4 = 0$ when $x = -2$.
Critical points: $-4, -2, 0$
$f(-4) = 4, f(-2) = 0, f(0) = 4$
Maximum value = 4, minimum value = 0

7. $\Psi'(x) = 2x + 3;\ 2x + 3 = 0$ when $x = -\frac{3}{2}$.

Critical points: $-2, -\frac{3}{2}, 1$

$\Psi(-2) = -2, \Psi\left(-\frac{3}{2}\right) = -\frac{9}{4}, \Psi(1) = 4$

Maximum value = 4, minimum value = $-\frac{9}{4}$

9. $f'(x) = 3x^2 - 3; 3x^2 - 3 = 0$ when $x = -1, 1$.
Critical points: $-1, 1$
$f(-1) = 3, f(1) = -1$
No maximum value, minimum value = -1

(See graph.)

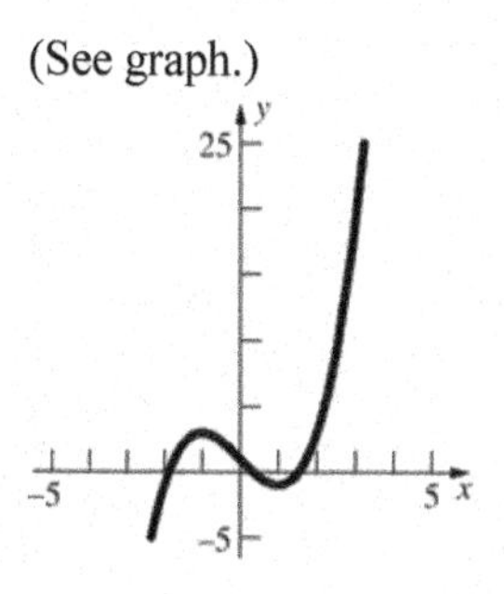

11. $h'(r) = -\frac{1}{r^2}$; $h'(r)$ is never 0; $h'(r)$ is not defined when $r = 0$, but $r = 0$ is not in the domain on $[-1, 3]$ since $h(0)$ is not defined.
Critical points: $-1, 3$
Note that $\lim_{x\to 0^-} h(r) = -\infty$ and $\lim_{x\to 0^+} h(x) = \infty$.
No maximum value, no minimum value.

13. $f'(x) = 4x^3 - 4x$
$= 4x(x^2 - 1)$
$= 4x(x-1)(x+1)$
$4x(x-1)(x+1) = 0$ when $x = 0, 1, -1$.
Critical points: $-2, -1, 0, 1, 2$
$f(-2) = 10$; $f(-1) = 1$; $f(0) = 2$; $f(1) = 1$;
$f(2) = 10$
Maximum value: 10
Minimum value: 1

15. $g'(x) = -\frac{2x}{(1+x^2)^2}$; $-\frac{2x}{(1+x^2)^2} = 0$ when $x = 0$.
Critical point: 0
$g(0) = 1$
As $x \to \infty, g(x) \to 0^+$; as $x \to -\infty, g(x) \to 0^+$.
Maximum value = 1, no minimum value
(See graph.)

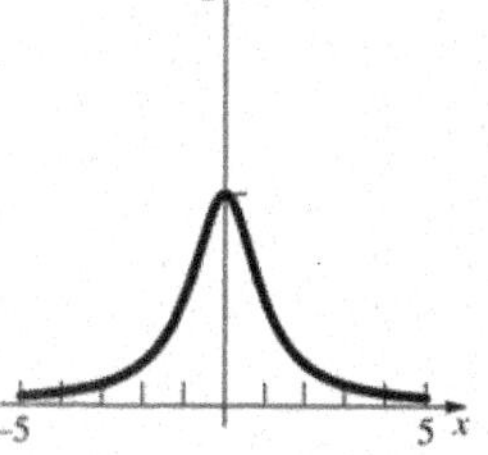

17. $r'(\theta) = \cos\theta$; $\cos\theta = 0$ when $\theta = \frac{\pi}{2} + k\pi$

Critical points: $-\frac{\pi}{4}, \frac{\pi}{6}$

$r\left(-\frac{\pi}{4}\right) = -\frac{1}{\sqrt{2}},\ r\left(\frac{\pi}{6}\right) = \frac{1}{2}$

Maximum value $= \frac{1}{2}$, minimum value $= -\frac{1}{\sqrt{2}}$

19. $a'(x) = \dfrac{x-1}{|x-1|}$; $a'(x)$ does not exist when $x = 1$.

Critical points: 0, 1, 3
$a(0) = 1$, $a(1) = 0$, $a(3) = 2$
Maximum value = 2, minimum value = 0

21. $g'(x) = \dfrac{1}{3x^{2/3}}$; $f'(x)$ does not exist when $x = 0$.

Critical points: –1, 0, 27
$g(-1) = -1$, $g(0) = 0$, $g(27) = 3$
Maximum value = 3, minimum value = –1

23. $H'(t) = -\sin t$

$-\sin t = 0$ when
$t = 0, \pi, 2\pi, 3\pi, 4\pi, 5\pi, 6\pi, 7\pi, 8\pi$
Critical points: $0, \pi, 2\pi, 3\pi, 4\pi, 5\pi, 6\pi, 7\pi, 8\pi$
$H(0) = 1$; $H(\pi) = -1$; $H(2\pi) = 1$;
$H(3\pi) = -1$; $H(4\pi) = 1$; $H(5\pi) = -1$;
$H(6\pi) = 1$; $H(7\pi) = -1$; $H(8\pi) = 1$
Maximum value: 1
Minimum value: –1

25. $g'(\theta) = \theta^2(\sec\theta\tan\theta) + 2\theta\sec\theta$
$= \theta\sec\theta(\theta\tan\theta + 2)$

$\theta\sec\theta(\theta\tan\theta + 2) = 0$ when $\theta = 0$.

Consider the graph:

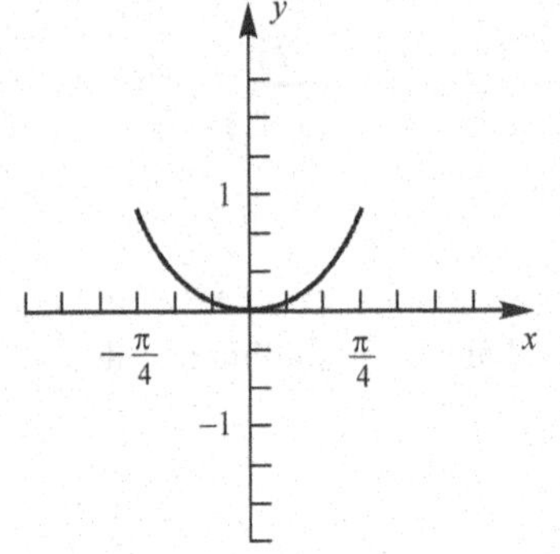

Critical points: $-\dfrac{\pi}{4}, 0, \dfrac{\pi}{4}$

$g\left(-\dfrac{\pi}{4}\right) = \dfrac{\pi^2\sqrt{2}}{16}$; $g(0) = 0$; $g\left(\dfrac{\pi}{4}\right) = \dfrac{\pi^2\sqrt{2}}{16}$

Maximum value: $\dfrac{\pi^2\sqrt{2}}{16}$; Minimum value: 0

27. a. $f'(x) = 3x^2 - 12x + 1$; $3x^2 - 12x + 1 = 0$

when $x = 2 - \dfrac{\sqrt{33}}{3}$ and $x = 2 + \dfrac{\sqrt{33}}{3}$.

Critical points: $-1, 2 - \dfrac{\sqrt{33}}{3}, 2 + \dfrac{\sqrt{33}}{3}, 5$

$f(-1) = -6$, $f\left(2 - \dfrac{\sqrt{33}}{3}\right) \approx 2.04$,

$f\left(2 + \dfrac{\sqrt{33}}{3}\right) \approx -26.04$, $f(5) = -18$

Maximum value ≈ 2.04;
minimum value ≈ -26.04

b. $g'(x) = \dfrac{(x^3 - 6x^2 + x + 2)(3x^2 - 12x + 1)}{|x^3 - 6x^2 + x + 2|}$;

$g'(x) = 0$ when $x = 2 - \dfrac{\sqrt{33}}{3}$ and

$x = 2 + \dfrac{\sqrt{33}}{3}$. $g'(x)$ does not exist when

$f(x) = 0$; on $[-1, 5]$, $f(x) = 0$ when
$x \approx -0.4836$ and $x \approx 0.7172$

Critical points: $-1, -0.4836, 2 - \dfrac{\sqrt{33}}{3}$,

$0.7172, 2 + \dfrac{\sqrt{33}}{3}, 5$

$g(-1) = 6$, $g(-0.4836) = 0$,

$g\left(2 - \dfrac{\sqrt{33}}{3}\right) \approx 2.04$, $g(0.7172) = 0$,

$g\left(2 + \dfrac{\sqrt{33}}{3}\right) \approx 26.04$, $g(5) = 18$

Maximum value ≈ 26.04,
minimum value = 0

29. Answers will vary. One possibility:

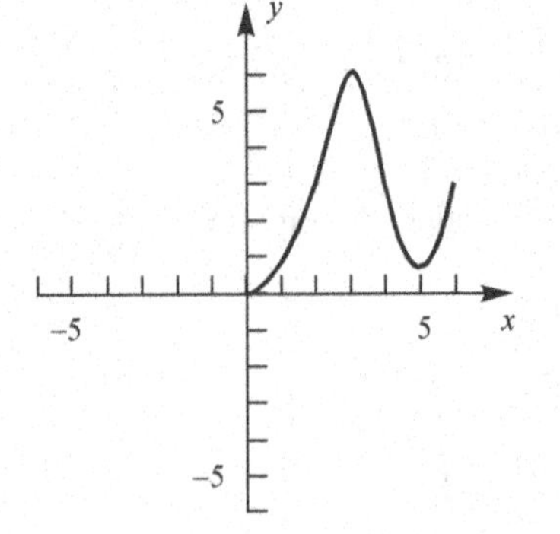

31. Answers will vary. One possibility:

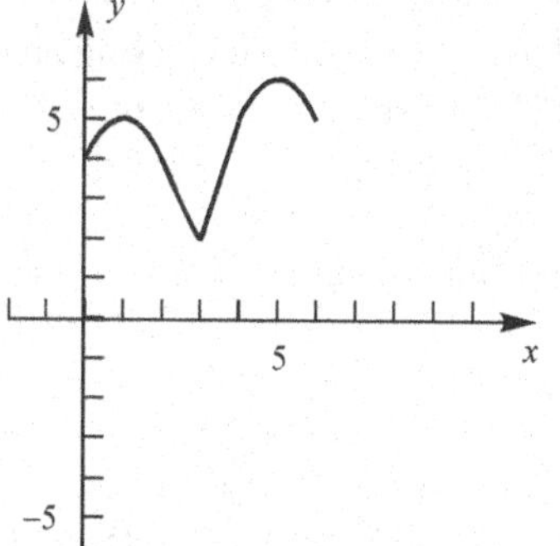

33. Answers will vary. One possibility:

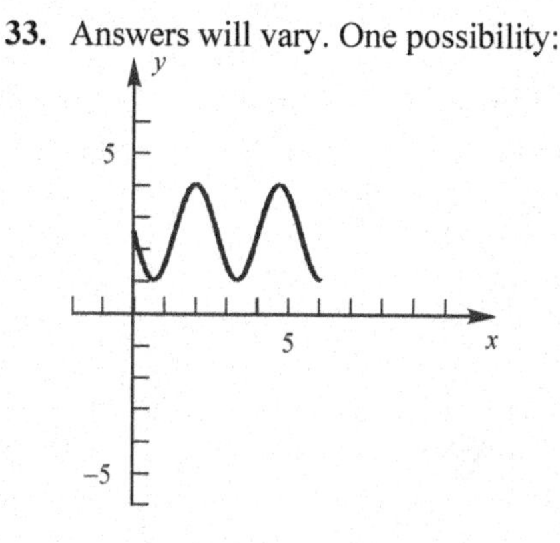

35. Answers will vary. One possibility:

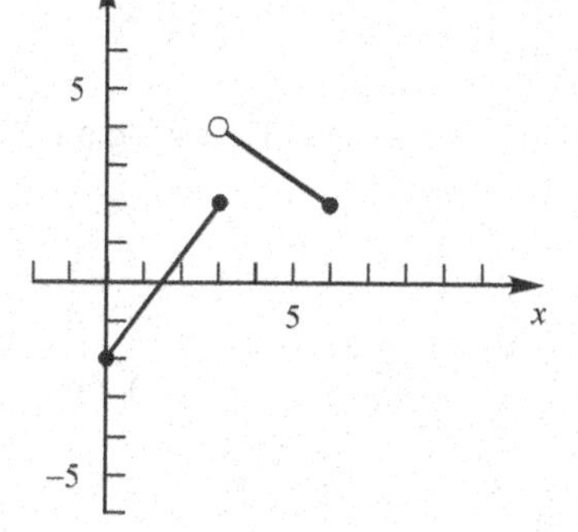

3.2 Concepts Review

1. Increasing; concave up

3. An inflection point

Problem Set 3.2

1. $f'(x) = 3;\ 3 > 0$ for all x. $f(x)$ is increasing for all x.

3. $h'(t) = 2t + 2;\ 2t + 2 > 0$ when $t > -1$. $h(t)$ is increasing on $[-1, \infty)$ and decreasing on $(-\infty, -1]$.

5. $G'(x) = 6x^2 - 18x + 12 = 6(x-2)(x-1)$

Split the x-axis into the intervals $(-\infty, 1)$, $(1, 2)$, $(2, \infty)$.

Test points: $x = 0, \frac{3}{2}, 3;\ G'(0) = 12,\ G'\left(\frac{3}{2}\right) = -\frac{3}{2},$

$G'(3) = 12$

$G(x)$ is increasing on $(-\infty, 1] \cup [2, \infty)$ and decreasing on $[1, 2]$.

7. $h'(z) = z^3 - 2z^2 = z^2(z-2)$

Split the x-axis into the intervals $(-\infty, 0)$, $(0, 2)$, $(2, \infty)$.

Test points: $z = -1, 1, 3;\ h'(-1) = -3,\ h'(1) = -1,$

$h'(3) = 9$

$h(z)$ is increasing on $[2, \infty)$ and decreasing on $(-\infty, 2]$.

9. $H'(t) = \cos t;\ H'(t) > 0$ when $0 \le t < \frac{\pi}{2}$ and

$\frac{3\pi}{2} < t \le 2\pi.$

$H(t)$ is increasing on $\left[0, \frac{\pi}{2}\right] \cup \left[\frac{3\pi}{2}, 2\pi\right]$ and

decreasing on $\left[\frac{\pi}{2}, \frac{3\pi}{2}\right]$.

11. $f''(x) = 2;\ 2 > 0$ for all x. $f(x)$ is concave up for all x; no inflection points.

13. $T''(t) = 18t;\ 18t > 0$ when $t > 0$. $T(t)$ is concave up on $(0, \infty)$ and concave down on $(-\infty, 0)$; $(0, 0)$ is the only inflection point.

15. $q''(x) = 12x^2 - 36x - 48$; $q''(x) > 0$ when $x < -1$ and $x > 4$.
$q(x)$ is concave up on $(-\infty, -1) \cup (4, \infty)$ and concave down on $(-1, 4)$; inflection points are $(-1, -19)$ and $(4, -499)$.

17. $F''(x) = 2\sin^2 x - 2\cos^2 x + 4 = 6 - 4\cos^2 x$;
$6 - 4\cos^2 x > 0$ for all x since $0 \le \cos^2 x \le 1$.
$F(x)$ is concave up for all x; no inflection points.

19. $f'(x) = 3x^2 - 12$; $3x^2 - 12 > 0$ when $x < -2$ or $x > 2$.
$f(x)$ is increasing on $(-\infty, -2] \cup [2, \infty)$ and decreasing on $[-2, 2]$.
$f''(x) = 6x$; $6x > 0$ when $x > 0$. $f(x)$ is concave up on $(0, \infty)$ and concave down on $(-\infty, 0)$.

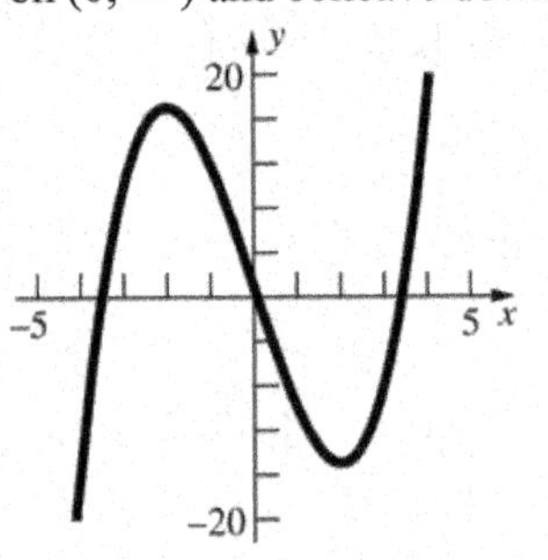

21. $g'(x) = 12x^3 - 12x^2 = 12x^2(x-1)$; $g'(x) > 0$ when $x > 1$. $g(x)$ is increasing on $[1, \infty)$ and decreasing on $(-\infty, 1]$.
$g''(x) = 36x^2 - 24x = 12x(3x-2)$; $g''(x) > 0$ when $x < 0$ or $x > \frac{2}{3}$. $g(x)$ is concave up on $(-\infty, 0) \cup \left(\frac{2}{3}, \infty\right)$ and concave down on $\left(0, \frac{2}{3}\right)$.

23. $G'(x) = 15x^4 - 15x^2 = 15x^2(x^2 - 1)$; $G'(x) > 0$ when $x < -1$ or $x > 1$. $G(x)$ is increasing on $(-\infty, -1] \cup [1, \infty)$ and decreasing on $[-1, 1]$.
$G''(x) = 60x^3 - 30x = 30x(2x^2 - 1)$;

Split the x-axis into the intervals $\left(-\infty, -\frac{1}{\sqrt{2}}\right)$, $\left(-\frac{1}{\sqrt{2}}, 0\right)$, $\left(0, \frac{1}{\sqrt{2}}\right)$, $\left(\frac{1}{\sqrt{2}}, \infty\right)$.

Test points: $x = -1, -\frac{1}{2}, \frac{1}{2}, 1$; $G''(-1) = -30$, $G''\left(-\frac{1}{2}\right) = \frac{15}{2}$, $G''\left(\frac{1}{2}\right) = -\frac{15}{2}$, $G''(1) = 30$.

$G(x)$ is concave up on $\left(-\frac{1}{\sqrt{2}}, 0\right) \cup \left(\frac{1}{\sqrt{2}}, \infty\right)$ and concave down on $\left(-\infty, -\frac{1}{\sqrt{2}}\right) \cup \left(0, \frac{1}{\sqrt{2}}\right)$.

25. $f'(x) = \frac{\cos x}{2\sqrt{\sin x}}$; $f'(x) > 0$ when $0 < x < \frac{\pi}{2}$. $f(x)$ is increasing on $\left[0, \frac{\pi}{2}\right]$ and decreasing on $\left[\frac{\pi}{2}, \pi\right]$.

$f''(x) = \frac{-\cos^2 x - 2\sin^2 x}{4\sin^{3/2} x}$; $f''(x) < 0$ for all x in $(0, \infty)$. $f(x)$ is concave down on $(0, \pi)$.

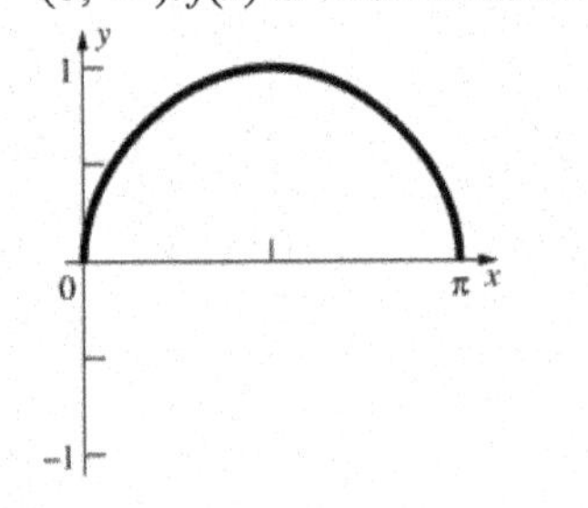

27. $f'(x)=\frac{2-5x}{3x^{1/3}}$; $2-5x>0$ when $x<\frac{2}{5}$, $f'(x)$ does not exist at $x = 0$.

Split the x-axis into the intervals $(-\infty, 0)$, $\left(0,\frac{2}{5}\right),\left(\frac{2}{5},\infty\right)$.

Test points: $-1,\frac{1}{5},1$; $f'(-1)=-\frac{7}{3}$, $f'\left(\frac{1}{5}\right)=\frac{\sqrt[3]{5}}{3}$, $f'(1)=-1$.

$f(x)$ is increasing on $\left[0,\frac{2}{5}\right]$ and decreasing on $(-\infty,0]\cup\left[\frac{2}{5},\infty\right)$.

$f''(x)=\frac{-2(5x+1)}{9x^{4/3}}$; $-2(5x+1)>0$ when $x<-\frac{1}{5}$, $f''(x)$ does not exist at $x = 0$.

Test points: $-1,-\frac{1}{10},1$; $f''(-1)=\frac{8}{9}$, $f''\left(-\frac{1}{10}\right)=-\frac{10^{4/3}}{9}$, $f(1)=-\frac{4}{3}$.

$f(x)$ is concave up on $\left(-\infty,-\frac{1}{5}\right)$ and concave down on $\left(-\frac{1}{5},0\right)\cup(0,\infty)$.

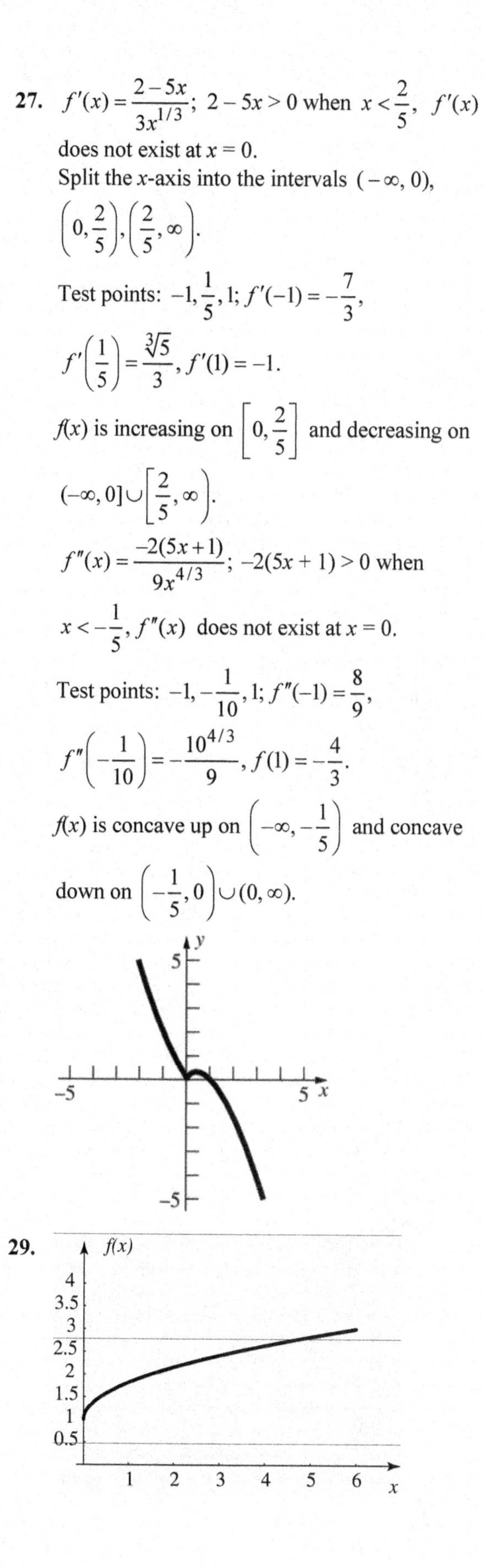

29.

31.

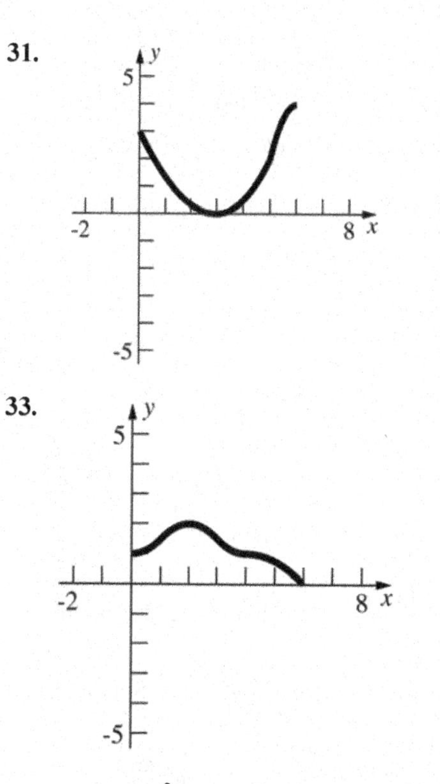

33.

35. $f(x)=ax^2+bx+c$; $f'(x)=2ax+b$; $f''(x)=2a$

An inflection point would occur where $f''(x)=0$, or $2a = 0$. This would only occur when $a = 0$, but if $a = 0$, the equation is not quadratic. Thus, quadratic functions have no points of inflection.

37. Suppose that there are points x_1 and x_2 in I where $f'(x_1)>0$ and $f'(x_2)<0$. Since f' is continuous on I, the Intermediate Value Theorem says that there is some number c between x_1 and x_2 such that $f'(c)=0$, which is a contradiction. Thus, either $f'(x)>0$ for all x in I and f is increasing throughout I or $f'(x)<0$ for all x in I and f is decreasing throughout I.

39. a. Let $f(x) = x^2$ and let $I = [0, a], a > y$.
$f'(x) = 2x > 0$ on I. Therefore, $f(x)$ is increasing on I, so $f(x) < f(y)$ for $x < y$.

b. Let $f(x) = \sqrt{x}$ and let $I = [0, a], a > y$.
$f'(x) = \frac{1}{2\sqrt{x}} > 0$ on I. Therefore, $f(x)$ is increasing on I, so $f(x) < f(y)$ for $x < y$.

c. Let $f(x) = \frac{1}{x}$ and let $I = [0, a]$, $a > y$.
$f'(x) = -\frac{1}{x^2} < 0$ on I. Therefore $f(x)$ is decreasing on I, so $f(x) > f(y)$ for $x < y$.

41. $f''(x) = \frac{3b - ax}{4x^{5/2}}$. If (4, 13) is an inflection point then $13 = 2a + \frac{b}{2}$ and $\frac{3b - 4a}{4 \cdot 32} = 0$. Solving these equations simultaneously, $a = \frac{39}{8}$ and $b = \frac{13}{2}$.

43. a. $[f(x) + g(x)]' = f'(x) + g'(x)$.
Since $f'(x) > 0$ and $g'(x) > 0$ for all x, $f'(x) + g'(x) > 0$ for all x. No additional conditions are needed.

b. $[f(x) \cdot g(x)]' = f(x)g'(x) + f'(x)g(x)$.
$f(x)g'(x) + f'(x)g(x) > 0$ if
$f(x) > -\frac{f'(x)}{g'(x)} g(x)$ for all x.

c. $[f(g(x))]' = f'(g(x))g'(x)$.
Since $f'(x) > 0$ and $g'(x) > 0$ for all x, $f'(g(x))g'(x) > 0$ for all x. No additional conditions are needed.

45. a.

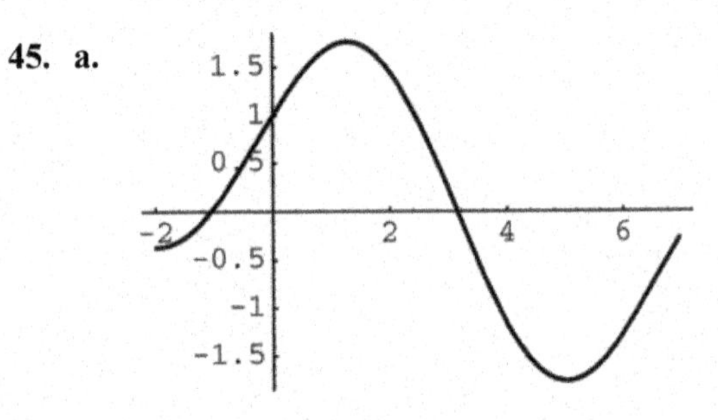

b. $f'(x) < 0 : (1.3, 5.0)$

c. $f''(x) < 0 : (-0.25, 3.1) \cup (6.5, 7]$

d. $f'(x) = \cos x - \frac{1}{2}\sin\frac{x}{2}$

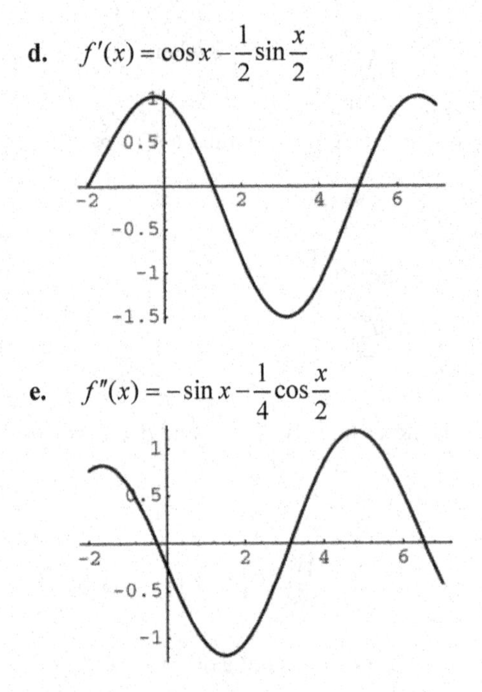

e. $f''(x) = -\sin x - \frac{1}{4}\cos\frac{x}{2}$

47. $f'(x) > 0$ on (–0.598, 0.680)
f is increasing on [–0.598, 0.680].

49. Let s be the distance traveled. Then $\frac{ds}{dt}$ is the speed of the car.

a. $\frac{ds}{dt} = ks$, k a constant

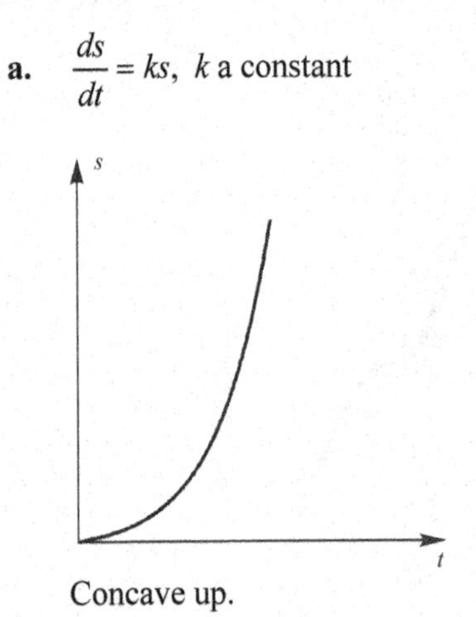

Concave up.

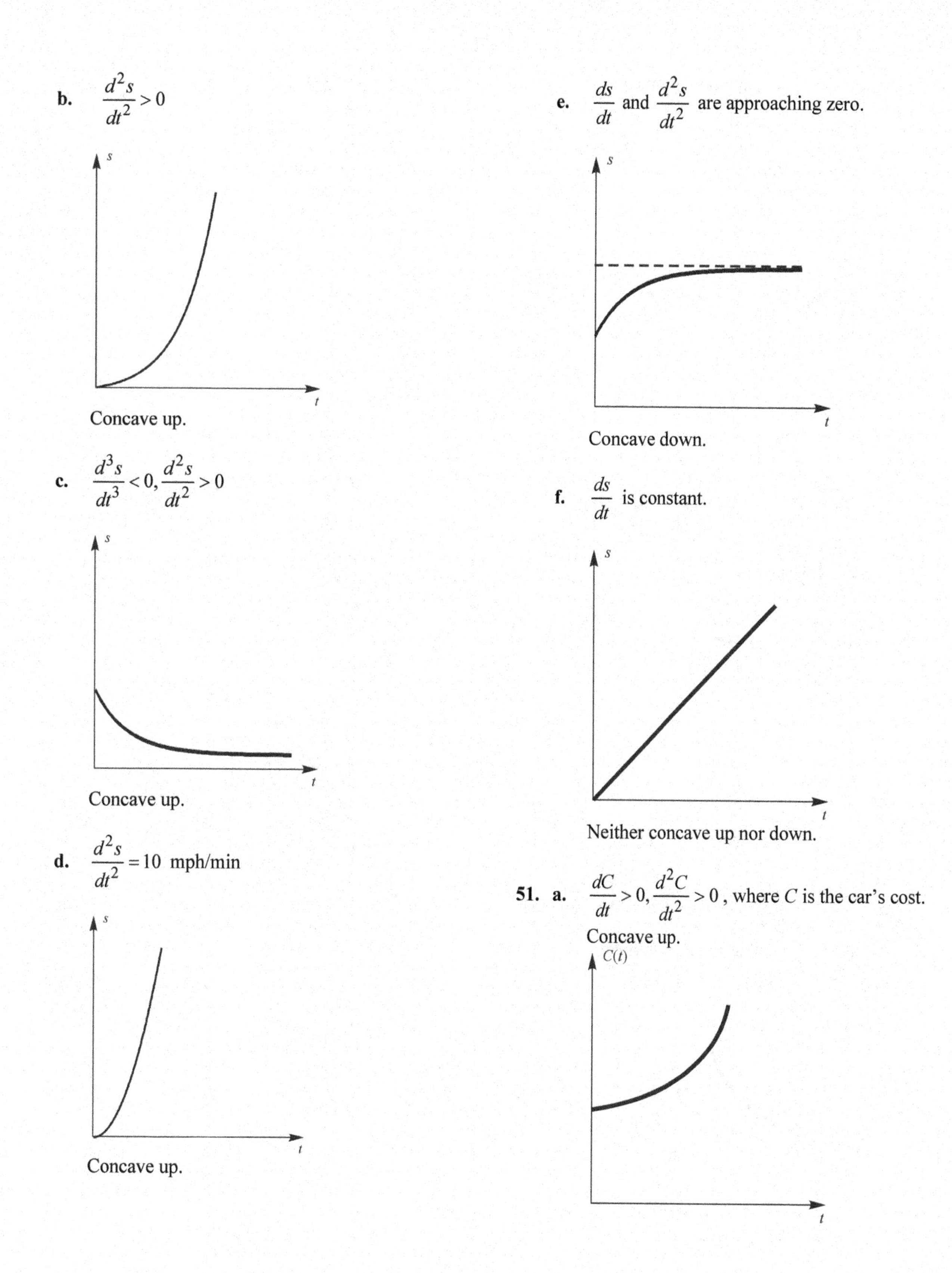

b. $\frac{d^2s}{dt^2} > 0$

Concave up.

c. $\frac{d^3s}{dt^3} < 0, \frac{d^2s}{dt^2} > 0$

Concave up.

d. $\frac{d^2s}{dt^2} = 10$ mph/min

Concave up.

e. $\frac{ds}{dt}$ and $\frac{d^2s}{dt^2}$ are approaching zero.

Concave down.

f. $\frac{ds}{dt}$ is constant.

Neither concave up nor down.

51. a. $\frac{dC}{dt} > 0, \frac{d^2C}{dt^2} > 0$, where C is the car's cost.

Concave up.

b. $f(t)$ is oil consumption at time t.

$\frac{df}{dt} < 0, \frac{d^2 f}{dt^2} > 0$

Concave up.

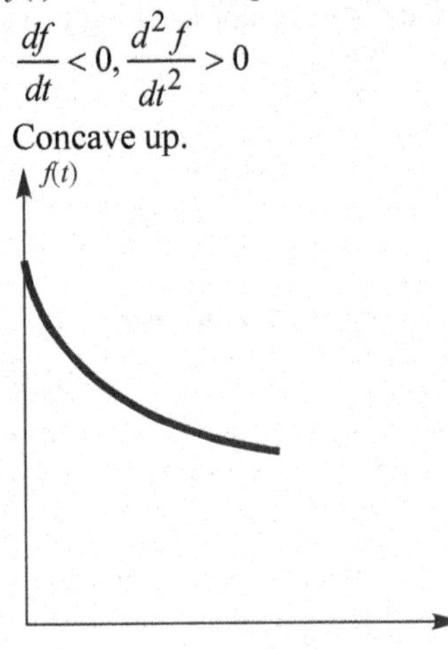

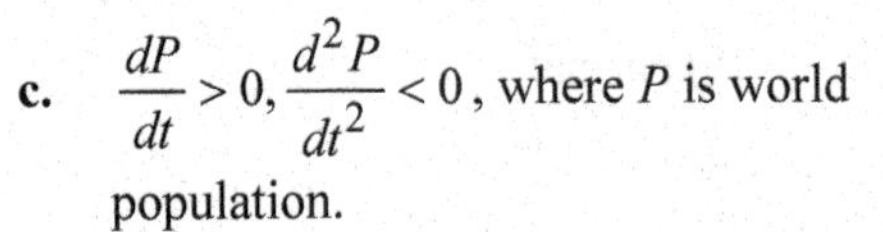

c. $\frac{dP}{dt} > 0, \frac{d^2 P}{dt^2} < 0$, where P is world population.

Concave down.

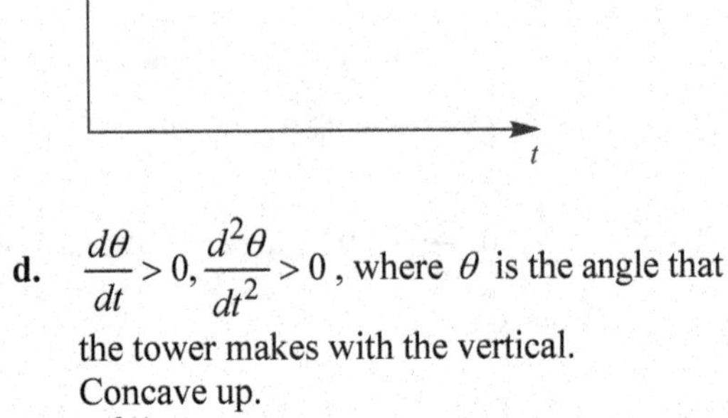

d. $\frac{d\theta}{dt} > 0, \frac{d^2\theta}{dt^2} > 0$, where θ is the angle that the tower makes with the vertical.

Concave up.

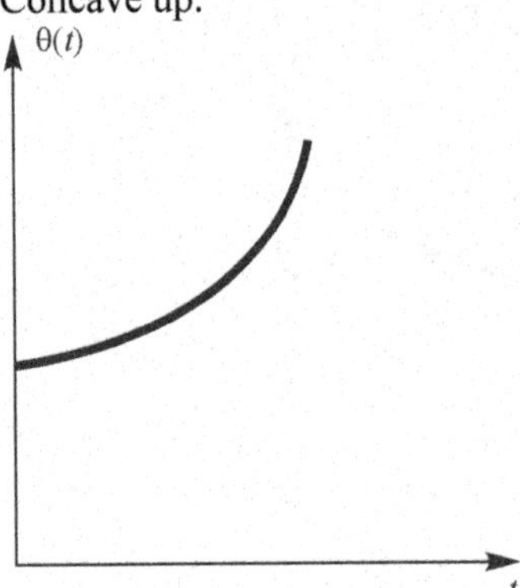

e. $P = f(t)$ is profit at time t.

$\frac{dP}{dt} > 0, \frac{d^2 P}{dt^2} < 0$

Concave down.

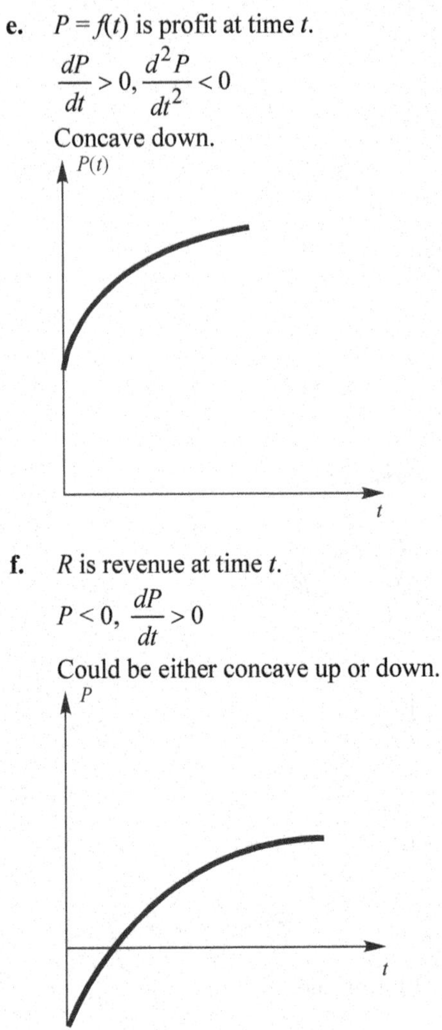

f. R is revenue at time t.

$P < 0, \frac{dP}{dt} > 0$

Could be either concave up or down.

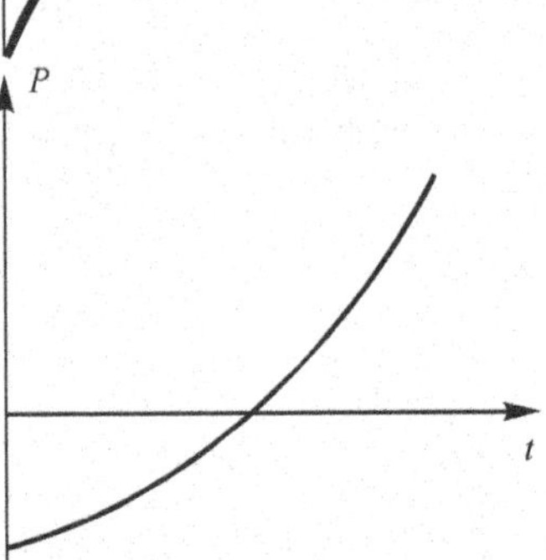

53. $\frac{dV}{dt} = 2 \text{ in}^3/\text{sec}$

The cup is a portion of a cone with the bottom cut off. If we let x represent the height of the missing cone, we can use similar triangles to show that

$$\frac{x}{3} = \frac{x+5}{3.5}$$

$$3.5x = 3x + 15$$

$$0.5x = 15$$

$$x = 30$$

Similar triangles can be used again to show that, at any given time, the radius of the cone at water level is

$$r = \frac{h+30}{20}$$

Therefore, the volume of water can be expressed as

$$V = \frac{\pi(h+30)^3}{1200} - \frac{45\pi}{2}.$$

We also know that $V = 2t$ from above. Setting the two volume equations equal to each other and solving for h gives $h = \sqrt[3]{\frac{2400}{\pi}t + 27000} - 30$.

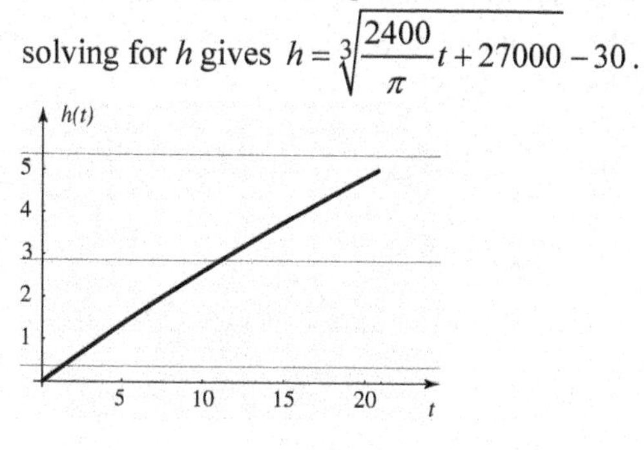

55. $V = 3t,\ 0 \le t \le 8$. The height is always increasing, so $h'(t) > 0$. The rate of change of the height decreases from time $t = 0$ until time t_1 when the water reaches the middle of the rounded bottom part. The rate of change then increases until time t_2 when the water reaches the middle of the neck. Then the rate of change decreases until $t = 8$ and the vase is full. Thus, $h''(t) > 0$ for $t_1 < t < t_2$ and $h''(t) < 0$ for $t_2 < t < 8$.

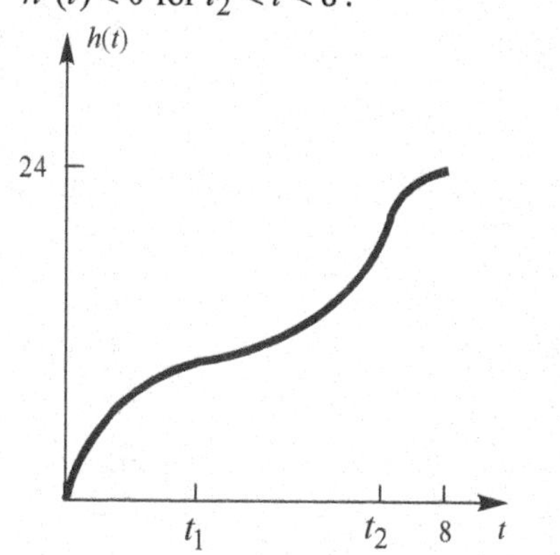

57. a. The cross-sectional area of the vase is approximately equal to ΔV and the corresponding radius is $r = \sqrt{\Delta V/\pi}$. The table below gives the approximate values for r. The vase becomes slightly narrower as you move above the base, and then gets wider as you near the top.

Depth	V	$A \approx \Delta V$	$r = \sqrt{\Delta V/\pi}$
1	4	4	1.13
2	8	4	1.13
3	11	3	0.98
4	14	3	0.98
5	20	6	1.38
6	28	8	1.60

b. Near the base, this vase is like the one in part (a), but just above the base it becomes larger. Near the middle of the vase it becomes very narrow. The top of the vase is similar to the one in part (a).

Depth	V	$A \approx \Delta V$	$r = \sqrt{\Delta V/\pi}$
1	4	4	1.13
2	9	5	1.26
3	12	3	0.98
4	14	2	0.80
5	20	6	1.38
6	28	8	1.60

3.3 Concepts Review

1. maximum

3. maximum

Problem Set 3.3

1. $f'(x) = 3x^2 - 12x = 3x(x-4)$
   Critical points: 0, 4
   $f'(x) > 0$ on $(-\infty, 0)$, $f'(x) < 0$ on $(0, 4)$,
   $f'(x) > 0$ on $(4, \infty)$
   $f''(x) = 6x - 12$; $f''(0) = -12$, $f''(4) = 12$.
   Local minimum at $x = 4$;
   local maximum at $x = 0$

3. $f'(\theta) = 2\cos 2\theta$; $2\cos 2\theta \neq 0$ on $\left(0, \frac{\pi}{4}\right)$
   No critical points; no local maxima or minima on $\left(0, \frac{\pi}{4}\right)$.

5. $\Psi'(\theta) = 2\sin\theta\cos\theta$
   $-\frac{\pi}{2} < \theta < \frac{\pi}{2}$
   Critical point: 0
   $\Psi'(\theta) < 0$ on $\left(-\frac{\pi}{2}, 0\right)$, $\Psi'(\theta) > 0$ on $\left(0, \frac{\pi}{2}\right)$,
   $\Psi''(\theta) = 2\cos^2\theta - 2\sin^2\theta$; $\Psi''(0) = 2$
   Local minimum at $x = 0$

7. $f'(x) = \frac{(x^2+4)\cdot 1 - x(2x)}{(x^2+4)^2} = \frac{4-x^2}{(x^2+4)^2}$
   Critical points: $-2, 2$
   $f'(x) < 0$ on $(-\infty, -2)$ and $(2, \infty)$;
   $f'(x) > 0$ on $(-2, 2)$
   $f''(x) = \frac{2x(x^2-12)}{(x^2+4)^3}$
   $f''(-2) = \frac{1}{16}$; $f''(2) = -\frac{1}{16}$
   Local minima at $x = -2$; Local maxima at $x = 2$

9. $h'(y) = 2y + \frac{1}{y^2}$
   Critical point: $-\frac{\sqrt[3]{4}}{2}$
   $h'(y) < 0$ on $\left(-\infty, -\frac{\sqrt[3]{4}}{2}\right)$
   $h'(y) > 0$ on $\left(-\frac{\sqrt[3]{4}}{2}, 0\right)$ and $(0, \infty)$
   $h''(y) = 2 - \frac{2}{y^3}$
   $h\left(-\frac{\sqrt[3]{4}}{2}\right) = 2 - \frac{2}{\left(-\frac{\sqrt[3]{4}}{2}\right)^3} = 2 + \frac{16}{4} = 6$
   Local minima at $-\frac{\sqrt[3]{4}}{2}$

11. $f'(x) = 3x^2 - 3 = 3(x^2 - 1)$
   Critical points: -1, 1
   $f''(x) = 6x$; $f''(-1) = -6$, $f''(1) = 6$
   Local minimum value $f(1) = -2$;
   local maximum value $f(-1) = 2$

13. $H'(x) = 4x^3 - 6x^2 = 2x^2(2x-3)$
   Critical points: $0, \frac{3}{2}$
   $H''(x) = 12x^2 - 12x = 12x(x-1)$; $H''(0) = 0$,
   $H''\left(\frac{3}{2}\right) = 9$
   $H'(x) < 0$ on $(-\infty, 0)$, $H'(x) < 0$ on $\left(0, \frac{3}{2}\right)$
   Local minimum value $H\left(\frac{3}{2}\right) = -\frac{27}{16}$; no local maximum values ($x = 0$ is neither a local minimum nor maximum)

15. $g'(t) = -\frac{2}{3(t-2)^{1/3}}$; $g'(t)$ does not exist at $t = 2$.
   Critical point: 2
   $g'(1) = \frac{2}{3}$, $g'(3) = -\frac{2}{3}$
   No local minimum values; local maximum value $g(2) = \pi$.

17. $f'(t) = 1 + \frac{1}{t^2}$
   No critical points
   No local minimum or maximum values

19. $\Lambda'(\theta) = -\frac{1}{1+\sin\theta}$; $\Lambda'(\theta)$ does not exist at $\theta = \frac{3\pi}{2}$, but $\Lambda(\theta)$ does not exist at that point either.
No critical points
No local minimum or maximum values

21. $f'(x) = 4(\sin 2x)(\cos 2x)$
$4(\sin 2x)(\cos 2x) = 0$ when $x = \frac{(2k-1)\pi}{4}$ or $x = \frac{k\pi}{2}$ where k is an integer.
Critical points: $0, \frac{\pi}{4}, \frac{\pi}{2}, 2$
$f(0) = 0$; $f\left(\frac{\pi}{4}\right) = 1$; $f\left(\frac{\pi}{2}\right) = 0$;
$f(2) \approx 0.5728$
Minimum value: $f(0) = f\left(\frac{\pi}{2}\right) = 0$
Maximum value: $f\left(\frac{\pi}{4}\right) = 1$

23. $g'(x) = \frac{-x(x^3-64)}{(x^3+32)^2}$
$g'(x) = 0$ when $x = 0$ or $x = 4$.
Critical points: 0, 4
$g(0) = 0$; $g(4) = \frac{1}{6}$
As x approaches ∞, the value of g approaches 0 but never actually gets there.
Maximum value: $g(4) = \frac{1}{6}$
Minimum value: $g(0) = 0$

25. $F'(x) = \frac{3}{\sqrt{x}} - 4$; $\frac{3}{\sqrt{x}} - 4 = 0$ when $x = \frac{9}{16}$
Critical points: $0, \frac{9}{16}, 4$
$F(0) = 0$, $F\left(\frac{9}{16}\right) = \frac{9}{4}$, $F(4) = -4$
Minimum value $F(4) = -4$; maximum value $F\left(\frac{9}{16}\right) = \frac{9}{4}$

27. $f'(x) = 64(-1)(\sin x)^{-2}\cos x$
$+27(-1)(\cos x)^{-2}(-\sin x)$
$= -\frac{64\cos x}{\sin^2 x} + \frac{27\sin x}{\cos^2 x}$
$= \frac{(3\sin x - 4\cos x)(9\sin^2 x + 12\cos x\sin x + 16\cos^2 x)}{\sin^2 x\cos^2 x}$
On $\left(0, \frac{\pi}{2}\right)$, $f'(x) = 0$ only where $3\sin x = 4\cos x$;
$\tan x = \frac{4}{3}$;
$x = \tan^{-1}\frac{4}{3} \approx 0.9273$
Critical point: 0.9273
For $0 < x < 0.9273$, $f'(x) < 0$, while for $0.9273 < x < \frac{\pi}{2}$, $f'(x) > 0$
Minimum value $f\left(\tan^{-1}\frac{4}{3}\right) = \frac{64}{\frac{4}{5}} + \frac{27}{\frac{3}{5}} = 125$;
no maximum value

29. $H'(x) = \frac{2x(x^2-1)}{|x^2-1|}$
$H'(x) = 0$ when $x = 0$.
$H'(x)$ is undefined when $x = -1$ or $x = 1$
Critical points: $-2, -1, 0, 1, 2$
$H(-2) = 3$; $H(-1) = 0$; $H(0) = 1$; $H(1) = 0$;
$H(2) = 3$
Minimum value: $H(-1) = H(1) = 0$
Maximum value: $H(-2) = H(2) = 3$

31. $f'(x) = 0$ when $x = 0$ and $x = 1$. On the interval $(-\infty, 0)$ we get $f'(x) < 0$. On $(0, \infty)$, we get $f'(x) > 0$. Thus there is a local min at $x = 0$ but no local max.

33. $f'(x) = 0$ at $x = 1, 2, 3, 4$; $f'(x)$ is negative on $(3, 4)$ and positive on $(-\infty, 1) \cup (1, 2) \cup (2, 3) \cup (4, \infty)$ Thus, the function has a local minimum at $x = 4$ and a local maximum at $x = 3$.

35. Since $f'(x) \geq 0$ for all x, the function is always increasing. Therefore, there are no local extrema.

37. Answers will vary. One possibility:

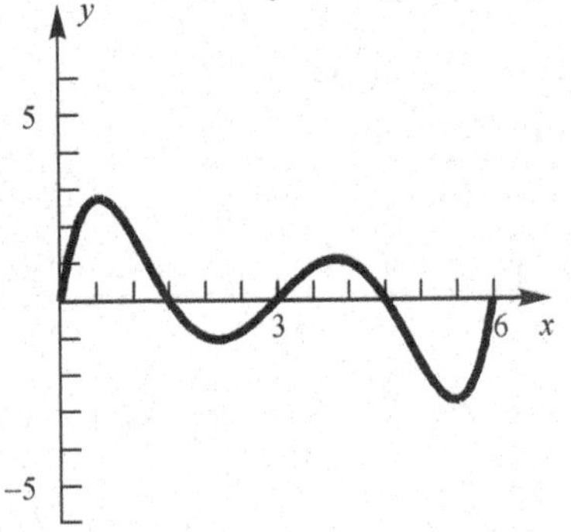

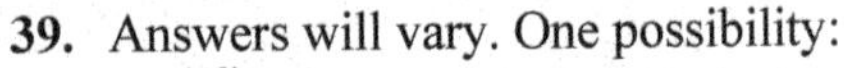

39. Answers will vary. One possibility:

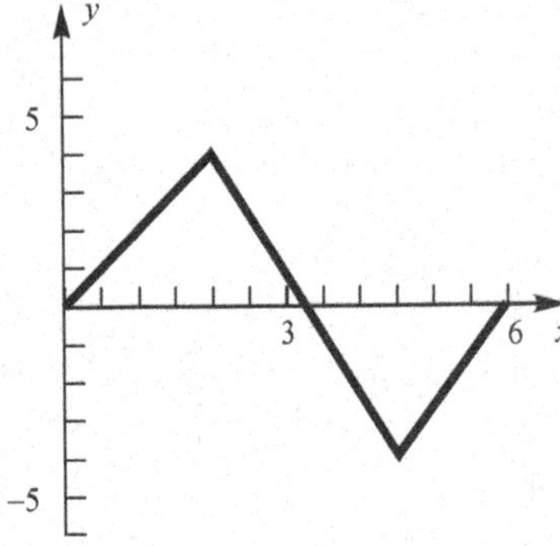

41. Answers will vary. One possibility:

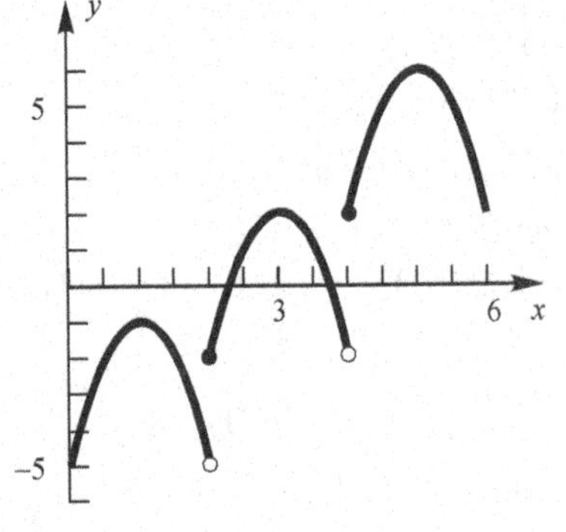

43. The graph of f is a parabola which opens up.

$f'(x) = 2Ax + B = 0 \rightarrow x = -\frac{B}{2A}$

$f''(x) = 2A$

Since $A > 0$, the graph of f is always concave up. There is exactly one critical point which yields the minimum of the graph.

$$f\left(-\frac{B}{2A}\right) = A\left(-\frac{B}{2A}\right)^2 + B\left(-\frac{B}{2A}\right) + C$$

$$= \frac{B^2}{4A} - \frac{B^2}{2A} + C = \frac{B^2 - 2B^2 + 4AC}{4A}$$

$$= \frac{4AC - B^2}{4A} = -\frac{B^2 - 4AC}{4A}$$

If $f(x) \geq 0$ with $A > 0$, then $-(B^2 - 4AC) \geq 0$,

or $B^2 - 4AC \leq 0$.

If $B^2 - 4AC \leq 0$, then we get $f\left(-\frac{B}{2A}\right) \geq 0$

Since $0 \leq f\left(-\frac{B}{2A}\right) \leq f(x)$ for all x, we get

$f(x) \geq 0$ for all x.

45. $f'''(c) > 0$ implies that f'' is increasing at c, so f is concave up to the right of c (since $f''(x) > 0$ to the right of c) and concave down to the left of c (since $f''(x) < 0$ to the left of c). Therefore f has a point of inflection at c.

3.4 Concepts Review

1. $0 < x < \infty$

3. $S = \sum_{i=1}^{n} (y_i - bx_i)^2$

Problem Set 3.4

1. Let x be one number, y be the other, and Q be the sum of the squares.

$xy = -16$

$y = -\frac{16}{x}$

The possible values for x are in $(-\infty, 0)$ or $(0, \infty)$.

$Q = x^2 + y^2 = x^2 + \frac{256}{x^2}$

$\frac{dQ}{dx} = 2x - \frac{512}{x^3}$

$2x - \frac{512}{x^3} = 0$

$x^4 = 256$

$x = \pm 4$

The critical points are –4, 4.

$\frac{dQ}{dx} < 0$ on $(-\infty, -4)$ and $(0, 4)$. $\frac{dQ}{dx} > 0$ on $(-4, 0)$ and $(4, \infty)$.

When $x = -4$, $y = 4$ and when $x = 4$, $y = -4$. The two numbers are –4 and 4.

3. Let x be the number.

$Q = \sqrt[4]{x} - 2x$

x will be in the interval $(0, \infty)$.

$$\frac{dQ}{dx} = \frac{1}{4}x^{-3/4} - 2$$

$$\frac{1}{4}x^{-3/4} - 2 = 0$$

$$x^{-3/4} = 8$$

$$x = \frac{1}{16}$$

$\frac{dQ}{dx} > 0$ on $\left(0, \frac{1}{16}\right)$ and $\frac{dQ}{dx} < 0$ on $\left(\frac{1}{16}, \infty\right)$

Q attains its maximum value at $x = \frac{1}{16}$.

5. Let Q be the square of the distance between (x, y) and $(0, 5)$.

$$Q = (x-0)^2 + (y-5)^2 = x^2 + (x^2-5)^2$$
$$= x^4 - 9x^2 + 25$$

$$\frac{dQ}{dx} = 4x^3 - 18x$$

$$4x^3 - 18x = 0$$

$$2x(2x^2 - 9) = 0$$

$$x = 0, \pm\frac{3}{\sqrt{2}}$$

$\frac{dQ}{dx} < 0$ on $\left(-\infty, -\frac{3}{\sqrt{2}}\right)$ and $\left(0, \frac{3}{\sqrt{2}}\right)$.

$\frac{dQ}{dx} > 0$ on $\left(-\frac{3}{\sqrt{2}}, 0\right)$ and $\left(\frac{3}{\sqrt{2}}, \infty\right)$.

When $x = -\frac{3}{\sqrt{2}}, y = \frac{9}{2}$ and when $x = \frac{3}{\sqrt{2}}$, $y = \frac{9}{2}$.

The points are $\left(-\frac{3}{\sqrt{2}}, \frac{9}{2}\right)$ and $\left(\frac{3}{\sqrt{2}}, \frac{9}{2}\right)$.

7. $x \ge x^2$ if $0 \le x \le 1$

$f(x) = x - x^2;\ f'(x) = 1 - 2x;$

$f'(x) = 0$ when $x = \frac{1}{2}$

Critical points: $0, \frac{1}{2}, 1$

$f(0) = 0, f(1) = 0,\ f\left(\frac{1}{2}\right) = \frac{1}{4}$; therefore, $\frac{1}{2}$ exceeds its square by the maximum amount.

9. Let x be the width of the square to be cut out and V the volume of the resulting open box.

$$V = x(24-2x)^2 = 4x^3 - 96x^2 + 576x$$

$$\frac{dV}{dx} = 12x^2 - 192x + 576 = 12(x-12)(x-4);$$

$12(x - 12)(x - 4) = 0$; $x = 12$ or $x = 4$.

Critical points: 0, 4, 12

At $x = 0$ or 12, $V = 0$; at $x = 4$, $V = 1024$.

The volume of the largest box is 1024 in.^{3}

11. Let x be the width of each pen, then the length along the barn is $80 - 4x$.

$$A = x(80-4x) = 80x - 4x^2; \frac{dA}{dx} = 80 - 8x;$$

$\frac{dA}{dx} = 0$ when $x = 10$.

Critical points: 0, 10, 20

At $x = 0$ or 20, $A = 0$; at $x = 10$, $A = 400$.

The area is largest with width 10 ft and length 40 ft.

13. $xy = 900;\ y = \frac{900}{x}$

The possible values for x are in $(0, \infty)$.

$$Q = 4x + 3y = 4x + 3\left(\frac{900}{x}\right) = 4x + \frac{2700}{x}$$

$$\frac{dQ}{dx} = 4 - \frac{2700}{x^2}$$

$$4 - \frac{2700}{x^2} = 0$$

$$x^2 = 675$$

$$x = \pm 15\sqrt{3}$$

$x = 15\sqrt{3}$ is the only critical point in $(0, \infty)$.

$\frac{dQ}{dx} < 0$ on $(0, 15\sqrt{3})$ and

$\frac{dQ}{dx} > 0$ on $(15\sqrt{3}, \infty)$.

When $x = 15\sqrt{3}, y = \frac{900}{15\sqrt{3}} = 20\sqrt{3}$.

Q has a minimum when $x = 15\sqrt{3} \approx 25.98$ ft and $y = 20\sqrt{3} \approx 34.64$ ft.

15. $xy = 300;\ y = \dfrac{300}{x}$

The possible values for x are in $(0, \infty)$.

$Q = 3(6x + 2y) + 2(2y) = 18x + 10y = 18x + \dfrac{3000}{x}$

$\dfrac{dQ}{dx} = 18 - \dfrac{3000}{x^2}$

$18 - \dfrac{3000}{x^2} = 0$

$x^2 = \dfrac{500}{3}$

$x = \pm\dfrac{10\sqrt{5}}{\sqrt{3}}$

$x = \dfrac{10\sqrt{5}}{\sqrt{3}}$ is the only critical point in $(0, \infty)$.

$\dfrac{dQ}{dx} < 0$ on $\left(0, \dfrac{10\sqrt{5}}{\sqrt{3}}\right)$ and

$\dfrac{dQ}{dx} > 0$ on $\left(\dfrac{10\sqrt{5}}{\sqrt{3}}, \infty\right)$.

When $x = \dfrac{10\sqrt{5}}{\sqrt{3}},\ y = \dfrac{300}{\frac{10\sqrt{5}}{\sqrt{3}}} = 6\sqrt{15}$

Q has a minimum when $x = \dfrac{10\sqrt{5}}{\sqrt{3}} \approx 12.91$ ft and

$y = 6\sqrt{15} \approx 23.24$ ft.

17. Let D be the square of the distance.

$D = (x-0)^2 + (y-4)^2 = x^2 + \left(\dfrac{x^2}{4} - 4\right)^2$

$= \dfrac{x^4}{16} - x^2 + 16$

$\dfrac{dD}{dx} = \dfrac{x^3}{4} - 2x;\ \dfrac{x^3}{4} - 2x = 0;\ x(x^2 - 8) = 0$

$x = 0, x = \pm 2\sqrt{2}$

Critical points: $0, 2\sqrt{2}, 2\sqrt{3}$

Since D is continuous and we are considering a closed interval for x, there is a maximum and minimum value of D on the interval. These extrema must occur at one of the critical points. At $x = 0$, $y = 0$, and $D = 16$. At $x = 2\sqrt{2}, y = 2,$ and $D = 12$. At $x = 2\sqrt{3}$, $y = 3$, and $D = 13$.

Therefore, the point on $y = \dfrac{x^2}{4}$ closest to $(0,4)$ is $P(2\sqrt{2}, 2)$ and the point farthest from $(0,4)$ is $Q(0,0)$.

19. Let x be the distance from P to where the woman lands the boat. She must row a distance of $\sqrt{x^2+4}$ miles and walk $10 - x$ miles. This will take her $T(x) = \dfrac{\sqrt{x^2+4}}{3} + \dfrac{10-x}{4}$ hours;

$0 \le x \le 10$. $T'(x) = \dfrac{x}{3\sqrt{x^2+4}} - \dfrac{1}{4}; T'(x) = 0$

when $x = \dfrac{6}{\sqrt{7}}$.

$T(0) = \dfrac{19}{6}$ hr $= 3$ hr 10 min ≈ 3.17 hr ,

$T\left(\dfrac{6}{\sqrt{7}}\right) = \dfrac{15+\sqrt{7}}{6} \approx 2.94$ hr,

$T(10) = \dfrac{\sqrt{104}}{3} \approx 3.40$ hr

She should land the boat $\dfrac{6}{\sqrt{7}} \approx 2.27$ mi down the shore from P.

21. $T(x) = \dfrac{\sqrt{x^2+4}}{20} + \dfrac{10-x}{4}, 0 \le x \le 10.$

$T'(x) = \dfrac{x}{20\sqrt{x^2+4}} - \dfrac{1}{4}; T'(x) = 0$ has no solution.

$T(0) = \dfrac{2}{20} + \dfrac{10}{4} = \dfrac{13}{5}$ hr $= 2$ hr, 36 min

$T(10) = \dfrac{\sqrt{104}}{20} \approx 0.5$ hr

She should take the boat all the way to town.

23. Let the coordinates of the first ship at 7:00 a.m. be (0, 0). Thus, the coordinates of the second ship at 7:00 a.m. are (–60, 0). Let t be the time in hours since 7:00 a.m. The coordinates of the first and second ships at t are $(-20t, 0)$ and $\left(-60 + 15\sqrt{2}t, -15\sqrt{2}t\right)$ respectively. Let D be the square of the distances at t.

$D = \left(-20t + 60 - 15\sqrt{2}t\right)^2 + \left(0 + 15\sqrt{2}t\right)^2$

$= \left(1300 + 600\sqrt{2}\right)t^2 - \left(2400 + 1800\sqrt{2}\right)t + 3600$

$\dfrac{dD}{dt} = 2\left(1300 + 600\sqrt{2}\right)t - \left(2400 + 1800\sqrt{2}\right)$

$2\left(1300 + 600\sqrt{2}\right)t - \left(2400 + 1800\sqrt{2}\right) = 0$ when

$t = \dfrac{12 + 9\sqrt{2}}{13 + 6\sqrt{2}} \approx 1.15$ hrs or 1 hr, 9 min

D is the minimum at $t = \dfrac{12 + 9\sqrt{2}}{13 + 6\sqrt{2}}$ since $\dfrac{d^2D}{dt^2} > 0$ for all t.

The ships are closest at 8:09 A.M.

25. Let x be the radius of the base of the cylinder and h the height.

$$V = \pi x^2 h; r^2 = x^2 + \left(\frac{h}{2}\right)^2; x^2 = r^2 - \frac{h^2}{4}$$

$$V = \pi\left(r^2 - \frac{h^2}{4}\right)h = \pi hr^2 - \frac{\pi h^3}{4}$$

$$\frac{dV}{dh} = \pi r^2 - \frac{3\pi h^2}{4}; V' = 0 \text{ when } h = \pm\frac{2\sqrt{3}r}{3}$$

Since $\frac{d^2V}{dh^2} = -\frac{3\pi h}{2}$, the volume is maximized when $h = \frac{2\sqrt{3}r}{3}$.

$$V = \pi\left(\frac{2\sqrt{3}}{3}r\right)r^2 - \frac{\pi\left(\frac{2\sqrt{3}}{3}r\right)^3}{4}$$

$$= \frac{2\pi\sqrt{3}}{3}r^3 - \frac{2\pi\sqrt{3}}{9}r^3 = \frac{4\pi\sqrt{3}}{9}r^3$$

27. Let x be the radius of the cylinder, r the radius of the sphere, and h the height of the cylinder.

$$A = 2\pi xh;\ r^2 = x^2 + \frac{h^2}{4}; x = \sqrt{r^2 - \frac{h^2}{4}}$$

$$A = 2\pi\sqrt{r^2 - \frac{h^2}{4}}h = 2\pi\sqrt{h^2r^2 - \frac{h^4}{4}}$$

$$\frac{dA}{dh} = \frac{\pi\left(2r^2h - h^3\right)}{\sqrt{h^2r^2 - \frac{h^4}{4}}}; A' = 0 \text{ when } h = 0, \pm\sqrt{2}r$$

$\frac{dA}{dh} > 0$ on $(0, \sqrt{2}r)$ and $\frac{dA}{dh} < 0$ on $(\sqrt{2}r, 2r)$,

so A is a maximum when $h = \sqrt{2}r$.

The dimensions are $h = \sqrt{2}r, x = \frac{r}{\sqrt{2}}$.

29. Let x be the length of a side of the square, so $\frac{100-4x}{3}$ is the side of the triangle, $0 \le x \le 25$

$$A = x^2 + \frac{1}{2}\left(\frac{100-4x}{3}\right)\frac{\sqrt{3}}{2}\left(\frac{100-4x}{3}\right)$$

$$= x^2 + \frac{\sqrt{3}}{4}\left(\frac{10{,}000 - 800x + 16x^2}{9}\right)$$

$$\frac{dA}{dx} = 2x - \frac{200\sqrt{3}}{9} + \frac{8\sqrt{3}}{9}x$$

$A'(x) = 0$ when $x = \frac{300\sqrt{3}}{11} - \frac{400}{11} \approx 10.874$.

Critical points: $x = 0, 10.874, 25$

At $x = 0, A \approx 481$; at $x = 10.874, A \approx 272$; at $x = 25, A = 625$.

a. For minimum area, the cut should be approximately $4(10.874) \approx 43.50$ cm from one end and the shorter length should be bent to form the square.

b. For maximum area, the wire should not be cut; it should be bent to form a square.

31. Let r be the radius of the cylinder and h the height of the cylinder.

$$V = \pi r^2 h + \frac{2}{3}\pi r^3; h = \frac{V - \frac{2}{3}\pi r^3}{\pi r^2} = \frac{V}{\pi r^2} - \frac{2}{3}r$$

Let k be the cost per square foot of the cylindrical wall. The cost is

$$C = k(2\pi rh) + 2k(2\pi r^2)$$

$$= k\left(2\pi r\left(\frac{V}{\pi r^2} - \frac{2}{3}r\right) + 4\pi r^2\right) = k\left(\frac{2V}{r} + \frac{8\pi r^2}{3}\right)$$

$$\frac{dC}{dr} = k\left(-\frac{2V}{r^2} + \frac{16\pi r}{3}\right); k\left(-\frac{2V}{r^2} + \frac{16\pi r}{3}\right) = 0$$

when $r^3 = \frac{3V}{8\pi}, r = \frac{1}{2}\left(\frac{3V}{\pi}\right)^{1/3}$

$$h = \frac{4V}{\pi\left(\frac{3V}{\pi}\right)^{2/3}} - \frac{1}{3}\left(\frac{3V}{\pi}\right)^{1/3} = \left(\frac{3V}{\pi}\right)^{1/3}$$

For a given volume V, the height of the cylinder is $\left(\frac{3V}{\pi}\right)^{1/3}$ and the radius is $\frac{1}{2}\left(\frac{3V}{\pi}\right)^{1/3}$.

33. $A = \frac{r^2\theta}{2};\ \theta = \frac{2A}{r^2}$

The perimeter is

$$Q = 2r + r\theta = 2r + \frac{2Ar}{r^2} = 2r + \frac{2A}{r}$$

$$\frac{dQ}{dr} = 2 - \frac{2A}{r^2}; Q' = 0 \text{ when } r = \sqrt{A}$$

$$\theta = \frac{2A}{(\sqrt{A})^2} = 2$$

$\frac{d^2Q}{dr^2} = \frac{4A}{r^3} > 0$, so this minimizes the perimeter.

35. x is limited by $0 \le x \le \sqrt{12}$.

$$A = 2x(12 - x^2) = 24x - 2x^3; \frac{dA}{dx} = 24 - 6x^2;$$

$$24 - 6x^2 = 0; x = -2, 2$$

Critical points: $0, 2, \sqrt{12}$.

When $x = 0$ or $\sqrt{12}$, $A = 0$.

When $x = 2$, $y = 12 - (2)^2 = 8$.

The dimensions are $2x = 2(2) = 4$ by 8.

37. If the end of the cylinder has radius r and h is the height of the cylinder, the surface area is

$A = 2\pi r^2 + 2\pi rh$ so $h = \frac{A}{2\pi r} - r.$

The volume is

$V = \pi r^2 h = \pi r^2\left(\frac{A}{2\pi r} - r\right) = \frac{Ar}{2} - \pi r^3.$

$V'(r) = \frac{A}{2} - 3\pi r^2; V'(r) = 0$ when $r = \sqrt{\frac{A}{6\pi}},$

$V''(r) = -6\pi r,$ so the volume is maximum when

$r = \sqrt{\frac{A}{6\pi}}.$

$h = \frac{A}{2\pi r} - r = 2\sqrt{\frac{A}{6\pi}} = 2r$

39. If the rectangle has length l and width w, the diagonal is $d = \sqrt{l^2 + w^2}$, so $l = \sqrt{d^2 - w^2}$. The area is $A = lw = w\sqrt{d^2 - w^2}$.

$A'(w) = \sqrt{d^2 - w^2} - \frac{w^2}{\sqrt{d^2 - w^2}} = \frac{d^2 - 2w^2}{\sqrt{d^2 - w^2}};$

$A'(w) = 0$ when $w = \frac{d}{\sqrt{2}}$ and so

$l = \sqrt{d^2 - \frac{d^2}{2}} = \frac{d}{\sqrt{2}}$. $A'(w) > 0$ on $\left(0, \frac{d}{\sqrt{2}}\right)$ and

$A'(w) < 0$ on $\left(\frac{d}{\sqrt{2}}, d\right)$. Maximum area is for a square.

41. The carrying capacity of the gutter is maximized when the area of the vertical end of the gutter is maximized. The height of the gutter is $3\sin\theta$. The area is

$A = 3(3\sin\theta) + 2\left(\frac{1}{2}\right)(3\cos\theta)(3\sin\theta)$

$= 9\sin\theta + 9\cos\theta\sin\theta.$

$\frac{dA}{d\theta} = 9\cos\theta + 9(-\sin\theta)\sin\theta + 9\cos\theta\cos\theta$

$= 9(\cos\theta - \sin^2\theta + \cos^2\theta)$

$= 9(2\cos^2\theta + \cos\theta - 1)$

$2\cos^2\theta + \cos\theta - 1 = 0;\ \cos\theta = -1, \frac{1}{2}; \theta = \pi, \frac{\pi}{3}$

Since $0 \le \theta \le \frac{\pi}{2}$, the critical points are

$0, \frac{\pi}{3}$, and $\frac{\pi}{2}$.

When $\theta = 0$, $A = 0$.

When $\theta = \frac{\pi}{3}$, $A = \frac{27\sqrt{3}}{4} \approx 11.7.$

When $\theta = \frac{\pi}{2}$, $A = 9.$

The carrying capacity is maximized when $\theta = \frac{\pi}{3}$.

43. Let V be the volume. $y = 4 - x$ and $z = 5 - 2x$. x is limited by $0 \le x \le 2.5$.

$V = x(4 - x)(5 - 2x) = 20x - 13x^2 + 2x^3$

$\frac{dV}{dx} = 20 - 26x + 6x^2; 2(3x^2 - 13x + 10) = 0;$

$2(3x - 10)(x - 1) = 0;$

$x = 1, \frac{10}{3}$

Critical points: 0, 1, 2.5
At $x = 0$ or 2.5, $V = 0$. At $x = 1$, $V = 9$.
Maximum volume when $x = 1$, $y = 4 - 1 = 3$, and $z = 5 - 2(1) = 3$.

45. Consider the figure below.

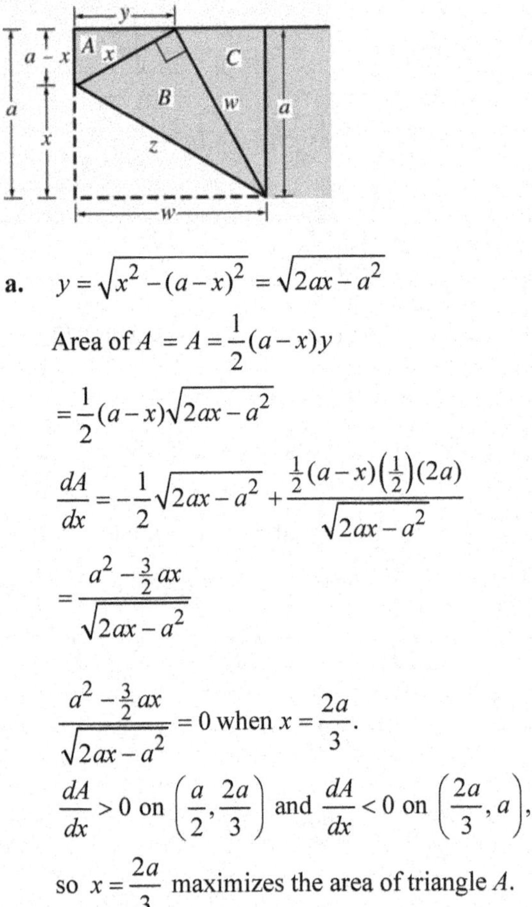

a. $y = \sqrt{x^2 - (a - x)^2} = \sqrt{2ax - a^2}$

Area of A = $A = \frac{1}{2}(a - x)y$

$= \frac{1}{2}(a - x)\sqrt{2ax - a^2}$

$\frac{dA}{dx} = -\frac{1}{2}\sqrt{2ax - a^2} + \frac{\frac{1}{2}(a - x)\left(\frac{1}{2}\right)(2a)}{\sqrt{2ax - a^2}}$

$= \frac{a^2 - \frac{3}{2}ax}{\sqrt{2ax - a^2}}$

$\frac{a^2 - \frac{3}{2}ax}{\sqrt{2ax - a^2}} = 0$ when $x = \frac{2a}{3}$.

$\frac{dA}{dx} > 0$ on $\left(\frac{a}{2}, \frac{2a}{3}\right)$ and $\frac{dA}{dx} < 0$ on $\left(\frac{2a}{3}, a\right)$,

so $x = \frac{2a}{3}$ maximizes the area of triangle A.

b. Triangle A is similar to triangle C, so

$$w = \frac{ax}{y} = \frac{ax}{\sqrt{2ax - a^2}}$$

Area of $B = B = \frac{1}{2}xw = \frac{ax^2}{2\sqrt{2ax - a^2}}$

$$\frac{dB}{dx} = \frac{a}{2}\left(\frac{2x\sqrt{2ax - a^2} - x^2 \frac{a}{\sqrt{2ax-a^2}}}{2ax - a^2}\right)$$

$$= \frac{a}{2}\left(\frac{2x(2ax - a^2) - ax^2}{(2ax - a^2)^{3/2}}\right) = \frac{a}{2}\left(\frac{3ax^2 - 2xa^2}{(2ax - a^2)^{3/2}}\right)$$

$$\frac{a^2}{2}\left(\frac{3x^2 - 2xa}{(2ax - a^2)^{3/2}}\right) = 0 \text{ when } x = 0, \frac{2a}{3}$$

Since $x = 0$ is not possible, $x = \frac{2a}{3}$.

$\frac{dB}{dx} < 0$ on $\left(\frac{a}{2}, \frac{2a}{3}\right)$ and $\frac{dB}{dx} > 0$ on $\left(\frac{2a}{3}, a\right)$,

so $x = \frac{2a}{3}$ minimizes the area of triangle B.

c. $z = \sqrt{x^2 + w^2} = \sqrt{x^2 + \frac{a^2x^2}{2ax - a^2}}$

$$= \sqrt{\frac{2ax^3}{2ax - a^2}}$$

$$\frac{dz}{dx} = \frac{1}{2}\sqrt{\frac{2ax - a^2}{2ax^3}}\left(\frac{6ax^2(2ax - a^2) - 2ax^3(2a)}{(2ax - a^2)^2}\right)$$

$$= \frac{4a^2x^3 - 3a^3x^2}{\sqrt{2ax^3(2ax - a^2)^3}}$$

$\frac{dz}{dx} = 0$ when $x = 0, \frac{3a}{4} \rightarrow x = \frac{3a}{4}$

$\frac{dz}{dx} < 0$ on $\left(\frac{a}{2}, \frac{3a}{4}\right)$ and $\frac{dz}{dx} > 0$ on $\left(\frac{3a}{4}, a\right)$,

so $x = \frac{3a}{4}$ minimizes length z.

47. a. $L'(\theta) = 15(9 + 25 - 30\cos\theta)^{-1/2}\sin\theta = 15(34 - 30\cos\theta)^{-1/2}\sin\theta$

$$L''(\theta) = -\frac{15}{2}(34 - 30\cos\theta)^{-3/2}(30\sin\theta)\sin\theta + 15(34 - 30\cos\theta)^{-1/2}\cos\theta$$

$$= -225(34 - 30\cos\theta)^{-3/2}\sin^2\theta + 15(34 - 30\cos\theta)^{-1/2}\cos\theta$$

$$= 15(34 - 30\cos\theta)^{-3/2}[-15\sin^2\theta + (34 - 30\cos\theta)\cos\theta]$$

$$= 15(34 - 30\cos\theta)^{-3/2}[-15\sin^2\theta + 34\cos\theta - 30\cos^2\theta]$$

$$= 15(34 - 30\cos\theta)^{-3/2}[-15 + 34\cos\theta - 15\cos^2\theta]$$

$$= -15(34 - 30\cos\theta)^{-3/2}[15\cos^2\theta - 34\cos\theta + 15]$$

$L'' = 0$ when $\cos\theta = \frac{34 \pm \sqrt{(34)^2 - 4(15)(15)}}{2(15)} = \frac{5}{3}, \frac{3}{5}$

$$\theta = \cos^{-1}\left(\frac{3}{5}\right)$$

$$L'\left(\cos^{-1}\left(\frac{3}{5}\right)\right) = 15\left(9 + 25 - 30\left(\frac{3}{5}\right)\right)^{-1/2}\left(\frac{4}{5}\right) = 3$$

$$L\left(\cos^{-1}\left(\frac{3}{5}\right)\right) = \left(9 + 25 - 30\left(\frac{3}{5}\right)\right)^{1/2} = 4$$

$\phi = 90°$ since the resulting triangle is a 3-4-5 right triangle.

b. $L'(\theta) = 65(25+169-130\cos\theta)^{-1/2}\sin\theta = 65(194-130\cos\theta)^{-1/2}\sin\theta$

$$L''(\theta) = -\frac{65}{2}(194-130\cos\theta)^{-3/2}(130\sin\theta)\sin\theta + 65(194-130\cos\theta)^{-1/2}\cos\theta$$

$$= -4225(194-130\cos\theta)^{-3/2}\sin^2\theta + 65(194-130\cos\theta)^{-1/2}\cos\theta$$

$$= 65(194-130\cos\theta)^{-3/2}[-65\sin^2\theta + (194-130\cos\theta)\cos\theta]$$

$$= 65(194-130\cos\theta)^{-3/2}[-65\sin^2\theta + 194\cos\theta - 130\cos^2\theta]$$

$$= 65(194-130\cos\theta)^{-3/2}[-65\cos^2\theta + 194\cos\theta - 65]$$

$$= -65(194-130\cos\theta)^{-3/2}[65\cos^2\theta - 194\cos\theta + 65]$$

$L'' = 0$ when $\cos\theta = \dfrac{194 \pm \sqrt{(194)^2 - 4(65)(65)}}{2(65)} = \dfrac{13}{5}, \dfrac{5}{13}$

$$\theta = \cos^{-1}\left(\frac{5}{13}\right)$$

$$L'\left(\cos^{-1}\left(\frac{5}{13}\right)\right) = 65\left(25+169-130\left(\frac{5}{13}\right)\right)^{1/2}\left(\frac{12}{13}\right) = 5$$

$$L\left(\cos^{-1}\left(\frac{5}{13}\right)\right) = \left(25+169-130\left(\frac{5}{13}\right)\right)^{1/2} = 12$$

$\phi = 90°$ since the resulting triangle is a 5-12-13 right triangle.

c. When the tips are separating most rapidly, $\phi = 90°$, $L = \sqrt{m^2 - h^2}$, $L' = h$

d. $L'(\theta) = hm(h^2 + m^2 - 2hm\cos\theta)^{-1/2}\sin\theta$

$$L''(\theta) = -h^2m^2(h^2+m^2-2hm\cos\theta)^{-3/2}\sin^2\theta + hm(h^2+m^2-2hm\cos\theta)^{-1/2}\cos\theta$$

$$= hm(h^2+m^2-2hm\cos\theta)^{-3/2}[-hm\sin^2\theta + (h^2+m^2)\cos\theta - 2hm\cos^2\theta]$$

$$= hm(h^2+m^2-2hm\cos\theta)^{-3/2}[-hm\cos^2\theta + (h^2+m^2)\cos\theta - hm]$$

$$= -hm(h^2+m^2-2hm\cos\theta)^{-3/2}[hm\cos^2\theta - (h^2+m^2)\cos\theta + hm]$$

$L'' = 0$ when $hm\cos^2\theta - (h^2+m^2)\cos\theta + hm = 0$

$(h\cos\theta - m)(m\cos\theta - h) = 0$

$$\cos\theta = \frac{m}{h}, \frac{h}{m}$$

Since $h < m$, $\cos\theta = \dfrac{h}{m}$ so $\theta = \cos^{-1}\left(\dfrac{h}{m}\right)$.

$$L'\left(\cos^{-1}\left(\frac{h}{m}\right)\right) = hm\left(h^2+m^2-2hm\left(\frac{h}{m}\right)\right)^{-1/2}\frac{\sqrt{m^2-h^2}}{m} = hm(m^2-h^2)^{-1/2}\frac{\sqrt{m^2-h^2}}{m} = h$$

$$L\left(\cos^{-1}\left(\frac{h}{m}\right)\right) = \left(h^2+m^2-2hm\left(\frac{h}{m}\right)\right)^{1/2} = \sqrt{m^2-h^2}$$

Since $h^2 + L^2 = m^2$, $\phi = 90°$.

49. Here we are interested in minimizing the distance between the earth and the asteroid. Using the coordinates P and Q for the two bodies, we can use the distance formula to obtain a suitable equation. However, for simplicity, we will minimize the squared distance to find the critical points. The squared distance between the objects is given by

$$D(t) = (93\cos(2\pi t) - 60\cos[2\pi(1.51t-1)])^2$$
$$+(93\sin(2\pi t) - 120\sin[2\pi(1.51t-1)])^2$$

The first derivative is

$$D'(t) \approx -34359[\cos(2\pi t)][\sin(9.48761t)]$$
$$+[\cos(9.48761t)][(204932\sin(9.48761t)$$
$$-141643\sin(2\pi t))]$$

Plotting the function and its derivative reveal a periodic relationship due to the orbiting of the objects. Careful examination of the graphs reveals that there is indeed a minimum squared distance (and hence a minimum distance) that occurs only once. The critical value for this occurrence is $t \approx 13.82790355$. This value gives a squared distance between the objects of ≈ 0.0022743 million miles. The actual distance is ≈ 0.047851 million miles $\approx 47{,}851$ miles.

51. Consider the following sketch.

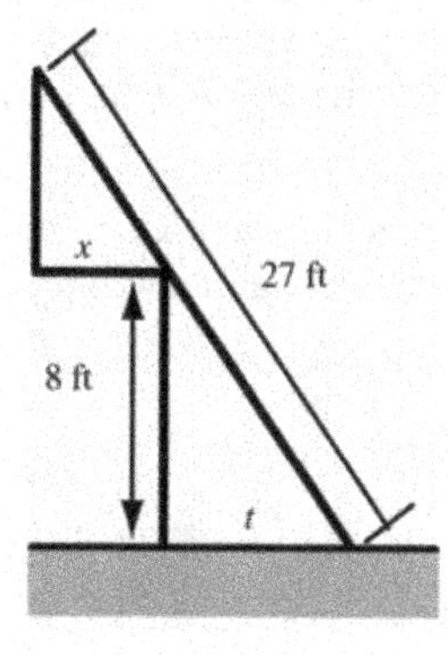

By similar triangles, $\dfrac{x}{27-\sqrt{t^2+64}} = \dfrac{t}{\sqrt{t^2+64}}$.

$$x = \frac{27t}{\sqrt{t^2+64}} - t$$

$$\frac{dx}{dt} = \frac{27\sqrt{t^2+64} - \frac{27t^2}{\sqrt{t^2+64}}}{t^2+64} - 1 = \frac{1728}{(t^2+64)^{3/2}} - 1$$

$$\frac{1728}{(t^2+64)^{3/2}} - 1 = 0 \text{ when } t = 4\sqrt{5}$$

$$\frac{d^2x}{dt^2} = \frac{-5184t}{(t^2+64)^{5/2}};\ \left.\frac{d^2x}{dt^2}\right|_{t=4\sqrt{5}} < 0$$

Therefore

$$x = \frac{27(4\sqrt{5})}{\sqrt{(4\sqrt{5})^2+64}} - 4\sqrt{5} = 5\sqrt{5} \approx 11.18 \text{ ft is the}$$

maximum horizontal overhang.

53. a.

$$\frac{dS}{db} = \frac{d}{db}\sum_{i=1}^{n}[y_i - (5+bx_i)]^2$$
$$= \sum_{i=1}^{n}\frac{d}{db}[y_i - (5+bx_i)]^2$$
$$= \sum_{i=1}^{n}2(y_i - 5 - bx_i)(-x_i)$$
$$= 2\left[\sum_{i=1}^{n}\left(-x_iy_i + 5x_i + bx_i^2\right)\right]$$
$$= -2\sum_{i=1}^{n}x_iy_i + 10\sum_{i=1}^{n}x_i + 2b\sum_{i=1}^{n}x_i^2$$

Setting $\dfrac{dS}{db} = 0$ gives

$$0 = -2\sum_{i=1}^{n}x_iy_i + 10\sum_{i=1}^{n}x_i + 2b\sum_{i=1}^{n}x_i^2$$
$$0 = -\sum_{i=1}^{n}x_iy_i + 5\sum_{i=1}^{n}x_i + b\sum_{i=1}^{n}x_i^2$$
$$b\sum_{i=1}^{n}x_i^2 = \sum_{i=1}^{n}x_iy_i - 5\sum_{i=1}^{n}x_i$$
$$b = \frac{\sum_{i=1}^{n}x_iy_i - 5\sum_{i=1}^{n}x_i}{\sum_{i=1}^{n}x_i^2}$$

You should check that this is indeed the value of b that minimizes the sum. Taking the second derivative yields

$$\frac{d^2S}{db^2} = 2\sum_{i=1}^{n}x_i^2$$

which is always positive (unless all the x values are zero). Therefore, the value for b above does minimize the sum as required.

b. Using the formula from **a.**, we get that

$$b = \frac{(2037) - 5(52)}{590} \approx 3.0119$$

c. The Least Squares Regression line is $y = 5 + 3.0119x$

Using this line, the predicted total number of labor hours to produce a lot of 15 brass bookcases is

$y = 5 + 3.0119(15) \approx 50.179$ hours

55. $n = 100 + 10\frac{250 - p(n)}{5}$ so $p(n) = 300 - \frac{n}{2}$

$R(n) = np(n) = 300n - \frac{n^2}{2}$

57.

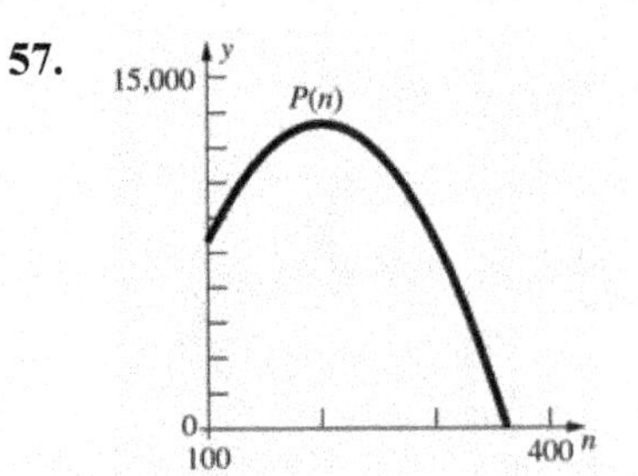

Estimate $n \approx 200$

$P'(n) = 200 - n; 200 - n = 0$ when $n = 200$.

$P''(n) = -1$, so profit is maximum at $n = 200$.

59. $\frac{C(n)}{n} = \frac{1000}{n} + \frac{n}{1200}$

When $n = 800$, $\frac{C(n)}{n} \approx 1.9167$ or \$1.92 per unit.

$\frac{dC}{dn} = \frac{n}{600}$

$C'(800) \approx 1.333$ or \$1.33

61. a. $R(x) = xp(x) = 20x + 4x^2 - \frac{x^3}{3}$

$\frac{dR}{dx} = 20 + 8x - x^2 = (10 - x)(x + 2)$

b. Increasing when $\frac{dR}{dx} > 0$

$20 + 8x - x^2 > 0$ on $[0, 10)$

Total revenue is increasing if $0 \le x \le 10$.

c. $\frac{d^2R}{dx^2} = 8 - 2x; \frac{d^2R}{dx^2} = 0$ when $x = 4$

$\frac{d^3R}{dx^3} = -2; \frac{dR}{dx}$ is maximum at $x = 4$.

63. $R(x) = \frac{800x}{x+3} - 3x$

$\frac{dR}{dx} = \frac{(x+3)(800) - 800x}{(x+3)^2} - 3 = \frac{2400}{(x+3)^2} - 3;$

$\frac{dR}{dx} = 0$ when $x = 20\sqrt{2} - 3 \approx 25$

$x_1 = 25; R(25) \approx 639.29$

At $x_1, \frac{dR}{dx} = 0$.

65. The revenue function would be $R(x) = x \cdot p(x) = 200x - 0.15x^2$. This, together with the cost function yields the following profit function:

$$P(x) = \begin{cases} -5000 + 194x - 0.148x^2 & \text{if } 0 \le x \le 500 \\ -9000 + 194x - 0.148x^2 & \text{if } 500 < x \le 750 \end{cases}$$

a. The only difference in the two pieces of the profit function is the constant. Since the derivative of a constant is 0, we can say that on the interval $0 < x < 750$,

$\frac{dP}{dx} = 194 - 0.296x$

There are no singular points in the given interval. To find stationary points, we solve

$$\frac{dP}{dx} = 0$$

$194 - 0.296x = 0$

$-0.296x = -194$

$x \approx 655$

Thus, the critical points are 0, 500, 655, and 750.

$P(0) = -5000$; $P(500) = 55{,}000$;

$P(655) = 54{,}574.30$; $P(750) = 53{,}250$

The profit is maximized if the company produces 500 chairs. The current machine can handle this work, so they should not buy the new machine.

b. Without the new machine, a production level of 500 chairs would yield a maximum profit of \$55,000.

67. $R(x) = 10x - 0.001x^2; 0 \le x \le 300$

$P(x) = (10x - 0.001x^2) - (200 + 4x - 0.01x^2)$

$= -200 + 6x + 0.009x^2$

$\frac{dP}{dx} = 6 + 0.018x; \frac{dP}{dx} = 0$ when $x \approx -333$

Critical numbers: $x = 0, 300$; $P(0) = -200$; $P(300) = 2410$; Maximum profit is \$2410 at $x = 300$.

69. a. $ab \le \left(\frac{a+b}{2}\right)^2 = \frac{a^2 + 2ab + b^2}{4}$

$= \frac{a^2}{4} + \frac{1}{2}ab + \frac{b^2}{4}$

This is true if

$0 \le \frac{a^2}{4} - \frac{1}{2}ab + \frac{b^2}{4} = \left(\frac{a}{2} - \frac{b}{2}\right)^2 = \left(\frac{a-b}{2}\right)^2$

Since a square can never be negative, this is always true.

b. $F(b)=\dfrac{a^2+2ab+b^2}{4b}$

As $b\to 0^+$, $a^2+2ab+b^2\to a^2$ while $4b\to 0^+$, thus $\lim\limits_{b\to 0^+} F(b)=\infty$ which is not close to a.

$$\lim_{b\to\infty}\frac{a^2+2ab+b^2}{4b}=\lim_{b\to\infty}\frac{\frac{a^2}{b}+2a+b}{4}=\infty,$$

so when b is very large, $F(b)$ is not close to a.

$$F'(b)=\frac{2(a+b)(4b)-4(a+b)^2}{16b^2}$$

$$=\frac{4b^2-4a^2}{16b^2}=\frac{b^2-a^2}{4b^2};$$

$F'(b)=0$ when $b^2=a^2$ or $b=a$ since a and b are both positive.

$$F(a)=\frac{(a+a)^2}{4a}=\frac{4a^2}{4a}=a$$

Thus $a\le\dfrac{(a+b)^2}{4b}$ for all $b>0$ or

$ab\le\dfrac{(a+b)^2}{4}$ which leads to $\sqrt{ab}\le\dfrac{a+b}{2}.$

c. Let $F(b)=\dfrac{1}{b}\left(\dfrac{a+b+c}{3}\right)^3=\dfrac{(a+b+c)^3}{27b}$

$$F'(b)=\frac{3(a+b+c)^2(27b)-27(a+b+c)^3}{27^2b^2}$$

$$=\frac{(a+b+c)^2[3b-(a+b+c)]}{27b^2}$$

$$=\frac{(a+b+c)^2(2b-a-c)}{27b^2};$$

$F'(b)=0$ when $b=\dfrac{a+c}{2}.$

$$F\left(\frac{a+c}{2}\right)=\frac{2}{a+c}\cdot\left(\frac{a+c}{3}+\frac{a+c}{6}\right)^3$$

$$=\frac{2}{a+c}\left(\frac{3(a+c)}{6}\right)^3=\frac{2}{a+c}\left(\frac{a+c}{2}\right)^3=\left(\frac{a+c}{2}\right)^2$$

Thus $\left(\dfrac{a+c}{2}\right)^2\le\dfrac{1}{b}\left(\dfrac{a+b+c}{3}\right)^3$ for all $b>0$.

From (b), $ac\le\left(\dfrac{a+c}{2}\right)^2$, thus

$ac\le\dfrac{1}{b}\left(\dfrac{a+b+c}{3}\right)^3$ or $abc\le\left(\dfrac{a+b+c}{3}\right)^3$

which gives the desired result

$(abc)^{1/3}\le\dfrac{a+b+c}{3}.$

3.5 Concepts Review

1. $f(x)$; $-f(x)$

3. $x=-1, x=2, x=3; y=1$

Problem Set 3.5

1. Domain: $(-\infty,\infty)$; range: $(-\infty,\infty)$
Neither an even nor an odd function.
y-intercept: 5; x-intercept: ≈ -2.3
$f'(x)=3x^2-3;\ 3x^2-3=0$ when $x=-1, 1$
Critical points: -1, 1
$f'(x)>0$ when $x<-1$ or $x>1$
$f(x)$ is increasing on $(-\infty,-1]\cup[1,\infty)$ and decreasing on $[-1, 1]$.
Local minimum $f(1)=3$;
local maximum $f(-1)=7$
$f''(x)=6x;\ f''(x)>0$ when $x>0$.
$f(x)$ is concave up on $(0,\infty)$ and concave down on $(-\infty, 0)$; inflection point $(0, 5)$.

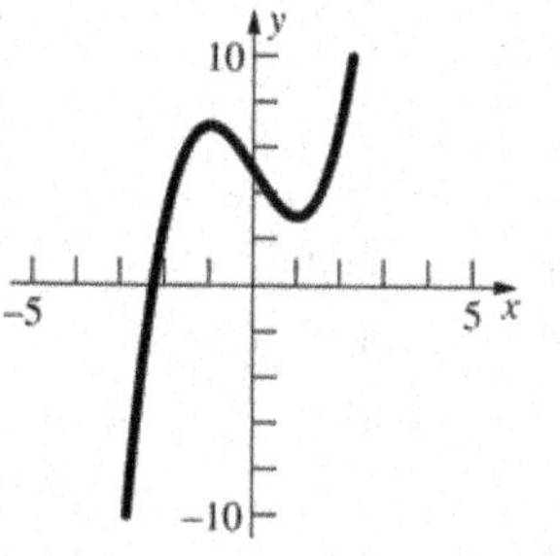

3. Domain: $(-\infty,\infty)$; range: $(-\infty,\infty)$
Neither an even nor an odd function.
y-intercept: 3; x-intercepts: $\approx -2.0, 0.2, 3.2$
$f'(x)=6x^2-6x-12=6(x-2)(x+1);$
$f'(x)=0$ when $x=-1, 2$
Critical points: -1, 2
$f'(x)>0$ when $x<-1$ or $x>2$
$f(x)$ is increasing on $(-\infty,-1]\cup[2,\infty)$ and decreasing on $[-1, 2]$.
Local minimum $f(2)=-17$;
local maximum $f(-1)=10$
$f''(x)=12x-6=6(2x-1);\ f''(x)>0$ when $x>\dfrac{1}{2}.$
$f(x)$ is concave up on $\left(\dfrac{1}{2},\infty\right)$ and concave down on $\left(-\infty,\dfrac{1}{2}\right)$; inflection point: $\left(\dfrac{1}{2},-\dfrac{7}{2}\right)$

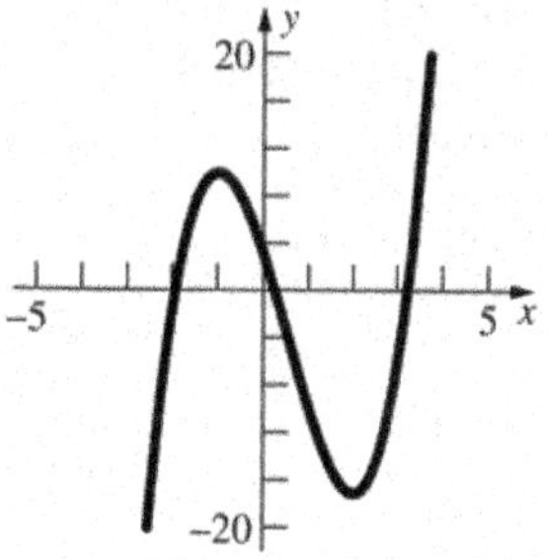

5. Domain: $(-\infty, \infty)$; range: $[0, \infty)$
Neither an even nor an odd function.
y-intercept: 1; x-intercept: 1
$G'(x) = 4(x-1)^3; G'(x) = 0$ when $x = 1$
Critical point: 1
$G'(x) > 0$ for $x > 1$
$G(x)$ is increasing on $[1, \infty)$ and decreasing on $(-\infty, 1]$.
Global minimum $f(1) = 0$; no local maxima
$G''(x) = 12(x-1)^2; G''(x) > 0$ for all $x \neq 1$
$G(x)$ is concave up on $(-\infty, 1) \cup (1, \infty)$; no inflection points

7. Domain: $(-\infty, \infty)$; range: $(-\infty, \infty)$
Neither an even nor an odd function.
y-intercept: 10; x-intercept: $1 - 11^{1/3} \approx -1.2$
$f'(x) = 3x^2 - 6x + 3 = 3(x-1)^2; f'(x) = 0$ when $x = 1$.
Critical point: 1
$f'(x) > 0$ for all $x \neq 1$.
$f(x)$ is increasing on $(-\infty, \infty)$ and decreasing nowhere.
No local maxima or minima
$f''(x) = 6x - 6 = 6(x-1); f''(x) > 0$ when $x > 1$.
$f(x)$ is concave up on $(1, \infty)$ and concave down on $(-\infty, 1)$; inflection point $(1, 11)$

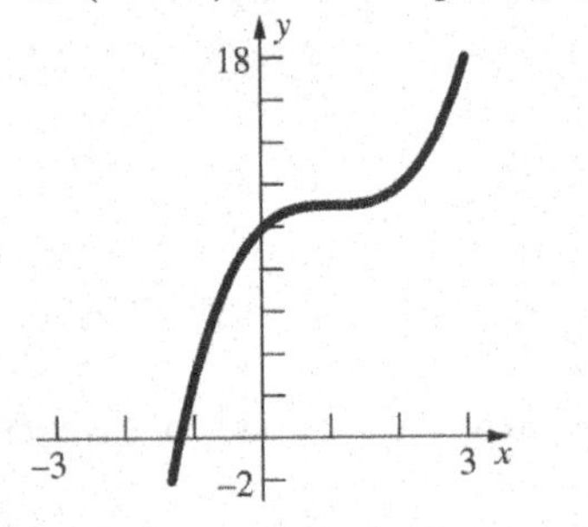

9. Domain: $(-\infty, -1) \cup (-1, \infty)$;
range: $(-\infty, 1) \cup (1, \infty)$
Neither an even nor an odd function
y-intercept: 0; x-intercept: 0
$g'(x) = \frac{1}{(x+1)^2}$; $g'(x)$ is never 0.
No critical points
$g'(x) > 0$ for all $x \neq -1$.
$g(x)$ is increasing on $(-\infty, -1) \cup (-1, \infty)$.
No local minima or maxima
$g''(x) = -\frac{2}{(x+1)^3}$; $g''(x) > 0$ when $x < -1$.
$g(x)$ is concave up on $(-\infty, -1)$ and concave down on $(-1, \infty)$; no inflection points (-1 is not in the domain of g).

$$\lim_{x\to\infty} \frac{x}{x+1} = \lim_{x\to\infty} \frac{1}{1+\frac{1}{x}} = 1;$$

$$\lim_{x\to-\infty} \frac{x}{x+1} = \lim_{x\to-\infty} \frac{1}{1+\frac{1}{x}} = 1;$$

horizontal asymptote: $y = 1$

As $x \to -1^-, x+1 \to 0^-$ so $\lim_{x\to-1^-} \frac{x}{x+1} = \infty$;

as $x \to -1^+, x+1 \to 0^+$ so $\lim_{x\to-1^+} \frac{x}{x+1} = -\infty$;

vertical asymptote: $x = -1$

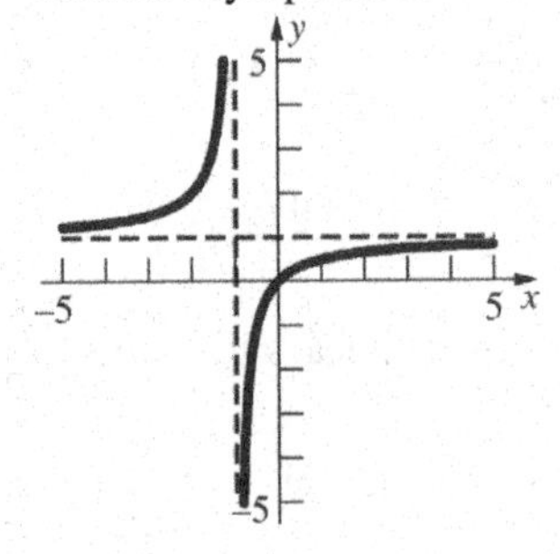

11. Domain: $(-\infty, \infty)$; range: $\left[-\frac{1}{4}, \frac{1}{4}\right]$

$f(-x) = \frac{-x}{(-x)^2+4} = -\frac{x}{x^2+4} = -f(x)$; odd function; symmetric with respect to the origin.
y-intercept: 0; x-intercept: 0
$f'(x) = \frac{4-x^2}{(x^2+4)^2}$; $f'(x) = 0$ when $x = -2, 2$
Critical points: $-2, 2$
$f'(x) > 0$ for $-2 < x < 2$
$f(x)$ is increasing on $[-2, 2]$ and decreasing on $(-\infty, -2] \cup [2, \infty)$.
Global minimum $f(-2) = -\frac{1}{4}$; global maximum $f(2) = \frac{1}{4}$

$f''(x) = \frac{2x(x^2-12)}{(x^2+4)^3}$; $f''(x) > 0$ when $-2\sqrt{3} < x < 0$ or $x > 2\sqrt{3}$
$f(x)$ is concave up on $(-2\sqrt{3}, 0) \cup (2\sqrt{3}, \infty)$ and concave down on $(-\infty, -2\sqrt{3}) \cup (0, 2\sqrt{3})$;
inflection points $\left(-2\sqrt{3}, -\frac{\sqrt{3}}{8}\right), (0, 0),$

$\left(2\sqrt{3}, \frac{\sqrt{3}}{8}\right)$

$$\lim_{x\to\infty} \frac{x}{x^2+4} = \lim_{x\to\infty} \frac{\frac{1}{x}}{1+\frac{4}{x^2}} = 0;$$

$$\lim_{x\to-\infty} \frac{x}{x^2+4} = \lim_{x\to-\infty} \frac{\frac{1}{x}}{1+\frac{4}{x^2}} = 0;$$

$y = 0$ is a horizontal asymptote.
No vertical asymptotes

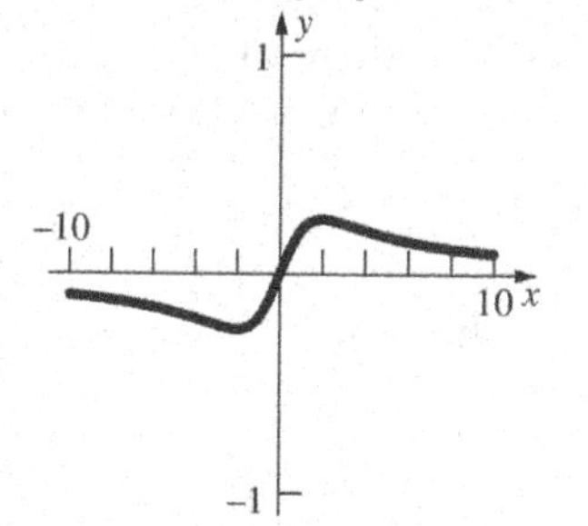

13. Domain: $(-\infty, 1) \cup (1, \infty)$;
range $(-\infty, 1) \cup (1, \infty)$
Neither an even nor an odd function
y-intercept: 0; x-intercept: 0

$h(x) = -\frac{1}{(x-1)^2}$; $h'(x)$ is never 0.

No critical points
$h'(x) < 0$ for all $x \neq 1$.
$h(x)$ is increasing nowhere and decreasing on $(-\infty, 1) \cup (1, \infty)$.
No local maxima or minima

$h''(x) = \frac{2}{(x-1)^3}$; $h''(x) > 0$ when $x > 1$

$h(x)$ is concave up on $(1, \infty)$ and concave down on $(-\infty, 1)$; no inflection points (1 is not in the domain of $h(x)$)

$$\lim_{x\to\infty} \frac{x}{x-1} = \lim_{x\to\infty} \frac{1}{1-\frac{1}{x}} = 1;$$

$$\lim_{x\to-\infty} \frac{x}{x-1} = \lim_{x\to-\infty} \frac{1}{1-\frac{1}{x}} = 1;$$

$y = 1$ is a horizontal asymptote.

As $x \to 1^-, x-1 \to 0^-$ so $\lim_{x\to 1^-} \frac{x}{x-1} = -\infty$;

as $x \to 1^+, x-1 \to 0^+$ so $\lim_{x\to 1^+} \frac{x}{x-1} = \infty$;

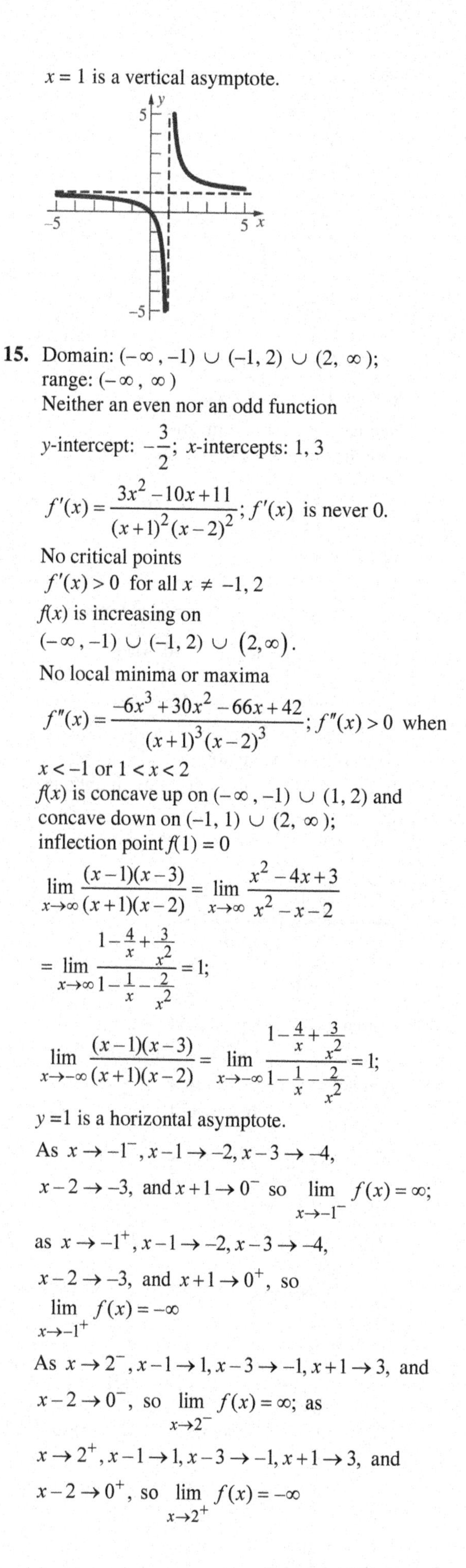

$x = 1$ is a vertical asymptote.

15. Domain: $(-\infty, -1) \cup (-1, 2) \cup (2, \infty)$;
range: $(-\infty, \infty)$
Neither an even nor an odd function

y-intercept: $-\frac{3}{2}$; x-intercepts: 1, 3

$f'(x) = \frac{3x^2 - 10x + 11}{(x+1)^2(x-2)^2}$; $f'(x)$ is never 0.

No critical points
$f'(x) > 0$ for all $x \neq -1, 2$
$f(x)$ is increasing on
$(-\infty, -1) \cup (-1, 2) \cup (2, \infty)$.
No local minima or maxima

$f''(x) = \frac{-6x^3 + 30x^2 - 66x + 42}{(x+1)^3(x-2)^3}$; $f''(x) > 0$ when

$x < -1$ or $1 < x < 2$
$f(x)$ is concave up on $(-\infty, -1) \cup (1, 2)$ and concave down on $(-1, 1) \cup (2, \infty)$;
inflection point $f(1) = 0$

$$\lim_{x\to\infty} \frac{(x-1)(x-3)}{(x+1)(x-2)} = \lim_{x\to\infty} \frac{x^2-4x+3}{x^2-x-2}$$

$$= \lim_{x\to\infty} \frac{1-\frac{4}{x}+\frac{3}{x^2}}{1-\frac{1}{x}-\frac{2}{x^2}} = 1;$$

$$\lim_{x\to-\infty} \frac{(x-1)(x-3)}{(x+1)(x-2)} = \lim_{x\to-\infty} \frac{1-\frac{4}{x}+\frac{3}{x^2}}{1-\frac{1}{x}-\frac{2}{x^2}} = 1;$$

$y = 1$ is a horizontal asymptote.

As $x \to -1^-, x-1 \to -2, x-3 \to -4,$

$x-2 \to -3,$ and $x+1 \to 0^-$ so $\lim_{x\to -1^-} f(x) = \infty$;

as $x \to -1^+, x-1 \to -2, x-3 \to -4,$

$x-2 \to -3,$ and $x+1 \to 0^+$, so
$\lim_{x\to -1^+} f(x) = -\infty$

As $x \to 2^-, x-1 \to 1, x-3 \to -1, x+1 \to 3,$ and

$x-2 \to 0^-$, so $\lim_{x\to 2^-} f(x) = \infty$; as

$x \to 2^+, x-1 \to 1, x-3 \to -1, x+1 \to 3,$ and

$x-2 \to 0^+$, so $\lim_{x\to 2^+} f(x) = -\infty$

$x = -1$ and $x = 2$ are vertical asymptotes.

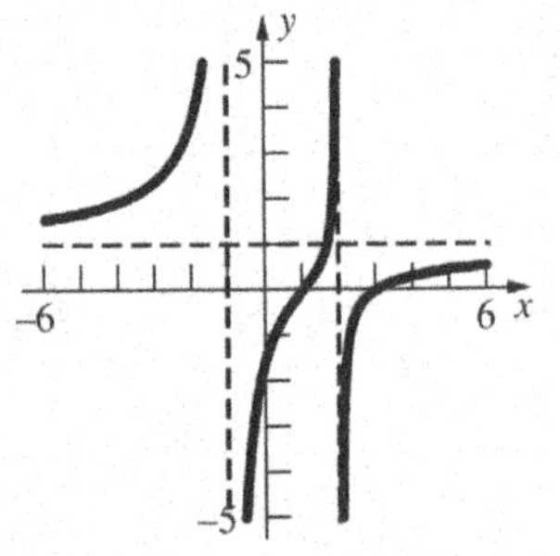

17. Domain: $(-\infty,1)\cup(1,\infty)$

Range: $(-\infty,\infty)$

Neither even nor odd function.

y-intercept: $y = 6$; x-intercept: $x = -3, 2$

$g'(x) = \dfrac{x^2 - 2x + 5}{(x-1)^2}$; $g'(x)$ is never zero. No critical points.

$g'(x) > 0$ over the entire domain so the function is always increasing. No local extrema.

$f''(x) = \dfrac{-8}{(x-1)^3}$; $f''(x) > 0$ when $x < 1$ (concave up) and $f''(x) < 0$ when $x > 1$ (concave down); no inflection points.

No horizontal asymptote; $x = 1$ is a vertical asymptote; the line $y = x + 2$ is an oblique (or slant) asymptote.

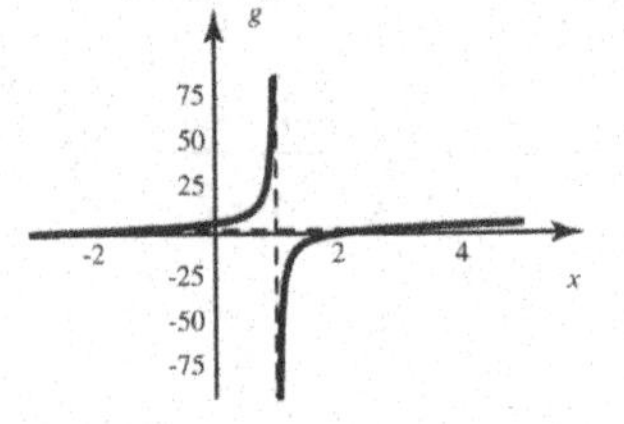

19. Domain: $(-\infty, \infty)$; range: $(-\infty, \infty)$

$R(-z) = -z|-z| = -z|z| = -R(z)$; odd function; symmetric with respect to the origin.

y-intercept: 0; *z*-intercept: 0

$R'(z) = |z| + \dfrac{z^2}{|z|} = 2|z|$ since $z^2 = |z|^2$ for all *z*;

$R'(z) = 0$ when $z = 0$

Critical point: 0

$R'(z) > 0$ when $z \neq 0$

R(*z*) is increasing on $(-\infty, \infty)$ and decreasing nowhere.

No local minima or maxima

$R''(z) = \dfrac{2z}{|z|}$; $R''(z) > 0$ when $z > 0$.

R(*z*) is concave up on $(0, \infty)$ and concave down on $(-\infty, 0)$; inflection point (0, 0).

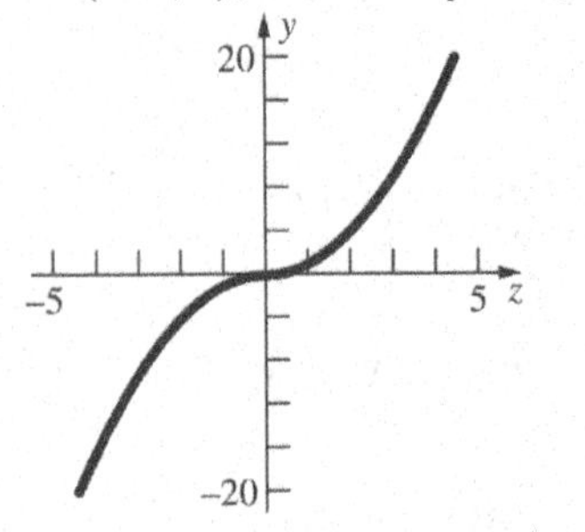

21. Domain: $(-\infty, \infty)$; range: $[0, \infty)$

Neither an even nor an odd function.

Note that for $x \le 0$, $|x| = -x$ so $|x| + x = 0$, while for $x > 0$, $|x| = x$ so $\dfrac{|x| + x}{2} = x$.

$$g(x) = \begin{cases} 0 & \text{if } x \le 0 \\ 3x^2 + 2x & \text{if } x > 0 \end{cases}$$

y-intercept: 0; *x*-intercepts: $(-\infty, 0]$

$$g'(x) = \begin{cases} 0 & \text{if } x \le 0 \\ 6x + 2 & \text{if } x > 0 \end{cases}$$

No critical points for $x > 0$.

g(*x*) is increasing on $[0, \infty)$ and decreasing nowhere.

$$g''(x) = \begin{cases} 0 & \text{if } x \le 0 \\ 6 & \text{if } x > 0 \end{cases}$$

g(*x*) is concave up on $(0, \infty)$; no inflection points

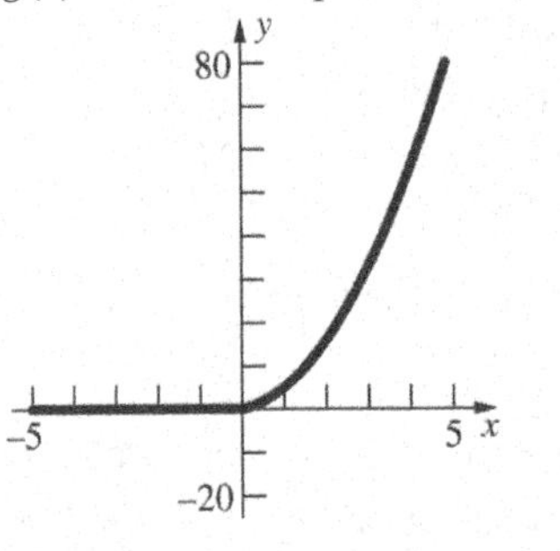

23. Domain: $(-\infty, \infty)$; range: $[0, 1]$

$f(-x) = |\sin(-x)| = |-\sin x| = |\sin x| = f(x)$; even function; symmetric with respect to the *y*-axis.

y-intercept: 0; *x*-intercepts: $k\pi$ where *k* is any integer.

$f'(x) = \dfrac{\sin x}{|\sin x|}\cos x$; $f'(x) = 0$ when $x = \dfrac{\pi}{2} + k\pi$ and $f'(x)$ does not exist when $x = k\pi$, where *k* is any integer.

Critical points: $\dfrac{k\pi}{2}$ and $k\pi + \dfrac{\pi}{2}$, where *k* is any integer; $f'(x) > 0$ when $\sin x$ and $\cos x$ are either both positive or both negative.

f(*x*) is increasing on $\left[k\pi, k\pi + \dfrac{\pi}{2}\right]$ and decreasing on $\left[k\pi+\frac{\pi}{2},(k+1)\pi\right]$ where k is any integer.

Global minima $f(k\pi)=0$; global maxima $f\left(k\pi+\frac{\pi}{2}\right)=1$, where k is any integer.

$$f''(x)=\frac{\cos^2 x}{|\sin x|}-\frac{\sin^2 x}{|\sin x|}$$

$$+\sin x\cos x\left(-\frac{1}{|\sin x|^2}\right)\left(\frac{\sin x}{|\sin x|}\right)(\cos x)$$

$$=\frac{\cos^2 x}{|\sin x|}-\frac{\sin^2 x}{|\sin x|}-\frac{\cos^2 x}{|\sin x|}=-\frac{\sin^2 x}{|\sin x|}=-|\sin x|$$

$f''(x)<0$ when $x\neq k\pi$, k any integer

$f(x)$ is never concave up and concave down on $(k\pi,(k+1)\pi)$ where k is any integer.

No inflection points

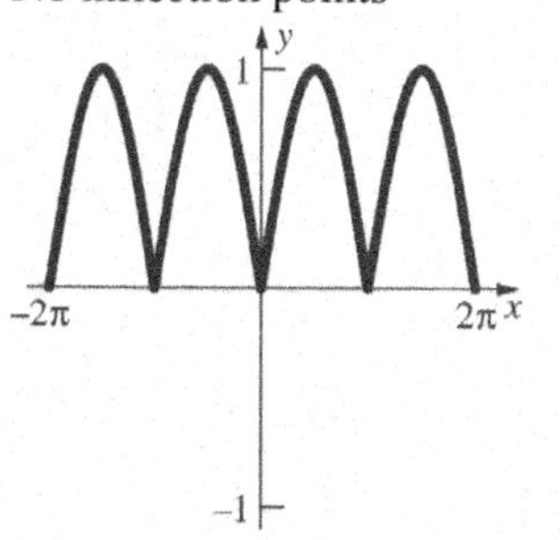

25. Domain: $(-\infty,\infty)$

Range: $[0,1]$

Even function since $h(-t)=\cos^2(-t)=\cos^2 t=h(t)$ so the function is symmetric with respect to the y-axis.

y-intercept: $y=1$; t-intercepts: $x=\frac{\pi}{2}+k\pi$ where k is any integer.

$h'(t)=-2\cos t\sin t$; $h'(t)=0$ at $t=\frac{k\pi}{2}$.

Critical points: $t=\frac{k\pi}{2}$

$h'(t)>0$ when $k\pi+\frac{\pi}{2}<t<(k+1)\pi$. The function is increasing on the intervals $[k\pi+(\pi/2),(k+1)\pi]$ and decreasing on the intervals $[k\pi,k\pi+(\pi/2)]$.

Global maxima $h(k\pi)=1$

Global minima $h\left(\frac{\pi}{2}+k\pi\right)=0$

$h''(t)=2\sin^2 t-2\cos^2 t=-2(\cos 2t)$

$h''(t)<0$ on $\left(k\pi-\frac{\pi}{4},k\pi+\frac{\pi}{4}\right)$ so h is concave down, and $h''(t)>0$ on $\left(k\pi+\frac{\pi}{4},k\pi+\frac{3\pi}{4}\right)$ so h is concave up.

Inflection points: $\left(\frac{k\pi}{2}+\frac{\pi}{4},\frac{1}{2}\right)$

No vertical asymptotes; no horizontal asymptotes.

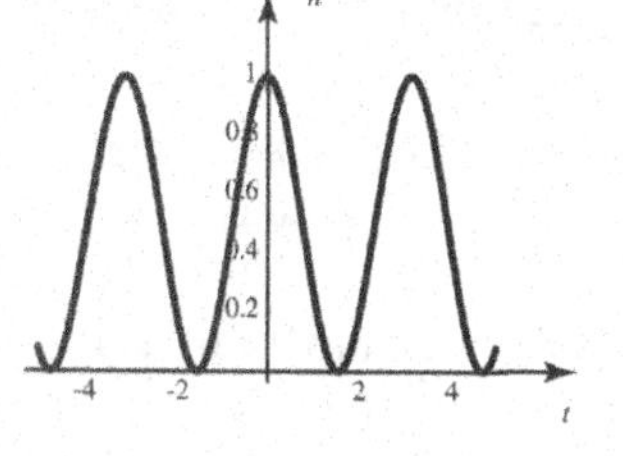

27. Domain: $\approx(-\infty,0.44)\cup(0.44,\infty)$;

range: $(-\infty,\infty)$

Neither an even nor an odd function

y-intercept: 0; x-intercepts: 0, ≈ 0.24

$$f'(x)=\frac{74.6092x^3-58.2013x^2+7.82109x}{(7.126x-3.141)^2};$$

$f'(x)=0$ when $x=0$, ≈ 0.17, ≈ 0.61

Critical points: 0, ≈ 0.17, ≈ 0.61

$f'(x)>0$ when $0<x<0.17$ or $0.61<x$

$f(x)$ is increasing on $\approx[0,0.17]\cup[0.61,\infty)$

and decreasing on

$(-\infty,0]\cup[0.17,0.44)\cup(0.44,0.61]$

Local minima $f(0)=0$, $f(0.61)\approx 0.60$; local maximum $f(0.17)\approx 0.01$

$$f''(x)=\frac{531.665x^3-703.043x^2+309.887x-24.566}{(7.126x-3.141)^3};$$

$f''(x)>0$ when $x<0.10$ or $x>0.44$

$f(x)$ is concave up on $(-\infty,0.10)\cup(0.44,\infty)$

and concave down on $(0.10, 0.44)$;

inflection point $\approx(0.10, 0.003)$

$$\lim_{x\to\infty}\frac{5.235x^3-1.245x^2}{7.126x-3.141}=\lim_{x\to\infty}\frac{5.235x^2-1.245x}{7.126-\frac{3.141}{x}}=\infty$$

so $f(x)$ does not have a horizontal asymptote.

As $x \to 0.44^-, 5.235x^3 - 1.245x^2 \to 0.20$ while

$7.126x - 3.141 \to 0^-$, so $\lim\limits_{x \to 0.44^-} f(x) = -\infty$;

as $x \to 0.44^+, 5.235x^3 - 1.245x^2 \to 0.20$ while

$7.126x - 3.141 \to 0^+$, so $\lim\limits_{x \to 0.44^+} f(x) = \infty$;

$x \approx 0.44$ is a vertical asymptote of $f(x)$.

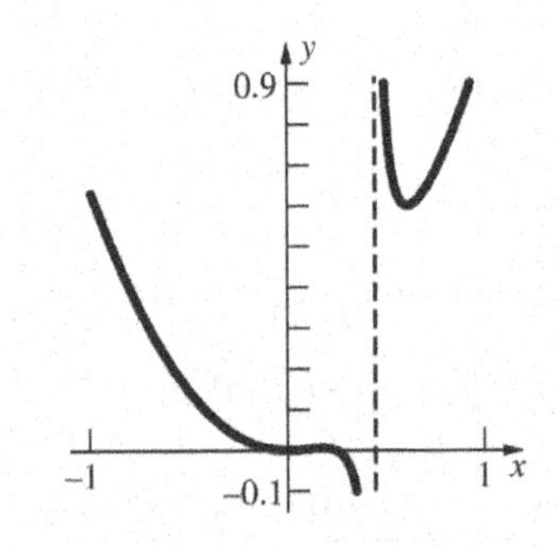

29.

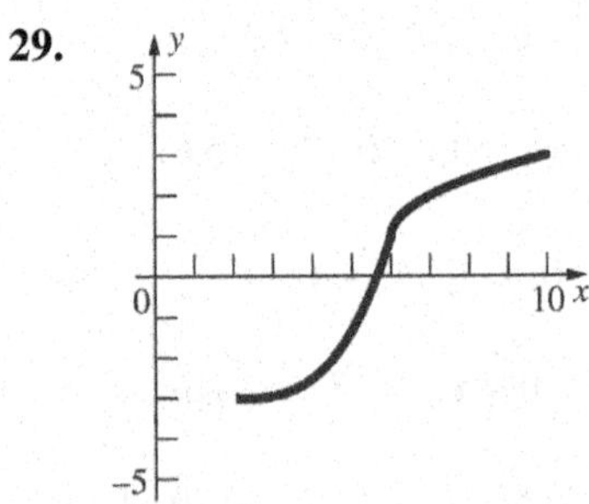

31.

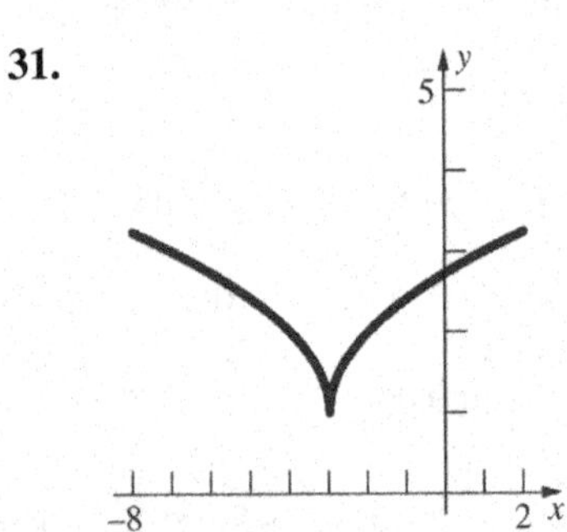

33.

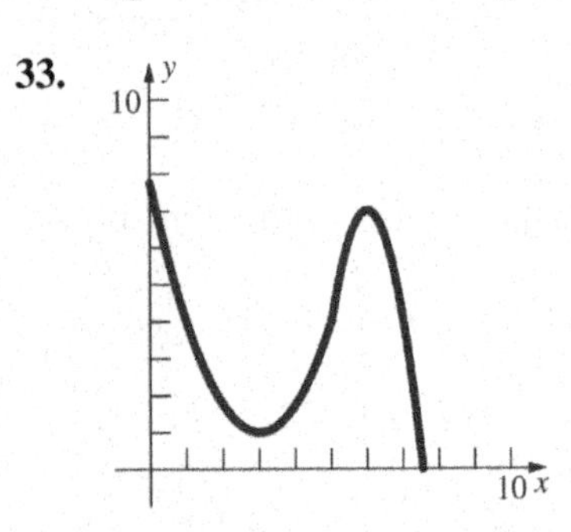

35.

37.

39.

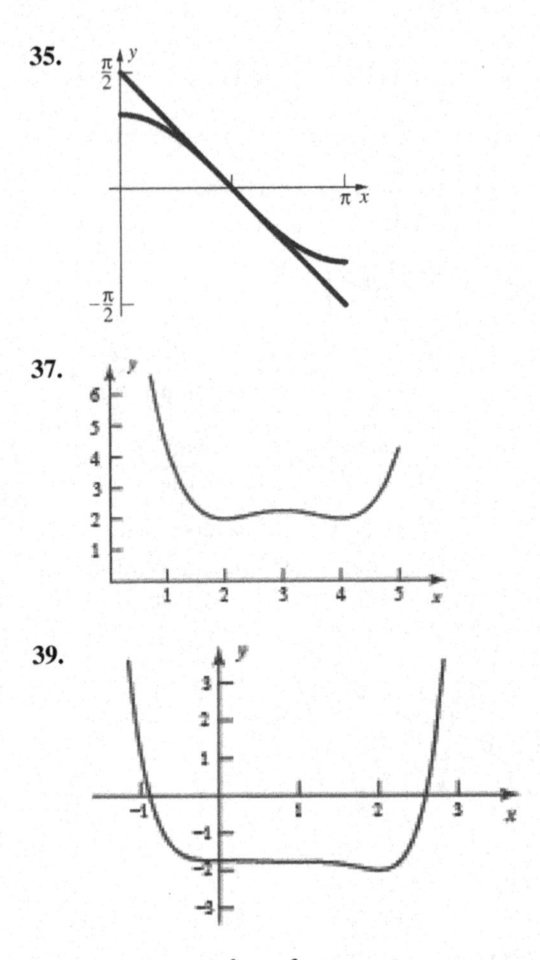

41. Let $f(x) = ax^3 + bx^2 + cx + d$, then $f'(x) = 3ax^2 + 2bx + c$ and $f''(x) = 6ax + 2b$. As long as $a \neq 0$, $f''(x)$ will be positive on one side of $x = \dfrac{b}{3a}$ and negative on the other side. $x = \dfrac{b}{3a}$ is the only inflection point.

43. Since the c term is squared, the only difference occurs when $c = 0$. When $c = 0$, $y = x^2\sqrt{x^2} = |x|^3$ which has domain $(-\infty, \infty)$ and range $[0, \infty)$. When $c \neq 0$, $y = x^2\sqrt{x^2 - c^2}$ has domain $(-\infty, -|c|] \cup [|c|, \infty)$ and range $[0, \infty)$.

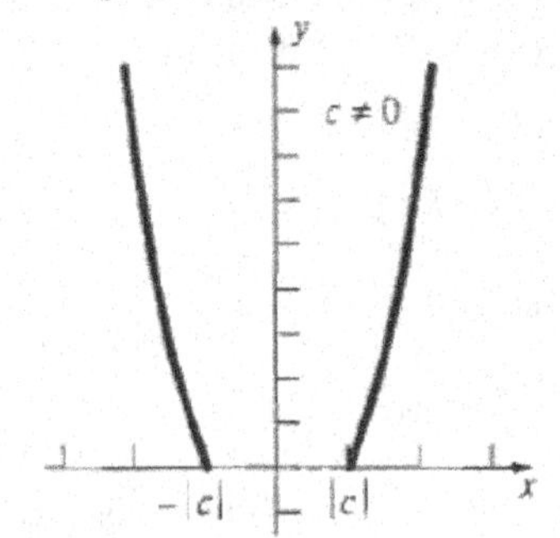

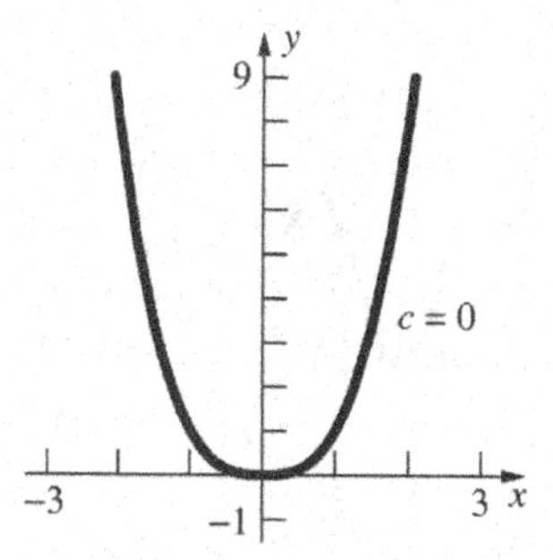

The only extremum points are $\pm|c|$. For $c = 0$, there is one minimum, for $c \neq 0$ there are two. No maxima, independent of c. No inflection points, independent of c.

45. $f(x) = \dfrac{1}{(cx^2-4)^2 + cx^2}$, then

$$f'(x) = \frac{2cx(7-2cx^2)}{[(cx^2-4)^2+cx^2]^2};$$

If $c > 0$, $f'(x) = 0$ when $x = 0, \pm\sqrt{\frac{7}{2c}}$.

If $c < 0$, $f'(x) = 0$ when $x = 0$.

Note that $f(x) = \frac{1}{16}$ (a horizontal line) if $c = 0$.

If $c > 0$, $f'(x) > 0$ when $x < -\sqrt{\frac{7}{2c}}$ and $0 < x < \sqrt{\frac{7}{2c}}$, so $f(x)$ is increasing on $\left(-\infty, -\sqrt{\frac{7}{2c}}\right] \cup \left[0, \sqrt{\frac{7}{2c}}\right]$ and decreasing on $\left[-\sqrt{\frac{7}{2c}}, 0\right] \cup \left[\sqrt{\frac{7}{2c}}, \infty\right)$. Thus, $f(x)$ has local maxima $f\left(-\sqrt{\frac{7}{2c}}\right) = \frac{4}{15}$, $f\left(\sqrt{\frac{7}{2c}}\right) = \frac{4}{15}$ and local minimum $f(0) = \frac{1}{16}$. If $c < 0$, $f'(x) > 0$ when $x < 0$, so $f(x)$ is increasing on $(-\infty, 0]$ and decreasing on $[0, \infty)$. Thus, $f(x)$ has a local maximum $f(0) = \frac{1}{16}$. Note that $f(x) > 0$ and has horizontal asymptote $y = 0$.

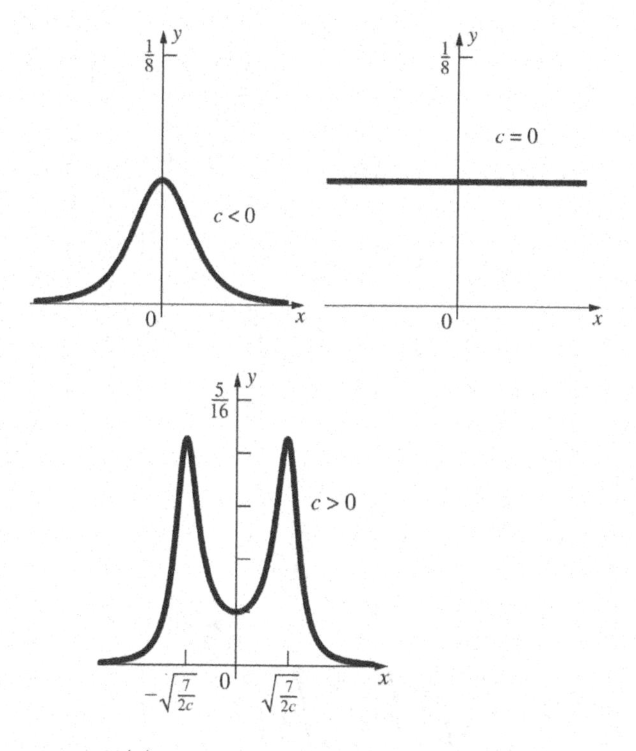

47. $f(x) = c + \sin cx$.

Since c is constant for all x and $\sin cx$ is continuous everywhere, the function $f(x)$ is continuous everywhere.

$f'(x) = c \cdot \cos cx$

$f'(x) = 0$ when $cx = \left(k + \frac{1}{2}\right)\pi$ or $x = \left(k + \frac{1}{2}\right)\frac{\pi}{c}$ where k is an integer.

$f''(x) = -c^2 \cdot \sin cx$

$f''\left(\left(k+\frac{1}{2}\right)\frac{\pi}{c}\right) = -c^2 \cdot \sin\left(c \cdot \left(k+\frac{1}{2}\right)\frac{\pi}{c}\right) = -c^2 \cdot (-1)^k$

In general, the graph of f will resemble the graph of $y = \sin x$. The period will decrease as $|c|$ increases and the graph will shift up or down depending on whether c is positive or negative.

If $c = 0$, then $f(x) = 0$.

If $c < 0$:

$f(x)$ is decreasing on $\left[\frac{(4k+1)\pi}{2c}, \frac{(4k-1)\pi}{2c}\right]$

$f(x)$ is increasing on $\left[\frac{(4k-1)\pi}{2c}, \frac{(4k-3)\pi}{2c}\right]$

$f(x)$ has local minima at $x = \frac{(4k-1)}{2c}\pi$ and local maxima at $x = \frac{(4k-3)\pi}{2c}$ where k is an integer.

If $c = 0$, $f(x) = 0$ and there are no extrema.

If $c > 0$:

$f(x)$ is decreasing on $\left[\frac{(4k-3)\pi}{2c}, \frac{(4k-1)\pi}{2c}\right]$

$f(x)$ is increasing on $\left[\frac{(4k-1)\pi}{2c}, \frac{(4k+1)\pi}{2c}\right]$

$f(x)$ has local minima at $x = \frac{(4k-1)}{2c}\pi$ and local maxima at $x = \frac{(4k-3)\pi}{2c}$ where k is an integer.

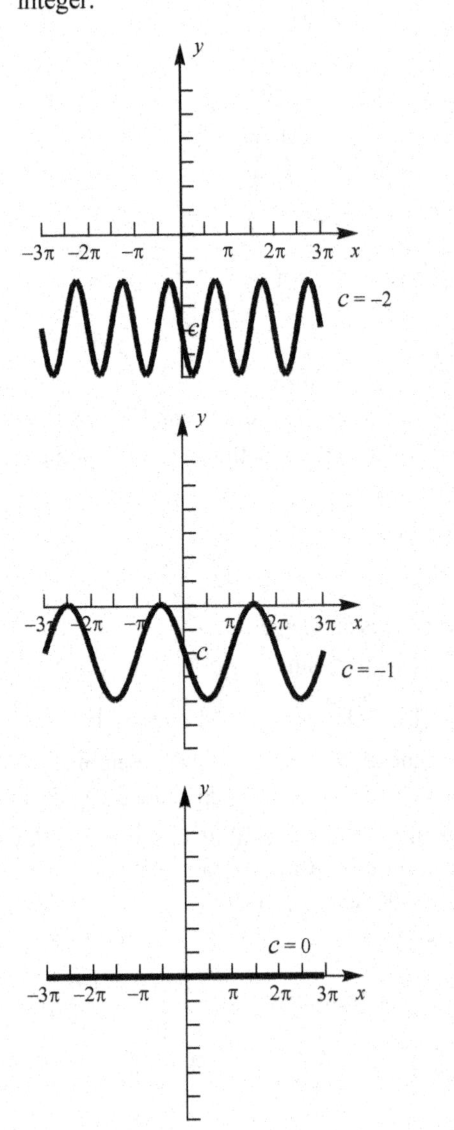

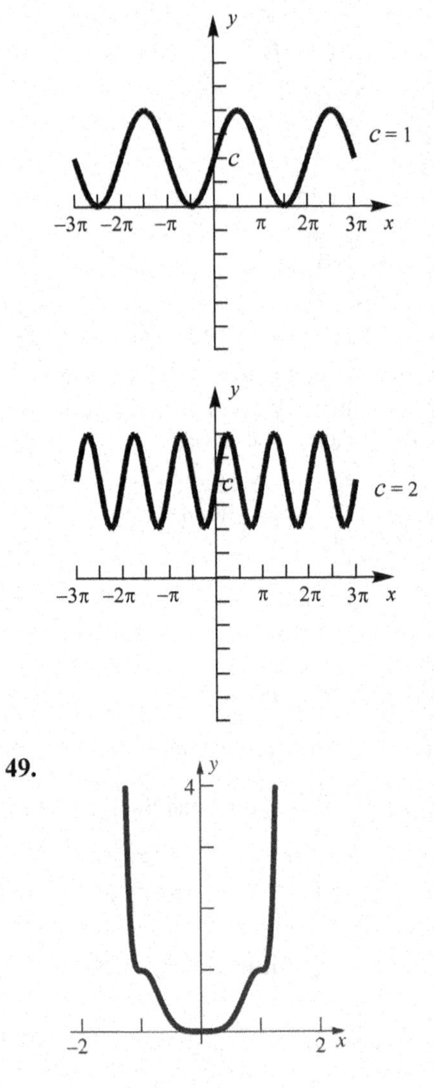

49.

Justification:

$f(1) = g(1) = 1$

$f(-x) = g((-x)^4) = g(x^4) = f(x)$

f is an even function; symmetric with respect to the y-axis.

$f'(x) = g'(x^4)4x^3$

$f'(x) > 0$ for x on $(0,1) \cup (1,\infty)$

$f'(x) < 0$ for x on $(-\infty,-1) \cup (-1,0)$

$f'(x) = 0$ for $x = -1, 0, 1$ since f' is continuous.

$f''(x) = g''(x^4)16x^6 + g'(x)12x^2$

$f''(x) = 0$ for $x = -1, 0, 1$

$f''(x) > 0$ for x on $(0, x_0) \cup (1,\infty)$

$f''(x) < 0$ for x on $(x_0, 1)$

Where x_0 is a root of $f''(x) = 0$ (assume that there is only one root on (0, 1)).

51. a. Not possible; $F'(x) > 0$ means that $F(x)$ is increasing. $F''(x) > 0$ means that the rate at which $F(x)$ is increasing never slows down. Thus the values of F must eventually become positive.

b. Not possible; If $F(x)$ is concave down for all x, then $F(x)$ cannot always be positive.

c.

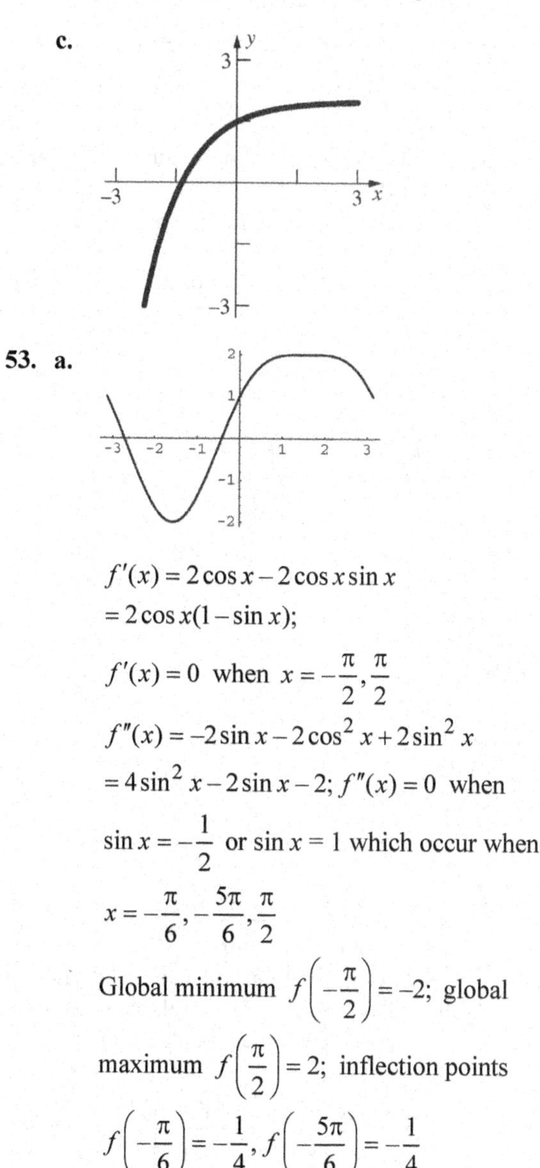

53. a.

$f'(x) = 2\cos x - 2\cos x \sin x$
$= 2\cos x(1 - \sin x);$

$f'(x) = 0$ when $x = -\frac{\pi}{2}, \frac{\pi}{2}$

$f''(x) = -2\sin x - 2\cos^2 x + 2\sin^2 x$
$= 4\sin^2 x - 2\sin x - 2;\ f''(x) = 0$ when

$\sin x = -\frac{1}{2}$ or $\sin x = 1$ which occur when

$x = -\frac{\pi}{6}, -\frac{5\pi}{6}, \frac{\pi}{2}$

Global minimum $f\left(-\frac{\pi}{2}\right) = -2$; global maximum $f\left(\frac{\pi}{2}\right) = 2$; inflection points $f\left(-\frac{\pi}{6}\right) = -\frac{1}{4}, f\left(-\frac{5\pi}{6}\right) = -\frac{1}{4}$

b.

$f'(x) = 2\cos x + 2\sin x \cos x$
$= 2\cos x(1 + \sin x);\ f'(x) = 0$ when
$x = -\frac{\pi}{2}, \frac{\pi}{2}$

$f''(x) = -2\sin x + 2\cos^2 x - 2\sin^2 x$
$= -4\sin^2 x - 2\sin x + 2;\ f''(x) = 0$ when

$\sin x = -1$ or $\sin x = \frac{1}{2}$ which occur when

$x = -\frac{\pi}{2}, \frac{\pi}{6}, \frac{5\pi}{6}$

Global minimum $f\left(-\frac{\pi}{2}\right) = -1$; global maximum $f\left(\frac{\pi}{2}\right) = 3$; inflection points $f\left(\frac{\pi}{6}\right) = \frac{5}{4}, f\left(\frac{5\pi}{6}\right) = \frac{5}{4}$.

c.

$f'(x) = -2\sin 2x + 2\sin x$
$= -4\sin x \cos x + 2\sin x = 2\sin x(1 - 2\cos x);$

$f'(x) = 0$ when $x = -\pi, -\frac{\pi}{3}, 0, \frac{\pi}{3}, \pi$

$f''(x) = -4\cos 2x + 2\cos x;\ f''(x) = 0$ when
$x \approx -2.206, -0.568, 0.568, 2.206$

Global minimum $f\left(-\frac{\pi}{3}\right) = f\left(\frac{\pi}{3}\right) = -1.5;$
Global maximum $f(-\pi) = f(\pi) = 3;$
Inflection points: $\approx (-2.206, 0.890)$, $(-0.568, -1.265)$, $(0.568, -1.265)$, $(2.206, 0.890)$

d.

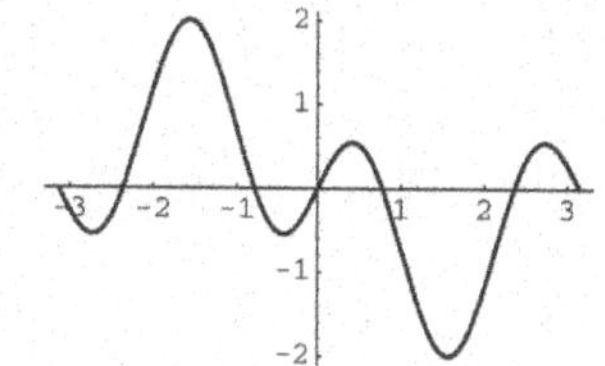

$f'(x) = 3\cos 3x - \cos x$; $f'(x) = 0$ when $3\cos 3x = \cos x$ which occurs when

$x = -\frac{\pi}{2}, \frac{\pi}{2}$ and when

$x \approx -2.7, -0.4, 0.4, 2.7$

$f''(x) = -9\sin 3x + \sin x$ which occurs when

$x = -\pi, 0, \pi$ and when

$x \approx -2.126, -1.016, 1.016, 2.126$

Global minimum $f\left(\frac{\pi}{2}\right) = -2$;

global maximum $f\left(-\frac{\pi}{2}\right) = 2$;

Inflection points: $\approx (-2.126, 0.755)$, $(-1.016, 0.755)$, $(0, 0)$, $(1.016, -0.755)$, $(2.126, -0.755)$

e.

$f'(x) = 2\cos 2x + 3\sin 3x$

Using the graphs, $f(x)$ has a global minimum at $f(2.17) \approx -1.9$ and a global maximum at $f(0.97) \approx 1.9$

$f''(x) = -4\sin 2x + 9\cos 3x$; $f''(x) = 0$ when

$x = -\frac{\pi}{2}, \frac{\pi}{2}$ and when

$x \approx -2.469, -0.673, 0.413, 2.729.$

Inflection points: $\left(-\frac{\pi}{2}, 0\right), \left(\frac{\pi}{2}, 0\right)$, $\approx (-2.469, 0.542), (-0.673, -0.542)$, $(0.413, 0.408)$, $(2.729, -0.408)$

55.

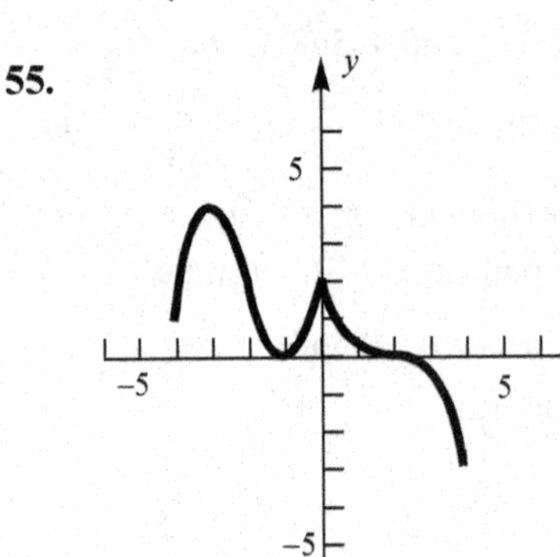

a. f is increasing on the intervals $(-\infty, -3]$ and $[-1, 0]$.
f is decreasing on the intervals $[-3, -1]$ and $[0, \infty)$.

b. f is concave down on the intervals $(-\infty, -2)$ and $(2, \infty)$.
f is concave up on the intervals $(-2, 0)$ and $(0, 2)$.

c. f attains a local maximum at $x = -3$ and $x = 0$.
f attains a local minimum at $x = -1$.

d. f has a point of inflection at $x = -2$ and $x = 2$.

57.

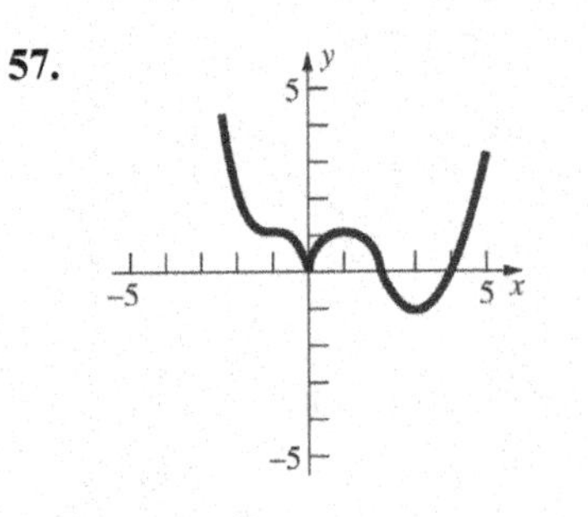

59. a.

$$f'(x) = \frac{2x^2 - 9x + 40}{\sqrt{x^2 - 6x + 40}};\ f'(x) \text{ is never 0,}$$

and always positive, so $f(x)$ is increasing for all x. Thus, on $[-1, 7]$, the global minimum is $f(-1) \approx -6.9$ and the global maximum if $f(7) \approx 48.0$.

$$f''(x) = \frac{2x^3 - 18x^2 + 147x - 240}{(x^2 - 6x + 40)^{3/2}};\ f''(x) = 0$$

when $x \approx 2.02$; inflection point $f(2.02) \approx 11.4$

b.

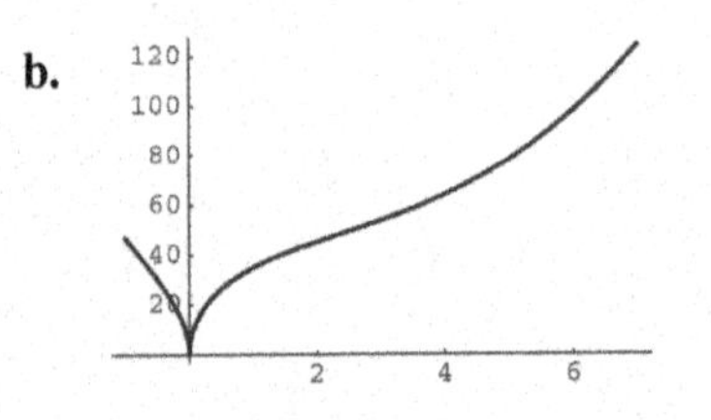

Global minimum $f(0) = 0$; global maximum $f(7) \approx 124.4$; inflection point at $x \approx 2.34$, $f(2.34) \approx 48.09$

c.

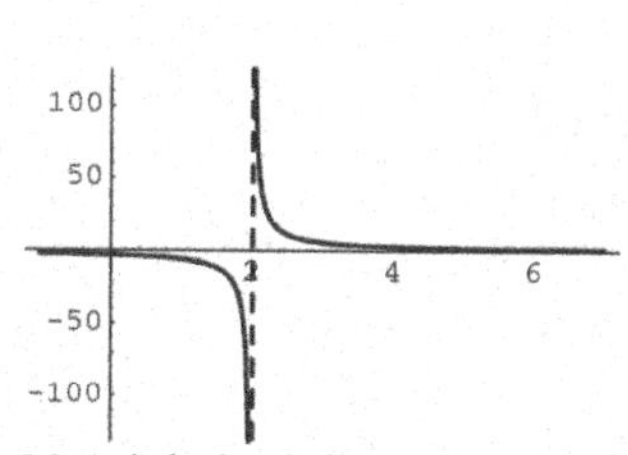

No global minimum or maximum; no inflection points

d.

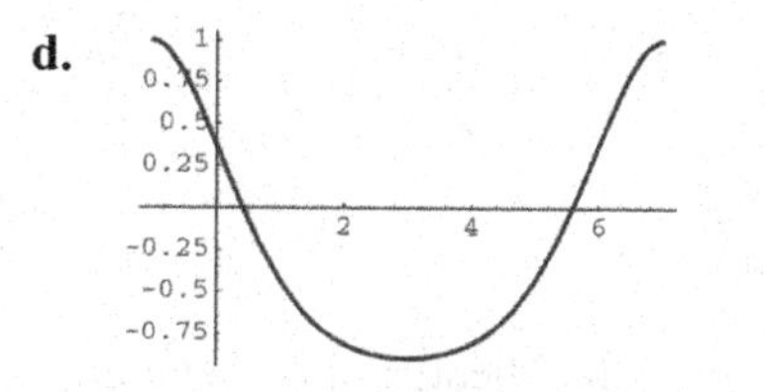

Global minimum $f(3) \approx -0.9$;
global maximum $f(-1) \approx 1.0$ or $f(7) \approx 1.0$;
Inflection points at $x \approx 0.05$ and $x \approx 5.9$,
$f(0.05) \approx 0.3, f(5.9) \approx 0.3$.

3.6 Concepts Review

1. continuous; (a, b); $f(b) - f(a) = f'(c)(b - a)$

3. $F(x) = G(x) + C$

Problem Set 3.6

1. $f'(x) = \dfrac{x}{|x|}$

$$\frac{f(2)-f(1)}{2-1} = \frac{2-1}{1} = 1$$

$\dfrac{c}{|c|} = 1$ for all $c > 0$, hence for all c in (1, 2)

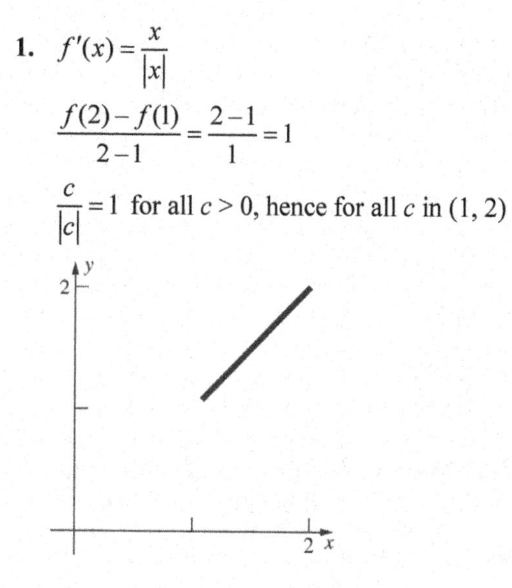

3. $f'(x) = 2x + 1$

$$\frac{f(2)-f(-2)}{2-(-2)} = \frac{6-2}{4} = 1$$

$2c + 1 = 1$ when $c = 0$

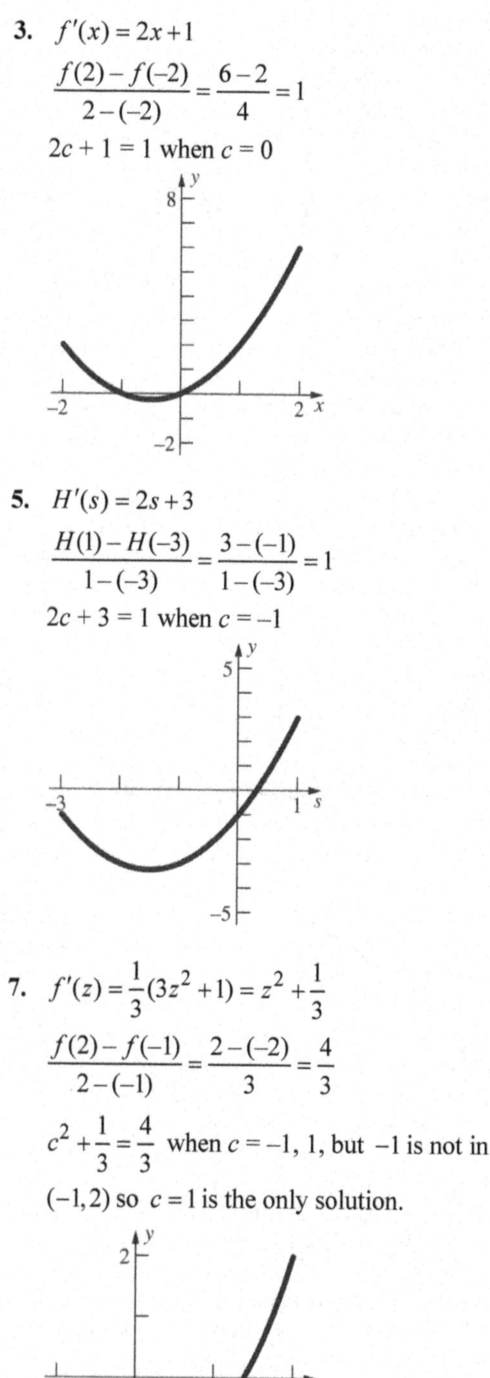

5. $H'(s) = 2s + 3$

$$\frac{H(1)-H(-3)}{1-(-3)} = \frac{3-(-1)}{1-(-3)} = 1$$

$2c + 3 = 1$ when $c = -1$

7. $f'(z) = \frac{1}{3}(3z^2 + 1) = z^2 + \frac{1}{3}$

$$\frac{f(2)-f(-1)}{2-(-1)} = \frac{2-(-2)}{3} = \frac{4}{3}$$

$c^2 + \frac{1}{3} = \frac{4}{3}$ when $c = -1$, 1, but -1 is not in $(-1, 2)$ so $c = 1$ is the only solution.

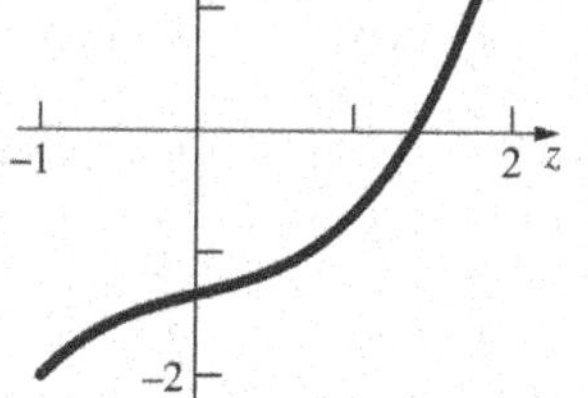

9. $h'(x) = -\dfrac{3}{(x-3)^2}$

$\dfrac{h(2)-h(0)}{2-0} = \dfrac{-2-0}{2} = -1$

$-\dfrac{3}{(c-3)^2} = -1$ when $c = 3 \pm \sqrt{3}$,

$c = 3 - \sqrt{3} \approx 1.27$ $(3+\sqrt{3}$ is not in $(0, 2).)$

11. $h'(t) = \dfrac{2}{3t^{1/3}}$

$\dfrac{h(2)-h(0)}{2-0} = \dfrac{2^{2/3}-0}{2} = 2^{-1/3}$

$\dfrac{2}{3c^{1/3}} = 2^{-1/3}$ when $c = \dfrac{16}{27} \approx 0.59$

13. $g'(x) = \dfrac{5}{3}x^{2/3}$

$\dfrac{g(1)-g(0)}{1-0} = \dfrac{1-0}{1} = 1$

$\dfrac{5}{3}c^{2/3} = 1$ when $c = \pm\left(\dfrac{3}{5}\right)^{3/2}$,

$c = \left(\dfrac{3}{5}\right)^{3/2} \approx 0.46$, $\left(-\left(\dfrac{3}{5}\right)^{3/2} \text{ is not in } (0, 1).\right)$

15. $S'(\theta) = \cos\theta$

$\dfrac{S(\pi)-S(-\pi)}{\pi-(-\pi)} = \dfrac{0-0}{2\pi} = 0$

$\cos c = 0$ when $c = \pm\dfrac{\pi}{2}$.

17. The Mean Value Theorem does not apply because $T(\theta)$ is not continuous at $\theta = \dfrac{\pi}{2}$.

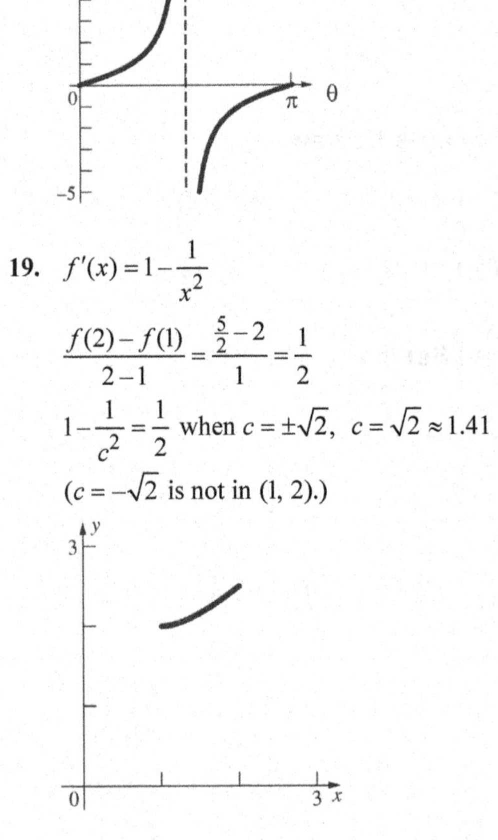

19. $f'(x) = 1 - \dfrac{1}{x^2}$

$\dfrac{f(2)-f(1)}{2-1} = \dfrac{\frac{5}{2}-2}{1} = \dfrac{1}{2}$

$1 - \dfrac{1}{c^2} = \dfrac{1}{2}$ when $c = \pm\sqrt{2}$, $c = \sqrt{2} \approx 1.41$

$(c = -\sqrt{2}$ is not in $(1, 2).)$

21. The Mean Value Theorem does not apply because f is not differentiable at $x = 0$.

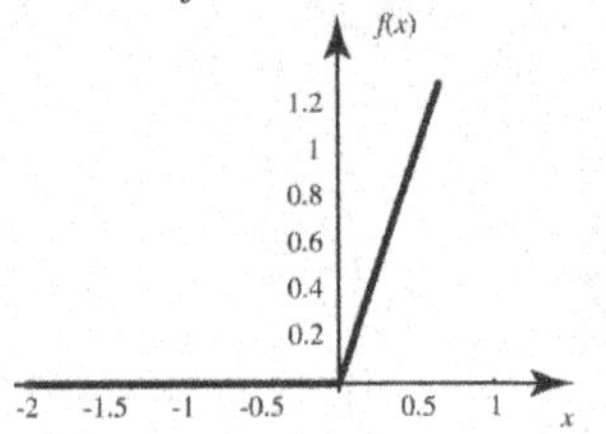

23. $\frac{f(8)-f(0)}{8-0}=-\frac{1}{4}$

There are three values for c such that $f'(c)=-\frac{1}{4}$.

They are approximately 1.5, 3.75, and 7.

25. By the Monotonicity Theorem, f is increasing on the intervals (a, x_0) and (x_0, b).

To show that $f(x_0) > f(x)$ for x in (a, x_0), consider f on the interval $(a, x_0]$.

f satisfies the conditions of the Mean Value Theorem on the interval $[x, x_0]$ for x in (a, x_0).

So for some c in (x, x_0),

$f(x_0)-f(x)=f'(c)(x_0-x)$.

Because

$f'(c)>0$ and $x_0-x>0$, $f(x_0)-f(x)>0$,

so $f(x_0)>f(x)$.

Similar reasoning shows that

$f(x)>f(x_0)$ for x in (x_0, b).

Therefore, f is increasing on (a, b).

27. $s(t)$ is defined in any interval not containing $t = 0$.

$s'(c)=-\frac{1}{c^2}<0$ for all $c \neq 0$. For any a, b with $a<b$ and both either positive or negative, the Mean Value Theorem says

$s(b)-s(a)=s'(c)(b-a)$ for some c in (a, b).

Since $a<b$, $b-a>0$ while $s'(c)<0$, hence $s(b)-s(a)<0$, or $s(b)<s(a)$.

Thus, $s(t)$ is decreasing on any interval not containing $t = 0$.

29. $F'(x)=0$ and $G(x)=0; G'(x)=0$.

By Theorem B,

$F(x) = G(x) + C$, so $F(x) = 0 + C = C$.

31. Let $G(x)=Dx; F'(x)=D$ and $G'(x)=D$.

By Theorem B, $F(x) = G(x) + C$; $F(x) = Dx + C$.

33. Since $f(a)$ and $f(b)$ have opposite signs, 0 is between $f(a)$ and $f(b)$. $f(x)$ is continuous on $[a, b]$, since it has a derivative. Thus, by the Intermediate Value Theorem, there is at least one point c,

$a<c<b$ with $f(c) = 0$.

Suppose there are two points, c and c', $c<c'$ in (a, b) with $f(c)=f(c')=0$. Then by Rolle's Theorem, there is at least one number d in (c, c') with $f'(d)=0$. This contradicts the given information that $f'(x)\neq 0$ for all x in $[a, b]$, thus there cannot be more than one x in $[a, b]$ where $f(x) = 0$.

35. Suppose there is more than one zero between successive distinct zeros of f'. That is, there are a and b such that $f(a) = f(b) = 0$ with a and b between successive distinct zeros of f'. Then by Rolle's Theorem, there is a c between a and b such that $f'(c)=0$. This contradicts the supposition that a and b lie between successive distinct zeros.

37. $f(x)$ is a polynomial function so it is continuous on $[0, 4]$ and $f''(x)$ exists for all x on $(0, 4)$.

$f(1) = f(2) = f(3) = 0$, so by Problem 36, there are at least two values of x in $[0, 4]$ where $f'(x)=0$ and at least one value of x in $[0, 4]$ where $f''(x)=0$.

39. $f'(x)=2\cos 2x; |f'(x)|\leq 2$

$\frac{|f(x_2)-f(x_1)|}{|x_2-x_1|}=|f'(x)|; \frac{|f(x_2)-f(x_1)|}{|x_2-x_1|}\leq 2$

$|f(x_2)-f(x_1)|\leq 2|x_2-x_1|;$

$|\sin 2x_2-\sin 2x_1|\leq 2|x_2-x_1|$

41. Suppose $f'(x)\geq 0$. Let a and b lie in the interior of I such that $b > a$. By the Mean Value Theorem, there is a point c between a and b such that

$f'(c)=\frac{f(b)-f(a)}{b-a}; \frac{f(b)-f(a)}{b-a}\geq 0$.

Since $a < b$, $f(b) \geq f(a)$, so f is nondecreasing.

Suppose $f'(x)\leq 0$. Let a and b lie in the interior of I such that $b > a$. By the Mean Value Theorem, there is a point c between a and b such that

$f'(c)=\frac{f(b)-f(a)}{b-a}; \frac{f(b)-f(a)}{b-a}\leq 0$. Since $a<b$, $f(a)\geq f(b)$, so f is nonincreasing.

43. Let $f(x) = h(x) - g(x)$.

$f'(x)=h'(x)-g'(x); f'(x)\geq 0$ for all x in (a, b) since $g'(x)\leq h'(x)$ for all x in (a, b), so f is nondecreasing on (a, b) by Problem 41. Thus

$x_1<x_2 \Rightarrow f(x_1)\leq f(x_2);$

$h(x_1)-g(x_1)\leq h(x_2)-g(x_2);$

$g(x_2)-g(x_1)\leq h(x_2)-h(x_1)$ for all x_1 and x_2 in (a, b).

45. Let $f(x) = \sin x$. $f'(x) = \cos x$, so
$|f'(x)| = |\cos x| \le 1$ for all x.
By the Mean Value Theorem,
$$\frac{f(x)-f(y)}{x-y} = f'(c) \text{ for some } c \text{ in } (x, y).$$
Thus, $\dfrac{|f(x)-f(y)|}{|x-y|} = |f'(c)| \le 1$;
$|\sin x - \sin y| \le |x - y|$.

47. Let s be the difference in speeds between horse A and horse B as function of time t.
Then s' is the difference in accelerations.
Let t_2 be the time in Problem 46 at which the horses had the same speeds and let t_1 be the finish time of the race.
$s(t_2) = s(t_1) = 0$
By the Mean Value Theorem,
$$\frac{s(t_1)-s(t_2)}{t_1-t_2} = s'(c) \text{ for some } c \text{ in } (t_2, t_1).$$
Therefore $s'(c) = 0$ for some c in (t_2, t_1).

49. Fix an arbitrary x.
$$f'(x) = \lim_{y \to x} \frac{f(y)-f(x)}{y-x} = 0, \text{ since}$$
$$\left|\frac{f(y)-f(x)}{y-x}\right| \le M|y-x|.$$
So, $f' \equiv 0 \to f = \text{constant}$.

51. Let $f(t)$ be the distance traveled at time t.
$$\frac{f(2)-f(0)}{2-0} = \frac{112-0}{2} = 56$$
By the Mean Value Theorem, there is a time c such that $f'(c) = 56$.
At some time during the trip, Johnny must have gone 56 miles per hour.

53. Since the car is stationary at $t = 0$, and since v is continuous, there exists a δ such that $v(t) < \frac{1}{2}$ for all t in the interval $[0, \delta]$. $v(t)$ is therefore less than $\frac{1}{2}$ and $s(\delta) < \delta \cdot \frac{1}{2} = \frac{\delta}{2}$. By the Mean Value Theorem, there exists a c in the interval $(\delta, 20)$ such that
$$v(c) = s'(c) = \frac{\left(20 - \frac{\delta}{2}\right)}{(20-\delta)}$$
$$> \frac{20-\delta}{20-\delta}$$
$= 1$ mile per minute
$= 60$ miles per hour

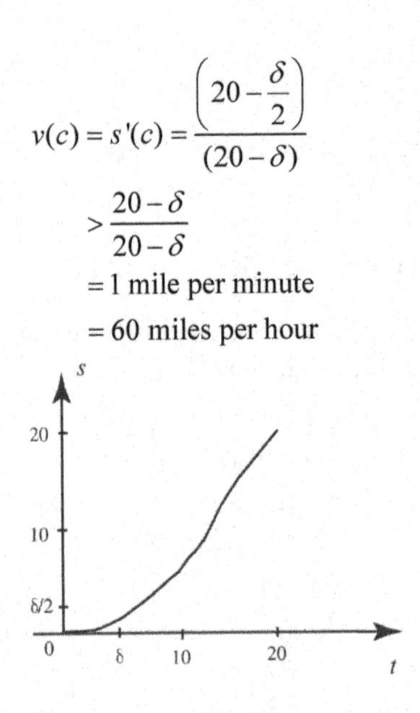

3.7 Concepts Review

1. slowness of convergence

3. algorithms

Problem Set 3.7

1. Let $f(x) = x^3 + 2x - 6$.
$f(1) = -3, f(2) = 6$

n	h_n	m_n	$f(m_n)$
1	0.5	1.5	0.375
2	0.25	1.25	–1.546875
3	0.125	1.375	–0.650391
4	0.0625	1.4375	–0.154541
5	0.03125	1.46875	0.105927
6	0.015625	1.45312	–0.0253716
7	0.0078125	1.46094	0.04001
8	0.00390625	1.45703	0.00725670
9	0.00195312	1.45508	–0.00907617

$r \approx 1.46$

3. Let $f(x) = 2\cos x - \sin x$.

$f(1) \approx 0.23913$; $f(2) \approx -1.74159$

n	h_n	m_n	$f(m_n)$
1	0.5	1.5	–0.856021
2	0.25	1.25	–0.318340
3	0.125	1.125	–0.039915
4	0.0625	1.0625	0.998044
5	0.03125	1.09375	0.029960
6	0.01563	1.109375	–0.004978

$r \approx 1.11$

5. Let $f(x) = x^3 + 6x^2 + 9x + 1 = 0$.

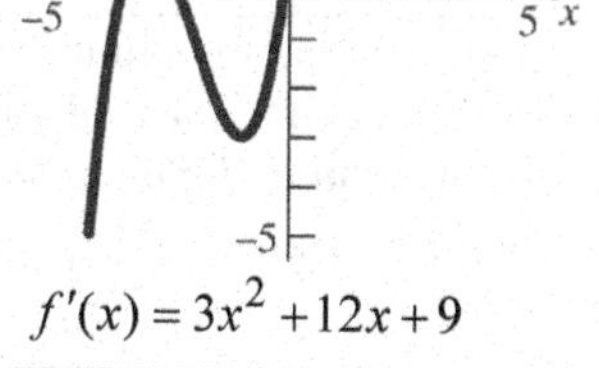

$f'(x) = 3x^2 + 12x + 9$

n	x_n
1	0
2	–0.1111111
3	–0.1205484
4	–0.1206148
5	–0.1206148

$r \approx -0.12061$

7. Let $f(x) = x - 2 + 2\cos x$.

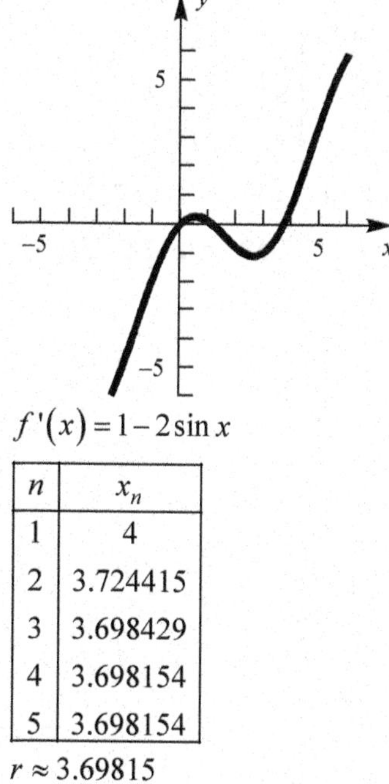

$f'(x) = 1 - 2\sin x$

n	x_n
1	4
2	3.724415
3	3.698429
4	3.698154
5	3.698154

$r \approx 3.69815$

9. Let $f(x) = \cos x - 2x$.

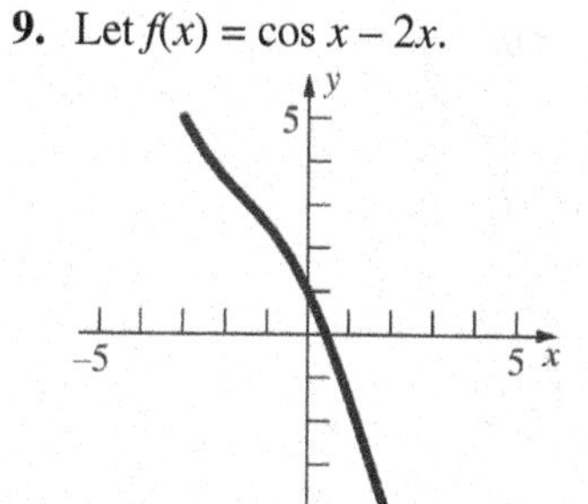

$f'(x) = -\sin x - 2$

n	x_n
1	0.5
2	0.4506267
3	0.4501836
4	0.4501836

$r \approx 0.45018$

11. Let $f(x) = x^4 - 8x^3 + 22x^2 - 24x + 8$.

$f'(x) = 4x^3 - 24x^2 + 44x - 24$

Note that $f(2) = 0$.

n	x_n
1	0.5
2	0.575
3	0.585586
4	0.585786

n	x_n
1	3.5
2	3.425
3	3.414414
4	3.414214
5	3.414214

$r = 2$, $r \approx 0.58579$, $r \approx 3.41421$

13. Let $f(x) = 2x^2 - \sin x$.

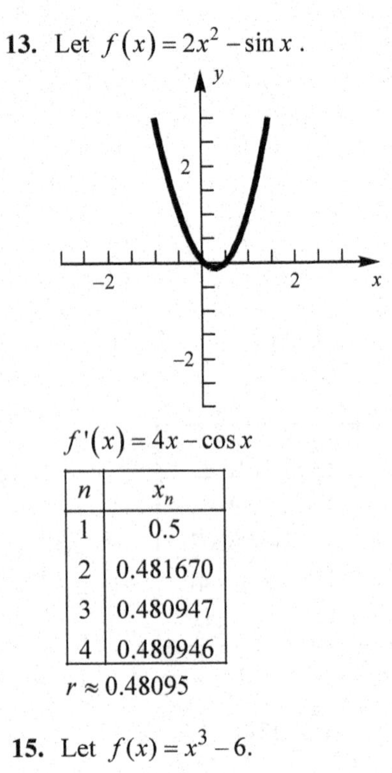

$f'(x) = 4x - \cos x$

n	x_n
1	0.5
2	0.481670
3	0.480947
4	0.480946

$r \approx 0.48095$

15. Let $f(x) = x^3 - 6$.

$f'(x) = 3x^2$

n	x_n
1	1.5
2	1.888889
3	1.819813
4	1.817125
5	1.817121
6	1.817121

$\sqrt[3]{6} \approx 1.81712$

17. $f(x) = x^4 + x^3 + x^2 + x$ is continuous on the given interval.

WINDOW
Xmin=-1
Xmax=1
Xscl=1
Ymin=-2
Ymax=6
Yscl=1
Xres=1

From the graph of *f*, we see that the maximum value of the function on the interval occurs at the right endpoint. The minimum occurs at a stationary point within the interval. To find where the minimum occurs, we solve $f'(x) = 0$ on the interval $[-1,1]$.

$f'(x) = 4x^3 + 3x^2 + 2x + 1 = g(x)$

Using Newton's Method to solve $g(x) = 0$, we get:

n	x_n
1	0
2	−0.5
3	−0.625
4	−0.60638
5	−0.60583
6	−0.60583

Minimum: $f(-0.60583) \approx -0.32645$

Maximum: $f(1) = 4$

19. $f(x) = \dfrac{\sin x}{x}$ is continuous on the given interval.

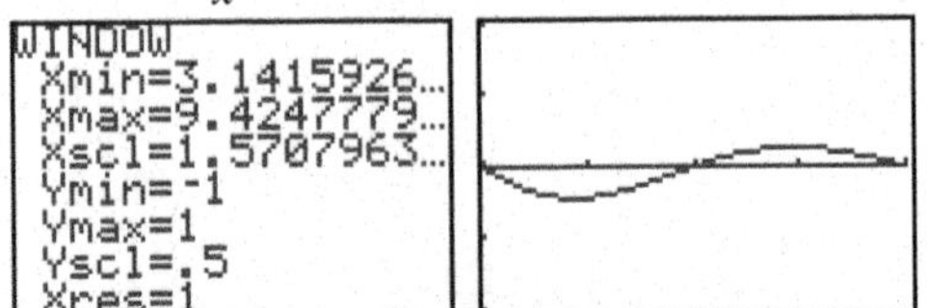

From the graph of *f*, we see that the minimum value and maximum value on the interval will occur at stationary points within the interval. To find these points, we need to solve $f'(x) = 0$ on the interval.

$f'(x) = \dfrac{x\cos x - \sin x}{x^2} = g(x)$

Using Newton's method to solve $g(x) = 0$ on the interval, we use the starting values of $\dfrac{3\pi}{2}$ and $\dfrac{5\pi}{2}$.

n	x_n
1	4.712389
2	4.479179
3	4.793365
4	4.493409
5	4.493409

n	x_n
1	7.853982
2	7.722391
3	7.725251
4	7.725252
5	7.725252

Minimum: $f(4.493409) \approx -0.21723$

Maximum: $f(7.725252) \approx 0.128375$

21. Graph $y = x$ and $y = 0.8 + 0.2 \sin x$.

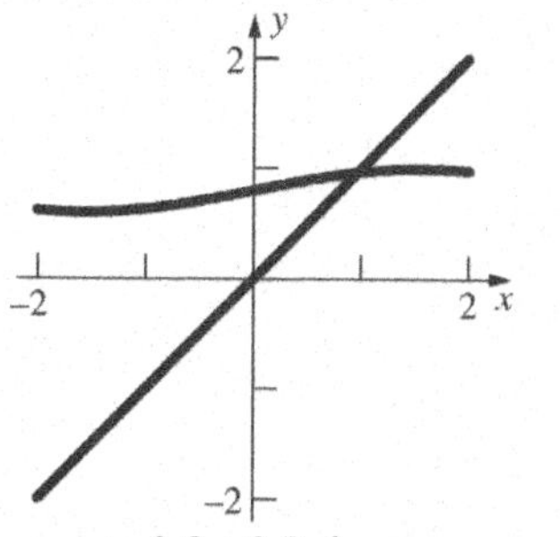

$x_{n+1} = 0.8 + 0.2 \sin x_n$

Let $x_1 = 1$.

n	x_n
1	1
2	0.96829
3	0.96478
4	0.96439
5	0.96434
6	0.96433
7	0.96433

$x \approx 0.9643$

23. a. For Tom's car, $P = 2000$, $R = 100$, and $k = 24$, thus

$$2000 = \frac{100}{i}\left[1 - \frac{1}{(1+i)^{24}}\right] \text{ or}$$

$$20i = 1 - \frac{1}{(1+i)^{24}}, \text{ which is equivalent to}$$

$$20i(1+i)^{24} - (1+i)^{24} + 1 = 0.$$

b. Let

$$f(i) = 20i(1+i)^{24} - (1+i)^{24} + 1$$
$$= (1+i)^{24}(20i - 1) + 1.$$

Then

$$f'(i) = 20(1+i)^{24} + 480i(1+i)^{23} - 24(1+i)^{23}$$
$$= (1+i)^{23}(500i - 4), \text{ so}$$

$$i_{n+1} = i_n - \frac{f(i_n)}{f'(i_n)} = i_n - \frac{(1+i_n)^{24}(20i_n - 1) + 1}{(1+i_n)^{23}(500i_n - 4)}$$

$$= i_n - \left[\frac{20i_n^2 + 19i_n - 1 + (1+i_n)^{-23}}{500i_n - 4}\right].$$

c.

n	i_n
1	0.012
2	0.0165297
3	0.0152651
4	0.0151323
5	0.0151308
6	0.0151308

$i = 0.0151308$

$r = 18.157\%$

25. $x_{n+1} = \dfrac{x_n + 1.5 \cos x_n}{2}$

n	x_n
1	1
2	0.905227
3	0.915744
4	0.914773

n	x_n
5	0.914864
6	0.914856
7	0.914857

$x \approx 0.91486$

27. $x_{n+1} = \sqrt{2.7 + x_n}$

n	x_n
1	1
2	1.923538
3	2.150241
4	2.202326
5	2.214120
6	2.216781
7	2.217382
8	2.217517
9	2.217548
10	2.217554
11	2.217556
12	2.217556

$x \approx 2.21756$

29. a.

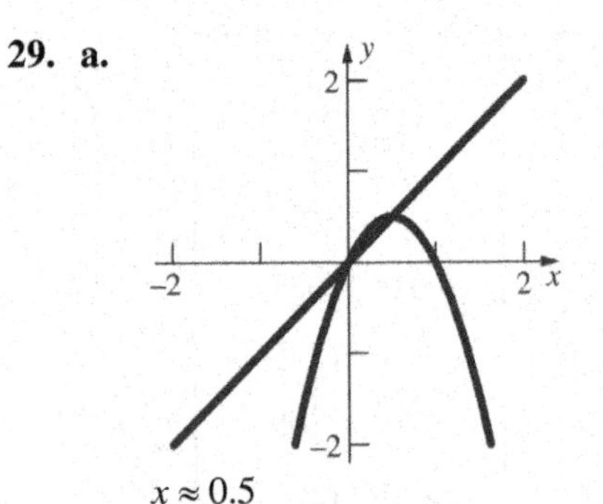

$x \approx 0.5$

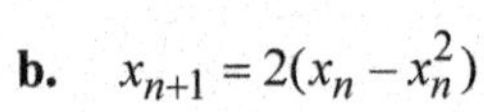

b. $x_{n+1} = 2(x_n - x_n^2)$

n	x_n
1	0.7
2	0.42
3	0.4872
4	0.4996723
5	0.4999998
6	0.5
7	0.5

c. $x = 2(x - x^2)$

$2x^2 - x = 0$

$x(2x - 1) = 0$

$x = 0,\ x = \dfrac{1}{2}$

31. a. $x_1 = 0$

$x_2 = \sqrt{1} = 1$

$x_3 = \sqrt{1+\sqrt{1}} = \sqrt{2} \approx 1.4142136$

$x_4 = \sqrt{1+\sqrt{1+\sqrt{1}}} \approx 1.553774$

$x_5 = \sqrt{1+\sqrt{1+\sqrt{1+\sqrt{1}}}} \approx 1.5980532$

b.

$$x = \sqrt{1+x}$$

$$x^2 = 1 + x$$

$$x^2 - x - 1 = 0$$

$$x = \frac{1 \pm \sqrt{1+4\cdot 1\cdot 1}}{2} = \frac{1 \pm \sqrt{5}}{2}$$

Taking the minus sign gives a negative solution for x, violating the requirement that $x \ge 0$. Hence, $x = \dfrac{1+\sqrt{5}}{2} \approx 1.618034$.

c. Let $x = \sqrt{1+\sqrt{1+\sqrt{1+\ldots}}}$. Then x satisfies the equation $x = \sqrt{1+x}$. From part (b) we know that x must equal $\left(1+\sqrt{5}\right)/2 \approx 1.618034$.

33. a. $x_1 = 1$

$x_2 = 1 + \dfrac{1}{1} = 2$

$x_3 = 1 + \dfrac{1}{1+\frac{1}{1}} = \dfrac{3}{2} = 1.5$

$x_4 = 1 + \dfrac{1}{1+\frac{1}{1+\frac{1}{1}}} = \dfrac{5}{3} \approx 1.6666667$

$x_5 = 1 + \dfrac{1}{1+\frac{1}{1+\frac{1}{1+\frac{1}{1}}}} = \dfrac{8}{5} = 1.6$

b.

$$x = 1 + \frac{1}{x}$$

$$x^2 = x + 1$$

$$x^2 - x - 1 = 0$$

$$x = \frac{1 \pm \sqrt{1+4\cdot 1\cdot 1}}{2} = \frac{1 \pm \sqrt{5}}{2}$$

Taking the minus sign gives a negative solution for x, violating the requirement that $x \ge 0$. Hence, $x = \dfrac{1+\sqrt{5}}{2} \approx 1.618034$.

c. Let

$$x = 1 + \cfrac{1}{1+\cfrac{1}{1+\cdots}}.$$

Then x satisfies the equation $x = 1 + \dfrac{1}{x}$. From part (b) we know that x must equal $\left(1+\sqrt{5}\right)/2 \approx 1.618034$.

35. a. The algorithm computes the root of $\dfrac{1}{x} - a = 0$ for x_1 close to $\dfrac{1}{a}$.

b. Let $f(x) = \dfrac{1}{x} - a$.

$$f'(x) = -\frac{1}{x^2}$$

$$\frac{f(x)}{f'(x)} = -x + ax^2$$

The recursion formula is

$$x_{n+1} = x_n - \frac{f(x_n)}{f'(x_n)} = 2x_n - a{x_n}^2.$$

37. The rod that barely fits around the corner will touch the outside walls as well as the inside corner.

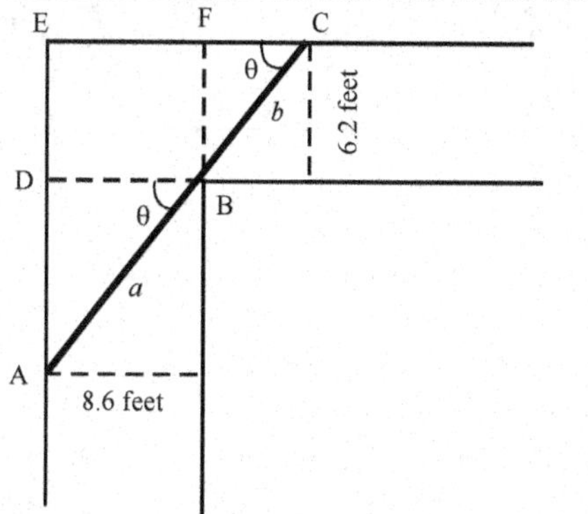

As suggested in the diagram, let a and b represent the lengths of the segments AB and BC, and let θ denote the angles $\angle DBA$ and $\angle FCB$. Consider the two similar triangles ΔADB and ΔBFC; these have hypotenuses a and b respectively. A little trigonometry applied to these angles gives

$a = \frac{8.6}{\cos\theta} = 8.6\sec\theta$ and $b = \frac{6.2}{\sin\theta} = 6.2\csc\theta$

Note that the angle θ determines the position of the rod. The total length of the rod is then

$L = a + b = 8.6\sec\theta + 6.2\csc\theta$

The domain for θ is the open interval $\left(0, \frac{\pi}{2}\right)$. The derivative of L is

$$L'(\theta) = \frac{8.6\sin^3\theta - 6.2\cos^3\theta}{\sin^2\theta \cdot \cos^2\theta}$$

Thus, $L'(\theta) = 0$ provided

$$8.6\sin^3\theta - 6.2\cos^3\theta = 0$$

$$8.6\sin^3\theta = 6.2\cos^3\theta$$

$$\frac{\sin^3\theta}{\cos^3\theta} = \frac{6.2}{8.6}$$

$$\tan^3\theta = \frac{6.2}{8.6}$$

$$\tan\theta = \sqrt[3]{\frac{6.2}{8.6}}$$

On $\left(0, \frac{\pi}{2}\right)$, there will only be one solution to this equation. We will use Newton's method to solve $\tan\theta - \sqrt[3]{\frac{6.2}{8.6}} = 0$ starting with $\theta_1 = \frac{\pi}{4}$.

n	θ_n
1	$\frac{\pi}{4} \approx 0.78540$
2	0.73373
3	0.73098
4	0.73097
5	0.73097

Note that $\theta \approx 0.73097$ minimizes the length of the rod that does *not* fit around the corner, which in turn maximizes the length of the rod that will fit around the corner (verify by using the Second Derivative Test).

$$L(0.73097) = 8.6\sec(0.73097) + 6.2\csc(0.73097)$$
$$\approx 20.84$$

Thus, the length of the longest rod that will fit around the corner is about 20.84 feet.

39. We can solve the equation $-\frac{2x^2}{25} + x + 42 = 0$ to find the value for x when the object hits the ground. We want the value to be positive, so we use the quadratic formula, keeping only the positive solution.

$$x = \frac{-1 - \sqrt{1^2 - 4(-0.08)(42)}}{2(-0.08)} = 30$$

We are interested in the global extrema for the distance of the object from the observer. We obtain the same extrema by considering the squared distance

$D(x) = (x-3)^2 + (42 + x - .08x^2)^2$

A graph of D will help us identify a starting point for our numeric approach.

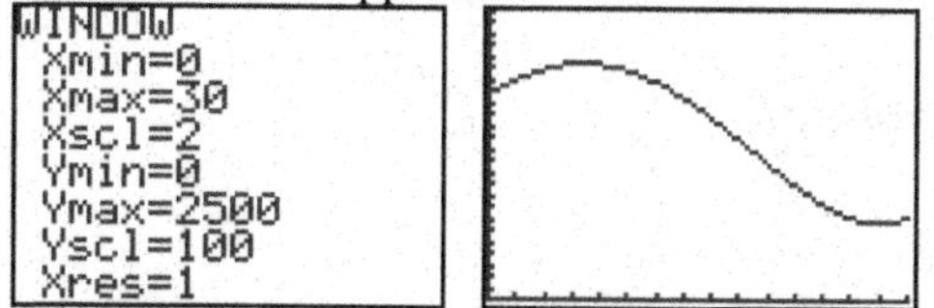

From the graph, it appears that D (and thus the distance from the observer) is maximized at about $x = 7$ feet and minimized just before the object hits the ground at about $x = 28$ feet.

The first derivative is given by

$$D'(x) = \frac{16}{625}x^3 - \frac{12}{25}x^2 - \frac{236}{25}x + 78.$$

a. We will use Newton's method to find the stationary point that yields the minimum distance, starting with $x_1 = 28$.

n	x_n
1	28
2	28.0280
3	28.0279
4	28.0279

$x \approx 28.0279$; $y \approx 7.1828$

The object is closest to the observer when it is at the point $(28.0279, 7.1828)$.

b. We will use Newton's method to find the stationary point that yields the maximum distance, starting with $x_1 = 7$.

n	x_n
1	7
2	6.7726
3	6.7728
4	6.7728

$x \approx 6.7728$; $y \approx 45.1031$

The object is closest to the observer when it is at the point $(6.7728, 45.1031)$.

3.8 Concepts Review

1. $rx^{r-1}; \frac{x^{r+1}}{r+1} + C, r \neq -1$

3. $u = x^4 + 3x^2 + 1, du = (4x^3 + 6x)dx$

$\int (x^4 + 3x^2 + 1)^8 (4x^3 + 6x)dx = \int u^8 du$

$= \frac{u^9}{9} + C = \frac{(x^4 + 3x^2 + 1)^9}{9} + C$

Problem Set 3.8

1. $\int 5dx = 5x + C$

3. $\int (x^2 + \pi)dx = \int x^2 dx + \pi \int 1 dx = \frac{x^3}{3} + \pi x + C$

5. $\int x^{5/4} dx = \frac{x^{9/4}}{\frac{9}{4}} + C = \frac{4}{9} x^{9/4} + C$

7. $\int \frac{1}{\sqrt[3]{x^2}} dx = \int x^{-2/3} dx = 3x^{1/3} + C = 3\sqrt[3]{x} + C$

9. $\int (x^2 - x)dx = \int x^2 dx - \int x\, dx = \frac{x^3}{3} - \frac{x^2}{2} + C$

11. $\int (4x^5 - x^3)dx = 4\int x^5 dx - \int x^3 dx$

$= 4\left(\frac{x^6}{6} + C_1\right) - \left(\frac{x^4}{4} + C_2\right)$

$= \frac{2x^6}{3} - \frac{x^4}{4} + C$

13. $\int (27x^7 + 3x^5 - 45x^3 + \sqrt{2}\, x)dx$

$= 27\int x^7 dx + 3\int x^5 dx - 45\int x^3 dx + \sqrt{2}\int x\, dx$

$= \frac{27x^8}{8} + \frac{x^6}{2} - \frac{45x^4}{4} + \frac{\sqrt{2}\, x^2}{2} + C$

15. $\int \left(\frac{3}{x^2} - \frac{2}{x^3}\right) dx = \int (3x^{-2} - 2x^{-3})\, dx$

$= 3\int x^{-2} dx - 2\int x^{-3} dx$

$= \frac{3x^{-1}}{-1} - \frac{2x^{-2}}{-2} + C$

$= -\frac{3}{x} + \frac{1}{x^2} + C$

17. $\int \frac{4x^6 + 3x^4}{x^3} dx = \int (4x^3 + 3x)\, dx$

$= 4\int x^3 dx + 3\int x\, dx$

$= x^4 + \frac{3x^2}{2} + C$

19. $\int (x^2 + x)\, dx = \int x^2 dx + \int x\, dx = \frac{x^3}{3} + \frac{x^2}{2} + C$

21. Let $u = x + 1$; then $du = dx$.

$\int (x+1)^2 dx = \int u^2 du = \frac{u^3}{3} + C = \frac{(x+1)^3}{3} + C$

23. $\int \frac{(z^2 + 1)^2}{\sqrt{z}} dz = \int \frac{z^4 + 2z^2 + 1}{\sqrt{z}} dz$

$= \int z^{7/2} dz + 2\int z^{3/2} + \int z^{-1/2} dz$

$= \frac{2}{9} z^{9/2} + \frac{4}{5} z^{5/2} + 2z^{1/2} + C$

25. $\int (\sin\theta - \cos\theta)d\theta = \int \sin\theta\, d\theta - \int \cos\theta\, d\theta$

$= -\cos\theta - \sin\theta + C$

27. Let $g(x) = \sqrt{2}\, x + 1$; then $g'(x) = \sqrt{2}$.

$\int \left(\sqrt{2}\, x + 1\right)^3 \sqrt{2}\, dx = \int [g(x)]^3 g'(x)dx$

$= \frac{[g(x)]^4}{4} + C = \frac{\left(\sqrt{2}\, x + 1\right)^4}{4} + C$

29. Let $u = 5x^3 + 3x - 8$; then $du = (15x^2 + 3)\,dx$.

$$\int (5x^2+1)(5x^3+3x-8)^6\,dx$$
$$= \int \frac{1}{3}(15x^2+3)(5x^3+3x-8)^6\,dx$$
$$= \frac{1}{3}\int u^6\,du = \frac{1}{3}\left(\frac{u^7}{7} + C_1\right)$$
$$= \frac{(5x^3+3x-8)^7}{21} + C$$

31. Let $u = 2t^2 - 11$; then $du = 4t\,dt$.

$$\int 3t\sqrt[3]{2t^2-11}\,dt = \int \frac{3}{4}(4t)(2t^2-11)^{1/3}\,dt$$
$$= \frac{3}{4}\int u^{1/3}du = \frac{3}{4}\left(\frac{3}{4}u^{4/3} + C_1\right)$$
$$= \frac{9}{16}(2t^2-11)^{4/3} + C$$
$$= \frac{9}{16}\sqrt[3]{(2t^2-11)^4} + C$$

33. Let $u = x^3 + 4$; then $du = 3x^2 dx$.

$$\int x^2\sqrt{x^3+4}\,dx = \int \frac{1}{3}3x^2\sqrt{x^3+4}\,dx$$
$$= \frac{1}{3}\int \sqrt{u}\,du = \frac{1}{3}\int u^{1/2}\,du$$
$$= \frac{1}{3}\left(\frac{2}{3}u^{3/2} + C_1\right)$$
$$= \frac{2}{9}\left(x^3+4\right)^{3/2} + C$$

35. Let $u = 1 + \cos x$; then $du = -\sin x\,dx$.

$$\int \sin x(1+\cos x)^4\,dx = -\int -\sin x(1+\cos x)^4\,dx$$
$$= -\int u^4\,du = -\left(\frac{1}{5}u^5 + C_1\right)$$
$$= -\frac{1}{5}(1+\cos x)^5 + C$$

37. $f'(x) = \int (3x+1)dx = \frac{3}{2}x^2 + x + C_1$

$$f(x) = \int\left(\frac{3}{2}x^2 + x + C_1\right)dx$$
$$= \frac{1}{2}x^3 + \frac{1}{2}x^2 + C_1x + C_2$$

39. $f'(x) = \int x^{1/2}dx = \frac{2}{3}x^{3/2} + C_1$

$$f(x) = \int\left(\frac{2}{3}x^{3/2} + C_1\right)dx$$
$$= \frac{4}{15}x^{5/2} + C_1x + C_2$$

41. $f''(x) = x + x^{-3}$

$$f'(x) = \int (x + x^{-3})dx = \frac{x^2}{2} - \frac{x^{-2}}{2} + C_1$$
$$f(x) = \int\left(\frac{1}{2}x^2 - \frac{1}{2}x^{-2} + C_1\right)dx$$
$$= \frac{1}{6}x^3 + \frac{1}{2}x^{-1} + C_1x + C_2$$
$$= \frac{1}{6}x^3 + \frac{1}{2x} + C_1x + C_2$$

43. The Product Rule for derivatives says

$$\frac{d}{dx}[f(x)g(x) + C] = f(x)g'(x) + f'(x)g(x).$$

Thus,

$$\int [f(x)g'(x) + f'(x)g(x)]dx = f(x)g(x) + C.$$

45. Let $f(x) = x^2$, $g(x) = \sqrt{x-1}$.

$$f'(x) = 2x,\; g'(x) = \frac{1}{2\sqrt{x-1}}$$
$$\int\left[\frac{x^2}{2\sqrt{x-1}} + 2x\sqrt{x-1}\right]dx$$
$$= \int [f(x)g'(x) + f'(x)g(x)]\,dx = f(x)g(x) + C$$
$$= x^2\sqrt{x-1} + C$$

47. $\int f''(x)dx = \int \frac{d}{dx}f'(x)dx = f'(x) + C$

$f'(x) = \sqrt{x^3+1} + \dfrac{3x^3}{2\sqrt{x^3+1}} = \dfrac{5x^3+2}{2\sqrt{x^3+1}}$ so

$$\int f''(x)dx = \frac{5x^3+2}{2\sqrt{x^3+1}} + C.$$

49. The Product Rule for derivatives says that

$$\frac{d}{dx}[f^m(x)g^n(x) + C]$$
$$= f^m(x)[g^n(x)]' + [f^m(x)]'g^n(x)$$
$$= f^m(x)[ng^{n-1}(x)g'(x)] + [mf^{m-1}(x)f'(x)]g^n(x)$$
$$= f^{m-1}(x)g^{n-1}(x)[nf(x)g'(x) + mg(x)f'(x)].$$

Thus,

$$\int f^{m-1}(x)g^{n-1}(x)[nf(x)g'(x) + mg(x)f'(x)]dx$$
$$= f^m(x)g^n(x) + C.$$

51. If $x \geq 0$, then $|x| = x$ and $\int |x|\,dx = \frac{1}{2}x^2 + C$.

If $x < 0$, then $|x| = -x$ and $\int |x|\,dx = -\frac{1}{2}x^2 + C$.

$$\int |x|\,dx = \begin{cases} \frac{1}{2}x^2 + C & \text{if } x \geq 0 \\ -\frac{1}{2}x^2 + C & \text{if } x < 0 \end{cases}$$

53. Different software may produce different, but equivalent answers. These answers were produced by Mathematica.

a. $\int 6\sin(3(x-2))\,dx = -2\cos(3(x-2)) + C$

b. $\int \sin^3\left(\frac{x}{6}\right)dx = \frac{1}{2}\cos\left(\frac{x}{2}\right) - \frac{9}{2}\cos\left(\frac{x}{6}\right) + C$

c. $\int (x^2\cos 2x + x\sin 2x)dx = \frac{x^2\sin 2x}{2} + C$

3.9 Concepts Review

1. differential equation

3. separate variables

Problem Set 3.9

1. $\frac{dy}{dx} = \frac{-2x}{2\sqrt{1-x^2}} = \frac{-x}{\sqrt{1-x^2}}$

$\frac{dy}{dx} + \frac{x}{y} = \frac{-x}{\sqrt{1-x^2}} + \frac{x}{\sqrt{1-x^2}} = 0$

3. $\frac{dy}{dx} = C_1\cos x - C_2\sin x;$

$\frac{d^2y}{dx^2} = -C_1\sin x - C_2\cos x$

$\frac{d^2y}{dx^2} + y$

$= (-C_1\sin x - C_2\cos x) + (C_1\sin x + C_2\cos x) = 0$

5. $\frac{dy}{dx} = x^2 + 1$

$dy = (x^2+1)\,dx$

$\int dy = \int (x^2+1)\,dx$

$y + C_1 = \frac{x^3}{3} + x + C_2$

$y = \frac{x^3}{3} + x + C$

At $x = 1$, $y = 1$:

$1 = \frac{1}{3} + 1 + C; C = -\frac{1}{3}$

$y = \frac{x^3}{3} + x - \frac{1}{3}$

7. $\frac{dy}{dx} = \frac{x}{y}$

$\int y\,dy = \int x\,dx$

$\frac{y^2}{2} + C_1 = \frac{x^2}{2} + C_2$

$y^2 = x^2 + C$

$y = \pm\sqrt{x^2 + C}$

At $x = 1$, $y = 1$:

$1 = \pm\sqrt{1+C}; C = 0$ and the square root is positive.

$y = \sqrt{x^2}$ or $y = x$

9. $\frac{dz}{dt} = t^2z^2$

$\int z^{-2}\,dz = \int t^2\,dt$

$-z^{-1} + C_1 = \frac{t^3}{3} + C_2$

$\frac{1}{z} = -\frac{t^3}{3} + C_3 = \frac{C - t^3}{3}$

$z = \frac{3}{C - t^3}$

At $t = 1$, $z = \frac{1}{3}$:

$\frac{1}{3} = \frac{3}{C-1}; C - 1 = 9; C = 10$

$z = \frac{3}{10 - t^3}$

11. $\frac{ds}{dt} = 16t^2 + 4t - 1$

$\int ds = \int (16t^2 + 4t - 1)\,dt$

$s + C_1 = \frac{16}{3}t^3 + 2t^2 - t + C_2$

$s = \frac{16}{3}t^3 + 2t^2 - t + C$

At $t = 0$, $s = 100$: $C = 100$

$s = \frac{16}{3}t^3 + 2t^2 - t + 100$

13. $\frac{dy}{dx} = (2x+1)^4$

$y = \int (2x+1)^4\,dx = \frac{1}{2}\int (2x+1)^4\,2\,dx$

$= \frac{1}{2}\frac{(2x+1)^5}{5} + C = \frac{(2x+1)^5}{10} + C$

At $x = 0$, $y = 6$:

$6 = \frac{1}{10} + C; C = \frac{59}{10}$

$y = \frac{(2x+1)^5}{10} + \frac{59}{10} = \frac{(2x+1)^5 + 59}{10}$

15. $\frac{dy}{dx} = 3x$

$y = \int 3x\,dx = \frac{3}{2}x^2 + C$

At (1, 2):

$2 = \frac{3}{2} + C$

$C = \frac{1}{2}$

$y = \frac{3}{2}x^2 + \frac{1}{2} = \frac{3x^2 + 1}{2}$

17. $v = \int t\,dt = \frac{t^2}{2} + v_0$

$v = \frac{t^2}{2} + 3$

$s = \int \left(\frac{t^2}{2} + 3\right) dt = \frac{t^3}{6} + 3t + s_0$

$s = \frac{t^3}{6} + 3t + 0 = \frac{t^3}{6} + 3t$

At $t = 2$:

$v = 5$ cm/s

$s = \frac{22}{3}$ cm

19. $v = \int (2t+1)^{1/3}\,dt = \frac{1}{2}\int (2t+1)^{1/3}\,2dt$

$= \frac{3}{8}(2t+1)^{4/3} + C_1$

$v_0 = 0 : 0 = \frac{3}{8} + C_1; C_1 = -\frac{3}{8}$

$v = \frac{3}{8}(2t+1)^{4/3} - \frac{3}{8}$

$s = \frac{3}{8}\int (2t+1)^{4/3}\,dt - \frac{3}{8}\int 1\,dt$

$= \frac{3}{16}\int (2t+1)^{4/3}\,2dt - \frac{3}{8}\int 1\,dt$

$= \frac{9}{112}(2t+1)^{7/3} - \frac{3}{8}t + C_2$

$s_0 = 10 : 10 = \frac{9}{112} + C_2; C_2 = \frac{1111}{112}$

$s = \frac{9}{112}(2t+1)^{7/3} - \frac{3}{8}t + \frac{1111}{112}$

At $t = 2$: $v = \frac{3}{8}(5)^{4/3} - \frac{3}{8} \approx 2.83$

$s = \frac{9}{112}(5)^{7/3} - \frac{6}{8} + \frac{1111}{112} \approx 12.6$

21. $v = -32t + 96$,

$s = -16t^2 + 96t + s_0 = -16t^2 + 96t$

$v = 0$ at $t = 3$

At $t = 3$, $s = -16(3^2) + 96(3) = 144$ ft

23. $\frac{dv}{dt} = -5.28$

$\int dv = -\int 5.28\,dt$

$v = \frac{ds}{dt} = -5.28t + v_0 = -5.28t + 56$

$\int ds = \int (-5.28t + 56)\,dt$

$s = -2.64t^2 + 56t + s_0 = -2.64t^2 + 56t + 1000$

When $t = 4.5$, $v = 32.24$ ft/s and $s = 1198.54$ ft

25. $\frac{dV}{dt} = -kS$

Since $V = \frac{4}{3}\pi r^3$ and $S = 4\pi r^2$,

$4\pi r^2 \frac{dr}{dt} = -k4\pi r^2$ so $\frac{dr}{dt} = -k$.

$\int dr = -\int k\,dt$

$r = -kt + C$

$2 = -k(0) + C$ and $0.5 = -k(10) + C$, so

$C = 2$ and $k = \frac{3}{20}$. Then, $r = -\frac{3}{20}t + 2$.

27. $v_{esc} = \sqrt{2gR}$

For the Moon, $v_{esc} \approx \sqrt{2(0.165)(32)(1080 \cdot 5280)}$
≈ 7760 ft/s ≈ 1.470 mi/s.

For Venus, $v_{esc} \approx \sqrt{2(0.85)(32)(3800 \cdot 5280)}$
$\approx 33{,}038$ ft/s ≈ 6.257 mi/s.

For Jupiter, $v_{esc} \approx 194{,}369$ ft/s ≈ 36.812 mi/s.

For the Sun, $v_{esc} \approx 2{,}021{,}752$ ft/s
≈ 382.908 mi/s.

29. $a = \frac{dv}{dt} = \frac{\Delta v}{\Delta t} = \frac{60-45}{10} = 1.5$ mi/h/s = 2.2 ft/s^2

31. For the first 10 s, $a = \frac{dv}{dt} = 6t$, $v = 3t^2$, and $s = t^3$. So $v(10) = 300$ and $s(10) = 1000$. After 10 s, $a = \frac{dv}{dt} = -10$, $v = -10(t - 10) + 300$, and $s = -5(t-10)^2 + 300(t-10) + 1000$. $v = 0$ at $t = 40$, at which time $s = 5500$ m.

33. a.

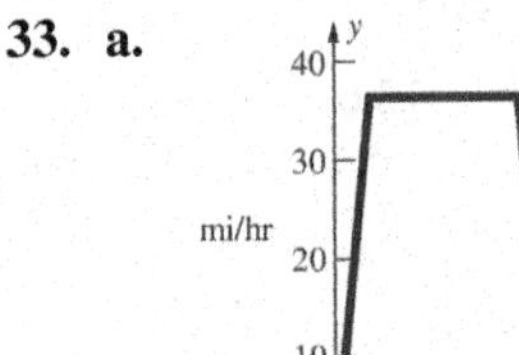

b. Since the trip that involves 1 min more travel time at speed v_m is 0.6 mi longer,
$v_m = 0.6$ mi/min
$= 36$ mi/h.

c. From part b, $v_m = 0.6$ mi/min. Note that the average speed during acceleration and deceleration is $\frac{v_m}{2} = 0.3$ mi/min. Let t be the time spent between stop C and stop D at the constant speed v_m, so
$0.6t + 0.3(4 - t) = 2$ miles. Therefore,
$t = 2\frac{2}{3}$ min and the time spent accelerating is $\frac{4-2\frac{2}{3}}{2} = \frac{2}{3}$ min.

$$a = \frac{0.6-0}{\frac{2}{3}} = 0.9 \text{ mi/min}^2.$$

35. a. $\frac{dV}{dt} = C_1\sqrt{h}$ where h is the depth of the water. Here, $V = \pi r^2 h = 100h$, so $h = \frac{V}{100}$.

Hence $\frac{dV}{dt} = C_1\frac{\sqrt{V}}{10}$, $V(0) = 1600$, $V(40) = 0$.

b. $\int 10V^{-1/2}dV = \int C_1 dt$; $20\sqrt{V} = C_1 t + C_2$;
$V(0) = 1600$: $C_2 = 20 \cdot 40 = 800$;
$V(40) = 0$: $C_1 = -\frac{800}{40} = -20$
$$V(t) = \frac{1}{400}(-20t + 800)^2 = (40 - t)^2$$

c. $V(10) = (40 - 10)^2 = 900$ cm^3

37. Initially, $v = -32t$ and $s = -16t^2 + 16$. $s = 0$ when $t = 1$. Later, the ball falls 9 ft in a time given by $0 = -16t^2 + 9$, or $\frac{3}{4}$ s, and on impact has a velocity of $-32\left(\frac{3}{4}\right) = -24$ ft/s. By symmetry, 24 ft/s must be the velocity right after the first bounce. So

a. $$v(t) = \begin{cases} -32t & \text{for } 0 \le t < 1 \\ -32(t-1) + 24 & \text{for } 1 < t \le 2.5 \end{cases}$$

b. $9 = -16t^2 + 16 \Rightarrow t \approx 0.66$ sec; s also equals 9 at the apex of the first rebound at $t = 1.75$ sec.

3.10 Chapter Review

Concepts Test

1. True: Max-Min Existence Theorem

3. True: For example, let $f(x) = \sin x$.

5. True: $f'(x) = 18x^5 + 16x^3 + 4x$;
$f''(x) = 90x^4 + 48x^2 + 4$, which is greater than zero for all x.

7. True: When $f'(x) > 0$, $f(x)$ is increasing.

9. True: $f(x) = ax^2 + bx + c$;
$f'(x) = 2ax + b$; $f''(x) = 2a$

11. False: $\tan^2 x$ has a minimum value of 0. This occurs whenever $x = k\pi$ where k is an integer.

13. True: $\lim_{x \to \frac{\pi}{2}^-} (2x^3 + x + \tan x) = \infty$ while
$\lim_{x \to -\frac{\pi}{2}^+} (2x^3 + x + \tan x) = -\infty$.

15. True: $\lim_{x \to \infty} \frac{x^2 + 1}{1 - x^2} = \lim_{x \to \infty} \frac{1 + \frac{1}{x^2}}{\frac{1}{x^2} - 1}$
$= \frac{1}{-1} = -1$ and
$\lim_{x \to -\infty} \frac{x^2 + 1}{1 - x^2} = \lim_{x \to -\infty} \frac{1 + \frac{1}{x^2}}{\frac{1}{x^2} - 1}$
$= \frac{1}{-1} = -1.$

17. True: The function is differentiable on (0, 2).

19. False: There are two points: $x = -\frac{\sqrt{3}}{3}, \frac{\sqrt{3}}{3}$.

21. False: For example if $f(x) = x^4$, $f'(0) = f''(0) = 0$ but f has a minimum at $x = 0$.

23. False: The rectangle will have *minimum* perimeter if it is a square.
$A = xy = K;\ y = \frac{K}{x}$
$P = 2x + \frac{2K}{x};\ \frac{dP}{dx} = 2 - \frac{2K}{x^2};\ \frac{d^2P}{dx^2} = \frac{4K}{x^3}$
$\frac{dP}{dx} = 0$ and $\frac{d^2P}{dx^2} > 0$
when $x = \sqrt{K}, y = \sqrt{K}$.

25. True: If $f(x_1) < f(x_2)$ and $g(x_1) < g(x_2)$ for $x_1 < x_2$,
$f(x_1) + g(x_1) < f(x_2) + g(x_2)$, so $f + g$ is increasing.

27. True: Since $f''(x) > 0$, $f'(x)$ is increasing for $x \ge 0$. Therefore, $f'(x) > 0$ for x in $[0, \infty)$, so $f(x)$ is increasing.

29. True: If the function is nondecreasing, $f'(x)$ must be greater than or equal to zero, and if $f'(x) \ge 0$, f is nondecreasing. This can be seen using the Mean Value Theorem.

31. False: For example, let $f(x) = e^x$.
$\lim_{x \to -\infty} e^x = 0$, so $y = 0$ is a horizontal asymptote.

33. True: $f'(x) = 3ax^2 + 2bx + c;\ f'(x) = 0$
when $x = \frac{-b \pm \sqrt{b^2 - 3ac}}{3a}$ by the Quadratic Formula. $f''(x) = 6ax + 2b$
so
$f''\left(\frac{-b \pm \sqrt{b^2 - 3ac}}{3a}\right) = \pm 2\sqrt{b^2 - 3ac}.$
Thus, if $b^2 - 3ac > 0$, one critical point is a local maximum and the other is a local minimum.
(If $b^2 - 3ac = 0$ the only critical point is an inflection point while if $b^2 - 3ac < 0$ there are no critical points.)
On an open interval, no local maxima can come from endpoints, so there can be at most one local maximum in an open interval.

35. True: Intermediate Value Theorem

37. False: $x_{n+1} = x_n - \frac{f(x_n)}{f'(x_n)} = -2x_n.$

39. True: From the Fixed-point Theorem, if g is continuous on $[a,b]$ and $a \le g(x) \le b$ whenever $a \le x \le b$, then there is at least one fixed point on $[a,b]$. The given conditions satisfy these criteria.

41. True: Theorem 3.8.C

43. True: $(-\sin x)^2 = \sin^2 x = 1 - \cos^2 x$

45. False: $f(x) = x^2 + 2x + 1$ and $g(x) = x^2 + 7x - 5$ are a counter-example.

47. True: At any given height, speed on the downward trip is the negative of speed on the upward.

Sample Test Problems

1. $f'(x) = 2x - 2;\ 2x - 2 = 0$ when $x = 1$.
Critical points: 0, 1, 4
$f(0) = 0, f(1) = -1, f(4) = 8$
Global minimum $f(1) = -1$;
global maximum $f(4) = 8$

3. $f'(z) = -\frac{2}{z^3}; -\frac{2}{z^3}$ is never 0.

Critical points: $-2, -\frac{1}{2}$

$f(-2) = \frac{1}{4}, f\left(-\frac{1}{2}\right) = 4$

Global minimum $f(-2) = \frac{1}{4}$;

global maximum $f\left(-\frac{1}{2}\right) = 4$.

5. $f'(x) = \frac{x}{|x|}; f'(x)$ does not exist at $x = 0$.

Critical points: $-\frac{1}{2}, 0, 1$

$f\left(-\frac{1}{2}\right) = \frac{1}{2}, f(0) = 0, f(1) = 1$

Global minimum $f(0) = 0$;
global maximum $f(1) = 1$

7. $f'(x) = 12x^3 - 12x^2 = 12x^2(x-1); f'(x) = 0$ when $x = 0, 1$
Critical points: $-2, 0, 1, 3$
$f(-2) = 80, f(0) = 0, f(1) = -1, f(3) = 135$
Global minimum $f(1) = -1$;
global maximum $f(3) = 135$

9. $f'(x) = 10x^4 - 20x^3 = 10x^3(x-2)$;
$f'(x) = 0$ when $x = 0, 2$
Critical points: $-1, 0, 2, 3$
$f(-1) = 0, f(0) = 7, f(2) = -9, f(3) = 88$
Global minimum $f(2) = -9$;
global maximum $f(3) = 88$

11. $f'(\theta) = \cos\theta; f'(\theta) = 0$ when $\theta = \frac{\pi}{2}$ in $\left[\frac{\pi}{4}, \frac{4\pi}{3}\right]$

Critical points: $\frac{\pi}{4}, \frac{\pi}{2}, \frac{4\pi}{3}$

$f\left(\frac{\pi}{4}\right) = \frac{1}{\sqrt{2}} \approx 0.71, f\left(\frac{\pi}{2}\right) = 1,$

$f\left(\frac{4\pi}{3}\right) = -\frac{\sqrt{3}}{2} \approx -0.87$

Global minimum $f\left(\frac{4\pi}{3}\right) \approx -0.87$;

global maximum $f\left(\frac{\pi}{2}\right) = 1$

13. $f'(x) = 3 - 2x; f'(x) > 0$ when $x < \frac{3}{2}$.
$f''(x) = -2; f''(x)$ is always negative.
$f(x)$ is increasing on $\left(-\infty, \frac{3}{2}\right]$ and concave down on $(-\infty, \infty)$.

15. $f'(x) = 3x^2 - 3 = 3(x^2 - 1); f'(x) > 0$ when $x < -1$ or $x > 1$.
$f''(x) = 6x; f''(x) < 0$ when $x < 0$.
$f(x)$ is increasing on $(-\infty, -1] \cup [1, \infty)$ and concave down on $(-\infty, 0)$.

17. $f'(x) = 4x^3 - 20x^4 = 4x^3(1 - 5x); f'(x) > 0$ when $0 < x < \frac{1}{5}$.

$f''(x) = 12x^2 - 80x^3 = 4x^2(3 - 20x); f''(x) < 0$ when $x > \frac{3}{20}$.

$f(x)$ is increasing on $\left[0, \frac{1}{5}\right]$ and concave down on $\left(\frac{3}{20}, \infty\right)$.

19. $f'(x) = 3x^2 - 4x^3 = x^2(3-4x)$; $f'(x) > 0$ when $x < \frac{3}{4}$.

$f''(x) = 6x - 12x^2 = 6x(1-2x)$; $f''(x) < 0$ when $x < 0$ or $x > \frac{1}{2}$.

$f(x)$ is increasing on $\left(-\infty, \frac{3}{4}\right]$ and concave down on $(-\infty, 0) \cup \left(\frac{1}{2}, \infty\right)$.

21. $f'(x) = 2x(x-4) + x^2 = 3x^2 - 8x = x(3x-8)$;

$f'(x) > 0$ when $x < 0$ or $x > \frac{8}{3}$

$f(x)$ is increasing on $(-\infty, 0] \cup \left[\frac{8}{3}, \infty\right)$ and decreasing on $\left[0, \frac{8}{3}\right]$

Local minimum $f\left(\frac{8}{3}\right) = -\frac{256}{27} \approx -9.48$;
local maximum $f(0) = 0$

$f''(x) = 6x - 8$; $f''(x) > 0$ when $x > \frac{4}{3}$.

$f(x)$ is concave up on $\left(\frac{4}{3}, \infty\right)$ and concave down on $\left(-\infty, \frac{4}{3}\right)$; inflection point $\left(\frac{4}{3}, -\frac{128}{27}\right)$

23. $f'(x) = 4x^3 - 2$; $f'(x) = 0$ when $x = \frac{1}{\sqrt[3]{2}}$.

$f''(x) = 12x^2$; $f''(x) = 0$ when $x = 0$.

$f''\left(\frac{1}{\sqrt[3]{2}}\right) = \frac{12}{2^{2/3}} > 0$, so

$f\left(\frac{1}{\sqrt[3]{2}}\right) = \frac{1}{2^{4/3}} - \frac{2}{2^{1/3}} = -\frac{3}{2^{4/3}}$ is a global minimum.

$f''(x) > 0$ for all $x \neq 0$; no inflection points
No horizontal or vertical asymptotes

25. $f'(x) = \frac{3x-6}{2\sqrt{x-3}}$; $f'(x) = 0$ when $x = 2$, but $x = 2$ is not in the domain of $f(x)$. $f'(x)$ does not exist when $x = 3$.

$f''(x) = \frac{3(x-4)}{4(x-3)^{3/2}}$; $f''(x) = 0$ when $x = 4$.

Global minimum $f(3) = 0$; no local maxima
Inflection point (4, 4)
No horizontal or vertical asymptotes.

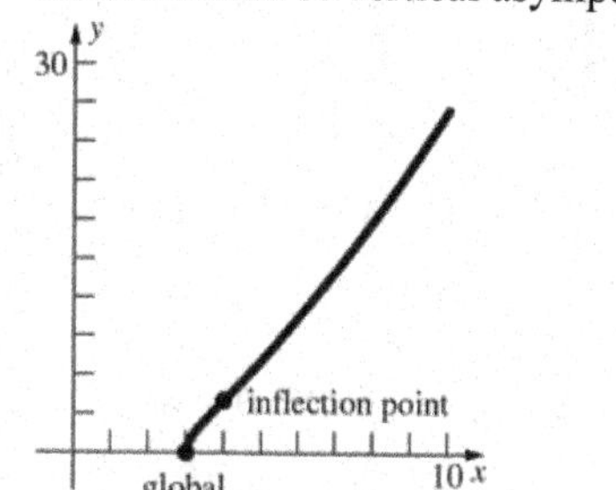

27. $f'(x) = 12x^3 - 12x^2 = 12x^2(x-1)$; $f'(x) = 0$ when $x = 0, 1$.

$f''(x) = 36x^2 - 24x = 12x(3x-2)$; $f''(x) = 0$ when $x = 0, \frac{2}{3}$.

$f''(1) = 12$, so $f(1) = -1$ is a minimum.
Global minimum $f(1) = -1$; no local maxima
Inflection points (0, 0), $\left(\frac{2}{3}, -\frac{16}{27}\right)$
No horizontal or vertical asymptotes.

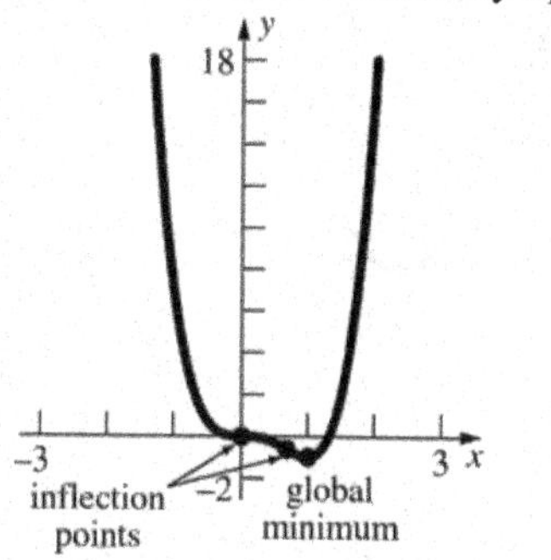

29. $f'(x) = 3 + \frac{1}{x^2}$; $f'(x) > 0$ for all $x \neq 0$.

$f''(x) = -\frac{2}{x^3}$; $f''(x) > 0$ when $x < 0$ and

$f''(x) < 0$ when $x > 0$

No local minima or maxima

No inflection points

$f(x) = 3x - \frac{1}{x}$, so

$\lim_{x\to\infty} [f(x) - 3x] = \lim_{x\to\infty} \left(-\frac{1}{x}\right) = 0$ and $y = 3x$ is an oblique asymptote.

Vertical asymptote $x = 0$

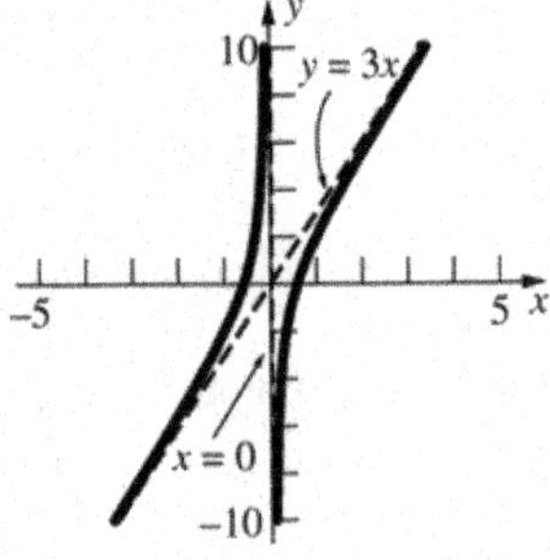

31. $f'(x) = -\sin x - \cos x$; $f'(x) = 0$ when

$x = -\frac{\pi}{4}, \frac{3\pi}{4}$.

$f''(x) = -\cos x + \sin x$; $f''(x) = 0$ when

$x = -\frac{3\pi}{4}, \frac{\pi}{4}$.

$f''\left(-\frac{\pi}{4}\right) = -\sqrt{2}$, $f''\left(\frac{3\pi}{4}\right) = \sqrt{2}$

Global minimum $f\left(\frac{3\pi}{4}\right) = -\sqrt{2}$;

global maximum $f\left(-\frac{\pi}{4}\right) = \sqrt{2}$

Inflection points $\left(-\frac{3\pi}{4}, 0\right), \left(\frac{\pi}{4}, 0\right)$

33. $f'(x) = x\sec^2 x + \tan x$; $f'(x) = 0$ when $x = 0$

$f''(x) = 2\sec^2 x(1 + x\tan x)$; $f''(x)$ is never 0 on $\left(-\frac{\pi}{2}, \frac{\pi}{2}\right)$.

$f''(0) = 2$; Global minimum $f(0) = 0$

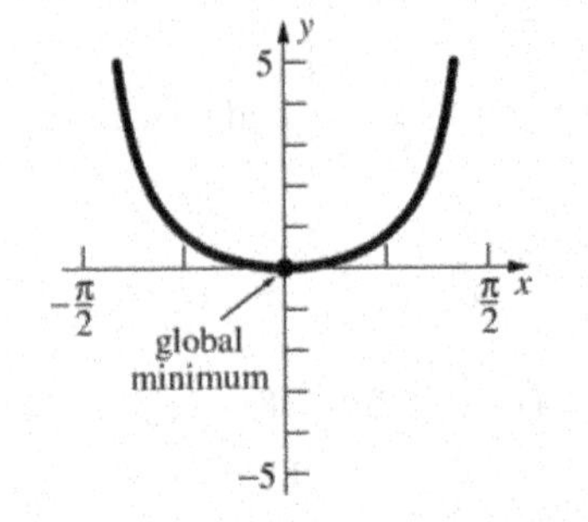

35. $f'(x) = \cos x - 2\cos x \sin x = \cos x(1 - 2\sin x)$;

$f'(x) = 0$ when $x = -\frac{\pi}{2}, \frac{\pi}{6}, \frac{\pi}{2}, \frac{5\pi}{6}$

$f''(x) = -\sin x + 2\sin^2 x - 2\cos^2 x$; $f''(x) = 0$ when $x \approx -2.51, -0.63, 1.00, 2.14$

$f''\left(-\frac{\pi}{2}\right) = 3$, $f''\left(\frac{\pi}{6}\right) = -\frac{3}{2}$, $f''\left(\frac{\pi}{2}\right) = 1$,

$f''\left(\frac{5\pi}{6}\right) = -\frac{3}{2}$

Global minimum $f\left(-\frac{\pi}{2}\right) = -2$,

local minimum $f\left(\frac{\pi}{2}\right) = 0$;

global maxima $f\left(\frac{\pi}{6}\right) = \frac{1}{4}$, $f\left(\frac{5\pi}{6}\right) = \frac{1}{4}$

Inflection points (–2.51, –0.94), (–0.63, –0.94), (1.00, 0.13), (2.14, 0.13)

37.

39.

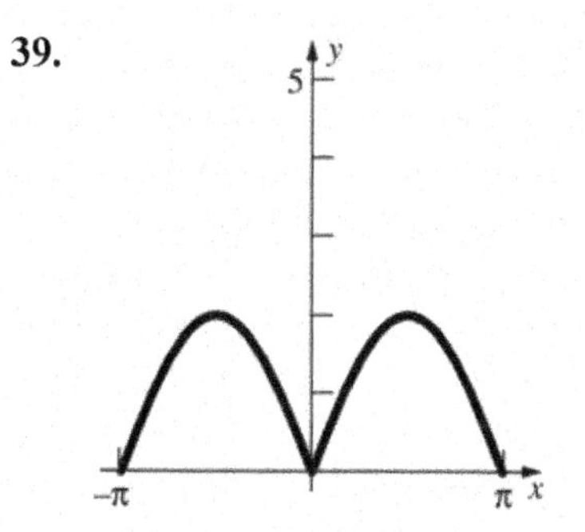

41. Let p be the length of the plank and let x be the distance from the fence to where the plank touches the ground.
See the figure below.

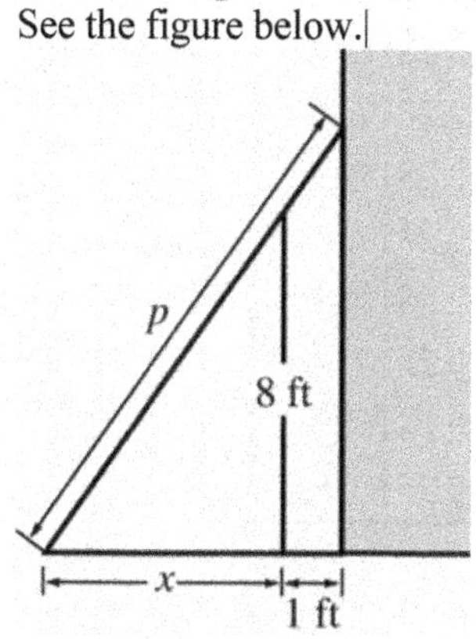

By properties of similar triangles,

$$\frac{p}{x+1} = \frac{\sqrt{x^2+64}}{x}$$

$$p = \left(1+\frac{1}{x}\right)\sqrt{x^2+64}$$

Minimize p:

$$\frac{dp}{dx} = -\frac{1}{x^2}\sqrt{x^2+64} + \left(1+\frac{1}{x}\right)\frac{x}{\sqrt{x^2+64}}$$

$$= \frac{1}{x^2\sqrt{x^2+64}}\left(-(x^2+64)+\left(1+\frac{1}{x}\right)x^3\right)$$

$$= \frac{x^3-64}{x^2\sqrt{x^2+64}}$$

$$\frac{x^3-64}{x^2\sqrt{x^2+64}} = 0; x = 4$$

$$\frac{dp}{dx} < 0 \text{ if } x < 4, \frac{dp}{dx} > 0 \text{ if } x > 4$$

When $x = 4$, $p = \left(1+\frac{1}{4}\right)\sqrt{16+64} \approx 11.18$ ft.

43. $\frac{1}{2}\pi r^2 h = 128\pi$

$$h = \frac{256}{r^2}$$

Let S be the surface area of the trough.

$$S = \pi r^2 + \pi r h = \pi r^2 + \frac{256\pi}{r}$$

$$\frac{dS}{dr} = 2\pi r - \frac{256\pi}{r^2}$$

$$2\pi r - \frac{256\pi}{r^2} = 0; r^3 = 128, r = 4\sqrt[3]{2}$$

Since $\frac{d^2S}{dr^2} > 0$ when $r = 4\sqrt[3]{2}$, $r = 4\sqrt[3]{2}$ minimizes S.

$$h = \frac{256}{\left(4\sqrt[3]{2}\right)^2} = 8\sqrt[3]{2}$$

45. a. $f'(x) = x^2$

$$\frac{f(3)-f(-3)}{3-(-3)} = \frac{9+9}{6} = 3$$

$$c^2 = 3; c = -\sqrt{3}, \sqrt{3}$$

b. The Mean Value Theorem does not apply because $F'(0)$ does not exist.

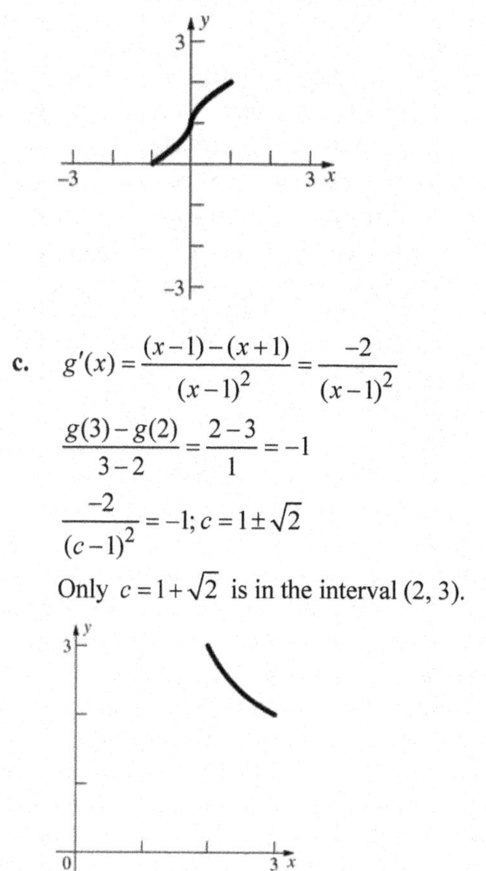

c. $g'(x) = \frac{(x-1)-(x+1)}{(x-1)^2} = \frac{-2}{(x-1)^2}$

$$\frac{g(3)-g(2)}{3-2} = \frac{2-3}{1} = -1$$

$$\frac{-2}{(c-1)^2} = -1; c = 1 \pm \sqrt{2}$$

Only $c = 1+\sqrt{2}$ is in the interval (2, 3).

47.

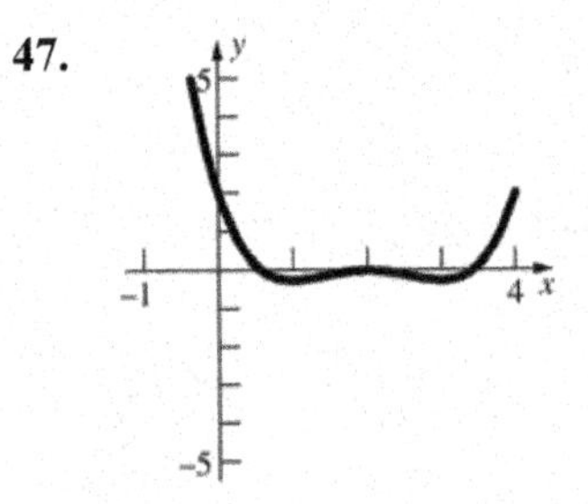

49. Let $f(x) = 3x - \cos 2x$; $a_1 = 0$, $b_1 = 1$.

$f(0) = -1$; $f(1) \approx 3.4161468$

n	h_n	m_n	$f(m_n)$
1	0.5	0.5	0.9596977
2	0.25	0.25	−0.1275826
3	0.125	0.375	0.3933111
4	0.0625	0.3125	0.1265369
5	0.03125	0.28125	−0.0021745
6	0.015625	0.296875	0.0617765
7	0.0078125	0.2890625	0.0296988
8	0.0039063	0.2851563	0.0137364
9	0.0019532	0.2832031	0.0057745
10	0.0009766	0.2822266	0.0017984
11	0.0004883	0.2817383	−0.0001884
12	0.0002442	0.2819824	0.0008049
13	0.0001221	0.2818604	0.0003082
14	0.0000611	0.2817994	0.0000600
15	0.0000306	0.2817689	−0.0000641
16	0.0000153	0.2817842	−0.0000018
17	0.0000077	0.2817918	0.0000293
18	0.0000039	0.2817880	0.0000138
19	0.0000020	0.2817861	0.0000061
20	0.0000010	0.2817852	0.0000022
21	0.0000005	0.2817847	0.0000004
22	0.0000003	0.2817845	−0.0000006
23	0.0000002	0.2817846	−0.0000000

$x \approx 0.281785$

51. $x_{n+1} = \dfrac{\cos 2x_n}{3}$

n	x_n
1	0.5
2	0.18010
3	0.311942
4	0.270539
5	0.285718
6	0.280375
7	0.282285
8	0.281606
9	0.281848
10	0.281762
11	0.281793
12	0.281782
13	0.281786
14	0.281784
15	0.281785
16	0.281785

$x \approx 0.2818$

53. $\int \left(x^3 - 3x^2 + 3\sqrt{x}\right) dx$

$= \int \left(x^3 - 3x^2 + 3x^{1/2}\right) dx$

$= \frac{1}{4}x^4 - x^3 + 3 \cdot \frac{2}{3}x^{3/2} + C$

$= \frac{1}{4}x^4 - x^3 + 2x^{3/2} + C$

55. $\int \frac{y^3 - 9y\sin y + 26y^{-1}}{y} dy$

$= \int \left(y^2 - 9\sin y + 26\right) dy$

$= \frac{1}{3}y^3 + 9\cos y + 26y + C$

57. Let $u = 2z^2 - 3$; then $du = 4z\,dz$ or $\frac{1}{4}du = z\,dz$.

$\int z\left(2z^2 - 3\right)^{1/3} dz = \int u^{1/3} \cdot \frac{1}{4} du$

$= \frac{1}{4} \int u^{1/3}\, du$

$= \frac{1}{4} \cdot \frac{3}{4} u^{4/3} + C$

$= \frac{3}{16}\left(2z^2 - 3\right)^{4/3} + C$

59. $u = \tan(3x^2 + 6x), du = (6x+6)\sec^2(3x^2 + 6x)$

$$\int (x+1)\tan^2\left(3x^2+6x\right)\sec^2\left(3x^2+6x\right)dx$$

$$= \frac{1}{6}\int u^2 du = \frac{1}{18}u^3 + C$$

$$= \frac{1}{18}\tan^3\left(3x^2+6x\right) + C$$

61. Let $u = t^5 + 5$; then $du = 5t^4 dt$ or $\frac{1}{5}du = t^4 dt$.

$$\int t^4\left(t^5+5\right)^{2/3} dt = \int \frac{1}{5}u^{2/3}du$$

$$= \frac{1}{5}\int u^{2/3}du$$

$$= \frac{1}{5}\cdot\frac{3}{5}u^{5/3} + C$$

$$= \frac{3}{25}\left(t^5+5\right)^{5/3} + C$$

63. Let $u = x^3 + 9$; then $du = 3x^2 dx$ or $\frac{1}{3}du = x^2 dx$.

$$\int \frac{x^2}{\sqrt{x^3+9}}dx = \frac{1}{3}\int \frac{du}{\sqrt{u}}$$

$$= \frac{1}{3}\int u^{-1/2}du$$

$$= \frac{1}{3}\cdot 2u^{1/2} + C$$

$$= \frac{2}{3}\sqrt{x^3+9} + C$$

65. Let $u = 2y - 1$; then $du = 2dy$.

$$\int \frac{2}{(2y-1)^3}dy = \int \frac{du}{u^3}$$

$$= \int u^{-3}du$$

$$= -\frac{1}{2}u^{-2} + C$$

$$= -\frac{1}{2(2y-1)^2} + C$$

67. $u = 2y^3 + 3y^2 + 6y, du = (6y^2 + 6y + 6)\,dy$

$$\frac{1}{6}\int u^{-1/5}du = \frac{5}{24}(2y^3+3y^2+6y)^{4/5} + C$$

69. $\int dy = \int \frac{1}{\sqrt{x+1}}dx$

$$y = 2\sqrt{x+1} + C$$

$$y = 2\sqrt{x+1} + 14$$

71. $\int dy = \int \sqrt{2t-1}\,dt$

$$y = \frac{1}{3}(2t-1)^{3/2} + C$$

$$y = \frac{1}{3}(2t-1)^{3/2} - 1$$

73. $\int 2y\,dy = \int (6x - x^3)dx$

$$y^2 = 3x^2 - \frac{1}{4}x^4 + C$$

$$y^2 = 3x^2 - \frac{1}{4}x^4 + 9$$

$$y = \sqrt{3x^2 - \frac{1}{4}x^4 + 9}$$

75. $s(t) = -16t^2 + 48t + 448$; $s = 0$ at $t = 7$;

$v(t) = s'(t) = -32t + 48$

when $t = 7$, $v = -32(7) + 48 = -176$ ft/s

Review and Preview Problems

1. $A_{\text{region}} = \frac{1}{2}bh = \frac{1}{2}aa\sin 60^\circ = \frac{\sqrt{3}}{4}a^2$

3. $A_{\text{region}} = 10\left(\frac{1}{2}\text{base}\times\text{height}\right) = 5\frac{a^2}{4}\cot 36^\circ$

$$= \frac{5}{4}a^2\cot 36^\circ$$

5. $A_{\text{region}} = A_{\text{rect}} + A_{\text{semic.}} = 3.6\cdot 5.8 + \frac{1}{2}\pi(1.8)^2$

$$\approx 25.97$$

7. $A_{\text{region}} = 0.5(1 + 1.5 + 2 + 2.5) = 3.5$

9. $A_{\text{region}} = A_{\text{rect}} + A_{\text{tri}} = 1x + \frac{1}{2}x\cdot x = \frac{1}{2}x^2 + x$

11. $y = 5 - x$; $A_{\text{region}} = A_{\text{rect}} + A_{\text{tri}}$

$$= 2(2) + \frac{1}{2}(2)(2) = 6$$

CHAPTER 4	The Definite Integral

4.1 Concepts Review

1. $2 \cdot \frac{5(6)}{2} = 30; 2(5) = 10$

3. inscribed; circumscribed

Problem Set 4.1

1. $\sum_{k=1}^{6}(k-1) = \sum_{k=1}^{6} k - \sum_{k=1}^{6} 1 = \frac{6(7)}{2} - 6(1) = 15$

3. $\sum_{k=1}^{7} \frac{1}{k+1} = \frac{1}{1+1} + \frac{1}{2+1} + \frac{1}{3+1}$
$+\frac{1}{4+1} + \frac{1}{5+1} + \frac{1}{6+1} + \frac{1}{7+1}$
$= \frac{1}{2} + \frac{1}{3} + \frac{1}{4} + \frac{1}{5} + \frac{1}{6} + \frac{1}{7} + \frac{1}{8} = \frac{1443}{840} = \frac{481}{280}$

5. $\sum_{m=1}^{8}(-1)^m 2^{m-2}$
$= (-1)^1 2^{-1} + (-1)^2 2^0 + (-1)^3 2^1$
$+(-1)^4 2^2 + (-1)^5 2^3 + (-1)^6 2^4$
$+(-1)^7 2^5 + (-1)^8 2^6$
$= -\frac{1}{2} + 1 - 2 + 4 - 8 + 16 - 32 + 64 = \frac{85}{2}$

7. $\sum_{n=1}^{6} n\cos(n\pi) = \sum_{n=1}^{6}(-1)^n \cdot n$
$= -1 + 2 - 3 + 4 - 5 + 6 = 3$

9. $1 + 2 + 3 + \cdots + 41 = \sum_{i=1}^{41} i$

11. $1 + \frac{1}{2} + \frac{1}{3} + \cdots + \frac{1}{100} = \sum_{i=1}^{100} \frac{1}{i}$

13. $a_1 + a_3 + a_5 + a_7 + \cdots + a_{99} = \sum_{i=1}^{50} a_{2i-1}$

15. $\sum_{i=1}^{10}(a_i + b_i) = \sum_{i=1}^{10} a_i + \sum_{i=1}^{10} b_i = 40 + 50 = 90$

17. $\sum_{p=0}^{9}(a_{p+1} - b_{p+1}) = \sum_{p=1}^{10} a_p - \sum_{p=1}^{10} b_p = 40 - 50 = -10$

19. $\sum_{i=1}^{100}(3i-2) = 3\sum_{i=1}^{100} i - \sum_{i=1}^{100} 2 = 3(5050) - 2(100)$
$= 14,950$

21. $\sum_{k=1}^{10}(k^3 - k^2) = \sum_{k=1}^{10} k^3 - \sum_{k=1}^{10} k^2 = 3025 - 385$
$= 2640$

23. $\sum_{i=1}^{n}(2i^2 - 3i + 1) = 2\sum_{i=1}^{n} i^2 - 3\sum_{i=1}^{n} i + \sum_{i=1}^{n} 1$
$= \frac{2n(n+1)(2n+1)}{6} - \frac{3n(n+1)}{2} + n$
$= \frac{2n^3 + 3n^2 + n}{3} - \frac{3n^2 + 3n}{2} + n = \frac{4n^3 - 3n^2 - n}{6}$

25. $S = 1 + 2 + 3 + \cdots + (n-2) + (n-1) + n$
$+S = n + (n-1) + (n-2) + \cdots + 3 + 2 + 1$
$2S = (n+1) + (n+1) + (n+1) + \cdots + (n+1) + (n+1) + (n+1)$
$2S = n(n+1)$
$S = \frac{n(n+1)}{2}$

27. a. $\sum_{k=0}^{10}\left(\frac{1}{2}\right)^k = \frac{1-\left(\frac{1}{2}\right)^{11}}{\frac{1}{2}} = 2 - \left(\frac{1}{2}\right)^{10}$, so
$\sum_{k=1}^{10}\left(\frac{1}{2}\right)^k = 1 - \left(\frac{1}{2}\right)^{10} = \frac{1023}{1024}$.

b. $\sum_{k=0}^{10} 2^k = \frac{1-2^{11}}{-1} = 2^{11} - 1$, so
$\sum_{k=1}^{10} 2^k = 2^{11} - 2 = 2046$.

29. $$(i+1)^3 - i^3 = 3i^2 + 3i + 1$$

$$\sum_{i=1}^{n}\left[(i+1)^3 - i^3\right] = \sum_{i=1}^{n}\left(3i^2 + 3i + 1\right)$$

$$(n+1)^3 - 1^3 = 3\sum_{i=1}^{n} i^2 + 3\sum_{i=1}^{n} i + \sum_{i=1}^{n} 1$$

$$n^3 + 3n^2 + 3n = 3\sum_{i=1}^{n} i^2 + 3\frac{n(n+1)}{2} + n$$

$$2n^3 + 6n^2 + 6n = 6\sum_{i=1}^{n} i^2 + 3n^2 + 3n + 2n$$

$$\frac{2n^3 + 3n^2 + n}{6} = \sum_{i=1}^{n} i^2$$

$$\frac{n(n+1)(2n+1)}{6} = \sum_{i=1}^{n} i^2$$

31. $$(i+1)^5 - i^5 = 5i^4 + 10i^3 + 10i^2 + 5i + 1$$

$$\sum_{i=1}^{n}\left[(i+1)^5 - i^5\right] = 5\sum_{i=1}^{n} i^4 + 10\sum_{i=1}^{n} i^3 + 10\sum_{i=1}^{n} i^2 + 5\sum_{i=1}^{n} i + \sum_{i=1}^{n} 1$$

$$(n+1)^5 - 1^5 = 5\sum_{i=1}^{n} i^4 + 10\frac{n^2(n+1)^2}{4} + 10\frac{n(n+1)(2n+1)}{6} + 5\frac{n(n+1)}{2} + n$$

$$n^5 + 5n^4 + 10n^3 + 10n^2 + 5n = 5\sum_{i=1}^{n} i^4 + \tfrac{5}{2}n^2(n+1)^2 + \tfrac{10}{6}n(n+1)(2n+1) + \frac{5}{2}n(n+1) + n$$

Solving for $\sum_{i=1}^{n} i^4$ yields

$$\sum_{i=1}^{n} i^4 = \tfrac{1}{5}\left[n^5 + \tfrac{5}{2}n^4 + \tfrac{5}{3}n^3 - \tfrac{1}{6}n\right]$$

$$= \frac{n(n+1)(2n+1)(3n^2 + 3n - 1)}{30}$$

33. $$\overline{x} = \frac{1}{7}(2+5+7+8+9+10+14) = \frac{55}{7} \approx 7.86$$

$$s^2 = \frac{1}{7}\left[\left(2-\frac{55}{7}\right)^2 + \left(5-\frac{55}{7}\right)^2 + \left(7-\frac{55}{7}\right)^2 + \left(8-\frac{55}{7}\right)^2 + \left(9-\frac{55}{7}\right)^2 + \left(10-\frac{55}{7}\right)^2 + \left(14-\frac{55}{7}\right)^2\right]$$

$$= \frac{608}{49} \approx 12.4$$

35. **a.** $\sum_{i=1}^{n}(x_i - \overline{x}) = \sum_{i=1}^{n} x_i - \sum_{i=1}^{n} \overline{x} = n\overline{x} - n\overline{x} = 0$

b. $s^2 = \frac{1}{n}\sum_{i=1}^{n}(x_i - \overline{x})^2 = \frac{1}{n}\sum_{i=1}^{n}(x_i^2 - 2\overline{x}\,x_i + \overline{x}^2)$

$= \frac{1}{n}\sum_{i=1}^{n} x_i^2 - \frac{2\overline{x}}{n}\sum_{i=1}^{n} x_i + \frac{1}{n}\sum_{i=1}^{n}\overline{x}^2$

$= \frac{1}{n}\sum_{i=1}^{n} x_i^2 - \frac{2\overline{x}}{n}(n\overline{x}) + \frac{1}{n}(n\overline{x}^2)$

$= \left(\frac{1}{n}\sum_{i=1}^{n} x_i^2\right) - 2\overline{x}^2 + \overline{x}^2 = \left(\frac{1}{n}\sum_{i=1}^{n} x_i^2\right) - \overline{x}^2$

37. Let $S(c) = \sum_{i=1}^{n}(x_i - c)^2$. Then

$S'(c) = \frac{d}{dc}\sum_{i=1}^{n}(x_i - c)^2$

$= \sum_{i=1}^{n}\frac{d}{dc}(x_i - c)^2$

$= \sum_{i=1}^{n} 2(x_i - c)(-1)$

$= -2\sum_{i=1}^{n} x_i + 2nc$

$S''(c) = 2n$

Set $S'(c) = 0$ and solve for c:

$-2\sum_{i=1}^{n} x_i + 2nc = 0$

$c = \frac{1}{n}\sum_{i=1}^{n} x_i = \overline{x}$

Since $S''(\overline{x}) = 2n > 0$ we know that $\overline{x}$ minimizes $S(c)$.

39. The bottom layer contains $10 \cdot 16 = 160$ oranges, the next layer contains $9 \cdot 15 = 135$ oranges, the third layer contains $8 \cdot 14 = 112$ oranges, and so on, up to the top layer, which contains $1 \cdot 7 = 7$ oranges. The stack contains $1 \cdot 7 + 2 \cdot 8 + \cdots + 9 \cdot 15 + 10 \cdot 16$

$= \sum_{i=1}^{10} i(6+i) = 715$ oranges.

41. For a general stack whose base is *m* rows of *n* oranges with $m \le n$, the stack contains

$$\sum_{i=1}^{m} i(n-m+i) = (n-m)\sum_{i=1}^{m} i + \sum_{i=1}^{m} i^2 = (n-m)\frac{m(m+1)}{2} + \frac{m(m+1)(2m+1)}{6} = \frac{m(m+1)(3n-m+1)}{6}$$

43. $A = \frac{1}{2}\left[1 + \frac{3}{2} + 2 + \frac{5}{2}\right] = \frac{7}{2}$

45. $A = \frac{1}{2}\left[\frac{3}{2} + 2 + \frac{5}{2} + 3\right] = \frac{9}{2}$

47. $A = \frac{1}{2}\left[\left(\frac{1}{2}\cdot 0^2 + 1\right) + \left(\frac{1}{2}\cdot\left(\frac{1}{2}\right)^2 + 1\right) + \left(\frac{1}{2}\cdot 1^2 + 1\right) + \left(\frac{1}{2}\cdot\left(\frac{3}{2}\right)^2 + 1\right)\right] = \frac{1}{2}\left(1 + \frac{9}{8} + \frac{3}{2} + \frac{17}{8}\right) = \frac{23}{8}$

49.

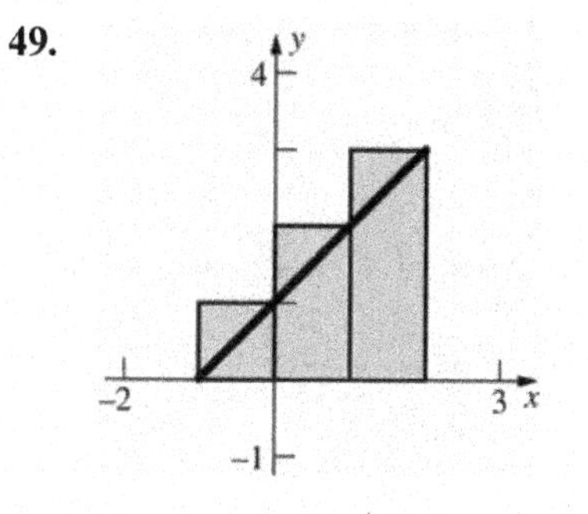

$A = 1(1 + 2 + 3) = 6$

51.

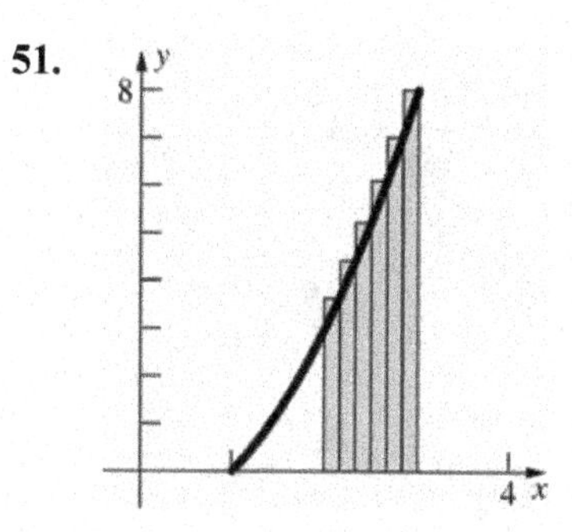

$$A = \frac{1}{6}\left[\left(\left(\frac{13}{6}\right)^2 - 1\right) + \left(\left(\frac{7}{3}\right)^2 - 1\right) + \left(\left(\frac{5}{2}\right)^2 - 1\right) + \left(\left(\frac{8}{3}\right)^2 - 1\right) + \left(\left(\frac{17}{6}\right)^2 - 1\right) + (3^2 - 1)\right]$$

$$= \frac{1}{6}\left(\frac{133}{36} + \frac{40}{9} + \frac{21}{4} + \frac{55}{9} + \frac{253}{36} + 8\right) = \frac{1243}{216}$$

53. $\Delta x = \frac{1}{n},\ x_i = \frac{i}{n}$

$f(x_i)\Delta x = \left(\frac{i}{n} + 2\right)\left(\frac{1}{n}\right) = \frac{i}{n^2} + \frac{2}{n}$

$A(S_n) = \left[\left(\frac{1}{n^2} + \frac{2}{n}\right) + \left(\frac{2}{n^2} + \frac{2}{n}\right) + \cdots + \left(\frac{n}{n^2} + \frac{2}{n}\right)\right] = \frac{1}{n^2}(1 + 2 + 3 + \cdots + n) + 2 = \frac{n(n+1)}{2n^2} + 2 = \frac{1}{2n} + \frac{5}{2}$

$\lim_{n\to\infty} A(S_n) = \lim_{n\to\infty}\left(\frac{1}{2n} + \frac{5}{2}\right) = \frac{5}{2}$

55. $\Delta x = \frac{2}{n}, x_i = -1 + \frac{2i}{n}$

$$f(x_i)\Delta x = \left[2\left(-1+\frac{2i}{n}\right)+2\right]\left(\frac{2}{n}\right) = \frac{8i}{n^2}$$

$$A(S_n) = \left[\left(\frac{8}{n^2}\right)+\left(\frac{16}{n^2}\right)+\cdots+\left(\frac{8n}{n^2}\right)\right]$$

$$= \frac{8}{n^2}(1+2+3+\cdots+n) = \frac{8}{n^2}\left[\frac{n(n+1)}{2}\right]$$

$$= 4\left[\frac{n^2+n}{n^2}\right] = 4+\frac{4}{n}$$

$$\lim_{n\to\infty} A(S_n) = \lim_{n\to\infty}\left(4+\frac{4}{n}\right) = 4$$

57. $\Delta x = \frac{1}{n}, x_i = \frac{i}{n}$

$$f(x_i)\Delta x = \left(\frac{i}{n}\right)^3\left(\frac{1}{n}\right) = \frac{i^3}{n^4}$$

$$A(S_n) = \left[\frac{1}{n^4}(1^3)+\frac{1}{n^4}(2^3)+\cdots+\frac{1}{n^4}(n^3)\right]$$

$$= \frac{1}{n^4}(1^3+2^3+\cdots+n^3) = \frac{1}{n^4}\left[\frac{n(n+1)}{2}\right]^2$$

$$= \frac{1}{n^4}\left[\frac{n^4+2n^3+n^2}{4}\right] = \frac{1}{4}\left[1+\frac{2}{n}+\frac{1}{n^2}\right]$$

$$\lim_{n\to\infty} A(S_n) = \lim_{n\to\infty}\frac{1}{4}\left[1+\frac{2}{n}+\frac{1}{n^2}\right] = \frac{1}{4}$$

59. $f(t_i)\Delta t = \left[\frac{i}{n}+2\right]\frac{1}{n} = \frac{i}{n^2}+\frac{2}{n}$

$$A(S_n) = \sum_{i=1}^{n}\left(\frac{i}{n^2}+\frac{2}{n}\right) = \frac{1}{n^2}\sum_{i=1}^{n} i + \sum_{i=1}^{n}\frac{2}{n}$$

$$= \frac{1}{n^2}\left[\frac{n(n+1)}{2}\right]+2$$

$$= \left[\frac{n^2+n}{2n^2}\right]+2$$

$$= \left(\frac{1}{2}+\frac{1}{2n}\right)+2$$

$$\lim_{n\to\infty} A(S_n) = \frac{1}{2}+2 = \frac{5}{2}$$

The object traveled $2\frac{1}{2}$ ft.

61. a. $f(x_i)\Delta x = \left(\frac{ib}{n}\right)^2\left(\frac{b}{n}\right) = \frac{b^3 i^2}{n^3}$

$$A_0^b = \frac{b^3}{n^3}\sum_{i=1}^{n} i^2 = \frac{b^3}{n^3}\left[\frac{n(n+1)(2n+1)}{6}\right]$$

$$= \frac{b^3}{6}\left[2+\frac{3}{n}+\frac{1}{n^2}\right]$$

$$\lim_{n\to\infty} A_0^b = \frac{2b^3}{6} = \frac{b^3}{3}$$

b. Since $a \ge 0$, $A_0^b = A_0^a + A_a^b$, or

$$A_a^b = A_0^b - A_0^a = \frac{b^3}{3} - \frac{a^3}{3}.$$

63. a. $A_0^5 = \frac{5^3}{3} = \frac{125}{3}$

b. $A_1^4 = \frac{4^3}{3} - \frac{1^3}{3} = \frac{63}{3} = 21$

c. $A_2^5 = \frac{5^3}{3} - \frac{2^3}{3} = \frac{117}{3} = 39$

65. a. $A_0^2(x^3) = \frac{2^{3+1}}{3+1} = 4$

b. $A_1^2(x^3) = \frac{2^{3+1}}{3+1} - \frac{1^{3+1}}{3+1} = 4 - \frac{1}{4} = \frac{15}{4}$

c. $A_1^2(x^5) = \frac{2^{5+1}}{5+1} - \frac{1^{5+1}}{5+1} = \frac{32}{3} - \frac{1}{6} = \frac{63}{6}$

$$= \frac{21}{2} = 10.5$$

d. $A_0^2(x^9) = \frac{2^{9+1}}{9+1} = \frac{1024}{10} = 102.4$

4.2 Concepts Review

1. Riemann sum

3. $A_{\text{up}} - A_{\text{down}}$

Problem Set 4.2

1. $R_P = f(2)(2.5-1) + f(3)(3.5-2.5) + f(4.5)(5-3.5) = 4(1.5) + 3(1) + (-2.25)(1.5) = 5.625$

3. $R_P = \sum_{i=1}^{5} f(\bar{x}_i)\Delta x_i = f(3)(3.75-3) + f(4)(4.25-3.75) + f(4.75)(5.5-4.25) + f(6)(6-5.5) + f(6.5)(7-6)$
$= 2(0.75) + 3(0.5) + 3.75(1.25) + 5(0.5) + 5.5(1) = 15.6875$

5. $R_P = \sum_{i=1}^{8} f(\bar{x}_i)\Delta x_i = [f(-1.75) + f(-1.25) + f(-0.75) + f(-0.25) + f(0.25) + f(0.75) + f(1.25) + f(1.75)](0.5)$
$= [-0.21875 - 0.46875 - 0.46875 - 0.21875 + 0.28125 + 1.03125 + 2.03125 + 3.28125](0.5) = 2.625$

7. $\int_1^3 x^3\, dx$

9. $\int_{-1}^1 \frac{x^2}{1+x}\, dx$

11. $\Delta x = \frac{2}{n}, \bar{x}_i = \frac{2i}{n}$

$f(\bar{x}_i) = \bar{x}_i + 1 = \frac{2i}{n} + 1$

$\sum_{i=1}^{n} f(\bar{x}_i)\Delta x = \sum_{i=1}^{n}\left[1 + i\left(\frac{2}{n}\right)\right]\frac{2}{n}$

$= \frac{2}{n}\sum_{i=1}^{n} 1 + \frac{4}{n^2}\sum_{i=1}^{n} i = \frac{2}{n}(n) + \frac{4}{n^2}\left[\frac{n(n+1)}{2}\right]$

$= 2 + 2\left(1 + \frac{1}{n}\right)$

$\int_0^2 (x+1)dx = \lim_{n\to\infty}\left[2 + 2\left(1 + \frac{1}{n}\right)\right] = 4$

13. $\Delta x = \frac{3}{n}, \bar{x}_i = -2 + \frac{3i}{n}$

$f(\bar{x}_i) = 2\left(-2 + \frac{3i}{n}\right) + \pi = \pi - 4 + \frac{6i}{n}$

$\sum_{i=1}^{n} f(\bar{x}_i)\Delta x = \sum_{i=1}^{n}\left[\pi - 4 + \frac{6i}{n}\right]\frac{3}{n}$

$= \frac{3}{n}\sum_{i=1}^{n}(\pi - 4) + \frac{18}{n^2}\sum_{i=1}^{n} i = 3(\pi - 4) + \frac{18}{n^2}\left[\frac{n(n+1)}{2}\right]$

$= 3\pi - 12 + 9\left(1 + \frac{1}{n}\right)$

$\int_{-2}^1 (2x + \pi)\, dx = \lim_{n\to\infty}\left[3\pi - 12 + 9\left(1 + \frac{1}{n}\right)\right]$
$= 3\pi - 3$

15. $\Delta x = \frac{5}{n}, \bar{x}_i = \frac{5i}{n}$

$f(\bar{x}_i) = 1 + \frac{5i}{n}$

$\sum_{i=1}^{n} f(\bar{x}_i)\Delta x = \sum_{i=1}^{n}\left[1 + i\left(\frac{5}{n}\right)\right]\frac{5}{n}$

$= \frac{5}{n}\sum_{i=1}^{n} 1 + \frac{25}{n^2}\sum_{i=1}^{n} i = 5 + \frac{25}{n^2}\left[\frac{n(n+1)}{2}\right]$

$= 5 + \frac{25}{2}\left(1 + \frac{1}{n}\right)$

$\int_0^5 (x+1)\, dx = \lim_{n\to\infty}\left[5 + \frac{25}{2}\left(1 + \frac{1}{n}\right)\right] = \frac{35}{2}$

17.

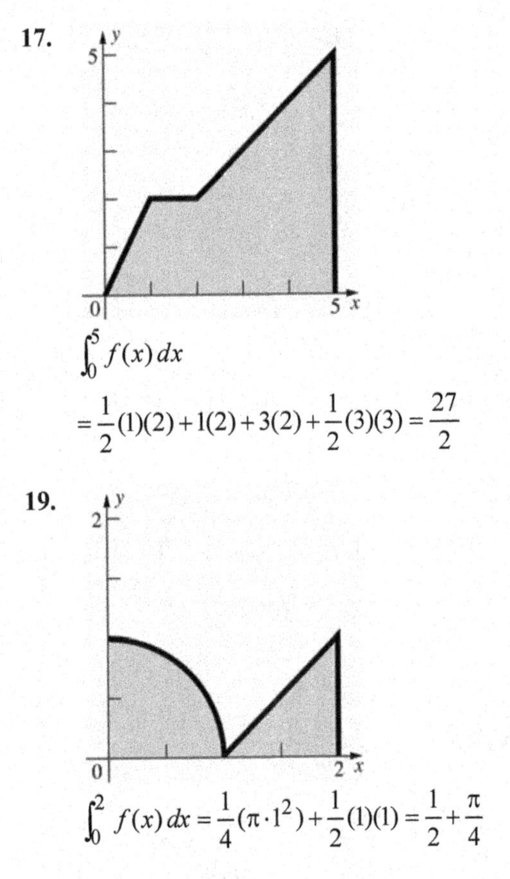

$\int_0^5 f(x)\,dx$

$= \frac{1}{2}(1)(2) + 1(2) + 3(2) + \frac{1}{2}(3)(3) = \frac{27}{2}$

19.

$\int_0^2 f(x)\,dx = \frac{1}{4}(\pi \cdot 1^2) + \frac{1}{2}(1)(1) = \frac{1}{2} + \frac{\pi}{4}$

21. The area under the curve is equal to the area of a semi-circle: $\int_{-A}^{A} \sqrt{A^2 - x^2}\,dx = \frac{1}{2}\pi A^2$.

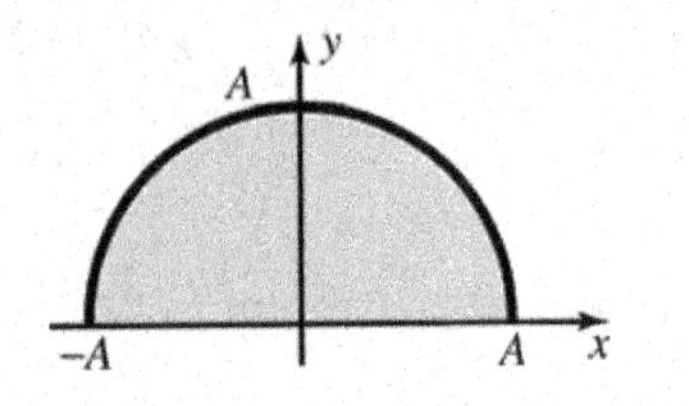

23. $s(4) = \int_0^4 v(t)\,dt = \frac{1}{2}4\left(\frac{4}{60}\right) = \frac{2}{15}$

25. $s(4) = \int_0^4 v(t)\,dt = \frac{1}{2}2(1) + 2(1) = 3$

27.

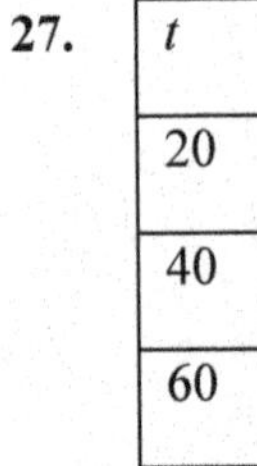

t	$s(t)$
20	40
40	80
60	120
80	160
100	200
120	240

29.

t	$s(t)$
20	20
40	80
60	160
80	240
100	320
120	400

31. a. $\int_{-3}^{3} [\![x]\!]\,dx = (-3-2-1+0+1+2)(1) = -3$

b. $\int_{-3}^{3} [\![x]\!]^2\,dx = [(-3)^2 + (-2)^2 + (-1)^2 + 0 + 1 + 4](1) = 19$

c. $\int_{-3}^{3} (x - [\![x]\!])\,dx = 6\left[\frac{1}{2}(1)(1)\right] = 3$

d. $\int_{-3}^{3} (x - [\![x]\!])^2\,dx = 6\int_0^1 x^2\,dx = 6 \cdot \frac{1^3}{3} = 2$

e. $\int_{-3}^{3} |x|\,dx = \frac{1}{2}(3)(3) + \frac{1}{2}(3)(3) = 9$

f. $\int_{-3}^{3} x|x|\,dx = \frac{(-3)^3}{3} + \frac{(3)^3}{3} = 0$

g. $\int_{-1}^{2} |x|[\![x]\!]\,dx = -\int_{-1}^{0} |x|\,dx + 0\int_0^1 |x|\,dx + \int_1^2 |x|\,dx$

$= -\frac{1}{2}(1)(1) + 1(1) + \frac{1}{2}(1)(1) = 1$

h. $\int_{-1}^{2} x^2[\![x]\!]\,dx = -\int_{-1}^{0} x^2\,dx + 0\int_0^1 x^2\,dx + \int_1^2 x^2\,dx$

$= -\frac{1^3}{3} + \left(\frac{2^3}{3} - \frac{1^3}{3}\right) = 2$

33. $R_P = \frac{1}{2}\sum_{i=1}^{n}(x_i + x_{i-1})(x_i - x_{i-1})$

$= \frac{1}{2}\sum_{i=1}^{n}\left(x_i^2 - x_{i-1}^2\right)$

$= \frac{1}{2}\Big[(x_1^2 - x_0^2) + (x_2^2 - x_1^2) + (x_3^2 - x_2^2) + \cdots + (x_n^2 - x_{n-1}^2)\Big]$

$= \frac{1}{2}(x_n^2 - x_0^2)$

$= \frac{1}{2}(b^2 - a^2)$

$\lim_{n\to\infty} \frac{1}{2}(b^2 - a^2) = \frac{1}{2}(b^2 - a^2)$

35. Left: $\int_0^2 (x^3 + 1)\,dx = 5.24$

Right: $\int_0^2 (x^3 + 1)\,dx = 6.84$

Midpoint: $\int_0^2 (x^3 + 1)\,dx = 5.98$

37. Left: $\int_0^1 \cos x\,dx \approx 0.8638$

Right: $\int_0^1 \cos x\,dx \approx 0.8178$

Midpoint: $\int_0^1 \cos x\,dx \approx 0.8418$

39. Partition [0, 1] into n regular intervals, so $\|P\| = \frac{1}{n}$.

If $\overline{x}_i = \frac{i}{n} + \frac{1}{2n}$, $f(\overline{x}_i) = 1$.

$\lim_{\|P\|\to 0} \sum_{i=1}^{n} f(\overline{x}_i)\Delta x_i = \lim_{n\to\infty} \sum_{i=1}^{n} \frac{1}{n} = 1$

If $\overline{x}_i = \frac{i}{n} + \frac{1}{\pi n}$, $f(\overline{x}_i) = 0$.

$\lim_{\|P\|\to 0} \sum_{i=1}^{n} f(\overline{x}_i)\Delta x_i = \lim_{n\to\infty} \sum_{i=1}^{n} 0 = 0$

Thus f is not integrable on [0, 1].

4.3 Concepts Review

1. $4(4 - 2) = 8$; $16(4 - 2) = 32$

3. $\int_1^4 f(x)\,dx$; $\int_2^5 \sqrt{x}\,dx$

Problem Set 4.3

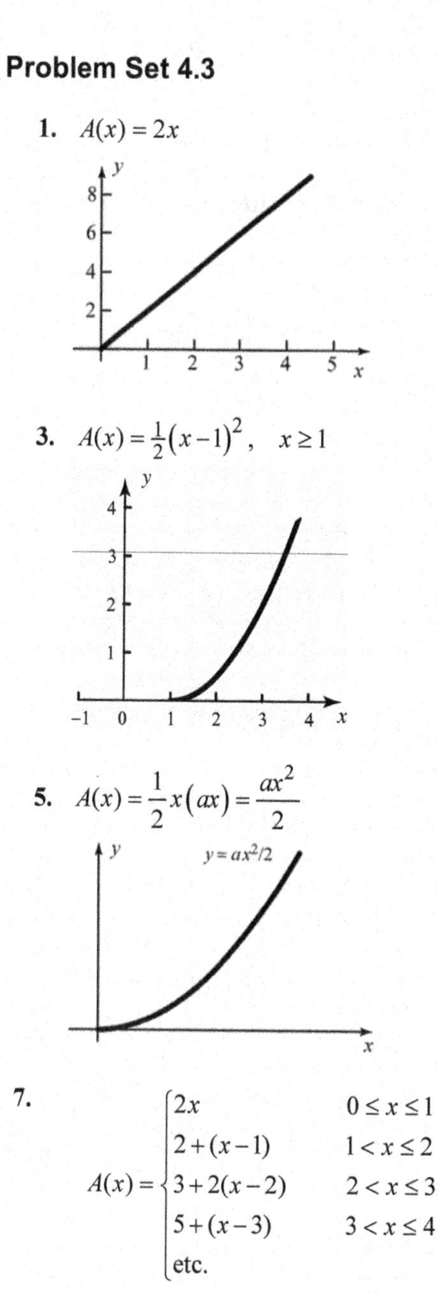

1. $A(x) = 2x$

3. $A(x) = \frac{1}{2}(x-1)^2,\quad x \ge 1$

5. $A(x) = \frac{1}{2}x(ax) = \frac{ax^2}{2}$

7.
$$A(x) = \begin{cases} 2x & 0 \le x \le 1 \\ 2 + (x-1) & 1 < x \le 2 \\ 3 + 2(x-2) & 2 < x \le 3 \\ 5 + (x-3) & 3 < x \le 4 \\ \text{etc.} & \end{cases}$$

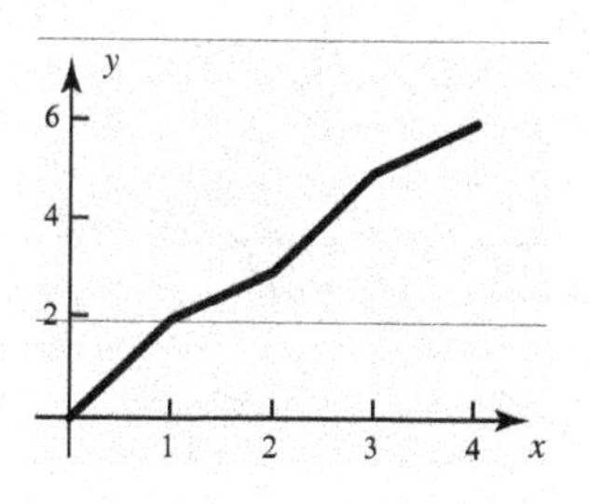

9. $\int_1^2 2f(x)\,dx = 2\int_1^2 f(x)\,dx = 2(3) = 6$

11. $\int_0^2 [2f(x)+g(x)]\,dx = 2\int_0^2 f(x)\,dx + \int_0^2 g(x)\,dx$

$= 2\left[\int_0^1 f(x)\,dx + \int_1^2 f(x)\,dx\right] + \int_0^2 g(x)\,dx$

$= 2(2 + 3) + 4 = 14$

13. $\int_2^1 [2f(s)+5g(s)]\,ds = -2\int_1^2 f(s)\,ds - 5\int_1^2 g(s)\,ds$

$= -2(3) - 5\left[\int_0^2 g(s)\,ds - \int_0^1 g(s)\,ds\right]$

$= -6 - 5[4 + 1] = -31$

15. $\int_0^2 [3f(t)+2g(t)]\,dt$

$= 3\left[\int_0^1 f(t)\,dt + \int_1^2 f(t)\,dt\right] + 2\int_0^2 g(t)\,dt$

$= 3(2 + 3) + 2(4) = 23$

17. $G'(x) = D_x\left[\int_1^x 2t\,dt\right] = 2x$

19. $G'(x) = D_x\left[\int_0^x \left(2t^2 + \sqrt{t}\right)dt\right] = 2x^2 + \sqrt{x}$

21. $G'(x) = D_x\left[\int_x^{\pi/4} (s-2)\cot(2s)\,ds\right]$

$= D_x\left[-\int_{\pi/4}^x (s-2)\cot(2s)\,ds\right]$

$= -(x-2)\cot(2x)$

23. $G'(x) = D_x\left[\int_1^{x^2} \sin t\,dt\right] = 2x\sin(x^2)$

25.
$$G(x) = \int_{-x^2}^{x} \frac{t^2}{1+t^2}\,dt$$
$$= \int_{-x^2}^{0} \frac{t^2}{1+t^2}\,dt + \int_0^x \frac{t^2}{1+t^2}\,dt$$
$$= -\int_0^{-x^2} \frac{t^2}{1+t^2}\,dt + \int_0^x \frac{t^2}{1+t^2}\,dt$$
$$G'(x) = -\frac{\left(-x^2\right)^2}{1+\left(-x^2\right)^2}(-2x) + \frac{x^2}{1+x^2}$$
$$= \frac{2x^5}{1+x^4} + \frac{x^2}{1+x^2}$$

27. $f'(x) = \dfrac{x}{\sqrt{1+x^2}};\ f''(x) = \dfrac{1}{\left(x^2+1\right)^{3/2}}$

So, $f(x)$ is increasing on $[0,\infty)$ and concave up on $(0,\infty)$.

29. $f'(x) = \cos x;\ \ f''(x) = -\sin x$

So, $f(x)$ is increasing on $\left[0,\frac{\pi}{2}\right],\left[\frac{3\pi}{2},\frac{5\pi}{2}\right],\ldots$ and concave up on $(\pi,2\pi),(3\pi,4\pi),\ldots$.

31. $f'(x) = \dfrac{1}{x};\ \ f''(x) = -\dfrac{1}{x^2}$

So, $f(x)$ is increasing on $(0,\infty)$ and never concave up.

33.

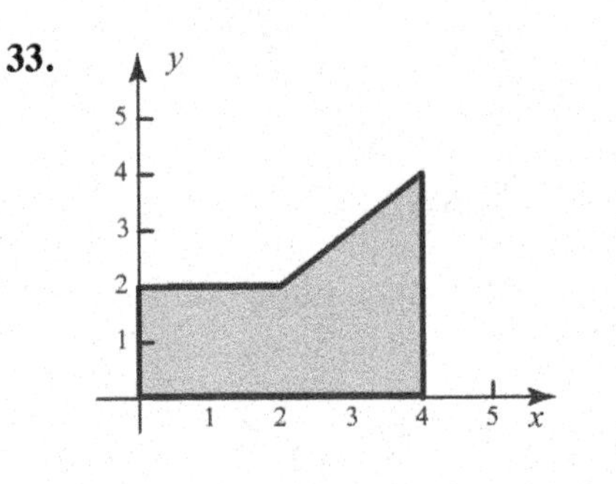

$\int_0^4 f(x)\,dx = \int_0^2 2\,dx + \int_2^4 x\,dx = 4 + 6 = 10$

35.

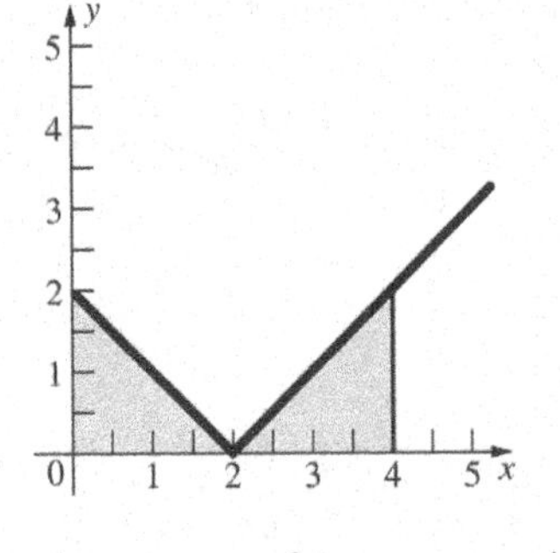

$\int_0^4 f(x)\,dx = \int_0^2 (2-x)\,dx + \int_2^4 (x-2)\,dx$

$= 2 + 2 = 4$

37. **a.** Local minima at 0, ≈ 3.8, ≈ 5.8, ≈ 7.9, ≈ 9.9;
local maxima at ≈ 3.1, ≈ 5, ≈ 7.1, ≈ 9, 10

b. Absolute minimum at 0, absolute maximum at ≈ 9

c. $\approx (0.7, 1.5), (2.5, 3.5), (4.5, 5.5), (6.5, 7.5), (8.5, 9.5)$

d.

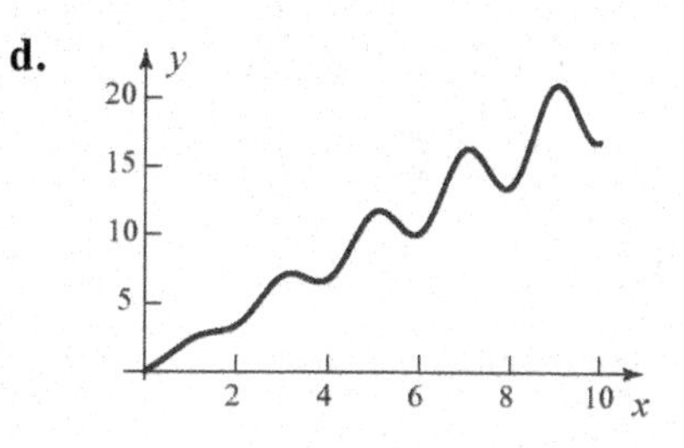

39. **a.** $F(0) = \int_0^0 (t^4 + 1)\, dt = 0$

b. $y = F(x)$

$$\frac{dy}{dx} = F'(x) = x^4 + 1$$

$$dy = (x^4 + 1)\, dx$$

$$y = \tfrac{1}{5}x^5 + x + C$$

c. Now apply the initial condition $y(0) = 0$:

$$0 = \tfrac{1}{5}0^5 + 0 + C$$

$$C = 0$$

Thus $y = F(x) = \tfrac{1}{5}x^5 + x$

41. For $t \ge 1$, $\sqrt{t} \le t$. Since $1 + x^4 \ge 1$ for all x, $1 \le \sqrt{1+x^4} \le 1 + x^4$.

$$\int_0^1 dx \le \int_0^1 \sqrt{1+x^4}\, dx \le \int_0^1 (1+x^4)\, dx$$

By problem 39d, $1 \le \int_0^1 \sqrt{1+x^4}\, dx \le \frac{6}{5}$

43. $5 \le f(x) \le 69$ so

$$4 \cdot 5 \le \int_0^4 (5 + x^3)\, dx \le 4 \cdot 69$$

$$20 \le \int_0^4 (5 + x^3)\, dx \le 276$$

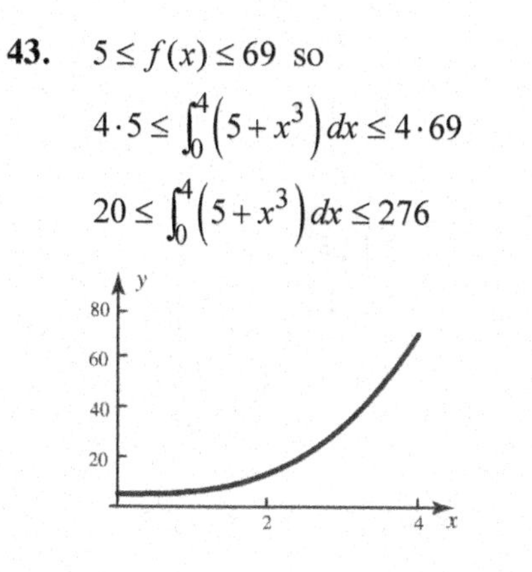

45. On [1,5],

$$3 + \frac{2}{5} \le 3 + \frac{2}{x} \le 3 + \frac{2}{1}$$

$$4\left(\frac{17}{5}\right) \le \int_1^5 \left(3 + \frac{2}{x}\right) dx \le 4 \cdot 5$$

$$\frac{68}{5} \le \int_1^5 \left(3 + \frac{2}{x}\right) dx \le 20$$

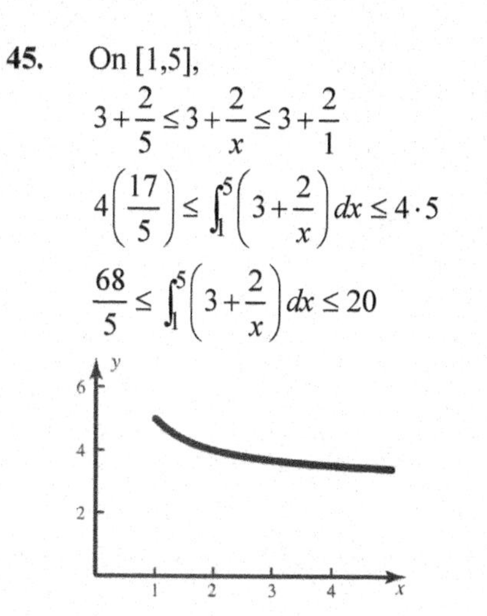

47. On $[4\pi, 8\pi]$

$$5 \le 5 + \tfrac{1}{20}\sin^2 x \le 5 + \tfrac{1}{20}$$

$$(4\pi)(5) \le \int_{4\pi}^{8\pi} \left(5 + \tfrac{1}{20}\sin^2 x\right) dx \le (4\pi)\left(5 + \tfrac{1}{20}\right)$$

$$20\pi \le \int_{4\pi}^{8\pi} \left(5 + \tfrac{1}{20}\sin^2 x\right) dx \le \frac{101}{5}\pi$$

49. Let $F(x) = \int_0^x \frac{1+t}{2+t}\, dt$. Then

$$\lim_{x \to 0} \frac{1}{x} \int_0^x \frac{1+t}{2+t}\, dt = \lim_{x \to 0} \frac{F(x) - F(0)}{x - 0}$$

$$= F'(0) = \frac{1+0}{2+0} = \frac{1}{2}$$

51. $\int_1^x f(t)\, dt = 2x - 2$

Differentiate both sides with respect to x:

$$\frac{d}{dx} \int_1^x f(t)\, dt = \frac{d}{dx}(2x - 2)$$

$$f(x) = 2$$

If such a function exists, it must satisfy $f(x) = 2$, but both sides of the first equality may differ by a constant yet still have equal derivatives. When $x = 1$ the left side is $\int_1^1 f(t)\, dt = 0$ and the right side is $2 \cdot 1 - 2 = 0$.

Thus the function $f(x) = 2$ satisfies

$$\int_1^x f(t)\, dt = 2x - 2.$$

53. $\int_0^{x^2} f(t)\,dt = \frac{1}{3}x^3$

Differentiate both sides with respect to x:

$$\frac{d}{dx}\int_0^{x^2} f(t)\,dt = \frac{d}{dx}\left(\frac{1}{3}x^3\right)$$

$$f\left(x^2\right)(2x) = x^2$$

$$f\left(x^2\right) = \frac{x}{2}$$

$$f(x) = \frac{\sqrt{x}}{2}$$

55. True; by Theorem B (Comparison Property)

57. False. $a = -1$, $b = 1$, $f(x) = x$ is a counterexample.

59. True. $\int_a^b f(x)dx - \int_a^b g(x)dx$
$= \int_a^b [f(x) - g(x)]dx$

61. $v(t) = \begin{cases} 2+(t-2), & t \le 2 \\ 2-(t-2), & t > 2 \end{cases}$
$= \begin{cases} t, & t \le 2 \\ 4-t, & t > 2 \end{cases}$

$s(t) = \int_0^t v(u)\,du$

$= \begin{cases} \int_0^t u\,du, & 0 \le t \le 2 \\ \int_0^2 u\,du + \int_2^t (4-u)\,du, & t > 2 \end{cases}$

$= \begin{cases} \frac{t^2}{2}, & 0 \le t \le 2 \\ 2+\left[4t - \frac{t^2}{2}\right], & t > 2 \end{cases}$

$= \begin{cases} \frac{t^2}{2}, & 0 \le t \le 2 \\ -4+4t-\frac{t^2}{2} & t > 2 \end{cases}$

$\frac{t^2}{2} - 4t + 4 = 0;\ t = 4 + 2\sqrt{2} \approx 6.83$

63. $-|f(x)| \le f(x) \le |f(x)|$, so

$\int_a^b -|f(x)|\,dx \le \int_a^b f(x)\,dx \Rightarrow$

$\int_a^b |f(x)|\,dx \ge -\int_a^b f(x)\,dx$

and combining this with

$\int_a^b |f(x)|\,dx \ge \int_a^b f(x)\,dx,$

we can conclude that

$\left|\int_a^b f(x)\,dx\right| \le \int_a^b |f(x)|\,dx$

4.4 Concepts Review

1. antiderivative; $F(b) - F(a)$

3. $F(d) - F(c)$

Problem Set 4.4

1. $\int_0^2 x^3\,dx = \left[\frac{x^4}{4}\right]_0^2 = 4 - 0 = 4$

3. $\int_{-1}^2 (3x^2 - 2x + 3)\,dx = \left[x^3 - x^2 + 3x\right]_{-1}^2$
$= (8 - 4 + 6) - (-1 - 1 - 3) = 15$

5. $\int_1^4 \frac{1}{w^2}\,dw = \left[-\frac{1}{w}\right]_1^4 = \left(-\frac{1}{4}\right) - (-1) = \frac{3}{4}$

7. $\int_0^4 \sqrt{t}\,dt = \left[\frac{2}{3}t^{3/2}\right]_0^4 = \left(\frac{2}{3}\cdot 8\right) - 0 = \frac{16}{3}$

9. $\int_{-4}^{-2}\left(y^2 + \frac{1}{y^3}\right)dy = \left[\frac{y^3}{3} - \frac{1}{2y^2}\right]_{-4}^{-2}$
$= \left(-\frac{8}{3} - \frac{1}{8}\right) - \left(-\frac{64}{3} - \frac{1}{32}\right) = \frac{1783}{96}$

11. $\int_0^{\pi/2} \cos x\,dx = [\sin x]_0^{\pi/2} = 1 - 0 = 1$

13. $\int_0^1 (2x^4 - 3x^2 + 5)\,dx = \left[\frac{2}{5}x^5 - x^3 + 5x\right]_0^1$
$= \left(\frac{2}{5} - 1 + 5\right) - 0 = \frac{22}{5}$

15. $u = 3x + 2$, $du = 3\,dx$
$\int \sqrt{u}\cdot\frac{1}{3}\,du = \frac{2}{9}u^{3/2} + C = \frac{2}{9}(3x+2)^{3/2} + C$

17. $u = 3x + 2, du = 3\,dx$

$$\int \cos(u) \cdot \frac{1}{3}\,du = \frac{1}{3}\sin u + C = \frac{1}{3}\sin(3x+2) + C$$

19. $u = 6x - 7, du = 6dx$

$$\int \sin u \cdot \frac{1}{6}\,du = -\frac{1}{6}\cos u + C$$

$$= -\frac{1}{6}\cos(6x-7) + C$$

21. $u = x^2 + 4, du = 2x\,dx$

$$\int \sqrt{u} \cdot \frac{1}{2}\,du = \frac{1}{3}u^{3/2} + C = \frac{1}{3}(x^2+4)^{3/2} + C$$

23. $u = x^2 + 3, du = 2x\,dx$

$$\int u^{-12/7} \cdot \frac{1}{2}\,du = -\frac{7}{10}u^{-5/7} + C$$

$$= -\frac{7}{10}(x^2+3)^{-5/7} + C$$

25. $u = x^2 + 4, du = 2x\,dx$

$$\int \sin(u) \cdot \frac{1}{2}\,du = -\frac{1}{2}\cos u + C$$

$$= -\frac{1}{2}\cos(x^2+4) + C$$

27. $u = \sqrt{x^2+4}, du = \dfrac{x}{\sqrt{x^2+4}}\,dx$

$$\int \sin u\,du = -\cos u + C = -\cos\sqrt{x^2+4} + C$$

29. $u = (x^3+5)^9$,

$du = 9(x^3+5)^8(3x^2)dx = 27x^2(x^3+5)^8\,dx$

$$\int \cos u \cdot \frac{1}{27}\,du = \frac{1}{27}\sin u + C$$

$$= \frac{1}{27}\sin\left[(x^3+5)^9\right] + C$$

31. $u = \sin(x^2+4), du = 2x\cos(x^2+4)\,dx$

$$\int \sqrt{u} \cdot \frac{1}{2}\,du = \frac{1}{3}u^{3/2} + C$$

$$= \frac{1}{3}\left[\sin(x^2+4)\right]^{3/2} + C$$

33. $u = \cos(x^3+5), du = -3x^2\sin(x^3+5)\,dx$

$$\int u^9 \cdot \left(-\frac{1}{3}\right)du = -\frac{1}{30}u^{10} + C$$

$$= -\frac{1}{30}\cos^{10}(x^3+5) + C$$

35. $u = x^2 + 1, du = 2x\,dx$

$$\int_0^1 (x^2+1)^{10}(2x)dx = \int_1^2 u^{10}\,du = \left[\frac{u^{11}}{11}\right]_1^2$$

$$= \left[\frac{1}{11}(2)^{11}\right] - \left[\frac{1}{11}(1)^{11}\right] = \frac{2047}{11}$$

37. $u = t + 2, du = dt$

$$\int_{-1}^{3} \frac{1}{(t+2)^2}\,dt = \int_1^5 u^{-2}\,du = \left[-\frac{1}{u}\right]_1^5$$

$$= \left[-\frac{1}{5}\right] - [-1] = \frac{4}{5}$$

39. $u = 3x + 1, du = 3\,dx$

$$\int_5^8 \sqrt{3x+1}\,dx = \frac{1}{3}\int_5^8 \sqrt{3x+1} \cdot 3dx = \frac{1}{3}\int_{16}^{25} \sqrt{u}\,du$$

$$= \left[\frac{2}{9}u^{3/2}\right]_{16}^{25} = \left[\frac{2}{9}(125)\right] - \left[\frac{2}{9}(64)\right] = \frac{122}{9}$$

41. $u = 7 + 2t^2, du = 4t\,dt$

$$\int_{-3}^{3} \sqrt{7+2t^2}\,(8t)\,dt = 2\int_{-3}^{3} \sqrt{7+2t^2} \cdot (4t)\,dt$$

$$= 2\int_{25}^{25} \sqrt{u}\,du = \left[\frac{4}{3}u^{3/2}\right]_{25}^{25}$$

$$= \left[\frac{4}{3}(125)\right] - \left[\frac{4}{3}(125)\right] = 0$$

43. $u = \cos x, du = -\sin x\,dx$

$$\int_0^{\pi/2} \cos^2 x \sin x\,dx = -\int_0^{\pi/2} \cos^2 x(-\sin x)\,dx$$

$$= -\int_1^0 u^2\,du = \left[-\frac{u^3}{3}\right]_1^0$$

$$= 0 - \left(-\frac{1}{3}\right) = \frac{1}{3}$$

45. $u = x^2 + 2x, du = (2x+2)\,dx = 2(x+1)\,dx$

$$\int_0^1 (x+1)(x^2+2x)^2\,dx$$

$$= \int_0^1 \frac{1}{2}(x^2+2x)^2 2(x+1)\,dx$$

$$= \frac{1}{2}\int_0^3 u^2\,du = \left[\frac{u^3}{6}\right]_0^3 = \frac{9}{2}$$

47. $u = \sin\theta, du = \cos\theta\,d\theta$

$$\int_0^{1/2} u^3\,du = \left[\frac{u^4}{4}\right]_0^{1/2} = \frac{1}{64} - 0 = \frac{1}{64}$$

49. $u = 3x - 3,\ du = 3dx$

$$\frac{1}{3}\int_{-3}^{0} \cos u\, du = \frac{1}{3}[\sin u]_{-3}^{0} = \frac{1}{3}(0 - \sin(-3))$$

$$= \frac{\sin 3}{3}$$

51. $u = \pi x^2,\ du = 2\pi x\, dx$

$$\frac{1}{2\pi}\int_{0}^{\pi} \sin u\, du = -\frac{1}{2\pi}[\cos u]_{0}^{\pi} = -\frac{1}{2\pi}(-1-1)$$

$$= \frac{1}{\pi}$$

53. $u = 2x,\ du = 2dx$

$$\frac{1}{2}\int_{0}^{\pi/2} \cos u\, du + \frac{1}{2}\int_{0}^{\pi/2} \sin u\, du$$

$$= \frac{1}{2}[\sin u]_{0}^{\pi/2} - \frac{1}{2}[\cos u]_{0}^{\pi/2}$$

$$= \frac{1}{2}(1-0) - \frac{1}{2}(0-1) = 1$$

55. $u = \cos x,\ du = -\sin x\, dx$

$$-\int_{1}^{0} \sin u\, du = [\cos u]_{1}^{0} = 1 - \cos 1$$

57. $u = \cos(x^2),\ du = -2x\sin(x^2)dx$

$$-\frac{1}{2}\int_{1}^{\cos 1} u^3\, du = -\frac{1}{2}\left[\frac{u^4}{4}\right]_{1}^{\cos 1} = -\frac{\cos^4 1}{8} + \frac{1}{8}$$

$$= \frac{1 - \cos^4 1}{8}$$

59. **a.** Between 0 and 3, $f(x) > 0$. Thus,

$$\int_0^3 f(x)\, dx > 0.$$

b. Since f is an antiderivative of f',

$$\int_0^3 f'(x)\, dx = f(3) - f(0)$$
$$= 0 - 2 = -2 < 0$$

c. $$\int_0^3 f''(x)\, dx = f'(3) - f'(0)$$
$$= -1 - 0 = -1 < 0$$

d. Since f is concave down at 0, $f''(0) < 0$.

$$\int_0^3 f'''(x)\, dx = f''(3) - f''(0)$$
$$= 0 - (\text{negative number}) > 0$$

61. $V(t) = \int V'(t) = \int (20 - t)\, dt = 20t - \frac{1}{2}t^2 + C$

$V(0) = C = 0$ since no water has leaked out at time $t = 0$. Thus, $V(t) = 20t - \frac{1}{2}t^2$, so

$V(20) - V(10) = 200 - 150 = 50$ gallons.

Time to drain: $20t - \frac{1}{2}t^2 = 200;\ t = 20$ hours.

63. Use a midpoint Riemann sum with $n = 12$ partitions.

$$V = \sum_{i=1}^{12} f(x_i)\Delta x_i$$

$$\approx 1(5.4 + 6.3 + 6.4 + 6.5 + 6.9 + 7.5 + 8.4 + 8.4 + 8.0 + 7.5 + 7.0 + 6.5)$$

$$= 84.8$$

65. Use a midpoint Riemann sum with $n = 12$ partitions.

$$E = \sum_{i=0}^{12} P(t_i)\Delta t_i$$

$$\approx 2(3.0 + 3.0 + 3.8 + 5.8 + 7.8 + 6.9 + 6.5 + 6.3 + 7.2 + 8.2 + 8.7 + 5.4)$$

$$= 145.2$$

67. **a.** $\int_a^b x^n dx = B_n;\ \int_{a^n}^{b^n} \sqrt[n]{y}\, dy = A_n$

Using Figure 3 of the text,

$(a)(a^n) + A_n + B_n = (b)(b^n)$ or

$B_n + A_n = b^{n+1} - a^{n+1}$. Thus

$$\int_a^b x^n dx + \int_{a^n}^{b^n} \sqrt[n]{y}\, dy = b^{n+1} - a^{n+1}$$

b. $$\int_a^b x^n\, dx + \int_{a^n}^{b^n} \sqrt[n]{y}\, dy$$

$$= \left[\frac{x^{n+1}}{n+1}\right]_a^b + \left[\frac{n}{n+1}y^{(n+1)/n}\right]_{a^n}^{b^n}$$

$$= \left(\frac{b^{n+1}}{n+1} - \frac{a^{n+1}}{n+1}\right) + \left(\frac{n}{n+1}b^{n+1} - \frac{n}{n+1}a^{n+1}\right)$$

$$= \frac{(n+1)b^{n+1} - (n+1)a^{n+1}}{n+1} = b^{n+1} - a^{n+1}$$

c. $B_n = \int_a^b x^n\,dx = \frac{1}{n+1}\left[x^{n+1}\right]_a^b$

$= \frac{1}{n+1}(b^{n+1} - a^{n+1})$

$A_n = \int_{a^n}^{b^n} \sqrt[n]{y}\,dy = \left[\frac{n}{n+1}y^{(n+1)/n}\right]_{a^n}^{b^n}$

$= \frac{n}{n+1}\left(b^{n+1} - a^{n+1}\right)$

$nB_n = \frac{n}{n+1}(b^{n+1} - a^{n+1}) = A_n$

69. $\int_0^3 x^2\,dx = \left[\frac{x^3}{3}\right]_0^3 = 9 - 0 = 9$

71. $\int_0^\pi \sin x\,dx = [-\cos x]_0^\pi = 1 + 1 = 2$

73. The right-endpoint Riemann sum is
$\sum_{i=1}^{n}\left(0+\frac{1-0}{n}i\right)^2\left(\frac{1}{n}\right) = \frac{1}{n^3}\sum_{i=1}^{n} i^2$, which for $n = 10$ equals $\frac{77}{200} = 0.385$.

$\int_0^1 x^2\,dx = \left[\frac{1}{3}x^3\right]_0^1 = \frac{1}{3} = 0.\overline{333}$

75. $\frac{d}{dx}\left(\frac{1}{2}x|x|\right) = \frac{1}{2}x\left(\frac{|x|}{x}\right) + \frac{|x|}{2} = |x|$

$\int_a^b |x|\,dx = \left[\frac{1}{2}x|x|\right]_a^b = \frac{1}{2}(b|b| - a|a|)$

77. **a.** Let c be in (a,b). Then $G'(c) = f(c)$ by the First Fundamental Theorem of Calculus. Since G is differentiable at c, G is continuous there. Now suppose $c = a$.

Then $\lim_{x\to c} G(x) = \lim_{x\to a}\int_a^x f(t)\,dt$. Since f is continuous on $[a,b]$, there exist (by the Min-Max Existence Theorem) m and M such that $f(m) \le f(x) \le f(M)$ for all x in $[a,b]$.
Then

$\int_a^x f(m)\,dt \le \int_a^x f(t)\,dt \le \int_a^x f(M)\,dt$

$(x-a)f(m) \le G(x) \le (x-a)f(M)$

By the Squeeze Theorem

$\lim_{x\to a^+}(x-a)f(m) \le \lim_{x\to a^+} G(x)$

$\le \lim_{x\to a^+}(x-a)f(M)$

Thus,

$\lim_{x\to a^+} G(x) = 0 = \int_a^a f(t)\,dt = G(a)$

Therefore G is right-continuous at $x = a$.
Now, suppose $c = b$. Then

$\lim_{x\to b^-} G(x) = \lim_{x\to b^-}\int_x^b f(t)\,dt$

As before,
$(b-x)f(m) \le G(x) \le (b-x)f(M)$ so we can apply the Squeeze Theorem again to obtain

$\lim_{x\to b^-}(b-x)f(m) \le \lim_{x\to b^-} G(x)$

$\le \lim_{x\to b^-}(b-x)f(M)$

Thus

$\lim_{x\to b^-} G(x) = 0 = \int_b^b f(t)\,dt = G(b)$

Therefore, G is left-continuous at $x = b$.

b. Let F be any antiderivative of f. Note that G is also an antiderivative of f. Thus, $F(x) = G(x) + C$. We know from part (a) that $G(x)$ is continuous on $[a,b]$. Thus $F(x)$, being equal to $G(x)$ plus a constant, is also continuous on $[a,b]$.

4.5 Concepts Review

1. $\frac{1}{b-a}\int_a^b f(x)\,dx$

3. 0; $2\int_0^2 f(x)\,dx$

Problem Set 4.5

1. $\frac{1}{3-1}\int_1^3 4x^3\,dx = \frac{1}{2}\left[x^4\right]_1^3 = 40$

3. $\frac{1}{3-0}\int_0^3 \frac{x}{\sqrt{x^2+16}}\,dx = \frac{1}{3}\left[\sqrt{x^2+16}\right]_0^3 = \frac{1}{3}$

5. $\frac{1}{1+2}\int_{-2}^1 (2+|x|)\,dx$

$= \frac{1}{3}\left[\int_{-2}^0 (2-x)\,dx + \int_0^1 (2+x)\,dx\right]$

$= \frac{1}{3}\left\{\left[2x-\frac{1}{2}x^2\right]_{-2}^0 + \left[2x+\frac{1}{2}x^2\right]_0^1\right\}$

$= \frac{1}{3}\left(-2(-2)+\frac{1}{2}(-2)^2+2+\frac{1}{2}\right) = \frac{17}{6}$

7. $\frac{1}{\pi}\int_0^\pi \cos x\,dx = \frac{1}{\pi}[\sin x]_0^\pi$

$= \frac{1}{\pi}[\sin\pi - \sin 0] = 0$

9. $\frac{1}{\sqrt{\pi}-0}\int_0^{\sqrt{\pi}} x\cos x^2\,dx = \frac{1}{\sqrt{\pi}}\left(\frac{1}{2}\sin x^2\right)_0^{\sqrt{\pi}}$

$= \frac{1}{\sqrt{\pi}}(0-0) = 0$

11. $\frac{1}{2-1}\int_1^2 y\left(1+y^2\right)^3\,dy = \left[\frac{1}{8}\left(1+y^2\right)^4\right]_1^2$

$= \frac{625}{8} - 2 = \frac{609}{8} = 76.125$

13. $\frac{1}{\pi/4}\int_{\pi/4}^{\pi/2} \frac{\sin\sqrt{z}}{\sqrt{z}}\,dz = \frac{4}{\pi}\left[-2\cos\sqrt{z}\right]_{\pi/4}^{\pi/2}$

$= \frac{8}{\pi}\left(\cos\sqrt{\pi/4} - \cos\sqrt{\pi/2}\right) \approx 0.815$

15. $\int_0^3 \sqrt{x+1}\,dx = \sqrt{c+1}(3-0)$

$\left[\frac{2}{3}(x+1)^{3/2}\right]_0^3 = 3\sqrt{c+1}$

$14/3 = 3\sqrt{c+1};\ c = \frac{115}{81} \approx 1.42$

17. $\int_{-4}^3 \left(1-x^2\right)dx = \left(1-c^2\right)(3+4)$

$\left[x-\frac{1}{3}x^3\right]_{-4}^3 = 7-7c^2$

$c = \pm\frac{\sqrt{39}}{3} \approx \pm 2.08$

19. $\int_0^2 |x|\,dx = |c|(2-0);\ \left[\frac{x|x|}{2}\right]_0^2 = 2|c|;\ c = 1$

21. $\int_{-\pi}^{\pi} \sin z\,dz = \sin c(\pi+\pi)$

$[-\cos z]_{-\pi}^{\pi} = 2\pi\sin c;\quad c = 0$

23. $\int_0^2 \left(v^2-v\right)dv = \left(c^2-c\right)(2-0)$

$\left[\frac{1}{3}v^3 - \frac{1}{2}v^2\right]_0^2 = 2c^2 - 2c$

$c = \frac{\sqrt{21}+3}{6} \approx 1.26$

25. $\int_1^4 (ax+b)\,dx = (ac+b)(4-1)$

$\left[\frac{a}{2}x^2 + bx\right]_1^4 = 3ac+3b;\ c = \frac{5}{2}$

27. $\frac{\int_A^B (ax+b)\,dx}{B-A} = f(c)$

$\frac{\left[\frac{a}{2}x^2+bx\right]_A^B}{B-A} = ac+b$

$\frac{\frac{a}{2}(B-A)(B+A)+b(B-A)}{B-A} = ac+b$

$\frac{a}{2}B + \frac{a}{2}A + b = ac+b;$

$c = \frac{1}{2}B + \frac{1}{2}A = (A+B)/2$

29.. Using $c = \pi$ yields $2\pi(5)^4 = 1250\pi \approx 3927$

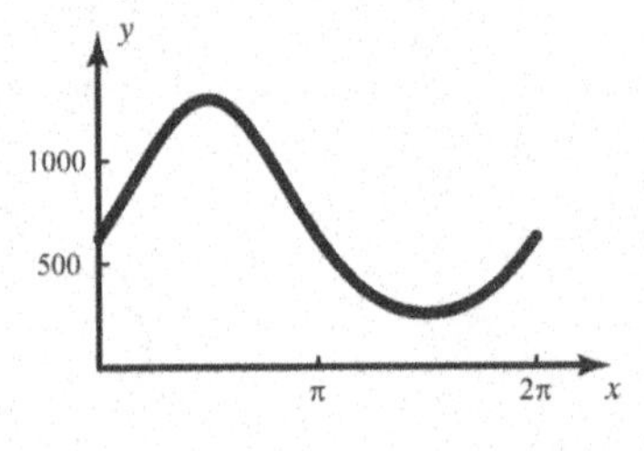

31. Using $c = 0.5$ yields $2\dfrac{2}{1+0.5^2} = 3.2$

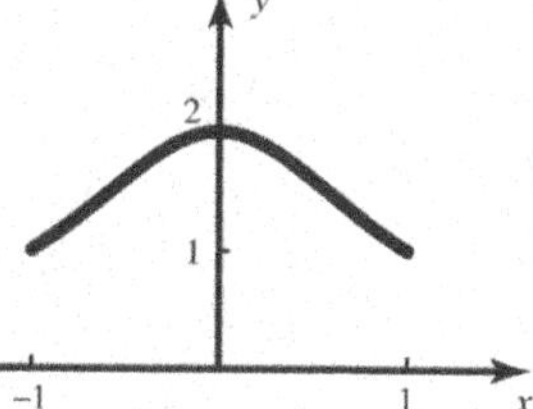

33. A rectangle with height 25 and width 7 has approximately the same area as that under the curve. Thus

$$\frac{1}{7}\int_0^7 H(t)\,dt \approx 25$$

35. $\int_{-\pi}^{\pi} (\sin x + \cos x)\,dx = \int_{-\pi}^{\pi} \sin x\,dx + 2\int_0^{\pi} \cos x\,dx$

$$= 0 + 2[\sin x]_0^{\pi} = 0$$

37. $\int_{-\pi/2}^{\pi/2} \frac{\sin x}{1+\cos x}dx = 0$, since the integrand is odd.

39. $\int_{-\pi}^{\pi} (\sin x + \cos x)^2\,dx$

$$= \int_{-\pi}^{\pi} (\sin^2 x + 2\sin x\cos x + \cos^2 x)\,dx$$

$$= \int_{-\pi}^{\pi} (1 + 2\sin x\cos x)\,dx = \int_{-\pi}^{\pi} dx + \int_{-\pi}^{\pi} \sin 2x\,dx$$

$$= 2\int_0^{\pi} dx + 0 = 2[x]_0^{\pi} = 2\pi$$

41. $\int_{-1}^{1} (1 + x + x^2 + x^3)\,dx$

$$= \int_{-1}^{1} dx + \int_{-1}^{1} x\,dx + \int_{-1}^{1} x^2\,dx + \int_{-1}^{1} x^3\,dx$$

$$= 2[x]_0^1 + 0 + 2\left[\frac{x^3}{3}\right]_0^1 + 0 = \frac{8}{3}$$

43. $\int_{-1}^{1} \left(\left|x^3\right| + x^3\right)dx = 2\int_0^1 \left|x^3\right|dx + \int_{-1}^{1} x^3\,dx$

$$= 2\left[\frac{x^4}{4}\right]_0^1 + 0 = \frac{1}{2}$$

45. $\int_{-b}^{-a} f(x)\,dx = \int_a^b f(x)\,dx$ when *f* is even.

$\int_{-b}^{-a} f(x)\,dx = -\int_a^b f(x)\,dx$ when *f* is odd.

47. $\int_0^{4\pi} |\cos x|\,dx = 8\int_0^{\pi/2} |\cos x|\,dx$

$$= 8[\sin x]_0^{\pi/2} = 8$$

49. $\int_1^{1+\pi} |\sin x|\,dx = \int_0^{\pi} |\sin x|\,dx = \int_0^{\pi} \sin x\,dx$

$$= [-\cos x]_0^{\pi} = 2$$

51. $\int_1^{1+\pi} |\cos x|\,dx = \int_0^{\pi} |\cos x|\,dx = 2\int_0^{\pi/2} \cos x\,dx$

$$= 2[\sin x]_0^{\pi/2} = 2(1-0) = 2$$

53. All the statements are true.

a. $\overline{u} + \overline{v} = \frac{1}{b-a}\int_a^b u\,dx + \frac{1}{b-a}\int_a^b v\,dx$

$$= \frac{1}{b-a}\int_a^b (u+v)\,dx = \overline{u+v}$$

b. $k\overline{u} = \frac{k}{b-a}\int_a^b u\,dx = \frac{1}{b-a}\int_a^b ku\,dx = \overline{ku}$

c. Note that

$\overline{u} = \frac{1}{b-a}\int_a^b u(x)\,dx = \frac{1}{a-b}\int_b^a u(x)\,dx$, so we can assume $a < b$.

$$\overline{u} = \frac{1}{b-a}\int_a^b u\,dx \le \frac{1}{b-a}\int_a^b v\,dx = \overline{v}$$

55. Since *f* is continuous on a closed interval $[a,b]$ there exist (by the Min-Max Existence Theorem) an *m* and *M* in $[a,b]$ such that $f(m) \le f(x) \le f(M)$ for all x in $[a,b]$. Thus

$$\int_a^b f(m)\,dx \le \int_a^b f(x)\,dx \le \int_a^b f(M)\,dx$$

$$(b-a)f(m) \le \int_a^b f(x)\,dx \le (b-a)f(M)$$

$$f(m) \le \frac{1}{b-a}\int_a^b f(x)\,dx \le f(M)$$

Since *f* is continuous, we can apply the Intermediate Value Theorem and say that *f* takes on every value between $f(m)$ and $f(M)$. Since $\frac{1}{b-a}\int_a^b f(x)\,dx$ is between $f(m)$ and $f(M)$, there exists a *c* in $[a,b]$ such that

$$f(c) = \frac{1}{b-a}\int_a^b f(x)\,dx.$$

57. **a.** Even

b. 2π

c. On $[0,\pi]$, $|\sin x| = \sin x$.

$u = \cos x$, $du = -\sin x\, dx$

$$\int f(x)\,dx = \int \sin x \cdot \sin(\cos x)\,dx = -\int \sin u\, du = \cos u + C = \cos(\cos x) + C$$

Likewise, on $[\pi, 2\pi]$, $\int f(x)\,dx = -\cos(\cos x) + C$

$$\int_0^{\pi/2} f(x)\,dx = 1 - \cos 1 \approx 0.46$$

$$\int_{-\pi/2}^{\pi/2} f(x)\,dx = 2\int_0^{\pi/2} f(x)\,dx = 2(1 - \cos 1) \approx 0.92$$

$$\int_0^{3\pi/2} f(x)\,dx = \int_0^{\pi} f(x)\,dx + \int_{\pi}^{3\pi/2} f(x)\,dx = \cos 1 - 1 \approx -0.46$$

$$\int_{-3\pi/2}^{3\pi/2} f(x)\,dx = 2\int_0^{3\pi/2} f(x)\,dx = 2(\cos 1 - 1) \approx -0.92$$

$$\int_0^{2\pi} f(x)\,dx = 0$$

$$\int_{\pi/6}^{4\pi/3} f(x)\,dx = 2\cos 1 - \cos\left(\frac{\sqrt{3}}{2}\right) + \cos\left(\frac{1}{2}\right) \approx -0.44$$

$$\int_{13\pi/6}^{10\pi/3} f(x)\,dx = \int_{\pi/6}^{4\pi/3} f(x)\,dx \approx -0.44$$

59. **a.** Written response.

b. $A = \int_0^a g(x)\,dx = \int_0^a \frac{a}{c} f\left(\frac{c}{a}x\right)dx = \int_0^c \frac{a}{c} f(x) \frac{a}{c}\,dx = \frac{a^2}{c^2}\int_0^c f(x)\,dx$

$B = \int_0^b h(x)\,dx = \int_0^b \frac{b}{c} f\left(\frac{c}{b}x\right)dx = \int_0^c \frac{b}{c} f(x) \frac{b}{c}\,dx = \frac{b^2}{c^2}\int_0^c f(x)\,dx$

Thus, $\int_0^a g(x)\,dx + \int_0^b h(x)\,dx = \frac{a^2}{c^2}\int_0^c f(x)\,dx + \frac{b^2}{c^2}\int_0^c f(x)\,dx = \frac{a^2 + b^2}{c^2}\int_0^c f(x)\,dx = \int_0^c f(x)\,dx$ since

$a^2 + b^2 = c^2$ from the triangle.

4.6 Concepts Review

1. 1, 2, 2, 2, …, 2, 1

3. n^4

Problem Set 4.6

1. $f(x)=\frac{1}{x^2}; h=\frac{3-1}{8}=0.25$

$x_0=1.00$	$f(x_0)=1$	$x_5=2.25$	$f(x_5)\approx 0.1975$
$x_1=1.25$	$f(x_1)=0.64$	$x_6=2.50$	$f(x_6)=0.16$
$x_2=1.50$	$f(x_2)\approx 0.4444$	$x_7=2.75$	$f(x_7)\approx 0.1322$
$x_3=1.75$	$f(x_3)\approx 0.3265$	$x_8=3.00$	$f(x_8)\approx 0.1111$
$x_4=2.00$	$f(x_4)=0.25$		

Left Riemann Sum: $\int_1^3 \frac{1}{x^2}dx \approx 0.25[f(x_0)+f(x_1)+\ldots+f(x_7)]\approx 0.7877$

Right Riemann Sum: $\int_1^3 \frac{1}{x^2}dx \approx 0.25[f(x_1)+f(x_2)+\ldots+f(x_8)]\approx 0.5655$

Trapezoidal Rule: $\int_1^3 \frac{1}{x^2}dx \approx \frac{0.25}{2}[f(x_0)+2f(x_1)+\ldots+2f(x_7)+f(x_8)]\approx 0.6766$

Parabolic Rule: $\int_1^3 \frac{1}{x^2}dx \approx \frac{0.25}{3}[f(x_0)+4f(x_1)+2f(x_2)+\ldots+4f(x_7)+f(x_8)]\approx 0.6671$

Fundamental Theorem of Calculus: $\int_1^3 \frac{1}{x^2}dx = \left[-\frac{1}{x}\right]_1^3 = -\frac{1}{3}+1=\frac{2}{3}\approx 0.6667$

3. $f(x)=\sqrt{x}; h=\frac{2-0}{8}=0.25$

$x_0=0.00$	$f(x_0)=0$	$x_5=1.25$	$f(x_5)\approx 1.1180$
$x_1=0.25$	$f(x_1)=0.5$	$x_6=1.50$	$f(x_6)\approx 1.2247$
$x_2=0.50$	$f(x_2)\approx 0.7071$	$x_7=1.75$	$f(x_7)\approx 1.3229$
$x_3=0.75$	$f(x_3)\approx 0.8660$	$x_8=2.00$	$f(x_8)\approx 1.4142$
$x_4=1.00$	$f(x_4)=1$		

Left Riemann Sum: $\int_0^2 \sqrt{x}\,dx \approx 0.25[f(x_0)+f(x_1)+\ldots+f(x_7)]\approx 1.6847$

Right Riemann Sum: $\int_0^2 \sqrt{x}\,dx \approx 0.25[f(x_1)+f(x_2)+\ldots+f(x_8)]\approx 2.0383$

Trapezoidal Rule: $\int_0^2 \sqrt{x}dx \approx \frac{0.25}{2}[f(x_0)+2f(x_1)+\ldots+2f(x_7)+f(x_8)]\approx 1.8615$

Parabolic Rule: $\int_0^2 \sqrt{x}dx \approx \frac{0.25}{3}[f(x_1)+4f(x_2)+2f(x_3)+\ldots+4f(x_7)+f(x_8)]\approx 1.8755$

Fundamental Theorem of Calculus: $\int_0^2 \sqrt{x}dx = \left[\frac{2}{3}x^{3/2}\right]_0^2 = \frac{4\sqrt{2}}{3}\approx 1.8856$

5. $f(x)=x\left(x^2+1\right)^5$; $h=\frac{1-0}{8}=0.125$

$x_0=0.00$ $f(x_0)=0$

$x_1=0.125$ $f(x_1)\approx 0.1351$

$x_2=0.250$ $f(x_2)\approx 0.3385$

$x_3=0.375$ $f(x_3)\approx 0.7240$

$x_4=0.500$ $f(x_4)\approx 1.5259$

$x_5=0.625$ $f(x_5)\approx 3.2504$

$x_6=0.750$ $f(x_6)\approx 6.9849$

$x_7=0.875$ $f(x_7)\approx 15.0414$

$x_8=1.000$ $f(x_8)=32$

Left Riemann Sum: $\int_0^1 x\left(x^2+1\right)^5 dx \approx 0.125[f(x_0)+f(x_1)+\cdots+f(x_7)]\approx 3.4966$

Right Riemann Sum: $\int_0^1 x\left(x^2+1\right)^5 dx \approx 0.125[f(x_1)+f(x_2)+\ldots+f(x_8)]\approx 7.4966$

Trapezoidal Rule: $\int_0^1 x\left(x^2+1\right)^5 dx \approx \frac{0.125}{2}[f(x_0)+2f(x_1)+\cdots+2f(x_7)+f(x_8)]\approx 5.4966$

Parabolic Rule: $\int_0^1 x\left(x^2+1\right)^5 dx \approx \frac{0.125}{3}[f(x_0)+4f(x_1)+2f(x_2)+\cdots+4f(x_7)+f(x_8)]\approx 5.2580$

Fundamental Theorem of Calculus: $\int_0^1 x\left(x^2+1\right)^5 dx = \left[\frac{1}{12}\left(x^2+1\right)^6\right]_0^1 = 5.25$

7.

	LRS	RRS	MRS	Trap	Parabolic
$n=4$	0.5728	0.3728	0.4590	0.4728	0.4637
$n=8$	0.5159	0.4159	0.4625	0.4659	0.4636
$n=16$	0.4892	0.4392	0.4634	0.4642	0.4636

9.

	LRS	RRS	MRS	Trap	Parabolic
$n=4$	2.6675	3.2855	2.9486	2.9765	2.9580
$n=8$	2.8080	3.1171	2.9556	2.9625	2.9579
$n=16$	2.8818	3.0363	2.9573	2.9591	2.9579

11. $f'(x)=-\frac{1}{x^2}$; $f''(x)=\frac{2}{x^3}$

The largest that $|f''(c)|$ can be on $[1,3]$ occurs when $c=1$, and $|f''(1)|=2$

$\frac{(3-1)^3}{12n^2}(2)\le 0.01$; $n\ge\sqrt{\frac{400}{3}}$ Round up: $n=12$

$\int_1^3 \frac{1}{x}dx \approx \frac{0.167}{2}[f(x_0)+2f(x_1)+\cdots+2f(x_{11})+f(x_{12})]\approx 1.1007$

13. $f'(x) = \frac{1}{2\sqrt{x}}$; $f''(x) = -\frac{1}{4x^{3/2}}$

The largest that $|f''(c)|$ can be on $[1, 4]$ occurs when $c = 1$, and $|f''(1)| = \frac{1}{4}$.

$\frac{(4-1)^3}{12n^2}\left(\frac{1}{4}\right) \le 0.01$; $n \ge \sqrt{\frac{900}{16}}$ Round up: $n = 8$

$\int_1^4 \sqrt{x}\, dx \approx \frac{0.375}{2}[f(x_0) + 2f(x_1) + \cdots + 2f(x_7) + f(x_8)] \approx 4.6637$

15. $f'(x) = -\frac{1}{x^2}$; $f''(x) = \frac{2}{x^3}$; $f'''(x) = -\frac{6}{x^4}$; $f^{(4)}(x) = \frac{24}{x^5}$

The largest that $\left|f^{(4)}(c)\right|$ can be on $[1, 3]$ occurs when $c = 1$, and $\left|f^{(4)}(1)\right| = 24$.

$\frac{(4-1)^5}{180n^4}(24) \le 0.01$; $n \approx 4.545$ Round up to even: $n = 6$

$\int_1^3 \frac{1}{x} dx \approx \frac{0.333}{3}[f(x_0) + 4f(x_1) + \ldots + 4f(x_5) + f(x_6)] \approx 1.0989$

17. $\int_{m-h}^{m+h} (ax^2 + bx + c)dx = \left[\frac{a}{3}x^3 + \frac{b}{2}x^2 + cx\right]_{m-h}^{m+h}$

$= \frac{a}{3}(m+h)^3 + \frac{b}{2}(m+h)^2 + c(m+h) - \frac{a}{3}(m-h)^3 - \frac{b}{2}(m-h)^2 - c(m-h)$

$= \frac{a}{3}(6m^2h + 2h^3) + \frac{b}{2}(4mh) + c(2h) = \frac{h}{3}[a(6m^2 + 2h^2) + b(6m) + 6c]$

$\frac{h}{3}[f(m-h) + 4f(m) + f(m+h)]$

$= \frac{h}{3}[a(m-h)^2 + b(m-h) + c + 4am^2 + 4bm + 4c + a(m+h)^2 + b(m+h) + c]$

$= \frac{h}{3}[a(6m^2 + 2h^2) + b(6m) + 6c]$

19. The left Riemann sum will be smaller than $\int_a^b f(x)\, dx$.

If the function is increasing, then $f(x_i) < f(x_{i+1})$ on the interval $[x_i, x_{i+1}]$. Therefore, the left Riemann sum will underestimate the value of the definite integral. The following example illustrates this behavior:

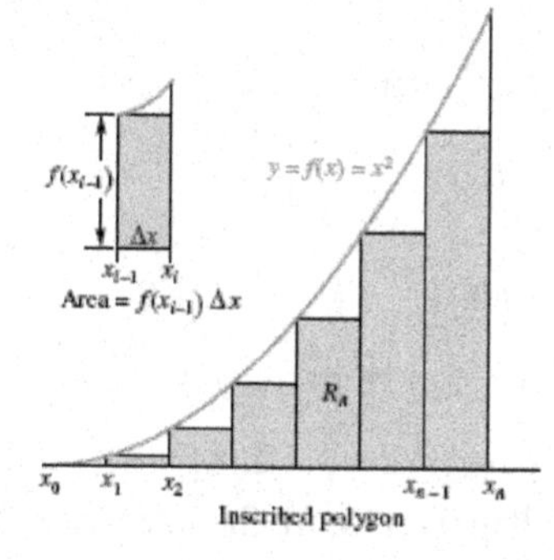

If f is increasing, then $f'(c) > 0$ for any $c \in (a, b)$. Thus, the error $E_n = \frac{(b-a)^2}{2n} f'(c) > 0$. Since the error is positive, then the Riemann sum must be less than the integral.

21. The midpoint Riemann sum will be larger than $\int_a^b f(x)\,dx$.

If f is concave down, then $f''(c) < 0$ for any $c \in (a,b)$. Thus, the error $E_n = \frac{(b-a)^3}{24n^2} f''(c) < 0$. Since the error is negative, then the Riemann sum must be greater than the integral.

23. Let $n = 2$.

$f(x) = x^k;\ h = a$

$x_0 = -a \qquad f(x_0) = -a^k$

$x_1 = 0 \qquad f(x_1) = 0$

$x_2 = a \qquad f(x_2) = a^k$

$$\int_{-a}^{a} x^k\,dx \approx \frac{a}{2}[-a^k + 2\cdot 0 + a^k] = 0$$

$$\int_{-a}^{a} x^k\,dx = \left[\frac{1}{k+1}x^{k+1}\right]_{-a}^{a} = \frac{1}{k+1}[a^{k+1} - (-a)^{k+1}] = \frac{1}{k+1}[a^{k+1} - a^{k+1}] = 0$$

A corresponding argument works for all n.

25. The integrand is increasing and concave down. By problems 19-22,
LRS < TRAP < MRS < RRS.

27. $A \approx \frac{10}{2}[75 + 2\cdot 71 + 2\cdot 60 + 2\cdot 45 + 2\cdot 45 + 2\cdot 52 + 2\cdot 57 + 2\cdot 60 + 59] = 4570\ \text{ft}^2$

29. $A \approx \frac{20}{3}[0 + 4\cdot 7 + 2\cdot 12 + 4\cdot 18 + 2\cdot 20 + 4\cdot 20 + 2\cdot 17 + 4\cdot 10 + 0] = 2120\ \text{ft}^2$

4 mi/h = 21,120 ft/h

(2120)(21,120)(24) = 1,074,585,600 ft^3

31. Using a right-Riemann sum,

$$\text{Water Usage} = \int_0^{120} F(t)\,dt \approx \sum_{i=1}^{10} F(t_i)\Delta t = 12(71 + 68 + \cdots + 148) = 13{,}740\ \text{gallons}$$

4.7 Chapter Review

Concepts Test

1. True: Theorem 4.3.D

3. True: If $F(x) = \int f(x)\,dx$, $f(x)$ is a derivative of $F(x)$.

5. False: The two sides will in general differ by a constant term.

7. True:
$$a_1 + a_0 + a_2 + a_1 + a_3 + a_2 + \cdots + a_{n-1} + a_{n-2} + a_n + a_{n-1}$$
$$= a_0 + 2a_1 + 2a_2 + \cdots + 2a_{n-1} + a_n$$

9. True:
$$\sum_{i=1}^{10}(a_i + 1)^2 = \sum_{i=1}^{10} a_i^2 + 2\sum_{i=1}^{10} a_i + \sum_{i=1}^{100} 1$$
$$= 100 + 2(20) + 10 = 150$$

11. True: The area of a vertical line segment is 0.

13. False: A counterexample is
$$f(x) = \begin{cases} 0, & x \neq 0 \\ 1, & x = 0 \end{cases}$$
with $\int_{-1}^{1} [f(x)]^2\,dx = 0$.

If $f(x)$ is continuous, then $[f(x)]^2 \geq 0$, and if $[f(x)]^2$ is greater than 0 on $[a, b]$, the integral will be also.

15. True: $\sin x + \cos x$ has period 2π, so $\int_x^{x+2\pi}(\sin x + \cos x)\,dx$ is independent of x.

17. True: $\sin^{13} x$ is an odd function.

19. False: The statement is not true if $c > d$.

21. True: Both sides equal 4.

23. True: If f is odd, then the accumulation function $F(x) = \int_0^x f(t)\,dt$ is even, and so is $F(x) + C$ for any C.

25. False: $f(x) = x^2$ is a counterexample.

27. False: $f(x) = x^2$, $v(x) = 2x + 1$ is a counterexample.

29. False: $f(x) = \sqrt{x}$ is a counterexample.

31. True: $F(b) - F(a) = \int_a^b F'(x)\,dx$
$= \int_a^b G'(x)\,dx = G(b) - G(a)$

33. False: $z(t) = t^2$ is a counterexample.

35. True: Odd-exponent terms cancel themselves out over the interval, since they are odd.

37. False: $a = 0, b = 1, f(x) = -1, g(x) = 0$ is a counterexample.

39. True: Note that $-|f(x)| \le f(x) \le |f(x)|$ and use Theorem 4.3.B.

41. True: Definition of Definite Integral

43. True. Right Riemann sum always bigger.

45. False. Trapeziod rule overestimates integral.

Sample Test Problems

1. $\left[\frac{1}{4}x^4 - x^3 + 2x^{3/2}\right]_0^1 = \frac{5}{4}$

3. $\left[\frac{1}{3}y^3 + 9\cos y - \frac{26}{y}\right]_1^{\pi} = \frac{50}{3} - \frac{26}{\pi} + \frac{\pi^3}{3} - 9\cos(1)$

5. $\left[\frac{3}{16}(2z^2 - 3)^{4/3}\right]_2^8 = \frac{-15\left(-125 + \sqrt[3]{5}\right)}{16}$

7. $u = \tan(3x^2 + 6x),\ du = (6x + 6)\sec^2(3x^2 + 6x)$
$\frac{1}{6}\int u^2\,du = \frac{1}{18}u^3 + C$
$\frac{1}{18}\left[\tan^3(3x^2 + 6x)\right]_0^{\pi} = \frac{1}{18}\tan^3(3\pi^2 + 6\pi)$

9. $\frac{1}{5}\left[\frac{3}{5}(t^5 + 5)^{5/3}\right]_1^2 = \frac{3}{25}\left[37^{5/3} - 6^{5/3}\right] \approx 46.9$

11. $\int (x+1)\sin\left(x^2 + 2x + 3\right)dx$
$= \frac{1}{2}\int \sin\left(x^2 + 2x + 3\right)(2x + 2)\,dx$
$= \frac{1}{2}\int \sin u\,du$
$= -\frac{1}{2}\cos\left(x^2 + 2x + 3\right) + C$

13. $\sum_{i=1}^{4}\left[\left(\frac{i}{2}\right)^2 - 1\right]\left(\frac{1}{2}\right) = \frac{7}{4}$

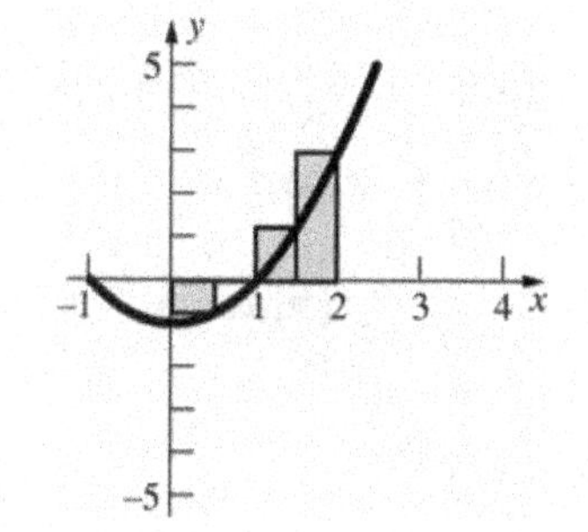

15. $\int_0^3 (2 - \sqrt{x+1})^2\,dx$
$= \int_0^3\left(x + 5 - 4\sqrt{x+1}\right)dx$
$= \left[\frac{1}{2}x^2 + 5x - \frac{8}{3}(x+1)^{3/2}\right]_0^3 = \frac{5}{6}$

17. $\int_2^4\left(5 - \frac{1}{x^2}\right)dx = \left[5x + \frac{1}{x}\right]_2^4 = \frac{39}{4}$

19. $\sum_{i=1}^{10}(6i^2 - 8i) = 6\sum_{i=1}^{10} i^2 - 8\sum_{i=1}^{10} i$
$= 6\left[\frac{10(11)(21)}{6}\right] - 8\left[\frac{10(11)}{2}\right] = 1870$

21. a. $\sum_{n=2}^{78} \frac{1}{n}$

b. $\sum_{n=1}^{50} nx^{2n}$

23. a. $\int_1^2 f(x)\,dx = \int_1^0 f(x)\,dx + \int_0^2 f(x)\,dx$
$= -4 + 2 = -2$

b. $\int_1^0 f(x)\,dx = -\int_0^1 f(x)\,dx = -4$

c. $\int_0^2 3f(u)\,du = 3\int_0^2 f(u)\,du = 3(2) = 6$

d. $\int_0^2 [2g(x) - 3f(x)]\,dx$
$= 2\int_0^2 g(x) - 3\int_0^2 f(x)\,dx$
$= 2(-3) - 3(2) = -12$

e. $\int_0^{-2} f(-x)\,dx = -\int_0^2 f(x)\,dx = -2$

25. a. $\int_{-2}^2 f(x)\,dx = 2\int_0^2 f(x)\,dx = 2(-4) = -8$

b. Since $f(x) \le 0$, $|f(x)| = -f(x)$ and
$\int_{-2}^2 |f(x)|\,dx = -\int_{-2}^2 f(x)\,dx$
$= -2\int_0^2 f(x)\,dx = 8$

c. $\int_{-2}^2 g(x)\,dx = 0$

d. $\int_{-2}^2 [f(x) + f(-x)]\,dx$
$= 2\int_0^2 f(x)\,dx + 2\int_0^2 f(x)\,dx$
$= 4(-4) = -16$

e. $\int_0^2 [2g(x) + 3f(x)]\,dx$
$= 2\int_0^2 g(x)\,dx + 3\int_0^2 f(x)\,dx$
$= 2(5) + 3(-4) = -2$

f. $\int_{-2}^0 g(x)\,dx = -\int_0^2 g(x)\,dx = -5$

27. $\int_{-4}^{-1} 3x^2\,dx = 3c^2(-1+4)$
$\left[x^3\right]_{-4}^{-1} = 9c^2$
$c^2 = 7$
$c = -\sqrt{7} \approx -2.65$

29. a. $G'(x) = \sin^2 x$

b. $G'(x) = f(x+1) - f(x)$

c. $G'(x) = -\frac{1}{x^2}\int_0^x f(z)\,dz + \frac{1}{x} f(x)$

d. $G'(x) = \int_0^x f(t)\,dt$

e. $G(x) = \int_0^{g(x)} \frac{dg(u)}{du}\,du = [g(u)]_0^{g(x)}$
$= g(g(x)) - g(0)$
$G'(x) = g'(g(x))g'(x)$

f. $G(x) = \int_0^{-x} f(-t)\,dt = \int_0^x f(u)(-du)$
$= -\int_0^x f(u)\,du$
$G'(x) = -f(x)$

31. $f(x) = \int_{2x}^{5x} \frac{1}{t}\,dt = \int_1^{5x} \frac{1}{t}\,dt - \int_1^{2x} \frac{1}{t}\,dt$
$f'(x) = \frac{1}{5x} \cdot 5 - \frac{1}{2x} \cdot 2 = 0$

33. $\int_1^2 \frac{1}{1+x^4}\,dx \approx \frac{0.125}{2}[f(x_0) + 2f(x_1) + \ldots + 2f(x_7) + f(x_8)] \approx 0.2043$

$$|f''(c)| = \left|\frac{4c^2(5c^4-3)}{(1+c^4)^3}\right| \le \frac{(4)(2^2)\left((5)(2^4)-3\right)}{\left(1+1^4\right)^3} = 154$$

$$|E_n| = \left|-\frac{(2-1)^3}{(12)8^2} f''(c)\right| = \frac{1}{(12)(64)}|f''(c)| \le \frac{154}{768} \approx 0.2005$$

Remark: A plot of f'' shows that in fact $|f''(c)| < 1.5$, so $|E_n| < 0.002$.

35. $|f''(c)| = \left|\dfrac{4c^2(5c^4-3)}{(1+c^4)^3}\right| \le \dfrac{(4)(2^2)\left((5)(2^4)+3\right)}{\left(1+1^4\right)^3} = 166$

$|E_n| = \left|-\dfrac{(2-1)^3}{12n^2} f''(c)\right| = \dfrac{1}{12n^2}|f''(c)| \le \dfrac{166}{12n^2} < 0.0001$

$n^2 > \dfrac{166}{(12)(0.0001)} \approx 138{,}333$ so $n > \sqrt{138{,}333} \approx 371.9$ Round up to $n = 372$.

Remark: A plot of f'' shows that in fact $|f''(c)| < 1.5$ which leads to $n = 36$.

37. The integrand is decreasing and concave up. Therefore, we get:
Midpoint Rule, Trapezoidal rule, Left Riemann Sum

Review and Preview Problems

1. $\dfrac{1}{2} - \left(\dfrac{1}{2}\right)^2 = \dfrac{1}{2} - \dfrac{1}{4} = \dfrac{1}{4}$

3. the distance between $(1,4)$ and $(\sqrt[3]{4},4)$ is $\sqrt[3]{4}$ -1

5. the distance between (2,4) and (1,1) is $\sqrt{(2-1)^2+(4-1)^2} = \sqrt{10}$

7. $V = (\pi \cdot 2^2)0.4 = 1.6\pi$

9. $V = [\pi(r_2^2 - r_1^2)]\Delta x$

11. $\int_{-1}^{2}\left(x^4 - 2x^3 + 2\right)dx = \left[\dfrac{x^5}{5} - \dfrac{x^4}{2} + 2x\right]_{-1}^{2}$

$= \dfrac{12}{5} - \left(-\dfrac{27}{10}\right) = \dfrac{51}{10}$

13. $\int_0^2\left(1 - \dfrac{x^2}{2} + \dfrac{x^4}{16}\right)dx = \left[x - \dfrac{x^3}{6} + \dfrac{x^5}{80}\right]_0^2 = \dfrac{16}{15}$

CHAPTER 5

Applications of the Integral

5.1 Concepts Review

1. $\int_a^b f(x)dx; -\int_a^b f(x)dx$

3. $g(x)-f(x); f(x)=g(x)$

Problem Set 5.1

1. Slice vertically.

$\Delta A \approx (x^2+1)\Delta x$

$A=\int_{-1}^{2}(x^2+1)dx=\left[\frac{1}{3}x^3+x\right]_{-1}^{2}$

$=\left(\frac{8}{3}+2\right)-\left(-\frac{1}{3}-1\right)=6$

3. Slice vertically.

$\Delta A \approx \left[(x^2+2)-(-x)\right]\Delta x=(x^2+x+2)\Delta x$

$A=\int_{-2}^{2}(x^2+x+2)dx=\left[\frac{1}{3}x^3+\frac{1}{2}x^2+2x\right]_{-2}^{2}$

$=\left(\frac{8}{3}+2+4\right)-\left(-\frac{8}{3}+2-4\right)=\frac{40}{3}$

5. To find the intersection points, solve $2-x^2=x$.

$x^2+x-2=0$
$(x+2)(x-1)=0$
$x=-2, 1$
Slice vertically.

$\Delta A \approx \left[(2-x^2)-x\right]\Delta x=(-x^2-x-2)\Delta x$

$A=\int_{-2}^{1}(-x^2-x+2)dx=\left[-\frac{1}{3}x^3-\frac{1}{2}x^2+2x\right]_{-2}^{1}$

$=\left(-\frac{1}{3}-\frac{1}{2}+2\right)-\left(\frac{8}{3}-2-4\right)=\frac{9}{2}$

7. Solve $x^3-x^2-6x=0$.

$x(x^2-x-6)=0$
$x(x+2)(x-3)=0$
$x=-2, 0, 3$
Slice vertically.

$\Delta A_1 \approx (x^3-x^2-6x)\Delta x$

$\Delta A_2 \approx -(x^3-x^2-6x)\Delta x=(-x^3+x^2+6x)\Delta x$

$A=A_1+A_2$

$=\int_{-2}^{0}(x^3-x^2-6x)dx+\int_0^3(-x^3+x^2+6x)dx$

$=\left[\frac{1}{4}x^4-\frac{1}{3}x^3-3x^2\right]_{-2}^{0}$

$+\left[-\frac{1}{4}x^4+\frac{1}{3}x^3+3x^2\right]_0^3$

$=\left[0-\left(4+\frac{8}{3}-12\right)\right]+\left[-\frac{81}{4}+9+27-0\right]$

$=\frac{16}{3}+\frac{63}{4}=\frac{253}{12}$

9. To find the intersection points, solve $y+1=3-y^2$.

$y^2+y-2=0$
$(y+2)(y-1)=0$
$y=-2, 1$
Slice horizontally.

$\Delta A \approx \left[(3-y^2)-(y+1)\right]\Delta y=(-y^2-y+2)\Delta y$

$A=\int_{-2}^{1}(-y^2-y+2)dy=\left[-\frac{1}{3}y^3-\frac{1}{2}y^2+2y\right]_{-2}^{1}$

$=\left(-\frac{1}{3}-\frac{1}{2}+2\right)-\left(\frac{8}{3}-2-4\right)=\frac{9}{2}$

11.

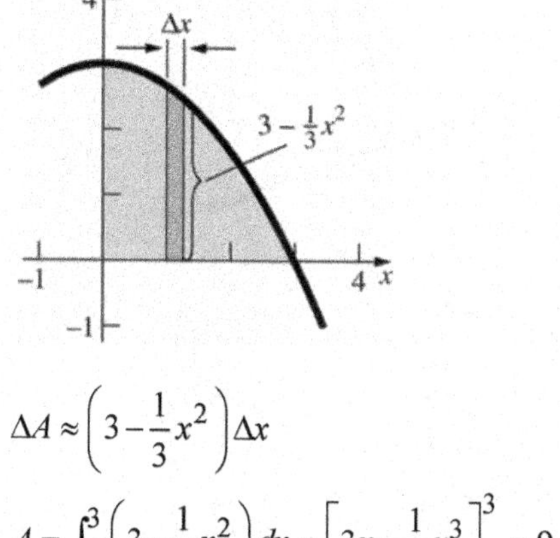

$$\Delta A \approx \left(3 - \frac{1}{3}x^2\right)\Delta x$$

$$A = \int_0^3 \left(3 - \frac{1}{3}x^2\right)dx = \left[3x - \frac{1}{9}x^3\right]_0^3 = 9 - 3 = 6$$

Estimate the area to be (3)(2) = 6.

13.

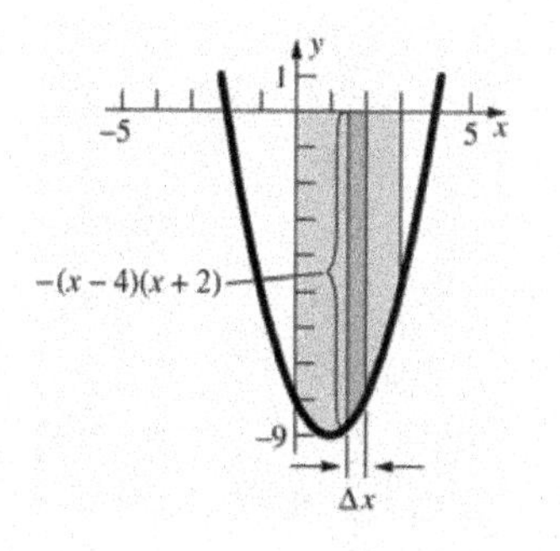

$$\Delta A \approx -(x-4)(x+2)\Delta x = (-x^2 + 2x + 8)\Delta x$$

$$A = \int_0^3 (-x^2 + 2x + 8)dx = \left[-\frac{1}{3}x^3 + x^2 + 8x\right]_0^3$$

$$= -9 + 9 + 24 = 24$$

Estimate the area to be (3)(8) = 24.

15.

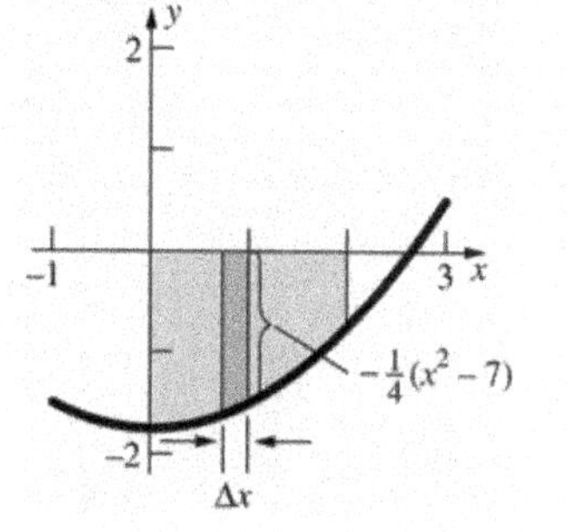

$$\Delta A \approx -\frac{1}{4}(x^2 - 7)\Delta x$$

$$A = \int_0^2 -\frac{1}{4}(x^2 - 7)dx = -\frac{1}{4}\left[\frac{1}{3}x^3 - 7x\right]_0^2$$

$$= -\frac{1}{4}\left(\frac{8}{3} - 14\right) = \frac{17}{6} \approx 2.83$$

Estimate the area to be $(2)\left(1\frac{1}{2}\right) = 3$.

17.

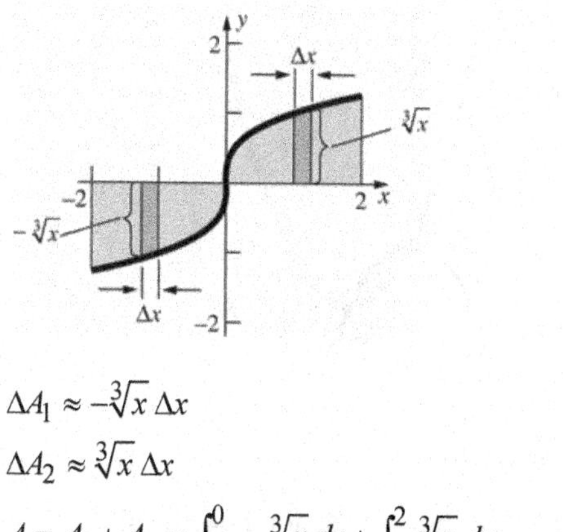

$$\Delta A_1 \approx -\sqrt[3]{x}\,\Delta x$$

$$\Delta A_2 \approx \sqrt[3]{x}\,\Delta x$$

$$A = A_1 + A_2 = \int_{-2}^0 -\sqrt[3]{x}\,dx + \int_0^2 \sqrt[3]{x}\,dx$$

$$= \left[-\frac{3}{4}x^{4/3}\right]_{-2}^0 + \left[\frac{3}{4}x^{4/3}\right]_0^2 = \left(\frac{3\sqrt[3]{2}}{2}\right) + \left(\frac{3\sqrt[3]{2}}{2}\right)$$

$$= 3\sqrt[3]{2} \approx 3.78$$

Estimate the area to be (2)(1) + (2)(1) = 4.

19.

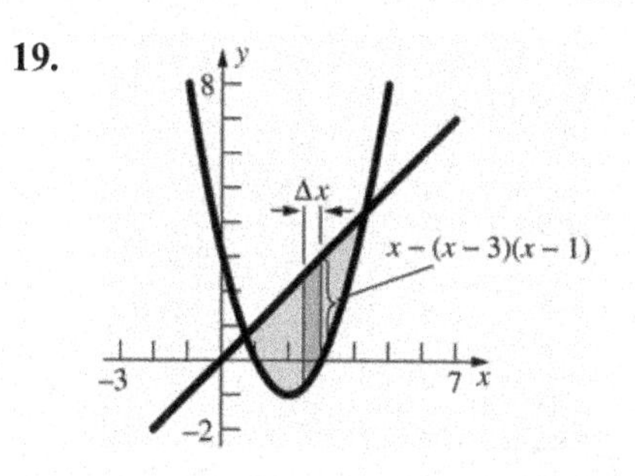

$$\Delta A \approx [x - (x-3)(x-1)]\Delta x$$

$$= \left[x - (x^2 - 4x + 3)\right]\Delta x = (-x^2 + 5x - 3)\Delta x$$

To find the intersection points, solve $x = (x - 3)(x - 1)$.

$$x^2 - 5x + 3 = 0$$

$$x = \frac{5 \pm \sqrt{25 - 12}}{2}$$

$$x = \frac{5 \pm \sqrt{13}}{2}$$

$$A = \int_{\frac{5-\sqrt{13}}{2}}^{\frac{5+\sqrt{13}}{2}} (-x^2 + 5x - 3)dx$$

$$= \left[-\frac{1}{3}x^3 + \frac{5}{2}x^2 - 3x\right]_{\frac{5-\sqrt{13}}{2}}^{\frac{5+\sqrt{13}}{2}} = \frac{13\sqrt{13}}{6} \approx 7.81$$

Estimate the area to be $\frac{1}{2}(4)(4) = 8$.

21.

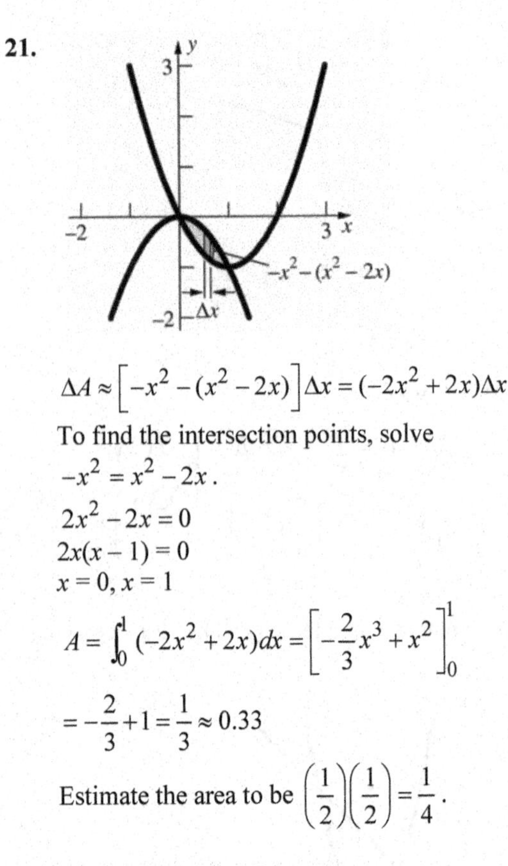

$$\Delta A \approx \left[-x^2-(x^2-2x)\right]\Delta x = (-2x^2+2x)\Delta x$$

To find the intersection points, solve

$-x^2 = x^2 - 2x$.

$2x^2 - 2x = 0$

$2x(x-1) = 0$

$x = 0, x = 1$

$$A = \int_0^1 (-2x^2+2x)\,dx = \left[-\frac{2}{3}x^3 + x^2\right]_0^1$$

$$= -\frac{2}{3}+1 = \frac{1}{3} \approx 0.33$$

Estimate the area to be $\left(\frac{1}{2}\right)\left(\frac{1}{2}\right) = \frac{1}{4}$.

23.

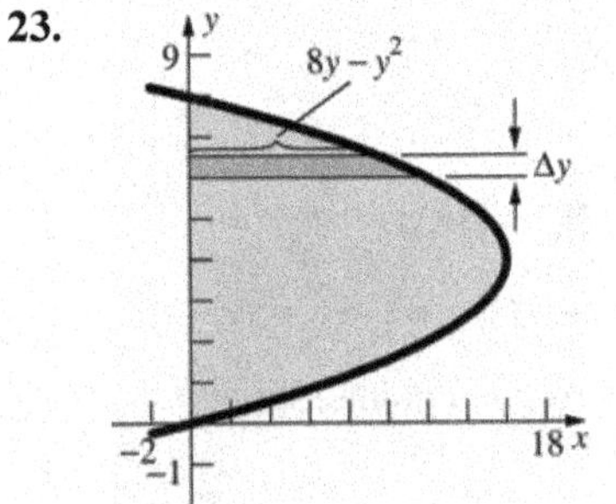

$$\Delta A \approx (8y-y^2)\Delta y$$

To find the intersection points, solve

$8y - y^2 = 0$.

$y(8-y) = 0$

$y = 0, 8$

$$A = \int_0^8 (8y-y^2)\,dy = \left[4y^2 - \frac{1}{3}y^3\right]_0^8$$

$$= 256 - \frac{512}{3} = \frac{256}{3} \approx 85.33$$

Estimate the area to be $(16)(5) = 80$.

25.

$$\Delta A \approx \left[(-6y^2+4y)-(2-3y)\right]\Delta y$$

$$= (-6y^2+7y-2)\Delta y$$

To find the intersection points, solve

$-6y^2 + 4y = 2 - 3y$.

$6y^2 - 7y + 2 = 0$

$(2y-1)(3y-2) = 0$

$y = \frac{1}{2}, \frac{2}{3}$

$$A = \int_{1/2}^{2/3} (-6y^2+7y-2)\,dy$$

$$= \left[-2y^3 + \frac{7}{2}y^2 - 2y\right]_{1/2}^{2/3}$$

$$= \left(-\frac{16}{27}+\frac{14}{9}-\frac{4}{3}\right) - \left(-\frac{1}{4}+\frac{7}{8}-1\right)$$

$$= \frac{1}{216} \approx 0.0046$$

Estimate the area to be

$\frac{1}{2}\left(\frac{1}{2}\right)\left(\frac{1}{5}\right) - \frac{1}{2}\left(\frac{1}{2}\right)\left(\frac{1}{6}\right) = \frac{1}{120}$.

27.

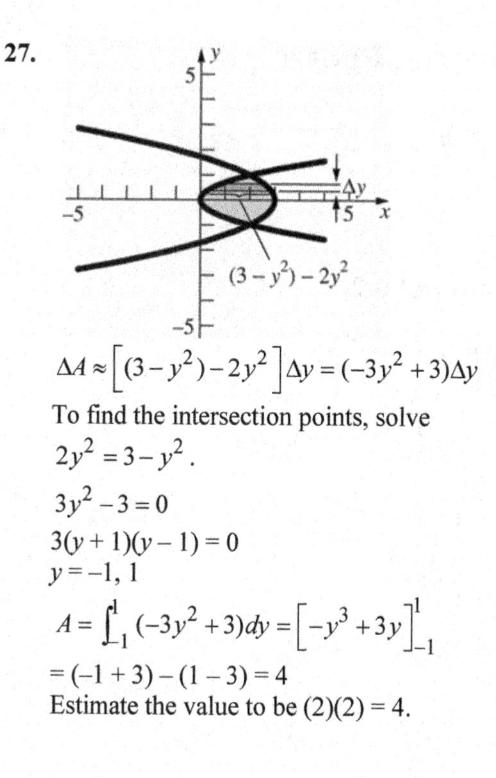

$\Delta A \approx \left[(3-y^2)-2y^2\right]\Delta y = (-3y^2+3)\Delta y$

To find the intersection points, solve

$2y^2 = 3 - y^2$.

$3y^2 - 3 = 0$

$3(y+1)(y-1) = 0$

$y = -1, 1$

$A = \int_{-1}^{1}(-3y^2+3)dy = \left[-y^3+3y\right]_{-1}^{1}$

$= (-1+3) - (1-3) = 4$

Estimate the value to be (2)(2) = 4.

29.

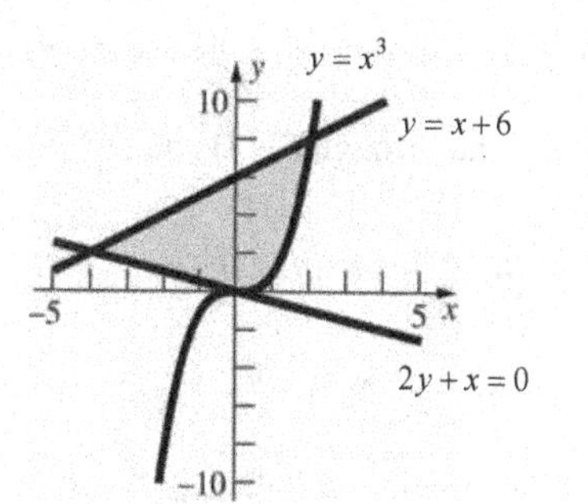

Let R_1 be the region bounded by $2y + x = 0$, $y = x + 6$, and $x = 0$.

$A(R_1) = \int_{-4}^{0}\left[(x+6) - \left(-\frac{1}{2}x\right)\right]dx$

$= \int_{-4}^{0}\left(\frac{3}{2}x+6\right)dx$

Let R_2 be the region bounded by $y = x + 6$, $y = x^3$, and $x = 0$.

$A(R_2) = \int_{0}^{2}\left[(x+6)-x^3\right]dx = \int_{0}^{2}(-x^3+x+6)dx$

$A(R) = A(R_1) + A(R_2)$

$= \int_{-4}^{0}\left(\frac{3}{2}x+6\right)dx + \int_{0}^{2}(-x^3+x+6)dx$

$= \left[\frac{3}{4}x^2+6x\right]_{-4}^{0} + \left[-\frac{1}{4}x^4+\frac{1}{2}x^2+6x\right]_{0}^{2}$

$= 12 + 10 = 22$

31. $\int_{-1}^{9}(3t^2-24t+36)dt = \left[t^3-12t^2+36t\right]_{-1}^{9} = (729-972+324)-(-1-12-36) = 130$

The displacement is 130 ft. Solve $3t^2-24t+36=0 \Rightarrow 3(t-2)(t-6)=0 \Rightarrow t=2, 6$

$|V(t)| = \begin{cases} 3t^2-24t+36 & t \le 2, t \ge 6 \\ -3t^2+24t-36 & 2 < t < 6 \end{cases}$

$\int_{-1}^{9}\left|3t^2-24t+36\right|dt = \int_{-1}^{2}(3t^2-24t+36)\,dt + \int_{2}^{6}(-3t^2+24t-36)\,dt + \int_{6}^{9}(3t^2-24t+36)\,dt$

$= \left[t^3-12t^2+36t\right]_{-1}^{2} + \left[-t^3+12t^2-36t\right]_{2}^{6} + \left[t^3-12t^2+36t\right]_{6}^{9} = 81+32+81 = 194$

The total distance traveled is 194 feet.

33. $s(t) = \int v(t)dt = \int(2t-4)dt = t^2-4t+C$

Since $s(0) = 0$, $C = 0$ and $s(t) = t^2 - 4t$. $s = 12$ when $t = 6$, so it takes the object 6 seconds to get $s = 12$.

$|2t-4| = \begin{cases} 4-2t & 0 \le t < 2 \\ 2t-4 & 2 \le t \end{cases}$

$\int_{0}^{2}|2t-4|\,dt = \left[-t^2+4t\right]_{0}^{2} = 4$, so the object travels a distance of 4 cm in the first two seconds.

$\int_{2}^{x}|2t-4|\,dt = \left[t^2-4t\right]_{2}^{x} = x^2-4x+4$

$x^2-4x+4 = 8$ when $x = 2+2\sqrt{2}$, so the object takes $2+2\sqrt{2} \approx 4.83$ seconds to travel a total distance of 12 centimeters.

35. Equation of line through (–2, 4) and (3, 9):
$y = x + 6$
Equation of line through (2, 4) and (–3, 9):
$y = -x + 6$

$$A(A) = \int_{-3}^{0}[9-(-x+6)]dx + \int_{0}^{3}[9-(x+6)]dx$$

$$= \int_{-3}^{0}(3+x)dx + \int_{0}^{3}(3-x)dx$$

$$= \left[3x+\frac{1}{2}x^2\right]_{-3}^{0} + \left[3x-\frac{1}{2}x^2\right]_{0}^{3} = \frac{9}{2}+\frac{9}{2} = 9$$

$$A(B) = \int_{-3}^{-2}[(-x+6)-x^2]dx$$

$$+ \int_{-2}^{0}[(-x+6)-(x+6)]dx$$

$$= \int_{-3}^{-2}(-x^2-x+6)dx + \int_{-2}^{0}(-2x)dx$$

$$= \left[-\frac{1}{3}x^3-\frac{1}{2}x^2+6x\right]_{-3}^{-2} + \left[-x^2\right]_{-2}^{0} = \frac{37}{6}$$

$$A(C) = A(B) = \frac{37}{6} \text{ (by symmetry)}$$

$$A(D) = \int_{-2}^{0}[(x+6)-x^2]dx + \int_{0}^{2}[(-x+6)-x^2]dx$$

$$= \left[-\frac{1}{3}x^3+\frac{1}{2}x^2+6x\right]_{-2}^{0} + \left[-\frac{1}{3}x^3-\frac{1}{2}x^2+6x\right]_{0}^{2}$$

$$= \frac{44}{3}$$

$$A(A)+A(B)+A(C)+A(D) = 36$$

$$A(A+B+C+D) = \int_{-3}^{3}(9-x^2)dx = \left[9x-\frac{1}{3}x^3\right]_{-3}^{3}$$

$$= 36$$

37. The height of the triangular region is given by for $0 \le x \le 1$. We need only show that the height of the second region is the same in order to apply Cavalieri''s Principle. The height of the second region is

$$h_2 = (x^2-2x+1)-(x^2-3x+1)$$

$$= x^2-2x+1-x^2+3x-1$$

$$= x \text{ for } 0 \le x \le 1.$$

Since $h_1 = h_2$ over the same closed interval, we can conclude that their areas are equal.

5.2 Concepts Review

1. $\pi r^2 h$

3. $\pi x^4 \Delta x$

Problem Set 5.2

1. Slice vertically.

$$\Delta V \approx \pi(x^2+1)^2\Delta x = \pi(x^4+2x^2+1)\Delta x$$

$$V = \pi\int_0^2 (x^4+2x^2+1)dx$$

$$= \pi\left[\frac{1}{5}x^5+\frac{2}{3}x^3+x\right]_0^2 = \pi\left(\frac{32}{5}+\frac{16}{3}+2\right) = \frac{206\pi}{15}$$

$$\approx 43.14$$

3. a. Slice vertically.

$$\Delta V \approx \pi(4-x^2)^2\Delta x = \pi(16-8x^2+x^4)\Delta x$$

$$V = \pi\int_0^2 (16-8x^2+x^4)dx$$

$$= \frac{256\pi}{15} \approx 53.62$$

b. Slice horizontally.

$x = \sqrt{4-y}$

Note that when $x = 0$, $y = 4$.

$$\Delta V \approx \pi\left(\sqrt{4-y}\right)^2 \Delta y = \pi(4-y)\Delta y$$

$$V = \pi\int_0^4 (4-y)dy = \pi\left[4y-\frac{1}{2}y^2\right]_0^4$$

$$= \pi(16-8) = 8\pi \approx 25.13$$

5.

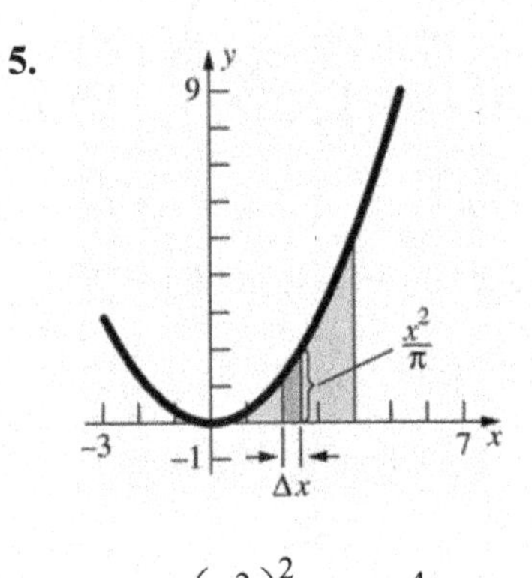

$$\Delta V \approx \pi\left(\frac{x^2}{\pi}\right)^2 \Delta x = \frac{x^4}{\pi}\Delta x$$

$$V = \int_0^4 \frac{x^4}{\pi}dx = \frac{1}{\pi}\left[\frac{1}{5}x^5\right]_0^4 = \frac{1024}{5\pi} \approx 65.19$$

7.

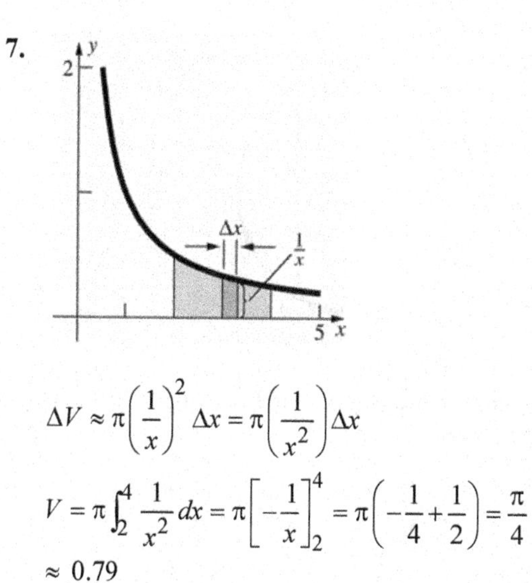

$$\Delta V \approx \pi\left(\frac{1}{x}\right)^2 \Delta x = \pi\left(\frac{1}{x^2}\right)\Delta x$$

$$V = \pi\int_2^4 \frac{1}{x^2}dx = \pi\left[-\frac{1}{x}\right]_2^4 = \pi\left(-\frac{1}{4}+\frac{1}{2}\right) = \frac{\pi}{4}$$

≈ 0.79

9. 

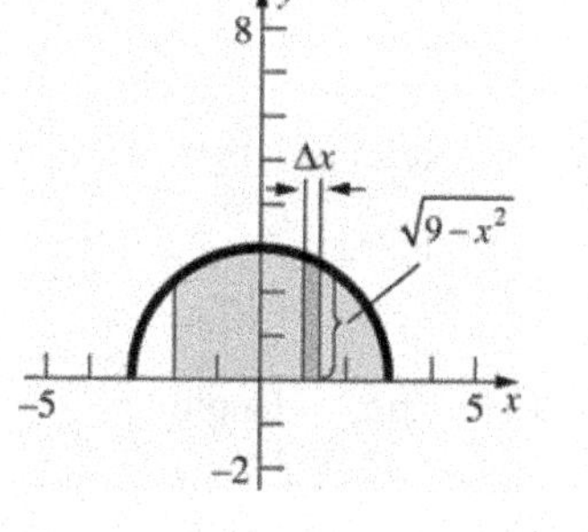

$$\Delta V \approx \pi\left(\sqrt{9-x^2}\right)^2 \Delta x = \pi(9-x^2)\Delta x$$

$$V = \pi\int_{-2}^{3}(9-x^2)dx = \pi\left[9x - \frac{1}{3}x^3\right]_{-2}^{3}$$

$$= \pi\left[(27-9)-\left(-18+\frac{8}{3}\right)\right] = \frac{100\pi}{3} \approx 104.72$$

11.

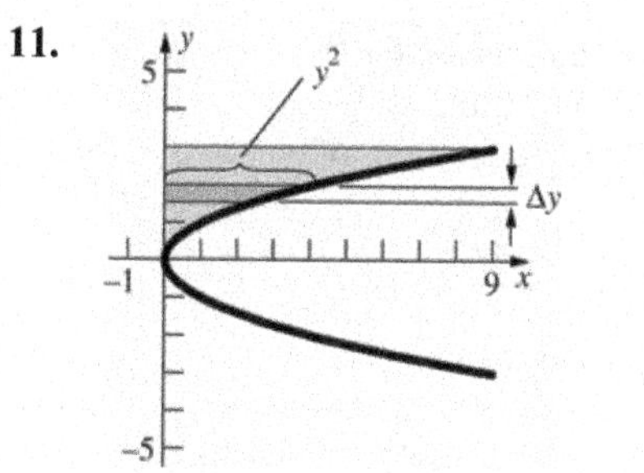

$$\Delta V \approx \pi(y^2)^2\Delta y = \pi y^4 \Delta y$$

$$V = \pi\int_0^3 y^4 dy = \pi\left[\frac{1}{5}y^5\right]_0^3 = \frac{243\pi}{5} \approx 152.68$$

13.

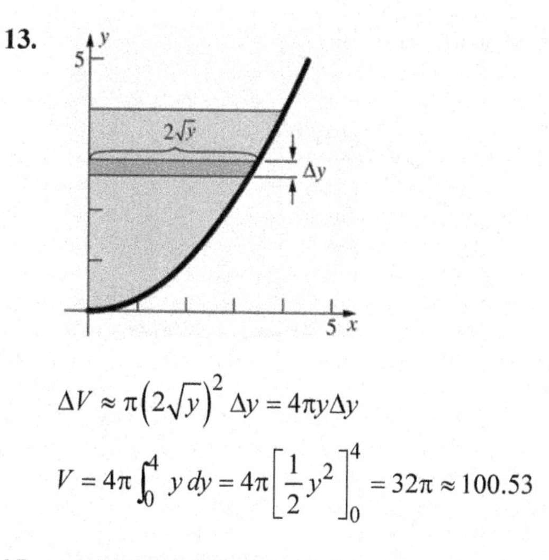

$$\Delta V \approx \pi\left(2\sqrt{y}\right)^2 \Delta y = 4\pi y \Delta y$$

$$V = 4\pi\int_0^4 y\,dy = 4\pi\left[\frac{1}{2}y^2\right]_0^4 = 32\pi \approx 100.53$$

15.

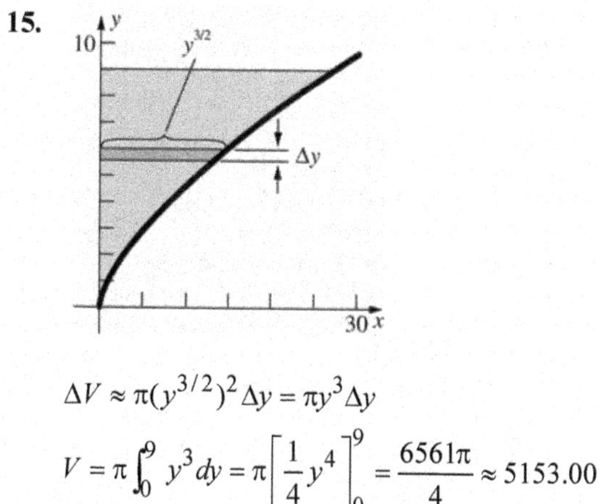

$$\Delta V \approx \pi(y^{3/2})^2\Delta y = \pi y^3 \Delta y$$

$$V = \pi\int_0^9 y^3 dy = \pi\left[\frac{1}{4}y^4\right]_0^9 = \frac{6561\pi}{4} \approx 5153.00$$

17. The equation of the upper half of the ellipse is

$$y = b\sqrt{1-\frac{x^2}{a^2}} \text{ or } y = \frac{b}{a}\sqrt{a^2-x^2}.$$

$$V = \pi\int_{-a}^{a}\frac{b^2}{a^2}(a^2-x^2)dx$$

$$= \frac{b^2\pi}{a^2}\left[a^2x - \frac{x^3}{3}\right]_{-a}^{a}$$

$$= \frac{b^2\pi}{a^2}\left[\left(a^3-\frac{a^3}{3}\right)-\left(-a^3+\frac{a^3}{3}\right)\right] = \frac{4}{3}ab^2\pi$$

19. Sketch the region.

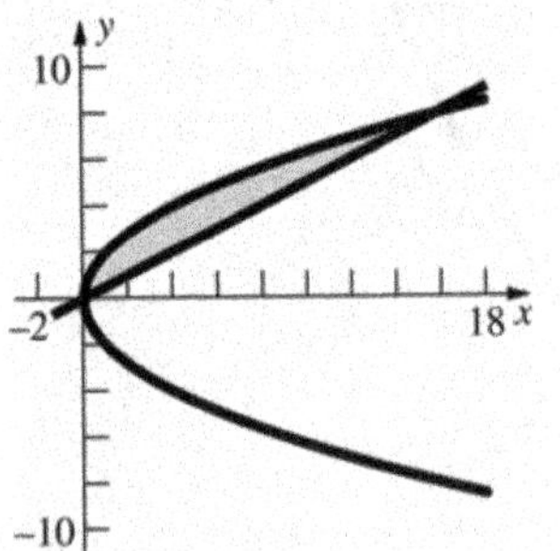

To find the intersection points, solve $\frac{x}{2} = 2\sqrt{x}$.

$$\frac{x^2}{4} = 4x$$
$$x^2 - 16x = 0$$
$$x(x-16) = 0$$
$$x = 0, 16$$

$$\Delta V \approx \pi\left[\left(2\sqrt{x}\right)^2 - \left(\frac{x}{2}\right)^2\right]\Delta x = \pi\left(4x - \frac{x^2}{4}\right)\Delta x$$

$$V = \pi\int_0^{16}\left(4x - \frac{x^2}{4}\right)dx = \pi\left[2x^2 - \frac{x^3}{12}\right]_0^{16}$$

$$= \pi\left(512 - \frac{1024}{3}\right) = \frac{512\pi}{3} \approx 536.17$$

21. Sketch the region.

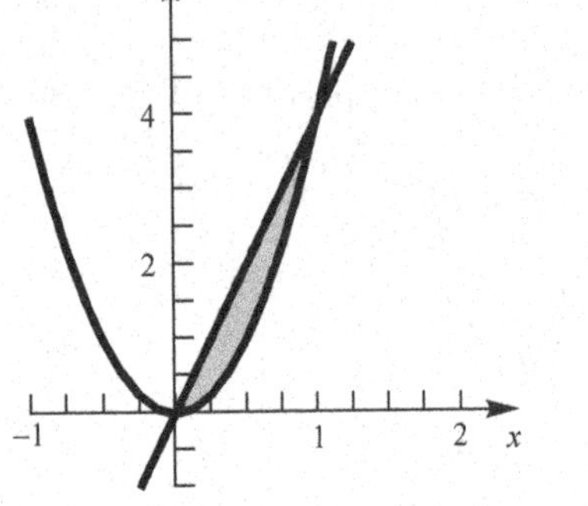

To find the intersection points, solve $\frac{y}{4} = \frac{\sqrt{y}}{2}$.

$$\frac{y^2}{16} = \frac{y}{4}$$
$$y^2 - 4y = 0$$
$$y(y-4) = 0$$
$$y = 0, 4$$

$$\Delta V \approx \pi\left[\left(\frac{\sqrt{y}}{2}\right)^2 - \left(\frac{y}{4}\right)^2\right]\Delta y = \pi\left(\frac{y}{4} - \frac{y^2}{16}\right)\Delta y$$

$$V = \pi\int_0^4\left(\frac{y}{4} - \frac{y^2}{16}\right)dy = \pi\left[\frac{y^2}{8} - \frac{y^3}{48}\right]_0^4$$

$$= \frac{2\pi}{3} \approx 2.0944$$

23.

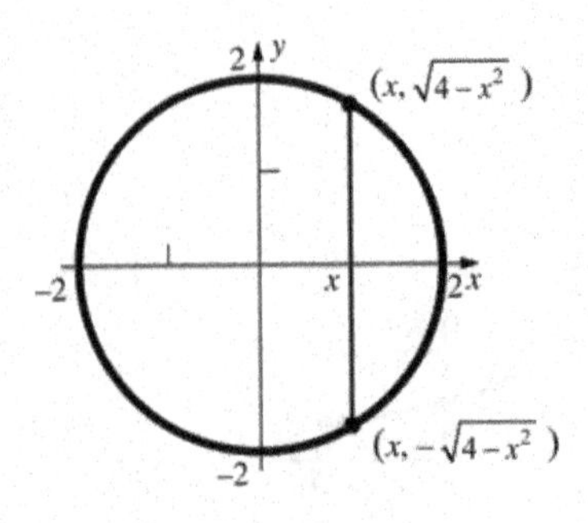

The square at x has sides of length $2\sqrt{4-x^2}$, as shown.

$$V = \int_{-2}^{2}\left(2\sqrt{4-x^2}\right)^2 dx = \int_{-2}^{2} 4(4-x^2)dx$$

$$= 4\left[4x - \frac{x^3}{3}\right]_{-2}^{2} = 4\left[\left(8 - \frac{8}{3}\right) - \left(-8 + \frac{8}{3}\right)\right] = \frac{128}{3}$$

$$\approx 42.67$$

25. The square at x has sides of length $\sqrt{\cos x}$.

$$V = \int_{-\pi/2}^{\pi/2} \cos x\,dx = [\sin x]_{-\pi/2}^{\pi/2} = 2$$

27. The square at x has sides of length $\sqrt{1-x^2}$.

$$V = \int_0^1 (1-x^2)dx = \left[x - \frac{x^3}{3}\right]_0^1 = \frac{2}{3} \approx 0.67$$

29. Using the result from Problem 28, the volume of one octant of the common region in the "+" is

$$\int_0^r (r^2 - y^2)dy = r^2 y - \frac{1}{3}y^2 \Big|_0^r$$

$$= r^3 - \frac{1}{3}r^3 = \frac{2}{3}r^3$$

Thus, the volume inside the "+" for two cylinders of radius r and length L is

$V =$ vol. of cylinders - vol. of common region

$$= 2(\pi r^2 L) - 8\left(\frac{2}{3}r^3\right)$$

$$= 2\pi r^2 L - \frac{16}{3}r^3$$

31. From Problem 30, the general form for the volume of a "T" formed by two cylinders with the same radius is

$V =$ vol. of cylinders - vol. of common region

$$= (\pi r^2)(L_1 + L_2) - 4\left(\frac{2}{3}r^3\right)$$

$$= \pi r^2(L_1 + L_2) - \frac{8}{3}r^3$$

33. Sketch the region.

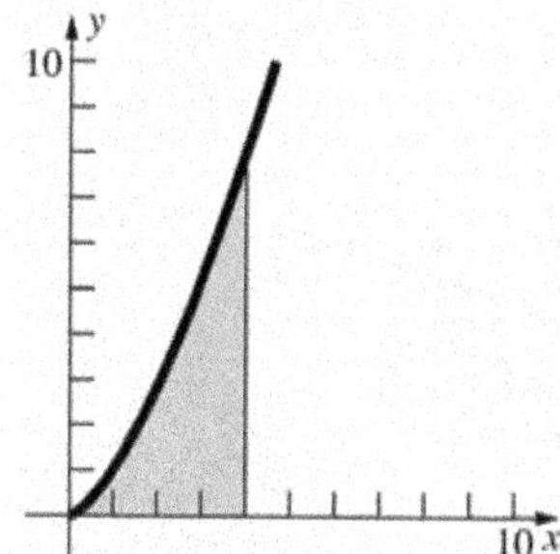

a. Revolving about the line $x = 4$, the radius of the disk at y is $4-\sqrt[3]{y^2} = 4 - y^{2/3}$.

$V = \pi\int_0^8 (4 - y^{2/3})^2\, dy$

$= \pi\int_0^8 (16 - 8y^{2/3} + y^{4/3})dy$

$= \pi\left[16y - \frac{24}{5}y^{5/3} + \frac{3}{7}y^{7/3}\right]_0^8$

$= \pi\left(128 - \frac{768}{5} + \frac{384}{7}\right)$

$= \frac{1024\pi}{35} \approx 91.91$

b. Revolving about the line $y = 8$, the inner radius of the disk at x is $8-\sqrt{x^3} = 8 - x^{3/2}$.

$V = \pi\int_0^4\left[8^2 - (8 - x^{3/2})^2\right]dx$

$= \pi\int_0^4 (16x^{3/2} - x^3)dx$

$= \pi\left[\frac{32}{5}x^{5/2} - \frac{1}{4}x^4\right]_0^4 = \pi\left(\frac{1024}{5} - 64\right)$

$= \frac{704\pi}{5} \approx 442.34$

35. The area of a quarter circle with radius 2 is $\int_0^2 \sqrt{4-y^2}\, dy = \pi$.

$\int_0^2 \left[2\sqrt{4-y^2} + 4 - y^2\right]dy$

$= 2\int_0^2 \sqrt{4-y^2}\, dy + \int_0^2 (4-y^2)dy$

$= 2\pi + \left[4y - \frac{1}{3}y^3\right]_0^2 = 2\pi + \left(8 - \frac{8}{3}\right)$

$= 2\pi + \frac{16}{3} \approx 11.62$

37. Let the x-axis lie on the base perpendicular to the diameter through the center of the base. The slice at x is a rectangle with base of length $2\sqrt{r^2 - x^2}$ and height $x\tan\theta$.

$V = \int_0^r 2x\tan\theta\sqrt{r^2 - x^2}\, dx$

$= \left[-\frac{2}{3}\tan\theta(r^2 - x^2)^{3/2}\right]_0^r$

$= \frac{2}{3}r^3\tan\theta$

39. Let A lie on the xy-plane. Suppose $\Delta A = f(x)\Delta x$ where $f(x)$ is the length at x, so $A = \int f(x)dx$. Slice the general cone at height z parallel to A. The slice of the resulting region is A_z and ΔA_z is a region related to $f(x)$ and Δx by similar triangles:

$\Delta A_z = \left(1 - \frac{z}{h}\right)f(x)\cdot\left(1 - \frac{z}{h}\right)\Delta x$

$= \left(1 - \frac{z}{h}\right)^2 f(x)\Delta x$

Therefore, $A_z = \left(1 - \frac{z}{h}\right)^2 \int f(x)dx = \left(1 - \frac{z}{h}\right)^2 A.$

$\Delta V \approx A_z\Delta z = A\left(1 - \frac{z}{h}\right)^2 \Delta z$ $V = A\int_0^h \left(1 - \frac{z}{h}\right)^2 dz$

$= A\left[-\frac{h}{3}\left(1 - \frac{z}{h}\right)^3\right]_0^h = \frac{1}{3}Ah.$

a. $A = \pi r^2$

$V = \frac{1}{3}Ah = \frac{1}{3}\pi r^2 h$

b. A face of a regular tetrahedron is an equilateral triangle. If the side of an equilateral triangle has length r, then the area is $A = \frac{1}{2}r\cdot\frac{\sqrt{3}}{2}r = \frac{\sqrt{3}}{4}r^2$.

The center of an equilateral triangle is $\frac{2}{3}\cdot\frac{\sqrt{3}}{2}r = \frac{1}{\sqrt{3}}r$ from a vertex. Then the height of a regular tetrahedron is

$h = \sqrt{r^2 - \left(\frac{1}{\sqrt{3}}r\right)^2} = \sqrt{\frac{2}{3}r^2} = \frac{\sqrt{2}}{\sqrt{3}}r$.

$V = \frac{1}{3}Ah = \frac{\sqrt{2}}{12}r^3$

41. First we examine the cross-sectional areas of each shape.
Hemisphere: cross-sectional shape is a circle.
The radius of the circle at height y is $\sqrt{r^2-y^2}$.
Therefore, the cross-sectional area for the hemisphere is
$A_h = \pi(\sqrt{r^2-y^2})^2 = \pi(r^2-y^2)$
Cylinder w/o cone: cross-sectional shape is a washer. The outer radius is a constant, r. The inner radius at height y is equal to y. Therefore, the cross-sectional area is
$A_2 = \pi r^2 - \pi y^2 = \pi(r^2-y^2)$.
Since both cross-sectional areas are the same, we can apply Cavaleri's Principle. The volume of the hemisphere of radius r is
$V =$ vol. of cylinder - vol. of cone

$$= \pi r^2 h - \frac{1}{3}\pi r^2 h$$

$$= \frac{2}{3}\pi r^2 h$$

With the height of the cylinder and cone equal to r, the volume of the hemisphere is

$$V = \frac{2}{3}\pi r^2(r) = \frac{2}{3}\pi r^3$$

5.3 Concepts Review

1. $2\pi x\, f(x)\Delta x$

3. $2\pi \int_0^2 (1+x)x\, dx$

Problem Set 5.3

1. a, b.

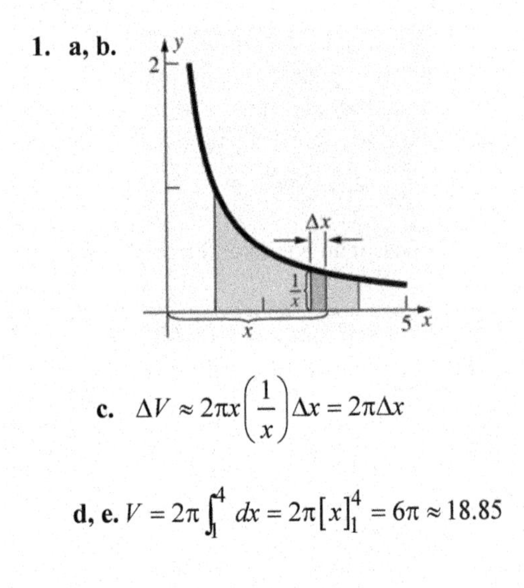

c. $\Delta V \approx 2\pi x\left(\frac{1}{x}\right)\Delta x = 2\pi\Delta x$

d, e. $V = 2\pi \int_1^4 dx = 2\pi[x]_1^4 = 6\pi \approx 18.85$

3. a, b.

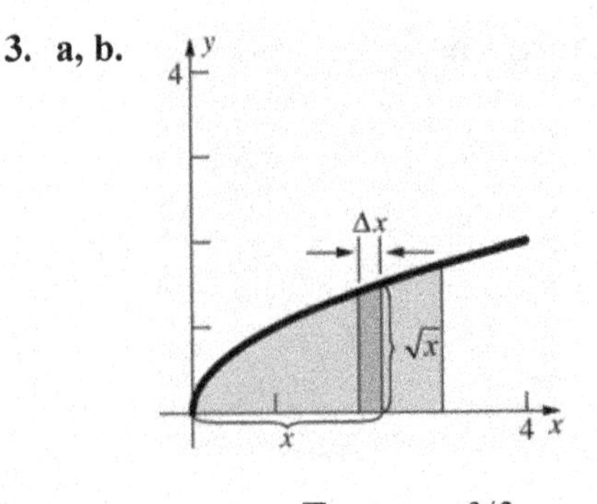

c. $\Delta V \approx 2\pi x\sqrt{x}\,\Delta x = 2\pi x^{3/2}\Delta x$

d, e. $V = 2\pi \int_0^3 x^{3/2} dx = 2\pi\left[\frac{2}{5}x^{5/2}\right]_0^3$

$$= \frac{36\sqrt{3}}{5}\pi \approx 39.18$$

5. a, b.

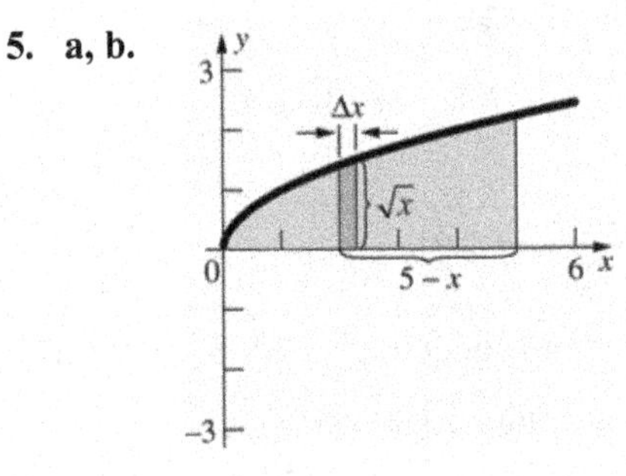

c. $\Delta V \approx 2\pi(5-x)\sqrt{x}\,\Delta x$
$= 2\pi(5x^{1/2} - x^{3/2})\Delta x$

d, e. $V = 2\pi \int_0^5 (5x^{1/2} - x^{3/2})dx$

$$= 2\pi\left[\frac{10}{3}x^{3/2} - \frac{2}{5}x^{5/2}\right]_0^5$$

$$= 2\pi\left(\frac{50\sqrt{5}}{3} - 10\sqrt{5}\right) = \frac{40\sqrt{5}}{3}\pi \approx 93.66$$

7. a, b.

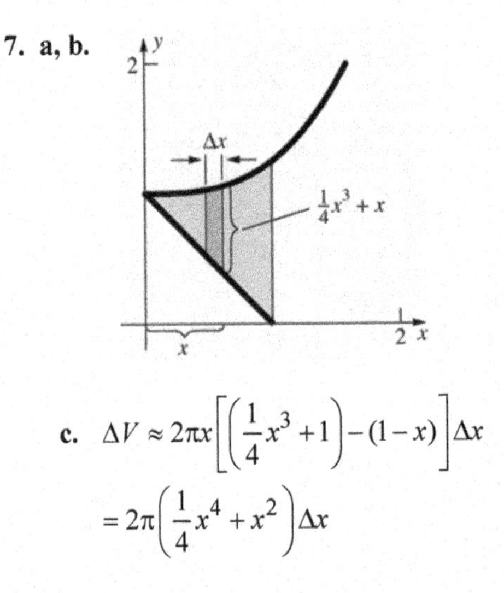

c. $\Delta V \approx 2\pi x\left[\left(\frac{1}{4}x^3 + 1\right) - (1-x)\right]\Delta x$
$= 2\pi\left(\frac{1}{4}x^4 + x^2\right)\Delta x$

d, e. $V = 2\pi \int_0^1 \left(\frac{1}{4}x^4 + x^2\right) dx$

$= 2\pi\left[\frac{1}{20}x^5 + \frac{1}{3}x^3\right]_0^1 = 2\pi\left(\frac{1}{20} + \frac{1}{3}\right)$

$= \frac{23\pi}{30} \approx 2.41$

9. a, b.

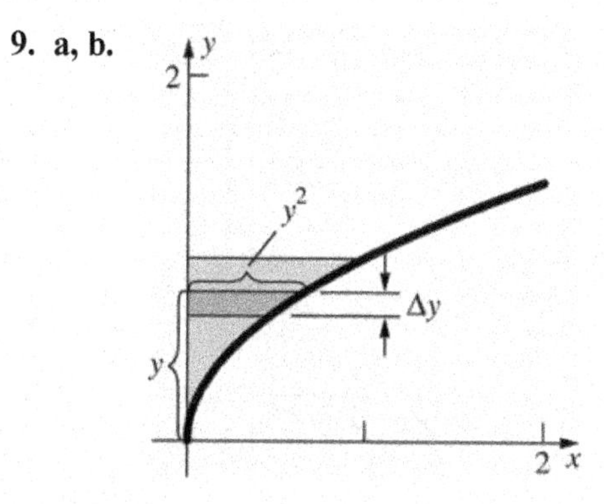

c. $\Delta V \approx 2\pi y(y^2)\Delta y = 2\pi y^3 \Delta y$

d, e. $V = 2\pi \int_0^1 y^3 dy = 2\pi\left[\frac{1}{4}y^4\right]_0^1 = \frac{\pi}{2} \approx 1.57$

11. a, b.

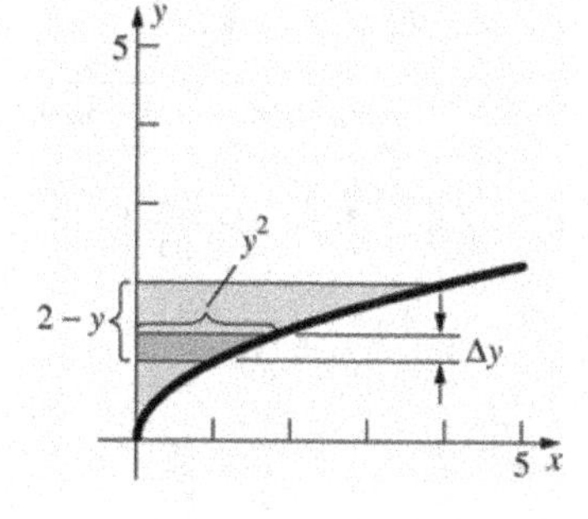

c. $\Delta V \approx 2\pi(2 - y)y^2 \Delta y = 2\pi(2y^2 - y^3)\Delta y$

d, e. $V = 2\pi \int_0^2 (2y^2 - y^3) dy = 2\pi\left[\frac{2}{3}y^3 - \frac{1}{4}y^4\right]_0^2$

$= 2\pi\left(\frac{16}{3} - 4\right) = \frac{8\pi}{3} \approx 8.38$

13. a. $\pi \int_a^b \left[f(x)^2 - g(x)^2\right] dx$

b. $2\pi \int_a^b x\left[f(x) - g(x)\right] dx$

c. $2\pi \int_a^b (x - a)\left[f(x) - g(x)\right] dx$

d. $2\pi \int_a^b (b - x)\left[f(x) - g(x)\right] dx$

15.

a. $A = \int_1^3 \frac{1}{x^3} dx$

b. $V = 2\pi \int_1^3 x\left(\frac{1}{x^3}\right) dx = 2\pi \int_1^3 \frac{1}{x^2} dx$

c. $V = \pi \int_1^3 \left[\left(\frac{1}{x^3} + 1\right)^2 - (-1)^2\right] dx$

$= \pi \int_1^3 \left(\frac{1}{x^6} + \frac{2}{x^3}\right) dx$

d. $V = 2\pi \int_1^3 (4 - x)\left(\frac{1}{x^3}\right) dx$

$= 2\pi \int_1^3 \left(\frac{4}{x^3} - \frac{1}{x^2}\right) dx$

17. To find the intersection point, solve $\sqrt{y} = \frac{y^3}{32}$.

$y = \frac{y^6}{1024}$

$y^6 - 1024y = 0$

$y(y^5 - 1024) = 0$

$y = 0, 4$

$V = 2\pi \int_0^4 y\left(\sqrt{y} - \frac{y^3}{32}\right) dy$

$= 2\pi \int_0^4 \left(y^{3/2} - \frac{y^4}{32}\right) dy$

$= 2\pi\left[\frac{2}{5}y^{5/2} - \frac{y^5}{160}\right]_0^4 = 2\pi\left(\frac{64}{5} - \frac{32}{5}\right) = \frac{64\pi}{5}$

≈ 40.21

19. Let R be the region bounded by $y=\sqrt{b^2-x^2}$, $y=-\sqrt{b^2-x^2}$, and $x=a$. When R is revolved about the y-axis, it produces the desired solid.

$$V=2\pi\int_a^b x\left(\sqrt{b^2-x^2}+\sqrt{b^2-x^2}\right)dx$$

$$=4\pi\int_a^b x\sqrt{b^2-x^2}\,dx=4\pi\left[-\frac{1}{3}(b^2-x^2)^{3/2}\right]_a^b$$

$$=4\pi\left[\frac{1}{3}(b^2-a^2)^{3/2}\right]=\frac{4\pi}{3}(b^2-a^2)^{3/2}$$

21. To find the intersection point, solve $\sin(x^2)=\cos(x^2)$.

$$\tan(x^2)=1$$

$$x^2=\frac{\pi}{4}$$

$$x=\frac{\sqrt{\pi}}{2}$$

$$V=2\pi\int_0^{\sqrt{\pi}/2} x\left[\cos(x^2)-\sin(x^2)\right]dx$$

$$=2\pi\int_0^{\sqrt{\pi}/2}\left[x\cos(x^2)-x\sin(x^2)\right]dx$$

$$=2\pi\left[\frac{1}{2}\sin(x^2)+\frac{1}{2}\cos(x^2)\right]_0^{\sqrt{\pi}/2}$$

$$=2\pi\left[\left(\frac{1}{2\sqrt{2}}+\frac{1}{2\sqrt{2}}\right)-\frac{1}{2}\right]=\pi\left(\sqrt{2}-1\right)\approx 1.30$$

23. a. The curves intersect when $x=0$ and $x=1$.

$$V=\pi\int_0^1[x^2-(x^2)^2]dx=\pi\int_0^1(x^2-x^4)dx$$

$$=\pi\left[\frac{1}{3}x^3-\frac{1}{5}x^5\right]_0^1=\pi\left(\frac{1}{3}-\frac{1}{5}\right)=\frac{2\pi}{15}\approx 0.42$$

b. $V=2\pi\int_0^1 x(x-x^2)dx=2\pi\int_0^1(x^2-x^3)dx$

$$=2\pi\left[\frac{1}{3}x^3-\frac{1}{4}x^4\right]_0^1=2\pi\left(\frac{1}{3}-\frac{1}{4}\right)=\frac{\pi}{6}$$

$$\approx 0.52$$

c. Slice perpendicular to the line $y=x$. At (a, a), the perpendicular line has equation $y=-(x-a)+a=-x+2a$. Substitute $y=-x+2a$ into $y=x^2$ and solve for $x\ge 0$.

$$x^2+x-2a=0$$

$$x=\frac{-1\pm\sqrt{1+8a}}{2}$$

$$x=\frac{-1+\sqrt{1+8a}}{2}$$

Substitute into $y=-x+2a$, so $y=\frac{1+4a-\sqrt{1+8a}}{2}$. Find an expression for r^2, the square of the distance from (a, a) to $\left(\frac{-1+\sqrt{1+8a}}{2},\frac{1+4a-\sqrt{1+8a}}{2}\right)$.

$$r^2=\left[a-\frac{-1+\sqrt{1+8a}}{2}\right]^2+\left[a-\frac{1+4a-\sqrt{1+8a}}{2}\right]^2$$

$$=\left[\frac{2a+1-\sqrt{1+8a}}{2}\right]^2+\left[-\frac{2a+1-\sqrt{1+8a}}{2}\right]^2$$

$$=2\left[\frac{2a+1-\sqrt{1+8a}}{2}\right]^2$$

$$=2a^2+6a+1-2a\sqrt{1+8a}-\sqrt{1+8a}$$

$$\Delta V\approx\pi r^2\Delta a$$

$$V=\pi\int_0^1(2a^2+6a+1-2a\sqrt{1+8a}-\sqrt{1+8a})\,da$$

$$=\pi\left[\frac{2}{3}a^3+3a^2+a-\frac{1}{12}(1+8a)^{3/2}\right]_0^1-\pi\int_0^1 2a\sqrt{1+8a}\,da$$

$$=\pi\left[\left(\frac{2}{3}+3+1-\frac{9}{4}\right)-\left(-\frac{1}{12}\right)\right]-\pi\int_0^1 2a\sqrt{1+8a}\,da$$

$$=\frac{5\pi}{2}-\pi\int_0^1 2a\sqrt{1+8a}\,da$$

To integrate $\int_0^1 2a\sqrt{1+8a}\,da$, use the substitution $u=1+8a$.

$$\int_0^1 2a\sqrt{1+8a}\,da=\int_1^9\frac{1}{4}(u-1)\sqrt{u}\frac{1}{8}du$$

$$=\frac{1}{32}\int_1^9(u^{3/2}-u^{1/2})du$$

$$=\frac{1}{32}\left[\frac{2}{5}u^{5/2}-\frac{2}{3}u^{3/2}\right]_1^9$$

$$=\frac{1}{32}\left[\left(\frac{486}{5}-18\right)-\left(\frac{2}{5}-\frac{2}{3}\right)\right]=\frac{149}{60}$$

$$V=\frac{5\pi}{2}-\frac{149\pi}{60}=\frac{\pi}{60}\approx 0.052$$

25. $\Delta V \approx \frac{x^2}{r^2} S\Delta x$

$$V = \frac{S}{r^2}\int_0^r x^2 dx = \frac{S}{r^2}\left[\frac{1}{3}x^3\right]_0^r = \frac{1}{3}rS$$

5.4 Concepts Review

1. Circle
$x^2 + y^2 = 16\cos^2 t + 16\sin^2 t = 16$

3. $\int_a^b \sqrt{[f'(t)]^2 + [g'(t)]^2}\, dt$

Problem Set 5.4

1. $f(x) = 4x^{3/2}, f'(x) = 6x^{1/2}$

$$L = \int_{1/3}^5 \sqrt{1+(6x^{1/2})^2}\, dx = \int_{1/3}^5 \sqrt{1+36x}\, dx$$
$$= \left[\frac{1}{36}\cdot\frac{2}{3}(1+36x)^{3/2}\right]_{1/3}^5$$
$$= \frac{1}{54}\left(181\sqrt{181} - 13\sqrt{13}\right) \approx 44.23$$

3. $f(x) = (4 - x^{2/3})^{3/2}$,

$$f'(x) = \frac{3}{2}(4 - x^{2/3})^{1/2}\left(-\frac{2}{3}x^{-1/3}\right)$$
$$= -x^{-1/3}(4 - x^{2/3})^{1/2}$$
$$L = \int_1^8 \sqrt{1+\left[-x^{-1/3}(4 - x^{2/3})^{1/2}\right]^2}\, dx$$
$$= \int_1^8 \sqrt{4x^{-2/3}}\, dx = \int_1^8 2x^{-1/3} dx$$
$$= 2\left[\frac{3}{2}x^{2/3}\right]_1^8 = 3(4-1) = 9$$

5. $g(y) = \frac{y^4}{16} + \frac{1}{2y^2}, g'(y) = \frac{y^3}{4} - \frac{1}{y^3}$

$$L = \int_{-3}^{-2} \sqrt{1+\left(\frac{y^3}{4} - \frac{1}{y^3}\right)^2}\, dy$$
$$= \int_{-3}^{-2} \sqrt{\frac{y^6}{16} + \frac{1}{2} + \frac{1}{y^6}}\, dy = \int_{-3}^{-2} \sqrt{\left(\frac{y^3}{4} + \frac{1}{y^3}\right)^2}\, dy$$
$$= \int_{-3}^{-2} -\left(\frac{y^3}{4} + \frac{1}{y^3}\right) dy = -\left[\frac{y^4}{16} - \frac{1}{2y^2}\right]_{-3}^{-2}$$
$$= -\left[\left(1 - \frac{1}{8}\right) - \left(\frac{81}{16} - \frac{1}{18}\right)\right] = \frac{595}{144} \approx 4.13$$

7.

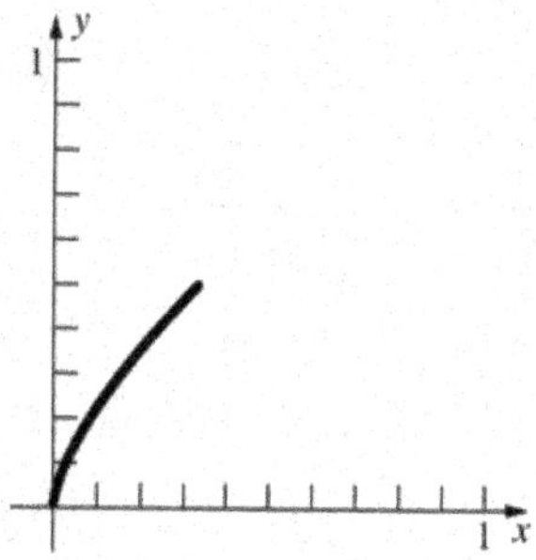

$$\frac{dx}{dt} = t^2, \frac{dy}{dt} = t$$
$$L = \int_0^1 \sqrt{(t^2)^2 + (t)^2}\, dt = \int_0^1 \sqrt{t^4 + t^2}\, dt$$
$$= \int_0^1 t\sqrt{t^2+1}\, dt = \left[\frac{1}{3}(t^2+1)^{3/2}\right]_0^1 = \frac{1}{3}\left(2\sqrt{2} - 1\right)$$
$$\approx 0.61$$

9.

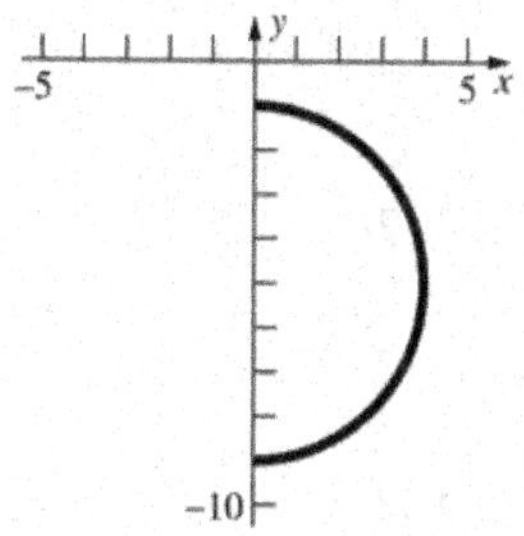

$$\frac{dx}{dt} = 4\cos t, \frac{dy}{dt} = -4\sin t$$
$$L = \int_0^\pi \sqrt{(4\cos t)^2 + (-4\sin t)^2}\, dt$$
$$= \int_0^\pi \sqrt{16\cos^2 t + 16\sin^2 t}\, dt = \int_0^\pi 4dt$$
$$= 4\pi \approx 12.57$$

11. $f(x) = 2x + 3, f'(x) = 2$

$$L = \int_1^3 \sqrt{1+(2)^2}\, dx = \sqrt{5}\int_1^3 dx = 2\sqrt{5}$$

At $x = 1, y = 2(1) + 3 = 5$.
At $x = 3, y = 2(3) + 3 = 9$.

$$d = \sqrt{(3-1)^2 + (9-5)^2} = \sqrt{20} = 2\sqrt{5}$$

13. $\frac{dx}{dt}=1, \frac{dy}{dt}=2t$

$$L=\int_0^2 \sqrt{1^2+(2t)^2}\,dt=\int_0^2 \sqrt{1+4t^2}\,dt$$

Let $f(t)=\sqrt{1+4t^2}$. Using the Parabolic Rule

with n = 8,

$$L \approx \frac{2-0}{3\times 8}\left[f(0)+4f\left(\frac{1}{4}\right)+2f\left(\frac{1}{2}\right)+4f\left(\frac{3}{4}\right)\right.$$
$$\left.+2f(1)+4f\left(\frac{5}{4}\right)+2f\left(\frac{3}{2}\right)+4f\left(\frac{7}{4}\right)+f(2)\right]$$
$$\approx \frac{1}{12}[1+4\times 1.118+2\times 1.4142$$
$$+4\times 1.8028+2\times 2.2361$$
$$+4\times 2.6926+2\times 3.1623$$
$$+4\times 3.6401+4.1231]\approx 4.6468$$

15. $\frac{dx}{dt}=\cos t, \frac{dy}{dt}=-2\sin 2t$

$$L=\int_0^{\pi/2}\sqrt{(\cos t)^2+(-2\sin 2t)^2}\,dt$$
$$=\int_0^{\pi/2}\sqrt{\cos^2 t+4\sin^2 2t}\,dt$$

Let $f(t)=\sqrt{\cos^2 t+4\sin^2 2t}$. Using the Parabolic Rule with n = 8,

$$L\approx \frac{\pi/2-0}{3\times 8}\left[f(0)+4f\left(\frac{\pi}{16}\right)+2f\left(\frac{2\pi}{16}\right)\right.$$
$$+4f\left(\frac{3\pi}{16}\right)+2f\left(\frac{4\pi}{16}\right)+4f\left(\frac{5\pi}{16}\right)+2f\left(\frac{6\pi}{16}\right)$$
$$\left.+4f\left(\frac{7\pi}{16}\right)+f\left(\frac{\pi}{2}\right)\right]\approx \frac{\pi}{48}[1+4\times 1.2441$$
$$+2\times 1.6892+4\times 2.0262+2\times 2.1213+4\times 1.9295$$
$$+2\times 1.4651+4\times 0.7898+0)\approx 2.3241$$

17.

$\frac{dx}{dt}=3a\cos t\sin^2 t, \frac{dy}{dt}=-3a\sin t\cos^2 t$

The first quadrant length is L

$$=\int_0^{\pi/2}\sqrt{(3a\cos t\sin^2 t)^2+(-3a\sin t\cos^2 t)^2}\,dt$$
$$=\int_0^{\pi/2}\sqrt{9a^2\cos^2 t\sin^4 t+9a^2\sin^2 t\cos^4 t}\,dt$$
$$=\int_0^{\pi/2}\sqrt{9a^2\cos^2 t\sin^2 t(\sin^2 t+\cos^2 t)}\,dt$$
$$=\int_0^{\pi/2}3a\cos t\sin t\,dt=3a\left[-\frac{1}{2}\cos^2 t\right]_0^{\pi/2}=\frac{3a}{2}$$

(The integral can also be evaluated as

$3a\left[\frac{1}{2}\sin^2 t\right]_0^{\pi/2}$ with the same result.)

The total length is $6a$.

19. From Problem 18,

$x=a(\theta-\sin\theta), y=a(1-\cos\theta)$

$\frac{dx}{d\theta}=a(1-\cos\theta), \frac{dy}{d\theta}=a\sin\theta$ so

$$\left(\frac{dx}{d\theta}\right)^2+\left(\frac{dy}{d\theta}\right)^2=[a(1-\cos\theta)]^2+[a\sin\theta]^2$$
$$=a^2-2a^2\cos\theta+a^2\cos^2\theta+a^2\sin^2\theta$$
$$=2a^2-2a^2\cos\theta=2a^2(1-\cos\theta)$$
$$=4a^2\frac{1-\cos\theta}{2}=4a^2\sin^2\left(\frac{\theta}{2}\right).$$

The length of one arch of the cycloid is

$$\int_0^{2\pi}\sqrt{4a^2\sin^2\left(\frac{\theta}{2}\right)}\,d\theta=\int_0^{2\pi}2a\sin\left(\frac{\theta}{2}\right)d\theta$$
$$=2a\left[-2\cos\frac{\theta}{2}\right]_0^{2\pi}=2a(2+2)=8a$$

21. **a.** $\frac{dy}{dx}=\sqrt{x^3-1}$

$$L=\int_1^2\sqrt{1+x^3-1}\,dx=\int_1^2 x^{3/2}dx$$
$$=\left[\frac{2}{5}x^{5/2}\right]_1^2=\frac{2}{5}\left(4\sqrt{2}-1\right)\approx 1.86$$

b. $f'(t)=1-\cos t, g'(t)=\sin t$

$$L=\int_0^{4\pi}\sqrt{2-2\cos t}\,dt=\int_0^{4\pi}2\left|\sin\left(\frac{t}{2}\right)\right|dt$$

$\sin\left(\frac{t}{2}\right)$ is positive for $0<t<2\pi$, and by symmetry, we can double the integral from 0 to 2π.

$$L=4\int_0^{2\pi}\sin\left(\frac{t}{2}\right)dt=\left[-8\cos\frac{t}{2}\right]_0^{2\pi}$$
$$=8+8=16$$

23. $f(x) = 6x,\ f'(x) = 6$

$$A = 2\pi \int_0^1 6x\sqrt{1+36}\,dx = 12\sqrt{37}\pi \int_0^1 x\,dx$$

$$= 12\sqrt{37}\pi\left[\frac{1}{2}x^2\right]_0^1 = 6\sqrt{37}\pi \approx 114.66$$

25. $f(x) = \frac{x^3}{3},\ f'(x) = x^2$

$$A = 2\pi \int_1^{\sqrt{7}} \frac{x^3}{3}\sqrt{1+x^4}\,dx$$

$$= 2\pi\left[\frac{1}{18}(1+x^4)^{3/2}\right]_1^{\sqrt{7}} = \frac{\pi}{9}\left(250\sqrt{2} - 2\sqrt{2}\right)$$

$$= \frac{248\pi\sqrt{2}}{9} \approx 122.43$$

27. $\frac{dx}{dt} = 1, \frac{dy}{dt} = 3t^2$

$$A = 2\pi \int_0^1 t^3\sqrt{1+9t^4}\,dt$$

$$= 2\pi\left[\frac{1}{54}(1+9t^4)^{3/2}\right]_0^1 = \frac{\pi}{27}\left(10\sqrt{10} - 1\right)$$

$$\approx 3.56$$

29. $y = f(x) = \sqrt{r^2 - x^2}$

$$f'(x) = -x(r^2 - x^2)^{-1/2}$$

$$A = 2\pi \int_{-r}^{r} \sqrt{r^2 - x^2}\sqrt{1 + \left[-x(r^2 - x^2)^{-1/2}\right]^2}\,dx$$

$$= 2\pi \int_{-r}^{r} \sqrt{r^2 - x^2}\sqrt{1 + x^2(r^2 - x^2)^{-1}}\,dx$$

$$= 2\pi \int_{-r}^{r} \sqrt{\left(r^2 - x^2\right)\left(1 + x^2(r^2 - x^2)^{-1}\right)}\,dx$$

$$= 2\pi \int_{-r}^{r} \sqrt{r^2 - x^2 + x^2}\,dx$$

$$= 2\pi \int_{-r}^{r} \sqrt{r^2}\,dx = 2\pi \int_{-r}^{r} r\,dx = 2\pi r x\,\Big|_{-r}^{r} = 4\pi r^2$$

31. a. The base circumference is equal to the arc length of the sector, so $2\pi r = \theta l$. Therefore, $\theta = \frac{2\pi r}{l}$.

b. The area of the sector is equal to the lateral surface area. Therefore, the lateral surface area is $\frac{1}{2}l^2\theta = \frac{1}{2}l^2\left(\frac{2\pi r}{l}\right) = \pi r l$.

c. Assume $r_2 > r_1$. Let l_1 and l_2 be the slant heights for r_1 and r_2, respectively. Then $A = \pi r_2 l_2 - \pi r_1 l_1 = \pi r_2(l_1 + l) - \pi r_1 l_1$.

From part a, $\theta = \frac{2\pi r_2}{l_2} = \frac{2\pi r_2}{l_1 + l} = \frac{2\pi r_1}{l_1}$.

Solve for $l_1: l_1 r_2 = l_1 r_1 + l r_1$

$$l_1(r_2 - r_1) = l r_1$$

$$l_1 = \frac{l r_1}{r_2 - r_1}$$

$$A = \pi r_2\left(\frac{l r_1}{r_2 - r_1} + l\right) - \pi r_1\left(\frac{l r_1}{r_2 - r_1}\right)$$

$$= \pi(l r_1 + l r_2) = 2\pi\left[\frac{r_1 + r_2}{2}\right]l$$

33. a. $\frac{dx}{dt} = a(1 - \cos t), \frac{dy}{dt} = a\sin t$

$$A = 2\pi \int_0^{2\pi} a(1 - \cos t)\cdot \sqrt{a^2(1 - \cos t)^2 + a^2\sin^2 t}\;dt$$

$$= 2\pi a \int_0^{2\pi} (1 - \cos t)\sqrt{2a^2 - 2a^2\cos t}\,dt$$

$$= 2\sqrt{2}\pi a^2 \int_0^{2\pi} (1 - \cos t)^{3/2}\,dt$$

b. $1 - \cos t = 2\sin^2\left(\frac{t}{2}\right)$, so

$$A = 2\sqrt{2}\pi a^2 \int_0^{2\pi} 2^{3/2}\sin^3\left(\frac{t}{2}\right)dt$$

$$= 8\pi a^2 \int_0^{2\pi} \sin\left(\frac{t}{2}\right)\sin^2\left(\frac{t}{2}\right)dt$$

$$= 8\pi a^2 \int_0^{2\pi} \sin\left(\frac{t}{2}\right)\left[1 - \cos^2\left(\frac{t}{2}\right)\right]dt$$

$$= 8\pi a^2\left[-2\cos\left(\frac{t}{2}\right) + \frac{2}{3}\cos^3\left(\frac{t}{2}\right)\right]_0^{2\pi}$$

$$= 8\pi a^2\left[\left(2 - \frac{2}{3}\right) - \left(-2 + \frac{2}{3}\right)\right] = \frac{64}{3}\pi a^2$$

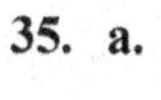

35. a.

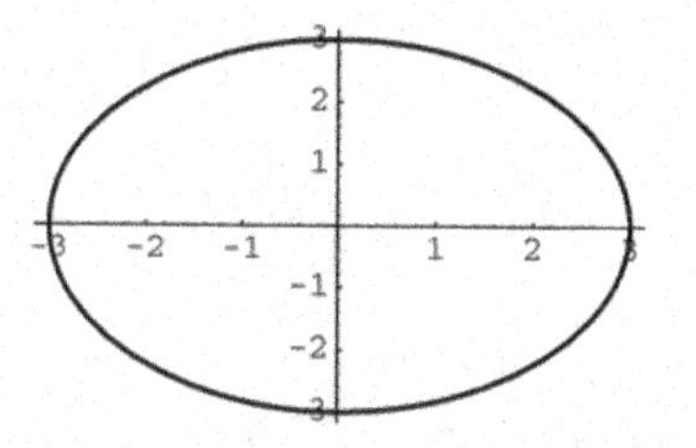

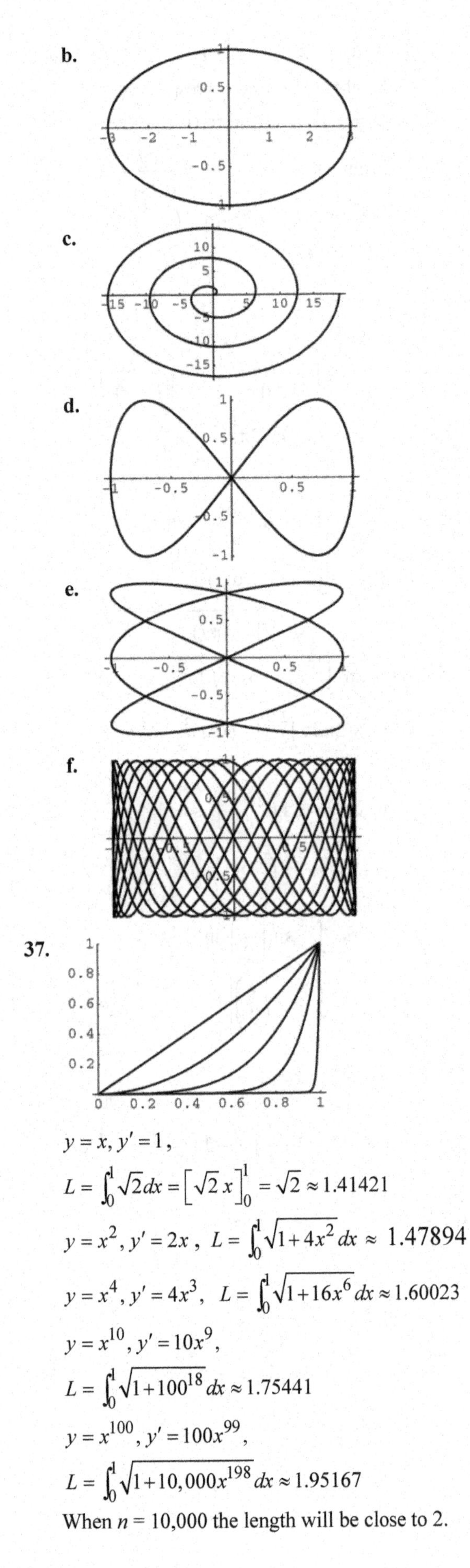

$y = x, y' = 1,$

$L = \int_0^1 \sqrt{2}dx = \left[\sqrt{2}\,x\right]_0^1 = \sqrt{2} \approx 1.41421$

$y = x^2, y' = 2x,\ L = \int_0^1 \sqrt{1+4x^2}\,dx \approx 1.47894$

$y = x^4, y' = 4x^3,\ L = \int_0^1 \sqrt{1+16x^6}\,dx \approx 1.60023$

$y = x^{10}, y' = 10x^9,$

$L = \int_0^1 \sqrt{1+100^{18}}\,dx \approx 1.75441$

$y = x^{100}, y' = 100x^{99},$

$L = \int_0^1 \sqrt{1+10{,}000x^{198}}\,dx \approx 1.95167$

When $n = 10{,}000$ the length will be close to 2.

5.5 Concepts Review

1. $F \cdot (b-a);\ \int_a^b F(x)\,dx$

3. the depth of that part of the surface

Problem Set 5.5

1. $F\left(\frac{1}{2}\right) = 6; k \cdot \frac{1}{2} = 6, k = 12$

$F(x) = 12x$

$W = \int_0^{1/2} 12x\,dx = \left[6x^2\right]_0^{1/2} = \frac{3}{2} = 1.5$ ft-lb

3. $F(0.01) = 0.6; k = 60$

$F(x) = 60x$

$W = \int_0^{0.02} 60x\,dx = \left[30x^2\right]_0^{0.02} = 0.012$ Joules

5. $W = \int_0^d kx dx = \left[\frac{1}{2}kx^2\right]_0^d = \frac{1}{2}k(d^2 - 0) = \frac{1}{2}kd^2$

7. $W = \int_0^2 9s\,ds = 9\left[\frac{1}{2}s^2\right]_0^2 = 18$ ft-lb

9. A slab of thickness Δy at height y has width $4 - \frac{4}{5}y$ and length 10. The slab will be lifted a distance $10 - y$.

$\Delta W \approx \delta \cdot 10 \cdot \left(4 - \frac{4}{5}y\right)\Delta y(10 - y)$

$= 8\delta(y^2 - 15y + 50)\Delta y$

$W = \int_0^5 8\delta(y^2 - 15y + 50)\,dy$

$= 8(62.4)\left[\frac{1}{3}y^3 - \frac{15}{2}y^2 + 50y\right]_0^5$

$= 8(62.4)\left(\frac{125}{3} - \frac{375}{2} + 250\right) = 52{,}000$ ft-lb

11. A slab of thickness Δy at height y has width $\frac{3}{4}y+3$ and length 10. The slab will be lifted a distance $9-y$. $\Delta W \approx \delta \cdot 10 \cdot \left(\frac{3}{4}y+3\right)\Delta y(9-y)$

$= \frac{15}{2}\delta(36+5y-y^2)\Delta y$

$W = \int_0^4 \frac{15}{2}\delta(36+5y-y^2)dy$

$= \frac{15}{2}(62.4)\left[36y+\frac{5}{2}y^2-\frac{1}{3}y^3\right]_0^4$

$= \frac{15}{2}(62.4)\left(144+40-\frac{64}{3}\right) = 76{,}128$ ft-lb

13. The volume of a disk with thickness Δy is $16\pi\Delta y$. If it is at height y, it will be lifted a distance $10-y$.

$\Delta W \approx \delta 16\pi\Delta y(10-y) = 16\pi\delta(10-y)\Delta y$

$W = \int_0^{10} 16\pi\delta(10-y)dy = 16\pi(50)\left[10y-\frac{1}{2}y^2\right]_0^{10}$

$= 16\pi(50)(100-50) \approx 125{,}664$ ft-lb

15. The total force on the face of the piston is $A \cdot f(x)$ if the piston is x inches from the cylinder head. The work done by moving the piston from x_1 to x_2 is $W = \int_{x_1}^{x_2} A \cdot f(x)dx = A\int_{x_1}^{x_2} f(x)dx$. This is the work done by the gas in moving the piston. The work done by the piston to compress the gas is the opposite of this or $A\int_{x_2}^{x_1} f(x)dx$.

17. $c = 40(16)^{1.4}$

$A = 2;\ p(v) = cv^{-1.4}$

$f(x) = c(2x)^{-1.4}$

$x_1 = \frac{16}{2} = 8,\ x_2 = \frac{2}{2} = 1$

$W = 2\int_1^8 c(2x)^{-1.4}dx = 2c\left[-1.25(2x)^{-0.4}\right]_1^8$

$= 80(16)^{1.4}(-1.25)(16^{-0.4}-2^{-0.4})$

≈ 2075.83 in.-lb

19. The total work is equal to the work W_1 to haul the load by itself and the work W_2 to haul the rope by itself.

$W_1 = 200 \cdot 500 = 100{,}000$ ft-lb

Let $y = 0$ be the bottom of the shaft. When the rope is at y, $\Delta W_2 \approx 2\Delta y(500-y)$.

$W_2 = \int_0^{500} 2(500-y)dy = 2\left[500y-\frac{1}{2}y^2\right]_0^{500}$

$= 2(250{,}000-125{,}000) = 250{,}000$ ft-lb

$W = W_1 + W_2 = 100{,}000 + 250{,}000$

$= 350{,}000$ ft-lb

21. $f(x) = \frac{k}{x^2};\ f(4000) = 5000$

$\frac{k}{4000^2} = 5000,\ k = 80{,}000{,}000{,}000$

$W = \int_{4000}^{4200} \frac{80{,}000{,}000{,}000}{x^2}dx$

$= 80{,}000{,}000{,}000\left[-\frac{1}{x}\right]_{4000}^{4200}$

$= \frac{20{,}000{,}000}{21} \approx 952{,}381$ mi-lb

23. The relationship between the height of the bucket and time is $y = 2t$, so $t = \frac{1}{2}y$. When the bucket is a height y, the sand has been leaking out of the bucket for $\frac{1}{2}y$ seconds. The weight of the bucket and sand is $100+500-3\left(\frac{1}{2}y\right) = 600-\frac{3}{2}y$.

$\Delta W \approx \left(600-\frac{3}{2}y\right)\Delta y$

$W = \int_0^{80}\left(600-\frac{3}{2}y\right)dy = \left[600y-\frac{3}{4}y^2\right]_0^{80}$

$= 48{,}000-4800 = 43{,}200$ ft-lb

25. Let y measure the height of a narrow rectangle with $0 \le y \le 3$. The force against this rectangle at depth $3-y$ is $\Delta F \approx \delta(3-y)(6)\Delta y$. Thus,

$F = \int_0^3 \delta(3-y)(6)\,dy = 6\delta\left[3y-\frac{y^2}{2}\right]_0^3$

$= 6 \cdot 62.4(4.5) = 1684.8$ pounds

27. Place the equilateral triangle in the coordinate system such that the vertices are $(-3,0),(3,0)$ and $(0,-3\sqrt{3})$.

The equation of the line in Quadrant I is

$y=\sqrt{3}\cdot x-3\sqrt{3}$ or $x=\frac{y}{\sqrt{3}}+3.$

$\Delta F\approx\delta(-y)\left(2\left(\frac{y}{\sqrt{3}}+3\right)\right)\Delta y$ and

$$F=\int_{-3\sqrt{3}}^{0}\delta(-y)\left(2\left(\frac{y}{\sqrt{3}}+3\right)\right)dy$$

$$=-2\delta\int_{-3\sqrt{3}}^{0}\left(\frac{y^2}{\sqrt{3}}+3y\right)dy$$

$$=-2\delta\left[\frac{y^3}{3\sqrt{3}}+\frac{3y^2}{2}\right]_{-3\sqrt{3}}^{0}=-2\cdot 62.4(0-13.5)$$

$=1684.8$ pounds

29. $\Delta F\approx\delta(1-y)\left(\sqrt{y}\right)\Delta y;\ F=\int_0^1\delta(1-y)\left(\sqrt{y}\right)dy$

$$=\delta\int_0^1\left(y^{1/2}-y^{3/2}\right)dy$$

$$=\delta\left[\frac{2}{3}y^{3/2}-\frac{2}{5}y^{5/2}\right]_0^1=62.4\left(\frac{4}{15}\right)$$

$=16.64$ pounds

31. Place a rectangle in the coordinate system such that the vertices are (0,0), (0,b), (a,0) and (a,b). The equation of the diagonal from (0,0) to (a,b) is $y=\frac{b}{a}x$ or $x=\frac{a}{b}y.$ For the upper left triangle I,

$\Delta F\approx\delta(b-y)\left(\frac{a}{b}y\right)\Delta y$ and

$$F=\int_0^b\delta(b-y)\left(\frac{a}{b}y\right)dy$$

$$=\delta\int_0^b\left(y-\frac{a}{b}y^2\right)dy=\delta\left[\frac{ay^2}{2}-\frac{ay^3}{3b}\right]_0^b$$

$$=\delta\left(\frac{ab^2}{2}-\frac{ab^2}{3}\right)=\delta\frac{ab^2}{6}$$

For the lower right triangle II,

$\Delta F\approx\delta(b-y)\left(a-\frac{a}{b}y\right)dy$ and

$$F=\int_0^b\delta(b-y)\left(a-\frac{a}{b}y\right)dy$$

$$=\int_0^b\delta\left(ab-2ay+\frac{a}{b}y^2\right)dy$$

$$=\delta\left[aby-ay^2+\frac{ay^3}{3b}\right]_0^b=\delta\left(ab^2-ab^2+\frac{ab^2}{3}\right)$$

$$=\delta\frac{ab^2}{3}$$

The total force on one half of the dam is twice the total force on the other half since $\frac{\delta\frac{ab^2}{3}}{\delta\frac{ab^2}{6}}=2.$

33. We can position the x-axis along the bottom of the pool as shown:

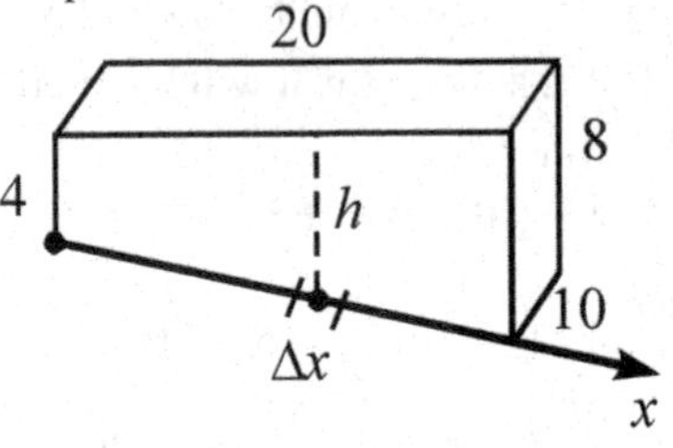

From the diagram, we let h = the depth of an arbitrary slice along the width of the bottom of the pool.

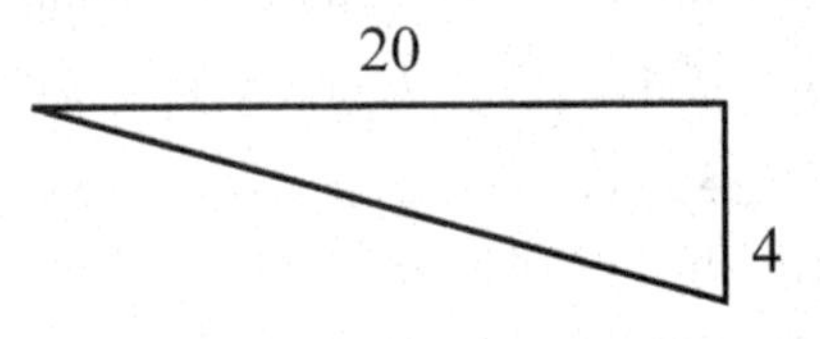

Using the Pythagorean Theorem, we can find that the length of the bottom of the pool is $\sqrt{20^2+4^2}=\sqrt{416}=4\sqrt{26}$

Next, we need to get h in terms of x. This can be done by using similar triangles to set up a proportion.

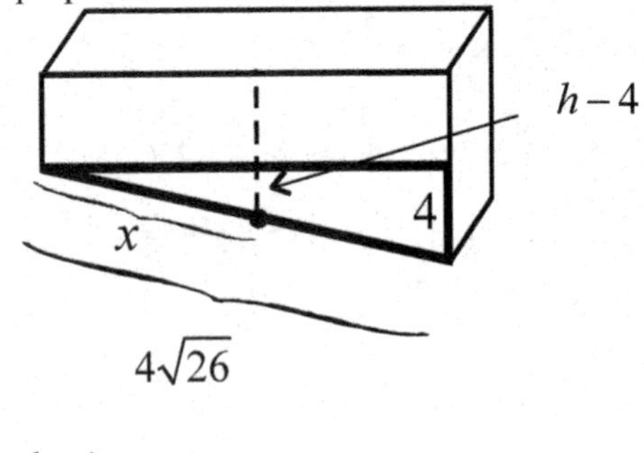

$$\frac{h-4}{4}=\frac{x}{4\sqrt{26}}\ \rightarrow\ h=4+\frac{x}{\sqrt{26}}$$

$$\Delta F = \delta \cdot h \cdot \Delta A$$

$$F = \int_0^{4\sqrt{26}} \delta\left(4 + \frac{x}{\sqrt{26}}\right)(10)\,dx$$

$$= \int_0^{4\sqrt{26}} 62.4\left(4 + \frac{x}{\sqrt{26}}\right)(10)\,dx$$

$$= 624\int_0^{4\sqrt{26}} \left(4 + \frac{x}{\sqrt{26}}\right)dx$$

$$= 624\left[4x + \frac{x^2}{2\sqrt{26}}\right]_0^{4\sqrt{26}}$$

$$= 624\left(16\sqrt{26} + 8\sqrt{26}\right) = 624\left(24\sqrt{26}\right)$$

$$= 14{,}976\sqrt{26} \text{ lb} \quad (\approx 76{,}362.92 \text{ lb})$$

35. Let W_1 be the work to lift V to the surface and W_2 be the work to lift V from the surface to 15 feet above the surface. The volume displaced by the buoy y feet above its original position is

$$\frac{1}{3}\pi\left(a - \frac{a}{h}y\right)^2 (h - y) = \frac{1}{3}\pi a^2 h\left(1 - \frac{y}{h}\right)^3.$$

The weight displaced is $\frac{\delta}{3}\pi a^2 h\left(1 - \frac{y}{h}\right)^3$.

Note by Archimede's Principle $m = \frac{\delta}{3}\pi a^2 h$ or

$a^2 h = \frac{3m}{\delta\pi}$, so the displaced weight is

$$m\left(1 - \frac{y}{h}\right)^3.$$

$$\Delta W_1 \approx \left(m - m\left(1 - \frac{y}{h}\right)^3\right)\Delta y = m\left(1 - \left(1 - \frac{y}{h}\right)^3\right)\Delta y$$

$$W_1 = m\int_0^h \left(1 - \left(1 - \frac{y}{h}\right)^3\right)dy$$

$$= m\left[y + \frac{h}{4}\left(1 - \frac{y}{h}\right)^4\right]_0^h = \frac{3mh}{4}$$

$$W_2 = m \cdot 15 = 15m$$

$$W = W_1 + W_2 = \frac{3mh}{4} + 15m$$

37. Since $\delta\left(\frac{1}{3}\pi a^2\right)(8) = 300$, $a = \sqrt{\frac{225}{2\pi\delta}}$.

When the buoy is at z feet $(0 \le z \le 2)$ below floating position, the radius r at the water level is

$$r = \left(\frac{8+z}{8}\right)a = \sqrt{\frac{225}{2\pi\delta}}\left(\frac{8+z}{8}\right).$$

$$F = \delta\left(\frac{1}{3}\pi r^2\right)(8 + z) - 300$$

$$= \frac{75}{128}(8+z)^3 - 300$$

$$W = \int_0^2 \left[\frac{75}{128}(8+z)^3 - 300\right]dz$$

$$= \left[\frac{75}{512}(8+z)^4 - 300z\right]_0^2$$

$$= \left(\frac{46{,}875}{32} - 600\right) - (600 - 0)$$

$$= \frac{8475}{32} \approx 264.84 \text{ ft-lb}$$

5.6 Concepts Review

1. right; $\frac{4 \cdot 1 + 6 \cdot 3}{4 + 6} = 2.2$

3. 1; 3

Problem Set 5.6

1. $\bar{x} = \frac{2 \cdot 5 + (-2) \cdot 7 + 1 \cdot 9}{5 + 7 + 9} = \frac{5}{21}$

3. $\bar{x} = \frac{\int_0^7 x\sqrt{x}\,dx}{\int_0^7 \sqrt{x}\,dx} = \frac{\left[\frac{2}{5}x^{5/2}\right]_0^7}{\left[\frac{2}{3}x^{3/2}\right]_0^7} = \frac{\frac{2}{5}\left(49\sqrt{7}\right)}{\frac{2}{3}\left(7\sqrt{7}\right)} = \frac{21}{5}$

5. $M_y = 1 \cdot 2 + 7 \cdot 3 + (-2) \cdot 4 + (-1) \cdot 6 + 4 \cdot 2 = 17$

$M_x = 1 \cdot 2 + 1 \cdot 3 + (-5) \cdot 4 + 0 \cdot 6 + 6 \cdot 2 = -3$

$m = 2 + 3 + 4 + 6 + 2 = 17$

$\bar{x} = \frac{M_y}{m} = 1, \bar{y} = \frac{M_x}{m} = -\frac{3}{17}$

7. Consider two regions R_1 and R_2 such that R_1 is bounded by *f(x)* and the *x*-axis, and R_2 is bounded by *g(x)* and the *x*-axis. Let R_3 be the region formed by $R_1 - R_2$. Make a regular partition of the homogeneous region R_3 such that each sub-region is of width , Δx and let *x* be the distance from the *y*-axis to the center of mass of a sub-region. The heights of R_1 and R_2 at *x* are approximately *f(x)* and *g(x)* respectively. The mass of R_3 is approximately

$$\begin{aligned}\Delta m &= \Delta m_1 - \Delta m_2 \\ &\approx \delta f(x)\Delta x - \delta g(x)\Delta x \\ &= \delta[f(x) - g(x)]\Delta x\end{aligned}$$

where δ is the density. The moments for R_3 are approximately

$$\begin{aligned}M_x &= M_x(R_1) - M_x(R_2) \\ &\approx \frac{\delta}{2}[f(x)]^2\,\Delta x - \frac{\delta}{2}[g(x)]^2\,\Delta x \\ &= \frac{\delta}{2}\left[(f(x))^2 - (g(x))^2\right]\Delta x\end{aligned}$$

$$\begin{aligned}M_y &= M_y(R_1) - M_y(R_2) \\ &\approx x\delta f(x)\Delta x - x\delta g(x)\Delta x \\ &= x\delta[f(x) - g(x)]\Delta x\end{aligned}$$

Taking the limit of the regular partition as $\Delta x \to 0$ yields the resulting integrals in Figure 10.

9.

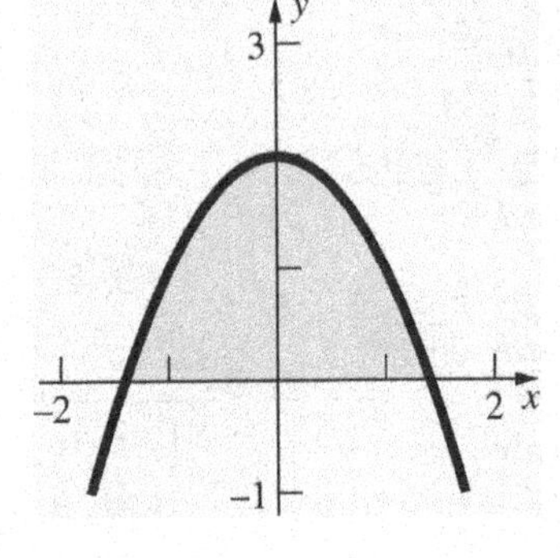

$\bar{x} = 0$ (by symmetry)

$$\bar{y} = \frac{\frac{1}{2}\int_{-\sqrt{2}}^{\sqrt{2}}(2-x^2)^2\,dx}{\int_{-\sqrt{2}}^{\sqrt{2}}(2-x^2)dx} = \frac{\frac{1}{2}\int_{-\sqrt{2}}^{\sqrt{2}}(4-4x^2+x^4)dx}{\left[2x - \frac{1}{3}x^3\right]_{-\sqrt{2}}^{\sqrt{2}}}$$

$$= \frac{\frac{1}{2}\left[4x - \frac{4}{3}x^3 + \frac{1}{5}x^5\right]_{-\sqrt{2}}^{\sqrt{2}}}{\frac{8\sqrt{2}}{3}} = \frac{\frac{32\sqrt{2}}{15}}{\frac{8\sqrt{2}}{3}} = \frac{4}{5}$$

11.

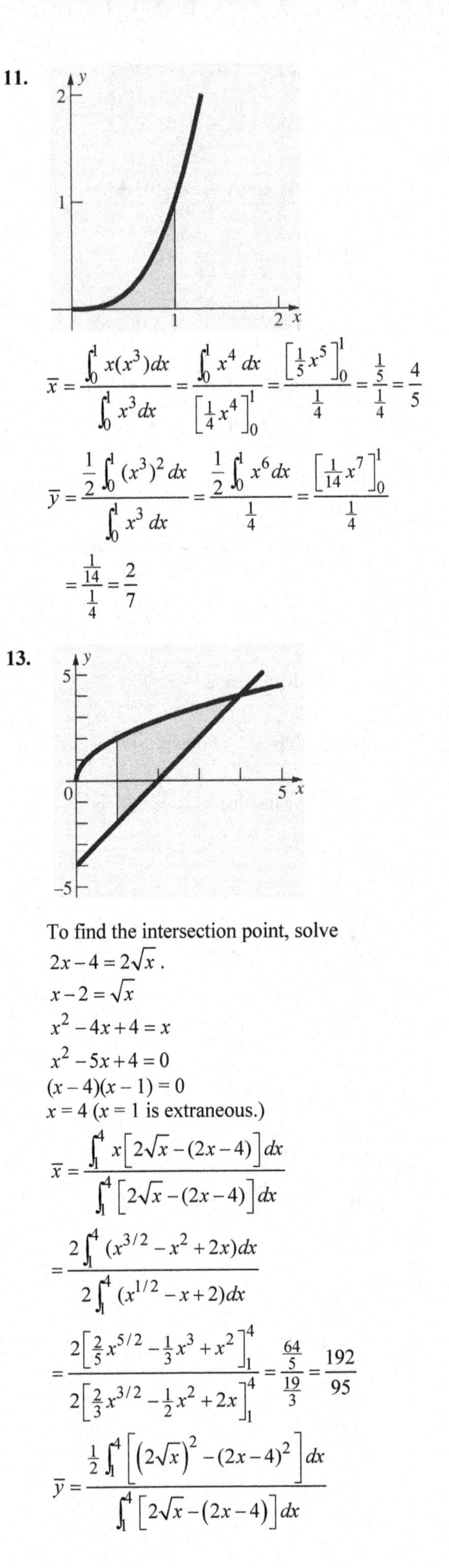

$$\bar{x} = \frac{\int_0^1 x(x^3)dx}{\int_0^1 x^3\,dx} = \frac{\int_0^1 x^4\,dx}{\left[\frac{1}{4}x^4\right]_0^1} = \frac{\left[\frac{1}{5}x^5\right]_0^1}{\frac{1}{4}} = \frac{\frac{1}{5}}{\frac{1}{4}} = \frac{4}{5}$$

$$\bar{y} = \frac{\frac{1}{2}\int_0^1 (x^3)^2\,dx}{\int_0^1 x^3\,dx} = \frac{\frac{1}{2}\int_0^1 x^6\,dx}{\frac{1}{4}} = \frac{\left[\frac{1}{14}x^7\right]_0^1}{\frac{1}{4}}$$

$$= \frac{\frac{1}{14}}{\frac{1}{4}} = \frac{2}{7}$$

13.

To find the intersection point, solve

$2x - 4 = 2\sqrt{x}$.

$x - 2 = \sqrt{x}$

$x^2 - 4x + 4 = x$

$x^2 - 5x + 4 = 0$

$(x-4)(x-1) = 0$

$x = 4$ ($x = 1$ is extraneous.)

$$\bar{x} = \frac{\int_1^4 x\left[2\sqrt{x} - (2x-4)\right]dx}{\int_1^4\left[2\sqrt{x} - (2x-4)\right]dx}$$

$$= \frac{2\int_1^4 (x^{3/2} - x^2 + 2x)dx}{2\int_1^4 (x^{1/2} - x + 2)dx}$$

$$= \frac{2\left[\frac{2}{5}x^{5/2} - \frac{1}{3}x^3 + x^2\right]_1^4}{2\left[\frac{2}{3}x^{3/2} - \frac{1}{2}x^2 + 2x\right]_1^4} = \frac{\frac{64}{5}}{\frac{19}{3}} = \frac{192}{95}$$

$$\bar{y} = \frac{\frac{1}{2}\int_1^4\left[\left(2\sqrt{x}\right)^2 - (2x-4)^2\right]dx}{\int_1^4\left[2\sqrt{x} - (2x-4)\right]dx}$$

$$= \frac{2\int_1^4 (-x^2 + 5x - 4)dx}{\frac{19}{3}}$$

$$= \frac{2\left[-\frac{1}{3}x^3 + \frac{5}{2}x^2 - 4x\right]_1^4}{\frac{19}{3}} = \frac{9}{\frac{19}{3}} = \frac{27}{19}$$

15.

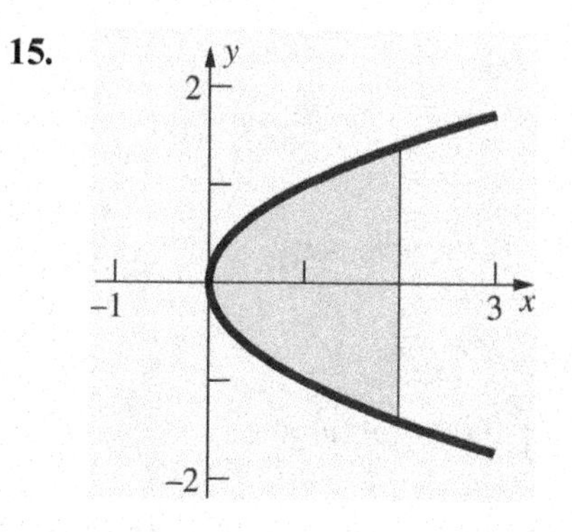

To find the intersection points, solve $y^2 = 2$.

$y = \pm\sqrt{2}$

$$\overline{x} = \frac{\frac{1}{2}\int_{-\sqrt{2}}^{\sqrt{2}}\left[2^2 - (y^2)^2\right]dy}{\int_{-\sqrt{2}}^{\sqrt{2}}(2 - y^2)dy} = \frac{\frac{1}{2}\int_{-\sqrt{2}}^{\sqrt{2}}(4 - y^4)dy}{\left[2y - \frac{1}{3}y^3\right]_{-\sqrt{2}}^{\sqrt{2}}}$$

$$= \frac{\frac{1}{2}\left[4y - \frac{1}{5}y^5\right]_{-\sqrt{2}}^{\sqrt{2}}}{\frac{8\sqrt{2}}{3}} = \frac{\frac{16\sqrt{2}}{5}}{\frac{8\sqrt{2}}{3}} = \frac{6}{5}$$

$\overline{y} = 0$ (by symmetry)

17. We let δ be the density of the regions and A_i be the area of region i.

Region R_1:

$$m(R_1) = \delta A_1 = \delta(1/2)(1)(1) = \frac{1}{2}\delta$$

$$\overline{x}_1 = \frac{\int_0^1 x(x)dx}{\int_0^1 xdx} = \frac{\frac{1}{3}x^3\Big|_0^1}{\frac{1}{2}x^2\Big|_0^1} = \frac{\frac{1}{3}}{\frac{1}{2}} = \frac{2}{3}$$

Since R_1 is symmetric about the line $y = 1 - x$, the centroid must lie on this line. Therefore, $\overline{y}_1 = 1 - \overline{x}_1 = 1 - \frac{2}{3} = \frac{1}{3}$; and we have

$$M_y(R_1) = \overline{x}_2 \cdot m(R_1) = \frac{1}{3}\delta$$

$$M_x(R_1) = \overline{y}_2 \cdot m(R_1) = \frac{1}{6}\delta$$

Region R_2:

$m(R_2) = \delta A_2 = \delta(2)(1) = 2\delta$

By symmetry we get

$\overline{x}_2 = 2$ and $\overline{y}_2 = \frac{1}{2}$.

Thus,

$M_y(R_2) = \overline{x}_2 \cdot m(R_2) = 4\delta$

$M_x(R_2) = \overline{y}_2 \cdot m(R_2) = \delta$

19. $m(R_1) = \delta\int_a^b (g(x) - f(x))dx$

$m(R_2) = \delta\int_b^c (g(x) - f(x))dx$

$M_x(R_1) = \frac{\delta}{2}\int_a^b ((g(x))^2 - (f(x))^2)dx$

$M_x(R_2) = \frac{\delta}{2}\int_b^c ((g(x))^2 - (f(x))^2)dx$

$M_y(R_1) = \delta\int_a^b x(g(x) - f(x))dx$

$M_y(R_2) = \delta\int_b^c x(g(x) - f(x))dx$

Now,

$$m(R_3) = \delta\int_a^c (g(x) - f(x))dx$$
$$= \delta\int_a^b (g(x) - f(x))dx + \delta\int_b^c (g(x) - f(x))dx$$
$$= m(R_1) + m(R_2)$$

$$M_x(R_3) = \frac{\delta}{2}\int_a^c ((g(x))^2 - (f(x))^2)dx$$
$$= \frac{\delta}{2}\int_a^b ((g(x))^2 - (f(x))^2)dx$$
$$+ \frac{\delta}{2}\int_b^c ((g(x))^2 - (f(x))^2)dx$$
$$= M_x(R_1) + M_x(R_2)$$

$$M_y(R_3) = \delta\int_a^c x(g(x) - f(x))dx$$
$$= \delta\int_a^b x(g(x) - f(x))dx$$
$$+ \delta\int_b^c x(g(x) - f(x))dx$$
$$= M_y(R_1) + M_y(R_2)$$

21. Let region 1 be the region bounded by $x = -2$, $x = 2$, $y = 0$, and $y = 1$, so $m_1 = 4 \cdot 1 = 4$.

By symmetry, $\bar{x}_1 = 0$ and $\bar{y}_1 = \frac{1}{2}$. Therefore $M_{1y} = \bar{x}_1 m_1 = 0$ and $M_{1x} = \bar{y}_1 m_1 = 2$.

Let region 2 be the region bounded by $x = -2$, $x = 1$, $y = -1$, and $y = 0$, so $m_2 = 3 \cdot 1 = 3$.

By symmetry, $\bar{x}_2 = -\frac{1}{2}$ and $\bar{y}_2 = -\frac{1}{2}$. Therefore

$M_{2y} = \bar{x}_2 m_2 = -\frac{3}{2}$ and $M_{2x} = \bar{y}_2 m_2 = -\frac{3}{2}$.

$$\bar{x} = \frac{M_{1y} + M_{2y}}{m_1 + m_2} = \frac{-\frac{3}{2}}{7} = -\frac{3}{14}$$

$$\bar{y} = \frac{M_{1x} + M_{2x}}{m_1 + m_2} = \frac{\frac{1}{2}}{7} = \frac{1}{14}$$

23. Let region 1 be the region bounded by $x = -2$, $x = 2$, $y = 2$, and $y = 4$, so $m_1 = 4 \cdot 2 = 8$. By symmetry, $\bar{x}_1 = 0$ and $\bar{y}_1 = 3$. Therefore, $M_{1y} = \bar{x}_1 m_1 = 0$ and $M_{1x} = \bar{y}_1 m_1 = 24$. Let region 2 be the region bounded by $x = -1$, $x = 2$, $y = 0$, and $y = 2$, so $m_2 = 3 \cdot 2 = 6$. By symmetry, $\bar{x}_2 = \frac{1}{2}$ and $\bar{y}_2 = 1$. Therefore, $M_{2y} = \bar{x}_2 m_2 = 3$ and $M_{2x} = \bar{y}_2 m_2 = 6$. Let region 3 be the region bounded by $x = 2$, $x = 4$, $y = 0$, and $y = 1$, so $m_3 = 2 \cdot 1 = 2$. By symmetry, $\bar{x}_3 = 3$ and $\bar{y}_2 = \frac{1}{2}$. Therefore, $M_{3y} = \bar{x}_3 m_3 = 6$ and $M_{3x} = \bar{y}_3 m_3 = 1$.

$$\bar{x} = \frac{M_{1y} + M_{2y} + M_{3y}}{m_1 + m_2 + m_3} = \frac{9}{16}$$

$$\bar{y} = \frac{M_{1x} + M_{2x} + M_{3x}}{m_1 + m_2 + m_3} = \frac{31}{16}$$

25. $A = \int_0^1 x^3\, dx = \left[\frac{1}{4}x^4\right]_0^1 = \frac{1}{4}$

From Problem 11, $\bar{x} = \frac{4}{5}$.

$V = A(2\pi\bar{x}) = \frac{1}{4}\left(2\pi \cdot \frac{4}{5}\right) = \frac{2\pi}{5}$

Using cylindrical shells:

$V = 2\pi \int_0^1 x \cdot x^3 dx = 2\pi \int_0^1 x^4 dx = 2\pi\left[\frac{1}{5}x^5\right]_0^1 = \frac{2\pi}{5}$

27. The volume of a sphere of radius a is $\frac{4}{3}\pi a^3$. If the semicircle $y = \sqrt{a^2 - x^2}$ is revolved about the x-axis the result is a sphere of radius a. The centroid of the region travels a distance of $2\pi\bar{y}$.

The area of the region is $\frac{1}{2}\pi a^2$. Pappus's Theorem says that

$(2\pi\bar{y})\left(\frac{1}{2}\pi a^2\right) = \pi^2 a^2 \bar{y} = \frac{4}{3}\pi a^3$.

$\bar{y} = \frac{4a}{3\pi}$, $\bar{x} = 0$ (by symmetry)

29. a. $\Delta V \approx 2\pi(K - y)w(y)\Delta y$

$V = 2\pi \int_c^d (K - y)w(y)dy$

b. $\Delta m \approx w(y)\Delta y$, so $m = \int_c^d w(y)dy = A$.

$\Delta M_x \approx yw(y)\Delta y$, so $M_x = \int_c^d yw(y)dy$.

$$\bar{y} = \frac{\int_c^d yw(y)dy}{A}$$

The distance traveled by the centroid is $2\pi(K - \bar{y})$.

$2\pi(K - \bar{y})A = 2\pi(KA - M_x)$

$= 2\pi\left(\int_c^d Kw(y)dy - \int_c^d yw(y)dy\right)$

$= 2\pi \int_c^d (K - y)w(y)dy$

Therefore, $V = 2\pi(K - \bar{y})A$.

31. a. The area of a regular polygon P of $2n$ sides is $2r^2 n \sin\frac{\pi}{2n}\cos\frac{\pi}{2n}$. (To find this consider the isosceles triangles with one vertex at the center of the polygon and the other vertices on adjacent corners of the polygon. Each such triangle has base of length $2r\sin\frac{\pi}{2n}$ and height $r\cos\frac{\pi}{2n}$.) Since P is a regular polygon the centroid is at its center. The distance from the centroid to any side is $r\cos\frac{\pi}{2n}$, so the centroid travels a distance of $2\pi r\cos\frac{\pi}{2n}$. Thus, by Pappus's Theorem, the volume of the resulting solid is

$\left(2\pi r\cos\frac{\pi}{2n}\right)\left(2r^2 n\sin\frac{\pi}{2n}\cos\frac{\pi}{2n}\right)$

$= 4\pi r^3 n\sin\frac{\pi}{2n}\cos^2\frac{\pi}{2n}$.

b. $\lim_{n\to\infty} 4\pi r^3 n \sin\frac{\pi}{2n}\cos^2\frac{\pi}{2n}$

$$\lim_{n\to\infty} \frac{\sin\frac{\pi}{2n}}{\frac{\pi}{2n}} 2\pi^2 r^3 \cos^2\frac{\pi}{2n} = 2\pi^2 r^3$$

As $n \to \infty$, the regular polygon approaches a circle. Using Pappus's Theorem on the circle of area πr^2 whose centroid (= center) travels a distance of $2\pi r$, the volume of the solid is $(\pi r^2)(2\pi r) = 2\pi^2 r^3$ which agrees with the results from the polygon.

33. Consider the region $S - R$.

$$\overline{y}_{S-R} = \frac{\frac{1}{2}\int_0^1 \left[g^2(x) - f^2(x)\right]dx}{S-R} \geq \overline{y}_R$$

$$= \frac{\frac{1}{2}\int_0^1 f^2(x)dx}{R}$$

$$\frac{1}{2}R\int_0^1 \left[g^2(x) - f^2(x)\right]dx \geq \frac{1}{2}(S-R)\int_0^1 f^2(x)dx$$

$$\frac{1}{2}R\int_0^1 \left[g^2(x) - f^2(x)\right]dx + \frac{1}{2}R\int_0^1 f^2(x)dx$$

$$\geq \frac{1}{2}(S-R)\int_0^1 f^2(x)dx + \frac{1}{2}R\int_0^1 f^2(x)dx$$

$$\frac{1}{2}R\int_0^1 g^2(x)dx \geq \frac{1}{2}S\int_0^1 f^2(x)dx$$

$$\frac{\frac{1}{2}\int_0^1 g^2(x)\,dx}{S} \geq \frac{\frac{1}{2}\int_0^1 f^2(x)dx}{R}$$

$$\overline{y}_S \geq \overline{y}_R$$

35. First we place the lamina so that the origin is centered inside the hole. We then recompute the centroid of Problem 34 (in this position) as

$$\overline{x} \approx \frac{\sum_{i=1}^{8} x_i h_i}{\sum_{i=1}^{8} h_i}$$

$$= \frac{(-25)(6.5) + (-15)(8) + \cdots + (5)(10) + (10)(8)}{6.5 + 8 + \cdots + 10 + 8}$$

$$= \frac{-480}{72.5} \approx -6.62$$

$$\overline{y} \approx \frac{\frac{1}{2}\sum_{i=1}^{8}((h_i - 4)^2 - (-4)^2)}{\sum_{i=1}^{8} h_i}$$

$$= \frac{(1/2)((2.5^2 - (-4)^2) + \cdots + (4^2 - (-4)^2))}{6.5 + 8 + \cdots + 10 + 8}$$

$$= \frac{45.875}{72.5} \approx 0.633$$

A quick computation will show that these values agree with those in Problem 34 (using a different reference point).

Now consider the whole lamina as R_3, the circular hole as R_2, and the remaining lamina as R_1. We can find the centroid of R_1 by noting that

$$M_x(R_1) = M_x(R_3) - M_x(R_2)$$

and similarly for $M_y(R_1)$.

From symmetry, we know that the centroid of a circle is at the center. Therefore, both $M_x(R_2)$ and $M_y(R_2)$ must be zero in our case.

This leads to the following equations

$$\overline{x} = \frac{M_y(R_3) - M_y(R_2)}{m(R_3) - m(R_2)}$$

$$= \frac{\delta\Delta x(-480)}{\delta\Delta x(72.5) - \delta\pi(2.5)^2}$$

$$= \frac{-2400}{342.87} \approx -7$$

$$\overline{y} = \frac{M_x(R_3) - M_x(R_2)}{m(R_3) - m(R_2)}$$

$$= \frac{\delta\Delta x(45.875)}{\delta\Delta x(72.5) - \delta\pi(2.5)^2}$$

$$= \frac{229.375}{342.87} \approx 0.669$$

Thus, the centroid is 7 cm above the center of the hole and 0.669 cm to the right of the center of the hole.

5.7 Concepts Review

1. discrete, continuous

3. $\int_0^5 f(x)\,dx$

Problem Set 5.7

1. a. $P(X \ge 2) = P(2) + P(3) = 0.05 + 0.05 = 0.1$

b. $E(X) = \sum_{i=1}^{4} x_i p_i$
$= 0(0.8) + 1(0.1) + 2(0.05) + 3(0.05)$
$= 0.35$

3. a. $P(X \ge 2) = P(2) = 0.2$

b. $E(X) = -2(0.2) + (-1)(0.2) + 0(0.2)$
$+ 1(0.2) + 2(0.2)$
$= 0$

5. a. $P(X \ge 2) = P(2) + P(3) + P(4)$
$= 0.2 + 0.2 + 0.2$
$= 0.6$

b. $E(X) = 1(0.4) + 2(0.2) + 3(0.2) + 4(0.2)$
$= 2.2$

7. a. $P(X \ge 2) = P(2) + P(3) + P(4)$
$= \frac{3}{10} + \frac{2}{10} + \frac{1}{10} = \frac{6}{10} = 0.6$

b. $E(X) = 1(0.4) + 2(0.3) + 3(0.2) + 4(0.1) = 2$

9. a. $P(X \ge 2) = \int_2^{20} \frac{1}{20}\,dx = \frac{1}{20} \cdot 18 = 0.9$

b. $E(X) = \int_0^{20} x \cdot \frac{1}{20}\,dx = \left[\frac{x^2}{40}\right]_0^{20} = 10$

c. For x between 0 and 20,
$F(x) = \int_0^x \frac{1}{20}\,dt = \frac{1}{20} \cdot x = \frac{x}{20}$

11. a. $P(X \ge 2) = \int_2^8 \frac{3}{256} x(8-x)\,dx$
$= \frac{3}{256}\left[4x^2 - \frac{x^3}{3}\right]_2^8 = \frac{3}{256} \cdot 72 = \frac{27}{32}$

b. $E(X) = \int_0^8 x \cdot \frac{3}{256} x(8-x)\,dx$
$= \frac{3}{256}\int_0^8 \left(8x^2 - x^3\right) dx$
$= \frac{3}{256}\left[\frac{8x^3}{3} - \frac{x^4}{4}\right]_0^8 = 4$

c. For $0 \le x \le 8$
$F(x) = \int_0^x \frac{3}{256} t(8-t)\,dt = \frac{3}{256}\left[4t^2 - \frac{t^3}{3}\right]_0^x$
$= \frac{3}{64}x^2 - \frac{1}{256}x^3$

13. a. $P(X \ge 2) = \int_2^4 \frac{3}{64} x^2(4-x)\,dx$
$= \frac{3}{64}\left[\frac{4x^3}{3} - \frac{x^4}{4}\right]_2^4 = 0.6875$

b. $E(X) = \int_0^4 x \cdot \frac{3}{64} x^2(4-x)\,dx$
$= \frac{3}{64}\int_0^4 \left(4x^3 - x^4\right) dx = \frac{3}{64}\left[x^4 - \frac{x^5}{5}\right]_0^4 = 2.4$

c. For $0 \le x \le 4$
$F(x) = \int_0^x \frac{3}{64} t^2(4-t)\,dt = \frac{3}{64}\left[\frac{4t^3}{3} - \frac{t^4}{4}\right]_0^x$
$= \frac{1}{16}x^3 - \frac{3}{256}x^4$

15. a. $P(X \ge 2) = \int_2^4 \frac{\pi}{8} \sin\left(\frac{\pi x}{4}\right) dx$
$= \frac{\pi}{8}\left[-\frac{4}{\pi}\cos\frac{\pi x}{4}\right]_2^4 = -\frac{1}{2}(-1-0) = \frac{1}{2}$

b. $E(X) = \int_0^4 x \cdot \frac{\pi}{8} \sin\left(\frac{\pi x}{4}\right) dx$
Using integration by parts or a CAS, $E(X) = 2$.

c. For $0 \le x \le 4$
$F(x) = \int_0^x \frac{\pi}{8} \sin\left(\frac{\pi t}{4}\right) dt = \frac{\pi}{8}\left[\frac{-4}{\pi}\cos\frac{\pi t}{4}\right]_0^x$
$= -\frac{1}{2}\left(\cos\frac{\pi x}{4} - 1\right) = -\frac{1}{2}\cos\frac{\pi x}{4} + \frac{1}{2}$

17. **a.** $P(X \ge 2) = \int_2^4 \frac{4}{3x^2}\,dx = \left[-\frac{4}{3x}\right]_2^4 = \frac{1}{3}$

b. $E(X) = \int_1^4 x \cdot \frac{4}{3x^2}\,dx = \left[\frac{4}{3}\ln x\right]_1^4$

$= \frac{4}{3}\ln 4 \approx 1.85$

c. For $1 \le x \le 4$

$F(x) = \int_1^x \frac{4}{3t^2}\,dt = \left[-\frac{4}{3t}\right]_1^1 = \frac{-4}{3x} + \frac{4}{3}$

$= \frac{4x-4}{3x}$

19. Proof of $F'(x) = f(x)$:

By definition, $F(x) = \int_A^x f(t)\,dt$. By the First Fundamental Theorem of Calculus, $F'(x) = f(x)$.

Proof of $F(A) = 0$ and $F(B) = 1$:

$F(A) = \int_A^A f(x)\,dx = 0;$

$F(B) = \int_A^B f(x)\,dx = 1$

Proof of $P(a \le X \le b) = F(b) - F(a)$:

$P(a \le X \le b) = \int_a^b f(x)\,dx = F(b) - F(a)$ due to the Second Fundamental Theorem of Calculus.

21. The median will be the solution to the equation $\int_a^{x_0} \frac{1}{b-a}\,dx = 0.5$.

$\frac{1}{b-a}(x_0 - a) = 0.5$

$x_0 - a = \frac{b-a}{2}$

$x_0 = \frac{a+b}{2}$

23. Since the PDF must integrate to one, solve $\int_0^5 kx(5-x)\,dx = 1$.

$\left[\frac{5kx^2}{2} - \frac{kx^3}{3}\right]_0^5 = 1$

$\frac{125k}{2} - \frac{125k}{3} = 1$

$375k - 250k = 6$

$k = \frac{6}{125}$

25. **a.** Solve $\int_0^4 k\left(2-|x-2|\right)dx = 1$

Due to the symmetry about the line x = 2, the solution can be found by solving

$2\int_0^2 kx\,dx = 1$

$k \cdot x^2\Big|_0^2 = 1$

$4k = 1$

$k = \frac{1}{4}$

b. $P(3 \le X \le 4) = \int_3^4 \frac{1}{4}\left(2-|x-2|\right)dx$

$= \int_3^4 \frac{1}{4}(2-(x-2))\,dx = \frac{1}{4}\int_3^4 (4-x)\,dx$

$= \frac{1}{4}\left[4x - \frac{x^2}{2}\right]_3^4 = \frac{1}{8}$

c. $E(X) = \int_0^4 x \cdot \frac{1}{4}(2-|x-2|)\,dx$

$= \int_0^2 x \cdot \frac{1}{4}(2+(x-2))\,dx + \int_2^4 x \cdot \frac{1}{4}(2-(x-2))\,dx$

$= \frac{1}{4}\int_0^2 x^2\,dx + \frac{1}{4}\int_2^4 (4x - x^2)\,dx$

$= \frac{1}{12}x^3\Big|_0^2 + \frac{1}{4}\left[2x^2 - \frac{x^3}{3}\right]_2^4 = \frac{2}{3} + \frac{4}{3} = 2$

d. If $0 \le x \le 2$, $F(x) = \int_0^x \frac{1}{4}t\,dt = \left[\frac{t^2}{8}\right]_0^x = \frac{x^2}{8}$

If $2 < x \le 4$, $F(x) = \int_0^2 \frac{1}{4}x\,dx + \int_2^x \frac{1}{4}(4-t)\,dt$

$= \frac{x^2}{8}\Big|_0^2 + \frac{1}{4}\left[4t - \frac{t^2}{2}\right]_2^x = \frac{1}{2} + \left(x - \frac{x^2}{8} - \frac{3}{2}\right)$

$= -\frac{x^2}{8} + x - 1$

$$F(x) = \begin{cases} 0 & \text{if } x < 0 \\ \frac{x^2}{8} & \text{if } 0 \le x \le 2 \\ -\frac{x^2}{8} + x - 1 & \text{if } 2 < x \le 4 \\ 1 & \text{if } x > 4 \end{cases}$$

e. Using a similar procedure as shown in part (a), the PDF for Y is

$$f(y) = \frac{1}{14,400}(120 - |y - 120|)$$

If $0 \le y < 120$, $F(y) = \int_0^y \frac{1}{14,400} t\,dt$

$$= \left[\frac{t^2}{28,800}\right]_0^y = \frac{y^2}{28,800}$$

If $120 < y \le 240$,

$$F(y) = \frac{1}{2} + \int_{120}^y \frac{1}{14,400}(240 - t)\,dt$$

$$= \frac{1}{2} + \frac{1}{14,400}\left[240t - \frac{t^2}{2}\right]_{120}^y$$

$$= \frac{1}{2} + \frac{y}{60} - \frac{y^2}{28,800} - \frac{3}{2} = -\frac{y^2}{28,800} + \frac{y}{60} - 1$$

$$F(x) = \begin{cases} 0 & \text{if } y < 0 \\ \frac{y^2}{28,800} & \text{if } 0 \le y \le 120 \\ -\frac{y^2}{28,800} + \frac{y}{60} - 1 & \text{if } 120 < y \le 240 \\ 1 & \text{if } y > 240 \end{cases}$$

27. a. Solve $\int_0^{0.6} kx^6(0.6 - x)^8\,dx = 1$.

$$k\int_0^{0.6} x^6(0.6 - x)^8\,dx = 1$$

Using a CAS, $k \approx 95,802,719$

b. The probability that a unit is scrapped is $1 - P(0.35 \le X \le 0.45)$

$$= 1 - k\int_{0.35}^{0.45} x^6(0.6 - x)^8\,dx$$

≈ 0.884 using a CAS

c. $E(X) = \int_0^{0.6} x \cdot kx^6(0.6 - x)^8\,dx$

$$= k\int_0^{0.6} x^7(0.6 - x)^8\,dx$$

≈ 0.2625

d. $F(x) = \int_0^x 95,802,719t^6(0.6 - t)^8\,dt$

Using a CAS,

$$F(x) \approx 6,386,850x^7(x^8 - 5.14286x^7 + 11.6308x^6 - 15.12x^5 + 12.3709x^4 - 6.53184x^3 + 2.17728x^2 - 0.419904x + 0.36)$$

e. If X = measurement in mm, and Y = measurement in inches, then $Y = X/25.4$. Thus,

$$F_Y(y) = P(Y \le y) = P(X/25.4 \le y) = P(X \le 25.4y) = F(25.4y)$$

where $F(x)$ is given in part (d).

Alternatively, we can proceed as follows:

Solve $\int_0^{3/127} k \cdot y^6\left(\frac{3}{127} - y\right)^8 dy = 1$ using a CAS.

$k \approx 1.132096857 \times 10^{29}$

$$F_Y(y) = \int_0^y k \cdot t^6\left(\frac{3}{127} - t\right)^8 dt$$

Using a CAS,

$$F_Y(y) \approx (7.54731 \times 10^{27})y^7(y^8 - 0.202475y^7 + 0.01802y^6 - 0.000923y^5 + 0.00003y^4 - (6.17827 \times 10^{-7})y^3 + (8.108 \times 10^{-9})y^2 - (6.156 \times 10^{-11})y + 2.07746 \times 10^{-13})$$

29. The PDF for the random variable X is

$$f(x) = \begin{cases} 1 & \text{if } 0 \le x \le 1 \\ 0 & \text{otherwise} \end{cases}$$

From Problem 20, the CDF for X is $F(x) = x$

Y is the distance from $(1, X)$ to the origin, so

$$Y = \sqrt{(1-0)^2 + (X-0)^2} = \sqrt{1 + X^2}$$

Here we have a one-to-one transformation from the set $\{x : 0 \le x \le 1\}$ to $\{y : 1 \le y \le \sqrt{2}\}$. For every $1 < a < b < \sqrt{2}$, the event $a < Y < b$ will occur when, and only when, $\sqrt{a^2 - 1} < X < \sqrt{b^2 - 1}$. If we let $a = 1$ and $b = y$, we can obtain the CDF for Y.

$$P(1 \le Y \le y) = P\left(\sqrt{1^2 - 1} \le X \le \sqrt{y^2 - 1}\right) = P\left(0 \le X \le \sqrt{y^2 - 1}\right) = F\left(\sqrt{y^2 - 1}\right) = \sqrt{y^2 - 1}$$

To find the PDF, we differentiate the CDF with respect to y.

$$PDF = \frac{d}{dy}\sqrt{y^2 - 1} = \frac{1}{2} \cdot \frac{1}{\sqrt{y^2 - 1}} \cdot 2y = \frac{y}{\sqrt{y^2 - 1}}$$

Therefore, for $0 \le y \le \sqrt{2}$ the PDF and CDF are respectively

$g(y) = \frac{y}{\sqrt{y^2 - 1}}$ and $G(y) = \sqrt{y^2 - 1}$.

31. By the defintion of a complement of a set, $A \cup A^c = S$, where S denotes the sample space.

Since $P(S) = 1$, $P(A \cup A^c) = 1$.

Since $P(A \cup A^c) = P(A) + P(A^c)$,

$P(A) + P(A^c) = 1$ and $P(A^c) = 1 - P(A)$.

33. $$F(x) = \begin{cases} 0 & \text{if } x < 0 \\ 0.8 & \text{if } 0 \le x < 1 \\ 0.9 & \text{if } 1 \le x < 2 \\ 0.95 & \text{if } 2 \le x < 3 \\ 1 & \text{if } x > 3 \end{cases}$$

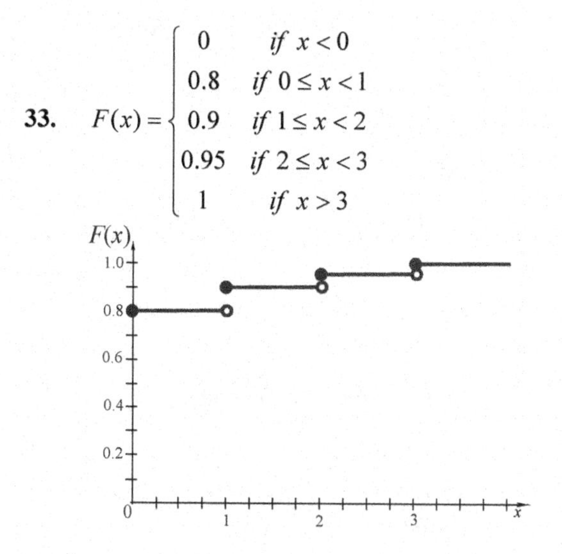

35. **a.** $P(Y < 2) = P(Y \le 2) = F(2) = 1$

b. $P(0.5 < Y < 0.6) = F(0.6) - F(0.5)$

$$= \frac{1.2}{1.6} - \frac{1}{1.5} = \frac{1}{12}$$

c. $f(y) = F'(y) = \dfrac{2}{(y+1)^2}, 0 \le y \le 1$

d. $E(Y) = \int_0^1 y \cdot \dfrac{2}{(y+1)^2} dy \approx 0.38629$

37. $E(X) = \int_0^4 x \cdot \dfrac{15}{512} x^2 (4-x)^2 \, dx = 2$

and $E(X^2) = \int_0^4 x^2 \cdot \dfrac{15}{512} x^2 (4-x)^2 \, dx$

$= \dfrac{32}{7} \approx 4.57$ using a CAS

39. $V(X) = E\left[(X-\mu)^2\right]$, where $\mu = E(X) = 2$

$V(X) = \int_0^4 (x-2)^2 \cdot \dfrac{15}{512} x^2 (4-x)^2 \, dx = \dfrac{4}{7}$

41. $E\left[(X-\mu)^2\right] = E(X^2 - 2X\mu + \mu^2)$

$= E(X^2) - E(2X\mu) + E(\mu^2)$

$= E(X^2) - 2\mu \cdot E(X) + \mu^2$

$= E(X^2) - 2\mu^2 + \mu^2$ since $E(X) = \mu$

$= E(X^2) - \mu^2$

For Problem 37, $V(X) = E(X^2) - \mu^2$ and using previous results, $V(X) = \dfrac{32}{7} - 2^2 = \dfrac{4}{7}$

5.8 Chapter Review

Concepts Test

1. False: $\int_0^{\pi} \cos x \, dx = 0$ because half of the area lies above the x-axis and half below the x-axis.

3. False: The statement would be true if either $f(x) \ge g(x)$ or $g(x) \ge f(x)$ for $a \le x \le b$. Consider Problem 1 with $f(x) = \cos x$ and $g(x) = 0$.

5. True: Since the cross sections in all planes parallel to the bases have the same area, the integrals used to compute the volumes will be equal.

7. False: Using the method of shells, $V = 2\pi \int_0^1 x(-x^2 + x)dx$. To use the method of washers we need to solve $y = -x^2 + x$ for x in terms of y.

9. False: Consider the curve given by $x = \dfrac{\cos t}{t}$, $y = \dfrac{\sin t}{t}, 2 \le t < \infty$.

11. False: If the cone-shaped tank is placed with the point downward, then the amount of water that needs to be pumped from near the bottom of the tank is much less than the amount that needs to be pumped from near the bottom of the cylindrical tank.

13. True: This is the definition of the center of mass.

15. True: By symmetry, the centroid is on the line $x = \frac{\pi}{2}$, so the centroid travels a distance of $2\pi\left(\frac{\pi}{2}\right) = \pi^2$.

17. True: Since the density is proportional to the square of the distance from the midpoint, equal masses are on either side of the midpoint.

19. True: A discrete random variable takes on a finite number of possible values, or an infinite set of possible outcomes provided that these outcomes can be put in a list such as $\{x_1, x_2, \ldots\}$.

21. True: $E(X) = 5 \cdot 1 = 5$

23. True: $P(X = 1) = P(1 \le X \le 1) = \int_1^1 f(x)\,dx = 0$

Sample Test Problems

1. $A = \int_0^1 (x - x^2)dx = \left[\frac{1}{2}x^2 - \frac{1}{3}x^3\right]_0^1 = \frac{1}{6}$

3. $V = 2\pi \int_0^1 x(x - x^2)dx = 2\pi \int_0^1 (x^2 - x^3)dx$

$= 2\pi\left[\frac{1}{3}x^3 - \frac{1}{4}x^4\right]_0^1 = \frac{\pi}{6}$

5. $V = 2\pi \int_0^1 (3 - x)(x - x^2)dx$

$= 2\pi \int_0^1 (x^3 - 4x^2 + 3x)\,dx$

$= 2\pi\left[\frac{1}{4}x^4 - \frac{4}{3}x^3 + \frac{3}{2}x^2\right]_0^1 = \frac{5\pi}{6}$

7. From Problem 1, $A = \frac{1}{6}$.

From Problem 6, $\overline{x} = \frac{1}{2}$ and $\overline{y} = \frac{1}{10}$.

$V(S_1) = 2\pi\left(\frac{1}{10}\right)\left(\frac{1}{6}\right) = \frac{\pi}{30}$

$V(S_2) = 2\pi\left(\frac{1}{2}\right)\left(\frac{1}{6}\right) = \frac{\pi}{6}$

$V(S_3) = 2\pi\left(\frac{1}{10} + 2\right)\left(\frac{1}{6}\right) = \frac{7\pi}{10}$

$V(S_4) = 2\pi\left(3 - \frac{1}{2}\right)\left(\frac{1}{6}\right) = \frac{5\pi}{6}$

9. $W = \int_0^6 (62.4)(5^2)\pi(10 - y)dy$

$= 1560\pi \int_0^6 (10 - y)dy = 1560\pi\left[10y - \frac{1}{2}y^2\right]_0^6$

$= 65{,}520\pi \approx 205{,}837$ ft-lb

11. a. To find the intersection points, solve $4x = x^2$.

$x^2 - 4x = 0$

$x(x - 4) = 0$

$x = 0, 4$

$A = \int_0^4 (4x - x^2)dx = \left[2x^2 - \frac{1}{3}x^3\right]_0^4$

$= \left(32 - \frac{64}{3}\right) = \frac{32}{3}$

b. To find the intersection points, solve $\frac{y}{4} = \sqrt{y}$.

$\frac{y^2}{16} = y$

$y^2 - 16y = 0$

$y(y - 16) = 0$

$y = 0, 16$

$A = \int_0^{16}\left(\sqrt{y} - \frac{y}{4}\right)dy = \left[\frac{2}{3}y^{3/2} - \frac{1}{8}y^2\right]_0^{16}$

$= \left(\frac{128}{3} - 32\right) = \frac{32}{3}$

13. $V = \pi \int_0^4 \left[(4x)^2 - (x^2)^2\right]dx$

$= \pi \int_0^4 (16x^2 - x^4)\,dx$

$= \pi\left[\frac{16}{3}x^3 - \frac{1}{5}x^5\right]_0^4 = \frac{2048\pi}{15}$

Using Pappus's Theorem:

From Problem 11, $A = \frac{32}{3}$.

From Problem 12, $\overline{y} = \frac{32}{5}$.

$V = 2\pi\overline{y} \cdot A = 2\pi\left(\frac{32}{5}\right)\left(\frac{32}{3}\right) = \frac{2048\pi}{15}$

15. $\frac{dy}{dx} = x^2 - \frac{1}{4x^2}$

$$L = \int_1^3 \sqrt{1+\left(x^2 - \frac{1}{4x^2}\right)^2}\, dx$$

$$= \int_1^3 \sqrt{x^4 + \frac{1}{2} + \frac{1}{16x^4}}\, dx = \int_1^3 \left(x^2 + \frac{1}{4x^2}\right) dx$$

$$= \left[\frac{1}{3}x^3 - \frac{1}{4x}\right]_1^3 = \left(9 - \frac{1}{12}\right) - \left(\frac{1}{3} - \frac{1}{4}\right) = \frac{53}{6}$$

17. $V = \int_{-3}^{3} \left(\sqrt{9-x^2}\right)^2 dx = \int_{-3}^{3} (9-x^2)\, dx$

$$= \left[9x - \frac{1}{3}x^3\right]_{-3}^{3} = (27-9) - (-27+9) = 36$$

19. $V = \pi \int_a^b \left[f^2(x) - g^2(x)\right] dx$

21. $M_y = \delta \int_a^b x[f(x) - g(x)]\, dx$

$$M_x = \frac{\delta}{2} \int_a^b \left[f^2(x) - g^2(x)\right] dx$$

23. $A_1 = 2\pi \int_a^b f(x)\sqrt{1+[f'(x)]^2}\, dx$

$$A_2 = 2\pi \int_a^b g(x)\sqrt{1+[g'(x)]^2}\, dx$$

$$A_3 = \pi\left[f^2(a) - g^2(a)\right]$$

$$A_4 = \pi\left[f^2(b) - g^2(b)\right]$$

Total surface area $= A_1 + A_2 + A_3 + A_4$.

25. a. $P(X \le 3) = F(3) = 1 - \frac{(6-3)^2}{36} = \frac{3}{4}$

b. $f(x) = F'(x) = 2 \cdot \frac{1}{36}(6-x) = \frac{6-x}{18}$, $0 \le x \le 6$

c. $E(X) = \int_0^6 x \cdot \left(\frac{6-x}{18}\right) dx$

$$= \frac{1}{18}\left[3x^2 - \frac{x^3}{3}\right]_0^6 = 2$$

Review and Preview Problems

1. By the Power Rule

$$\int \frac{1}{x^2}\, dx = \int x^{-2}\, dx = \frac{x^{-2+1}}{-2+1} = \frac{x^{-1}}{-1} = -\frac{1}{x} + C$$

3. By the Power Rule

$$\int \frac{1}{x^{1.01}}\, dx = \int x^{-1.01}\, dx = \frac{x^{-1.01+1}}{-1.01+1}$$

$$= \frac{x^{-0.01}}{-0.01} = -\frac{100}{x^{0.01}} + C$$

5. $F(1) = \int_1^1 \frac{1}{t}\, dt = 0$

7. Let $g(x) = x^2$; then by the Chain Rule and problem 6,

$$D_x F(x^2) = D_x F(g(x)) = F'(g(x))g'(x)$$

$$= \left(\frac{1}{x^2}\right)(2x) = \frac{2}{x}$$

9. a. $(1+1)^{1/1} = 2^1 = 2$

b. $\left(1+\frac{1}{5}\right)^{1/\frac{1}{5}} = \left(\frac{6}{5}\right)^5 = 2.48832$

c. $\left(1+\frac{1}{10}\right)^{1/\frac{1}{10}} = \left(\frac{11}{10}\right)^{10} \approx 2.593742$

d. $\left(1+\frac{1}{50}\right)^{1/\frac{1}{50}} = \left(\frac{51}{50}\right)^{50} \approx 2.691588$

e. $\left(1+\frac{1}{100}\right)^{1/\frac{1}{100}} = \left(\frac{101}{100}\right)^{100} \approx 2.704814$

11. a. $\left(1+\frac{1}{2}\right)^{2/1} = \left(\frac{3}{2}\right)^2 = 2.25$

b. $\left(1+\frac{\frac{1}{5}}{2}\right)^{2/\frac{1}{5}} = \left(1+\frac{1}{10}\right)^{10} = \left(\frac{11}{10}\right)^{10} \approx 2.593742$

c. $\left(1+\frac{\frac{1}{10}}{2}\right)^{2/\frac{1}{10}} = \left(1+\frac{1}{20}\right)^{20} = \left(\frac{21}{20}\right)^{20} \approx 2.6533$

d. $\left(1+\frac{\frac{1}{50}}{2}\right)^{2/\frac{1}{50}} = \left(1+\frac{1}{100}\right)^{100} = \left(\frac{101}{100}\right)^{100}$

≈ 2.70481

e. $\left(1+\frac{\frac{1}{100}}{2}\right)^{2/\frac{1}{100}} = \left(1+\frac{1}{200}\right)^{200} = \left(\frac{201}{200}\right)^{200}$

≈ 2.71152

13. We know from trigonometry that, for any x and any integer k, $\sin(x+2k\pi)=\sin(x)$. Since

$$\sin\left(\frac{\pi}{6}\right)=\frac{1}{2} \quad\text{and}\quad \sin\left(\frac{5\pi}{6}\right)=\frac{1}{2},$$

$$\sin(x)=\frac{1}{2} \text{ if } x=\frac{\pi}{6}+2k\pi=\frac{12k+1}{6}\pi$$

$$\text{or } x=\frac{5\pi}{6}+2k\pi=\frac{12k+5}{6}\pi$$

where k is any integer.

15. We know from trigonometry that, for any x and any integer k, $\tan(x+k\pi)=\tan(x)$. Since

$$\tan\left(\frac{\pi}{4}\right)=1,\ \tan(x)=1 \text{ if } x=\frac{\pi}{4}+k\pi=\frac{4k+1}{4}\pi$$

where k is any integer.

17. In the triangle, relative to θ,

$opp=\sqrt{x^2-1}$, $adj=1$, $hypot=x$ so that

$$\sin\theta=\frac{\sqrt{x^2-1}}{x} \quad \cos\theta=\frac{1}{x} \quad \tan\theta=\sqrt{x^2-1}$$

$$\cot\theta=\frac{1}{\sqrt{x^2-1}} \quad \sec\theta=x \quad \csc\theta=\frac{x}{\sqrt{x^2-1}}$$

19. In the triangle, relative to θ,

$opp=1$, $adj=x$, $hypot=\sqrt{1+x^2}$ so that

$$\sin\theta=\frac{1}{\sqrt{1+x^2}} \quad \cos\theta=\frac{x}{\sqrt{1+x^2}} \quad \tan\theta=\frac{1}{x}$$

$$\cot\theta=x \quad \sec\theta=\frac{\sqrt{1+x^2}}{x} \quad \csc\theta=\sqrt{1+x^2}$$

21.

$$y'=xy^2 \rightarrow dy=xy^2dx$$

$$\frac{1}{y^2}dy=xdx$$

$$\int\frac{dy}{y^2}=\int xdx$$

$$-\frac{1}{y}=\frac{1}{2}x^2+C$$

When $x=0$ and $y=1$ we get $C=-1$. Thus,

$$-\frac{1}{y}=\frac{1}{2}x^2-1=\frac{x^2-2}{2}$$

$$y=-\frac{2}{x^2-2}$$

CHAPTER 6 Transcendental Functions

6.1 Concepts Review

1. $\int_1^x \frac{1}{t}dt$; $(0, \infty)$; $(-\infty, \infty)$

3. $\frac{1}{x}$; $\ln|x| + C$

Problem Set 6.1

1. a. $\ln 6 = \ln(2 \cdot 3) = \ln 2 + \ln 3 = 0.693 + 1.099 = 1.792$

   b. $\ln 1.5 = \ln\left(\frac{3}{2}\right) = \ln 3 - \ln 2 = 1.099 - 0.693 = 0.406$

   c. $\ln 81 = \ln 3^4 = 4\ln 3 = 4(1.099) = 4.396$

   d. $\ln\sqrt{2} = \ln 2^{1/2} = \frac{1}{2}\ln 2 = \frac{1}{2}(0.693) = 0.3465$

   e. $\ln\left(\frac{1}{36}\right) = -\ln 36 = -\ln(2^2 \cdot 3^2) = -2\ln 2 - 2\ln 3 = -2(0.693) - 2(1.099) = -3.584$

   f. $\ln 48 = \ln(2^4 \cdot 3) = 4\ln 2 + \ln 3 = 4(0.693) + 1.099 = 3.871$

3. $D_x \ln(x^2 + 3x + \pi) = \frac{1}{x^2 + 3x + \pi} \cdot D_x(x^2 + 3x + \pi) = \frac{2x+3}{x^2 + 3x + \pi}$

5. $D_x \ln(x-4)^3 = D_x 3\ln(x-4) = 3 \cdot \frac{1}{x-4} D_x(x-4) = \frac{3}{x-4}$

7. $\frac{dy}{dx} = 3 \cdot \frac{1}{x} = \frac{3}{x}$

9. $z = x^2 \ln x^2 + (\ln x)^3 = x^2 \cdot 2\ln x + (\ln x)^3$

   $\frac{dz}{dx} = x^2 \cdot \frac{2}{x} + 2x \cdot 2\ln x + 3(\ln x)^2 \cdot \frac{1}{x} = 2x + 4x\ln x + \frac{3}{x}(\ln x)^2$

11. $g'(x) = \frac{1}{x + \sqrt{x^2+1}}\left[1 + \frac{1}{2}(x^2+1)^{-1/2} \cdot 2x\right] = \frac{1}{\sqrt{x^2+1}}$

13. $f(x) = \ln\sqrt[3]{x} = \frac{1}{3}\ln x$

    $f'(x) = \frac{1}{3} \cdot \frac{1}{x} = \frac{1}{3x}$

    $f'(81) = \frac{1}{3 \cdot 81} = \frac{1}{243}$

15. Let $u = 2x + 1$ so $du = 2\,dx$.

    $\int \frac{1}{2x+1}dx = \frac{1}{2}\int \frac{1}{u}du = \frac{1}{2}\ln|u| + C = \frac{1}{2}\ln|2x+1| + C$

17. Let $u = 3v^2 + 9v$ so $du = 6v + 9$.

    $\int \frac{6v+9}{3v^2+9v}dv = \int \frac{1}{u}du = \ln|u| + C = \ln\left|3v^2 + 9v\right| + C$

19. Let $u = \ln x$ so $du = \frac{1}{x}dx$

    $\int \frac{2\ln x}{x}dx = 2\int u\,du = u^2 + C = (\ln x)^2 + C$

21. Let $u = 2x^5 + \pi$ so $du = 10x^4\,dx$.

$$\int \frac{x^4}{2x^5+\pi}\,dx = \frac{1}{10}\int \frac{1}{u}\,du$$

$$= \frac{1}{10}\ln|u| + C = \frac{1}{10}\ln\left|2x^5+\pi\right| + C$$

$$\int_0^3 \frac{x^4}{2x^5+\pi}\,dx = \left[\frac{1}{10}\ln\left|2x^5+\pi\right|\right]_0^3$$

$$= \frac{1}{10}[\ln(486+\pi) - \ln\pi] = \ln\sqrt[10]{\frac{486+\pi}{\pi}} \approx 0.5048$$

23. By long division, $\frac{x^2}{x-1} = x+1+\frac{1}{x-1}$

so $\int \frac{x^2}{x-1}\,dx = \int x\,dx + \int 1\,dx + \int \frac{1}{x-1}\,dx$

$$= \frac{x^2}{2} + x + \ln|x-1| + C$$

25. By long division,

$\frac{x^4}{x+4} = x^3 - 4x^2 + 16x - 64 + \frac{256}{x+4}$ so

$\int \frac{x^4}{x+4}\,dx =$

$\int x^3\,dx - \int 4x^2\,dx + \int 16x\,dx - \int 64\,dx + 256\int \frac{1}{x+4}\,dx$

$$= \frac{x^4}{4} - \frac{4x^3}{3} + 8x^2 - 64x + 256\ln|x+4| + C$$

27. $2\ln(x+1) - \ln x = \ln(x+1)^2 - \ln x = \ln\frac{(x+1)^2}{x}$

29. $\ln(x-2) - \ln(x+2) + 2\ln x$

$$= \ln(x-2) - \ln(x+2) + \ln x^2$$

$$= \ln\frac{x^2(x-2)}{x+2}$$

31. $\ln y = \ln(x+11) - \frac{1}{2}\ln(x^3-4)$

$$\frac{1}{y}\frac{dy}{dx} = \frac{1}{x+11}\cdot 1 - \frac{1}{2}\cdot\frac{1}{x^3-4}\cdot 3x^2$$

$$= \frac{1}{x+11} - \frac{3x^2}{2(x^3-4)}$$

$$\frac{dy}{dx} = y\cdot\left[\frac{1}{x+11} - \frac{3x^2}{2(x^3-4)}\right]$$

$$= \frac{x+11}{\sqrt{x^3-4}}\left[\frac{1}{x+11} - \frac{3x^2}{2(x^3-4)}\right]$$

$$= -\frac{x^3+33x^2+8}{2(x^3-4)^{3/2}}$$

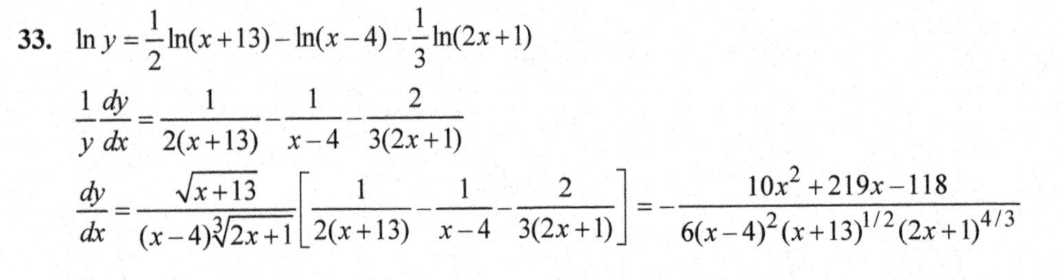

33. $\ln y = \frac{1}{2}\ln(x+13) - \ln(x-4) - \frac{1}{3}\ln(2x+1)$

$$\frac{1}{y}\frac{dy}{dx} = \frac{1}{2(x+13)} - \frac{1}{x-4} - \frac{2}{3(2x+1)}$$

$$\frac{dy}{dx} = \frac{\sqrt{x+13}}{(x-4)\sqrt[3]{2x+1}}\left[\frac{1}{2(x+13)} - \frac{1}{x-4} - \frac{2}{3(2x+1)}\right] = -\frac{10x^2+219x-118}{6(x-4)^2(x+13)^{1/2}(2x+1)^{4/3}}$$

35.

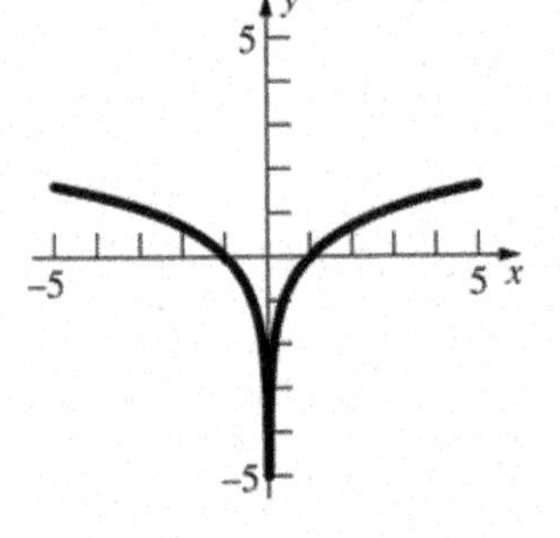

$y = \ln x$ is reflected across the y-axis.

37.

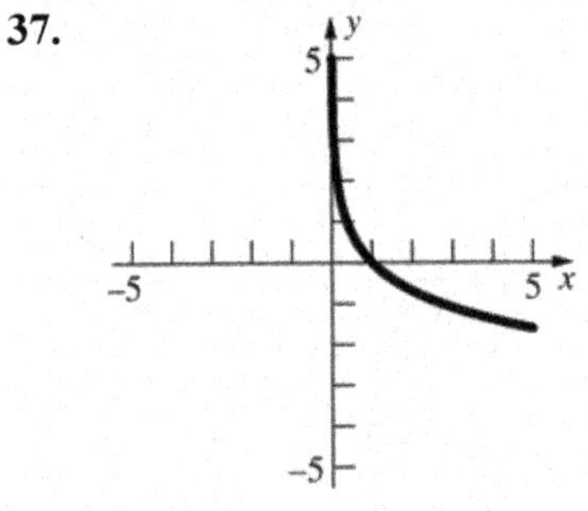

$y = \ln x$ is reflected across the x-axis since $\ln\left(\frac{1}{x}\right) = -\ln x$.

39. 

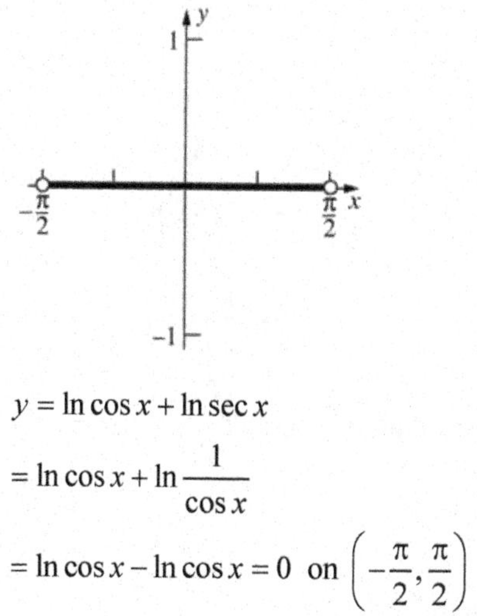

$$y = \ln \cos x + \ln \sec x$$
$$= \ln \cos x + \ln \frac{1}{\cos x}$$
$$= \ln \cos x - \ln \cos x = 0 \text{ on } \left(-\frac{\pi}{2}, \frac{\pi}{2}\right)$$

41. The domain is $(0, \infty)$.

$$f'(x) = 4x \ln x + 2x^2\left(\frac{1}{x}\right) - 2x = 4x \ln x$$

$f'(x) = 0$ if $\ln x = 0$, or $x = 1$.

$f'(x) < 0$ for $x < 1$ and $f'(x) > 0$ for $x > 1$

so $f(1) = -1$ is a minimum.

43. $\ln 4 > 1$

so $\ln 4^m = m \ln 4 > m \cdot 1 = m$

Thus $x > 4^m \Rightarrow \ln x > m$

so $\lim_{x \to \infty} \ln x = \infty$

45. $\int_{1/3}^{x} \frac{1}{t} dt = 2 \int_{1}^{x} \frac{1}{t} dt$

$$\int_{1/3}^{1} \frac{1}{t} dt + \int_{1}^{x} \frac{1}{t} dt = 2 \int_{1}^{x} \frac{1}{t} dt$$
$$\int_{1/3}^{1} \frac{1}{t} dt = \int_{1}^{x} \frac{1}{t} dt$$
$$-\int_{1}^{1/3} \frac{1}{t} dt = \int_{1}^{x} \frac{1}{t} dt$$
$$-\ln \frac{1}{3} = \ln x$$

$\ln 3 = \ln x$

$x = 3$

47. $\lim_{n \to \infty} \left[\frac{1}{n+1} + \frac{1}{n+2} + \cdots + \frac{1}{2n}\right]$

$$= \lim_{n \to \infty} \left[\frac{1}{1+\frac{1}{n}} + \frac{1}{1+\frac{2}{n}} + \cdots + \frac{1}{1+\frac{n}{n}}\right] \cdot \frac{1}{n}$$
$$= \lim_{n \to \infty} \sum_{i=1}^{n} \left(\frac{1}{1+\frac{i}{n}}\right) \cdot \frac{1}{n} = \int_{1}^{2} \frac{1}{x} dx = \ln 2 \approx 0.693$$

49. a. $f(x) = \ln\left(\frac{ax-b}{ax+b}\right)^c = c \ln\left(\frac{ax-b}{ax+b}\right)$

$$= \frac{a^2-b^2}{2ab}[\ln(ax-b) - \ln(ax+b)]$$
$$f'(x) = \frac{a^2-b^2}{2ab}\left[\frac{a}{ax-b} - \frac{a}{ax+b}\right]$$
$$= \frac{a^2-b^2}{2ab}\left[\frac{2ab}{(ax-b)(ax+b)}\right] = \frac{a^2-b^2}{a^2x^2-b^2}$$
$$f'(1) = \frac{a^2-b^2}{a^2-b^2} = 1$$

b. $f'(x) = \cos^2 u \cdot \frac{du}{dx}$

$$= \cos^2[\ln(x^2+x-1)] \cdot \frac{2x+1}{x^2+x-1}$$
$$f'(1) = \cos^2[\ln(1^2+1-1)] \cdot \frac{2 \cdot 1+1}{1^2+1-1}$$
$$= 3\cos^2(0) = 3$$

51. $\int_{\pi/4}^{\pi/3} \sec x \csc x \, dx = \left[-\ln|\cos x| + \ln|\sin x|\right]_{\pi/4}^{\pi/3}$

$$= \left[\ln|\tan x|\right]_{\pi/4}^{\pi/3} = \ln|\tan \tfrac{\pi}{3}| - \ln|\tan \tfrac{\pi}{4}|$$
$$= \ln(\sqrt{3}) - \ln 1 = 0.5493 - 0 = 0.5493$$

53. $V = 2\pi \int_{1}^{4} x f(x) dx = \int_{1}^{4} \frac{2\pi x}{x^2+4} dx$

Let $u = x^2 + 4$ so $du = 2x\, dx$.

$$\int \frac{2\pi x}{x^2+4} dx = \pi \int \frac{1}{u} du = \pi \ln|u| + C$$
$$= \pi \ln|x^2+4| + C$$
$$\int_{1}^{4} \frac{2\pi x}{x^2+4} dx = \left[\pi \ln|x^2+4|\right]_{1}^{4}$$
$$= \pi \ln 20 - \pi \ln 5 = \pi \ln 4 \approx 4.355$$

55.

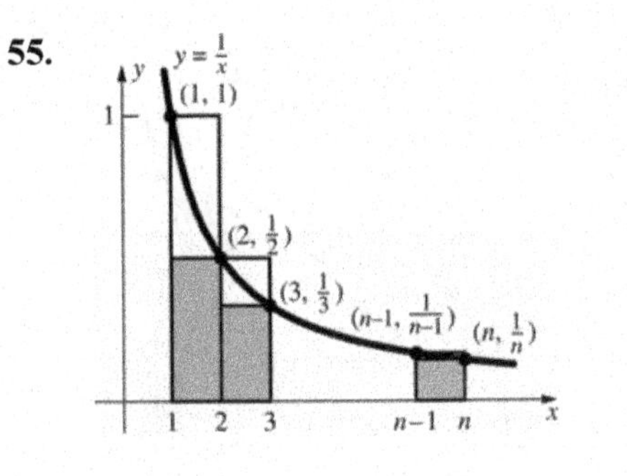

$\frac{1}{2}+\frac{1}{3}+\cdots+\frac{1}{n}=$ the lower approximate area

$1+\frac{1}{2}+\cdots+\frac{1}{n-1}=$ the upper approximate area

$\ln n =$ the exact area under the curve

Thus,

$$\frac{1}{2}+\frac{1}{3}+\cdots+\frac{1}{n}<\ln n<1+\frac{1}{2}+\frac{1}{3}+\cdots+\frac{1}{n-1}.$$

57. a. $f'(x)=\frac{1}{1.5+\sin x}\cdot\cos x=\frac{\cos x}{1.5+\sin x}$

$f'(x)=0$ when $\cos x = 0$.

Critical points: $0, \frac{\pi}{2}, \frac{3\pi}{2}, \frac{5\pi}{2}, 3\pi$

$f(0) \approx 0.405$,

$f\left(\frac{\pi}{2}\right)\approx 0.916, f\left(\frac{3\pi}{2}\right)\approx -0.693,$

$f\left(\frac{5\pi}{2}\right)\approx 0.916, f(3\pi)\approx 0.405.$

On $[0, 3\pi]$, the maximum value points are $\left(\frac{\pi}{2}, 0.916\right), \left(\frac{5\pi}{2}, 0.916\right)$ and the minimum value point is $\left(\frac{3\pi}{2}, -0.693\right)$.

b. $f''(x)=-\frac{1+1.5\sin x}{(1.5+\sin x)^2}$

On $[0, 3\pi]$, $f''(x)=0$ when $x \approx 3.871$, 5.553.

Inflection points are (3.871, –0.182), (5.553, –0.182).

c. $\int_0^{3\pi}\ln(1.5+\sin x)dx \approx 4.042$

59.

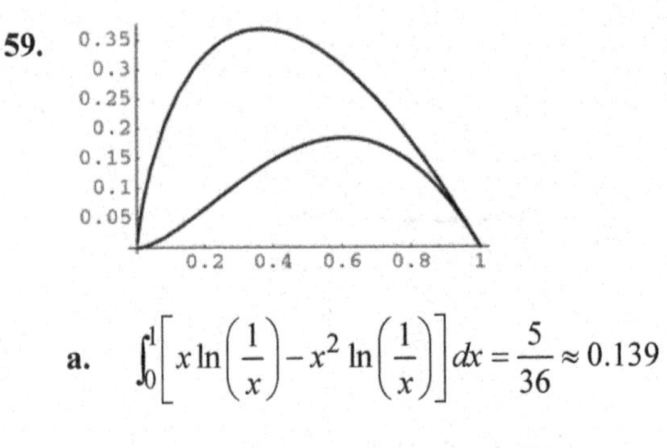

a. $\int_0^1\left[x\ln\left(\frac{1}{x}\right)-x^2\ln\left(\frac{1}{x}\right)\right]dx=\frac{5}{36}\approx 0.139$

b. Maximum of ≈ 0.260 at $x \approx 0.236$

6.2 Concepts Review

1. $f(x_1)\neq f(x_2)$

3. monotonic; strictly increasing; strictly decreasing

Problem Set 6.2

1. $f(x)$ is one-to-one, so it has an inverse.
Since $f(4)=2, f^{-1}(2)=4$.

3. $f(x)$ is not one-to-one, so it does not have an inverse.

5. $f(x)$ is one-to-one, so it has an inverse.
Since $f(-1.3) \approx 2$, $f^{-1}(2)\approx -1.3$.

7. $f'(x)=-5x^4-3x^2=-(5x^4+3x^2)<0$ for all $x \neq 0$. $f(x)$ is strictly decreasing at $x = 0$ because $f(x) > 0$ for $x < 0$ and $f(x) < 0$ for $x > 0$. Therefore $f(x)$ is strictly decreasing for x and so it has an inverse.

9. $f'(\theta)=-\sin\theta<0$ for $0<\theta<\pi$
$f(\theta)$ is decreasing at $\theta = 0$ because $f(0) = 1$ and $f(\theta) < 1$ for $0 < \theta < \pi$. $f(\theta)$ is decreasing at $\theta = \pi$ because $f(\pi) = -1$ and $f(\theta) > -1$ for $0 < \theta < \pi$. Therefore $f(\theta)$ is strictly decreasing on $0 \le \theta \le \pi$ and so it has an inverse.

11. $f'(z)=2(z-1)>0$ for $z > 1$
$f(z)$ is increasing at $z = 1$ because $f(1) = 0$ and $f(z) > 0$ for $z > 1$. Therefore, $f(z)$ is strictly increasing on $z \ge 1$ and so it has an inverse.

13. $f'(x)=\sqrt{x^4+x^2+10}>0$ for all real x. $f(x)$ is strictly increasing and so it has an inverse.

15. Step 1:
$y = x + 1$
$x = y - 1$
Step 2: $f^{-1}(y) = y - 1$
Step 3: $f^{-1}(x) = x - 1$
Check:
$f^{-1}(f(x)) = (x+1) - 1 = x$
$f(f^{-1}(x)) = (x-1) + 1 = x$

17. Step 1:
$y = \sqrt{x+1}$ (note that $y \geq 0$)
$x + 1 = y^2$
$x = y^2 - 1, y \geq 0$
Step 2: $f^{-1}(y) = y^2 - 1, y \geq 0$
Step 3: $f^{-1}(x) = x^2 - 1, x \geq 0$
Check:
$f^{-1}(f(x)) = (\sqrt{x+1})^2 - 1 = (x+1) - 1 = x$
$f(f^{-1}(x)) = \sqrt{(x^2-1)+1} = \sqrt{x^2} = |x| = x$

19. Step 1:
$$y = -\frac{1}{x-3}$$
$$x - 3 = -\frac{1}{y}$$
$$x = 3 - \frac{1}{y}$$
Step 2: $f^{-1}(y) = 3 - \frac{1}{y}$
Step 3: $f^{-1}(x) = 3 - \frac{1}{x}$
Check:
$$f^{-1}(f(x)) = 3 - \frac{1}{-\frac{1}{x-3}} = 3 + (x-3) = x$$
$$f(f^{-1}(x)) = -\frac{1}{\left(3 - \frac{1}{x}\right) - 3} = -\frac{1}{-\frac{1}{x}} = x$$

21. Step 1:
$y = 4x^2, x \leq 0$ (note that $y \geq 0$)
$$x^2 = \frac{y}{4}$$
$$x = -\sqrt{\frac{y}{4}} = -\frac{\sqrt{y}}{2}, \text{ negative since } x \leq 0$$
Step 2: $f^{-1}(y) = -\frac{\sqrt{y}}{2}$
Step 3: $f^{-1}(x) = -\frac{\sqrt{x}}{2}$
Check:
$$f^{-1}(f(x)) = -\frac{\sqrt{4x^2}}{2} = -\sqrt{x^2} = -|x| = -(-x) = x$$
$$f(f^{-1}(x)) = 4\left(-\frac{\sqrt{x}}{2}\right)^2 = 4 \cdot \frac{x}{4} = x$$

23. Step 1:
$y = (x-1)^3$
$x - 1 = \sqrt[3]{y}$
$x = 1 + \sqrt[3]{y}$
Step 2: $f^{-1}(y) = 1 + \sqrt[3]{y}$
Step 3: $f^{-1}(x) = 1 + \sqrt[3]{x}$
Check: $f^{-1}(f(x)) = 1 + \sqrt[3]{(x-1)^3} = 1 + (x-1) = x$
$f(f^{-1}(x)) = [(1 + \sqrt[3]{x}) - 1]^3 = (\sqrt[3]{x})^3 = x$

25. Step 1:
$$y = \frac{x-1}{x+1}$$
$xy + y = x - 1$
$x - xy = 1 + y$
$$x = \frac{1+y}{1-y}$$
Step 2: $f^{-1}(y) = \frac{1+y}{1-y}$
Step 3: $f^{-1}(x) = \frac{1+x}{1-x}$
Check:
$$f^{-1}(f(x)) = \frac{1 + \frac{x-1}{x+1}}{1 - \frac{x-1}{x+1}} = \frac{x+1+x-1}{x+1-x+1} = \frac{2x}{2} = x$$
$$f(f^{-1}(x)) = \frac{\frac{1+x}{1-x} - 1}{\frac{1+x}{1-x} + 1} = \frac{1+x-1+x}{1+x+1-x} = \frac{2x}{2} = x$$

27. Step 1:

$$y = \frac{x^3+2}{x^3+1}$$

$$x^3y + y = x^3 + 2$$

$$x^3y - x^3 = 2 - y$$

$$x^3 = \frac{2-y}{y-1}$$

$$x = \left(\frac{2-y}{y-1}\right)^{1/3}$$

Step 2: $f^{-1}(y) = \left(\frac{2-y}{y-1}\right)^{1/3}$

Step 3: $f^{-1}(x) = \left(\frac{2-x}{x-1}\right)^{1/3}$

Check:

$$f^{-1}(f(x)) = \left(\frac{2-\frac{x^3+2}{x^3+1}}{\frac{x^3+2}{x^3+1}-1}\right)^{1/3} = \left(\frac{2x^3+2-x^3-2}{x^3+2-x^3-1}\right)^{1/3}$$

$$= \left(\frac{x^3}{1}\right)^{1/3} = x$$

$$f(f^{-1}(x)) = \frac{\left[\left(\frac{2-x}{x-1}\right)^{1/3}\right]^3+2}{\left[\left(\frac{2-x}{x-1}\right)^{1/3}\right]^3+1} = \frac{\frac{2-x}{x-1}+2}{\frac{2-x}{x-1}+1}$$

$$= \frac{2-x+2x-2}{2-x+x-1} = \frac{x}{1} = x$$

29. By similar triangles, $\frac{r}{h} = \frac{4}{6}$. Thus, $r = \frac{2h}{3}$

This gives

$$V = \frac{\pi r^2 h}{3} = \frac{\pi\left(4h^2/9\right)h}{3} = \frac{4\pi h^3}{27}$$

$$h^3 = \frac{27V}{4\pi}$$

$$h = 3\sqrt[3]{\frac{V}{4\pi}}$$

31. $f'(x) = 4x+1$; $f'(x) > 0$ when $x > -\frac{1}{4}$ and

$f'(x) < 0$ when $x < -\frac{1}{4}$.

The function is decreasing on $\left(-\infty, -\frac{1}{4}\right]$ and increasing on $\left[-\frac{1}{4}, \infty\right)$. Restrict the domain to $\left(-\infty, -\frac{1}{4}\right]$ or restrict it to $\left[-\frac{1}{4}, \infty\right)$.

Then $f^{-1}(x) = \frac{1}{4}(-1-\sqrt{8x+33})$ or

$f^{-1}(x) = \frac{1}{4}(-1+\sqrt{8x+33})$.

33.

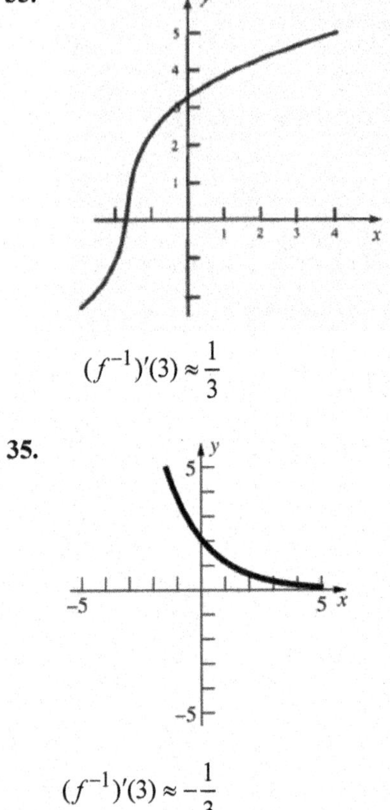

$(f^{-1})'(3) \approx \frac{1}{3}$

35.

$(f^{-1})'(3) \approx -\frac{1}{3}$

37. $f'(x) = 15x^4 + 1$ and $y = 2$ corresponds to $x = 1$,

so $(f^{-1})'(2) = \frac{1}{f'(1)} = \frac{1}{15+1} = \frac{1}{16}$.

39. $f'(x) = 2\sec^2 x$ and $y = 2$ corresponds to $x = \frac{\pi}{4}$,

so $(f^{-1})'(2) = \frac{1}{f'\left(\frac{\pi}{4}\right)} = \frac{1}{2\sec^2\left(\frac{\pi}{4}\right)} = \frac{1}{2}\cos^2\left(\frac{\pi}{4}\right)$

$= \frac{1}{4}$.

41. $(g^{-1} \circ f^{-1})(h(x)) = (g^{-1} \circ f^{-1})(f(g(x)))$

$= g^{-1} \circ [f^{-1}(f(g(x)))] = g^{-1} \circ [g(x)] = x$

Similarly,

$h((g^{-1} \circ f^{-1})(x)) = f(g((g^{-1} \circ f^{-1})(x)))$

$= f(g(g^{-1}(f^{-1}(x)))) = f(f^{-1}(x)) = x$

Thus $h^{-1} = g^{-1} \circ f^{-1}$

43. f has an inverse because it is monotonic (increasing): $f'(x) = \sqrt{1+\cos^2 x} > 0$

a. $(f^{-1})'(A) = \dfrac{1}{f'\left(\frac{\pi}{2}\right)} = \dfrac{1}{\sqrt{1+\cos^2\left(\frac{\pi}{2}\right)}} = 1$

b. $(f^{-1})'(B) = \dfrac{1}{f'\left(\frac{5\pi}{6}\right)} = \dfrac{1}{\sqrt{1+\cos^2\left(\frac{5\pi}{6}\right)}} = \dfrac{1}{\sqrt{\frac{7}{4}}}$

$= \dfrac{2}{\sqrt{7}}$

c. $(f^{-1})'(0) = \dfrac{1}{f'(0)} = \dfrac{1}{\sqrt{1+\cos^2(0)}} = \dfrac{1}{\sqrt{2}}$

45.

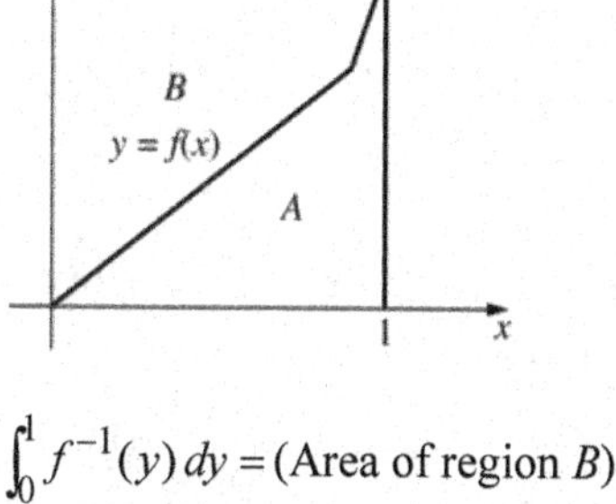

$\int_0^1 f^{-1}(y)\,dy = $ (Area of region B)

$= 1 -$ (Area of region A)

$= 1 - \int_0^1 f(x)\,dx = 1 - \dfrac{2}{5} = \dfrac{3}{5}$

47. Given $p > 1, q > 1, \dfrac{1}{p} + \dfrac{1}{q} = 1$, and $f(x) = x^{p-1}$,

solving $\dfrac{1}{p} + \dfrac{1}{q} = 1$ for p gives $p = \dfrac{q}{q-1}$, so

$\dfrac{1}{p-1} = \dfrac{1}{\frac{q}{q-1} - 1} = \dfrac{1}{\left[\frac{q-(q-1)}{q-1}\right]} = \dfrac{q-1}{1} = q-1.$

Thus, if $y = x^{p-1}$ then $x = y^{\frac{1}{p-1}} = y^{q-1}$, so $f^{-1}(y) = y^{q-1}$.

By Problem 44, since $f(x) = x^{p-1}$ is strictly increasing for $p > 1$, $ab \le \int_0^a x^{p-1}dx + \int_0^b y^{q-1}dy$

$ab \le \left[\dfrac{x^p}{p}\right]_0^a + \left[\dfrac{y^q}{q}\right]_0^b$

$ab \le \dfrac{a^p}{p} + \dfrac{b^q}{q}$

6.3 Concepts Review

1. increasing; exp

3. x; x

Problem Set 6.3

1. a. 20.086

b. 8.1662

c. $e^{\sqrt{2}} \approx e^{1.41} \approx 4.1$

d. $e^{\cos(\ln 4)} \approx e^{0.18} \approx 1.20$

3. $e^{3\ln x} = e^{\ln x^3} = x^3$

5. $\ln e^{\cos x} = \cos x$

7. $\ln(x^3 e^{-3x}) = \ln x^3 + \ln e^{-3x} = 3\ln x - 3x$

9. $e^{\ln 3 + 2\ln x} = e^{\ln 3} \cdot e^{2\ln x} = 3 \cdot e^{\ln x^2} = 3x^2$

11. $D_x e^{x+2} = e^{x+2} D_x(x+2) = e^{x+2}$

13. $D_x e^{\sqrt{x+2}} = e^{\sqrt{x+2}} D_x \sqrt{x+2} = \dfrac{e^{\sqrt{x+2}}}{2\sqrt{x+2}}$

15. $D_x e^{2\ln x} = D_x e^{\ln x^2} = D_x x^2 = 2x$

17. $D_x(x^3 e^x) = x^3 D_x e^x + e^x D_x(x^3)$

$= x^3 e^x + e^x \cdot 3x^2 = x^2 e^x (x+3)$

19. $D_x[\sqrt{e^{x^2}} + e^{\sqrt{x^2}}] = D_x (e^{x^2})^{1/2} + D_x e^{\sqrt{x^2}}$

$= \dfrac{1}{2}(e^{x^2})^{-1/2} D_x e^{x^2} + e^{\sqrt{x^2}} D_x \sqrt{x^2}$

$= \dfrac{1}{2}(e^{x^2})^{-1/2} e^{x^2} D_x x^2 + e^{\sqrt{x^2}} \cdot \dfrac{x}{\sqrt{x^2}}$

$= \dfrac{1}{2}(e^{x^2})^{1/2} 2x + e^{\sqrt{x^2}} \cdot \dfrac{x}{|x|}$

$= x\sqrt{e^{x^2}} + \dfrac{xe^{\sqrt{x^2}}}{|x|}$

21. $D_x[e^{xy} + xy] = D_x[2]$

$e^{xy}(xD_xy + y) + (xD_xy + y) = 0$

$xe^{xy}D_xy + ye^{xy} + xD_xy + y = 0$

$xe^{xy}D_xy + xD_xy = -ye^{xy} - y$

$$D_xy = \frac{-ye^{xy} - y}{xe^{xy} + x} = -\frac{y(e^{xy}+1)}{x(e^{xy}+1)} = -\frac{y}{x}$$

23. a.

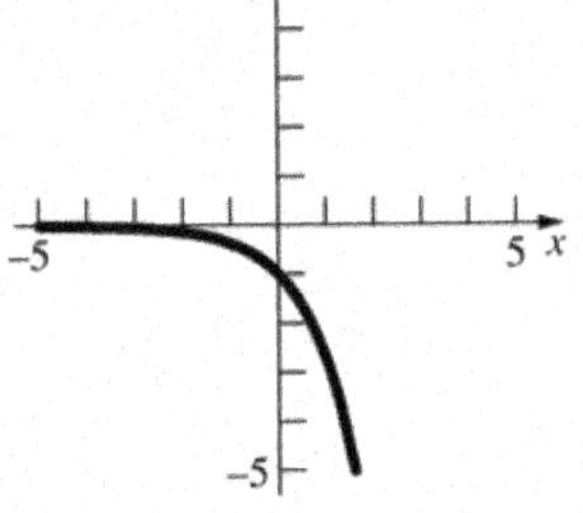

The graph of $y = e^x$ is reflected across the x-axis.

b.

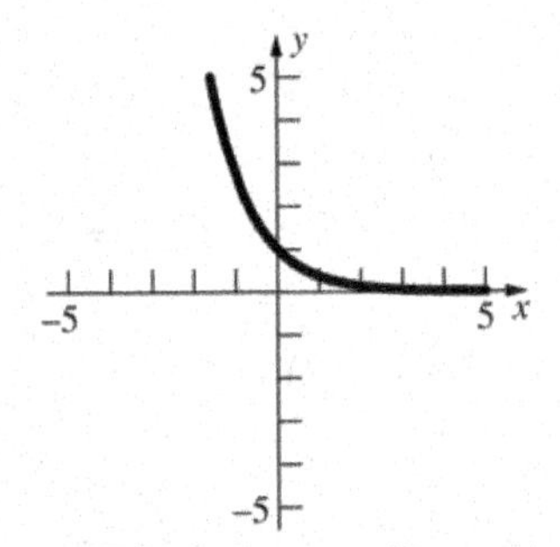

The graph of $y = e^x$ is reflected across the y-axis.

25. $f(x) = e^{2x}$ Domain $= (-\infty, \infty)$

$f'(x) = 2e^{2x}$, $f''(x) = 4e^{2x}$

Since $f'(x) > 0$ for all x, f is increasing on $(-\infty, \infty)$.

Since $f''(x) > 0$ for all x, f is concave upward on $(-\infty, \infty)$.

Since f and f' are both monotonic, there are no extreme values or points of inflection.

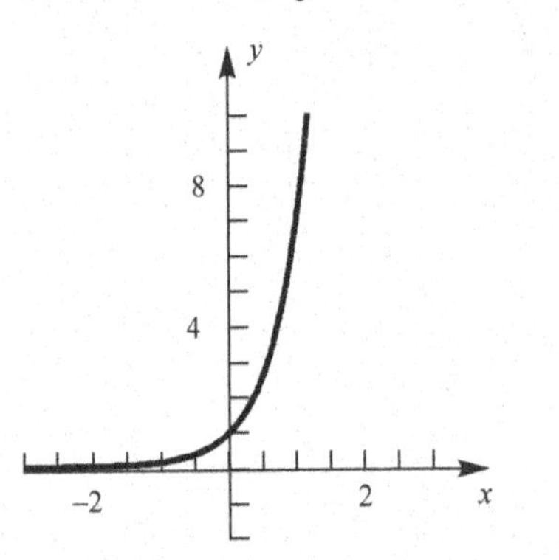

27. $f(x) = xe^{-x}$ Domain $= (-\infty, \infty)$

$f'(x) = (1-x)e^{-x}$, $f''(x) = (x-2)e^{-x}$

x	$(-\infty,1)$	1	$(1,2)$	2	$(2,\infty)$
f'	+	0	−	−	−
f''	−	−	−	0	+

f is increasing on $(-\infty, 1]$ and decreasing on $[1, \infty)$. f has a maximum at $(1, \frac{1}{e})$

f is concave up on $(2, \infty)$ and concave down on $(-\infty, 2)$. f has a point of inflection at $(2, \frac{2}{e^2})$

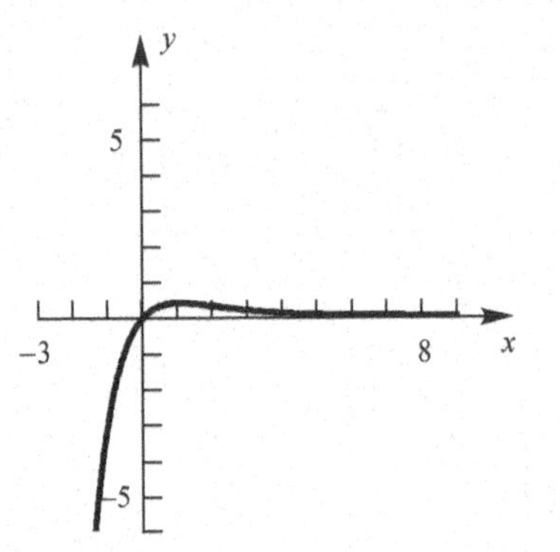

29. $f(x) = \ln(x^2 + 1)$ Since $x^2 + 1 > 0$ for all x, domain $= (-\infty, \infty)$

$$f'(x) = \frac{2x}{x^2+1}, \quad f''(x) = \frac{-2(x^2-1)}{(x^2+1)^2}$$

x	$(-\infty,-1)$	-1	$(-1,0)$	0	$(0,1)$	1	$(1,\infty)$
f'	−	−	−	0	+	+	+
f''	−	0	+	+	+	0	−

f is increasing on $(0, \infty)$ and decreasing on $(-\infty, 0)$. f has a minimum at $(0, 0)$

f is concave up on $(-1, 1)$ and concave down on $(-\infty, -1) \cup (1, \infty)$. f has points of inflection at $(-1, \ln 2)$ and $(1, \ln 2)$

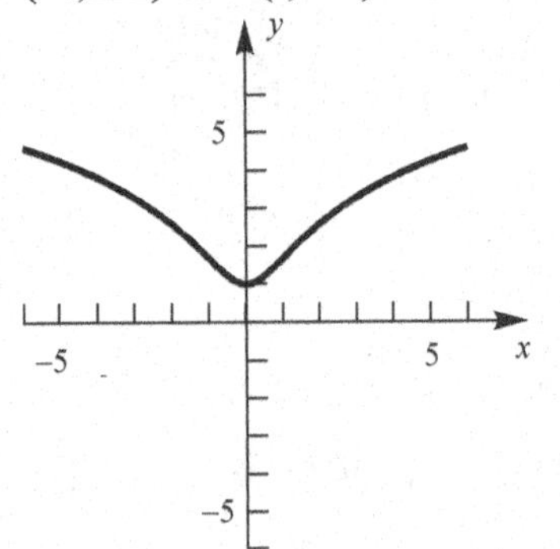

31. $f(x) = \ln(1+e^x)$ Since $1+e^x > 0$ for all x, domain $= (-\infty, \infty)$

$$f'(x) = \frac{e^x}{1+e^x}, \quad f''(x) = \frac{e^x}{(1+e^x)^2}$$

Since $f'(x) > 0$ for all x, f is increasing on $(-\infty, \infty)$.

Since $f''(x) > 0$ for all x, f is concave upward on $(-\infty, \infty)$.

Since f and f' are both monotonic, there are no extreme values or points of inflection.

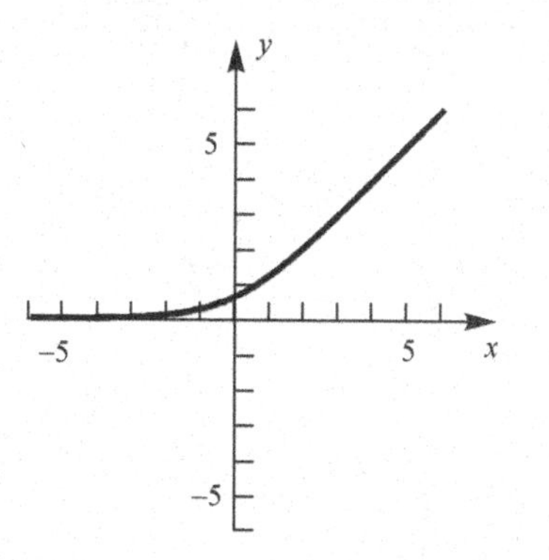

33. $f(x) = e^{-(x-2)^2}$ Domain $= (-\infty, \infty)$

$f'(x) = (4-2x)e^{-(x-2)^2}$,

$f''(x) = (4x^2 - 16x + 14)e^{-(x-2)^2}$

Note that $4x^2 - 16x + 14 = 0$ when

$$x = \frac{4 \pm \sqrt{2}}{2} \approx 2 \pm 0.707$$

x	$(-\infty,1.293)$	≈ 1.293	$(1.293,2)$	2	$(2,2.707)$	≈ 2.707	$(2.707,\infty)$
f'	+	+	+	0	−	−	−
f''	+	0	−	−	−	0	+

f is increasing on $(-\infty, 2]$ and decreasing on $[2, \infty)$. f has a maximum at $(2,1)$

f is concave up on $(-\infty, \frac{4-\sqrt{2}}{2}) \cup (\frac{4+\sqrt{2}}{2}, \infty)$ and concave down on $(\frac{4-\sqrt{2}}{2}, \frac{4+\sqrt{2}}{2})$. f has points of inflection at $(\frac{4-\sqrt{2}}{2}, \frac{1}{\sqrt{e}})$ and $(\frac{4+\sqrt{2}}{2}, \frac{1}{\sqrt{e}})$.

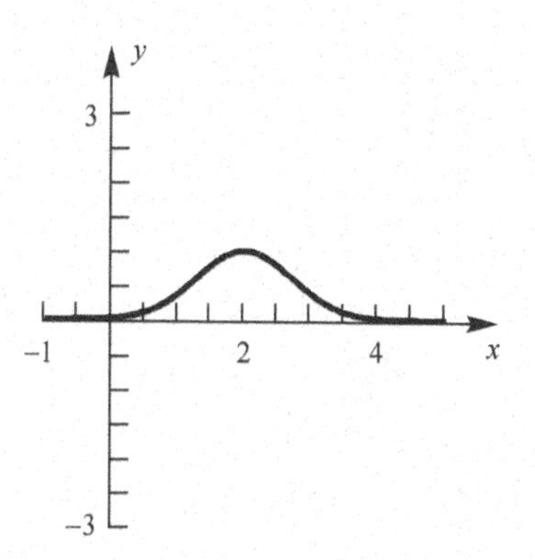

35. $f(x) = \int_0^x e^{-t^2}\,dt$ Domain $= (-\infty, \infty)$

$f'(x) = e^{-x^2}$, $f''(x) = -2xe^{-x^2}$

x	$(-\infty,0)$	0	$(0,\infty)$
f'	+	+	+
f''	+	0	−

f is increasing on $(-\infty, \infty)$ and so has no extreme values. f is concave up on $(-\infty, 0)$ and concave down on $(0, \infty)$. f has a point of inflection at $(0,0)$

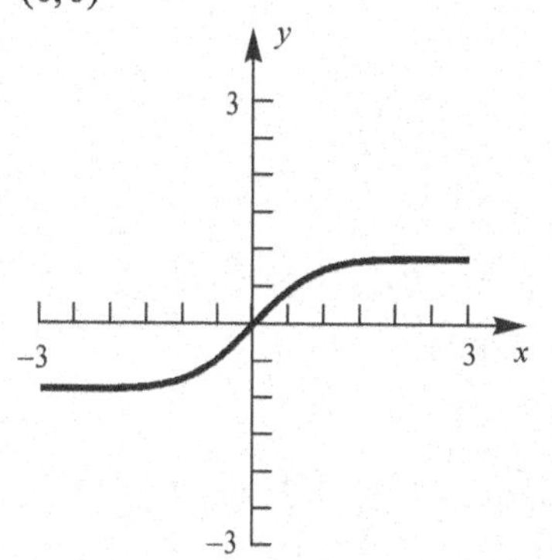

37. Let $u = 3x + 1$, so $du = 3dx$.

$$\int e^{3x+1}dx = \frac{1}{3}\int e^{3x+1}3dx = \frac{1}{3}\int e^u du = \frac{1}{3}e^u + C$$

$$= \frac{1}{3}e^{3x+1} + C$$

39. Let $u = x^2 + 6x$, so $du = (2x + 6)dx$.

$$\int (x+3)e^{x^2+6x}dx = \frac{1}{2}\int e^u du = \frac{1}{2}e^u + C$$

$$= \frac{1}{2}e^{x^2+6x} + C$$

41. Let $u = -\frac{1}{x}$, so $du = \frac{1}{x^2}dx$.

$$\int \frac{e^{-1/x}}{x^2}dx = \int e^u du = e^u + C = e^{-1/x} + C$$

43. Let $u = 2x + 3$, so $du = 2dx$

$$\int e^{2x+3}dx = \frac{1}{2}\int e^u du = \frac{1}{2}e^u + C = \frac{1}{2}e^{2x+3} + C$$

$$\int_0^1 e^{2x+3}dx = \left[\frac{1}{2}e^{2x+3}\right]_0^1 = \frac{1}{2}e^5 - \frac{1}{2}e^3$$

$$= \frac{1}{2}e^3(e^2 - 1) \approx 64.2$$

45. $V = \pi\int_0^{\ln 3}(e^x)^2 dx = \pi\int_0^{\ln 3} e^{2x}dx$

$$= \pi\left[\frac{1}{2}e^{2x}\right]_0^{\ln 3} = \pi\left(\frac{1}{2}e^{2\ln 3} - \frac{1}{2}e^0\right) = 4\pi \approx 12.57$$

47. The line through (0, 1) and $\left(1, \frac{1}{e}\right)$ has slope

$$\frac{\frac{1}{e} - 1}{1 - 0} = \frac{1}{e} - 1 = \frac{1-e}{e} \Rightarrow y - 1 = \frac{1-e}{e}(x - 0);$$

$$y = \frac{1-e}{e}x + 1$$

$$\int_0^1\left[\left(\frac{1-e}{e}x + 1\right) - e^{-x}\right]dx = \left[\frac{1-e}{2e}x^2 + x + e^{-x}\right]_0^1$$

$$= \frac{1-e}{2e} + 1 + \frac{1}{e} - 1 = \frac{3-e}{2e} \approx 0.052$$

49. a. Exact:

$$10! = 10\cdot 9\cdot 8\cdot 7\cdot 6\cdot 5\cdot 4\cdot 3\cdot 2\cdot 1$$
$$= 3{,}628{,}800$$

Approximate:

$$10! \approx \sqrt{20\pi}\left(\frac{10}{e}\right)^{10} \approx 3{,}598{,}696$$

b. $60! \approx \sqrt{120\pi}\left(\frac{60}{e}\right)^{60} \approx 8.31\times 10^{81}$

51. $x = e^t \sin t$, so $dx = (e^t \sin t + e^t \cos t)dt$

$y = e^t \cos t$, so $dy = (e^t \cos t - e^t \sin t)dt$

$$ds = \sqrt{dx^2 + dy^2}$$

$$= e^t\sqrt{(\sin t + \cos t)^2 + (\cos t - \sin t)^2}\,dt$$

$$= e^t\sqrt{2\sin^2 t + 2\cos^2 t}\,dt = \sqrt{2}e^t\,dt$$

The length of the curve is

$$\int_0^{\pi}\sqrt{2}e^t dt = \sqrt{2}\left[e^t\right]_0^{\pi} = \sqrt{2}(e^{\pi} - 1) \approx 31.312$$

53. a. $\lim_{x\to 0^+}\frac{\ln x}{1 + (\ln x)^2}$ is of the form $\frac{\infty}{\infty}$.

$$= \lim_{x\to 0^+}\frac{D_x \ln x}{D_x[1 + (\ln x)^2]} = \lim_{x\to 0^+}\frac{\frac{1}{x}}{2\ln x\cdot\frac{1}{x}}$$

$$= \lim_{x\to 0^+}\frac{1}{2\ln x} = 0$$

$$\lim_{x\to\infty}\frac{\ln x}{1 + (\ln x)^2} = \lim_{x\to\infty}\frac{1}{2\ln x} = 0$$

b. $$f'(x) = \frac{[1 + (\ln x)^2]\cdot\frac{1}{x} - \ln x\cdot 2\ln x\cdot\frac{1}{x}}{[1 + (\ln x)^2]^2}$$

$$= \frac{1 - (\ln x)^2}{x[1 + (\ln x)^2]^2}$$

$f'(x) = 0$ when $\ln x = \pm 1$ so $x = e^1 = e$

or $x = e^{-1} = \frac{1}{e}$

$$f(e) = \frac{\ln e}{1 + (\ln e)^2} = \frac{1}{1 + 1^2} = \frac{1}{2}$$

$$f\left(\frac{1}{e}\right) = \frac{\ln\frac{1}{e}}{1 + \left(\ln\frac{1}{e}\right)^2} = \frac{-1}{1 + (-1)^2} = -\frac{1}{2}$$

Maximum value of $\frac{1}{2}$ at $x = e$; minimum value of $-\frac{1}{2}$ at $x = e^{-1}$.

c. $$F(x) = \int_1^{x^2}\frac{\ln t}{1 + (\ln t)^2}dt$$

$$F'(x) = \frac{\ln x^2}{1 + (\ln x^2)^2}\cdot 2x$$

$$F'(\sqrt{e}) = \frac{\ln(\sqrt{e})^2}{1 + [\ln(\sqrt{e})^2]^2}\cdot 2\sqrt{e} = \frac{1}{1 + 1^2}\cdot 2\sqrt{e}$$

$$= \sqrt{e} \approx 1.65$$

55. a. $$\int_{-3}^{3}\exp\left(-\frac{1}{x^2}\right)dx = 2\int_0^3\exp\left(-\frac{1}{x^2}\right)dx \approx 3.11$$

b. $$\int_0^{8\pi}e^{-0.1x}\sin x\,dx \approx 0.910$$

57. $f(x) = e^{-x^2}$

$f'(x) = -2xe^{-x^2}$

$f''(x) = -2e^{-x^2} + 4x^2e^{-x^2} = 2e^{-x^2}(2x^2 - 1)$

$y = f(x)$ and $y = f''(x)$ intersect when

$e^{-x^2} = 2e^{-x^2}(2x^2 - 1); 1 = 4x^2 - 2;$

$4x^2 - 3 = 0, x = \pm\frac{\sqrt{3}}{2}$

Both graphs are symmetric with respect to the y-axis so the area is

$$2\left\{\int_0^{\frac{\sqrt{3}}{2}}[e^{-x^2} - 2e^{x^2}(2x^2 - 1)]\,dx + \int_{\frac{\sqrt{3}}{2}}^{3}[2e^{-x^2}(2x^2 - 1) - e^{-x^2}]\,dx\right\}$$

≈ 4.2614

59. $\lim_{x\to-\infty} \ln(x^2 + e^{-x}) = \infty$ (behaves like $-x$)

$\lim_{x\to\infty} \ln(x^2 + e^{-x}) = \infty$ (behaves like $2\ln x$)

6.4 Concepts Review

1. $e^{\sqrt{3}\ln\pi}$; $e^{x\ln a}$

3. $\frac{\ln x}{\ln a}$

Problem Set 6.4

1. $2^x = 8 = 2^3$; $x = 3$

3. $x = 4^{3/2} = 8$

5. $\log_9\left(\frac{x}{3}\right) = \frac{1}{2}$

$\frac{x}{3} = 9^{1/2} = 3$

$x = 9$

7. $\log_2(x+3) - \log_2 x = 2$

$\log_2 \frac{x+3}{x} = 2$

$\frac{x+3}{x} = 2^2 = 4$

$x + 3 = 4x$

$x = 1$

9. $\log_5 12 = \frac{\ln 12}{\ln 5} \approx 1.544$

11. $\log_{11}(8.12)^{1/5} = \frac{1}{5}\frac{\ln 8.12}{\ln 11} \approx 0.1747$

13. $x \ln 2 = \ln 17$

$x = \frac{\ln 17}{\ln 2} \approx 4.08746$

15. $(2s - 3)\ln 5 = \ln 4$

$2s - 3 = \frac{\ln 4}{\ln 5}$

$s = \frac{1}{2}\left(3 + \frac{\ln 4}{\ln 5}\right) \approx 1.9307$

17. $D_x(6^{2x}) = 6^{2x}\ln 6 \cdot D_x(2x) = 2\cdot 6^{2x}\ln 6$

19. $D_x \log_3 e^x = \frac{1}{e^x \ln 3}\cdot D_x e^x$

$= \frac{e^x}{e^x \ln 3} = \frac{1}{\ln 3} \approx 0.9102$

Alternate method:

$D_x \log_3 e^x = D_x(x\log_3 e) = \log_3 e$

$= \frac{\ln e}{\ln 3} = \frac{1}{\ln 3} \approx 0.9102$

21. $D_z[3^z \ln(z+5)]$

$= 3^z \cdot \frac{1}{z+5}(1) + \ln(z+5)\cdot 3^z \ln 3$

$= 3^z\left[\frac{1}{z+5} + \ln(z+5)\ln 3\right]$

23. Let $u = x^2$ so $du = 2x\,dx$.

$\int x\cdot 2^{x^2}\,dx = \frac{1}{2}\int 2^u\,du = \frac{1}{2}\cdot\frac{2^u}{\ln 2} + C$

$= \frac{2^{x^2}}{2\ln 2} + C = \frac{2^{x^2-1}}{\ln 2} + C$

25. Let $u = \sqrt{x}$, so $du = \frac{1}{2\sqrt{x}}dx$.

$\int \frac{5^{\sqrt{x}}}{\sqrt{x}}dx = 2\int 5^u\,du = 2\cdot\frac{5^u}{\ln 5} + C$

$= \frac{2\cdot 5^{\sqrt{x}}}{\ln 5} + C$

$\int_1^4 \frac{5^{\sqrt{x}}}{\sqrt{x}}dx = 2\left[\frac{5^{\sqrt{x}}}{\ln 5}\right]_1^4 = 2\left(\frac{25}{\ln 5} - \frac{5}{\ln 5}\right)$

$= \frac{40}{\ln 5} \approx 24.85$

27. $\frac{d}{dx}10^{(x^2)} = 10^{(x^2)}\ln 10\frac{d}{dx}x^2 = 10^{(x^2)}2x\ln 10$

$\frac{d}{dx}(x^2)^{10} = \frac{d}{dx}x^{20} = 20x^{19}$

$\frac{dy}{dx} = \frac{d}{dx}[10^{(x^2)} + (x^2)^{10}]$

$= 10^{(x^2)}2x\ln 10 + 20x^{19}$

29. $\frac{d}{dx}x^{\pi+1} = (\pi+1)x^{\pi}$

$\frac{d}{dx}(\pi+1)^x = (\pi+1)^x\ln(\pi+1)$

$\frac{dy}{dx} = \frac{d}{dx}[x^{\pi+1} + (\pi+1)^x]$

$= (\pi+1)x^{\pi} + (\pi+1)^x\ln(\pi+1)$

31. $y = (x^2+1)^{\ln x} = e^{(\ln x)\ln(x^2+1)}$

$\frac{dy}{dx} = e^{(\ln x)\ln(x^2+1)}\frac{d}{dx}[(\ln x)\ln(x^2+1)]$

$= e^{(\ln x)\ln(x^2+1)}\left[\frac{1}{x}\ln(x^2+1) + \ln x\frac{2x}{x^2+1}\right]$

$= (x^2+1)^{\ln x}\left(\frac{\ln(x^2+1)}{x} + \frac{2x\ln x}{x^2+1}\right)$

33. $f(x) = x^{\sin x} = e^{\sin x\ln x}$

$f'(x) = e^{\sin x\ln x}\frac{d}{dx}(\sin x\ln x)$

$= e^{\sin x\ln x}\left[(\sin x)\left(\frac{1}{x}\right) + (\cos x)(\ln x)\right]$

$= x^{\sin x}\left(\frac{\sin x}{x} + \cos x\ln x\right)$

$f'(1) = 1^{\sin 1}\left(\frac{\sin 1}{1} + \cos 1\ln 1\right) = \sin 1 \approx 0.8415$

35. $f(x) = 2^{-x} = e^{(\ln 2)(-x)}$ Domain $= (-\infty, \infty)$

$f'(x) = (-\ln 2)2^{-x}$, $f''(x) = (\ln 2)^2 2^{-x}$

Since $f'(x) < 0$ for all x, f is decreasing on $(-\infty, \infty)$.

Since $f''(x) > 0$ for all x, f is concave upward on $(-\infty, \infty)$.

Since f and f' are both monotonic, there are no extreme values or points of inflection.

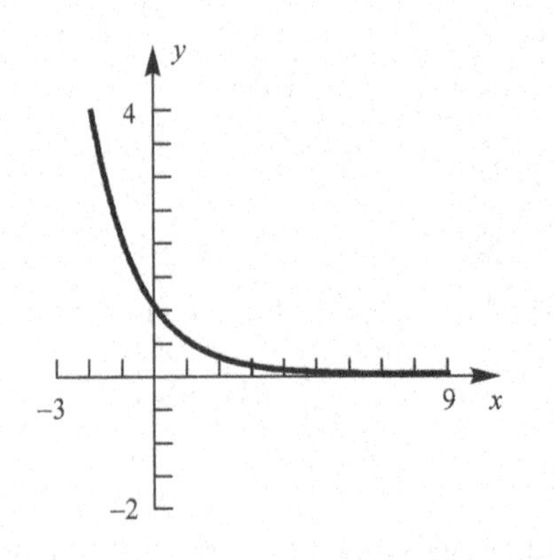

37. $f(x) = \log_2(x^2+1) = \frac{\ln(x^2+1)}{\ln 2}$. Since $x^2+1 > 0$ for all x, domain $= (-\infty, \infty)$

$f'(x) = \left(\frac{2}{\ln 2}\right)\left(\frac{x}{x^2+1}\right)$, $f''(x) = \left(\frac{2}{\ln 2}\right)\left(\frac{1-x^2}{(x^2+1)^2}\right)$

x	$(-\infty,-1)$	-1	$(-1,0)$	0	$(0,1)$	1	$(1,\infty)$
f'	–	–	–	0	+	+	+
f''	–	0	+	+	+	0	–

f is increasing on $[0, \infty)$ and decreasing on $(-\infty, 0]$. f has a minimum at $(0, 0)$

f is concave up on $(-1, 1)$ and concave down on $(-\infty, -1) \cup (1, \infty)$. f has points of inflection at $(-1, 1)$ and $(1, 1)$

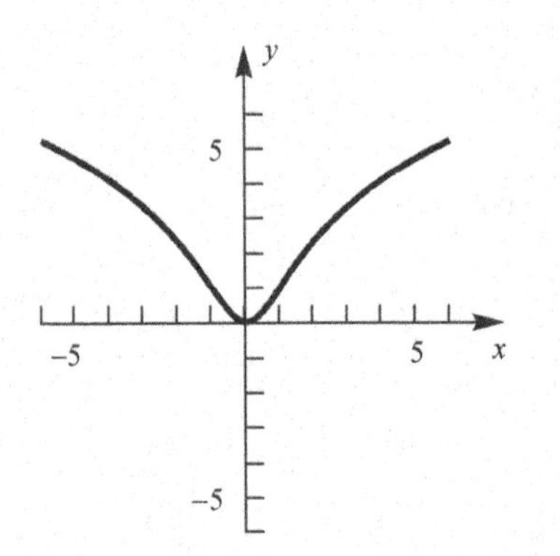

39. $f(x)=\int_1^x 2^{-t^2}\,dt$ Domain = $(-\infty,\infty)$

$f'(x)=2^{-x^2}$, $f''(x)=-2(\ln 2)x2^{-x^2}$

x	$(-\infty,0)$	0	$(0,\infty)$
f'	+	+	+
f''	+	0	–

f is increasing on $(-\infty,\infty)$ and so has no extreme values.

f is concave up on $(-\infty,0)$ and concave down on $(0,\infty)$. *f* has a point of inflection at $(0,\int_1^0 2^{-t^2}\,dt)\approx(0,-0.81)$

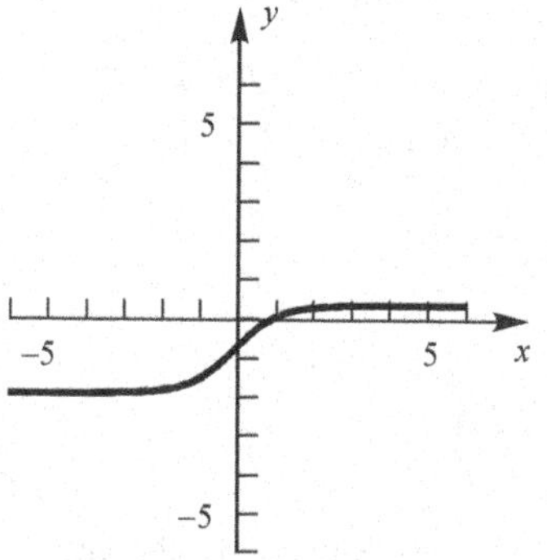

41. $\log_{1/2}x=\dfrac{\ln x}{\ln\frac{1}{2}}=\dfrac{\ln x}{-\ln 2}=-\log_2 x$

43. $M=0.67\log_{10}(0.37E)+1.46$

$\log_{10}(0.37E)=\dfrac{M-1.46}{0.67}$

$E=\dfrac{10^{\frac{M-1.46}{0.67}}}{0.37}$

Evaluating this expression for $M=7$ and $M=8$ gives $E\approx 5.017\times 10^8$ kW-h and $E\approx 1.560\times 10^{10}$ kW-h, respectively.

45. If *r* is the ratio between the frequencies of successive notes, then the frequency of $\overline{C}=r^{12}$ (the frequency of *C*). Since $\overline{C}$ has twice the frequency of *C*, $r=2^{1/12}\approx 1.0595$

Frequency of $\overline{C}=440(2^{1/12})^3=440\sqrt[4]{2}\approx 523.25$

47. If $y=A\cdot b^x$, then $\ln y=\ln A+x\ln b$, so the $\ln y$ vs. x plot will be linear.

If $y=C\cdot x^d$, then $\ln y=\ln C+d\ln x$, so the $\ln y$ vs. $\ln x$ plot will be linear.

49. $f(x)=(x^x)^x=x^{(x^2)}\neq x^{(x^x)}=g(x)$

$f(x)=x^{(x^2)}=e^{x^2\ln x}$

$f'(x)=e^{x^2\ln x}\dfrac{d}{dx}(x^2\ln x)$

$=e^{x^2\ln x}\left(2x\ln x+x^2\cdot\dfrac{1}{x}\right)$

$=x^{(x^2)}(2x\ln x+x)$

$g(x)=x^{(x^x)}=e^{x^x\ln x}$

Using the result from Example 5

$\left(\dfrac{d}{dx}x^x=x^x(1+\ln x)\right)$:

$g'(x)=e^{x^x\ln x}\dfrac{d}{dx}(x^x\ln x)$

$=e^{x^x\ln x}\left[x^x(1+\ln x)\ln x+x^x\cdot\dfrac{1}{x}\right]$

$=x^{(x^x)}x^x\left[(1+\ln x)\ln x+\dfrac{1}{x}\right]$

$=x^{x^x+x}\left[\ln x+(\ln x)^2+\dfrac{1}{x}\right]$

51. a. Let $g(x)=\ln f(x)=\ln\left(\dfrac{x^a}{a^x}\right)=a\ln x-x\ln a$.

$g'(x)=\left(\dfrac{a}{x}\right)-\ln a$

$g'(x)<0$ when $x>\dfrac{a}{\ln a}$, so as $x\to\infty$ $g(x)$ is decreasing. $g''(x)=-\dfrac{a}{x^2}$, so $g(x)$ is concave down. Thus, $\lim_{x\to\infty}g(x)=-\infty$, so $\lim_{x\to\infty}f(x)=\lim_{x\to\infty}e^{g(x)}=0$.

b. Again let $g(x)=\ln f(x)=a\ln x-x\ln a$. Since $y=\ln x$ is an increasing function, $f(x)$ is maximized when $g(x)$ is maximized.

$g'(x)=\left(\dfrac{a}{x}\right)-\ln a$, so $g'(x)>0$ on $\left(0,\dfrac{a}{\ln a}\right)$ and $g'(x)<0$ on $\left(\dfrac{a}{\ln a},\infty\right)$.

Therefore, $g(x)$ (and hence $f(x)$) is maximized at $x_0=\dfrac{a}{\ln a}$.

c. Note that $x^a = a^x$ is equivalent to $g(x) = 0$.

By part b., $g(x)$ is maximized at $x_0 = \frac{a}{\ln a}$.

If $a = e$, then

$g(x_0) = g\left(\frac{e}{\ln e}\right) = g(e) = e\ln e - e\ln e = 0.$

Since $g(x) < g(x_0) = 0$ for all $x \neq x_0$, the equation $g(x) = 0$ (and hence $x^a = a^x$) has just one positive solution. If $a \neq e$, then

$g(x_0) = g\left(\frac{a}{\ln a}\right) = a\ln\left(\frac{a}{\ln a}\right) - \frac{a}{\ln a}(\ln a)$

$= a\left[\ln\left(\frac{a}{\ln a}\right) - 1\right].$

Now $\frac{a}{\ln a} > e$ (justified below), so

$g(x_0) = a\left[\ln\frac{a}{\ln a} - 1\right] > a(\ln e - 1) = 0.$ Since $g'(x) > 0$ on $(0, x_0)$, $g(x_0) > 0$, and $\lim_{x\to 0} g(x) = -\infty$, $g(x) = 0$ has exactly one solution on $(0, x_0)$.

Since $g'(x) < 0$ on (x_0, ∞), $g(x_0) > 0$, and $\lim_{x\to\infty} g(x) = -\infty$, $g(x) = 0$ has exactly one solution on (x_0, ∞). Therefore, the equation $g(x) = 0$ (and hence $x^a = a^x$) has exactly two positive solutions.

To show that $\frac{a}{\ln a} > e$ when $a \neq e$:

Consider the function $h(x) = \frac{x}{\ln x}$, for $x > 1$.

$h'(x) = \frac{\ln(x)(1) - x\left(\frac{1}{x}\right)}{(\ln x)^2} = \frac{\ln x - 1}{(\ln x)^2}$

Note that $h'(x) < 0$ on $(1, e)$ and $h'(x) > 0$ on (e, ∞), so $h(x)$ has its minimum at (e, e).

Therefore $\frac{x}{\ln x} > e$ for all $x \neq e, x > 1$.

d. For the case $a = e$, part c. shows that $g(x) = e\ln x - x\ln e < 0$ for $x \neq e$.

Therefore, when $x \neq e$, $\ln x^e < \ln e^x$, which implies $x^e < e^x$. In particular, $\pi^e < e^\pi$.

53. $f(x) = x^x = e^{x\ln x}$

Let $g(x) = x\ln x$.

Using L'Hôpital's Rule,

$\lim_{x\to 0^+} g(x) = \lim_{x\to 0^+} \frac{\ln x}{\frac{1}{x}} = \lim_{x\to 0^+} \frac{\frac{1}{x}}{-\frac{1}{x^2}}$

$= \lim_{x\to 0^+} (-x) = 0$

Therefore, $\lim_{x\to 0^+} x^x = e^0 = 1$.

$g'(x) = 1 + \ln x$

Since $g'(x) < 0$ on $(0, 1/e)$ and $g'(x) > 0$ on $(1/e, \infty)$, $g(x)$ has its minimum at $x = \frac{1}{e}$.

Therefore, $f(x)$ has its minimum at $(e^{-1}, e^{-1/e})$.

Note: this point could also be written as

$\left(\frac{1}{e}, \left(\frac{1}{e}\right)^{\frac{1}{e}}\right).$

55. $\int_0^{4\pi} x^{\sin x}\,dx \approx 20.2259$

57. a. In order of increasing slope, the graphs represent the curves $y = 2^x$, $y = 3^x$, and $y = 4^x$.

b. $\ln y$ is linear with respect to x, and at $x = 0$, $y = 1$ since $C = 1$.

c. The graph passes through the points (0.2, 4) and (0.6, 8). Thus, $4 = Cb^{0.2}$ and $8 = Cb^{0.6}$. Dividing the second equation by the first, gets $2 = b^{0.4}$ so $b = 2^{5/2}$.

Therefore $C = 2^{3/2}$.

6.5 Concepts Review

1. ky; $ky(L - y)$

3. half-life

Problem Set 6.5

1. $k = -6$, $y_0 = 4$, so $y = 4e^{-6t}$

3. $k = 0.005$, so $y = y_0 e^{0.005t}$

$y(10) = y_0 e^{0.005(10)} = y_0 e^{0.05}$

$y(10) = 2 \Rightarrow y_0 = \frac{2}{e^{0.05}}$

$y = \frac{2}{e^{0.05}} e^{0.005t} = 2e^{0.005t - 0.05} = 2e^{0.005(t-10)}$

5. $y_0 = 10,000,\ y(10) = 20,000$

$20,000 = 10,000e^{k(10)}$

$2 = e^{10k}$

$\ln 2 = 10k;\quad k = \frac{\ln 2}{10}$

$y = 10,000e^{((\ln 2)/10)t} = 10,000 \cdot 2^{t/10}$

After 25 days, $y = 10,000 \cdot 2^{2.5} \approx 56,568.$

7. $3y_0 = y_0 e^{((\ln 2)/10)t}$

$3 = e^{((\ln 2)/10)t}$

$\ln 3 = \frac{\ln 2}{10}t$

$t = \frac{10 \ln 3}{\ln 2} \approx 15.8$ days

9. 1 year: (4.5 million) $(1.032) \approx 4.64$ million

2 years: (4.5 million) $(1.032)^2 \approx 4.79$ million

10 years: (4.5 million) $(1.032)^{10} \approx 6.17$ million

100 years: (4.5 million) $(1.032)^{100} \approx 105$ million

11. The formula to use is $y = y_0 e^{kt}$, where y = population after t years, y_0 =population at time t = 0, and k is the rate of growth. We are given

$235,000 = y_0 e^{k(12)}$ and

$164,000 = y_0 e^{k(5)}$

Dividing one equation by the other yields

$1.43293 = e^{12k-5k} = e^{7k}$ or

$k = \frac{\ln(1.43293)}{7} \approx 0.0513888$

Thus $y_0 = \frac{235,000}{e^{12(0.0513888)}} = 126,839.$

13. $\frac{1}{2} = e^{k(700)}$ and $y_0 = 10$

$-\ln 2 = 700k$

$k = -\frac{\ln 2}{700} \approx -0.00099$

$y = 10e^{-0.00099t}$

At $t = 300,\ y = 10e^{-0.00099 \cdot 300} \approx 7.43.$

After 300 years there will be about 7.43 g.

15. The basic formula is $y = y_0 e^{kt}$. If t_* denotes the half-life of the material, then (see Example 3)

$\frac{1}{2} = e^{kt_*}$ or $k = \frac{\ln(0.5)}{t_*}$. Thus

$k_C = \frac{-0.693}{30.22} = -0.0229$ and $k_S = \frac{-0.693}{28.8} = -0.0241$

To find when 1% of each material will remain, we use $0.01y_0 = y_0 e^{kt}$ or $t = \frac{\ln(0.01)}{k}$. Thus

$t_C = \frac{-4.6052}{-0.0229} \approx 201$ years (2187) and

$t_S = \frac{-4.6052}{-0.0241} \approx 191$ years (2177)

17. $\frac{1}{2} = e^{5730k}$

$k = \frac{\ln\left(\frac{1}{2}\right)}{5730} \approx -1.210 \times 10^{-4}$

$0.7y_0 = y_0 e^{(-1.210 \times 10^{-4})t}$

$t = \frac{\ln 0.7}{-1.210 \times 10^{-4}} \approx 2950$

The fort burned down about 2950 years ago.

19. From Example 4, $T(t) = T_1 + (T_0 - T_1)e^{kt}$. In this problem, $200 = T(0.5) = 75 + (300 - 75)e^{k(0.5)}$ so

$k = \frac{\ln\left(\frac{125}{225}\right)}{0.5} = -1.1756$ and

$T(3) = 75 + 225e^{(-1.1756)(3)} = 81.6^\circ$ F

21. From Example 4, $T(t) = T_1 + (T_0 - T_1)e^{kt}$. In this problem, $70 = T(5) = 90 + (26 - 90)e^{k(5)}$ so

$k = \frac{\ln\left(\frac{-20}{-64}\right)}{5} = -0.2326$ and

$T(10) = 90 - 64e^{(-0.2326)(10)} = 90 - 64(0.0977) = 83.7^\circ$ C

23. From Example 4, $T(t) = T_1 + (T_0 - T_1)e^{kt}$.

Let w = the time of death; then

$82 = T(10 - w) = 70 + (98.6 - 70)e^{k(10-w)}$

$76 = T(11 - w) = 70 + (98.6 - 70)e^{k(11-w)}$

or $\quad 12 = 28.6e^{k(10-w)}$

$\quad 6 = 28.6e^{k(11-w)}$

Dividing: $2 = e^{k(-1)}$ or $k = \ln(0.5) = -0.693$

To find w:

$12 = 28.6e^{-0.693(10-w)}$ so $10 - w = \frac{\ln\left(\frac{12}{28.6}\right)}{-0.693} = 1.25$

Therefore $w = 10 - 1.25 = 8.75 = 8:45$ pm .

25. a. $(\$375)(1.035)^2 \approx \$401.71$

b. $(\$375)\left(1+\frac{0.035}{12}\right)^{24} \approx \$402.15$

c. $(\$375)\left(1+\frac{0.035}{365}\right)^{730} \approx \$402.19$

d. $(\$375)e^{0.035 \cdot 2} \approx \$402.19$

27. a. $\left(1+\frac{0.06}{12}\right)^{12t} = 2$

$1.005^{12t} = 2$

$12t = \frac{\ln 2}{\ln 1.005}$ so $t = \frac{\ln 2}{12 \ln 1.005} \approx 11.58$

It will take about 11.58 years or 11 years, 6 months, 29 days.

b. $e^{0.06t} = 2 \Rightarrow t = \frac{\ln 2}{0.06} \approx 11.55$

It will take about 11.55 years or 11 years, 6 months, and 18 days.

29. 1626 to 2000 is 374 years.

$y = 24e^{0.06 \cdot 374} \approx \$133.6$ billion

31. $1000e^{(0.05)(1)} = \$1051.27$

33. If t is the doubling time, then

$\left(1+\frac{p}{100}\right)^t = 2$

$t \ln\left(1+\frac{p}{100}\right) = \ln 2$

$t = \frac{\ln 2}{\ln\left(1+\frac{p}{100}\right)} \approx \frac{\ln 2}{\frac{p}{100}} = \frac{100 \ln 2}{p} \approx \frac{70}{p}$

35. $y = \frac{16(6.4)}{6.4+(16-6.4)e^{-16(0.00186)t}}$

$= \frac{102.4}{6.4+9.6e^{-0.02976t}}$

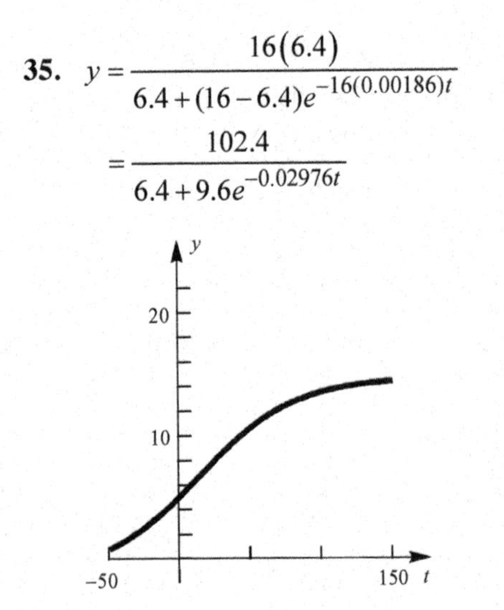

37. a. $\lim_{x \to 0}(1-x)^{1/x} = \lim_{x \to 0}\frac{1}{[1+(-x)]^{1/(-x)}} = \frac{1}{e}$

b. $\lim_{x \to 0}(1+3x)^{1/x} = \lim_{x \to 0}\left[(1+3x)^{\frac{1}{3x}}\right]^3 = e^3$

c. $\lim_{n \to \infty}\left(\frac{n+2}{n}\right)^n = \lim_{n \to \infty}\left(1+\frac{2}{n}\right)^n$

$= \lim_{x \to 0^+}(1+2x)^{1/x}$

$= \lim_{x \to 0^+}\left[(1+2x)^{\frac{1}{2x}}\right]^2 = e^2$

d. $\lim_{n \to \infty}\left(\frac{n-1}{n}\right)^{2n} = \lim_{n \to \infty}\left(1-\frac{1}{n}\right)^{2n}$

$= \lim_{x \to 0^+}(1-x)^{2/x}$

$= \lim_{x \to 0^+}\left[(1-x)^{\frac{1}{-x}}\right]^{-2} = \frac{1}{e^2}$

39. Let y = population in millions, $t = 0$ in 1985, $a = 0.012, b = 0.06$, $y_0 = 10$

$\frac{dy}{dt} = 0.012y + 0.06$

$y = \left(10+\frac{0.06}{0.012}\right)e^{0.012t} - \frac{0.06}{0.012} = 15e^{0.012t} - 5$

From 1985 to 2010 is 25 years. At $t = 25$, $y = 15e^{0.012 \cdot 25} - 5 \approx 15.25$. The population in 2010 will be about 15.25 million.

41. If $f(t) = e^{kt}$, then $\frac{f'(t)}{f(t)} = \frac{ke^{kt}}{e^{kt}} = k$.

43. $\frac{f'(x)}{f(x)} = k > 0$ can be written as

$\frac{1}{y}\frac{dy}{dx} = k$ where $y = f(x)$.

$\frac{dy}{y} = k\,dx$ has the solution $y = Ce^{kx}$.

Thus, the equation $f(x) = Ce^{kx}$ represents exponential growth since $k > 0$.

45. Maximum population:

$$13,500,000 \text{ mi}^2 \cdot \frac{640 \text{ acres}}{1 \text{ mi}^2} \cdot \frac{1 \text{ person}}{\frac{1}{2} \text{ acre}}$$

$= 1.728 \times 10^{10}$ people

Let $t = 0$ be in 2004.

$(6.4 \times 10^9)e^{0.0132t} = 1.728 \times 10^{10}$

$$t = \frac{\ln\left(\frac{1.728 \cdot 10^{10}}{6.4 \cdot 10^9}\right)}{0.0132} \approx 75.2 \text{ years from 2004, or}$$

sometime in the year 2079.

47. a. $k = 0.0132 - 0.0001t$

b. $y' = (0.0132 - 0.0001t)y$

c. $\frac{dy}{dt} = (0.0132 - 0.0001t)y$

$\frac{dy}{y} = (0.0132 - 0.0001t)\,dt$

$\ln y = 0.0132t - 0.00005t^2 + C_0$

$y = C_1 e^{0.0132t - 0.00005t^2}$

The initial condition $y(0) = 6.4$ implies that

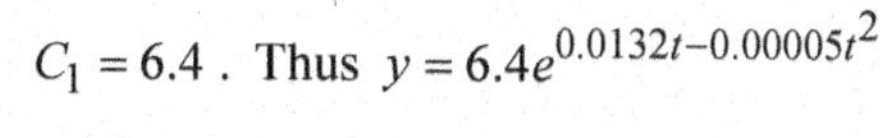

$C_1 = 6.4$. Thus $y = 6.4e^{0.0132t - 0.00005t^2}$

d.

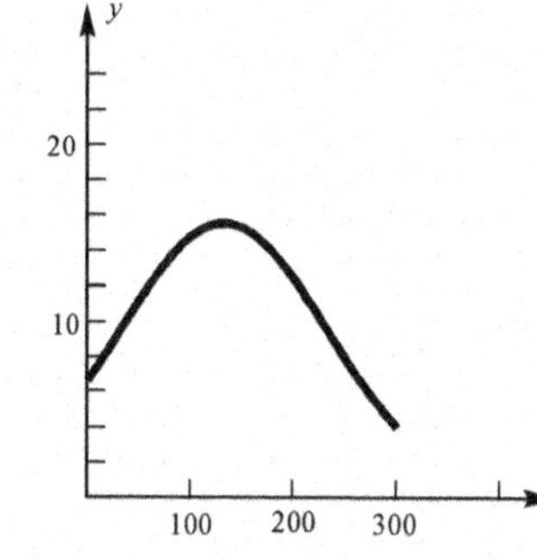

e. The maximum population will occur when

$\frac{d}{dt}\left(0.0132t - 0.00005t^2\right) = 0$

$0.0132 = 0.0001t$

$t = 0.0132 / 0.0001 = 132$

$t = 132$, which is year 2136

The population will equal the 2004 value of 6.4 billion when $0.0132t - 0.00005t^2 = 0$

$t = 0$ or $t = 264$.

The model predicts that the population will return to the 2004 level in year 2268.

49.

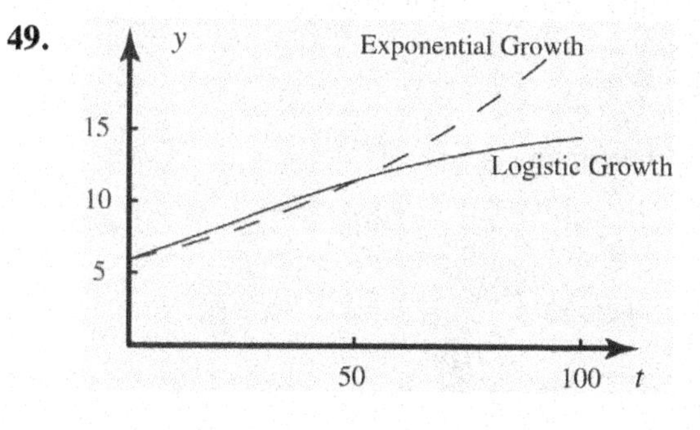

Exponential growth:
In 2010 ($t = 6$): 6.93 billion
In 2040 ($t = 36$): 10.29 billion
In 2090 ($t = 86$): 19.92 billion
Logistic growth:
In 2010 ($t = 6$): 7.13 billion
In 2040 ($t = 36$): 10.90 billion
In 2090 ($t = 86$): 15.15 billion

6.6 Concepts Review

1. $\exp\left(\int P(x)dx\right)$

3. $\frac{1}{x}; \frac{d}{dx}\left(\frac{y}{x}\right) = 1; x^2 + Cx$

Problem Set 6.6

1. Integrating factor is e^x.

$D(ye^x) = 1$

$y = e^{-x}(x + C)$

3. $y' + \frac{x}{1-x^2}y = \frac{ax}{1-x^2}$

Integrating factor:

$\exp \int \frac{x}{1-x^2}dx = \exp\left[\ln(1-x^2)^{-1/2}\right]$

$= (1-x^2)^{-1/2}$

$D[y(1-x^2)^{-1/2}] = ax(1-x^2)^{-3/2}$

Then $y(1-x^2)^{-1/2} = a(1-x^2)^{-1/2} + C$, so $y = a + C(1-x^2)^{1/2}$.

5. Integrating factor is $\frac{1}{x}$.

$D\left[\frac{y}{x}\right] = e^x$

$y = xe^x + Cx$

7. Integrating factor is x. $D[yx] = 1$; $y = 1 + Cx^{-1}$

9. $y' + f(x)y = f(x)$

Integrating factor: $e^{\int f(x)dx}$

$D\left[ye^{\int f(x)dx}\right] = f(x)e^{\int f(x)dx}$

Then $ye^{\int f(x)dx} = e^{\int f(x)dx} + C$, so

$y = 1 + Ce^{-\int f(x)dx}$.

11. Integrating factor is $\frac{1}{x}$. $D\left[\frac{y}{x}\right] = 3x^2$; $y = x^4 + Cx$

$y = x^4 + 2x$ goes through $(1, 3)$.

13. Integrating factor: xe^x

$d[yxe^x] = 1$; $y = e^{-x}(1 + Cx^{-1})$; $y = e^{-x}(1 - x^{-1})$ goes through $(1, 0)$.

15. Let y denote the number of pounds of chemical A after t minutes.

$\frac{dy}{dt} = \left(2\frac{\text{lbs}}{\text{gal}}\right)\left(3\frac{\text{gal}}{\text{min}}\right) - \left(\frac{y \text{ lbs}}{20 \text{ gal}}\right)\left(\frac{3 \text{ gal}}{\text{min}}\right)$

$= 6 - \frac{3y}{20}$ lb/min

$y' + \frac{3}{20}y = 6$

Integrating factor: $e^{\int (3/20)dt} = e^{3t/20}$

$D[ye^{3t/20}] = 6e^{3t/20}$

Then $ye^{3t/20} = 40e^{3t/20} + C$. $t = 0, y = 10$

$\Rightarrow C = -30$.

Therefore, $y(t) = 40 - 30e^{-3t/20}$, so

$y(20) = 40 - 30e^{-3} \approx 38.506$ lb.

17. $\frac{dy}{dt} = 4 - \left[\frac{y}{(120 - 2t)}\right](6)$ or $y' + \left[\frac{3}{(60 - t)}\right]y = 4$

Integrating factor is $(60 - t)^{-3}$.

$D[y(60 - t)^{-3}] = 4(60 - t)^{-3}$

$y(t) = 2(60 - t) + C(60 - t)^3$

$y(t) = 2(60 - t) - \left(\frac{1}{1800}\right)(60 - t)^3$ goes through $(0, 0)$.

19. $I' + 10^6 I = 1$

Integrating factor $= \exp(10^6 t)$

$D[I \exp(10^6 t)] = \exp(10^6 t)$

$I(t) = 10^{-6} + C\exp(-10^6 t)$

$I(t) = 10^{-6}[1 - \exp(-10^6 t)]$ goes through $(0, 0)$.

21. $1000\,I = 120 \sin 377t$

$I(t) = 0.12 \sin 377t$

23. Let y be the number of gallons of pure alcohol in the tank at time t.

a. $y' = \frac{dy}{dt} = 5(0.25) - \left(\frac{5}{100}\right)y = 1.25 - 0.05y$

Integrating factor is $e^{0.05t}$.

$y(t) = 25 + Ce^{-0.05t}$; $y = 100, t = 0, C = 75$

$y(t) = 25 + 75e^{-0.05t}$; $y = 50, t = T$,

$T = 20(\ln 3) \approx 21.97$ min

b. Let A be the number of gallons of pure alcohol drained away.

$(100 - A) + 0.25A = 50 \Rightarrow A = \frac{200}{3}$

It took $\frac{\frac{200}{3}}{5}$ minutes for the draining and the same amount of time to refill, so

$T = \frac{2\left(\frac{200}{3}\right)}{5} = \frac{80}{3} \approx 26.67$ min.

c. c would need to satisfy

$\frac{\frac{200}{3}}{5} + \frac{\frac{200}{3}}{c} < 20(\ln 3)$.

$c > \frac{10}{(3\ln 3 - 2)} \approx 7.7170$

d. $y' = 4(0.25) - 0.05y = 1 - 0.05y$

Solving for y, as in part a, yields $y = 20 + 80e^{-0.05t}$. The drain is closed when $t = 0.8T$. We require that

$(20 + 80e^{-0.05 \cdot 0.8T}) + 4 \cdot 0.25 \cdot 0.2T = 50$,

or $400e^{-0.04T} + T = 150$.

25. a. $v_\infty = -\frac{32}{0.05} = -640$

$v(t) = [120-(-640)]e^{-0.05t} + (-640) = 0$ if $t = 20\ln\left(\frac{19}{16}\right)$.

$y(t) = 0 + (-640)t + \left(\frac{1}{0.05}\right)[120-(-640)](1-e^{-0.05t})$

$= -640t + 15{,}200(1-e^{-0.05t})$

Therefore, the maximum altitude is

$$y\left(20\ln\left(\frac{19}{16}\right)\right) = -12{,}800\ln\left(\frac{19}{16}\right) + \frac{45{,}600}{19} \approx 200.32 \text{ ft}$$

b. $-640T + 15{,}200(1-e^{-0.05T}) = 0;$

$95 - 4T - 95e^{-0.05T} = 0$

27. a. $e^{-\ln x + C}\left(\frac{dy}{dx} - \frac{y}{x}\right) = x^2 e^{-\ln x + C}$

$e^{-\ln x}e^C\left(\frac{dy}{dx} - \frac{y}{x}\right) = x^2 e^C e^{-\ln x}$

$\frac{1}{x}e^C\frac{dy}{dx} - ye^C\frac{1}{x^2} = x^2 e^C \frac{1}{x}$

$\frac{d}{dx}\left(e^C\frac{1}{x}y\right) = xe^C$

b. $e^C\frac{y}{x} = e^C\int x\,dx$

$\frac{y}{x} = \frac{x^2}{2} + C_1$

$y = \frac{x^3}{2} + C_1 x$

6.7 Concepts Review

1. slope field

3. $y_{n-1} + hf(x_{n-1}, y_{n-1})$

Problem Set 6.7

1.

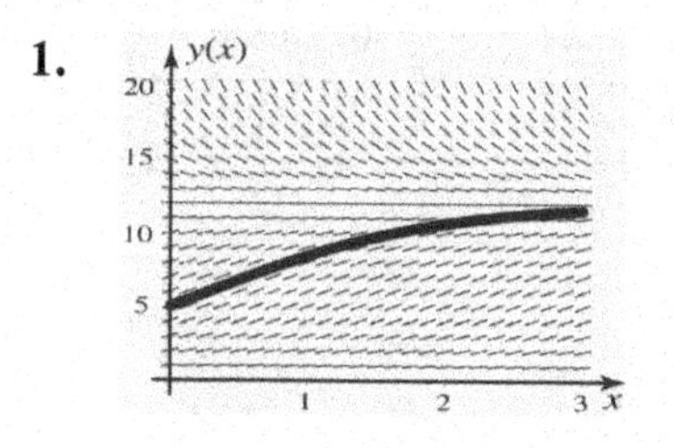

$\lim_{x\to\infty} y(x) = 12$ and $y(2) \approx 10.5$

3.

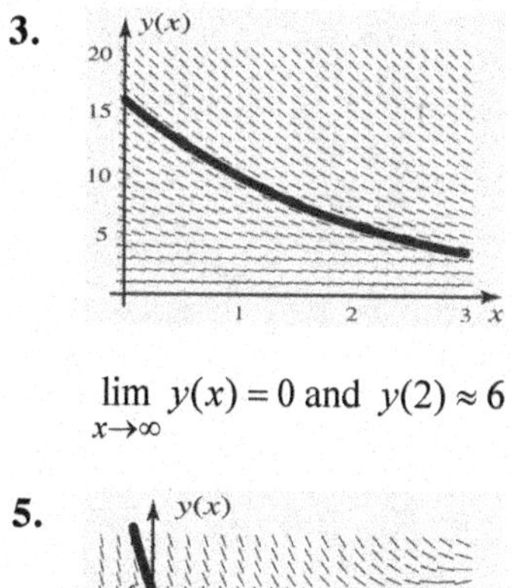

$\lim_{x\to\infty} y(x) = 0$ and $y(2) \approx 6$

5.

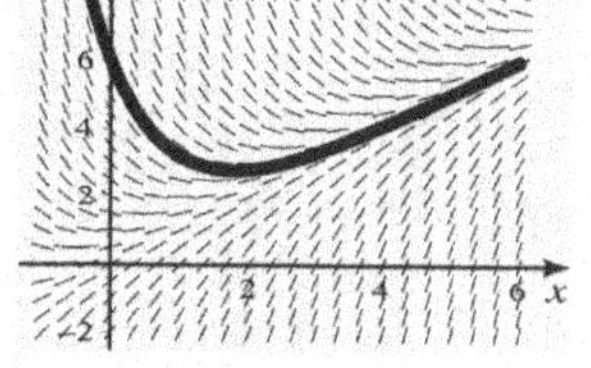

The oblique asymptote is $y = x$.

7.

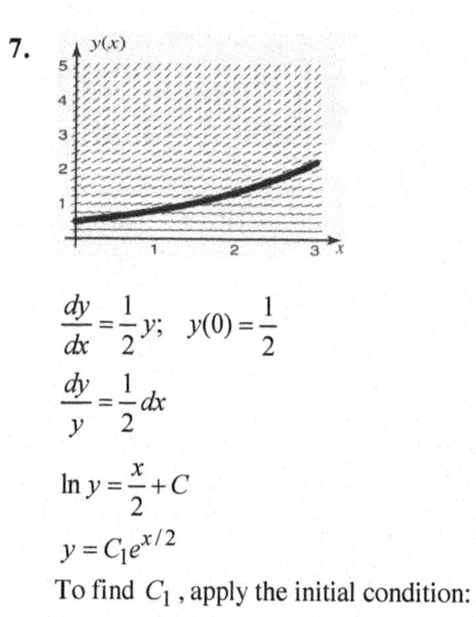

$\frac{dy}{dx} = \frac{1}{2}y;\quad y(0) = \frac{1}{2}$

$\frac{dy}{y} = \frac{1}{2}dx$

$\ln y = \frac{x}{2} + C$

$y = C_1 e^{x/2}$

To find C_1, apply the initial condition:

$\frac{1}{2} = y(0) = C_1 e^0 = C_1$

$y = \frac{1}{2}e^{x/2}$

1.0	0.65536	0.74148

9. 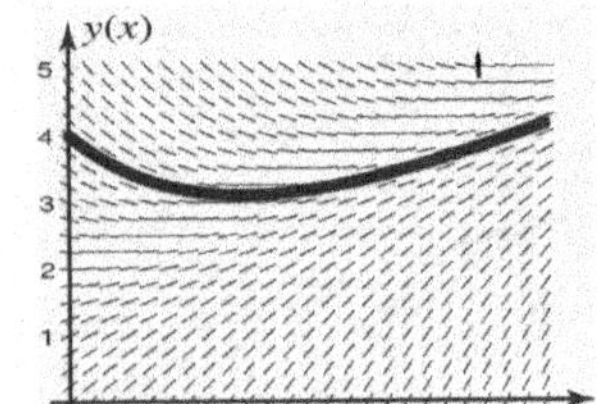

$y' + y = x + 2$

The integrating factor is $e^{\int 1dx} = e^x$.

$e^x y' + ye^x = e^x(x+2)$

$\frac{d}{dx}\left(e^x y\right) = (x+2)e^x$

$e^x y = \int (x+2)e^x\, dx$

Integrate by parts: let $u = x+2, \quad dv = e^x\, dx$.

Then $du = dx$ and $v = e^x$. Thus

$e^x\, y = (x+2)e^x - \int e^x\, dx$

$e^x\, y = (x+2)e^x - e^x + C$

$y = x + 2 - 1 + Ce^{-x}$

To find C, apply the initial condition:

$4 = y(0) = 0 + 1 + Ce^{-0} = 1 + C \quad \rightarrow \quad C = 3$

Thus, $y = x + 1 + 3e^{-x}$.

Note: Solutions to Problems 22-28 are given along with the corresponding solutions to 11-16.

11., 22.

x_n	Euler's Method y_n	Improved Euler Method y_n
0.0	3.0	3.0
0.2	4.2	4.44
0.4	5.88	6.5712
0.6	8.232	9.72538
0.8	11.5248	14.39356
1.0	16.1347	21.30246

12., 23.

x_n	Euler's Method y_n	Improved Euler Method y_n
0.0	2.0	2.0
0.2	1.6	1.64
0.4	1.28	1.3448
0.6	1.024	1.10274
0.8	0.8195	0.90424

13., 24.

x_n	Euler's Method y_n	Improved Euler Method y_n
0.0	0.0	0.0
0.2	0.0	0.02
0.4	0.04	0.08
0.6	0.12	0.18
0.8	0.24	0.32
1.0	0.40	0.50

14., 25.

x_n	Euler's Method y_n	Improved Euler Method y_n
0.0	0.0	0.0
0.2	0.0	0.004
0.4	0.008	0.024
0.6	0.040	0.076
0.8	0.112	0.176
1.0	0.240	0.340

15., 26.

x_n	Euler's Method y_n	Improved Euler Method y_n
1.0	1.0	1.0
1.2	1.2	1.244
1.4	1.488	1.60924
1.6	1.90464	2.16410
1.8	2.51412	3.02455
2.0	3.41921	4.391765

16., 27.

x_n	Euler's Method y_n	Improved Euler Method y_n
1.0	2.0	2.0
1.2	1.2	1.312
1.4	0.624	0.80609
1.6	0.27456	0.46689
1.8	0.09884	0.25698
2.0	0.02768	0.13568

17. a. $y_0 = 1$

$y_1 = y_0 + hf(x_0, y_0)$
$= y_0 + hy_0 = (1+h)y_0$

$y_2 = y_1 + hf(x_1, y_1) = y_1 + hy_1$
$= (1+h)y_1 = (1+h)^2 y_0$

$y_3 = y_2 + hf(x_2, y_2) = y_2 + hy_2$
$= (1+h)y_2 = (1+h)^3 y_0$

$\vdots$

$y_n = y_{n-1} + hf(x_{n-1}, y_{n-1}) = y_{n-1} + hy_{n-1}$
$= (1+h)y_{n-1} = (1+h)^n y_0 = (1+h)^n$

b. Let $N = 1/h$. Then y_N is an approximation to the solution at $x = Nh = (1/h)h = 1$. The exact solution is $y(1) = e$. Thus, $(1+1/N)^N \approx e$ for large N. From Chapter 7, we know that $\lim_{N\to\infty}(1+1/N)^N = e$.

19. a. $\int_{x_0}^{x_1} y'(x)dx = \int_{x_0}^{x_1} \sin x^2 dx$

$y(x_1) - y(x_0) \approx (x_1 - x_0)\sin x_0^2$
$y(x_1) - y(0) = h\sin x_0^2$
$y(x_1) - 0 \approx 0.1\sin 0^2$
$y(x_1) \approx 0$

b. $\int_{x_0}^{x_2} y'(x)dx = \int_{x_0}^{x_2} \sin x^2 dx$

$y(x_2) - y(x_0) \approx (x_1 - x_0)\sin x_0^2 + (x_2 - x_1)\sin x_1^2$
$y(x_2) - y(0) = h\sin x_0^2 + h\sin x_1^2$
$y(x_2) - 0 \approx 0.1\sin 0^2 + 0.1\sin 0.1^2$
$y(x_2) \approx 0.00099998$

c. $\int_{x_0}^{x_3} y'(x)dx = \int_{x_0}^{x_3} \sin x^2 dx$

$y(x_3) - y(x_0) \approx (x_1 - x_0)\sin x_0^2 + (x_2 - x_1)\sin x_1^2 + (x_3 - x_2)\sin x_1^2$
$y(x_3) - y(0) = h\sin x_0^2 + h\sin x_1^2 + h\sin x_2^2$
$y(x_3) - 0 \approx 0.1\sin 0^2 + 0.1\sin 0.1^2 + 0.1\sin 0.2^2$
$y(x_3) \approx 0.004999$

Continuing in this fashion, we have

$\int_{x_0}^{x_n} y'(x)dx = \int_{x_0}^{x_n} \sin x^2 dx$

$y(x_n) - y(x_0) \approx \sum_{i=0}^{n-1}(x_{i+1} - x_i)\sin x_i^2$

$y(x_n) \approx h\sum_{i=0}^{n-1} f(x_{i-1})$

When $n = 10$, this becomes

$y(x_{10}) = y(1) \approx 0.269097$

d. The result $y(x_n) \approx h\sum_{i=0}^{n-1} f(x_{i-1})$ is the same as that given in Problem 18. Thus, when $f(x, y)$ depends only on x, then the two methods (1) Euler's method for approximating the solution to $y' = f(x)$ at x_n, and (2) the left-endpoint Riemann sum for approximating $\int_0^{x_n} f(x)\,dx$, are equivalent.

21. a. $\frac{\Delta y}{\Delta x} = \frac{1}{2}[f(x_0, y_0) + f(x_1 + \hat{y}_1)]$

b. $\frac{y_1 - y_0}{h} = \frac{\Delta y}{\Delta x} = \frac{1}{2}[f(x_0, y_0) + f(x_1 + \hat{y}_1)] \Rightarrow$

$2(y_1 - y_0) = h[f(x_0, y_0) + f(x_1 + \hat{y}_1)] \Rightarrow$

$y_1 - y_0 = \frac{h}{2}[f(x_0, y_0) + f(x_1 + \hat{y}_1)] \Rightarrow$

$y_1 = y_0 + \frac{h}{2}[f(x_0, y_0) + f(x_1 + \hat{y}_1)]$

c. 1. $x_{n-1} + h$

2. $y_{n-1} + hf(x_{n-1}, y_{n-1})$

3. $y_{n-1} + \frac{h}{2}[f(x_{n-1}, y_{n-1}) + f(x_n, \hat{y}_n)]$

22-27. See problems 11-16

6.8 Concepts Review

1. $\left[-\frac{\pi}{2}, \frac{\pi}{2}\right]$; arcsin

3. 1

Problem Set 6.8

1. $\arccos\left(\frac{\sqrt{2}}{2}\right) = \frac{\pi}{4}$ since $\cos\frac{\pi}{4} = \frac{\sqrt{2}}{2}$

3. $\sin^{-1}\left(-\frac{\sqrt{3}}{2}\right) = -\frac{\pi}{3}$ since $\sin\left(-\frac{\pi}{3}\right) = -\frac{\sqrt{3}}{2}$

5. $\arctan(\sqrt{3}) = \frac{\pi}{3}$ since $\tan\left(\frac{\pi}{3}\right) = \sqrt{3}$

7. $\arcsin\left(-\frac{1}{2}\right) = -\frac{\pi}{6}$ since $\sin\left(-\frac{\pi}{6}\right) = -\frac{1}{2}$

9. $\sin(\sin^{-1} 0.4567) = 0.4567$ by definition

11. $\sin^{-1}(0.1113) \approx 0.1115$

13. $\cos(\text{arccot}\, 3.212) = \cos\left(\arctan \frac{1}{3.212}\right)$

$\approx \cos 0.3018 \approx 0.9548$

15. $\sec^{-1}(-2.222) = \cos^{-1}\left(\frac{1}{-2.222}\right) \approx 2.038$

17. $\cos(\sin(\tan^{-1} 2.001)) \approx 0.6259$

19. $\theta = \sin^{-1} \frac{x}{8}$

21. $\theta = \sin^{-1} \frac{5}{x}$

23. Let θ_1 be the angle opposite the side of length 3, and $\theta_2 = \theta_1 - \theta$, so $\theta = \theta_1 - \theta_2$. Then $\tan \theta_1 = \frac{3}{x}$ and $\tan \theta_2 = \frac{1}{x}$. $\theta = \tan^{-1} \frac{3}{x} - \tan^{-1} \frac{1}{x}$.

25. $\cos\left[2 \sin^{-1}\left(-\frac{2}{3}\right)\right] = 1 - 2\sin^2\left[\sin^{-1}\left(-\frac{2}{3}\right)\right]$

$= 1 - 2\left(-\frac{2}{3}\right)^2 = \frac{1}{9}$

27. $\sin\left[\cos^{-1}\left(\frac{3}{5}\right) + \cos^{-1}\left(\frac{5}{13}\right)\right] = \sin\left[\cos^{-1}\left(\frac{3}{5}\right)\right]\cos\left[\cos^{-1}\left(\frac{5}{13}\right)\right] + \cos\left[\cos^{-1}\left(\frac{3}{5}\right)\right]\sin\left[\cos^{-1}\left(\frac{5}{13}\right)\right]$

$= \sqrt{1 - \left(\frac{3}{5}\right)^2} \cdot \frac{5}{13} + \frac{3}{5}\sqrt{1 - \left(\frac{5}{13}\right)^2} = \frac{56}{65}$

29. $\tan(\sin^{-1} x) = \frac{\sin(\sin^{-1} x)}{\cos(\sin^{-1} x)} = \frac{x}{\sqrt{1 - x^2}}$

31. $\cos(2\sin^{-1} x) = 1 - 2\sin^2(\sin^{-1} x) = 1 - 2x^2$

33. a. $\lim_{x \to \infty} \tan^{-1} x = \frac{\pi}{2}$ since $\lim_{\theta \to \pi/2^-} \tan \theta = \infty$

b. $\lim_{x \to -\infty} \tan^{-1} x = -\frac{\pi}{2}$ since $\lim_{\theta \to -\pi/2^+} \tan \theta = -\infty$

35. a. Let $L = \lim_{x \to 1^-} \sin^{-1} x$. Since

$\sin(\sin^{-1} x) = x$,

$\lim_{x \to 1^-} \sin(\sin^{-1} x) = \lim_{x \to 1^-} x = 1$.

Thus, since sin is continuous, the Composite Limit Theorem gives us

$\lim_{x \to 1^-} \sin(\sin^{-1} x) = \lim_{x \to 1^-} \sin(L)$; hence

$\sin L = 1$ and since the range of $\sin^{-1}$ is $\left[-\frac{\pi}{2}, \frac{\pi}{2}\right]$, $L = \frac{\pi}{2}$.

b. Let $L = \lim_{x \to -1^+} \sin^{-1} x$. Since

$\sin(\sin^{-1} x) = x$,

$\lim_{x \to -1^+} \sin(\sin^{-1} x) = \lim_{x \to -1^+} x = -1$.

Thus, since sin is continuous, the Composite Limit Theorem gives us

$\lim_{x \to -1^+} \sin(\sin^{-1} x) = \lim_{x \to -1^+} \sin(L)$; hence

$\sin L = -1$ and since the range of $\sin^{-1}$ is $\left[-\frac{\pi}{2}, \frac{\pi}{2}\right]$, $L = -\frac{\pi}{2}$.

37. Let $f(x) = y = \sin^{-1} x$; then the slope of the tangent line to the graph of y at c is

$f'(c) = \frac{1}{\sqrt{1 - c^2}}$. Hence, $\lim_{c \to 1^-} f'(c) = \infty$ so

that the tangent lines approach the vertical.

39. $y = \ln(2 + \sin x)$. Let $u = 2 + \sin x$; then $y = \ln u$ so by the Chain Rule

$$\frac{dy}{dx} = \frac{dy}{du}\frac{du}{dx} = \left(\frac{1}{u}\right)\frac{du}{dx} = \left(\frac{1}{2+\sin x}\right)\cdot \cos x$$
$$= \frac{\cos x}{2+\sin x}$$

41. $$\frac{d}{dx}\ln(\sec x + \tan x) = \frac{\sec x \tan x + \sec^2 x}{\sec x + \tan x}$$
$$= \frac{(\sec x)(\tan x + \sec x)}{\sec x + \tan x} = \sec x$$

43. $$\frac{d}{dx}\sin^{-1}(2x^2) = \frac{1}{\sqrt{1-(2x^2)^2}}\cdot 4x = \frac{4x}{\sqrt{1-4x^4}}$$

45. $$\frac{d}{dx}[x^3 \tan^{-1}(e^x)] = x^3 \cdot \frac{e^x}{1+(e^x)^2} + 3x^2 \tan^{-1}(e^x)$$
$$= x^2\left[\frac{xe^x}{1+e^{2x}} + 3\tan^{-1}(e^x)\right]$$

47. $$\frac{d}{dx}(\tan^{-1} x)^3 = 3(\tan^{-1} x)^2 \cdot \frac{1}{1+x^2} = \frac{3(\tan^{-1} x)^2}{1+x^2}$$

49. $$\frac{d}{dx}\sec^{-1}(x^3) = \frac{1}{|x^3|\sqrt{(x^3)^2 - 1}}\cdot 3x^2 = \frac{3}{|x|\sqrt{x^6 - 1}}$$

51. $$\frac{d}{dx}(1+\sin^{-1} x)^3 = 3(1+\sin^{-1} x)^2 \cdot \frac{1}{\sqrt{1-x^2}}$$
$$= \frac{3(1+\sin^{-1} x)^2}{\sqrt{1-x^2}}$$

53. $y = \tan^{-1}\left(\ln x^2\right)$

Let $u = x^2$, $v = \ln u$; then $y = \tan^{-1}\left(v(u(x))\right)$ so by the Chain Rule:

$$\frac{dy}{dx} = \frac{dy}{dv}\frac{dv}{du}\frac{du}{dx} = \frac{1}{1+v^2}\cdot\frac{1}{u}\cdot 2x =$$
$$\frac{1}{1+(\ln x^2)^2}\cdot\frac{1}{x^2}\cdot 2x =$$
$$\frac{2}{x[1+(\ln x^2)^2]}$$

55. $\int \cos 3x \, dx$

Let $u = 3x$, $du = 3dx$; then

$$\int \cos 3x \, dx = \frac{1}{3}\int \cos 3x(3dx) =$$
$$\frac{1}{3}\int \cos u \, du = \frac{1}{3}\sin u + C = \frac{1}{3}\sin 3x + C$$

57. Let $u = \sin 2x$, so $du = 2\cos 2x \, dx$.

$$\int \sin 2x \cos 2x \, dx = \frac{1}{2}\int \sin 2x(2\cos 2x)dx$$
$$= \frac{1}{2}\int u \, du$$
$$= \frac{u^2}{4} + C = \frac{1}{4}\sin^2 2x + C$$

59. Let $u = e^{2x}$, so $du = 2e^{2x}dx$.

$$\int e^{2x}\cos(e^{2x})dx = \frac{1}{2}\int \cos(e^{2x})(2e^{2x})dx$$
$$= \frac{1}{2}\int \cos u \, du$$
$$= \frac{1}{2}\sin u + C = \frac{1}{2}\sin(e^{2x}) + C$$
$$\int_0^1 e^{2x}\cos(e^{2x})\,dx = \left[\frac{1}{2}\sin(e^{2x})\right]_0^1$$
$$= \left[\frac{1}{2}\sin(e^2) - \frac{1}{2}\sin(e^0)\right]$$
$$= \frac{\sin e^2 - \sin 1}{2} \approx 0.0262$$

61. $$\int_0^{\sqrt{2}/2}\frac{1}{\sqrt{1-x^2}}dx = [\arcsin x]_0^{\sqrt{2}/2}$$
$$= \arcsin\frac{\sqrt{2}}{2} - \arcsin 0 = \frac{\pi}{4}$$

63. $$\int_{-1}^{1}\frac{1}{1+x^2}dx = \left[\tan^{-1} x\right]_{-1}^{1} = \tan^{-1} 1 - \tan^{-1}(-1)$$
$$= \frac{\pi}{4} - \left(-\frac{\pi}{4}\right) = \frac{\pi}{2}$$

65. Let $u = 2x$, so $du = 2\,dx$.

$$\int \frac{1}{1+4x^2}dx = \frac{1}{2}\int \frac{1}{1+(2x)^2}2dx$$
$$= \frac{1}{2}\int \frac{1}{1+u^2}du = \frac{1}{2}\arctan u + C$$
$$= \frac{1}{2}\arctan 2x + C$$

67. $\int \frac{1}{\sqrt{12-9x^2}}dx = \int \frac{1}{\sqrt{12\left(1-\frac{3}{4}x^2\right)}}dx$

$= \frac{1}{2\sqrt{3}}\int \frac{1}{\sqrt{1-\left(\frac{\sqrt{3}}{2}x\right)^2}}dx$

Let $u = \frac{\sqrt{3}}{2}x,\ du = \frac{\sqrt{3}}{2}dx$; then

$\frac{1}{2\sqrt{3}}\int \frac{1}{\sqrt{1-\left(\frac{\sqrt{3}}{2}x\right)^2}}dx = \frac{1}{2\sqrt{3}}\left(\frac{2}{\sqrt{3}}\right)\int \frac{1}{\sqrt{1-u^2}}du$

$= \frac{1}{3}\sin^{-1}u + C = \frac{1}{3}\sin^{-1}\left(\frac{\sqrt{3}}{2}x\right) + C$

69. $\int \frac{1}{x^2-6x+13}dx = \int \frac{1}{(x^2-6x+9)+4}dx$

$= \int \frac{1}{(x-3)^2+4}dx$

Let $u = x-3,\ du = dx,\ a = 2;$ then

$\int \frac{1}{(x-3)^2+4}dx = \int \frac{1}{u^2+a^2}du = \frac{1}{a}\tan^{-1}\left(\frac{u}{a}\right) + C$

$= \frac{1}{2}\tan^{-1}\left(\frac{x-3}{2}\right) + C$

71. $\int \frac{1}{x\sqrt{4x^2-9}}dx$. Let $u = 2x,\ du = 2\,dx, a = 3$; then

$\int \frac{1}{x\sqrt{4x^2-9}}dx = \int \frac{1}{2x\sqrt{4x^2-9}}(2\,dx) =$

$\int \frac{1}{u\sqrt{u^2-a^2}}du = \frac{1}{a}\sec^{-1}\left(\frac{|u|}{a}\right) + C =$

$\frac{1}{3}\sec^{-1}\left(\frac{2|x|}{3}\right) + C$

73. The top of the picture is 7.6 ft above eye level, and the bottom of the picture is 2.6 ft above eye level. Let θ_1 be the angle between the viewer's line of sight to the top of the picture and the horizontal. Then call $\theta_2 = \theta_1 - \theta$, so $\theta = \theta_1 - \theta_2$.

$\tan\theta_1 = \frac{7.6}{b}$; $\tan\theta_2 = \frac{2.6}{b}$;

$\theta = \tan^{-1}\frac{7.6}{b} - \tan^{-1}\frac{2.6}{b}$

If $b = 12.9$, $\theta \approx 0.3335$ or $19.1°$.

75. $\tan\left[2\tan^{-1}\left(\frac{1}{4}\right)\right] = \frac{2\tan\left[\tan^{-1}\left(\frac{1}{4}\right)\right]}{1-\tan^2\left[\tan^{-1}\left(\frac{1}{4}\right)\right]}$

$= \frac{2\cdot\frac{1}{4}}{1-\left(\frac{1}{4}\right)^2} = \frac{8}{15}$

$\tan\left[3\tan^{-1}\left(\frac{1}{4}\right)\right] = \tan\left[2\tan^{-1}\left(\frac{1}{4}\right) + \tan^{-1}\left(\frac{1}{4}\right)\right]$

$= \frac{\tan\left[2\tan^{-1}\left(\frac{1}{4}\right)\right] + \tan\left[\tan^{-1}\left(\frac{1}{4}\right)\right]}{1-\tan\left[2\tan^{-1}\left(\frac{1}{4}\right)\right]\tan\left[\tan^{-1}\left(\frac{1}{4}\right)\right]}$

$= \frac{\frac{8}{15}+\frac{1}{4}}{1-\frac{8}{15}\cdot\frac{1}{4}} = \frac{47}{52}$

$\tan\left[3\tan^{-1}\left(\frac{1}{4}\right) + \tan^{-1}\left(\frac{5}{99}\right)\right]$

$= \frac{\tan\left[3\tan^{-1}\left(\frac{1}{4}\right)\right] + \tan\left[\tan^{-1}\left(\frac{5}{99}\right)\right]}{1-\tan\left[3\tan^{-1}\left(\frac{1}{4}\right)\right]\tan\left[\tan^{-1}\left(\frac{5}{99}\right)\right]}$

$= \frac{\frac{47}{52}+\frac{5}{99}}{1-\frac{47}{52}\cdot\frac{5}{99}} = \frac{4913}{4913} = 1 = \tan\frac{\pi}{4}$

Thus, $3\tan^{-1}\left(\frac{1}{4}\right) + \tan^{-1}\left(\frac{5}{99}\right) = \tan^{-1}(1) = \frac{\pi}{4}$.

77.

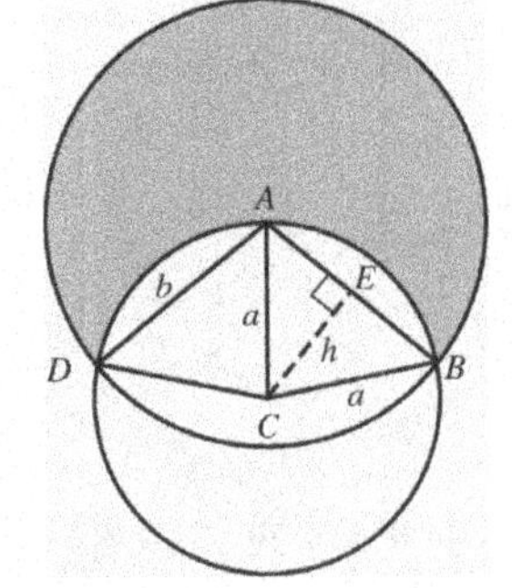

Let θ represent $\angle DAB$, then $\angle CAB$ is $\frac{\theta}{2}$. Since ΔABC is isosceles, $|AE| = \frac{b}{2}$, $\cos\frac{\theta}{2} = \frac{\frac{b}{2}}{a} = \frac{b}{2a}$ and $\theta = 2\cos^{-1}\frac{b}{2a}$. Thus sector ADB has area $\frac{1}{2}\left(2\cos^{-1}\frac{b}{2a}\right)b^2 = b^2\cos^{-1}\frac{b}{2a}$. Let ϕ represent $\angle DCB$, then $\angle ACB$ is $\frac{\phi}{2}$ and $\angle ECA$ is $\frac{\phi}{4}$, so $\sin\frac{\phi}{4} = \frac{\frac{b}{2}}{a} = \frac{b}{2a}$ and $\phi = 4\sin^{-1}\frac{b}{2a}$. Thus sector DCB

has area $\frac{1}{2}\left(4\sin^{-1}\frac{b}{2a}\right)a^2 = 2a^2\sin^{-1}\frac{b}{2a}$. These sectors overlap on the triangles ΔDAC and ΔCAB, each of which has area

$$\frac{1}{2}|AB|h = \frac{1}{2}b\sqrt{a^2 - \left(\frac{b}{2}\right)^2} = \frac{1}{2}b\frac{\sqrt{4a^2-b^2}}{2}.$$

The large circle has area πb^2, hence the shaded region has area

$$\pi b^2 - b^2\cos^{-1}\frac{b}{2a} - 2a^2\sin^{-1}\frac{b}{2a} + \frac{1}{2}b\sqrt{4a^2-b^2}$$

79.

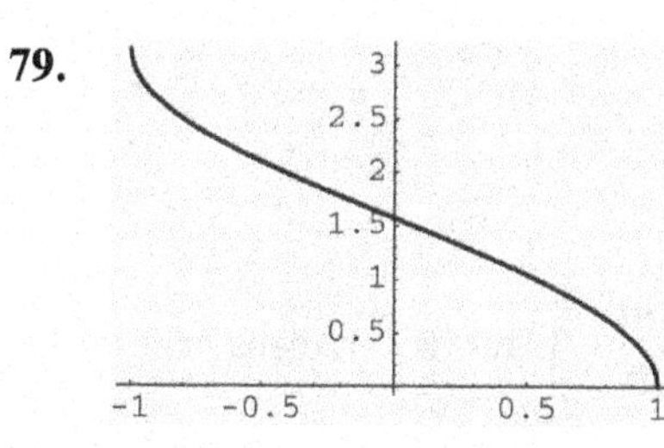

It is the same graph as $y = \arccos x$.

Conjecture: $\frac{\pi}{2} - \arcsin x = \arccos x$

Proof: Let $\theta = \frac{\pi}{2} - \arcsin x$

Then $x = \sin\left(\frac{\pi}{2} - \theta\right) = \cos\theta$

so $\theta = \arccos x$.

81. $$\int\frac{dx}{\sqrt{a^2-x^2}} = \int\frac{dx}{\sqrt{a^2\left[1-\left(\frac{x}{a}\right)^2\right]}}$$

$$= \int\frac{1}{|a|}\cdot\frac{dx}{\sqrt{1-\left(\frac{x}{a}\right)^2}} = \int\frac{1}{a}\cdot\frac{dx}{\sqrt{1-\left(\frac{x}{a}\right)^2}} \text{ since } a > 0$$

Let $u = \frac{x}{a}$, so $du = \frac{1}{a}dx$.

$$\int\frac{1}{a}\cdot\frac{dx}{\sqrt{1-\left(\frac{x}{a}\right)^2}} = \int\frac{du}{\sqrt{1-u^2}} = \sin^{-1}u + C$$

$$= \sin^{-1}\frac{x}{a} + C$$

83. Let $u = \frac{x}{a}$, so $du = \frac{1}{a}dx$

$$\int\frac{dx}{a^2+x^2} = \frac{1}{a}\int\frac{1}{1+\left(\frac{x}{a}\right)^2}\frac{1}{a}dx$$

$$= \frac{1}{a}\int\frac{1}{1+u^2}du = \frac{1}{a}\tan^{-1}u + C$$

$$= \frac{1}{a}\tan^{-1}\left(\frac{x}{a}\right) + C$$

85. Note that $\frac{d}{dx}\sin^{-1}\left(\frac{x}{a}\right) = \frac{1}{\sqrt{a^2-x^2}}$ (See Problem 67).

$$\frac{d}{dx}\left[\frac{x}{2}\sqrt{a^2-x^2} + \frac{a^2}{2}\sin^{-1}\frac{x}{a} + C\right]$$

$$= \frac{1}{2}\sqrt{a^2-x^2} + \frac{x}{2}\frac{1}{2\sqrt{a^2-x^2}}(-2x)$$

$$+ \frac{a^2}{2}\frac{1}{\sqrt{a^2-x^2}} + 0$$

$$= \frac{1}{2}\sqrt{a^2-x^2} + \frac{1}{2}\frac{-x^2+a^2}{\sqrt{a^2-x^2}} = \sqrt{a^2-x^2}$$

87. Let θ be the angle subtended by viewer's eye.

$$\theta = \tan^{-1}\left(\frac{12}{b}\right) - \tan^{-1}\left(\frac{2}{b}\right)$$

$$\frac{d\theta}{db} = \frac{1}{1+\left(\frac{12}{b}\right)^2}\left(-\frac{12}{b^2}\right) - \frac{1}{1+\left(\frac{2}{b}\right)^2}\left(-\frac{2}{b^2}\right)$$

$$= \frac{2}{b^2+4} - \frac{12}{b^2+144} = \frac{10(24-b^2)}{(b^2+4)(b^2+144)}$$

Since $\frac{d\theta}{db} > 0$ for b in $\left[0, 2\sqrt{6}\right)$

and $\frac{d\theta}{db} < 0$ for b in $\left(2\sqrt{6}, \infty\right)$, the angle is maximized for $b = 2\sqrt{6} \approx 4.899$.
The ideal distance is about 4.9 ft from the wall.

89. Let $h(t)$ represent the height of the elevator (the number of feet above the spectator's line of sight) t seconds after the line of sight passes horizontal, and let $\theta(t)$ denote the angle of elevation.

Then $h(t) = 15t$, so $\theta(t) = \tan^{-1}\left(\frac{15t}{60}\right) = \tan^{-1}\left(\frac{t}{4}\right)$.

$$\frac{d\theta}{dt} = \frac{1}{1+\left(\frac{t}{4}\right)^2}\left(\frac{1}{4}\right) = \frac{4}{16+t^2}$$

At $t = 6$, $\frac{d\theta}{dt} = \frac{4}{16+6^2} = \frac{1}{13}$ radians per second or about 4.41° per second.

91. Let x represent the position on the shoreline and let θ represent the angle of the beam ($x = 0$ and $\theta = 0$ when the light is pointed at P). Then $\theta = \tan^{-1}\left(\frac{x}{2}\right)$, so $\frac{d\theta}{dt} = \frac{1}{1+\left(\frac{x}{2}\right)^2}\frac{1}{2}\frac{dx}{dt} = \frac{2}{4+x^2}\frac{dx}{dt}$ When $x = 1$,

$\frac{dx}{dt} = 5\pi$, so $\frac{d\theta}{dt} = \frac{2}{4+1^2}(5\pi) = 2\pi$ The beacon revolves at a rate of 2π radians per minute or 1 revolution per minute.

93. Let x represent the distance to the *center* of the earth and let θ represent the angle subtended by the earth. Then

$\theta = 2\sin^{-1}\left(\frac{6376}{x}\right)$, so

$$\frac{d\theta}{dt} = 2\frac{1}{\sqrt{1-\left(\frac{6376}{x}\right)^2}}\left(-\frac{6376}{x^2}\right)\frac{dx}{dt} = -\frac{12,752}{x\sqrt{x^2-6376^2}}\frac{dx}{dt}$$

When she is 3000 km from the surface

$x = 3000 + 6376 = 9376$ and $\frac{dx}{dt} = -2$. Substituting these values, we obtain $\frac{d\theta}{dt} \approx 3.96\times 10^{-4}$ radians per second.

6.9 Concepts Review

1. $\frac{e^x - e^{-x}}{2}; \frac{e^x + e^{-x}}{2}$

3. the graph of $x^2 - y^2 = 1$, a hyperbola

Problem Set 6.9

1. $\cosh x + \sinh x = \frac{e^x + e^{-x}}{2} + \frac{e^x - e^{-x}}{2} = \frac{2e^x}{2} = e^x$

3. $\cosh x - \sinh x = \frac{e^x + e^{-x}}{2} - \frac{e^x - e^{-x}}{2} = \frac{2e^{-x}}{2} = e^{-x}$

5. $\sinh x \cosh y + \cosh x \sinh y = \frac{e^x - e^{-x}}{2}\cdot\frac{e^y + e^{-y}}{2} + \frac{e^x + e^{-x}}{2}\cdot\frac{e^y - e^{-y}}{2}$

$$= \frac{e^{x+y} + e^{x-y} - e^{-x+y} - e^{-x-y}}{4} + \frac{e^{x+y} - e^{x-y} + e^{-x+y} - e^{-x-y}}{4} = \frac{2e^{x+y} - 2e^{-(x+y)}}{4} = \frac{e^{x+y} - e^{-(x+y)}}{2} = \sinh(x+y)$$

7. $\cosh x \cosh y + \sinh x \sinh y = \frac{e^x + e^{-x}}{2}\cdot\frac{e^y + e^{-y}}{2} + \frac{e^x - e^{-x}}{2}\cdot\frac{e^y - e^{-y}}{2}$

$$= \frac{e^{x+y} + e^{x-y} + e^{-x+y} + e^{-x-y}}{4} + \frac{e^{x+y} - e^{x-y} - e^{-x+y} + e^{-x-y}}{4} = \frac{2e^{x+y} + 2e^{-x-y}}{4} = \frac{e^{x+y} + e^{-(x+y)}}{2} = \cosh(x+y)$$

9. $$\frac{\tanh x + \tanh y}{1 + \tanh x \tanh y} = \frac{\frac{\sinh x}{\cosh x} + \frac{\sinh y}{\cosh y}}{1 + \frac{\sinh x}{\cosh x}\cdot\frac{\sinh y}{\cosh y}} = \frac{\sinh x \cosh y + \cosh x \sinh y}{\cosh x \cosh y + \sinh x \sinh y} = \frac{\sinh(x+y)}{\cosh(x+y)} = \tanh(x+y)$$

11. $2\sinh x\cosh x = \sinh x\cosh x + \cosh x\sinh x = \sinh(x+x) = \sinh 2x$

13. $D_x \sinh^2 x = 2\sinh x\cosh x = \sinh 2x$

15. $D_x(5\sinh^2 x) = 10\sinh x \cdot \cosh x = 5\sinh 2x$

17. $D_x \cosh(3x+1) = \sinh(3x+1)\cdot 3 = 3\sinh(3x+1)$

19. $D_x \ln(\sinh x) = \frac{1}{\sinh x}\cdot \cosh x = \frac{\cosh x}{\sinh x} = \coth x$

21. $D_x(x^2 \cosh x) = x^2 \cdot \sinh x + \cosh x \cdot 2x = x^2 \sinh x + 2x\cosh x$

23. $D_x(\cosh 3x \sinh x) = \cosh 3x \cdot \cosh x + \sinh x \cdot \sinh 3x \cdot 3 = \cosh 3x \cosh x + 3\sinh 3x \sinh x$

25. $D_x(\tanh x \sinh 2x) = \tanh x \cdot \cosh 2x \cdot 2 + \sinh 2x \cdot \operatorname{sech}^2 x = 2\tanh x \cosh 2x + \sinh 2x \operatorname{sech}^2 x$

27. $D_x \sinh^{-1}(x^2) = \frac{1}{\sqrt{(x^2)^2+1}}\cdot 2x = \frac{2x}{\sqrt{x^4+1}}$

29. $D_x \tanh^{-1}(2x-3) = \frac{1}{1-(2x-3)^2}\cdot 2 = \frac{2}{1-(4x^2-12x+9)} = \frac{2}{-4x^2+12x-8} = -\frac{1}{2(x^2-3x+2)}$

31. $D_x[x\cosh^{-1}(3x)] = x\cdot\frac{1}{\sqrt{(3x)^2-1}}\cdot 3 + \cosh^{-1}(3x)\cdot 1 = \frac{3x}{\sqrt{9x^2-1}} + \cosh^{-1} 3x$

33. $D_x \ln(\cosh^{-1} x) = \frac{1}{\cosh^{-1} x}\cdot\frac{1}{\sqrt{x^2-1}}$

$= \frac{1}{\sqrt{x^2-1}\cosh^{-1} x}$

35. $D_x \tanh(\cot x) = \operatorname{sech}^2(\cot x)\cdot(-\csc^2 x)$

$= -\csc^2 x \operatorname{sech}^2(\cot x)$

37. Area $= \int_0^{\ln 3} \cosh 2x\,dx = \left[\frac{1}{2}\sinh 2x\right]_0^{\ln 3}$

$= \frac{1}{2}\left(\frac{e^{2\ln 3} - e^{-2\ln 3}}{2} - \frac{e^0 - e^{-0}}{2}\right)$

$= \frac{1}{4}(e^{\ln 9} - e^{\ln\frac{1}{9}}) = \frac{1}{4}\left(9 - \frac{1}{9}\right) = \frac{20}{9}$

39. Let $u = \pi x^2 + 5$, so $du = 2\pi x\,dx$.

$\int x\cosh(\pi x^2+5)dx = \frac{1}{2\pi}\int \cosh u\,du$

$= \frac{1}{2\pi}\sinh u + C = \frac{1}{2\pi}\sinh(\pi x^2+5) + C$

41. Let $u = 2z^{1/4}$, so $du = \frac{1}{4}\cdot 2z^{-3/4}dz = \frac{1}{2\sqrt[4]{z^3}}dz.$

$\int \frac{\sinh(2z^{1/4})}{\sqrt[4]{z^3}}dz = 2\int \sinh u\,du = 2\cosh u + C$

$= 2\cosh(2z^{1/4}) + C$

43. Let $u = \sin x$, so $du = \cos x\,dx$

$\int \cos x \sinh(\sin x)dx = \int \sinh u\,du = \cosh u + C$

$= \cosh(\sin x) + C$

45. Let $u = \ln(\sinh x^2)$, so

$du = \frac{1}{\sinh x^2}\cdot \cosh x^2 \cdot 2x\,dx = 2x\coth x^2\,dx$.

$\int x\coth x^2 \ln(\sinh x^2)dx = \frac{1}{2}\int u\,du = \frac{1}{2}\cdot\frac{u^2}{2} + C$

$= \frac{1}{4}[\ln(\sinh x^2)]^2 + C$

47. Note that the graphs of $y = \sinh x$ and $y = 0$ intersect at the origin.

Area $= \int_0^{\ln 2} \sinh x\, dx = [\cosh x]_0^{\ln 2}$

$= \frac{e^{\ln 2} + e^{-\ln 2}}{2} - \frac{e^0 + e^0}{2} = \frac{1}{2}\left(2 + \frac{1}{2}\right) - 1 = \frac{1}{4}$

49. Volume $= \int_0^1 \pi \cosh^2 x\, dx = \frac{\pi}{2}\int_0^1 (1 + \cosh 2x)dx$

$= \frac{\pi}{2}\left[x + \frac{\sinh 2x}{2}\right]_0^1$

$= \frac{\pi}{2}\left(1 + \frac{\sinh 2}{2} - 0\right)$

$= \frac{\pi}{2} + \frac{\pi \sinh 2}{4} \approx 4.42$

51. Note that $1 + \sinh^2 x = \cosh^2 x$ and $\cosh^2 x = \frac{1 + \cosh 2x}{2}$

Surface area $= \int_0^1 2\pi y\sqrt{1 + \left(\frac{dy}{dx}\right)^2}\, dx$

$= \int_0^1 2\pi \cosh x\sqrt{1 + \sinh^2 x}\, dx$

$= \int_0^1 2\pi \cosh x \cosh x\, dx$

$= \int_0^1 \pi(1 + \cosh 2x)dx$

$= \left[\pi x + \frac{\pi}{2}\sinh 2x\right]_0^1 = \pi + \frac{\pi}{2}\sinh 2 \approx 8.84$

53. $y = a\cosh\left(\frac{x}{a}\right) + C$

$\frac{dy}{dx} = \sinh\left(\frac{x}{a}\right)$

$\frac{d^2y}{dx^2} = \frac{1}{a}\cosh\left(\frac{x}{a}\right)$

We need to show that $\frac{d^2y}{dx^2} = \frac{1}{a}\sqrt{1 + \left(\frac{dy}{dx}\right)^2}$.

Note that $1 + \sinh^2\left(\frac{x}{a}\right) = \cosh^2\left(\frac{x}{a}\right)$ and $\cosh\left(\frac{x}{a}\right) > 0$. Therefore,

$\frac{1}{a}\sqrt{1 + \left(\frac{dy}{dx}\right)^2} = \frac{1}{a}\sqrt{1 + \sinh^2\left(\frac{x}{a}\right)} = \frac{1}{a}\sqrt{\cosh^2\left(\frac{x}{a}\right)} = \frac{1}{a}\cosh\left(\frac{x}{a}\right) = \frac{d^2y}{dx^2}$

55. a.

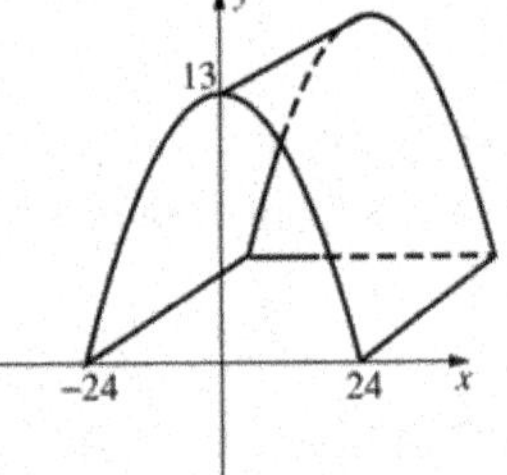

b. Area under the curve is

$\int_{-24}^{24}\left[37 - 24\cosh\left(\frac{x}{24}\right)\right]dx = \left[37x - 576\sinh\left(\frac{x}{24}\right)\right]_{-24}^{24} \approx 422$

Volume is about (422)(100) = 42,200 ft^3.

c. Length of the curve is

$\int_{-24}^{24}\sqrt{1 + \left(\frac{dy}{dx}\right)^2}\, dx = \int_{-24}^{24}\sqrt{1 + \sinh^2\left(\frac{x}{24}\right)}dx = \int_{-24}^{24}\cosh\left(\frac{x}{24}\right)dx = \left[24\sinh\left(\frac{x}{24}\right)\right]_{-24}^{24} = 48\sinh 1 \approx 56.4$

Surface area $\approx$ (56.4)(100) = 5640 ft^2

57. a. $(\sinh x + \cosh x)^r = \left(\frac{e^x - e^{-x}}{2} + \frac{e^x + e^{-x}}{2}\right)^r = \left(\frac{2e^x}{2}\right)^r = e^{rx}$

$\sinh rx + \cosh rx = \frac{e^{rx} - e^{-rx}}{2} + \frac{e^{rx} + e^{-rx}}{2} = \frac{2e^{rx}}{2} = e^{rx}$

b. $(\cosh x - \sinh x)^r = \left(\frac{e^x + e^{-x}}{2} - \frac{e^x - e^{-x}}{2}\right)^r = \left(\frac{2e^{-x}}{2}\right)^r = e^{-rx}$

$\cosh rx - \sinh rx = \frac{e^{rx} + e^{-rx}}{2} - \frac{e^{rx} - e^{-rx}}{2} = \frac{2e^{-rx}}{2} = e^{-rx}$

c. $(\cos x + i \sin x)^r = \left(\frac{e^{ix} + e^{-ix}}{2} + i\frac{e^{ix} - e^{-ix}}{2i}\right)^r = \left(\frac{2e^{ix}}{2}\right)^r = e^{irx}$

$\cos rx + i \sin rx = \frac{e^{irx} + e^{-irx}}{2} + i\frac{e^{irx} - e^{-irx}}{2i} = \frac{2e^{irx}}{2} = e^{irx}$

d. $(\cos x - i \sin x)^r = \left(\frac{e^{ix} + e^{-ix}}{2} - i\frac{e^{ix} - e^{-ix}}{2i}\right)^r = \left(\frac{2e^{-ix}}{2}\right)^r = e^{-irx}$

$\cos rx - i \sin rx = \frac{e^{irx} + e^{-irx}}{2} - i\frac{e^{irx} - e^{-irx}}{2i} = \frac{2e^{-irx}}{2} = e^{-irx}$

59. Area $= \int_0^x \cosh t \, dt = [\sinh t]_0^x = \sinh x$

Arc length =

$\int_0^x \sqrt{1 + [D_t \cosh t]^2} \, dt = \int_0^x \sqrt{1 + \sinh^2 t} \, dt$

$= \int_0^x \cosh t \, dt = [\sinh t]_0^x = \sinh x$

61.

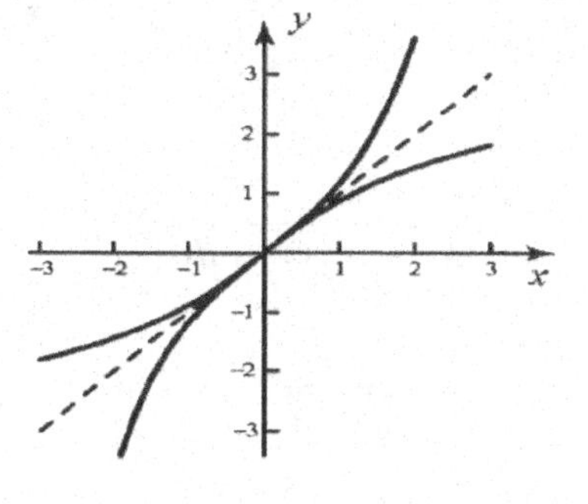

The functions $y = \sinh x$ and $y = \ln(x + \sqrt{x^2 + 1})$ are inverse functions.

6.10 Chapter Review

Concepts Test

1. False: $\ln 0$ is undefined.

3. True: $\int_1^{e^3} \frac{1}{t} dt = \left[\ln|t|\right]_1^{e^3} = \ln e^3 - \ln 1 = 3$

5. True: The range of $y = \ln x$ is the set of all real numbers.

7. False: $4 \ln x = \ln(x^4)$

9. True: $f(g(x)) = 4 + e^{\ln(x-4)}$
$= 4 + (x - 4) = x$
and
$g(f(x)) = \ln(4 + e^x - 4) = \ln e^x = x$

11. True: $\ln x$ is an increasing function.

13. True: $e^z > 0$ for all z.

15. True: $\lim_{x \to 0^+} (\ln \sin x - \ln x)$

$= \lim_{x \to 0^+} \ln\left(\frac{\sin x}{x}\right) = \ln 1 = 0$

17. False: $\ln \pi$ is a constant so $\frac{d}{dx} \ln \pi = 0$.

19. True: e is a number.

21. False: $D_x(x^x) = x^x(1+\ln x)$

23. True: The integrating factor is
$e^{\int 4/x\,dx} = e^{4\ln x} = \left(e^{\ln x}\right)^4 = x^4$

25. False: The solution is $y(x) = e^{2x}$, so $y'(x) = 2e^{2x}$. In general, Euler's method will underestimate the solution if the slope of the solution is increasing as it is in this case.

27. False: $\arcsin(\sin 2\pi) = \arcsin 0 = 0$

29. False: $\cosh x$ is not increasing.

31. True: $|\sinh x| \le \frac{1}{2}e^{|x|}$ is equivalent to $\left|e^x - e^{-x}\right| \le e^{|x|}$. When $x = 0$, $\sinh x = 0 < \frac{1}{2}e^0 = \frac{1}{2}$. If $x > 0$, $e^x > 1$ and $e^{-x} < 1 < e^x$, thus
$\left|e^x - e^{-x}\right| = e^x - e^{-x} < e^x = e^{|x|}$.
If $x < 0$, $e^{-x} > 1$ and $e^x < 1 < e^{-x}$, thus
$\left|e^x - e^{-x}\right| = -(e^x - e^{-x})$
$= e^{-x} - e^x < e^{-x} = e^{|x|}$.

33. False: $\cosh(\ln 3) = \frac{e^{\ln 3} + e^{-\ln 3}}{2}$
$= \frac{1}{2}\left(3 + \frac{1}{3}\right) = \frac{5}{3}$

35. True: $\lim_{x\to-\infty} \tan^{-1} x = -\frac{\pi}{2}$, since $\lim_{x\to-\frac{\pi}{2}^+} \tan x = -\infty$.

37. True: $\tanh x = \frac{\sinh x}{\cosh x}$; $\sinh x$ is an odd function and $\cosh x$ is an even function.

39. True: $\ln 3^{100} = 100\ln 3 > 100 \cdot 1$ since $\ln 3 > 1$.

41. True: y triples every time t increases by t_1.

43. True: $(y(t) + z(t))' = y'(t) + z'(t)$
$= ky(t) + kz(t) = k(y(t) + z(t))$

45. False: Use the substitution $u = -h$.
$\lim_{h\to 0}(1-h)^{-1/h} = \lim_{u\to 0}(1+u)^{1/u} = e$
by Theorem 6.5.A.

47. True: If $D_x(a^x) = a^x \ln a = a^x$, then $\ln a = 1$, so $a = e$.

Sample Test Problems

1. $\ln\frac{x^4}{2} = 4\ln x - \ln 2$

$$\frac{d}{dx}\ln\frac{x^4}{2} = \frac{d}{dx}(4\ln x - \ln 2) = \frac{4}{x}$$

3. $\frac{d}{dx}e^{x^2-4x} = e^{x^2-4x}\frac{d}{dx}(x^2 - 4x)$

$= (2x-4)e^{x^2-4x}$

5. $\frac{d}{dx}\tan(\ln e^x) = \frac{d}{dx}\tan x = \sec^2 x$

7. $\frac{d}{dx}2\tanh\sqrt{x} = 2\operatorname{sech}^2\sqrt{x}\frac{d}{dx}\sqrt{x} = \frac{\operatorname{sech}^2\sqrt{x}}{\sqrt{x}}$

9. $\frac{d}{dx}\sinh^{-1}(\tan x) = \frac{1}{\sqrt{\tan^2 x + 1}}\frac{d}{dx}\tan x$

$= \frac{\sec^2 x}{\sqrt{\tan^2 x + 1}} = \frac{\sec^2 x}{\sqrt{\sec^2 x}} = |\sec x|$

11. $\frac{d}{dx}\sec^{-1} e^x = \frac{1}{\left|e^x\right|\sqrt{(e^x)^2 - 1}}\frac{d}{dx}e^x$

$= \frac{e^x}{e^x\sqrt{e^{2x} - 1}} = \frac{1}{\sqrt{e^{2x} - 1}}$

13. $\frac{d}{dx}3\ln(e^{5x} + 1) = \frac{3}{e^{5x} + 1}(5e^{5x}) = \frac{15e^{5x}}{e^{5x} + 1}$

15. $\frac{d}{dx}\cos e^{\sqrt{x}} = -\sin e^{\sqrt{x}}\frac{d}{dx}e^{\sqrt{x}}$

$= (-\sin e^{\sqrt{x}})e^{\sqrt{x}}\frac{d}{dx}\sqrt{x}$

$= -\frac{e^{\sqrt{x}}\sin e^{\sqrt{x}}}{2\sqrt{x}}$

17. $\frac{d}{dx}2\cos^{-1}\sqrt{x} = \frac{-2}{\sqrt{1-(\sqrt{x})^2}}\frac{d}{dx}\sqrt{x}$

$= \frac{-2}{\sqrt{1-x}}\frac{1}{2\sqrt{x}} = -\frac{1}{\sqrt{x-x^2}}$

19. $\frac{d}{dx}2\csc e^{\ln\sqrt{x}} = \frac{d}{dx}2\csc\sqrt{x}$

$= -2\csc\sqrt{x}\cot\sqrt{x}\frac{d}{dx}\sqrt{x}$

$= -\frac{\csc\sqrt{x}\cot\sqrt{x}}{\sqrt{x}}$

21. $\frac{d}{dx}4\tan 5x\sec 5x$

$= 20\sec^2 5x\sec 5x + 20\tan 5x\sec 5x\tan 5x$

$= 20\sec 5x(\sec^2 5x + \tan^2 5x)$

$= 20\sec 5x(2\sec^2 5x - 1)$

23. $\frac{d}{dx}x^{1+x} = \frac{d}{dx}e^{(1+x)\ln x}$

$= e^{(1+x)\ln x}\frac{d}{dx}[(1+x)\ln x]$

$= x^{1+x}\left[(1)(\ln x) + (1+x)\left(\frac{1}{x}\right)\right]$

$= x^{1+x}\left(\ln x + 1 + \frac{1}{x}\right)$

25. Let $u = 3x - 1$, so $du = 3\,dx$.

$\int e^{3x-1}dx = \frac{1}{3}\int e^{3x-1}3dx = \frac{1}{3}\int e^u\,du$

$= \frac{1}{3}e^u + C = \frac{1}{3}e^{3x-1} + C$

Check:

$\frac{d}{dx}\left(\frac{1}{3}e^{3x-1} + C\right) = \frac{1}{3}e^{3x-1}\frac{d}{dx}(3x-1) = e^{3x-1}$

27. Let $u = e^x$, so $du = e^x dx$.

$\int e^x\sin e^x dx = \int\sin u\,du = -\cos u + C$

$= -\cos e^x + C$

Check:

$\frac{d}{dx}(-\cos e^x + C) = (\sin e^x)\frac{d}{dx}e^x = e^x\sin e^x$

29. Let $u = e^{x+3} + 1$, so $du = e^{x+3}dx$.

$\int\frac{e^{x+2}}{e^{x+3}+1}dx = \frac{1}{e}\int\frac{1}{e^{x+3}+1}e^{x+3}dx = \frac{1}{e}\int\frac{1}{u}du$

$= \frac{1}{e}\ln|u| + C = \frac{\ln(e^{x+3}+1)}{e} + C$

Check:

$\frac{d}{dx}\left(\frac{\ln(e^{x+3}+1)}{e} + C\right) = \frac{1}{e}\frac{1}{e^{x+3}+1}\frac{d}{dx}(e^{x+3}+1)$

$= \frac{e^{x+3}e^{-1}}{e^{x+3}+1} = \frac{e^{x+2}}{e^{x+3}+1}$

31. Let $u = 2x$, so $du = 2\,dx$.

$\int\frac{4}{\sqrt{1-4x^2}}dx = 2\int\frac{1}{\sqrt{1-(2x)^2}}2dx$

$= 2\int\frac{1}{\sqrt{1-u^2}}du$

$= 2\sin^{-1}u + C = 2\sin^{-1}2x + C$

Check:

$\frac{d}{dx}(2\sin^{-1}2x + C) = 2\left(\frac{1}{\sqrt{1-(2x)^2}}\right)\frac{d}{dx}2x$

$= \frac{4}{\sqrt{1-4x^2}}$

33. Let $u = \ln x$, so $du = \frac{1}{x}dx$.

$\int\frac{-1}{x + x(\ln x)^2}dx = -\int\frac{1}{1+(\ln x)^2}\cdot\frac{1}{x}dx$

$= -\int\frac{1}{1+u^2}du = -\tan^{-1}u + C = -\tan^{-1}(\ln x) + C$

Check:

$\frac{d}{dx}[-\tan^{-1}(\ln x) + C] = -\frac{1}{1+(\ln x)^2}\frac{d}{dx}\ln x$

$= \frac{-1}{x + x(\ln x)^2}$

35. $f'(x) = \cos x - \sin x$; $f'(x) = 0$ when $\tan x = 1$,

$x = \frac{\pi}{4}$

$f'(x) > 0$ when $\cos x > \sin x$ which occurs when

$-\frac{\pi}{2} \le x < \frac{\pi}{4}$.

$f''(x) = -\sin x - \cos x$; $f''(x) = 0$ when

$\tan x = -1$, $x = -\frac{\pi}{4}$

$f''(x) > 0$ when $\cos x < -\sin x$ which occurs

when $-\frac{\pi}{2} \le x < -\frac{\pi}{4}$.

Increasing on $\left[-\frac{\pi}{2}, \frac{\pi}{4}\right]$

Decreasing on $\left[\frac{\pi}{4}, \frac{\pi}{2}\right]$

Concave up on $\left(-\frac{\pi}{2}, -\frac{\pi}{4}\right)$

Concave down on $\left(-\frac{\pi}{4}, \frac{\pi}{2}\right)$

Inflection point at $\left(-\frac{\pi}{4}, 0\right)$

Global maximum at $\left(\frac{\pi}{4}, \sqrt{2}\right)$

Global minimum at $\left(-\frac{\pi}{2}, -1\right)$

37. a. $f'(x) = 5x^4 + 6x^2 + 4 \ge 4 > 0$ for all x, so $f(x)$ is increasing.

b. $f(1) = 7$, so $g(7) = f^{-1}(7) = 1$.

c. $g'(7) = \frac{1}{f'(1)} = \frac{1}{15}$

39.

x_n	y_n
1.0	2.0
1.2	2.4
1.4	2.976
1.6	3.80928
1.8	5.02825
2.0	6.83842

41. $y = (\cos x)^{\sin x} = e^{\sin x \ln(\cos x)}$

$\frac{dy}{dx} = e^{\sin x \ln(\cos x)} \frac{d}{dx}[\sin x \ln(\cos x)]$

$= e^{\sin x \ln(\cos x)}\left[\cos x \ln(\cos x) + (\sin x)\left(\frac{1}{\cos x}\right)(-\sin x)\right]$

$= (\cos x)^{\sin x}\left[\cos x \ln(\cos x) - \frac{\sin^2 x}{\cos x}\right]$

At $x = 0$, $\frac{dy}{dx} = 1^0(1 \ln 1 - 0) = 0$.

The tangent line has slope 0, so it is horizontal: $y = 1$.

43. Integrating factor is x. $D[yx] = 0$; $y = Cx^{-1}$

45. (Linear first-order) $y' + 2xy = 2x$

Integrating factor: $e^{\int 2x dx} = e^{x^2}$

$D[ye^{x^2}] = 2xe^{x^2}$; $ye^{x^2} = e^{x^2} + C$;

$y = 1 + Ce^{-x^2}$

If $x = 0$, $y = 3$, then $3 = 1 + C$, so $C = 2$.

Therefore, $y = 1 + 2e^{-x^2}$.

47. Integrating factor is e^{-2x}.

$D[ye^{-2x}] = e^{-x}$; $y = -e^x + Ce^{2x}$

Review and Preview Problems

1. $\int \sin 2x \, dx \underset{\substack{u=2x \\ du=2dx}}{=} \frac{1}{2}\int \sin u \, du = -\frac{1}{2}\cos u + C =$

$-\frac{1}{2}\cos 2x + C$

3. $\int x \sin x^2 \, dx \underset{\substack{u=x^2 \\ du=2x dx}}{=} \frac{1}{2}\int \sin u \, du = -\frac{1}{2}\cos u + C =$

$-\frac{1}{2}\cos x^2 + C$

5. $\displaystyle\int_{\substack{u=\cos t\\ du=-\sin t\,dt}} \frac{\sin t}{\cos t}\,dt = -\int \frac{1}{u}\,du = -\ln|u| + C =$

$\ln\left|\frac{1}{u}\right| + C = \ln\left|\frac{1}{\cos t}\right| + C = \ln|\sec t| + C$

7. $\displaystyle\int_{\substack{u=x^2+2\\ du=2x\,dx}} x\sqrt{x^2+2}\,dx = \frac{1}{2}\int \sqrt{u}\,du = \frac{1}{3}u^{3/2} + C$

$= \frac{1}{3}\left(x^2+2\right)^{3/2} + C$

9. $f'(x) = \left[x\left(\frac{1}{x}\right) + (\ln x)(1)\right] - 1 = \ln x$

11. $f'(x) = \left[(-2x)(\cos x) + (-x^2)(-\sin x)\right] +$

$\left[(2)(\sin x) + (2x)(\cos x)\right] + \left[2(-\sin x)\right]$

$= x^2 \sin x$

13. $\cos 2x = 1 - 2\sin^2 x$; thus $\sin^2 x = \frac{1-\cos 2x}{2}$

15. $\cos^4 x = (\cos^2 x)^2 = \left(\frac{1+\cos 2x}{2}\right)^2$

17. $\cos u \cos v = \frac{\cos(u+v) + \cos(u-v)}{2} \Rightarrow$

$\cos 3x \cos 5x = \frac{\cos(8x) + \cos(-2x)}{2}$

$= \frac{\cos 8x + \cos 2x}{2}$

19. $\sqrt{a^2 - (a\sin t)^2} = \sqrt{a^2(1-\sin^2 t)} =$

$|a|\sqrt{\cos^2 t} = |a|\cos t$

21. $\sqrt{(a\sec t)^2 - a^2 -} = \sqrt{a^2(\sec^2 t - 1)} =$

$|a|\sqrt{\tan^2 t} = |a| \cdot |\tan t|$

23. $\frac{1}{1-x} - \frac{1}{x} = \frac{x-(1-x)}{(1-x)x} = \frac{2x-1}{x(1-x)}$

25. $-\frac{1}{x} - \frac{1}{2(x+1)} + \frac{3}{2(x-3)}$

$= \frac{-2(x+1)(x-3) - x(x-3) + 3x(x+1)}{2x(x+1)(x-3)}$

$= \frac{-2(x^2-2x-3) - (x^2-3x) + (3x^2+3x)}{2x(x+1)(x-3)}$

$= \frac{10x+6}{2x(x+1)(x-3)} = \frac{2(5x+3)}{2x(x+1)(x-3)}$

$= \frac{(5x+3)}{x(x+1)(x-3)}$

CHAPTER 7 Techniques of Integration

7.1 Concepts Review

1. elementary function

3. e^x

Problem Set 7.1

1. $\int (x-2)^5\,dx = \frac{1}{6}(x-2)^6 + C$

3. $u = x^2+1,\ du = 2x\,dx$
When $x = 0,\ u = 1$ and when $x = 2,\ u = 5$.
$$\int_0^2 x(x^2+1)^5\,dx = \frac{1}{2}\int_0^2 (x^2+1)^5(2x\,dx)$$
$$= \frac{1}{2}\int_1^5 u^5\,du$$
$$= \left[\frac{u^6}{12}\right]_1^5 = \frac{5^6-1^6}{12} = \frac{15624}{12} = 1302$$

5. $\int \frac{dx}{x^2+4} = \frac{1}{2}\tan^{-1}\left(\frac{x}{2}\right) + C$

7. $u = x^2+4,\ du = 2x\,dx$
$$\int \frac{x}{x^2+4}\,dx = \frac{1}{2}\int \frac{du}{u}$$
$$= \frac{1}{2}\ln|u| + C = \frac{1}{2}\ln\left|x^2+4\right| + C$$
$$= \frac{1}{2}\ln(x^2+4) + C$$

9. $u = 4+z^2,\ du = 2z\,dz$
$$\int 6z\sqrt{4+z^2}\,dz = 3\int \sqrt{u}\,du$$
$$= 2u^{3/2} + C = 2(4+z^2)^{3/2} + C$$

11. $\int \frac{\tan z}{\cos^2 z}\,dz = \int \tan z \sec^2 z\,dz$
$u = \tan z,\ du = \sec^2 z\,dz$
$$\int \tan z\sec^2 z\,dz = \int u\,du = \frac{1}{2}u^2 + C$$
$$= \frac{1}{2}\tan^2 z + C$$

13. $u = \sqrt{t},\ du = \frac{1}{2\sqrt{t}}\,dt$
$$\int \frac{\sin\sqrt{t}}{\sqrt{t}}\,dt = 2\int \sin u\,du$$
$$= -2\cos u + C = -2\cos\sqrt{t} + C$$

15. $u = \sin x,\ du = \cos x\,dx$
$$\int_0^{\pi/4} \frac{\cos x}{1+\sin^2 x}\,dx = \int_0^{\sqrt{2}/2} \frac{du}{1+u^2}$$
$$= [\tan^{-1} u]_0^{\sqrt{2}/2} = \tan^{-1}\frac{\sqrt{2}}{2} \approx 0.6155$$

17. $\int \frac{3x^2+2x}{x+1}\,dx = \int (3x-1)\,dx + \int \frac{1}{x+1}\,dx$
$$= \frac{3}{2}x^2 - x + \ln|x+1| + C$$

19. $u = \ln 4x^2,\ du = \frac{2}{x}\,dx$
$$\int \frac{\sin(\ln 4x^2)}{x}\,dx = \frac{1}{2}\int \sin u\,du$$
$$= -\frac{1}{2}\cos u + C = -\frac{1}{2}\cos(\ln 4x^2) + C$$

21. $u = e^x,\ du = e^x\,dx$
$$\int \frac{6e^x}{\sqrt{1-e^{2x}}}\,dx = 6\int \frac{du}{\sqrt{1-u^2}}\,du = 6\sin^{-1} u + C$$
$$= 6\sin^{-1}(e^x) + C$$

23. $u = 1-e^{2x},\ du = -2e^{2x}\,dx$
$$\int \frac{3e^{2x}}{\sqrt{1-e^{2x}}}\,dx = -\frac{3}{2}\int \frac{du}{\sqrt{u}}$$
$$= -3\sqrt{u} + C = -3\sqrt{1-e^{2x}} + C$$

25. $\int_0^1 t\,3^{t^2}\,dt = \frac{1}{2}\int_0^1 2t\,3^{t^2}\,dt$
$$= \left[\frac{3^{t^2}}{2\ln 3}\right]_0^1 = \frac{3}{2\ln 3} - \frac{1}{2\ln 3} = \frac{1}{\ln 3} \approx 0.9102$$

27. $\int \frac{\sin x - \cos x}{\sin x} dx = \int \left(1 - \frac{\cos x}{\sin x}\right) dx$

$u = \sin x,\ du = \cos x\, dx$

$\int \frac{\sin x - \cos x}{\sin x} dx = x - \int \frac{du}{u}$

$= x - \ln|u| + C$

$= x - \ln|\sin x| + C$

29. $u = e^x,\ du = e^x dx$

$\int e^x \sec e^x dx = \int \sec u\, du$

$= \ln|\sec u + \tan u| + C$

$= \ln\left|\sec e^x + \tan e^x\right| + C$

31. $\int \frac{\sec^3 x + e^{\sin x}}{\sec x} dx = \int (\sec^2 x + e^{\sin x} \cos x)\, dx$

$= \tan x + \int e^{\sin x} \cos x\, dx$

$u = \sin x,\ du = \cos x\, dx$

$\tan x + \int e^{\sin x} \cos x\, dx = \tan x + \int e^u\, du$

$= \tan x + e^u + C = \tan x + e^{\sin x} + C$

33. $u = t^3 - 2,\ du = 3t^2 dt$

$\int \frac{t^2 \cos(t^3 - 2)}{\sin^2(t^3 - 2)} dt = \frac{1}{3} \int \frac{\cos u}{\sin^2 u} du$

$v = \sin u,\ dv = \cos u\, du$

$\frac{1}{3} \int \frac{\cos u}{\sin^2 u} du = \frac{1}{3} \int v^{-2} dv = -\frac{1}{3} v^{-1} + C$

$= -\frac{1}{3\sin u} + C$

$= -\frac{1}{3\sin(t^3 - 2)} + C.$

35. $u = t^3 - 2,\ du = 3t^2 dt$

$\int \frac{t^2 \cos^2(t^3 - 2)}{\sin^2(t^3 - 2)} dt = \frac{1}{3} \int \frac{\cos^2 u}{\sin^2 u} du$

$= \frac{1}{3} \int \cot^2 u\, du = \frac{1}{3} \int (\csc^2 u - 1) du$

$= \frac{1}{3}[-\cot u - u] + C_1$

$= \frac{1}{3}[-\cot(t^3 - 2) - (t^3 - 2)] + C_1$

$= -\frac{1}{3}[\cot(t^3 - 2) + t^3] + C$

37. $u = \tan^{-1} 2t,\ du = \frac{2}{1 + 4t^2} dt$

$\int \frac{e^{\tan^{-1} 2t}}{1 + 4t^2} dt = \frac{1}{2} \int e^u du$

$= \frac{1}{2} e^u + C = \frac{1}{2} e^{\tan^{-1} 2t} + C$

39. $u = 3y^2,\ du = 6y\, dy$

$\int \frac{y}{\sqrt{16 - 9y^4}} dy = \frac{1}{6} \int \frac{1}{\sqrt{4^2 - u^2}} du$

$= \frac{1}{6} \sin^{-1}\left(\frac{u}{4}\right) + C$

$= \frac{1}{6} \sin^{-1}\left(\frac{3y^2}{4}\right) + C$

41. $u = x^3,\ du = 3x^2 dx$

$\int x^2 \sinh x^3 dx = \frac{1}{3} \int \sinh u\, du$

$= \frac{1}{3} \cosh u + C$

$= \frac{1}{3} \cosh x^3 + C$

43. $u = e^{3t},\ du = 3e^{3t} dt$

$\int \frac{e^{3t}}{\sqrt{4 - e^{6t}}} dt = \frac{1}{3} \int \frac{1}{\sqrt{2^2 - u^2}} du$

$= \frac{1}{3} \sin^{-1}\left(\frac{u}{2}\right) + C$

$= \frac{1}{3} \sin^{-1}\left(\frac{e^{3t}}{2}\right) + C$

45. $u = \cos x,\ du = -\sin x\, dx$

$\int_0^{\pi/2} \frac{\sin x}{16 + \cos^2 x} dx = -\int_1^0 \frac{1}{16 + u^2} du$

$= \int_0^1 \frac{1}{16 + u^2} du$

$= \left[\frac{1}{4} \tan^{-1}\left(\frac{u}{4}\right)\right]_0^1 = \left[\frac{1}{4} \tan^{-1}\left(\frac{1}{4}\right) - \frac{1}{4} \tan^{-1} 0\right]$

$= \frac{1}{4} \tan^{-1}\left(\frac{1}{4}\right) \approx 0.0612$

47. $\int \frac{1}{x^2+2x+5}dx = \int \frac{1}{x^2+2x+1+4}dx$

$= \int \frac{1}{(x+1)^2+2^2}d(x+1)$

$= \frac{1}{2}\tan^{-1}\left(\frac{x+1}{2}\right)+C$

49. $\int \frac{dx}{9x^2+18x+10} = \int \frac{dx}{9x^2+18x+9+1}$

$= \int \frac{dx}{(3x+3)^2+1^2}$

$u = 3x + 3$, $du = 3\,dx$

$\int \frac{dx}{(3x+3)^2+1^2} = \frac{1}{3}\int \frac{du}{u^2+1^2}$

$= \frac{1}{3}\tan^{-1}(3x+3)+C$

51. $\int \frac{x+1}{9x^2+18x+10}dx = \frac{1}{18}\int \frac{18x+18}{9x^2+18x+10}dx$

$= \frac{1}{18}\ln\left|9x^2+18x+10\right|+C$

$= \frac{1}{18}\ln\left(9x^2+18x+10\right)+C$

53. $u = \sqrt{2}t$, $du = \sqrt{2}dt$

$\int \frac{dt}{t\sqrt{2t^2-9}} = \int \frac{du}{u\sqrt{u^2-3^2}}$

$= \frac{1}{3}\sec^{-1}\left(\frac{\left|\sqrt{2}t\right|}{3}\right)+C$

55. The length is given by

$L = \int_a^b \sqrt{1+\left(\frac{dy}{dx}\right)^2}\,dx$

$= \int_0^{\pi/4}\sqrt{1+\left[\frac{1}{\cos x}(-\sin x)\right]^2}\,dx$

$= \int_0^{\pi/4}\sqrt{1+\tan^2 x}\,dx = \int_0^{\pi/4}\sqrt{\sec^2 x}\,dx$

$= \int_0^{\pi/4}\sec x\,dx = \left[\ln\left|\sec x+\tan x\right|\right]_0^{\pi/4}$

$= \ln\left|\sqrt{2}+1\right| - \ln|1| = \ln\left|\sqrt{2}+1\right| \approx 0.881$

57. $u = x - \pi$, $du = dx$

$\int_0^{2\pi}\frac{x|\sin x|}{1+\cos^2 x}dx = \int_{-\pi}^{\pi}\frac{(u+\pi)|\sin(u+\pi)|}{1+\cos^2(u+\pi)}du$

$= \int_{-\pi}^{\pi}\frac{(u+\pi)|\sin u|}{1+\cos^2 u}du$

$= \int_{-\pi}^{\pi}\frac{u|\sin u|}{1+\cos^2 u}du + \int_{-\pi}^{\pi}\frac{\pi|\sin u|}{1+\cos^2 u}du$

$\int_{-\pi}^{\pi}\frac{u|\sin u|}{1+\cos^2 u}du = 0$ by symmetry.

$\int_{-\pi}^{\pi}\frac{\pi|\sin u|}{1+\cos^2 u}du = 2\int_0^{\pi}\frac{\pi\sin u}{1+\cos^2 u}du$

$v = \cos u$, $dv = -\sin u\,du$

$-2\int_1^{-1}\frac{\pi}{1+v^2}dv = 2\pi\int_{-1}^{1}\frac{1}{1+v^2}dv$

$= 2\pi[\tan^{-1} v]_{-1}^{1} = 2\pi\left[\frac{\pi}{4}-\left(-\frac{\pi}{4}\right)\right]$

$= 2\pi\left(\frac{\pi}{2}\right) = \pi^2$

7.2 Concepts Review

1. $uv - \int v\,du$

3. 1

Problem Set 7.2

1. $u = x \qquad dv = e^x dx$

$du = dx \qquad v = e^x$

$\int xe^x dx = xe^x - \int e^x dx = xe^x - e^x + C$

3. $u = t \qquad dv = e^{5t+\pi}dt$

$du = dt \qquad v = \frac{1}{5}e^{5t+\pi}$

$\int te^{5t+\pi}dt = \frac{1}{5}te^{5t+\pi} - \int \frac{1}{5}e^{5t+\pi}dt$

$= \frac{1}{5}te^{5t+\pi} - \frac{1}{25}e^{5t+\pi} + C$

5. $u = x \qquad dv = \cos x\,dx$

$du = dx \qquad v = \sin x$

$\int x\cos x\,dx = x\sin x - \int \sin x\,dx$

$= x\sin x + \cos x + C$

7. $u = t - 3 \qquad dv = \cos(t-3)dt$

$du = dt \qquad v = \sin(t-3)$

$\int (t-3)\cos(t-3)dt = (t-3)\sin(t-3) - \int \sin(t-3)dt$

$= (t-3)\sin(t-3) + \cos(t-3) + C$

9. $u = t \qquad dv = \sqrt{t+1}\,dt$

$du = dt \qquad v = \frac{2}{3}(t+1)^{3/2}$

$\int t\sqrt{t+1}\,dt = \frac{2}{3}t(t+1)^{3/2} - \int \frac{2}{3}(t+1)^{3/2}dt$

$= \frac{2}{3}t(t+1)^{3/2} - \frac{4}{15}(t+1)^{5/2} + C$

11. $u = \ln 3x \qquad dv = dx$

$du = \frac{1}{x}dx \qquad v = x$

$\int \ln 3x\,dx = x\ln 3x - \int x\frac{1}{x}dx \;= x\ln 3x - x + C$

13. $u = \arctan x \qquad dv = dx$

$du = \frac{1}{1+x^2}dx \qquad v = x$

$\int \arctan x = x\arctan x - \int \frac{x}{1+x^2}dx$

$= x\arctan x - \frac{1}{2}\int \frac{2x}{1+x^2}dx$

$= x\arctan x - \frac{1}{2}\ln(1+x^2) + C$

15. $u = \ln x \qquad dv = \frac{dx}{x^2}$

$du = \frac{1}{x}dx \qquad v = -\frac{1}{x}$

$\int \frac{\ln x}{x^2}dx = -\frac{\ln x}{x} - \int -\frac{1}{x}\left(\frac{1}{x}\right)dx$

$= -\frac{\ln x}{x} - \frac{1}{x} + C$

17. $u = \ln t \qquad dv = \sqrt{t}\,dt$

$du = \frac{1}{t}dt \qquad v = \frac{2}{3}t^{3/2}$

$\int_1^e \sqrt{t}\ln t\,dt = \left[\frac{2}{3}t^{3/2}\ln t\right]_1^e - \int_1^e \frac{2}{3}t^{1/2}dt$

$= \frac{2}{3}e^{3/2}\ln e - \frac{2}{3}\cdot 1\ln 1 - \left[\frac{4}{9}t^{3/2}\right]_1^e$

$= \frac{2}{3}e^{3/2} - 0 - \frac{4}{9}e^{3/2} + \frac{4}{9} = \frac{2}{9}e^{3/2} + \frac{4}{9} \approx 1.4404$

19. $u = \ln z \qquad dv = z^3dz$

$du = \frac{1}{z}dz \qquad v = \frac{1}{4}z^4$

$\int z^3\ln z\,dz = \frac{1}{4}z^4\ln z - \int \frac{1}{4}z^4\cdot\frac{1}{z}dz$

$= \frac{1}{4}z^4\ln z - \frac{1}{4}\int z^3dz$

$= \frac{1}{4}z^4\ln z - \frac{1}{16}z^4 + C$

21. $u = \arctan\left(\frac{1}{t}\right) \qquad dv = dt$

$du = -\frac{1}{1+t^2}dt \qquad v = t$

$\int \arctan\left(\frac{1}{t}\right)dt = t\arctan\left(\frac{1}{t}\right) + \int \frac{t}{1+t^2}dt$

$= t\arctan\left(\frac{1}{t}\right) + \frac{1}{2}\ln(1+t^2) + C$

23. $u = x \qquad dv = \csc^2 x\,dx$

$du = dx \qquad v = -\cot x$

$\int_{\pi/6}^{\pi/2} x\csc^2 x\,dx = [-x\cot x]_{\pi/6}^{\pi/2} + \int_{\pi/6}^{\pi/2}\cot x\,dx \;= [-x\cot x + \ln|\sin x|]_{\pi/6}^{\pi/2}$

$= -\frac{\pi}{2}\cdot 0 + \ln 1 + \frac{\pi}{6}\sqrt{3} - \ln\frac{1}{2} = \frac{\pi}{2\sqrt{3}} + \ln 2 \approx 1.60$

25. $u = x^3 \qquad dv = x^2\sqrt{x^3+4}\,dx$

$du = 3x^2dx \qquad v = \frac{2}{9}(x^3+4)^{3/2}$

$\int x^5\sqrt{x^3+4}\,dx = \frac{2}{9}x^3(x^3+4)^{3/2} - \int \frac{2}{3}x^2(x^3+4)^{3/2}dx \;= \frac{2}{9}x^3(x^3+4)^{3/2} - \frac{4}{45}(x^3+4)^{5/2} + C$

27. $u = t^4$ $\quad dv = \dfrac{t^3}{(7-3t^4)^{3/2}}dt$

$du = 4t^3\,dt$ $\quad v = \dfrac{1}{6(7-3t^4)^{1/2}}$

$$\int \frac{t^7}{(7-3t^4)^{3/2}}dt = \frac{t^4}{6(7-3t^4)^{1/2}} - \frac{2}{3}\int \frac{t^3}{(7-3t^4)^{1/2}}dt = \frac{t^4}{6(7-3t^4)^{1/2}} + \frac{1}{9}(7-3t^4)^{1/2} + C$$

29. $u = z^4$ $\quad dv = \dfrac{z^3}{(4-z^4)^2}dz$

$du = 4z^3\,dz$ $\quad v = \dfrac{1}{4(4-z^4)}$

$$\int \frac{z^7}{(4-z^4)^2}dz = \frac{z^4}{4(4-z^4)} - \int \frac{z^3}{4-z^4}dz = \frac{z^4}{4(4-z^4)} + \frac{1}{4}\ln\left|4-z^4\right| + C$$

31. $u = x$ $\quad dv = \sinh x\,dx$

$du = dx$ $\quad v = \cosh x$

$$\int x\sinh x\,dx = x\cosh x - \int \cosh x\,dx = x\cosh x - \sinh x + C$$

33. $u = x$ $\quad dv = (3x+10)^{49}\,dx$

$du = dx$ $\quad v = \dfrac{1}{150}(3x+10)^{50}$

$$\int x(3x+10)^{49}dx = \frac{x}{150}(3x+10)^{50} - \frac{1}{150}\int (3x+10)^{50}dx = \frac{x}{150}(3x+10)^{50} - \frac{1}{22{,}950}(3x+10)^{51} + C$$

35. $u = x$ $\quad dv = 2^x\,dx$

$du = dx$ $\quad v = \dfrac{1}{\ln 2}2^x$

$$\int x2^x\,dx = \frac{x}{\ln 2}2^x - \frac{1}{\ln 2}\int 2^x\,dx$$

$$= \frac{x}{\ln 2}2^x - \frac{1}{(\ln 2)^2}2^x + C$$

37. $u = x^2$ $\quad dv = e^x\,dx$

$du = 2x\,dx$ $\quad v = e^x$

$$\int x^2e^x\,dx = x^2e^x - \int 2xe^x\,dx$$

$u = x$ $\quad dv = e^x\,dx$

$du = dx$ $\quad v = e^x$

$$\int x^2e^x\,dx = x^2e^x - 2\left(xe^x - \int e^x\,dx\right)$$

$$= x^2e^x - 2xe^x + 2e^x + C$$

39. $u = \ln^2 z$ $\quad dv = dz$

$du = \dfrac{2\ln z}{z}dz$ $\quad v = z$

$$\int \ln^2 z\,dz = z\ln^2 z - 2\int \ln z\,dz$$

$u = \ln z$ $\quad dv = dz$

$du = \dfrac{1}{z}dz$ $\quad v = z$

$$\int \ln^2 z\,dz = z\ln^2 z - 2\left(z\ln z - \int dz\right)$$

$$= z\ln^2 z - 2z\ln z + 2z + C$$

41. $u = e^t$ $\quad dv = \cos t\,dt$

$du = e^t\,dt$ $\quad v = \sin t$

$$\int e^t\cos t\,dt = e^t\sin t - \int e^t\sin t\,dt$$

$u = e^t$ $\quad dv = \sin t\,dt$

$du = e^t\,dt$ $\quad v = -\cos t$

$$\int e^t\cos t\,dt = e^t\sin t - \left[-e^t\cos t + \int e^t\cos t\,dt\right]$$

$$\int e^t\cos t\,dt = e^t\sin t + e^t\cos t - \int e^t\cos t\,dt$$

$$2\int e^t\cos t\,dt = e^t\sin t + e^t\cos t + C$$

$$\int e^t\cos t\,dt = \frac{1}{2}e^t(\sin t + \cos t) + C$$

43. $u = x^2 \quad dv = \cos x\, dx$

$du = 2x\, dx \quad v = \sin x$

$\int x^2 \cos x\, dx = x^2 \sin x - \int 2x \sin x\, dx$

$u = 2x \quad dv = \sin x\, dx$

$du = 2dx \quad v = -\cos x$

$\int x^2 \cos x\, dx = x^2 \sin x - \left(-2x\cos x + \int 2\cos x\, dx\right) = x^2 \sin x + 2x\cos x - 2\sin x + C$

45. $u = \sin(\ln x) \qquad dv = dx$

$du = \cos(\ln x) \cdot \frac{1}{x} dx \qquad v = x$

$\int \sin(\ln x) dx = x\sin(\ln x) - \int \cos(\ln x)\, dx$

$u = \cos\,(\ln x) \qquad dv = dx$

$du = -\sin(\ln x) \cdot \frac{1}{x} dx \qquad v = x$

$\int \sin(\ln x) dx = x\sin(\ln x) - \left[x\cos(\ln x) - \int -\sin(\ln x) dx\right]$

$\int \sin(\ln x) dx = x\sin(\ln x) - x\cos(\ln x) - \int \sin(\ln x) dx$

$2\int \sin(\ln x) dx = x\sin(\ln x) - x\cos(\ln x) + C$

$\int \sin(\ln x) dx = \frac{x}{2}[\sin(\ln x) - \cos(\ln x)] + C$

47. $u = (\ln x)^3 \qquad dv = dx$

$du = \frac{3\ln^2 x}{x} dx \qquad v = x$

$\int (\ln x)^3\, dx = x(\ln x)^3 - 3\int \ln^2 x\, dx$

$= x\ln^3 x - 3(x\ln^2 x - 2x\ln x + 2x + C)$

$= x\ln^3 x - 3x\ln^2 x + 6x\ln x - 6x + C$

49. $u = \sin x \qquad dv = \sin(3x) dx$

$du = \cos x\, dx \qquad v = -\frac{1}{3}\cos(3x)$

$\int \sin x \sin(3x) dx = -\frac{1}{3}\sin x\cos(3x) + \frac{1}{3}\int \cos x\cos(3x) dx$

$u = \cos x \qquad dv = \cos(3x) dx$

$du = -\sin x\, dx \qquad v = \frac{1}{3}\sin(3x)$

$\int \sin x\sin(3x) dx = -\frac{1}{3}\sin x\cos(3x) + \frac{1}{3}\left[\frac{1}{3}\cos x\sin(3x) + \frac{1}{3}\int \sin x\sin(3x) dx\right]$

$= -\frac{1}{3}\sin x\cos(3x) + \frac{1}{9}\cos x\sin(3x) + \frac{1}{9}\int \sin x\sin(3x) dx$

$\frac{8}{9}\int \sin x\sin(3x) dx = -\frac{1}{3}\sin x\cos(3x) + \frac{1}{9}\cos x\sin(3x) + C$

$\int \sin x\sin(3x) dx = -\frac{3}{8}\sin x\cos(3x) + \frac{1}{8}\cos x\sin(3x) + C$

51. $u = e^{\alpha z}$ $\qquad dv = \sin \beta z\, dz$

$du = \alpha e^{\alpha z} dz \qquad v = -\frac{1}{\beta}\cos \beta z$

$$\int e^{\alpha z} \sin \beta z\, dz = -\frac{1}{\beta} e^{\alpha z} \cos \beta z + \frac{\alpha}{\beta} \int e^{\alpha z} \cos \beta z\, dz$$

$u = e^{\alpha z}$ $\qquad dv = \cos \beta z\, dz$

$du = \alpha e^{\alpha z} dz \qquad v = \frac{1}{\beta}\sin \beta z$

$$\int e^{\alpha z} \sin \beta z\, dz = -\frac{1}{\beta} e^{\alpha z} \cos \beta z + \frac{\alpha}{\beta}\left[\frac{1}{\beta} e^{\alpha z} \sin \beta z - \frac{\alpha}{\beta} \int e^{\alpha z} \sin \beta z\, dz\right]$$

$$= -\frac{1}{\beta} e^{\alpha z} \cos \beta z + \frac{\alpha}{\beta^2} e^{\alpha z} \sin \beta z - \frac{\alpha^2}{\beta^2} \int e^{\alpha z} \sin \beta z\, dz$$

$$\frac{\beta^2 + \alpha^2}{\beta^2} \int e^{\alpha z} \sin \beta z\, dz = -\frac{1}{\beta} e^{\alpha z} \cos \beta z + \frac{\alpha}{\beta^2} e^{\alpha z} \sin \beta z + C$$

$$\int e^{\alpha z} \sin \beta z\, dz = \frac{-\beta}{\alpha^2 + \beta^2} e^{\alpha z} \cos \beta z + \frac{\alpha}{\alpha^2 + \beta^2} e^{\alpha z} \sin \beta z + C = \frac{e^{\alpha z}(\alpha \sin \beta z - \beta \cos \beta z)}{\alpha^2 + \beta^2} + C$$

53. $u = \ln x$ $\qquad dv = x^{\alpha} dx$

$du = \frac{1}{x} dx \qquad v = \frac{x^{\alpha+1}}{\alpha+1},\ \alpha \neq -1$

$$\int x^{\alpha} \ln x\, dx = \frac{x^{\alpha+1}}{\alpha+1} \ln x - \frac{1}{\alpha+1} \int x^{\alpha} dx = \frac{x^{\alpha+1}}{\alpha+1} \ln x - \frac{x^{\alpha+1}}{(\alpha+1)^2} + C, \alpha \neq -1$$

55. $u = x^{\alpha}$ $\qquad dv = e^{\beta x} dx$

$du = \alpha x^{\alpha-1} dx \qquad v = \frac{1}{\beta} e^{\beta x}$

$$\int x^{\alpha} e^{\beta x} dx = \frac{x^{\alpha} e^{\beta x}}{\beta} - \frac{\alpha}{\beta} \int x^{\alpha-1} e^{\beta x} dx$$

57. $u = x^{\alpha}$ $\qquad dv = \cos \beta x\, dx$

$du = \alpha x^{\alpha-1} dx \qquad v = \frac{1}{\beta}\sin \beta x$

$$\int x^{\alpha} \cos \beta x\, dx = \frac{x^{\alpha} \sin \beta x}{\beta} - \frac{\alpha}{\beta} \int x^{\alpha-1} \sin \beta x\, dx$$

59. $u = (a^2 - x^2)^{\alpha}$ $\qquad dv = dx$

$du = -2\alpha x(a^2 - x^2)^{\alpha-1} dx \qquad v = x$

$$\int (a^2 - x^2)^{\alpha} dx = x(a^2 - x^2)^{\alpha} + 2\alpha \int x^2 (a^2 - x^2)^{\alpha-1} dx$$

61. $u = \cos^{\alpha-1} \beta x \qquad dv = \cos \beta x\, dx$

$du = -\beta(\alpha-1)\cos^{\alpha-2} \beta x \sin \beta x\, dx \qquad v = \frac{1}{\beta}\sin \beta x$

$$\int \cos^{\alpha} \beta x\, dx = \frac{\cos^{\alpha-1} \beta x \sin \beta x}{\beta} + (\alpha-1)\int \cos^{\alpha-2} \beta x \sin^2 \beta x\, dx$$

$$= \frac{\cos^{\alpha-1} \beta x \sin \beta x}{\beta} + (\alpha-1)\int \cos^{\alpha-2} \beta x(1-\cos^2 \beta x)\, dx$$

$$= \frac{\cos^{\alpha-1} \beta x \sin \beta x}{\beta} + (\alpha-1)\int \cos^{\alpha-2} \beta x\, dx - (\alpha-1)\int \cos^{\alpha} \beta x\, dx$$

$$\alpha \int \cos^{\alpha} \beta x = \frac{\cos^{\alpha-1} \beta x \sin \beta x}{\beta} + (\alpha-1)\int \cos^{\alpha-2} \beta x\, dx$$

$$\int \cos^{\alpha} \beta x\, dx = \frac{\cos^{\alpha-1} \beta x \sin \beta x}{\alpha\beta} + \frac{\alpha-1}{\alpha}\int \cos^{\alpha-2} \beta x\, dx$$

63. $\int x^4 \cos 3x\, dx = \frac{1}{3}x^4 \sin 3x - \frac{4}{3}\int x^3 \sin 3x\, dx = \frac{1}{3}x^4 \sin 3x - \frac{4}{3}\left[-\frac{1}{3}x^3 \cos 3x + \int x^2 \cos 3x\, dx\right]$

$$= \frac{1}{3}x^4 \sin 3x + \frac{4}{9}x^3 \cos 3x - \frac{4}{3}\left[\frac{1}{3}x^2 \sin 3x - \frac{2}{3}\int x \sin 3x\, dx\right]$$

$$= \frac{1}{3}x^4 \sin 3x + \frac{4}{9}x^3 \cos 3x - \frac{4}{9}x^2 \sin 3x + \frac{8}{9}\left[-\frac{1}{3}x \cos 3x + \frac{1}{3}\int \cos 3x\, dx\right]$$

$$= \frac{1}{3}x^4 \sin 3x + \frac{4}{9}x^3 \cos 3x - \frac{4}{9}x^2 \sin 3x - \frac{8}{27}x \cos 3x + \frac{8}{81}\sin 3x + C$$

65. First make a sketch.

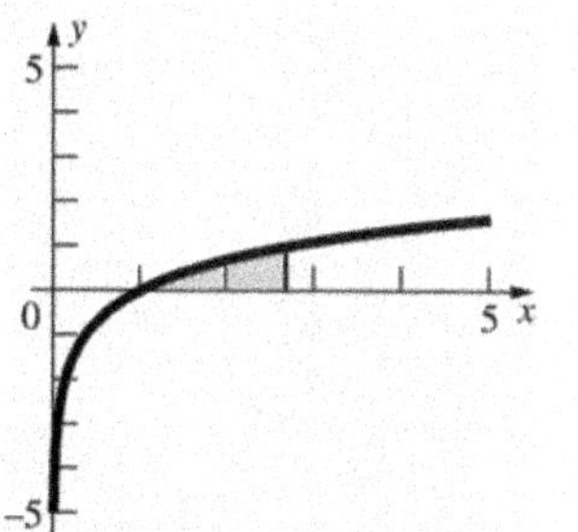

From the sketch, the area is given by

$$\int_1^e \ln x\, dx$$

$u = \ln x \qquad dv = dx$

$du = \frac{1}{x}dx \qquad v = x$

$$\int_1^e \ln x\, dx = [x \ln x]_1^e - \int_1^e dx = [x \ln x - x]_1^e = (e-e) - (1 \cdot 0 - 1) = 1$$

67.

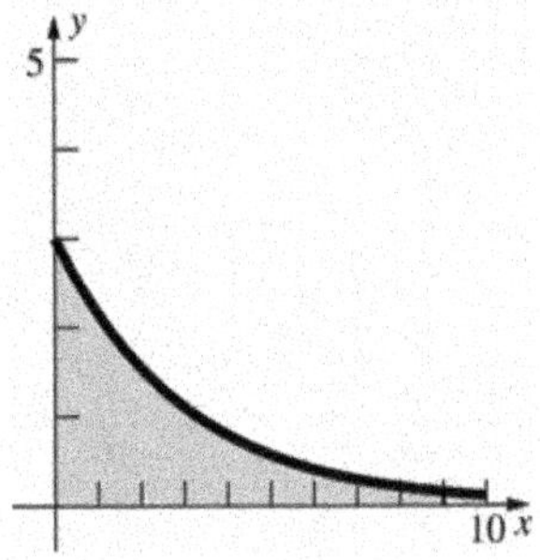

$$\int_0^9 3e^{-x/3}dx = -9\int_0^9 e^{-x/3}\left(-\frac{1}{3}dx\right) = -9[e^{-x/3}]_0^9 = -\frac{9}{e^3}+9 \approx 8.55$$

69. $\int_0^{\pi/4}(x\cos x - x\sin x)dx = \int_0^{\pi/4} x\cos x\,dx - \int_0^{\pi/4} x\sin x\,dx$

$$= \left([x\sin x]_0^{\pi/4} - \int_0^{\pi/4}\sin x\,dx\right) - \left([-x\cos x]_0^{\pi/4} + \int_0^{\pi/4}\cos x\,dx\right)$$

$$= [x\sin x + \cos x + x\cos x - \sin x]_0^{\pi/4} = \frac{\sqrt{2}\pi}{4} - 1 \approx 0.11$$

Use Problems 60 and 61 for $\int x\sin x\,dx$ and $\int x\cos x\,dx$.

71. $\int_1^e \ln x^2 dx = 2\int_1^e \ln x\,dx$

$u = \ln x \qquad dv = dx$

$du = \frac{1}{x}dx \qquad v = x$

$$2\int_1^e \ln x\,dx = 2\left([x\ln x]_1^e - \int_1^e dx\right) = 2\left(e - [x]_1^e\right) = 2$$

$\int_1^e x\ln x^2 dx = 2\int_1^e x\ln x\,dx$

$u = \ln x \qquad dv = x\,dx$

$du = \frac{1}{x}dx \qquad v = \frac{1}{2}x^2$

$$2\int_1^e x\ln x\,dx = 2\left(\left[\frac{1}{2}x^2\ln x\right]_1^e - \int_1^e \frac{1}{2}x\,dx\right) = 2\left(\frac{1}{2}e^2 - \left[\frac{1}{4}x^2\right]_1^e\right) = \frac{1}{2}(e^2+1)$$

$\frac{1}{2}\int_1^e (\ln x)^2 dx$

$u = (\ln x)^2 \qquad dv = dx$

$du = \frac{2\ln x}{x}dx \qquad v = x$

$$\frac{1}{2}\int_1^e (\ln x)^2 dx = \frac{1}{2}\left([x(\ln x)^2]_1^e - 2\int_1^e \ln x\,dx\right) = \frac{1}{2}(e-2)$$

$$\overline{x} = \frac{\frac{1}{2}(e^2+1)}{2} = \frac{e^2+1}{4}$$

$$\overline{y} = \frac{\frac{1}{2}(e-2)}{2} = \frac{e-2}{4}$$

73. a. $p(x) = x^3 - 2x$

$g(x) = e^x$

All antiderivatives of $g(x) = e^x$

$\int (x^3 - 2x)e^x dx = (x^3 - 2x)e^x - (3x^2 - 2)e^x + 6xe^x - 6e^x + C$

b. $p(x) = x^2 - 3x + 1$

$g(x) = \sin x$

$G_1(x) = -\cos x$

$G_2(x) = -\sin x$

$G_3(x) = \cos x$

$\int (x^2 - 3x + 1)\sin x\, dx = (x^2 - 3x + 1)(-\cos x) - (2x - 3)(-\sin x) + 2\cos x + C$

75. $u = f(x)$ $\quad dv = \sin nx\, dx$

$du = f'(x)dx$ $\quad v = -\frac{1}{n}\cos nx$

$$a_n = \frac{1}{\pi}\left[\underbrace{\left[-\frac{1}{n}\cos(nx)f(x) \right]_{-\pi}^{\pi}}_{\text{Term 1}} + \underbrace{\frac{1}{n}\int_{-\pi}^{\pi} \cos(nx)f'(x)dx}_{\text{Term 2}} \right]$$

$\text{Term 1} = \frac{1}{n}\cos(n\pi)(f(-\pi) - f(\pi)) = \pm\frac{1}{n}(f(-\pi) - f(\pi))$

Since $f'(x)$ is continuous on $[-\infty, \infty]$, it is bounded. Thus, $\int_{-\pi}^{\pi} \cos(nx)f'(x)dx$ is bounded so

$\lim_{n\to\infty} a_n = \lim_{n\to\infty} \frac{1}{\pi n}\left[\pm(f(-\pi) - f(\pi)) + \int_{-\pi}^{\pi} \cos(nx)f'(x)\,dx \right] = 0.$

77. The proof fails to consider the constants when integrating $\frac{1}{t}$.

The symbol $\int (1/t)\,dt$ is a *family* of functions, all of who whom have derivative $\frac{1}{t}$. We know that any two of these functions will differ by a constant, so it is perfectly correct (notationally) to write $\int (1/t)\,dt = \int (1/t)\,dt + 1$

79. $u = f(x)$ $\quad dv = dx$

$du = f'(x)dx$ $\quad v = x$

$\int_a^b f(x)dx = [xf(x)]_a^b - \int_a^b xf'(x)dx$

Starting with the same integral,

$u = f(x)$ $\quad dv = dx$

$du = f'(x)dx$ $\quad v = x - a$

$\int_a^b f(x)\,dx = [(x-a)f(x)]_a^b - \int_a^b (x-a)f'(x)dx$

81. Use proof by induction.

$n = 1$: $f(a) + f'(a)(t-a) + \int_a^t (t-x)f''(x)dx = f(a) + f'(a)(t-a) + [f'(x)(t-x)]_a^t + \int_a^t f'(x)dx$

$= f(a) + f'(a)(t-a) - f'(a)(t-a) + [f(x)]_a^t = f(t)$

Thus, the statement is true for $n = 1$. Note that integration by parts was used with $u = (t - x)$, $dv = f''(x)dx$.

Suppose the statement is true for n.

$$f(t) = f(a) + \sum_{i=1}^{n} \frac{f^{(i)}(a)}{i!}(t-a)^i + \int_a^t \frac{(t-x)^n}{n!} f^{(n+1)}(x)dx$$

Integrate $\int_a^t \frac{(t-x)^n}{n!} f^{(n+1)}(x)dx$ by parts.

$u = f^{(n+1)}(x) \qquad dv = \frac{(t-x)^n}{n!}dx$

$du = f^{(n+2)}(x) \qquad v = -\frac{(t-x)^{n+1}}{(n+1)!}$

$$\int_a^t \frac{(t-x)^n}{n!} f^{(n+1)}(x)dx = \left[-\frac{(t-x)^{n+1}}{(n+1)!} f^{(n+1)}(x)\right]_a^t + \int_a^t \frac{(t-x)^{n+1}}{(n+1)!} f^{(n+2)}(x)dx$$

$$= \frac{(t-a)^{n+1}}{(n+1)!} f^{(n+1)}(a) + \int_a^t \frac{(t-x)^{n+1}}{(n+1)!} f^{(n+2)}(x)dx$$

Thus $f(t) = f(a) + \sum_{i=1}^{n} \frac{f^{(i)}(a)}{i!}(t-a)^i + \frac{(t-a)^{n+1}}{(n+1)!} f^{(n+1)}(a) + \int_a^t \frac{(t-x)^{n+1}}{(n+1)!} f^{(n+2)}(x)dx$

$$= f(a) + \sum_{i=1}^{n+1} \frac{f^{(i)}(a)}{i!}(t-a)^i + \int_a^t \frac{(t-x)^{n+1}}{(n+1)!} f^{(n+2)}(x)dx$$

Thus, the statement is true for $n + 1$.

83. $u = f(t) \qquad dv = f''(t)dt$

$du = f'(t)dt \qquad v = f'(t)$

$$\int_a^b f''(t)f(t)dt = [f(t)f'(t)]_a^b - \int_a^b [f'(t)]^2\, dt$$

$$= f(b)f'(b) - f(a)f'(a) - \int_a^b [f'(t)]^2 dt = -\int_a^b [f'(t)]^2\, dt$$

$[f'(t)]^2 \geq 0$, so $-\int_a^b [f'(t)]^2 \leq 0$.

85. Let $I = \int_0^x \int_0^{t_1} \cdots \int_0^{t_{n-1}} f(t_n)\, dt_n ... dt_2 dt_1$ be the iterated integral. Note that for $i \geq 2$, the limits of integration of the integral with respect to t_i are 0 to t_{i-1} so that any change of variables in an outer integral affects the limits, and hence the variables in all interior integrals. We use induction on n, noting that the case $n = 2$ is solved in the previous problem.

Assume we know the formula for $n-1$, and we want to show it for n.

$$I = \int_0^x \int_0^{t_1} \int_0^{t_2} \cdots \int_0^{t_{n-1}} f(t_n)\, dt_n ... dt_3 dt_2\, dt_1 = \int_0^{t_1} \int_0^{t_2} \cdots \int_0^{t_{n-2}} F(t_{n-1})\, dt_{n-1} ... dt_3 dt_2 dt_1$$

where $F(t_{n-1}) = \int_0^{t_{n-1}} f(t_n)\, dn$.

By induction,

$$I = \frac{1}{(n-2)!} \int_0^x F(t_1)(x - t_1)^{n-2}\, dt_1$$

$u = F(t_1) = \int_0^{t_1} f(t_n)\, dt_n, \quad dv = (x - t_1)^{n-2}; \qquad du = f(t_1)\, dt_1, \quad v = -\frac{1}{n-1}(x - t_1)^{n-1}$

$$I = \frac{1}{(n-2)!} \left\{ \left[-\frac{1}{n-1}(x - t_1)^{n-1} \int_0^{t_1} f(t_n)\, dt_n \right]_{t_1=0}^{t_1=x} + \frac{1}{n-1} \int_0^x f(t_1)(x - t_1)^{n-1}\, dt_1 \right\}.$$

$$= \frac{1}{(n-1)!} \int_0^x f(t_1)(x - t_1)^{n-1}\, dt_1$$

(note: that the quantity in square brackets equals 0 when evaluated at the given limits)

87. $\int (3x^4 + 2x^2)e^x\, dx = e^x \sum_{j=0}^{4} (-1)^j \frac{d^j(3x^4 + 2x^2)}{dx^j} = e^x[3x^4 + 2x^2 - 12x^3 - 4x + 36x^2 + 4 - 72x + 72]$

$$= e^x(3x^4 - 12x^3 + 38x^2 - 76x + 76)$$

7.3 Concepts Review

1. $\int \frac{1+\cos 2x}{2}\,dx$

3. $\int \sin^2 x(1-\sin^2 x)\cos x\,dx$

Problem Set 7.3

1. $\int \sin^2 x\,dx = \int \frac{1-\cos 2x}{2}\,dx$

$= \frac{1}{2}\int dx - \frac{1}{2}\int \cos 2x\,dx$

$= \frac{1}{2}x - \frac{1}{4}\sin 2x + C$

3. $\int \sin^3 x\,dx = \int \sin x(1-\cos^2 x)dx$

$= \int \sin x\,dx - \int \sin x \cos^2 x\,dx$

$= -\cos x + \frac{1}{3}\cos^3 x + C$

5. $\int_0^{\pi/2} \cos^5 \theta\,d\theta = \int_0^{\pi/2} (1-\sin^2\theta)^2 \cos\theta\,d\theta$

$= \int_0^{\pi/2} (1-2\sin^2\theta+\sin^4\theta)\cos\theta\,d\theta$

$= \left[\sin\theta - \frac{2}{3}\sin^3\theta + \frac{1}{5}\sin^5\theta\right]_0^{\pi/2}$

$= \left(1-\frac{2}{3}+\frac{1}{5}\right) - 0 = \frac{8}{15}$

7. $\int \sin^5 4x\cos^2 4x\,dx = \int (1-\cos^2 4x)^2 \cos^2 4x \sin 4x\,dx = \int (1-2\cos^2 4x+\cos^4 4x)\cos^2 4x\sin 4x\,dx$

$= -\frac{1}{4}\int (\cos^2 4x - 2\cos^4 4x + \cos^6 4x)(-4\sin 4x)dx = -\frac{1}{12}\cos^3 4x + \frac{1}{10}\cos^5 4x - \frac{1}{28}\cos^7 4x + C$

9. $\int \cos^3 3\theta \sin^{-2} 3\theta\,d\theta = \int (1-\sin^2 3\theta)\sin^{-2} 3\theta\cos 3\theta\,d\theta = \frac{1}{3}\int (\sin^{-2} 3\theta - 1)3\cos 3\theta\,d\theta$

$= -\frac{1}{3}\csc 3\theta - \frac{1}{3}\sin 3\theta + C$

11. $\int \sin^4 3t\cos^4 3t\,dt = \int \left(\frac{1-\cos 6t}{2}\right)^2\left(\frac{1+\cos 6t}{2}\right)^2 dt = \frac{1}{16}\int (1-2\cos^2 6t+\cos^4 6t)dt$

$= \frac{1}{16}\int\left[1-(1+\cos 12t)+\frac{1}{4}(1+\cos 12t)^2\right]dt = -\frac{1}{16}\int \cos 12t\,dt + \frac{1}{64}\int (1+2\cos 12t+\cos^2 12t)dt$

$= -\frac{1}{192}\int 12\cos 12t\,dt + \frac{1}{64}\int dt + \frac{1}{384}\int 12\cos 12t\,dt + \frac{1}{128}\int (1+\cos 24t)dt$

$= -\frac{1}{192}\sin 12t + \frac{1}{64}t + \frac{1}{384}\sin 12t + \frac{1}{128}t + \frac{1}{3072}\sin 24t + C = \frac{3}{128}t - \frac{1}{384}\sin 12t + \frac{1}{3072}\sin 24t + C$

13. $\int \sin 4y\cos 5y\,dy = \frac{1}{2}\int [\sin 9y + \sin(-y)]dy = \frac{1}{2}\int (\sin 9y - \sin y)dy$

$= \frac{1}{2}\left(-\frac{1}{9}\cos 9y + \cos y\right) + C = \frac{1}{2}\cos y - \frac{1}{18}\cos 9y + C$

15. $\int \sin^4\left(\frac{w}{2}\right)\cos^2\left(\frac{w}{2}\right)dw = \int \left(\frac{1-\cos w}{2}\right)^2\left(\frac{1+\cos w}{2}\right)dw = \frac{1}{8}\int (1-\cos w-\cos^2 w+\cos^3 w)dw$

$= \frac{1}{8}\int\left[1-\cos w-\frac{1}{2}(1+\cos 2w)+(1-\sin^2 w)\cos w\right]dw = \frac{1}{8}\int\left[\frac{1}{2}-\frac{1}{2}\cos 2w - \sin^2 w\cos w\right]dw$

$= \frac{1}{16}w - \frac{1}{32}\sin 2w - \frac{1}{24}\sin^3 w + C$

17. $\int x\cos^2 x\sin x\,dx$

$u = x \qquad du = 1\,dx$

$dv = \cos^2 x\sin x\;dx$

$v = -\underset{t=\cos x}{\int}(\cos x)^2(-\sin x)\,dx = -\frac{1}{3}\cos^3 x$

Thus

$\int x\cos^2 x\sin x\,dx = x(-\frac{1}{3}\cos^3 x) - \int(1)(-\frac{1}{3}\cos^3 x)\,dx = \frac{1}{3}\left[-x\cos^3 x + \int\cos^3 x\,dx\right]$

$= \frac{1}{3}\left[-x\cos^3 x + \int\cos x(1-\sin^2 x)\,dx\right] = \frac{1}{3}\left[-x\cos^3 x + \underset{t=\sin x}{\int}(\cos x - \cos x\sin^2 x)\;dx\right] = \frac{1}{3}\left[-x\cos^3 x + \sin x - \frac{1}{3}\sin^3 x\right] + C$

19. $\int\tan^4 x\;dx = \int\left(\tan^2 x\right)\left(\tan^2 x\right)\;dx = \int\left(\tan^2 x\right)(\sec^2 x - 1)\;dx = \int\left(\tan^2 x\sec^2 x - \tan^2 x\right)dx$

$= \int\tan^2 x\sec^2 x\;dx - \int(\sec^2 x - 1)dx = \frac{1}{3}\tan^3 x - \tan x + x + C$

21. $\tan^3 x = \int(\tan x)\left(\tan^2 x\right)dx = \int(\tan x)\left(\sec^2 x - 1\right)dx = \frac{1}{2}\tan^2 x + \ln\left|\cos x\right| + C$

23. $\int\tan^5\left(\frac{\theta}{2}\right)d\theta$

$u = \left(\frac{\theta}{2}\right);\; du = \frac{d\theta}{2}$

$\int\tan^5\left(\frac{\theta}{2}\right)d\theta = 2\int\tan^5 u\;du = 2\int\left(\tan^3 u\right)\left(\sec^2 u - 1\right)du = 2\int\tan^3 u\sec^2 u\;du - 2\int\tan^3 u\;du$

$= 2\int\tan^3 u\sec^2 u\;du - 2\int\tan u\left(\sec^2 u - 1\right)du = 2\int\tan^3 u\sec^2 u\;du - 2\int\tan u\sec^2 u\;du + 2\int\tan u\;\mathrm{du}$

$= \frac{1}{2}\tan^4\left(\frac{\theta}{2}\right) - \tan^2\left(\frac{\theta}{2}\right) - 2\ln\left|\cos\frac{\theta}{2}\right| + C$

25. $\int\tan^{-3} x\sec^4 x dx = \int\left(\tan^{-3} x\right)\left(\sec^2 x\right)\left(\sec^2 x\right)dx = \int\left(\tan^{-3} x\right)\left(1 + \tan^2 x\right)\left(\sec^2 x\right)dx$

$= \int\tan^{-3} x\sec^2 x\;\mathrm{dx} + \int(\tan x)^{-1}\sec^2 x\;dx = -\frac{1}{2}\tan^{-2} x + \ln\left|\tan x\right| + C$

27. $\int\tan^3 x\sec^2 x\;dx$

Let $u = \tan x$. Then $du = \sec^2 x\,dx$.

$\int\tan^3 x\sec^2 x\;dx = \int u^3 du = \frac{1}{4}u^4 + C = \frac{1}{4}\tan^4 x + C$

29. $\int_{-\pi}^{\pi}\cos mx\cos nx\,dx = \frac{1}{2}\int_{-\pi}^{\pi}(\cos[(m+n)x] + \cos[(m-n)x])dx = \frac{1}{2}\left[\frac{1}{m+n}\sin[(m+n)x] + \frac{1}{m-n}\sin[(m-n)x]\right]_{-\pi}^{\pi}$

$= 0$ for $m \neq n$, since $\sin k\pi = 0$ for all integers k.

31. $\int_0^{\pi}\pi(x+\sin x)^2\,dx = \pi\int_0^{\pi}(x^2 + 2x\sin x + \sin^2 x)\,dx = \pi\int_0^{\pi}x^2\,dx + 2\pi\int_0^{\pi}x\sin x\,dx + \frac{\pi}{2}\int_0^{\pi}(1-\cos 2x)dx$

$= \pi\left[\frac{1}{3}x^3\right]_0^{\pi} + 2\pi\left[\sin x - x\cos x\right]_0^{\pi} + \frac{\pi}{2}\left[x - \frac{1}{2}\sin 2x\right]_0^{\pi} = \frac{1}{3}\pi^4 + 2\pi(0+\pi-0) + \frac{\pi}{2}(\pi-0-0) = \frac{1}{3}\pi^4 + \frac{5}{2}\pi^2 \approx 57.1437$

Use Formula 40 with $u = x$ for $\int x\sin x\,dx$

33. a. $\frac{1}{\pi}\int_{-\pi}^{\pi} f(x)\sin(mx)dx = \frac{1}{\pi}\int_{-\pi}^{\pi}\left(\sum_{n=1}^{N} a_n \sin(nx)\right)\sin(mx)dx = \frac{1}{\pi}\sum_{n=1}^{N} a_n \int_{-\pi}^{\pi}\sin(nx)\sin(mx)dx$

From Example 6,

$\int_{-\pi}^{\pi}\sin(nx)\sin(mx)dx = \begin{cases} 0 \text{ if } n \neq m \\ \pi \text{ if } n = m \end{cases}$ so every term in the sum is 0 except for when $n = m$.

If $m > N$, there is no term where $n = m$, while if $m \le N$, then $n = m$ occurs. When $n = m$

$a_n \int_{-\pi}^{\pi}\sin(nx)\sin(mx)\,dx = a_m\pi$ so when $m \le N$, $\frac{1}{\pi}\int_{-\pi}^{\pi} f(x)\sin(mx)\,dx = \frac{1}{\pi}\cdot a_m \cdot \pi = a_m$.

b. $\frac{1}{\pi}\int_{-\pi}^{\pi} f^2(x)dx = \frac{1}{\pi}\int_{-\pi}^{\pi}\left(\sum_{n=1}^{N} a_n \sin(nx)\right)\left(\sum_{m=1}^{N} a_m \sin(mx)\right)dx = \frac{1}{\pi}\sum_{n=1}^{N} a_n \sum_{m=1}^{N} a_m \int_{-\pi}^{\pi}\sin(nx)\sin(mx)\,dx$

From Example 6, the integral is 0 except when $m = n$. When $m = n$, we obtain

$\frac{1}{\pi}\sum_{n=1}^{N} a_n(a_n\pi) = \sum_{n=1}^{N} a_n^2$.

35. Using the half-angle identity $\cos\frac{x}{2} = \sqrt{\frac{1+\cos x}{2}}$, we see that since

$\cos\frac{\pi}{4} = \cos\frac{\frac{\pi}{2}}{2} = \frac{\sqrt{2}}{2}$

$\cos\frac{\pi}{8} = \cos\frac{\frac{\pi}{2}}{4} = \sqrt{\frac{1+\frac{\sqrt{2}}{2}}{2}} = \frac{\sqrt{2+\sqrt{2}}}{2}$,

$\cos\frac{\pi}{16} = \cos\frac{\frac{\pi}{2}}{8} = \sqrt{\frac{1+\frac{\sqrt{2+\sqrt{2}}}{2}}{2}} = \frac{\sqrt{2+\sqrt{2+\sqrt{2}}}}{2}$, etc.

Thus,

$\frac{\sqrt{2}}{2}\cdot\frac{\sqrt{2+\sqrt{2}}}{2}\cdot\frac{\sqrt{2+\sqrt{2+\sqrt{2}}}}{2}\cdots = \cos\left(\frac{\frac{\pi}{2}}{2}\right)\cos\left(\frac{\frac{\pi}{2}}{4}\right)\cos\left(\frac{\frac{\pi}{2}}{8}\right)\cdots = \lim_{n\to\infty}\cos\left(\frac{\frac{\pi}{2}}{2}\right)\cos\left(\frac{\frac{\pi}{2}}{4}\right)\cdots\cos\left(\frac{\frac{\pi}{2}}{2^n}\right) = \frac{\sin\left(\frac{\pi}{2}\right)}{\frac{\pi}{2}} = \frac{2}{\pi}$

7.4 Concepts Review

1. $\sqrt{x-3}$

3. $2 \tan t$

Problem Set 7.4

1. $u = \sqrt{x+1}, u^2 = x+1, 2u\,du = dx$

$\int x\sqrt{x+1}dx = \int(u^2-1)u(2u\,du)$

$= \int(2u^4 - 2u^2)du = \frac{2}{5}u^5 - \frac{2}{3}u^3 + C$

$= \frac{2}{5}(x+1)^{5/2} - \frac{2}{3}(x+1)^{3/2} + C$

3. $u = \sqrt{3t+4}, u^2 = 3t+4,\ 2u\,du = 3\,dt$

$\int\frac{t\,dt}{\sqrt{3t+4}} = \int\frac{\frac{1}{3}(u^2-4)\frac{2}{3}u\,du}{u} = \frac{2}{9}\int(u^2-4)du$

$= \frac{2}{27}u^3 - \frac{8}{9}u + C$

$= \frac{2}{27}(3t+4)^{3/2} - \frac{8}{9}(3t+4)^{1/2} + C$

5. $u = \sqrt{t}, u^2 = t,\ 2u\,du = dt$

$\int_1^2\frac{dt}{\sqrt{t}+e} = \int_1^{\sqrt{2}}\frac{2u\,du}{u+e} = 2\int_1^{\sqrt{2}}\frac{u+e-e}{u+e}du$

$= 2\int_1^{\sqrt{2}}du - 2\int_1^{\sqrt{2}}\frac{e}{u+e}du$

$= 2[u]_1^{\sqrt{2}} - 2e\left[\ln|u+e|\right]_1^{\sqrt{2}}$

$= 2(\sqrt{2}-1) - 2e[\ln(\sqrt{2}+e) - \ln(1+e)]$

$= 2\sqrt{2} - 2 - 2e\ln\left(\frac{\sqrt{2}+e}{1+e}\right)$

7. $u = (3t+2)^{1/2}, u^2 = 3t+2, 2u\,du = 3dt$

$$\int t(3t+2)^{3/2}dt = \int \frac{1}{3}(u^2-2)u^3\left(\frac{2}{3}u\,du\right)$$

$$= \frac{2}{9}\int (u^6 - 2u^4)du = \frac{2}{63}u^7 - \frac{4}{45}u^5 + C$$

$$= \frac{2}{63}(3t+2)^{7/2} - \frac{4}{45}(3t+2)^{5/2} + C$$

9. $x = 2\sin t$, $dx = 2\cos t\,dt$

$$\int \frac{\sqrt{4-x^2}}{x}dx = \int \frac{2\cos t}{2\sin t}(2\cos t\,dt)$$

$$= 2\int \frac{1-\sin^2 t}{\sin t}dt = 2\int \csc t\,dt - 2\int \sin t\,dt$$

$$= 2\ln|\csc t - \cot t| + 2\cos t + C$$

$$= 2\ln\left|\frac{2-\sqrt{4-x^2}}{x}\right| + \sqrt{4-x^2} + C$$

11. $x = 2\tan t$, $dx = 2\sec^2 t\,dt$

$$\int \frac{dx}{(x^2+4)^{3/2}} = \int \frac{2\sec^2 t\,dt}{(4\sec^2 t)^{3/2}} = \frac{1}{4}\int \cos t\,dt$$

$$= \frac{1}{4}\sin t + C = \frac{x}{4\sqrt{x^2+4}} + C$$

13. $t = \sec x$, $dt = \sec x\tan x\,dx$

Note that $\frac{\pi}{2} < x \le \pi$.

$$\sqrt{t^2-1} = |\tan x| = -\tan x$$

$$\int_{-2}^{-3} \frac{\sqrt{t^2-1}}{t^3}dt = \int_{2\pi/3}^{\sec^{-1}(-3)} \frac{-\tan x}{\sec^3 x}\sec x\tan x\,dx$$

$$= \int_{2\pi/3}^{\sec^{-1}(-3)} -\sin^2 x\,dx = \int_{2\pi/3}^{\sec^{-1}(-3)}\left(\frac{1}{2}\cos 2x - \frac{1}{2}\right)dx$$

$$= \left[\frac{1}{4}\sin 2x - \frac{1}{2}x\right]_{2\pi/3}^{\sec^{-1}(-3)}$$

$$= \left[\frac{1}{2}\sin x\cos x - \frac{1}{2}x\right]_{2\pi/3}^{\sec^{-1}(-3)}$$

$$= -\frac{\sqrt{2}}{9} - \frac{1}{2}\sec^{-1}(-3) + \frac{\sqrt{3}}{8} + \frac{\pi}{3} \approx 0.151252$$

15. $z = \sin t$, $dz = \cos t\,dt$

$$\int \frac{2z-3}{\sqrt{1-z^2}}dz = \int (2\sin t - 3)dt$$

$$= -2\cos t - 3t + C$$

$$= -2\sqrt{1-z^2} - 3\sin^{-1} z + C$$

17. $x^2+2x+5 = x^2+2x+1+4 = (x+1)^2+4$

$u = x+1$, $du = dx$

$$\int \frac{dx}{\sqrt{x^2+2x+5}} = \int \frac{du}{\sqrt{u^2+4}}$$

$u = 2\tan t$, $du = 2\sec^2 t\,dt$

$$\int \frac{du}{\sqrt{u^2+4}} = \int \sec t\,dt = \ln|\sec t + \tan t| + C_1$$

$$= \ln\left|\frac{\sqrt{u^2+4}}{2} + \frac{u}{2}\right| + C_1$$

$$= \ln\left|\frac{\sqrt{x^2+2x+5}+x+1}{2}\right| + C_1$$

$$= \ln\left|\sqrt{x^2+2x+5}+x+1\right| + C$$

19. $x^2+2x+5 = x^2+2x+1+4 = (x+1)^2+4$

$u = x+1$, $du = dx$

$$\int \frac{3x}{\sqrt{x^2+2x+5}}dx = \int \frac{3u-3}{\sqrt{u^2+4}}du$$

$$= 3\int \frac{u}{\sqrt{u^2+4}}du - 3\int \frac{du}{\sqrt{u^2+4}}$$

(Use the result of Problem 17.)

$$= 3\sqrt{u^2+4} - 3\ln\left|\sqrt{u^2+4}+u\right| + C$$

$$= 3\sqrt{x^2+2x+5} - 3\ln\left|\sqrt{x^2+2x+5}+x+1\right| + C$$

21. $5-4x-x^2 = 9-(4+4x+x^2) = 9-(x+2)^2$

$u = x+2$, $du = dx$

$$\int \sqrt{5-4x-x^2}\,dx = \int \sqrt{9-u^2}\,du$$

$u = 3\sin t$, $du = 3\cos t\,dt$

$$\int \sqrt{9-u^2}\,du = 9\int \cos^2 t\,dt = \frac{9}{2}\int (1+\cos 2t)dt$$

$$= \frac{9}{2}\left(t + \frac{1}{2}\sin 2t\right) + C = \frac{9}{2}(t + \sin t\cos t) + C$$

$$= \frac{9}{2}\sin^{-1}\left(\frac{u}{3}\right) + \frac{1}{2}u\sqrt{9-u^2} + C$$

$$= \frac{9}{2}\sin^{-1}\left(\frac{x+2}{3}\right) + \frac{x+2}{2}\sqrt{5-4x-x^2} + C$$

23. $4x-x^2=4-(4-4x+x^2)=4-(x-2)^2$

$u=x-2,\ du=dx$

$$\int\frac{dx}{\sqrt{4x-x^2}}=\int\frac{du}{\sqrt{4-u^2}}$$

$u=2\sin t,\ du=2\cos t\,dt$

$$\int\frac{du}{\sqrt{4-u^2}}=\int dt=t+C=\sin^{-1}\left(\frac{u}{2}\right)+C$$

$$=\sin^{-1}\left(\frac{x-2}{2}\right)+C$$

25. $x^2+2x+2=x^2+2x+1+1=(x+1)^2+1$

$u=x+1,\ du=dx$

$$\int\frac{2x+1}{x^2+2x+2}dx=\int\frac{2u-1}{u^2+1}du$$

$$=\int\frac{2u}{u^2+1}du-\int\frac{du}{u^2+1}$$

$$=\ln\left|u^2+1\right|-\tan^{-1}u+C$$

$$=\ln\left(x^2+2x+2\right)-\tan^{-1}(x+1)+C$$

27. $$V=\pi\int_0^1\left(\frac{1}{x^2+2x+5}\right)^2dx$$

$$=\pi\int_0^1\left[\frac{1}{(x+1)^2+4}\right]^2dx$$

$x+1=2\tan t,\ dx=2\sec^2 t\,dt$

$$V=\pi\int_{\tan^{-1}(1/2)}^{\pi/4}\left(\frac{1}{4\sec^2 t}\right)^2 2\sec^2 t\,dt$$

$$=\frac{\pi}{8}\int_{\tan^{-1}(1/2)}^{\pi/4}\frac{1}{\sec^2 t}dt=\frac{\pi}{8}\int_{\tan^{-1}(1/2)}^{\pi/4}\cos^2 t\,dt$$

$$=\frac{\pi}{8}\int_{\tan^{-1}(1/2)}^{\pi/4}\left(\frac{1}{2}+\frac{1}{2}\cos 2t\right)dt$$

$$=\frac{\pi}{8}\left[\frac{1}{2}t+\frac{1}{4}\sin 2t\right]_{\tan^{-1}(1/2)}^{\pi/4}$$

$$=\frac{\pi}{8}\left[\frac{1}{2}t+\frac{1}{2}\sin t\cos t\right]_{\tan^{-1}(1/2)}^{\pi/4}$$

$$=\frac{\pi}{8}\left[\left(\frac{\pi}{8}+\frac{1}{4}\right)-\left(\frac{1}{2}\tan^{-1}\frac{1}{2}+\frac{1}{5}\right)\right]$$

$$=\frac{\pi}{16}\left(\frac{1}{10}+\frac{\pi}{4}-\tan^{-1}\frac{1}{2}\right)\approx 0.082811$$

29. a. $u=x^2+9,\ du=2x\,dx$

$$\int\frac{x\,dx}{x^2+9}=\frac{1}{2}\int\frac{du}{u}=\frac{1}{2}\ln|u|+C$$

$$=\frac{1}{2}\ln\left|x^2+9\right|+C=\frac{1}{2}\ln\left(x^2+9\right)+C$$

b. $x=3\tan t,\ dx=3\sec^2 t\,dt$

$$\int\frac{x\,dx}{x^2+9}=\int\tan t\,dt=-\ln|\cos t|+C$$

$$=-\ln\left|\frac{3}{\sqrt{x^2+9}}\right|+C_1=-\ln\left(\frac{3}{\sqrt{x^2+9}}\right)+C_1$$

$$=\ln\left(\sqrt{x^2+9}\right)-\ln 3+C_1$$

$$=\ln\left((x^2+9)^{1/2}\right)+C=\frac{1}{2}\ln\left(x^2+9\right)+C$$

31. a. $u=\sqrt{4-x^2},\ u^2=4-x^2,\ 2u\,du=-2x\,dx$

$$\int\frac{\sqrt{4-x^2}}{x}dx=\int\frac{\sqrt{4-x^2}}{x^2}x\,dx=-\int\frac{u^2du}{4-u^2}$$

$$=\int\frac{-4+4-u^2}{4-u^2}du=-4\int\frac{1}{4-u^2}du+\int du$$

$$=-4\cdot\frac{1}{4}\ln\left|\frac{u+2}{u-2}\right|+u+C$$

$$=-\ln\left|\frac{\sqrt{4-x^2}+2}{\sqrt{4-x^2}-2}\right|+\sqrt{4-x^2}+C$$

b. $x=2\sin t,\ dx=2\cos t\,dt$

$$\int\frac{\sqrt{4-x^2}}{x}dx=2\int\frac{\cos^2 t}{\sin t}dt$$

$$=2\int\frac{(1-\sin^2 t)}{\sin t}dt$$

$$=2\int\csc t\,dt-2\int\sin t\,dt$$

$$=2\ln|\csc t-\cot t|+2\cos t+C$$

$$=2\ln\left|\frac{2}{x}-\frac{\sqrt{4-x^2}}{x}\right|+\sqrt{4-x^2}+C$$

$$=2\ln\left|\frac{2-\sqrt{4-x^2}}{x}\right|+\sqrt{4-x^2}+C$$

To reconcile the answers, note that

$$-\ln\left|\frac{\sqrt{4-x^2}+2}{\sqrt{4-x^2}-2}\right|=\ln\left|\frac{\sqrt{4-x^2}-2}{\sqrt{4-x^2}+2}\right|$$

$$=\ln\left|\frac{(\sqrt{4-x^2}-2)^2}{(\sqrt{4-x^2}+2)(\sqrt{4-x^2}-2)}\right|$$

$$=\ln\left|\frac{(2-\sqrt{4-x^2})^2}{4-x^2-4}\right|=\ln\left|\frac{(2-\sqrt{4-x^2})^2}{-x^2}\right|$$

$$=\ln\left|\left(\frac{2-\sqrt{4-x^2}}{x}\right)^2\right|=2\ln\left|\frac{2-\sqrt{4-x^2}}{x}\right|.$$

33. a. The coordinate of C is $(0, -a)$. The lower arc of the lune lies on the circle given by the equation $x^2+(y+a)^2=2a^2$ or

$y=\pm\sqrt{2a^2-x^2}-a.$ The upper arc of the lune lies on the circle given by the equation

$x^2+y^2=a^2$ or $y=\pm\sqrt{a^2-x^2}.$

$$A=\int_{-a}^{a}\sqrt{a^2-x^2}\,dx-\int_{-a}^{a}\left(\sqrt{2a^2-x^2}-a\right)dx$$

$$=\int_{-a}^{a}\sqrt{a^2-x^2}\,dx-\int_{-a}^{a}\sqrt{2a^2-x^2}\,dx+2a^2$$

Note that $\int_{-a}^{a}\sqrt{a^2-x^2}\,dx$ is the area of a semicircle with radius a, so

$$\int_{-a}^{a}\sqrt{a^2-x^2}\,dx=\frac{\pi a^2}{2}.$$

For $\int_{-a}^{a}\sqrt{2a^2-x^2}\,dx,$ let

$x=\sqrt{2}a\sin t,\ dx=\sqrt{2}a\cos t\,dt$

$$\int_{-a}^{a}\sqrt{2a^2-x^2}\,dx=\int_{-\pi/4}^{\pi/4}2a^2\cos^2 t\,dt$$

$$=a^2\int_{-\pi/4}^{\pi/4}(1+\cos 2t)dt=a^2\left[t+\frac{1}{2}\sin 2t\right]_{-\pi/4}^{\pi/4}$$

$$=\frac{\pi a^2}{2}+a^2$$

$$A=\frac{\pi a^2}{2}-\left(\frac{\pi a^2}{2}+a^2\right)+2a^2=a^2$$

Thus, the area of the lune is equal to the area of the square

b. Without using calculus, consider the following labels on the figure.

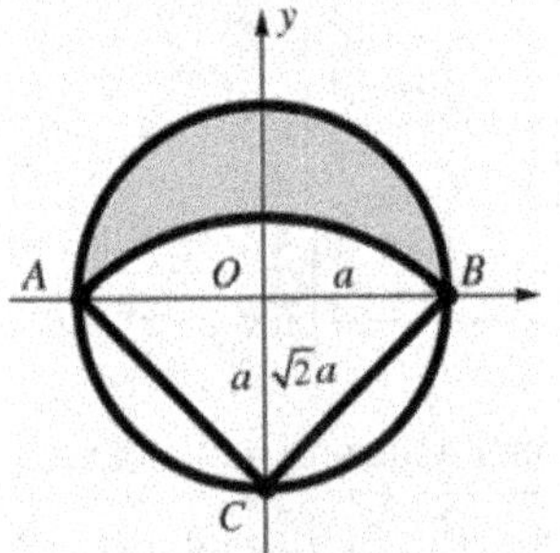

Area of the lune = Area of the semicircle of radius a at O + Area (ΔABC) – Area of the sector ABC.

$$A=\frac{1}{2}\pi a^2+a^2-\frac{1}{2}\left(\frac{\pi}{2}\right)(\sqrt{2}a)^2$$

$$=\frac{1}{2}\pi a^2+a^2-\frac{1}{2}\pi a^2=a^2$$

Note that since BC has length $\sqrt{2}a,$ the measure of angle OCB is $\frac{\pi}{4},$ so the measure of angle ACB is $\frac{\pi}{2}.$

35. $\frac{dy}{dx}=-\frac{\sqrt{a^2-x^2}}{x};\quad y=\int -\frac{\sqrt{a^2-x^2}}{x}dx$

$x=a\sin t,\ dx=a\cos t\,dt$

$$y=\int-\frac{a\cos t}{a\sin t}a\cos t\,dt=-a\int\frac{\cos^2 t}{\sin t}dt$$

$$=-a\int\frac{1-\sin^2 t}{\sin t}dt=a\int(\sin t-\csc t)dt$$

$$=a(-\cos t-\ln|\csc t-\cot t|)+C$$

$$\cos t=\frac{\sqrt{a^2-x^2}}{a},\ \csc t=\frac{a}{x},\ \cot t=\frac{\sqrt{a^2-x^2}}{x}$$

$$y=a\left(-\frac{\sqrt{a^2-x^2}}{a}-\ln\left|\frac{a}{x}-\frac{\sqrt{a^2-x^2}}{x}\right|\right)+C$$

$$=-\sqrt{a^2-x^2}-a\ln\left|\frac{a-\sqrt{a^2-x^2}}{x}\right|+C$$

Since $y=0$ when $x=a$,
$0=0-a\ln 1+C$, so $C=0$.

$$y=-\sqrt{a^2-x^2}-a\ln\left|\frac{a-\sqrt{a^2-x^2}}{x}\right|$$

7.5 Concepts Review

1. proper

3. $a=2;\ b=3;\ c=-1$

Problem Set 7.5

1. $\frac{1}{x(x+1)}=\frac{A}{x}+\frac{B}{x+1}$

$1=A(x+1)+Bx$

$A=1,\ B=-1$

$$\int\frac{1}{x(x+1)}dx=\int\frac{1}{x}dx-\int\frac{1}{x+1}dx$$

$$=\ln|x|-\ln|x+1|+C$$

3. $\frac{3}{x^2-1}=\frac{3}{(x+1)(x-1)}=\frac{A}{x+1}+\frac{B}{x-1}$

$3=A(x-1)+B(x+1)$

$A=-\frac{3}{2},\ B=\frac{3}{2}$

$$\int\frac{3}{x^2-1}dx=-\frac{3}{2}\int\frac{1}{x+1}dx+\frac{3}{2}\int\frac{1}{x-1}dx$$

$$=-\frac{3}{2}\ln|x+1|+\frac{3}{2}\ln|x-1|+C$$

5. $\frac{x-11}{x^2+3x-4}=\frac{x-11}{(x+4)(x-1)}=\frac{A}{x+4}+\frac{B}{x-1}$

$x-11=A(x-1)+B(x+4)$

$A=3, B=-2$

$\int\frac{x-11}{x^2+3x-4}dx=3\int\frac{1}{x+4}dx-2\int\frac{1}{x-1}dx=3\ln|x+4|-2\ln|x-1|+C$

7. $\frac{3x-13}{x^2+3x-10}=\frac{3x-13}{(x+5)(x-2)}=\frac{A}{x+5}+\frac{B}{x-2}$

$3x-13=A(x-2)+B(x+5)$

$A=4, B=-1$

$\int\frac{3x-13}{x^2+3x-10}dx=4\int\frac{1}{x+5}dx-\int\frac{1}{x-2}dx=4\ln|x+5|-\ln|x-2|+C$

9. $\frac{2x+21}{2x^2+9x-5}=\frac{2x+21}{(2x-1)(x+5)}=\frac{A}{2x-1}+\frac{B}{x+5}$

$2x+21=A(x+5)+B(2x-1)$

$A=4, B=-1$

$\int\frac{2x+21}{2x^2+9x-5}dx=\int\frac{4}{2x-1}dx-\int\frac{1}{x+5}dx=2\ln|2x-1|-\ln|x+5|+C$

11. $\frac{17x-3}{3x^2+x-2}=\frac{17x-3}{(3x-2)(x+1)}=\frac{A}{3x-2}+\frac{B}{x+1}$

$17x-3=A(x+1)+B(3x-2)$

$A=5, B=4$

$\int\frac{17x-3}{3x^2+x-2}dx=\int\frac{5}{3x-2}dx+\int\frac{4}{x+1}dx=\frac{5}{3}\ln|3x-2|+4\ln|x+1|+C$

13. $\frac{2x^2+x-4}{x^3-x^2-2x}=\frac{2x^2+x-4}{x(x+1)(x-2)}=\frac{A}{x}+\frac{B}{x+1}+\frac{C}{x-2}$

$2x^2+x-4=A(x+1)(x-2)+Bx(x-2)+Cx(x+1)$

$A=2, B=-1, C=1$

$\int\frac{2x^2+x-4}{x^3-x^2-2x}dx=\int\frac{2}{x}dx-\int\frac{1}{x+1}dx+\int\frac{1}{x-2}dx=2\ln|x|-\ln|x+1|+\ln|x-2|+C$

15. $\frac{6x^2+22x-23}{(2x-1)(x^2+x-6)}=\frac{6x^2+22x-23}{(2x-1)(x+3)(x-2)}=\frac{A}{2x-1}+\frac{B}{x+3}+\frac{C}{x-2}$

$6x^2+22x-23=A(x+3)(x-2)+B(2x-1)(x-2)+C(2x-1)(x+3)$

$A=2, B=-1, C=3$

$\int\frac{6x^2+22x-23}{(2x-1)(x^2+x-6)}dx=\int\frac{2}{2x-1}dx-\int\frac{1}{x+3}dx+\int\frac{3}{x-2}dx=\ln|2x-1|-\ln|x+3|+3\ln|x-2|+C$

17. $\frac{x^3}{x^2+x-2} = x-1+\frac{3x-2}{x^2+x-2}$

$\frac{3x-2}{x^2+x-2} = \frac{3x-2}{(x+2)(x-1)} = \frac{A}{x+2}+\frac{B}{x-1}$

$3x-2 = A(x-1)+B(x+2)$

$A = \frac{8}{3}, B = \frac{1}{3}$

$\int \frac{x^3}{x^2+x-2}dx = \int (x-1)\,dx + \frac{8}{3}\int \frac{1}{x+2}dx + \frac{1}{3}\int \frac{1}{x-1}dx = \frac{1}{2}x^2 - x + \frac{8}{3}\ln|x+2| + \frac{1}{3}\ln|x-1| + C$

19. $\frac{x^4+8x^2+8}{x^3-4x} = x+\frac{12x^2+8}{x(x+2)(x-2)}$

$\frac{12x^2+8}{x(x+2)(x-2)} = \frac{A}{x}+\frac{B}{x+2}+\frac{C}{x-2}$

$12x^2+8 = A(x+2)(x-2)+Bx(x-2)+Cx(x+2)$

$A = -2, B = 7, C = 7$

$\int \frac{x^4+8x^2+8}{x^3-4x}dx = \int x\,dx - 2\int \frac{1}{x}dx + 7\int \frac{1}{x+2}dx + 7\int \frac{1}{x-2}dx = \frac{1}{2}x^2 - 2\ln|x| + 7\ln|x+2| + 7\ln|x-2| + C$

21. $\frac{x+1}{(x-3)^2} = \frac{A}{x-3}+\frac{B}{(x-3)^2}$

$x+1 = A(x-3)+B$

$A = 1, B = 4$

$\int \frac{x+1}{(x-3)^2}dx = \int \frac{1}{x-3}dx + \int \frac{4}{(x-3)^2}dx = \ln|x-3| - \frac{4}{x-3} + C$

23. $\frac{3x+2}{x^3+3x^2+3x+1} = \frac{3x+2}{(x+1)^3} = \frac{A}{x+1}+\frac{B}{(x+1)^2}+\frac{C}{(x+1)^3}$

$3x+2 = A(x+1)^2+B(x+1)+C$

$A = 0, B = 3, C = -1$

$\int \frac{3x+2}{x^3+3x^2+3x+1}dx = \int \frac{3}{(x+1)^2}dx - \int \frac{1}{(x+1)^3}dx = -\frac{3}{x+1} + \frac{1}{2(x+1)^2} + C$

25. $\frac{3x^2-21x+32}{x^3-8x^2+16x} = \frac{3x^2-21x+32}{x(x-4)^2} = \frac{A}{x}+\frac{B}{x-4}+\frac{C}{(x-4)^2}$

$3x^2-21x+32 = A(x-4)^2+Bx(x-4)+Cx$

$A = 2, B = 1, C = -1$

$\int \frac{3x^2-21x+32}{x^3-8x^2+16}dx = \int \frac{2}{x}dx + \int \frac{1}{x-4}dx - \int \frac{1}{(x-4)^2}dx = 2\ln|x| + \ln|x-4| + \frac{1}{x-4} + C$

27. $\frac{2x^2+x-8}{x^3+4x} = \frac{2x^2+x-8}{x(x^2+4)} = \frac{A}{x}+\frac{Bx+C}{x^2+4}$

$A = -2, B = 4, C = 1$

$\int \frac{2x^2+x-8}{x^3+4x}dx = -2\int \frac{1}{x}dx + \int \frac{4x+1}{x^2+4}dx = -2\int \frac{1}{x}dx + 2\int \frac{2x}{x^2+4}dx + \int \frac{1}{x^2+4}dx$

$= -2\ln|x| + 2\ln|x^2+4| + \frac{1}{2}\tan^{-1}\left(\frac{x}{2}\right) + C$

29. $\frac{2x^2-3x-36}{(2x-1)(x^2+9)}=\frac{A}{2x-1}+\frac{Bx+C}{x^2+9}$

$A=-4, B=3, C=0$

$\int\frac{2x^2-3x-36}{(2x-1)(x^2+9)}dx=-4\int\frac{1}{2x-1}dx+\int\frac{3x}{x^2+9}dx=-2\ln|2x-1|+\frac{3}{2}\ln\left|x^2+9\right|+C$

31. $\frac{1}{(x-1)^2(x+4)^2}=\frac{A}{x-1}+\frac{B}{(x-1)^2}+\frac{C}{x+4}+\frac{D}{(x+4)^2}$

$A=-\frac{2}{125}, B=\frac{1}{25}, C=\frac{2}{125}, D=\frac{1}{25}$

$\int\frac{1}{(x-1)^2(x+4)^2}dx=-\frac{2}{125}\int\frac{1}{x-1}dx+\frac{1}{25}\int\frac{1}{(x-1)^2}dx+\frac{2}{125}\int\frac{1}{x+4}dx+\frac{1}{25}\int\frac{1}{(x+4)^2}dx$

$=-\frac{2}{125}\ln|x-1|-\frac{1}{25(x-1)}+\frac{2}{125}\ln|x+4|-\frac{1}{25(x+4)}+C$

33. $x=\sin t$, $dx=\cos t\,dt$

$\int\frac{(\sin^3 t-8\sin^2 t-1)\cos t}{(\sin t+3)(\sin^2 t-4\sin t+5)}dt=\int\frac{x^3-8x^2-1}{(x+3)(x^2-4x+5)}dx$

$=x-\frac{50}{13}\ln|x+3|-\frac{68}{13}\tan^{-1}(x-2)-\frac{41}{26}\ln\left|x^2-4x+5\right|+C$

which is the result of Problem 32.

$\int\frac{(\sin^3 t-8\sin^2 t-1)\cos t}{(\sin t+3)(\sin^2 t-4\sin t+5)}dt=\sin t-\frac{50}{13}\ln|\sin t+3|-\frac{68}{13}\tan^{-1}(\sin t-2)-\frac{41}{26}\ln\left|\sin^2 t-4\sin t+5\right|+C$

35. $\frac{x^3-4x}{(x^2+1)^2}=\frac{Ax+B}{x^2+1}+\frac{Cx+D}{(x^2+1)^2}$

$A=1, B=0, C=-5, D=0$

$\int\frac{x^3-4x}{(x^2+1)^2}dx=\int\frac{x}{x^2+1}dx-5\int\frac{x}{(x^2+1)^2}dx=\frac{1}{2}\ln\left|x^2+1\right|+\frac{5}{2(x^2+1)}+C$

37. $\frac{2x^3+5x^2+16x}{x^5+8x^3+16x}=\frac{x(2x^2+5x+16)}{x(x^4+8x^2+16)}=\frac{2x^2+5x+16}{(x^2+4)^2}=\frac{Ax+B}{x^2+4}+\frac{Cx+D}{(x^2+4)^2}$

$A=0, B=2, C=5, D=8$

$\int\frac{2x^3+5x^2+16x}{x^5+8x^3+16x}dx=\int\frac{2}{x^2+4}dx+\int\frac{5x+8}{(x^2+4)^2}dx=\int\frac{2}{x^2+4}dx+\int\frac{5x}{(x^2+4)^2}dx+\int\frac{8}{(x^2+4)^2}dx$

To integrate $\int\frac{8}{(x^2+4)^2}dx$, let $x=2\tan\theta$, $dx=2\sec^2\theta\,d\theta$.

$\int\frac{8}{(x^2+4)^2}dx=\int\frac{16\sec^2\theta}{16\sec^4\theta}d\theta=\int\cos^2\theta\,d\theta=\int\left(\frac{1}{2}+\frac{1}{2}\cos 2\theta\right)d\theta$

$=\frac{1}{2}\theta+\frac{1}{4}\sin 2\theta+C=\frac{1}{2}\theta+\frac{1}{2}\sin\theta\cos\theta+C=\frac{1}{2}\tan^{-1}\frac{x}{2}+\frac{x}{x^2+4}+C$

$\int\frac{2x^3+5x^2+16x}{x^5+8x^3+16x}dx=\tan^{-1}\frac{x}{2}-\frac{5}{2(x^2+4)}+\frac{1}{2}\tan^{-1}\frac{x}{2}+\frac{x}{x^2+4}+C=\frac{3}{2}\tan^{-1}\frac{x}{2}+\frac{2x-5}{2(x^2+4)}+C$

39. $u = \sin\theta$, $du = \cos\theta\, d\theta$

$$\int_0^{\pi/4} \frac{\cos\theta}{(1-\sin^2\theta)(\sin^2\theta+1)^2}d\theta = \int_0^{1/\sqrt{2}} \frac{1}{(1-u^2)(u^2+1)^2}du = \int_0^{1/\sqrt{2}} \frac{1}{(1-u)(1+u)(u^2+1)^2}du$$

$$\frac{1}{(1-u^2)(u^2+1)^2} = \frac{A}{1-u} + \frac{B}{1+u} + \frac{Cu+D}{u^2+1} + \frac{Eu+F}{(u^2+1)^2}$$

$$A = \frac{1}{8}, B = \frac{1}{8}, C = 0, D = \frac{1}{4}, E = 0, F = \frac{1}{2}$$

$$\int_0^{1/\sqrt{2}} \frac{1}{(1-u^2)(u^2+1)^2}du = \frac{1}{8}\int_0^{1/\sqrt{2}} \frac{1}{1-u}du + \frac{1}{8}\int_0^{1/\sqrt{2}} \frac{1}{1+u}du + \frac{1}{4}\int_0^{1/\sqrt{2}} \frac{1}{u^2+1}du + \frac{1}{2}\int_0^{1/\sqrt{2}} \frac{1}{(u^2+1)^2}du$$

$$= \left[-\frac{1}{8}\ln|1-u| + \frac{1}{8}\ln|1+u| + \frac{1}{4}\tan^{-1}u + \frac{1}{4}\left(\tan^{-1}u + \frac{u}{u^2+1}\right)\right]_0^{1/\sqrt{2}}$$

$$= \left[\frac{1}{8}\ln\left|\frac{1+u}{1-u}\right| + \frac{1}{2}\tan^{-1}u + \frac{u}{4(u^2+1)}\right]_0^{1/\sqrt{2}}$$

$$= \frac{1}{8}\ln\left|\frac{\sqrt{2}+1}{\sqrt{2}-1}\right| + \frac{1}{2}\tan^{-1}\frac{1}{\sqrt{2}} + \frac{1}{6\sqrt{2}} \approx 0.65$$

(To integrate $\int \frac{1}{(u^2+1)^2}du$, let $u = \tan t$.)

41. $\frac{dy}{dt} = y(1-y)$ so that $\int \frac{1}{y(1-y)}dy = \int 1\, dt = t + C_1$

a. Using partial fractions:

$$\frac{1}{y(1-y)} = \frac{A}{y} + \frac{B}{1-y} = \frac{A(1-y)+By}{y(1-y)} \Rightarrow$$

$$A + (B-A)y = 1 + 0y \Rightarrow A = 1,\ B - A = 0 \Rightarrow A = 1,\ B = 1 \Rightarrow \frac{1}{y(1-y)} = \frac{1}{y} + \frac{1}{1-y}$$

Thus: $t + C_1 = \int\left(\frac{1}{y} + \frac{1}{1-y}\right)dy = \ln y - \ln(1-y) = \ln\left(\frac{y}{1-y}\right)$ so that

$$\frac{y}{1-y} = e^{t+C_1} \underset{(C=e^{C_1})}{=} Ce^t \quad \text{or} \quad y(t) = \frac{e^t}{\frac{1}{C}+e^t}$$

Since $y(0) = 0.5$, $0.5 = \frac{1}{\frac{1}{C}+1}$ or $C = 1$; thus $y(t) = \frac{e^t}{1+e^t}$

b. $y(3) = \frac{e^3}{1+e^3} \approx 0.953$

43. $\frac{dy}{dt} = 0.0003y(8000-y)$ so that

$$\int \frac{1}{y(8000-y)}dy = \int 0.0003\,dt = 0.0003t + C_1$$

a. Using partial fractions:

$$\frac{1}{y(8000-y)} = \frac{A}{y} + \frac{B}{8000-y} = \frac{A(8000-y)+By}{y(8000-y)}$$

$$\Rightarrow 8000A + (B-A)y = 1 + 0y$$

$$\Rightarrow 8000A = 1,\ B - A = 0$$

$$\Rightarrow A = \frac{1}{8000},\ B = \frac{1}{8000}$$

$$\Rightarrow \frac{1}{y(8000-y)} = \frac{1}{8000}\left[\frac{1}{y} + \frac{1}{(8000-y)}\right]$$

Thus:

$$0.0003t + C_1 = \frac{1}{8000}\int\left(\frac{1}{y} + \frac{1}{(8000-y)}\right)dy =$$

$$\frac{1}{8000}[\ln y - \ln(8000-y)] = \frac{1}{8000}\ln\left(\frac{y}{8000-y}\right)$$

so that

$$\frac{y}{8000-y} = e^{2.4t+8000C_1} \underset{(C=e^{8000C_1})}{=} Ce^{2.4t} \text{ or}$$

$$y(t) = \frac{8000e^{2.4t}}{\frac{1}{C}+e^{2.4t}}$$

Since $y(0) = 1000,\ 1000 = \frac{8000}{\frac{1}{C}+1}$ or $C = \frac{1}{7}$;

thus $y(t) = \frac{8000e^{2.4t}}{7+e^{2.4t}}$

b. $y(3) = \frac{8000e^{7.2}}{7+e^{7.2}} \approx 7958.4$

45. $\frac{dy}{dt} = ky(L-y)$ so that

$$\int \frac{1}{y(L-y)}dy = \int k\,dt = kt + C_1$$

Using partial fractions:

$$\frac{1}{y(L-y)} = \frac{A}{y} + \frac{B}{L-y} = \frac{A(L-y)+By}{y(L-y)} \Rightarrow$$

$$LA + (B-A)y = 1 + 0y \Rightarrow LA = 1,\ B - A = 0 \Rightarrow$$

$$A = \frac{1}{L},\ B = \frac{1}{L} \Rightarrow \frac{1}{y(L-y)} = \frac{1}{L}\left[\frac{1}{y} + \frac{1}{L-y}\right]$$

Thus: $kt + C_1 = \frac{1}{L}\int\left(\frac{1}{y} + \frac{1}{L-y}\right)dy =$

$\frac{1}{L}[\ln y - \ln(L-y)] = \frac{1}{L}\ln\left(\frac{y}{L-y}\right)$ so that

$$\frac{y}{L-y} = e^{kLt+LC_1} \underset{(C=e^{LC_1})}{=} Ce^{kLt} \text{ or } y(t) = \frac{Le^{kLt}}{\frac{1}{C}+e^{kLt}}$$

If $y_0 = y(0) = \frac{L}{\frac{1}{C}+1}$ then $\frac{1}{C} = \frac{L-y_0}{y_0}$; so our final formula is $y(t) = \frac{Le^{kLt}}{\left(\frac{L-y_0}{y_0}\right)+e^{kLt}}$.

(Note: if $y_0 < L$, then $u = \frac{L-y_0}{y_0} > 0$ and $\frac{e^{kLt}}{u+e^{kLt}} < 1$; thus $y(t) < L$ for all t)

47. If $y_0 < L$, then $y'(0) = ky_0(L-y_0) > 0$ and the population is increasing initially.

49. a. $\frac{dy}{dt} = ky(16-y)$

$$\frac{dy}{y(16-y)} = kdt$$

$$\int \frac{dy}{y(16-y)} = \int kdt$$

$$\frac{1}{16}\int\left(\frac{1}{y} + \frac{1}{16-y}\right)dy = kt + C$$

$$\frac{1}{16}(\ln|y| - \ln|16-y|) = kt + C$$

$$\ln\left|\frac{y}{16-y}\right| = 16kt + C$$

$$\frac{y}{16-y} = Ce^{16kt}$$

$$y(0) = 2: \frac{1}{7} = C; \frac{y}{16-y} = \frac{1}{7}e^{16kt}$$

$$y(50) = 4: \frac{1}{3} = \frac{1}{7}e^{800k},\ \text{so } k = \frac{1}{800}\ln\frac{7}{3}$$

$$\frac{y}{16-y} = \frac{1}{7}e^{\left(\frac{1}{50}\ln\frac{7}{3}\right)t}$$

$$7y = 16e^{\left(\frac{1}{50}\ln\frac{7}{3}\right)t} - ye^{\left(\frac{1}{50}\ln\frac{7}{3}\right)t}$$

$$y = \frac{16e^{\left(\frac{1}{50}\ln\frac{7}{3}\right)t}}{7+e^{\left(\frac{1}{50}\ln\frac{7}{3}\right)t}} = \frac{16}{1+7e^{-\left(\frac{1}{50}\ln\frac{7}{3}\right)t}}$$

b. $y(90) = \frac{16}{1+7e^{-\left(\frac{1}{50}\ln\frac{7}{3}\right)90}} \approx 6.34$ billion

c. $$9 = \frac{16}{1+7e^{-\left(\frac{1}{50}\ln\frac{7}{3}\right)t}}$$

$$7e^{-\left(\frac{1}{50}\ln\frac{7}{3}\right)t} = \frac{16}{9} - 1$$

$$e^{-\left(\frac{1}{50}\ln\frac{7}{3}\right)t} = \frac{1}{9}$$

$$-\left(\tfrac{1}{50}\ln\tfrac{7}{3}\right)t = \ln\frac{1}{9}$$

$$t = -50\left(\frac{\ln\frac{1}{9}}{\ln\frac{7}{3}}\right) \approx 129.66$$

The population will be 9 billion in 2055.

51. a. Separating variables, we obtain

$$\frac{dx}{(a-x)(b-x)} = k\,dt$$

$$\frac{1}{(a-x)(b-x)} = \frac{A}{a-x} + \frac{B}{b-x}$$

$$A = -\frac{1}{a-b},\ B = \frac{1}{a-b}$$

$$\int \frac{dx}{(a-x)(b-x)}$$

$$= \frac{1}{a-b}\int\left(-\frac{1}{a-x} + \frac{1}{b-x}\right)dx = \int k\,dt$$

$$\frac{\ln|a-x| - \ln|b-x|}{a-b} = kt + C$$

$$\frac{1}{a-b}\ln\left|\frac{a-x}{b-x}\right| = kt + C$$

$$\frac{a-x}{b-x} = Ce^{(a-b)kt}$$

Since $x = 0$ when $t = 0$, $C = \frac{a}{b}$, so

$$a - x = (b-x)\frac{a}{b}e^{(a-b)kt}$$

$$a\left(1 - e^{(a-b)kt}\right) = x\left(1 - \frac{a}{b}e^{(a-b)kt}\right)$$

$$x(t) = \frac{a(1-e^{(a-b)kt})}{1-\frac{a}{b}e^{(a-b)kt}} = \frac{ab(1-e^{(a-b)kt})}{b-ae^{(a-b)kt}}$$

b. Since $b > a$ and $k > 0$, $e^{(a-b)kt} \to 0$ as $t \to \infty$. Thus,

$$x \to \frac{ab(1)}{b-0} = a.$$

c. $$x(t) = \frac{8(1-e^{-2kt})}{4-2e^{-2kt}}$$

$x(20) = 1$, so $4 - 2e^{-40k} = 8 - 8e^{-40k}$

$$6e^{-40k} = 4$$

$$k = -\frac{1}{40}\ln\frac{2}{3}$$

$$e^{-2kt} = e^{t/20\ln 2/3} = e^{\ln(2/3)^{t/20}} = \left(\frac{2}{3}\right)^{t/20}$$

$$x(t) = \frac{4\left(1-\left(\frac{2}{3}\right)^{t/20}\right)}{2-\left(\frac{2}{3}\right)^{t/20}}$$

$$x(60) = \frac{4\left(1-\left(\frac{2}{3}\right)^{3}\right)}{2-\left(\frac{2}{3}\right)^{3}} = \frac{38}{23} \approx 1.65 \text{ grams}$$

d. If $a = b$, the differential equation is, after separating variables

$$\frac{dx}{(a-x)^2} = k\,dt$$

$$\int \frac{dx}{(a-x)^2} = \int k\,dt$$

$$\frac{1}{a-x} = kt + C$$

$$\frac{1}{kt+C} = a - x$$

$$x(t) = a - \frac{1}{kt+C}$$

Since $x = 0$ when $t = 0$, $C = \frac{1}{a}$, so

$$x(t) = a - \frac{1}{kt+\frac{1}{a}} = a - \frac{a}{akt+1}$$

$$= a\left(1 - \frac{1}{akt+1}\right) = a\left(\frac{akt}{akt+1}\right).$$

53. Separating variables, we obtain

$$\frac{dy}{(A-y)(B+y)} = k\,dt$$

$$\frac{1}{(A-y)(B+y)} = \frac{C}{A-y} + \frac{D}{B+y}$$

$$C = \frac{1}{A+B},\ D = \frac{1}{A+B}$$

$$\int \frac{dy}{(A-y)(B+y)} = \frac{1}{A+B}\int\left(\frac{1}{A-y} + \frac{1}{B+y}\right)dy$$

$$= \int k\,dt$$

$$\frac{-\ln(A-y) + \ln(B+y)}{A+B} = kt + C$$

$$\frac{1}{A+B}\ln\left|\frac{B+y}{A-y}\right| = kt + C$$

$$\frac{B+y}{A-y} = Ce^{(A+B)kt}$$

$$B + y = (A-y)Ce^{(A+B)kt}$$

$$y(1 + Ce^{(A+B)kt}) = ACe^{(A+B)kt} - B$$

$$y(t) = \frac{ACe^{(A+B)kt} - B}{1 + Ce^{(A+B)kt}}$$

7.6 Concepts Review

1. substitution

3. approximation

Problem Set 7.6

Note: Throughout this section, the notation Fxxx refers to integration formula number xxx in the back of the book.

1. Integration by parts.

$u = x \qquad dv = e^{-5x}$

$du = 1\,dx \qquad v = -\frac{1}{5}e^{-5x}$

$$\int xe^{-5x}\,dx = -\frac{1}{5}xe^{-5x} - \int -\frac{1}{5}e^{-5x}\,dx$$

$$= -\frac{1}{5}xe^{-5x} - \frac{1}{25}e^{-5x} + C$$

$$= -\frac{1}{5}e^{-5x}\left(x + \frac{1}{5}\right) + C$$

3. Substitution

$$\int_1^2 \frac{\ln x}{x}\,dx = \int_0^{\ln 2} u\,du = \left[\frac{u^2}{2}\right]_0^{\ln 2} = \frac{(\ln 2)^2}{2} \approx 0.2402$$

$u = \ln x$, $du = \frac{1}{x}dx$

5. Trig identity $\cos^2 u = \frac{1 + \cos 2u}{2}$ and substitution.

$$\int \cos^4 2x\,dx = \int\left(\frac{1+\cos 4x}{2}\right)^2 dx =$$

$$\frac{1}{4}\left[\int 1 + 2\cos 4x + \cos^2 4x\right]dx =$$

$u = 4x$, $du = 4\,dx$.

$$\frac{1}{4}\left[x + \frac{1}{2}\sin 4x + \int\left(\frac{1+\cos 8x}{2}\right)dx\right] =$$

$v = 8x$, $dv = 8dx$

$$\frac{1}{4}\left[x + \frac{1}{2}\sin 4x + \frac{1}{2}x + \frac{1}{16}\sin 8x\right] + C =$$

$$\frac{1}{64}[24x + 8\sin 4x + \sin 8x] + C$$

7. Partial fractions

$$\int \frac{1}{x^2 + 6x + 8}\,dx = \int \frac{1}{(x+4)(x+2)}\,dx$$

$$\frac{1}{(x+4)(x+2)} = \frac{A}{(x+4)} + \frac{B}{(x+2)} =$$

$$\frac{A(x+2) + B(x+4)}{(x+4)(x+2)} = \frac{(A+B)x + (2A+4B)}{(x+4)(x+2)} \Rightarrow$$

$$A + B = 0,\ 2A + 4B = 1 \Rightarrow A = -\frac{1}{2}, B = \frac{1}{2}$$

$$\int_1^2 \frac{1}{x^2+6x+8} = \frac{1}{2}\int_1^2\left(\frac{1}{x+2} - \frac{1}{x+4}\right)dx$$

$$= \frac{1}{2}\left[\ln|x+2| - \ln|x+4|\right]_1^2 = \frac{1}{2}\left[\ln\left|\frac{(x+2)}{(x+4)}\right|\right]_1^2$$

$$= \frac{1}{2}\left(\ln\frac{4}{6} - \ln\frac{3}{5}\right) = \frac{1}{2}\ln\frac{10}{9} \approx 0.0527$$

9. Substitution

$$\int_0^5 x\sqrt{x+2}\,dx = \int_{\sqrt{2}}^{\sqrt{7}} (u^2-2)(u)2u\,du =$$

$u=\sqrt{x+2}$
$u^2=x+2$
$2u\,du=dx$

$$\int_{\sqrt{2}}^{\sqrt{7}} 2u^4-4u^2\,du = 2\left[\frac{u^5}{5}-\frac{2u^3}{3}\right]_{\sqrt{2}}^{\sqrt{7}} =$$

$$\frac{2}{15}\left[3u^5-10u^3\right]_{\sqrt{2}}^{\sqrt{7}} = \frac{2}{15}\left[77\sqrt{7}+8\sqrt{2}\right] \approx 28.67$$

11. Use of symmetry; this is an odd function, so

$$\int_{-\pi/2}^{\pi/2} \cos^2 x \sin x\,dx = 0$$

13. a. Formula 96

$$\int x\sqrt{3x+1}\,dx \underset{\substack{F96\\a=3,b=1}}{=} \frac{2}{135}(9x-2)(3x+1)^{3/2}+C$$

b. Substitution; Formula 96

$$\int e^x\sqrt{3e^x+1}\;dx = \int u\sqrt{3u+1}\,du \underset{\substack{F96\\a=3,b=1}}{=}$$

$u=e^x,\ du=e^x\,dx$

$$\frac{2}{135}\left(9e^x-2\right)\left(3e^x+1\right)^{3/2}+C$$

15. a. Substitution, Formula 18

$$\int \frac{dx}{9-16x^2} = \frac{1}{4}\int \frac{du}{9-u^2} \underset{\substack{F18\\a=3}}{=}$$

$u=4x,\ du=4dx$

$$\frac{1}{4}\left[\frac{1}{6}\ln\left|\frac{u+3}{u-3}\right|\right]+C = \frac{1}{24}\ln\left|\frac{4x+3}{4x-3}\right|+C$$

b. Substitution, Formula 18

$$\int \frac{e^x}{9-16e^{2x}}\,dx = \frac{1}{4}\int\frac{du}{9-u^2} \underset{\text{part a.}}{=}$$

$u=4e^x,\ du=4e^x dx$

$$\frac{1}{24}\ln\left|\frac{4e^x+3}{4e^x-3}\right|+C$$

17. a. Substitution, Formula 57

$$\int x^2\sqrt{9-2x^2}\,dx = \frac{\sqrt{2}}{4}\int u^2\sqrt{9-u^2}\,du \underset{\substack{F57\\a=3}}{=}$$

$u=\sqrt{2}x$
$du=\sqrt{2}\,dx$

$$\frac{1}{16}\left(x(4x^2-9)\sqrt{9-2x^2}\right)+\frac{81\sqrt{2}}{32}\sin^{-1}\left(\frac{\sqrt{2}x}{3}\right)+C$$

b. Substitution, Formula 57

$$\int \sin^2 x\cos x\sqrt{9-2\sin^2 x}\,dx = \frac{\sqrt{2}}{4}\int u^2\sqrt{9-u^2}\,du$$

$u=\sqrt{2}\sin x$
$du=\sqrt{2}\cos x\,dx$

$$\underset{\substack{F57\\a=3}}{=} \frac{1}{16}\left(\sin x(4\sin^2 x-9)\sqrt{9-2\sin^2 x}\right)$$

$$+\frac{81\sqrt{2}}{32}\sin^{-1}\left(\frac{\sqrt{2}\sin x}{3}\right)+C$$

19. a. Substitution, Formula 45

$$\int\frac{dx}{\sqrt{5+3x^2}} = \frac{\sqrt{3}}{3}\int\frac{du}{\sqrt{5+u^2}} \underset{\substack{F45\\a=\sqrt{5}}}{=}$$

$u=\sqrt{3}x$
$du=\sqrt{3}\,dx$

$$\frac{\sqrt{3}}{3}\ln\left|\sqrt{3}x+\sqrt{5+3x^2}\right|+C$$

b. Substitution, Formula 45

$$\int\frac{x}{\sqrt{5+3x^4}}\,dx = \frac{\sqrt{3}}{6}\int\frac{du}{\sqrt{5+u^2}} \underset{\substack{F45\\a=\sqrt{5}}}{=}$$

$u=\sqrt{3}x^2$
$du=2\sqrt{3}\,x\,dx$

$$\frac{\sqrt{3}}{6}\ln\left|\sqrt{3}x^2+\sqrt{5+3x^4}\right|+C$$

21. a. Complete the square; substitution; Formula 45.

$$\int\frac{dt}{\sqrt{t^2+2t-3}} = \int\frac{dt}{\sqrt{(t+1)^2-4}} = \int\frac{du}{\sqrt{u^2-4}} \underset{\substack{F45\\a=2}}{=}$$

$u=t+1$
$du=dt$

$$\ln\left|(t+1)+\sqrt{t^2+2t-3}\right|+C$$

b. Complete the square; substitution; Formula 45.

$$\int\frac{dt}{\sqrt{t^2+3t-5}} = \int\frac{dt}{\sqrt{(t+\frac{3}{2})^2-\frac{29}{4}}} =$$

$u=t+3/2$
$du=dt$

$$\int\frac{du}{\sqrt{u^2-\frac{29}{4}}} \underset{\substack{F45\\a=\sqrt{29}/2}}{=} \ln\left|(t+\tfrac{3}{2})+\sqrt{t^2+3t-5}\right|+C$$

23. a. Formula 98

$$\int \frac{y}{\sqrt{3y+5}}\,dy \underset{\substack{F98\\a=3,b=5}}{=} \frac{2}{27}(3y-10)\sqrt{3y+5}+C$$

b. Substitution, Formula 98

$$\underset{\substack{u=\sin t\\du=\cos t\,dt}}{\int \frac{\sin t\cos t}{\sqrt{3\sin t+5}}} = \int \frac{u}{\sqrt{3u+5}}\,du \underset{\substack{F98\\a=3,b=5}}{=}$$

$$\frac{2}{27}(3\sin t-10)\sqrt{3\sin t+5}+C$$

25. Substitution; Formula 84

$$\underset{\substack{u=3t\\du=3dt}}{\int \sinh^2 3t\,dt} = \frac{1}{3}\int \sinh^2 u\,du \underset{F84}{=}$$

$$\frac{1}{3}\left(\frac{1}{4}\sinh 6t-\frac{3}{2}t\right)+C=\frac{1}{12}(\sinh 6t-6t)+C$$

27. Substitution; Formula 98

$$\underset{\substack{u=\cos t\\du=-\sin t\,dt}}{\int \frac{\cos t\sin t}{\sqrt{2\cos t+1}}\,dt} = -\int \frac{u}{\sqrt{2u+1}}\,du \underset{\substack{F98\\a=2\\b=1}}{=}$$

$$-\frac{1}{6}(2\cos t-2)\sqrt{2\cos t+1}+C=$$

$$\frac{1}{3}(1-\cos t)\sqrt{2\cos t+1}+C$$

29. Substitution; Formula 99, Formula 98

$$\underset{\substack{u=\cos t\\du=-\sin t\,dt}}{\int \frac{\cos^2 t\sin t}{\sqrt{\cos t+1}}\,dt} = -\int \frac{u^2}{\sqrt{u+1}}\,du \underset{\substack{F99\\n=2\\a=1\\b=1}}{=}$$

$$-\frac{2}{5}\left[u^2\sqrt{u+1}-2\int \frac{u}{\sqrt{u+1}}\,du\right] \underset{F98}{=}$$

$$-\frac{2}{5}\left[u^2\sqrt{u+1}-2\left(\frac{2}{3}(u-2)\sqrt{u+1}\right)\right]+C=$$

$$-\frac{2}{5}\sqrt{\cos t+1}\left[\cos^2 t-\frac{4}{3}(\cos t-2)\right]+C$$

31. Using a CAS, we obtain:

$$\int_0^{\pi} \frac{\cos^2 x}{1+\sin x}\,dx=\pi-2\approx 1.14159$$

33. Using a CAS, we obtain:

$$\int_0^{\pi/2} \sin^{12} x\,dx=\frac{231\pi}{2048}\approx 0.35435$$

35. Using a CAS, we obtain:

$$\int_1^4 \frac{\sqrt{t}}{1+t^8}\,dt\approx 0.11083$$

37. Using a CAS, we obtain:

$$\int_0^{\pi/2} \frac{1}{1+2\cos^5 x}\,dx\approx 1.10577$$

39. Using a CAS, we obtain:

$$\int_2^3 \frac{x^2+2x-1}{x^2-2x+1}\,dx=4\ln(2)+2\approx 4.77259$$

41. $\int_0^c \frac{1}{x+1}\,dx \underset{F3}{=} \left[\ln|x+1|\right]_0^c=\ln(c+1)$

$\ln(c+1)=1 \Rightarrow c+1=e \Rightarrow$

$c=e-1\approx 1.71828$

43. Substitution; Formula 65

$$\underset{\substack{u=x+1\\du=dx}}{\int \ln(x+1)\,dx} = \int \ln u\,du \underset{F65}{=}$$

$(x+1)[\ln(x+1)-1]$. Thus

$\int_0^c \ln(x+1)\,dx=(x+1)[\ln(x+1)-1]_0^c=$

$(c+1)\ln(c+1)-c$ and

$(c+1)\ln(c+1)-c=1 \Rightarrow \ln(c+1)=1 \Rightarrow$

$c+1=e \Rightarrow c=e-1\approx 1.71828$

45. There is no antiderivative that can be expressed in terms of elementary functions; an approximation for the integral, as well as a process such as Newton's Method, must be used. Several approaches are possible. $c\approx 0.59601$

47. There is no antiderivative that can be expressed in terms of elementary functions; an approximation for the integral, as well as a process such as Newton's Method, must be used. Several approaches are possible. $c\approx 0.16668$

49. There is no antiderivative that can be expressed in terms of elementary functions; an approximation for the integral, as well as a process such as Newton's Method, must be used. Several approaches are possible. $c\approx 9.2365$

51. $f(x)=8-x \quad g(x)=cx \quad a=0 \quad b=\frac{8}{c+1}$

a. $\int_a^b x(f(x)-g(x))\,dx=\int_0^{8/(c+1)} 8x-(c+1)x^2\,dx=$

$$\left[4x^2-\left(\frac{c+1}{3}\right)x^3\right]_0^{8/(c+1)}=\frac{256}{(c+1)^2}-\frac{512}{3(c+1)^2}=$$

$$\frac{256}{3(c+1)^2}$$

b. $\int_0^{c+1}(f(x)-g(x))\,dx = \int_0^{8/c+1} 8-(c+1)x\,dx =$

$$\left[8x-\left(\frac{c+1}{2}\right)x^2\right]_0^{8/c+1} = \frac{64}{(c+1)} - \frac{32}{(c+1)} =$$

$$\frac{32}{(c+1)}$$

c. $\bar{x} = \left(\frac{256}{3(c+1)^2}\right)\left(\frac{c+1}{32}\right) = \frac{8}{3(c+1)}$

$$\bar{x} = 2 \Rightarrow \frac{8}{3(c+1)} = 2 \Rightarrow c = \frac{1}{3}$$

53. $f(x) = 6e^{-x/3}$ $g(x) = 0$ $a = 0$ $b = c$

a. $\int_a^b x(f(x)-g(x))\,dx = 6\int_0^c xe^{-x/3}\,dx =$

$u = x$, $dv = e^{-x/3}$, $du = dx$, $v = -3e^{-x/3}$

$$6\left[-3xe^{-x/3}\right]_0^c + 18\int_0^c e^{-x/3}\,dx =$$

$$\left[-18e^{-x/3}(x+3)\right]_0^c = -18e^{-c/3}(c+3)+54$$

b. $\int_a^b (f(x)-g(x))\,dx = \int_0^c 6e^{-x/3}\,dx =$

$$-18\left(e^{-c/3}-1\right)$$

c. For notational convenience, let

$u = -18e^{-c/3}$; then

$$\bar{x} = \frac{u(c+3)+54}{u+18} = \frac{cu}{u+18} + \frac{3(u+18)}{u+18} =$$

$$\frac{cu}{u+18}+3$$

$$\bar{x} = 2 \Rightarrow \frac{cu}{u+18} = -1 \Rightarrow \frac{c}{1+\frac{18}{u}} = -1 \Rightarrow$$

$$c = \frac{1}{e^{-c/3}} - 1 \Rightarrow \frac{1}{c+1} = e^{-c/3}$$

Let

$$h(c) = \frac{1}{c+1} - e^{-c/3},\quad h'(c) = \frac{1}{3}e^{-c/3} - \frac{1}{(c+1)^2}$$

and apply Newton's Method

n	1	2	3	4	5	6
a_n	2.0000	5.0000	5.6313	5.7103	5.7114	5.7114

$c \approx 5.7114$

55. a. $erf(x) = \frac{2}{\sqrt{\pi}}\int_0^x e^{-t^2}\,dt$

$$\therefore \frac{d}{dx}erf(x) = \frac{2}{\sqrt{\pi}}e^{-x^2}$$

b. $Si(x) = \int_0^x \frac{\sin t}{t}\,dt$

$$\therefore \frac{d}{dx}Si(x) = \frac{\sin x}{x}$$

57. a. (See problem 55 a.). Since $erf'(x) > 0$ for all x, $erf(x)$ is increasing on $(0,\infty)$.

b. $erf''(x) = \frac{-4x}{\sqrt{\pi}}e^{-x^2}$ which is negative on $(0,\infty)$, so $erf(x)$ is not concave up anywhere on the interval.

59. a. (See problem 56 b.) Since $C'(x) = \cos\left(\frac{\pi}{2}x^2\right)$, $C'(x) > 0$ when $0 < \frac{\pi}{2}x^2 < \frac{\pi}{2}$ or $\frac{3\pi}{2} < \frac{\pi}{2}x^2 < 2\pi$; thus $C(x)$ is increasing on $(0,1)\cup(\sqrt{3},2)$.

b. Since $C''(x) = -\pi x\sin\left(\frac{\pi}{2}x^2\right)$, $C''(x) > 0$ when $\pi < \frac{\pi}{2}x^2 < 2\pi$. Thus $C(x)$ is concave up on $(\sqrt{2},2)$.

7.7 Chapter Review

Concepts Test

1. True: The resulting integrand will be of the form $\sin u$.

3. False: Try the substitution $u = x^4, du = 4x^3\,dx$

5. True: The resulting integrand will be of the form $\frac{1}{a^2+u^2}$.

7. True: This integral is most easily solved with a partial fraction decomposition.

9. True: Because both exponents are even positive integers, half-angle formulas are used.

11. False: Use the substitution $u = -x^2-4x, du = (-2x-4)dx$

13. True: Then expand and use the substitution $u = \sin x$, $du = \cos x\,dx$

15. True: Let $u = \ln x$ $\quad dv = x^2 dx$

$du = \frac{1}{x}dx \quad v = \frac{1}{3}x^3$

17. False: $\frac{x^2}{x^2-1} = 1 + \frac{1}{2(x-1)} - \frac{1}{2(x+1)}$

19. True: $\frac{x^2+2}{x(x^2+1)} = \frac{2}{x} + \frac{-x}{x^2+1}$

21. False: To complete the square, add $\frac{b^2}{4a}$.

23. True: Polynomials with the same values for all x will have identical coefficients for like degree terms.

25. False: It can, however, be solved by the substitution $u = 25 - 4x^2$; then $du = -8x\,dx$ and

$\int x\sqrt{25-4x^2}\,dx = -\frac{1}{8}\int\sqrt{u}\,du =$

$-\frac{1}{12}(25-4x^2)^{3/2} + C$

27. True: by the First Fundamental Theorem of Calculus.

Sample Test Problems

1. $\int_0^4 \frac{t}{\sqrt{9+t^2}}dt = \left[\sqrt{9+t^2}\right]_0^4 = 5 - 3 = 2$

3. $\int_0^{\pi/2} e^{\cos x}\sin x\,dx = \left[-e^{\cos x}\right]_0^{\pi/2} = e - 1 \approx 1.718$

5. $\int\frac{y^3+y}{y+1}dy = \int\left(y^2 - y + 2 - \frac{2}{1+y}\right)dy$

$= \frac{1}{3}y^3 - \frac{1}{2}y^2 + 2y - 2\ln|1+y| + C$

7. $\int\frac{y-2}{y^2-4y+2}dy = \frac{1}{2}\int\frac{2y-4}{y^2-4y+2}dy$

$= \frac{1}{2}\ln\left|y^2-4y+2\right| + C$

9. $\int\frac{e^{2t}}{e^t-2}dt = e^t + 2\ln\left|e^t-2\right| + C$

(Use the substitution $u = e^t - 2$, $du = e^t\,dt$ which gives the integral $\int\frac{u+2}{u}du.$)

11. $\int\frac{dx}{\sqrt{16+4x-2x^2}} = \frac{1}{\sqrt{2}}\sin^{-1}\left(\frac{x-1}{3}\right) + C$

(Complete the square.)

13. $y = \sqrt{\frac{2}{3}}\tan t$, $dy = \sqrt{\frac{2}{3}}\sec^2 t\,dt$

$\int\frac{dy}{\sqrt{2+3y^2}} = \int\frac{\sqrt{\frac{2}{3}}\sec^2 t}{\sqrt{2}\sec t}dt$

$= \frac{1}{\sqrt{3}}\int\sec t\,dt = \frac{1}{\sqrt{3}}\ln|\sec t + \tan t| + C_1$

$= \frac{1}{\sqrt{3}}\ln\left|\frac{\sqrt{y^2+\frac{2}{3}}}{\sqrt{\frac{2}{3}}} + \frac{y}{\sqrt{\frac{2}{3}}}\right| + C_1$

$= \frac{1}{\sqrt{3}}\ln\left|\frac{\sqrt{y^2+\frac{2}{3}}+y}{\sqrt{\frac{2}{3}}}\right| + C_1$

$= \frac{1}{\sqrt{3}}\ln\left|\sqrt{y^2+\frac{2}{3}}+y\right| + C$

Note that $\tan t = \frac{y}{\sqrt{\frac{2}{3}}}$, so $\sec t = \frac{\sqrt{y^2+\frac{2}{3}}}{\sqrt{\frac{2}{3}}}$.

15. $\int\frac{\tan x}{\ln|\cos x|}dx = -\ln\left|\ln|\cos x|\right| + C$

Use the substitution $u = \ln|\cos x|$.

17. $\int\sinh x\,dx = \cosh x + C$

19. $u = x \quad dv = \cot^2 x\,dx$

$du = dx \quad v = -\cot x - x$

$\int x\cot^2 x\,dx = -x\cot x - x^2 - \int(-\cot x - x)dx$

$= -x\cot x - \frac{1}{2}x^2 + \ln|\sin x| + C$

Use $\cot^2 x = \csc^2 x - 1$ for $\int\cot^2 x\,dx$.

21. $u = \ln t^2$, $du = \frac{2}{t}dt$

$\int\frac{\ln t^2}{t}dt = \frac{[\ln(t^2)]^2}{4} + C$

23. $\int e^{t/3}\sin 3t\,dt = \dfrac{-3e^{t/3}(9\cos 3t - \sin 3t)}{82} + C$

Use integration by parts twice.

25. $\int \sin\dfrac{3x}{2}\cos\dfrac{x}{2}\,dx = -\dfrac{\cos x}{2} - \dfrac{\cos 2x}{4} + C$

Use a product identity.

27. $\int \tan^3 2x\sec 2x\,dx = \dfrac{1}{2}\int(\sec^2 2x - 1)d(\sec 2x)$

$= \dfrac{1}{6}\sec^3(2x) - \dfrac{1}{2}\sec(2x) + C$

29. $\int \tan^{3/2} x\sec^4 x\,dx = \int \tan^{3/2} x(1+\tan^2 x)\sec^2 x\,dx = \int \tan^{3/2} x\sec^2 x\,dx + \int \tan^{7/2} x\sec^2 x\,dx$

$= \dfrac{2}{5}\tan^{5/2} x + \dfrac{2}{9}\tan^{9/2} x + C$

31. $u = 9 - e^{2y},\ du = -2e^{2y}\,dy$

$\int \dfrac{e^{2y}}{\sqrt{9-e^{2y}}}dy = -\dfrac{1}{2}\int u^{-1/2}du = -\sqrt{u} + C = -\sqrt{9-e^{2y}} + C$

33. $\int e^{\ln(3\cos x)}dx = \int 3\cos x\,dx = 3\sin x + C$

35. $u = e^{4x},\ du = 4e^{4x}dx$

$\int \dfrac{e^{4x}}{1+e^{8x}}dx = \dfrac{1}{4}\int\dfrac{du}{1+u^2} = \dfrac{1}{4}\tan^{-1}(e^{4x}) + C$

37. $u = \sqrt{w+5},\ u^2 = w+5,\ 2u\,du = dw$

$\int \dfrac{w}{\sqrt{w+5}}dw = 2\int(u^2-5)du = \dfrac{2}{3}u^3 - 10u + C$

$= \dfrac{2}{3}(w+5)^{3/2} - 10(w+5)^{1/2} + C$

39. $u = \cos^2 y,\ du = -2\cos y\sin y\,dy$

$\int \dfrac{\sin y\cos y}{9+\cos^4 y}dy = -\dfrac{1}{2}\int\dfrac{du}{9+u^2}$

$= -\dfrac{1}{6}\tan^{-1}\left(\dfrac{\cos^2 y}{3}\right) + C$

41. $\dfrac{4x^2+3x+6}{x^2(x^2+3)} = \dfrac{A}{x} + \dfrac{B}{x^2} + \dfrac{Cx+D}{x^2+3}$

$A = 1,\ B = 2,\ C = -1,\ D = 2$

$\int \dfrac{4x^2+3x+6}{x^2(x^2+3)}dx = \int\dfrac{1}{x}dx + 2\int\dfrac{1}{x^2}dx + \int\dfrac{-x+2}{x^2+3}dx$

$= \int\dfrac{1}{x}dx + 2\int\dfrac{1}{x^2}dx - \dfrac{1}{2}\int\dfrac{2x}{x^2+3}dx + 2\int\dfrac{1}{x^2+3}dx$

$= \ln|x| - \dfrac{2}{x} - \dfrac{1}{2}\ln\left|x^2+3\right| + \dfrac{2}{\sqrt{3}}\tan^{-1}\left(\dfrac{x}{\sqrt{3}}\right) + C$

43. **a.** $\dfrac{3-4x^2}{(2x+1)^3} = \dfrac{A}{2x+1} + \dfrac{B}{(2x+1)^2} + \dfrac{C}{(2x+1)^3}$

b. $\dfrac{7x-41}{(x-1)^2(2-x)^3} = \dfrac{A}{x-1} + \dfrac{B}{(x-1)^2} + \dfrac{C}{2-x} + \dfrac{D}{(2-x)^2} + \dfrac{E}{(2-x)^3}$

c. $\dfrac{3x+1}{(x^2+x+10)^2} = \dfrac{Ax+B}{x^2+x+10} + \dfrac{Cx+D}{(x^2+x+10)^2}$

d. $\dfrac{(x+1)^2}{(x^2-x+10)^2(1-x^2)^2} = \dfrac{A}{1-x} + \dfrac{B}{(1-x)^2} + \dfrac{C}{1+x} + \dfrac{D}{(1+x)^2} + \dfrac{Ex+F}{x^2-x+10} + \dfrac{Gx+H}{(x^2-x+10)^2}$

e. $\dfrac{x^5}{(x+3)^4(x^2+2x+10)^2} = \dfrac{A}{x+3} + \dfrac{B}{(x+3)^2} + \dfrac{C}{(x+3)^3} + \dfrac{D}{(x+3)^4} + \dfrac{Ex+F}{x^2+2x+10} + \dfrac{Gx+H}{(x^2+2x+10)^2}$

f. $\dfrac{(3x^2+2x-1)^2}{(2x^2+x+10)^3} = \dfrac{Ax+B}{2x^2+x+10} + \dfrac{Cx+D}{(2x^2+x+10)^2} + \dfrac{Ex+F}{(2x^2+x+10)^3}$

45. $y = \frac{x^2}{16}, y' = \frac{x}{8}$

$$L = \int_0^4 \sqrt{1+\left(\frac{x}{8}\right)^2}\,dx = \int_0^4 \sqrt{1+\frac{x^2}{64}}\,dx$$

$x = 8\tan t,\ dx = 8\sec^2 t$

$$L = \int_0^{\tan^{-1}\frac{1}{2}} \sec t \cdot 8\sec^2 t\,dt = 8\int_0^{\tan^{-1}\frac{1}{2}} \sec^3 t\,dt = 4\left[\sec t\tan t + \ln|\sec t + \tan t|\right]_0^{\tan^{-1}\frac{1}{2}}$$

$$= 4\left[\left(\frac{\sqrt{5}}{2}\right)\left(\frac{1}{2}\right) + \ln\left|\frac{1}{2}+\frac{\sqrt{5}}{2}\right|\right] = \sqrt{5} + 4\ln\left(\frac{1+\sqrt{5}}{2}\right) \approx 4.1609$$

Note: Use Formula 28 for $\int \sec^3 t\,dt$.

47. $V = 2\pi\int_0^3 \frac{x}{x^2+5x+6}\,dx$

$$\frac{x}{x^2+5x+6} = \frac{A}{x+2} + \frac{B}{x+3}$$

$A = -2, B = 3$

$$V = 2\pi\int_0^3 \left[-\frac{2}{x+2} + \frac{3}{x+3}\right]dx = 2\pi\left[-2\ln(x+2) + 3\ln(x+3)\right]_0^3$$

$$= 2\pi[(-2\ln 5 + 3\ln 6) - (-2\ln 2 + 3\ln 3)] = 2\pi\left(3\ln 2 + 2\ln\frac{2}{5}\right) = 2\pi\ln\frac{32}{25} \approx 1.5511$$

49. $V = 2\pi\int_0^{\ln 3} 2(e^x - 1)(\ln 3 - x)dx = 4\pi\int_0^{\ln 3} [(\ln 3)e^x - xe^x - \ln 3 + x]dx$

Note that $\int xe^x dx = xe^x - \int e^x dx = xe^x - e^x + C$ by using integration by parts.

$$V = 4\pi\left[(\ln 3)e^x - xe^x + e^x - (\ln 3)x + \frac{1}{2}x^2\right]_0^{\ln 3} = 4\pi\left[\left(3\ln 3 - 3\ln 3 + 3 - (\ln 3)^2 + \frac{1}{2}(\ln 3)^2\right) - (\ln 3 + 1)\right]$$

$$= 4\pi\left[2 - \ln 3 - \frac{1}{2}(\ln 3)^2\right] \approx 3.7437$$

51. $A = -\int_{-6}^0 \frac{t}{(t-1)^2}\,dt$

$$\frac{t}{(t-1)^2} = \frac{A}{(t-1)} + \frac{B}{(t-1)^2}$$

$A = 1, B = 1$

$$A = -\int_{-6}^0 \left[\frac{1}{t-1} + \frac{1}{(t-1)^2}\right]dt = -\left[\ln|t-1| - \frac{1}{t-1}\right]_{-6}^0 = -\left[(0+1) - \left(\ln 7 + \frac{1}{7}\right)\right] = \ln 7 - \frac{6}{7} \approx 1.0888$$

53. The length is given by

$$\int_{\pi/6}^{\pi/3} \sqrt{1+[f'(x)]^2}\,dx = \int_{\pi/6}^{\pi/3} \sqrt{1+\frac{\cos^2 x}{\sin^2 x}}\,dx = \int_{\pi/6}^{\pi/3} \sqrt{\frac{\sin^2 x + \cos^2 x}{\sin^2 x}}\,dx = \int_{\pi/6}^{\pi/3} \frac{1}{\sin x}\,dx = \int_{\pi/6}^{\pi/3} \csc x\,dx$$

$$= \left[\ln|\csc x - \cot x|\right]_{\pi/6}^{\pi/3} = \ln\left|\frac{2}{\sqrt{3}} - \frac{1}{\sqrt{3}}\right| - \ln\left|2-\sqrt{3}\right| = \ln\left(\frac{1}{\sqrt{3}}\right) - \ln(2-\sqrt{3}) = \ln\left(\frac{2\sqrt{3}+3}{3}\right) \approx 0.768$$

55. a. First substitute $u = \sin x,\ du = \cos x\,dx$ to obtain $\int \cos x\sqrt{\sin^2 x + 4}\,dx = \int \sqrt{u^2+4}\,du$, then use Formula 44:

$$\int \cos x\sqrt{\sin^2 x + 4}\,dx = \frac{\sin x}{2}\sqrt{\sin^2 x + 4} + 2\ln\left|\sin x + \sqrt{\sin^2 x + 4}\right| + C$$

b. First substitute $u = 2x$, $du = 2dx$ to obtain $\int \frac{1}{1-4x^2}dx = \frac{1}{2}\int \frac{du}{1-u^2}$.

Then use Formula 18: $\int \frac{1}{1-4x^2}dx = \frac{1}{4}\ln\left|\frac{2x+1}{2x-1}\right| + C$.

57. Using partial fractions (see Section 7.6, prob 46 b.):

$$\frac{1}{1+x^3} = \frac{1}{(x+1)(x^2-x+1)} = \frac{A}{x+1} + \frac{Bx+C}{x^2-x+1} = \frac{(A+B)x^2 + (B+C-A)x + (A+C)}{(x+1)(x^2-x+1)} \Rightarrow$$

$A + C = 1 \quad B + C = A \quad A = -B \quad \Rightarrow A = \frac{1}{3} \quad B = -\frac{1}{3} \quad C = \frac{2}{3}.$

Therefore:

$$\int \frac{1}{1+x^3}dx = \frac{1}{3}\left[\int \frac{1}{x+1}dx - \int \frac{x-2}{x^2-x+1}dx\right] = \frac{1}{3}\left[\ln|x+1| - \underbrace{\int \frac{x-2}{(x-\frac{1}{2})^2 + \frac{3}{4}}dx}_{u = x-\frac{1}{2},\ du = dx}\right]$$

$$= \left[\ln|x+1| - \int \frac{u - \frac{3}{2}}{u^2 + \frac{3}{4}}du\right] \underset{F17}{=} \frac{1}{3}\left[\ln\left|\frac{x+1}{\sqrt{x^2-x+1}}\right| + \sqrt{3}\tan^{-1}\left(\frac{2}{\sqrt{3}}(x - \tfrac{1}{2})\right)\right]$$

so $\int_0^c \frac{1}{1+x^3}dx = \frac{1}{3}\left[\ln\left|\frac{c+1}{\sqrt{c^2-c+1}}\right| + \sqrt{3}\left[\tan^{-1}\left(\frac{2}{\sqrt{3}}(c - \tfrac{1}{2})\right) + \frac{\pi}{6}\right]\right]$.

Letting $G(c) = \frac{1}{3}\left[\ln\left|\frac{c+1}{\sqrt{c^2-c+1}}\right| + \sqrt{3}\left[\tan^{-1}\left(\frac{2}{\sqrt{3}}(c - \tfrac{1}{2})\right) + \frac{\pi}{6}\right]\right] - 0.5$ and $G'(c) = \frac{1}{1+c^3}$ we apply Newton's Method to find the value of c such that $\int_0^c \frac{1}{1+x^3}dx = 0.5$:

n	1	2	3	4	5	6
a_n	1.0000	0.3287	0.5090	0.5165	0.5165	0.5165

Thus $c \approx 0.5165$.

Review and Preview Problems

1. $\lim_{x\to 2} \frac{x^2+1}{x^2-1} = \frac{2^2+1}{2^2-1} = \frac{5}{3}$

3. $\lim_{x\to 3} \frac{x^2-9}{x-3} = \lim_{x\to 3} \frac{(x+3)(x-3)}{x-3} =$
$\lim_{x\to 3}(x+3) = 3+3 = 6$

5. $\lim_{x\to 0} \frac{\sin 2x}{x} = \lim_{x\to 0} \frac{2\sin x\cos x}{x} =$
$\lim_{x\to 0} 2\left(\frac{\sin x}{x}\right)\cos x = 2(1)(1) = 2$

7. $\lim_{x\to\infty} \frac{x^2+1}{x^2-1} = \lim_{x\to\infty} \frac{1+\frac{1}{x^2}}{1-\frac{1}{x^2}} = \frac{1+0}{1-0} = 1$ or:

$\lim_{x\to\infty} \frac{x^2+1}{x^2-1} = \lim_{x\to\infty} 1 + \frac{2}{x^2-1} = 1+0 = 1$

9. $\lim_{x\to\infty} e^{-x} = \lim_{x\to\infty} \frac{1}{e^x} = 0$

11. $\lim_{x\to\infty} e^{2x} = \infty$ (has no finite value)

13. $\lim_{x\to\infty} \tan^{-1} x = \frac{\pi}{2}$

15. $f(x) = xe^{-x}$

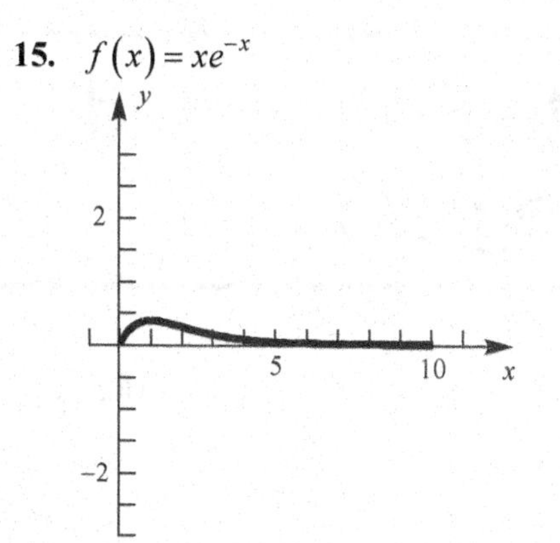

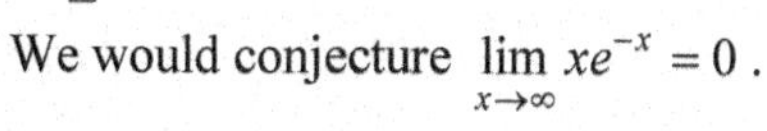

We would conjecture $\lim_{x\to\infty} xe^{-x} = 0$.

17. $f(x) = x^3e^{-x}$

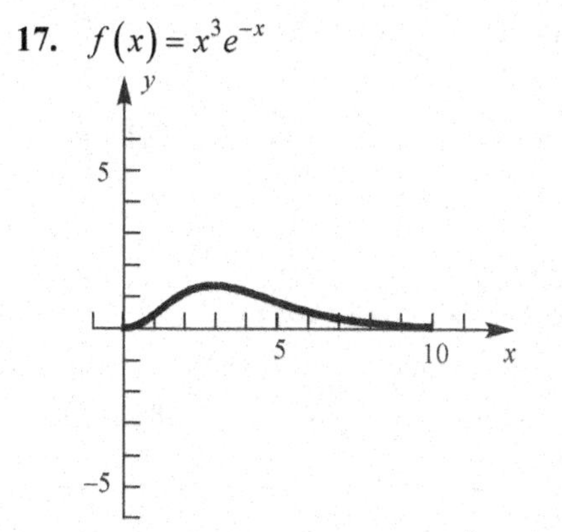

We would conjecture $\lim_{x\to\infty} x^3e^{-x} = 0$.

19. $y = x^{10}e^{-x}$

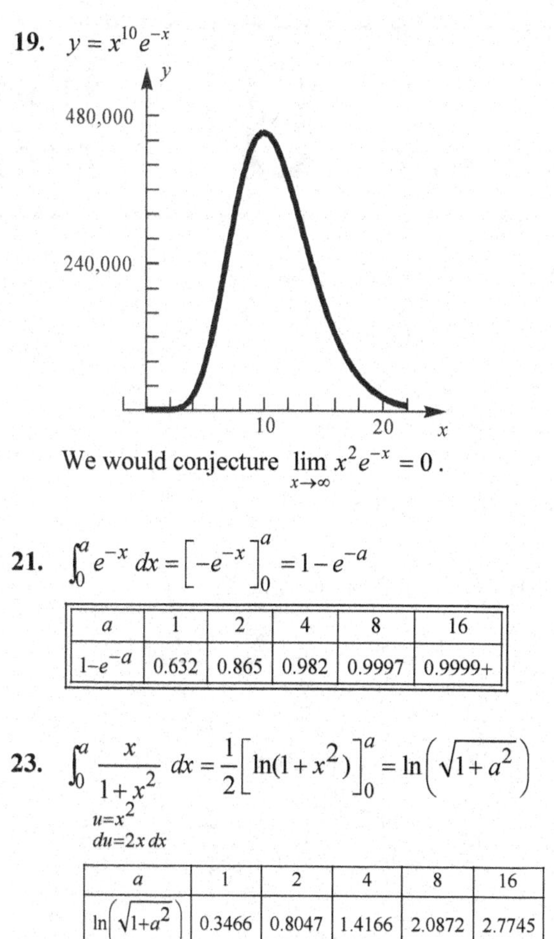

We would conjecture $\lim_{x\to\infty} x^2e^{-x} = 0$.

21. $\int_0^a e^{-x}\,dx = \left[-e^{-x}\right]_0^a = 1 - e^{-a}$

a	1	2	4	8	16
$1-e^{-a}$	0.632	0.865	0.982	0.9997	0.9999+

23. $\int_0^a \frac{x}{1+x^2}\,dx = \frac{1}{2}\left[\ln(1+x^2)\right]_0^a = \ln\left(\sqrt{1+a^2}\right)$

$u = x^2$
$du = 2x\,dx$

a	1	2	4	8	16
$\ln\left(\sqrt{1+a^2}\right)$	0.3466	0.8047	1.4166	2.0872	2.7745

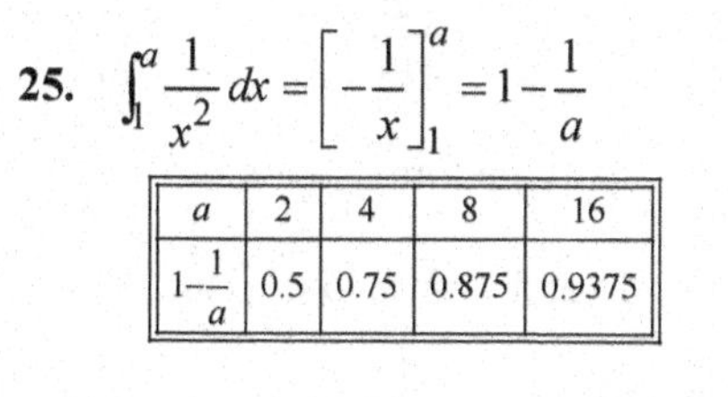

25. $\int_1^a \frac{1}{x^2}\,dx = \left[-\frac{1}{x}\right]_1^a = 1 - \frac{1}{a}$

a	2	4	8	16
$1-\frac{1}{a}$	0.5	0.75	0.875	0.9375

27. $\int_a^4 \frac{1}{\sqrt{x}}\,dx = \left[2\sqrt{x}\right]_a^4 = 4 - 2\sqrt{a}$

a	1	$\frac{1}{2}$	$\frac{1}{4}$	$\frac{1}{8}$	$\frac{1}{16}$
$4-2\sqrt{a}$	2	2.58579	3	3.29289	3.5

CHAPTER 8	Indeterminate Forms and Improper Integrals

8.1 Concepts Review

1. $\lim_{x\to a} f(x);\ \lim_{x\to a} g(x)$

3. $\sec^2 x$; 1; $\lim_{x\to 0} \cos x \neq 0$

Problem Set 8.1

1. The limit is of the form $\frac{0}{0}$.

$$\lim_{x\to 0}\frac{2x-\sin x}{x}=\lim_{x\to 0}\frac{2-\cos x}{1}=1$$

3. The limit is of the form $\frac{0}{0}$.

$$\lim_{x\to 0}\frac{x-\sin 2x}{\tan x}=\lim_{x\to 0}\frac{1-2\cos 2x}{\sec^2 x}=\frac{1-2}{1}=-1$$

5. The limit is of the form $\frac{0}{0}$.

$$\lim_{x\to -2}\frac{x^2+6x+8}{x^2-3x-10}=\lim_{x\to -2}\frac{2x+6}{2x-3}$$
$$=\frac{2}{-7}=-\frac{2}{7}$$

7. The limit is not of the form $\frac{0}{0}$.

As $x\to 1^-, x^2-2x+2\to 1$, and $x^2-1\to 0^-$ so

$$\lim_{x\to 1^-}\frac{x^2-2x+2}{x^2+1}=-\infty$$

9. The limit is of the form $\frac{0}{0}$.

$$\lim_{x\to\pi/2}\frac{\ln(\sin x)^3}{\pi/2-x}=\lim_{x\to\pi/2}\frac{\frac{1}{\sin^3 x}3\sin^2 x\cos x}{-1}$$
$$=\frac{0}{-1}=0$$

11. The limit is of the form $\frac{0}{0}$.

$$\lim_{t\to 1}\frac{\sqrt{t}-t^2}{\ln t}=\lim_{t\to 1}\frac{\frac{1}{2\sqrt{t}}-2t}{\frac{1}{t}}=\frac{-\frac{3}{2}}{1}=-\frac{3}{2}$$

13. The limit is of the form $\frac{0}{0}$. (Apply l'Hôpital's Rule twice.)

$$\lim_{x\to 0}\frac{\ln\cos 2x}{7x^2}=\lim_{x\to 0}\frac{\frac{-2\sin 2x}{\cos 2x}}{14x}=\lim_{x\to 0}\frac{-2\sin 2x}{14x\cos 2x}$$
$$=\lim_{x\to 0}\frac{-4\cos 2x}{14\cos 2x-28x\sin 2x}=\frac{-4}{14-0}=-\frac{2}{7}$$

15. The limit is of the form $\frac{0}{0}$. (Apply l'Hôpital's Rule three times.)

$$\lim_{x\to 0}\frac{\tan x-x}{\sin 2x-2x}=\lim_{x\to 0}\frac{\sec^2 x-1}{2\cos 2x-2}$$
$$=\lim_{x\to 0}\frac{2\sec^2 x\tan x}{-4\sin 2x}=\lim_{x\to 0}\frac{2\sec^4 x+4\sec^2 x\tan^2 x}{-8\cos 2x}$$
$$=\frac{2+0}{-8}=-\frac{1}{4}$$

17. The limit is of the form $\frac{0}{0}$. (Apply l'Hôpital's Rule twice.)

$$\lim_{x\to 0^+}\frac{x^2}{\sin x-x}=\lim_{x\to 0^+}\frac{2x}{\cos x-1}=\lim_{x\to 0^+}\frac{2}{-\sin x}$$

This limit is not of the form $\frac{0}{0}$. As $x\to 0^+, 2\to 2$, and $-\sin x\to 0^-$, so

$$\lim_{x\to 0^+}\frac{2}{\sin x}=-\infty.$$

19. The limit is of the form $\frac{0}{0}$. (Apply l'Hôpital's Rule twice.)

$$\lim_{x\to 0}\frac{\tan^{-1}x-x}{8x^3}=\lim_{x\to 0}\frac{\frac{1}{1+x^2}-1}{24x^2}=\lim_{x\to 0}\frac{\frac{-2x}{(1+x^2)^2}}{48x}$$

$$=\lim_{x\to 0}-\frac{1}{24(1+x^2)^2}=-\frac{1}{24}$$

21. The limit is of the form $\frac{0}{0}$. (Apply l'Hôpital's Rule twice.)

$$\lim_{x\to 0^+}\frac{1-\cos x-x\sin x}{2-2\cos x-\sin^2 x}$$

$$=\lim_{x\to 0^+}\frac{-x\cos x}{2\sin x-2\cos x\sin s}$$

$$=\lim_{x\to 0^+}\frac{x\sin x-\cos x}{2\cos x-2\cos^2 x+2\sin^2 x}$$

This limit is not of the form $\frac{0}{0}$.

As $x\to 0^+, x\sin x-\cos x\to -1$ and $2\cos x-2\cos^2 x+2\sin^2 x\to 0^+$, so

$$\lim_{x\to 0^+}\frac{x\sin x-\cos x}{2\cos x-2\cos^2 x+2\sin^2 x}=-\infty$$

23. The limit is of the form $\frac{0}{0}$.

$$\lim_{x\to 0}\frac{\int_0^x\sqrt{1+\sin t}\,dt}{x}=\lim_{x\to 0}\sqrt{1+\sin x}=1$$

25. It would not have helped us because we proved $\lim_{x\to 0}\frac{\sin x}{x}=1$ in order to find the derivative of $\sin x$.

27. a. $\overline{OB}=\cos t, \overline{BC}=\sin t$ and $\overline{AB}=1-\cos t$, so the area of triangle ABC is $\frac{1}{2}\sin t(1-\cos t)$.

The area of the sector COA is $\frac{1}{2}t$ while the area of triangle COB is $\frac{1}{2}\cos t\sin t$, thus the area of the curved region ABC is $\frac{1}{2}(t-\cos t\sin t)$.

$$\lim_{t\to 0^+}\frac{\text{area of triangle } ABC}{\text{area of curved region } ABC}=\lim_{t\to 0^+}\frac{\frac{1}{2}\sin t(1-\cos t)}{\frac{1}{2}(t-\cos t\sin t)}$$

$$=\lim_{t\to 0^+}\frac{\sin t(1-\cos t)}{t-\cos t\sin t}=\lim_{t\to 0^+}\frac{\cos t-\cos^2 t+\sin^2 t}{1-\cos^2 t+\sin^2 t}=\lim_{t\to 0^+}\frac{4\sin t\cos t-\sin t}{4\cos t\sin t}=\lim_{t\to 0^+}\frac{4\cos t-1}{4\cos t}=\frac{3}{4}$$

(L'Hôpital's Rule was applied twice.)

b. The area of the sector BOD is $\frac{1}{2}t\cos^2 t$, so the area of the curved region BCD is $\frac{1}{2}\cos t\sin t-\frac{1}{2}t\cos^2 t$.

$$\lim_{t\to 0^+}\frac{\text{area of curved region } BCD}{\text{area of curved region } ABC}=\lim_{t\to 0^+}\frac{\frac{1}{2}\cos t(\sin t-t\cos t)}{\frac{1}{2}(t-\cos t\sin t)}$$

$$=\lim_{t\to 0^+}\frac{\cos t(\sin t-t\cos t)}{t-\sin t\cos t}=\lim_{t\to 0^+}\frac{\sin t(2t\cos t-\sin t)}{1-\cos^2 t+\sin^2 t}=\lim_{t\to 0^+}\frac{2t(\cos^2 t-\sin^2 t)}{4\cos t\sin t}=\lim_{t\to 0^+}\frac{t(\cos^2 t-\sin^2 t)}{2\cos t\sin t}$$

$$=\lim_{t\to 0^+}\frac{\cos^2 t-4t\cos t\sin t-\sin^2 t}{2\cos^2 t-2\sin^2 t}=\frac{1-0-0}{2-0}=\frac{1}{2}$$

(L'Hôpital's Rule was applied three times.)

29. By l'Hôpital's Rule$\left(\frac{0}{0}\right)$, we have $\lim_{x\to 0^+} f(x) = \lim_{x\to 0^+} \frac{e^x - 1}{x} = \lim_{x\to 0^+} \frac{e^x}{1} = 1$ and

$\lim_{x\to 0^-} f(x) = \lim_{x\to 0^-} \frac{e^x - 1}{x} = \lim_{x\to 0^-} \frac{e^x}{1} = 1$ so we define $f(0) = 1$.

31. A should approach $4\pi b^2$, the surface area of a sphere of radius b.

$$\lim_{a\to b^+}\left[2\pi b^2 + \frac{2\pi a^2 b \arcsin \frac{\sqrt{a^2-b^2}}{a}}{\sqrt{a^2-b^2}}\right] = 2\pi b^2 + 2\pi b \lim_{a\to b^+} \frac{a^2 \arcsin \frac{\sqrt{a^2-b^2}}{a}}{\sqrt{a^2-b^2}}$$

Focusing on the limit, we have

$$\lim_{a\to b^+} \frac{a^2 \arcsin \frac{\sqrt{a^2-b^2}}{a}}{\sqrt{a^2-b^2}} = \lim_{a\to b^+} \frac{2a \arcsin \frac{\sqrt{a^2-b^2}}{a} + a^2\left(\frac{b}{a\sqrt{a^2-b^2}}\right)}{\frac{a}{\sqrt{a^2-b^2}}} = \lim_{a\to b^+}\left(2\sqrt{a^2-b^2}\arcsin\frac{\sqrt{a^2-b^2}}{a} + b\right) = b.$$

Thus, $\lim_{a\to b^+} A = 2\pi b^2 + 2\pi b(b) = 4\pi b^2$.

33. If $f'(a)$ and $g'(a)$ both exist, then f and g are both continuous at a. Thus, $\lim_{x\to a} f(x) = 0 = f(a)$ and $\lim_{x\to a} g(x) = 0 = g(a)$.

$$\lim_{x\to a} \frac{f(x)}{g(x)} = \lim_{x\to a} \frac{f(x) - f(a)}{g(x) - g(a)}$$

$$\lim_{x\to a} \frac{\frac{f(x)-f(a)}{x-a}}{\frac{g(x)-g(a)}{x-a}} = \frac{\lim_{x\to a} \frac{f(x)-f(a)}{x-a}}{\lim_{x\to a} \frac{g(x)-g(a)}{x-a}} = \frac{f'(a)}{g'(a)}$$

35. $\lim_{x\to 0} \frac{e^x - 1 - x - \frac{x^2}{2} - \frac{x^3}{6}}{x^4} = \frac{1}{24}$

37. $\lim_{x\to 0} \frac{\tan x - x}{\arcsin x - x} = \lim_{x\to 0} \frac{\sec^2 x - 1}{\frac{1}{\sqrt{1-x^2}} - 1} = 2$

39.

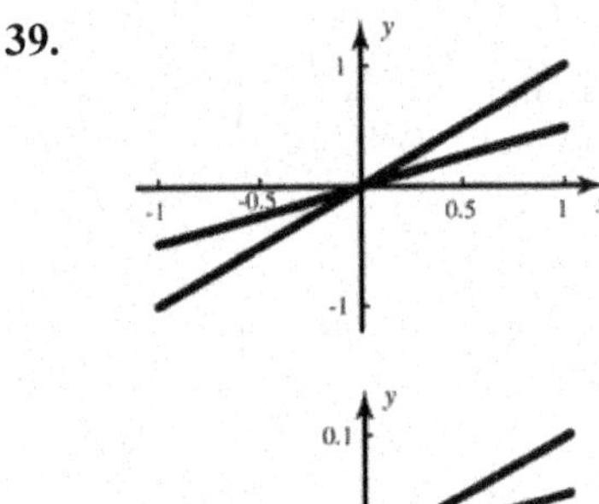

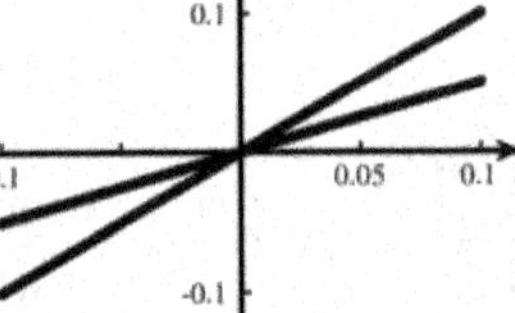

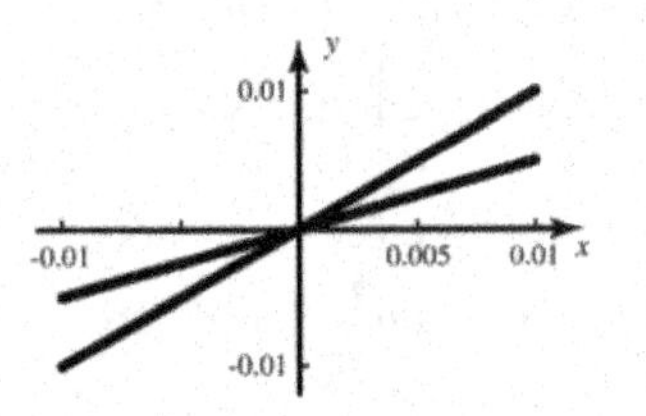

The slopes are approximately $0.005/0.01 = 1/2$ and $0.01/0.01 = 1$. The ratio of the slopes is therefore $1/2$, indicating that the limit of the ratio should be about $1/2$. An application of l'Hopital's Rule confirms this.

41.

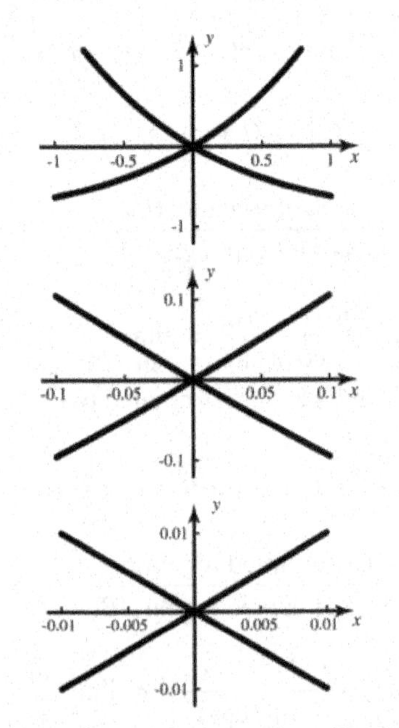

The slopes are approximately $0.01/0.01 = 1$ and $-0.01/0.01 = 1$. The ratio of the slopes is therefore $-1/1 = -1$, indicating that the limit of the ratio should be about -1. An application of l'Hopital's Rule confirms this.

8.2 Concepts Review

1. $\frac{f'(x)}{g'(x)}$

3. $\infty-\infty, 0^\circ, \infty^\circ, 1^\infty$

Problem Set 8.2

1. The limit is of the form $\frac{\infty}{\infty}$.

$$\lim_{x\to\infty}\frac{\ln x^{1000}}{x}=\lim_{x\to\infty}\frac{\frac{1}{x^{1000}}1000x^{999}}{1}$$
$$=\lim_{x\to\infty}\frac{1000}{x}=0$$

3. $\lim_{x\to\infty}\frac{x^{10000}}{e^x}=0$ (See Example 2).

5. The limit is of the form $\frac{\infty}{\infty}$.

$$\lim_{x\to\frac{\pi}{2}}\frac{3\sec x+5}{\tan x}=\lim_{x\to\frac{\pi}{2}}\frac{3\sec x\tan x}{\sec^2 x}$$
$$=\lim_{x\to\frac{\pi}{2}}\frac{3\tan x}{\sec x}=\lim_{x\to\frac{\pi}{2}}3\sin x=3$$

7. The limit is of the form $\frac{\infty}{\infty}$.

$$\lim_{x\to\infty}\frac{\ln(\ln x^{1000})}{\ln x}=\lim_{x\to\infty}\frac{\frac{1}{\ln x^{1000}}\left(\frac{1}{x^{1000}}1000x^{999}\right)}{\frac{1}{x}}$$
$$=\lim_{x\to\infty}\frac{1000}{x\ln x^{1000}}=0$$

9. The limit is of the form $\frac{\infty}{\infty}$.

$$\lim_{x\to0^+}\frac{\cot x}{\sqrt{-\ln x}}=\lim_{x\to0^+}\frac{-\csc^2 x}{-\frac{1}{2x\sqrt{-\ln x}}}$$
$$=\lim_{x\to0^+}\frac{2x\sqrt{-\ln x}}{\sin^2 x}$$
$$=\lim_{x\to0^+}\left[\frac{2x}{\sin x}\csc x\sqrt{-\ln x}\right]=\infty$$

since $\lim_{x\to0^+}\frac{x}{\sin x}=1$ while $\lim_{x\to0^+}\csc x=\infty$ and $\lim_{x\to0^+}\sqrt{-\ln x}=\infty$.

11. $\lim_{x\to0}(x\ln x^{1000})=\lim_{x\to0}\frac{\ln x^{1000}}{\frac{1}{x}}$

The limit is of the form $\frac{\infty}{\infty}$.

$$\lim_{x\to0}\frac{\ln x^{1000}}{\frac{1}{x}}=\lim_{x\to0}\frac{\frac{1}{x^{1000}}1000x^{999}}{-\frac{1}{x^2}}$$
$$=\lim_{x\to0}-1000x=0$$

13. $\lim_{x\to0}(\csc^2 x-\cot^2 x)=\lim_{x\to0}\frac{1-\cos^2 x}{\sin^2 x}$

$$=\lim_{x\to0}\frac{\sin^2 x}{\sin^2 x}=1$$

15. The limit is of the form 0^0.

Let $y=(3x)^{x^2}$, then $\ln y=x^2\ln 3x$

$$\lim_{x\to0^+}x^2\ln 3x=\lim_{x\to0^+}\frac{\ln 3x}{\frac{1}{x^2}}$$

The limit is of the form $\frac{\infty}{\infty}$.

$$\lim_{x\to0^+}\frac{\ln 3x}{\frac{1}{x^2}}=\lim_{x\to0^+}\frac{\frac{1}{3x}\cdot 3}{-\frac{2}{x^3}}=\lim_{x\to0^+}-\frac{x^2}{2}=0$$

$$\lim_{x\to0^+}(3x)^{x^2}=\lim_{x\to0^+}e^{\ln y}=1$$

17. The limit is of the form 0^∞, which is not an indeterminate form. $\lim_{x\to(\pi/2)^-}(5\cos x)^{\tan x}=0$

19. The limit is of the form 1^∞.

Let $y=(x+e^{x/3})^{3/x}$, then $\ln y=\frac{3}{x}\ln(x+e^{x/3})$.

$$\lim_{x\to0}\frac{3}{x}\ln(x+e^{x/3})=\lim_{x\to0}\frac{3\ln(x+e^{x/3})}{x}$$

The limit is of the form $\frac{0}{0}$.

$$\lim_{x\to0}\frac{3\ln(x+e^{x/3})}{x}=\lim_{x\to0}\frac{\frac{3}{x+e^{x/3}}\left(1+\frac{1}{3}e^{x/3}\right)}{1}$$
$$=\lim_{x\to0}\frac{3+e^{x/3}}{x+e^{x/3}}=\frac{4}{1}=4$$

$$\lim_{x\to0}(x+e^{x/3})^{3/x}=\lim_{x\to0}e^{\ln y}=e^4$$

21. The limit is of the form 1^0, which is not an indeterminate form.

$$\lim_{x\to\frac{\pi}{2}} (\sin x)^{\cos x} = 1$$

23. The limit is of the form ∞^0. Let $y = x^{1/x}$, then $\ln y = \frac{1}{x}\ln x$.

$$\lim_{x\to\infty} \frac{1}{x}\ln x = \lim_{x\to\infty} \frac{\ln x}{x}$$

The limit is of the form $\frac{-\infty}{\infty}$.

$$\lim_{x\to\infty} \frac{\ln x}{x} = \lim_{x\to\infty} \frac{\frac{1}{x}}{1} = \lim_{x\to\infty} \frac{1}{x} = 0$$

$$\lim_{x\to\infty} x^{1/x} = \lim_{x\to\infty} e^{\ln y} = 1$$

25. The limit is of the form 0^∞, which is not an indeterminate form.

$$\lim_{x\to 0^+} (\tan x)^{2/x} = 0$$

27. The limit is of the form 0^0. Let $y = (\sin x)^x$, then $\ln y = x\ln(\sin x)$.

$$\lim_{x\to 0^+} x\ln(\sin x) = \lim_{x\to 0^+} \frac{\ln(\sin x)}{\frac{1}{x}}$$

The limit is of the form $\frac{-\infty}{\infty}$.

$$\lim_{x\to 0^+} \frac{\ln(\sin x)}{\frac{1}{x}} = \lim_{x\to 0^+} \frac{\frac{1}{\sin x}\cos x}{-\frac{1}{x^2}}$$

$$= \lim_{x\to 0^+} \left[\frac{x}{\sin x}(-x\cos x)\right] = 1\cdot 0 = 0$$

$$\lim_{x\to 0^+} (\sin x)^x = \lim_{x\to 0^+} e^{\ln y} = 1$$

29. The limit is of the form $\infty - \infty$.

$$\lim_{x\to 0}\left(\csc x - \frac{1}{x}\right) = \lim_{x\to 0}\left(\frac{1}{\sin x} - \frac{1}{x}\right) = \lim_{x\to 0} \frac{x - \sin x}{x\sin x}$$

The limit is of the form $\frac{0}{0}$. (Apply l'Hôpital's Rule twice.)

$$\lim_{x\to 0} \frac{x - \sin x}{x\sin x} = \lim_{x\to 0} \frac{1 - \cos x}{\sin x + x\cos x}$$

$$= \lim_{x\to 0} \frac{\sin x}{2\cos x - x\sin x} = \frac{0}{2} = 0$$

31. The limit is of the form 3^∞, which is not an indeterminate form.

$$\lim_{x\to 0^+} (1 + 2e^x)^{1/x} = \infty$$

33. The limit is of the form 1^∞.

Let $y = (\cos x)^{1/x}$, then $\ln y = \frac{1}{x}\ln(\cos x)$.

$$\lim_{x\to 0} \frac{1}{x}\ln(\cos x) = \lim_{x\to 0} \frac{\ln(\cos x)}{x}$$

The limit is of the form $\frac{0}{0}$.

$$\lim_{x\to 0} \frac{\ln(\cos x)}{x} = \lim_{x\to 0} \frac{\frac{1}{\cos x}(-\sin x)}{1} = \lim_{x\to 0} -\frac{\sin x}{\cos x} = 0$$

$$\lim_{x\to 0} (\cos x)^{1/x} = \lim_{x\to 0} e^{\ln y} = 1$$

35. Since $\cos x$ oscillates between -1 and 1 as $x \to \infty$, this limit is not of an indeterminate form previously seen.
Let $y = e^{\cos x}$, then $\ln y = (\cos x)\ln e = \cos x$

$\lim_{x\to\infty} \cos x$ does not exist, so $\lim_{x\to\infty} e^{\cos x}$ does not exist.

37. The limit is of the form $\frac{0}{-\infty}$, which is not an indeterminate form.

$$\lim_{x\to 0^+} \frac{x}{\ln x} = 0$$

39. $\sqrt{1+e^{-t}} > 1$ for all t, so $\int_1^x \sqrt{1+e^{-t}}\,dt > \int_1^x dt = x - 1$.

The limit is of the form $\frac{\infty}{\infty}$.

$$\lim_{x\to\infty} \frac{\int_1^x \sqrt{1+e^{-t}}\,dt}{x} = \lim_{x\to\infty} \frac{\sqrt{1+e^{-x}}}{1} = 1$$

41. a. Let $y = \sqrt[n]{a}$, then $\ln y = \frac{1}{n}\ln a$.

$$\lim_{n\to\infty} \frac{1}{n}\ln a = 0$$

$$\lim_{n\to\infty} \sqrt[n]{a} = \lim_{n\to\infty} e^{\ln y} = 1$$

b. The limit is of the form ∞^0.

Let $y = \sqrt[n]{n}$, then $\ln y = \frac{1}{n}\ln n$.

$$\lim_{n\to\infty} \frac{1}{n}\ln n = \lim_{n\to\infty} \frac{\ln n}{n}$$

This limit is of the form $\frac{\infty}{\infty}$.

$$\lim_{n\to\infty} \frac{\ln n}{n} = \lim_{n\to\infty} \frac{\frac{1}{n}}{1} = 0$$

$$\lim_{n\to\infty} \sqrt[n]{n} = \lim_{n\to\infty} e^{\ln y} = 1$$

c. $\lim_{n\to\infty} n\left(\sqrt[n]{a} - 1\right) = \lim_{n\to\infty} \frac{\sqrt[n]{a} - 1}{\frac{1}{n}}$

This limit is of the form $\frac{0}{0}$,

since $\lim_{n\to\infty} \sqrt[n]{a} = 1$ by part a.

$$\lim_{n\to\infty} \frac{\sqrt[n]{a} - 1}{\frac{1}{n}} = \lim_{n\to\infty} \frac{-\frac{1}{n^2}\sqrt[n]{a}\ln a}{-\frac{1}{n^2}}$$

$$= \lim_{n\to\infty} \sqrt[n]{a}\ln a = \ln a$$

d. $\lim_{n\to\infty} n\left(\sqrt[n]{n} - 1\right) = \lim_{n\to\infty} \frac{\sqrt[n]{n} - 1}{\frac{1}{n}}$

This limit is of the form $\frac{0}{0}$,

since $\lim_{n\to\infty} \sqrt[n]{n} = 1$ by part b.

$$\lim_{n\to\infty} \frac{\sqrt[n]{n} - 1}{\frac{1}{n}} = \lim_{n\to\infty} \frac{\sqrt[n]{n}\left(\frac{1}{n^2}\right)(1 - \ln n)}{-\frac{1}{n^2}}$$

$$= \lim_{n\to\infty} \sqrt[n]{n}(\ln n - 1) = \infty$$

43.

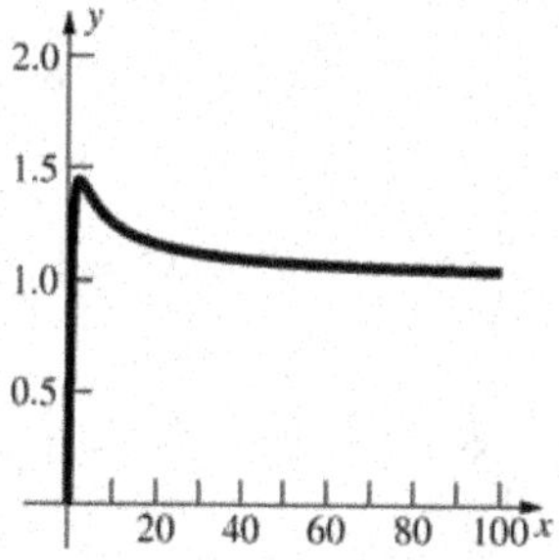

$$\ln y = \frac{\ln x}{x}$$

$$\lim_{x\to 0^+} \frac{\ln x}{x} = -\infty, \text{ so } \lim_{x\to 0^+} x^{1/x} = \lim_{x\to 0^+} e^{\ln y} = 0$$

$$\lim_{x\to\infty} \frac{\ln x}{x} = \lim_{x\to\infty} \frac{\frac{1}{x}}{1} = 0, \text{ so } \lim_{x\to\infty} x^{1/x} = \lim_{x\to\infty} e^{\ln y} = 1$$

$$y = x^{1/x} = e^{\frac{1}{x}\ln x}$$

$$y' = \left(\frac{1}{x^2} - \frac{\ln x}{x^2}\right)e^{\frac{1}{x}\ln x}$$

$y' = 0$ when $x = e$.

y is maximum at $x = e$ since $y' > 0$ on $(0, e)$ and $y' < 0$ on (e, ∞). When $x = e$, $y = e^{1/e}$.

45. $\lim_{n\to\infty} \frac{1^k + 2^k + \cdots + n^k}{n^{k+1}}$

$$= \lim_{n\to\infty} \frac{1}{n}\left[\left(\frac{1}{n}\right)^k + \left(\frac{2}{n}\right)^k + \cdots \left(\frac{n}{n}\right)^k\right]$$

$$= \lim_{n\to\infty} \sum_{i=1}^{n} \frac{1}{n}\cdot\left(\frac{i}{n}\right)^k$$

The summation has the form of a Reimann sum for $f(x) = x^k$ on the interval $[0,1]$ using a regular partition and evaluating the function at each right endpoint. Thus, $\Delta x_i = \frac{1}{n}$, $\overline{x}_i = \frac{i}{n}$, and $f(\overline{x}_i) = \left(\frac{i}{n}\right)^k$. Therefore,

$$\lim_{n\to\infty} \frac{1^k + 2^k + \cdots + n^k}{n^{k+1}} = \lim_{n\to\infty} \sum_{i=1}^{n} \frac{1}{n}\cdot\left(\frac{i}{n}\right)^k$$

$$= \int_0^1 x^k\,dx = \left[\frac{1}{k+1}x^{k+1}\right]_0^1$$

$$= \frac{1}{k+1}$$

47. a. $\lim_{t\to 0^+}\left(\frac{1}{2}2^t+\frac{1}{2}5^t\right)^{1/t}=\sqrt{2}\sqrt{5}\approx 3.162$

b. $\lim_{t\to 0^+}\left(\frac{1}{5}2^t+\frac{4}{5}5^t\right)^{1/t}=\sqrt[5]{2}\cdot\sqrt[5]{5^4}\approx 4.163$

c. $\lim_{t\to 0^+}\left(\frac{1}{10}2^t+\frac{9}{10}5^t\right)^{1/t}=\sqrt[10]{2}\cdot\sqrt[10]{5^9}\approx 4.562$

49. Note $f(x) > 0$ on $[0, \infty)$.

$$\lim_{x\to\infty} f(x)=\lim_{x\to\infty}\left(\frac{x^{25}}{e^x}+\frac{x^3}{e^x}+\left(\frac{2}{e}\right)^x\right)=0$$

Therefore there is no absolute minimum.

$$f'(x)=(25x^{24}+3x^2+2^x\ln 2)e^{-x}$$
$$-(x^{25}+x^3+2^x)e^{-x}$$
$$=(-x^{25}+25x^{24}-x^3+3x^2-2^x+2^x\ln 2)e^{-x}$$

Solve for x when $f'(x)=0$. Using a numerical method, $x\approx 25$.
A graph using a computer algebra system verifies that an absolute maximum occurs at about $x = 25$.

8.3 Concepts Review

1. converge

3. $\int_{-\infty}^{0} f(x)dx;\ \int_{0}^{\infty} f(x)dx$

Problem Set 8.3

In this section and the chapter review, it is understood that $[g(x)]_a^{\infty}$ means $\lim_{b\to\infty}[g(x)]_a^b$ and likewise for similar expressions.

1. $\int_{100}^{\infty} e^x dx=\left[e^x\right]_{100}^{\infty}=\infty-e^{100}=\infty$
The integral diverges.

3. $\int_{1}^{\infty} 2xe^{-x^2}dx=\left[-e^{-x^2}\right]_1^{\infty}=0-(-e^{-1})=\frac{1}{e}$

5. $\int_{9}^{\infty}\frac{x\,dx}{\sqrt{1+x^2}}=\left[\sqrt{1+x^2}\right]_9^{\infty}=\infty-\sqrt{82}=\infty$
The integral diverges.

7. $\int_{1}^{\infty}\frac{dx}{x^{1.00001}}=\left[-\frac{1}{0.00001x^{0.00001}}\right]_1^{\infty}$

$$=0-\left(-\frac{1}{0.00001}\right)=\frac{1}{0.00001}=100{,}000$$

9. $\int_{1}^{\infty}\frac{dx}{x^{0.99999}}=\left[\frac{x^{0.00001}}{0.00001}\right]_1^{\infty}=\infty-100{,}000=\infty$
The integral diverges.

11. $\int_{e}^{\infty}\frac{1}{x\ln x}dx=[\ln(\ln x)]_e^{\infty}=\infty-0=\infty$
The integral diverges.

13. Let $u=\ln x,\ du=\frac{1}{x}dx,\ dv=\frac{1}{x^2}dx,\ v=-\frac{1}{x}.$

$$\int_2^{\infty}\frac{\ln x}{x^2}dx=\lim_{b\to\infty}\int_2^b\frac{\ln x}{x^2}dx$$
$$=\lim_{b\to\infty}\left[-\frac{\ln x}{x}\right]_2^b+\lim_{b\to\infty}\int_2^b\frac{1}{x^2}dx$$
$$=\lim_{b\to\infty}\left[-\frac{\ln x}{x}-\frac{1}{x}\right]_2^b=\frac{\ln 2+1}{2}$$

15. $\int_{-\infty}^{1}\frac{dx}{(2x-3)^3}=\left[-\frac{1}{4(2x-3)^2}\right]_{-\infty}^{1}$

$$=-\frac{1}{4}-(-0)=-\frac{1}{4}$$

17. $\int_{-\infty}^{\infty} \frac{x}{\sqrt{x^2+9}} dx = \int_{-\infty}^{0} \frac{x}{\sqrt{x^2+9}} dx + \int_{0}^{\infty} \frac{x}{\sqrt{x^2+9}} dx = \left[\sqrt{x^2+9}\right]_{-\infty}^{0} + \left[\sqrt{x^2+9}\right]_{0}^{\infty} = (3-\infty)+(\infty-3)$

The integral diverges since both $\int_{-\infty}^{0} \frac{x}{\sqrt{x^2+9}} dx$ and $\int_{0}^{\infty} \frac{x}{\sqrt{x^2+9}} dx$ diverge.

19. $\int_{-\infty}^{\infty} \frac{1}{x^2+2x+10} dx = \int_{-\infty}^{\infty} \frac{1}{(x+1)^2+9} dx = \int_{-\infty}^{0} \frac{1}{(x+1)^2+9} dx + \int_{0}^{\infty} \frac{1}{(x+1)^2+9} dx$

$\int \frac{1}{(x+1)^2+9} dx = \frac{1}{3}\tan^{-1}\frac{x+1}{3}$ by using the substitution $x + 1 = 3 \tan \theta$.

$\int_{-\infty}^{0} \frac{1}{(x+1)^2+9} dx = \left[\frac{1}{3}\tan^{-1}\frac{x+1}{3}\right]_{-\infty}^{0} = \frac{1}{3}\tan^{-1}\frac{1}{3} - \frac{1}{3}\left(-\frac{\pi}{2}\right) = \frac{1}{6}\left(\pi + 2\tan^{-1}\frac{1}{3}\right)$

$\int_{0}^{\infty} \frac{1}{(x+1)^2+9} dx = \left[\frac{1}{3}\tan^{-1}\frac{x+1}{3}\right]_{0}^{\infty} = \frac{1}{3}\left(\frac{\pi}{2}\right) - \frac{1}{3}\tan^{-1}\frac{1}{3} = \frac{1}{6}\left(\pi - 2\tan^{-1}\frac{1}{3}\right)$

$\int_{-\infty}^{\infty} \frac{1}{x^2+2x+10} dx = \frac{1}{6}\left(\pi + 2\tan^{-1}\frac{1}{3}\right) + \frac{1}{6}\left(\pi - 2\tan^{-1}\frac{1}{3}\right) = \frac{\pi}{3}$

21. $\int_{-\infty}^{\infty} \text{sech}\, x\, dx = \int_{-\infty}^{0} \text{sech}\, x\, dx = \int_{0}^{\infty} \text{sech}\, x\, dx$

$= [\tan^{-1}(\sinh x)]_{-\infty}^{0} + [\tan^{-1}(\sinh x)]_{0}^{\infty}$

$= \left[0 - \left(-\frac{\pi}{2}\right)\right] + \left[\frac{\pi}{2} - 0\right] = \pi$

23. $\int_{0}^{\infty} e^{-x} \cos x\, dx = \left[\frac{1}{2e^x}(\sin x - \cos x)\right]_{0}^{\infty}$

$= 0 - \frac{1}{2}(0-1) = \frac{1}{2}$

(Use Formula 68 with $a = -1$ and $b = 1$.)

25. The area is given by

$\int_{1}^{\infty} \frac{2}{4x^2-1} dx = \int_{1}^{\infty} \left(\frac{1}{2x-1} - \frac{1}{2x+1}\right) dx$

$= \frac{1}{2}\left[\ln|2x-1| - \ln|2x+1|\right]_{1}^{\infty} = \frac{1}{2}\left[\ln\left|\frac{2x-1}{2x+1}\right|\right]_{1}^{\infty}$

$= \frac{1}{2}\left(0 - \ln\left(\frac{1}{3}\right)\right) = \frac{1}{2}\ln 3$

Note:. $\lim_{x\to\infty} \ln = \left|\frac{2x-1}{2x+1}\right| = 0$ since

$\lim_{x\to\infty} \left(\frac{2x-1}{2x+1}\right) = 1$ ·

27. The integral would take the form

$k\int_{3960}^{\infty} \frac{1}{x} dx = [k \ln x]_{3960}^{\infty} = \infty$

which would make it impossible to send anything out of the earth's gravitational field.

29. $FP = \int_{0}^{\infty} e^{-rt} f(t)\, dt = \int_{0}^{\infty} 100{,}000e^{-0.08t}$

$= \left[-\frac{1}{0.08} 100{,}000e^{-0.08t}\right]_{0}^{\infty} = 1{,}250{,}000$

The present value is \$1,250,000.

31. **a.** $\int_{-\infty}^{\infty} f(x)\,dx = \int_{-\infty}^{a} 0\,dx + \int_{a}^{b} \frac{1}{b-a}\,dx + \int_{b}^{\infty} 0\,dx = 0 + \frac{1}{b-a}[x]_a^b + 0 = \frac{1}{b-a}(b-a)$

b. $\mu = \int_{-\infty}^{\infty} x\,f(x)\,dx = \int_{-\infty}^{a} x\cdot 0\,dx + \int_{a}^{b} x\frac{1}{b-a}\,dx + \int_{b}^{\infty} x\cdot 0\,dx = 0 + \frac{1}{b-a}\left[\frac{x^2}{2}\right]_a^b + 0$

$$= \frac{b^2-a^2}{2(b-a)} = \frac{(b+a)(b-a)}{2(b-a)} = \frac{a+b}{2}$$

$$\sigma^2 = \int_{-\infty}^{\infty} (x-\mu)^2\,dx = \int_{-\infty}^{a} (x-\mu)^2\cdot 0\,dx + \int_{a}^{b} (x-\mu)^2 \frac{1}{b-a}\,dx + \int_{b}^{\infty} (x-\mu)^2\cdot 0\,dx$$

$$= 0 + \frac{1}{b-a}\left[\frac{(x-\mu)^3}{3}\right]_a^b + 0 = \frac{1}{b-a}\frac{(b-\mu)^3-(a-\mu)^3}{3} = \frac{1}{b-a}\frac{b^3-3b^2\mu+3b\mu^2-a^3+3a^2\mu-3a\mu^2}{3}$$

Next, substitute $\mu = (a+b)/2$ to obtain

$$\sigma^2 = \frac{1}{3(b-a)}\left[\tfrac{1}{4}b^3 - \tfrac{3}{4}b^2a + \tfrac{3}{4}ba^2 - \tfrac{1}{4}a^3\right] = \frac{1}{12(b-a)}(b-a)^3 = \frac{(b-a)^2}{12}$$

c. $P(X<2) = \int_{-\infty}^{2} f(x)\,dx = \int_{-\infty}^{0} 0\,dx + \int_{0}^{2} \frac{1}{10-0}\,dx = \frac{2}{10} = \frac{1}{5}$

33.

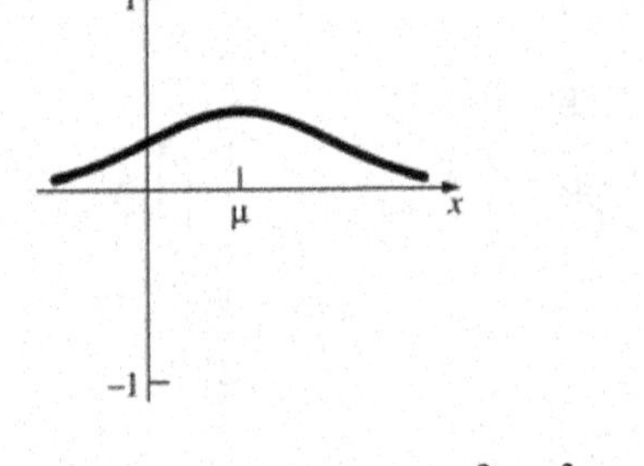

$$f'(x) = -\frac{x-\mu}{\sigma^3\sqrt{2\pi}}e^{-(x-\mu)^2/2\sigma^2}$$

$$f''(x) = -\frac{1}{\sigma^3\sqrt{2\pi}}e^{-(x-\mu)^2/2\sigma^2} + \frac{(x-\mu)^2}{\sigma^5\sqrt{2\pi}}e^{-(x-\mu)^2/2\sigma^2}$$

$$= \left(\frac{(x-\mu)^2}{\sigma^5\sqrt{2\pi}} - \frac{1}{\sigma^3\sqrt{2\pi}}\right)e^{-(x-\mu)^2/2\sigma^2} =$$

$$\frac{1}{\sigma^5\sqrt{2\pi}}[(x-\mu)^2-\sigma^2]e^{-(x-\mu)^2/2\sigma^2}$$

$f''(x) = 0$ when $(x-\mu)^2 = \sigma^2$ so $x = \mu \pm \sigma$ and the distance from μ to each inflection point is σ.

35. We use the results from problem 34:

a. To have a probability density function (34 a.) we need $C = k$; so $C = 3$. Also, $\mu = \frac{kM}{k-1}$ (34 b.) and since, in our problem, $\mu = 20,000$ and $k = 3$, we have $20000 = \frac{3}{2}M$ or $M = \frac{4\times 10^4}{3}$.

b. By 34 c., $\sigma^2 = \frac{kM^2}{(k-2)(k-1)^2}$ so that

$$\sigma^2 = \frac{3}{4}\left(\frac{4\times 10^4}{3}\right)^2 = \frac{4\times 10^8}{3}$$

c. $\int_{10^5}^{\infty} f(x)\,dx = \left(\frac{4\times 10^4}{3}\right)^3 \lim_{t\to\infty}\int_{10^5}^{t} \frac{3}{x^4}\,dx =$

$$-\left(\frac{4\times 10^4}{3}\right)^3 \lim_{t\to\infty}\left[\frac{1}{x^3}\right]_{10^5}^{t}$$

$$= \left(\frac{4\times 10^4}{3}\right)^3 \lim_{t\to\infty}\left[\frac{1}{10^{15}} - \frac{1}{t^3}\right] = \frac{64}{27\times 10^3}$$

≈ 0.0024

Thus $6/25$ of one percent earn over \$100,000.

37. a. $\int_{-\infty}^{\infty} \sin x\,dx = \int_{-\infty}^{0} \sin x\,dx + \int_{0}^{\infty} \sin x\,dx$

$= \lim_{a\to\infty}[-\cos x]_0^a + \lim_{a\to -\infty}[-\cos x]_a^0$

Both do not converge since $-\cos x$ is oscillating between –1 and 1, so the integral diverges.

b. $\lim_{a\to\infty}\int_{-a}^{a} \sin x\,dx = \lim_{a\to\infty}[-\cos x]_{-a}^{a}$

$= \lim_{a\to\infty}[-\cos a + \cos(-a)]$

$= \lim_{a\to\infty}[-\cos a + \cos a] = \lim_{a\to\infty} 0 = 0$

39. For example, the region under the curve $y = \frac{1}{x}$ to the right of $x = 1$.
Rotated about the x-axis the volume is $\pi\int_1^{\infty}\frac{1}{x^2}dx = \pi$. Rotated about the y-axis, the volume is $2\pi\int_1^{\infty} x\cdot\frac{1}{x}dx$ which diverges.

41. $\int_1^{100}\frac{1}{x^2}dx = \left[-\frac{1}{x}\right]_1^{100} = 0.99$

$\int_1^{100}\frac{1}{x^{1.1}}dx = \left[-\frac{1}{0.1x^{0.1}}\right]_1^{100} \approx 3.69$

$\int_1^{100}\frac{1}{x^{1.01}}dx = \left[-\frac{1}{0.01x^{0.01}}\right]_1^{100} \approx 4.50$

$\int_1^{100}\frac{1}{x}dx = [\ln x]_1^{100} = \ln 100 \approx 4.61$

$\int_1^{100}\frac{1}{x^{0.99}}dx = \left[\frac{x^{0.01}}{0.01}\right]_1^{100} \approx 4.71$

43. $\int_0^1 \frac{1}{\sqrt{2\pi}}\exp(-0.5x^2)dx \approx 0.3413$

$\int_0^2 \frac{1}{\sqrt{2\pi}}\exp(-0.5x^2)dx \approx 0.4772$

$\int_0^3 \frac{1}{\sqrt{2\pi}}\exp(-0.5x^2)dx \approx 0.4987$

$\int_0^4 \frac{1}{\sqrt{2\pi}}\exp(-0.5x^2)dx \approx 0.5000$

8.4 Concepts Review

1. unbounded

3. $\lim_{b\to 4^-}\int_0^b \frac{1}{\sqrt{4-x}}dx$

Problem Set 8.4

1. $\int_1^3 \frac{dx}{(x-1)^{1/3}} = \lim_{b\to 1^+}\left[\frac{3(x-1)^{2/3}}{2}\right]_b^3$

$= \frac{3}{2}\sqrt[3]{2^2} - \lim_{b\to 1^+}\frac{3(b-1)^{2/3}}{2} = \frac{3}{\sqrt[3]{2}} - 0 = \frac{3}{\sqrt[3]{2}}$

3. $\int_3^{10}\frac{dx}{\sqrt{x-3}} = \lim_{b\to 3^+}\left[2\sqrt{x-3}\right]_b^{10}$

$= 2\sqrt{7} - \lim_{b\to 3^+} 2\sqrt{b-3} = 2\sqrt{7}$

5. $\int_0^1 \frac{dx}{\sqrt{1-x^2}} = \lim_{b\to 1^-}\left[\sin^{-1}x\right]_0^b$

$= \lim_{b\to 1^-}\sin^{-1}b - \sin^{-1}0 = \frac{\pi}{2} - 0 = \frac{\pi}{2}$

7. $\int_{-1}^{3}\frac{1}{x^3}dx = \lim_{b\to 0^-}\int_{-1}^{b}\frac{1}{x^3}dx + \lim_{b\to 0^+}\int_{b}^{3}\frac{1}{x^3}dx$

$= \lim_{b\to 0^-}\left[-\frac{1}{2x^2}\right]_{-1}^{b} + \lim_{b\to 0^+}\left[-\frac{1}{2x^2}\right]_{b}^{3}$

$= \left(\lim_{b\to 0^-} -\frac{1}{2b^2}+\frac{1}{2}\right)+\left(-\frac{1}{18}+\lim_{b\to 0^+}\frac{1}{2b^2}\right)$

$= \left(-\infty+\frac{1}{2}\right)+\left(-\frac{1}{8}+\infty\right)$

The integral diverges.

9. $\int_{-1}^{128} x^{-5/7}dx$

$= \lim_{b\to 0^-}\int_{-1}^{b} x^{-5/7}dx + \lim_{b\to 0^+}\int_{b}^{128} x^{-5/7}dx$

$= \lim_{b\to 0^-}\left[\frac{7}{2}x^{2/7}\right]_{-1}^{b} + \lim_{b\to 0^+}\left[\frac{7}{2}x^{2/7}\right]_{b}^{128}$

$= \lim_{b\to 0^-}\frac{7}{2}b^{2/7} - \frac{7}{2}(-1)^{2/7} + \frac{7}{2}(128)^{2/7} - \lim_{b\to 0^+}\frac{7}{2}b^{2/7}$

$= 0 - \frac{7}{2} + \frac{7}{2}(4) - 0 = \frac{21}{2}$

11. $\int_{0}^{4}\frac{dx}{(2-3x)^{1/3}} = \lim_{b\to \frac{2}{3}^-}\int_{0}^{b}\frac{dx}{(2-3x)^{1/3}} + \lim_{b\to \frac{2}{3}^+}\int_{b}^{4}\frac{dx}{(2-3x)^{1/3}} = \lim_{b\to \frac{2}{3}^-}\left[-\frac{1}{2}(2-3x)^{2/3}\right]_{0}^{b} + \lim_{b\to \frac{2}{3}^+}\left[-\frac{1}{2}(2-3x)^{2/3}\right]_{b}^{4}$

$= \lim_{b\to \frac{2}{3}^-} -\frac{1}{2}(2-3b)^{2/3} + \frac{1}{2}(2)^{2/3} - \frac{1}{2}(-10)^{2/3} + \lim_{b\to \frac{2}{3}^+}\frac{1}{2}(2-3b)^{2/3}$

$= 0 + \frac{1}{2}2^{2/3} - \frac{1}{2}10^{2/3} + 0 = \frac{1}{2}(2^{2/3} - 10^{2/3})$

13. $\int_{0}^{-4}\frac{x}{16-2x^2}dx = \lim_{b\to -\sqrt{8}^+}\int_{0}^{b}\frac{x}{16-2x^2}dx + \lim_{b\to -\sqrt{8}^-}\int_{b}^{-4}\frac{x}{16-2x^2}dx$

$= \lim_{b\to -\sqrt{8}^+}\left[-\frac{1}{4}\ln\left|16-2x^2\right|\right]_{0}^{b} + \lim_{b\to -\sqrt{8}^-}\left[-\frac{1}{4}\ln\left|16-2x^2\right|\right]_{b}^{-4}$

$= \lim_{b\to -\sqrt{8}^+} -\frac{1}{4}\ln\left|16-2b^2\right| + \frac{1}{4}\ln 16 - \frac{1}{4}\ln 16 + \lim_{b\to -\sqrt{8}^-}\frac{1}{4}\ln\left|16-2b^2\right|$

$= \left[-(-\infty)+\frac{1}{4}\ln 16\right]+\left[-\frac{1}{4}\ln 16+(-\infty)\right]$

The integral diverges.

15. $\int_{-2}^{-1}\frac{dx}{(x+1)^{4/3}} = \lim_{b\to -1^-}\left[-\frac{3}{(x+1)^{1/3}}\right]_{-2}^{b} = \lim_{b\to -1^-} -\frac{3}{(b+1)^{1/3}} + \frac{3}{(-1)^{1/3}} = -(-\infty) - 3$

The integral diverges.

17. Note that $\frac{1}{x^3-x^2-x+1} = \frac{1}{2(x-1)^2} - \frac{1}{4(x-1)} + \frac{1}{4(x+1)}$

$\int_{0}^{3}\frac{dx}{x^3-x^2-x+1} = \lim_{b\to 1^-}\int_{0}^{b}\frac{dx}{x^3-x^2-x+1} + \lim_{b\to 1^+}\int_{b}^{3}\frac{dx}{x^3-x^2-x+1}$

$= \lim_{b\to 1^-}\left[-\frac{1}{2(x-1)} - \frac{1}{4}\ln|x-1| + \frac{1}{4}\ln|x+1|\right]_{0}^{b} + \lim_{b\to 1^+}\left[-\frac{1}{2(x-1)} - \frac{1}{4}\ln|x-1| + \frac{1}{4}\ln|x+1|\right]_{b}^{3}$

$\lim_{b\to 1^-}\left[\left(-\frac{1}{2(b-1)} + \frac{1}{4}\ln\left|\frac{b+1}{b-1}\right|\right) + \left(-\frac{1}{2}+0\right)\right] + \lim_{b\to 1^+}\left[-\frac{1}{4}+\frac{1}{4}\ln 2 - \left(-\frac{1}{2(b-1)} + \frac{1}{4}\ln\left|\frac{b+1}{b-1}\right|\right)\right]$

$= \left(\infty+\infty-\frac{1}{2}\right)+\left(-\frac{1}{4}+\frac{1}{4}\ln 2+\infty-\infty\right)$

The integral diverges.

19. $\int_0^{\pi/4} \tan 2x dx = \lim_{b\to\frac{\pi}{4}^-} \left[-\frac{1}{2}\ln|\cos 2x|\right]_0^b$

$= \lim_{b\to\frac{\pi}{4}^-} -\frac{1}{2}\ln|\cos 2b| + \frac{1}{2}\ln 1 = -(-\infty) + 0$

The integral diverges.

21. $\int_0^{\pi/2} \frac{\sin x}{1-\cos x} dx = \lim_{b\to 0^+} \left[\ln|1-\cos x|\right]_b^{\pi/2}$

$= \ln 1 - \lim_{b\to 0^+} \ln|1-\cos b| = 0 - (-\infty)$

The integral diverges.

23. $\int_0^{\pi/2} \tan^2 x \sec^2 x \, dx = \lim_{b\to\frac{\pi}{2}^-} \left[\frac{1}{3}\tan^3 x\right]_0^b$

$= \lim_{b\to\frac{\pi}{2}^-} \frac{1}{3}\tan^3 b - \frac{1}{3}(0)^3 = \infty$

The integral diverges.

25. Since $\frac{1-\cos x}{2} = \sin^2 \frac{x}{2}$,

$\frac{1}{\cos x - 1} = -\frac{1}{2}\csc^2 \frac{x}{2}$.

$\int_0^{\pi} \frac{dx}{\cos x - 1} = \lim_{b\to 0^+} \left[\cot \frac{x}{2}\right]_b^{\pi}$

$= \cot\frac{\pi}{2} - \lim_{b\to 0^+} \cot\frac{b}{2} = 0 - \infty$

The integral diverges.

27. $\int_0^{\ln 3} \frac{e^x dx}{\sqrt{e^x - 1}} = \lim_{b\to 0^+} \left[2\sqrt{e^x - 1}\right]_b^{\ln 3}$

$= 2\sqrt{3-1} - \lim_{b\to 0^+} 2\sqrt{e^b - 1} = 2\sqrt{2} - 0 = 2\sqrt{2}$

29. $\int_1^e \frac{dx}{x \ln x} = \lim_{b\to 1^+} [\ln(\ln x)]_b^e = \ln(\ln e) - \lim_{b\to 1^+} \ln(\ln b) = \ln 1 - \ln 0 = 0 + \infty$

The integral diverges.

31. $\int_{2c}^{4c} \frac{dx}{\sqrt{x^2 - 4c^2}} = \lim_{b\to 2c^+} \left[\ln\left|x + \sqrt{x^2 - 4c^2}\right|\right]_b^{4c} = \ln\left[(4+2\sqrt{3})c\right] - \lim_{b\to 2c^+} \ln\left|b + \sqrt{b^2 - 4c^2}\right|$

$= \ln\left[(4+2\sqrt{3})c\right] - \ln 2c = \ln(2+\sqrt{3})$

33. For $0 < c < 1$, $\frac{1}{\sqrt{x}(1+x)}$ is continuous. Let $u = \frac{1}{1+x}$, $du = -\frac{1}{(1+x)^2} dx$.

$dv = \frac{1}{\sqrt{x}} dx, v = 2\sqrt{x}$.

$\int_c^1 \frac{1}{\sqrt{x}(1+x)} dx = \left[\frac{2\sqrt{x}}{1+x}\right]_c^1 + 2\int_c^1 \frac{\sqrt{x}dx}{(1+x)^2} = \frac{2}{2} - \frac{2\sqrt{c}}{1+c} + 2\int_c^1 \frac{\sqrt{x}dx}{(1+x)^2} = 1 - \frac{2\sqrt{c}}{1+c} + 2\int_c^1 \frac{\sqrt{x}dx}{(1+x)^2}$

Thus, $\lim_{c\to 0} \int_c^1 \frac{1}{\sqrt{x}(1+x)} dx = \lim_{c\to 0}\left[1 - \frac{2\sqrt{c}}{1+c} + 2\int_c^1 \frac{\sqrt{x}dx}{(1+x)^2}\right] = 1 - 0 + 2\int_0^1 \frac{\sqrt{x}dx}{(1+x)^2}$

This last integral is a proper integral.

35. $\int_{-3}^3 \frac{x}{\sqrt{9-x^2}} dx = \int_{-3}^0 \frac{x}{\sqrt{9-x^2}} dx + \int_0^3 \frac{x}{\sqrt{9-x^2}} dx = \lim_{b\to -3^+} \left[-\sqrt{9-x^2}\right]_b^0 + \lim_{b\to 3^-} \left[-\sqrt{9-x^2}\right]_0^b$

$= -\sqrt{9} + \lim_{b\to -3^+} \sqrt{9-b^2} - \lim_{b\to 3^-} \sqrt{9-b^2} + \sqrt{9} = -3 + 0 - 0 + 3 = 0$

37. $\int_{-4}^{4}\frac{1}{16-x^2}dx=\int_{-4}^{0}\frac{1}{16-x^2}dx+\int_{0}^{4}\frac{1}{16-x^2}dx=\lim_{b\to -4^+}\left[\frac{1}{8}\ln\left|\frac{x+4}{x-4}\right|\right]_b^0+\lim_{b\to 4^-}\left[\frac{1}{8}\ln\left|\frac{x+4}{x-4}\right|\right]_0^b$

$=\frac{1}{8}\ln 1-\lim_{b\to -4^+}\frac{1}{8}\ln\left|\frac{b+4}{b-4}\right|+\lim_{b\to 4^-}\frac{1}{8}\ln\left|\frac{b+4}{b-4}\right|-\frac{1}{8}\ln 1=(0+\infty)+(\infty-0)$

The integral diverges.

39. $\int_0^\infty \frac{1}{x^p}dx=\int_0^1\frac{1}{x^p}dx+\int_1^\infty\frac{1}{x^p}dx$

If $p>1$, $\int_0^1\frac{1}{x^p}dx=\left[\frac{1}{-p+1}x^{-p+1}\right]_0^1$ diverges since $\lim_{x\to 0^+}x^{-p+1}=\infty$.

If $p<1$ and $p\neq 0$, $\int_1^\infty\frac{1}{x^p}dx=\left[\frac{1}{-p+1}x^{-p+1}\right]_1^\infty$ diverges since $\lim_{x\to\infty}x^{-p+1}=\infty$.

If $p=0$, $\int_0^\infty dx=\infty$.

If $p=1$, both $\int_0^1\frac{1}{x}dx$ and $\int_1^\infty\frac{1}{x}dx$ diverge.

41. $\int_0^8 (x-8)^{-2/3}dx=\lim_{b\to 8^-}\left[3(x-8)^{1/3}\right]_0^b$
$=3(0)-3(-2)=6$

43. a. $\int_0^1 x^{-2/3}dx=\lim_{b\to 0^+}\left[3x^{1/3}\right]_b^1=3$

b. $V=\pi\int_0^1 x^{-4/3}dx=\lim_{b\to 0^+}\pi\left[-3x^{-1/3}\right]_b^1$
$=-3\pi+3\pi\lim_{b\to 0}b^{-1/3}$
The limit tends to infinity as $b\to 0$, so the volume is infinite.

45. $\int_0^1\frac{\sin x}{x}dx$ is not an improper integral since $\frac{\sin x}{x}$ is bounded in the interval $0\le x\le 1$.

47. For $x\ge 1$, $x^2\ge x$ so $-x^2\le -x$, thus $e^{-x^2}\le e^{-x}$.

$\int_1^\infty e^{-x}dx=\lim_{b\to\infty}[-e^{-x}]_1^b=-\lim_{b\to\infty}\frac{1}{e^b}+e^{-1}$
$=-0+\frac{1}{e}=\frac{1}{e}$

Thus, by the Comparison Test, $\int_1^\infty e^{-x^2}dx$ converges.

49. Since $x^2\ln(x+1)\ge x^2$, we know that $\frac{1}{x^2\ln(x+1)}\le\frac{1}{x^2}$. Since $\int_1^\infty\frac{1}{x^2}dx=\left[-\frac{1}{x}\right]_1^\infty=1$ we can apply the Comparison Test of Problem 46 to conclude that $\int_1^\infty\frac{1}{x^2\ln(x+1)}dx$ converges.

51. a. From Example 2 of Section 8.2, $\lim_{x\to\infty}\frac{x^a}{e^x}=0$ for a any positive real number.
Thus $\lim_{x\to\infty}\frac{x^{n+1}}{e^x}=0$ for any positive real number n, hence there is a number M such that $0<\frac{x^{n+1}}{e^x}\le 1$ for $x\ge M$. Divide the inequality by x^2 to get that $0<\frac{x^{n-1}}{e^x}\le\frac{1}{x^2}$ for $x\ge M$.

b. $\int_1^\infty\frac{1}{x^2}dx=\lim_{b\to\infty}\left[-\frac{1}{x}\right]_1^b=-\lim_{b\to\infty}\frac{1}{b}+\frac{1}{1}$
$=-0+1=1$
$\int_1^\infty x^{n-1}e^{-x}dx=\int_1^M x^{n-1}e^{-x}dx+\int_M^\infty x^{n-1}e^{-x}dx$
$\le\int_1^M x^{n-1}e^{-x}dx+\int_1^\infty\frac{1}{x^2}dx$
$=1+\int_1^M x^{n-1}e^{-x}dx$
by part a and Problem 46. The remaining integral is finite, so $\int_1^\infty x^{n-1}e^{-x}dx$ converges.

53. a. $\Gamma(1)=\int_0^\infty x^0 e^{-x}dx=\left[-e^{-x}\right]_0^\infty=1$

b. $\Gamma(n+1)=\int_0^\infty x^n e^{-x}dx$

Let $u=x^n$, $dv=e^{-x}dx$, $du=nx^{n-1}dx, v=-e^{-x}$.

$$\Gamma(n+1)=[-x^n e^{-x}]_0^\infty+\int_0^\infty nx^{n-1}e^{-x}dx=0+n\int_0^\infty x^{n-1}e^{-x}dx=n\Gamma(n)$$

c. From parts a and b,
$\Gamma(1)=1, \Gamma(2)=1\cdot\Gamma(1)=1,$
$\Gamma(3)=2\cdot\Gamma(2)=2\cdot 1=2!.$
Suppose $\Gamma(n)=(n-1)!$, then by part b, $\Gamma(n+1)=n\Gamma(n)=n[(n-1)!]=n!$.

55. a. $\int_{-\infty}^\infty f(x)dx=\int_0^\infty Cx^{\alpha-1}e^{-\beta x}dx$

Let $y=\beta x$, so $x=\frac{y}{\beta}$ and $dx=\frac{1}{\beta}dy$.

$$\int_0^\infty Cx^{\alpha-1}e^{-\beta x}dx=\int_0^\infty C\left(\frac{y}{\beta}\right)^{\alpha-1}e^{-y}\frac{1}{\beta}dy=\frac{C}{\beta^\alpha}\int_0^\infty y^{\alpha-1}e^{-y}dy=C\beta^{-\alpha}\Gamma(\alpha)$$

$C\beta^{-\alpha}\Gamma(\alpha)=1$ when $C=\frac{1}{\beta^{-\alpha}\Gamma(\alpha)}=\frac{\beta^\alpha}{\Gamma(\alpha)}$.

b. $\mu=\int_{-\infty}^\infty xf(x)dx=\int_0^\infty x\frac{\beta^\alpha}{\Gamma(\alpha)}x^{\alpha-1}e^{-\beta x}dx=\frac{\beta^\alpha}{\Gamma(\alpha)}\int_0^\infty x^\alpha e^{-\beta x}dx$

Let $y=\beta x$, so $x=\frac{y}{\beta}$ and $dx=\frac{1}{\beta}dy$.

$$\mu=\frac{\beta^\alpha}{\Gamma(\alpha)}\int_0^\infty\left(\frac{y}{\beta}\right)^\alpha e^{-y}\frac{1}{\beta}dy=\frac{1}{\beta\Gamma(\alpha)}\int_0^\infty y^\alpha e^{-y}dy=\frac{1}{\beta\Gamma(\alpha)}\Gamma(\alpha+1)=\frac{1}{\beta\Gamma(\alpha)}\alpha\Gamma(\alpha)=\frac{\alpha}{\beta}$$

(Recall that $\Gamma(\alpha+1)=\alpha\Gamma(\alpha)$ for $\alpha>0$.)

c. $\sigma^2=\int_{-\infty}^\infty(x-\mu)^2 f(x)dx=\int_0^\infty\left(x-\frac{\alpha}{\beta}\right)^2\frac{\beta^\alpha}{\Gamma(\alpha)}x^{\alpha-1}e^{-\beta x}dx=\frac{\beta^\alpha}{\Gamma(\alpha)}\int_0^\infty\left(x^2-\frac{2\alpha}{\beta}x+\frac{\alpha^2}{\beta^2}\right)x^{\alpha-1}e^{-\beta x}dx$

$$=\frac{\beta^\alpha}{\Gamma(\alpha)}\int_0^\infty x^{\alpha+1}e^{-\beta x}dx-\frac{2\alpha\beta^{\alpha-1}}{\Gamma(\alpha)}\int_0^\infty x^\alpha e^{-\beta x}dx+\frac{\alpha^2\beta^{\alpha-2}}{\Gamma(\alpha)}\int_0^\infty x^{\alpha-1}e^{-\beta x}dx$$

In all three integrals, let $y=\beta x$, so $x=\frac{y}{\beta}$ and $dx=\frac{1}{\beta}dy$.

$$\sigma^2=\frac{\beta^\alpha}{\Gamma(\alpha)}\int_0^\infty\left(\frac{y}{\beta}\right)^{\alpha+1}e^{-y}\frac{1}{\beta}dy-\frac{2\alpha\beta^{\alpha-1}}{\Gamma(\alpha)}\int_0^\infty\left(\frac{y}{\beta}\right)^\alpha e^{-y}\frac{1}{\beta}dy+\frac{\alpha^2\beta^{\alpha-2}}{\Gamma(\alpha)}\int_0^\infty\left(\frac{y}{\beta}\right)^{\alpha-1}e^{-y}\frac{1}{\beta}dy$$

$$=\frac{1}{\beta^2\Gamma(\alpha)}\int_0^\infty y^{\alpha+1}e^{-y}dy-\frac{2\alpha}{\beta^2\Gamma(\alpha)}\int_0^\infty y^\alpha e^{-y}dy+\frac{\alpha^2}{\beta^2\Gamma(\alpha)}\int_0^\infty y^{\alpha-1}e^{-y}dy$$

$$=\frac{1}{\beta^2\Gamma(\alpha)}\Gamma(\alpha+2)-\frac{2\alpha}{\beta^2\Gamma(\alpha)}\Gamma(\alpha+1)+\frac{\alpha^2}{\beta^2\Gamma(\alpha)}\Gamma(\alpha)=\frac{1}{\beta^2\Gamma(\alpha)}(\alpha+1)\alpha\Gamma(\alpha)-\frac{2\alpha}{\beta^2\Gamma(\alpha)}\alpha\Gamma(\alpha)+\frac{\alpha^2}{\beta^2}$$

$$=\frac{\alpha^2+\alpha}{\beta^2}-\frac{2\alpha^2}{\beta^2}+\frac{\alpha^2}{\beta^2}=\frac{\alpha}{\beta^2}$$

57. a. The integral is the area between the curve $y^2 = \frac{1-x}{x}$ and the x-axis from $x = 0$ to $x = 1$.

$y^2 = \frac{1-x}{x};\ xy^2 = 1-x;\ x(y^2+1) = 1$

$x = \frac{1}{y^2+1}$

As $x \to 0,\ y = \sqrt{\frac{1-x}{x}} \to \infty,$ while

when $x = 1,\ y = \sqrt{\frac{1-1}{1}} = 0,$ thus the area is

$\int_0^{\infty} \frac{1}{y^2+1} dy = \lim_{b\to\infty} [\tan^{-1} y]_0^b$

$= \lim_{b\to\infty} \tan^{-1} b - \tan^{-1} 0 = \frac{\pi}{2}$

b. The integral is the area between the curve $y^2 = \frac{1+x}{1-x}$ and the x-axis from $x = -1$ to $x = 1$.

$y^2 = \frac{1+x}{1-x};\ y^2 - xy^2 = 1+x;\ y^2 - 1 = x(y^2+1);$

$x = \frac{y^2-1}{y^2+1}$

When $x = -1,\ y = \sqrt{\frac{1+(-1)}{1-(-1)}} = \sqrt{\frac{0}{2}} = 0,$ while

as $x \to 1,\ y = \sqrt{\frac{1+x}{1-x}} \to \infty.$

The area in question is the area to the right of the curve $y = \sqrt{\frac{1+x}{1-x}}$ and to the left of the line $x = 1$. Thus, the area is

$\int_0^{\infty} \left(1 - \frac{y^2-1}{y^2+1}\right) dy = \int_0^{\infty} \frac{2}{y^2+1} dy$

$= \lim_{b\to\infty} \left[2\tan^{-1} y\right]_0^b$

$\lim_{b\to\infty} 2\tan^{-1} b - 2\tan^{-1} 0 = 2\left(\frac{\pi}{2}\right) = \pi$

8.5 Chapter Review

Concepts Test

1. True: See Example 2 of Section 8.2.

3. False: $\lim_{x\to\infty} \frac{1000x^4+1000}{0.001x^4+1} = \frac{1000}{0.001} = 10^6$

5. False: For example, if $f(x) = x$ and $g(x) = e^x$,

$\lim_{x\to\infty} \frac{x}{e^x} = 0.$

7. True: Take the inner limit first.

9. True: Since $\lim_{x\to a} f(x) = -1 \neq 0,$ it serves only to affect the sign of the limit of the product.

11. False: Consider $f(x) = 3x^2$ and $g(x) = x^2+1,$ then

$\lim_{x\to\infty} \frac{f(x)}{g(x)} = \lim_{x\to\infty} \frac{3x^2}{x^2+1}$

$= \lim_{x\to\infty} \frac{3}{1+\frac{1}{x^2}} = 3,$ but

$\lim_{x\to\infty} [f(x) - 3g(x)]$

$= \lim_{x\to\infty} [3x^2 - 3(x^2+1)]$

$= \lim_{x\to\infty} [-3] = -3$

13. True: See Example 7 of Section 8.2.

15. True: Use repeated applications of l'Hôpital's Rule.

17. False: Consider $f(x) = 3x^2 + x + 1$ and $g(x) = 4x^3 + 2x + 3;\ f'(x) = 6x+1$

$g'(x) = 12x^2 + 2,$ and so

$\lim_{x\to 0} \frac{f'(x)}{g'(x)} = \lim_{x\to 0} \frac{6x+1}{12x^2+2} = \frac{1}{2}$ while

$\lim_{x\to 0} \frac{f(x)}{g(x)} = \lim_{x\to 0} \frac{3x^2+x+1}{4x^3+2x+3} = \frac{1}{3}$

19. True: $\int_0^\infty \frac{1}{x^p}dx = \int_0^1 \frac{1}{x^p}dx + \int_1^\infty \frac{1}{x^p}dx;$

$\int_0^1 \frac{1}{x^p}dx$ diverges for $p \geq 1$ and

$\int_1^\infty \frac{1}{x^p}dx$ diverges for $p \leq 1$.

21. True: $\int_{-\infty}^\infty f(x)dx = \int_{-\infty}^0 f(x)dx + \int_0^\infty f(x)dx$

If f is an even function, then $f(-x) = f(x)$ so

$\int_{-\infty}^0 f(x)dx = \int_0^\infty f(x)dx.$

Thus, both integrals making up $\int_{-\infty}^\infty f(x)dx$ converge so their sum converges.

23. True: $\int_0^\infty f'(x)dx = \lim_{b\to\infty}\int_0^b f'(x)dx$

$= \lim_{b\to\infty}[f(x)]_0^b = \lim_{b\to\infty} f(b) - f(0)$

$= 0 - f(0) = -f(0).$

$f(0)$ must exist and be finite since $f'(x)$ is continuous on $[0, \infty)$.

25. False: The integrand is bounded on the interval $\left[0, \frac{\pi}{4}\right]$.

Sample Test Problems

1. The limit is of the form $\frac{0}{0}$.

$$\lim_{x\to 0}\frac{4x}{\tan x} = \lim_{x\to 0}\frac{4}{\sec^2 x} = 4$$

3. The limit is of the form $\frac{0}{0}$. (Apply l'Hôpital's Rule twice.)

$$\lim_{x\to 0}\frac{\sin x - \tan x}{\frac{1}{3}x^2} = \lim_{x\to 0}\frac{\cos x - \sec^2 x}{\frac{2}{3}x}$$

$$= \lim_{x\to 0}\frac{-\sin x - 2\sec x(\sec x \tan x)}{\frac{2}{3}} = 0$$

5. $\lim_{x\to 0} 2x\cot x = \lim_{x\to 0}\frac{2x\cos x}{\sin x}$

The limit is of the form $\frac{0}{0}$.

$$\lim_{x\to 0}\frac{2x\cos x}{\sin x} = \lim_{x\to 0}\frac{2\cos x - 2x\sin x}{\cos x} = \frac{2-0}{1} = 2$$

7. The limit is of the form $\frac{\infty}{\infty}$.

$$\lim_{t\to\infty}\frac{\ln t}{t^2} = \lim_{t\to\infty}\frac{\frac{1}{t}}{2t} = \lim_{t\to\infty}\frac{1}{2t^2} = 0$$

9. As $x \to 0, \sin x \to 0$, and $\frac{1}{x} \to \infty$. A number less than 1, raised to a large power, is a very small number $\left(\left(\frac{1}{2}\right)^{32} = 2.328\times 10^{-10}\right)$ so

$$\lim_{x\to 0^+}(\sin x)^{1/x} = 0.$$

11. The limit is of the form 0^0.

Let $y = x^x$, then $\ln y = x \ln x$.

$$\lim_{x\to 0^+} x\ln x = \lim_{x\to 0^+}\frac{\ln x}{\frac{1}{x}}$$

The limit is of the form $\frac{\infty}{\infty}$.

$$\lim_{x\to 0^+}\frac{\ln x}{\frac{1}{x}} = \lim_{x\to 0^+}\frac{\frac{1}{x}}{-\frac{1}{x^2}} = \lim_{x\to 0^+} -x = 0$$

$$\lim_{x\to 0^+} x^x = \lim_{x\to 0^+} e^{\ln y} = 1$$

13. $\lim_{x\to 0^+}\sqrt{x}\ln x = \lim_{x\to 0^+}\frac{\ln x}{\frac{1}{\sqrt{x}}}$

The limit is of the form $\frac{\infty}{\infty}$.

$$\lim_{x\to 0^+}\frac{\ln x}{\frac{1}{\sqrt{x}}} = \lim_{x\to 0^+}\frac{\frac{1}{x}}{-\frac{1}{2x^{3/2}}} = \lim_{x\to 0^+} -2\sqrt{x} = 0$$

15. $\lim_{x\to 0^+}\left(\frac{1}{\sin x} - \frac{1}{x}\right) = \lim_{x\to 0^+}\frac{x - \sin x}{x\sin x}$

The limit is of the form $\frac{0}{0}$. (Apply l'Hôpital's Rule twice.)

$$\lim_{x\to 0^+}\frac{x - \sin x}{x\sin x} = \lim_{x\to 0^+}\frac{1 - \cos x}{\sin x + x\cos x}$$

$$= \lim_{x\to 0^+}\frac{\sin x}{2\cos x - x\sin x} = \frac{0}{2} = 0$$

17. The limit is of the form 1^{∞}.

Let $y = (\sin x)^{\tan x}$, then $\ln y = \tan x \ln(\sin x)$.

$$\lim_{x\to\frac{\pi}{2}} \tan x \ln(\sin x) = \lim_{x\to\frac{\pi}{2}} \frac{\sin x \ln(\sin x)}{\cos x}$$

The limit is of the form $\frac{0}{0}$.

$$\lim_{x\to\frac{\pi}{2}} \frac{\sin x \ln(\sin x)}{\cos x} = \lim_{x\to\frac{\pi}{2}} \frac{\cos x \ln(\sin x) + \frac{\sin x}{\sin x}\cos x}{\sin x}$$

$$= \lim_{x\to\frac{\pi}{2}} \frac{\cos x(1+\ln(\sin x))}{\sin x} = \frac{0}{1} = 0$$

$$\lim_{x\to\frac{\pi}{2}} (\sin x)^{\tan x} = \lim_{x\to\frac{\pi}{2}} e^{\ln y} = 1$$

19. $\int_0^{\infty} \frac{dx}{(x+1)^2} = \left[-\frac{1}{x+1}\right]_0^{\infty} = 0+1=1$

21. $\int_{-\infty}^{1} e^{2x}\,dx = \left[\frac{1}{2}e^{2x}\right]_{-\infty}^{1} = \frac{1}{2}e^2 - 0 = \frac{1}{2}e^2$

23. $\int_0^{\infty} \frac{dx}{x+1} = [\ln(x+1)]_0^{\infty} = \infty - 0 = \infty$

The integral diverges.

25. $\int_1^{\infty} \frac{dx}{x^2+x^4} = \int_1^{\infty}\left(\frac{1}{x^2} - \frac{1}{1+x^2}\right)dx = \left[-\frac{1}{x} - \tan^{-1} x\right]_1^{\infty} = 0 - \frac{\pi}{2} + 1 + \tan^{-1} 1 = 1 + \frac{\pi}{4} - \frac{\pi}{2} = 1 - \frac{\pi}{4}$

27. $\int_{-2}^{0} \frac{dx}{2x+3} = \lim_{b\to -\frac{3}{2}^-} \int_{-2}^{b} \frac{dx}{2x+3} + \lim_{b\to -\frac{3}{2}^+} \int_{b}^{0} \frac{dx}{2x+3} = \lim_{b\to -\frac{3}{2}^-} \left[\frac{1}{2}\ln|2x+3|\right]_{-2}^{b} + \lim_{b\to -\frac{3}{2}^+} \left[\frac{1}{2}\ln|2x+3|\right]_{b}^{0}$

$$= \left(\lim_{b\to -\frac{3}{2}^-} \frac{1}{2}\ln|2b+3| - \frac{1}{2}(0)\right) + \left(\frac{1}{2}\ln 3 - \lim_{b\to -\frac{3}{2}^+} \frac{1}{2}\ln|2b+3|\right) = (-\infty) + \left(\frac{1}{2}\ln 3 + \infty\right)$$

The integral diverges.

29. $\int_2^{\infty} \frac{dx}{x(\ln x)^2} = \left[-\frac{1}{\ln x}\right]_2^{\infty} = -0 + \frac{1}{\ln 2} = \frac{1}{\ln 2}$

31. $\int_3^5 \frac{dx}{(4-x)^{2/3}} = \lim_{b\to 4^-} \int_3^b \frac{dx}{(4-x)^{2/3}} + \lim_{b\to 4^+} \int_b^5 \frac{dx}{(4-x)^{2/3}} = \lim_{b\to 4^-} \left[-3(4-x)^{1/3}\right]_3^b + \lim_{b\to 4^+} \left[-3(4-x)^{1/3}\right]_b^5$

$$= \lim_{b\to 4^-} -3(4-b)^{1/3} + 3(1)^{1/3} - 3(-1)^{1/3} + \lim_{b\to 4^+} 3(4-b)^{1/3} = 0+3+3+0 = 6$$

33. $\int_{-\infty}^{\infty} \frac{x}{x^2+1}dx = \int_{-\infty}^{0} \frac{x}{x^2+1}dx + \int_0^{\infty} \frac{x}{x^2+1}dx$

$$= \frac{1}{2}\left[\ln(x^2+1)\right]_{-\infty}^{0} + \frac{1}{2}\left[\ln(x^2+1)\right]_0^{\infty} =$$

$(0+\infty)+(\infty-0)$

The integral diverges.

35. $\frac{e^x}{e^{2x}+1} = \frac{e^x}{(e^x)^2+1}$

Let $u = e^x$, $du = e^x\,dx$

$$\int_0^{\infty} \frac{e^x}{e^{2x}+1}dx = \int_1^{\infty} \frac{1}{u^2+1}du = \left[\tan^{-1} u\right]_1^{\infty} = \frac{\pi}{2} - \tan^{-1} 1 = \frac{\pi}{2} - \frac{\pi}{4} = \frac{\pi}{4}$$

37. $\int_{-3}^{3} \frac{x}{\sqrt{9-x^2}}dx = 0$

See Problem 35 in Section 8.4.

39. For $p \neq 1, p \neq 0$, $\int_1^\infty \frac{1}{x^p}dx = \left[-\frac{1}{(p-1)x^{p-1}}\right]_1^\infty = \lim_{b\to\infty}\frac{1}{(1-p)b^{p-1}} + \frac{1}{p-1}$

$\lim_{b\to\infty}\frac{1}{b^{p-1}} = 0$ when $p-1>0$ or $p>1$, and $\lim_{b\to\infty}\frac{1}{b^{p-1}} = \infty$ when $p<1, p\neq 0$.

When $p=1$, $\int_1^\infty \frac{1}{x}dx = [\ln x]_1^\infty = \infty - 0$. The integral diverges.

When $p=0$, $\int_1^\infty 1dx = [x]_1^\infty = \infty - 1$. The integral diverges.

$\int_1^\infty \frac{1}{x^p}dx$ converges when $p>1$ and diverges when $p \leq 1$.

41. For $x \geq 1$, $x^6 + x > x^6$, so $\sqrt{x^6+x} > \sqrt{x^6} = x^3$ and $\frac{1}{\sqrt{x^6+x}} < \frac{1}{x^3}$. Hence, $\int_1^\infty \frac{1}{\sqrt{x^6+x}}dx < \int_1^\infty \frac{1}{x^3}dx$ which converges since $3>1$ (see Problem 39). Thus $\int_1^\infty \frac{1}{\sqrt{x^6+x}}dx$ converges.

43. For $x>3$, $\ln x > 1$, so $\frac{\ln x}{x} > \frac{1}{x}$. Hence, $\int_3^\infty \frac{\ln x}{x}dx > \int_3^\infty \frac{1}{x}dx = [\ln x]_3^\infty = \infty - \ln 3$.

The integral diverges, thus $\int_3^\infty \frac{\ln x}{x}dx$ also diverges.

Review and Preview Problems

1. Original: If $x>0$, then $x^2>0$ (AT)

Converse: If $x^2>0$, then $x>0$

Contrapositive: If $x^2 \leq 0$, then $x \leq 0$ (AT)

3. Original:
f differentiable at $c \Rightarrow f$ continuous at c (AT)
Converse:
f continuous at $c \Rightarrow f$ differentiable at c
Contrapositive:
f discontinuous at $c \Rightarrow f$ non-differentiable at c (AT)

5. Original:
f right continuous at $c \Rightarrow f$ continuous at c
Converse:
f continuous at $c \Rightarrow f$ right continuous at c (AT)
Contrapositive:
f discontinuous at $c \Rightarrow f$ not right continuous at c

7. Original: $f(x)=x^2 \Rightarrow f'(x)=2x$ (AT)

Converse: $f'(x)=2x \Rightarrow f(x)=x^2$

(Could have $f(x)=x^2+3$)

Contrapositive: $f'(x) \neq 2x \Rightarrow f(x) \neq x^2$ (AT)

9. $1+\frac{1}{2}+\frac{1}{4} = \frac{4}{4}+\frac{2}{4}+\frac{1}{4} = \frac{7}{4}$

11. $\sum_{i=1}^4 \frac{1}{i} = \frac{1}{1}+\frac{1}{2}+\frac{1}{3}+\frac{1}{4} = \frac{12+6+4+3}{12} = \frac{25}{12}$

13. By L'Hopital's Rule $\left(\frac{\infty}{\infty}\right)$:

$\lim_{x\to\infty}\frac{x}{2x+1} = \lim_{x\to\infty}\frac{1}{2} = \frac{1}{2}$

15. By L'Hopital's Rule $\left(\frac{\infty}{\infty}\right)$ twice:

$\lim_{x\to\infty}\frac{x^2}{e^x} = \lim_{x\to\infty}\frac{2x}{e^x} = \lim_{x\to\infty}\frac{2}{e^x} = 0$

17. $\int_1^\infty \frac{1}{x}dx = \lim_{t\to\infty}\int_1^t \frac{1}{x}dx =$

$\lim_{t\to\infty}[\ln x]_1^t = \lim_{t\to\infty}[\ln t] = \infty$

Integral does not converge.

19. $\int_1^\infty \frac{1}{x^{1.001}}dx = \lim_{t\to\infty}\int_1^t \frac{1}{x^{1.001}}dx =$

$\lim_{t\to\infty}\left[-\frac{1000}{x^{0.001}}\right]_1^t = \lim_{t\to\infty}\left[1000 - \frac{1000}{t^{0.001}}\right] = 1000$

Integral converges.

21. $\int_1^\infty \frac{x}{x^2+1}dx = \lim_{t\to\infty}\int_1^t \frac{x}{x^2+1}dx = \frac{1}{2}\ln\left(x^2+1\right)\Big|_1^\infty = \infty$

Integral does not converge.

CHAPTER 9

Infinite Series

9.1 Concepts Review

1. a sequence

3. bounded above

Problem Set 9.1

1. $a_1 = \frac{1}{2}, a_2 = \frac{2}{5}, a_3 = \frac{3}{8}, a_4 = \frac{4}{11}, a_5 = \frac{5}{14}$

$\lim_{n\to\infty} \frac{n}{3n-1} = \lim_{n\to\infty} \frac{1}{3-\frac{1}{n}} = \frac{1}{3}$; converges

3. $a_1 = \frac{6}{3} = 2, a_2 = \frac{18}{9} = 2, a_3 = \frac{38}{17},$

$a_4 = \frac{66}{27} = \frac{22}{9}, a_5 = \frac{102}{39} = \frac{34}{13}$

$\lim_{n\to\infty} \frac{4n^2+2}{n^2+3n-1} = \lim_{n\to\infty} \frac{4+\frac{2}{n^2}}{1+\frac{3}{n}-\frac{1}{n^2}} = 4$; converges

5. $a_1 = \frac{7}{8}, a_2 = \frac{26}{27}, a_3 = \frac{63}{64}, \quad a_4 = \frac{124}{125}, a_5 = \frac{215}{216}$

$\lim_{n\to\infty} \frac{n^3+3n^2+3n}{(n+1)^3} = \lim_{n\to\infty} \frac{n^3+3n^2+3n}{n^3+3n^2+3n+1}$

$= \lim_{n\to\infty} \frac{1+\frac{3}{n}+\frac{3}{n^2}}{1+\frac{3}{n}+\frac{3}{n^2}+\frac{1}{n^3}} = 1$

7. $a_1 = -\frac{1}{3}, a_2 = \frac{2}{4} = \frac{1}{2}, a_3 = -\frac{3}{5}, a_4 = \frac{4}{6} = \frac{2}{3},$

$a_5 = -\frac{5}{7}$

$\lim_{n\to\infty} \frac{n}{n+2} = \lim_{n\to\infty} \frac{1}{1+\frac{2}{n}} = 1$, but since it alternates between positive and negative, the sequence diverges.

9. $a_1 = -1, a_2 = \frac{1}{2}, a_3 = -\frac{1}{3}, a_4 = \frac{1}{4}, a_5 = -\frac{1}{5}$

$\cos(n\pi) = (-1)^n$, so $-\frac{1}{n} \le \frac{\cos(n\pi)}{n} \le \frac{1}{n}$.

$\lim_{n\to\infty} -\frac{1}{n} = \lim_{n\to\infty} \frac{1}{n} = 0$, so by the Squeeze Theorem, the sequence converges to 0.

11. $a_1 = \frac{e^2}{3} \approx 2.4630, a_2 = \frac{e^4}{9} \approx 6.0665,\ a_3 = \frac{e^6}{17} \approx 23.7311, a_4 = \frac{e^8}{27} \approx 110.4059,\ a_5 = \frac{e^{10}}{39} \approx 564.7812$

Consider $\lim_{x\to\infty} \frac{e^{2x}}{x^2+3x-1} = \lim_{x\to\infty} \frac{2e^{2x}}{2x+3} = \lim_{x\to\infty} \frac{4e^{2x}}{2} = \infty$ by using l'Hôpital's Rule twice. The sequence diverges.

13. $a_1 = -\frac{\pi}{5} \approx -0.6283, a_2 = \frac{\pi^2}{25} \approx 0.3948,\ a_3 = -\frac{\pi^3}{125} \approx -0.2481, a_4 = \frac{\pi^4}{625} \approx 0.1559,\ a_5 = -\frac{\pi^5}{3125} \approx -0.0979$

$\frac{(-\pi)^n}{5^n} = \left(-\frac{\pi}{5}\right)^n, -1 < -\frac{\pi}{5} < 1$, thus the sequence converges to 0.

15. $a_1 = 2.99, a_2 = 2.9801, a_3 \approx 2.9703,\ a_4 \approx 2.9606, a_5 \approx 2.9510$

$(0.99)^n$ converges to 0 since $-1 < 0.99 < 1$, thus $2 + (0.99)^n$ converges to 2.

17. $a_1 = \frac{\ln 1}{\sqrt{1}} = 0, a_2 = \frac{\ln 2}{\sqrt{2}} \approx 0.4901,\ a_3 = \frac{\ln 3}{\sqrt{3}} \approx 0.6343, a_4 = \frac{\ln 4}{2} \approx 0.6931,$

$a_5 = \frac{\ln 5}{\sqrt{5}} \approx 0.7198$

Consider $\lim_{x\to\infty} \frac{\ln x}{\sqrt{x}} = \lim_{x\to\infty} \frac{\frac{1}{x}}{\frac{1}{2\sqrt{x}}} = \lim_{x\to\infty} \frac{2}{\sqrt{x}} = 0$ by using l'Hôpital's Rule. Thus, $\lim_{n\to\infty} \frac{\ln n}{\sqrt{n}} = 0$; converges.

19. $a_1 = \left(1+\frac{2}{1}\right)^{1/2} = \sqrt{3} \approx 1.7321,$

$a_2 = \left(1+\frac{2}{2}\right)^{2/2} = 2,\ a_3 = \left(1+\frac{2}{3}\right)^{3/2} = \left(\frac{5}{3}\right)^{3/2} \approx 2.1517,\ a_4 = \left(1+\frac{2}{4}\right)^{4/2} = \left(\frac{3}{2}\right)^2 = \frac{9}{4},$

$a_5 = \left(1+\frac{2}{5}\right)^{5/2} = \left(\frac{7}{5}\right)^{5/2} \approx 2.3191$

Let $\frac{2}{n} = h$, then as $n \to \infty, h \to 0$ and $\lim_{n\to\infty}\left(1+\frac{2}{n}\right)^{n/2} = \lim_{h\to 0}(1+h)^{1/h} = e$ by Theorem 6.5A; converges

21. $a_n = \frac{n}{n+1}$ or $a_n = 1 - \frac{1}{n+1};\ \lim_{n\to\infty}\left(1-\frac{1}{n+1}\right) = 1 - \lim_{n\to\infty}\frac{1}{n+1} = 1$; converges

23. $a_n = (-1)^n \frac{n}{2n-1};\ \lim_{n\to\infty} \frac{n}{2n-1}$

$= \lim_{n\to\infty} \frac{1}{2-\frac{1}{n}} = \frac{1}{2}$, but due to $(-1)^n$, the terms of the sequence alternate between positive and negative, so the sequence diverges.

25. $a_n = \frac{n}{n^2-(n-1)^2} = \frac{n}{n^2-(n^2-2n+1)} = \frac{n}{2n-1}$;

$\lim_{n\to\infty} \frac{n}{2n-1} = \lim_{n\to\infty} \frac{1}{2-\frac{1}{n}} = \frac{1}{2}$; converges

27. $a_n = n\sin\frac{1}{n}$; $\lim_{n\to\infty} n\sin\frac{1}{n} = \lim_{n\to\infty} \frac{\sin\frac{1}{n}}{\frac{1}{n}} = 1$ since

$\lim_{x\to 0}\frac{\sin x}{x} = 1$; converges

29. $a_n = \frac{2^n}{n^2}$;

$\lim_{n\to\infty}\frac{2^n}{n^2} = \lim_{n\to\infty}\frac{2^n \ln 2}{2n} = \lim_{n\to\infty}\frac{2^n(\ln 2)^2}{2} = \infty$;
diverges

31. $a_1 = \frac{1}{2}, a_2 = \frac{5}{4}, a_3 = \frac{9}{8}, a_4 = \frac{13}{16}$

a_n is positive for all n, and $a_{n+1} < a_n$ for all $n \ge 2$ since $a_{n+1} - a_n = -\frac{4n-7}{2^{n+1}}$, so $\{a_n\}$ converges to a limit $L \ge 0$.

33. $a_2 = \frac{3}{4}; a_3 = \left(\frac{3}{4}\right)\left(\frac{8}{9}\right) = \frac{2}{3}$;

$a_4 = \left(\frac{3}{4}\right)\left(\frac{8}{9}\right)\left(\frac{15}{16}\right) = \frac{5}{8}$;

$a_5 = \left(\frac{3}{4}\right)\left(\frac{8}{9}\right)\left(\frac{15}{16}\right)\left(\frac{24}{25}\right) = \frac{3}{5}$

$a_n > 0$ for all n and $a_{n+1} < a_n$ since

$a_{n+1} = a_n\left(1-\frac{1}{(n+1)^2}\right)$ and $1-\frac{1}{(n+1)^2} < 1$, so $\{a_n\}$ converges to a limit $L \ge 0$.

35. $a_1 = 1, a_2 = 1+\frac{1}{2}(1) = \frac{3}{2}, a_3 = 1+\frac{1}{2}\left(\frac{3}{2}\right) = \frac{7}{4}$,

$a_4 = 1+\frac{1}{2}\left(\frac{7}{4}\right) = \frac{15}{8}$

Suppose that $1 < a_n < 2$, then $\frac{1}{2} < \frac{1}{2}a_n < 1$, so $\frac{3}{2} < 1+\frac{1}{2}a_n < 2$, or $\frac{3}{2} < a_{n+1} < 2$. Thus, since $1 < a_2 < 2$, every subsequent term is between $\frac{3}{2}$ and 2.

$a_n < 2$ thus $\frac{1}{2}a_n < 1$, so $a_n < 1+\frac{1}{2}a_n = a_{n+1}$ and the sequence is nondecreasing, so $\{a_n\}$ converges to a limit $L \le 2$.

37.

n	u_n
1	1.73205
2	2.17533
3	2.27493
4	2.29672
5	2.30146
6	2.30249
7	2.30271
8	2.30276
9	2.30277
10	2.30278
11	2.30278

$\lim_{n\to\infty} u_n \approx 2.3028$

39. If $u = \lim_{n\to\infty} u_n$, then $u = \sqrt{3+u}$ or $u^2 = 3+u$;

$u^2 - u - 3 = 0$ when $u = \frac{1}{2}\left(1 \pm \sqrt{13}\right)$ so

$u = \frac{1}{2}\left(1+\sqrt{13}\right) \approx 2.3028$ since $u > 0$ and

$\frac{1}{2}\left(1-\sqrt{13}\right) < 0$.

41.

n	u_n
1	0
2	1
3	1.1
4	1.11053
5	1.11165
6	1.11177
7	1.11178
8	1.11178

$\lim_{n\to\infty} u_n \approx 1.1118$

43. As $n \to \infty, \frac{k}{n} \to 0$; using $\Delta x = \frac{1}{n}$, an equivalent definite integral is

$\int_0^1 \sin x\, dx = [-\cos x]_0^1 = -\cos 1 + \cos 0 = 1 - \cos 1$
≈ 0.4597

45. $\left|\frac{n}{n+1} - 1\right| = \left|\frac{n-(n+1)}{n+1}\right| = \left|\frac{-1}{n+1}\right| = \frac{1}{n+1}$;

$\frac{1}{n+1} < \varepsilon$ is the same as $\frac{1}{\varepsilon} < n+1$. For any given $\varepsilon > 0$, choose $N > \frac{1}{\varepsilon} - 1$ then

$n \ge N \Rightarrow \left|\frac{n}{n+1} - 1\right| < \varepsilon$.

47. Recall that every rational number can be written as either a terminating or a repeating decimal. Thus if the sequence 1, 1.4, 1.41, 1.414, … has a limit within the rational numbers, the terms of the sequence would eventually either repeat or terminate, which they do not since they are the decimal approximations to $\sqrt{2}$, which is irrational. Within the real numbers, the least upper bound is $\sqrt{2}$.

49. If $\{b_n\}$ is bounded, there are numbers N and M with $N \le |b_n| \le M$ for all n. Then

$|a_n N| \le |a_n b_n| \le |a_n M|$.

$\lim_{n\to\infty} |a_n N| = |N| \lim_{n\to\infty} |a_n| = 0$ and

$\lim_{n\to\infty} |a_n M| = |M| \lim_{n\to\infty} |a_n| = 0$, so $\lim_{n\to\infty} |a_n b_n| = 0$

by the Squeeze Theorem, and by Theorem C, $\lim_{n\to\infty} a_n b_n = 0$.

51. No. Consider $a_n = (-1)^n$ and $b_n = (-1)^{n+1}$. Both $\{a_n\}$ and $\{b_n\}$ diverge, but $a_n + b_n = (-1)^n + (-1)^{n+1} = (-1)^n(1+(-1)) = 0$ so $\{a_n + b_n\}$ converges.

53.

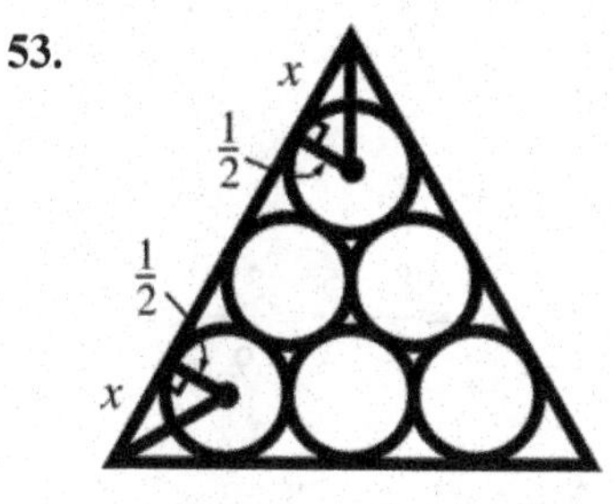

From the figure shown, the sides of the triangle have length $n - 1 + 2x$. The small right triangles marked are 30-60-90 right triangles, so $x = \frac{\sqrt{3}}{2}$; thus the sides of the large triangle have lengths $n-1+\sqrt{3}$ and $B_n = \frac{\sqrt{3}}{4}\left(n-1+\sqrt{3}\right)^2$

$= \frac{\sqrt{3}}{4}\left(n^2 + 2\sqrt{3}n - 2n - 2\sqrt{3} + 4\right)$ while

$$A_n = \frac{n(n+1)}{2}\pi\left(\frac{1}{2}\right)^2 = \frac{\pi}{8}(n^2+n)$$

$$\lim_{n\to\infty} \frac{A_n}{B_n} = \lim_{n\to\infty} \frac{\frac{\pi}{8}(n^2+n)}{\frac{\sqrt{3}}{4}(n^2+2\sqrt{3}n-2n-2\sqrt{3}+4)}$$

$$= \lim_{n\to\infty} \frac{\pi\left(1+\frac{1}{n}\right)}{2\sqrt{3}\left(1+\frac{2\sqrt{3}}{n}-\frac{2}{n}-\frac{2\sqrt{3}}{n^2}+\frac{4}{n^2}\right)} = \frac{\pi}{2\sqrt{3}}$$

55. Let $f(x) = \left(1+\frac{1}{2x}\right)^x$.

$$\lim_{x\to\infty}\left(1+\frac{1}{2x}\right)^x = \lim_{x\to 0^+}\left(1+\frac{x}{2}\right)^{1/x}$$

$$= \lim_{x\to 0^+}\left[\left(1+\frac{x}{2}\right)^{2/x}\right]^{1/2} = e^{1/2}, \text{ so}$$

$$\lim_{n\to\infty}\left(1+\frac{1}{2n}\right)^n = e^{1/2}.$$

57. Let $f(x) = \left(\frac{x-1}{x+1}\right)^x$.

Using the fact that $\lim_{x\to\infty} f(x) = \lim_{x\to 0^+} f\left(\frac{1}{x}\right)$, we can write

$$\lim_{x\to\infty}\left(\frac{x-1}{x+1}\right)^x = \lim_{x\to 0^+}\left(\frac{\frac{1}{x}-1}{\frac{1}{x}+1}\right)^{1/x} = \lim_{x\to 0^+}\left(\frac{\frac{1-x}{x}}{\frac{1+x}{x}}\right)^{1/x}$$

$= \lim_{x\to 0^+}\left(\frac{1-x}{1+x}\right)^{1/x}$ which leads to the indeterminate form 1^∞.

Let $y = \left(\frac{1-x}{1+x}\right)^{1/x}$. Then,

$$\ln y = \ln\left(\frac{1-x}{1+x}\right)^{1/x}$$

$$\ln y = \frac{1}{x}\ln\left(\frac{1-x}{1+x}\right)$$

$$\lim_{x\to 0^+} \ln y = \lim_{x\to 0^+} \frac{1}{x}\ln\left(\frac{1-x}{1+x}\right)$$

$$\ln\left[\lim_{x\to 0^+} y\right] = \lim_{x\to 0^+} \frac{\ln\left(\frac{1-x}{1+x}\right)}{x}$$

$$= \lim_{x\to 0^+} \frac{-2}{1-x^2} \quad \text{(l'Hopital's Rule)}$$

$$-2$$

This gives us,

$$\ln\left[\lim_{x\to 0^+} y\right] = -2$$

$$\lim_{x\to 0^+} y = e^{-2} \quad \text{or} \quad \lim_{x\to 0^+}\left(\frac{1-x}{1+x}\right)^{1/x} = e^{-2}$$

Thus, $\lim_{n\to\infty}\left(\frac{n-1}{n+1}\right)^n = e^{-2}$.

59. Let $f(x) = \left(\frac{2+x^2}{3+x^2}\right)^{x^2}$

Using the fact that $\lim_{x\to\infty} f(x) = \lim_{x\to 0^+} f\left(\frac{1}{x}\right)$, we can write

$$\lim_{x\to\infty}\left(\frac{2+x^2}{3+x^2}\right)^{x^2} = \lim_{x\to 0^+}\left(\frac{2+\frac{1}{x^2}}{3+\frac{1}{x^2}}\right)^{1/x^2}$$

$$= \lim_{x\to 0^+}\left(\frac{\frac{2x^2+1}{x^2}}{\frac{3x^2+1}{x^2}}\right)^{1/x^2} = \lim_{x\to 0^+}\left(\frac{2x^2+1}{3x^2+1}\right)^{1/x^2}$$ which

leads to the indeterminate form 1^∞.

Let $y = \left(\frac{2x^2+1}{3x^2+1}\right)^{1/x^2}$. Then,

$$\ln y = \ln\left(\frac{2x^2+1}{3x^2+1}\right)^{1/x^2}$$

$$\lim_{x\to 0^+} \ln y = \lim_{x\to 0^+} \frac{1}{x^2}\ln\left(\frac{2x^2+1}{3x^2+1}\right)$$

$$\ln\left[\lim_{x\to 0^+} y\right] = \lim_{x\to 0^+} \frac{\ln\left(\frac{2x^2+1}{3x^2+1}\right)}{x^2}$$

$$= \lim_{x\to 0^+}\left[\frac{-1}{(2x^2+1)(3x^2+1)}\right]$$ (l'Hopital's Rule)

$= -1$

This gives us,

$$\ln\left[\lim_{x\to 0^+} y\right] = -1$$

$$\lim_{x\to 0^+} y = e^{-1} \quad \text{or} \quad \lim_{x\to 0^+}\left(\frac{1-x}{1+x}\right)^{1/x^2} = e^{-1}$$

Thus,

$$\lim_{n\to\infty}\left(\frac{2+n^2}{3+n^2}\right)^{n^2} = e^{-1}.$$

9.2 Concepts Review

1. an infinite series

3. $|r| < 1; \frac{a}{1-r}$

Problem Set 9.2

1. $\sum_{k=1}^{\infty}\left(\frac{1}{7}\right)^k = \frac{1}{7} + \frac{1}{7}\cdot\frac{1}{7} + \frac{1}{7}\left(\frac{1}{7}\right)^2 + \ldots$; a geometric series with $a = \frac{1}{7}, r = \frac{1}{7}$; $S = \frac{\frac{1}{7}}{1-\frac{1}{7}} = \frac{\frac{1}{7}}{\frac{6}{7}} = \frac{1}{6}$

3. $\sum_{k=0}^{\infty} 2\left(\frac{1}{4}\right)^k = 2 + 2\cdot\frac{1}{4} + 2\left(\frac{1}{4}\right)^2 + \ldots$; a geometric series with $a = 2$, $r = \frac{1}{4}$; $S = \frac{2}{1-\frac{1}{4}} = \frac{2}{\frac{3}{4}} = \frac{8}{3}$.

$\sum_{k=0}^{\infty} 3\left(-\frac{1}{5}\right)^k = 3 - 3\cdot\frac{1}{5} + 3\left(\frac{1}{5}\right)^2 - \ldots$; a geometric series with $a = 3$, $r = -\frac{1}{5}$;

$$S = \frac{3}{1-\left(-\frac{1}{5}\right)} = \frac{3}{\frac{6}{5}} = \frac{5}{2}$$

Thus, by Theorem B,

$$\sum_{k=0}^{\infty}\left[2\left(\frac{1}{4}\right)^k + 3\left(-\frac{1}{5}\right)^k\right] = \frac{8}{3} + \frac{5}{2} = \frac{31}{6}$$

5. $\sum_{k=1}^{\infty}\frac{k-5}{k+2} = -\frac{4}{3} - \frac{3}{4} - \frac{2}{5} - \frac{1}{6} + 0 + \frac{1}{8} + \frac{2}{9} + \ldots$;

$\lim_{k\to\infty}\frac{k-5}{k+2} = \lim_{k\to\infty}\frac{1-\frac{5}{k}}{1+\frac{2}{k}} = 1 \neq 0$; the series diverges.

7. $\sum_{k=2}^{\infty}\left(\frac{1}{k} - \frac{1}{k-1}\right) = \left(\frac{1}{2} - \frac{1}{1}\right) + \left(\frac{1}{3} - \frac{1}{2}\right) + \left(\frac{1}{4} - \frac{1}{3}\right) + \ldots$;

$$S_n = \left(\frac{1}{2} - 1\right) + \left(\frac{1}{3} - \frac{1}{2}\right) + \ldots + \left(\frac{1}{n-1} - \frac{1}{n-2}\right) + \left(\frac{1}{n} - \frac{1}{n-1}\right) = -1 + \frac{1}{n};$$

$\lim_{n\to\infty} S_n = \lim_{n\to\infty} -1 + \frac{1}{n} = -1$, so $\sum_{k=2}^{\infty}\left(\frac{1}{k} - \frac{1}{k-1}\right) = -1$

9. $\sum_{k=1}^{\infty}\frac{k!}{100^k}=\frac{1}{100}+\frac{2}{10,000}+\frac{6}{1,000,000}+\ldots$

Consider $\{a_n\}$, where $a_{n+1}=\frac{n+1}{100}a_n, a_1=\frac{1}{100}$. $a_n>0$ for all n, and for $n>99$, $a_{n+1}>a_n$, so the sequence is eventually an increasing sequence, hence $\lim_{n\to\infty} a_n \neq 0$. The sequence can also be described by

$a_n=\frac{n!}{100^n}$, hence $\sum_{k=1}^{\infty}\frac{k!}{100^k}$ diverges.

11. $\sum_{k=1}^{\infty}\left(\frac{e}{\pi}\right)^{k+1}=\left(\frac{e}{\pi}\right)^2+\left(\frac{e}{\pi}\right)^2\cdot\frac{e}{\pi}+\left(\frac{e}{\pi}\right)^2\left(\frac{e}{\pi}\right)^2+\ldots$; a geometric series with $a=\left(\frac{e}{\pi}\right)^2, r=\frac{e}{\pi}<1$;

$$S=\frac{\left(\frac{e}{\pi}\right)^2}{1-\frac{e}{\pi}}=\frac{\left(\frac{e}{\pi}\right)^2}{\frac{\pi-e}{\pi}}=\frac{e^2}{\pi(\pi-e)}\approx 5.5562$$

13. $\sum_{k=2}^{\infty}\left(\frac{3}{(k-1)^2}-\frac{3}{k^2}\right)=\left(\frac{3}{1}-\frac{3}{4}\right)+\left(\frac{3}{4}-\frac{3}{9}\right)+\left(\frac{3}{9}-\frac{3}{16}\right)+\ldots;$

$$S_n=\left(3-\frac{3}{4}\right)+\left(\frac{3}{4}-\frac{1}{3}\right)+\left(\frac{1}{3}-\frac{3}{16}\right)+\ldots+\left(\frac{3}{(n-2)^2}-\frac{3}{(n-1)^2}\right)+\left(\frac{3}{(n-1)^2}-\frac{3}{n^2}\right)$$

$=3-\frac{3}{n^2}$; $\lim_{n\to\infty}S_n=3-\lim_{n\to\infty}\frac{3}{n^2}=3$, so

$$\sum_{k=2}^{\infty}\left(\frac{3}{(k-1)^2}-\frac{3}{k^2}\right)=3.$$

15. $0.22222\ldots=\sum_{k=1}^{\infty}\frac{2}{10}\left(\frac{1}{10}\right)^{k-1}$

$$=\frac{\frac{2}{10}}{1-\frac{1}{10}}=\frac{2}{9}$$

17. $0.013013013\ldots=\sum_{k=1}^{\infty}\frac{13}{1000}\left(\frac{1}{1000}\right)^{k-1}$

$$=\frac{\frac{13}{1000}}{1-\frac{1}{1000}}=\frac{13}{999}$$

19. $0.4999\ldots=\frac{4}{10}+\sum_{k=1}^{\infty}\frac{9}{100}\left(\frac{1}{10}\right)^{k-1}$

$$=\frac{4}{10}+\frac{\frac{9}{100}}{1-\frac{1}{10}}=\frac{1}{2}$$

21. Let $s=1-r$, so $r=1-s$. Since $0<r<2$, $-1<1-r<1$, so

$|s|<1$, and $\sum_{k=0}^{\infty}r(1-r)^k=\sum_{k=0}^{\infty}(1-s)s^k$

$$=\sum_{k=1}^{\infty}(1-s)s^{k-1}=\frac{1-s}{1-s}=1$$

23. $\ln\frac{k}{k+1} = \ln k - \ln(k+1)$

$S_n = (\ln 1 - \ln 2) + (\ln 2 - \ln 3) + (\ln 3 - \ln 4) + \ldots + (\ln(n-1) - \ln n) + (\ln n - \ln(n+1)) = \ln 1 - \ln(n+1) = -\ln(n+1)$

$\lim_{n\to\infty} S_n = \lim_{n\to\infty} -\ln(n+1) = -\infty$, thus $\sum_{k=1}^{\infty} \ln\frac{k}{k+1}$ diverges.

25. The ball drops 100 feet, rebounds up $100\left(\frac{2}{3}\right)$ feet, drops $100\left(\frac{2}{3}\right)$ feet, rebounds up $100\left(\frac{2}{3}\right)^2$ feet, drops $100\left(\frac{2}{3}\right)^2$, etc. The total distance it travels is

$$100 + 200\left(\frac{2}{3}\right) + 200\left(\frac{2}{3}\right)^2 + 200\left(\frac{2}{3}\right)^3 + \ldots = -100 + 200 + 200\left(\frac{2}{3}\right) + 200\left(\frac{2}{3}\right)^2 + 200\left(\frac{2}{3}\right)^3 + \ldots$$

$$= -100 + \sum_{k=1}^{\infty} 200\left(\frac{2}{3}\right)^{k-1} = -100 + \frac{200}{1-\frac{2}{3}} = 500 \text{ feet}$$

27. \$1 billion + 75% of \$1 billion + 75% of 75% of \$1 billion + ... = $\sum_{k=1}^{\infty} (\$1\text{ billion})0.75^{k-1} = \frac{\$1\text{ billion}}{1-0.75} = \$4\text{ billion}$

29. As the midpoints of the sides of a square are connected, a new square is formed. The new square has sides $\frac{1}{\sqrt{2}}$ times the sides of the old square. Thus, the new square has area $\frac{1}{2}$ the area of the old square. Then in the next step, $\frac{1}{8}$ of each new square is shaded.

$$\text{Area} = \frac{1}{8}\cdot 1 + \frac{1}{8}\cdot\frac{1}{2} + \frac{1}{8}\cdot\frac{1}{4} + \ldots = \sum_{k=1}^{\infty} \frac{1}{8}\left(\frac{1}{2}\right)^{k-1} = \frac{\frac{1}{8}}{1-\frac{1}{2}} = \frac{1}{4}$$

The area will be $\frac{1}{4}$.

31. $$\frac{3}{4} + \frac{3}{4}\left(\frac{1}{4}\cdot\frac{1}{4}\right) + \frac{3}{4}\left(\frac{1}{4}\cdot\frac{1}{4}\right)\left(\frac{1}{4}\cdot\frac{1}{4}\right) + \ldots = \sum_{k=1}^{\infty} \frac{3}{4}\left(\frac{1}{16}\right)^{k-1} = \frac{\frac{3}{4}}{1-\frac{1}{16}} = \frac{4}{5}$$

The original does not need to be equilateral since each smaller triangle will have $\frac{1}{4}$ area of the previous larger triangle.

33. a. We first note that, at each stage, the number of sides is four times the number in the previous stage and the length of each side is one-third the length in the previous stage. Summarizing:

Stage	# of sides	length/ side (in.)	perimeter p_n
0	3	9	27
1	3(4)	$9\left(\frac{1}{3}\right)$	36
⋮	⋮	⋮	⋮
n	$3(4^n)$	$9\left(\frac{1}{3^n}\right)$	$27\left(\frac{4}{3}\right)^n$

The perimeter of the Koch snowflake is $\lim_{n\to\infty} p_n = \lim_{n\to\infty} 27\left(\frac{4}{3}\right)^n$ which is infinite since $\frac{4}{3} > 1$.

b. We note the following:

1. The area of an equilateral triangle of side s is $\frac{\sqrt{3}}{4}s^2$

2. The number of new triangles added at each stage is equal to the number of sides the figure had at the previous stage and

3. the area of each new triangle at a given stage is $\frac{\sqrt{3}}{4}(\text{side length at that stage})^2$. Using results from part a. we can summarize:

Stage	Additional triangles (col 2, part *a.*)	Area of each new Δ (see col 3, part *a.*)	Additional area, A_n
0	original	$\frac{\sqrt{3}}{4}(9^2)$	$\frac{\sqrt{3}}{4}(9^2)$
1	3	$\frac{\sqrt{3}}{4}(3^2)$	36
⋮	⋮	⋮	⋮
n	$3(4^{n-1})$	$\frac{\sqrt{3}}{4}\left(\frac{9}{3^n}\right)^2$	$3\sqrt{3}\left(\frac{4}{9}\right)^{n-2}$

Thus the area of the Koch snowflake is

$$\sum_{n=0}^{\infty} A_n = \frac{81\sqrt{3}}{4} + \frac{27\sqrt{3}}{4} + \sum_{n=1}^{\infty} 3\sqrt{3}\left(\frac{4}{9}\right)^{n-1}$$

$$= \frac{81\sqrt{3}}{4} + \frac{27\sqrt{3}}{4} + \left(\frac{3\sqrt{3}}{\left(1-\frac{4}{9}\right)}\right)$$

$$= \frac{81\sqrt{3}}{4} + \frac{1}{3}\left(\frac{81\sqrt{3}}{4}\right) + \frac{4}{15}\left(\frac{81\sqrt{3}}{4}\right) = \frac{8}{5}\left(\frac{81\sqrt{3}}{4}\right)$$

Note: By generalizing the above argument it can be shown that, no matter what the size of the original equilateral triangle, the area of the Koch snowflake constructed from it will be $\frac{8}{5}$ times the area of the original triangle.

35. Both Achilles and the tortoise will have moved.

$$100 + 10 + 1 + \frac{1}{10} + \frac{1}{100} + \ldots = \sum_{k=1}^{\infty} 100\left(\frac{1}{10}\right)^{k-1}$$

$$= \frac{100}{1-\frac{1}{10}} = 111\frac{1}{9} \text{ yards}$$

Also, one can see this by the following reasoning. In the time it takes the tortoise to run $\frac{d}{10}$ yards, Achilles will run d yards. Solve $d = 100 + \frac{d}{10}$. $d = \frac{1000}{9} = 111\frac{1}{9}$ yards

37. Note that:

1. If we let t_n be the probability that Peter wins on his nth flip, then the total probability that Peter wins is $T = \sum_{n=1}^{\infty} t_n$

2. The probability that neither man wins in their first k flips is $\left(\frac{2}{3}\cdot\frac{2}{3}\right)^k = \left(\frac{4}{9}\right)^k$.

3. The probability that Peter wins on his nth flip requires that (i) he gets a head on the nth flip, and (ii) neither he nor Paul gets a head on their previous n-1 flips. Thus:

$$t_n = \left(\frac{1}{3}\right)\left(\frac{4}{9}\right)^{n-1} \text{ and } T = \sum_{n=1}^{\infty}\left(\frac{1}{3}\right)\left(\frac{4}{9}\right)^{n-1} =$$

$$\left(\frac{1/3}{(1-4/9)}\right) = \frac{1}{3}\cdot\frac{9}{5} = \frac{3}{5}$$

39. Let X = *number of rolls needed to get first 6* For X to equal n, two things must occur:

1. Mary must get a non-6 (probability $= \frac{5}{6}$) on each of her first n-1 rolls, and

2. Mary rolls a 6 (probability $= \frac{1}{6}$) on the nth roll. Thus,

$$\Pr(X = n) = \left(\frac{5}{6}\right)^{n-1}\left(\frac{1}{6}\right) \text{ and}$$

$$\sum_{n=1}^{\infty}\left(\frac{5}{6}\right)^{n-1}\left(\frac{1}{6}\right) = \frac{1/6}{(1-5/6)} = 1$$

41. (Proof by contradiction) Assume $\sum_{k=1}^{\infty} ca_k$ converges, and $c \neq 0$. Then $\frac{1}{c}$ is defined, so $\sum_{k=1}^{\infty} a_k = \sum_{k=1}^{\infty}\frac{1}{c}ca_k = \frac{1}{c}\sum_{k=1}^{\infty} ca_k$ would also converge, by Theorem B(i).

43. **a.** The top block is supported *exactly* at its center of mass. The location of the center of mass of the top n blocks is the average of the locations of their individual centers of mass, so the nth block moves the center of mass left by $\frac{1}{n}$ of the location of its center of mass, that is, $\frac{1}{n}\cdot\frac{1}{2}$ or $\frac{1}{2n}$ to the left. But this is exactly how far the (n + 1)st block underneath it is offset.

b. Since $\frac{1}{2}+\frac{1}{4}+\frac{1}{6}+\ldots = \frac{1}{2}\sum_{k=1}^{\infty}\frac{1}{k}$, which diverges, there is no limit to how far the top block can protrude.

45. (Proof by contradiction) Assume $\sum_{k=1}^{\infty}(a_k + b_k)$ converges. Since $\sum_{k=1}^{\infty} b_k$ converges, so would $\sum_{k=1}^{\infty} a_k = \sum_{k=1}^{\infty}(a_k + b_k) + (-1)\sum_{k=1}^{\infty} b_k$, by Theorem B(ii).

47. Taking vertical strips, the area is

$$1\cdot 1 + 1\cdot\frac{1}{2} + 1\cdot\frac{1}{4} + 1\cdot\frac{1}{8} + \cdots = \sum_{k=1}^{\infty}\left(\frac{1}{2}\right)^{k-1}.$$

Taking horizontal strips, the area is

$$\frac{1}{2}\cdot 1 + \frac{1}{4}\cdot 2 + \frac{1}{8}\cdot 3 + \frac{1}{16}\cdot 4 + \cdots = \sum_{k=1}^{\infty}\frac{k}{2^k}.$$

a. $\sum_{k=1}^{\infty}\frac{k}{2^k} = \sum_{k=1}^{\infty}\left(\frac{1}{2}\right)^{k-1} = \frac{1}{1-\frac{1}{2}} = 2$

b. The moment about $x = 0$ is

$$\sum_{k=0}^{\infty}\left(\frac{1}{2}\right)^k \cdot (1)k = \sum_{k=0}^{\infty}\frac{k}{2^k} = \sum_{k=1}^{\infty}\frac{k}{2^k} = 2.$$

$$\bar{x} = \frac{\text{moment}}{\text{area}} = \frac{2}{2} = 1$$

49. **a.** $A = \sum_{n=0}^{\infty} Ce^{-nkt} = \sum_{n=1}^{\infty} C\left(\frac{1}{e^{kt}}\right)^{n-1}$

$$= \frac{C}{1-\frac{1}{e^{kt}}} = \frac{Ce^{kt}}{e^{kt}-1}$$

b. $\frac{1}{2} = e^{-kt} = e^{-6k} \Rightarrow k = \frac{\ln 2}{6} \Rightarrow A = \frac{4}{3}C;$

if $C = 2$ mg, then $A = \frac{8}{3}$ mg.

51. $\frac{1}{f_k f_{k+1}} - \frac{1}{f_{k+1} f_{k+2}} = \frac{f_{k+2} - f_k}{f_k f_{k+1} f_{k+2}} = \frac{1}{f_k f_{k+2}}$

since $f_{k+2} = f_{k+1} + f_k$. Thus,

$\sum_{k=1}^{\infty} \frac{1}{f_k f_{k+2}} = \sum_{k=1}^{\infty} \left(\frac{1}{f_k f_{k+1}} - \frac{1}{f_{k+1} f_{k+2}} \right)$ and

$$S_n = \left(\frac{1}{f_1 f_2} - \frac{1}{f_2 f_3} \right) + \left(\frac{1}{f_2 f_3} - \frac{1}{f_3 f_4} \right) + \cdots + \left(\frac{1}{f_{n-1} f_n} - \frac{1}{f_n f_{n+1}} \right) + \left(\frac{1}{f_n f_{n+1}} - \frac{1}{f_{n+1} f_{n+2}} \right)$$

$$= \frac{1}{f_1 f_2} - \frac{1}{f_{n+1} f_{n+2}} = \frac{1}{1 \cdot 1} - \frac{1}{f_{n+1} f_{n+2}} = 1 - \frac{1}{f_{n+1} f_{n+2}}$$

The terms of the Fibonacci sequence increase without bound, so

$$\lim_{n \to \infty} S_n = 1 - \lim_{n \to \infty} \frac{1}{f_{n+1} f_{n+2}} = 1 - 0 = 1$$

9.3 Concepts Review

1. bounded above

3. convergence or divergence

Problem Set 9.3

1. $\frac{1}{x+3}$ is continuous, positive, and nonincreasing on $[0, \infty)$.

$$\int_0^{\infty} \frac{1}{x+3} dx = \left[\ln|x+3| \right]_0^{\infty} = \infty - \ln 3 = \infty$$

The series diverges.

3. $\frac{x}{x^2+3}$ is continuous, positive, and nonincreasing on $[2, \infty)$.

$$\int_2^{\infty} \frac{x}{x^2+3} dx = \left[\frac{1}{2} \ln\left|x^2+3\right| \right]_2^{\infty} = \infty - \frac{1}{2} \ln 7 = \infty$$

The series diverges.

5. $\frac{2}{\sqrt{x+2}}$ is continuous, positive, and nonincreasing on $[1, \infty)$.

$$\int_1^{\infty} \frac{2}{\sqrt{x+2}} dx = \left[4\sqrt{x+2} \right]_1^{\infty} = \infty - 4\sqrt{3} = \infty$$

Thus $\sum_{k=1}^{\infty} \frac{2}{\sqrt{k+2}}$ diverges, hence

$\sum_{k=1}^{\infty} \frac{-2}{\sqrt{k+2}} = -\sum_{k=1}^{\infty} \frac{2}{\sqrt{k+2}}$ also diverges.

7. $\frac{7}{4x+2}$ is continuous, positive, and nonincreasing on. $[2, \infty)$

$$\int_2^{\infty} \frac{7}{4x+2} dx = \left[\frac{7}{4} \ln|4x+2| \right]_2^{\infty} = \infty - \frac{7}{4} \ln 10 = \infty$$

The series diverges.

9. $\frac{3}{(4+3x)^{7/6}}$ is continuous, positive, and nonincreasing on $[1, \infty)$.

$$\int_1^{\infty} \frac{3}{(4+3x)^{7/6}} dx = \left[-\frac{6}{(4+3x)^{1/6}} \right]_1^{\infty}$$

$$= 0 + \frac{6}{7^{1/6}} = 6 \cdot 7^{-1/6} < \infty$$

The series converges.

11. xe^{-3x^2} is continuous, positive, and nonincreasing on $[1, \infty)$.

$$\int_1^{\infty} xe^{-3x^2} dx = \left[-\frac{1}{6} e^{-3x^2} \right]_1^{\infty} = 0 + \frac{1}{6} e^{-3}$$

$$= \frac{1}{6e^3} < \infty$$

The series converges.

13. $\lim_{k \to \infty} \frac{k^2+1}{k^2+5} = \lim_{k \to \infty} \frac{1 + \frac{1}{k^2}}{1 + \frac{5}{k^2}} = 1 \neq 0$, so the series diverges.

15. $\sum_{k=1}^{\infty}\left(\frac{1}{2}\right)^k$ is a geometric series with $r=\frac{1}{2}; \left|\frac{1}{2}\right|<1$ so the series converges.

In $\sum_{k=1}^{\infty}\frac{k-1}{2k+1}$, $\lim_{k\to\infty}\frac{k-1}{2k+1}=\lim_{k\to\infty}\frac{1-\frac{1}{k}}{2+\frac{1}{k}}=\frac{1}{2}\neq 0$, so the series diverges. Thus, the sum of the series diverges.

17. $\sin\left(\frac{k\pi}{2}\right)=\begin{cases}1 & k=4j+1\\ -1 & k=4j+3,\\ 0 & k \text{ is even}\end{cases}$

where j is any nonnegative integer.

Thus $\lim_{k\to\infty}\left|\sin\left(\frac{k\pi}{2}\right)\right|$ does not exist, hence

$\lim_{k\to\infty}\left|\sin\left(\frac{k\pi}{2}\right)\right|\neq 0$ and the series diverges.

19. $x^2e^{-x^3}$ is continuous, positive, and nonincreasing on $[1,\infty)$.

$\int_1^{\infty}x^2e^{-x^3}dx=\left[-\frac{1}{3}e^{-x^3}\right]_1^{\infty}=0+\frac{1}{3}e^{-1}<\infty$, so the series converges.

21. $\frac{\tan^{-1}x}{1+x^2}$ is continuous, positive, and nonincreasing on $[1,\infty)$.

$\int_1^{\infty}\frac{\tan^{-1}x}{1+x^2}dx=\left[\frac{1}{2}(\tan^{-1}x)^2\right]_1^{\infty}$

$=\frac{1}{2}\left(\frac{\pi}{2}\right)^2-\frac{1}{2}\left(\frac{\pi}{4}\right)^2=\frac{3\pi^2}{32}<\infty$, so the series converges.

23. $\frac{x}{e^x}$ is continuous, positive, and nonincreasing on $[5,\infty)$.

$E=\sum_{k=6}^{\infty}\frac{k}{e^k}\le\int_5^{\infty}\frac{x}{e^x}dx=[-xe^{-x}]_5^{\infty}+\int_5^{\infty}e^{-x}dx$

$=[-xe^{-x}-e^{-x}]_5^{\infty}=0+5e^{-5}+e^{-5}=6e^{-5}$

≈ 0.0404

25. $\frac{1}{1+x^2}$ is continuous, positive, and nonincreasing on $[5,\infty)$.

$E=\sum_{k=6}^{\infty}\frac{1}{1+k^2}\le\int_5^{\infty}\frac{1}{1+x^2}dx=[\tan^{-1}x]_5^{\infty}$

$=\frac{\pi}{2}-\tan^{-1}5\approx 0.1974$

27. $E_n=\sum_{k=n+1}^{\infty}\frac{1}{k^2}<\int_n^{\infty}\frac{1}{x^2}dx=\lim_{A\to\infty}\int_n^{A}\frac{1}{x^2}dx=$

$\lim_{A\to\infty}\left[\frac{1}{n}-\frac{1}{A}\right]=\frac{1}{n}$

$\frac{1}{n}<0.0002\Rightarrow n>5000$

29. $E_n=\sum_{k=n+1}^{\infty}\frac{1}{1+k^2}<\int_n^{\infty}\frac{1}{1+x^2}dx$

$=\lim_{A\to\infty}\int_n^{A}\frac{1}{1+x^2}dx=\lim_{A\to\infty}\left[\tan^{-1}A-\tan^{-1}n\right]$

$=\frac{\pi}{2}-\tan^{-1}n$

$\frac{\pi}{2}-\tan^{-1}n<0.0002\Rightarrow\tan^{-1}n>\frac{\pi}{2}-0.0002$

$\Rightarrow n>\tan\left(\frac{\pi}{2}-0.0002\right)\approx 5000$

31. $E_n=\sum_{k=n+1}^{\infty}\frac{k}{1+k^4}<\int_n^{\infty}\frac{x}{1+x^4}dx=\lim_{A\to\infty}\frac{1}{2}\int_{n^2}^{A}\frac{du}{1+u^2}$

$u=x^2$
$du=2x\,dx$

$=\frac{1}{2}\lim_{A\to\infty}\left[\tan^{-1}A-\tan^{-1}n^2\right]=\frac{1}{2}\left[\frac{\pi}{2}-\tan^{-2}\left(n^2\right)\right]$

$\frac{1}{2}\left[\frac{\pi}{2}-\tan^{-2}\left(n^2\right)\right]<0.0002$

$\Rightarrow\frac{\pi}{2}-\tan^{-2}\left(n^2\right)<0.0004\Rightarrow\tan^{-1}\left(n^2\right)>1.5703963$

$\Rightarrow n>\sqrt{\tan(1.5703963)}\approx 50$

33. Consider $\int_2^{\infty}\frac{1}{x(\ln x)^p}dx$. Let $u=\ln x$,

$du=\frac{1}{x}dx.$

$\int_2^{\infty}\frac{1}{x(\ln x)^p}dx=\int_{\ln 2}^{\infty}\frac{1}{u^p}du$ which converges for $p>1$.

35. 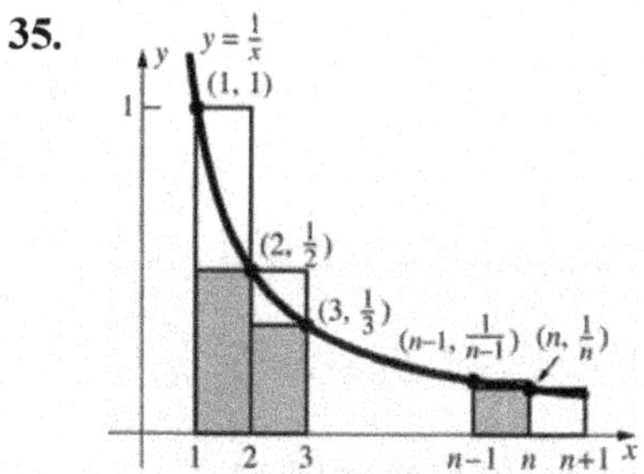

The upper rectangles, which extend to $n + 1$ on the right, have area $1+\frac{1}{2}+\frac{1}{3}+\ldots+\frac{1}{n}$. These rectangles are above the curve $y=\frac{1}{x}$ from $x = 1$ to $x = n + 1$. Thus,

$$\int_1^{n+1}\frac{1}{x}dx = [\ln x]_1^{n+1} = \ln(n+1) - \ln 1 = \ln(n+1)$$
$$< 1+\frac{1}{2}+\frac{1}{3}+\ldots+\frac{1}{n}.$$

The lower (shaded) rectangles have area $\frac{1}{2}+\frac{1}{3}+\ldots+\frac{1}{n}$. These rectangles lie below the curve $y=\frac{1}{x}$ from $x = 1$ to $x = n$. Thus

$$\frac{1}{2}+\frac{1}{3}+\ldots+\frac{1}{n} < \int_1^n \frac{1}{x}dx = \ln n, \text{ so}$$
$$1+\frac{1}{2}+\frac{1}{3}+\ldots+\frac{1}{n} < 1+\ln n.$$

37. $\{B_n\}$ is a nondecreasing sequence that is bounded above, thus by the Monotonic Sequence Theorem (Theorem D of Section 9.1), $\lim_{n\to\infty} B_n$ exists. The rationality of γ is a famous unsolved problem.

39. $\gamma + \ln(n+1) > 20 \Rightarrow \ln(n+1) > 20-\gamma \approx 19.4228$
$\Rightarrow n+1 > e^{19.4228} \approx 272{,}404{,}867$
$\Rightarrow n > 272{,}404{,}866$

41. Every time n is incremented by 1, a positive amount of area is added, thus $\{A_n\}$ is an increasing sequence. Each curved region has horizontal width 1, and can be moved into the heavily outlined triangle without any overlap. This can be done by shifting the nth shaded region, which goes from $(n, f(n))$ to $(n + 1, f(n + 1))$, as follows:
shift $(n + 1, f(n + 1))$ to $(2, f(2))$ and $(n, f(n))$ to $(1, f(2)-[f(n + 1) - f(n)])$.
The slope of the line forming the bottom of the shaded region between $x = n$ and $x = n + 1$ is

$$\frac{f(n+1)-f(n)}{(n+1)-n} = f(n+1)-f(n) > 0$$

since f is increasing.
By the Mean Value Theorem,
$f(n+1)-f(n) = f'(c)$ for some c in $(n, n + 1)$.
Since f is concave down, $n < c < n+1$ means that $f'(c) < f'(b)$ for all b in $[1, n]$. Thus, the nth shaded region will not overlap any other shaded region when shifted into the heavily outlined triangle. Thus, the area of all of the shaded regions is less than or equal to the area of the heavily outlined triangle, so $\lim_{n\to\infty} A_n$ exists.

43. (Refer to *fig 2* in the text). Let $b_k = \int_k^{k+1} f(x)\,dx$; then from *fig 2*, it is clear that $a_k \ge b_k$ for $k = 1, 2, \ldots, n, \ldots$

Therefore $\sum_{k=n+1}^{t} a_k \ge \sum_{k=n+1}^{t} b_k = \int_{n+1}^{t} f(x)\,dx$ so that

$$E_n = \sum_{k=n+1}^{\infty} a_k = \lim_{t\to\infty}\sum_{k=n+1}^{t} a_k \ge \lim_{t\to\infty}\int_{n+1}^{t} f(x)\,dx = \int_{n+1}^{\infty} f(x)\,dx .$$

9.4 Concepts Review

1. $0 \le a_k \le b_k$

3. $\rho < 1; \rho > 1; \rho = 1$

Problem Set 9.4

1. $a_n = \frac{n}{n^2+2n+3}; b_n = \frac{1}{n}$

$$\lim_{n\to\infty} \frac{a_n}{b_n} = \lim_{n\to\infty} \frac{n^2}{n^2+2n+3} = \lim_{n\to\infty} \frac{1}{1+\frac{2}{n}+\frac{3}{n^2}} = 1;$$

$0 < 1 < \infty$

$\sum_{n=1}^{\infty} b_n$ diverges $\Rightarrow \sum_{n=1}^{\infty} a_n$ diverges.

3. $a_n = \frac{1}{n\sqrt{n+1}} = \frac{1}{\sqrt{n^3+n^2}}; \; b_n = \frac{1}{n^{3/2}}$

$$\lim_{n\to\infty} \frac{a_n}{b_n} = \lim_{n\to\infty} \frac{n^{3/2}}{\sqrt{n^3+n^2}} = \lim_{n\to\infty} \sqrt{\frac{n^3}{n^3+n^2}}$$

$$= \lim_{n\to\infty} \sqrt{\frac{1}{1+\frac{1}{n}}} = 1; \; 0 < 1 < \infty$$

$\sum_{n=1}^{\infty} b_n$ converges $\Rightarrow \sum_{n=1}^{\infty} a_n$ converges

5. $\lim_{n\to\infty} \frac{a_{n+1}}{a_n} = \lim_{n\to\infty} \frac{8^{n+1} n!}{(n+1)! 8^n} = \lim_{n\to\infty} \frac{8}{n+1} = 0 < 1$

The series converges.

7. $\lim_{n\to\infty} \frac{a_{n+1}}{a_n} = \lim_{n\to\infty} \frac{(n+1)! n^{100}}{(n+1)^{100} n!} = \lim_{n\to\infty} \frac{n^{100}}{(n+1)^{99}}$

$$= \lim_{n\to\infty} \frac{n}{\left(\frac{n+1}{n}\right)^{99}} = \infty \text{ since } \lim_{n\to\infty} \left(\frac{n+1}{n}\right)^{99} = 1$$

The series diverges.

9. $\lim_{n\to\infty} \frac{a_{n+1}}{a_n} = \lim_{n\to\infty} \frac{(n+1)^3 (2n)!}{(2n+2)! n^3}$

$$= \lim_{n\to\infty} \frac{(n+1)^3}{(2n+2)(2n+1)n^3} = \lim_{n\to\infty} \frac{n^3+3n^2+3n+1}{4n^5+6n^4+2n^3}$$

$$= \lim_{n\to\infty} \frac{\frac{1}{n^2}+\frac{3}{n^3}+\frac{3}{n^4}+\frac{1}{n^5}}{4+\frac{6}{n}+\frac{2}{n^2}} = 0 < 1$$

The series converges.

11. $\lim_{n\to\infty} \frac{n}{n+200} = \lim_{n\to\infty} \frac{1}{1+\frac{200}{n}} = 1 \ne 0$

The series diverges; nth-Term Test

13. $a_n = \frac{n+3}{n^2\sqrt{n}}; b_n = \frac{1}{n^{3/2}}$

$$\lim_{n\to\infty} \frac{a_n}{b_n} = \lim_{n\to\infty} \frac{n^{5/2}+3n^{3/2}}{n^{5/2}} = \lim_{n\to\infty} \frac{1+\frac{3}{n}}{1} = 1;$$

$0 < 1 < \infty$. $\sum_{n=1}^{\infty} b_n$ converges $\Rightarrow \sum_{n=1}^{\infty} a_n$ converges; Limit Comparison Test

15. $\lim_{n\to\infty} \frac{a_{n+1}}{a_n} = \lim_{n\to\infty} \frac{(n+1)^2 n!}{(n+1)! n^2} = \lim_{n\to\infty} \frac{n^2+2n+1}{(n+1)n^2}$

$$= \lim_{n\to\infty} \frac{n^2+2n+1}{n^3+n^2} = \lim_{n\to\infty} \frac{\frac{1}{n}+\frac{2}{n^2}+\frac{1}{n^3}}{1+\frac{1}{n}} = 0 < 1$$

The series converges; Ratio Test

17. $a_n = \frac{4n^3+3n}{n^5-4n^2+1}; b_n = \frac{1}{n^2}$

$$\lim_{n\to\infty} \frac{a_n}{b_n} = \lim_{n\to\infty} \frac{4n^5+3n^3}{n^5-4n^2+1} = \lim_{n\to\infty} \frac{4+\frac{3}{n^2}}{1-\frac{4}{n^3}+\frac{1}{n^5}} = 4;$$

$0 < 4 < \infty$

$\sum_{n=1}^{\infty} b_n$ converges $\Rightarrow \sum_{n=1}^{\infty} a_n$ converges; Limit Comparison Test

19. $a_n = \frac{1}{n(n+1)} = \frac{1}{n^2+n}; b_n = \frac{1}{n^2}$

$$\lim_{n\to\infty} \frac{a_n}{b_n} = \lim_{n\to\infty} \frac{n^2}{n^2+n} = \lim_{n\to\infty} \frac{1}{1+\frac{1}{n}} = 1;$$

$0 < 1 < \infty$

$\sum_{n=1}^{\infty} b_n$ converges $\Rightarrow \sum_{n=1}^{\infty} a_n$ converges; Limit Comparison Test

21. $a_n = \frac{n+1}{n(n+2)(n+3)} = \frac{n+1}{n^3+5n^2+6n}; b_n = \frac{1}{n^2}$

$$\lim_{n\to\infty} \frac{a_n}{b_n} = \lim_{n\to\infty} \frac{n^3+n^2}{n^3+5n^2+6n} = \lim_{n\to\infty} \frac{1+\frac{1}{n}}{1+\frac{5}{n}+\frac{6}{n^2}} = 1;$$

$0 < 1 < \infty$

$\sum_{n=1}^{\infty} b_n$ converges $\Rightarrow \sum_{n=1}^{\infty} a_n$ converges; Limit Comparison Test

23. $a_n = \frac{n}{3^n}$; $\lim_{n\to\infty} \frac{a_{n+1}}{a_n} = \lim_{n\to\infty} \frac{(n+1)3^n}{3^{n+1}n}$

$= \lim_{n\to\infty} \frac{n+1}{3n} = \lim_{n\to\infty} \frac{1+\frac{1}{n}}{3} = \frac{1}{3} < 1$

The series converges; Ratio Test

25. $a_n = \frac{1}{n\sqrt{n}} = \frac{1}{n^{3/2}}$; $\frac{1}{x^{3/2}}$ is continuous, positive, and nonincreasing on $[1, \infty)$.

$$\int_1^\infty \frac{1}{x^{3/2}}dx = \left[-\frac{2}{\sqrt{x}}\right]_1^\infty = 0 + 2 = 2 < \infty$$

The series converges; Integral Test

27. $0 \le \sin^2 n \le 1$ for all n, so

$2 \le 2 + \sin^2 n \le 3 \Rightarrow \frac{1}{2} \ge \frac{1}{2+\sin^2 n} \ge \frac{1}{3}$ for all n.

Thus, $\lim_{n\to\infty} \frac{1}{2+\sin^2 n} \ne 0$ and the series diverges; nth-Term Test

29. $-1 \le \cos n \le 1$ for all n, so

$3 \le 4 + \cos n \le 5 \Rightarrow \frac{3}{n^3} \le \frac{4+\cos n}{n^3} \le \frac{5}{n^3}$ for all n.

$\sum_{n=1}^{\infty} \frac{5}{n^3}$ converges $\Rightarrow \sum_{n=1}^{\infty} \frac{4+\cos n}{n^3}$ converges;

Comparison Test

31. $\lim_{n\to\infty} \frac{a_{n+1}}{a_n} = \lim_{n\to\infty} \frac{(n+1)^{n+1}(2n)!}{(2n+2)!n^n}$

$= \lim_{n\to\infty} \frac{(n+1)^{n+1}}{(2n+2)(2n+1)n^n} = \lim_{n\to\infty} \frac{(n+1)^{n+1}}{2(n+1)(2n+1)n^n}$

$= \lim_{n\to\infty} \frac{(n+1)^n}{2(2n+1)n^n} = \lim_{n\to\infty} \left[\frac{1}{4n+2}\left(\frac{n+1}{n}\right)^n\right]$

$= \left[\lim_{n\to\infty} \frac{1}{4n+2}\right]\left[\lim_{n\to\infty}\left(\frac{n+1}{n}\right)^n\right] = 0 \cdot e = 0 < 1$

(The limits can be separated since both limits exist.) The series converges; Ratio Test

33. $\lim_{n\to\infty} \frac{a_{n+1}}{a_n} = \lim_{n\to\infty} \frac{(4^{n+1}+n+1)n!}{(n+1)!(4^n+n)}$

$= \lim_{n\to\infty} \frac{4^{n+1}+n+1}{(n+1)(4^n+n)} = \lim_{n\to\infty} \frac{4+\frac{n}{4^n}+\frac{1}{4^n}}{(n+1)\left(1+\frac{n}{4^n}\right)}$

$= \lim_{n\to\infty} \frac{4+\frac{n}{4^n}+\frac{1}{4^n}}{1+n+\frac{n}{4^n}+\frac{n^2}{4^n}} = 0$

since $\lim_{n\to\infty} \frac{n^2}{4^n} = 0$, $\lim_{n\to\infty} \frac{n}{4^n} = 0$, and

$\lim_{n\to\infty} \frac{1}{4^n} = 0$. The series converges; Ratio Test

35. Since $\sum a_n$ converges, $\lim_{n\to\infty} a_n = 0$. Thus, there is some positive integer N such that $0 < a_n < 1$ for all $n \ge N$. $a_n < 1 \Rightarrow a_n{}^2 < a_n$, thus $\sum_{n=N}^{\infty} a_n{}^2 < \sum_{n=N}^{\infty} a_n$. Hence $\sum_{n=N}^{\infty} a_n{}^2$ converges, and $\sum a_n{}^2$ also converges, since adding a finite number of terms does not affect the convergence or divergence of a series.

37. If $\lim_{n\to\infty} \frac{a_n}{b_n} = 0$ then there is some positive integer N such that $0 \le \frac{a_n}{b_n} < 1$ for all $n \ge N$. Thus, for $n \ge N$, $a_n < b_n$. By the Comparison Test, since $\sum_{n=N}^{\infty} b_n$ converges, $\sum_{n=N}^{\infty} a_n$ also converges. Thus, $\sum a_n$ converges since adding a finite number of terms will not affect the convergence or divergence of a series.

39. If $\lim_{n\to\infty} na_n = 1$ then there is some positive integer N such that $a_n \ge 0$ for all $n \ge N$. Let $b_n = \frac{1}{n}$, so $\lim_{n\to\infty} \frac{a_n}{b_n} = \lim_{n\to\infty} na_n = 1 < \infty$.

Since $\sum_{n=N}^{\infty} \frac{1}{n}$ diverges, $\sum_{n=N}^{\infty} a_n$ diverges by the Limit Comparison Test.

Thus $\sum a_n$ diverges since adding a finite number of terms will not affect the convergence or divergence of a series.

41. Suppose that $\lim_{n\to\infty} (a_n)^{1/n} = R$ where $a_n > 0$.
If $R < 1$, there is some number r with $R < r < 1$ and some positive integer N such that $\left|(a_n)^{1/n} - R\right| < r - R$ for all $n \geq N$. Thus,

$R - r < (a_n)^{1/n} - R < r - R$ or

$-r < (a_n)^{1/n} < r < 1$. Since $a_n > 0$,

$0 < (a_n)^{1/n} < r$ and $0 < a_n < r^n$ for all $n \geq N$.

Thus, $\sum_{n=N}^{\infty} a_n < \sum_{n=N}^{\infty} r^n$, which converges since

$|r| < 1$. Thus, $\sum_{n=N}^{\infty} a_n$ converges so $\sum a_n$ also converges.

If $R > 1$, there is some number r with $1 < r < R$ and some positive integer N such that $\left|(a_n)^{1/n} - R\right| < R - r$ for all $n \geq N$. Thus,

$r - R < (a_n)^{1/n} - R < R - r$ or

$r < (a_n)^{1/n} < 2R - r$ for all $n \geq N$. Hence

$r^n < a_n$ for all $n \geq N$, so $\sum_{n=N}^{\infty} r^n < \sum_{n=N}^{\infty} a_n$, and

since $\sum_{n=N}^{\infty} r^n$ diverges $(r > 1)$, $\sum_{n=N}^{\infty} a_n$ also

diverges, so $\sum a_n$ diverges.

43. a. $\ln\left(1+\frac{1}{n}\right) = \ln\left(\frac{n+1}{n}\right) = \ln(n+1) - \ln n$

$S_n = (\ln 2 - \ln 1) + (\ln 3 - \ln 2) + \ldots$
$+ (\ln n - \ln(n-1)) + (\ln(n+1) - \ln n)$
$= -\ln 1 + \ln(n+1) = \ln(n+1)$

$\lim_{n\to\infty} S_n = \lim_{n\to\infty} \ln(n+1) = \infty$

Since the partial sums are unbounded, the series diverges.

b. $\ln\frac{(n+1)^2}{n(n+2)} = 2\ln(n+1) - \ln n - \ln(n+2)$

$S_n = (2\ln 2 - \ln 1 - \ln 3) + (2\ln 3 - \ln 2 - \ln 4)$
$+ (2\ln 4 - \ln 3 - \ln 5) + \ldots$
$+ (2\ln n - \ln(n-1) - \ln(n+1))$
$+ (2\ln(n+1) - \ln n - \ln(n+2))$
$= \ln 2 - \ln 1 + \ln(n+1) - \ln(n+2)$
$= \ln 2 + \ln\frac{n+1}{n+2}$

$\lim_{n\to\infty} S_n = \ln 2 + \lim_{n\to\infty} \ln\frac{n+1}{n+2} = \ln 2$

Since the partial sums converge, the series converges.

c. $\left(\frac{1}{\ln x}\right)^{\ln x}$ is continuous, positive, and

nonincreasing on $[2, \infty)$, thus $\sum_{n=2}^{\infty}\left(\frac{1}{\ln n}\right)^{\ln n}$

converges if and only if $\int_2^{\infty}\left(\frac{1}{\ln x}\right)^{\ln x} dx$

converges.

Let $u = \ln x$, so $x = e^u$ and $dx = e^u\,du$.

$$\int_2^{\infty}\left(\frac{1}{\ln x}\right)^{\ln x} dx = \int_{\ln 2}^{\infty}\left(\frac{1}{u}\right)^u e^u\,du = \int_{\ln 2}^{\infty}\left(\frac{e}{u}\right)^u du$$

This integral converges if and only if the associated series, $\sum_{n=1}^{\infty}\left(\frac{e}{n}\right)^n$ converges. With

$a_n = \left(\frac{e}{n}\right)^n$, the Root Test (Problem 41)

gives $\lim_{n\to\infty}(a_n)^{1/n} = \lim_{n\to\infty}\left[\left(\frac{e}{n}\right)^n\right]^{1/n}$

$= \lim_{n\to\infty}\frac{e}{n} = 0 < 1$

Thus, $\sum_{n=1}^{\infty}\left(\frac{e}{n}\right)^n$ converges, so $\int_{\ln 2}^{\infty}\left(\frac{e}{u}\right)^u du$

converges, whereby $\sum_{n=2}^{\infty}\frac{1}{(\ln n)^{\ln n}}$

converges.

d. $\left(\frac{1}{\ln(\ln x)}\right)^{\ln x}$ is continuous, positive, and

nonincreasing on $[3, \infty)$, thus

$\sum_{n=3}^{\infty}\left(\frac{1}{\ln(\ln n)}\right)^{\ln n}$ converges if and only if

$\int_3^{\infty}\left(\frac{1}{\ln(\ln x)}\right)^{\ln x} dx$ converges.

Let $u = \ln x$, so $x = e^u$ and $dx = e^u\,du$.

$$\int_3^{\infty}\left(\frac{1}{\ln(\ln x)}\right)^{\ln x} dx$$

$$= \int_{\ln 3}^{\infty}\left(\frac{1}{\ln u}\right)^u e^u\,du = \int_{\ln 3}^{\infty}\left(\frac{e}{\ln u}\right)^u du.$$

This integral converges if and only if the

associated series, $\sum_{n=2}^{\infty}\left(\frac{e}{\ln n}\right)^n$ converges.

With $a_n = \left(\frac{e}{\ln n}\right)^n$, the Root Test (Problem 41) gives

$$\lim_{n\to\infty}(a_n)^{1/n} = \lim_{n\to\infty}\left[\left(\frac{e}{\ln n}\right)^n\right]^{1/n}.$$

$$= \lim_{n\to\infty}\frac{e}{\ln n} = 0 < 1$$

Thus, $\sum_{n=2}^{\infty}\left(\frac{e}{\ln n}\right)^n$ converges, so

$\int_{\ln 3}^{\infty}\left(\frac{e}{\ln u}\right)^u du$ converges, whereby

$\sum_{n=3}^{\infty}\frac{1}{(\ln(\ln n))^{\ln n}}$ converges.

e. $a_n = 1/n;\ b_n = 1/(\ln n)^4$

$$\lim_{n\to\infty}\frac{a_n}{b_n} = \lim_{n\to\infty}\frac{1/n}{1/(\ln n)^4} = \lim_{n\to\infty}\frac{(\ln n)^4}{n}$$

$$= \lim_{n\to\infty}\frac{4(\ln n)^3(1/n)}{1} = \lim_{n\to\infty}\frac{4(\ln n)^3}{n}$$

$$= \lim_{n\to\infty}\frac{12(\ln n)^2(1/n)}{1} = \lim_{n\to\infty}\frac{12(\ln n)^2}{n}$$

$$= \lim_{n\to\infty}\frac{24(\ln n)}{n} = \lim_{n\to\infty}\frac{24(1/n)}{1}$$

$$= \lim_{n\to\infty}\frac{24}{n} = 0$$

$\sum_{n=2}^{\infty}\frac{1}{n}$ diverges $\Rightarrow \sum_{n=2}^{\infty}\frac{1}{(\ln n)^4}$ diverges

f. $\left(\frac{\ln x}{x}\right)^2$ is continuous, positive, and nonincreasing on $[3, \infty)$. Using integration by parts twice,

$$\int_3^{\infty}\left(\frac{\ln x}{x}\right)^2 dx = \left[-\frac{(\ln x)^2}{x}\right]_3^{\infty} + \int_3^{\infty}\frac{2\ln x}{x^2}dx$$

$$= \left[-\frac{(\ln x)^2}{x}\right]_3^{\infty} + \left[-\frac{2\ln x}{x}\right]_3^{\infty} + \int_3^{\infty}\frac{2}{x^2}dx$$

$$= \left[-\frac{(\ln x)^2}{x} - \frac{2\ln x}{x} - \frac{2}{x}\right]_3^{\infty} \approx 1.8 < \infty$$

Thus, $\sum_{n=3}^{\infty}\left(\frac{\ln x}{x}\right)^2$ converges.

45. Let $a_n = \frac{1}{n^p}\left(1 + \frac{1}{2^p} + \frac{1}{3^p} + \ldots + \frac{1}{n^p}\right)$ and $b_n = \frac{1}{n^p}$. Then $\lim_{n\to\infty}\frac{a_n}{b_n} = \lim_{n\to\infty}\left(1 + \frac{1}{2^p} + \frac{1}{3^p} + \ldots + \frac{1}{n^p}\right) = \sum_{n=1}^{\infty}\frac{1}{n^p}$ which converges if $p > 1$. Thus, by the Limit Comparison Test, if $\sum_{n=1}^{\infty} b_n$ converges for $p > 1$, so does $\sum_{n=1}^{\infty} a_n$. Since $\sum_{n=1}^{\infty} b_n = \sum_{n=1}^{\infty}\frac{1}{n^p}$ converges for $p > 1$, $\sum_{n=1}^{\infty}\frac{1}{n^p}\left(1 + \frac{1}{2^p} + \frac{1}{3^p} + \cdots + \frac{1}{n^p}\right)$ also converges. For $p \le 1$, since $1 + \frac{1}{2^p} + \ldots + \frac{1}{n^p} > 1$, $\frac{1}{n^p}\left(1 + \frac{1}{2^p} + \ldots + \frac{1}{n^p}\right) > \frac{1}{n^p}$. Hence, since $\sum_{n=1}^{\infty}\frac{1}{n^p}$ diverges for $p \le 1$, $\sum_{n=1}^{\infty}\frac{1}{n^p}\left(1 + \frac{1}{2^p} + \ldots + \frac{1}{n^p}\right)$ also diverges. The series converges for $p > 1$ and diverges for $p \le 1$.

9.5 Concepts Review

1. $\lim_{n \to \infty} a_n = 0$

3. the alternating harmonic series

Problem Set 9.5

1. $a_n = \frac{2}{3n+1}; \frac{2}{3n+1} > \frac{2}{3n+4}$, so $a_n > a_{n+1}$;

$\lim_{n \to \infty} \frac{2}{3n+1} = 0$. $S_9 \approx 0.363$. The error made by using S_9 is not more than $a_{10} \approx 0.065$.

3. $a_n = \frac{1}{\ln(n+1)}; \frac{1}{\ln(n+1)} > \frac{1}{\ln(n+2)}$, so $a_n > a_{n+1}$;

$\lim_{n \to \infty} \frac{1}{\ln(n+1)} = 0$. $S_9 \approx 1.137$. The error made by using S_9 is not more than $a_{10} \approx 0.417$.

5. $a_n = \frac{\ln n}{n}; \frac{\ln n}{n} > \frac{\ln(n+1)}{n+1}$ is equivalent to

$\ln \frac{n^{n+1}}{(n+1)^n} > 0$ or $\frac{n^{n+1}}{(n+1)^n} > 1$ which is true for $n > 2$. $S_9 \approx -0.041$. The error made by using S_9 is not more than $a_{10} \approx 0.230$.

7. $\frac{|u_{n+1}|}{|u_n|} = \frac{\left|\left(-\frac{3}{4}\right)^{n+1}\right|}{\left|\left(-\frac{3}{4}\right)^n\right|} = \frac{3}{4} < 1$, so the series converges absolutely.

9. $\frac{|u_{n+1}|}{|u_n|} = \frac{\frac{n+1}{2^{n+1}}}{\frac{n}{2^n}} = \frac{n+1}{2n}$; $\lim_{n \to \infty} \frac{n+1}{2n} = \frac{1}{2} < 1$, so the series converges absolutely.

11. $n(n+1) = n^2 + n > n^2$ for all $n > 0$, thus

$\frac{1}{n(n+1)} < \frac{1}{n^2}$, so $\sum_{n=1}^{\infty} |u_n| = \sum_{n=1}^{\infty} \frac{1}{n(n+1)} < \sum_{n=1}^{\infty} \frac{1}{n^2}$ which converges since $2 > 1$, thus

$\sum_{n=1}^{\infty} (-1)^{n+1} \frac{1}{n(n+1)}$ converges absolutely.

13. $\sum_{n=1}^{\infty} (-1)^{n+1} \frac{1}{5n} = \frac{1}{5} \sum_{n=1}^{\infty} \frac{(-1)^{n+1}}{n}$ which converges since $\sum_{n=1}^{\infty} \frac{(-1)^{n+1}}{n}$ converges. The series is conditionally convergent since $\frac{1}{5} \sum_{n=1}^{\infty} \frac{1}{n}$ diverges.

15. $\lim_{n \to \infty} \frac{n}{10n+1} = \frac{1}{10} \neq 0$. Thus the sequence of partial sums does not converge; the series diverges.

17. $\lim_{n \to \infty} \frac{1}{n \ln n} = 0; \frac{1}{n \ln n} > \frac{1}{(n+1)\ln(n+1)}$ is equivalent to $(n+1)^{n+1} > n^n$ which is true for all $n > 0$ so $a_n > a_{n+1}$. The alternating series converges.

$\sum_{n=2}^{\infty} |u_n| = \sum_{n=2}^{\infty} \frac{1}{n \ln n}; \frac{1}{x \ln x}$ is continuous, positive, and nonincreasing on $[2, \infty)$.

Using $u = \ln x$, $du = \frac{1}{x} dx$,

$\int_2^{\infty} \frac{1}{x \ln x} dx = \int_{\ln 2}^{\infty} \frac{1}{u} du = \left[\ln|u|\right]_{\ln 2}^{\infty} = \infty$. Thus,

$\sum_{n=2}^{\infty} \frac{1}{n \ln n}$ diverges and $\sum_{n=2}^{\infty} (-1)^n \frac{1}{n \ln n}$ is conditionally convergent.

19. $\frac{|u_{n+1}|}{|u_n|} = \frac{\frac{(n+1)^4}{2^{n+1}}}{\frac{n^4}{2^n}} = \frac{(n+1)^4}{2n^4}$;

$\lim_{n \to \infty} \frac{(n+1)^4}{2n^4} = \frac{1}{2} < 1$.

The series is absolutely convergent.

21. $a_n = \frac{n}{n^2+1}; \frac{n}{n^2+1} > \frac{n+1}{(n+1)^2+1}$ is equivalent to $n^2 + n - 1 > 0$, which is true for $n > 1$, so $a_n > a_{n+1}$; $\lim_{n \to \infty} \frac{n}{n^2+1} = 0$, hence the alternating series converges. Let $b_n = \frac{1}{n}$, then

$\lim_{n \to \infty} \frac{a_n}{b_n} = \lim_{n \to \infty} \frac{n^2}{n^2+1} = 1; 0 < 1 < \infty$. Thus, since

$\sum_{n=1}^{\infty} b_n = \sum_{n=1}^{\infty} \frac{1}{n}$ diverges, $\sum_{n=1}^{\infty} a_n = \sum_{n=1}^{\infty} \frac{n}{n^2+1}$ also diverges. The series is conditionally convergent.

23. $\cos n\pi = (-1)^n = \frac{1}{(-1)}(-1)^{n+1}$ so the series is

$-1\sum_{n=1}^{\infty}\frac{(-1)^{n+1}}{n}$, -1 times the alternating harmonic series. The series is conditionally convergent.

25. $|\sin n| \le 1$ for all n, so

$\sum_{n=1}^{\infty}|u_n| = \sum_{n=1}^{\infty}\frac{|\sin n|}{n\sqrt{n}} \le \sum_{n=1}^{\infty}\frac{1}{n^{3/2}}$ which converges

since $\frac{3}{2} > 1$. Thus the series is absolutely convergent.

27. $a_n = \frac{1}{\sqrt{n(n+1)}}$; $\frac{1}{\sqrt{n(n+1)}} > \frac{1}{\sqrt{(n+1)(n+2)}}$ and

$\lim_{n\to\infty}\frac{1}{\sqrt{n(n+1)}} = 0$ so the alternating series converges.

Let $b_n = \frac{1}{n}$, then

$\lim_{n\to\infty}\frac{a_n}{b_n} = \lim_{n\to\infty}\frac{n}{\sqrt{n^2+n}} = \lim_{n\to\infty}\frac{1}{\sqrt{1+\frac{1}{n}}} = 1$;

$0 < 1 < \infty$.

Thus, since $\sum_{n=1}^{\infty}b_n = \sum_{n=1}^{\infty}\frac{1}{n}$ diverges,

$\sum_{n=1}^{\infty}a_n = \sum_{n=1}^{\infty}\frac{1}{\sqrt{n(n+1)}}$ also diverges.

The series is conditionally convergent.

29. $\sum_{n=1}^{\infty}\frac{(-3)^{n+1}}{n^2} = \sum_{n=1}^{\infty}(-1)^{n+1}\frac{3^{n+1}}{n^2}$; $\lim_{n\to\infty}\frac{3^{n+1}}{n^2} \ne 0$, so the series diverges.

31. Suppose $\sum|a_n|$ converges. Thus, $\sum 2|a_n|$ converges, so $\sum(|a_n| + a_n)$ converges since $0 \le |a_n| + a_n \le 2|a_n|$. By the linearity of convergent series, $\sum a_n = \sum(|a_n| + a_n) - \sum|a_n|$ converges, which is a contradiction.

33. The positive-term series is

$1 + \frac{1}{3} + \frac{1}{5} + \frac{1}{7} + \ldots = \sum_{n=1}^{\infty}\frac{1}{2n-1}$.

$\sum_{n=1}^{\infty}\frac{1}{2n-1} > \frac{1}{2}\sum_{n=1}^{\infty}\frac{1}{n}$ which diverges since the harmonic series diverges.

Thus, $\sum_{n=1}^{\infty}\frac{1}{2n-1}$ diverges.

The negative-term series is

$-\frac{1}{2} - \frac{1}{4} - \frac{1}{6} - \frac{1}{8} - \ldots = -\frac{1}{2}\sum_{n=1}^{\infty}\frac{1}{n}$ which diverges, since the harmonic series diverges.

35. a. $1 + \frac{1}{3} \approx 1.33$

b. $1 + \frac{1}{3} - \frac{1}{2} \approx 0.833$

c. $1 + \frac{1}{3} - \frac{1}{2} + \frac{1}{5} + \frac{1}{7} + \frac{1}{9} + \frac{1}{11} \approx 1.38$

$1 + \frac{1}{3} - \frac{1}{2} + \frac{1}{5} + \frac{1}{7} + \frac{1}{9} + \frac{1}{11} - \frac{1}{4} \approx 1.13$

37. Written response. Consider the partial sum of the positive terms of the series, and the partial sum of the negative terms. If both partial sums were bounded, the series would be absolutely convergent. Therefore, at least one of the partial sums must sum to ∞ (or $-\infty$). If the series of positive terms summed to ∞ and the series of negative terms summed to a finite number, the original series would not be convergent (similarly for the series of negative terms). Therefore, the positive terms sum to ∞ and the negative terms sum to $-\infty$. We can then rearrange the terms to make the original series sum to any value we wish.

39. Consider $1 - 1 + \frac{1}{2} - \frac{1}{4} + \frac{1}{3} - \frac{1}{9} + \ldots$

It is clear that $\lim_{n\to\infty}a_n = 0$. Pairing successive terms, we obtain $\frac{1}{n} - \frac{1}{n^2} = \frac{n-1}{n^2} > 0$ for $n > 1$.

Let $a_n = \frac{n-1}{n^2}$ and $b_n = \frac{1}{n}$. Then

$\lim_{n\to\infty}\frac{a_n}{b_n} = \lim_{n\to\infty}\frac{n^2-n}{n^2} = 1$; $0 < 1 < \infty$.

Thus, since $\sum_{n=1}^{\infty}b_n = \sum_{n=1}^{\infty}\frac{1}{n}$ diverges,

$\sum_{n=1}^{\infty}a_n = \sum_{n=1}^{\infty}\left(\frac{1}{n} - \frac{1}{n^2}\right)$ also diverges.

41. Note that $(a_k + b_k)^2 \ge 0$ and $(a_k - b_k)^2 \ge 0$ for all *k*. Thus, $a_k^2 \pm 2a_k b_k + b_k^2 \ge 0$, or $a_k^2 + b_k^2 \ge \pm 2a_k b_k$ for all *k*, and $a_k^2 + b_k^2 \ge 2|a_k b_k|$. Since $\sum_{k=1}^{\infty} a_k^2$ and $\sum_{k=1}^{\infty} b_k^2$ both converge, $\sum_{k=1}^{\infty}(a_k^2 + b_k^2)$ also converges, and by the Comparison Test, $\sum_{k=1}^{\infty} 2|a_k b_k|$ converges. Hence, $\sum_{k=1}^{\infty}|a_k b_k| = \frac{1}{2}\sum_{k=1}^{\infty} 2|a_k b_k|$ converges, i.e., $\sum_{k=1}^{\infty} a_k b_k$ converges absolutely.

43. Consider the graph of $\frac{|\sin x|}{x}$ on the interval $[k\pi, (k+1)\pi]$.

Note that for $k\pi + \frac{\pi}{6} \le x \le k\pi + \frac{5\pi}{6}$, $\frac{1}{2} \le |\sin x|$ while $\frac{1}{\left(k+\frac{5}{6}\right)\pi} \le \frac{1}{x}$. Thus on $\left[\left(k+\frac{1}{6}\right)\pi, \left(k+\frac{5}{6}\right)\pi\right]$

$$\frac{1}{2\left(k+\frac{5}{6}\right)\pi} = \frac{1}{\left(2k+\frac{5}{3}\right)\pi} \le \frac{|\sin x|}{x}, \text{ so } \int_{k\pi}^{(k+1)\pi} \frac{|\sin x|}{x} dx \ge \int_{(k+1/6)\pi}^{(k+5/6)\pi} \frac{|\sin x|}{x} dx \ge \frac{1}{\left(2k+\frac{5}{3}\right)\pi} \int_{(k+1/6)\pi}^{(k+5/6)\pi} dx = \frac{1}{3k+\frac{5}{2}}.$$

Hence, $\int_{\pi}^{\infty} \frac{|\sin x|}{x} dx \ge \sum_{k=1}^{\infty} \frac{1}{3k+\frac{5}{2}}$. Let $a_k = \frac{1}{3k+\frac{5}{2}}$ and $b_k = \frac{1}{k}$.

$\lim_{k\to\infty} \frac{a_k}{b_k} = \lim_{k\to\infty} \frac{k}{3k+\frac{5}{2}} = \lim_{k\to\infty} \frac{1}{3+\frac{5}{2k}} = \frac{1}{3}$; $0 < \frac{1}{3} < \infty$. Thus, since $\sum_{k=1}^{\infty} b_k = \sum_{k=1}^{\infty} \frac{1}{k}$ diverges, $\sum_{k=1}^{\infty} a_k = \sum_{k=1}^{\infty} \frac{1}{3k+\frac{5}{2}}$ also diverges. Hence, $\int_{\pi}^{\infty} \frac{|\sin x|}{x} dx$ also diverges and adding $\int_0^{\pi} \frac{|\sin x|}{x} dx$ will not affect its divergence.

45. $\frac{1}{n+1} + \frac{1}{n+2} + \cdots + \frac{1}{2n} = \left[\frac{1}{1+\frac{1}{n}} + \frac{1}{1+\frac{2}{n}} + \cdots + \frac{1}{1+\frac{n}{n}}\right]\left(\frac{1}{n}\right)$

This is a Riemann sum for the function $f(x) = \frac{1}{x}$ from $x = 1$ to 2 where $\Delta x = \frac{1}{n}$.

$$\lim_{n\to\infty} \sum_{k=1}^{n} \left[\frac{1}{1+\frac{k}{n}}\left(\frac{1}{n}\right)\right] = \int_1^2 \frac{1}{x} dx = \ln 2$$

9.6 Concepts Review

1. power series

3. interval; endpoints

Problem Set 9.6

1. $\sum_{n=1}^{\infty}\frac{x^n}{(n-1)!};\ \rho=\lim_{n\to\infty}\left|\frac{(n-1)!x^{n+1}}{n!x^n}\right|=\lim_{n\to\infty}\left|\frac{x}{n}\right|=0$. Series converges for all x.

3. $\sum_{n=1}^{\infty}\frac{x^n}{n^2};\ \rho=\lim_{n\to\infty}\left|\frac{n^2x^{n+1}}{(n+1)^2x^n}\right|=\lim_{n\to\infty}\left|\frac{x}{(1+2/n+1/n^2)}\right|=|x|$; convergence on $(-1,1)$.

   For $x=1,\ a_n=\frac{1}{n^2}$ (*p*-series, *p*=2) and the series converges.

   For $x=-1,\ a_n=\frac{(-1)^n}{n^2}$ (alternating *p*-series, *p*=2) and the series converges. by the Absolute Convergence Test.
   Series converges on $[-1,1]$

5. This is the alternating series for problem 3; thus it converges on $[-1,1]$ by the Absolute Convergence Test.

7. Let $u=x-2$; then, from problem 6, the series converges when $u\in(-1,1]$; that is when $x\in(1,3]$.

9. $\sum_{n=1}^{\infty}\frac{(-1)^{n+1}x^n}{n(n+1)};\ \rho=\lim_{n\to\infty}\left|\frac{x^{n+1}}{(n+1)(n+2)}\div\frac{x^n}{n(n+1)}\right|=\lim_{n\to\infty}|x|\left|\frac{n}{n+2}\right|=|x|$

   When $x=1$, the series is $\sum_{n=1}^{\infty}(-1)^{n+1}\frac{1}{n(n+1)}$ which converges absolutely by comparison with the series $\sum_{n=1}^{\infty}\frac{1}{n^2}$.

   When $x=-1$, the series is $\sum_{n=1}^{\infty}(-1)^{n+1}\frac{(-1)^n}{n(n+1)}=\sum_{n=1}^{\infty}(-1)^{2n+1}\frac{1}{n(n+1)}$

   $=\sum_{n=1}^{\infty}(-1)\frac{1}{n(n+1)}=(-1)\sum_{n=1}^{\infty}\frac{1}{n(n+1)}$ which converges since $\sum_{n=1}^{\infty}\frac{1}{n(n+1)}$ converges.
   The series converges on $-1\le x\le 1$.

11. $\sum_{n=1}^{\infty}\frac{(-1)^{n+1}x^{2n-1}}{(2n-1)!};\ \rho=\lim_{n\to\infty}\left|\frac{x^{2n+1}}{(2n+1)!}\div\frac{x^{2n-1}}{(2n-1)!}\right|=\lim_{n\to\infty}\left|x^2\right|\left|\frac{1}{2n(2n+1)}\right|=0$
    The series converges for all x.

13. $\sum_{n=1}^{\infty}nx^n;\ \rho=\lim_{n\to\infty}\left|\frac{(n+1)x^{n+1}}{nx^n}\right|$

    $=\lim_{n\to\infty}|x|\left|\frac{n+1}{n}\right|=|x|$

    When $x=1$, the series is $\sum_{n=1}^{\infty}n$ which clearly diverges.

    When $x=-1$, the series is $\sum_{n=1}^{\infty}n(-1)^n;\ a_n=n$; $\lim_{n\to\infty}a_n\neq 0$, thus the series diverges.
    The series converges on $-1<x<1$.

15. $1+\sum_{n=1}^{\infty}\frac{(-1)^n x^n}{n}$; $\rho=\lim_{n\to\infty}\left|\frac{x^{n+1}}{n+1}\div\frac{x^n}{n}\right|$

$=\lim_{n\to\infty}|x|\left|\frac{n}{n+1}\right|=|x|$

When $x = 1$, the series is $1+\sum_{n=1}^{\infty}(-1)^n\frac{1}{n}$, which is 1 added to the alternating harmonic series multiplied by –1, which converges.
When $x = -1$, the series is

$1+\sum_{n=1}^{\infty}(-1)^n\frac{(-1)^n}{n}=1+\sum_{n=1}^{\infty}\frac{1}{n}$, which diverges.

The series converges on $-1 < x \le 1$.

17. $1+\sum_{n=1}^{\infty}\frac{(-1)^n x^n}{n(n+2)}$;

$\rho=\lim_{n\to\infty}\left|\frac{x^{n+1}}{(n+1)(n+3)}\div\frac{x^n}{n(n+2)}\right|$

$=\lim_{n\to\infty}|x|\left|\frac{n^2+2n}{n^2+4n+3}\right|=|x|$

When $x = 1$ the series is $1+\sum_{n=1}^{\infty}(-1)^n\frac{1}{n(n+2)}$ which converges absolutely by comparison with the series $\sum_{n=1}^{\infty}\frac{1}{n^2}$.

When $x = -1$, the series is

$1+\sum_{n=1}^{\infty}(-1)^n\frac{(-1)^n}{n(n+2)}=1+\sum_{n=1}^{\infty}\frac{1}{n(n+2)}$ which

converges by comparison with $\sum_{n=1}^{\infty}\frac{1}{n^2}$.

The series converges on $-1 \le x \le 1$.

19. $\sum_{n=0}^{\infty}\frac{(-1)^n x^n}{2^n}$; $\rho=\lim_{n\to\infty}\left|\frac{x^{n+1}}{2^{n+1}}\div\frac{x^n}{2^n}\right|=\lim_{n\to\infty}\left|\frac{x}{2}\right|$

$=\left|\frac{x}{2}\right|$; $\left|\frac{x}{2}\right|<1$ when $-2 < x < 2$.

When $x = 2$, the series is $\sum_{n=0}^{\infty}(-1)^n\frac{2^n}{2^n}=\sum_{n=0}^{\infty}(-1)^n$

which diverges.
When $x = -2$, the series is

$\sum_{n=0}^{\infty}(-1)^n\frac{(-2)^n}{2^n}=\sum_{n=0}^{\infty}(-1)^n(-1)^n=\sum_{n=0}^{\infty}1$ which

diverges. The series converges on $-2 < x < 2$.

21. $\sum_{n=0}^{\infty}\frac{2^n x^n}{n!}$; $\rho=\lim_{n\to\infty}\left|\frac{2^{n+1}x^{n+1}}{(n+1)!}\div\frac{2^n x^n}{n!}\right|$

$=\lim_{n\to\infty}|2x|\left|\frac{1}{n+1}\right|=0$.

The series converges for all x.

23. $\sum_{n=1}^{\infty}\frac{(x-1)^n}{n}$; $\rho=\lim_{n\to\infty}\left|\frac{(x-1)^{n+1}}{n+1}\div\frac{(x-1)^n}{n}\right|$

$=\lim_{n\to\infty}|x-1|\left|\frac{n}{n+1}\right|=|x-1|$; $|x-1|<1$ when $0 < x < 2$.

When $x = 0$, the series is $\sum_{n=1}^{\infty}\frac{(-1)^n}{n}$ which converges.

When $x = 2$, the series is $\sum_{n=1}^{\infty}\frac{1}{n}$ which diverges.

The series converges on $0 \le x < 2$.

25. $\sum_{n=0}^{\infty}\frac{(x+1)^n}{2^n}$; $\rho=\lim_{n\to\infty}\left|\frac{(x+1)^{n+1}}{2^{n+1}}\div\frac{(x+1)^n}{2^n}\right|$

$=\lim_{n\to\infty}\left|\frac{x+1}{2}\right|=\left|\frac{x+1}{2}\right|$; $\left|\frac{x+1}{2}\right|<1$ when $-3 < x < 1$.

When $x = -3$, the series is $\sum_{n=0}^{\infty}\frac{(-2)^n}{2^n}=\sum_{n=0}^{\infty}(-1)^n$

which diverges.

When $x = 1$, the series is $\sum_{n=0}^{\infty}\frac{2^n}{2^n}=\sum_{n=0}^{\infty}1$ which

diverges.
The series converges on $-3 < x < 1$.

27. $\sum_{n=1}^{\infty}\frac{(x+5)^n}{n(n+1)}$; $\rho=\lim_{n\to\infty}\left|\frac{(x+5)^{n+1}}{(n+1)(n+2)}\div\frac{(x+5)^n}{n(n+1)}\right|$

$=\lim_{n\to\infty}|x+5|\left|\frac{n}{n+2}\right|=|x+5|$; $|x+5|<1$ when $-6 < x < -4$.

When $x = -4$, the series is $\sum_{n=1}^{\infty}\frac{1}{n(n+1)}$ which

converges by comparison with $\sum_{n=1}^{\infty}\frac{1}{n^2}$.

When $x = -6$, the series is $\sum_{n=1}^{\infty}\frac{(-1)^n}{n(n+1)}$ which

converges absolutely since $\sum_{n=1}^{\infty}\frac{1}{n(n+1)}$

converges.
The series converges on $-6 \le x \le -4$.

29. If for some x_0, $\lim_{n\to\infty} \frac{x_0^n}{n!} \neq 0$, then $\sum \frac{x_0^n}{n!}$ could not converge.

31. The Absolute Ratio Test gives

$$\rho = \lim_{n\to\infty}\left|\frac{(n+1)!x^{2n+3}}{1\cdot 3\cdot 5\cdots(2n+1)} \div \frac{n!x^{2n+1}}{1\cdot 3\cdot 5\cdots(2n-1)}\right|$$

$$= \lim_{n\to\infty}\left|x^2\right|\left|\frac{n+1}{2n+1}\right| = \left|\frac{x^2}{2}\right|;\ \left|\frac{x^2}{2}\right| < 1 \text{ when } |x| < \sqrt{2}.$$

The radius of convergence is $\sqrt{2}$.

33. This is a geometric series, so it converges for $|x-3| < 1$, $2 < x < 4$. For these values of x, the series converges to $\frac{1}{1-(x-3)} = \frac{1}{4-x}$.

35. a. $\rho = \lim_{n\to\infty}\left|\frac{(3x+1)^{n+1}}{(n+1)\cdot 2^{n+1}} \div \frac{(3x+1)^n}{n\cdot 2^n}\right| = \lim_{n\to\infty}|3x+1|\left|\frac{n}{2n+2}\right| = \frac{1}{2}|3x+1|;\ \frac{1}{2}|3x+1| < 1$ when $-1 < x < \frac{1}{3}$.

When $x = -1$, the series is $\sum_{n=1}^{\infty}\frac{(-2)^n}{n\cdot 2^n} = \sum_{n=1}^{\infty}(-1)^n\frac{1}{n}$, which converges.

When $x = \frac{1}{3}$, the series is $\sum_{n=1}^{\infty}\frac{2^n}{n\cdot 2^n} = \sum_{n=1}^{\infty}\frac{1}{n}$, which diverges. The series converges on $-1 \le x < \frac{1}{3}$.

b. $\rho = \lim_{n\to\infty}\left|\frac{(-1)^{n+1}(2x-3)^{n+1}}{4^{n+1}\sqrt{n+1}} \div \frac{(-1)^n(2x-3)^n}{4^n\sqrt{n}}\right| = \lim_{n\to\infty}|2x-3|\left|\frac{\sqrt{n}}{4\sqrt{n+1}}\right| = \frac{1}{4}|2x-3|;$

$\frac{1}{4}|2x-3| < 1$ when $-\frac{1}{2} < x < \frac{7}{2}$.

When $x = -\frac{1}{2}$, the series is $\sum_{n=1}^{\infty}(-1)^n\frac{(-4)^n}{4^n\sqrt{n}} = \sum_{n=1}^{\infty}\frac{1}{\sqrt{n}}$ which diverges since $\frac{1}{2} < 1$.

When $x = \frac{7}{2}$, the series is $\sum_{n=1}^{\infty}(-1)^n\frac{4^n}{4^n\sqrt{n}} = \sum_{n=1}^{\infty}(-1)^n\frac{1}{\sqrt{n}};\ a_n = \frac{1}{\sqrt{n}};\ \frac{1}{\sqrt{n}} > \frac{1}{\sqrt{n+1}}$, so $a_n > a_{n+1}$;

$\lim_{n\to\infty}\frac{1}{\sqrt{n}} = 0$, so $\sum_{n=1}^{\infty}(-1)^n\frac{1}{\sqrt{n}}$ converges. The series converges on $-\frac{1}{2} < x \le \frac{7}{2}$.

37. If $a_{n+3} = a_n$, then $a_0 = a_3 = a_6 = a_{3n}$, $a_1 = a_4 = a_7 = a_{3n+1}$, and $a_2 = a_5 = a_8 = a_{3n+2}$. Thus,

$$\sum_{n=0}^{\infty}a_nx^n = a_0 + a_1x + a_2x^2 + a_0x^3 + a_1x^4 + a_2x^5 + \cdots = (a_0 + a_1x + a_2x^2)(1 + x^3 + x^6 + \cdots)$$

$$= (a_0 + a_1x + a_2x^2)\sum_{n=0}^{\infty}x^{3n} = (a_0 + a_1x + a_2x^2)\sum_{n=0}^{\infty}(x^3)^n.$$

$a_0 + a_1x + a_2x^2$ is a polynomial, which will converge for all x.

$\sum_{n=0}^{\infty}(x^3)^n$ is a geometric series which, converges for $\left|x^3\right| < 1$, or, equivalently, $|x| < 1$.

Since $\sum_{n=0}^{\infty}(x^3)^n = \frac{1}{1-x^3}$ for $|x| < 1$, $S(x) = \frac{a_0 + a_1x + a_2x^2}{1-x^3}$ for $|x| < 1$.

9.7 Concepts Review

1. integrated; interior

3. $1+x^2+\frac{x^4}{2}+\frac{x^6}{6}$

Problem Set 9.7

1. From the geometric series for $\frac{1}{1-x}$ with x replaced by $-x$, we get
$\frac{1}{1+x}=1-x+x^2-x^3+x^4-x^5+\cdots$,
radius of convergence 1.

3. $\frac{d}{dx}\left(\frac{1}{1-x}\right)=\frac{1}{(1-x)^2};\ \frac{d}{dx}\left(\frac{1}{(1-x)^2}\right)=\frac{2}{(1-x)^3}$,
so $\frac{1}{(1-x)^3}$ is $\frac{1}{2}$ of the second derivative of $\frac{1}{1-x}$. Thus,
$\frac{1}{(1-x)^3}=1+3x+6x^2+10x^3+\cdots$;
radius of convergence 1.

5. From the geometric series for $\frac{1}{1-x}$ with x replaced by $\frac{3}{2}x$, we get
$$\frac{1}{2-3x}=\frac{1}{2}+\frac{3x}{4}+\frac{9x^2}{8}+\frac{27x^3}{16}+\cdots;$$
radius of convergence $\frac{2}{3}$.

7. From the geometric series for $\frac{1}{1-x}$ with x replaced by x^4, we get
$\frac{x^2}{1-x^4}=x^2+x^6+x^{10}+x^{14}+\cdots$;
radius of convergence 1.

9. From the geometric series for $\ln(1+x)$ with x replaced by t, we get
$\int_0^x \ln(1+t)dt=\frac{x^2}{2}-\frac{x^3}{6}+\frac{x^4}{12}-\frac{x^5}{20}+\ldots$;
radius of convergence 1.

11. $\ln(1+x)=x-\frac{x^2}{2}+\frac{x^3}{3}-\frac{x^4}{4}+\ldots, -1<x\le 1$

$\ln(1-x)=-x-\frac{x^2}{2}-\frac{x^3}{3}-\frac{x^4}{4}+\ldots, -1\le x<1$

$\ln\frac{1+x}{1-x}=\ln(1+x)-\ln(1-x)=2x+\frac{2x^3}{3}+\frac{2x^5}{5}+\cdots$; radius of convergence 1.

13. Substitute $-x$ for x in the series for e^x to get:
$e^{-x}=1-x+\frac{x^2}{2!}-\frac{x^3}{3!}+\frac{x^4}{4!}-\frac{x^5}{5!}+\cdots$.

15. Add the result of Problem 13 to the series for e^x to get:
$e^x+e^{-x}=2+\frac{2x^2}{2!}+\frac{2x^4}{4!}+\frac{2x^6}{6!}+\cdots$.

17. $e^{-x}\cdot\frac{1}{1-x}=\left(1-x+\frac{x^2}{2!}-\frac{x^3}{3!}+\cdots\right)(1+x+x^2+\cdots)=1+\frac{x^2}{2}+\frac{x^3}{3}+\frac{3x^4}{8}+\frac{11x^5}{30}+\cdots$

19. $\frac{\tan^{-1}x}{e^x}=e^{-x}\tan^{-1}x=\left(1-x+\frac{x^2}{2!}-\frac{x^3}{3!}+\cdots\right)\left(x-\frac{x^3}{3}+\frac{x^5}{5}-\frac{x^7}{7}+\cdots\right)=x-x^2+\frac{x^3}{6}+\frac{x^4}{6}+\frac{3x^5}{40}+\cdots$

21. $(\tan^{-1}x)(1+x^2+x^4)=\left(x-\frac{x^3}{3}+\frac{x^5}{5}-\frac{x^7}{7}+\cdots\right)(1+x^2+x^4)=x+\frac{2x^3}{3}+\frac{13x^5}{15}-\frac{29x^7}{105}+\cdots$

23. The series representation of $\frac{e^x}{1+x}$ is $1+\frac{x^2}{2}-\frac{x^3}{3}+\frac{3x^4}{8}-\frac{11x^5}{30}+\cdots$, so $\int_0^x \frac{e^t}{1+t}dt = x+\frac{1}{6}x^3-\frac{1}{12}x^4+\frac{3}{40}x^5-\cdots$.

25. **a.** $\frac{1}{1+x}=1-x+x^2-x^3+x^4-x^5+\cdots$, so $\frac{x}{1+x}=x-x^2+x^3-x^4+x^5-\cdots$.

b. $e^x=1+x+\frac{x^2}{2!}+\frac{x^3}{3!}+\frac{x^4}{4!}+\frac{x^5}{5!}+\cdots$, so $\frac{e^x-(1+x)}{x^2}=\frac{1}{2!}+\frac{x}{3!}+\frac{x^2}{4!}+\frac{x^3}{5!}+\cdots$.

c. $-\ln(1-x)=x+\frac{x^2}{2}+\frac{x^3}{3}+\frac{x^4}{4}+\frac{x^5}{5}+\cdots$, so $-\ln(1-2x)=2x+\frac{4x^2}{2}+\frac{8x^3}{3}+\frac{16x^4}{4}+\cdots$.

27. Differentiating the series for $\frac{1}{1-x}$ yields $\frac{1}{(1-x)^2}=1+2x+3x^2+4x^3+\cdots$ multiplying this series by x gives

$\frac{x}{(1-x)^2}=x+2x^2+3x^3+4x^4+\cdots$, hence $\sum_{n=1}^{\infty}nx^n=\frac{x}{(1-x)^2}$ for $-1<x<1$.

29. **a.** $\tan^{-1}(e^x-1)=(e^x-1)-\frac{(e^x-1)^3}{3}+\frac{(e^x-1)^5}{5}-\cdots=\left(x+\frac{x^2}{2!}+\frac{x^3}{3!}+\cdots\right)-\frac{1}{3}\left(x+\frac{x^2}{2!}+\cdots\right)^3+\cdots$

$=x+\frac{x^2}{2}-\frac{x^3}{6}-\cdots$

b. $e^{e^x-1}=1+(e^x-1)+\frac{(e^x-1)^2}{2!}+\frac{(e^x-1)^3}{3!}+\cdots$

$=1+\left(x+\frac{x^2}{2!}+\frac{x^3}{3!}+\cdots\right)+\frac{1}{2!}\left(x+\frac{x^2}{2!}+\cdots\right)^2+\frac{1}{3!}\left(x+\frac{x^2}{2!}+\cdots\right)^3$

$=1+\left(x+\frac{x^2}{2!}+\frac{x^3}{3!}+\cdots\right)+\frac{1}{2!}\left(x^2+2\frac{x^3}{2!}+\cdots\right)+\frac{1}{3!}\left(x^3+3\frac{x^4}{2!}+\cdots\right)=1+x+x^2+\frac{5x^3}{6}+\cdots$

31. $\frac{x}{x^2-3x+2}=\frac{x}{(x-2)(x-1)}=\frac{2}{x-2}-\frac{1}{x-1}=-\frac{1}{1-\frac{x}{2}}+\frac{1}{1-x}=-\left(1+\frac{x}{2}+\frac{x^2}{4}+\frac{x^3}{8}+\cdots\right)+\left(1+x+x^2+x^3+\cdots\right)$

$=\frac{x}{2}+\frac{3x^2}{4}+\frac{7x^3}{8}+\cdots=\sum_{n=1}^{\infty}\frac{(2^n-1)x^n}{2^n}$

33. $F(x)-xF(x)-x^2F(x)=(f_0+f_1x+f_2x^2+f_3x^3+\cdots)-(f_0x+f_1x^2+f_2x^3+\cdots)-(f_0x^2+f_1x^3+f_2x^4+\cdots)$

$=f_0+(f_1-f_0)x+(f_2-f_1-f_0)x^2+(f_3-f_2-f_1)x^3+\cdots$

$=f_0+(f_1-f_0)x+\sum_{n=2}^{\infty}(f_n-f_{n-1}-f_{n-2})x^n=0+x+\sum_{n=0}^{\infty}(f_{n+2}-f_{n+1}-f_n)x^{n+2}$

Since $f_{n+2}=f_{n+1}+f_n$, $f_{n+2}-f_{n+1}-f_n=0$. Thus $F(x)-xF(x)-x^2F(x)=x$.

$F(x)=\frac{x}{1-x-x^2}$

35. $\pi\approx 16\left(\frac{1}{5}-\frac{1}{375}+\frac{1}{15{,}625}-\frac{1}{546{,}875}+\frac{1}{17{,}578{,}125}\right)-4\left(\frac{1}{239}\right)\approx 3.14159$

9.8 Concepts Review

1. $\dfrac{f^{(k)}(0)}{k!}$

3. $-\infty$; ∞

Problem Set 9.8

1. $\tan x = \dfrac{\sin x}{\cos x} = \dfrac{x - \frac{x^3}{3!} + \frac{x^5}{5!} - \frac{x^7}{7!} + \ldots}{1 - \frac{x^2}{2!} + \frac{x^4}{4!} - \frac{x^6}{6!} + \ldots} = x + \dfrac{x^3}{3} + \dfrac{2x^5}{15} + \ldots$

3. $e^x \sin x = \left(1 + x + \dfrac{x^2}{2!} + \dfrac{x^3}{3!} + \dfrac{x^4}{4!} + \ldots\right)\left(x - \dfrac{x^3}{3!} + \dfrac{x^5}{5!} - \ldots\right) = x + x^2 + \dfrac{x^3}{3} - \dfrac{x^5}{30} - \ldots$

5. $\cos x \ln(1+x) = \left(1 - \dfrac{x^2}{2!} + \dfrac{x^4}{4!} - \ldots\right)\left(x - \dfrac{x^2}{2} + \dfrac{x^3}{3} - \dfrac{x^4}{4} + \ldots\right) = x - \dfrac{x^2}{2} - \dfrac{x^3}{6} + \dfrac{3x^5}{40} - \ldots$

7. $e^x + x + \sin x = x + \left(1 + x + \dfrac{x^2}{2!} + \ldots\right) + \left(x - \dfrac{x^3}{3!} + \dfrac{x^5}{5!} - \ldots\right) = 1 + 3x + \dfrac{x^2}{2!} + \dfrac{x^4}{4!} + \dfrac{2x^5}{5!} + \ldots$

9. $\dfrac{1}{1-x}\cosh x = (1 + x + x^2 + x^3 + \ldots)\left(1 + \dfrac{x^2}{2!} + \dfrac{x^4}{4!} + \ldots\right) = 1 + x + \dfrac{3x^2}{2} + \dfrac{3x^3}{2} + \dfrac{37x^4}{24} + \dfrac{37x^5}{24} + \ldots,\ -1 < x < 1$

11. $\dfrac{1}{1+x+x^2} = \dfrac{1}{1+x+x^2} \cdot \dfrac{1-x}{1-x} = \dfrac{1}{1-x^3}(1-x) = (1-x)\displaystyle\sum_{n=0}^{\infty} x^{3n} = 1 - x + x^3 - x^4 + \cdots,\ |x| < 1$

13. $\sin^3 x = \left(x - \dfrac{x^3}{3!} + \dfrac{x^5}{5!} - \ldots\right)^2 \left(x - \dfrac{x}{3!} + \dfrac{x^5}{5!} - \ldots\right) = \left(x^2 - 2\dfrac{x^4}{3!} + \ldots\right)\left(x - \dfrac{x^3}{3!} + \dfrac{x^5}{5!} - \ldots\right) = x^3 - \dfrac{x^5}{2} + \ldots$

15. $x\sec(x^2) + \sin x = \dfrac{x}{\cos(x^2)} + \sin x = \dfrac{x}{1 - \frac{x^4}{2!} + \frac{x^8}{4!} - \ldots} + \left(x - \dfrac{x^3}{3!} + \dfrac{x^5}{5!} - \ldots\right)$

$= \left(x + \dfrac{x^5}{2} + \ldots\right) + \left(x - \dfrac{x^3}{3!} + \dfrac{x^5}{5!} - \ldots\right) = 2x - \dfrac{x^3}{3!} + \dfrac{61x^5}{120} + \ldots$

17. $(1+x)^{3/2} = 1 + \dfrac{3x}{2} + \dfrac{3x^2}{8} - \dfrac{x^3}{16} + \dfrac{3x^4}{128} - \dfrac{3x^5}{256} + \ldots,\ -1 < x < 1$

19. $f^{(n)}(x) = e^x$ for all n. $f(1) = f'(1) = f''(1) = f'''(1) = e$

$e^x \approx e + e(x-1) + \dfrac{e}{2}(x-1)^2 + \dfrac{e}{6}(x-1)^3$

21. $f\left(\dfrac{\pi}{3}\right) = \dfrac{1}{2}; f'\left(\dfrac{\pi}{3}\right) = -\dfrac{\sqrt{3}}{2}; f''\left(\dfrac{\pi}{3}\right) = -\dfrac{1}{2}; f'''\left(\dfrac{\pi}{3}\right) = \dfrac{\sqrt{3}}{2}$; $\cos x \approx \dfrac{1}{2} - \dfrac{\sqrt{3}}{2}\left(x - \dfrac{\pi}{3}\right) - \dfrac{1}{4}\left(x - \dfrac{\pi}{3}\right)^2 + \dfrac{\sqrt{3}}{12}\left(x - \dfrac{\pi}{3}\right)^3$

23. $f(1) = 3$; $f'(1) = 2 + 3 = 5$;

$f''(1) = 2 + 6 = 8$; $f'''(1) = 6$

$1 + x^2 + x^3 = 3 + 5(x-1) + 4(x-1)^2 + (x-1)^3$

This is exact since $f^{(n)}(x) = 0$ for $n \ge 4$.

25. The derivative of an even function is an odd function and the derivative of an odd function is an even function. (Problem 50 of Section 3.2).

Since $f(x) = \sum a_n x^n$ is an even function, $f'(x)$ is an odd function, so $f''(x)$ is an even function, hence $f'''(x)$ is an odd function, etc.

Thus $f^{(n)}(x)$ is an even function when n is even and an odd function when n is odd.

By the Uniqueness Theorem, if $f(x) = \sum a_n x^n$, then $a_n = \frac{f^{(n)}(0)}{n!}$. If $g(x)$ is an odd function, $g(0) = 0$, thence $a_n = 0$ for all odd n since $f^{(n)}(x)$ is an odd function for odd n.

27. $\frac{1}{\sqrt{1-t^2}} = [1 + (-t^2)]^{-1/2}$

$= 1 - \frac{1}{2}(-t^2) + \frac{3}{8}(-t^2)^2 - \frac{5}{16}(-t^2)^3 + \cdots$

$= 1 + \frac{t^2}{2} + \frac{3t^4}{8} + \frac{5t^6}{16} + \cdots$

Thus, $\sin^{-1} x = \int_0^x \frac{1}{\sqrt{1-t^2}} dt$

$= \int_0^x \left(1 + \frac{t^2}{2} + \frac{3t^4}{8} + \frac{5t^6}{16} + \cdots\right) dt$

$= \left[t + \frac{t^3}{6} + \frac{3t^5}{40} + \frac{5t^7}{112} + \cdots\right]_0^x$

$= x + \frac{x^3}{6} + \frac{3x^5}{40} + \frac{5x^7}{112} + \cdots$

29. $\cos(x^2) = 1 - \frac{x^4}{2!} + \frac{x^8}{4!} - \frac{x^{12}}{6!} + \frac{x^{16}}{8!} - \cdots$

$\int_0^1 \cos(x^2) dx = \int_0^1 \left(1 - \frac{x^4}{2!} + \frac{x^8}{4!} - \frac{x^{12}}{6!} + \frac{x^{16}}{8!} - \cdots\right) dx$

$= \left[x - \frac{x^5}{10} + \frac{x^9}{216} - \frac{x^{13}}{9360} + \frac{x^{17}}{685,440} - \cdots\right]_0^1$

$= 1 - \frac{1}{10} + \frac{1}{216} - \frac{1}{9360} + \frac{1}{685,440} - \cdots \approx 0.90452$

31. $\frac{1}{x} = \frac{1}{1-(1-x)} = 1 + (1-x) + (1-x)^2 + (1-x)^3 + \cdots = 1 - (x-1) + (x-1)^2 - (x-1)^3 + \cdots$

for $-1 < 1 - x < 1$, or $0 < x < 2$.

33. a. $f(x) = 1 + (x + x^2) + \frac{(x+x^2)^2}{2!} + \frac{(x+x^2)^3}{3!} + \frac{(x+x^2)^4}{4!} + \ldots$

$= 1 + (x + x^2) + \frac{1}{2}(x^2 + 2x^3 + x^4) + \frac{1}{6}(x^3 + 3x^4 + 3x^5 + x^6) + \frac{1}{24}(x^4 + 4x^5 + 6x^6 + 4x^7 + x^8) + \ldots$

$= 1 + x + \frac{3x^2}{2} + \frac{7x^3}{6} + \frac{25x^4}{24} + \ldots = \sum_{n=0}^{\infty} \frac{f^{(n)}(0)}{n!} x^n$

Thus $\frac{f^{(4)}(0)}{4!} = \frac{25}{24}$ so $f^{(4)}(0) = \frac{25}{24} 4! = 25$.

b. $f(x)=1+\sin x+\frac{\sin^2 x}{2!}+\frac{\sin^3 x}{3!}+\frac{\sin^4 x}{4!}+\ldots$

$$=1+\left(x-\frac{x^3}{3!}+\ldots\right)+\frac{1}{2}\left(x-\frac{x^3}{3!}+\ldots\right)^2+\frac{1}{6}\left(x-\frac{x^3}{3!}+\ldots\right)^3+\frac{1}{24}\left(x-\frac{x^3}{3!}+\ldots\right)^4$$

$$=1+\left(x-\frac{x^3}{3!}+\ldots\right)+\frac{1}{2}\left(x^2-2\frac{x^4}{3!}+\ldots\right)+\frac{1}{6}(x^3-\ldots)+\frac{1}{24}(x^4-\ldots)=1+x+\frac{x^2}{2}-\frac{x^4}{8}-\ldots=\sum_{n=0}^{\infty}\frac{f^{(n)}(0)}{n!}x^n$$

Thus, $\frac{f^{(4)}(0)}{4!}=-\frac{1}{8}$ so $f^{(4)}(0)=-\frac{1}{8}4!=-3.$

c. $e^{t^2}-1=-1+\left(1+t^2+\frac{t^4}{2!}+\frac{t^6}{3!}+\frac{t^8}{4!}+\ldots\right)=t^2+\frac{t^4}{2!}+\frac{t^6}{3!}+\frac{t^8}{4!}+\ldots$

so $\frac{e^{t^2}-1}{t^2}=1+\frac{t^2}{2}+\frac{t^4}{6}+\frac{t^6}{24}+\ldots$

$$f(x)=\int_0^x\left(1+\frac{t^2}{2}+\frac{t^4}{6}+\frac{t^6}{24}+\ldots\right)dt=\left[t+\frac{t^3}{6}+\frac{t^5}{30}+\ldots\right]_0^x=x+\frac{x^3}{6}+\frac{x^5}{30}+\ldots=\sum_{n=0}^{\infty}\frac{f^{(n)}(0)}{n!}x^n$$

Thus, $\frac{f^{(4)}(0)}{4!}=0$ so $f^{(4)}(0)=0.$

d. $e^{\cos x-1}=1+(\cos x-1)+\frac{(\cos x-1)^2}{2!}+\frac{(\cos x-1)^3}{3!}+\ldots$

$$=1+\left(-\frac{x^2}{2!}+\frac{x^4}{4!}-\ldots\right)+\frac{1}{2}\left(-\frac{x^2}{2!}+\frac{x^4}{4!}-\ldots\right)^2+\frac{1}{6}\left(-\frac{x^2}{2!}+\frac{x^4}{4!}-\ldots\right)^3+\ldots$$

$$=1+\left(-\frac{x^2}{2}+\frac{x^4}{24}-\ldots\right)+\frac{1}{2}\left(\frac{x^4}{4}-\ldots\right)+\frac{1}{6}\left(-\frac{x^6}{8}+\ldots\right)=1-\frac{x^2}{2}+\frac{x^4}{6}-\ldots$$

Hence $f(x)=e-\frac{e}{2}x^2+\frac{e}{6}x^4-\ldots=\sum_{n=0}^{\infty}\frac{f^{(n)}(0)}{n!}x^n$

Thus, $\frac{f^{(4)}(0)}{4!}=\frac{e}{6}$ so $f^{(4)}(0)=\frac{e}{6}4!=4e.$

e. Observe that $\ln(\cos^2 x)=\ln(1-\sin^2 x)$.

$$\sin^2 x=\left(x-\frac{x^3}{3!}+\frac{x^5}{5!}-\cdots\right)^2=x^2-\frac{x^4}{3}+\frac{2x^6}{45}-\cdots$$

$$\ln(1-\sin^2 x)=-\sin^2 x-\frac{\sin^4 x}{2}-\frac{\sin^6 x}{3}-\cdots$$

$$=-\left(x^2-\frac{x^4}{3}+\cdots\right)-\frac{1}{2}\left(x^2-\frac{x^4}{3}+\cdots\right)^2-\frac{1}{3}\left(x^2-\frac{x^4}{3}+\cdots\right)^3$$

$$=-\left(x^2-\frac{x^4}{3}+\frac{2x^6}{45}-\cdots\right)-\frac{1}{2}\left(x^4-\frac{2x^6}{3}+\cdots\right)-\frac{1}{3}(x^6-\cdots)=-x^2-\frac{x^4}{6}-\frac{2x^6}{45}-\cdots$$

Hence $f(x)=-x^2-\frac{1}{6}x^4-\frac{2}{45}x^6-\cdots=\sum_{n=0}^{\infty}\frac{f^{(n)}(0)}{n!}x^n.$

Thus, $\frac{f^{(4)}(0)}{4!}=-\frac{1}{6}$ so $f^{(4)}(0)=-\frac{1}{6}4!=-4.$

35. $\tanh x = \dfrac{\sinh x}{\cosh x} = a_0 + a_1 x + a_2 x^2 + \ldots$

so $\sinh x = \cosh x(a_0 + a_1 x + a_2 x^2 + \ldots)$

or $x + \dfrac{x^3}{6} + \dfrac{x^5}{120} + \ldots = \left(1 + \dfrac{x^2}{2} + \dfrac{x^4}{24} + \ldots\right)\left(a_0 + a_1 x + a_2 x^2 + \ldots\right)$

$= a_0 + a_1 x + \left(a_2 + \dfrac{a_0}{2}\right)x^2 + \left(a_3 + \dfrac{a_1}{2}\right)x^3 + \left(a_4 + \dfrac{a_2}{2} + \dfrac{a_0}{24}\right)x^4 + \left(a_5 + \dfrac{a_3}{2} + \dfrac{a_1}{24}\right)x^5 + \ldots$

Thus $a_0 = 0, a_1 = 1, a_2 + \dfrac{a_0}{2} = 0, a_3 + \dfrac{a_1}{2} = \dfrac{1}{6},$

$a_4 + \dfrac{a_2}{2} + \dfrac{a_0}{24} = 0, a_5 + \dfrac{a_3}{2} + \dfrac{a_1}{24} = \dfrac{1}{120},$ so

$a_0 = 0, a_1 = 1, a_2 = 0, a_3 = -\dfrac{1}{3}, a_4 = 0, a_5 = \dfrac{2}{15}$ and therefore

$\tanh x = x - \dfrac{1}{3}x^3 + \dfrac{2}{15}x^5 - \ldots$

37. a. First define $R_3(x)$ by

$$R_3(x) = f(x) - f(a) - f'(a)(x-a) - \frac{f''(a)}{2!}(x-a)^2 - \frac{f'''(a)}{3!}(x-a)^3$$

For any t in the interval $[a, x]$ we define

$$g(t) = f(x) - f(t) - f'(t)(x-t) - \frac{f''(t)}{2!}(x-t)^2 - \frac{f'''(t)}{3!}(x-t)^3 - R_3(x)\frac{(x-t)^4}{(x-a)^4}$$

Next we differentiate with respect to t using the Product and Power Rules:

$$g'(t) = 0 - f'(t) - [-f'(t) + f''(t)(x-t)] - \frac{1}{2!}\left[-2f''(t)(x-t) + f'''(t)(x-t)^2\right]$$

$$-\frac{1}{3!}\left[-3f'''(t)(x-t)^2 + f^{(4)}(t)(x-t)^3\right] + R_3(x)\frac{4(x-t)^3}{(x-a)^4}$$

$$= -\frac{f^{(4)}(t)(x-t)^3}{3!} + 4R_3(x)\frac{(x-t)^3}{(x-a)^4}$$

Since $g(x) = 0$, $g(a) = R_3(x) - R_3(x) = 0$, and $g(t)$ is continuous on $[a, x]$, we can apply the Mean Value Theorem for Derivatives. There exists, therefore, a number c between a and x such that $g'(c) = 0$. Thus,

$$0 = g'(c) = -\frac{f^{(4)}(c)(x-c)^3}{3!} + 4R_3(x)\frac{(x-c)^3}{(x-a)^4}$$

which leads to:

$$R_3(x) = \frac{f^{(4)}(c)}{4!}(x-a)^4$$

b. Like the previous part, first define $R_n(x)$ by

$$R_n(x) = f(x) - f(a) - f'(a)(x-a) - \frac{f''(a)}{2!}(x-a)^2 - \cdots - \frac{f^{(n)}(a)}{n!}(x-a)^n$$

For any t in the interval $[a, x]$ we define

$$g(t) = f(x) - f(t) - f'(t)(x-t) - \frac{f''(t)}{2!}(x-t)^2 - \cdots - \frac{f^{(n)}(t)}{n!}(x-t)^n - R_n(x)\frac{(x-t)^{n+1}}{(x-a)^{n+1}}$$

Next we differentiate with respect to t using the Product and Power Rules:

$$g'(t) = 0 - f'(t) - \left[-f'(t) + f''(t)(x-t)\right] - \frac{1}{2!}\left[-2f''(t)(x-t) + f'''(t)(x-t)^2\right] - \cdots$$

$$-\frac{1}{n!}\left[-nf^{(n)}(t)(x-t)^{n-1} + f^{(n+1)}(t)(x-t)^n\right] + R_n(x)\frac{(n+1)(x-t)^n}{(x-a)^{n+1}}$$

$$= -\frac{f^{(n+1)}(t)(x-t)^n}{n!} + (n+1)R_n(x)\frac{(x-t)^n}{(x-a)^{n+1}}$$

Since $g(x) = 0$, $g(a) = R_n(x) - R_n(x) = 0$, and $g(t)$ is continuous on $[a, x]$, we can apply the Mean Value Theorem for Derivatives. There exists, therefore, a number c between a and x such that $g'(c) = 0$. Thus,

$$0 = g'(c) = -\frac{f^{(n+1)}(c)(x-c)^n}{n!} + (n+1)R_n(x)\frac{(x-c)^n}{(x-a)^{n+1}}$$

which leads to:

$$R_n(x) = \frac{f^{(n+1)}(c)}{(n+1)!}(x-a)^{n+1}$$

39. $f'(t) = \begin{cases} 0 & \text{if } t < 0 \\ 4t^3 & \text{if } t \ge 0 \end{cases}$,

$f''(t) = \begin{cases} 0 & \text{if } t < 0 \\ 12t^2 & \text{if } t \ge 0 \end{cases}$,

$f'''(t) = \begin{cases} 0 & \text{if } t < 0 \\ 24t & \text{if } t \ge 0 \end{cases}$,

$f^{(4)}(t) = \begin{cases} 0 & \text{if } t < 0 \\ 24 & \text{if } t \ge 0 \end{cases}$

$\lim_{t \to 0^+} f^{(4)}(t) = 24$ while $\lim_{t \to 0^-} f^{(4)}(t) = 0$, thus $f^{(4)}(0)$ does not exist, and $f(t)$ cannot be represented by a Maclaurin series.

Suppose that $g(t)$ as described in the text is represented by a Maclaurin series, so

$g(t) = a_0 + a_1 t + a_2 t^2 + \ldots = \sum_{n=0}^{\infty} \frac{g^{(n)}(0)}{n!} t^n$ for all t in $(-R, R)$ for some $R > 0$. It is clear that, for $t \le 0$, $g(t)$ is represented by

$g(t) = 0 + 0t + 0t^2 + \ldots$. However, this will not represent $g(t)$ for any $t > 0$ since the car is moving for $t > 0$. Similarly, any series that represents $g(t)$ for $t > 0$ cannot be 0 everywhere, so it will not represent $g(t)$ for $t < 0$. Thus, $g(t)$ cannot be represented by a Maclaurin series.

41. $\sin x = x - \frac{x^3}{6} + \frac{x^5}{120} - \frac{x^7}{5040} + \cdots$

43. $3\sin x - 2\exp x = -2 + x - x^2 - \frac{5x^3}{6} - \cdots$

$3\sin x = 3x - \frac{x^3}{2} + \frac{x^5}{40} - \frac{x^7}{1680} + \cdots$

$-2\exp x = -2 - 2x - x^2 - \frac{x^3}{3} - \cdots$

Thus, $3\sin x - 2\exp x = -2 + x - x^2 - \frac{5x^3}{6} - \cdots$

45. $\sin(\exp x-1)=x+\frac{x^2}{2}-\frac{5x^4}{24}-\frac{23x^5}{120}-\cdots$

$\exp x-1=x+\frac{x^2}{2}+\frac{x^3}{6}+\frac{x^4}{24}+\cdots$

$$\sin(\exp x-1)=\left(x+\frac{x^2}{2}+\frac{x^3}{6}+\frac{x^4}{24}+\frac{x^5}{120}+\ldots\right)-\frac{1}{6}\left(x+\frac{x^2}{2}+\frac{x^3}{6}+\ldots\right)^3+\frac{1}{120}\left(x+\frac{x^2}{2}+\ldots\right)^5-\cdots$$

$$=\left(x+\frac{x^2}{2}+\frac{x^3}{6}+\frac{x^4}{24}+\frac{x^5}{120}+\cdots\right)-\frac{1}{6}\left(x^3+\frac{3x^4}{2}+\frac{5x^5}{4}+\cdots\right)+\frac{1}{120}(x^5+\cdots)-\cdots$$

$$=\left(x+\frac{x^2}{2}+\frac{x^3}{6}+\frac{x^4}{24}+\frac{x^5}{120}+\ldots\right)-\left(\frac{x^3}{6}+\frac{x^4}{4}+\frac{5x^5}{24}+\ldots\right)+\left(\frac{x^5}{120}+\ldots\right)-\cdots=x+\frac{x^2}{2}-\frac{5x^4}{24}-\frac{23x^5}{120}-\cdots$$

47. $(\sin x)(\exp x)=x+x^2+\frac{x^3}{3}-\frac{x^5}{30}-\cdots$

$$(\sin x)(\exp x)=\left(x-\frac{x^3}{6}+\frac{x^5}{120}-\cdots\right)\left(1+x+\frac{x^2}{2}+\frac{x^3}{6}+\frac{x^4}{24}+\cdots\right)$$

$$=\left(x-\frac{x^3}{6}+\frac{x^5}{120}+\cdots\right)+\left(x^2-\frac{x^4}{6}+\frac{x^6}{120}-\cdots\right)+\left(\frac{x^3}{2}-\frac{x^5}{12}+\cdots\right)+\left(\frac{x^4}{6}-\frac{x^6}{36}+\cdots\right)+\left(\frac{x^5}{24}-\frac{x^7}{144}+\cdots\right)+\cdots$$

$$=x+x^2+\frac{x^3}{3}-\frac{x^5}{30}-\cdots$$

9.9 Concepts Review

1. $f(1)$; $f'(1)$; $f''(1)$

3. error of the method; error of calculation

Problem Set 9.9

1. $f(x)=e^{2x}\quad f(0)=1$

$f'(x)=2e^{2x}\ f'(0)=2;\quad f''(x)=4e^{2x}\qquad f''(0)=4$

$f^{(3)}(x)=8e^{2x}\ f^{(3)}(0)=8;\quad f^{(4)}(x)=16e^{2x}\ f^{(4)}(0)=16$

$f(x)\approx 1+2x+\frac{4}{2!}x^2+\frac{8}{3!}x^3+\frac{16}{4!}x^4=1+2x+2x^2+\frac{4}{3}x^3+\frac{2}{3}x^4$

$f(0.12)\approx 1+2(0.12)+2(0.12)^2+\frac{4}{3}(0.12)^3+\frac{2}{3}(0.12)^4\approx 1.2712$

3. $f(x)=\sin 2x\ f(0)=0$

$f'(x)=2\cos 2x\ f'(0)=2$

$f''(x)=-4\sin 2x\ f''(0)=0$

$f^{(3)}(x)=-8\cos 2x\ f^{(3)}(0)=-8$

$f^{(4)}(x)=16\sin 2x\ f^{(4)}(0)=0$

$f(x)\approx 2x-\frac{8}{3!}x^3=2x-\frac{4}{3}x^3$

$f(0.12)\approx 2(0.12)-\frac{4}{3}(0.12)^3\approx 0.2377$

5. $f(x) = \ln(1+x) \qquad f(0) = 0$

$f'(x) = \frac{1}{1+x} \quad f'(0) = 1$

$f''(x) = -\frac{1}{(1+x)^2} \quad f''(0) = -1$

$f^{(3)}(x) = \frac{2}{(1+x)^3} \quad f^{(3)}(0) = 2$

$f^{(4)}(x) = -\frac{6}{(1+x)^4} \quad f^{(4)}(0) = -6$

$f(x) \approx x - \frac{1}{2!}x^2 + \frac{2}{3!}x^3 - \frac{6}{4!}x^4 = x - \frac{1}{2}x^2 + \frac{1}{3}x^3 - \frac{1}{4}x^4$

$f(0.12) \approx 0.12 - \frac{1}{2}(0.12)^2 + \frac{1}{3}(0.12)^3 - \frac{1}{4}(0.12)^4$

≈ 0.1133

7. $f(x) = \tan^{-1} x \qquad f(0) = 0$

$f'(x) = \frac{1}{1+x^2} \qquad f'(0) = 1$

$f''(x) = -\frac{2x}{(1+x^2)^2} \qquad f''(0) = 0$

$f'''(x) = \frac{6x^2 - 2}{(1+x^2)^3} \qquad f'''(0) = -2$

$f^{(4)}(x) = \frac{-24x^3 + 24x}{(1+x^2)^4} \qquad f^{(4)}(0) = 0$

$f(x) \approx x - \frac{2}{3!}x^3 = x - \frac{1}{3}x^3$

$f(0.12) \approx 0.12 - \frac{1}{3}(0.12)^3 \approx 0.1194$

9. $f(x) = e^x \qquad f(1) = e$

$f'(x) = e^x \quad f'(1) = e$

$f''(x) = e^x \quad f''(1) = e$

$f'''(x) = e^x \quad f'''(1) = e$

$P_3(x) = e + e(x-1) + \frac{e}{2}(x-1)^2 + \frac{e}{6}(x-1)^3$

11. $f(x) = \tan x; \quad f\left(\frac{\pi}{6}\right) = \frac{\sqrt{3}}{3}$

$f'(x) = \sec^2 x; \quad f'\left(\frac{\pi}{6}\right) = \frac{4}{3}$

$f''(x) = 2\sec^2 x \tan x; \quad f''\left(\frac{\pi}{6}\right) = \frac{8\sqrt{3}}{9}$

$f'''(x) = 2\sec^4 x + 4\sec^2 x \tan^2 x; \quad f'''\left(\frac{\pi}{6}\right) = \frac{16}{3}$

$P_3(x) = \frac{\sqrt{3}}{3} + \frac{4}{3}\left(x - \frac{\pi}{6}\right) + \frac{4\sqrt{3}}{9}\left(x - \frac{\pi}{6}\right)^2 + \frac{8}{9}\left(x - \frac{\pi}{6}\right)^3$

13. $f(x) = \cot^{-1} x; \quad f(1) = \frac{\pi}{4}$

$f'(x) = -\frac{1}{1+x^2}; \quad f'(1) = -\frac{1}{2}$

$f''(x) = \frac{2x}{(1+x^2)^2}; \quad f''(1) = \frac{1}{2}$

$f'''(x) = \frac{-6x^2 + 2}{(1+x^2)^3}; \quad f'''(1) = -\frac{1}{2}$

$P_3(x) = \frac{\pi}{4} - \frac{1}{2}(x-1) + \frac{1}{4}(x-1)^2 - \frac{1}{12}(x-1)^3$

15. $f(x) = x^3 - 2x^2 + 3x + 5; \quad f(1) = 7$

$f'(x) = 3x^2 - 4x + 3; \quad f'(1) = 2$

$f''(x) = 6x - 4; \quad f''(1) = 2$

$f^{(3)}(x) = 6; \quad f^{(3)}(1) = 6$

$P_3(x) = 7 + 2(x-1) + (x-1)^2 + (x-1)^3$

$= 5 + 3x - 2x^2 + x^3 = f(x)$

17. $f(x) = \frac{1}{1-x}; \quad f(0) = 1$

$f'(x) = \frac{1}{(1-x)^2}; \quad f'(0) = 1$

$f''(x) = \frac{2}{(1-x)^3}; \quad f''(0) = 2$

$f^{(3)}(x) = \frac{6}{(1-x)^4}; \quad f^{(3)}(0) = 6$

$f^{(4)}(x) = \frac{24}{(1-x)^5}; \quad f^{(4)}(0) = 24$

$f^{(n)}(x) = \frac{n!}{(1-x)^{n+1}}; \quad f^{(n)}(0) = n!$

$f(x) \approx 1 + x + \frac{2}{2!}x^2 + \frac{6}{3!}x^3 + \ldots + \frac{n!}{n!}x^n$

$= 1 + x + x^2 + x^3 + \ldots + x^n$

Using $n = 4$, $f(x) \approx 1 + x + x^2 + x^3 + x^4$

a. $f(0.1) \approx 1.1111$

b. $f(0.5) \approx 1.9375$

c. $f(0.9) \approx 4.0951$

d. $f(2) \approx 31$

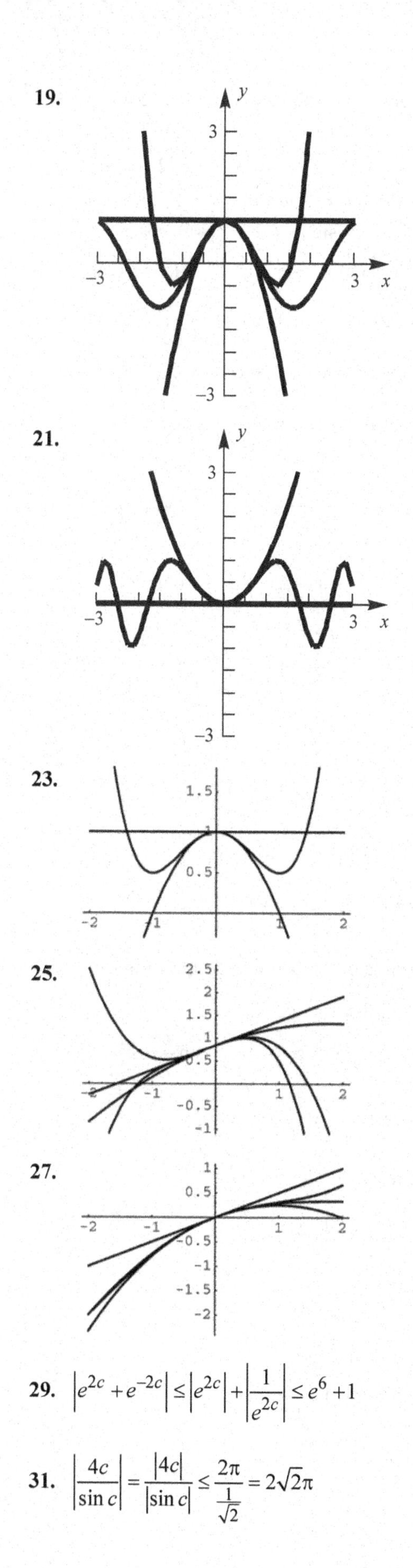

29. $\left|e^{2c}+e^{-2c}\right| \le \left|e^{2c}\right| + \left|\dfrac{1}{e^{2c}}\right| \le e^6+1$

31. $\left|\dfrac{4c}{\sin c}\right| = \dfrac{|4c|}{|\sin c|} \le \dfrac{2\pi}{\frac{1}{\sqrt{2}}} = 2\sqrt{2}\pi$

33. $\left|\dfrac{e^c}{c+5}\right| = \dfrac{\left|e^c\right|}{|c+5|} \le \dfrac{e^4}{3}$

35. $\left|\dfrac{c^2+\sin c}{10\ln c}\right| = \dfrac{\left|c^2+\sin c\right|}{|10\ln c|} \le \dfrac{\left|c^2\right|+|\sin c|}{|10\ln c|}$

$\le \dfrac{16+1}{10\ln 2} = \dfrac{17}{10\ln 2}$

37. $f(x)=\ln(2+x);\ f'(x)=\dfrac{1}{2+x};$

$f''(x) = -\dfrac{1}{(2+x)^2};\ f^{(3)}(x) = \dfrac{2}{(2+x)^3};$

$f^{(4)}(x) = -\dfrac{6}{(2+x)^4};\ f^{(5)}(x) = \dfrac{24}{(2+x)^5};$

$f^{(6)}(x) = -\dfrac{120}{(2+x)^6};\ f^{(7)}(x) = \dfrac{720}{(2+x)^7}$

$R_6(x) = \dfrac{1}{7!}\cdot\dfrac{720}{(2+c)^7}x^7 = \dfrac{x^7}{7(2+c)^7}$

$|R_6(0.5)| \le \left|\dfrac{0.5^7}{7\cdot 2^7}\right| \approx 8.719\times 10^{-6}$

39. $f(x)=\sin x;\ f^{(7)}(x) = -\cos x$

$R_6(x) = \dfrac{-\cos c}{7!}\left(x-\dfrac{\pi}{4}\right)^7 = \dfrac{-\cos c\left(x-\frac{\pi}{4}\right)^7}{5040}$

$|R_6(0.5)| \le \left|\dfrac{\cos 0.5\left(0.5-\frac{\pi}{4}\right)^7}{5040}\right| \approx 2.685\times 10^{-8}$

41. If $f(x)=\dfrac{1}{x}$, it is easily verified that

$f^{(n)}(x) = \dfrac{(-1)^n n!}{x^{(n+1)}}$. Thus for $a=1$

$R_6(x) = \dfrac{-(x-1)^7}{c^8}$, where c is between x and 1 .

Thus, $|R_6(0.5)| = \dfrac{(0.5)^7}{c^8}$, where $c\in(0.5,1)$.

Therefore, $|R_6(0.5)| \le \dfrac{(0.5)^7}{(0.5)^8} = 2$

43. $R_n(x) = \frac{e^c}{(n+1)!}x^{n+1}$

Note that $e^1 < 3$.

$|R_n(1)| < \frac{3}{(n+1)!}$

$\frac{3}{(n+1)!} < 0.000005$ or $600000 < (n+1)!$ when $n \geq 9$.

45. This is a Binomial Series ($p = \frac{1}{2}$), so the third-order Maclaurin polynomial is (see section 9.8, Thm D and example 6) $P_3(x) = 1 + \frac{x}{2} - \frac{x^2}{8} + \frac{x^3}{16}$; further,

$R_3(x) = \frac{-5x^4}{128(1+c)^{7/2}}$. Now if $x \in [-0.5, 0.5]$ and c is between 0 and x, then

$1 + c > 0.5$ and $x^4 \leq \left(\frac{1}{2}\right)^4 = \frac{1}{16}$ so that, for all x,

$$|R_3(x)| \leq \frac{5\left(\frac{1}{16}\right)}{128(0.5)^{7/2}} = \frac{5\sqrt{2}}{256} \approx 0.0276$$

47. $f(x) = (1+x)^{-1/2}$ $\qquad f(0) = 1$

$f'(x) = -\frac{1}{2}(1+x)^{-3/2}$ $\qquad f'(0) = -\frac{1}{2}$

$f''(x) = \frac{3}{4}(1+x)^{-5/2}$ $\qquad f''(0) = \frac{3}{4}$

$f^{(3)}(x) = -\frac{15}{8}(1+x)^{-7/2}$ $\qquad f^{(3)}(0) = -\frac{15}{8}$

$f^{(4)}(x) = \frac{105}{16}(1+x)^{-9/2}$ $\qquad f^{(4)}(c) = \frac{105}{16}(1+c)^{-9/2}$

$(1+x)^{-1/2} \approx 1 - \frac{1}{2}x + \frac{3}{8}x^2 - \frac{5}{16}x^3$

$R_3(x) = \frac{35}{128}(1+c)^{-9/2}x^4$

$|R_3(x)| \leq \left|\frac{35}{128}(0.95)^{-9/2}(0.05)^4\right| \approx 2.15 \times 10^{-6}$

49. $R_4(x) = \frac{\cos c}{5!}x^5$

$|R_4(x)| \leq \frac{(0.5)^5}{5!} \approx 0.00026042 \leq 0.0002605$

$\int_0^{0.5} \sin x\, dx \approx \int_0^{0.5}\left(x - \frac{1}{6}x^3\right)dx$

$= \left[\frac{1}{2}x^2 - \frac{1}{24}x^4\right]_0^{0.5} \approx 0.1224$

Error $\leq 0.0002605(0.5 - 0) = 0.00013025$

51. Assume n is odd; that is $n = 2m+1$ for $m \geq 0$.

Then, $R_{n+1}(x) = R_{2m+2}(x) = \frac{f^{(2m+3)}(c)}{(2m+3)!}x^{2m+3}$.

Note that, for all m,

$f^{(4m)}(x) = \sin x,\ f^{(4m+1)}(x) = \cos x$

$f^{(4m+2)}(x) = -\sin x,\ f^{(4m+3)}(x) = -\cos x$;

therefore, $|R_{n+1}(x)| = \frac{\cos c}{(2m+3)!}x^{2m+3}$ where c is between 0 and x. For $x \in [0, \pi/2]$, $c \in (0, x)$ so that $\cos c < 1$ and $x \leq \frac{\pi}{2}$; hence

$|R_{n+1}(x)| \leq \frac{\left(\frac{\pi}{2}\right)^{2m+3}}{(2m+3)!}$. Now, for

$k = 2, 3, \ldots, 2m+3,\ \frac{\frac{\pi}{2}}{k} \leq \frac{\pi}{4}$ so that

$|R_{n+1}(x)| \leq \frac{\left(\frac{\pi}{2}\right)^{2m+3}}{(2m+3)!} \leq \frac{\pi}{2}\left(\frac{\pi}{4}\right)^{2m+2}$ for all $x \in [0, \frac{\pi}{2}]$. Now

$\frac{\pi}{2}\left(\frac{\pi}{4}\right)^{2m+2} \leq 0.00005 \Rightarrow$

$(2m+2)\ln\left(\frac{\pi}{4}\right) \leq \ln\left(\frac{2(0.00005)}{\pi}\right) \Rightarrow$

$2m+2 \geq 42.8666 \Rightarrow n = 2m+1 > 42$

53. The area of the sector with angle t is $\frac{1}{2}tr^2$. The area of the triangle is

$\frac{1}{2}\left(r\sin\frac{t}{2}\right)\left(2r\cos\frac{t}{2}\right) = r^2\sin\frac{t}{2}\cos\frac{t}{2} = \frac{1}{2}r^2\sin t$

$A = \frac{1}{2}tr^2 - \frac{1}{2}r^2\sin t$

Using $n = 3$, $\sin t \approx t - \frac{1}{6}t^3$.

$A \approx \frac{1}{2}tr^2 - \frac{1}{2}r^2\left(t - \frac{1}{6}t^3\right) = \frac{1}{12}r^2t^3$

55. a. $\ln\left(1 + \frac{r}{12}\right)^{12n} = \ln 2$

$12n\ln\left(1 + \frac{r}{12}\right) = \ln 2$

$n = \frac{\ln 2}{12\ln\left(1 + \frac{r}{12}\right)}$

b. $f(x) = \ln(1+x)$; $f(0) = 0$

$$f'(x) = \frac{1}{1+x};\quad f'(0) = 1$$

$$f''(x) = -\frac{1}{(1+x)^2};\quad f''(0) = -1$$

$$\ln(1+x) \approx x - \frac{x^2}{2}$$

$$n \approx \frac{\ln 2}{r - \frac{r^2}{24}} = \left[\frac{24}{r(24-r)}\right]\ln 2$$

$$= \frac{\ln 2}{r} + \frac{\ln 2}{24-r}$$

$$\approx \frac{\ln 2}{r} + \frac{\ln 2}{24} \approx \frac{0.693}{r} + 0.029$$

We let $24 - r \approx 24$ since the interest rate r is going to be close to 0.

c.

r	n (exact)	n (approx.)	n (rule 72)
0.05	13.8918	13.889	14.4
0.10	6.9603	6.959	7.2
0.15	4.6498	4.649	4.8
0.20	3.4945	3.494	3.6

57. $f(x) = x^4 - 3x^3 + 2x^2 + x - 2$; $f(1) = -1$

$f'(x) = 4x^3 - 9x^2 + 4x + 1$; $f'(1) = 0$

$f''(x) = 12x^2 - 18x + 4$; $f''(1) = -2$

$f^{(3)}(x) = 24x - 18$; $f^{(3)}(1) = 6$

$f^{(4)}(x) = 24$; $f^{(4)}(1) = 24$

$f^{(5)}(x) = 0$

Since $f^{(5)}(x) = 0$, $R_5(x) = 0$.

$$x^4 - 3x^3 + 2x^2 + x - 2$$
$$= -1 - (x-1)^2 + (x-1)^3 + (x-1)^4$$

59. $f(x) = \sin x$; $f\left(\frac{\pi}{4}\right) = \frac{\sqrt{2}}{2}$

$f'(x) = \cos x$; $f'\left(\frac{\pi}{4}\right) = \frac{\sqrt{2}}{2}$

$f''(x) = -\sin x$; $f''\left(\frac{\pi}{4}\right) = -\frac{\sqrt{2}}{2}$

$f^{(3)}(x) = -\cos x$; $f^{(3)}\left(\frac{\pi}{4}\right) = -\frac{\sqrt{2}}{2}$

$f^{(4)}(x) = \sin x$; $f^{(4)}(c) = \sin c$

$43° = \frac{\pi}{4} - \frac{\pi}{90}$ radians

$$\sin x = \frac{\sqrt{2}}{2} + \frac{\sqrt{2}}{2}\left(x - \frac{\pi}{4}\right) - \frac{\sqrt{2}}{4}\left(x - \frac{\pi}{4}\right)^2 - \frac{\sqrt{2}}{12}\left(x - \frac{\pi}{4}\right)^3 + R_3(x)$$

$$\sin\left(\frac{\pi}{4} - \frac{\pi}{90}\right) = \frac{\sqrt{2}}{2} + \frac{\sqrt{2}}{2}\left(-\frac{\pi}{90}\right) - \frac{\sqrt{2}}{4}\left(-\frac{\pi}{90}\right)^2 - \frac{\sqrt{2}}{12}\left(-\frac{\pi}{90}\right)^3 + R_3\left(\frac{\pi}{4} - \frac{\pi}{90}\right)$$

$$\approx 0.681998 + R_3$$

$$|R_3| = \left|\frac{\sin c}{4!}\left(-\frac{\pi}{90}\right)^4\right| < \frac{1}{24}\left(\frac{\pi}{90}\right)^4 \approx 6.19 \times 10^{-8}$$

61. $|R_9(x)| \le \frac{1}{10!}x^{10} \le \frac{1}{10!}\left(\frac{\pi}{2}\right)^{10} \approx 2.5202 \times 10^{-5}$

63. The kth derivative of $h(x)f(x)$ is $\sum_{i=0}^{k}\binom{k}{i}h^{(i)}(x)f^{(k-i)}(x)$. If $h(x) = x^{n+1}$,

$$h^{(i)}(x) = \frac{(n+1)!}{(n+1-i)!}x^{n+1-i}.$$

Thus for $i \le n+1$, $h^{(i)}(0) = 0$. Let $q(x) = x^{n+1}f(x)$. Then

$$q^{(k)}(0) = \sum_{i=0}^{k}\binom{k}{i}h^{(i)}(0)f^{(k-i)}(0) = 0$$

for $k \le n+1$.

$g^{(k)}(x) = p^{(k)}(x) + q^{(k)}(x)$, so

$g^{(k)}(0) = p^{(k)}(0) + q^{(k)}(0) = p^{(k)}(0)$

for $k \le n+1$.

The Maclaurin polynomial of order n for g is

$p(0) + p'(0)x + \frac{p''(0)}{2!}x^2 + \ldots + \frac{p^{(n)}(0)}{n!}x^n$ which is the Maclaurin polynomial of order n for $p(x)$. Since $p(x)$ is a polynomial of degree at most n, the remainder $R_n(x)$ of Maclaurin's Formula for $p(x)$ is 0, so the Maclaurin polynomial of order n for $g(x)$ is $p(x)$.

9.10 Chapter Review

Concepts Test

1. False: If $b_n = 100$ and $a_n = 50 + (-1)^n$ then since $a_n = \begin{cases} 51 & \text{if } n \text{ is even} \\ 49 & \text{if } n \text{ is odd} \end{cases}$, $0 \le a_n \le b_n$ for all n and $\lim_{n\to\infty} b_n = 100$ while $\lim_{n\to\infty} a_n$ does not exist.

3. True: If $\lim_{n\to\infty} a_n = L$ then for any $\varepsilon > 0$ there is a number $M > 0$ such that $|a_n - L| < \varepsilon$ for all $n \ge M$. Thus, for the same ε, $|a_{3n+4} - L| < \varepsilon$ for $3n + 4 \ge$ M or $n \ge \frac{M-4}{3}$. Since ε was arbitrary, $\lim_{n\to\infty} a_{3n+4} = L$.

5. False: Let a_n be given by
$$a_n = \begin{cases} 1 & \text{if } n \text{ is prime} \\ 0 & \text{if } n \text{ is composite} \end{cases}$$
Then $a_{mn} = 0$ for all mn since $m \ge 2$, hence $\lim_{n\to\infty} a_{mn} = 0$ for $m \ge 2$.
$\lim_{n\to\infty} a_n$ does not exist since for any $M > 0$ there will be a_n's with $a_n = 1$ since there are infinitely many prime numbers.

7. False: Let $a_n = 1 + \frac{1}{2} + \cdots + \frac{1}{n}$. Then $a_n - a_{n+1} = -\frac{1}{n+1}$ so $\lim_{n\to\infty} (a_n - a_{n+1}) = 0$ but $\lim_{n\to\infty} a_n$ is not finite since $\lim_{n\to\infty} a_n = \sum_{k=1}^{\infty} \frac{1}{k}$, which diverges.

9. True: If $\{a_n\}$ converges, then for some N, there are numbers m and M with $m \le a_n \le M$ for all $n \ge N$. Thus $\frac{m}{n} \le \frac{a_n}{n} \le \frac{M}{n}$ for all $n \ge N$. Since $\left\{\frac{m}{n}\right\}$ and $\left\{\frac{M}{n}\right\}$ both converge to 0, $\left\{\frac{a_n}{n}\right\}$ must also converge to 0.

11. True: The series converges by the Alternating Series Test.
$S_1 = a_1, S_2 = a_1 - a_2, S_3 = a_1 - a_2 + a_3$, $S_4 = a_1 - a_2 + a_3 - a_4$, etc.
$0 < a_2 < a_1 \Rightarrow 0 < a_1 - a_2 = S_2 < a_1$;
$0 < a_3 < a_2 \Rightarrow -a_2 < -a_2 + a_3 < 0$ so
$0 < a_1 - a_2 < a_1 - a_2 + a_3 = S_3 < a_1$;
$0 < a_4 < a_3 \Rightarrow 0 < a_3 - a_4 < a_3$, so
$-a_2 < -a_2 + a_3 - a_4 < -a_2 + a_3 < 0$,
hence
$0 < a_1 - a_2 < a_1 - a_2 + a_3 - a_4 = S_4$
$< a_1 - a_2 + a_3 < a_1$; etc.
For each even n, $0 < S_{n-1} - a_n$ while for each odd n, $n > 1$, $S_{n-1} + a_n < a_1$.

13. False: $\sum_{n=1}^{\infty} (-1)^n$ diverges but the partial sums are bounded ($S_n = -1$ for odd n and $S_n = 0$ for even n.)

15. True: $\rho = \lim_{n\to\infty} \frac{a_{n+1}}{a_n} = 1$, Ratio Test is inconclusive. (See the discussion before Example 5 in Section 9.4.)

17. False: $\lim_{n\to\infty} \left(1 - \frac{1}{n}\right)^n = \frac{1}{e} \ne 0$ so the series cannot converge.

19. True:
$$\sum_{n=2}^{\infty} \frac{n+1}{(n \ln n)^2}$$
$$= \sum_{n=2}^{\infty} \left[\frac{n}{(n \ln n)^2} + \frac{1}{(n \ln n)^2}\right]$$
$$= \sum_{n=2}^{\infty} \left[\frac{1}{n(\ln n)^2} + \frac{1}{n^2(\ln n)^2}\right]$$
$\frac{1}{x(\ln x)^2}$ is continuous, positive, and nonincreasing on $[2, \infty)$. Using

$u = \ln x,\ du = \frac{1}{x}dx,$

$$\int_2^\infty \frac{1}{x(\ln x)^2}dx = \int_{\ln 2}^\infty \frac{1}{u^2}du$$

$$= \left[-\frac{1}{u}\right]_{\ln 2}^\infty = 0 + \frac{1}{\ln 2} < \infty \text{ so}$$

$\sum_{n=2}^\infty \frac{1}{n(\ln n)^2}$ converges.

For $n \ge 3$, $\ln n > 1$, so $(\ln n)^2 > 1$ and

$\frac{1}{n^2(\ln n)^2} < \frac{1}{n^2}$. Thus

$\sum_{n=3}^\infty \frac{1}{n^2(\ln n)^2} < \sum_{n=3}^\infty \frac{1}{n^2}$ so

$\sum_{n=3}^\infty \frac{1}{n^2(\ln n)^2}$ converges by the Comparison Test. Since both series converge, so does their sum.

21. True: If $0 \le a_{n+100} \le b_n$ for all n in N, then $\sum_{n=101}^\infty a_n \le \sum_{n=1}^\infty b_n$ so $\sum_{n=1}^\infty a_n$ also converges, since adding a finite number of terms does not affect the convergence or divergence of a series.

23. True: $\frac{1}{3} + \left(\frac{1}{3}\right)^2 + \left(\frac{1}{3}\right)^3 + \ldots = \sum_{n=1}^\infty \left(\frac{1}{3}\right)^n$

$$= \sum_{n=1}^\infty \frac{1}{3}\left(\frac{1}{3}\right)^{n-1} = \frac{\frac{1}{3}}{1-\frac{1}{3}} = \frac{\frac{1}{3}}{\frac{2}{3}} = \frac{1}{2},$$ so the sum of the first thousand terms is less than $\frac{1}{2}$.

25. True: If $b_n \le a_n \le 0$ for all n in N then $0 \le -a_n \le -b_n$ for all n in N.

$\sum_{n=1}^\infty -b_n = (-1)\sum_{n=1}^\infty b_n$ which converges since $\sum_{n=1}^\infty b_n$ converges.

Thus, by the Comparison Test, $\sum_{n=1}^\infty -a_n$ converges, hence $\sum_{n=1}^\infty a_n = (-1)\sum_{n=1}^\infty (-a_n)$ also converges.

27. True: $\left|\sum_{n=1}^\infty (-1)^{n+1}\frac{1}{n} - \sum_{n=1}^{99}(-1)^{n+1}\frac{1}{n}\right|$

$$= \left|-\frac{1}{100} + \frac{1}{101} - \frac{1}{102} - \cdots\right| < \frac{1}{100} = 0.01$$

29. True: $|3-(-1.1)| = 4.1$, so the radius of convergence of the series is at least 4.1.
$|3-7| = 4 < 4.1$ so $x = 7$ is within the interval of convergence.

31. True: The radius of convergence is at least 1.5, so 1 is within the interval of convergence.

Thus $\int_0^1 f(x)dx = \left[\sum_{n=0}^\infty \frac{a_n x^{n+1}}{n+1}\right]_0^1$

$$= \sum_{n=0}^\infty \frac{a_n}{n+1}.$$

33. False: Consider the function

$$f(x) = \begin{cases} e^{-\frac{1}{x^2}} & x \ne 0 \\ 0 & x = 0 \end{cases}.$$

The Maclaurin series for this function represents the function only at $x = 0$.

35. True: $\sum_{n=0}^\infty \frac{(-1)^n x^n}{n!} = e^{-x}$, $\frac{d}{dx}e^{-x} + e^{-x} = 0$

37. True: If $p(x)$ and $q(x)$ are polynomials of degree less than or equal to n, satisfying $p(a) = q(a) = f(a)$ and $p^{(k)}(a) = q^{(k)}(a) = f^{(k)}(a)$ for $k \le n$, then $p(x) = q(x)$.

39. True: After simplifying, $P_3(x) = f(x)$.

41. True: The Maclaurin polynomial of an even function involves only even powers of x, so $f'(0) = 0$ if $f(x)$ is an even function.

Sample Test Problems

1. $\lim_{n\to\infty} \frac{9n}{\sqrt{9n^2+1}} = \lim_{n\to\infty} \frac{9}{\sqrt{9+\frac{1}{n^2}}} = 3$

The sequence converges to 3.

3. $\lim_{n\to\infty}\left(1+\frac{4}{n}\right)^n = \lim_{n\to\infty}\left(\left(1+\frac{4}{n}\right)^{n/4}\right)^4 = e^4$

The sequence converges to e^4.

5. Let $y = \sqrt[n]{n} = n^{1/n}$ then $\ln y = \frac{1}{n}\ln n$.

$\lim_{n\to\infty} \frac{1}{n}\ln n = \lim_{n\to\infty} \frac{\ln n}{n} = \lim_{n\to\infty} \frac{\frac{1}{n}}{1} = \lim_{n\to\infty} \frac{1}{n} = 0$ by using l'Hôpital's Rule. Thus, $\lim_{n\to\infty} \sqrt[n]{n} = \lim_{n\to\infty} e^{\ln y} = 1$. The sequence converges to 1.

7. $a_n \ge 0$; $\lim_{n\to\infty} \frac{\sin^2 n}{\sqrt{n}} \le \lim_{n\to\infty} \frac{1}{\sqrt{n}} = 0$

The sequence converges to 0.

9. $S_n = \left(\frac{1}{\sqrt{1}} - \frac{1}{\sqrt{2}}\right) + \left(\frac{1}{\sqrt{2}} - \frac{1}{\sqrt{3}}\right) + \ldots + \left(\frac{1}{\sqrt{n-1}} - \frac{1}{\sqrt{n}}\right) + \left(\frac{1}{\sqrt{n}} - \frac{1}{\sqrt{n+1}}\right) = 1 - \frac{1}{\sqrt{n+1}}$, so

$\lim_{n\to\infty} S_n = \lim_{n\to\infty}\left(1 - \frac{1}{\sqrt{n+1}}\right) = 1$. The series converges to 1.

11. $\ln\frac{1}{2} + \ln\frac{2}{3} + \ln\frac{3}{4} + \ldots = \sum_{n=1}^{\infty} \ln\frac{n}{n+1} = \sum_{n=1}^{\infty}[\ln n - \ln(n+1)]$

$S_n = (\ln 1 - \ln 2) + (\ln 2 - \ln 3) + \ldots + (\ln(n-1) - \ln n) + (\ln n - \ln(n+1)) = \ln 1 - \ln(n+1) = \ln\frac{1}{n+1}$

As $n \to \infty$, $\frac{1}{n+1} \to 0$ so $\lim_{n\to\infty} S_n = \lim_{n\to\infty} \ln\frac{1}{n+1} = -\infty$.
The series diverges.

13. $\sum_{k=0}^{\infty} e^{-2k} = \sum_{k=0}^{\infty}\left(\frac{1}{e^2}\right)^k = \frac{1}{1-\frac{1}{e^2}} = \frac{e^2}{e^2-1} \approx 1.1565$

since $\frac{1}{e^2} < 1$.

15. $\sum_{k=1}^{\infty} 91\left(\frac{1}{100}\right)^k = \frac{91}{1-\frac{1}{100}} - 91 = \frac{9100}{99} - 91 = \frac{91}{99}$

The series converges since $\left|\frac{1}{100}\right| < 1$.

17. $\cos x = 1 - \frac{x^2}{2!} + \frac{x^4}{4!} - \frac{x^6}{6!} + \cdots$, so

$1 - \frac{2^2}{2!} + \frac{2^4}{4!} - \frac{2^6}{6!} + \cdots$ converges to $\cos 2 \approx -0.41615$.

19. Let $a_n = \frac{n}{1+n^2}$ and $b_n = \frac{1}{n}$.

$\lim_{n\to\infty} \frac{a_n}{b_n} = \lim_{n\to\infty} \frac{n^2}{1+n^2} = \lim_{n\to\infty} \frac{1}{\frac{1}{n^2}+1} = 1$;

$0 < 1 < \infty$.

By the Limit Comparison Test, since $\sum_{n=1}^{\infty} b_n = \sum_{n=1}^{\infty} \frac{1}{n}$ diverges, $\sum_{n=1}^{\infty} a_n = \sum_{n=1}^{\infty} \frac{n}{1+n^2}$ also diverges.

21. Since the series alternates, $\frac{1}{\sqrt[3]{n}} > \frac{1}{\sqrt[3]{n+1}} > 0$, and

$\lim_{n\to\infty} \frac{1}{\sqrt[3]{n}} = 0$, the series converges by the Alternating Series Test.

23. $\sum_{n=1}^{\infty} \frac{2^n+3^n}{4^n} = \sum_{n=1}^{\infty}\left(\left(\frac{1}{2}\right)^n + \left(\frac{3}{4}\right)^n\right)$

$= \left(\frac{1}{1-\frac{1}{2}} - 1\right) + \left(\frac{1}{1-\frac{3}{4}} - 1\right) = 1 + 3 = 4$

The series converges to 4. The 1's must be subtracted since the index starts with $n = 1$.

25. $\lim_{n\to\infty} \frac{n+1}{10n+12} = \frac{1}{10} \neq 0$, so the series diverges.

27. $\rho = \lim_{n\to\infty}\left|\frac{(n+1)^2}{(n+1)!} \div \frac{n^2}{n!}\right| = \lim_{n\to\infty}\left|\frac{n+1}{n^2}\right| = 0 < 1$, so the series converges.

29. $\rho = \lim_{n\to\infty}\left|\frac{2^{n+1}(n+1)!}{(n+3)!} \div \frac{2^n n!}{(n+2)!}\right|$

$= \lim_{n\to\infty}\left|\frac{2(n+1)}{n+3}\right| = 2 > 1$

The series diverges.

31. $\rho = \lim_{n\to\infty}\left|\frac{(n+1)^2\left(\frac{2}{3}\right)^{n+1}}{n^2\left(\frac{2}{3}\right)^n}\right| = \lim_{n\to\infty}\left|\frac{2}{3}\right|\left|\frac{(n+1)^2}{n^2}\right|$

$= \frac{2}{3} < 1$, so the series converges.

33. $a_n = \frac{1}{3n-1}; \frac{1}{3n-1} > \frac{1}{3n+2}$ so $a_n > a_{n+1}$;

$\lim_{n\to\infty} a_n = \lim_{n\to\infty} \frac{1}{3n-1} = 0$, so the series

$\sum_{n=1}^{\infty}(-1)^n \frac{1}{3n-1}$ converges by the Alternating Series Test.

Let $b_n = \frac{1}{n}$, then

$\lim_{n\to\infty}\frac{a_n}{b_n} = \lim_{n\to\infty}\frac{n}{3n-1} = \lim_{n\to\infty}\frac{1}{3-\frac{1}{n}} = \frac{1}{3}$;

$0 < \frac{1}{3} < \infty$. By the Limit Comparison Test, since

$\sum_{n=1}^{\infty} b_n = \sum_{n=1}^{\infty}\frac{1}{n}$ diverges, $\sum_{n=1}^{\infty} a_n = \sum_{n=1}^{\infty}\frac{1}{3n-1}$ also diverges.

The series is conditionally convergent.

35. $\frac{3^n}{2^{n+8}} = \frac{1}{2^8}\left(\frac{3}{2}\right)^n$;

$\lim_{n\to\infty}\frac{1}{2^8}\left(\frac{3}{2}\right)^n = \frac{1}{2^8}\lim_{n\to\infty}\left(\frac{3}{2}\right)^n = \infty$ since $\frac{3}{2} > 1$.

The series is divergent.

37. $\rho = \lim_{n\to\infty}\left|\frac{x^{n+1}}{(n+1)^3+1} \div \frac{x^n}{n^3+1}\right|$

$= \lim_{n\to\infty}|x|\left|\frac{n^3+1}{(n+1)^3+1}\right| = |x|$

When $x = 1$, the series is

$\sum_{n=0}^{\infty}\frac{1}{n^3+1} = 1 + \sum_{n=1}^{\infty}\frac{1}{n^3+1} \le 1 + \sum_{n=1}^{\infty}\frac{1}{n^3}$, which converges.

When $x = -1$, the series is $\sum_{n=0}^{\infty}\frac{(-1)^n}{n^3+1}$ which

converges absolutely since $\sum_{n=0}^{\infty}\frac{1}{n^3+1}$ converges.

The series converges on $-1 \le x \le 1$.

39. $\rho = \lim_{n\to\infty}\left|\frac{(x-4)^{n+1}}{n+2} \div \frac{(x-4)^n}{n+1}\right|$

$= \lim_{n\to\infty}|x-4|\left|\frac{n+1}{n+2}\right| = |x-4|$; $|x-4| < 1$ when $3 < x < 5$.

When $x = 5$, the series is

$\sum_{n=0}^{\infty}\frac{(-1)^n(1)^n}{n+1} = \sum_{n=0}^{\infty}\frac{(-1)^n}{n+1}$.

$a_n = \frac{1}{n+1}; \frac{1}{n+1} > \frac{1}{n+2}$, so $a_n > a_{n+1}$;

$\lim_{n\to\infty}\frac{1}{n+1} = 0$ so $\sum_{n=0}^{\infty}\frac{(-1)^n}{n+1}$ converges by the Alternating Series Test.

When $x = 3$, the series is

$\sum_{n=0}^{\infty}\frac{(-1)^n(-1)^n}{n+1} = \sum_{n=0}^{\infty}\frac{1}{n+1}$. $a_n = \frac{1}{n+1}$, let

$b_n = \frac{1}{n}$ then $\lim_{n\to\infty}\frac{a_n}{b_n} = \lim_{n\to\infty}\frac{n}{n+1} = 1$; $0 < 1 < \infty$

hence since $\sum_{n=1}^{\infty} b_n = \sum_{n=1}^{\infty}\frac{1}{n}$ diverges, $\sum_{n=0}^{\infty}\frac{1}{n+1}$ also diverges.

The series converges on $3 < x \le 5$.

41. $\rho = \lim_{n\to\infty}\left|\frac{(x-3)^{n+1}}{2^{n+1}+1} \div \frac{(x-3)^n}{2^n+1}\right|$

$= \lim_{n\to\infty}|x-3|\left|\frac{1+\frac{1}{2^n}}{2+\frac{1}{2^n}}\right| = \frac{|x-3|}{2}; \ \frac{|x-3|}{2} < 1$

when $1 < x < 5$.
When $x = 5$, the series is

$\sum_{n=0}^{\infty}\frac{2^n}{2^n+1} = \sum_{n=0}^{\infty}\frac{1}{1+\left(\frac{1}{2}\right)^n}; \ \lim_{n\to\infty}\frac{1}{1+\left(\frac{1}{2}\right)^n} = 1 \neq 0$

so the series diverges.

When $x = 1$, the series is

$\sum_{n=0}^{\infty}\frac{(-2)^n}{2^n+1} = \sum_{n=0}^{\infty}\frac{(-1)^n}{1+\left(\frac{1}{2}\right)^n}; \ \lim_{n\to\infty}\frac{1}{1+\left(\frac{1}{2}\right)^n} = 1 \neq 0$

so the series diverges.
The series converges on $1 < x < 5$.

43. $\frac{1}{1+x} = 1 - x + x^2 - x^3 + \ldots$ for $-1 < x < 1$.

If $f(x) = \frac{1}{1+x}$, then $f'(x) = -\frac{1}{(1+x)^2}$. Thus, differentiating the series for $\frac{1}{1+x}$ and multiplying by -1 yields

$\frac{1}{(1+x)^2} = 1 - 2x + 3x^2 - 4x^3 + \cdots$. The series converges on $-1 < x < 1$.

45. $\sin^2 x = \left(x - \frac{x^3}{3!} + \frac{x^5}{5!} - \frac{x^7}{7!} + \cdots\right)^2$

$= x^2 - \frac{x^4}{3} + \frac{2x^6}{45} - \frac{x^8}{315} + \cdots$

Since the series for $\sin x$ converges for all x, so does the series for $\sin^2 x$.

47. $\sin x + \cos x = 1 + x - \frac{x^2}{2!} - \frac{x^3}{3!} + \frac{x^4}{4!} + \frac{x^5}{5!} - \cdots$

Since the series for $\sin x$ and $\cos x$ converge for all x, so does the series for $\sin x + \cos x$.

49. Let $a_k = \frac{k}{e^{k^2}}$ and define $f(x) = \frac{x}{e^{x^2}}$; then $f(k) = a_k$ and f is positive, continuous and non-increasing (since $f'(x) = \frac{1-2x^2}{e^{x^2}} < 0$) on $[1,\infty)$. Thus, by the Integral Test, $E_n < \int_n^{\infty}\frac{x}{e^{x^2}}dx = \lim_{A\to\infty}\left[-\frac{1}{2e^{x^2}}\right]_n^A =$

$\frac{1}{2}\left[-\lim_{A\to\infty}\frac{1}{2e^{A^2}} + \frac{1}{2e^{n^2}}\right] = \frac{1}{4e^{n^2}}$. Now $\frac{1}{4e^{n^2}} \le 0.000005 \Rightarrow e^{n^2} \ge 50{,}000 \Rightarrow n^2 \ge \ln(50{,}000) \approx 10.82 \Rightarrow \ n > 3$.

51. **a.** From the Maclaurin series for $\frac{1}{1-x}$, we have $\frac{1}{1-x^3} = 1 + x^3 + x^6 + \cdots$.

b. In Example 6 of Section 9.8 it is shown that $\sqrt{1+x} = 1 + \frac{1}{2}x - \frac{1}{8}x^2 + \frac{1}{16}x^3 - \frac{5}{128}x^4 + \cdots$ so

$\sqrt{1+x^2} = 1 + \frac{1}{2}x^2 - \frac{1}{8}x^4 + \cdots$.

c. $e^{-x} = 1 - x + \frac{x^2}{2!} - \frac{x^3}{3!} + \frac{x^4}{4!} - \frac{x^5}{5!} + \cdots$, so $e^{-x} - 1 + x = \frac{x^2}{2!} - \frac{x^3}{3!} + \frac{x^4}{4!} - \cdots$.

d. Using division with the Maclaurin series for $\cos x$, we get $\sec x = 1 + \frac{x^2}{2} + \frac{5x^4}{4!} + \frac{61x^6}{6!} + \cdots$.

Thus, $x\sec x = x + \frac{x^3}{2} + \frac{5x^5}{4!} + \cdots$.

e. $e^{-x}\sin x = \left(1 - x + \frac{x^2}{2!} - \frac{x^3}{3!} + \cdots\right)\left(x - \frac{x^3}{3!} + \frac{x^5}{5!} - \frac{x^7}{7!} + \cdots\right) = x - x^2 + \frac{x^3}{3} - \cdots$

f. $1 + \sin x = 1 + x - \frac{x^3}{3!} + \frac{x^5}{5!} - \frac{x^7}{7!} + \cdots$; Using division, we get $\frac{1}{1+\sin x} = 1 - x + x^2 - \cdots$.

53. $f(0)=0$

$f'(x)=\cos^2 x-2x^2\sin^2 x$

$f'(0)=1$

$p(x)=x;\ p(0.2)=0.2;\ f(0.2)=0.1998$

55. $g(x)=x^3-2x^2+5x-7 \qquad g(2)=3$

$g'(x)=3x^2-4x+5 \qquad g'(2)=9$

$g''(x)=6x-4 \qquad g''(2)=8$

$g^{(3)}(x)=6 \qquad g^{(3)}(2)=6$

Since $g^{(4)}(x)=0,\ R_3(x)=0,$ so the Taylor polynomial of order 3 based at 2 is an exact representation.

$g(x)=P_4(x)=3+9(x-2)+4(x-2)^2+(x-2)^3$

57. $f(x)=\frac{1}{x+1} \qquad f(1)=\frac{1}{2}$

$f'(x)=-\frac{1}{(x+1)^2} \qquad f'(1)=-\frac{1}{4}$

$f''(x)=\frac{2}{(x+1)^3} \qquad f''(1)=\frac{1}{4}$

$f^{(3)}(x)=-\frac{6}{(x+1)^4} \quad f^{(3)}(1)=-\frac{3}{8}$

$f^{(4)}(x)=\frac{24}{(x+1)^5} \qquad f^{(4)}(1)=\frac{3}{4}$

$$f(x)\approx\frac{1}{2}-\frac{1}{4}(x-1)+\frac{1}{8}(x-1)^2-\frac{1}{16}(x-1)^3+\frac{1}{32}(x-1)^4$$

59. $f(x)=\frac{1}{2}(1-\cos 2x) \qquad f(0)=0$

$f'(x)=\sin 2x \qquad f'(0)=0$

$f''(x)=2\cos 2x \qquad f''(0)=2$

$f^{(3)}(x)=-4\sin 2x \qquad f^{(3)}(0)=0$

$f^{(4)}(x)=-8\cos 2x \qquad f^{(4)}(0)=-8$

$f^{(5)}(x)=16\sin 2x \qquad f^{(5)}(0)=0$

$f^{(6)}(x)=32\cos 2x \qquad f^{(6)}(c)=32\cos 2c$

$$\sin^2 x\approx\frac{2}{2!}x^2-\frac{8}{4!}x^4=x^2-\frac{1}{3}x^4$$

$$|R_4(x)|=|R_5(x)|=\left|\frac{32}{6!}(\cos 2c)x^6\right|\le\frac{2}{45}(0.2)^6$$

$<2.85\times10^{-6}$

61. From Problem 60,

$$\ln x\approx(x-1)-\frac{1}{2}(x-1)^2+\frac{1}{3}(x-1)^3-\frac{1}{4}(x-1)^4+\frac{1}{5}(x-1)^5.$$

$$|R_5(x)|=\left|\frac{1}{6c^6}(x-1)^6\right|\le\frac{0.2^6}{6\cdot0.8^6}<4.07\times10^{-5}$$

$$\int_{0.8}^{1.2}\ln x\,dx\approx\int_{0.8}^{1.2}\left[(x-1)-\frac{1}{2}(x-1)^2+\frac{1}{3}(x-1)^3-\frac{1}{4}(x-1)^4+\frac{1}{5}(x-1)^5\right]dx$$

$$=\left[\frac{1}{2}(x-1)^2-\frac{1}{6}(x-1)^3+\frac{1}{12}(x-1)^4-\frac{1}{20}(x-1)^5+\frac{1}{30}(x-1)^6\right]_{0.8}^{1.2}$$

≈ -0.00269867

Error $\le(1.2-0.8)4.07\times10^{-5}<1.63\times10^{-5}$

Review and Preview Problems

1. $f(x)=\frac{x^2}{4}$ so that $f'(x)=\frac{x}{2}$ and $f'(2)=1$

a. The tangent line will be the line through the point $(2,1)$ having slope $m=1$. Using the point slope formula: $(y-1)=1(x-2)$ or $y=x-1$.

b. The normal line will be the line through the point $(2,1)$ having slope $m=-\frac{1}{1}=-1$. Using the point slope formula: $(y-1)=-1(x-2)$ or $y=-x+3$ or $x+y=3$.

3. Solving equation 1 for y^2: $y^2=9-\frac{9}{16}x^2$ and putting this result into equation 2:

$$\frac{x^2}{9}+\frac{1}{16}\left(9-\frac{9}{16}x^2\right)=1$$

$175x^2=1008 \quad x^2=5.76 \quad x=\pm2.4$

Putting these values into equation 1 we get

$y^2=9-\frac{9}{16}(5.76)=9-3.24=5.76$ so $y=\pm2.4$ also. Thus the points of intersection are $(2.4,2.4),\ (-2.4,2.4),\ (2.4,-2.4),\ (-2.4,-2.4)$

5. Since we are given a point, all we need is the slope to determine the equation of our tangent line.

$$\frac{d}{dx}\left(x^2+\frac{y^2}{4}\right)=\frac{d}{dx}(1)$$

$$2x+\frac{y}{2}\frac{dy}{dx}=0$$

$$\frac{y}{2}\frac{dy}{dx}=-2x$$

$$\frac{dy}{dx}=\frac{-4x}{y}$$

At the point $\left(-\frac{\sqrt{3}}{2},1\right)$, we get

$$\frac{dy}{dx}=\frac{-4\left(-\sqrt{3}/2\right)}{1}=2\sqrt{3}=m_{\tan}$$

Therefore, the equation of the tangent line to the curve at $\left(-\frac{\sqrt{3}}{2},1\right)$ is given by

$$y-1=2\sqrt{3}\left(x-\left(-\frac{\sqrt{3}}{2}\right)\right)$$

$$y-1=2\sqrt{3}\,x+3$$

$$y=2\sqrt{3}\,x+4$$

7. Denote the curves as

$$C_1:\frac{x^2}{100}+\frac{y^2}{64}=1 \quad\text{and}\quad C_2:\frac{x^2}{9}-\frac{y^2}{27}=1$$

a. From C_2, $3x^2-27=y^2$ and so, from C_1,

$$16x^2+25\left(3x^2-27\right)=1600$$

$$91x^2=2275,\ x^2=25,\ x=\pm 5$$

$$y^2=3(25)-27,\ y^2=48,\ y=\pm 4\sqrt{3}$$

Thus the point of intersection in the first quadrant is $P=(5,4\sqrt{3})$.

b. Slope m_1 of the line tangent to C_1 at P:

$$C_1':\ \frac{x}{50}+\frac{y}{32}y'=0,\qquad y'=-\frac{16x}{25y}\ \text{so}$$

$$m_1=-\frac{80}{100\sqrt{3}}=-\frac{4\sqrt{3}}{15}$$

The line T_1 tangent to C_1 at P is

$$T_1:\ (y-4\sqrt{3})=-\frac{4\sqrt{3}}{15}(x-5)\ \text{or}$$

$$4\sqrt{3}\,x+15y=80\sqrt{3}$$

Slope m_2 of the line tangent to C_2 at P:

$$C_2':\ \frac{2}{9}x-\frac{2}{27}yy'=0,\qquad y'=\frac{3x}{y}\ \text{so}$$

$$m_2=\frac{15}{4\sqrt{3}}=\frac{5\sqrt{3}}{4}.$$

The line T_2 tangent to C_2 at P is

$$T_2:\ (y-4\sqrt{3})=\frac{5\sqrt{3}}{4}(x-5)\ \text{or}$$

$$5\sqrt{3}\,x-4y=9\sqrt{3}$$

c. To find the angles between the tangent lines, you can use problem 40 of section 0.7 or consider the diagram below:

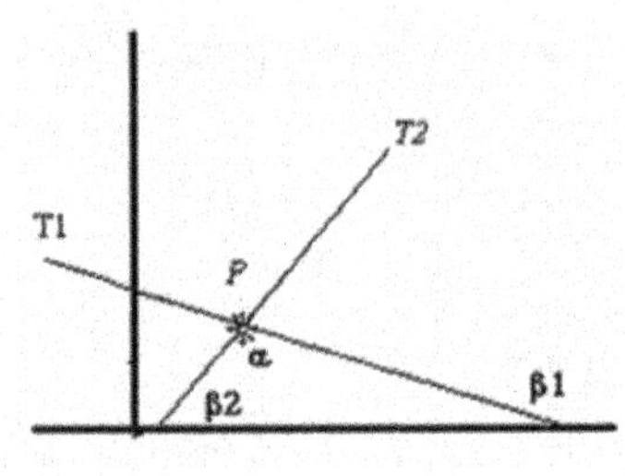

Note that

$\alpha+\beta 2+(180-\beta 1)=180$ or $\alpha=\beta 1-\beta 2$; further

$$\beta 1=180-\tan^{-1}|m_1|=180-\tan^{-1}\left(\frac{4\sqrt{3}}{15}\right)=155.2^\circ$$

$$\beta 2=\tan^{-1}m_2=\tan^{-1}\left(\frac{5\sqrt{3}}{4}\right)=65.2$$

Thus $\alpha=155.2-65.2=90^\circ$ so the tangent lines are perpendicular. This can be verified by noting that $-\frac{1}{m_1}=\frac{15}{4\sqrt{3}}=\frac{15\sqrt{3}}{12}=\frac{5\sqrt{3}}{4}=m_2$

9. By the Pythagorean Theorem, $r^2=3^2+4^2$ or $r=\sqrt{9+16}=5$. Since

$$\sin\theta=\frac{3}{r}=\frac{3}{5},\quad \theta=\sin^{-1}(0.6)=36.9^\circ$$

11. Since the triangle is an isosceles right triangle, $x=y$ and $x^2+y^2=8^2$. Thus $2x^2=64$ and $x=y=\sqrt{32}=4\sqrt{2}$

CHAPTER 10 Conics and Polar Coordinates

10.1 Concepts Review

1. $e < 1; e = 1; e > 1$

3. $(0, 1); y = -1$

Problem Set 10.1

1. $y^2 = 4(1)x$
Focus at $(1, 0)$
Directrix: $x = -1$

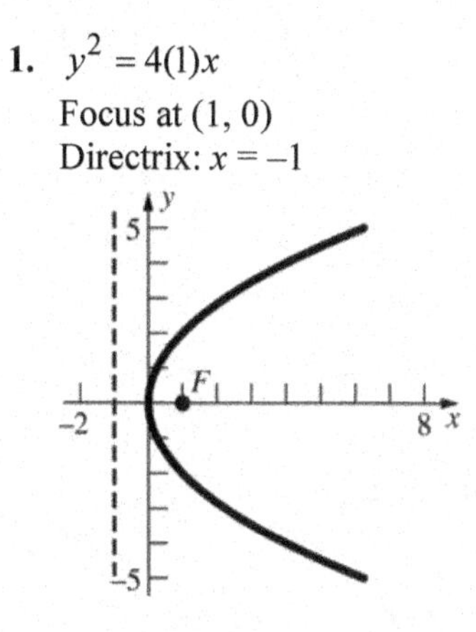

3. $x^2 = -4(3)y$
Focus at $(0, -3)$
Directrix: $y = 3$

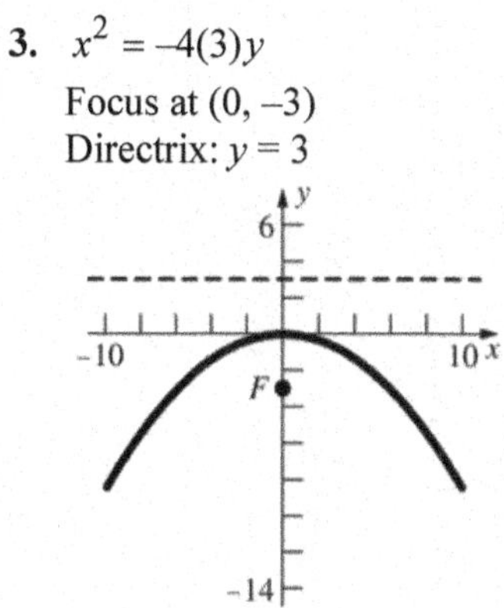

5. $y^2 = 4\left(\frac{1}{4}\right)x$

Focus at $\left(\frac{1}{4}, 0\right)$

Directrix: $x = -\frac{1}{4}$

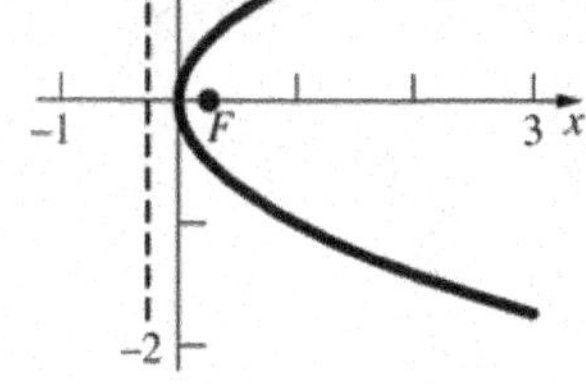

7. $2x^2 = 6y$

$x^2 = 4\left(\frac{3}{4}\right)y$

Focus at $\left(0, \frac{3}{4}\right)$

Directrix: $y = -\frac{3}{4}$

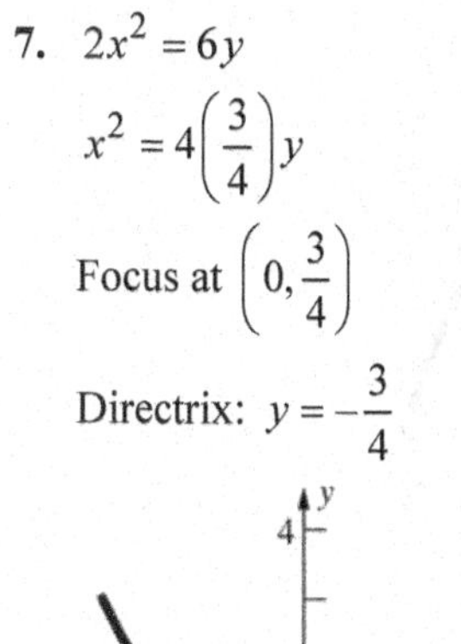
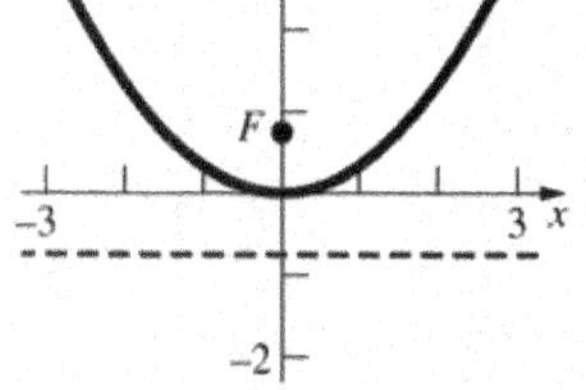

9. The parabola opens to the right, and $p = 2$.
$y^2 = 8x$

11. The parabola opens downward, and $p = 2$.
$x^2 = -8y$

13. The parabola opens to the left, and $p = 4$.
$y^2 = -16x$

15. The equation has the form $y^2 = cx$, so

$(-1)^2 = 3c.$

$c = \frac{1}{3}$

$y^2 = \frac{1}{3}x$

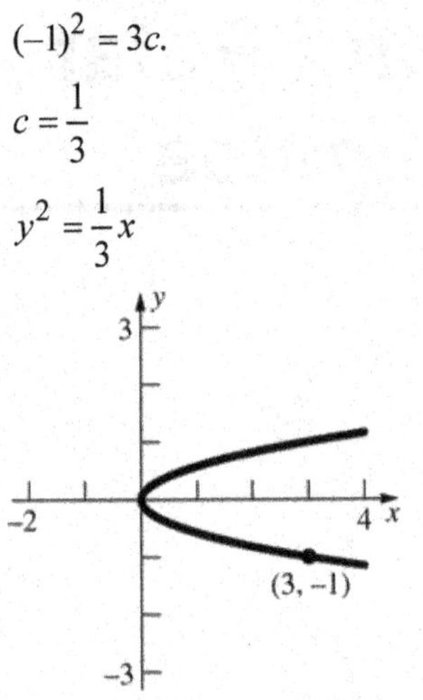

17. The equation has the form $x^2 = cy$, so

$(6)^2 = -5c.$

$c = -\frac{36}{5}$

$x^2 = -\frac{36}{5}y$

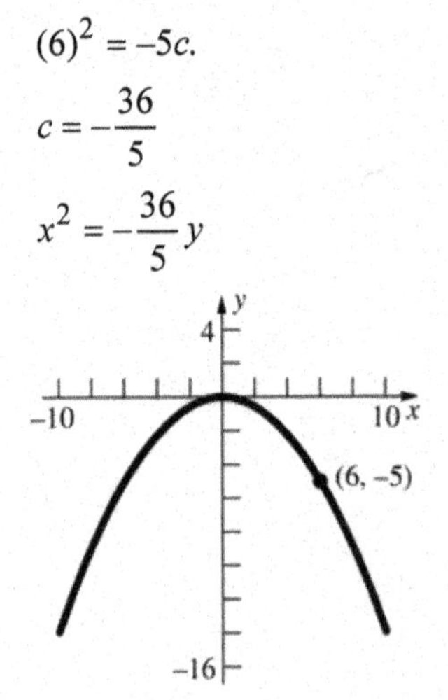

19. $y^2 = 16x$

$2yy' = 16$

$y' = \frac{16}{2y}$

At $(1, -4)$, $y' = -2$.

Tangent: $y + 4 = -2(x - 1)$ or $y = -2x - 2$

Normal: $y + 4 = \frac{1}{2}(x - 1)$ or $y = \frac{1}{2}x - \frac{9}{2}$

21. $x^2 = 2y$

$2x = 2y'$

$y' = x$

At $(4, 8)$, $y' = 4$.

Tangent: $y - 8 = 4(x - 4)$ or $y = 4x - 8$

Normal: $y - 8 = -\frac{1}{4}(x - 4)$ or $y = -\frac{1}{4}x + 9$

23. $y^2 = -15x$

$2yy' = -15$

$y' = -\frac{15}{2y}$

At $\left(-3, -3\sqrt{5}\right)$, $y' = \frac{\sqrt{5}}{2}$.

Tangent: $y + 3\sqrt{5} = \frac{\sqrt{5}}{2}(x + 3)$ or

$y = \frac{\sqrt{5}}{2}x - \frac{3\sqrt{5}}{2}$

Normal: $y + 3\sqrt{5} = -\frac{2\sqrt{5}}{5}(x + 3)$ or

$y = -\frac{2\sqrt{5}}{5}x - \frac{21\sqrt{5}}{5}$

25. $x^2 = -6y$

$2x = -6y'$

$y' = -\frac{x}{3}$

At $\left(3\sqrt{2}, -3\right)$, $y' = -\sqrt{2}$.

Tangent: $y+3 = -\sqrt{2}\left(x - 3\sqrt{2}\right)$ or $y = -\sqrt{2}x + 3$

Normal: $y+3 = \frac{\sqrt{2}}{2}\left(x - 3\sqrt{2}\right)$ or $y = \frac{\sqrt{2}}{2}x - 6$

27. $y^2 = 5x$

$2yy' = 5$

$y' = \frac{5}{2y}$

$y' = \frac{\sqrt{5}}{4}$ when $y = 2\sqrt{5}$, so $x = 4$.

The point is $\left(4, 2\sqrt{5}\right)$.

29. The slope of the line is $\frac{3}{2}$.

$y^2 = -18x; 2yy' = -18$

$2y\left(\frac{3}{2}\right) = -18; y = -6$

$(-6)^2 = -18x; x = -2$

The equation of the tangent line is

$y + 6 = \frac{3}{2}(x+2)$ or $y = \frac{3}{2}x - 3$.

31. From Problem 30, if the parabola is described by the equation $y^2 = 4px$, the slopes of the tangent lines are $\frac{2p}{y_0}$ and $-\frac{y_0}{2p}$. Since they are negative reciprocals, the tangent lines are perpendicular.

33. Assume that the x- and y-axes are positioned such that the axis of the parabola is the y-axis with the vertex at the origin and opening upward. Then the equation of the parabola is $x^2 = 4py$ and $(0, p)$ is the focus. Let D be the distance from a point on the parabola to the focus.

$$D = \sqrt{(x-0)^2 + (y-p)^2} = \sqrt{x^2 + \left(\frac{x^2}{4p} - p\right)^2}$$

$$= \sqrt{\frac{x^4}{16p^2} + \frac{x^2}{2} + p^2} = \frac{x^2}{4p} + p$$

$D' = \frac{x}{2p}; \frac{x}{2p} = 0, x = 0$

$D'' > 0$ so at $x = 0$, D is minimum. $y = 0$

Therefore, the vertex (0, 0) is the point closest to the focus.

35. Let the y-axis be the axis of the parabola, so the Earth's coordinates are $(0, p)$ and the equation of the path is $x^2 = 4py$, where the coordinates are in millions of miles. When the line from Earth to the asteroid makes an angle of $75°$ with the axis of the parabola, the asteroid is at $(40\sin 75°, p + 40\cos 75°)$. (See figure.)

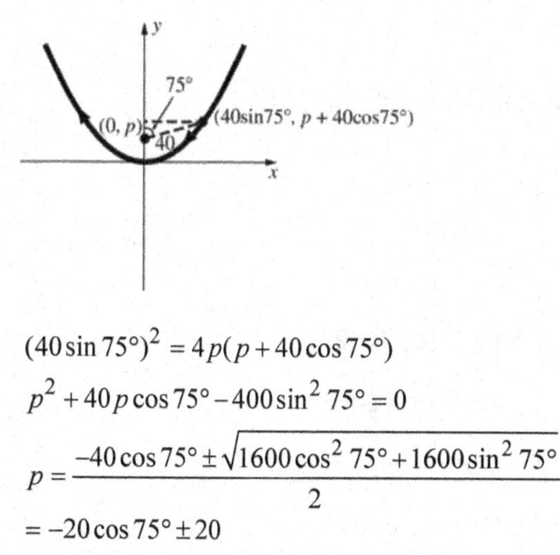

$(40\sin 75°)^2 = 4p(p + 40\cos 75°)$

$p^2 + 40p\cos 75° - 400\sin^2 75° = 0$

$$p = \frac{-40\cos 75° \pm \sqrt{1600\cos^2 75° + 1600\sin^2 75°}}{2}$$

$= -20\cos 75° \pm 20$

$p = 20 - 20\cos 75° \approx 14.8\ (p > 0)$

The closest point to Earth is (0, 0), so the asteroid will come within 14.8 million miles of Earth.

37. Let $|RL|$ be the distance from R to the directrix. Observe that the distance from the latus rectum to the directrix is $2p$ so $|RG| = 2p - |RL|$. From the definition of a parabola, $|RL| = |FR|$. Thus, $|FR| + |RG| = |RL| + 2p - |RL| = 2p$.

39. Let P_1 and P_2 denote (x_1, y_1) and (x_2, y_2), respectively. $|P_1P_2| = |P_1F| + |P_2F|$ since the focal chord passes through the focus. By definition of a parabola, $|P_1F| = p + x_1$ and $|P_2F| = p + x_2$. Thus, the length of the chord is $|P_1P_2| = x_1 + x_2 + 2p$.
$L = p + p + 2p = 4p$

41. $\frac{dy}{dx} = \frac{\delta x}{H}$

$y = \frac{\delta x^2}{2H} + C$

$y(0) = 0$ implies that $C = 0$. $y = \frac{\delta x^2}{2H}$
This is an equation for a parabola.

43.

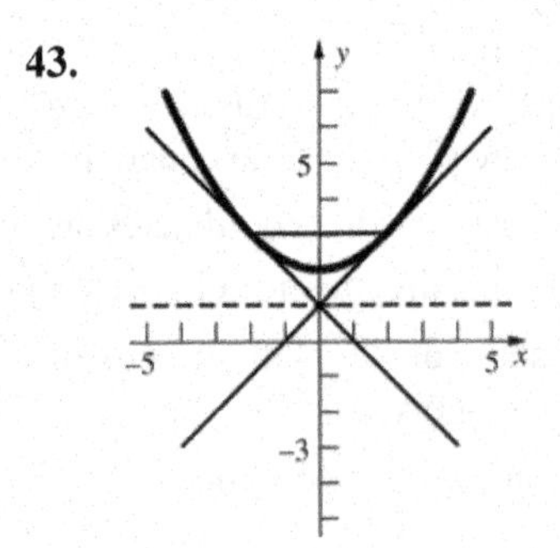

10.2 Concepts Review

1. $\frac{x^2}{a^2} + \frac{y^2}{b^2} = 1$

3. foci

Problem Set 10.2

1. Horizontal ellipse

3. Vertical hyperbola

5. Vertical parabola

7. Vertical ellipse

9. $\frac{x^2}{16} + \frac{y^2}{4} = 1$; horizontal ellipse
$a = 4, b = 2,\ c = 2\sqrt{3}$
Vertices: $(\pm 4, 0)$
Foci: $\left(\pm 2\sqrt{3}, 0\right)$

11. $\frac{y^2}{4} - \frac{x^2}{9} = 1$; vertical hyperbola
$a = 2, b = 3,\ c = \sqrt{13}$
Vertices: $(0, \pm 2)$
Foci: $\left(0, \pm\sqrt{13}\right)$

Asymptotes: $y = \pm\frac{2}{3}x$

13. $\frac{x^2}{2} + \frac{y^2}{8} = 1$; vertical ellipse
$a = 2\sqrt{2}, b = \sqrt{2}, c = \sqrt{6}$
Vertices: $(0, \pm 2\sqrt{2})$
Foci: $\left(0, \pm\sqrt{6}\right)$

15. $\frac{x^2}{10} - \frac{y^2}{4} = 1$; horizontal hyperbola

$a = \sqrt{10}, b = 2, c = \sqrt{14}$

Vertices: $\left(\pm\sqrt{10}, 0\right)$

Foci: $\left(\pm\sqrt{14}, 0\right)$

Asymptotes: $y = \pm\frac{2}{\sqrt{10}}x$

17. This is a horizontal ellipse with $a = 6$ and $c = 3$.

$b = \sqrt{36-9} = \sqrt{27}$

$\frac{x^2}{36} + \frac{y^2}{27} = 1$

19. This is a vertical ellipse with $c = 5$.

$a = \frac{c}{e} = \frac{5}{\frac{1}{3}} = 15, b = \sqrt{225-25} = \sqrt{200}$

$\frac{x^2}{200} + \frac{y^2}{225} = 1$

21. This is a horizontal ellipse with $a = 5$.

$\frac{x^2}{25} + \frac{y^2}{b^2} = 1$

$\frac{4}{25} + \frac{9}{b^2} = 1$

$b^2 = \frac{225}{21}$

$\frac{x^2}{25} + \frac{y^2}{\frac{225}{21}} = 1$

23. This is a vertical hyperbola with $a = 4$ and $c = 5$.

$b = \sqrt{25-16} = 3$

$\frac{y^2}{16} - \frac{x^2}{9} = 1$

25. This is a horizontal hyperbola with $a = 8$.

The asymptotes are $y = \pm\frac{1}{2}x$, so $\frac{b}{8} = \frac{1}{2}$ or $b = 4$.

$\frac{x^2}{64} - \frac{y^2}{16} = 1$

27. This is a horizontal ellipse with $c = 2$.

$8 = \frac{a}{e}, 8 = \frac{a}{\frac{c}{a}}$, so $a^2 = 8c = 16$.

$b = \sqrt{16-4} = \sqrt{12}$

$\frac{x^2}{16} + \frac{y^2}{12} = 1$

29. The asymptotes are $y = \pm\frac{1}{2}x$. If the hyperbola is horizontal $\frac{b}{a} = \frac{1}{2}$ or $a = 2b$. If the hyperbola is vertical, $\frac{a}{b} = \frac{1}{2}$ or $b = 2a$.

Suppose the hyperbola is horizontal.

$\frac{x^2}{4b^2} - \frac{y^2}{b^2} = 1$

$\frac{16}{4b^2} - \frac{9}{b^2} = 1$

$b^2 = -5$

This is not possible.

Suppose the hyperbola is vertical.

$\frac{y^2}{a^2} - \frac{x^2}{4a^2} = 1$

$\frac{9}{a^2} - \frac{16}{4a^2} = 1$

$a^2 = 5$

$\frac{y^2}{5} - \frac{x^2}{20} = 1$

31. This is an ellipse whose foci are (0, 9) and (0, - 9) and whose major diameter has length $2a = 26$. Since the foci are on the y-axis, it is the major axis of the ellipse so the equation has the form $\frac{y^2}{a^2} + \frac{x^2}{b^2} = 1$. Since $2a = 26$, $a^2 = 169$ and since $a^2 = b^2 + c^2$, $b^2 = 169 - (9)^2 = 88$. Thus the equation is $\frac{y^2}{169} + \frac{x^2}{88} = 1$

33. This is an hyperbola whose foci are (7, 0) and (- 7, 0) and whose axis is the x-axis. So the equation has the form $\frac{x^2}{a^2} - \frac{y^2}{b^2} = 1$. Since $2a = 12$, $a^2 = 36$ and $b^2 = c^2 - a^2 = (7^2) - 36 = 13$

Thus the equation is $\frac{x^2}{36} - \frac{y^2}{13} = 1$

35. Use implicit differentiation to find the slope: $\frac{2}{27}x+\frac{2}{9}yy'=0$. At the point $(3,\sqrt{6})$, $\frac{2}{9}+\frac{2\sqrt{6}}{9}y'=0$, or $y'=-\frac{\sqrt{6}}{6}$ so the equation of the tangent line is $(y-\sqrt{6})=-\frac{\sqrt{6}}{6}(x-3)$ or $x+\sqrt{6}y=9$.

37. Use implicit differentiation to find the slope: $\frac{2}{27}x+\frac{2}{9}yy'=0$. At the point $(3,-\sqrt{6})$, $\frac{2}{9}-\frac{2\sqrt{6}}{9}y'=0$, or $y'=\frac{\sqrt{6}}{6}$ so the equation of the tangent line is $(y+\sqrt{6})=-\frac{\sqrt{6}}{6}(x-3)$ or $x-\sqrt{6}y=9$.

39. Use implicit differentiation to find the slope: $2x+2yy'=0$. At the point $(5,12)$ $10+24y'=0$, or $y'=-\frac{5}{12}$ so the equation of the tangent line is $(y-12)=-\frac{5}{12}(x-5)$ or $5x+12y=169$.

41. Use implicit differentiation to find the slope: $\frac{1}{44}x+\frac{2}{169}yy'=0$. At the point $(0,13)$, $\frac{2}{13}y'=0$, or $y'=0$. The tangent line is horizontal and thus has equation $y=13$.

43. Let the y-axis run through the center of the arch and the x-axis lie on the floor. Thus $a = 5$ and $b = 4$ and the equation of the arch is $\frac{x^2}{25}+\frac{y^2}{16}=1$.

When $y = 2$, $\frac{x^2}{25}+\frac{(2)^2}{16}=1$, so $x=\pm\frac{5\sqrt{3}}{2}$.

The width of the box can at most be $5\sqrt{3}\approx 8.66$ ft.

45. The foci are at $(\pm c, 0)$.

$c=\sqrt{a^2-b^2}$

$\frac{a^2-b^2}{a^2}+\frac{y^2}{b^2}=1$

$y^2=\frac{b^4}{a^2}$, $y=\pm\frac{b^2}{a}$

Thus, the length of the latus rectum is $\frac{2b^2}{a}$.

47. $a = 18.09$, $b = 4.56$,

$c=\sqrt{(18.09)^2-(4.56)^2}\approx 17.51$

The comet's minimum distance from the sun is $18.09 - 17.51 \approx 0.58$ AU.

49. $a-c=4132$; $a+c=4583$

$2a=8715$; $a=4357.5$

$c=4357.5-4132=225.5$

$e=\frac{c}{a}=\frac{225.5}{4357.5}\approx 0.05175$

51. $\frac{x^2}{4}+\frac{y^2}{9}=1$

Equation of tangent line at (x_0, y_0):

$\frac{xx_0}{4}+\frac{yy_0}{9}=1$

At $(0, 6)$, $y_0=\frac{3}{2}$.

When $y=\frac{3}{2}$, $\frac{x^2}{4}+\frac{1}{4}=1$, $x=\pm\sqrt{3}$.

The points of tangency are $\left(\sqrt{3},\frac{3}{2}\right)$ and $\left(-\sqrt{3},\frac{3}{2}\right)$

53. $2x^2-7y^2-35=0; 4x-14yy'=0$

$y'=\frac{2x}{7y}; -\frac{2}{3}=\frac{2x}{7y}; x=-\frac{7y}{3}$

Substitute $x=-\frac{7y}{3}$ into the equation of the hyperbola.

$\frac{98}{9}y^2-7y^2-35=0, y=\pm 3$

The coordinates of the points of tangency are $(-7,3)$ and $(7,-3)$.

55. $y=\pm b\sqrt{1-\frac{x^2}{a^2}}$

$A=4b\int_0^a\sqrt{1-\frac{x^2}{a^2}}\,dx$

Let $x = a \sin t$ then $dx = a \cos t\, dt$. Then the limits are 0 and $\frac{\pi}{2}$.

$A=4b\int_0^a\sqrt{1-\frac{x^2}{a^2}}\,dx=4ab\int_0^{\pi/2}\cos^2 t\,dt$

$=2ab\int_0^{\pi/2}(1+\cos 2t)dt=2ab\left[t+\frac{\sin 2t}{2}\right]_0^{\pi/2}$

$=\pi ab$

57. $y = \pm b\sqrt{\frac{x^2}{a^2} - 1}$

The vertical line at one focus is $x = \sqrt{a^2 + b^2}$.

$$V = \pi \int_a^{\sqrt{a^2+b^2}} \left(b\sqrt{\frac{x^2}{a^2} - 1} \right)^2 dx$$

$$= b^2\pi \int_a^{\sqrt{a^2+b^2}} \left(\frac{x^2}{a^2} - 1 \right) dx = b^2\pi \left[\frac{x^3}{3a^2} - x \right]_a^{\sqrt{a^2+b^2}}$$

$$= b^2\pi \left[\frac{(a^2+b^2)^{3/2}}{3a^2} - \sqrt{a^2+b^2} + \frac{2}{3}a \right]$$

$$= \frac{\pi b^2}{3a^2}\left[(a^2+b^2)^{3/2} - 3a^2\sqrt{a^2+b^2} + 2a^3 \right]$$

59. If one corner of the rectangle is at (x, y) the sides have length $2x$ and $2y$.

$$x = \pm a\sqrt{1 - \frac{y^2}{b^2}}$$

$$A = 4xy = 4ya\sqrt{1 - \frac{y^2}{b^2}} = 4a\sqrt{y^2 - \frac{y^4}{b^2}}$$

$$\frac{dA}{dy} = \frac{2a\left(2y - \frac{4y^3}{b^2}\right)}{\sqrt{y^2 - \frac{y^4}{b^2}}}; \ \frac{dA}{dy} = 0 \text{ when}$$

$$y - \frac{2y^3}{b^2} = 0$$

$$y\left(1 - \frac{2y^2}{b^2}\right) = 0$$

$y = 0$ or $y = \pm\frac{b}{\sqrt{2}}$

The Second Derivative Test shows that $y = \frac{b}{\sqrt{2}}$ is a maximum.

$$x = a\sqrt{1 - \frac{\left(\frac{b}{\sqrt{2}}\right)^2}{b^2}} = \frac{a}{\sqrt{2}}$$

Therefore, the rectangle is $a\sqrt{2}$ by $b\sqrt{2}$.

61. Add the two equations to get $9y^2 = 675$.

$y = \pm 5\sqrt{3}$

Substitute $y = 5\sqrt{3}$ into either of the two equations and solve for $x \Rightarrow x = \pm 6$

The point in the first quadrant is $(6, 5\sqrt{3})$.

63.

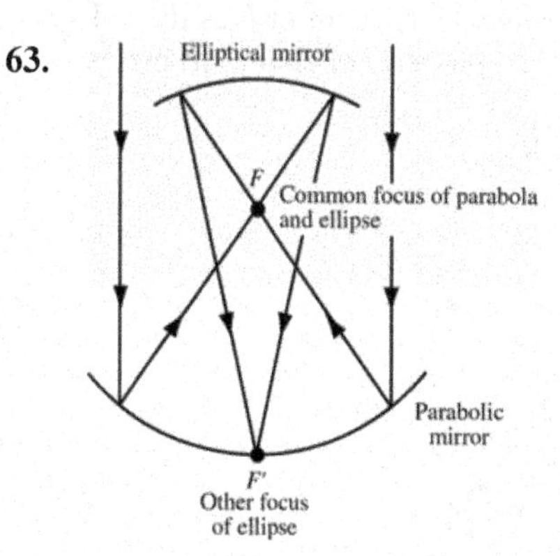

65. Written response. Possible answer: the ball will follow a path that does not go between the foci.

67. Possible answer: Attach one end of a string to F and attach one end of another string to F'. Place a spool at a vertex. Tightly wrap both strings in the same direction around the spool. Insert a pencil through the spool. Then trace out a branch of the hyperbola by unspooling the strings while keeping both strings taut.

69. Let $P(x, y)$ be the location of the explosion.

$3|AP| = 3|BP| + 12$

$|AP| - |BP| = 4$

Thus, P lies on the right branch of the horizontal hyperbola with $a = 2$ and $c = 8$, so $b = 2\sqrt{15}$.

$$\frac{x^2}{4} - \frac{y^2}{60} = 1$$

Since $|BP| = |CP|$, the y-coordinate of P is 5.

$$\frac{x^2}{4} - \frac{25}{60} = 1, x = \pm\sqrt{\frac{17}{3}}$$

P is at $\left(\sqrt{\frac{17}{3}}, 5\right)$.

71. $2a = p + q, \ 2c = |p - q|$

$$b^2 = a^2 - c^2 = \frac{(p+q)^2}{4} - \frac{(p-q)^2}{4} = pq$$

$b = \sqrt{pq}$

73. Let (x, y) be the coordinates of P as the ladder slides. Using a property of similar triangles,

$$\frac{x}{a} = \frac{\sqrt{b^2 - y^2}}{b}.$$

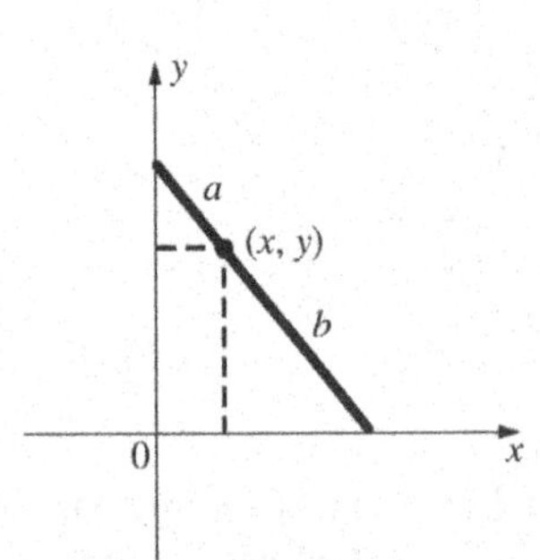

Square both sides to get

$$\frac{x^2}{a^2} = \frac{b^2 - y^2}{b^2} \text{ or } b^2x^2 + a^2y^2 = a^2b^2 \text{ or}$$

$$\frac{x^2}{a^2} + \frac{y^2}{b^2} = 1.$$

75. The equations of the hyperbolas are $\frac{x^2}{a^2} - \frac{y^2}{b^2} = 1$ and $\frac{y^2}{b^2} - \frac{x^2}{a^2} = 1.$

$$e = \frac{c}{a} = \frac{\sqrt{a^2 + b^2}}{a}$$

$$E = \frac{c}{b} = \frac{\sqrt{a^2 + b^2}}{b}$$

$$e^{-2} + E^{-2} = \frac{a^2}{a^2 + b^2} + \frac{b^2}{a^2 + b^2} = 1$$

77.

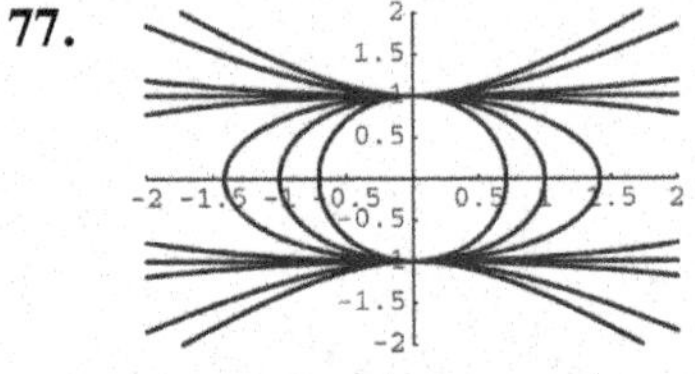

When $a < 0$, the conic is an ellipse. When $a > 0$, the conic is a hyperbola. When $a = 0$, the graph is two parallel lines.

10.3 Concepts Review

1. $\frac{a^2}{4}$

3. $\frac{A - C}{B}$

Problem Set 10.3

1. $x^2 + y^2 - 2x + 2y + 1 = 0$

$(x^2 - 2x + 1) + (y^2 + 2y + 1) = -1 + 1 + 1$

$(x - 1)^2 + (y + 1)^2 = 1$

This is a circle.

3. $9x^2 + 4y^2 + 72x - 16y + 124 = 0$

$9(x^2 + 8x + 16) + 4(y^2 - 4y + 4) = -124 + 144 + 16$

$9(x + 4)^2 + 4(y - 2)^2 = 36$

This is an ellipse.

5. $9x^2 + 4y^2 + 72x - 16y + 160 = 0$

$9(x^2 + 8x + 16) + 4(y^2 - 4y + 4) = -160 + 144 + 16$

$9(x + 4)^2 + 4(y - 2)^2 = 0$

This is a point.

7. $y^2 - 5x - 4y - 6 = 0$

$(y^2 - 4y + 4) = 5x + 6 + 4$

$(y - 2)^2 = 5(x + 2)$

This is a parabola.

9. $3x^2 + 3y^2 - 6x + 12y + 60 = 0$

$3(x^2 - 2x + 1) + 3(y^2 + 4y + 4) = -60 + 3 + 12$

$3(x - 1)^2 + 3(y + 2)^2 = -45$

This is the empty set.

11. $4x^2 - 4y^2 + 8x + 12y - 5 = 0$

$4(x^2 + 2x + 1) - 4\left(y^2 - 3y + \frac{9}{4}\right) = 5 + 4 - 9$

$4(x + 1)^2 - 4\left(y - \frac{3}{2}\right)^2 = 0$

This is two intersecting lines.

13. $4x^2 - 24x + 36 = 0$

$4(x^2 - 6x + 9) = -36 + 36$

$4(x - 3)^2 = 0$

This is a line.

15. $\dfrac{(x+3)^2}{4}+\dfrac{(y+2)^2}{16}=1$

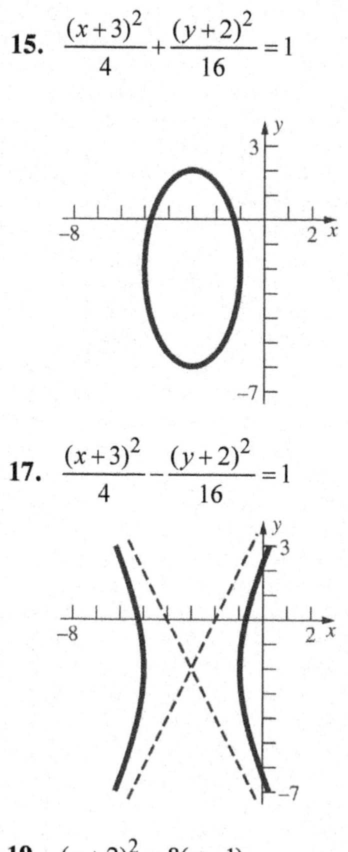

17. $\dfrac{(x+3)^2}{4}-\dfrac{(y+2)^2}{16}=1$

19. $(x+2)^2=8(y-1)$

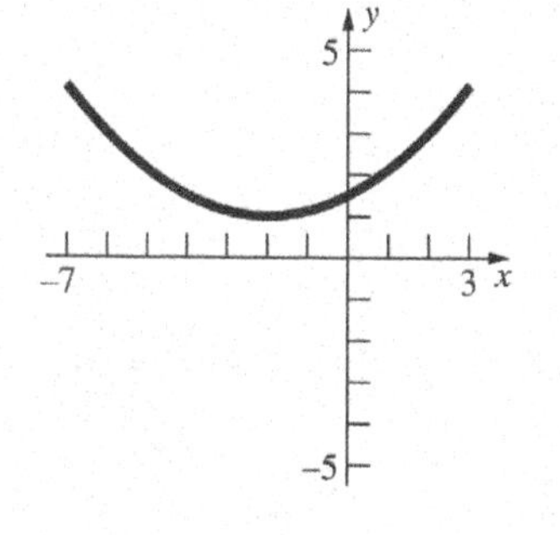

21. $(y-1)^2=16$

$y-1=\pm 4$

$y=5, y=-3$

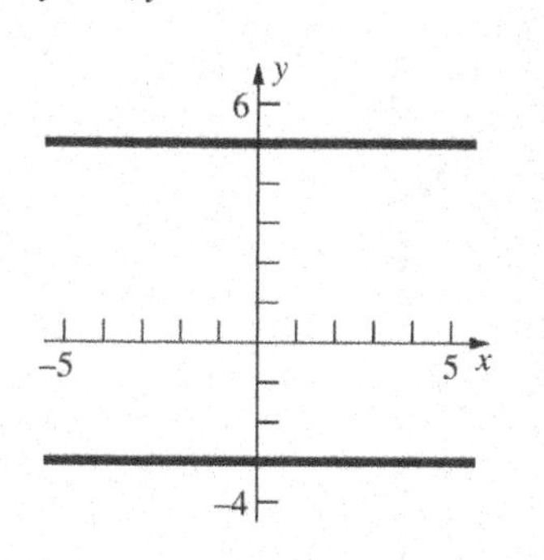

23. $x^2+4y^2-2x+16y+1=0$

$(x^2-2x+1)+4(y^2+4y+4)=-1+1+16$

$(x-1)^2+4(y+2)^2=16$

$\dfrac{(x-1)^2}{16}+\dfrac{(y+2)^2}{4}=1$

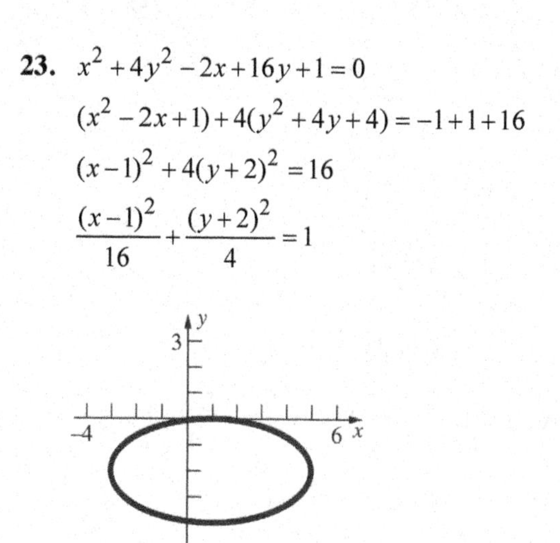

25. $9x^2-16y^2+54x+64y-127=0$

$9(x^2+6x+9)-16(y^2-4y+4)=127+81-64$

$9(x+3)^2-16(y-2)^2=144$

$\dfrac{(x+3)^2}{16}-\dfrac{(y-2)^2}{9}=1$

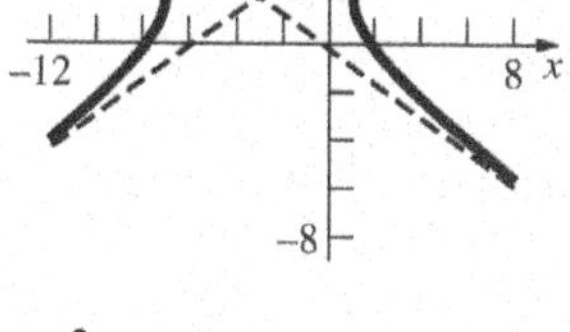

27. $4x^2+16x-16y+32=0$

$4(x^2+4x+4)=16y-32+16$

$4(x+2)^2=16(y-1)$

$(x+2)^2=4(y-1)$

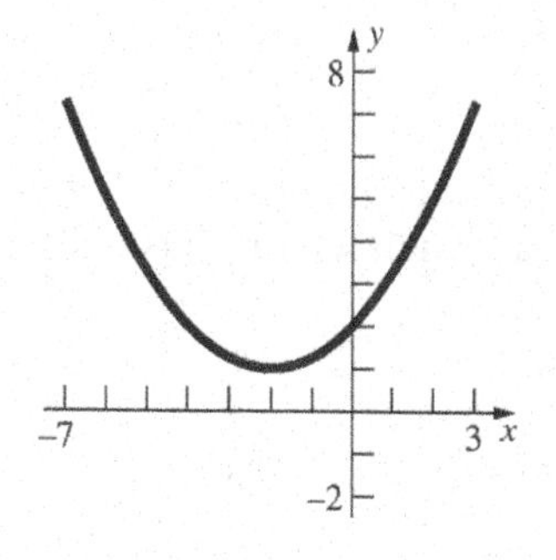

29. $2y^2 - 4y - 10x = 0$

$2(y^2 - 2y + 1) = 10x + 2$

$(y-1)^2 = 5\left(x + \frac{1}{5}\right)$

$(y-1)^2 = 4\left(\frac{5}{4}\right)\left(x + \frac{1}{5}\right)$

Horizontal parabola, $p = \frac{5}{4}$

Vertex $\left(-\frac{1}{5}, 1\right)$; Focus is at $\left(\frac{21}{20}, 1\right)$ and directrix is at $x = -\frac{29}{20}$.

31. $16(x-1)^2 + 25(y+2)^2 = 400$

$\frac{(x-1)^2}{25} + \frac{(y+2)^2}{16} = 1$

Horizontal ellipse, center $(1, -2)$, $a = 5, b = 4$,
$c = \sqrt{25-16} = 3$
Foci are at $(-2, -2)$ and $(4, -2)$.

33. $a = 5, b = 4$

$\frac{(x-5)^2}{25} + \frac{(y-1)^2}{16} = 1$

35. Vertical parabola, opens upward, $p = 5 - 3 = 2$

$(x-2)^2 = 4(2)(y-3)$

$(x-2)^2 = 8(y-3)$

37. Vertical hyperbola, center $(0, 3)$, $2a = 6, a = 3$,
$c = 5,\ b = \sqrt{25-9} = 4$

$\frac{(y-3)^2}{9} - \frac{x^2}{16} = 1$

39. Horizontal parabola, opens to the left

Vertex $(6, 5)$, $p = \frac{10-2}{2} = 4$

$(y-5)^2 = -4(4)(x-6)$

$(y-5)^2 = -16(x-6)$

41. Horizontal ellipse, center $(0, 2)$, $c = 2$
Since it passes through the origin and center is at $(0, 2)$, $b = 2$.
$a = \sqrt{4+4} = \sqrt{8}$

$\frac{x^2}{8} + \frac{(y-2)^2}{4} = 1$

43. $x^2 + xy + y^2 = 6$

$\cot 2\theta = 0$

$2\theta = \frac{\pi}{2}, \theta = \frac{\pi}{4}$

$x = u\frac{\sqrt{2}}{2} - v\frac{\sqrt{2}}{2} = \frac{\sqrt{2}}{2}(u - v)$

$y = u\frac{\sqrt{2}}{2} + v\frac{\sqrt{2}}{2} = \frac{\sqrt{2}}{2}(u + v)$

$\frac{1}{2}(u-v)^2 + \frac{1}{2}(u-v)(u+v) + \frac{1}{2}(u+v)^2 = 6$

$\frac{3}{2}u^2 + \frac{1}{2}v^2 = 6$

$\frac{u^2}{4} + \frac{v^2}{12} = 1$

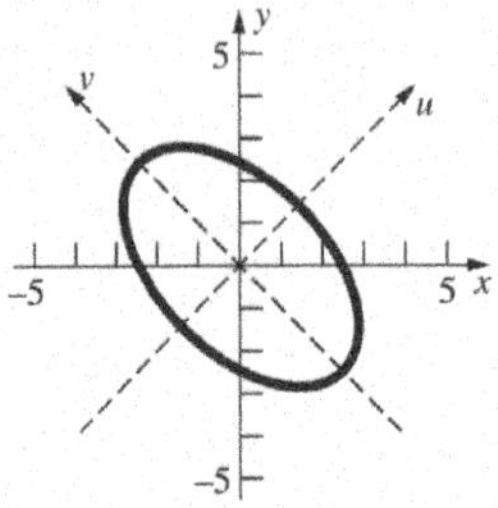

45. $4x^2 + xy + 4y^2 = 56$

$\cot 2\theta = 0,\ 2\theta = \frac{\pi}{2}, \theta = \frac{\pi}{4}$

$x = \frac{\sqrt{2}}{2}(u - v)$

$y = \frac{\sqrt{2}}{2}(u + v)$

$2(u-v)^2 + \frac{1}{2}(u-v)(u+v) + 2(u+v)^2 = 56$

$\frac{9}{2}u^2 + \frac{7}{2}v^2 = 56$

$\frac{u^2}{\frac{112}{9}} + \frac{v^2}{16} = 1$

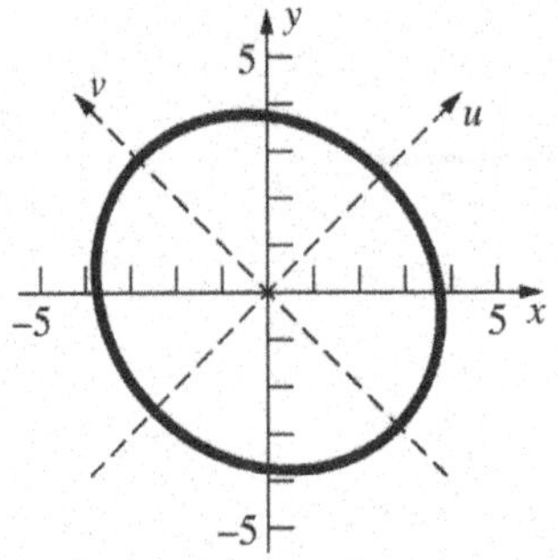

47. $-\frac{1}{2}x^2 + 7xy - \frac{1}{2}y^2 - 6\sqrt{2}x - 6\sqrt{2}y = 0$

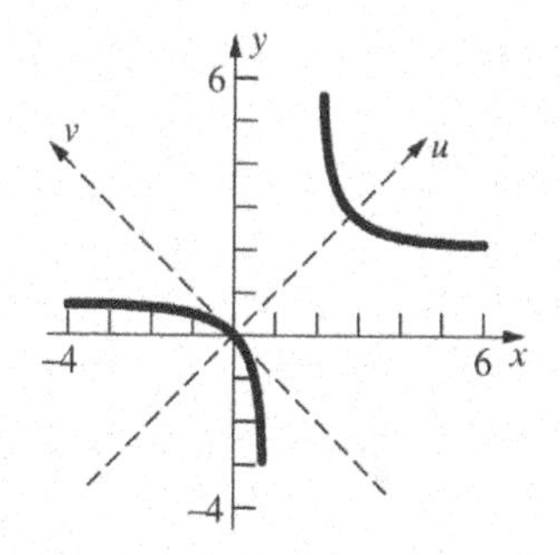

$\cot 2\theta = 0, 2\theta = \frac{\pi}{2}, \theta = \frac{\pi}{4}$

$x = \frac{\sqrt{2}}{2}(u - v);\quad y = \frac{\sqrt{2}}{2}(u + v)$

$-\frac{1}{4}(u-v)^2 + \frac{7}{2}(u-v)(u+v) - \frac{1}{4}(u+v)^2 - 6(u-v) - 6(u+v) = 0$

$3u^2 - 4v^2 - 12u = 0$

$3(u^2 - 4u + 4) - 4v^2 = 12$

$\frac{(u-2)^2}{4} - \frac{v^2}{3} = 1$

49. $A = 4,\ B = -3,\ C = D = E = 0,\ F = -18$

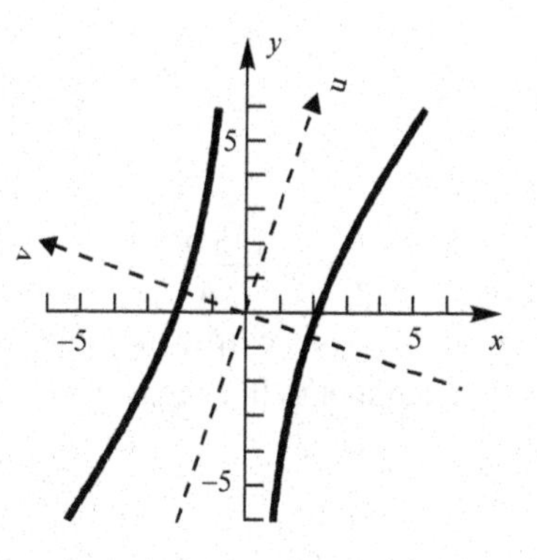

$\cot 2\theta = \frac{4-0}{-3} = -\frac{4}{3}$

Since $0 \le 2\theta \le \pi$, $\sin 2\theta$ is positive, so $\cos 2\theta$ is negative; using a 3-4-5 right triangle, we conclude $\cos 2\theta = -\frac{4}{5}$. Thus

$\sin\theta = \sqrt{\frac{1-\cos 2\theta}{2}} = \sqrt{\frac{1-(-4/5)}{2}} = \frac{3\sqrt{10}}{10}$ and

$\cos\theta = \sqrt{\frac{1+\cos 2\theta}{2}} = \sqrt{\frac{1+(-4/5)}{2}} = \frac{\sqrt{10}}{10}$. Rotating through the angle

$\theta = \frac{1}{2}\cos^{-1}(-0.8) = 71.6°$, we have

$4\left(\frac{\sqrt{10}}{10}u - \frac{3\sqrt{10}}{10}v\right)^2 - 3\left(\frac{\sqrt{10}}{10}u - \frac{3\sqrt{10}}{10}v\right)\left(\frac{3\sqrt{10}}{10}u + \frac{\sqrt{10}}{10}v\right) = 18$ or

$45v^2 - 5u^2 = 180$ or $\frac{v^2}{4} - \frac{u^2}{36} = 1$. This is a hyperbola in standard position in the uv-system; its axis is the v-axis, and $a = 2,\ b = 6$.

51. $34x^2 + 24xy + 41y^2 + 250y = -325$

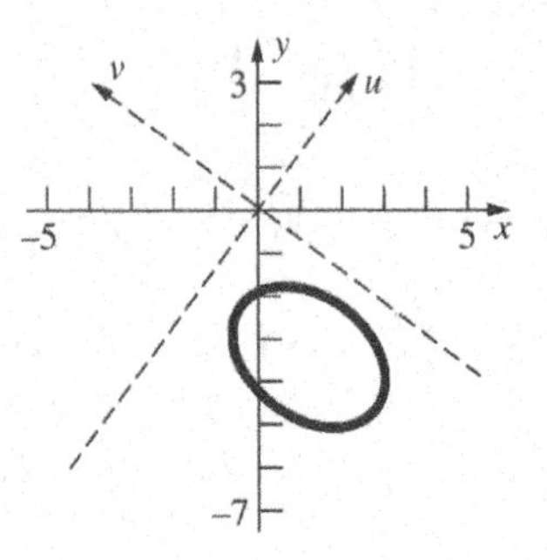

$\cot 2\theta = -\frac{7}{24}, r = 25$

$\cos 2\theta = -\frac{7}{25}$

$\cos\theta = \sqrt{\frac{1-\frac{7}{25}}{2}} = \frac{3}{5}$; $\sin\theta = \sqrt{\frac{1+\frac{7}{25}}{2}} = \frac{4}{5}$

$x = \frac{1}{5}(3u - 4v)$; $y = \frac{1}{5}(4u + 3v)$

$\frac{34}{25}(3u-4v)^2 + \frac{24}{25}(3u-4v)(4u+3v) + \frac{41}{25}(4u+3v)^2 + 50(4u+3v) = -325$

$50u^2 + 25v^2 + 200u + 150v = -325$

$2u^2 + v^2 + 8u + 6v = -13$

$2(u^2 + 4u + 4) + (v^2 + 6v + 9) = -13 + 8 + 9$

$2(u+2)^2 + (v+3)^2 = 4$

$\frac{(u+2)^2}{2} + \frac{(v+3)^2}{4} = 1$

This is an ellipse in standard position in the uv-system, with major axis parallel to the v-axis. Its center is $(u,v) = (-2,-3)$ and $a = 2$, $b = \sqrt{2}$.

53. a. If C is a vertical parabola, the equation for C can be written in the form $y = ax^2 + bx + c$. Substitute the three points into the equation.

$2 = a - b + c$
$0 = c$
$6 = 9a + 3b + c$

Solve the system to get $a = 1, b = -1, c = 0$.

$y = x^2 - x$

b. If C is a horizontal parabola, an equation for C can be written in the form $x = ay^2 + by + c$. Substitute the three points into the equation.

$-1 = 4a + 2b + c$
$0 = c$
$3 = 36a + 6b + c$

Solve the system to get $a = \frac{1}{4}, b = -1, c = 0$.

$x = \frac{1}{4}y^2 - y$

c. If C is a circle, an equation for C can be written in the form $(x-h)^2 + (y-k)^2 = r^2$. Substitute the three points into the equation.

$(-1-h)^2 + (2-k)^2 = r^2$

$h^2 + k^2 = r^2$

$(3-h)^2 + (6-k)^2 = r^2$

Solve the system to get $h = \frac{5}{2}, k = \frac{5}{2}$, and $r^2 = \frac{25}{2}$.

$\left(x - \frac{5}{2}\right)^2 + \left(y - \frac{5}{2}\right)^2 = \frac{25}{2}$

55. $y^2 = Lx + Kx^2$

$$K\left(x^2 + \frac{L}{K}x + \frac{L^2}{4K^2}\right) - y^2 = \frac{L^2}{4K}$$

$$K\left(x + \frac{L}{2K}\right)^2 - y^2 = \frac{L^2}{4K}$$

$$\frac{(x+\frac{L}{2K})^2}{\frac{L^2}{4K^2}} - \frac{y^2}{\frac{L^2}{4K}} = 1$$

If $K < -1$, the conic is a vertical ellipse. If $K = -1$, the conic is a circle. If $-1 < K < 0$, the conic is a horizontal ellipse. If $K = 0$, the original equation is $y^2 = Lx$, so the conic is a horizontal parabola. If $K > 0$, the conic is a horizontal hyperbola.

If $-1 < K < 0$ (a horizontal ellipse) the length of the latus rectum is (see problem 45, Section 10.2)

$$\frac{2b^2}{a} = 2\frac{L^2}{4|K|}\frac{1}{\frac{|L|}{2|K|}} = |L|$$

From general considerations, the result for a vertical ellipse is the same as the one just obtained.

For $K = -1$ (a circle) we have

$$\left(x - \frac{L}{2}\right)^2 + y^2 = \frac{L^2}{4} \Rightarrow 2\frac{|L|}{2} = |L|$$

If $K = 0$ (a horizontal parabola) we have $y^2 = Lx; y^2 = 4\frac{L}{4}x; p = \frac{L}{4}$, and the latus rectum is

$$2\sqrt{Lp} = 2\sqrt{L\frac{L}{4}} = |L|.$$

If $K > 0$ (a horizontal hyperbola) we can use the result of Problem 46, Section 10.2. The length of the latus rectum is $\frac{2b^2}{a}$, which is equal to $|L|$.

57. $x = u\cos\alpha - v\sin\alpha$

$y = u\sin\alpha + v\cos\alpha$

$(u\cos\alpha - v\sin\alpha)\cos\alpha + (u\sin\alpha + v\cos\alpha)\sin\alpha = d$

$u(\cos^2\alpha + \sin^2\alpha) = d$

$u = d$

Thus, the perpendicular distance from the origin is d.

59. $x = u\cos\theta - v\sin\theta;\quad y = u\sin\theta + v\cos\theta$

$$x(\cos\theta) + y(\sin\theta) = (u\cos^2\theta - v\cos\theta\sin\theta) + (u\sin^2\theta + v\cos\theta\sin\theta) = u$$

$$x(-\sin\theta) + y(\cos\theta) = (-u\cos\theta\sin\theta + v\sin^2\theta) + (u\cos\theta\sin\theta + v\cos^2\theta) = v$$

Thus, $u = x\cos\theta + y\sin\theta$ and $v = -x\sin\theta + y\cos\theta$.

61. Rotate to eliminate the xy-term.

$x^2 + 14xy + 49y^2 = 100$

$$\cot 2\theta = -\frac{24}{7};\quad \cos 2\theta = -\frac{24}{25}$$

$$\cos\theta = \sqrt{\frac{1-\frac{24}{25}}{2}} = \frac{1}{5\sqrt{2}};\quad \sin\theta = \sqrt{\frac{1+\frac{24}{25}}{2}} = \frac{7}{5\sqrt{2}};\quad x = \frac{1}{5\sqrt{2}}(u-7v);\quad y = \frac{1}{5\sqrt{2}}(7u+v)$$

$$\frac{1}{50}(u-7v)^2 + \frac{14}{50}(u-7v)(7u+v) + \frac{49}{50}(7u+v)^2 = 100$$

$50u^2 = 100$

$u^2 = 2$

$u = \pm\sqrt{2}$

Thus the points closest to the origin in uv-coordinates are $\left(\sqrt{2}, 0\right)$ and $\left(-\sqrt{2}, 0\right)$.

$$x = \frac{1}{5\sqrt{2}}\left(\sqrt{2}\right) = \frac{1}{5} \text{ or } x = \frac{1}{5\sqrt{2}}\left(-\sqrt{2}\right) = -\frac{1}{5};\quad y = \frac{1}{5\sqrt{2}}\left(7\sqrt{2}\right) = \frac{7}{5} \text{ or } y = \frac{1}{5\sqrt{2}}\left(-7\sqrt{2}\right) = -\frac{7}{5}$$

The points closest to the origin in xy-coordinates are $\left(\frac{1}{5}, \frac{7}{5}\right)$ and $\left(-\frac{1}{5}, -\frac{7}{5}\right)$.

63. From Problem 62, $a = A\cos^2\theta + B\cos\theta\sin\theta + C\sin^2\theta$,

$b = -2A\cos\theta\sin\theta + B(\cos^2\theta - \sin^2\theta) + 2C\cos\theta\sin\theta$, and

$c = A\sin^2\theta - B\cos\theta\sin\theta + C\cos^2\theta$.

$b^2 = B^2\cos^4\theta + 4(-AB + BC)\cos^3\theta\sin\theta + 2(2A^2 - B^2 - 4AC + 2C^2)\cos^2\theta\sin^2\theta$

$+4(AB - BC)\cos\theta\sin^3\theta + B^2\sin^4\theta$

$4ac = 4AC\cos^4\theta + 4(-AB + BC)\cos^3\theta\sin\theta + 4(A^2 - B^2 + C^2)\cos^2\theta\sin^2\theta + 4(AB - BC)\cos\theta\sin^3\theta + 4AC\sin^4\theta$

$b^2 - 4ac = (B^2 - 4AC)\cos^4\theta + 2(B^2 - 4AC)\cos^2\theta\sin^2\theta + (B^2 - 4AC)\sin^4\theta$

$= (B^2 - 4AC)(\cos^2\theta)(\cos^2\theta + \sin^2\theta) + (B^2 - 4AC)(\sin^2\theta)(\cos^2\theta + \sin^2\theta)$

$= (B^2 - 4AC)(\cos^2\theta + \sin^2\theta) = B^2 - 4AC$

65. a. From Problem 63, $-4ac = B^2 - 4AC = -\Delta$ or $\frac{1}{ac} = \frac{4}{\Delta}$.

b. From Problem 62, $a + c = A + C$.

$\frac{1}{a} + \frac{1}{c} = \frac{a+c}{ac} = \frac{4(A+C)}{\Delta}$

c. $\frac{2}{\Delta}\left(A + C \pm \sqrt{(A-C)^2 + B^2}\right)$

$= \frac{2}{\Delta}\left[\frac{\Delta}{4}\left(\frac{1}{a} + \frac{1}{c}\right) \pm \sqrt{A^2 + 2AC + C^2 + B^2 - 4AC}\right]$

$= \frac{1}{2}\left(\frac{1}{a} + \frac{1}{c}\right) \pm \frac{2}{\Delta}\sqrt{(A+C)^2 - \Delta}$

$= \frac{1}{2}\left(\frac{1}{a} + \frac{1}{c}\right) \pm \frac{2}{\Delta}\sqrt{\frac{\Delta^2}{16}\left(\frac{1}{a} + \frac{1}{c}\right)^2 - \Delta}$

$= \frac{1}{2}\left(\frac{1}{a} + \frac{1}{c}\right) \pm \frac{1}{2}\sqrt{\left(\frac{1}{a} + \frac{1}{c}\right)^2 - 4\left(\frac{4}{\Delta}\right)}$

$= \frac{1}{2}\left(\frac{1}{a} + \frac{1}{c}\right) \pm \sqrt{\frac{1}{a^2} + \frac{2}{ac} + \frac{1}{c^2} - 4\left(\frac{1}{ac}\right)}$

$= \frac{1}{2}\left(\frac{1}{a} + \frac{1}{c}\right) \pm \frac{1}{2}\sqrt{\left(\frac{1}{a} - \frac{1}{c}\right)^2}$

$= \frac{1}{2}\left(\frac{1}{a} + \frac{1}{c} \pm \left|\frac{1}{a} - \frac{1}{c}\right|\right)$

The two values are $\frac{1}{a}$ and $\frac{1}{c}$.

67. $\cot 2\theta = 0$, $\theta = \frac{\pi}{4}$

$x = \frac{\sqrt{2}}{2}(u - v)$

$y = \frac{\sqrt{2}}{2}(u + v)$

$\frac{1}{2}(u-v)^2 + \frac{B}{2}(u-v)(u+v) + \frac{1}{2}(u+v)^2 = 1$

$\frac{2+B}{2}u^2 + \frac{2-B}{2}v^2 = 1$

a. The graph is an ellipse if $\frac{2+B}{2} > 0$ and $\frac{2-B}{2} > 0$, so $-2 < B < 2$.

b. The graph is a circle if $\frac{2+B}{2} = \frac{2-B}{2}$, so $B = 0$.

c. The graph is a hyperbola if $\frac{2+B}{2} > 0$ and $\frac{2-B}{2} < 0$ or if $\frac{2+B}{2} < 0$ and $\frac{2-B}{2} > 0$, so $B < -2$ or $B > 2$.

d. The graph is two parallel lines if $\frac{2+B}{2} = 0$ or $\frac{2-B}{2} = 0$, so $B = \pm 2$.

69. From Figure 6 it is clear that $v = r\sin\phi$ and $u = r\cos\phi$.

Also noting that $y = r\sin(\theta + \phi)$ leads us to

$y = r\sin(\theta + \phi) = r\sin\theta\cos\phi + r\cos\theta\sin\phi$

$= (r\cos\phi)(\sin\theta) + (r\sin\phi)(\cos\theta)$

$= u\sin\theta + v\cos\theta$

10.4 Concepts Review

1. simple; closed; simple

3. cycloid

Problem Set 10.4

1. a.

t	x	y
–2	–6	–4
–1	–3	–2
0	0	0
1	3	2
2	6	4

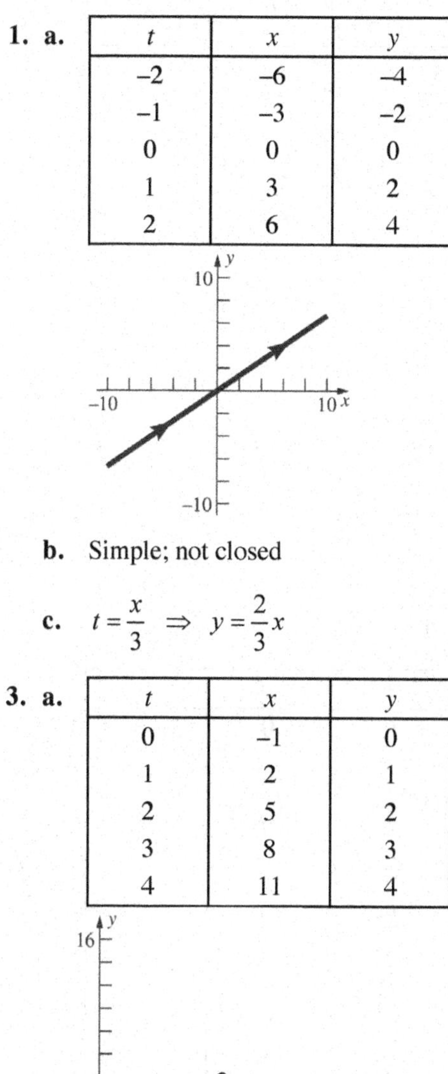

b. Simple; not closed

c. $t = \frac{x}{3} \Rightarrow y = \frac{2}{3}x$

3. a.

t	x	y
0	–1	0
1	2	1
2	5	2
3	8	3
4	11	4

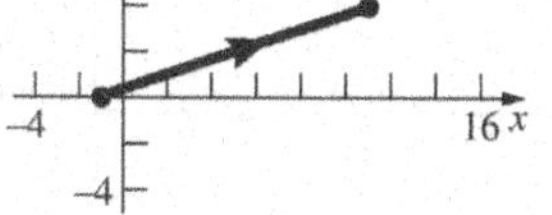

b. Simple; not closed

c. $t = \frac{1}{3}(x+1) \Rightarrow y = \frac{1}{3}(x+1)$

5. a.

t	x	y
0	4	0
1	3	1
2	2	$\sqrt{2}$
3	1	$\sqrt{3}$
4	0	2

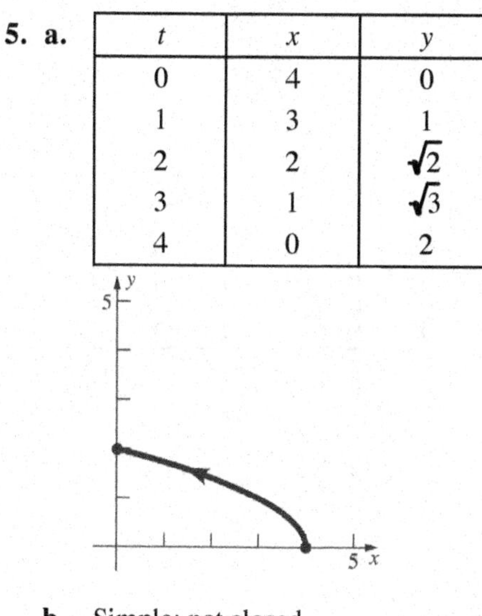

b. Simple; not closed

c. $t = 4 - x$

$y = \sqrt{4-x}$

7. a.

s	x	y
1	1	1
3	$\frac{1}{3}$	3
5	$\frac{1}{5}$	5
7	$\frac{1}{7}$	7
9	$\frac{1}{9}$	9

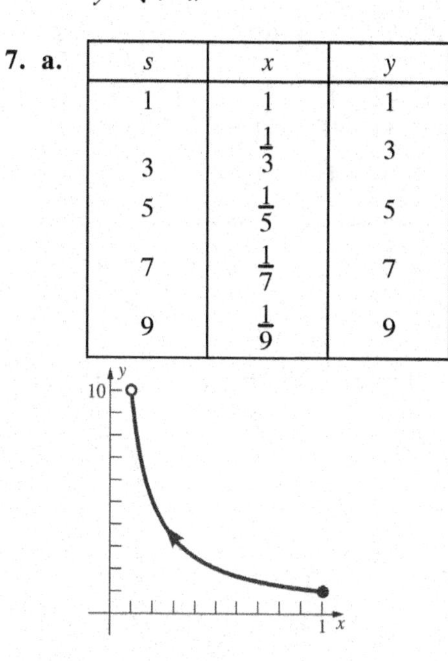

b. Simple; not closed

c. $s = \frac{1}{x}$

$y = \frac{1}{x}$

9. a.

t	x	y
–3	–15	5
–2	0	0
–1	3	–3
0	0	–4
1	–3	–3
2	0	0
3	15	5

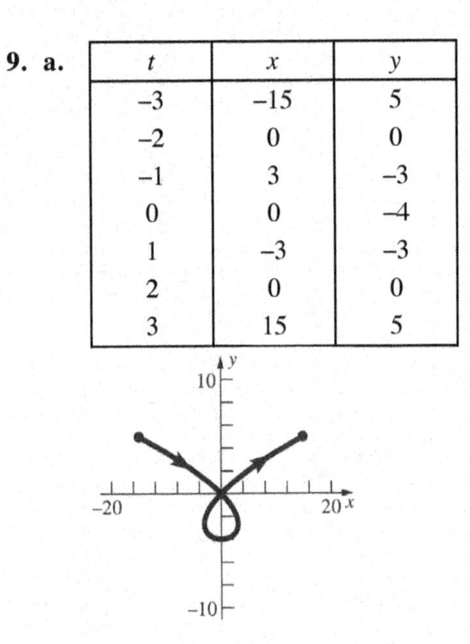

b. Not simple; not closed

c. $x^2 = t^6 - 8t^4 + 16t^2$

$t^2 = y + 4$

$x^2 = (y+4)^3 - 8(y+4)^2 + 16(y+4)$

$x^2 = y^3 + 4y^2$

11. a.

t	x	y
2	0	$3\sqrt{2}$
3	2	3
4	$2\sqrt{2}$	0

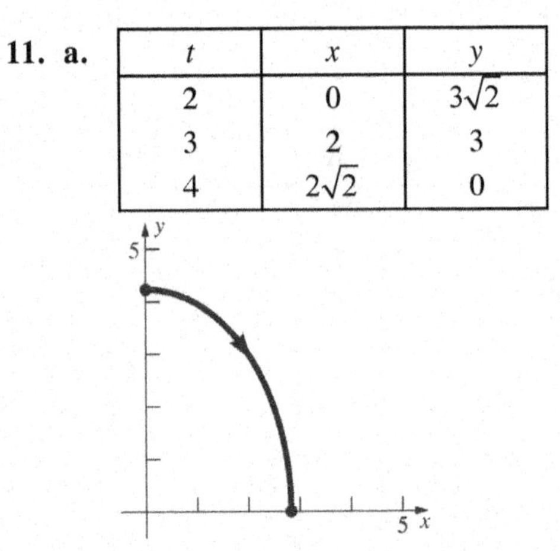

b. Simple; not closed

c. $t = \frac{1}{4}x^2 + 2$

$t = 4 - \frac{1}{9}y^2$

$\frac{1}{4}x^2 + 2 = 4 - \frac{1}{9}y^2$

$\frac{x^2}{8} + \frac{y^2}{18} = 1$

13. a.

t	x	y
0	0	3
$\frac{\pi}{2}$	2	0
π	0	–3
$\frac{3\pi}{2}$	–2	0
2π	0	3

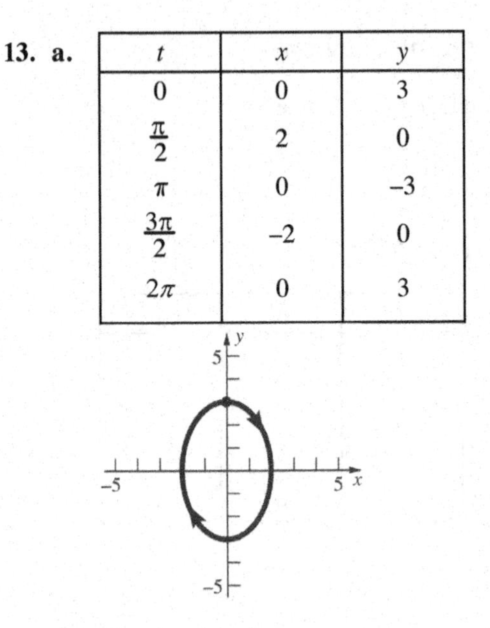

b. Simple; closed

c. $\sin^2 t = \frac{x^2}{4}$

$\cos^2 t = \frac{y^2}{9}$

$\frac{x^2}{4} + \frac{y^2}{9} = 1$

15. a.

r	x	y
0	0	–3
$\frac{\pi}{2}$	–2	0
π	0	3
$\frac{3\pi}{2}$	2	0
2π	0	–3
$\frac{5\pi}{2}$	–2	0
3π	0	3
$\frac{7\pi}{2}$	2	0
4π	0	–3

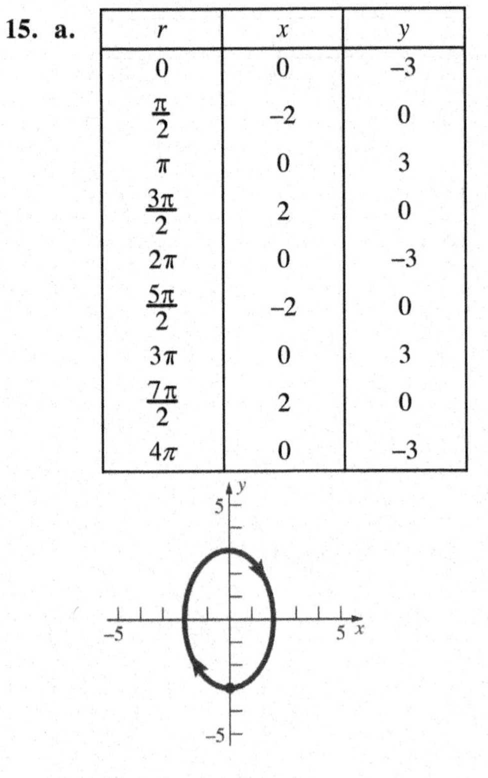

b. Not simple; closed

c. $\sin^2 r = \frac{x^2}{4}$

$\cos^2 r = \frac{y^2}{9}$

$\frac{x^2}{4} + \frac{y^2}{9} = 1$

17. a.

θ	x	y
0	0	9
$\frac{\pi}{4}$	$\frac{9}{2}$	$\frac{9}{2}$
$\frac{\pi}{2}$	0	9
$\frac{3\pi}{4}$	$\frac{9}{2}$	$\frac{9}{2}$
π	0	9

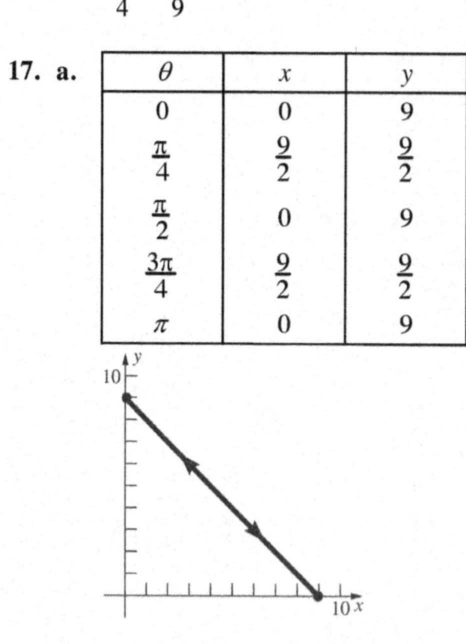

b. Not simple; closed

c. $\sin^2 \theta = \frac{x}{9}$

$\cos^2 \theta = \frac{y}{9}$

$\frac{x}{9} + \frac{y}{9} = 1$

$x + y = 9$

19. a.

θ	x	y
$0 + 2\pi n$	1	0
$\frac{\pi}{3} + 2\pi n$	$\frac{1}{2}$	$-\frac{3}{2}$
$\frac{\pi}{2} + 2\pi n$	0	0
$\frac{2\pi}{3} + 2\pi n$	$-\frac{1}{2}$	$-\frac{3}{2}$
$\pi + 2\pi n$	-1	0
$\frac{4\pi}{3} + 2\pi n$	$-\frac{1}{2}$	$-\frac{3}{2}$
$\frac{3\pi}{2} + 2\pi n$	0	0
$\frac{5\pi}{3} + 2\pi n$	$\frac{1}{2}$	$-\frac{3}{2}$

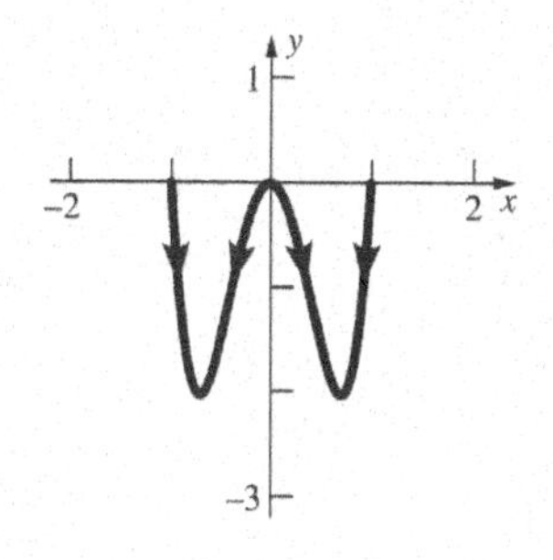

b. Not simple; not closed

c. $\cos \theta = x$

$\sin \theta = \sqrt{1 - x^2}$

$y = -8 \sin^2 \theta \cos^2 \theta$

$y = -8x^2(1 - x^2)$

21. $\frac{dx}{d\tau} = 6\tau$

$\frac{dy}{d\tau} = 12\tau^2$

$\frac{dy}{dx} = 2\tau$

$\frac{dy'}{d\tau} = 2$

$\frac{d^2y}{dx^2} = \frac{1}{3\tau}$

23. $\frac{dx}{d\theta} = 4\theta$

$\frac{dy}{d\theta} = 3\sqrt{5}\theta^2$

$\frac{dy}{dx} = \frac{3\sqrt{5}}{4}\theta$

$\frac{dy'}{d\theta} = \frac{3\sqrt{5}}{4}$

$\frac{d^2y}{dx^2} = \frac{3\sqrt{5}}{16\theta}$

25. $\frac{dx}{dt} = \sin t$

$\frac{dy}{dt} = \cos t$

$\frac{dy}{dx} = \cot t$

$\frac{dy'}{dt} = -\csc^2 t$

$\frac{d^2y}{dx^2} = -\csc^3 t$

27. $\frac{dx}{dt} = 3\sec^2 t$

$\frac{dy}{dt} = 5\sec t \tan t$

$\frac{dy}{dx} = \frac{5}{3}\sin t$

$\frac{dy'}{dt} = \frac{5}{3}\cos t$

$\frac{d^2y}{dx^2} = \frac{5}{9}\cos^3 t$

29. $\frac{dx}{dt} = -\frac{2t}{(1+t^2)^2}$

$\frac{dy}{dt} = \frac{2t-1}{t^2(1-t)^2}$

$\frac{dy}{dx} = \frac{(1-2t)(1+t^2)^2}{2t^3(1-t)^2}$

$\frac{dy'}{dt} = -\frac{3t^5+7t^4-6t^3+10t^2-9t+3}{2t^4(1-t)^3}$

$\frac{d^2y}{dx^2} = \frac{(3t^5+7t^4-6t^3+10t^2-9t+3)(1+t^2)^2}{4t^5(1-t)^3}$

31. $\frac{dx}{dt} = 2t, \frac{dy}{dt} = 3t^2$

$\frac{dy}{dx} = \frac{3}{2}t$

At $t = 2, x = 4, y = 8$, and $\frac{dy}{dx} = 3$.

Tangent line: $y - 8 = 3(x - 4)$ or $3x - y - 4 = 0$

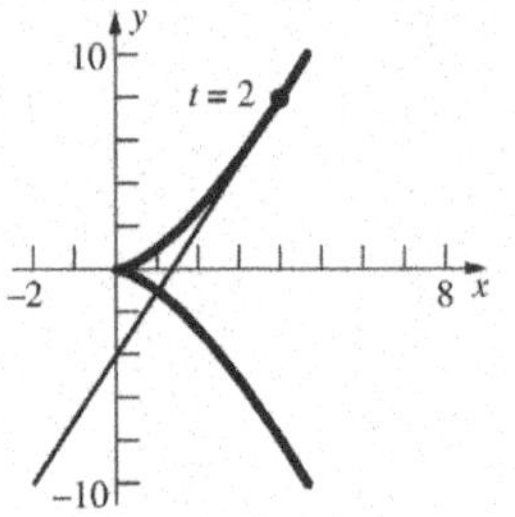

33. $\frac{dx}{dt} = 2\sec t \tan t, \frac{dy}{dt} = 2\sec^2 t$

$\frac{dy}{dx} = \csc t$

At $t = -\frac{\pi}{6}, x = \frac{4}{\sqrt{3}}, y = -\frac{2}{\sqrt{3}}$, and $\frac{dy}{dx} = -2$.

Tangent line: $y + \frac{2}{\sqrt{3}} = -2\left(x - \frac{4}{\sqrt{3}}\right)$ or

$2\sqrt{3}x + \sqrt{3}y - 6 = 0$

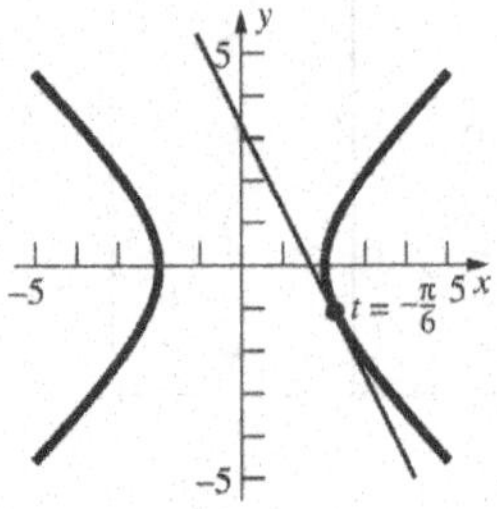

35. $\frac{dx}{dt} = 2, \frac{dy}{dt} = 3$

$L = \int_0^3 \sqrt{4+9}\,dt = \sqrt{13}\int_0^3 dt = \sqrt{13}[t]_0^3 = 3\sqrt{13}$

37. $\frac{dx}{dt} = 1, \frac{dy}{dt} = \frac{3}{2}t^{1/2}$

$L = \int_0^3 \sqrt{1+\frac{9}{4}t}\,dt$

$= \frac{1}{2}\int_0^3 \sqrt{4+9t}\,dt$

$= \frac{1}{18}\left[\frac{2}{3}(4+9t)^{3/2}\right]_0^3$

$= \frac{1}{27}(31^{3/2} - 8) = \frac{1}{27}(31\sqrt{31} - 8)$

39. $\frac{dx}{dt} = 6t, \frac{dy}{dt} = 3t^2$

$L = \int_0^2 \sqrt{36t^2 + 9t^4}\,dt = 3\int_0^2 t\sqrt{4+t^2}\,dt$

$3\left[\frac{1}{3}(4+t^2)^{3/2}\right]_0^2 = 16\sqrt{2} - 8$

41. $\frac{dx}{dt} = 2e^t, \frac{dy}{dt} = \frac{9}{2}e^{3t/2}$

$L = \int_{\ln 3}^{2\ln 3} \sqrt{4e^{2t} + \frac{81}{4}e^{3t}}\,dt = \int_{\ln 3}^{2\ln 3} e^t\sqrt{4 + \frac{81}{4}e^t}\,dt$

$= \left[\frac{8}{243}\left(4 + \frac{81}{4}e^t\right)^{3/2}\right]_{\ln 3}^{2\ln 3}$

$= \frac{745\sqrt{745} - 259\sqrt{259}}{243}$

43. $\frac{dx}{dt}=\frac{2}{\sqrt{t}}, \frac{dy}{dt}=2t-\frac{1}{2t^2}$

$$L=\int_{1/4}^{1}\sqrt{\frac{4}{t}+\left(4t^2-\frac{2}{t}+\frac{1}{4t^4}\right)}dt$$

$$=\int_{1/4}^{1}\sqrt{4t^2+\frac{2}{t}+\frac{1}{4t^4}}dt$$

$$=\int_{1/4}^{1}\sqrt{\left(2t+\frac{1}{2t^2}\right)^2}\,dt$$

$$=\int_{1/4}^{1}\left(2t+\frac{1}{2t^2}\right)dt=\left[t^2-\frac{1}{2t}\right]_{1/4}^{1}=\frac{39}{16}$$

45. $\frac{dx}{dt}=-\sin t,$

$$\frac{dy}{dt}=\frac{\sec t\tan t+\sec^2 t}{\sec t+\tan t}-\cos t=\sec t-\cos t$$

$$L=\int_0^{\pi/4}\sqrt{\sin^2 t+(\sec^2 t-2+\cos^2 t)}dt$$

$$=\int_0^{\pi/4}\tan t\,dt$$

$$=\left[-\ln|\cos t|\right]_0^{\pi/4}=-\ln\frac{1}{\sqrt{2}}=\frac{1}{2}\ln 2$$

47. a. $\frac{dx}{d\theta}=\cos\theta, \frac{dy}{d\theta}=-\sin\theta$

$$L=\int_0^{2\pi}\sqrt{\cos^2\theta+\sin^2\theta}\,d\theta=\int_0^{2\pi}d\theta$$

$$=[\theta]_0^{2\pi}=2\pi$$

b. $\frac{dx}{d\theta}=3\cos 3\theta, \frac{dy}{d\theta}=-3\sin 3\theta$

$$L=\int_0^{2\pi}\sqrt{9\cos^2 3\theta+9\sin^2 3\theta}\,d\theta$$

$$=3\int_0^{2\pi}d\theta=3[\theta]_0^{2\pi}=6\pi$$

c. The curve in part a goes around the unit circle once, while the curve in part b goes around the unit circle three times.

49. $\frac{dx}{dt}=-\sin t, \frac{dy}{dt}=\cos t$

$$S=\int_0^{2\pi}2\pi(1+\cos t)\sqrt{\sin^2 t+\cos^2 t}\,dt$$

$$=2\pi\int_0^{2\pi}(1+\cos t)dt=2\pi[t+\sin t]_0^{2\pi}=4\pi^2$$

51. $\frac{dx}{dt}=-\sin t, \frac{dy}{dt}=\cos t$

$$S=\int_0^{2\pi}2\pi(1+\sin t)\sqrt{\sin^2 t+\cos^2 t}\,dt$$

$$=2\pi\int_0^{2\pi}(1+\sin t)dt=2\pi[t-\cos t]_0^{2\pi}=4\pi^2$$

53. $\frac{dx}{dt}=1, \frac{dy}{dt}=t+\sqrt{7}$

$$S=\int_{-\sqrt{7}}^{\sqrt{7}}2\pi\left(t+\sqrt{7}\right)\sqrt{1+\left(t+\sqrt{7}\right)^2}\,dt$$

$$=2\pi\left[\frac{1}{3}\left(1+\left(t+\sqrt{7}\right)^2\right)^{3/2}\right]_{-\sqrt{7}}^{\sqrt{7}}$$

$$=\frac{2\pi}{3}\left(29\sqrt{29}-1\right)$$

55. $dx=dt$; when $x=0, t=-1$; when $x=1, t=0$.

$$\int_0^1(x^2-4y)dx=\int_{-1}^{0}[(t+1)^2-4(t^3+4)]dt$$

$$=\int_{-1}^{0}(-4t^3+t^2+2t-15)dt$$

$$=\left[-t^4+\frac{1}{3}t^3+t^2-15t\right]_{-1}^{0}=-\frac{44}{3}$$

57. $dx=2e^{2t}dt$

$$A=\int_1^{25}y\,dx=\int_0^{\ln 5}2e^t\,dt=[2e^t]_0^{\ln 5}=8$$

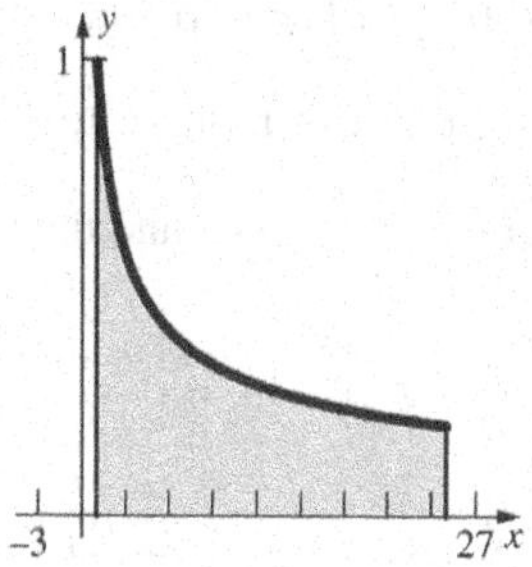

59.

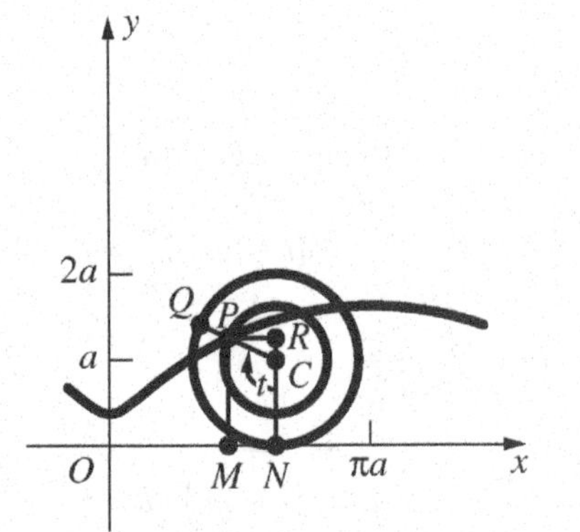

Let the wheel roll along the x-axis with P initially at $(0, a-b)$.

$|ON| = \text{arc } NQ = at$

$x = |OM| = |ON| - |MN| = at - b\sin t$

$y = |MP| = |RN| = |NC| + |CR| = a - b\cos t$

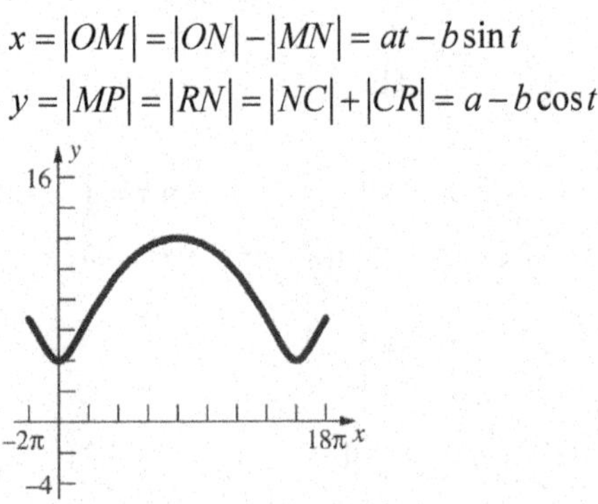

61. The x- and y-coordinates of the center of the circle of radius b are $(a-b)\cos t$ and $(a-b)\sin t$, respectively. The angle measure (in a clockwise direction) of arc BP is $\frac{a}{b}t$. The horizontal change from the center of the circle of radius b to P is

$b\cos\left(-\left(\frac{a}{b}t - t\right)\right) = b\cos\left(\frac{a-b}{b}t\right)$ and the vertical change is $b\sin\left(-\left(\frac{a}{b}t - t\right)\right) = -b\sin\left(\frac{a-b}{b}t\right)$.

Therefore, $x = (a-b)\cos t + b\cos\left(\frac{a-b}{b}t\right)$ and

$y = (a-b)\sin t - b\sin\left(\frac{a-b}{b}t\right)$.

63. Consider the following figure similar to the one in the text for Problem 61.

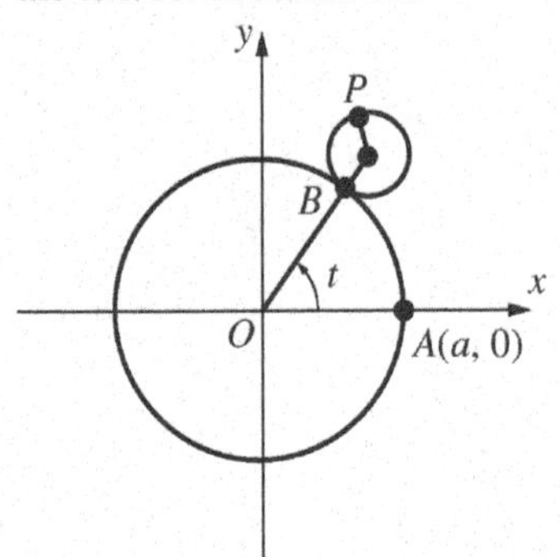

The x- and y-coordinates of the center of the circle of radius b are $(a+b)\cos t$ and $(a+b)\sin t$ respectively. The angle measure (in a counter-clockwise direction) of arc PB is $\frac{a}{b}t$. The horizontal change from the center of the circle of radius b to P is

$b\cos\left(\frac{a}{b}t + t + \pi\right) = -b\cos\left(\frac{a+b}{b}t\right)$ and the vertical change is

$b\sin\left(\frac{a}{b}t + t + \pi\right) = -b\sin\left(\frac{a+b}{b}t\right)$. Therefore,

$x = (a+b)\cos t - b\cos\left(\frac{a+b}{b}t\right)$ and

$y = (a+b)\sin t - b\sin\left(\frac{a+b}{b}t\right)$.

65. $\frac{dx}{dt} = \left(\frac{a}{3}\right)(-2\sin t - 2\sin 2t)$,

$\frac{dy}{dt} = \left(\frac{a}{3}\right)(2\cos t - 2\cos 2t)$

$\left(\frac{dx}{dt}\right)^2 = \left(\frac{a}{3}\right)^2 (4\sin^2 t + 8\sin t \sin 2t + 4\sin^2 2t)$

$\left(\frac{dy}{dt}\right)^2 = \left(\frac{a}{3}\right)^2 (4\cos^2 t - 8\cos t\cos 2t + 4\cos^2 2t)$

$\left(\frac{dx}{dt}\right)^2 + \left(\frac{dy}{dt}\right)^2 = \left(\frac{a}{3}\right)^2 (8 + 8\sin t\sin 2t - 8\cos t\cos 2t)$

$= \left(\frac{a}{3}\right)^2 (8 + 16\sin^2 t\cos t - 8\cos^3 t + 8\sin^2 t\cos t)$

$= \left(\frac{a}{3}\right)^2 (8 + 24\cos t\sin^2 t - 8\cos^3 t)$

$= \left(\frac{a}{3}\right)^2 (8 + 24\cos t - 32\cos^3 t)$

$L = 3\int_0^{2\pi/3} \sqrt{\left(\frac{dx}{dt}\right)^2 + \left(\frac{dy}{dt}\right)^2}\, dt$

$= a\int_0^{2\pi/3} \sqrt{8 + 24\cos t - 32\cos^3 t}\, dt$

Using a CAS to evaluate the length, $L = \frac{16a}{3}$.

67. a.

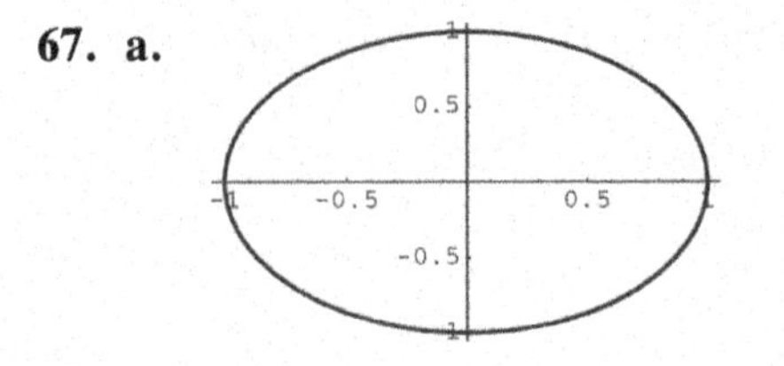

The curve touches a horizontal border twice and touches a vertical border twice.

b. 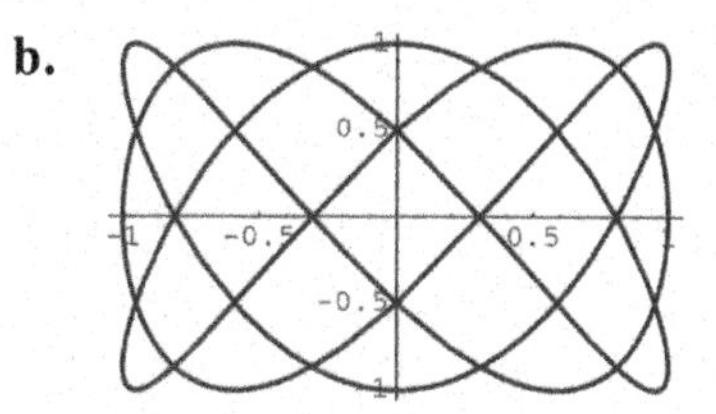

The curve touches a horizontal border five times and touches a vertical border three times.

c. 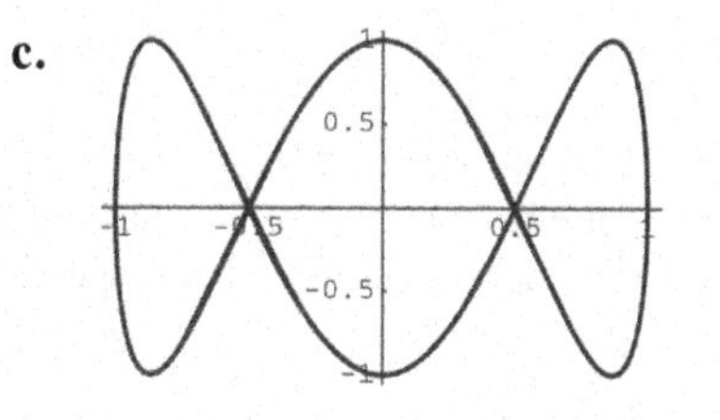

The curve touches a horizontal border six times and touches a vertical border twice. Note that the curve is traced out five times.

d. 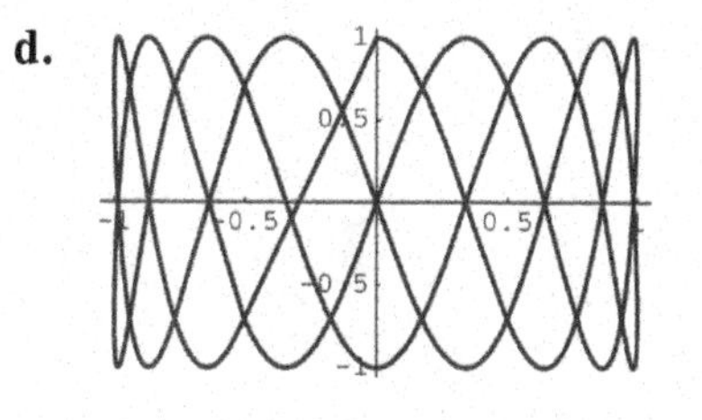

The curve touches a horizontal border 18 times and touches a vertical border four times.

69. a. $x = \cos t$; $y = \sin 2t$

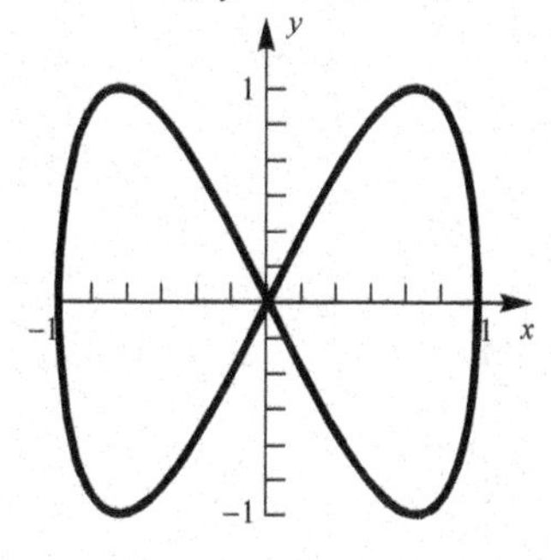

b. $x = \cos 4t$; $y = \sin 8t$

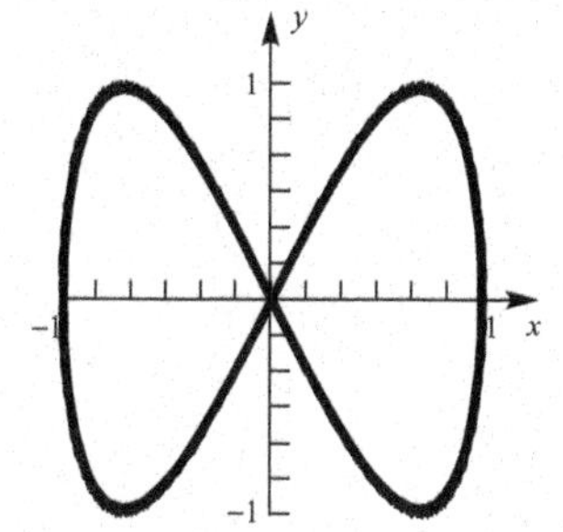

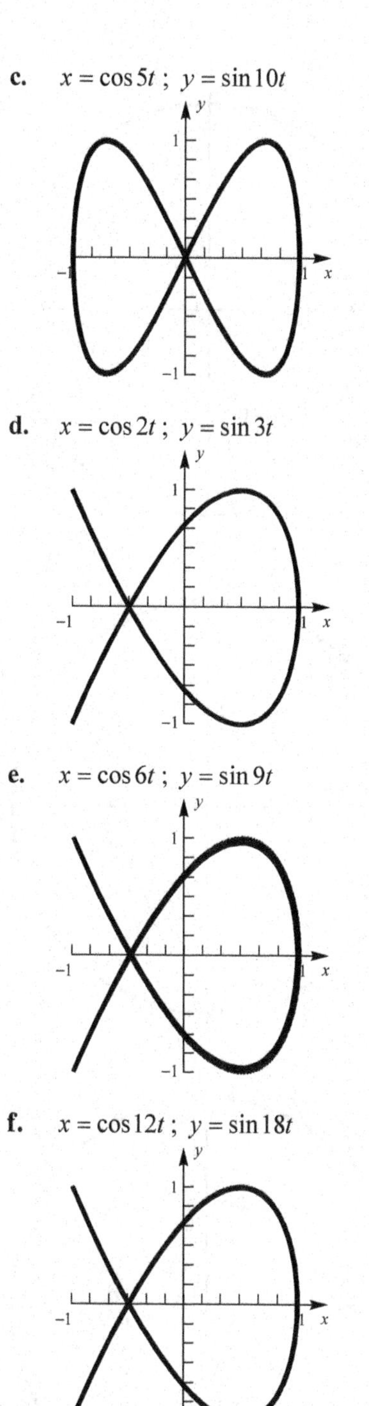

71. a.

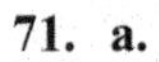

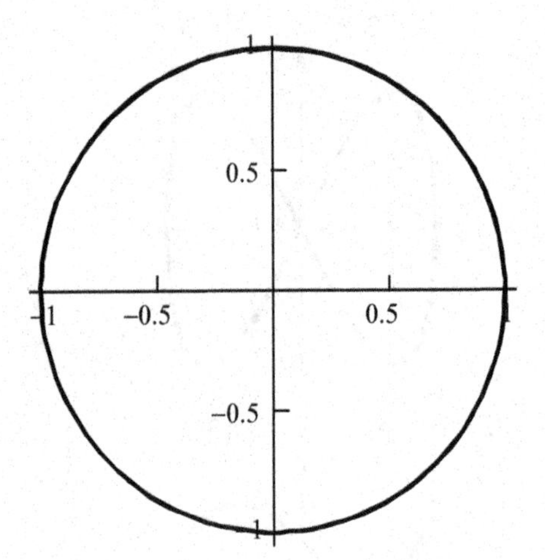

b.

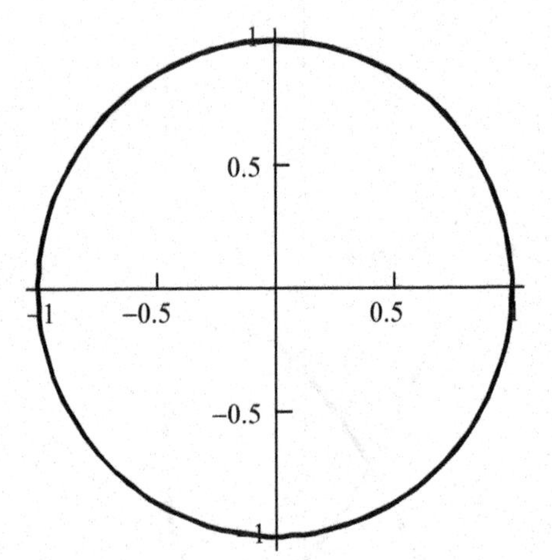

c.

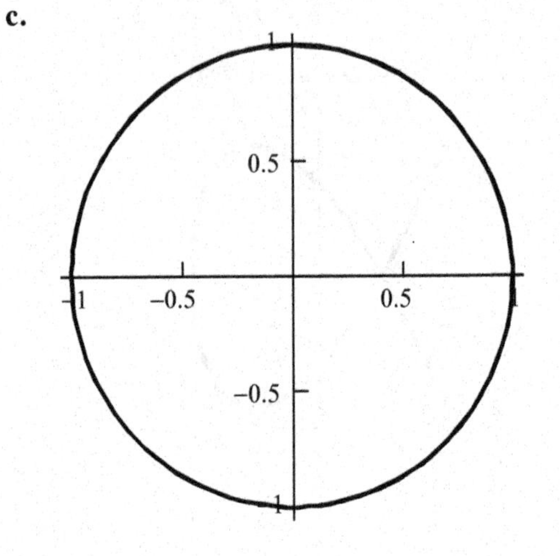

d.

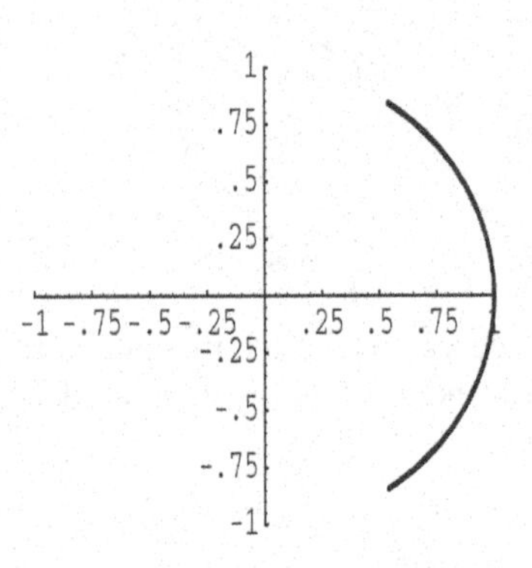

Given a parameterization of the form $x = \cos f(t)$ and $y = \sin f(t)$, the point moves around the curve (which is a circle of radius 1) at a speed of $|f'(t)|$. The point travels clockwise around the circle when $f(t)$ is decreasing and counterclockwise when $f(t)$ is increasing. Note that in part d, only part of the circle will be traced out since the range of $f(t) = \sin t$ is $[-1, 1]$.

73. a. $0 \le t \le 2\pi$

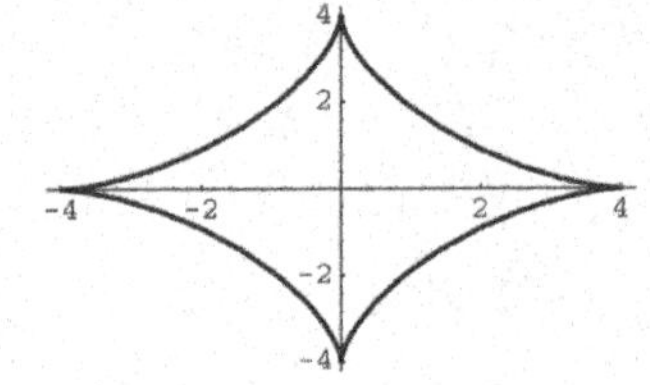

b. $0 \le t \le 2\pi$

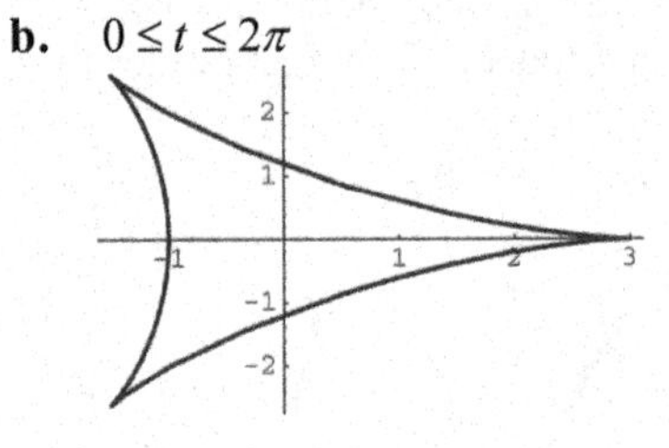

c. $0 \le t \le 4\pi$

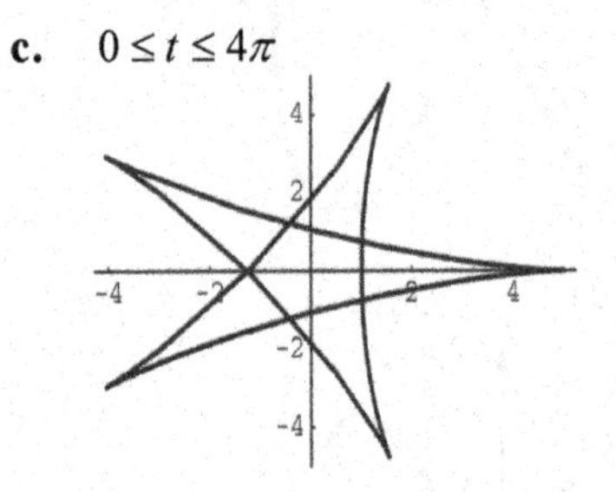

d. $0 \le t \le 8\pi$

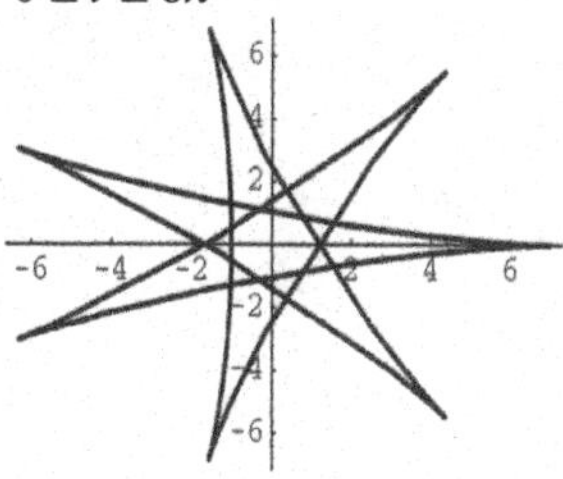

Let $\frac{p}{q} = \frac{a}{b}$ where $\frac{p}{q}$ is the reduced fraction of $\frac{a}{b}$. The length of the t-interval is $2q\pi$.

The number of times the graph would touch the circle of radius a during the t-interval is p.

If $\frac{a}{b}$ is irrational, the curve is not periodic.

75. $x = \frac{3t}{t^3+1}, y = \frac{3t^2}{t^3+1}$

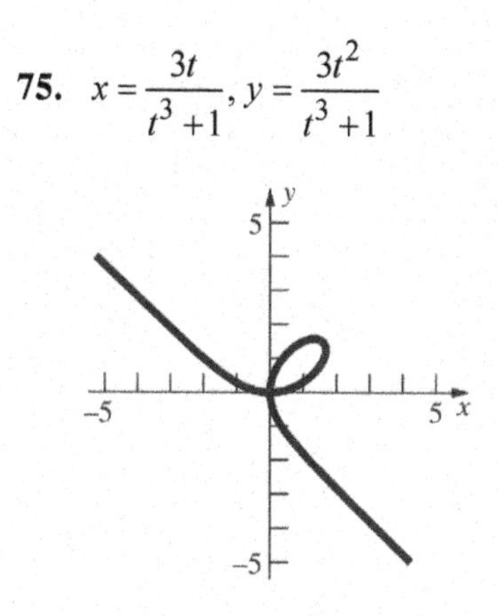

When $x > 0, t > 0$ or $t < -1$.
When $x < 0, -1 < t < 0$.
When $y > 0, t > -1$. When $y < 0, t < -1$.
Therefore the graph is in quadrant I for $t > 0$, quadrant II for $-1 < t < 0$, quadrant III for no t, and quadrant IV for $t < -1$.

10.5 Concepts Review

1. infinitely many

3. circle; line

Problem Set 10.5

1.

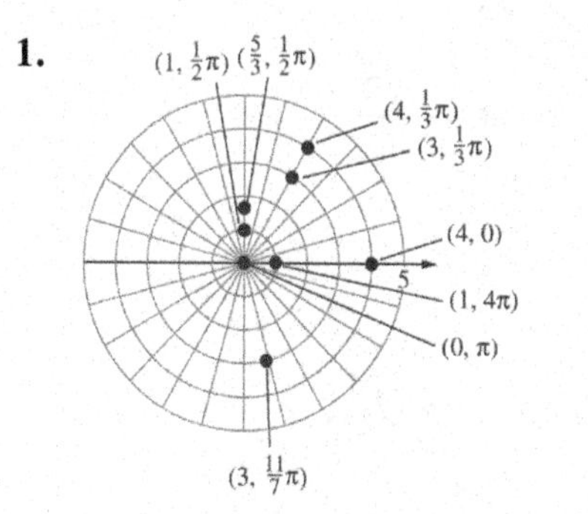

3.

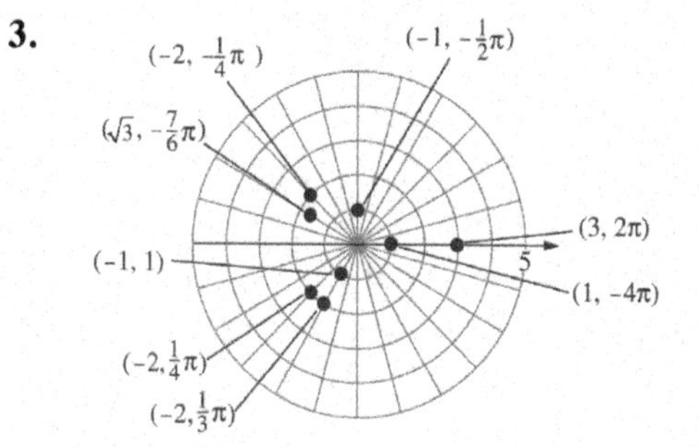

5.

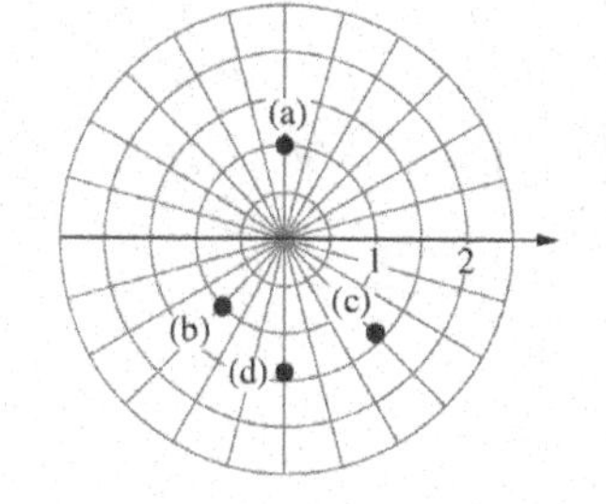

a. $\left(1, -\frac{3}{2}\pi\right), \left(1, \frac{5}{2}\pi\right), \left(-1, -\frac{1}{2}\pi\right), \left(-1, \frac{3}{2}\pi\right)$

b. $\left(1, -\frac{3}{4}\pi\right), \left(1, \frac{5}{4}\pi\right), \left(-1, -\frac{7}{4}\pi\right), \left(-1, \frac{9}{4}\pi\right)$

c. $\left(\sqrt{2}, -\frac{7}{3}\pi\right), \left(\sqrt{2}, \frac{5}{3}\pi\right), \left(-\sqrt{2}, -\frac{4}{3}\pi\right), \left(-\sqrt{2}, \frac{2}{3}\pi\right)$

d. $\left(\sqrt{2}, -\frac{1}{2}\pi\right), \left(\sqrt{2}, \frac{3}{2}\pi\right), \left(-\sqrt{2}, -\frac{3}{2}\pi\right), \left(-\sqrt{2}, \frac{1}{2}\pi\right)$

7. a. $x = 1\cos\frac{1}{2}\pi = 0$

$y = 1\sin\frac{1}{2}\pi = 1$

$(0, 1)$

b. $x = -1\cos\frac{1}{4}\pi = -\frac{\sqrt{2}}{2}$

$y = -1\sin\frac{1}{4}\pi = -\frac{\sqrt{2}}{2}$

$\left(-\frac{\sqrt{2}}{2}, -\frac{\sqrt{2}}{2}\right)$

c. $x = \sqrt{2}\cos\left(-\frac{1}{3}\pi\right) = \frac{\sqrt{2}}{2}$

$y = \sqrt{2}\sin\left(-\frac{1}{3}\pi\right) = -\frac{\sqrt{6}}{2}$

$\left(\frac{\sqrt{2}}{2}, -\frac{\sqrt{6}}{2}\right)$

d. $x = -\sqrt{2}\cos\frac{5}{2}\pi = 0$

$y = -\sqrt{2}\sin\frac{5}{2}\pi = -\sqrt{2}$

$\left(0, -\sqrt{2}\right)$

9. a. $r^2 = \left(3\sqrt{3}\right)^2 + 3^2 = 36,\ r = 6$

$\tan\theta = \frac{1}{\sqrt{3}}, \theta = \frac{\pi}{6}$

$\left(6, \frac{\pi}{6}\right)$

b. $r^2 = \left(-2\sqrt{3}\right)^2 + 2^2 = 16,\ r = 4$

$\tan\theta = \frac{2}{-2\sqrt{3}}, \theta = \frac{5\pi}{6}$

$\left(4, \frac{5\pi}{6}\right)$

c. $r^2 = \left(-\sqrt{2}\right)^2 + \left(-\sqrt{2}\right)^2 = 4,\ r = 2$

$\tan\theta = \frac{-\sqrt{2}}{-\sqrt{2}}, \theta = \frac{5\pi}{4}$

$\left(2, \frac{5\pi}{4}\right)$

d. $r^2 = 0^2 + 0^2 = 0,\ r = 0$

$\tan\theta = 0, \theta = 0$

$(0, 0)$

11. $x - 3y + 2 = 0$

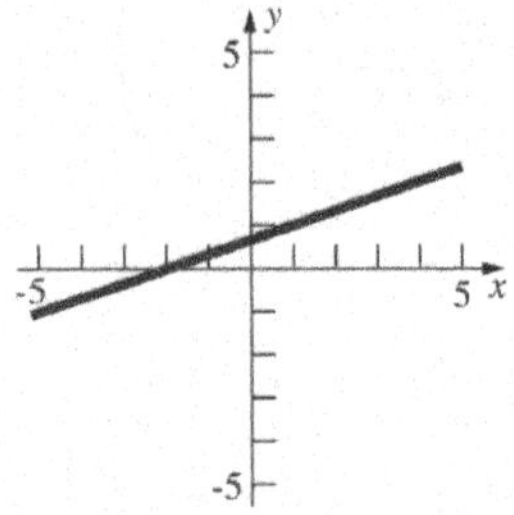

$r\cos\theta - 3r\sin\theta + 2 = 0$

$r = -\frac{2}{\cos\theta - 3\sin\theta}$

$r = \frac{2}{3\sin\theta - \cos\theta}$

13. $y = -2$

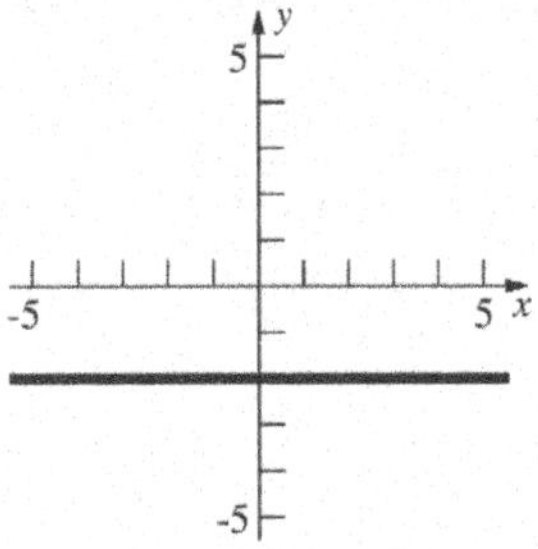

$r\sin\theta = -2$

$r = -\frac{2}{\sin\theta}$

$r = -2\csc\theta$

15. $x^2 + y^2 = 4$

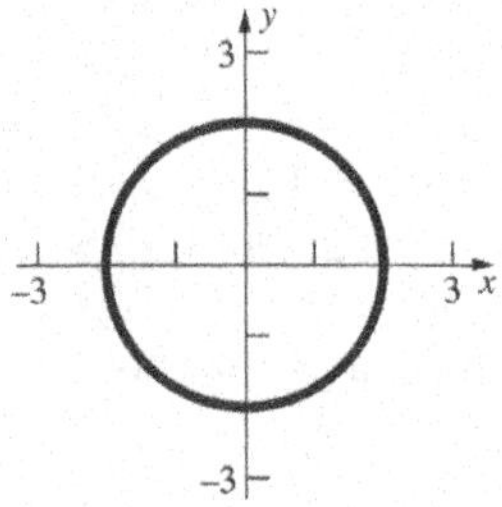

$(r\cos\theta)^2 + (r\sin\theta)^2 = 4$

$r^2 = 4$

$r = 2$

17. $\theta = \frac{\pi}{2}$

$\cot\theta = 0$

$\frac{x}{y} = 0$

$x = 0$

19. $r\cos\theta + 3 = 0$

$x + 3 = 0$

$x = -3$

21. $r \sin\theta - 1 = 0$
$y - 1 = 0$
$y = 1$

23. $r = 6$, circle

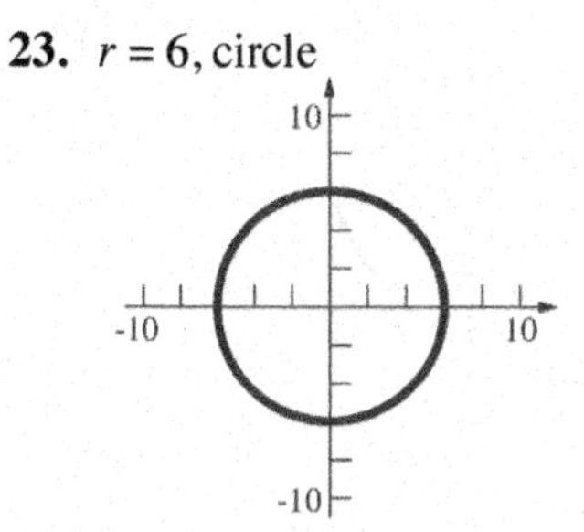

25. $r = \frac{3}{\sin\theta}$

$r = \frac{3}{\cos\left(\theta - \frac{\pi}{2}\right)}$, line

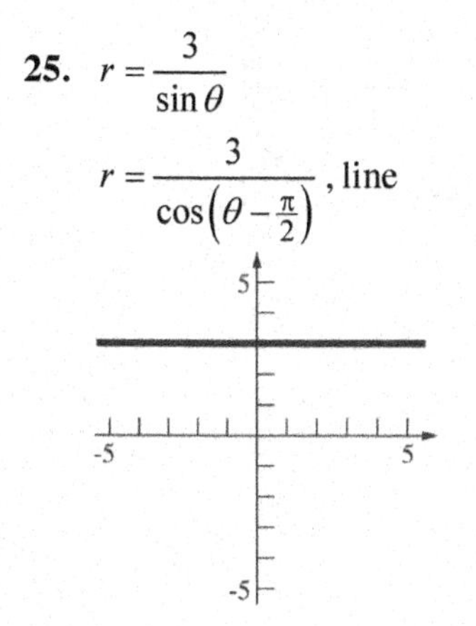

27. $r = 4\sin\theta$

$r = 2(2)\cos\left(\theta - \frac{\pi}{2}\right)$, circle

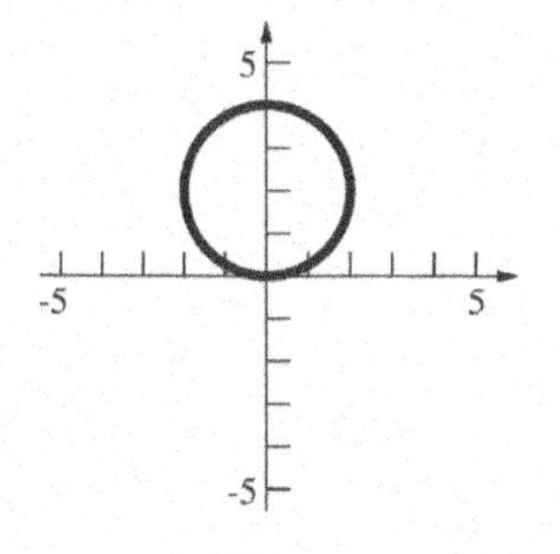

29. $r = \frac{4}{1+\cos\theta}$

$r = \frac{(1)(4)}{1+(1)\cos\theta}$, parabola

$e = 1$

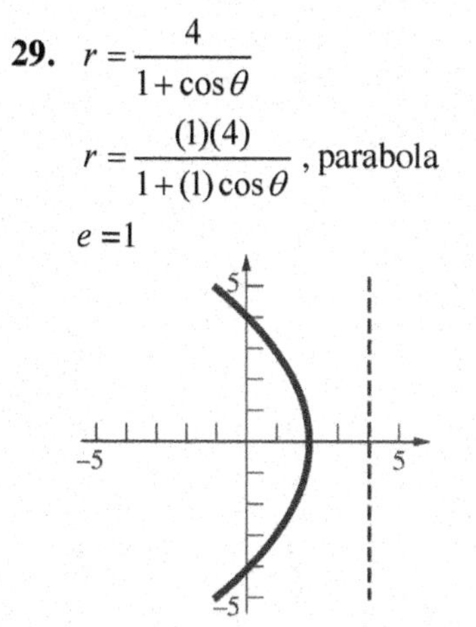

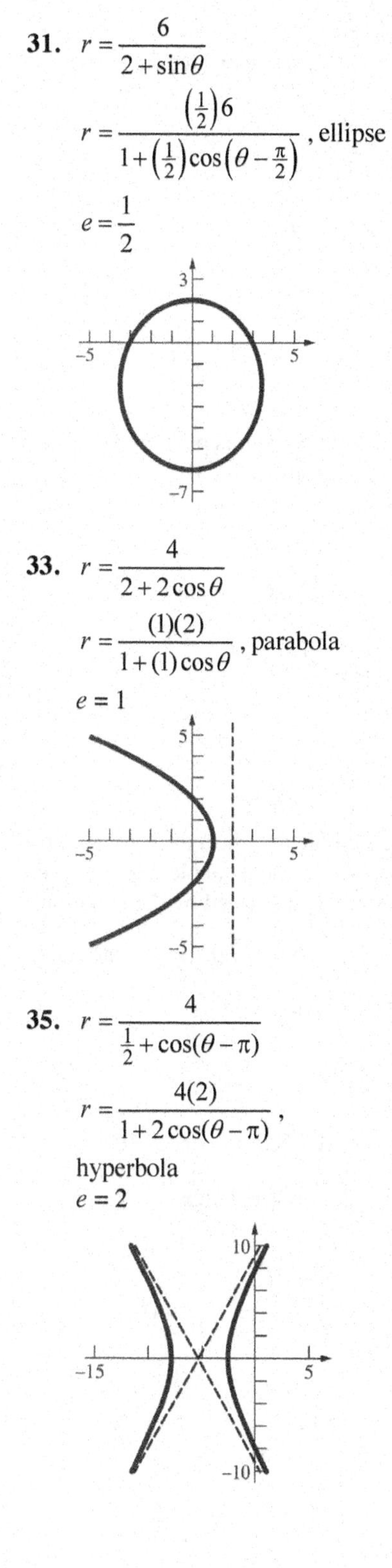

31. $r = \frac{6}{2+\sin\theta}$

$r = \frac{\left(\frac{1}{2}\right)6}{1+\left(\frac{1}{2}\right)\cos\left(\theta - \frac{\pi}{2}\right)}$, ellipse

$e = \frac{1}{2}$

33. $r = \frac{4}{2+2\cos\theta}$

$r = \frac{(1)(2)}{1+(1)\cos\theta}$, parabola

$e = 1$

35. $r = \frac{4}{\frac{1}{2}+\cos(\theta - \pi)}$

$r = \frac{4(2)}{1+2\cos(\theta - \pi)}$,

hyperbola

$e = 2$

37. By the Law of Cosines,
$a^2 = r^2 + c^2 - 2rc\cos(\theta - \alpha)$ (see figure below).

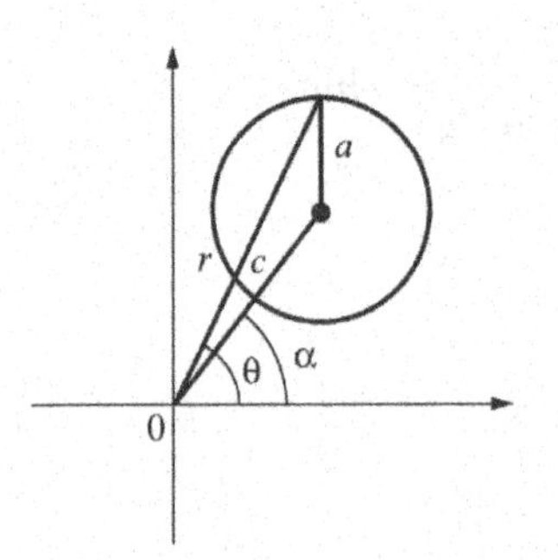

39. Recall that the latus rectum is perpendicular to the axis of the conic through a focus.

$$r\left(\theta_0 + \frac{\pi}{2}\right) = \frac{ed}{1 + e\cos\frac{\pi}{2}} = ed$$

Thus the length of the latus rectum is $2ed$.

41. $a + c = 183, a - c = 17$
$2a = 200, a = 100$
$2c = 166, c = 83$

$$e = \frac{c}{a} = 0.83$$

43. Let sun lie at the pole and the axis of the parabola lie on the pole so that the parabola opens to the left. Then the path is described by the equation $r = \frac{d}{1+\cos\theta}$. Substitute $(100, 120°)$ into the equation and solve for d.

$$100 = \frac{d}{1 + \cos 120°}$$

$d = 50$
The closest distance occurs when $\theta = 0°$.

$$r = \frac{50}{1 + \cos 0°} = 25 \text{ million miles}$$

45. $x = \frac{4e}{1 + e\cos t}\cos t, y = \frac{4e}{1 + e\cos t}\sin t$

$e = 0.1$

0.4
0.2
-0.4 -0.2 0.2
-0.2
-0.4

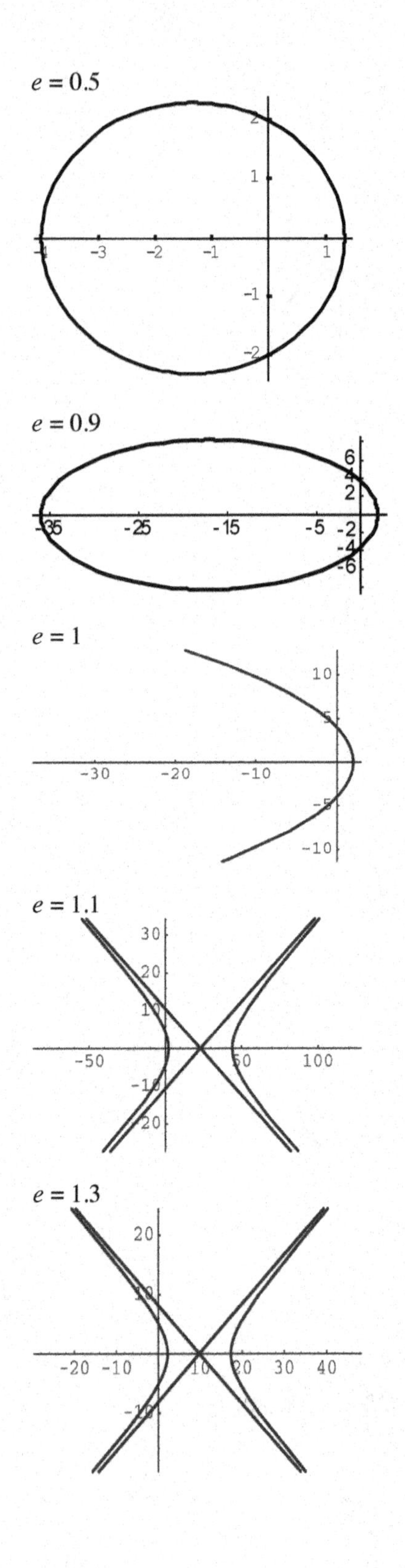

10.6 Concepts Review

1. limaçon

3. rose; odd; even

Problem Set 10.6

1. $\theta^2 - \frac{\pi^2}{16} = 0$

   $\theta = \pm\frac{\pi}{4}$

   Changing $\theta \to -\theta$ or $r \to -r$ yields an equivalent set of equations. Therefore all 3 tests are passed.

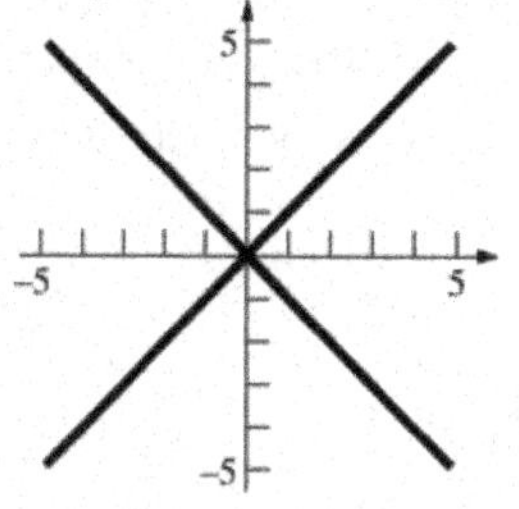

3. $r \sin\theta + 4 = 0$

   $r = -\frac{4}{\sin\theta}$

   Since $\sin(-\theta) = -\sin\theta$, test 2 is passed. The other two tests fail so the graph has only y-axis symmetry.

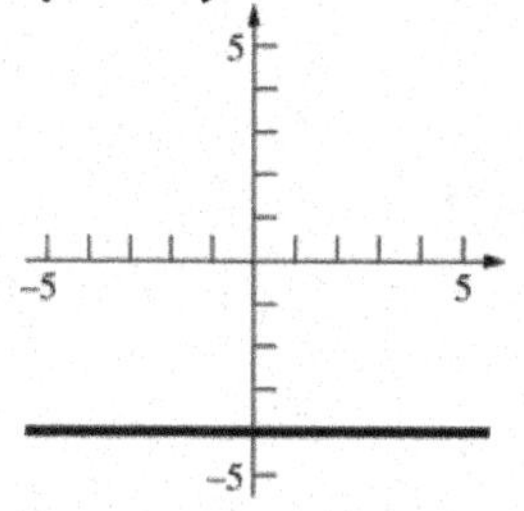

5. $r = 2\cos\theta$

   Since $\cos(-\theta) = \cos\theta$, the graph is symmetric about the x-axis. The other symmetry tests fail.

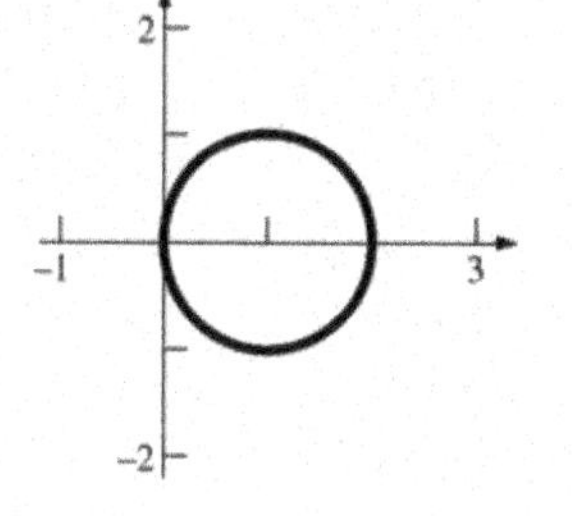

7. $r = \frac{2}{1-\cos\theta}$

   Since $\cos(-\theta) = \cos\theta$, the graph is symmetric about the x-axis. The other symmetry tests fail.

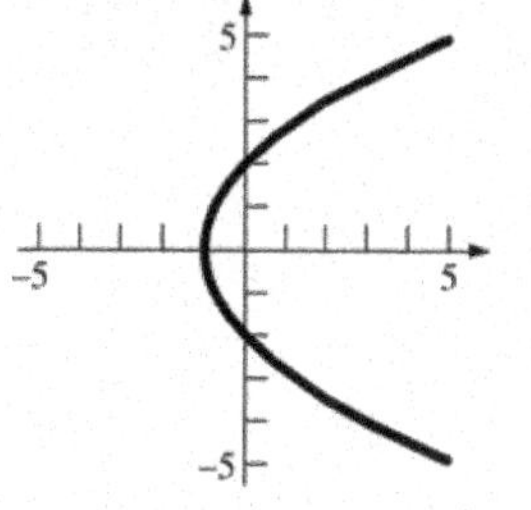

9. $r = 3 - 3\cos\theta$ (cardioid)

   Since $\cos(-\theta) = \cos\theta$, the graph is symmetric about the x-axis. The other symmetry tests fail.

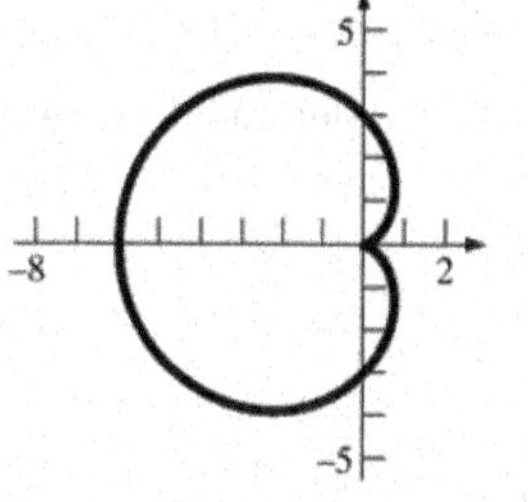

11. $r = 1 - 1\sin\theta$ (cardioid)

   Since $\sin(\pi - \theta) = \sin\theta$, the graph is symmetric about the y-axis. The other symmetry tests fail.

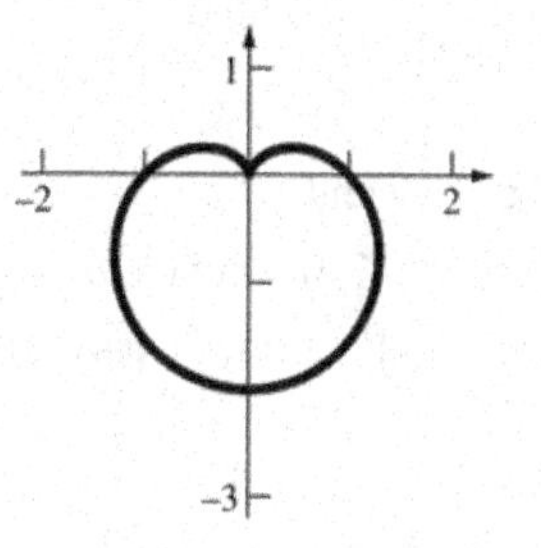

13. $r = 1 - 2\sin\theta$ (limaçon)

   Since $\sin(\pi - \theta) = \sin\theta$, the graph is symmetric about the y-axis. The other symmetry tests fail.

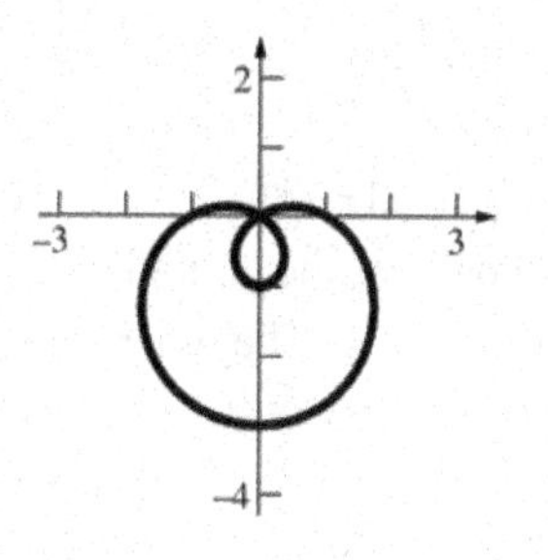

15. $r = 2 - 3\sin\theta$ (limaçon)

Since $\sin(\pi - \theta) = \sin\theta$, the graph is symmetric about the y-axis. The other symmetry tests fail.

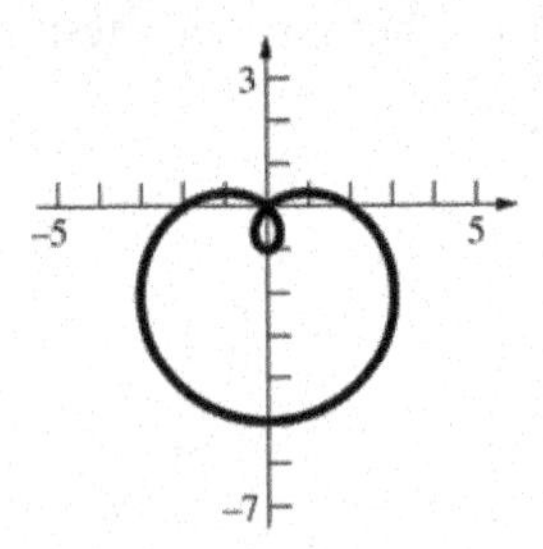

17. $r^2 = 4\cos 2\theta$ (lemniscate)

$r = \pm 2\sqrt{\cos 2\theta}$

Since $\cos(-2\theta) = \cos 2\theta$ and

$\cos(2(\pi - \theta)) = \cos(2\pi - 2\theta) = \cos(-2\theta) = \cos 2\theta$

the graph is symmetric about both axes and the origin.

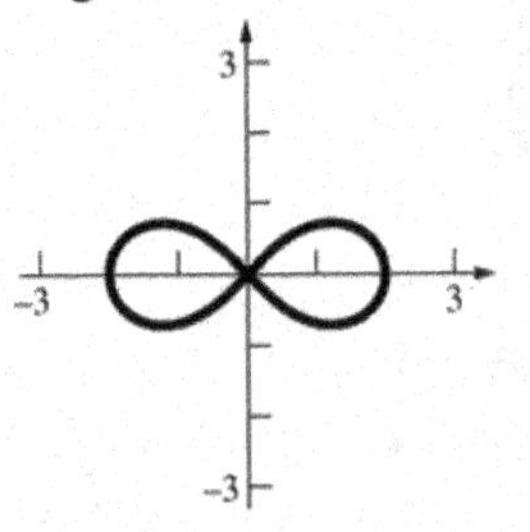

19. $r^2 = -9\cos 2\theta$ (lemniscate)

$r = \pm 3\sqrt{-\cos 2\theta}$

Since $\cos(-2\theta) = \cos 2\theta$ and

$\cos(2(\pi - \theta)) = \cos(2\pi - 2\theta) = \cos(-2\theta) = \cos 2\theta$

the graph is symmetric about both axes and the origin.

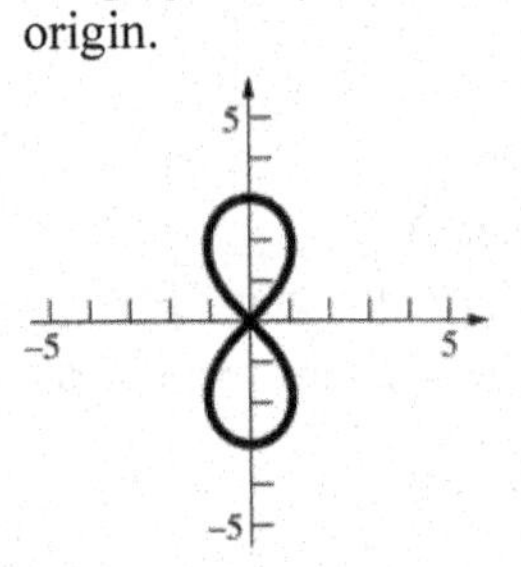

21. $r = 5\cos 3\theta$ (three-leaved rose)

Since $\cos(-3\theta) = \cos(3\theta)$, the graph is symmetric about the x-axis. The other symmetry tests fail.

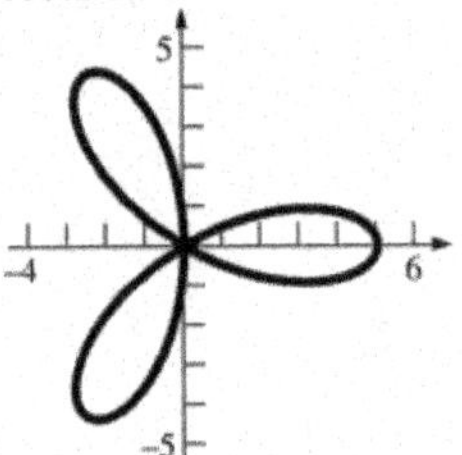

23. $r = 6\sin 2\theta$ (four-leaved rose)

Since

$\sin(2(\pi - \theta)) = \sin(2\pi - 2\theta)$

$= \sin(-2\theta) = -\sin(2\theta)$

and $\sin(-2\theta) = -\sin(2\theta)$, the graph is symmetric about both axes and the origin.

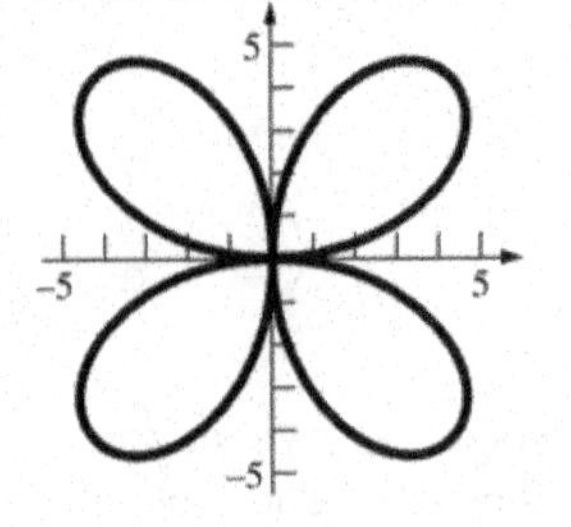

25. $r = 7\cos 5\theta$ (five-leaved rose)

Since $\cos(-5\theta) = \cos 5\theta$, the graph is symmetric about the x-axis. The other symmetry tests fail.

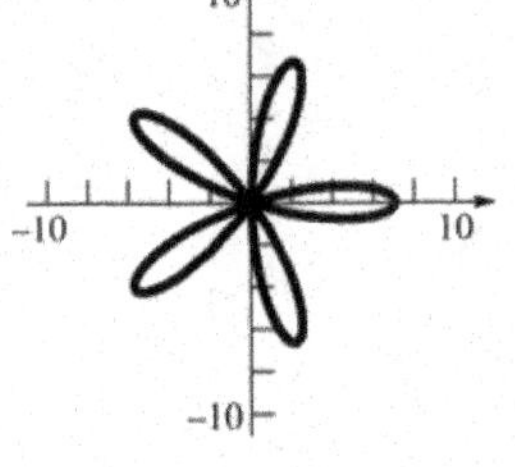

27. $r = \frac{1}{2}\theta, \theta \geq 0$ (spiral of Archimedes)

No symmetry. All three tests fail.

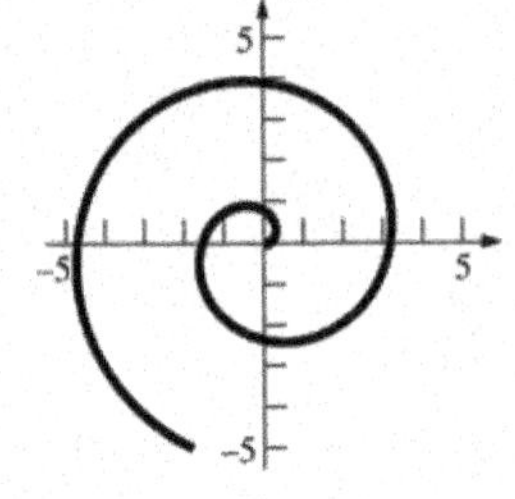

29. $r = e^{\theta}, \theta \geq 0$ (logarithmic spiral)

No symmetry. All three tests fail.

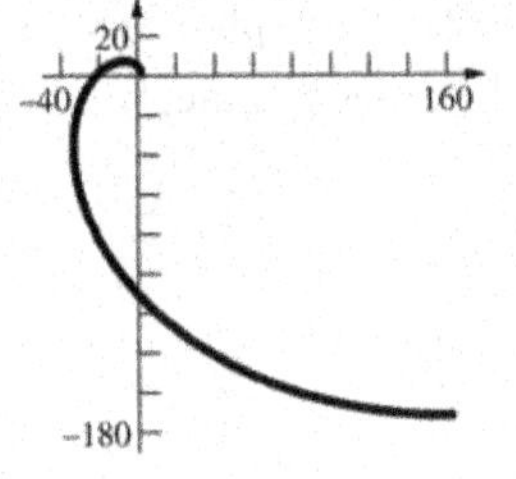

31. $r = \frac{2}{\theta}, \theta > 0$ (reciprocal spiral)

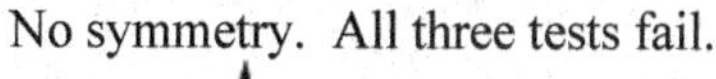

No symmetry. All three tests fail.

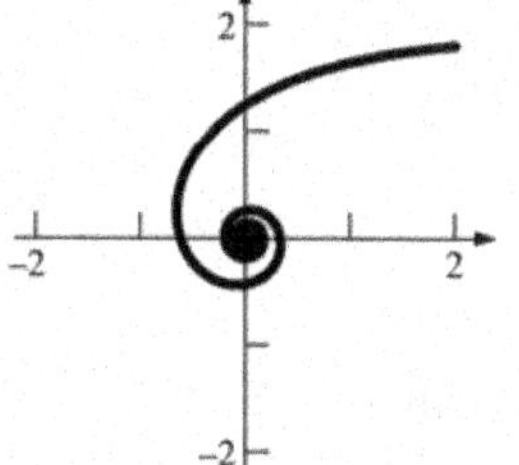

33. $r = 6,\ r = 4 + 4\cos\theta$

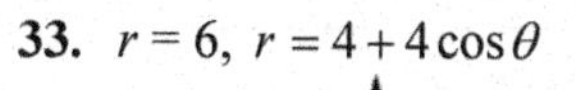

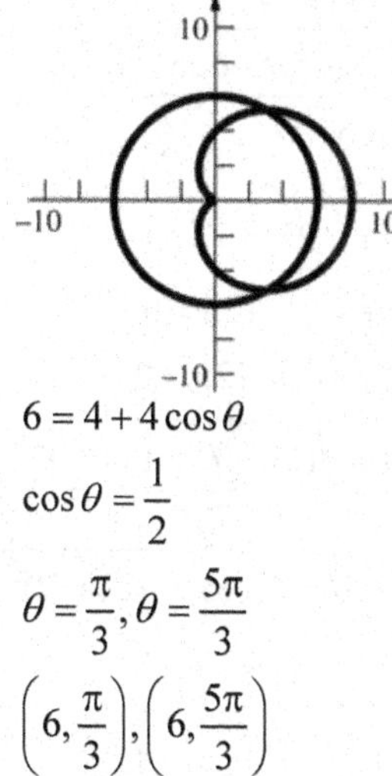

$6 = 4 + 4\cos\theta$

$\cos\theta = \frac{1}{2}$

$\theta = \frac{\pi}{3}, \theta = \frac{5\pi}{3}$

$\left(6, \frac{\pi}{3}\right), \left(6, \frac{5\pi}{3}\right)$

35. $r = 3\sqrt{3}\cos\theta, r = 3\sin\theta$

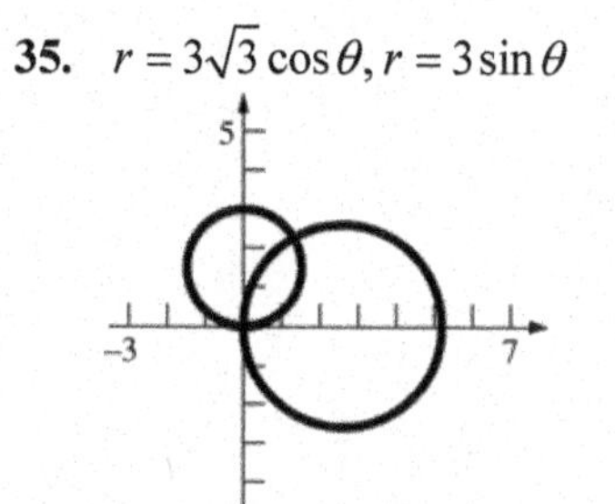

$3\sqrt{3}\cos\theta = 3\sin\theta$

$\tan\theta = \sqrt{3}$

$\theta = \frac{\pi}{3}, \theta = \frac{4\pi}{3}$

$\left(\frac{3\sqrt{3}}{2}, \frac{\pi}{3}\right) = \left(-\frac{3\sqrt{3}}{2}, \frac{4\pi}{3}\right)$

(0, 0) is also a solution since both graphs include the pole.

37. $r = 6\sin\theta, r = \frac{6}{1+2\sin\theta}$

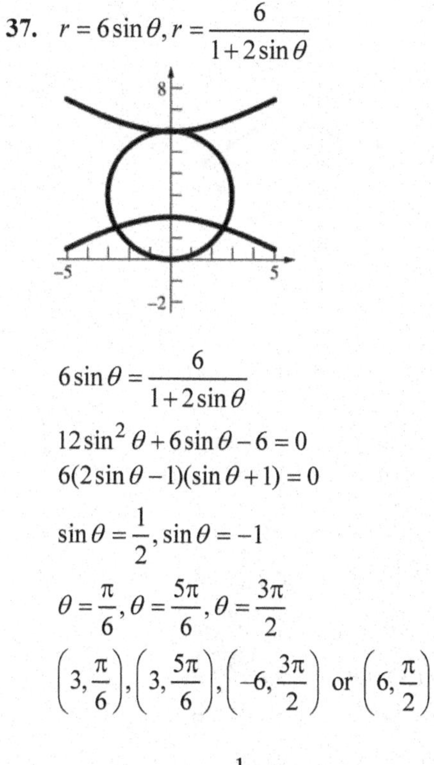

$6\sin\theta = \frac{6}{1+2\sin\theta}$

$12\sin^2\theta + 6\sin\theta - 6 = 0$

$6(2\sin\theta - 1)(\sin\theta + 1) = 0$

$\sin\theta = \frac{1}{2}, \sin\theta = -1$

$\theta = \frac{\pi}{6}, \theta = \frac{5\pi}{6}, \theta = \frac{3\pi}{2}$

$\left(3, \frac{\pi}{6}\right), \left(3, \frac{5\pi}{6}\right), \left(-6, \frac{3\pi}{2}\right)$ or $\left(6, \frac{\pi}{2}\right)$

39. Consider $r = \cos\frac{1}{2}\theta$.

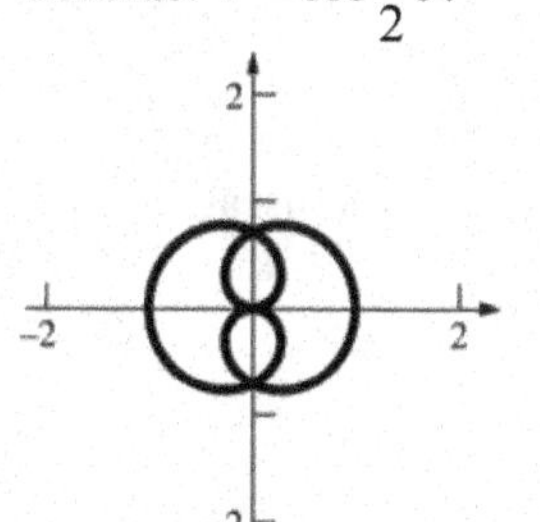

The graph is clearly symmetric with respect to the y-axis.

Substitute (r, θ) by $(-r, -\theta)$.

$-r = \cos\left(-\frac{1}{2}\theta\right) = \cos\frac{1}{2}\theta$

$r = -\cos\frac{1}{2}\theta$

Substitute (r, θ) by $(r, \pi - \theta)$

$r = \cos\frac{1}{2}(\pi - \theta) = \cos\frac{1}{2}\pi\cos\frac{1}{2}\theta + \sin\frac{1}{2}\pi\sin\frac{1}{2}\theta$

$= \sin\frac{1}{2}\theta$

$r = \sin\frac{1}{2}\theta$

41.

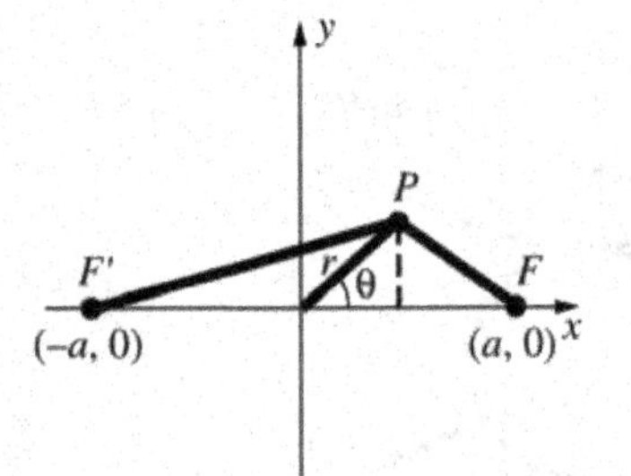

$$|PF| = \sqrt{(a - r\cos\theta)^2 + (r\sin\theta)^2}$$
$$= \sqrt{r^2 + a^2 - 2ar\cos\theta}$$
$$|PF'| = \sqrt{(a + r\cos\theta)^2 + (r\sin\theta)^2}$$
$$= \sqrt{r^2 + a^2 + 2ar\cos\theta}$$
$$|PF||PF'| = \sqrt{(r^2 + a^2)^2 - 4a^2r^2\cos^2\theta} = a^2$$
$$(r^4 + 2a^2r^2 + a^4) - 4a^2r^2\cos^2\theta = a^4$$
$$r^4 - 4a^2r^2\cos^2\theta + 2a^2r^2 = 0$$
$$r^2(r^2 - 2a^2(2\cos^2\theta - 1)) = 0$$
$$r^2 - 2a^2(2\cos^2\theta - 1) = 0$$
$$r^2 = 2a^2(1 + \cos 2\theta - 1)$$
$$r^2 = 2a^2\cos 2\theta$$

This is the equation of a lemniscate.

43. a. $y = 45$
$$r\sin\theta = 45$$
$$r = \frac{45}{\sin\theta}$$

b. $x^2 + y^2 = 36$
$$r^2 = 36$$
$$r = 6$$

c. $x^2 - y^2 = 1$
$$r^2\cos^2\theta - r^2\sin^2\theta = 1$$
$$r^2 = \frac{1}{\cos 2\theta}$$
$$r = \pm\frac{1}{\sqrt{\cos 2\theta}}$$

d. $4xy = 1$
$$4r^2\cos\theta\sin\theta = 1$$
$$r^2 = \frac{1}{2\sin 2\theta}$$
$$r = \pm\frac{1}{\sqrt{2\sin 2\theta}}$$

e. $y = 3x + 2$
$$r\sin\theta = 3r\cos\theta + 2$$
$$r(\sin\theta - 3\cos\theta) = 2$$
$$r = \frac{2}{\sin\theta - 3\cos\theta}$$

f. $3x^2 + 4y = 2$
$$3r^2\cos^2\theta + 4r\sin\theta = 2$$
$$(3\cos^2\theta)r^2 + (4\sin\theta)r - 2 = 0$$
$$r = \frac{-4\sin\theta \pm \sqrt{16\sin^2\theta + 24\cos^2\theta}}{6\cos^2\theta}$$
$$r = \frac{-2\sin\theta \pm \sqrt{4\sin^2\theta + 6\cos^2\theta}}{3\cos^2\theta}$$

g. $x^2 + 2x + y^2 - 4y - 25 = 0$
$$r^2 + 2r\cos\theta - 4r\sin\theta - 25 = 0$$
$$r^2 + (2\cos\theta - 4\sin\theta)r - 25 = 0$$
$$r = \frac{-2\cos\theta + 4\sin\theta \pm \sqrt{(2\cos\theta - 4\sin\theta)^2 + 100}}{2}$$
$$r = -\cos\theta + 2\sin\theta \pm \sqrt{(\cos\theta - 2\sin\theta)^2 + 25}$$

45. a. VII

b. I

c. VIII

d. III

e. V

f. II

g. VI

h. IV

47. $r = \cos\left(\frac{13\theta}{5}\right)$

49. $r = 1 + 3\cos\left(\frac{\theta}{3}\right)$

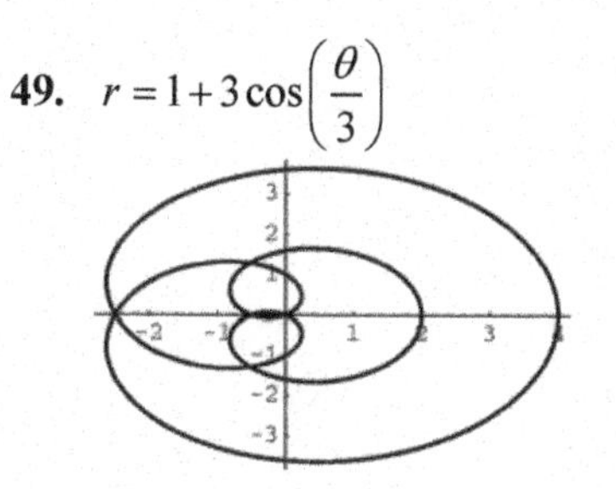

51. a. The graph for $\phi = 0$ is the graph for $\phi \neq 0$ rotated by ϕ counterclockwise about the pole.

 b. As n increases, the number of "leaves" increases.

 c. If $|a| > |b|$, the graph will not pass through the pole and will not "loop." If $|b| < |a|$, the graph will pass through the pole and will have $2n$ "loops" (n small "loops" and n large "loops"). If $|a| = |b|$, the graph passes through the pole and will have n "loops." If $ab \neq 0, n > 1$, and $\phi = 0$, the graph will be symmetric about $\theta = \frac{\pi}{n}k$, where $k = 0, n-1$.

53. The spiral will unwind clockwise for $c < 0$. The spiral will unwind counter-clockwise for $c > 0$.

55. a. III

 b. IV

 c. I

 d. II

 e. VI

 f. V

10.7 Concepts Review

1. $\frac{1}{2}r^2\theta$

3. $\frac{1}{2}\int_0^{2\pi}(2+2\cos\theta)^2\,d\theta$

Problem Set 10.7

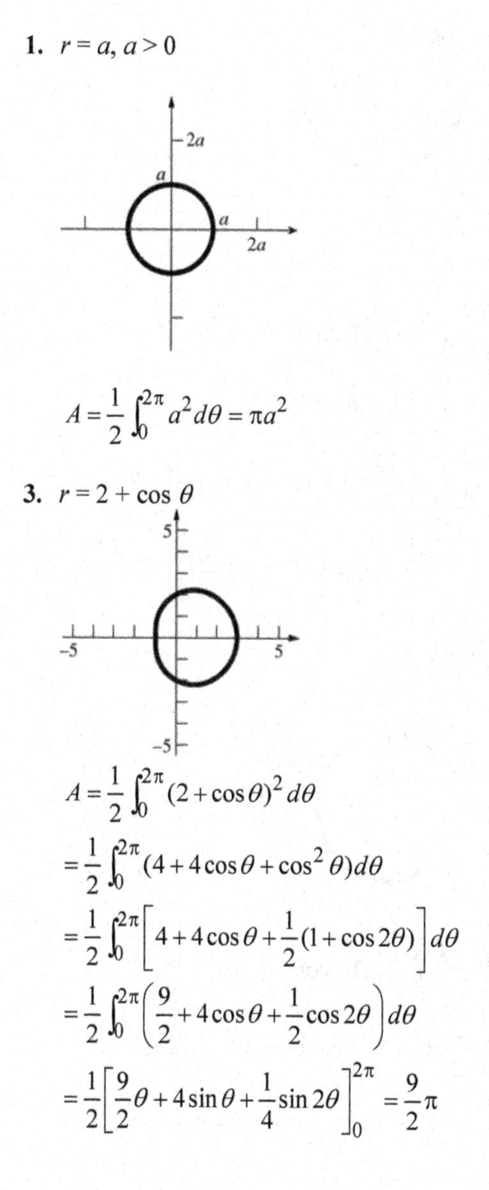

1. $r = a, a > 0$

$$A = \frac{1}{2}\int_0^{2\pi} a^2\,d\theta = \pi a^2$$

3. $r = 2 + \cos\theta$

$$A = \frac{1}{2}\int_0^{2\pi}(2+\cos\theta)^2\,d\theta$$

$$= \frac{1}{2}\int_0^{2\pi}(4 + 4\cos\theta + \cos^2\theta)\,d\theta$$

$$= \frac{1}{2}\int_0^{2\pi}\left[4 + 4\cos\theta + \frac{1}{2}(1+\cos 2\theta)\right]d\theta$$

$$= \frac{1}{2}\int_0^{2\pi}\left(\frac{9}{2} + 4\cos\theta + \frac{1}{2}\cos 2\theta\right)d\theta$$

$$= \frac{1}{2}\left[\frac{9}{2}\theta + 4\sin\theta + \frac{1}{4}\sin 2\theta\right]_0^{2\pi} = \frac{9}{2}\pi$$

5. $r = 3 - 3\sin\theta$

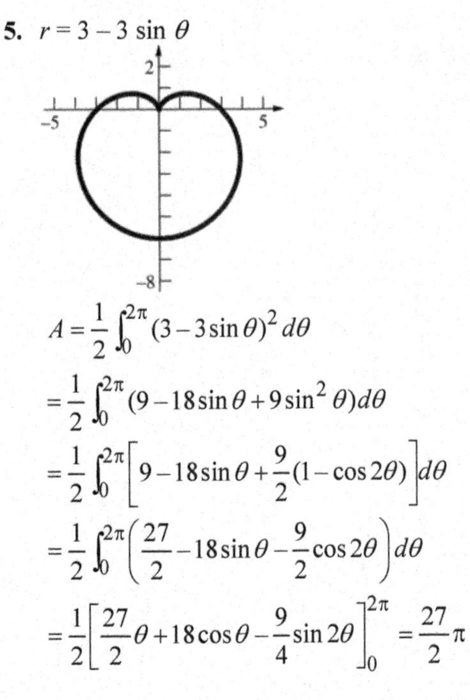

$$A = \frac{1}{2}\int_0^{2\pi}(3 - 3\sin\theta)^2\,d\theta$$

$$= \frac{1}{2}\int_0^{2\pi}(9 - 18\sin\theta + 9\sin^2\theta)d\theta$$

$$= \frac{1}{2}\int_0^{2\pi}\left[9 - 18\sin\theta + \frac{9}{2}(1 - \cos 2\theta)\right]d\theta$$

$$= \frac{1}{2}\int_0^{2\pi}\left(\frac{27}{2} - 18\sin\theta - \frac{9}{2}\cos 2\theta\right)d\theta$$

$$= \frac{1}{2}\left[\frac{27}{2}\theta + 18\cos\theta - \frac{9}{4}\sin 2\theta\right]_0^{2\pi} = \frac{27}{2}\pi$$

7. $r = a(1 + \cos\theta)$

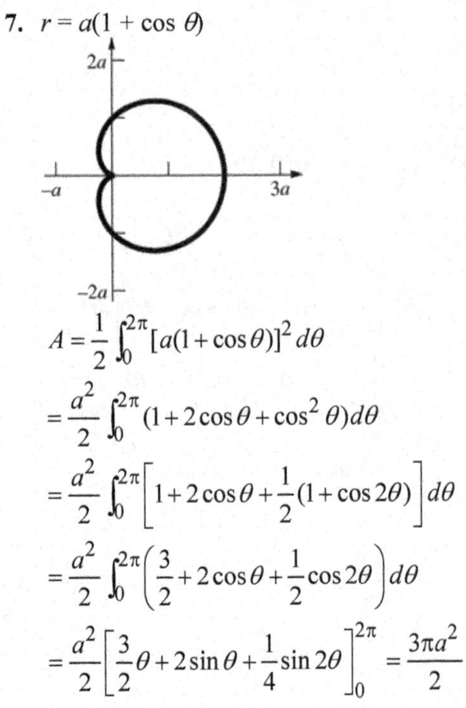

$$A = \frac{1}{2}\int_0^{2\pi}[a(1 + \cos\theta)]^2\,d\theta$$

$$= \frac{a^2}{2}\int_0^{2\pi}(1 + 2\cos\theta + \cos^2\theta)d\theta$$

$$= \frac{a^2}{2}\int_0^{2\pi}\left[1 + 2\cos\theta + \frac{1}{2}(1 + \cos 2\theta)\right]d\theta$$

$$= \frac{a^2}{2}\int_0^{2\pi}\left(\frac{3}{2} + 2\cos\theta + \frac{1}{2}\cos 2\theta\right)d\theta$$

$$= \frac{a^2}{2}\left[\frac{3}{2}\theta + 2\sin\theta + \frac{1}{4}\sin 2\theta\right]_0^{2\pi} = \frac{3\pi a^2}{2}$$

9. $r^2 = 9\sin 2\theta$

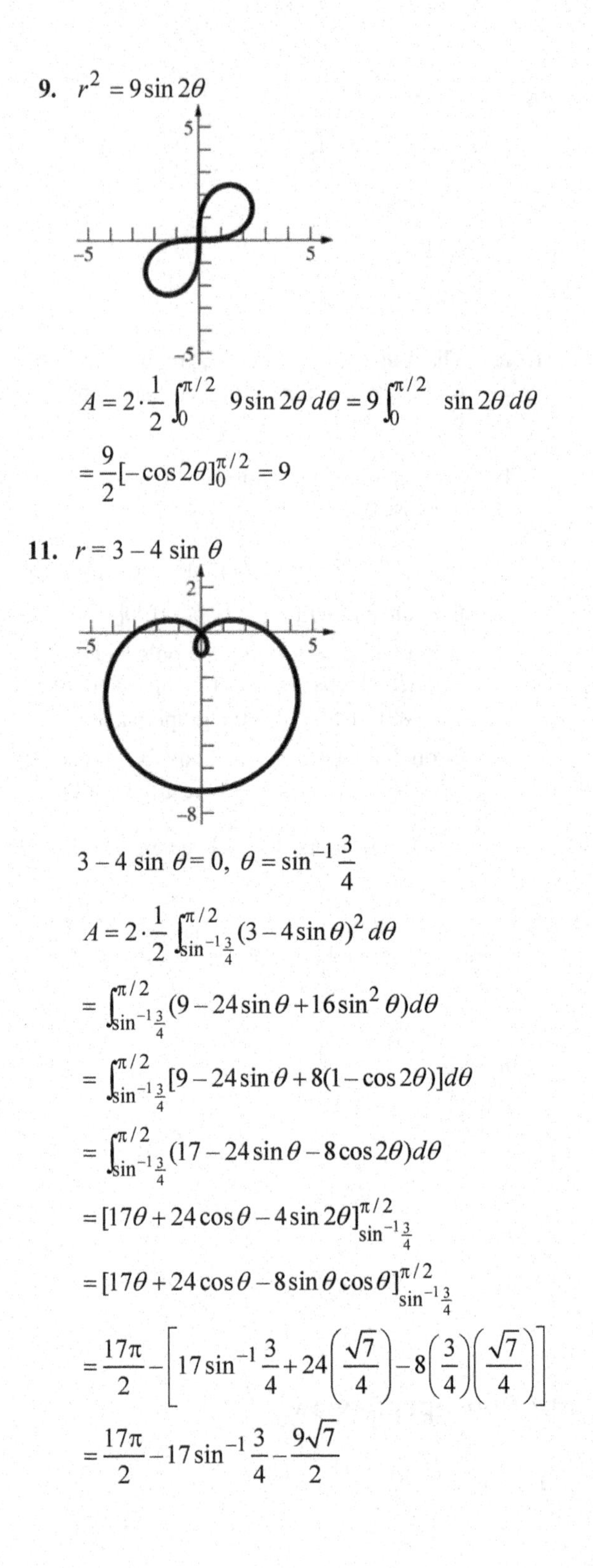

$$A = 2\cdot\frac{1}{2}\int_0^{\pi/2} 9\sin 2\theta\,d\theta = 9\int_0^{\pi/2}\sin 2\theta\,d\theta$$

$$= \frac{9}{2}[-\cos 2\theta]_0^{\pi/2} = 9$$

11. $r = 3 - 4\sin\theta$

$3 - 4\sin\theta = 0,\ \theta = \sin^{-1}\frac{3}{4}$

$$A = 2\cdot\frac{1}{2}\int_{\sin^{-1}\frac{3}{4}}^{\pi/2}(3 - 4\sin\theta)^2\,d\theta$$

$$= \int_{\sin^{-1}\frac{3}{4}}^{\pi/2}(9 - 24\sin\theta + 16\sin^2\theta)d\theta$$

$$= \int_{\sin^{-1}\frac{3}{4}}^{\pi/2}[9 - 24\sin\theta + 8(1 - \cos 2\theta)]d\theta$$

$$= \int_{\sin^{-1}\frac{3}{4}}^{\pi/2}(17 - 24\sin\theta - 8\cos 2\theta)d\theta$$

$$= [17\theta + 24\cos\theta - 4\sin 2\theta]_{\sin^{-1}\frac{3}{4}}^{\pi/2}$$

$$= [17\theta + 24\cos\theta - 8\sin\theta\cos\theta]_{\sin^{-1}\frac{3}{4}}^{\pi/2}$$

$$= \frac{17\pi}{2} - \left[17\sin^{-1}\frac{3}{4} + 24\left(\frac{\sqrt{7}}{4}\right) - 8\left(\frac{3}{4}\right)\left(\frac{\sqrt{7}}{4}\right)\right]$$

$$= \frac{17\pi}{2} - 17\sin^{-1}\frac{3}{4} - \frac{9\sqrt{7}}{2}$$

13. $r = 2 - 3\cos\theta$

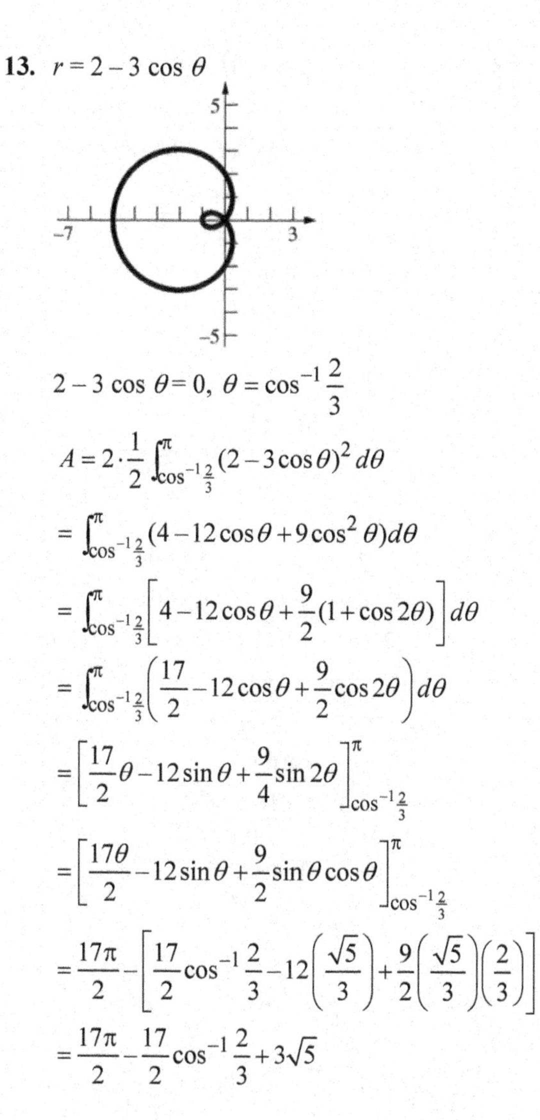

$2 - 3\cos\theta = 0,\ \theta = \cos^{-1}\frac{2}{3}$

$$A = 2 \cdot \frac{1}{2}\int_{\cos^{-1}\frac{2}{3}}^{\pi}(2-3\cos\theta)^2\, d\theta$$

$$= \int_{\cos^{-1}\frac{2}{3}}^{\pi}(4 - 12\cos\theta + 9\cos^2\theta)d\theta$$

$$= \int_{\cos^{-1}\frac{2}{3}}^{\pi}\left[4 - 12\cos\theta + \frac{9}{2}(1+\cos 2\theta)\right]d\theta$$

$$= \int_{\cos^{-1}\frac{2}{3}}^{\pi}\left(\frac{17}{2} - 12\cos\theta + \frac{9}{2}\cos 2\theta\right)d\theta$$

$$= \left[\frac{17}{2}\theta - 12\sin\theta + \frac{9}{4}\sin 2\theta\right]_{\cos^{-1}\frac{2}{3}}^{\pi}$$

$$= \left[\frac{17\theta}{2} - 12\sin\theta + \frac{9}{2}\sin\theta\cos\theta\right]_{\cos^{-1}\frac{2}{3}}^{\pi}$$

$$= \frac{17\pi}{2} - \left[\frac{17}{2}\cos^{-1}\frac{2}{3} - 12\left(\frac{\sqrt{5}}{3}\right) + \frac{9}{2}\left(\frac{\sqrt{5}}{3}\right)\left(\frac{2}{3}\right)\right]$$

$$= \frac{17\pi}{2} - \frac{17}{2}\cos^{-1}\frac{2}{3} + 3\sqrt{5}$$

15. $r = 4\cos 3\theta$

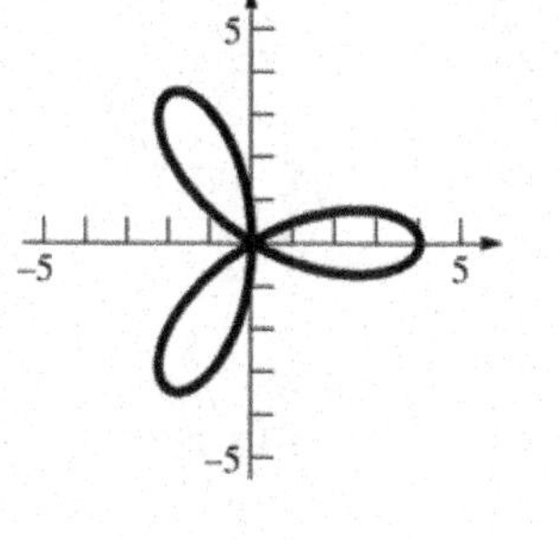

$$A = 6 \cdot \frac{1}{2}\int_0^{\pi/6}(4\cos 3\theta)^2\, d\theta = 48\int_0^{\pi/6}\cos^2 3\theta\, d\theta$$

$$= 24\int_0^{\pi/6}(1+\cos 6\theta)d\theta = 24\left[\theta + \frac{1}{6}\sin 6\theta\right]_0^{\pi/6} = 4\pi$$

17. $A = \frac{1}{2}\int_0^{2\pi}100\, d\theta - \frac{1}{2}\int_0^{2\pi}49\, d\theta = 51\pi$

19. $r = 2, r^2 = 8\cos 2\theta$

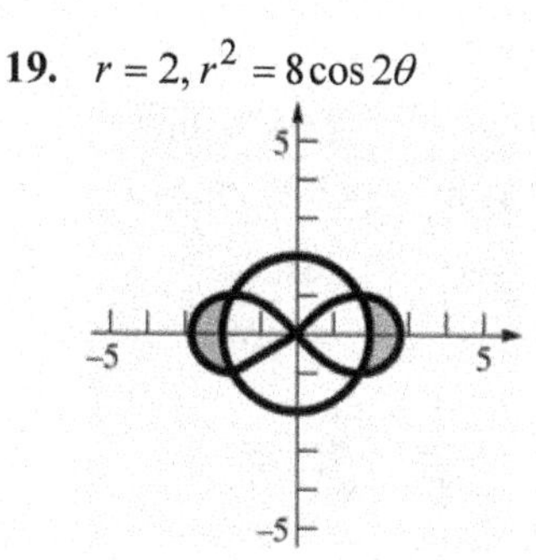

Solve for the θ-coordinate of the first intersection point.

$4 = 8\cos 2\theta$

$\cos 2\theta = \frac{1}{2}$

$2\theta = \frac{\pi}{3}$

$\theta = \frac{\pi}{6}$

$$A = 4 \cdot \frac{1}{2}\int_0^{\pi/6}(8\cos 2\theta - 4)d\theta$$

$$= 2[4\sin 2\theta - 4\theta]_0^{\pi/6}$$

$$= 4\sqrt{3} - \frac{4\pi}{3}$$

21. $r = 3 + 3\cos\theta, r = 3 + 3\sin\theta$

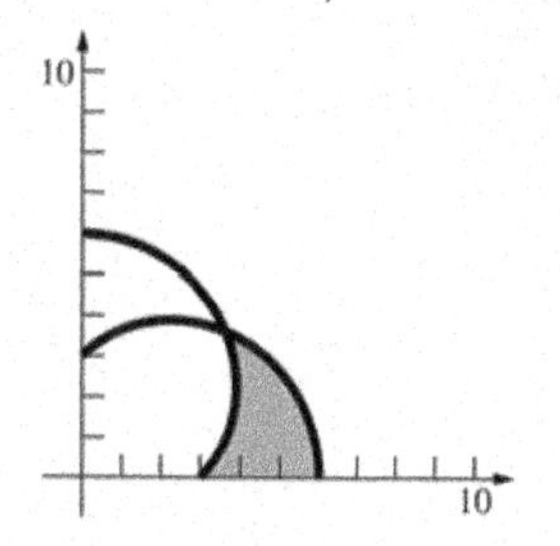

Solve for the θ-coordinate of the intersection point.

$3 + 3\cos\theta = 3 + 3\sin\theta$

$\tan\theta = 1$

$\theta = \frac{\pi}{4}$

$$A = \frac{1}{2}\int_0^{\pi/4}[(3+3\cos\theta)^2 - (3+3\sin\theta)^2]d\theta$$

$$= \frac{1}{2}\int_0^{\pi/4}(18\cos\theta + 9\cos^2\theta - 18\sin\theta - 9\sin^2\theta)d\theta$$

$$= \frac{1}{2}\int_0^{\pi/4}(18\cos\theta - 18\sin\theta + 9\cos 2\theta)d\theta$$

$$= \frac{1}{2}\left[18\sin\theta + 18\cos\theta + \frac{9}{2}\sin 2\theta\right]_0^{\pi/4}$$

$$= 9\sqrt{2} - \frac{27}{4}$$

23. a. $f(\theta) = 2\cos\theta, f'(\theta) = -2\sin\theta$

$$m = \frac{(2\cos\theta)\cos\theta + (-2\sin\theta)\sin\theta}{-(2\cos\theta)\sin\theta + (-2\sin\theta)\cos\theta}$$

$$= \frac{2\cos^2\theta - 2\sin^2\theta}{-4\cos\theta\sin\theta} = \frac{\cos 2\theta}{-\sin 2\theta}$$

At $\theta = \frac{\pi}{3}, m = \frac{-\frac{1}{2}}{-\frac{\sqrt{3}}{2}} = \frac{1}{\sqrt{3}}$.

b. $f(\theta) = 1 + \sin\theta, f'(\theta) = \cos\theta$

$$m = \frac{(1+\sin\theta)\cos\theta + (\cos\theta)\sin\theta}{-(1+\sin\theta)\sin\theta + (\cos\theta)\cos\theta}$$

$$= \frac{\cos\theta + 2\sin\theta\cos\theta}{\cos^2\theta - \sin^2\theta - \sin\theta} = \frac{\cos\theta + \sin 2\theta}{\cos 2\theta - \sin\theta}$$

At $\theta = \frac{\pi}{3}, m = \frac{\frac{1}{2} + \frac{\sqrt{3}}{2}}{-\frac{1}{2} - \frac{\sqrt{3}}{2}} = -1$.

c. $f(\theta) = \sin 2\theta, f'(\theta) = 2\cos 2\theta$

$$m = \frac{(\sin 2\theta)\cos\theta + (2\cos 2\theta)\sin\theta}{-(\sin 2\theta)\sin\theta + (2\cos 2\theta)\cos\theta}$$

At $\theta = \frac{\pi}{3}$, .

$$m = \frac{\left(\frac{\sqrt{3}}{2}\right)\left(\frac{1}{2}\right) + (-1)\left(\frac{\sqrt{3}}{2}\right)}{-\left(\frac{\sqrt{3}}{2}\right)\left(\frac{\sqrt{3}}{2}\right) + (-1)\left(\frac{1}{2}\right)} = \frac{-\frac{\sqrt{3}}{4}}{-\frac{5}{4}} = \frac{\sqrt{3}}{5}$$

d. $f(\theta) = 4 - 3\cos\theta, f'(\theta) = 3\sin\theta$

$$m = \frac{(4 - 3\cos\theta)\cos\theta + (3\sin\theta)\sin\theta}{-(4 - 3\cos\theta)\sin\theta + (3\sin\theta)\cos\theta}$$

$$= \frac{4\cos\theta - 3\cos^2\theta + 3\sin^2\theta}{-4\sin\theta + 6\sin\theta\cos\theta}$$

$$= \frac{4\cos\theta - 3\cos 2\theta}{-4\sin\theta + 3\sin 2\theta}$$

At $\theta = \frac{\pi}{3}$,

$$m = \frac{4\left(\frac{1}{2}\right) - 3\left(-\frac{1}{2}\right)}{-4\left(\frac{\sqrt{3}}{2}\right) + 3\left(\frac{\sqrt{3}}{2}\right)} = \frac{\frac{7}{2}}{-\frac{\sqrt{3}}{2}} = -\frac{7}{\sqrt{3}}.$$

25. $f(\theta) = 1 - 2\sin\theta, f'(\theta) = -2\cos\theta$

$$m = \frac{(1 - 2\sin\theta)\cos\theta + (-2\cos\theta)\sin\theta}{-(1 - 2\sin\theta)\sin\theta + (-2\cos\theta)\cos\theta}$$

$$= \frac{\cos\theta - 4\sin\theta\cos\theta}{-\sin\theta + 2\sin^2\theta - 2\cos^2\theta}$$

$$= \frac{\cos\theta(1 - 4\sin\theta)}{-\sin\theta + 2\sin^2\theta - 2\cos^2\theta}$$

$m = 0$ when $\cos\theta(1 - 4\sin\theta) = 0$

$\cos\theta = 0$, or $1 - 4\sin\theta = 0$

$\theta = \frac{\pi}{2}, \theta = \frac{3\pi}{2}, \theta = \sin^{-1}\left(\frac{1}{4}\right) \approx 0.25,$

$\theta = \pi - \sin^{-1}\left(\frac{1}{4}\right) \approx 2.89$

$f\left(\frac{\pi}{2}\right) = -1, f\left(\frac{3\pi}{2}\right) = 3, f\left(\sin^{-1}\left(\frac{1}{4}\right)\right) = \frac{1}{2},$

$f\left(\pi - \sin^{-1}\left(\frac{1}{4}\right)\right) = \frac{1}{2}$

$\left(-1, \frac{\pi}{2}\right), \left(3, \frac{3\pi}{2}\right), \left(\frac{1}{2}, 0.25\right), \left(\frac{1}{2}, 2.89\right)$

27. $f(\theta) = a(1 + \cos\theta), f'(\theta) = -a\sin\theta$

$$L = \int_0^{2\pi} \sqrt{[a(1+\cos\theta)]^2 + [-a\sin\theta]^2}\, d\theta = a\int_0^{2\pi} \sqrt{2 + 2\cos\theta}\, d\theta = 2a\int_0^{2\pi} \sqrt{\frac{1+\cos\theta}{2}}\, d\theta = 2a\int_0^{2\pi} \left|\cos\frac{\theta}{2}\right| d\theta$$

$$= 2a\left[\int_0^{\pi} \cos\frac{\theta}{2}\, d\theta - \int_{\pi}^{2\pi} \cos\frac{\theta}{2}\, d\theta\right] = 2a\left(\left[2\sin\frac{\theta}{2}\right]_0^{\pi} - \left[2\sin\frac{\theta}{2}\right]_{\pi}^{2\pi}\right) = 8a$$

29. If n is even, there are $2n$ leaves.

$$A = 2n\frac{1}{2}\int_{-\pi/2n}^{\pi/2n} (a\cos n\theta)^2\, d\theta = na^2\int_{-\pi/2n}^{\pi/2n} \cos^2 n\theta\, d\theta = na^2\int_{-\pi/2n}^{\pi/2n} \frac{1 + \cos 2n\theta}{2}\, d\theta$$

$$= na^2\left[\frac{1}{2}\theta + \frac{\sin 2n\theta}{4n}\right]_{-\pi/2n}^{\pi/2n} = \frac{1}{2}a^2\pi$$

If n is odd, there are n leaves.

$$A = n\cdot\frac{1}{2}\int_{-\pi/2n}^{\pi/2n} (a\cos n\theta)^2\, d\theta = \frac{na^2}{2}\int_{-\pi/2n}^{\pi/2n} \cos^2 n\theta\, d\theta = \frac{na^2}{2}\left[\frac{1}{2}\theta + \frac{\sin 2n\theta}{4n}\right]_{-\pi/2n}^{\pi/2n} = \frac{1}{4}a^2\pi$$

31. a. Sketch the graph.

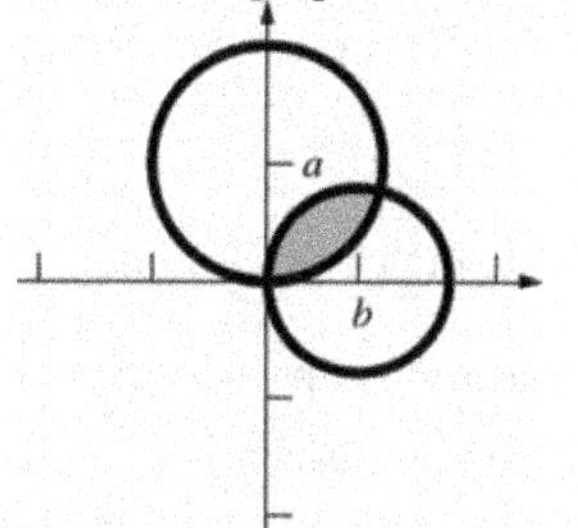

Solve for the θ-coordinate of the intersection.

$2a \sin\theta = 2b \cos\theta$

$\tan\theta = \frac{b}{a};\quad \theta = \tan^{-1}\left(\frac{b}{a}\right)$

Let $\theta_0 = \tan^{-1}\left(\frac{b}{a}\right)$.

$$A = \frac{1}{2}\int_0^{\theta_0}(2a\sin\theta)^2\,d\theta + \frac{1}{2}\int_{\theta_0}^{\pi/2}(2b\cos\theta)^2\,d\theta$$

$$= 2a^2\int_0^{\theta_0}\sin^2\theta\,d\theta + 2b^2\int_{\theta_0}^{\pi/2}\cos^2\theta\,d\theta$$

$$= a^2\int_0^{\theta_0}(1-\cos 2\theta)d\theta + b^2\int_{\theta_0}^{\pi/2}(1+\cos 2\theta)d\theta$$

$$= a^2\left[\theta - \frac{\sin 2\theta}{2}\right]_0^{\theta_0} + b^2\left[\theta + \frac{\sin 2\theta}{2}\right]_{\theta_0}^{\pi/2}$$

$$= a^2\theta_0 + b^2\left(\frac{\pi}{2} - \theta_0\right) - \frac{a^2+b^2}{2}\sin 2\theta_0$$

$$= a^2\theta_0 + b^2\left(\frac{\pi}{2} - \theta_0\right) - (a^2+b^2)\sin\theta_0\cos\theta_0$$

$$= a^2\tan^{-1}\left(\frac{b}{a}\right) + b^2\left(\frac{\pi}{2} - \tan^{-1}\left(\frac{b}{a}\right)\right) - ab.$$

Note that since

$\tan\theta = \frac{b}{a}$, $\cos\theta = \frac{a}{\sqrt{a^2+b^2}}$

and $\sin\theta = \frac{b}{\sqrt{a^2+b^2}}$.

b. Let m_1 be the slope of $r = 2a\sin\theta$.

$$m_1 = \frac{2a\sin\theta\cos\theta + 2a\cos\theta\sin\theta}{-2a\sin\theta\sin\theta + 2a\cos\theta\cos\theta}$$

$$= \frac{2\sin\theta\cos\theta}{\cos^2\theta - \sin^2\theta}$$

At $\theta = \tan^{-1}\left(\frac{b}{a}\right)$, $m_1 = \frac{2ab}{a^2-b^2}$.

At $\theta = 0$ (the pole), $m_1 = 0$.

Let m_2 be the slope of $r = 2b\cos\theta$.

$$m_2 = \frac{2b\cos\theta\cos\theta - 2b\sin\theta\sin\theta}{-2b\cos\theta\sin\theta - 2b\sin\theta\cos\theta}$$

$$= \frac{\cos^2\theta - \sin^2\theta}{-2\sin\theta\cos\theta}$$

At $\theta = \tan^{-1}\left(\frac{b}{a}\right)$, $m_2 = -\frac{a^2-b^2}{2ab}$.

At $\theta = \frac{\pi}{2}$ (the pole), m_2 is undefined.

Therefore the two circles intersect at right angles.

33. The edge of the pond is described by the equation $r = 2a\cos\theta$.
Solve for intersection points of the circles $r = ak$ and $r = 2a\cos\theta$.

$ak = 2a\cos\theta$

$\cos\theta = \frac{k}{2}, \theta = \cos^{-1}\left(\frac{k}{2}\right)$

Let A be the grazing area.

$$A = \frac{1}{2}\pi(ka)^2 + 2\cdot\frac{1}{2}\int_{\cos^{-1}\left(\frac{k}{2}\right)}^{\pi/2}[(ka)^2 - (2a\cos\theta)^2]d\theta = \frac{1}{2}k^2a^2\pi + a^2\int_{\cos^{-1}\left(\frac{k}{2}\right)}^{\pi/2}(k^2 - 4\cos^2\theta)d\theta$$

$$= \frac{1}{2}k^2a^2\pi + a^2\int_{\cos^{-1}\left(\frac{k}{2}\right)}^{\pi/2}((k^2-2) - 2\cos 2\theta)d\theta = \frac{1}{2}k^2a^2\pi + a^2\left[(k^2-2)\theta - \sin 2\theta\right]_{\cos^{-1}\left(\frac{k}{2}\right)}^{\pi/2}$$

$$= \frac{1}{2}k^2a^2\pi + a^2\left[k^2\theta - 2\theta - 2\sin\theta\cos\theta\right]_{\cos^{-1}\left(\frac{k}{2}\right)}^{\pi/2}$$

$$= \frac{1}{2}k^2a^2\pi + a^2\left[\frac{k^2\pi}{2} - \pi - k^2\cos^{-1}\left(\frac{k}{2}\right) + 2\cos^{-1}\left(\frac{k}{2}\right) + \frac{k\sqrt{4-k^2}}{2}\right]$$

$$= a^2\left[(k^2-1)\pi + (2-k^2)\cos^{-1}\left(\frac{k}{2}\right) + \frac{k\sqrt{4-k^2}}{2}\right]$$

35. The untethered goat has a grazing area of πa^2. From Problem 34, the tethered goat has a grazing area of $a^2\left(\frac{\pi k^2}{2}+\frac{k^3}{3}\right)$.

$$\pi a^2 = a^2\left(\frac{\pi k^2}{2}+\frac{k^3}{3}\right)$$

$$\pi = \frac{\pi k^2}{2}+\frac{k^3}{3}$$

$2k^3 + 3\pi k^2 - 6\pi = 0$

Using a numerical method or graphing calculator, $k \approx 1.26$. The length of the rope is approximately $1.26a$.

37. $A = 3\cdot\frac{1}{2}\int_0^{\pi/3}(4\sin 3\theta)^2\, d\theta = 24\int_0^{\pi/3}\sin^2 3\theta\, d\theta$

$= 12\int_0^{\pi/3}(1-\cos 6\theta)\, d\theta$

$= 12\left[\theta - \frac{\sin 6\theta}{6}\right]_0^{\pi/3} = 4\pi$

$f(\theta) = 4\sin 3\theta,\ f'(\theta) = 12\cos 3\theta$

$L = 3\int_0^{\pi/3}\sqrt{(4\sin 3\theta)^2 + (12\cos 3\theta)^2}\, d\theta$

$= 3\int_0^{\pi/3}\sqrt{16\sin^2 3\theta + 144\cos^2 3\theta}\, d\theta$

$= 12\int_0^{\pi/3}\sqrt{1+8\cos^2 3\theta}\, d\theta \approx 26.73$

39. $r = 4\sin\left(\frac{3\theta}{2}\right), 0 \le \theta \le 4\pi$

$f(\theta) = 4\sin\left(\frac{3\theta}{2}\right),\ f'(\theta) = 6\cos\left(\frac{3\theta}{2}\right)$

$$L = \int_0^{4\pi}\sqrt{\left[4\sin\left(\frac{3\theta}{2}\right)\right]^2 + \left[6\cos\left(\frac{3\theta}{2}\right)\right]^2}\, d\theta$$

$$= \int_0^{4\pi}\sqrt{16\sin^2\left(\frac{3\theta}{2}\right) + 36\cos^2\left(\frac{3\theta}{2}\right)}\, d\theta$$

$$= \int_0^{4\pi}\sqrt{16 + 20\cos^2\left(\frac{3\theta}{2}\right)}\, d\theta \approx 63.46$$

10.8 Chapter Review

Concepts Test

1. False: If $a = 0$, the graph is a line.

3. False: The defining condition of an ellipse is $|PF| = e|PL|$ where $0 < e < 1$. Hence the distance from the vertex to a directrix is a greater than the distance to a focus.

5. True: The asymptotes for both hyperbolas are $y = \pm\frac{b}{a}x$.

7. True: As e approaches 0, the ellipse becomes more circular.

9. False: The equation $x^2 - y^2 = 0$ represents the two lines $y = \pm x$.

11. True: If $k > 0$, the equation is a horizontal hyperbola; if $k < 0$, the equation is a vertical hyperbola.

13. False: If $b > a$, the distance is $2\sqrt{b^2 - a^2}$.

15. True: Since light from one focus reflects to the other focus, light emanating from a point between a focus and the nearest vertex will reflect beyond the other focus.

17. True: The equation is equivalent to

$$\left(x+\frac{C}{2}\right)^2 + \left(y+\frac{D}{2}\right)^2 = -F + \frac{C^2}{4} + \frac{D^2}{4}.$$

Thus, the graph is a circle if

$-F + \frac{C^2}{4} + \frac{D^2}{4} > 0$, a point if

$-F + \frac{C^2}{4} + \frac{D^2}{4} = 0$, or the empty set if

$-F^2 + \frac{C^2}{4} + \frac{D^2}{4} < 0$.

19. False: The limiting forms of two parallel lines and the empty set cannot be formed in such a manner.

21. False: For example, $xy = 1$ is a hyperbola with coordinates only in the first and third quadrants.

23. False: For example, $x = 0$, $y = t$, and $x = 0$, $y = -t$ both represent the line $x = 0$.

25. False: For example, the graph of $x = t^2, y = t$ does not represent y as a function of x. $y = \pm\sqrt{x}$, but $h(x) = \pm\sqrt{x}$ is not a function.

27. False: For example, if $x = t^3, y = t^3$ then $y = x$ so $\frac{d^2y}{dx^2} = 0$, but $\frac{g''(t)}{f''(t)} = 1$.

29. True: The graph of $r = 4\cos\theta$ is a circle of radius 2 centered at (2, 0). The graph $r = 4\cos\left(\theta - \frac{\pi}{3}\right)$ is the graph of $r = 4\cos\theta$ rotated $\frac{\pi}{3}$ counter-clockwise about the pole.

31. False: For example, if $f(\theta) = \cos\theta$ and $g(\theta) = \sin\theta$, solving the two equations simultaneously does not give the pole (which is $\left(0, \frac{\pi}{2}\right)$ for $f(\theta)$ and (0, 0) for $g(\theta)$).

33. True: Since f is even $f(-\theta) = f(\theta)$. Thus, if we replace (r, θ) by $(r, -\theta)$, the equation $r = f(-\theta)$ is $r = f(\theta)$. Therefore, the graph is symmetric about the x-axis.

Sample Test Problems

1. a. $x^2 - 4y^2 = 0; y = \pm\frac{x}{2}$

(5) Two intersecting lines

b. $x^2 - 4y^2 = 0.01; \frac{x^2}{0.01} - \frac{y^2}{0.0025} = 1$

(9) A hyperbola

c. $x^2 - 4 = 0; x = \pm 2$

(4) Two parallel lines

d. $x^2 - 4x + 4 = 0; x = 2$

(3) A single line

e. $x^2 + 4y^2 = 0; (0, 0)$

(2) A single point

f. $x^2 + 4y^2 = x; x^2 - x + \frac{1}{4} + 4y^2 = \frac{1}{4}$;

$$\frac{\left(x - \frac{1}{2}\right)^2}{\frac{1}{4}} + \frac{y^2}{\frac{1}{16}} = 1$$

(8) An ellipse

g. $x^2 + 4y^2 = -x; x^2 + x + \frac{1}{4} + 4y^2 = \frac{1}{4}$;

$$\frac{\left(x + \frac{1}{2}\right)^2}{\frac{1}{4}} + \frac{y^2}{\frac{1}{16}} = 1$$

(8) An ellipse

h. $x^2 + 4y^2 = -1$

(1) No graph

i. $(x^2 + 4y - 1)^2 = 0; x^2 + 4y - 1 = 0$

(7) A parabola

j. $3x^2 + 4y^2 = -x^2 + 1; x^2 + y^2 = \frac{1}{4}$

(6) A circle

3. $9x^2 + 4y^2 - 36 = 0; \frac{x^2}{4} + \frac{y^2}{9} = 1$

Vertical ellipse; $a = 3, b = 2, c = \sqrt{5}$

Foci are at $\left(0, \pm\sqrt{5}\right)$ and vertices are at $(0, \pm 3)$.

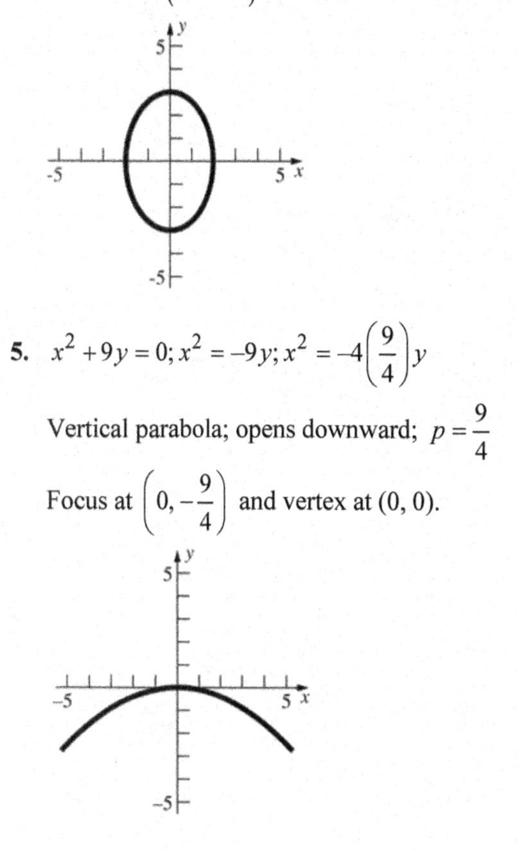

5. $x^2 + 9y = 0; x^2 = -9y; x^2 = -4\left(\frac{9}{4}\right)y$

Vertical parabola; opens downward; $p = \frac{9}{4}$

Focus at $\left(0, -\frac{9}{4}\right)$ and vertex at (0, 0).

7. $9x^2 + 25y^2 - 225 = 0; \frac{x^2}{25} + \frac{y^2}{9} = 1$

Horizontal ellipse, $a = 5$, $b = 3$, $c = 4$

Foci are at $(\pm 4, 0)$ and vertices are at $(\pm 5, 0)$.

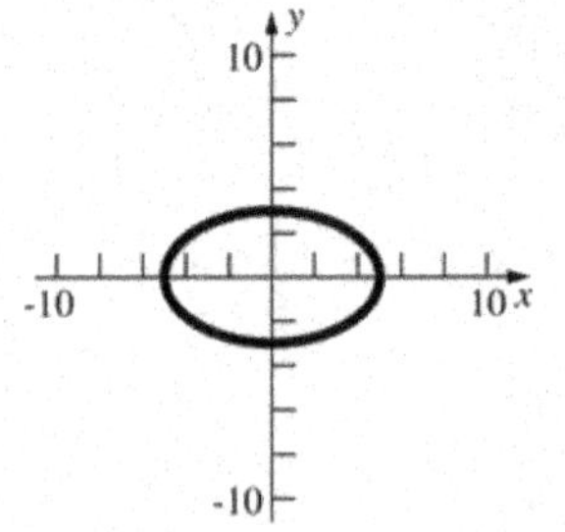

9. $r = \frac{5}{2 + 2\sin\theta} = \frac{\left(\frac{5}{2}\right)(1)}{1 + (1)\cos\left(\theta - \frac{\pi}{2}\right)}$

$e = 1$; parabola

Focus is at $(0, 0)$ and vertex is at $\left(0, \frac{5}{4}\right)$ (in Cartesian coordinates).

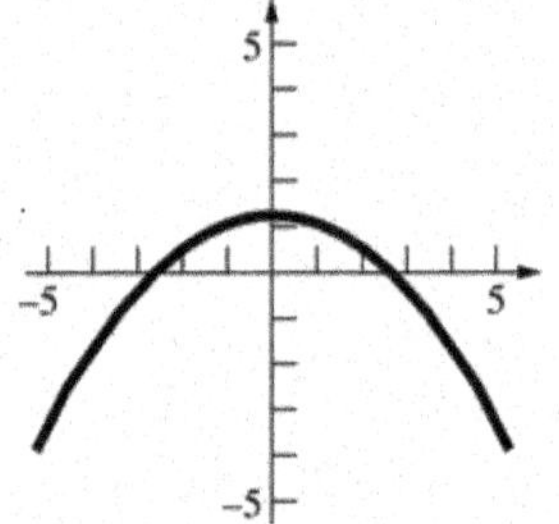

11. Horizontal ellipse; center at $(0, 0)$, $a = 4$,

$e = \frac{c}{a} = \frac{1}{2}, c = 2, b = \sqrt{16 - 4} = 2\sqrt{3}$

$\frac{x^2}{16} + \frac{y^2}{12} = 1$

13. Horizontal parabola;

$y^2 = ax, (3)^2 = a(-1), a = -9$

$y^2 = -9x$

15. Horizontal hyperbola, $a = 2$,

$x = \pm 2y, \frac{a}{b} = 2, b = 1$

$\frac{x^2}{4} - \frac{y^2}{1} = 1$

17. Horizontal ellipse; $2a = 10$, $a = 5$, $c = 4 - 1 = 3$,

$b = \sqrt{25 - 9} = 4$

$\frac{(x-1)^2}{25} + \frac{(y-2)^2}{16} = 1$

19. $4x^2 + 4y^2 - 24x + 36y + 81 = 0$

$4(x^2 - 6x + 9) + 4\left(y^2 + 9y + \frac{81}{4}\right) = -81 + 36 + 81$

$4(x-3)^2 + 4\left(y + \frac{9}{2}\right)^2 = 36$

$(x-3)^2 + \left(y + \frac{9}{2}\right)^2 = 9$; circle

21. $x^2 + 8x + 6y + 28 = 0$

$(x^2 + 8x + 16) = -6y - 28 + 16$

$(x+4)^2 = -6(y+2)$; parabola

23. $x = \frac{\sqrt{2}}{2}(u - v)$

$y = \frac{\sqrt{2}}{2}(u + v)$

$\frac{1}{2}(u-v)^2 + \frac{3}{2}(u-v)(u+v) + \frac{1}{2}(u+v)^2 = 10$

$\frac{5}{2}u^2 - \frac{1}{2}v^2 = 10$

$r = \frac{5}{2}, s = -\frac{1}{2}$

$\frac{u^2}{4} - \frac{v^2}{20} = 1$; hyperbola

$a = 2,\ b = 2\sqrt{5}, c = \sqrt{4 + 20} = 2\sqrt{6}$

The distance between foci is $4\sqrt{6}$.

25. $t = \frac{1}{6}(x-2)$

$y = \frac{1}{3}(x-2)$

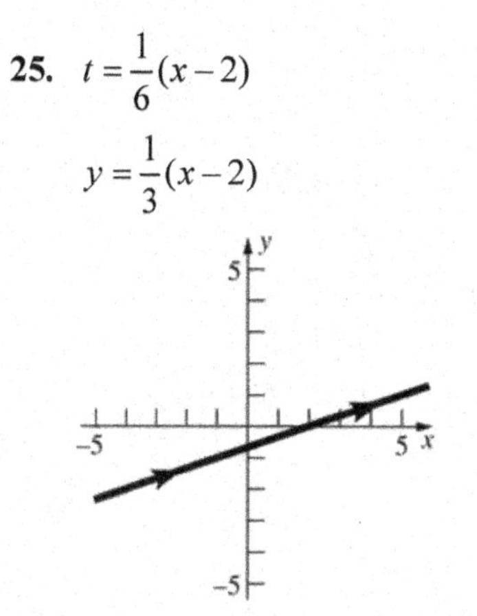

27. $\sin t = \frac{x+2}{4}, \cos t = \frac{y-1}{3}$

$$\frac{(x+2)^2}{16} + \frac{(y-1)^2}{9} = 1$$

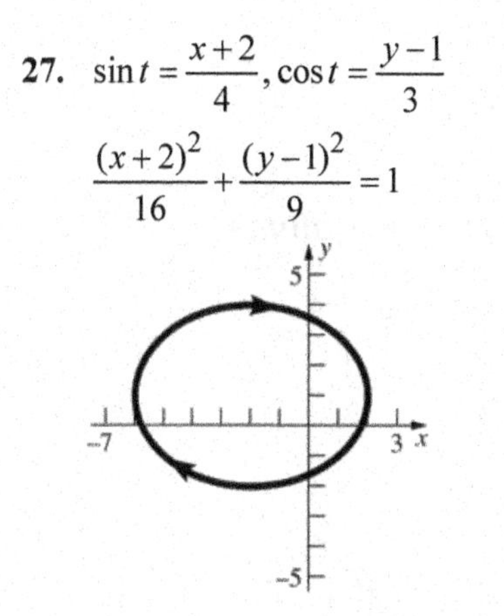

29. $\frac{dx}{dt} = 6t^2 - 4, \frac{dy}{dt} = 1 + \frac{1}{t+1} = \frac{t+2}{t+1}$

$$\frac{dy}{dx} = \frac{\frac{t+2}{t+1}}{6t^2-4} = \frac{t+2}{(t+1)(6t^2-4)}$$

At $t = 0, x = 7, y = 0$, and $\frac{dy}{dx} = -\frac{1}{2}$.

Tangent line: $y = -\frac{1}{2}(x-7)$ or $x + 2y - 7 = 0$

Normal line: $y = 2(x - 7)$ or $2x - y - 14 = 0$.

31. One approach is to use the arc length formula

$$L = \int_a^b \sqrt{\left(\frac{dx}{dt}\right)^2 + \left(\frac{dy}{dt}\right)^2}\, dt = \int_0^9 \sqrt{\frac{9t}{4} + \frac{9t}{4}}\, dt =$$

$$\frac{3\sqrt{2}}{2}\int_0^9 \sqrt{t}\, dt = \sqrt{2}\left[t^{3/2}\right]_0^9 = 27\sqrt{2}$$

Another way is to note that when $t = 0, (x, y) = (1, 2)$, when $t = 9$, $(x, y) = (28, 29)$, and $y = x + 1$, which is a straight line. Thus the curve length is simply the distance between the points $(1, 2)$ and $(28, 29)$ or $\sqrt{(28-1)^2 + (29-2)^2} = 27\sqrt{2}$

33 $r = 6\cos\theta$

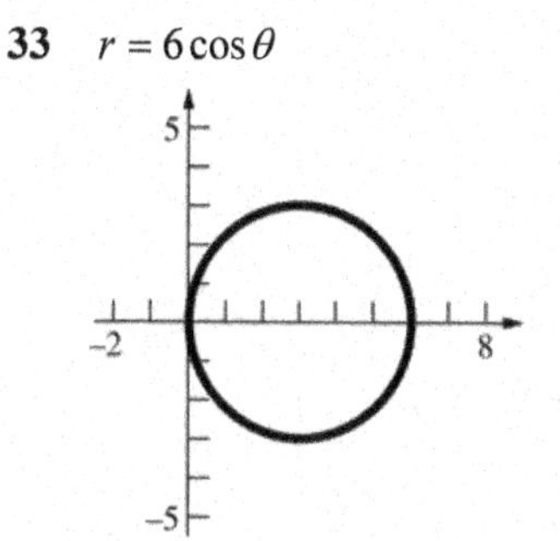

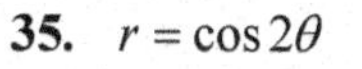

35. $r = \cos 2\theta$

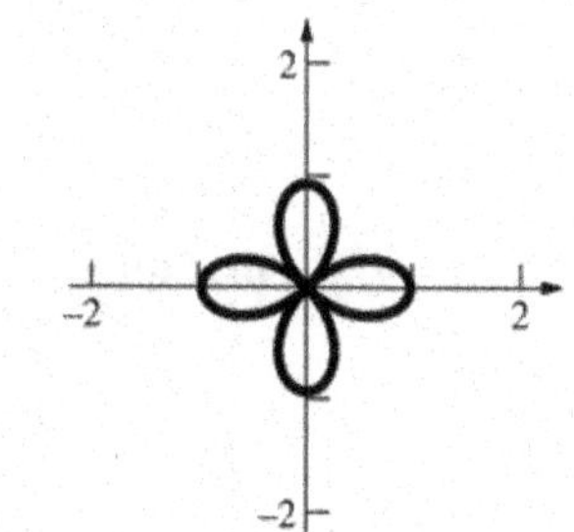

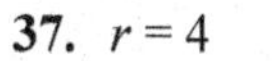

37. $r = 4$

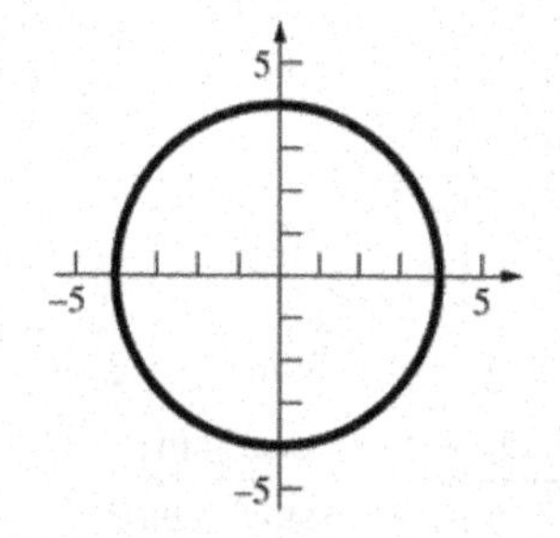

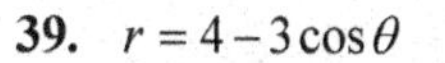

39. $r = 4 - 3\cos\theta$

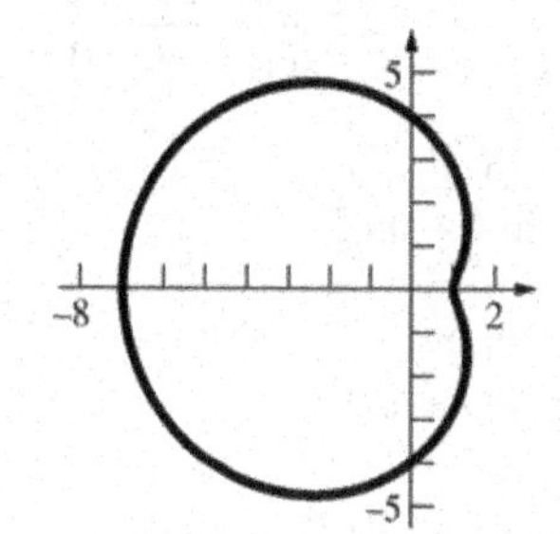

41. $\theta = \frac{2}{3}\pi$

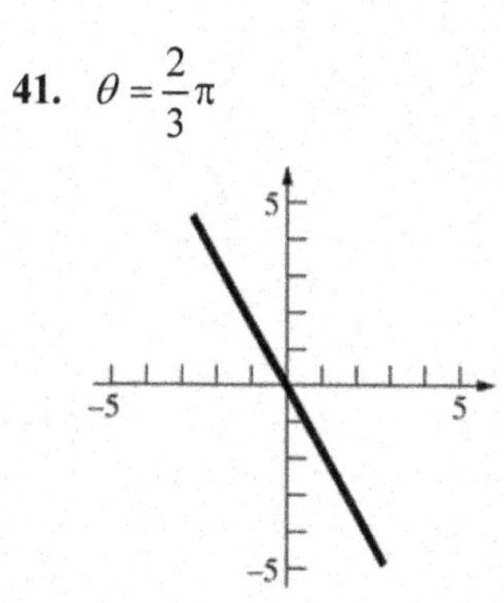

43. $r^2 = 16\sin 2\theta$

$r = \pm 4\sqrt{\sin 2\theta}$

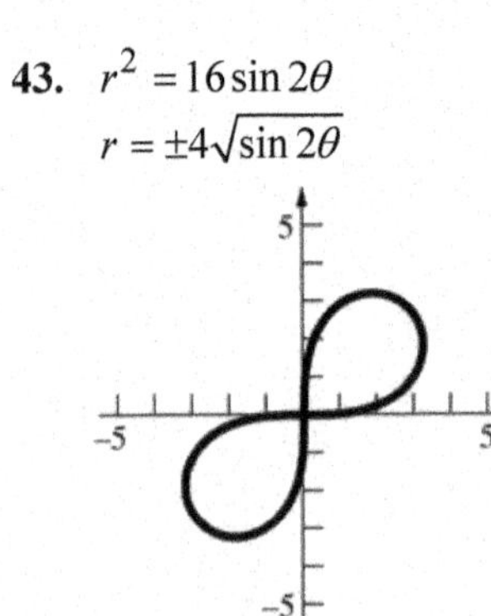

45. $r^2 - 6r(\cos\theta + \sin\theta) + 9 = 0$

$x^2 + y^2 - 6x - 6y + 9 = 0$

$(x^2 - 6x + 9) + (y^2 - 6y + 9) = -9 + 9 + 9$

$(x-3)^2 + (y-3)^2 = 9$

47. $f(\theta) = 3 + 3\cos\theta,\ f'(\theta) = -3\sin\theta$

$$m = \frac{(3+3\cos\theta)\cos\theta + (-3\sin\theta)\sin\theta}{-(3+3\cos\theta)\sin\theta + (-3\sin\theta)\cos\theta}$$

$$= \frac{\cos\theta + \cos^2\theta - \sin^2\theta}{-\sin\theta - 2\cos\theta\sin\theta} = \frac{\cos\theta + \cos 2\theta}{-\sin\theta - \sin 2\theta}$$

At $\theta = \frac{\pi}{6}, m = \frac{\cos\frac{\pi}{6} + \cos\frac{\pi}{3}}{-\sin\frac{\pi}{6} - \sin\frac{\pi}{3}} = -1$.

49. $A = 2 \cdot \frac{1}{2}\int_0^{\pi}(5 - 5\cos\theta)^2\, d\theta$

$= 25\int_0^{\pi}(1 - 2\cos\theta + \cos^2\theta)d\theta$

$= 25\int_0^{\pi}\left(\frac{3}{2} - 2\cos\theta + \frac{1}{2}\cos 2\theta\right)d\theta$

$= 25\left[\frac{3}{2}\theta - 2\sin\theta + \frac{1}{4}\sin 2\theta\right]_0^{\pi} = \frac{75\pi}{2}$

51. $\frac{x^2}{400} + \frac{y^2}{100} = 1; \frac{x}{200} + \frac{yy'}{50} = 0$

$y' = -\frac{x}{4y}; y' = -\frac{2}{3}$ at (16, 6)

Tangent line: $y - 6 = -\frac{2}{3}(x - 16)$

When $x = 14$, $y = -\frac{2}{3}(14 - 16) + 6 = \frac{22}{3}$.

$k = \frac{22}{3}$

53. **a.** I **b.** IV

c. III **d.** II

Review and Preview Problems

1. $x = 2t$, $y = t - 3$; $1 \le t \le 4$

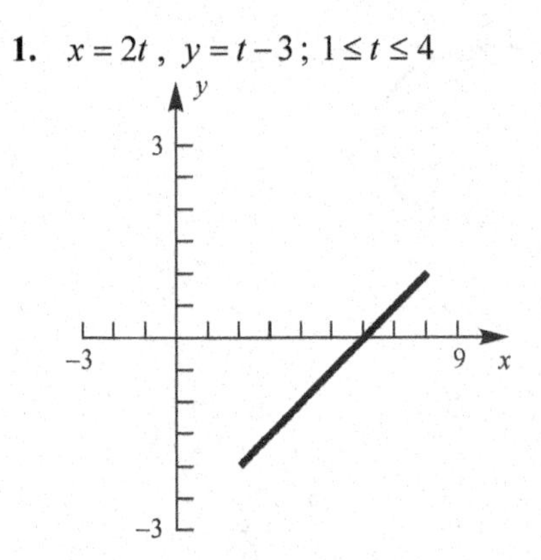

3. $x = 2\cos t$, $y = 2\sin t$; $0 \le t \le 2\pi$

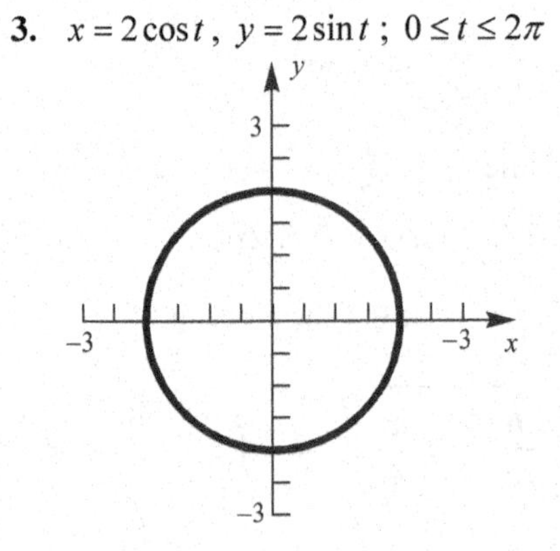

5. $x = t$, $y = \tan 2t$; $-\frac{\pi}{4} < t < \frac{\pi}{4}$

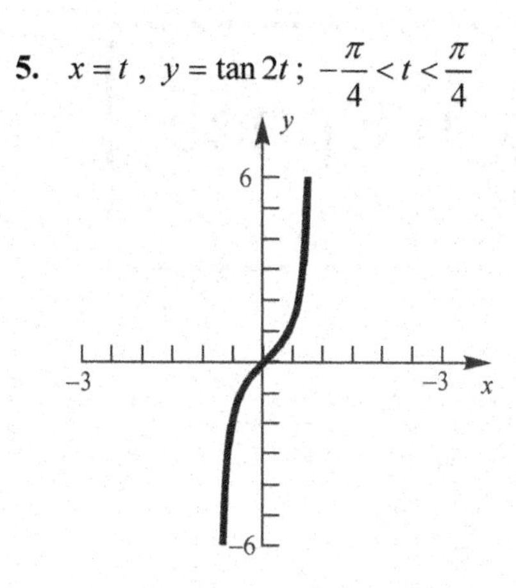

7. $x = h \cdot \cos\theta$
$y = h \cdot \sin\theta$

For problems 9-11, $L = \int_a^b \sqrt{\left(\frac{dx}{dt}\right)^2 + \left(\frac{dy}{dt}\right)^2}\, dt$

9. $\frac{dx}{dt} = 1$, $\frac{dy}{dt} = \frac{9}{2}\sqrt{t}$

$L = \int_0^4 \sqrt{1 + \frac{81}{4}t}\, dt = \int_0^4 \frac{1}{2}\sqrt{4 + 81t}\, dt =$ $\quad {\scriptstyle u = 4+81t,\ du = 81\,dt}$

$\frac{1}{162}\int_4^{328} \sqrt{u}\, du = \frac{2}{3} \cdot \frac{1}{162}\left[u^{3/2}\right]_4^{328} =$

$\frac{1}{243}\left[(328)^{3/2} - 8\right] \approx 24.4129$

11. $\frac{dx}{dt} = -2a\sin 2t$, $\frac{dy}{dt} = 2a\cos 2t$

$L = \int_0^{\pi/2} \sqrt{4a^2\sin^2 2t + 4a^2\cos^2 2t}\, dt =$

$\int_0^{\pi/2} 2|a|\sqrt{1}\, dt = 2|a|\left[t\right]_0^{\pi/2} = \pi|a|$

13. Let $(x, 2x+1)$ represent any point on the line $y = 2x + 1$; then the square of its distance from $(0,3)$ is $d(x) = x^2 + [(2x+1) - 3]^2 = 5x^2 - 8x + 4$ Now $d'(x) = 10x - 8$ so that $d'(0.8) = 0$; further, $d''(x) = 10 > 0$ so that the absolute minimum of the (square of the) distance occurs at the point $(0.8, 2.6)$ where the distance is

$\sqrt{d(0.8)} = \sqrt{5(0.8)^2 - 8(0.8) + 4} = \sqrt{0.8} \approx 0.894$

15. $s(t) = t^2 - 6t + 8$

a. $v(t) = s'(t) = 2t - 6$
$a(t) = v'(t) = 2$

b. The object is moving forward (in positive x-direction) when $v(t) > 0$ or $t > 3$.

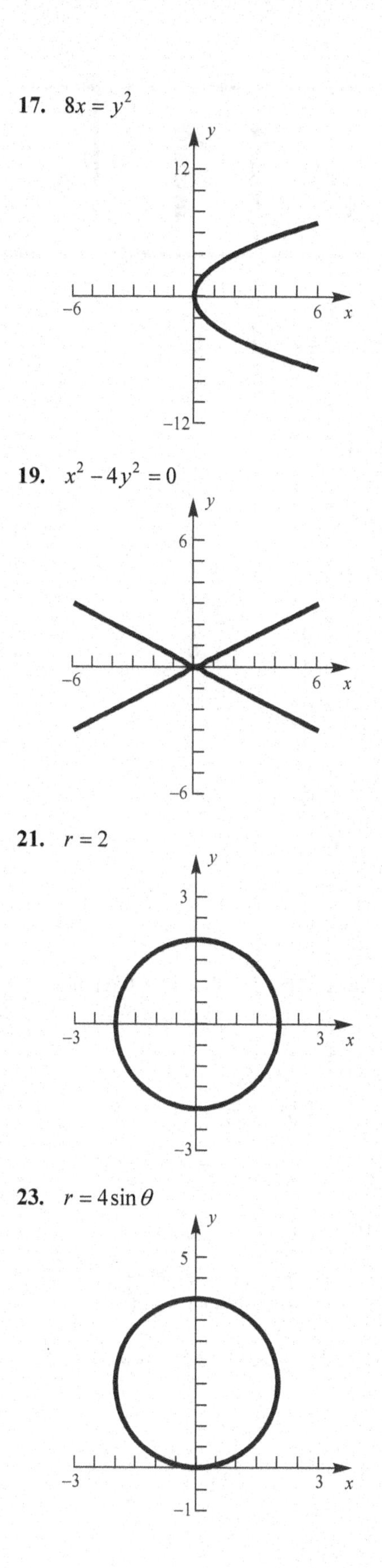

17. $8x = y^2$

19. $x^2 - 4y^2 = 0$

21. $r = 2$

23. $r = 4\sin\theta$

CHAPTER 11 Geometry in Space and Vectors

11.1 Concepts Review

1. coordinates

3. $(-1, 3, 5)$; 4

Problem Set 11.1

1. $A(1, 2, 3)$, $B(2, 0, 1)$, $C(-2, 4, 5)$, $D(0, 3, 0)$, $E(-1, -2, -3)$

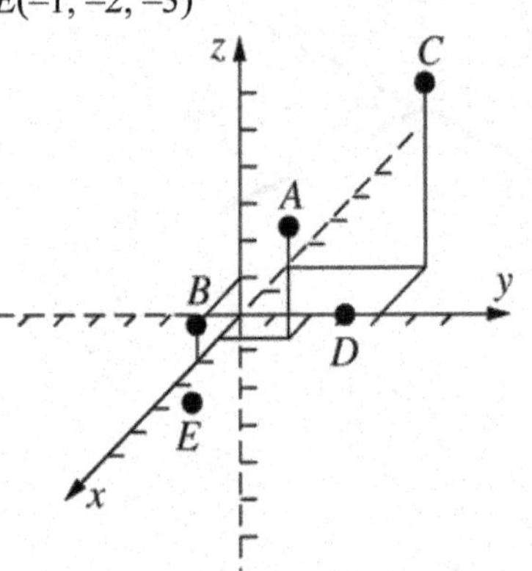

3. $x = 0$ in the yz-plane. $x = 0$ and $y = 0$ on the z-axis.

5. a. $\sqrt{(6-1)^2+(-1-2)^2+(0-3)^2}$
$= \sqrt{25+9+9} = \sqrt{43}$

b. $\sqrt{(-2-2)^2+(-2+2)^2+(0+3)^2}$
$= \sqrt{16+0+9} = 5$

c. $\sqrt{(e+\pi)^2+(\pi+4)^2+\left(0-\sqrt{3}\right)^2}$
$\sqrt{(e+\pi)^2+(\pi+4)^2+3} \approx 9.399$

7. $P(2, 1, 6)$, $Q(4, 7, 9)$, $R(8, 5, -6)$
$|PQ| = \sqrt{(2-4)^2+(1-7)^2+(6-9)^2}$
$= \sqrt{4+36+9} = 7$
$|PR| = \sqrt{(2-8)^2+(1-5)^2+(6+6)^2}$
$= \sqrt{36+16+144} = 14$
$|QR| = \sqrt{(4-8)^2+(7-5)^2+(9+6)^2}$
$= \sqrt{16+4+225} = \sqrt{245}$
$|PQ|^2+|PR|^2 = 49+196 = 245 = |QR|^2$, so the triangle formed by joining P, Q, and R is a right triangle, since it satisfies the Pythagorean Theorem.

9. Since the faces are parallel to the coordinate planes, the sides of the box are in the planes $x = 2, y = 3, z = 4, x = 6, y = -1$, and $z = 0$ and the vertices are at the points where 3 of these planes intersect. Thus, the vertices are $(2, 3, 4)$, $(2, 3, 0)$, $(2, -1, 4)$, $(2, -1, 0)$, $(6, 3, 4)$, $(6, 3, 0)$, $(6, -1, 4)$, and $(6, -1, 0)$

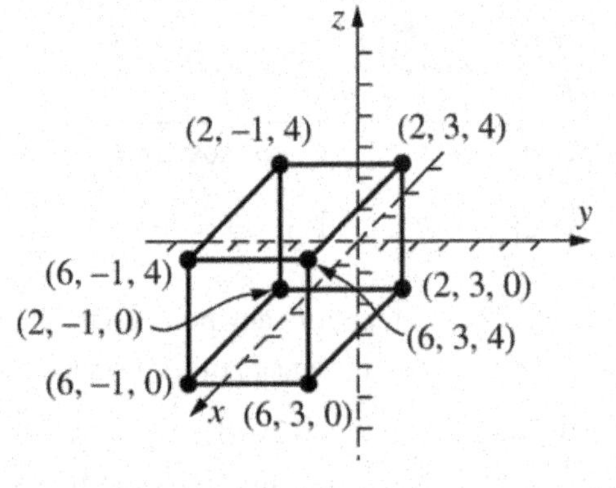

11. a. $(x-1)^2+(y-2)^2+(z-3)^2 = 25$

b. $(x+2)^2+(y+3)^2+(z+6)^2 = 5$

c. $(x-\pi)^2+(y-e)^2+\left(z-\sqrt{2}\right)^2 = \pi$

13. $(x^2 - 12x + 36) + (y^2 + 14y + 49) + (z^2 - 8z + 16) = -1 + 36 + 49 + 16$

$(x-6)^2 + (y+7)^2 + (z-4)^2 = 100$

Center: (6, –7, 4); radius 10

15. $x^2 + y^2 + z^2 - x + 2y + 4z = \frac{13}{4}$

$\left(x^2 - x + \frac{1}{4}\right) + (y^2 + 2y + 1) + (z^2 + 4z + 4) = \frac{13}{4} + \frac{1}{4} + 1 + 4$

$\left(x - \frac{1}{2}\right)^2 + (y+1)^2 + (z+2)^2 = \frac{17}{2}$

Center: $\left(\frac{1}{2}, -1, -2\right)$; radius $\sqrt{\frac{17}{2}} \approx 2.92$

17. x-intercept: $y = z = 0 \Rightarrow 2x = 12, x = 6$
y-intercept: $x = z = 0 \Rightarrow 6y = 12, y = 2$
z-intercept: $x = y = 0 \Rightarrow 3z = 12, z = 4$

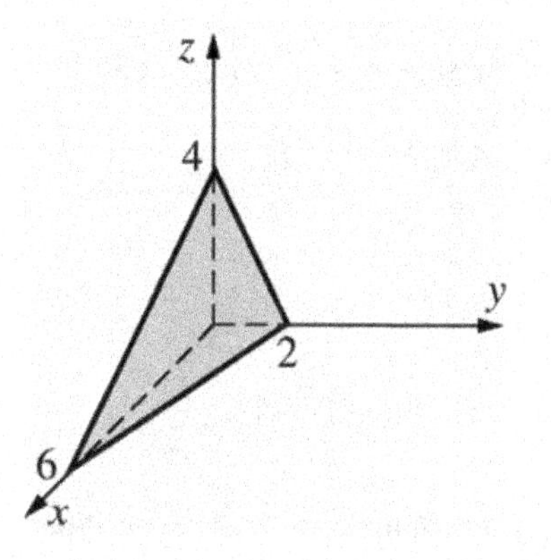

19. x-intercept: $y = z = 0 \Rightarrow x = 6$
y-intercept: $x = z = 0 \Rightarrow 3y = 6, y = 2$
z-intercept: $x = y = 0 \Rightarrow -z = 6, z = -6$

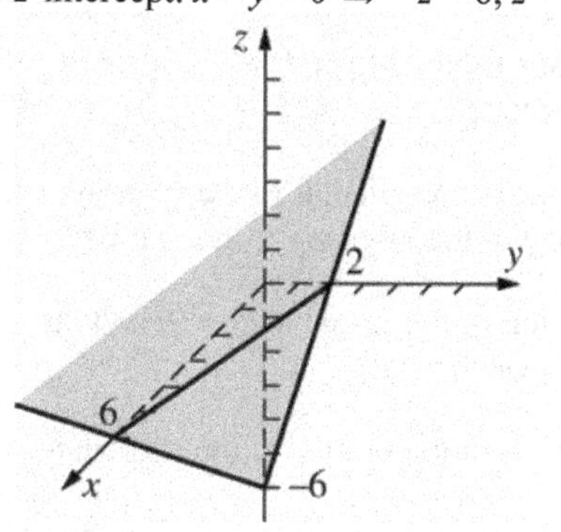

21. x and y cannot both be zero, so the plane is parallel to the z-axis.
x-intercept: $y = z = 0 \Rightarrow x = 8$

y-intercept: $x = z = 0 \Rightarrow 3y = 8,\ y = \frac{8}{3}$

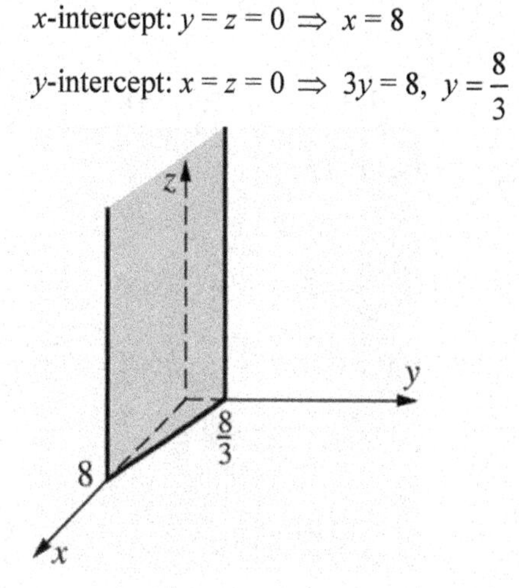

23. This is a sphere with center (0, 0, 0) and radius 3.

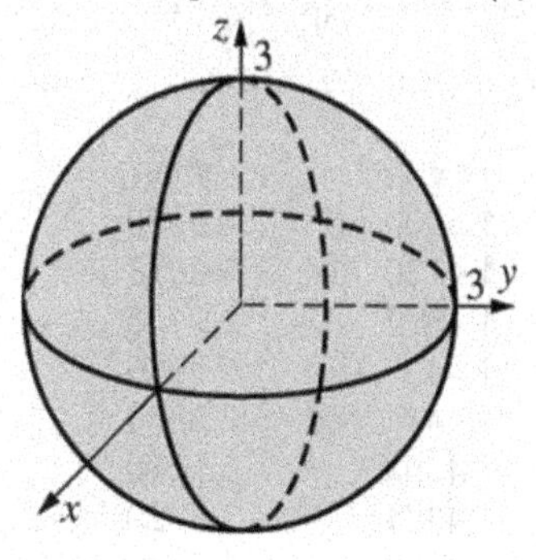

For problems 25-35, $L = \int_a^b \sqrt{\left(\frac{dx}{dt}\right)^2 + \left(\frac{dy}{dt}\right)^2 + \left(\frac{dz}{dt}\right)^2}\, dt$

25. $\frac{dx}{dt} = 1, \frac{dy}{dt} = 1, \frac{dz}{dt} = 2$

$L = \int_0^2 \sqrt{1^2 + 1^2 + 2^2}\, dt = \int_0^2 \sqrt{6}\, dt =$

$\left[\sqrt{6}t\right]_0^2 = 2\sqrt{6} \approx 4.899$

27. $\frac{dx}{dt} = \frac{3}{2}\sqrt{t}, \frac{dy}{dt} = 3, \frac{dz}{dt} = 4$

$L = \int_1^4 \sqrt{\left(\frac{9}{4}t\right) + 9 + 16}\, dt = \int_1^4 \frac{1}{2}\sqrt{9t + 100}\, dt =$ $\quad {\scriptstyle u = 9t+100,\ du = 9\,dt}$

$\frac{1}{18}\int_{109}^{136} \sqrt{u}\, du = \frac{1}{27}\left[u^{3/2}\right]_{109}^{136} \approx 16.59$

29. $\frac{dx}{dt} = 2t, \frac{dy}{dt} = 2\sqrt{t}, \frac{dz}{dt} = 1$

$L = \int_0^8 \sqrt{4t^2 + 4t + 1}\, dt = \int_0^8 \sqrt{(2t+1)^2}\, dt =$

$\int_0^8 (2t+1)\, dt = \left[t^2 + t\right]_0^8 = 72$

31. $\frac{dx}{dt} = -2\sin t, \frac{dy}{dt} = 2\cos t, \frac{dz}{dt} = 3$

$L = \int_{-\pi}^{\pi} \sqrt{4\sin^2 t + 4\cos^2 t + 9}\, dt = \int_{-\pi}^{\pi} \sqrt{13}\, dt =$

$\left[\sqrt{13}\,t\right]_{-\pi}^{\pi} = 2\pi\sqrt{13} \approx 22.654$

33. $\frac{dx}{dt} = \frac{1}{2\sqrt{t}}, \frac{dy}{dt} = 1, \frac{dz}{dt} = 1$

$L = \int_1^6 \sqrt{\left(\frac{1}{4t}\right) + 1 + 1}\, dt = \int_1^6 \sqrt{2 + \left(\frac{1}{4t}\right)}\, dt$

By the Parabolic Rule ($n = 10$):

i	x_i	$f(x_i)$	c_i	$c_i \cdot f(x_i)$
0	1	1.5000	1	1.5000
1	1.5	1.4720	4	5.8878
2	2	1.4577	2	2.9155
3	2.5	1.4491	4	5.7966
4	3	1.4434	2	2.8868
5	3.5	1.4392	4	5.7570
6	4	1.4361	2	2.8723
7	4.5	1.4337	4	5.7349
8	5	1.4318	2	2.8636
9	5.5	1.4302	4	5.7208
10	6	1.4289	1	1.4289
			approximation	**7.2273**

35. $\frac{dx}{dt} = -2\sin t, \frac{dy}{dt} = \cos t, \frac{dz}{dt} = 1$

$L = \int_0^{6\pi} \sqrt{4\sin^2 t + \cos^2 t + 1}\, dt =$

$\int_0^{6\pi} \sqrt{3\sin^2 t + 2}\, dt$

By the Parabolic Rule ($n = 10$):

i	x_i	$f(x_i)$	c_i	$c_i \cdot f(x_i)$
0	0	1.4142	1	1.4142
1	1.88	2.1711	4	8.6843
2	3.77	1.7425	2	3.4851
3	5.65	1.7425	4	6.9702
4	7.54	2.1711	2	4.3421
5	9.42	1.4142	4	5.6569
6	11.3	2.1711	2	4.3421
7	13.2	1.7425	4	6.9702
8	15.1	1.7425	2	3.4851
9	17	2.1711	4	8.6843
10	18.8	1.4142	1	1.4142
			approximation	**34.8394**

37. The center of the sphere is the midpoint of the diameter, so it is $\left(\frac{-2+4}{2}, \frac{3-1}{2}, \frac{6+5}{2}\right) = \left(1, 1, \frac{11}{2}\right)$. The radius is $\frac{1}{2}\sqrt{(-2-4)^2 + (3+1)^2 + (6-5)^2} = \frac{\sqrt{53}}{2}$. The equation is $(x-1)^2 + (y-1)^2 + \left(z - \frac{11}{2}\right)^2 = \frac{53}{4}$.

39. The center must be 6 units from each coordinate plane. Since it is in the first octant, the center is (6, 6, 6). The equation is $(x-6)^2 + (y-6)^2 + (z-6)^2 = 36$.

41.
- **a.** Plane parallel to and two units above the xy-plane
- **b.** Plane perpendicular to the xy-plane whose trace in the xy-plane is the line $x = y$.
- **c.** Union of the yz-plane ($x = 0$) and the xz-plane ($y = 0$)
- **d.** Union of the three coordinate planes
- **e.** Cylinder of radius 2, parallel to the z-axis
- **f.** Top half of the sphere with center (0, 0, 0) and radius 3

43. If $P(x, y, z)$ denotes the moving point, $\sqrt{(x-1)^2+(y-2)^2+(z+3)^2}=2\sqrt{(x-1)^2+(y-2)^2+(z-3)^2}$, which simplifies to $(x-1)^2+(y-2)^2+(z-5)^2=16$, is a sphere with radius 4 and center (1, 2, 5).

45. Note that the volume of a segment of height h in a hemisphere of radius r is $\pi h^2\left[r-\left(\frac{h}{3}\right)\right]$.

The resulting solid is the union of two segments, one for each sphere. Since the two spheres have the same radius, each segment will have the same value for h. h is the radius minus half the distance between the centers of the two spheres.

$$h=2-\frac{1}{2}\sqrt{(2-1)^2+(4-2)^2+(3-1)^2}=2-\frac{3}{2}=\frac{1}{2}$$

$$V=2\left[\pi\left(\frac{1}{2}\right)^2\left(2-\frac{1}{6}\right)\right]=\frac{11\pi}{12}$$

47. Plots will vary. We first note that the sign of c will influence the vertical direction an object moves (along the helix) with increasing time; if c is negative the object will spiral downward, whereas if c is positive it will spiral upward. The smaller $|c|$ is the "tighter" the spiral will be; that is the space between successive "coils" of the helix decreases as $|c|$ decreases.

11.2 Concepts Review

1. magnitude; direction

3. the tail of **u**; the head of **v**

Problem Set 11.2

1.

3.

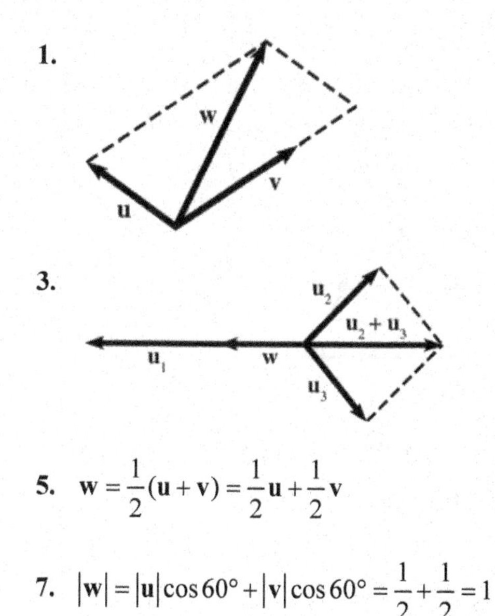

5. $\mathbf{w}=\frac{1}{2}(\mathbf{u}+\mathbf{v})=\frac{1}{2}\mathbf{u}+\frac{1}{2}\mathbf{v}$

7. $|\mathbf{w}|=|\mathbf{u}|\cos 60°+|\mathbf{v}|\cos 60°=\frac{1}{2}+\frac{1}{2}=1$

9. $\mathbf{u}+\mathbf{v}=\langle -1+3, 0+4\rangle=\langle 2,4\rangle$

$\mathbf{u}-\mathbf{v}=\langle -1-3, 0-4\rangle=\langle -4,-4\rangle$

$\|\mathbf{u}\|=\sqrt{(-1)^2+(0)^2}=\sqrt{1}=1$

$\|\mathbf{v}\|=\sqrt{(3)^2+(4)^2}=\sqrt{25}=5$

11. $\mathbf{u}+\mathbf{v}=\langle 12+(-2), 12+2\rangle=\langle 10,14\rangle$

$\mathbf{u}-\mathbf{v}=\langle 12-(-2), 12-2\rangle=\langle 14,10\rangle$

$\|\mathbf{u}\|=\sqrt{(12)^2+(12)^2}=\sqrt{288}=12\sqrt{2}$

$\|\mathbf{v}\|=\sqrt{(-2)^2+(2)^2}=\sqrt{8}=2\sqrt{2}$

13. $\mathbf{u}+\mathbf{v}=\langle -1+3, 0+4, 0+0\rangle=\langle 2,4,0\rangle$

$\mathbf{u}-\mathbf{v}=\langle -1-3, 0-4, 0-0\rangle=\langle -4,-4,0\rangle$

$\|\mathbf{u}\|=\sqrt{(-1)^2+(0)^2+(0)^2}=\sqrt{1}=1$

$\|\mathbf{v}\|=\sqrt{(3)^2+(4)^2+(0)^2}=\sqrt{25}=5$

15. $\mathbf{u}+\mathbf{v}=\langle 1+(-5), 0+0, 1+0\rangle=\langle -4,0,1\rangle$

$\mathbf{u}-\mathbf{v}=\langle 1-(-5), 0-0, 1-0\rangle=\langle 6,0,1\rangle$

$\|\mathbf{u}\|=\sqrt{(1)^2+(0)^2+(1)^2}=\sqrt{2}\approx 1.414$

$\|\mathbf{v}\|=\sqrt{(-5)^2+(0)^2+(0)^2}=\sqrt{25}=5$

17. Let θ be the angle of **w** measured clockwise from south.

$|\mathbf{w}|\cos\theta = |\mathbf{u}|\cos 30° + |\mathbf{v}|\cos 45° = 25\sqrt{3} + 25\sqrt{2}$

$= 25(\sqrt{3} + \sqrt{2})$

$|\mathbf{w}|\sin\theta = |\mathbf{v}|\sin 45° - |\mathbf{u}|\sin 30° = 25\sqrt{2} - 25$

$= 25(\sqrt{2} - 1)$

$|\mathbf{w}|^2 = |\mathbf{w}|^2\cos^2\theta + |\mathbf{w}|^2\sin^2\theta$

$= 625(\sqrt{3} + \sqrt{2})^2 + 625(\sqrt{2} - 1)^2$

$= 625(8 - 2\sqrt{2} + 2\sqrt{6})$

$|\mathbf{w}| = \sqrt{625(8 - 2\sqrt{2} + 2\sqrt{6})} = 25\sqrt{8 - 2\sqrt{2} + 2\sqrt{6}}$

≈ 79.34

$\tan\theta = \frac{|\mathbf{w}|\sin\theta}{|\mathbf{w}|\cos\theta} = \frac{\sqrt{2} - 1}{\sqrt{3} + \sqrt{2}}$

$\theta = \tan^{-1}\left(\frac{\sqrt{2} - 1}{\sqrt{3} + \sqrt{2}}\right) = 7.5°$

w has magnitude 79.34 lb in the direction S 7.5° W.

19. The force of 300 N parallel to the plane has magnitude 300 sin 30° = 150 N. Thus, a force of 150 N parallel to the plane will just keep the weight from sliding.

21. Let θ be the angle the plane makes from north, measured clockwise.

425 sin θ = 45 sin 20°

$\sin\theta = \frac{9}{85}\sin 20°$

$\theta = \sin^{-1}\left(\frac{9}{85}\sin 20°\right) \approx 2.08°$

Let x be the speed of airplane with respect to the ground.

$x = 45\cos 20° + 425\cos\theta \approx 467$

The plane flies in the direction N 2.08° E, flying 467 mi/h with respect to the ground.

23. Let x be the air speed.

$x\cos 60° = 40$

$x = \frac{40}{\cos 60°} = 80$

The air speed of the plane is 80 mi/hr

25. Let $\mathbf{u} = \langle u_1, u_2\rangle, \mathbf{v} = \langle v_1, v_2\rangle$, and $\mathbf{w} = \langle w_1, w_2\rangle$

a. $\mathbf{u} + \mathbf{v} = \langle u_1 + v_1, u_2 + v_2\rangle =$
$\langle v_1 + u_1, v_2 + u_2\rangle = \mathbf{v} + \mathbf{u}$

b. $(\mathbf{u} + \mathbf{v}) + \mathbf{w} = \langle u_1 + v_1, u_2 + v_2\rangle + \langle w_1, w_2\rangle =$
$\langle (u_1 + v_1) + w_1, (u_2 + v_2) + w_2\rangle =$
$\langle u_1 + (v_1 + w_1), u_2 + (v_2 + w_2)\rangle =$
$\langle u_1, u_2\rangle\langle v_1 + w_1, v_2 + w_2\rangle = \mathbf{u} + (\mathbf{v} + \mathbf{w})$

c. $\mathbf{u} + \mathbf{0} = \langle u_1 + 0, u_2 + 0\rangle = \langle u_1, u_2\rangle = \mathbf{u}$
$\mathbf{u} + \mathbf{0} = \mathbf{0} + \mathbf{u}$ by part a.

d. $\mathbf{u} + (\text{-}\mathbf{u}) = \langle u_1, u_2\rangle + \langle -u_1, -u_2\rangle =$
$\langle u_1 + (-u_1), u_2 + (-u_2)\rangle = \langle 0, 0\rangle = \mathbf{0}$

e. $a(b\mathbf{u}) = a(\langle bu_1, bu_2\rangle) = \langle a(bu_1), a(bu_2)\rangle =$
$\langle (ab)u_1, (ab)u_2\rangle = (ab)\mathbf{u}$

f. $a(\mathbf{u} + \mathbf{v}) = a\langle u_1 + v_1, u_2 + v_2\rangle =$
$\langle a(u_1 + v_1), a(u_2 + v_2)\rangle =$
$\langle au_1 + av_1, au_2 + av_2\rangle =$
$\langle au_1, au_2\rangle + \langle av_1, av_2\rangle = a\mathbf{u} + a\mathbf{v}$

g. $(a + b)\mathbf{u} = \langle (a + b)u_1, (a + b)u_2\rangle =$
$\langle au_1 + bu_1, au_2 + bu_2\rangle =$
$\langle au_1, au_2\rangle + \langle bu_1, bu_2\rangle = a\mathbf{u} + b\mathbf{u}$

h. $1\mathbf{u} = \langle 1\cdot u_1, 1\cdot u_2\rangle = \langle u_1, u_2\rangle = \mathbf{u}$

i. $\|a\mathbf{u}\| = \|\langle au_1, au_2\rangle\| = \sqrt{(au_1)^2 + (au_2)^2} =$
$\sqrt{a^2u_1^2 + a^2u_2^2} = \sqrt{a^2(u_1^2 + u_2^2)} =$
$\sqrt{a^2}\sqrt{u_1^2 + u_2^2} = |a|\cdot\|\mathbf{u}\|$

27. Given triangle ABC, let D be the midpoint of AB and E be the midpoint of BC. $\mathbf{u} = \overrightarrow{AB}$, $\mathbf{v} = \overrightarrow{BC}$, $\mathbf{w} = \overrightarrow{AC}$, $\mathbf{z} = \overrightarrow{DE}$

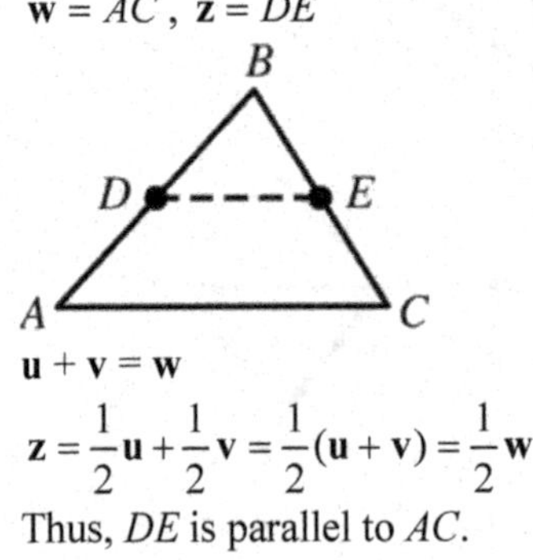

$\mathbf{u} + \mathbf{v} = \mathbf{w}$

$\mathbf{z} = \frac{1}{2}\mathbf{u} + \frac{1}{2}\mathbf{v} = \frac{1}{2}(\mathbf{u} + \mathbf{v}) = \frac{1}{2}\mathbf{w}$

Thus, DE is parallel to AC.

29. Let P_i be the tail of $\mathbf{v}_i$. Then

$\mathbf{v}_1 + \mathbf{v}_2 + \ldots + \mathbf{v}_n$

$= \overrightarrow{P_1P_2} + \overrightarrow{P_2P_3} + \cdots + \overrightarrow{P_nP_1}$

$= \overrightarrow{P_1P_1} = \mathbf{0}.$

31. The components of the forces along the lines containing AP, BP, and CP are in equilibrium; that is,

$W = W \cos \alpha + W \cos \beta$
$W = W \cos \beta + W \cos \gamma$
$W = W \cos \alpha + W \cos \gamma$

Thus, $\cos \alpha + \cos \beta = 1$, $\cos \beta + \cos \gamma = 1$, and $\cos \alpha + \cos \gamma = 1$. Solving this system of equations results in $\cos\alpha = \cos\beta = \cos\gamma = \frac{1}{2}$.

Hence $\alpha = \beta = \gamma = 60°$.
Therefore, $\alpha + \beta = \alpha + \gamma = \beta + \gamma = 120°$.

33. The components of the forces along the lines containing AP, BP, and CP are in equilibrium; that is,

$5w \cos \alpha + 4w \cos \beta = 3w$
$3w \cos \beta + 5w \cos \gamma = 4w$
$3w \cos \alpha + 4w \cos \gamma = 5w$

Thus,

$5 \cos \alpha + 4 \cos \beta = 3$
$3 \cos \beta + 5 \cos \gamma = 4$
$3 \cos \alpha + 4 \cos \gamma = 5.$

Solving this system of equations results in $\cos\alpha = \frac{3}{5}, \cos\beta = 0, \cos\gamma = \frac{4}{5}$, from which it follows that $\sin\alpha = \frac{4}{5}, \sin\beta = 1, \sin\gamma = \frac{3}{5}$.

Therefore, $\cos(\alpha+\beta) = -\frac{4}{5}, \cos(\alpha+\gamma) = 0,$

$\cos(\beta+\gamma) = -\frac{3}{5}$, so

$\alpha + \beta = \cos^{-1}\left(-\frac{4}{5}\right) \approx 143.13°, \quad \alpha + \gamma = 90°,$

$\beta + \gamma = \cos^{-1}\left(-\frac{3}{5}\right) \approx 126.87°.$

This problem can be modeled with three strings going through A, four strings through B, and five strings through C, with equal weights attached to the twelve strings. Then the quantity to be minimized is $3|AP| + 4|BP| + 5|CP|$.

35. By symmetry, the tension on each wire will be the same; denote it by $\|\mathbf{T}\|$ where $\mathbf{T}$ can be the tension vector along any of the wires. The chandelier exerts a force of 100 lbs. vertically downward. Each wire exerts a vertical tension of $\|\mathbf{T}\|\sin 45°$ upward. Since a state of equilibrium exists,

$4 \cdot \|\mathbf{T}\|\sin 45° = 100$ or $\|\mathbf{T}\| = \frac{100}{2\sqrt{2}} \approx 35.36$

The tension in each wire is approximately 35.36 lbs.

11.3 Concepts Review

1. $u_1v_1 + u_2v_2 + u_3v_3$; $\|\mathbf{u}\|\|\mathbf{v}\|\cos\theta$

3. $\mathbf{F} \bullet \mathbf{D}$

Problem Set 11.3

1. a. $2\mathbf{a} - 4\mathbf{b} = (-4\mathbf{i} + 6\mathbf{j}) + (-8\mathbf{i} + 12\mathbf{j})$
$= -12\mathbf{i} + 18\mathbf{j}$

b. $\mathbf{a} \cdot \mathbf{b} = (-2)(2) + (3)(-3) = -13$

c. $\mathbf{a} \cdot (\mathbf{b} + \mathbf{c}) = (-2\mathbf{i} + 3\mathbf{j}) \cdot (2\mathbf{i} - 8\mathbf{j})$
$= (-2)(2) + (3)(-8) = -28$

d. $(-2\mathbf{a} + 3\mathbf{b}) \cdot 5\mathbf{c} = 5[(10\mathbf{i} - 15\mathbf{j}) \cdot (-5\mathbf{j})]$
$= 5[(10)(0) + (-15)(-5)] = 375$

e. $\|\mathbf{a}\|\mathbf{c} \cdot \mathbf{a} = \sqrt{4+9}[(0)(-2) + (-5)(3)] = -15\sqrt{13}$

f. $\mathbf{b} \cdot \mathbf{b} - \|\mathbf{b}\| = (2)(2) + (-3)(-3) - \sqrt{4+9}$
$= 13 - \sqrt{13}$

3. a. $\cos\theta = \frac{\mathbf{a}\cdot\mathbf{b}}{\|\mathbf{a}\|\|\mathbf{b}\|} = \frac{(1)(-1) + (-3)(2)}{\left(\sqrt{10}\right)\left(\sqrt{5}\right)} = -\frac{7}{\sqrt{50}}$

$= -\frac{7}{5\sqrt{2}} \approx -0.9899$

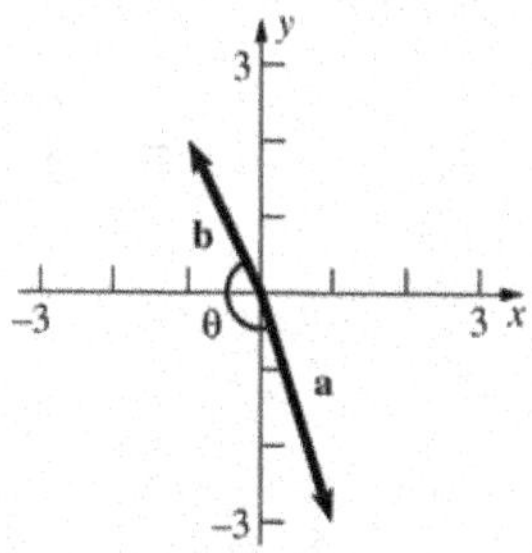

b. $\cos\theta = \dfrac{\mathbf{a}\cdot\mathbf{b}}{\|\mathbf{a}\|\|\mathbf{b}\|} = \dfrac{(-1)(6)+(-2)(0)}{(\sqrt{5})(6)} = -\dfrac{6}{6\sqrt{5}}$

$= -\dfrac{1}{\sqrt{5}} \approx -0.4472$

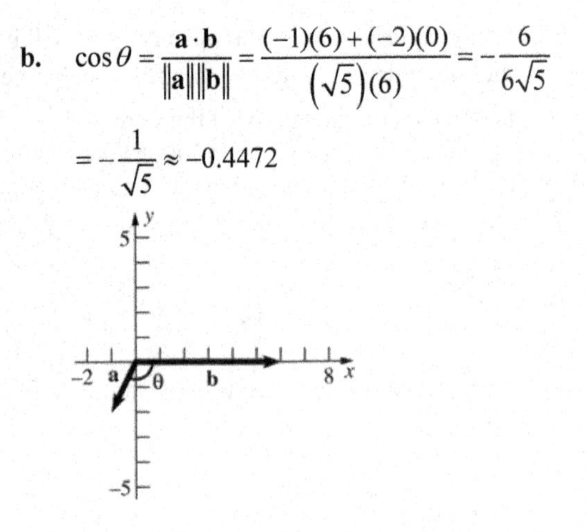

c. $\cos\theta = \dfrac{\mathbf{a}\cdot\mathbf{b}}{\|\mathbf{a}\|\|\mathbf{b}\|} = \dfrac{(2)(-2)+(-1)(-4)}{(\sqrt{5})(2\sqrt{5})} = \dfrac{0}{10} = 0$

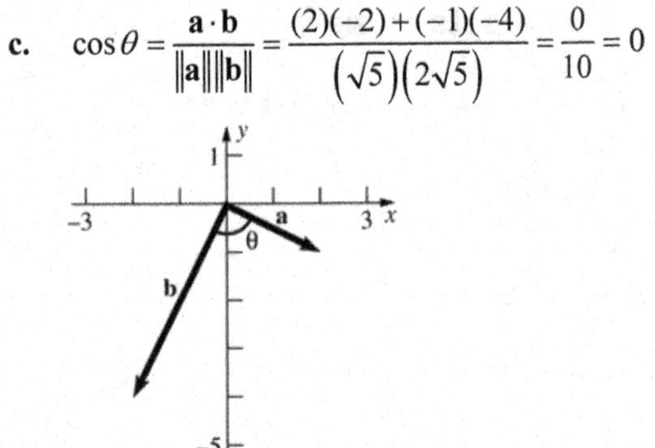

d. $\cos\theta = \dfrac{\mathbf{a}\cdot\mathbf{b}}{\|\mathbf{a}\|\|\mathbf{b}\|} = \dfrac{(4)(-8)+(-7)(10)}{(\sqrt{65})(2\sqrt{41})}$

$= \dfrac{-102}{2\sqrt{2665}} = -\dfrac{51}{\sqrt{2665}} \approx -0.9879$

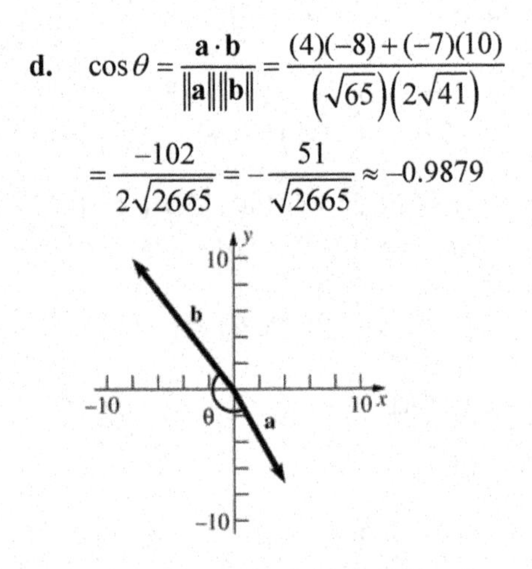

5. a. $\mathbf{a}\bullet\mathbf{b} = (1)(0)+(2)(1)+(-1)(1) = 1$

b. $(\mathbf{a}+\mathbf{c})\bullet\mathbf{b} = (3\mathbf{j}+\mathbf{k})\bullet(\mathbf{j}+\mathbf{k})$

$= (0)(0)+(3)(1)+(1)(1) = 4$

c. $\dfrac{\mathbf{a}}{\|\mathbf{a}\|} = \dfrac{1}{\sqrt{1^2+2^2+(-1)^2}}(\mathbf{i}+2\mathbf{j}-\mathbf{k})$

$= \dfrac{\sqrt{6}}{6}\mathbf{i} + \dfrac{\sqrt{6}}{3}\mathbf{j} - \dfrac{\sqrt{6}}{6}\mathbf{k}$

d. $(\mathbf{b}-\mathbf{c})\bullet\mathbf{a} = (\mathbf{i}-\mathbf{k})\bullet(\mathbf{i}+2\mathbf{j}-\mathbf{k})$

$= (1)(1)+(0)(2)+(-1)(-1) = 2$

e. $\dfrac{\mathbf{a}\bullet\mathbf{b}}{\|\mathbf{a}\|\|\mathbf{b}\|} = \dfrac{(1)(0)+(2)(1)+(-1)(1)}{\sqrt{1^2+2^2+(-1)^2}\sqrt{0^2+1^2+1^2}}$

$= \dfrac{1}{\sqrt{6}\sqrt{2}} = \dfrac{\sqrt{3}}{6}$

f. By Theorem A (5), $\mathbf{b}\bullet\mathbf{b} - \|\mathbf{b}\|^2 = 0$

7. The basic formula is $\cos\theta = \dfrac{\mathbf{u}\bullet\mathbf{v}}{\|\mathbf{u}\|\|\mathbf{v}\|}$

$\theta_{\mathbf{a},\mathbf{b}} = \cos^{-1}\left(\dfrac{\sqrt{2}-\sqrt{2}+0}{\sqrt{4}\cdot\sqrt{3}}\right) =$

$\cos^{-1} 0 = 90^\circ$

$\theta_{\mathbf{a},\mathbf{c}} = \cos^{-1}\left(\dfrac{-2\sqrt{2}+2\sqrt{2}+0}{\sqrt{4}\cdot\sqrt{9}}\right) =$

$\cos^{-1} 0 = 90^\circ$

$\theta_{\mathbf{b},\mathbf{c}} = \cos^{-1}\left(\dfrac{-2-2+1}{\sqrt{3}\cdot\sqrt{9}}\right) =$

$\cos^{-1}\left(-\dfrac{\sqrt{3}}{3}\right) \approx 125.26^\circ$

9. The basic formulae are

$\cos\alpha_{\mathbf{u}} = \dfrac{u_1}{\|\mathbf{u}\|} \quad \cos\beta_{\mathbf{u}} = \dfrac{u_2}{\|\mathbf{u}\|} \quad \cos\gamma_{\mathbf{u}} = \dfrac{u_3}{\|\mathbf{u}\|}$

a. $\mathbf{a} = \langle\sqrt{2},\sqrt{2},0\rangle \qquad \|\mathbf{a}\| = 2$

$\cos\alpha_{\mathbf{a}} = \dfrac{a_1}{\|\mathbf{a}\|} = \dfrac{\sqrt{2}}{2}, \quad \alpha_{\mathbf{a}} = 45^\circ$

$\cos\beta_{\mathbf{a}} = \dfrac{a_2}{\|\mathbf{a}\|} = \dfrac{\sqrt{2}}{2}, \quad \beta_{\mathbf{a}} = 45^\circ$

$\cos\gamma_{\mathbf{a}} = \dfrac{a_3}{\|\mathbf{a}\|} = \dfrac{0}{2} = 0, \quad \gamma_{\mathbf{a}} = 90^\circ$

b. $\mathbf{b} = \langle 1,-1,1\rangle \qquad \|\mathbf{b}\| = \sqrt{3}$

$\cos\alpha_{\mathbf{b}} = \dfrac{b_1}{\|\mathbf{b}\|} = \dfrac{1}{\sqrt{3}} \approx 0.577, \quad \alpha_{\mathbf{b}} \approx 54.74^\circ$

$\cos\beta_{\mathbf{b}} = \dfrac{b_2}{\|\mathbf{b}\|} = \dfrac{-1}{\sqrt{3}} \approx -0.577, \quad \beta_{\mathbf{b}} \approx 125.26^\circ$

$\cos\gamma_{\mathbf{b}} = \dfrac{b_3}{\|\mathbf{b}\|} = \dfrac{1}{\sqrt{3}} \approx 0.577, \quad \gamma_{\mathbf{b}} \approx 54.74^\circ$

c. $\mathbf{c} = \langle -2, 2, 1 \rangle \qquad \|\mathbf{c}\| = 3$

$$\cos\alpha_\mathbf{c} = \frac{c_1}{\|\mathbf{c}\|} = -\frac{2}{3}, \quad \alpha_\mathbf{c} \approx 131.81^\circ$$

$$\cos\beta_\mathbf{c} = \frac{c_2}{\|\mathbf{c}\|} = \frac{2}{3}, \quad \beta_\mathbf{c} \approx 48.19^\circ$$

$$\cos\gamma_\mathbf{c} = \frac{c_3}{\|\mathbf{c}\|} = \frac{1}{3}, \quad \gamma_\mathbf{c} \approx 70.53^\circ$$

11. $\langle 6,3\rangle \cdot \langle -1,2\rangle = (6)(-1) + (3)(2) = 0$
Therefore the vectors are orthogonal

13. $\mathbf{a}\cdot\mathbf{b} = (1)(1) + (-1)(1) + (0)(0) = 0$
$\mathbf{a}\cdot\mathbf{c} = (1)(0) + (-1)(0) + (0)(2) = 0$
$\mathbf{b}\cdot\mathbf{c} = (1)(0) + (1)(0) + (0)(2) = 0$
Therefore the vectors are mutually orthogonal.

15. If $x\mathbf{i} + y\mathbf{j} + z\mathbf{k}$ is perpendicular to $-4\mathbf{i} + 5\mathbf{j} + \mathbf{k}$ and $4\mathbf{i} + \mathbf{j}$, then $-4x + 5y + z = 0$ and $4x + y = 0$ since the dot product of perpendicular vectors is 0. Solving these equations yields $y = -4x$ and $z = 24x$. Hence, for any x, $x\mathbf{i} - 4x\mathbf{j} + 24x\mathbf{k}$ is perpendicular to the given vectors.

$$\|x\mathbf{i} - 4x\mathbf{j} + 24x\mathbf{k}\| = \sqrt{x^2 + 16x^2 + 576x^2}$$
$$= |x|\sqrt{593}$$

This length is 10 when $x = \pm\frac{10}{\sqrt{593}}$. The vectors are $\frac{10}{\sqrt{593}}\mathbf{i} - \frac{40}{\sqrt{593}}\mathbf{j} + \frac{240}{\sqrt{593}}\mathbf{k}$ and $-\frac{10}{\sqrt{593}}\mathbf{i} + \frac{40}{\sqrt{593}}\mathbf{j} - \frac{240}{\sqrt{593}}\mathbf{k}$.

17. A vector equivalent to $\overrightarrow{BA}$ is
$\mathbf{u} = \langle 1+4, 2-5, 3-6\rangle = \langle 5, -3, -3\rangle$.
A vector equivalent to $\overrightarrow{BC}$ is
$\mathbf{v} = \langle 1+4, 0-5, 1-6\rangle = \langle 5, -5, -5\rangle$.

$$\cos\theta = \frac{\mathbf{u}\cdot\mathbf{v}}{\|\mathbf{u}\|\|\mathbf{v}\|} = \frac{5\cdot 5 + (-3)(-5) + (-3)(-5)}{\sqrt{25+9+9}\sqrt{25+25+25}}$$
$$= \frac{55}{\sqrt{43}\sqrt{75}} = \frac{11}{\sqrt{129}}, \text{ so } \theta = \cos^{-1}\frac{11}{\sqrt{129}} \approx 14.4°.$$

19. $\langle c,6\rangle \cdot \langle c,-4\rangle = 0 \Rightarrow c^2 - 24 = 0 \Rightarrow$
$c^2 = 24 \Rightarrow c = \pm 2\sqrt{6}$

21. $(c\mathbf{i} + \mathbf{j} + \mathbf{k}) \cdot (0\mathbf{i} + 2\mathbf{j} + d\mathbf{k}) = 0 \Rightarrow 0c + 2 + d = 0 \Rightarrow$
c is any number, $d = -2$

For problems 23-33, the formula to use is

$$\text{proj}_\mathbf{a}\mathbf{b} = \left(\frac{\mathbf{b}\cdot\mathbf{a}}{\|\mathbf{a}\|^2}\right)\mathbf{a}$$

23. $\mathbf{u} = \langle 1,2\rangle$, $\mathbf{v} = \langle 2,-1\rangle$, $\mathbf{w} = \langle 1,5\rangle$

$$\text{proj}_\mathbf{v}\mathbf{u} = \frac{(1)(2) + (2)(-1)}{2^2 + (-1)^2}\langle 2,-1\rangle = \mathbf{0}$$

25. $\mathbf{u} = \langle 1,2\rangle$, $\mathbf{v} = \langle 2,-1\rangle$, $\mathbf{w} = \langle 1,5\rangle$

$$\text{proj}_\mathbf{u}\mathbf{w} = \frac{(1)(1) + (5)(2)}{1^2 + 2^2}\langle 1,2\rangle = \left\langle \frac{11}{5}, \frac{22}{5}\right\rangle$$

27. $\mathbf{u} = \langle 1,2\rangle$, $\mathbf{v} = \langle 2,-1\rangle$, $\mathbf{w} = \langle 1,5\rangle$

$$\text{proj}_\mathbf{j}\mathbf{u} = \frac{(1)(0) + (2)(1)}{0^2 + (1)^2}\langle 0,1\rangle = \langle 0,2\rangle$$

29. $\mathbf{u} = \langle 3,2,1\rangle$, $\mathbf{v} = \langle 2,0,-1\rangle$, $\mathbf{w} = \langle 1,5,-3\rangle$

$$\text{proj}_\mathbf{v}\mathbf{u} = \frac{(3)(2) + (2)(0) + (1)(-1)}{2^2 + 0^2 + (-1)^2}\langle 2,0,-1\rangle =$$
$$\langle 2,0,-1\rangle$$

31. $\mathbf{u} = \langle 3,2,1\rangle$, $\mathbf{v} = \langle 2,0,-1\rangle$, $\mathbf{w} = \langle 1,5,-3\rangle$

$$\text{proj}_\mathbf{u}\mathbf{w} = \frac{(1)(3) + (5)(2) + (-3)(1)}{3^2 + 2^2 + 1^2}\langle 3,2,1\rangle =$$
$$\left\langle \frac{15}{7}, \frac{10}{7}, \frac{5}{7}\right\rangle$$

33. $\mathbf{u} = \langle 3,2,1\rangle$, $\mathbf{v} = \langle 2,0,-1\rangle$, $\mathbf{w} = \langle 1,5,-3\rangle$

$$\text{proj}_\mathbf{k}\mathbf{u} = \frac{(3)(0) + (2)(0) + (1)(1)}{0^2 + 0^2 + (1)^2}\langle 0,0,1\rangle = \langle 0,0,1\rangle$$

35. a. $\text{proj}_\mathbf{u}\mathbf{u} = \left(\frac{\mathbf{u}\cdot\mathbf{u}}{\|\mathbf{u}\|^2}\right)\mathbf{u} = \left(\frac{\mathbf{u}\cdot\mathbf{u}}{\mathbf{u}\cdot\mathbf{u}}\right)\mathbf{u} = \mathbf{u}$

b. $\text{proj}_{-\mathbf{u}}\mathbf{u} = \left(\frac{\mathbf{u}\cdot(-\mathbf{u})}{\|-\mathbf{u}\|^2}\right)(-\mathbf{u}) = \left(\frac{\mathbf{u}\cdot\mathbf{u}}{\mathbf{u}\cdot\mathbf{u}}\right)\mathbf{u} = \mathbf{u}$

37. $\mathbf{u}\cdot\mathbf{v} = (-1)(-1) + 5\cdot 1 + 3(-1) = 3$
$\|\mathbf{v}\| = \sqrt{1+1+1} = \sqrt{3}$
$\frac{\mathbf{u}\cdot\mathbf{v}}{\|\mathbf{v}\|} = \frac{3}{\sqrt{3}} = \sqrt{3}$

39. $\|\mathbf{u}\| = (4 + 9 + z^2)^{1/2} = 5$ and $z > 0$, so
$z = 2\sqrt{3} \approx 3.4641$.

41. There are infinitely many such pairs. Note that $\langle -4, 2, 5\rangle \cdot \langle 1, 2, 0\rangle = -4+4+0=0$, so $\mathbf{u} = \langle 1, 2, 0\rangle$ is perpendicular to $\langle -4,2,5\rangle$. For any c, $\langle -2, 1, c\rangle \cdot \langle 1, 2, 0\rangle = -2+2+0=0$ so $\mathbf{v} = \langle -2, 1, c\rangle$ is a candidate.
$\langle -4, 2, 5\rangle \cdot \langle -2, 1, c\rangle = 8+2+5c$
$8 + 2 + 5c = 0 \Rightarrow c = -2$, so one pair is $\mathbf{u} = \langle 1, 2, 0\rangle$, $\mathbf{v} = \langle -2,1,-2\rangle$.

43. The following do not make sense.

a. $\mathbf{v} \cdot \mathbf{w}$ is not a vector.

b. $\mathbf{u} \cdot \mathbf{w}$ is not a vector.

45. $a\mathbf{u} + b\mathbf{u} = a\langle u_1, u_2\rangle + b\langle u_1, u_2\rangle$
$= \langle au_1, au_2\rangle + \langle bu_1, bu_2\rangle = \langle au_1 + bu_1, au_2 + bu_2\rangle$
$\langle (a+b)u_1, (a+b)u_2\rangle = (a+b)\langle u_1, u_2\rangle$
$= (a+b)\mathbf{u}$

47. $c(\mathbf{u} \cdot \mathbf{v}) = c(\langle u_1, u_2\rangle \cdot \langle v_1, v_2\rangle)$
$= c(u_1v_1 + u_2v_2) = c(u_1v_1) + c(u_2v_2)$
$= (cu_1)v_1 + (cu_2)v_2 = \langle cu_1, cu_2\rangle \cdot \langle v_1, v_2\rangle$
$= (c\langle u_1, u_2\rangle) \cdot \langle v_1, v_2\rangle = (c\mathbf{u}) \cdot \mathbf{v}$

49. $\mathbf{0} \cdot \mathbf{u} = 0u_1 + 0u_2 = 0$

51. $\mathbf{r} = k\mathbf{a} + m\mathbf{b} \Rightarrow 7 = k(3) + m(-3)$ and
$-8 = k(-2) + m(4)$
$3k - 3m = 7$
$-2k + 4m = -8$
Solve the system of equations to get
$k = \frac{2}{3}, m = -\frac{5}{3}$.

53. a and b cannot both be zero. If $a = 0$, then the line $ax + by = c$ is horizontal and $\mathbf{n} = b\mathbf{j}$ is vertical, so $\mathbf{n}$ is perpendicular to the line. Use a similar argument if $b = 0$. If $a \neq 0$ and $b \neq 0$, then $P_1\left(\frac{c}{a}, 0\right)$ and $P_2\left(0, \frac{c}{b}\right)$ are points on the line.

$$\mathbf{n} \cdot \overrightarrow{P_1P_2} = (a\mathbf{i} + b\mathbf{j}) \cdot \left(-\frac{c}{a}\mathbf{i} + \frac{c}{b}\mathbf{j}\right) = -c + c = 0$$

55. $\|\mathbf{u}+\mathbf{v}\|^2 - \|\mathbf{u}-\mathbf{v}\|^2 = (\mathbf{u}+\mathbf{v}) \cdot (\mathbf{u}+\mathbf{v})$
$-(\mathbf{u}-\mathbf{v}) \cdot (\mathbf{u}-\mathbf{v}) = [\mathbf{u} \cdot \mathbf{u} + 2(\mathbf{u} \cdot \mathbf{v}) + \mathbf{v} \cdot \mathbf{v}]$
$-[\mathbf{u} \cdot \mathbf{u} - 2(\mathbf{u} \cdot \mathbf{v}) + \mathbf{v} \cdot \mathbf{v}] = 4(\mathbf{u} \cdot \mathbf{v})$
so $\mathbf{u} \cdot \mathbf{v} = \frac{1}{4}\|\mathbf{u}+\mathbf{v}\|^2 - \frac{1}{4}\|\mathbf{u}-\mathbf{v}\|^2$

57. Place the box so that its corners are at the points (0, 0, 0), (4, 0, 0), (0, 6, 0), (4, 6, 0), (0, 0, 10), (4, 0, 10), (0, 6, 10), and (4, 6, 10). The main diagonals are (0, 0, 0) to (4, 6, 10), (4, 0, 0) to (0, 6, 10), (0, 6, 0) to (4, 0, 10), and (0, 0, 10) to (4, 6, 0). The corresponding vectors are $\langle 4, 6, 10\rangle, \langle -4,6,10\rangle$, $\langle 4,-6,10\rangle$, and $\langle 4,6,-10\rangle$.
All of these vectors have length $\sqrt{16+36+100} = \sqrt{152}$. Thus, the smallest angle θ between any pair, $\mathbf{u}$ and $\mathbf{v}$ of the diagonals is found from the largest value of $\mathbf{u} \cdot \mathbf{v}$, since

$$\cos\theta = \frac{\mathbf{u} \cdot \mathbf{v}}{\|\mathbf{u}\|\|\mathbf{v}\|} = \frac{\mathbf{u} \cdot \mathbf{v}}{152}.$$

There are six ways of pairing the four vectors. The largest value of $\mathbf{u} \cdot \mathbf{v}$ is 120 which occurs with
$\mathbf{u} = \langle 4, 6, 10\rangle$ and $\mathbf{v} = \langle -4,6,10\rangle$. Thus,

$$\cos\theta = \frac{120}{152} = \frac{15}{19} \text{ so } \theta = \cos^{-1}\frac{15}{19} \approx 37.86°.$$

59. Work $= \mathbf{F} \cdot \mathbf{D} = (3\mathbf{i} + 10\mathbf{j}) \cdot (10\mathbf{j})$
$= 0 + 100 = 100$ joules

61. $\mathbf{D} = 5\mathbf{i} + 8\mathbf{j}$
Work $= \mathbf{F} \cdot \mathbf{D} = (6)(5) + (8)(8) = 94$ ft-lb

63. $\mathbf{D} = (4-0)\mathbf{i} + (4-0)\mathbf{j} + (0-8)\mathbf{k} = 4 + 4\mathbf{j} - 8\,\mathbf{k}$
Thus, $W = \mathbf{F} \cdot \mathbf{D} = 0(4) + 0(4) - 4(-8) = 32$ joules.

65. $2(x-1) - 4(y-2) + 3(z+3) = 0$
$2x - 4y + 3z = -15$

67. $(x-1) + 4(y-2) + 4(z-1) = 0$
$x + 4y + 4z = 13$

69. The planes are $2x - 4y + 3z = -15$ and $3x - 2y - z = -4$. The normals to the planes are $\mathbf{u} = \langle 2,-4,3\rangle$ and $\mathbf{v} = \langle 3,-2,-1\rangle$. If θ is the angle between the planes,

$$\cos\theta = \frac{\mathbf{u} \cdot \mathbf{v}}{\|\mathbf{u}\|\|\mathbf{v}\|} = \frac{6+8-3}{\sqrt{29}\sqrt{14}} = \frac{11}{\sqrt{406}}, \text{ so}$$

$$\theta = \cos^{-1}\frac{11}{\sqrt{406}} \approx 56.91°.$$

71. a. Planes parallel to the xy-plane may be expressed as $z = D$, so $z = 2$ is an equation of the plane.

b. An equation of the plane is $2(x + 4) - 3(y + 1) - 4(z - 2) = 0$ or $2x - 3y - 4z = -13$.

73. Distance $= \frac{|(1)+3(-1)+(2)-7|}{\sqrt{1+9+1}} = \frac{7}{\sqrt{11}} \approx 2.1106$

75. (0, 0, 9) is on $-3x + 2y + z = 9$. The distance from (0, 0, 9) to $6x - 4y - 2z = 19$ is $\frac{|6(0)-4(0)-2(9)-19|}{\sqrt{36+16+4}} = \frac{37}{\sqrt{56}} \approx 4.9443$ is the distance between the planes.

77. The equation of the sphere in standard form is $(x+1)^2 + (y+3)^2 + (z-4)^2 = 26$, so its center is $(-1, -3, 4)$ and radius is $\sqrt{26}$. The distance from the sphere to the plane is the distance from the center to the plane minus the radius of the sphere or

$$\frac{|3(-1)+4(-3)+1(4)-15|}{\sqrt{9+16+1}} - \sqrt{26} = \sqrt{26} - \sqrt{26} = 0,$$

so the sphere is tangent to the plane.

79. $|\mathbf{u}\cdot\mathbf{v}| = |\cos\theta|\|\mathbf{u}\|\|\mathbf{v}\| \le \|\mathbf{u}\|\|\mathbf{v}\|$ since $|\cos\theta| \le 1$.

81. The 3 wires must offset the weight of the object, thus
$(3\mathbf{i} + 4\mathbf{j} + 15\mathbf{k}) + (-8\mathbf{i} - 2\mathbf{j} + 10\mathbf{k}) + (a\mathbf{i} + b\mathbf{j} + c\mathbf{k})$
$= 0\mathbf{i} + 0\mathbf{j} + 30\mathbf{k}$
Thus, $3 - 8 + a = 0$, so $a = 5$;
$4 - 2 + b = 0$, so $b = -2$;
$15 + 10 + c = 30$, so $c = 5$.

83. Let $\mathbf{x} = \langle x, y, z\rangle$, so

$$(\mathbf{x}-\mathbf{a})\cdot(\mathbf{x}-\mathbf{b}) =$$
$$\langle x-a_1, y-a_2, z-a_3\rangle \cdot \langle x-b_1, y-b_2, z-b_3\rangle$$
$$= x^2 - (a_1+b_1)x + a_1b_1 + y^2 - (a_2+b_2)y + a_2b_2$$
$$+ z^2 - (a_3+b_3)z + a_3b_3$$

Setting this equal to 0 and completing the squares yields

$$\left(x - \frac{a_1+b_1}{2}\right)^2 + \left(y - \frac{a_2+b_2}{2}\right)^2 + \left(z - \frac{a_3+b_3}{2}\right)^2$$
$$= \frac{1}{4}\left[(a_1-b_1)^2 + (a_2-b_2)^2 + (a_3-b_3)^2\right]$$

A sphere with center $\left(\frac{a_1+b_1}{2}, \frac{a_2+b_2}{2}, \frac{a_3+b_3}{2}\right)$

and radius

$$\frac{1}{4}\left[(a_1-b_1)^2 + (a_2-b_2)^2 + (a_3-b_3)^2\right] = \frac{1}{4}|\mathbf{a}-\mathbf{b}|^2$$

85. If **a**, **b**, and **c** are the position vectors of the vertices labeled *A*, *B*, and *C*, respectively, then the side *BC* is represented by the vector **c** – **b**. The position vector of the midpoint of *BC* is $\mathbf{b} + \frac{1}{2}(\mathbf{c}-\mathbf{b}) = \frac{1}{2}(\mathbf{b}+\mathbf{c})$. The segment from *A* to the midpoint of *BC* is $\frac{1}{2}(\mathbf{b}+\mathbf{c}) - \mathbf{a}$. Thus, the position vector of *P* is

$$\mathbf{a} + \frac{2}{3}\left[\frac{1}{2}(\mathbf{b}+\mathbf{c}) - \mathbf{a}\right] = \frac{\mathbf{a}+\mathbf{b}+\mathbf{c}}{3}$$

If the vertices are (2, 6, 5), (4, –1, 2), and (6, 1, 2), the corresponding position vectors are $\langle 2, 6, 5\rangle$, $\langle 4, -1, 2\rangle$, and $\langle 6, 1, 2\rangle$. The position vector of *P* is

$$\frac{1}{3}\langle 2+4+6,\ 6-1+1,\ 5+2+2\rangle = \frac{1}{3}\langle 12, 6, 9\rangle =$$

$\langle 4, 2, 3\rangle$. Thus *P* is (4, 2, 3).

87. After reflecting from the *xy*-plane, the ray has direction $a\mathbf{i} + b\mathbf{j} - c\mathbf{k}$. After reflecting from the *xz*-plane, the ray now has direction $a\mathbf{i} - b\mathbf{j} - c\mathbf{k}$. After reflecting from the *yz*-plane, the ray now has direction $-a\mathbf{i} - b\mathbf{j} - c\mathbf{k}$, the opposite of its original direction. ($a < 0, b < 0, c < 0$)

11.4 Concepts Review

1. $\begin{vmatrix} \mathbf{i} & \mathbf{j} & \mathbf{k} \\ -1 & 2 & 1 \\ 3 & 1 & -1 \end{vmatrix} = \begin{vmatrix} 2 & 1 \\ 1 & -1 \end{vmatrix}\mathbf{i} - \begin{vmatrix} -1 & 1 \\ 3 & -1 \end{vmatrix}\mathbf{j} + \begin{vmatrix} -1 & 2 \\ 3 & 1 \end{vmatrix}\mathbf{k}$
$= (-2-1)\mathbf{i} - (1-3)\mathbf{j} + (-1-6)\mathbf{k}$
$= -3\mathbf{i} + 2\mathbf{j} - 7\mathbf{k} = \langle -3, 2, -7\rangle$

3. $-(\mathbf{v} \times \mathbf{u})$

Problem Set 11.4

1. a. $\mathbf{a}\times\mathbf{b} = \begin{vmatrix} \mathbf{i} & \mathbf{j} & \mathbf{k} \\ -3 & 2 & -2 \\ -1 & 2 & -4 \end{vmatrix}$

$= \begin{vmatrix} 2 & -2 \\ 2 & -4 \end{vmatrix}\mathbf{i} - \begin{vmatrix} -3 & -2 \\ -1 & -4 \end{vmatrix}\mathbf{j} + \begin{vmatrix} -3 & 2 \\ -1 & 2 \end{vmatrix}\mathbf{k}$
$= (-8+4)\mathbf{i} - (12-2)\mathbf{j} + (-6+2)\mathbf{k}$
$= -4\mathbf{i} - 10\mathbf{j} - 4\mathbf{k}$

b. $\mathbf{b}+\mathbf{c}=6\mathbf{i}+5\mathbf{j}-8\mathbf{k}$, so

$$\mathbf{a}\times(\mathbf{b}+\mathbf{c})=\begin{vmatrix}\mathbf{i} & \mathbf{j} & \mathbf{k}\\ -3 & 2 & -2\\ 6 & 5 & -8\end{vmatrix}$$

$$=\begin{vmatrix}2 & -2\\ 5 & -8\end{vmatrix}\mathbf{i}-\begin{vmatrix}-3 & -2\\ 6 & -8\end{vmatrix}\mathbf{j}+\begin{vmatrix}-3 & 2\\ 6 & 5\end{vmatrix}\mathbf{k}$$

$=(-16+10)\mathbf{i}-(24+12)\mathbf{j}+(-15-12)\mathbf{k}$
$=-6\mathbf{i}-36\mathbf{j}-27\mathbf{k}$

c. $\mathbf{a}\cdot(\mathbf{b}+\mathbf{c})=-3(6)+2(5)-2(-8)=8$

d. $$\mathbf{b}\times\mathbf{c}=\begin{vmatrix}\mathbf{i} & \mathbf{j} & \mathbf{k}\\ -1 & 2 & -4\\ 7 & 3 & -4\end{vmatrix}$$

$$=\begin{vmatrix}2 & -4\\ 3 & -4\end{vmatrix}\mathbf{i}-\begin{vmatrix}-1 & -4\\ 7 & -4\end{vmatrix}\mathbf{j}+\begin{vmatrix}-1 & 2\\ 7 & 3\end{vmatrix}\mathbf{k}$$

$=(-8+12)\mathbf{i}-(4+28)\mathbf{j}+(-3-14)\mathbf{k}$
$=4\mathbf{i}-32\mathbf{j}-17\mathbf{k}$

$$\mathbf{a}\times(\mathbf{b}\times\mathbf{c})=\begin{vmatrix}\mathbf{i} & \mathbf{j} & \mathbf{k}\\ -3 & 2 & -2\\ 4 & -32 & -17\end{vmatrix}$$

$$=\begin{vmatrix}2 & -2\\ -32 & -17\end{vmatrix}\mathbf{i}-\begin{vmatrix}-3 & -2\\ 4 & -17\end{vmatrix}\mathbf{j}+\begin{vmatrix}-3 & 2\\ 4 & -32\end{vmatrix}\mathbf{k}$$

$=(-34-64)\mathbf{i}-(51+8)\mathbf{j}+(96-8)\mathbf{k}$
$=-98\mathbf{i}-59\mathbf{j}+88\mathbf{k}$

3. $$\mathbf{a}\times\mathbf{b}=\begin{vmatrix}\mathbf{i} & \mathbf{j} & \mathbf{k}\\ 1 & 2 & 3\\ -2 & 2 & -4\end{vmatrix}$$

$$=\begin{vmatrix}2 & 3\\ 2 & -4\end{vmatrix}\mathbf{i}-\begin{vmatrix}1 & 3\\ -2 & -4\end{vmatrix}\mathbf{j}+\begin{vmatrix}1 & 2\\ -2 & 2\end{vmatrix}\mathbf{k}$$

$=(-8-6)\mathbf{i}-(-4+6)\mathbf{j}+(2+4)\mathbf{k}=-14\mathbf{i}-2\mathbf{j}+6\mathbf{k}$
is perpendicular to both **a** and **b**. All perpendicular vectors will have the form $c(-14\mathbf{i}-2\mathbf{j}+6\mathbf{k})$ where c is a real number.

5. $\mathbf{u}=\langle 3-1,-1-3,2-5\rangle=\langle 2,-4,-3\rangle$ and
$\mathbf{v}=\langle 4-1,0-3,1-5\rangle=\langle 3,-3,-4\rangle$ are in the plane.

$$\mathbf{u}\times\mathbf{v}=\begin{vmatrix}\mathbf{i} & \mathbf{j} & \mathbf{k}\\ 2 & -4 & -3\\ 3 & -3 & -4\end{vmatrix}$$

$$=\begin{vmatrix}-4 & -3\\ -3 & -4\end{vmatrix}\mathbf{i}-\begin{vmatrix}2 & -3\\ 3 & -4\end{vmatrix}\mathbf{j}+\begin{vmatrix}2 & -4\\ 3 & -3\end{vmatrix}\mathbf{k}$$

$=(16-9)\mathbf{i}-(-8+9)\mathbf{j}+(-6+12)\mathbf{k}$
$=\langle 7,-1,6\rangle$ is perpendicular to the plane.

$$\pm\frac{1}{\sqrt{49+1+36}}\langle 7,-1,6\rangle=\pm\left\langle\frac{7}{\sqrt{86}},-\frac{1}{\sqrt{86}},\frac{6}{\sqrt{86}}\right\rangle$$

are the unit vectors perpendicular to the plane.

7. $$\mathbf{a}\times\mathbf{b}=\begin{vmatrix}\mathbf{i} & \mathbf{j} & \mathbf{k}\\ -1 & 1 & -3\\ 4 & 2 & -4\end{vmatrix}$$

$$=\begin{vmatrix}1 & -3\\ 2 & -4\end{vmatrix}\mathbf{i}-\begin{vmatrix}-1 & -3\\ 4 & -4\end{vmatrix}\mathbf{j}+\begin{vmatrix}-1 & 1\\ 4 & 2\end{vmatrix}\mathbf{k}$$

$=(-4+6)\mathbf{i}-(4+12)\mathbf{j}+(-2-4)\mathbf{k}$
$=2\mathbf{i}-16\mathbf{j}-6\mathbf{k}$

Area of parallelogram $=\|\mathbf{a}\times\mathbf{b}\|$
$=\sqrt{4+256+36}=2\sqrt{74}$

9. $\mathbf{a}=\langle 2-3,4-2,6-1\rangle=\langle -1,2,5\rangle$ and
$\mathbf{b}=\langle -1-3,2-2,5-1\rangle=\langle -4,0,4\rangle$ are adjacent sides of the triangle. The area of the triangle is half the area of the parallelogram with **a** and **b** as adjacent sides.

$$\mathbf{a}\times\mathbf{b}=\begin{vmatrix}\mathbf{i} & \mathbf{j} & \mathbf{k}\\ -1 & 2 & 5\\ -4 & 0 & 4\end{vmatrix}$$

$$=\begin{vmatrix}2 & 5\\ 0 & 4\end{vmatrix}\mathbf{i}-\begin{vmatrix}-1 & 5\\ -4 & 4\end{vmatrix}\mathbf{j}+\begin{vmatrix}-1 & 2\\ -4 & 0\end{vmatrix}\mathbf{k}$$

$=(8-0)\mathbf{i}-(-4+20)\mathbf{j}+(0+8)\mathbf{k}=\langle 8,-16,8\rangle$

Area of triangle $=\frac{1}{2}\sqrt{64+256+64}=\frac{1}{2}\left(8\sqrt{6}\right)$
$=4\sqrt{6}$

11. $\mathbf{u}=\langle 0-1,3-3,0-2\rangle=\langle -1,0,-2\rangle$ and
$\mathbf{v}=\langle 2-1,4-3,3-2\rangle=\langle 1,1,1\rangle$ are in the plane.

$$\mathbf{u}\times\mathbf{v}=\begin{vmatrix}\mathbf{i} & \mathbf{j} & \mathbf{k}\\ -1 & 0 & -2\\ 1 & 1 & 1\end{vmatrix}$$

$$=\begin{vmatrix}0 & -2\\ 1 & 1\end{vmatrix}\mathbf{i}-\begin{vmatrix}-1 & -2\\ 1 & 1\end{vmatrix}\mathbf{j}+\begin{vmatrix}-1 & 0\\ 1 & 1\end{vmatrix}\mathbf{k}$$

$=(0+2)\mathbf{i}-(-1+2)\mathbf{j}+(-1-0)\mathbf{k}=\langle 2,-1,-1\rangle$

The plane through (1, 3, 2) with normal $\langle 2,-1,-1\rangle$ has equation
$2(x-1)-1(y-3)-1(z-2)=0$ or
$2x-y-z=-3$.

13. $\mathbf{u} = \langle 0-7, 3-0, 0-0\rangle = \langle -7, 3, 0\rangle$ and $\mathbf{v} = \langle 0-7, 0-0, 5-0\rangle = \langle -7, 0, 5\rangle$ are in the plane.

$$\mathbf{u}\times\mathbf{v} = \begin{vmatrix} \mathbf{i} & \mathbf{j} & \mathbf{k} \\ -7 & 3 & 0 \\ -7 & 0 & 5 \end{vmatrix} =$$

$$\begin{vmatrix} 3 & 0 \\ 0 & 5 \end{vmatrix}\mathbf{i} - \begin{vmatrix} -7 & 0 \\ -7 & 5 \end{vmatrix}\mathbf{j} + \begin{vmatrix} -7 & 3 \\ -7 & 0 \end{vmatrix}\mathbf{k}$$

$= (15-0)\mathbf{i} - (-35-0)\mathbf{j} + (0+21)\mathbf{k} = \langle 15, 35, 21\rangle$

The plane through (7, 0, 0) with normal $\langle 15, 35, 21\rangle$ has equation $15(x-7) + 35(y-0) + 21(z-0) = 0$ or $15x + 35y + 21z = 105$.

15. An equation of the plane is $1(x-2) - 1(y-5) + 2(z-1) = 0$ or $x - y + 2z = -1$.

17. The plane's normals will be perpendicular to the normals of the other two planes. Then a normal is $\langle 1, -3, 2\rangle \times \langle 2, -2, -1\rangle = \langle 7, 5, 4\rangle$. An equation of the plane is $7(x+1) + 5(y+2) + 4(z-3) = 0$ or $7x + 5y + 4z = -5$.

19. $(4\mathbf{i} + 3\mathbf{j} - \mathbf{k}) \times (2\mathbf{i} - 5\mathbf{j} + 6\mathbf{k}) = 13\mathbf{i} - 26\mathbf{j} - 26\mathbf{k}$ $= 13(\mathbf{i} - 2\mathbf{j} - 2\mathbf{k})$ is normal to the plane. An equation of the plane is $1(x-2) - 2(y+3) - 2(z-2) = 0$ or $x - 2y - 2z = 4$.

21. Each vector normal to the plane we seek is parallel to the line of intersection of the given planes. Also, the cross product of vectors normal to the given planes will be parallel to both planes, hence parallel to the line of intersection. Thus, $\langle 4, -3, 2\rangle \times \langle 3, 2, -1\rangle = \langle -1, 10, 17\rangle$ is normal to the plane we seek. An equation of the plane is $-1(x-6) + 10(y-2) + 17(z+1) = 0$ or $-x + 10y + 17z = -3$.

23. $\text{Volume} = |\langle 2, 3, 4\rangle \cdot (\langle 0, 4, -1\rangle \times \langle 5, 1, 3\rangle)| = |\langle 2, 3, 4\rangle \cdot \langle 13, -5, -20\rangle| = |-69| = 69$

25. **a.** $\text{Volume} = |\mathbf{u}\cdot(\mathbf{v}\times\mathbf{w})| = |\langle 3, 2, 1\rangle \cdot \langle -3, -1, 2\rangle| = |-9| = 9$

b. $\text{Area} = \|\mathbf{u}\times\mathbf{v}\| = |\langle 3, -5, 1\rangle| = \sqrt{9+25+1} = \sqrt{35}$

c. Let θ be the angle. Then θ is the complement of the smaller angle between **u** and **v** × **w**.

$$\cos\left(\frac{\pi}{2} - \theta\right) = \sin\theta = \frac{|\mathbf{u}\cdot(\mathbf{v}\times\mathbf{w})|}{\|\mathbf{u}\|\|\mathbf{v}\times\mathbf{w}\|} = \frac{9}{\sqrt{9+4+1}\sqrt{9+1+4}} = \frac{9}{14},\ \theta \approx 40.01°$$

27. Choice (c) does not make sense because $(\mathbf{a}\cdot\mathbf{b})$ is a scalar and can't be crossed with a vector. Choice (d) does not make sense because $(\mathbf{a}\times\mathbf{b})$ is a vector and can't be added to a constant.

29. Let **b** and **c** determine the (triangular) base of the tetrahedron. Then the area of the base is $\frac{1}{2}\|\mathbf{b}\times\mathbf{c}\|$ which is half of the area of the parallelogram determined by **b** and **c**. Thus,

$$\frac{1}{3}(\text{area of base})(\text{height}) = \frac{1}{3}\left[\frac{1}{2}(\text{area of corresponding parallelogram})(\text{height})\right]$$

$$= \frac{1}{6}(\text{area of corresponding parallelpiped}) = \frac{1}{6}|\mathbf{a}\cdot(\mathbf{b}\times\mathbf{c})|$$

31. Let $\mathbf{u} = \langle u_1, u_2, u_3 \rangle$ and $\mathbf{v} = \langle v_1, v_2, v_3 \rangle$ then

$$\mathbf{u} \times \mathbf{v} = \langle u_2v_3 - u_3v_2, u_3v_1 - u_1v_3, u_1v_2 - u_2v_1 \rangle$$

$$\|\mathbf{u} \times \mathbf{v}\|^2 = u_2^2v_3^2 - 2u_2u_3v_2v_3 + u_3^2v_2^2 + u_3^2v_1^2 - 2u_1u_3v_1v_3 + u_1^2v_3^2 + u_1^2v_2^2 - 2u_1u_2v_1v_2 + u_2^2v_1^2$$

$$= u_1^2(v_1^2 + v_2^2 + v_3^2) - u_1^2v_1^2 + u_2^2(v_1^2 + v_2^2 + v_3^2) - u_2^2v_2^2 + u_3^2(v_1^2 + v_2^2 + v_3^2)$$

$$-u_3^2v_3^2 - 2u_2u_3v_2v_3 - 2u_1u_3v_1v_3 - 2u_1u_2v_1v_2$$

$$= (u_1^2 + u_2^2 + u_3^2)(v_1^2 + v_2^2 + v_3^2) - (u_1^2v_1^2 + u_2^2v_2^2 + u_3^2v_3^2 + 2u_2u_3v_2v_3 + 2u_1u_3v_1v_3 + 2u_1u_2v_1v_2)$$

$$= (u_1^2 + u_2^2 + u_3^2)(v_1^2 + v_2^2 + v_3^2) - (u_1v_1 + u_2v_2 + u_3v_3)^2 = \|\mathbf{u}\|^2\|\mathbf{v}\|^2 - (\mathbf{u} \cdot \mathbf{v})^2$$

33. $(\mathbf{v} + \mathbf{w}) \times \mathbf{u} = -[\mathbf{u} \times (\mathbf{v} + \mathbf{w})]$

$= -[(\mathbf{u} \times \mathbf{v}) + (\mathbf{u} \times \mathbf{w})] = -(\mathbf{u} \times \mathbf{v}) - (\mathbf{u} \times \mathbf{w})$

$= (\mathbf{v} \times \mathbf{u}) + (\mathbf{w} \times \mathbf{u})$

35. $\overrightarrow{PQ} = \langle -a, b, 0 \rangle$, $\overrightarrow{PR} = \langle -a, 0, c \rangle$,

The area of the triangle is half the area of the parallelogram with $\overrightarrow{PQ}$ and $\overrightarrow{PR}$ as adjacent sides, so area

$(\Delta PQR) = \frac{1}{2}\|\langle -a, b, 0 \rangle \times \langle -a, 0, c \rangle\| = \frac{1}{2}\|\langle bc, ac, ab \rangle\| = \frac{1}{2}\sqrt{b^2c^2 + a^2c^2 + a^2b^2}$.

37. From Problem 35, $D^2 = \frac{1}{4}(b^2c^2 + a^2c^2 + a^2b^2) = \left(\frac{1}{2}bc\right)^2 + \left(\frac{1}{2}ac\right)^2 + \left(\frac{1}{2}ab\right)^2 = A^2 + B^2 + C^2$.

39. The area of the triangle is $A = \frac{1}{2}\|\mathbf{a} \times \mathbf{b}\|$. Thus,

$$A^2 = \frac{1}{4}\|\mathbf{a} \times \mathbf{b}\|^2 = \frac{1}{4}\left(\|\mathbf{a}\|^2\|\mathbf{b}\|^2 - (\mathbf{a} \cdot \mathbf{b})^2\right) = \frac{1}{4}\left[\|\mathbf{a}\|^2\|\mathbf{b}\|^2 - \frac{1}{4}\left(\|\mathbf{a}\|^2 + \|\mathbf{b}\|^2 - \|\mathbf{a} - \mathbf{b}\|^2\right)^2\right]$$

$$= \frac{1}{16}\left[4a^2b^2 - (a^2 + b^2 - c^2)^2\right] = \frac{1}{16}(2a^2b^2 - a^4 + 2a^2c^2 - b^4 + 2b^2c^2 - c^4).$$

Note that $s - a = \frac{1}{2}(b + c - a)$,

$s - b = \frac{1}{2}(a + c - b)$, and $s - c = \frac{1}{2}(a + b - c)$.

$s(s - a)(s - b)(s - c)$

$= \frac{1}{16}(a + b + c)(b + c - a)(a + c - b)(a + b - c)$

$= \frac{1}{16}(2a^2b^2 - a^4 + 2a^2c^2 - b^4 + 2b^2c^2 - c^4)$

which is the same as was obtained above.

11.5 Concepts Review

1. a vector-valued function of a real variable

3. position

Problem Set 11.5

1. $\lim_{t\to1}[2t\mathbf{i}-t^2\mathbf{j}]=\lim_{t\to1}(2t)\mathbf{i}-\lim_{t\to1}(t^2)\mathbf{j}=2\mathbf{i}-\mathbf{j}$

3. $\lim_{t\to1}\left[\frac{t-1}{t^2-1}\mathbf{i}-\frac{t^2+2t-3}{t-1}\mathbf{j}\right]$

$=\lim_{t\to1}\left(\frac{t-1}{(t-1)(t+1)}\right)\mathbf{i}-\lim_{t\to1}\left(\frac{(t-1)(t+3)}{t-1}\right)\mathbf{j}$

$=\lim_{t\to1}\left(\frac{1}{t+1}\right)\mathbf{i}-\lim_{t\to1}(t+3)\mathbf{j}=\frac{1}{2}\mathbf{i}-4\mathbf{j}$

5. $\lim_{t\to0}\left[\frac{\sin t\cos t}{t}\mathbf{i}-\frac{7t^3}{e^t}\mathbf{j}+\frac{t}{t+1}\mathbf{k}\right]$

$=\lim_{t\to0}\left(\frac{\sin t\cos t}{t}\right)\mathbf{i}-\lim_{t\to0}\left(\frac{7t^3}{e^t}\right)\mathbf{j}$

$+\lim_{t\to0}\left(\frac{t}{t+1}\right)\mathbf{k}=\mathbf{i}$

7. $\lim_{t\to0^+}\left\langle\ln(t^3),t^2\ln t,t\right\rangle$ does not exist because $\lim_{t\to0^+}\ln(t^3)=-\infty.$

9. a. The domain of $f(t)=\frac{2}{t-4}$ is $(-\infty,4)\cup(4,\infty)$. The domain of $g(t)=\sqrt{3-t}$ is $(-\infty,3]$. The domain of $h(t)=\ln|4-t|$ is $(-\infty,4)\cup(4,\infty)$. Thus, the domain of **r** is $(-\infty,3]$ or $\{t\in\mathbb{R}:t\le3\}$.

b. The domain of $f(t)=[\![t^2]\!]$ is $(-\infty,\infty)$. The domain of $g(t)=\sqrt{20-t}$ is $(-\infty,20]$. The domain of $h(t)=3$ is $(-\infty,\infty)$. Thus, the domain of **r** is $(-\infty,20]$ or $\{t\in\mathbb{R}:t\le20\}$.

c. The domain of $f(t)=\cos t$ is $(-\infty,\infty)$. The domain of $g(t)=\sin t$ is also $(-\infty,\infty)$. The domain of $h(t)=\sqrt{9-t^2}$ is $[-3,3]$. Thus, the domain of **r** is $[-3,3]$, or $\{t\in\mathbb{R}:-3\le t\le3\}$.

11. a. $f(t)=\frac{2}{t-4}$ is continuous on $(-\infty,4)\cup(4,\infty)$. $g(t)=\sqrt{3-t}$ is continuous on $(-\infty,3]$. $h(t)=\ln|4-t|$ is continuous on $(-\infty,4)$ and on $(4,\infty)$. Thus, **r** is continuous on
$(-\infty,3]$ or $\{t\in\mathbb{R}:t\le3\}$.

b. $f(t)=[\![t^2]\!]$ is continuous on $\left(-\sqrt{n+1},-\sqrt{n}\right)\cup\left(\sqrt{n},\sqrt{n+1}\right)$ where n is a non-negative integer. $g(t)=\sqrt{20-t}$ is continuous on $(-\infty,20)$ or $\{t\in\mathbb{R}:t<20\}$. $h(t)=3$ is continuous on $(-\infty,\infty)$. Thus, **r** is continuous on $\left(-\sqrt{n+1},-\sqrt{n}\right)\cup\left(\sqrt{k},\sqrt{k+1}\right)$ where n and k are non-negative integers and $k<400$ or $\{t\in\mathbb{R}:t<20,t^2$ not an integer$\}$.

c. $f(t)=\cos t$ and $f(t)=\sin t$ are continuous on $(-\infty,\infty)$. $h(t)=\sqrt{9-t^2}$ is continuous on $[-3,3]$. Thus, **r** is continuous on $[-3,3]$.

13. a. $D_t\mathbf{r}(t)=9(3t+4)^2\mathbf{i}+2te^{t^2}\mathbf{j}+0\mathbf{k}$

$D_t^2\mathbf{r}(t)=54(3t+4)\mathbf{i}+2(2t^2+1)e^{t^2}\mathbf{j}$

b. $D_t\mathbf{r}(t)=\sin 2t\mathbf{i}-3\sin 3t\mathbf{j}+2t\mathbf{k}$

$D_t^2\mathbf{r}(t)=2\cos 2t\mathbf{i}-9\cos 3t\mathbf{j}+2\mathbf{k}$

15. $\mathbf{r}'(t)=-e^{-t}\mathbf{i}-\frac{2}{t}\mathbf{j};\ \mathbf{r}''(t)=e^{-t}\mathbf{i}+\frac{2}{t^2}\mathbf{j}$

$\mathbf{r}(t)\cdot\mathbf{r}''(t)=e^{-2t}-\frac{2}{t^2}\ln\left(t^2\right)$

$D_t[\mathbf{r}(t)\cdot\mathbf{r}''(t)]=-2e^{-2t}-\left[\frac{2}{t^2}\cdot\frac{1}{t^2}\cdot2t-\frac{4}{t^3}\ln\left(t^2\right)\right]$

$=-2e^{-2t}-\frac{4}{t^3}+\frac{4\ln\left(t^2\right)}{t^3}$

17. $h(t)\mathbf{r}(t)=e^{-3t}\sqrt{t-1}\mathbf{i}+e^{-3t}\ln(2t^2)\mathbf{j}$

$D_t[h(t)\mathbf{r}(t)]=-\frac{e^{-3t}}{2}\left(\frac{6t-7}{\sqrt{t-1}}\right)\mathbf{i}$

$+e^{-3t}\left(\frac{2}{t}-3\ln(2t^2)\right)\mathbf{j}$

19. $\mathbf{v}(t) = \mathbf{r}'(t) = 4\mathbf{i} + 10t\mathbf{j} + 2\mathbf{k}$
$\mathbf{a}(t) = \mathbf{r}''(t) = 10\mathbf{j}$
$\mathbf{v}(1) = 4\mathbf{i} + 10\mathbf{j} + 2\mathbf{k}$; $\mathbf{a}(1) = 10\mathbf{j}$;
$s(1) = \sqrt{16+100+4} = 2\sqrt{30} \approx 10.954$

21. $\mathbf{v}(t) = -\frac{1}{t^2}\mathbf{i} - \frac{2t}{(t^2-1)^2}\mathbf{j} + 5t^4\mathbf{k}$

$\mathbf{a}(t) = \frac{2}{t^3}\mathbf{i} + \frac{2+6t^2}{(t^2-1)^3}\mathbf{j} + 20t^3\mathbf{k}$

$\mathbf{v}(2) = -\frac{1}{4}\mathbf{i} - \frac{4}{9}\mathbf{j} + 80\mathbf{k}$; $\mathbf{a}(2) = \frac{1}{4}\mathbf{i} + \frac{26}{27}\mathbf{j} + 160\mathbf{k}$;

$s(2) = \sqrt{\frac{1}{16} + \frac{16}{81} + 6400} = \frac{\sqrt{8,294,737}}{36}$
≈ 80.002

23. $\mathbf{v}(t) = t^2\mathbf{j} + \frac{2}{3}t^{-1/3}\mathbf{k}$; $\mathbf{a}(t) = 2t\mathbf{j} - \frac{2}{9}t^{-4/3}\mathbf{k}$

$\mathbf{v}(2) = 4\mathbf{j} + \frac{2^{2/3}}{3}\mathbf{k}$; $\mathbf{a}(2) = 4\mathbf{j} - \frac{2^{-1/3}}{9}\mathbf{k}$

$s(2) = \sqrt{16 + \frac{2^{4/3}}{9}} \approx 4.035$

$\mathbf{v}(2) = 4\mathbf{j} + \frac{2^{2/3}}{3}\mathbf{k}$; $\mathbf{a}(2) = 4\mathbf{j} - \frac{1}{9\sqrt[3]{2}}\mathbf{k}$;

$s(2) = \sqrt{16 + \frac{2^{4/3}}{9}} \approx 4.035$

25. $\mathbf{v}(t) = -\sin t\,\mathbf{i} + \cos t\,\mathbf{j} + \mathbf{k}$
$\mathbf{a}(t) = -\cos t\,\mathbf{i} - \sin t\,\mathbf{j}$
$\mathbf{v}(\pi) = -\mathbf{j} + \mathbf{k}$; $\mathbf{a}(\pi) = \mathbf{i}$;
$s(\pi) = \sqrt{1+1} = \sqrt{2} \approx 1.414$

27. $\mathbf{v}(t) = \sec^2 t\mathbf{i} + 3e^t\mathbf{j} - 4\sin 4t\mathbf{k}$
$\mathbf{a}(t) = 2\sec^2 t \tan t\mathbf{i} + 3e^t\mathbf{j} - 16\cos 4t\mathbf{k}$

$\mathbf{v}\left(\frac{\pi}{4}\right) = 2\mathbf{i} + 3e^{\pi/4}\mathbf{j}$; $\mathbf{a}\left(\frac{\pi}{4}\right) = 4\mathbf{i} + 3e^{\pi/4}\mathbf{j} + 16\mathbf{k}$;

$s\left(\frac{\pi}{4}\right) = \sqrt{4 + 9e^{\pi/2}} \approx 6.877$

29. $\mathbf{v}(t) = (\pi t\cos\pi t + \sin\pi t)\mathbf{i}$
$+ (\cos\pi t - \pi t\sin\pi t)\mathbf{j} - e^{-t}\mathbf{k}$
$\mathbf{a}(t) = (2\pi\cos\pi t - \pi^2 t\sin\pi t)\mathbf{i}$
$+ (-2\pi\sin\pi t - \pi^2 t\cos\pi t)\mathbf{j} + e^{-t}\mathbf{k}$
$\mathbf{v}(2) = 2\pi\mathbf{i} + \mathbf{j} - e^{-2}\mathbf{k}$; $\mathbf{a}(2) = 2\pi\mathbf{i} - 2\pi^2\mathbf{j} + e^{-2}\mathbf{k}$;
$s(2) = \sqrt{4\pi^2 + 1 + e^{-4}} \approx 6.364$

31. If $\|\mathbf{v}\| = C$, then $\|\mathbf{v}\|^2 = \mathbf{v}\cdot\mathbf{v} = C$ Differentiate implicitly to get $D_t(\mathbf{v}\cdot\mathbf{v}) = 2\mathbf{v}\cdot\mathbf{v}' = 0$. Thus, $\mathbf{v}\cdot\mathbf{v}' = \mathbf{v}\cdot\mathbf{a} = 0$, so **a** is perpendicular to **v**.

33. $s = \int_0^2 \sqrt{1^2 + \cos^2 t + (-\sin t)^2}\,dt$
$= \int_0^2 \sqrt{1 + \cos^2 t + \sin^2 t}\,dt = \int_0^2 \sqrt{1+1}\,dt$
$= \sqrt{2}\int_0^2 dt = \sqrt{2}(2-0) = 2\sqrt{2}$

35. $s = \int_3^6 \sqrt{24t^2 + 4t^4 + 36}\,dt$
$= \int_3^6 2\sqrt{t^4 + 6t^2 + 9}\,dt$
$= \int_3^6 2(t^2+3)\,dt = 2\left[\frac{t^3}{3} + 3t\right]_3^6$
$= 2[72 + 18 - (9 + 9)] = 144$

37. $s = \int_0^1 \sqrt{9t^4 + 36t^4 + 324t^4}\,dt$
$= \int_0^1 3\sqrt{41}t^2\,dt = \left[\sqrt{41}t^3\right]_0^1 = \sqrt{41} \approx 6.403$

39. $\mathbf{f}'(u) = -\sin u\mathbf{i} + 3e^{3u}\mathbf{j}$; $g'(t) = 6t$
$\mathbf{F}'(t) = \mathbf{f}'(g(t))g'(t)$
$= -6t\sin(3t^2 - 4)\mathbf{i} + 18te^{9t^2-12}\mathbf{j}$

41. $\int_0^1 (e^t\mathbf{i} + e^{-t}\mathbf{j})\,dt = \left[e^t\mathbf{i} - e^{-t}\mathbf{j}\right]_0^1$
$= (e-1)\mathbf{i} + (1 - e^{-1})\mathbf{j}$

43. $\mathbf{r}(t) = 5\cos 6t\mathbf{i} + 5\sin 6t\mathbf{j}$
$\mathbf{v}(t) = -30\sin 6t\mathbf{i} + 30\cos 6t\mathbf{j}$
$|\mathbf{v}(t)| = \sqrt{900\sin^2 6t + 900\cos^2 6t} = 30$
$\mathbf{a}(t) = -180\cos 6t\mathbf{i} - 180\sin 6t\mathbf{j}$

45. a. The motion of the planet with respect to the sun can be given by $x = R_p \cos t$, $y = R_p \sin t$.

Assume that when $t = 0$, both the planet and the moon are on the x-axis.

Since the moon orbits the planet 10 times for every time the planet orbits the sun, the motion of the moon with respect to the planet can be given by $x = R_m \cos 10t$, $y = R_m \sin 10t$.

Combining these equations, the motion of the moon with respect to the sun is given by $x = R_p \cos t + R_m \cos 10t$, $y = R_p \sin t + R_m \sin 10t$.

b.

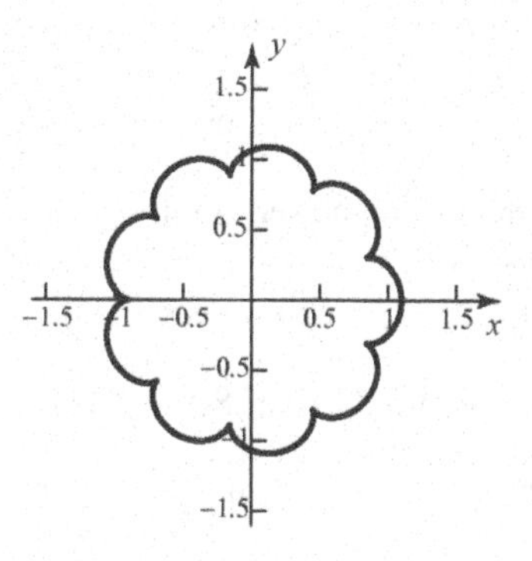

c. $x'(t) = -R_p \sin t - 10R_m \sin 10t$

$y'(t) = R_p \cos t + 10R_m \cos 10t$

The moon is motionless with respect to the sun when $x'(t)$ and $y'(t)$ are both 0.

Solve $x'(t) = 0$ for sin t and $y'(t) = 0$ for cos t to get $\sin t = -\frac{10R_m}{R_p} \sin 10t$,

$\cos t = -\frac{10R_m}{R_p} \cos 10t$.

Since $\sin^2 t + \cos^2 t = 1$

$1 = \frac{100R_m^2}{R_p^2} \sin^2 10t + \frac{100R_m^2}{R_p^2} \cos^2 10t$

$= \frac{100R_m^2}{R_p^2}$. Thus, $R_p^2 = 100R_m^2$ or

$R_p = 10R_m$. Substitute this into $x'(t) = 0$ and $y'(t) = 0$ to get $-R_p(\sin t + \sin 10t) = 0$ and $R_p(\cos t + \cos 10t) = 0$.

If $0 \le t \le \frac{\pi}{2}$, then to have $\sin t + \sin 10t = 0$ and $\cos t + \cos 10t = 0$ it must be that $10t = \pi + t$ or $t = \frac{\pi}{9}$.

Thus, when the radius of the planet's orbit around the sun is ten times the radius of the moon's orbit around the planet and $t = \frac{\pi}{9}$, the moon is motionless with respect to the sun.

47. a. Winding upward around the right circular cylinder $x = \sin t$, $y = \cos t$ as t increases.

b. Same as part a, but winding faster/slower by a factor of $3t^2$.

c. With standard orientation of the axes, the motion is winding to the right around the right circular cylinder $x = \sin t$, $z = \cos t$.

d. Spiraling upward, with increasing radius, along the spiral $x = t \sin t$, $y = t \cos t$.

11.6 Concepts Review

1. $1 + 4t$; $-3 - 2t$; $2 - t$

3. $2t\mathbf{i} - 3\mathbf{j} + 3t^2\mathbf{k}$

Problem Set 11.6

1. A parallel vector is $\mathbf{v} = \langle 4-1, 5+2, 6-3 \rangle = \langle 3, 7, 3 \rangle$.

$x = 1 + 3t$, $y = -2 + 7t$, $z = 3 + 3t$

3. A parallel vector is $\mathbf{v} = \langle 6-4, 2-2, -1-3 \rangle = \langle 2, 0, -4 \rangle$ or $\langle 1, 0, -2 \rangle$.

$x = 4 + t$, $y = 2$, $z = 3 - 2t$

5. $x = 4 + 3t$, $y = 5 + 2t$, $z = 6 + t$

$\frac{x-4}{3} = \frac{y-5}{2} = \frac{z-6}{1}$

7. Another parallel vector is $\langle 1, 10, 100 \rangle$.

$x = 1 + t$, $y = 1 + 10t$, $z = 1 + 100t$

$\frac{x-1}{1} = \frac{y-1}{10} = \frac{z-1}{100}$

9. Set $z=0$. Solving $4x+3y=1$ and $10x+6y=10$ yields $x=4, y=-5$. Thus $P_1(4,-5,0)$ is on the line. Set $y=0$. Solving $4x-7z=1$ and $10x-5z=10$ yields $x=\frac{13}{10}, z=\frac{3}{5}$. Thus $P_2\left(\frac{13}{10},0,\frac{3}{5}\right)$ is on the line.

$\overrightarrow{P_1P_2}=\left\langle\frac{13}{10}-4,0-(-5),\frac{3}{5}-0\right\rangle=\left\langle-\frac{27}{10},5,\frac{3}{5}\right\rangle$ is a direction vector. Thus, $\langle 27,-50,-6\rangle=-10\overrightarrow{P_1P_2}$ is also a direction vector. The symmetric equations are thus

$$\frac{x-4}{27}=\frac{y+5}{-50}=\frac{z}{-6}$$

11. $\mathbf{u}=\langle 1,4,-2\rangle$ and $\mathbf{v}=\langle 2,-1,-2\rangle$ are both perpendicular to the line, so $\mathbf{u}\times\mathbf{v}=\langle -10,-2,-9\rangle$, and hence $\langle 10,2,9\rangle$ is parallel to the line.
With $y=0, x-2z=13$ and $2x-2z=5$ yield $\left(-8,0,-\frac{21}{2}\right)$. The symmetric equations are

$$\frac{x+8}{10}=\frac{y}{2}=\frac{z+\frac{21}{2}}{9}$$

13. $\langle 1,-5,2\rangle$ is a vector in the direction of the line.

$$\frac{x-4}{1}=\frac{y}{-5}=\frac{z-6}{2}$$

15. The point of intersection on the z-axis is $(0,0,4)$. A vector in the direction of the line is $\langle 5-0,-3-0,4-4\rangle=\langle 5,-3,0\rangle$. Parametric equations are $x=5t, y=-3t, z=4$.

17. Using $t=0$ and $t=1$, two points on the first line are $(-2,1,2)$ and $(0,5,1)$. Using $t=0$, a point on the second line is $(2,3,1)$. Thus, two nonparallel vectors in the plane are
$\langle 0+2,5-1,1-2\rangle=\langle 2,4,1\rangle$ and
$\langle 2+2,3-1,1-2\rangle=\langle 4,2,-1\rangle$
Hence, $\langle 2,4,-1\rangle\times\langle 4,2,-1\rangle=\langle -2,-2,-12\rangle$ is a normal to the plane, and so is $\langle 1,1,6\rangle$. An equation of the plane is
$1(x+2)+1(y-1)+6(z-2)=0$ or
$x+y+6z=11$.

19. Using $t=0$, another point in the plane is $(1,-1,4)$ and $\langle 2,3,1\rangle$ is parallel to the plane. Another parallel vector is
$\langle 1-1,-1+1,5-4\rangle=\langle 0,0,1\rangle$. Thus,
$\langle 2,3,1\rangle\times\langle 0,0,1\rangle=\langle 3,-2,0\rangle$ is a normal to the plane. An equation of the plane is
$3(x-1)-2(y+1)+0(z-5)=0$ or $3x-2y=5$.

21. a. With $t=0$ in the first line, $x=2-0=2$, $y=3+4\cdot 0=3$, $z=2\cdot 0=0$, so $(2,3,0)$ is on the first line.

b. $\langle -1,4,2\rangle$ is parallel to the first line, while $\langle 1,0,2\rangle$ is parallel to the second line, so
$\langle -1,4,2\rangle\times\langle 1,0,2\rangle=\langle 8,4,-4\rangle=4\langle 2,1,-1\rangle$
is normal to both. Thus, π has equation
$2(x-2)+1(y-3)-1(z-0)=0$ or
$2x+y-z=7$, and contains the first line.

c. With $t=0$ in the second line,
$x=-1+0=-1, y=2, z=-1+2\cdot 0=-1$, so $Q(-1,2,-1)$ is on the second line.

d. From Example 10 of Section 11.3, the distance from Q to π is

$$\frac{|2(-1)+(2)-(-1)-7|}{\sqrt{4+1+1}}=\frac{6}{\sqrt{6}}=\sqrt{6}\approx 2.449.$$

23. $\mathbf{r}\left(\frac{\pi}{3}\right)=\mathbf{i}+3\sqrt{3}\mathbf{j}+\frac{\pi}{3}\mathbf{k}$, so $\left(1,3\sqrt{3},\frac{\pi}{3}\right)$ is on the tangent line.
$\mathbf{r}'(t)=-2\sin t\mathbf{i}+6\cos t\mathbf{j}+\mathbf{k}$, so
$\mathbf{r}'\left(\frac{\pi}{3}\right)=-\sqrt{3}\mathbf{i}+3\mathbf{j}+\mathbf{k}$ is parallel to the tangent line at $t=\frac{\pi}{3}$. The symmetric equations of the line are $\frac{x-1}{-\sqrt{3}}=\frac{y-3\sqrt{3}}{3}=\frac{z-\frac{\pi}{3}}{1}$.

25. The curve is given by $\mathbf{r}(t)=3t\mathbf{i}+2t^2\mathbf{j}+t^5\mathbf{k}$.
$\mathbf{r}(-1)=-3\mathbf{i}+2\mathbf{j}-\mathbf{k}$, so $(-3,2,-1)$ is on the plane.
$\mathbf{r}'(t)=3\mathbf{i}+4t\mathbf{j}+5t^4\mathbf{k}$, so $\mathbf{r}'(-1)=3\mathbf{i}-4\mathbf{j}+5\mathbf{k}$ is in the direction of the curve at $t=-1$, hence normal to the plane. An equation of the plane is
$3(x+3)-4(y-2)+5(z+1)=0$ or
$3x-4y+5z=-22$.

27. a. $[x(t)]^2+[y(t)]^2+[z(t)]^2$

$=(2t)^2+\left(\sqrt{7t}\right)^2+\left(\sqrt{9-7t-4t^2}\right)^2$

$=4t^2+7t+9-7t-4t^2=9$

Thus, the curve lies on the sphere $x^2+y^2+z^2=9$ whose center is at the origin.

b. $\mathbf{r}(t)=2t\,\mathbf{i}+\sqrt{7t}\,\mathbf{j}+\sqrt{9-7t-4t^2}\,\mathbf{k}$

$\mathbf{r}(1/4)=\frac{1}{2}\mathbf{i}+\sqrt{7/4}\,\mathbf{j}+\sqrt{9-7/4-1/4}\,\mathbf{k}$

$=\frac{1}{2}\mathbf{i}+\frac{\sqrt{7}}{2}\mathbf{j}+\sqrt{7}\,\mathbf{k}$

$\mathbf{r}'(t)=2\mathbf{i}+\frac{\sqrt{7}}{2\sqrt{t}}\mathbf{j}+\frac{-7-8t}{2\sqrt{9-7t-4t^2}}\mathbf{k}$

$\mathbf{r}'(1/4)=2\mathbf{i}+\sqrt{7}\,\mathbf{j}-\frac{9}{2\sqrt{7}}\mathbf{k}$

The tangent line is therefore

$x=\frac{1}{2}+2t$

$y=\frac{\sqrt{7}}{2}+\sqrt{7}\,t$

$z=\sqrt{7}-\frac{9}{2\sqrt{7}}t$

This line intersects the xz-plane when $y=0$, which occurs when $0=\frac{\sqrt{7}}{2}+\sqrt{7}t$, that is, when $t=-\frac{1}{2}$. For this value of t, $x=-\frac{1}{2}$, $y=0$, and $z=\frac{9}{4\sqrt{7}}+\sqrt{7}=\frac{37}{4\sqrt{7}}$. The point of intersection is therefore $\left(-\frac{1}{2},0,\frac{37}{4\sqrt{7}}\right)$.

29. a. $\mathbf{r}(t)=2t\,\mathbf{i}+t^2\mathbf{j}+(1-t^2)\mathbf{k}$. Notice that $x(t)+y(t)+z(t)=2t+t^2+1-t^2=2t+1$. Since $x=2t$, we have $x+y+z=x+1.$, so this curve lies on the plane with equation $y+z=1$.

b. Since $\mathbf{r}(2)=4\mathbf{i}+4\mathbf{j}-3\mathbf{k}$, $\mathbf{r}'(t)=2\,\mathbf{i}+2t\,\mathbf{j}-2t\,\mathbf{k}$ and $\mathbf{r}'(2)=2\mathbf{i}+4\mathbf{j}-4\mathbf{k}$, the equation of the tangent line is $x=4+2t$, $y=4+4t$, $z=-3-4t$. The line intersects the xy-plane when $z=0$, that is, when $t=-\frac{3}{4}$, which gives $x=\frac{5}{2}$, $y=1$, and $z=0$. The point of intersection is therefore $\left(\frac{5}{2},1,0\right)$.

31. Let $\overrightarrow{PR}$ be the scalar projection of $\overrightarrow{PQ}$ on $\mathbf{n}$.

Then $\left\|\overrightarrow{PQ}\right\|^2=\left\|\overrightarrow{PR}\right\|^2+d^2$ so

$$d^2=\left\|\overrightarrow{PQ}\right\|^2-\left\|\overrightarrow{PR}\right\|^2=\left\|\overrightarrow{PQ}\right\|^2-\frac{\left|\overrightarrow{PQ}\cdot\mathbf{n}\right|^2}{\|\mathbf{n}\|^2}$$

$$=\frac{\left\|\overrightarrow{PQ}\right\|^2\|\mathbf{n}\|^2-\left(\overrightarrow{PQ}\cdot\mathbf{n}\right)^2}{\|\mathbf{n}\|^2}=\frac{\left\|\overrightarrow{PQ}\times\mathbf{n}\right\|^2}{\|\mathbf{n}\|^2}$$

by Lagrange's Identity. Thus,

$$d=\frac{\left\|\overrightarrow{PQ}\times\mathbf{n}\right\|}{\|\mathbf{n}\|}.$$

a. $P(3,-2,1)$ is on the line, so

$\overrightarrow{PQ}=\langle 1-3,\ 0+2,\ -4-1\rangle=\langle -2,\ 2,\ -5\rangle$

while $\mathbf{n}=\langle 2,-2,1\rangle$, so

$d=\frac{\|\langle -2,\ 2,\ -5\rangle\times\langle 2,\ -2,\ 1\rangle\|}{\sqrt{4+4+1}}$

$=\frac{\|\langle -8,\ -8,\ 0\rangle\|}{3}=\frac{8\sqrt{2}}{3}\approx 3.771$

b. $P(1,-1,0)$ is on the line, so

$\overrightarrow{PQ}=\langle 2-1,\ -1+1,\ 3-0\rangle=\langle 1,\ 0,\ 3\rangle$ while $\mathbf{n}=\langle 2,3,-6\rangle$.

$d=\frac{\|\langle 1,\ 0,\ 3\rangle\times\langle 2,\ 3,\ -6\rangle\|}{\sqrt{4+9+36}}=\frac{\|\langle -9,\ 12,\ 3\rangle\|}{7}$

$=\frac{3\sqrt{26}}{7}\approx 2.185$

11.7 Concepts Review

1. $\frac{d\mathbf{T}}{ds}$

3. $\frac{d^2s}{dt^2};\left(\frac{ds}{dt}\right)^2\kappa$

Problem Set 11.7

1. $\mathbf{r}(t)=\langle t,t^2\rangle$

$\mathbf{v}(t)=\mathbf{r}'(t)=\langle 1,2t\rangle$

$\mathbf{v}(1)=\langle 1,2\rangle$

$\mathbf{a}(t)=\mathbf{v}'(t)=\langle 0,2\rangle$

$\mathbf{a}(1)=\langle 0,2\rangle$

$$\mathbf{T}(t)=\left\langle \frac{1}{\sqrt{1+4t^2}},\frac{2t}{\sqrt{1+4t^2}}\right\rangle$$

$$\mathbf{T}(1)=\left\langle \frac{1}{\sqrt{5}},\frac{2}{\sqrt{5}}\right\rangle$$

$$\mathbf{T}'(t)=\left\langle -\frac{4t}{\left(1+4t^2\right)^{3/2}},\frac{2}{\left(1+4t^2\right)^{3/2}}\right\rangle$$

$$\mathbf{T}'(1)=\left\langle -\frac{4}{5^{3/2}},\frac{2}{5^{3/2}}\right\rangle$$

$\|\mathbf{T}'(1)\|=\frac{2}{5}$

$\|\mathbf{v}(1)\|=\sqrt{5}$

$$\kappa=\frac{\|\mathbf{T}'(1)\|}{\|\mathbf{v}(1)\|}=\frac{2}{5\sqrt{5}}=\frac{2}{5^{3/2}}$$

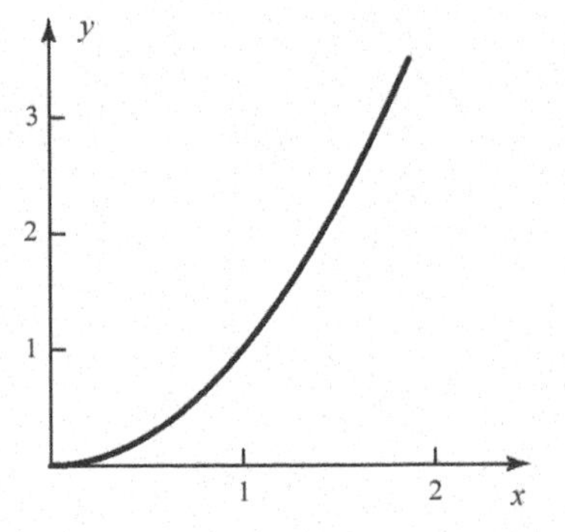

3. $\mathbf{r}(t)=\langle t,2\cos t,2\sin t\rangle$

$\mathbf{v}(t)=\mathbf{r}'(t)=\langle 1,-2\sin t,2\cos t\rangle$

$\mathbf{v}(\pi)=\langle 1,0,-2\rangle$

$\mathbf{a}(t)=\mathbf{v}'(t)=\langle 0,-2\cos t,-2\sin t\rangle$

$\mathbf{a}(\pi)=\langle 0,2,0\rangle$

$$\mathbf{T}(t)=\left\langle \frac{1}{\sqrt{5}},-\frac{2\sin t}{\sqrt{5}},\frac{2\cos t}{\sqrt{5}}\right\rangle$$

$$\mathbf{T}(\pi)=\left\langle \frac{1}{\sqrt{5}},0,-\frac{2}{\sqrt{5}}\right\rangle$$

$$\mathbf{T}'(t)=\left\langle 0,-\frac{2\cos t}{\sqrt{5}},-\frac{2\sin t}{\sqrt{5}}\right\rangle$$

$$\mathbf{T}'(\pi)=\left\langle 0,\frac{2}{\sqrt{5}},0\right\rangle$$

$\|\mathbf{T}'(\pi)\|=\frac{2}{\sqrt{5}}$

$\|\mathbf{v}(t)\|=\sqrt{5}$

$$\kappa=\frac{\|\mathbf{T}'(\pi)\|}{\|\mathbf{v}(\pi)\|}=\frac{2/\sqrt{5}}{\sqrt{5}}=\frac{2}{5}$$

5. $\mathbf{r}(t)=\left\langle \frac{t^2}{8},5\cos t,5\sin t\right\rangle$

$$\mathbf{v}(t)=\mathbf{r}'(t)=\left\langle \frac{t}{4},-5\sin t,5\cos t\right\rangle$$

$$\mathbf{v}(\pi)=\left\langle \frac{\pi}{4},0,-5\right\rangle$$

$$\mathbf{a}(t)=\mathbf{v}'(t)=\left\langle \frac{1}{4},-5\cos t,-5\sin t\right\rangle$$

$$\mathbf{a}(\pi)=\left\langle \frac{1}{4},5,0\right\rangle$$

$$\mathbf{T}(t)=\left\langle \frac{t}{\sqrt{400+t^2}},-\frac{20\sin t}{\sqrt{400+t^2}},\frac{20\cos t}{\sqrt{400+t^2}}\right\rangle$$

$$\mathbf{T}(\pi)=\left\langle \frac{\pi}{\sqrt{400+\pi^2}},0,-\frac{20}{\sqrt{400+\pi^2}}\right\rangle$$

$$\mathbf{T}'(t) = \left\langle \frac{400}{\left(400+t^2\right)^{3/2}}, \frac{-20\left((400+t^2)\cos t - t\sin t\right)}{\left(400+t^2\right)^{3/2}}, \frac{-20\left((400+t^2)\sin t + t\cos t\right)}{\left(400+t^2\right)^{3/2}} \right\rangle$$

$$\mathbf{T}'(\pi) = \left\langle \frac{400}{\left(400+\pi^2\right)^{3/2}}, \frac{-20\left((400+\pi^2)\cos\pi - \pi\sin\pi\right)}{\left(400+\pi^2\right)^{3/2}}, \frac{-20\left((400+\pi^2)\sin\pi + \pi\cos\pi\right)}{\left(400+\pi^2\right)^{3/2}} \right\rangle$$

$$= \left\langle \frac{400}{\left(400+\pi^2\right)^{3/2}}, \frac{-20}{\left(400+\pi^2\right)^{1/2}}, \frac{20\pi}{\left(400+\pi^2\right)^{3/2}} \right\rangle$$

$$\|\mathbf{T}'(\pi)\| = \frac{400^2}{\left(400+\pi^2\right)^3} + \frac{400}{\left(400+\pi^2\right)} + \frac{400\pi^2}{\left(400+\pi^2\right)^3}$$

$$\approx 0.989091$$

$$\|\mathbf{v}(t)\| = \sqrt{\frac{\pi^2}{16}+25}$$

$$\kappa = \frac{\|\mathbf{T}'(\pi)\|}{\|\mathbf{v}(\pi)\|} \approx 0.195422$$

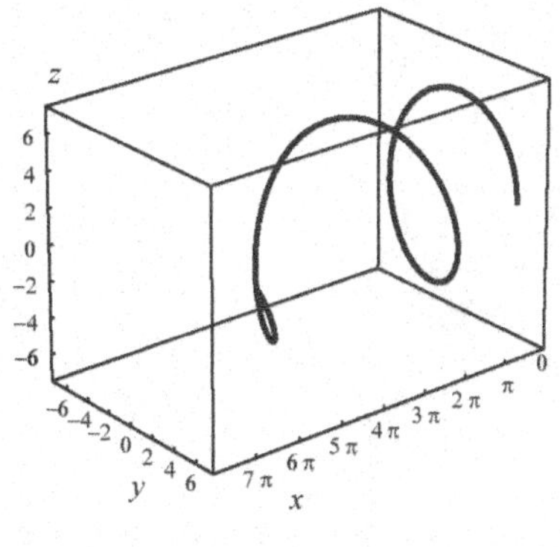

7. $\mathbf{u}'(t) = 8t\mathbf{i} + 4\mathbf{j}$

$$\|\mathbf{u}'(t)\| = 4\sqrt{4t^2+1}$$

$$\mathbf{T}(t) = \frac{\mathbf{u}'(t)}{\|\mathbf{u}'(t)\|} = \frac{2t}{\sqrt{4t^2+1}}\mathbf{i} + \frac{1}{\sqrt{4t^2+1}}\mathbf{j}$$

$$\mathbf{T}\left(\frac{1}{2}\right) = \frac{1}{\sqrt{2}}\mathbf{i} + \frac{1}{\sqrt{2}}\mathbf{j}$$

$x'(t) = 8t \qquad y'(t) = 4$

$x''(t) = 8 \qquad y''(t) = 0$

$$\kappa(t) = \frac{|x'y'' - y'x''|}{\left[x'^2 + y'^2\right]^{3/2}} = \frac{32}{64(4t^2+1)^{3/2}}$$

$$= \frac{1}{2(4t^2+1)^{3/2}}$$

$$\kappa\left(\frac{1}{2}\right) = \frac{1}{2(2)^{3/2}} = \frac{1}{4\sqrt{2}}$$

9. $\mathbf{z}'(t) = -3\sin t\mathbf{i} + 4\cos t\mathbf{j}$

$$\|\mathbf{z}'(t)\| = \sqrt{9\sin^2 t + 16\cos^2 t}$$

$$\mathbf{T}(t) = \frac{\mathbf{z}'(t)}{\|\mathbf{z}'(t)\|}$$

$$= -\frac{3\sin t}{\sqrt{9\sin^2 t + 16\cos^2 t}}\mathbf{i} + \frac{4\cos t}{\sqrt{9\sin^2 t + 16\cos^2 t}}\mathbf{j}$$

$$\mathbf{T}\left(\frac{\pi}{4}\right) = -\frac{3}{5}\mathbf{i} + \frac{4}{5}\mathbf{j}$$

$x'(t) = -3\sin t \qquad y'(t) = 4\cos t$

$x''(t) = -3\cos t \qquad y''(t) = -4\sin t$

$$\kappa(t) = \frac{|x'y'' - y'x''|}{\left[x'^2 + y'^2\right]^{3/2}} = \frac{12}{\left(9\sin^2 t + 16\cos^2 t\right)^{3/2}}$$

$$\kappa\left(\frac{\pi}{4}\right) = \frac{12}{\left(\frac{25}{2}\right)^{3/2}} = \frac{24\sqrt{2}}{125}$$

11. $x = 1 - t^2,\ \ y = 1 - t^3$

$x'(t) = -2t,\ \ y'(t) = -3t^2$

$x''(t) = -2,\ \ y''(t) = -6t$

$$\mathbf{r}(t) = \left(1-t^2\right)\mathbf{i} + \left(1-t^3\right)\mathbf{j}$$

$$\mathbf{r}'(t) = -2t\,\mathbf{i} - 3t^2\,\mathbf{j}$$

$$\mathbf{T}(t) = \frac{\mathbf{r}'(t)}{\|\mathbf{r}'(t)\|} = \frac{-2t\,\mathbf{i} - 3t^2\,\mathbf{j}}{\sqrt{4t^2+9t^4}}$$

$$\mathbf{T}(1) = \frac{-2(1)\mathbf{i} - 3(1)^2\mathbf{j}}{\sqrt{4(1)^2 + 9(1)^4}} = -\frac{2}{\sqrt{13}}\mathbf{i} - \frac{3}{\sqrt{13}}\mathbf{j}$$

$$\kappa(t) = \frac{|x'y'' - x''y'|}{\left((x')^2 + (y')^2\right)^{3/2}} = \frac{\left|(-2t)(-6t) - (-2)\left(-3t^2\right)\right|}{\left((-2t)^2 + \left(-3t^2\right)^2\right)^{3/2}}$$

$$= \frac{\left|12t^2 - 6t^2\right|}{\left(4t^2 + 9t^4\right)^{3/2}} = \frac{6t^2}{\left(4t^2 + 9t^4\right)^{3/2}}$$

When $t = 1$, the curvature is

$$\kappa = \frac{6}{(4+9)^{3/2}} = \frac{6}{13^{3/2}} \approx 0.128008$$

13. $\mathbf{r}'(t) = -(\cos t + \sin t)e^{-t}\mathbf{i} + (\cos t - \sin t)e^{t}\mathbf{j}$

$|\mathbf{r}'(t)| = \sqrt{2}e^{-t}$

$$\mathbf{T}(t) = \frac{\mathbf{r}'(t)}{\|\mathbf{r}'(t)\|} = -\frac{\cos t + \sin t}{\sqrt{2}}\mathbf{i} + \frac{\cos t - \sin t}{\sqrt{2}}\mathbf{j}$$

$$\mathbf{T}(0) = -\frac{1}{\sqrt{2}}\mathbf{i} + \frac{1}{\sqrt{2}}\mathbf{j}$$

$x'(t) = -(\cos t - \sin t)e^{-t}$ $\quad y'(t) = (\cos t - \sin t)e^{-t}$

$x''(t) = (2\sin t)e^{-t}$ $\quad y''(t) = (-2\cos t)e^{-t}$

$$\kappa(t) = \frac{|x'y'' - y'x''|}{\left[x'^2 + y'^2\right]^{3/2}} = \frac{2e^{-2t}}{2\sqrt{2}e^{-3t}} = \frac{e^{t}}{\sqrt{2}}$$

$$\kappa(0) = \frac{1}{\sqrt{2}}$$

15.

$y' = 4x,\ y'' = 4$

$$\kappa = \frac{4}{(1+16x^2)^{3/2}}$$

At $(1, 2)$, $\kappa = \dfrac{4}{17\sqrt{17}}$ and $R = \dfrac{17\sqrt{17}}{4}$.

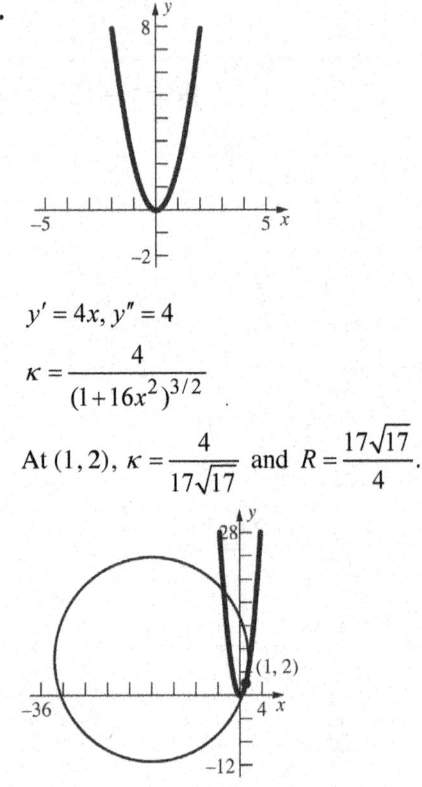

17.

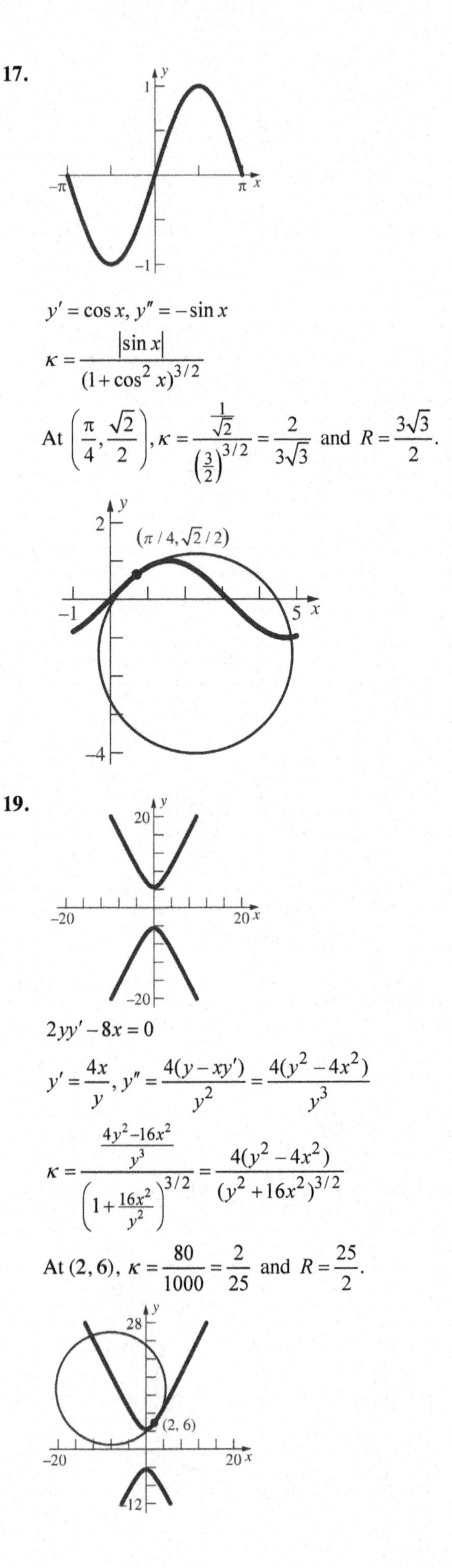

$y' = \cos x,\ y'' = -\sin x$

$$\kappa = \frac{|\sin x|}{(1+\cos^2 x)^{3/2}}$$

At $\left(\dfrac{\pi}{4}, \dfrac{\sqrt{2}}{2}\right)$, $\kappa = \dfrac{\frac{1}{\sqrt{2}}}{\left(\frac{3}{2}\right)^{3/2}} = \dfrac{2}{3\sqrt{3}}$ and $R = \dfrac{3\sqrt{3}}{2}$.

19.

$2yy' - 8x = 0$

$$y' = \frac{4x}{y},\ y'' = \frac{4(y - xy')}{y^2} = \frac{4(y^2 - 4x^2)}{y^3}$$

$$\kappa = \frac{\frac{4y^2 - 16x^2}{y^3}}{\left(1 + \frac{16x^2}{y^2}\right)^{3/2}} = \frac{4(y^2 - 4x^2)}{(y^2 + 16x^2)^{3/2}}$$

At $(2, 6)$, $\kappa = \dfrac{80}{1000} = \dfrac{2}{25}$ and $R = \dfrac{25}{2}$.

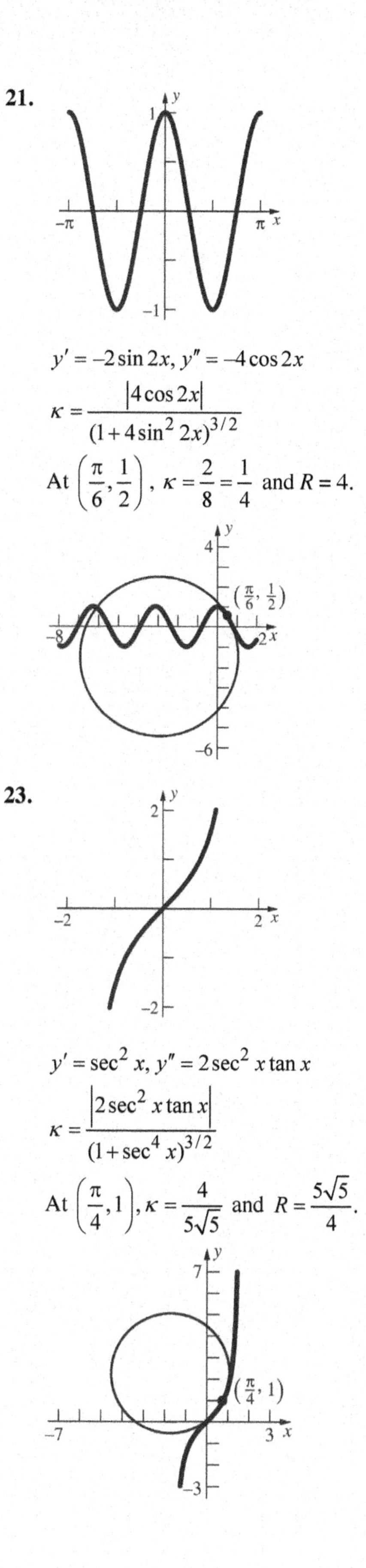

21.

$y' = -2\sin 2x,\ y'' = -4\cos 2x$

$$\kappa = \frac{|4\cos 2x|}{(1+4\sin^2 2x)^{3/2}}$$

At $\left(\frac{\pi}{6}, \frac{1}{2}\right)$, $\kappa = \frac{2}{8} = \frac{1}{4}$ and $R = 4$.

23.

$y' = \sec^2 x,\ y'' = 2\sec^2 x \tan x$

$$\kappa = \frac{|2\sec^2 x \tan x|}{(1+\sec^4 x)^{3/2}}$$

At $\left(\frac{\pi}{4}, 1\right)$, $\kappa = \frac{4}{5\sqrt{5}}$ and $R = \frac{5\sqrt{5}}{4}$.

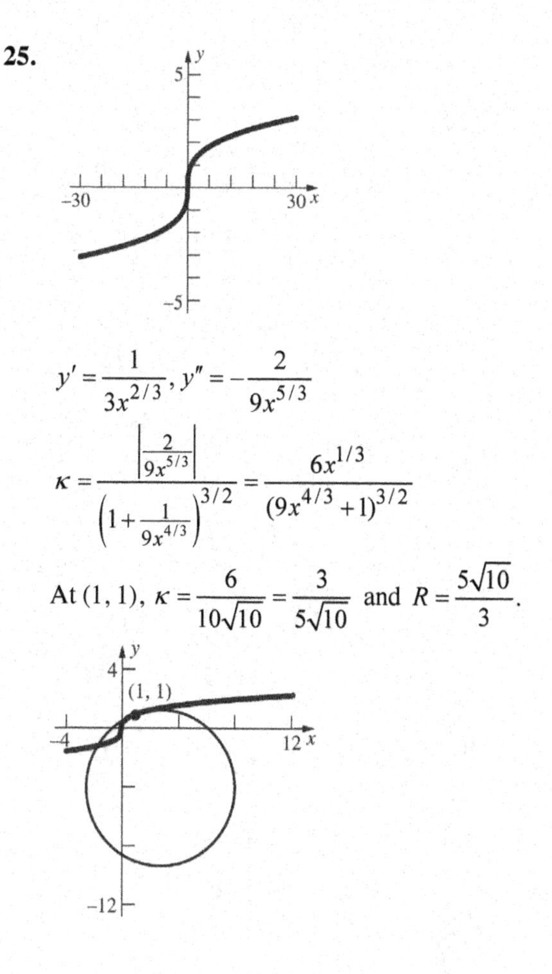

25.

$$y' = \frac{1}{3x^{2/3}},\ y'' = -\frac{2}{9x^{5/3}}$$

$$\kappa = \frac{\left|\frac{2}{9x^{5/3}}\right|}{\left(1+\frac{1}{9x^{4/3}}\right)^{3/2}} = \frac{6x^{1/3}}{(9x^{4/3}+1)^{3/2}}$$

At $(1, 1)$, $\kappa = \frac{6}{10\sqrt{10}} = \frac{3}{5\sqrt{10}}$ and $R = \frac{5\sqrt{10}}{3}$.

27. $\mathbf{r}'(t) = t\mathbf{i} + \mathbf{j} + t^2\mathbf{k}$

$\mathbf{r}''(t) = \mathbf{i} + 2t\mathbf{k}$

$$\mathbf{T}(2) = \frac{\mathbf{r}'(2)}{\|\mathbf{r}'(2)\|} = \frac{2\mathbf{i}+\mathbf{j}+4\mathbf{k}}{\sqrt{4+1+16}} = \frac{2}{\sqrt{21}}\mathbf{i} + \frac{1}{\sqrt{21}}\mathbf{j} + \frac{4}{\sqrt{21}}\mathbf{k}$$

$$a_T(2) = \frac{\mathbf{r}'(2)\cdot\mathbf{r}''(2)}{\|\mathbf{r}'(2)\|} = \frac{(2\mathbf{i}+\mathbf{j}+4\mathbf{k})\cdot(\mathbf{i}+4\mathbf{k})}{\sqrt{21}} = \frac{18}{\sqrt{21}}$$

$$a_N(2) = \frac{\|\mathbf{r}'(2)\times\mathbf{r}''(2)\|}{\|\mathbf{r}'(2)\|} = \frac{\|(2\mathbf{i}+\mathbf{j}+4\mathbf{k})\times(\mathbf{i}+4\mathbf{k})\|}{\sqrt{21}} = \frac{1}{\sqrt{21}}\|4\mathbf{i}-4\mathbf{j}-\mathbf{k}\| = \frac{\sqrt{33}}{\sqrt{21}} = \sqrt{\frac{11}{7}}$$

$$\mathbf{N}(2) = \frac{\mathbf{r}''(2) - a_T(2)\mathbf{T}(2)}{a_N(2)} = \frac{(\mathbf{i}+4\mathbf{k}) - \frac{18}{\sqrt{21}}\left(\frac{2}{\sqrt{21}}\mathbf{i} + \frac{1}{\sqrt{21}}\mathbf{j} + \frac{4}{\sqrt{21}}\mathbf{k}\right)}{\sqrt{\frac{11}{7}}} = \sqrt{\frac{7}{11}}\left(-\frac{15}{21}\mathbf{i} - \frac{18}{21}\mathbf{j} + \frac{12}{21}\mathbf{k}\right)$$

$$= \sqrt{\frac{7}{11}}\left(-\frac{5}{7}\mathbf{i} - \frac{6}{7}\mathbf{j} + \frac{4}{7}\mathbf{k}\right) = -\frac{5}{\sqrt{77}}\mathbf{i} - \frac{6}{\sqrt{77}}\mathbf{j} + \frac{4}{\sqrt{77}}\mathbf{k}$$

$$\kappa(2) = \frac{\|\mathbf{r}'(2)\times\mathbf{r}''(2)\|}{\|\mathbf{r}'(2)\|^3} = \frac{\sqrt{33}}{\left(\sqrt{21}\right)^3} = \sqrt{\frac{33}{9261}} = \sqrt{\frac{11}{3087}} = \frac{\sqrt{11}}{21\sqrt{7}}$$

$$\mathbf{B}(2) = \mathbf{T}(2)\times\mathbf{N}(2) = \left(\frac{2}{\sqrt{21}}\mathbf{i} + \frac{1}{\sqrt{21}}\mathbf{j} + \frac{4}{\sqrt{21}}\mathbf{k}\right)\times\left(-\frac{5}{\sqrt{77}}\mathbf{i} - \frac{6}{\sqrt{77}}\mathbf{j} + \frac{4}{\sqrt{77}}\mathbf{k}\right) = \frac{4}{\sqrt{33}}\mathbf{i} - \frac{4}{\sqrt{33}}\mathbf{j} - \frac{1}{\sqrt{33}}\mathbf{k}$$

29. $\mathbf{r}(t) = \langle 7\sin 3t,\ 7\cos 3t,\ 14t\rangle$

$\mathbf{r}'(t) = \langle 21\cos 3t,\ -21\sin 3t,\ 14\rangle$

$\mathbf{r}''(t) = \langle -63\sin 3t,\ -63\cos 3t,\ 0\rangle$

$$\mathbf{T}\left(\frac{\pi}{3}\right) = \frac{\mathbf{r}'\left(\frac{\pi}{3}\right)}{\left\|\mathbf{r}'\left(\frac{\pi}{3}\right)\right\|} = \frac{\langle -21, 0, 14\rangle}{\sqrt{441+196}} = \frac{1}{7\sqrt{13}}\langle -21, 0, 14\rangle = \left\langle -\frac{3}{\sqrt{13}}, 0, \frac{2}{\sqrt{13}}\right\rangle$$

$$a_T\left(\frac{\pi}{3}\right) = \frac{\mathbf{r}'\left(\frac{\pi}{3}\right)\cdot\mathbf{r}''\left(\frac{\pi}{3}\right)}{\left\|\mathbf{r}'\left(\frac{\pi}{3}\right)\right\|} = \frac{\langle -21, 0, 14\rangle\cdot\langle 0, -63, 0\rangle}{7\sqrt{13}} = 0$$

$$a_N\left(\frac{\pi}{3}\right) = \frac{\left\|\mathbf{r}'\left(\frac{\pi}{3}\right)\times\mathbf{r}''\left(\frac{\pi}{3}\right)\right\|}{\left\|\mathbf{r}'\left(\frac{\pi}{3}\right)\right\|} = \frac{|\langle -21, 0, 14\rangle\times\langle 0, -63, 0\rangle|}{7\sqrt{13}} = \frac{|\langle 882, 0, -1323\rangle|}{7\sqrt{13}} = \frac{441\sqrt{13}}{7\sqrt{13}} = 63$$

$$\mathbf{N}\left(\frac{\pi}{3}\right) = \frac{\mathbf{r}''\left(\frac{\pi}{3}\right) - a_T\left(\frac{\pi}{3}\right)\mathbf{T}\left(\frac{\pi}{3}\right)}{a_N\left(\frac{\pi}{3}\right)} = \frac{1}{63}\langle 0, 63, 0\rangle = \langle 0, 1, 0\rangle$$

$$\kappa\left(\frac{\pi}{3}\right) = \frac{\left\|\mathbf{r}\left(\frac{\pi}{3}\right)\times\mathbf{r}''\left(\frac{\pi}{3}\right)\right\|}{\left\|\mathbf{r}'\left(\frac{\pi}{3}\right)\right\|^3} = \frac{441\sqrt{13}}{\left(7\sqrt{13}\right)^3} = \frac{9}{91}$$

$$\mathbf{B}\left(\frac{\pi}{3}\right) = \mathbf{T}\left(\frac{\pi}{3}\right)\times\mathbf{N}\left(\frac{\pi}{3}\right) = \left\langle -\frac{3}{\sqrt{13}}, 0, \frac{2}{\sqrt{13}}\right\rangle\times\langle 0, 1, 0\rangle = \left\langle -\frac{2}{\sqrt{13}}, 0, -\frac{3}{\sqrt{13}}\right\rangle$$

31. $\mathbf{r}'(t) = \sinh\frac{t}{3}\mathbf{i} + \mathbf{j}$

$\mathbf{r}''(t) = \frac{1}{3}\cosh\frac{t}{3}\mathbf{i}$

$\mathbf{T}(1) = \frac{\mathbf{r}'(1)}{\|\mathbf{r}'(1)\|} = \frac{\sinh\frac{1}{3}\mathbf{i} + \mathbf{j}}{\sqrt{\sinh^2\frac{1}{3} + 1}} = \frac{\sinh\frac{1}{3}\mathbf{i} + \mathbf{j}}{\cosh\frac{1}{3}} = \tanh\frac{1}{3}\mathbf{i} + \operatorname{sech}\frac{1}{3}\mathbf{j}$

$a_T(1) = \frac{\mathbf{r}'(1)\cdot\mathbf{r}''(1)}{\|\mathbf{r}'(1)\|} = \frac{\left(\sinh\frac{1}{3}\mathbf{i} + \mathbf{j}\right)\cdot\left(\frac{1}{3}\cosh\frac{1}{3}\mathbf{i}\right)}{\cosh\frac{1}{3}} = \frac{1}{3}\sinh\frac{1}{3}$

$a_N(1) = \frac{\|\mathbf{r}'(1)\times\mathbf{r}''(1)\|}{\|\mathbf{r}'(1)\|} = \frac{\left\|\left(\sinh\frac{1}{3}\mathbf{i} + \mathbf{j}\right)\times\left(\frac{1}{3}\cosh\frac{1}{3}\mathbf{i}\right)\right\|}{\cosh\frac{1}{3}} = \frac{\left\|-\frac{1}{3}\cos\frac{1}{3}\mathbf{k}\right\|}{\cosh\frac{1}{3}} = \frac{1}{3}$

$\mathbf{N}(1) = \frac{\mathbf{r}''(1) - a_T(1)\mathbf{T}(1)}{a_N(1)} = 3\left[\frac{1}{3}\cosh\frac{1}{3}\mathbf{i} - \frac{1}{3}\sinh\frac{1}{3}\left(\tanh\frac{1}{3}\mathbf{i} + \operatorname{sech}\frac{1}{3}\mathbf{j}\right)\right] = \left(\cosh\frac{1}{3} - \frac{\sinh^2\frac{1}{3}}{\cosh\frac{1}{3}}\right)\mathbf{i} - \tanh\frac{1}{3}\mathbf{j}$

$= \operatorname{sech}\frac{1}{3}\mathbf{i} - \tanh\frac{1}{3}\mathbf{j}$

$\kappa(1) = \frac{|\mathbf{r}'(1)\times\mathbf{r}''(1)|}{|\mathbf{r}'(1)|^3} = \frac{\frac{1}{3}\cosh\frac{1}{3}}{\cosh^3\frac{1}{3}} = \frac{1}{3}\operatorname{sech}^2\frac{1}{3}$

$\mathbf{B}(1) = \mathbf{T}(1)\times\mathbf{N}(1) = \left(\tanh\frac{1}{3}\mathbf{i} + \operatorname{sech}\frac{1}{3}\mathbf{j}\right)\times\left(\operatorname{sech}\frac{1}{3}\mathbf{i} - \tanh\frac{1}{3}\mathbf{j}\right) = \left(-\operatorname{sech}^2\frac{1}{3} - \tanh^2\frac{1}{3}\right)\mathbf{k} = -\mathbf{k}$

33. $\mathbf{r}'(t) = -2e^{-2t}\mathbf{i} + 2e^{2t}\mathbf{j} + 2\sqrt{2}\mathbf{k}$

$\mathbf{r}''(t) = 4e^{-2t}\mathbf{i} + 4e^{2t}\mathbf{j}$

$\mathbf{T}(0) = \frac{\mathbf{r}'(0)}{\|\mathbf{r}'(0)\|} = \frac{-2\mathbf{i} + 2\mathbf{j} + 2\sqrt{2}\mathbf{k}}{\sqrt{4+4+8}} = -\frac{1}{2}\mathbf{i} + \frac{1}{2}\mathbf{j} + \frac{\sqrt{2}}{2}\mathbf{k}$

$a_T(0) = \frac{\mathbf{r}'(0)\cdot\mathbf{r}''(0)}{\|\mathbf{r}'(0)\|} = \frac{\left(-2\mathbf{i} + 2\mathbf{j} + 2\sqrt{2}\mathbf{k}\right)\cdot(4\mathbf{i} + 4\mathbf{j})}{4} = 0$

$a_N(0) = \frac{\|\mathbf{r}'(0)\times\mathbf{r}''(0)\|}{\|\mathbf{r}'(0)\|} = \frac{\left|-8\sqrt{2}\mathbf{i} + 8\sqrt{2}\mathbf{j} - 16\mathbf{k}\right|}{4} = \frac{16\sqrt{2}}{4} = 4\sqrt{2}$

$\mathbf{N}(0) = \frac{\mathbf{r}''(0) - a_T(0)\mathbf{T}(0)}{a_N(0)} = \frac{4\mathbf{i} + 4\mathbf{j}}{4\sqrt{2}} = \frac{1}{\sqrt{2}}\mathbf{i} + \frac{1}{\sqrt{2}}\mathbf{j}$

$\kappa(0) = \frac{\|\mathbf{r}'(0)\times\mathbf{r}''(0)\|}{\|\mathbf{r}'(0)\|^3} = \frac{16\sqrt{2}}{64} = \frac{\sqrt{2}}{4}$

$\mathbf{B}(0) = \mathbf{T}(0)\times\mathbf{N}(0) = \left(-\frac{1}{2}\mathbf{i} + \frac{1}{2}\mathbf{j} + \frac{\sqrt{2}}{2}\mathbf{k}\right)\times\left(\frac{1}{\sqrt{2}}\mathbf{i} + \frac{1}{\sqrt{2}}\mathbf{j}\right) = -\frac{1}{2}\mathbf{i} + \frac{1}{2}\mathbf{j} - \frac{1}{\sqrt{2}}\mathbf{k}$

35. $y' = \frac{1}{x}, y'' = -\frac{1}{x^2}$

$$\kappa = \frac{\left|\frac{1}{x^2}\right|}{\left(1+\frac{1}{x^2}\right)^{3/2}} = \frac{|x|}{(x^2+1)^{3/2}}$$

Since $0 < x < \infty$, $\kappa = \frac{x}{(x^2+1)^{3/2}}$.

$$\kappa' = \frac{(x^2+1)^{3/2} - 3x^2(x^2+1)^{1/2}}{(x^2+1)^3} = \frac{-2x^2+1}{(x^2+1)^{5/2}}$$

$\kappa' = 0$ when $x = \frac{1}{\sqrt{2}}$. Since $\kappa' > 0$ on $\left(0, \frac{1}{\sqrt{2}}\right)$ and $\kappa' < 0$ on $\left(\frac{1}{\sqrt{2}}, \infty\right)$, so κ is maximum when $x = \frac{1}{\sqrt{2}}, y = \ln\frac{1}{\sqrt{2}} = -\frac{\ln 2}{2}$. The point of maximum curvature is $\left(\frac{1}{\sqrt{2}}, -\frac{\ln 2}{2}\right)$.

37. $y' = \sinh x, y'' = \cosh x$

$$\kappa = \frac{\cosh x}{(1+\sinh^2 x)^{3/2}} = \operatorname{sech}^2 x$$

$\kappa' = -2\operatorname{sech}^2 x \tanh x$

$\kappa' = 0$ when $x = 0$. Since $\kappa' > 0$ on $(-\infty, 0)$ and $\kappa' < 0$ on $(0, \infty)$, so κ is maximum when $x = 0, y = 1$. The point of maximum curvature is $(0, 1)$.

39. $y' = e^x, y'' = e^x$

$$\kappa = \frac{e^x}{(1+e^{2x})^{3/2}}$$

$$\kappa' = \frac{e^x(1+e^{2x})^{3/2} - 3e^{3x}(1+e^{2x})^{1/2}}{(1+e^{2x})^3}$$

$$= \frac{e^x(1-2e^{2x})}{(1+e^{2x})^{5/2}}$$

$\kappa' = 0$ when $x = -\frac{1}{2}\ln 2$. Since $\kappa' > 0$ on $\left(-\infty, -\frac{1}{2}\ln 2\right)$ and $\kappa' < 0$ on $\left(-\frac{1}{2}\ln 2, \infty\right)$, so κ is maximum when $x = -\frac{1}{2}\ln 2$, $y = \frac{1}{\sqrt{2}}$. The point of maximum curvature is $\left(-\frac{1}{2}\ln 2, \frac{1}{\sqrt{2}}\right)$.

41. $\mathbf{r}'(t) = 3\mathbf{i} + 6t\mathbf{j}$

$\mathbf{r}''(t) = 6\mathbf{j}$

$$\frac{ds}{dt} = |\mathbf{r}'(t)| = 3\sqrt{1+4t^2}$$

$$a_T = \frac{d^2s}{dt^2} = \frac{12t}{\sqrt{1+4t^2}}$$

$$a_N^2 = |\mathbf{r}''(t)|^2 - a_T^2 = 36 - \frac{144t^2}{1+4t^2} = \frac{36}{1+4t^2}$$

$$a_N = \frac{6}{\sqrt{1+4t^2}}$$

At $t_1 = \frac{1}{3}$, $a_T = \frac{12}{\sqrt{13}}$ and $a_N = \frac{18}{\sqrt{13}}$.

43. $\mathbf{r}'(t) = 2\mathbf{i} + 2t\mathbf{j}$

$\mathbf{r}''(t) = 2\mathbf{j}$

$$\frac{ds}{dt} = \|\mathbf{r}'(t)\| = 2\sqrt{1+t^2}$$

$$a_T = \frac{d^2s}{dt^2} = \frac{2t}{\sqrt{1+t^2}}$$

$$a_N^2 = \|\mathbf{r}''(t)\|^2 - a_T^2 = 4 - \frac{4t^2}{1+t^2} = \frac{4}{1+t^2}$$

$$a_N = \frac{2}{\sqrt{1+t^2}}$$

At $t_1 = -1, a_T = -\sqrt{2}$ and $a_N = \sqrt{2}$.

45. $\mathbf{r}'(t) = a\sinh t\mathbf{i} + a\cosh t\mathbf{j}$

$\mathbf{r}''(t) = a\cosh t\mathbf{i} + a\sinh t\mathbf{j}$

$$\frac{ds}{dt} = \|\mathbf{r}'(t)\| = a\sqrt{\sinh^2 t + \cosh^2 t}$$

$$a_T = \frac{d^2s}{dt^2} = \frac{2a\cosh t\sinh t}{\sqrt{\sinh^2 t + \cosh^2 t}}$$

$$a_N^2 = \|\mathbf{r}''(t)\|^2\ a_T^2$$

$$= a^2(\cosh^2 t + \sinh^2 t) - \frac{4a^2\cosh^2 t\sinh^2 t}{\sinh^2 t + \cosh^2 t}$$

$$= \frac{a^2(\cosh^2 t - \sinh t^2)^2}{\sinh^2 t + \cosh^2 t}$$

$$a_N = \frac{a(\cosh^2 t - \sinh^2 t)}{\sqrt{\sinh^2 t + \cosh^2 t}}$$

At $t_1 = \ln 3, \cosh t = \frac{5}{3}, \sinh t = \frac{4}{3}$, and

$\sqrt{\sinh^2 t + \cosh^2 t} = \frac{\sqrt{41}}{3}$, so $a_T = \frac{40a}{3\sqrt{41}}$ and

$a_N = \frac{3a}{\sqrt{41}}$.

47. $\mathbf{r}'(t) = \mathbf{i} + 3\mathbf{j} + 2t\mathbf{k}$

$\mathbf{r}''(t) = 2\mathbf{k}$

$$a_T(t) = \frac{(\mathbf{i}+3\mathbf{j}+2t\mathbf{k})\cdot(2\mathbf{k})}{\sqrt{1+9+4t^2}} = \frac{4t}{\sqrt{10+4t^2}}$$

$$a_T(1) = \frac{4}{\sqrt{14}}$$

$$a_N(t) = \frac{|(\mathbf{i}+3\mathbf{j}+2t\mathbf{k})\times(2\mathbf{k})|}{\sqrt{10+4t^2}} = \frac{|6\mathbf{i}-2\mathbf{j}|}{\sqrt{10+4t^2}}$$

$$= \frac{\sqrt{36+4}}{\sqrt{10+4t^2}} = 2\sqrt{\frac{10}{10+4t^2}} = 2\sqrt{\frac{5}{5+2t^2}}$$

$$a_N(1) = 2\sqrt{\frac{5}{7}}$$

49. $\mathbf{r}(t) = \left\langle e^{-t}, 2t, e^t\right\rangle$; $\mathbf{r}'(t) = \left\langle -e^{-t}, 2, e^t\right\rangle$

$\mathbf{r}''(t) = \left\langle e^{-t}, 0, e^t\right\rangle$; $\mathbf{r}'(t)\cdot\mathbf{r}''(t) = -e^{-2t} + e^{2t}$

$$\|\mathbf{r}'(t)\| = \sqrt{e^{-2t} + 4 + e^{2t}}$$

$$\|\mathbf{r}'(t)\times\mathbf{r}''(t)\| = \left|\left\langle 2e^t, 2, -2e^{-t}\right\rangle\right|$$

$$= \sqrt{4e^{2t} + 4 + 4e^{-2t}}$$

$$= 2\sqrt{e^{2t} + 1 + e^{-2t}}$$

$$a_T(t) = \frac{e^{2t} - e^{-2t}}{\sqrt{e^{2t} + 4 + e^{-2t}}}$$

$$a_T(0) = 0$$

$$a_N(t) = 2\sqrt{\frac{e^{2t}+1+e^{-2t}}{e^{2t}+4+e^{-2t}}}$$

$$a_N(0) = 2\sqrt{\frac{3}{6}} = \sqrt{2}$$

51. $\mathbf{r}'(t) = (1-t^2)\mathbf{i} - (1+t^2)\mathbf{j} + \mathbf{k}$

$\mathbf{r}''(t) = -2t\mathbf{i} - 2t\mathbf{j}$

$\mathbf{r}'(t)\cdot\mathbf{r}''(t) = -2t(1-t^2) + 2t(1+t^2) = 4t^3$

$$\|\mathbf{r}'(t)\| = \sqrt{(1-t^2)^2 + (1+t^2)^2 + 1} = \sqrt{2t^4+3}$$

$$\|\mathbf{r}'(t)\times\mathbf{r}''(t)\| = \|2t\mathbf{i} - 2t\mathbf{j} - 4t\mathbf{k}\|$$

$$= \sqrt{4t^2 + 4t^2 + 16t^2}$$

$$= 2\sqrt{6}|t|$$

$$a_T(t) = \frac{4t^3}{\sqrt{2t^4+3}};\quad a_T(3) = 36\sqrt{\frac{3}{55}}$$

$$a_N(t) = \frac{2\sqrt{6}|t|}{\sqrt{2t^4+3}};\quad a_N(3) = 6\sqrt{\frac{2}{55}}$$

53.

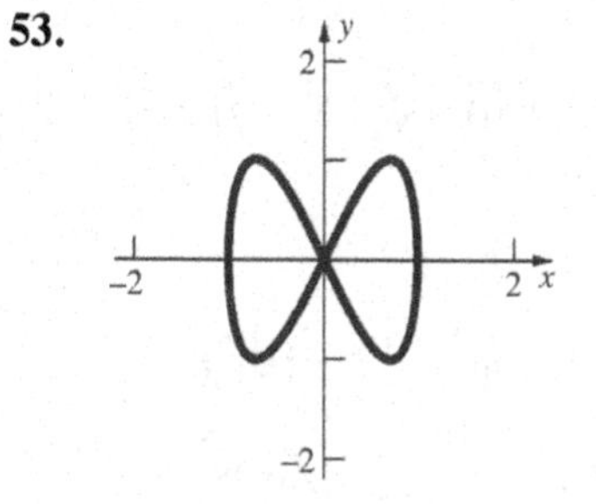

$\mathbf{v}(t) = \langle \cos t, 2\cos 2t\rangle$, $\mathbf{a}(t) = \langle -\sin t, -4\sin 2t\rangle$

$\mathbf{a}(t) = 0$ if and only if $-\sin t = 0$ and $-4\sin 2t = 0$, which occurs if and only if $t = 0$, π, 2π, so it occurs only at the origin.

$\mathbf{a}(t)$ points to the origin if and only if $\mathbf{a}(t) = -k\mathbf{r}(t)$ for some k and $\mathbf{r}(t)$ is not $\mathbf{0}$. This occurs if and only if $t = \frac{\pi}{2}, \frac{3\pi}{2}$, so it occurs only at $(1, 0)$ and $(-1, 0)$.

55. $s''(t) = a_T = 0 \Rightarrow$ speed $= s'(t) = c$ (a constant)

$$\kappa\left(\frac{ds}{dt}\right)^2 = a_N = 0 \Rightarrow \kappa = 0 \text{ or}$$

$$\frac{ds}{dt} = 0 \Rightarrow \kappa = 0$$

57. **v**(5) is tangent to the helix at the point where the particle is 12 meters above the ground. Its path is described by **v**(5) = cos 5**i** – sin 5**j** + 7**k**.

59. It is given that at (–12, 16), $s'(t) = 10$ ft/s and $s''(t) = 5$ ft/s^2. From Example 2, $\kappa = \frac{1}{20}$.

Therefore, $a_T = 5$ and $a_N = \left(\frac{1}{20}\right)(10)^2 = 5$, so **a** = 5**T** + 5**N**.

Let $\mathbf{r}(t) = \langle 20\cos t, 20\sin t\rangle$ describe the circle.

$\mathbf{r}(t) = \langle -12, 16\rangle \Rightarrow \cos t = -\frac{3}{5}$ and $\sin t = \frac{4}{5}$.

$\mathbf{v}(t) = \langle -20\sin t, 20\cos t\rangle$, so $|\mathbf{v}(t)| = 20$.

Then $\mathbf{T}(t) = \frac{\mathbf{v}(t)}{\|\mathbf{v}(t)\|} = \langle -\sin t, \cos t\rangle$.

Thus, at (–12, 16), $\mathbf{T} = \left\langle -\frac{4}{5}, -\frac{3}{5}\right\rangle$ and

$\mathbf{N} = \left\langle \frac{3}{5}, -\frac{4}{5}\right\rangle$ since **N** is a unit vector perpendicular to **T** and pointing to the concave side of the curve.

Therefore,

$$\mathbf{a} = 5\left\langle -\frac{4}{5}, -\frac{3}{5}\right\rangle + 5\left\langle \frac{3}{5}, -\frac{4}{5}\right\rangle = -\mathbf{i} - 7\mathbf{j}.$$

61. Let $\mu mg = \frac{mv_R^2}{R}$. Then $v_R = \sqrt{\mu gR}$. At the values given,

$v_R = \sqrt{(0.4)(32)(400)} = \sqrt{5120} \approx 71.55$ ft/s (about 49.79 mi/h).

63. $\tan\phi = y'$

$$(1 + y'^2)^{3/2} = (1 + \tan^2\phi)^{3/2} = \sec^3\phi$$

$$\kappa = \frac{|y''|}{\left[1 + y'^2\right]^{3/2}} = \frac{|y''|}{|\sec^3\phi|} = \left|y''\cos^3\phi\right|$$

65. $\mathbf{B} = \mathbf{T}\times\mathbf{N}$. Left-multiply by **T** and use Theorem 11.4C

$$\begin{aligned}\mathbf{T}\times\mathbf{B} &= \mathbf{T}\times(\mathbf{T}\times\mathbf{N})\\ &= (\mathbf{T}\cdot\mathbf{N})\mathbf{T} - (\mathbf{T}\cdot\mathbf{T})\mathbf{N}\\ &= 0\mathbf{T} - 1\mathbf{N}\\ &= -\mathbf{N}\end{aligned}$$

Thus, $\mathbf{N} = -\mathbf{T}\times\mathbf{B} = \mathbf{B}\times\mathbf{T}$.

To derive a result for **T** in terms of **N** and **B**, begin with $\mathbf{N} = \mathbf{B}\times\mathbf{T}$ and left multiply by **B**:

$$\begin{aligned}\mathbf{B}\times\mathbf{N} &= \mathbf{B}\times(\mathbf{B}\times\mathbf{T})\\ &= (\mathbf{B}\cdot\mathbf{T})\mathbf{B} - (\mathbf{B}\cdot\mathbf{B})\mathbf{T}\\ &= 0\mathbf{B} - 1\mathbf{T}\\ &= -\mathbf{T}\end{aligned}$$

Thus, $\mathbf{T} = -\mathbf{B}\times\mathbf{N} = \mathbf{N}\times\mathbf{B}$.

67. Let

$P_5(x) = a_0 + a_1x + a_2x^2 + a_3x^3 + a_4x^4 + a_5x^5$.

$P_5(0) = 0 \Rightarrow a_0 = 0$

$P_5'(x) = a_1 + 2a_2x + 3a_3x^2 + 4a_4x^3 + 5a_5x^4$, so

$P_5'(0) = 0 \Rightarrow a_1 = 0$.

$P''(x) = 2a_2 + 6a_3x + 12a_4x^2 + 20a_5x^3$, so

$P_5''(0) = 0 \Rightarrow a_2 = 0$.

Thus, $P_5(x) = a_3x^3 + a_4x^4 + a_5x^5$,

$P_5'(x) = 3a_3x^2 + 4a_4x^3 + 5a_5x^4$, and

$P_5''(x) = 6a_3x + 12a_4x^2 + 20a_5x^3$.

$P_5(1) = 1$, $P_5'(1) = 0$, and $P_5''(1) = 0 \Rightarrow$

$a_3 + a_4 + a_5 = 1$

$3a_3 + 4a_4 + 5a_5 = 0$

$6a_3 + 12a_4 + 20a_5 = 0$

The simultaneous solution to these equations is $a_3 = 10$, $a_4 = -15$, $a_5 = 6$, so

$P_5(x) = 10x^3 - 15x^4 + 6x^5$.

69. Let the polar coordinate equation of the curve be $r = f(\theta)$. Then the curve is parameterized by $x = r\cos\theta$ and $y = r\sin\theta$.

$x' = -r\sin\theta + r'\cos\theta \qquad y' = r\cos\theta + r'\sin\theta$

$x'' = -r\cos\theta - 2r'\sin\theta + r''\cos\theta \qquad y'' = -r\sin\theta + 2r'\cos\theta + r''\sin\theta$

By Theorem A, the curvature is

$$\kappa = \frac{|x'y'' - y'x''|}{\left|x'^2 + y'^2\right|^{3/2}}$$

$$= \frac{|(-r\sin\theta + r'\cos\theta)(-r\sin\theta + 2r'\cos\theta + r''\sin\theta) - (r\cos\theta + r'\sin\theta)(-r\cos\theta - 2r'\sin\theta + r''\cos\theta)|}{[(-r\sin\theta + r'\cos\theta)^2 + (r\cos\theta + r'\sin\theta)^2]^{3/2}}$$

$$= \frac{\left|r^2 + 2r'^2 - rr''\right|}{(r^2 + r'^2)^{3/2}}.$$

71. $r' = -\sin\theta, r'' = -\cos\theta$

At $\theta = 0, r = 2,\ r' = 0,$ and $r'' = -1.$

$$\kappa = \frac{|4+0+2|}{(4+0)^{3/2}} = \frac{6}{8} = \frac{3}{4}$$

73. $r' = -4\sin\theta, r'' = -4\cos\theta$

At $\theta = \frac{\pi}{2}, r = 4, r' = -4,$ and $r'' = 0.$

$$\kappa = \frac{|16+32-0|}{(16+16)^{3/2}} = \frac{48}{128\sqrt{2}} = \frac{3}{8\sqrt{2}}$$

75. $r' = 4\cos\theta, r'' = -4\sin\theta$

At $\theta = \frac{\pi}{2}, r = 8, r' = 0, r'' = -4.$

$$\kappa = \frac{|64+0+32|}{(64+0)^{3/2}} = \frac{96}{512} = \frac{3}{16}$$

77. $r = \sqrt{\cos 2\theta}$; $r' = -\frac{\sin 2\theta}{\sqrt{\cos 2\theta}}, r'' = -\frac{\cos^2 2\theta + 1}{(\cos 2\theta)^{3/2}}$

$$\kappa = \frac{\left|\cos 2\theta + \frac{2\sin^2 2\theta}{\cos 2\theta} + \frac{\cos^2 2\theta + 1}{\cos 2\theta}\right|}{\left(\cos 2\theta + \frac{\sin^2 2\theta}{\cos 2\theta}\right)^{3/2}} = \frac{\left|\frac{3}{\cos 2\theta}\right|}{\left(\frac{1}{\cos 2\theta}\right)^{3/2}}$$

$$= 3\left(\sqrt{\cos 2\theta}\right) = 3r$$

79.

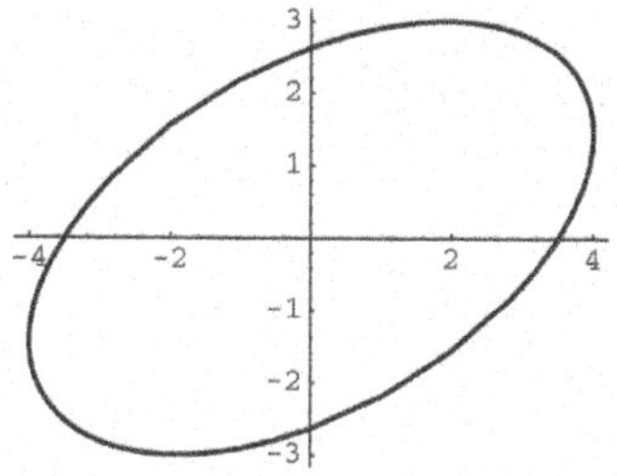

Maximum curvature ≈ 0.7606,
minimum curvature ≈ 0.1248

81. $\frac{d\mathbf{B}}{ds} = \frac{d}{ds}(\mathbf{T}\times\mathbf{N}) = \frac{d\mathbf{T}}{ds}\times\mathbf{N} + \mathbf{T}\times\frac{d\mathbf{N}}{ds}$

Since $\mathbf{N} = \frac{\frac{d\mathbf{T}}{ds}}{\left\|\frac{d\mathbf{T}}{ds}\right\|}, \frac{d\mathbf{T}}{ds}\times\mathbf{N} = 0,$ so $\frac{d\mathbf{B}}{ds} = \mathbf{T}\times\frac{d\mathbf{N}}{ds}.$

Thus $\mathbf{T}\cdot\frac{d\mathbf{B}}{ds} = \mathbf{T}\cdot\left(\mathbf{T}\times\frac{d\mathbf{N}}{ds}\right) = (\mathbf{T}\times\mathbf{T})\cdot\frac{d\mathbf{N}}{ds} = 0,$ so $\frac{d\mathbf{B}}{ds}$ is perpendicular to $\mathbf{T}$.

83. Let $ax + by + cz + d = 0$ be the equation of the plane containing the curve. Since **T** and **N** lie in the plane $\mathbf{B} = \pm\frac{a\mathbf{i} + b\mathbf{j} + c\mathbf{k}}{\sqrt{a^2 + b^2 + c^2}}$. Thus, **B** is a constant vector and $\frac{d\mathbf{B}}{ds} = \mathbf{0}$, so $\tau(s) = 0$, since **N** will not necessarily be **0** everywhere.

85. $\mathbf{r}(t) = 6\cos\pi t\mathbf{i} + 6\sin\pi t\mathbf{j}, + 2t\mathbf{k}, t > 0$

Let $(6\cos\pi t)^2 + (6\sin\pi t)^2 + (2t)^2 = 100.$

Then $36(\cos^2\pi t + \sin^2\pi t) + 4t^2 = 100;$

$4t^2 = 64;\ t = 4.$

$\mathbf{r}(4) = 6\mathbf{i} + 8\mathbf{k}$, so the fly will hit the sphere at the point (6, 0, 8).

$\mathbf{r}'(t) = -6\pi\sin\pi t\mathbf{i} + 6\pi\cos\pi t\mathbf{j} + 2\mathbf{k},$ so the fly will have traveled

$$\int_0^4 \sqrt{(-6\pi\sin\pi t)^2 + (6\pi\cos\pi t)^2 + (2)^2}\,dt$$

$$= \int_0^4 \sqrt{36\pi^2 + 4}\,dt$$

$$= \sqrt{36\pi^2 + 4}(4-0) = 8\sqrt{9\pi^2 + 1} \approx 75.8214$$

11.8 Concepts Review

1. traces; cross sections

3. ellipsoid

Problem Set 11.8

1. $\frac{x^2}{36}+\frac{y^2}{4}=1$
   Elliptic cylinder

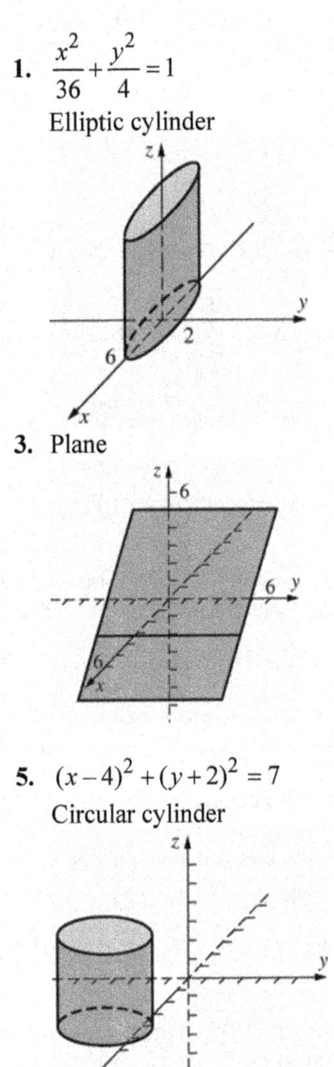

3. Plane

5. $(x-4)^2+(y+2)^2=7$
   Circular cylinder

7. $\frac{x^2}{441}+\frac{y^2}{196}+\frac{z^2}{36}=1$
   Ellipsoid

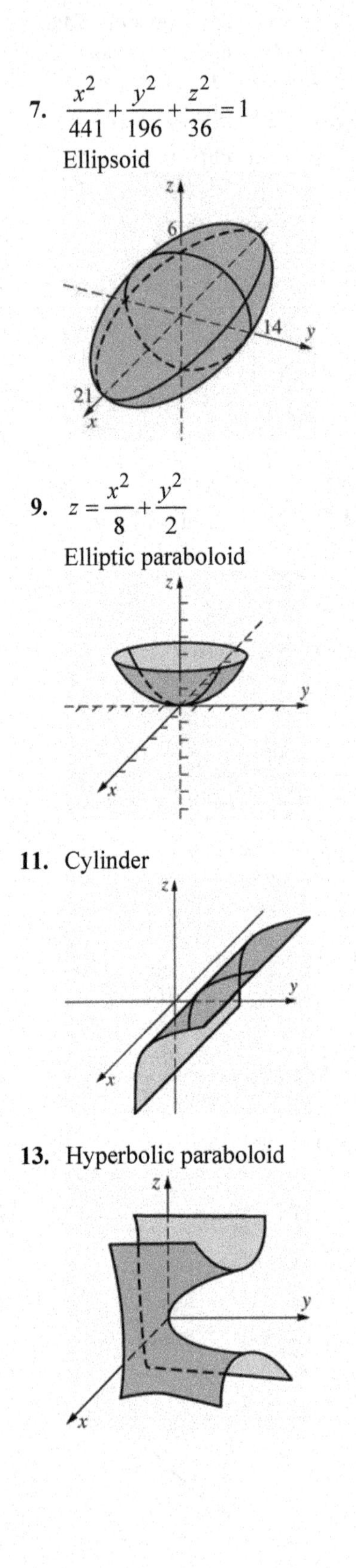

9. $z=\frac{x^2}{8}+\frac{y^2}{2}$
   Elliptic paraboloid

11. Cylinder

13. Hyperbolic paraboloid

15. $y = \frac{x^2}{4} + \frac{z^2}{9}$

Elliptic Paraboloid

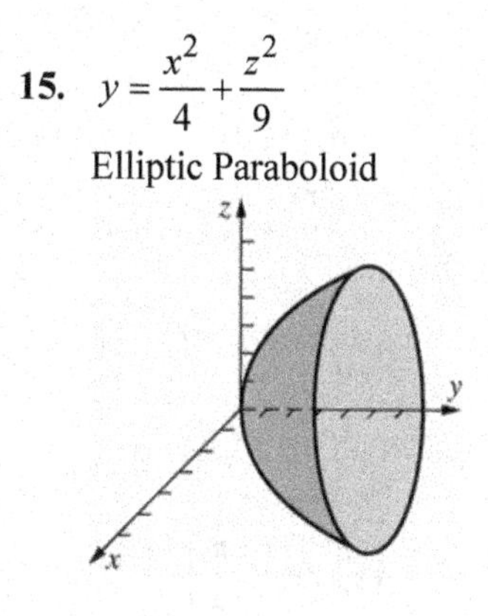

17. Plane

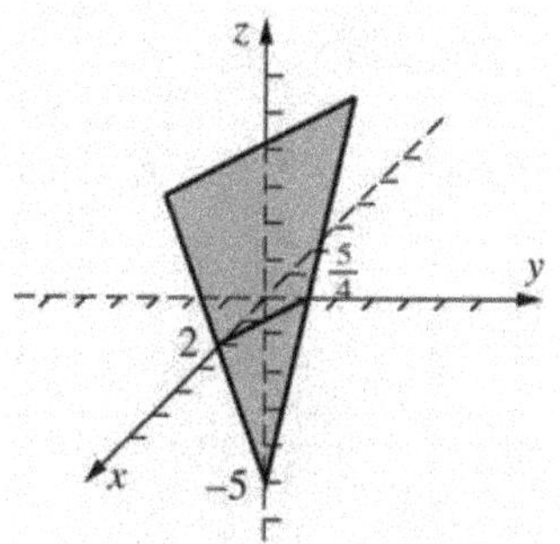

19. Hemisphere

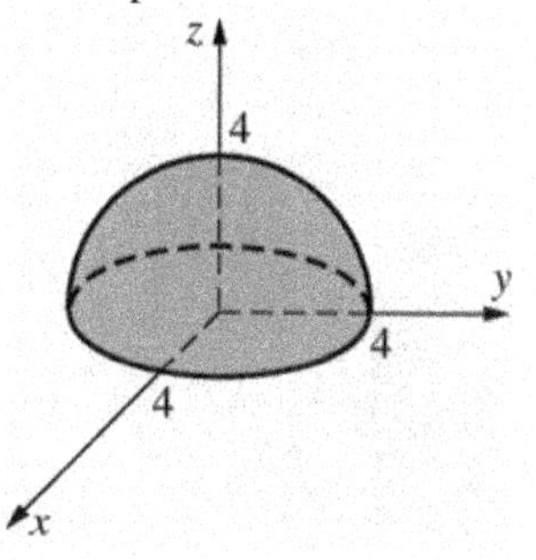

21. a. Replacing x by $-x$ results in an equivalent equation.

b. Replacing x by $-x$ and y by $-y$ results in an equivalent equation.

c. Replacing x by $-x$, y by $-y$, and z with $-z$, results in an equivalent equation.

23. All central ellipsoids are symmetric with respect to (a) the origin, (b) the x-axis, and the (c) xy-plane.

25. All central hyperboloids of two sheets are symmetric with respect to (a) the origin, (b) the z-axis, and (c) the yz-plane.

27. At $y = k$, the revolution generates a circle of radius $x = \sqrt{\frac{y}{2}} = \sqrt{\frac{k}{2}}$. Thus, the cross section in the plane $y = k$ is the circle $x^2 + z^2 = \frac{k}{2}$ or $2x^2 + 2z^2 = k$. The equation of the surface is $y = 2x^2 + 2z^2$.

29. At $y = k$, the revolution generates a circle of radius $x = \sqrt{3 - \frac{3}{4}y^2} = \sqrt{3 - \frac{3}{4}k^2}$. Thus, the cross section in the plane $y = k$ is the circle $x^2 + z^2 = 3 - \frac{3}{4}k^2$ or $12 - 4x^2 - 4z^2 = 3k^2$. The equation of the surface is $4x^2 + 3y^2 + 4z^2 = 12$.

31. When $z = 4$ the equation is $4 = \frac{x^2}{4} + \frac{y^2}{9}$ or $1 = \frac{x^2}{16} + \frac{y^2}{36}$, so $a^2 = 36, b^2 = 16$, and $c^2 = a^2 - b^2 = 20$, hence $c = \pm 2\sqrt{5}$. The major axis of the ellipse is on the y-axis so the foci are at $\left(0, \pm 2\sqrt{5}, 4\right)$.

33. When $z = h$, the equation is
$\frac{x^2}{a^2} + \frac{y^2}{b^2} + \frac{h^2}{c^2} = 1$ or $\frac{x^2}{a^2} + \frac{y^2}{b^2} = \frac{c^2 - h^2}{c^2}$
which is equivalent to

$\frac{x^2}{\frac{a^2(c^2-h^2)}{c^2}} + \frac{y^2}{\frac{b^2(c^2-h^2)}{c^2}} = 1$, which is

$\frac{x^2}{A^2} + \frac{y^2}{B^2} = 1$ with $A = \frac{a}{c}\sqrt{c^2 - h^2}$ and $B = \frac{b}{c}\sqrt{c^2 - h^2}$. Thus, the area is

$$\pi\left(\frac{a}{c}\sqrt{c^2 - h^2}\right)\left(\frac{b}{c}\sqrt{c^2 - h^2}\right) = \frac{\pi ab(c^2 - h^2)}{c^2}$$

35. Equating the expressions for y,

$4 - x^2 = x^2 + z^2$ or $1 = \frac{x^2}{2} + \frac{z^2}{4}$ which is the equation of an ellipse in the xz-plane with major diameter of $2\sqrt{4} = 4$ and minor diameter $2\sqrt{2}$.

37. $(t\cos t)^2 + (t\sin t)^2 - t^2$

$= t^2(\cos^2 t + \sin^2 t) - t^2 = t^2 - t^2 = 0,$

hence every point on the spiral is on the cone.

For $\mathbf{r} = 3t\cos t\mathbf{i} + t\sin t\mathbf{j} + t\mathbf{k}$, every point satisfies $x^2 + 9y^2 - 9z^2 = 0$ so the spiral lies on the elliptical cone.

11.9 Concepts Review

1. circular cylinder; sphere

3. $\rho^2 = r^2 + z^2$

Problem Set 11.9

1. Cylindrical to Spherical:

$\rho = \sqrt{r^2 + z^2}$

$\cos\phi = \frac{z}{\sqrt{r^2 + z^2}}$

$\theta = \theta$

Spherical to Cylindrical:

$r = \rho\sin\phi$

$z = \rho\cos\phi$

$\theta = \theta$

3. a. $x = 6\cos\left(\frac{\pi}{6}\right) = 3\sqrt{3}$

$y = 6\sin\left(\frac{\pi}{6}\right) = 3$

$z = -2$

b. $x = 4\cos\left(\frac{4\pi}{3}\right) = -2$

$y = 4\sin\left(\frac{4\pi}{3}\right) = -2\sqrt{3}$

$z = -8$

5. a. $\rho = \sqrt{x^2 + y^2 + z^2}$

$= \sqrt{4 + 12 + 16} = 4\sqrt{2}$

$\tan\theta = \frac{y}{x} = \frac{-2\sqrt{3}}{2} = -\sqrt{3}$ and (x, y) is in the 4th quadrant so $\theta = \frac{5\pi}{3}$.

$\cos\phi = \frac{z}{\rho} = \frac{4}{4\sqrt{2}} = \frac{\sqrt{2}}{2}$ so $\phi = \frac{\pi}{4}$.

Spherical: $\left(4\sqrt{2}, \frac{5\pi}{3}, \frac{\pi}{4}\right)$

b. $\rho = \sqrt{2 + 2 + 12} = 4$

$\tan\theta = \frac{\sqrt{2}}{-\sqrt{2}} = -1$ and (x, y) is in the 2nd quadrant so $\theta = \frac{3\pi}{4}$.

$\cos\phi = \frac{2\sqrt{3}}{4} = \frac{\sqrt{3}}{2}$ so $\frac{\pi}{6}$.

Spherical: $\left(4, \frac{3\pi}{4}, \frac{\pi}{6}\right)$

7. $r = 5$

Cylinder

9. $\phi = \frac{\pi}{6}$

Cone

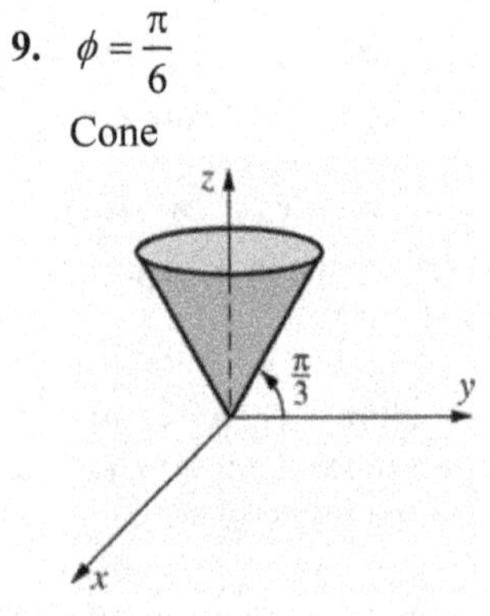

11. $r = 3\cos\theta$
Circular cylinder

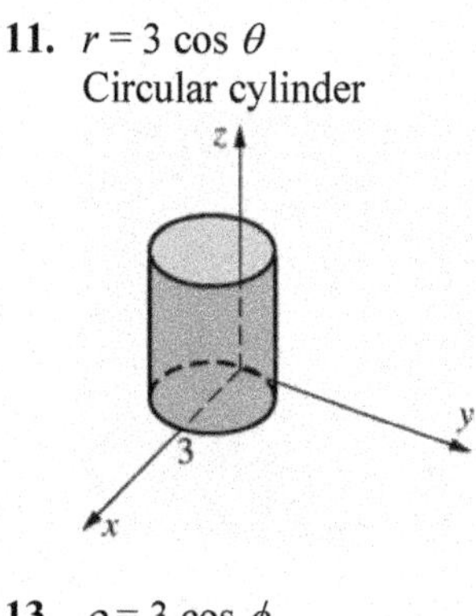

13. $\rho = 3\cos\phi$

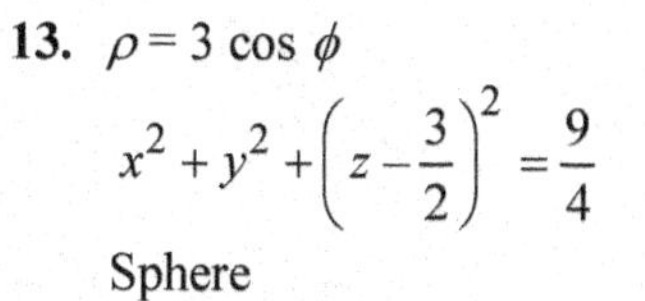

$$x^2 + y^2 + \left(z - \frac{3}{2}\right)^2 = \frac{9}{4}$$

Sphere

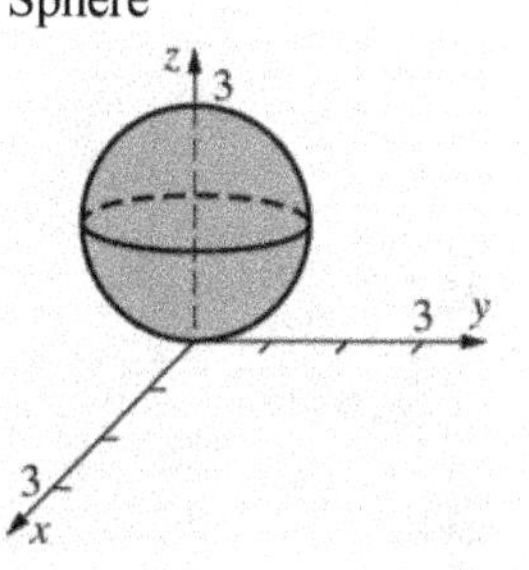

15. $r^2 + z^2 = 9$

$x^2 + y^2 + z^2 = 9$

Sphere

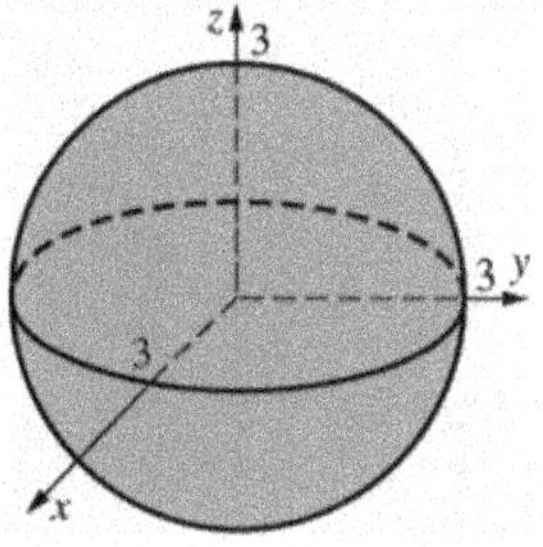

17. $x^2 + y^2 = 9;\ r^2 = 9;\ r = 3$

19. $r^2 + 4z^2 = 10$

21. $(x^2 + y^2 + z^2) - 3z^2 = 0;\ \rho^2 - 3\rho^2\cos^2\phi = 0;$

$\cos^2\phi = \frac{1}{3}$ (pole is not lost); $\cos^2\phi = \frac{1}{3}$ (or

$\sin^2\phi = \frac{2}{3}$ or $\tan^2\phi = 2$)

23. $(r^2 + z^2) + z^2 = 4;\ \rho^2 + \rho^2\cos^2\phi = 4;$

$$\rho^2 = \frac{4}{1 + \cos^2\phi}$$

25. $r\cos\theta + r\sin\theta = 4;\ r = \dfrac{4}{\sin\theta + \cos\theta}$

27. $(x^2 + y^2 + z^2) - z^2 = 9;$

$\rho^2 - \rho^2\cos^2\phi = 9;$

$\rho^2(1 - \cos^2\phi) = 9;\ \rho^2\sin^2\phi = 9$

$\rho\sin\phi = 3$

29. $r^2\cos 2\theta = z;\ r^2(\cos^2\theta - \sin^2\theta) = z;$

$(r\cos\theta)^2 - (r\sin\theta)^2 = z;\ x^2 - y^2 = z$

31. $z = 2x^2 + 2y^2 = 2(x^2 + y^2)$ (Cartesian);

$z = 2r^2$ (cylindrical)

33. For St. Paul:

$\rho = 3960,$

$\theta = 360° - 93.1° = 266.9° \approx 4.6583$ rad

$\phi = 90° - 45° = 45° = \dfrac{\pi}{4}$ rad

$$x = 3960\sin\frac{\pi}{4}\cos 4.6583 \approx -151.4$$

$$y = 3960\sin\frac{\pi}{4}\sin 4.6583 \approx -2796.0$$

$$z = 3960\cos\frac{\pi}{4} \approx 2800.1$$

For Oslo:

$\rho = 3960,\ \theta = 10.5° \approx 0.1833$ rad,

$\phi = 90° - 59.6° = 30.4° \approx 0.5306$ rad

$x = 3960 \sin 0.5306 \cos 0.1833 \approx 1970.4$

$y = 3960 \sin 0.5306 \sin 0.1833 \approx 365.3$

$z = 3960 \cos 0.5306 \approx 3415.5$

As in Example 7,

$$\cos\gamma \approx \frac{(-151.4)(1970.4) + (-2796.0)(365.3) + (2800.1)(3415.5)}{3960^2} \approx 0.5257$$

so $\gamma \approx 1.0173$ and the great-circle distance is $d \approx 3960(1.0173) \approx 4029$ mi

35. From Problem 33, the coordinates of St. Paul are $P(-151.4, -2796.0, 2800.1)$.

For Turin:

$\rho = 3960,\ \theta = 7.4° \approx 0.1292$ rad, $\phi = \dfrac{\pi}{4}$ rad

$$x = 3960\sin\frac{\pi}{4}\cos 0.1292 \approx 2776.8$$

$$y = 3960\sin\frac{\pi}{4}\sin 0.1292 \approx 360.8$$

$$z = 3960\cos\frac{\pi}{4} \approx 2800.1$$

$$\cos\gamma \approx \frac{(-151.4)(2776.8) + (-2796.0)(360.8) + (2800.1)(2800.1)}{3960^2} \approx 0.4088$$

so $\gamma \approx 1.1497$ and the great-circle distance is $d \approx 3960(1.1497) \approx 4553$ mi

37. Let St. Paul be at $P_1(-151.4, -2796.0, 2800.1)$ and Turin be at $P_2(2776.8, 360.8, 2800.1)$ and O be the center of the earth. Let β be the angle between the z-axis and the plane determined by O, P_1, and P_2. $\overrightarrow{OP_1} \times \overrightarrow{OP_2}$ is normal to the plane. The angle between the z-axis and $\overrightarrow{OP_1} \times \overrightarrow{OP_2}$ is complementary to β. Hence $\beta = \dfrac{\pi}{2} - \cos^{-1}\left(\dfrac{\left(\overrightarrow{OP_1} \times \overrightarrow{OP_2}\right)\cdot \mathbf{k}}{\left\|\overrightarrow{OP_1} \times \overrightarrow{OP_2}\right\|\|\mathbf{k}\|}\right) \approx \dfrac{\pi}{2} - \cos^{-1}\left(\dfrac{7.709\times 10^6}{1.431\times 10^7}\right) \approx 0.5689.$

The distance between the North Pole and the St. Paul-Turin great-circle is $3960(0.5689) \approx 2253$ mi

39. Let P_1 be (a_1, θ_1, ϕ_1) and P_2 be (a_2, θ_2, ϕ_2). If γ is the angle between $\overrightarrow{OP_1}$ and $\overrightarrow{OP_2}$ then the great-circle distance between P_1 and P_2 is $a\gamma$. $|OP_1| = |OP_2| = a$ while the straight-line distance between P_1 and P_2 is (from Problem 38)

$d^2 = (a-a)^2 + 2a^2[1 - \cos(\theta_1 - \theta_2)\sin\phi_1 \sin\phi_2 - \cos\phi_1 \cos\phi_2]$

$= 2a^2\{1 - [\cos(\theta_1 - \theta_2)\sin\phi_1 \sin\phi_2 + \cos\phi_1 \cos\phi_2]\}$.

Using the Law Of Cosines on the triangle OP_1P_2,

$d^2 = a^2 + a^2 - 2a^2\cos\gamma \ = 2a^2(1 - \cos\gamma)$.

Thus, γ is the central angle and $\cos\gamma = \cos(\theta_1 - \theta_2)\sin\phi_1 \sin\phi_2 + \cos\phi_1 \cos\phi_2$.

41. a. New York (–74°, 40.4°); Greenwich (0°, 51.3°)

$\cos\gamma = \cos(-74° - 0°)\cos(40.4°)\cos(51.3°) + \sin(40.4°)\sin(51.3°) \approx 0.637$

Then $\gamma \approx 0.880$ rad, so $d \approx 3960(0.8801) \approx 3485$ mi.

b. St. Paul (–93.1°, 45°); Turin (7.4°, 45°)

$\cos\gamma = \cos(-93.1° - 7.4°)\cos(45°)\cos(45°) + \sin(45°)\sin(45°) \approx 0.4089$

Then $\gamma \approx 1.495$ rad, so $d \approx 3960(1.1495) \approx 4552$ mi.

c. South Pole (7.4°, –90°); Turin (7.4°, 45°)

Note that any value of α can be used for the poles.

$\cos\gamma = \cos 0°\cos(-90°)\cos(45°) + \sin(-90°)\sin(45°) = -\frac{1}{\sqrt{2}}$

thus $\gamma = 135° = \frac{3\pi}{4}$ rad, so $d = 3960\left(\frac{3\pi}{4}\right) \approx 9331$ mi.

d. New York (–74°, 40.4°); Cape Town (18.4°, –33.9°)

$\cos\gamma = \cos(-74° - 18.4°)\cos(40.4°)\cos(-33.9°) + \sin(40.4°)\sin(-33.9°) \approx -0.3880$

Then $\gamma \approx 1.9693$ rad, so $d \approx 3960(1.9693) \approx 7798$ mi.

e. For these points $\alpha_1 = 100°$ and $\alpha_2 = -80°$ while $\beta_1 = \beta_2 = 0$, hence

$\cos\gamma = \cos 180°$ and $\gamma = \pi$ rad, so $d = 3960\pi \approx 12{,}441$ mi.

11.10 Chapter Review

Concepts Test

1. True: The coordinates are defined in terms of distances from the coordinate planes in such a way that they are unique.

3. True: See Section 11.2.

5. False: The distance between (0, 0, 3) and (0, 0, –3) (a point from each plane) is 6, so the distance between the planes is less than or equal to 6 units.

7. True: Let $t = \frac{1}{2}$.

9. True: $(2\mathbf{i} - 3\mathbf{j}) \cdot (6\mathbf{i} + 4\mathbf{j}) = 0$ if and only if the vectors are perpendicular.

11. False: The dot product for three vectors $(\mathbf{a} \cdot \mathbf{b}) \cdot \mathbf{c}$ is not defined.

13. True: If $\|\mathbf{u} \cdot \mathbf{v}\| = \|\mathbf{u}\|\|\mathbf{v}\|$, then $|\cos\theta| = 1$ since $\|\mathbf{u} \cdot \mathbf{v}\| = \|\mathbf{u}\|\|\mathbf{v}\||\cos\theta|$. Thus, **u** is a scalar multiple of **v**. If **u** is a scalar multiple of **v**,
$$\|\mathbf{u} \cdot \mathbf{v}\| = \|\mathbf{u} \cdot k\mathbf{u}\| = k\|\mathbf{u}\|^2$$
$$= \|\mathbf{u}\|\|k\mathbf{u}\| = \|\mathbf{u}\|\|\mathbf{v}\|.$$

15. True: $(\mathbf{u} + \mathbf{v}) \cdot (\mathbf{u} - \mathbf{v}) = \|\mathbf{u}\|^2 - \|\mathbf{v}\|^2 = 0$ so $\|\mathbf{u}\|^2 = \|\mathbf{v}\|^2$ or $|\mathbf{u}| = |\mathbf{v}|$.

17. True: Theorem 11.5A

19. True: $\| \|\mathbf{u}\| \mathbf{u} \| = \|\mathbf{u}\|\|\mathbf{u}\| = \|\mathbf{u}\|^2$

21. True: $\|\mathbf{u} \times \mathbf{v}\| = \|-\mathbf{v} \times \mathbf{u}\| = |-1|\|\mathbf{v} \times \mathbf{u}\|$
$$= \|\mathbf{v} \times \mathbf{u}\|$$

23. False: Obviously not true if **u** = **v**. (More generally, it is only true when **u** and **v** are also perpendicular.)

25. True: $\dfrac{\|\mathbf{u} \times \mathbf{v}\|}{(\mathbf{u} \cdot \mathbf{v})} = \dfrac{\|\mathbf{u}\|\|\mathbf{v}\|\sin\theta}{\|\mathbf{u}\|\|\mathbf{v}\|\cos\theta} = \tan\theta$

27. True: $|(2\mathbf{i} \times 2\mathbf{j}) \cdot (\mathbf{j} \times \mathbf{i})| = 4|(\mathbf{k}) \cdot (-\mathbf{k})|$
$$= 4(\mathbf{k} \cdot \mathbf{k}) = 4$$

29. True: Since $\langle b_1, b_2, b_3 \rangle$ is normal to the plane.

31. True: $\kappa = \dfrac{\|\mathbf{r}' \times \mathbf{r}''\|}{\|\mathbf{r}'\|^3} = 0 \Rightarrow \|\mathbf{r}' \times \mathbf{r}''\|$ $= \|\mathbf{r}'\|\|\mathbf{r}''\|\sin\theta = 0$. Thus, either $\mathbf{r}'$ and $\mathbf{r}''$ are parallel or either $\mathbf{r}'$ or $\mathbf{r}''$ is **0**, which implies that the path is a straight line.

33. False: κ depends only on the shape of the curve.

35. False: $x' = -2\sin t$ $\quad y' = 2\cos t$
$x'' = -2\cos t$ $\quad y'' = -2\sin t$
Thus,
$$\kappa = \frac{|x'y'' - y'x''|}{[x'^2 + y'^2]^{3/2}} = \frac{4}{8} = \frac{1}{2}.$$

37. False: Consider uniform circular motion:
$|dv/dt| = 0$ but $\|\mathbf{v}\| = a\omega$.

39. False: If $y'' = k$ then $y' = kx + C$ and
$$\kappa = \frac{|y''|}{[1 + y'^2]^{3/2}} = \frac{k}{[1 + (kx + C)^2]^{3/2}}$$
is not constant.

41. False: For example, if $\mathbf{r}(t) = \cos t^2\mathbf{i} + \sin t^2\mathbf{j}$, then $\mathbf{r}'(t) = -2t\sin t^2\mathbf{i} + 2t\cos t^2\mathbf{j}$, so $\|\mathbf{r}(t)\| = 1$ but $\|\mathbf{r}'(t)\| = 2t$.

43. True If $\mathbf{r}(t) = a\cos\omega t\,\mathbf{i} + a\sin\omega t\,\mathbf{j} + ct\mathbf{k}$, then $\mathbf{r}'(t) = -a\omega\sin\omega t\,\mathbf{i} + a\omega\cos\omega t\,\mathbf{j} + c\mathbf{k}$ so

$$\mathbf{T}(t) = \mathbf{r}'(t)/\|\mathbf{r}'(t)\| = \frac{-a\omega\sin\omega t\,\mathbf{i} + a\omega\cos\omega t\,\mathbf{j} + c\mathbf{k}}{\sqrt{a^2\omega^2 + c^2}}$$

$$\mathbf{T}'(t) = \frac{-a\omega^2\cos\omega t\,\mathbf{i} - a\omega^2\sin\omega t\,\mathbf{j}}{\sqrt{a\omega^2 + c^2}}$$

which points directly to the z-axis. Therefore $\mathbf{N}(t) = \mathbf{T}'(t)/\|\mathbf{T}'(t)\|$ points directly to the z-axis.

45. True: **T** depends only upon the shape of the curve, hence **N** and **B** also.

47. False: If $\mathbf{r}(t) = a\cos t\mathbf{i} + a\sin t\mathbf{j} + ct\mathbf{k}$ then **v** is perpendicular to **a**, but the path of motion is a circular helix, not a circle.

49. True: The curves are identical, although the motion of an object moving along the curves would be different.

51. True: The parameterization affects only the rate at which the curve is traced out.

53. False: For $\mathbf{r}(t) = \cos t\mathbf{i} + \sin t\,\mathbf{j}$, $\|\mathbf{r}(t)\| = 1$, but $\mathbf{r}'(t) \neq \mathbf{0}$.

55. False: The graph of $\rho = 0$ is the origin.

57. False: The origin, $\rho = 0$, has infinitely many spherical coordinates, since any value of θ and ϕ can be used.

Sample Test Problems

1. The center of the sphere is the midpoint $\left(\frac{-2+4}{2}, \frac{3+1}{2}, \frac{3+5}{2}\right) = (1, 2, 4)$ of the diameter. The radius is $r = \sqrt{(1+2)^2 + (2-3)^2 + (4-3)^2}$ $= \sqrt{9+1+1} = \sqrt{11}$. The equation of the sphere is $(x-1)^2 + (y-2)^2 + (z-4)^2 = 11$

3. **a.** $3\langle 2,-5\rangle - 2\langle 1,1\rangle = \langle 6,-15\rangle - \langle 2,2\rangle = \langle 4,-17\rangle$

b. $\langle 2,-5\rangle \cdot \langle 1,1\rangle = 2 + (-5) = -3$

c. $\langle 2,-5\rangle \cdot (\langle 1,1\rangle + \langle -6,0\rangle) = \langle 2,-5\rangle \cdot \langle -5,1\rangle = -10 + (-5) = -15$

d. $(4\langle 2,-5\rangle + 5\langle 1,1\rangle) \cdot 3\langle -6,0\rangle = \langle 13,-15\rangle \cdot \langle -18,0\rangle = -234 + 0 = -234$

e. $\sqrt{36+0}\langle -6,0\rangle \cdot \langle 1,-1\rangle = 6(-6+0) = -36$

f. $\langle -6,0\rangle \cdot \langle -6,0\rangle - \sqrt{36+0} = (36+0) - 6 = 30$

5. **a.** $\mathbf{a} + \mathbf{b} + \mathbf{c} = 2\mathbf{i} + \mathbf{j} + 4\mathbf{k}$

b. $\mathbf{b} \bullet \mathbf{c} = (0)(3) + (1)(-1) + (-2)(4) = -9$

c. $\mathbf{b} \times \mathbf{c} = \begin{vmatrix} \mathbf{i} & \mathbf{j} & \mathbf{k} \\ 0 & 1 & -2 \\ 3 & -1 & 4 \end{vmatrix} = 2\mathbf{i} - 6\mathbf{j} - 3\mathbf{k}$

$\mathbf{a} \cdot (\mathbf{b} \times \mathbf{c}) = (-\mathbf{i} + \mathbf{j} + 2\mathbf{k}) \cdot (2\mathbf{i} - 6\mathbf{j} - 3\mathbf{k}) = -2 - 6 - 6 = -14$

d. $\mathbf{b} \cdot \mathbf{c}$ is a scalar, and **a** crossed with a scalar doesn't exist.

e. $\|\mathbf{a} - \mathbf{b}\| = \|-\mathbf{i} + 0\mathbf{j} + 4\mathbf{k}\| = \sqrt{1^2 + 0^2 + 4^2} = \sqrt{17}$

f. From part (c), $\mathbf{b} \times \mathbf{c} = 2\mathbf{i} - 6\mathbf{j} - 3\mathbf{k}$ so $\|\mathbf{b} \times \mathbf{c}\| = \sqrt{2^2 + 6^2 + 3^2} = \sqrt{49} = 7$.

7.

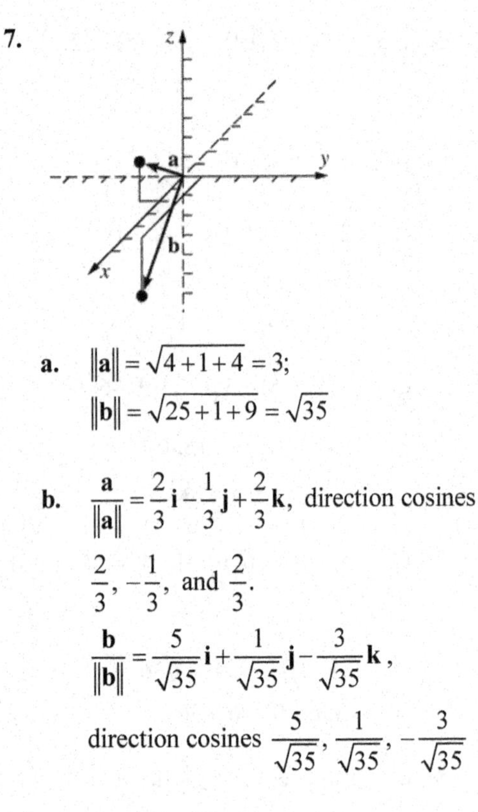

a. $\|\mathbf{a}\| = \sqrt{4+1+4} = 3;$
$\|\mathbf{b}\| = \sqrt{25+1+9} = \sqrt{35}$

b. $\frac{\mathbf{a}}{\|\mathbf{a}\|} = \frac{2}{3}\mathbf{i} - \frac{1}{3}\mathbf{j} + \frac{2}{3}\mathbf{k}$, direction cosines $\frac{2}{3}, -\frac{1}{3}$, and $\frac{2}{3}$.

$\frac{\mathbf{b}}{\|\mathbf{b}\|} = \frac{5}{\sqrt{35}}\mathbf{i} + \frac{1}{\sqrt{35}}\mathbf{j} - \frac{3}{\sqrt{35}}\mathbf{k}$,

direction cosines $\frac{5}{\sqrt{35}}, \frac{1}{\sqrt{35}}, -\frac{3}{\sqrt{35}}$

c. $\frac{2}{3}\mathbf{i} - \frac{1}{3}\mathbf{j} + \frac{2}{3}\mathbf{k}$

d. $\cos\theta = \frac{\mathbf{a}\cdot\mathbf{b}}{\|\mathbf{a}\|\|\mathbf{b}\|} = \frac{10-1-6}{3\sqrt{35}}$

$= \frac{3}{3\sqrt{35}} = \frac{1}{\sqrt{35}}$

$\theta = \cos^{-1}\frac{1}{\sqrt{35}} \approx 1.4010 \approx 80.27°$

9. $c\langle 3,3,-1\rangle \times \langle -1,-2,4\rangle = c\langle 10,-11,-3\rangle$ for any c in $\mathbb{R}$.

11. a. $y = 7$, since y must be a constant.

b. $x = -5$, since it is parallel to the yz-plane.

c. $z = -2$, since it is parallel to the xy-plane.

d. $3x - 4y + z = -45$, since it can be expressed as $3x - 4y + z = D$ and D must satisfy $3(-5) - 4(7) + (-2) = D$, so $D = -45$.

13. If the planes are perpendicular, their normals will also be perpendicular. Thus $0 = \langle 1, 5, C\rangle \cdot \langle 4, -1, 1\rangle = 4 - 5 + C$, so $C = 1$.

15. A vector in the direction of the line is $\langle 8,1,-8\rangle$. Parametric equations are $x = -2 + 8t$, $y = 1 + t$, $z = 5 - 8t$.

17. (0, 25, 16) and (–50, 0, 16) are on the line, so $\langle 50, 25, 0\rangle = 25\langle 2, 1, 0\rangle$ is in the direction of the line. Parametric equations are $x = 0 + 2t$, $y = 25 + 1t$, $z = 16 + 0t$ or $x = 2t$, $y = 25 + t$, $z = 16$.

19. $\langle 5,-4,-3\rangle$ is a vector in the direction of the line, and $\langle 2,-2,1\rangle$ is a position vector to the line. Then a vector equation of the line is $\mathbf{r}(t) = \langle 2,-2,1\rangle + t\langle 5,-4,-3\rangle$.

21. $\mathbf{r}'(t) = \langle 1, t, t^2\rangle$, $\mathbf{r}'(2) = \langle 1, 2, 4\rangle$ and $\mathbf{r}(2) = \left\langle 2, 2, \frac{8}{3}\right\rangle$. Symmetric equations for the tangent line are

$\frac{x-2}{1} = \frac{y-2}{2} = \frac{z-\frac{8}{3}}{4}$. Normal plane is

$1(x-2) + 2(y-2) + 4\left(z - \frac{8}{3}\right) = 0$ or

$3x + 6y + 12z = 50$.

23. $\mathbf{r}'(t) = e^t\langle \cos t + \sin t, -\sin t + \cos t, 1\rangle$

$\|\mathbf{r}'(t)\| = \sqrt{3}e^t$

Length is

$\int_1^5 \sqrt{3}e^t\,dt = \left[\sqrt{3}e^t\right]_1^5$

$= \sqrt{3}(e^5 - e) \approx 252.3509$.

25. Let the wind vector be

$\mathbf{w} = \langle 100\cos 30°, 100\sin 30° \rangle$

$= \langle 50\sqrt{3}, 50 \rangle.$

Let $\mathbf{p} = \langle p_1, p_2 \rangle$ be the plane's air velocity vector.

We want $\mathbf{w} + \mathbf{p} = 450\,\mathbf{j} = \langle 0, 450 \rangle$.

$\langle 50\sqrt{3}, 50 \rangle + \langle p_1, p_2 \rangle = \langle 0, 450 \rangle$

$\Rightarrow 50\sqrt{3} + p_1 = 0, 50 + p_2 = 450$

$\Rightarrow p_1 = -50\sqrt{3}, p_2 = 400$

Therefore, $\mathbf{p} = \langle -50\sqrt{3}, 400 \rangle$. The angle β formed with the vertical satisfies

$$\cos\beta = \frac{\mathbf{p}\cdot\mathbf{j}}{\|\mathbf{p}\|\|\mathbf{j}\|} = \frac{400}{\sqrt{167{,}500}};\ \beta \approx 12.22°.$$

Thus, the heading is N12.22°W. The air speed is $\|\mathbf{p}\| = \sqrt{167{,}500} \approx 409.27$ mi/h.

27. **a.** $\mathbf{r}'(t) = \left\langle \frac{1}{t}, -6t \right\rangle; \mathbf{r}''(t) = \left\langle -t^{-2}, -6 \right\rangle$

b. $\mathbf{r}'(t) = \langle \cos t, -2\sin 2t \rangle;$

$\mathbf{r}''(t) = \langle -\sin t, -4\cos 2t \rangle$

c. $\mathbf{r}'(t) = \left\langle \sec^2 t, -4t^3 \right\rangle;$

$\mathbf{r}''(t) = \left\langle 2\sec^2 t\tan t, -12t^2 \right\rangle$

29. $\mathbf{v}(t) = \left\langle 1, 2t, 3t^2 \right\rangle; \mathbf{a}(t) = \langle 0, 2, 6t \rangle$

$\mathbf{v}(1) = \langle 1, 2, 3 \rangle$

$\|\mathbf{v}(1)\| = \sqrt{14}$

$\mathbf{a}(1) = \langle 0, 2, 6 \rangle$

$$a_T = \frac{\mathbf{v}\cdot\mathbf{a}}{\|\mathbf{v}\|} = \frac{0+4+18}{\sqrt{14}} = \frac{22}{\sqrt{14}} \approx 5.880;$$

$$a_N = \frac{\|\mathbf{v}\times\mathbf{a}\|}{\|\mathbf{v}\|} = \frac{|\langle 6, -6, 2 \rangle|}{\sqrt{14}}$$

$$= \frac{2\sqrt{19}}{\sqrt{14}} \approx 2.330$$

31. Sphere

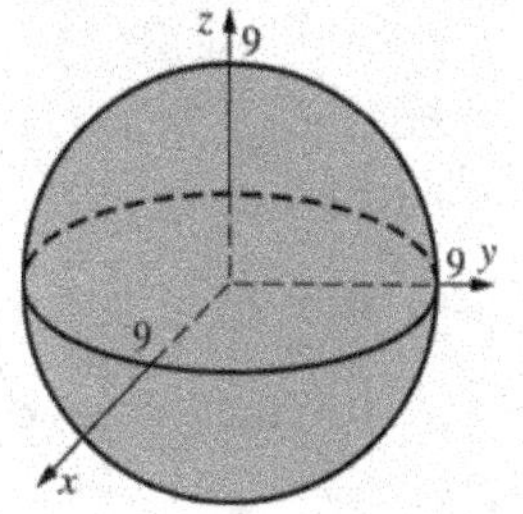

33. Circular paraboloid

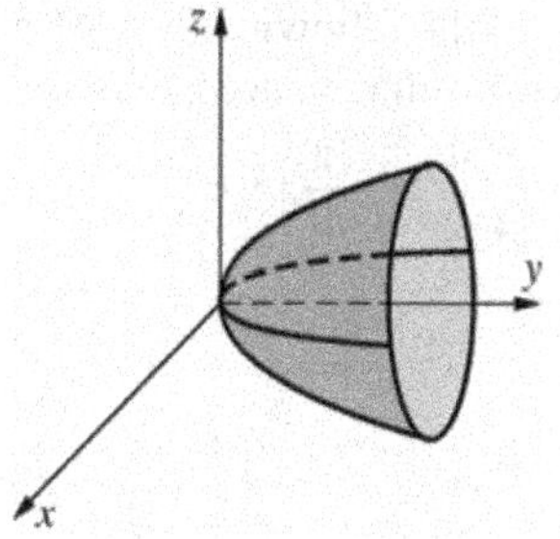

35. Plane

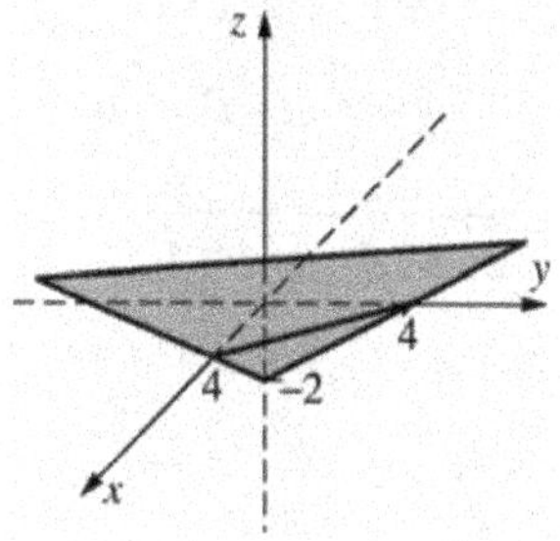

37. $\frac{x^2}{12} + \frac{y^2}{9} + \frac{z^2}{4} = 1$

Ellipsoid

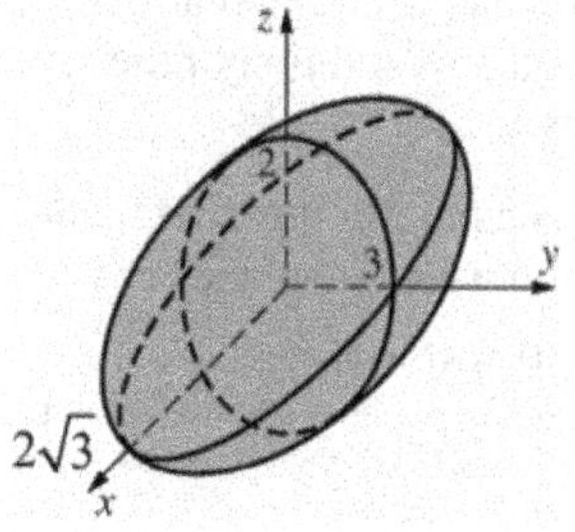

39. **a.** $r^2 = 9;\ r = 3$

b. $(x^2 + y^2) + 3y^2 = 16$

$r^2 + 3r^2\sin^2\theta = 16, r^2 = \frac{16}{1 + 3\sin^2\theta}$

c. $r^2 = 9z$

d. $r^2 + 4z^2 = 10$

41. a. $\rho^2 = 4; \rho = 2$

b. $x^2 + y^2 + z^2 - 2z^2 = 0;$

$\rho^2 - 2\rho^2 \cos^2 \phi = 0;$

$\rho^2(1 - 2\cos^2 \phi) = 0;$

$1 - 2\cos^2 \phi = 0;\ \cos^2 \phi = \frac{1}{2};\ \phi = \frac{\pi}{4}$

or $\phi = \frac{3\pi}{4}$.

Any of the following (as well as others) would be acceptable:

$\left(\phi - \frac{\pi}{4}\right)\left(\phi - \frac{3\pi}{4}\right) = 0$

$\cos^2 \phi = \frac{1}{2}$

$\sec^2 \phi = 2$

$\tan^2 \phi = 1$

c. $2x^2 - (x^2 + y^2 + z^2) = 1;$

$2\rho^2 \sin^2 \phi \cos^2 \theta - \rho^2 = 1;$

$\rho^2 = \frac{1}{2\sin^2 \phi \cos^2 \theta - 1}$

d. $x^2 + y^2 = z;$

$\rho^2 \sin^2 \phi \cos^2 \theta + \rho^2 \sin^2 \phi \sin^2 \theta = \rho \cos \phi$

$\rho^2 \sin^2 \phi(\cos^2 \theta + \sin^2 \theta) = \rho \cos \phi;$

$\rho \sin^2 \phi = \cos \phi;\ \rho = \cot \phi \csc \phi$

(Note that when we divided through by ρ in part c and d we did not lose the pole since it is also a solution of the resulting equations.)

43. (2, 0, 0) is a point of the first plane. The distance between the planes is

$$\frac{|2(2) - 3(0) + \sqrt{3}(0) - 9|}{\sqrt{4 + 9 + 3}} = \frac{5}{\sqrt{16}} = 1.25$$

45. If speed $= \frac{ds}{dt} = c$, a constant, then

$\mathbf{a} = \frac{d^2 s}{dt^2}\mathbf{T} + \left(\frac{ds}{dt}\right)^2 \kappa \mathbf{N} = c^2 \kappa \mathbf{N}$ since

$\frac{d^2 s}{dt^2} = 0$. **T** is in the direction of **v**, while **N** is perpendicular to **T** and hence to **v** also. Thus, $\mathbf{a} = c^2 \kappa \mathbf{N}$ is perpendicular to **v**.

Review and Preview

1. $x^2 + y^2 + z^2 = 64$

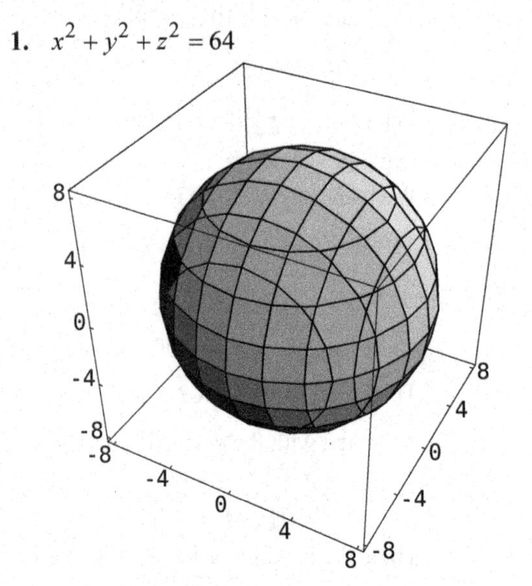

3. $z = x^2 + 4y^2$

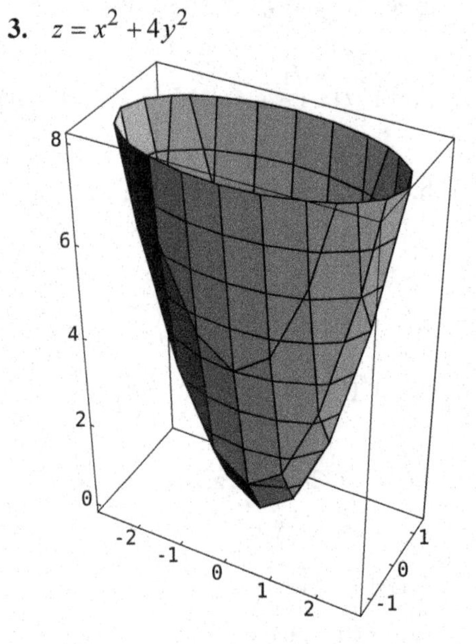

5. a. $\frac{d}{dx}2x^3 = 6x^2$

b. $\frac{d}{dx}5x^3 = 15x^2$

c. $\frac{d}{dx}kx^3 = 3kx^2$

d. $\frac{d}{dx}ax^3 = 3ax^2$

7. **a.** $\frac{d}{da}\sin 2a = 2\cos 2a$

b. $\frac{d}{da}\sin 17a = 17\cos 17a$

c. $\frac{d}{da}\sin ta = t\cos ta$

d. $\frac{d}{da}\sin sa = s\cos sa$

9. $f(x) = \frac{1}{x^2 - 1}$ is both continuous and differentiable at $x = 2$ since rational functions are continuous and differentiable at every real number in their domain.

11. $f(x) = |x - 4|$ is continuous at $x = 4$ since $\lim_{x \to 4} |x - 4| = f(4) = 0$. *f* is not differentiable at $x = 4$.

13. $f'(x) = 3 - 3(x-1)^2$; $f'(x) = 0$ when $x = 0, 2$;
$f''(x) = -6(x-1)$; Since $f''(2) = -6$, a local maximum occurs at $x = 2$. Since $f(0) = -1$,
$f(2) = 5$, and $f(4) = -15$, the maximum value of *f* on [0, 4] is 5 while the minimum value is -15.

15. $S = 2\pi r^2 + 2\pi rh$
Since the volume is to be 8 cubic feet, we have

$$V = 8$$
$$\pi r^2 h = 8$$
$$h = \frac{8}{\pi r^2}$$

Substituting for *h* in our surface area equation gives us

$$S = 2\pi r^2 + 2\pi r\left(\frac{8}{\pi r^2}\right) = 2\pi r^2 + \frac{16}{r}$$

Thus, we can write *S* as a function of *r*:

$$S(r) = 2\pi r^2 + \frac{16}{r}$$

CHAPTER 12 Derivatives for Functions of Two or More Variables

12.1 Concepts Review

1. real-valued function of two real variables

3. concentric circles

Problem Set 12.1

1. a. 5 b. 0

   c. 6 d. $a^6 + a^2$

   e. $2x^2,\ x \neq 0$ f. Undefined

   The natural domain is the set of all (x, y) such that y is nonnegative.

3. a. $\sin(2\pi) = 0$ b. $4\sin\left(\frac{\pi}{6}\right) = 2$

   c. $16\sin\left(\frac{\pi}{2}\right) = 16$

   d. $\pi^2 \sin(\pi^2) \approx -4.2469$

5. $F(t\cos t, \sec^2 t) = t^2 \cos^2 t \sec^2 t = t^2, \ \cos t \neq 0$

7. $z = 6$ is a plane.

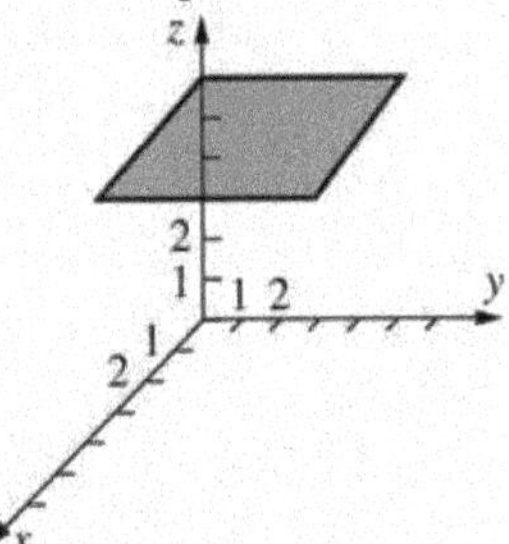

9. $x + 2y + z = 6$ is a plane.

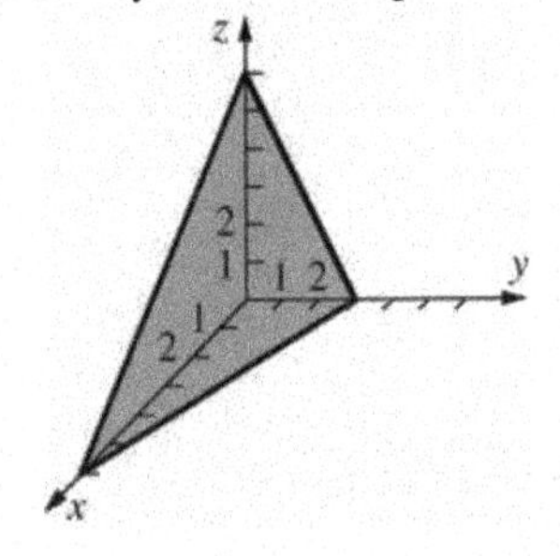

11. $x^2 + y^2 + z^2 = 16$, $z \geq 0$ is a hemisphere.

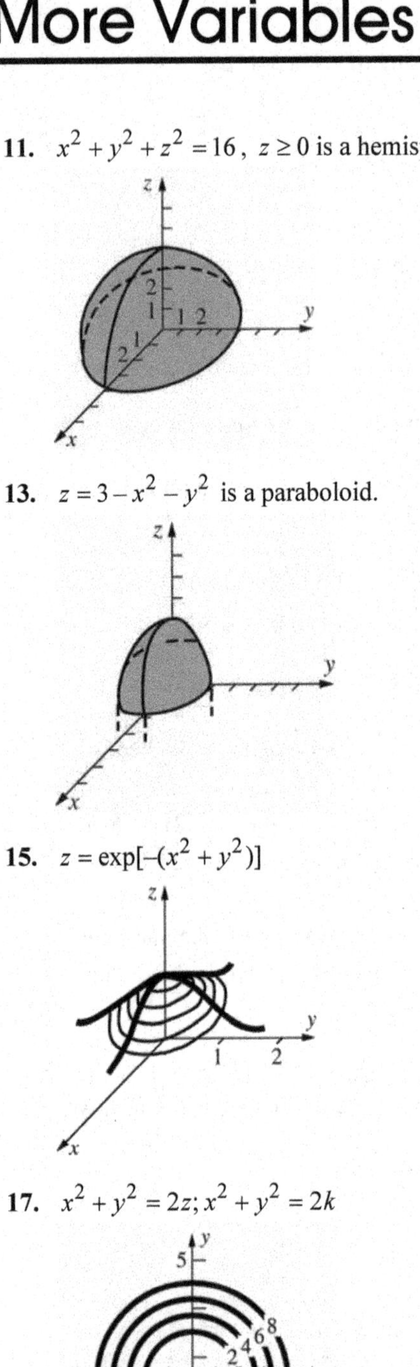

13. $z = 3 - x^2 - y^2$ is a paraboloid.

15. $z = \exp[-(x^2 + y^2)]$

17. $x^2 + y^2 = 2z; x^2 + y^2 = 2k$

19. $x^2 = zy, y \neq 0; x^2 = ky, y \neq 0$

21. $z = \dfrac{x^2+1}{x^2+y^2}, k = 1, 2, 4$

$k = 1$: $y^2 = 1$ or $y = \pm 1$;
two parallel lines

$k = 2$: $2x^2 + 2y^2 = x^2 + 1$

$\dfrac{x^2}{1} + \dfrac{y^2}{\frac{1}{2}} = 1$; ellipse

$k = 4$: $4x^2 + 4y^2 = x^2 + 1$

$\dfrac{x^2}{\frac{1}{3}} + \dfrac{y^2}{\frac{1}{4}} = 1$; ellipse

23. $x = 0$, if $T = 0$:

$y^2 = \left(\dfrac{1}{T} - 1\right)x^2$, if $y \neq 0$.

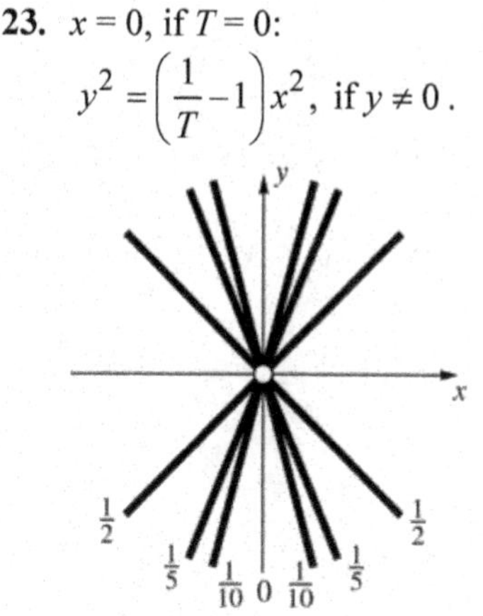

25. a. San Francisco and St. Louis had a temperature between 70 and 80 degrees Fahrenheit.

b. Drive northwest to get to cooler temperatures, and drive southeast to get warmer temperatures.

c. Since the level curve for 70 runs southwest to northeast, you could drive southwest or northeast and stay at about the same temperature.

27. $x^2 + y^2 + z^2 \geq 16$; the set of all points on and outside the sphere of radius 4 that is centered at the origin

29. $\dfrac{x^2}{9} + \dfrac{y^2}{16} + \dfrac{z^2}{1} \leq 1$; points inside and on the ellipsoid

31. Since the argument to the natural logarithm function must be positive, we must have $x^2 + y^2 + z^2 > 0$. This is true for all (x, y, z) except $(x, y, z) = (0, 0, 0)$. The domain consists all points in $\mathbb{R}^3$ except the origin.

33. $x^2 + y^2 + z^2 = k,\ k > 0$; set of all spheres centered at the origin

35. $\dfrac{x^2}{\frac{1}{16}} + \dfrac{y^2}{\frac{1}{4}} - \dfrac{z^2}{1} = k$; the elliptic cone

$\dfrac{x^2}{9} + \dfrac{y^2}{9} = \dfrac{z^2}{16}$ and all hyperboloids (one and two sheets) with z-axis for axis such that $a{:}b{:}c$ is $\left(\dfrac{1}{4}\right){:}\left(\dfrac{1}{4}\right){:}\left(\dfrac{1}{3}\right)$ or 3:3:4.

37. $4x^2 - 9y^2 = k,\ k$ in R; $\dfrac{x^2}{\frac{k}{4}} - \dfrac{y^2}{\frac{k}{9}} = 1$, if $k \neq 0$;

planes $y = \pm\dfrac{2x}{3}$ (for $k = 0$) and all hyperbolic cylinders parallel to the z-axis such that the ratio $a{:}b$ is $\left(\dfrac{1}{2}\right){:}\left(\dfrac{1}{3}\right)$ or 3:2 (where a is associated with the x-term)

39. a. All (w, x, y, z) except $(0, 0, 0, 0)$, which would cause division by 0.

b. All $(x_1, x_2, \ldots, x_n)$ in n-space.

c. All $(x_1, x_2, \ldots, x_n)$ that satisfy $x_1^2 + x_2^2 + \cdots + x_n^2 \leq 1$; other values of $(x_1, x_2, \ldots, x_n)$ would lead to the square root of a negative number.

41. a. AC is the least steep path and BC is the most steep path between A and C since the level curves are farthest apart along AC and closest together along BC.

b. $|AC| \approx \sqrt{(5750)^2 + (3000)^2} \approx 6490$ ft

$|BC| \approx \sqrt{(580)^2 + (3000)^2} \approx 3060$ ft

43.

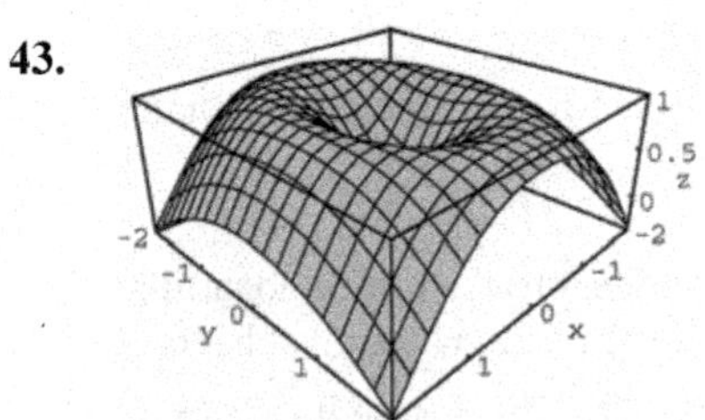

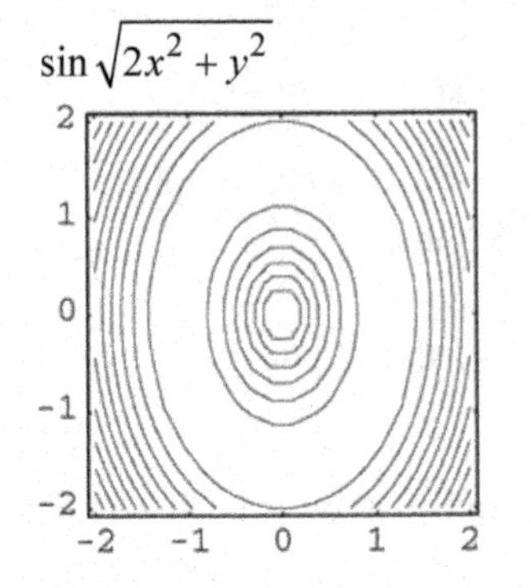

45.

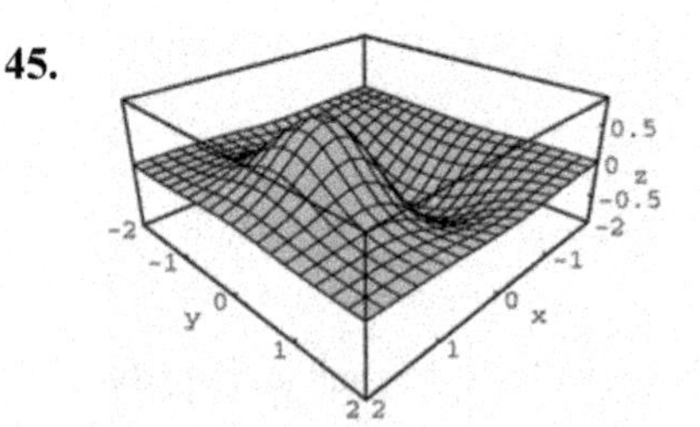

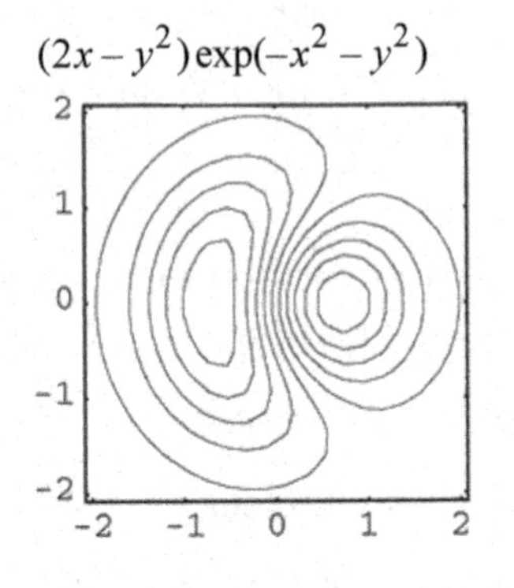

12.2 Concepts Review

1. $\lim_{h \to 0} \frac{[(f(x_0+h, y_0) - f(x_0, y_0)]}{h}$; partial derivative of f with respect to x

3. 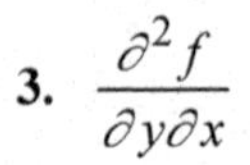
$\frac{\partial^2 f}{\partial y \partial x}$

Problem Set 12.2

1. $f_x(x, y) = 8(2x-y)^3$; $f_y(x, y) = -4(2x-y)^3$

3. $f_x(x, y) = \frac{(xy)(2x) - (x^2-y^2)(y)}{(xy)^2} = \frac{x^2+y^2}{x^2y}$

$f_y(x, y) = \frac{(xy)(-2y) - (x^2-y^2)(x)}{(xy)^2}$

$= -\frac{(x^2+y^2)}{xy^2}$

5. $f_x(x, y) = e^y \cos x$; $f_y(x, y) = e^y \sin x$

7. $f_x(x, y) = x(x^2-y^2)^{-1/2}$;
$f_y(x, y) = -y(x^2-y^2)^{-1/2}$

9. $g_x(x, y) = -ye^{-xy}$; $g_y(x, y) = -xe^{-xy}$

11. $f_x(x, y) = 4[1+(4x-7y)^2]^{-1}$;
$f_y(x, y) = -7[1+(4x-7y)^2]^{-1}$

13. $f_x(x, y) = -2xy\sin(x^2+y^2)$;
$f_y(x, y) = -2y^2\sin(x^2+y^2) + \cos(x^2+y^2)$

15. $F_x(x, y) = 2\cos x \cos y$; $F_y(x, y) = -2\sin x \sin y$

17. $f_x(x, y) = 4xy^3 - 3x^2y^5$;
$f_{xy}(x, y) = 12xy^2 - 15x^2y^4$
$f_y(x, y) = 6x^2y^2 - 5x^3y^4$;
$f_{yx}(x, y) = 12xy^2 - 15x^2y^4$

19. $f_x(x, y) = 6e^{2x}\cos y$; $f_{xy}(x, y) = -6e^{2x}\sin y$
$f_y(x, y) = -3e^{2x}\sin y$; $f_{yx}(x, y) = -6e^{2x}\sin y$

21. $F_x(x, y) = \frac{(xy)(2) - (2x-y)(y)}{(xy)^2} = \frac{y^2}{x^2y^2} = \frac{1}{x^2}$;

$F_x(3, -2) = \frac{1}{9}$

$F_y(x, y) = \frac{(xy)(-1) - (2x-y)(x)}{(xy)^2} = \frac{-2x^2}{x^2y^2} = -\frac{2}{x^2}$;

$F_y(3, -2) = -\frac{1}{2}$

23. $f_x(x, y) = -y^2(x^2 + y^4)^{-1};$

$f_x\left(\sqrt{5}, -2\right) = -\frac{4}{21} \approx -0.1905$

$f_y(x, y) = 2xy(x^2 + y^4)^{-1};$

$f_y\left(\sqrt{5}, -2\right) = -\frac{4\sqrt{5}}{21} \approx -0.4259$

25. Let $z = f(x, y) = \frac{x^2}{9} + \frac{y^2}{4}$.

$f_y(x, y) = \frac{y}{2}$

The slope is $f_y(3, 2) = 1$.

27. $z = f(x, y) = \left(\frac{1}{2}\right)(9x^2 + 9y^2 - 36)^{1/2}$

$f_x(x, y) = \frac{9x}{2(9x^2 + 9y^2 - 36)^{1/2}}$

$f_x(2, 1) = 3$

29. $V_r(r, h) = 2\pi rh;$

$V_r(6, 10) = 120\pi \approx 376.99 \text{ in.}^2$

31. $P(V, T) = \frac{kT}{V}$

$P_T(V, T) = \frac{k}{V};$

$P_T(100, 300) = \frac{k}{100}$ lb/in.^{2} per degree

33. $f_x(x, y) = 3x^2y - y^3;\ f_{xx}(x, y) = 6xy;$

$f_y(x, y) = x^3 - 3xy^2;\ f_{yy}(x, y) = -6xy$

Therefore, $f_{xx}(x, y) + f_{yy}(x, y) = 0$.

35. $F_y(x, y) = 15x^4y^4 - 6x^2y^2;$

$F_{yy}(x, y) = 60x^4y^3 - 12x^2y;$

$F_{yyy}(x, y) = 180x^4y^2 - 12x^2$

37. a. $\frac{\partial^3 f}{\partial y^3}$

b. $\frac{\partial^3 y}{\partial y \partial x^2}$

c. $\frac{\partial^4 y}{\partial y^3 \partial x}$

39. a. $f_x(x, y, z) = 6xy - yz$

b. $f_y(x, y, z) = 3x^2 - xz + 2yz^2;$

$f_y(0, 1, 2) = 8$

c. Using the result in a, $f_{xy}(x, y, z) = 6x - z$.

41. $f_x(x, y, x) = -yze^{-xyz} - y(xy - z^2)^{-1}$

43. If $f(x, y) = x^4 + xy^3 + 12,\ f_y(x, y) = 3xy^2;$ $f_y(1, -2) = 12$. Therefore, along the tangent line $\Delta y = 1 \Rightarrow \Delta z = 12$, so $\langle 0, 1, 12\rangle$ is a tangent vector (since $\Delta x = 0$). Then parametric equations of the tangent line are $\begin{Bmatrix} x = 1 \\ y = -2 + t \\ z = 5 + 12t \end{Bmatrix}$. Then the point of xy-plane at which the bee hits is $(1, 0, 29)$ [since $y = 0 \Rightarrow t = 2 \Rightarrow x = 1, z = 29$].

45. Domain: (Case $x < y$)
The lengths of the sides are then x, $y - x$, and $1 - y$. The sum of the lengths of any two sides must be greater than the length of the remaining side, leading to three inequalities:

$x + (y - x) > 1 - y \Rightarrow y > \frac{1}{2}$

$(y - x) + (1 - y) > x \Rightarrow x < \frac{1}{2}$

$x + (1 - y) > y - x \Rightarrow y < x + \frac{1}{2}$

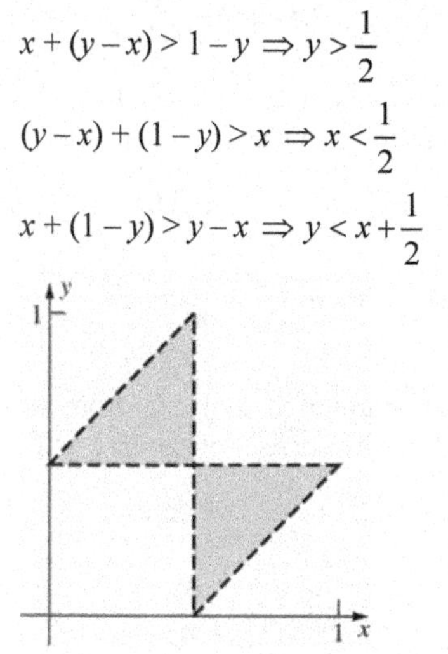

The case for $y < x$ yields similar inequalities (x and y interchanged). The graph of D_A, the domain of A is given above. In set notation it is

$D_A = \left\{(x, y) : x < \frac{1}{2}, y > \frac{1}{2}, y < x + \frac{1}{2}\right\}$

$\cup \left\{(x, y) : x > \frac{1}{2}, y < \frac{1}{2},\ x < y + \frac{1}{2}\right\}.$

Range: The area is greater than zero but can be arbitrarily close to zero since one side can be arbitrarily small and the other two sides are bounded above. It seems that the area would be largest when the triangle is equilateral. An equilateral triangle with sides equal to $\frac{1}{3}$ has area $\frac{\sqrt{3}}{36}$. Hence the range of A is $\left(0, \frac{\sqrt{3}}{36}\right]$. (In Sections 8 and 9 of this chapter methods will be presented which will make it easy to prove that the largest value of A will occur when the triangle is equilateral.)

47. a. Moving parallel to the y-axis from the point (1, 1) to the nearest level curve and approximating $\frac{\Delta z}{\Delta y}$, we obtain

$$f_y(1, 1) = \frac{4-5}{1.25-1} = -4.$$

b. Moving parallel to the x-axis from the point (–4, 2) to the nearest level curve and approximating $\frac{\Delta z}{\Delta x}$, we obtain

$$f_x(-4, 2) \approx \frac{1-0}{-2.5-(-4)} = \frac{2}{3}.$$

c. Moving parallel to the x-axis from the point (–5, –2) to the nearest level curve and approximately $\frac{\Delta z}{\Delta x}$, we obtain

$$f_x(-4, -5) \approx \frac{1-0}{-2.5-(-5)} = \frac{2}{5}.$$

d. Moving parallel to the y-axis from the point (0, –2) to the nearest level curve and approximating $\frac{\Delta z}{\Delta y}$, we obtain

$$f_y(0, 2) \approx \frac{0-1}{\frac{-19}{8}-(-2)} = \frac{8}{3}.$$

49. a. $f_y(x, y, z)$

$$= \lim_{\Delta y \to 0} \frac{f(x, y+\Delta y, z) - f(x, y, z)}{\Delta y}$$

b. $f_z(x, y, z)$

$$= \lim_{\Delta z \to 0} \frac{f(x, y, z+\Delta z) - f(x, y, z)}{\Delta z}$$

c. $G_x(w, x, y, z)$

$$= \lim_{\Delta x \to 0} \frac{G(w, x+\Delta x, y, z) - G(w, x, y, z)}{\Delta x}$$

d. $\frac{\partial}{\partial z}\lambda(x, y, z, t)$

$$= \lim_{\Delta z \to 0} \frac{\lambda(x, y, z+\Delta z, t) - \lambda(x, y, z, t)}{\Delta z}$$

e. $\frac{\partial}{\partial b_2} S(b_0, b_1, b_2, \ldots, b_n) =$

$$= \lim_{\Delta b_2 \to 0} \left(\frac{S(b_0, b_1, b_2 + \Delta b_2, \ldots, b_n) - S(b_0, b_1, b_2, \ldots, b_n)}{\Delta b_2} \right)$$

12.3 Concepts Review

1. 3; (x, y) approaches (1, 2).

3. contained in S

Problem Set 12.3

1. –18

3. $\lim_{(x, y) \to (2, \pi)} \left[x\cos^2 xy - \sin\left(\frac{xy}{3}\right) \right]$

$$= 2\cos^2 2\pi - \sin\left(\frac{2\pi}{3}\right) = 2 - \frac{\sqrt{3}}{2} \approx 1.1340$$

5. $-\frac{5}{2}$

7. 1

9. The limit does not exist since the function is not defined anywhere along the line $y = x$. That is, there is no neighborhood of the origin in which the function is defined everywhere except possibly at the origin.

11. Changing to polar coordinates,

$$\lim_{(x,y) \to (0,0)} \frac{xy}{\sqrt{x^2+y^2}} = \lim_{r \to 0} \frac{r\cos\theta \cdot r\sin\theta}{r}$$

$$= \lim_{r \to 0} r\cos\theta \cdot \sin\theta = 0$$

13. Use polar coordinates.

$$\frac{x^{7/3}}{x^2+y^2} = \frac{r^{7/3}(\cos\theta)^{7/3}}{r^2} = r^{1/3}(\cos\theta)^{7/3}$$

$r^{1/3}(\cos\theta)^{7/3} \to 0$ as $r \to 0$, so the limit is 0.

15. $$f(x, y) = \frac{r^4\cos^2\theta\sin^2\theta}{r^2\cos^2\theta + r^4\sin^4\theta}$$

$$= r^2\left(\frac{\cos^2\theta\sin^2\theta}{\cos^2\theta + r^2\sin^4\theta}\right)$$

If $\cos\theta = 0$, then $f(x, y) = 0$. If $\cos\theta \neq 0$, hen this converges to 0 as $r \to 0$. Thus the limit is 0.

17. $f(x, y)$ is continuous for all (x, y) since for all (x, y), $x^2 + y^2 + 1 \neq 0$.

19. Require $1 - x^2 - y^2 > 0; x^2 + y^2 < 1$. S is the interior of the unit circle centered at the origin.

21. Require $y - x^2 \neq 0$. S is the entire plane except the parabola $y = x^2$.

23. Require $x - y + 1 \geq 0; y \leq x + 1$. S is the region below and on the line $y = x + 1$.

25. $f(x, y, z)$ is continuous for all $(x, y, z) \neq (0,0,0)$ since for all $(x, y, z) \neq (0,0,0)$, $x^2 + y^2 + z^2 > 0$.

27. The boundary consists of the points that form the outer edge of the rectangle. The set is closed.

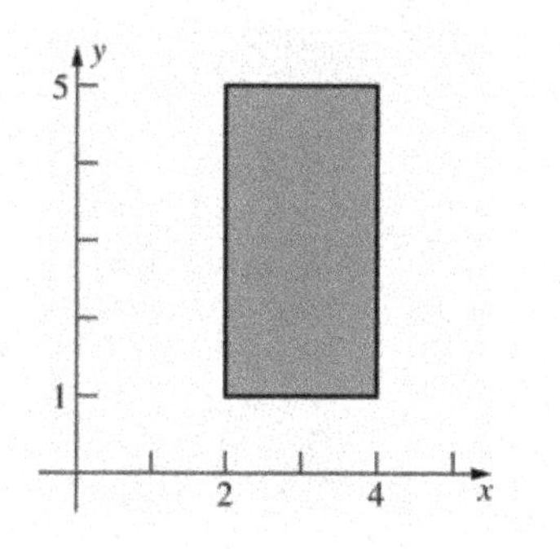

29. The boundary consists of the circle and the origin. The set is neither open (since, for example, (1, 0) is not an interior point), nor closed (since (0, 0) is not in the set).

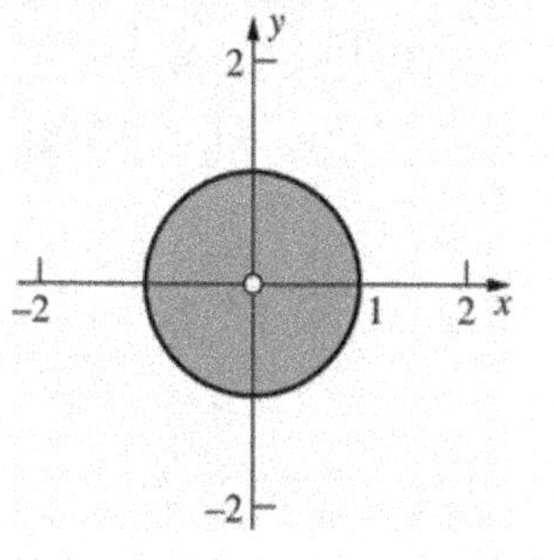

31. The boundary consists of the graph of $y = \sin\left(\frac{1}{x}\right)$ along with the part of the y-axis for which $y \leq 1$. The set is open.

33. $$\frac{x^2 - 4y^2}{x - 2y} = \frac{(x+2y)(x-2y)}{x-2y} = x + 2y \text{ (if } x \neq 2y\text{)}$$

If $x = 2y$, $x + 2y = 2x$. Take $g(x) = 2x$.

35. Along the x-axis ($y = 0$): $\lim_{(x, y)\to(0, 0)} \frac{0}{x^2+0} = 0$.

Along $y = x$:

$$\lim_{(x, y)\to(0, 0)} \frac{x^2}{2x^2} = \lim_{(x, y)\to(0, 0)} \frac{1}{2} = \frac{1}{2}.$$

Hence, the limit does not exist because for some points near the origin $f(x, y)$ is getting closer to 0, but for others it is getting closer to $\frac{1}{2}$.

37. a. $$\lim_{x\to 0} \frac{x^2(mx)}{x^4 + (mx)^2} = \lim_{x\to 0} \frac{mx^3}{x^4 + m^2x^2}$$

$$= \lim_{x\to 0} \frac{mx}{x^2 + m^2} = 0$$

b. $$\lim_{x\to 0} \frac{x^2(x^2)}{x^4 + (x^2)^2} = \lim_{x\to 0} \frac{x^4}{2x^4} = \lim_{x\to 0} \frac{1}{2} = \frac{1}{2}$$

c. $\lim_{(x, y)\to(0, 0)} \frac{x^2 y}{x^4 + y}$ does not exist.

39. a. $\{(x, y, z): x^2 + y^2 = 1, z \text{ in } [1, 2]\}$ [For $x^2 + y^2 < 1$, the particle hits the hemisphere and then slides to the origin (or bounds toward the origin); for $x^2 + y^2 = 1$, it bounces up; for $x^2 + y^2 > 1$, it falls straight down.]

b. $\{(x, y, z): x^2 + y^2 = 1, z = 1\}$ (As one moves at a level of $z = 1$ from the rim of the bowl toward any position away from the bowl there is a change from seeing all of the interior of the bowl to seeing none of it.)

c. $\{(x, y, z): z = 1\}$ [$f(x, y, z)$ is undefined (infinite) at $(x, y, 1)$.]

d. ϕ (Small changes in points of the domain result in small changes in the shortest path from the points to the origin.)

41. a. $f(x, y) = \begin{cases} (x^2 + y^2)^{1/2} + 1 & \text{if } y \neq 0 \\ |x - 1| & \text{if } y = 0 \end{cases}$. Check discontinuities where $y = 0$.

As $y = 0$, $(x^2 + y^2)^{1/2} + 1 = |x| + 1$, so f is continuous if $|x| + 1 = |x - 1|$. Squaring each side and simplifying yields $|x| = -x$, so f is continuous for $x \leq 0$. That is, f is discontinuous along the positive x-axis.

b. Let $P = (u, v)$ and $Q = (x, y)$.

$$f(u, v, x, y) = \begin{cases} |OP| + |OQ| & \text{if } P \text{ and } Q \text{ are not on same ray from the origin and neither is the origin} \\ |PQ| & \text{otherwise} \end{cases}.$$

This means that in the first case one travels from P to the origin and then to Q; in the second case one travels directly from P to Q without passing through the origin, so f is discontinuous on the set $\{(u, v, x, y): \langle u, v\rangle = k\langle x, y\rangle \text{ for some } k > 0, \langle u, v\rangle \neq \mathbf{0}, \langle x, y\rangle \neq \mathbf{0}\}$.

43.

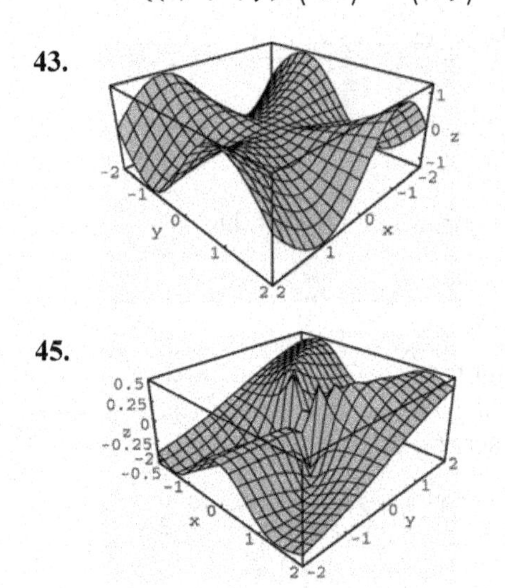

45.

47. If we approach the point $(0,0,0)$ along a straight path from the point (x,x,x), we have

$$\lim_{(x,x,x)\to(0,0,0)} \frac{x(x)(x)}{x^3 + x^3 + x^3} = \lim_{(x,x,x)\to(0,0,0)} \frac{x^3}{3x^3} = \frac{1}{3}$$

Since the limit does not equal to $f(0,0,0)$, the function is not continuous at the point $(0,0,0)$.

12.4 Concepts Review

1. gradient

3. $\frac{\partial f}{\partial x}(\mathbf{p})\mathbf{i}+\frac{\partial f}{\partial y}(\mathbf{p})\mathbf{j};\ \mathbf{y}^2\mathbf{i}+2xy\mathbf{j}$

Problem Set 12.4

1. $\left\langle 2xy+3y,\ x^2+3x\right\rangle$

3. $\nabla f(x,\ y)=\left\langle (x)(e^{xy}y)+(e^{xy})(1),\ xe^{xy}x\right\rangle = e^{xy}\left\langle xy+1,\ x^2\right\rangle$

5. $x(x+y)^{-2}\left\langle y(x+2),\ x^2\right\rangle$

7. $(x^2+y^2+z^2)^{-1/2}\langle x,\ y,\ z\rangle$

9. $\nabla f(x,\ y)=\left\langle (x^2y)(e^{x-z})+(e^{x-z})(2xy),\ x^2e^{x-z},\ x^2ye^{x-z}(-1)\right\rangle = xe^{x-z}\langle y(x+2), x, -xy\rangle$

11. $\nabla f(x,\ y)=\left\langle 2xy-y^2,\ x^2-2xy\right\rangle;\ \ \nabla f(-2,\ 3)=\langle -21,\ 16\rangle$

$z=f(-2,\ 3)+\langle -21,\ 16\rangle\cdot\langle x+2,\ y-3\rangle = 30+(-21x-42+16y-48)$

$z=-21x+16y-60$

13. $\nabla f(x,\ y)=\langle -\pi\sin(\pi x)\sin(\pi y),\ \pi\cos(\pi x)\cos(\pi y)+2\pi\cos(2\pi y)\rangle$

$\nabla f\left(-1,\ \frac{1}{2}\right)=\langle 0,\ -2\pi\rangle$

$z=f\left(-1,\ \frac{1}{2}\right)+\langle 0,\ -2\pi\rangle\cdot\left\langle x+1,\ y-\frac{1}{2}\right\rangle=-1+(0-2\pi y+\pi);$

$z=-2\pi y+(\pi-1)$

15. $\nabla f(x,\ y,\ z)=\left\langle 6x+z^2,\ -4y,\ 2xz\right\rangle$, so $\nabla f(1,\ 2,\ -1)=\langle 7,\ -8,\ -2\rangle$

Tangent hyperplane:

$w=f(1,\ 2,\ -1)+\nabla f(1,\ 2,\ -1)\cdot\langle x-1,\ y-2,\ z+1\rangle=-4+\langle 7,\ -8,\ -2\rangle\cdot\langle x-1,\ y-2,\ z+1\rangle$

$=-4+(7x-7-8y+16-2z-2)$

$w=7x-8y-2z+3$

17. $\nabla\left(\frac{f}{g}\right)=\frac{\langle gf_x-fg_x, gf_y-fg_y, gf_z-fg_z\rangle}{g^2}=\frac{g\langle f_x, f_y, f_z\rangle-f\langle g_x, g_y, g_z\rangle}{g^2}=\frac{g\nabla f-f\nabla g}{g^2}$

19. Let $F(x,y,z)=x^2-6x+2y^2-10y+2xy-z=0$

$\nabla F(x,y,z)=\langle 2x-6+2y, 4y-10+2x, -1\rangle$

The tangent plane will be horizontal if $\nabla F(x,y,z)=\langle 0,0,k\rangle$, where $k\neq 0$. Therefore, we have the following system of equations:

$2x+2y-6=0$

$2x+4y-10=0$

Solving this system yields $x=1$ and $y=2$. Thus, there is a horizontal tangent plane at $(x,y)=(1,2)$.

21. a. The point $(2,1,9)$ projects to $(2,1,0)$ on the xy plane. The equation of a plane containing this point and parallel to the x-axis is given by $y = 1$. The tangent plane to the surface at the point $(2,1,9)$ is given by

$$\begin{aligned} z &= f(2,1) + \nabla f(2,1) \cdot \langle x-2, y-1 \rangle \\ &= 9 + \langle 12,10 \rangle \langle x-2, y-1 \rangle \\ &= 12x + 10y - 25 \end{aligned}$$

The line of intersection of the two planes is the tangent line to the surface, passing through the point $(2,1,9)$, whose projection in the xy plane is parallel to the x-axis. This line of intersection is parallel to the cross product of the normal vectors for the planes. The normal vectors are $\langle 12,10,-1 \rangle$ and $\langle 0,1,0 \rangle$ for the tangent plane and vertical plane respectively. The cross product is given by $\langle 12,10,-1 \rangle \times \langle 0,1,0 \rangle = \langle 1,0,12 \rangle$

Thus, parametric equations for the desired tangent line are $x = 2 + t$

$$y = 1$$
$$z = 9 + 12t$$

b. Using the equation for the tangent plane from the previous part, we now want the vertical plane to be parallel to the y-axis, but still pass through the projected point $(2,1,0)$. The vertical plane now has equation $x = 2$. The normal equations are given by $\langle 12,10,-1 \rangle$ and $\langle 1,0,0 \rangle$ for the tangent and vertical planes respectively. Again we find the cross product of the normal vectors:

$\langle 12,10,-1 \rangle \times \langle 1,0,0 \rangle = \langle 0,10,10 \rangle$

Thus, parametric equations for the desired tangent line are $x = 2$

$$y = 1 + 10t$$
$$z = 9 + 10t$$

c. Using the equation for the tangent plane from the first part, we now want the vertical plane to be parallel to the line $y = x$, but still pass through the projected point $(2,1,0)$. The vertical plane now has equation $y - x + 1 = 0$. The normal equations are given by $\langle 12,10,-1 \rangle$ and $\langle 1,-1,0 \rangle$ for the tangent and vertical planes respectively. Again we find the cross product of the normal vectors:

$\langle 12,10,-1 \rangle \times \langle 1,-1,0 \rangle = \langle -1,-1,-22 \rangle$

Thus, parametric equations for the desired tangent line are $x = 2 - t$

$$y = 1 - t$$
$$z = 9 - 22t$$

23. $\nabla f(x, y) = \left\langle -10\left(\dfrac{1}{2\sqrt{|xy|}}\dfrac{|xy|}{xy}y\right), -10\left(\dfrac{1}{2\sqrt{|xy|}}\dfrac{|xy|}{xy}x\right)\right\rangle = \dfrac{-5xy}{|xy|^{3/2}}\langle y, x\rangle$ $\left[\text{Note that } \dfrac{|a|}{a} = \dfrac{a}{|a|}.\right]$

$\nabla f(1, -1) = \langle -5, 5 \rangle$

Tangent plane:

$z = f(1, -1) + \nabla f(1, -1) \cdot \langle x-1, y+1 \rangle = -10 + \langle -5, 5 \rangle \cdot \langle x-1, y+1 \rangle = -10 + (-5x + 5 + 5y + 5)$

$z = -5x + 5y$

25. $f(\mathbf{a}) - f(\mathbf{b}) = f\langle 2,1 \rangle - f\langle 0,0 \rangle = 4 - 9 = -5$

$\nabla f(x, y) = \langle -2x, -2y \rangle$; $\mathbf{b} - \mathbf{a} = \langle 2,1 \rangle$

The value $\mathbf{c} = \langle c_x, c_y \rangle$ will be a solution to

$-5 = \langle -2c_x, -2c_y \rangle \langle 2,1 \rangle$

$\mathbf{c} \in \{\langle c_x, c_y \rangle : 4c_x + 2c_y = 5\}$

In order for **c** to be between **a** and **b**, **c** must lie on the line $y = \frac{1}{2}x$. Consequently, **c** will be the solution to the following system of equations: $4c_x + 2c_y = 5$ and $c_y = \frac{1}{2}c_x$. The solution is $\mathbf{c} = \langle 1, \frac{1}{2} \rangle$.

27. $\nabla f(\mathbf{p}) = \nabla g(\mathbf{p}) \Rightarrow \nabla[f(\mathbf{p}) - g(\mathbf{p})] = \mathbf{0}$

$\Rightarrow f(\mathbf{p}) - g(\mathbf{p})$ is a constant.

29.

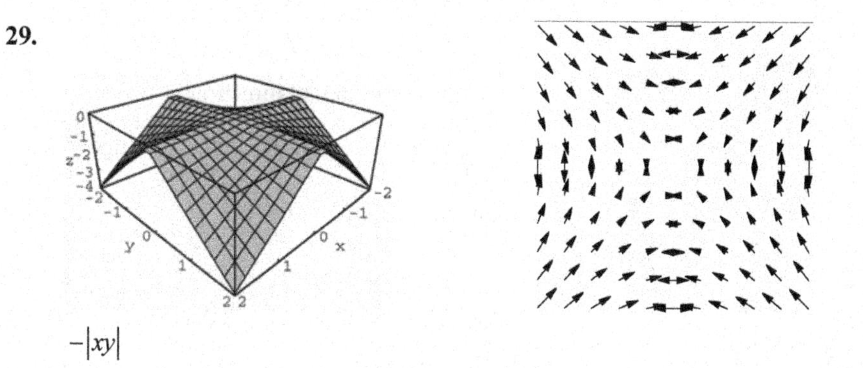

$-|xy|$

a. The gradient points in the direction of greatest increase of the function.

b. No. If it were, $|0+h|-|0| = 0+|h|\delta(h)$ where $\delta(h)\to 0$ as $h\to 0$, which is possible.

31. a. (i) $$\nabla[f+g] = \frac{\partial(f+g)}{\partial x}\mathbf{i}+\frac{\partial(f+g)}{\partial y}\mathbf{j}+\frac{\partial(f+g)}{\partial z}\mathbf{k} = \frac{\partial f}{\partial x}\mathbf{i}+\frac{\partial g}{\partial x}\mathbf{i}+\frac{\partial f}{\partial y}\mathbf{j}+\frac{\partial g}{\partial y}\mathbf{j}+\frac{\partial f}{\partial z}\mathbf{k}+\frac{\partial g}{\partial z}\mathbf{k}$$
$$= \frac{\partial f}{\partial x}\mathbf{i}+\frac{\partial f}{\partial y}\mathbf{j}+\frac{\partial f}{\partial z}\mathbf{k}+\frac{\partial g}{\partial x}\mathbf{i}+\frac{\partial g}{\partial y}\mathbf{j}+\frac{\partial g}{\partial z}\mathbf{k}$$
$$= \nabla f+\nabla g$$

(ii) $$\nabla[\alpha f] = \frac{\partial[\alpha f]}{\partial x}\mathbf{i}+\frac{\partial[\alpha f]}{\partial y}\mathbf{j}+\frac{\partial[\alpha f]}{\partial z}\mathbf{k} = \alpha\frac{\partial[f]}{\partial x}\mathbf{i}+\alpha\frac{\partial[f]}{\partial y}\mathbf{j}+\alpha\frac{\partial[f]}{\partial z}\mathbf{k} = \alpha\nabla f$$

(iii) $$\nabla[fg] = \frac{\partial(fg)}{\partial x}\mathbf{i}+\frac{\partial(fg)}{\partial y}\mathbf{j}+\frac{\partial(fg)}{\partial z}\mathbf{k}$$
$$= \left(f\frac{\partial g}{\partial x}+g\frac{\partial f}{\partial x}\right)\mathbf{i}+\left(f\frac{\partial g}{\partial y}+g\frac{\partial f}{\partial y}\right)\mathbf{j}+\left(f\frac{\partial g}{\partial z}+g\frac{\partial f}{\partial z}\right)\mathbf{k}$$
$$= f\left(\frac{\partial g}{\partial x}\mathbf{i}+\frac{\partial g}{\partial y}\mathbf{j}+\frac{\partial g}{\partial z}\mathbf{k}\right)+g\left(\frac{\partial f}{\partial x}\mathbf{i}+\frac{\partial f}{\partial y}\mathbf{j}+\frac{\partial f}{\partial z}\mathbf{k}\right) = f\nabla g+g\nabla f$$

b. (i) $$\nabla[f+g] = \frac{\partial(f+g)}{\partial x_1}\mathbf{i}_1+\frac{\partial(f+g)}{\partial x_2}\mathbf{i}_2+\cdots+\frac{\partial(f+g)}{\partial x_n}\mathbf{i}_n$$
$$= \frac{\partial f}{\partial x_1}\mathbf{i}_1+\frac{\partial g}{\partial x_1}\mathbf{i}_1+\frac{\partial f}{\partial x_2}\mathbf{i}_2+\frac{\partial g}{\partial x_2}\mathbf{i}_2+\cdots+\frac{\partial f}{\partial x_n}\mathbf{i}_n+\frac{\partial g}{\partial x_n}\mathbf{i}_n$$
$$= \frac{\partial f}{\partial x_1}\mathbf{i}_1+\frac{\partial f}{\partial x_2}\mathbf{i}_2+\cdots+\frac{\partial f}{\partial x_n}\mathbf{i}_n+\frac{\partial g}{\partial x_1}\mathbf{i}_1+\frac{\partial g}{\partial x_2}\mathbf{i}_2+\cdots+\frac{\partial g}{\partial x_n}\mathbf{i}_n$$
$$= \nabla f+\nabla g$$

(ii) $$\nabla[\alpha f] = \frac{\partial[\alpha f]}{\partial x_1}\mathbf{i}_1+\frac{\partial[\alpha f]}{\partial x_2}\mathbf{i}_2+\cdots+\frac{\partial[\alpha f]}{\partial x_n}\mathbf{i}_n = \alpha\frac{\partial[f]}{\partial x_1}\mathbf{i}_1+\alpha\frac{\partial[f]}{\partial x_2}\mathbf{i}_2+\cdots+\alpha\frac{\partial[f]}{\partial x_n}\mathbf{i}_n = \alpha\nabla f$$

(iii) $$\nabla[fg] = \frac{\partial(fg)}{\partial x_1}\mathbf{i}_1+\frac{\partial(fg)}{\partial x_2}\mathbf{i}_2+\cdots+\frac{\partial(fg)}{\partial x_n}\mathbf{i}_n$$
$$= \left(f\frac{\partial g}{\partial x_1}+g\frac{\partial f}{\partial x_1}\right)\mathbf{i}_1+\left(f\frac{\partial g}{\partial x_2}+g\frac{\partial f}{\partial x_2}\right)\mathbf{i}_2+\cdots+\left(f\frac{\partial g}{\partial x_n}+g\frac{\partial f}{\partial x_n}\right)\mathbf{i}_n$$
$$= f\left(\frac{\partial g}{\partial x_1}\mathbf{i}_1+\frac{\partial g}{\partial x_2}\mathbf{i}_2+\cdots+\frac{\partial g}{\partial x_n}\mathbf{i}_n\right)+g\left(\frac{\partial f}{\partial x_1}\mathbf{i}_1+\frac{\partial f}{\partial x_2}\mathbf{i}_2+\cdots+\frac{\partial f}{\partial x_n}\mathbf{i}_n\right)$$
$$= f\nabla g+g\nabla f$$

12.5 Concepts Review

1. $\dfrac{[f(\mathbf{p}+h\mathbf{u})-f(\mathbf{p})]}{h}$

3. greatest increase

Problem Set 12.5

1. $D_{\mathbf{u}}f(x, y) = \langle 2xy, x^2\rangle \cdot \left\langle \frac{3}{5}, -\frac{4}{5}\right\rangle$; $D_{\mathbf{u}}f(1, 2) = \frac{8}{5}$

3. $D_{\mathbf{u}}f(x, y) = f(x, y)\cdot \mathbf{u}$ $\left(\text{where } u = \dfrac{\mathbf{a}}{|\mathbf{a}|}\right)$

$= \langle 4x + y, x - 2y\rangle \cdot \dfrac{\langle 1, -1\rangle}{\sqrt{2}}$;

$D_{\mathbf{u}}f(3, -2) = \langle 10, 7\rangle \cdot \dfrac{\langle 1, -1\rangle}{\sqrt{2}} = \dfrac{3}{\sqrt{2}} \approx 2.1213$

5. $D_{\mathbf{u}}f(x, y) = e^x\langle \sin y, \cos y\rangle \cdot \left[\left(\frac{1}{2}\right)\langle 1, \sqrt{3}\rangle\right]$;

$D_{\mathbf{u}}f\left(0, \frac{\pi}{4}\right) = \dfrac{(\sqrt{2}+\sqrt{6})}{4} \approx 0.9659$

7. $D_{\mathbf{u}}f(x, y, z) = \mathbf{u}\cdot \nabla f(\mathbf{p})$

$= \langle 3x^2y, x^3 - 2yz^2, -2y^2z\rangle \cdot \left[\left(\frac{1}{3}\right)\langle 1, -2, 2\rangle\right]$;

$D_{\mathbf{u}}f(-2, 1, 3) = \dfrac{52}{3}$

9. f increases most rapidly in the direction of the gradient. $\nabla f(x, y) = \langle 3x^2, -5y^4\rangle$;

$\nabla f(2, -1) = \langle 12, -5\rangle$

$\dfrac{\langle 12, -5\rangle}{13}$ is the unit vector in that direction. The rate of change of $f(x, y)$ in that direction at that point is the magnitude of the gradient.

$|\langle 12, -5\rangle| = 13$

11. $\nabla f(x, y, z) = \langle 2xyz, x^2z, x^2y\rangle$;

$f(1, -1, 2) = \langle -4, 2, -1\rangle$

A unit vector in that direction is

$\left(\frac{1}{\sqrt{21}}\right)\langle -4, 2, -1\rangle$. The rate of change in that direction is $\sqrt{21} \approx 4.5826$.

13. $-\nabla f(x, y) = 2\langle x, y\rangle$; $-\nabla f(-1, 2) = 2\langle -1, 2\rangle$ is the direction of most rapid decrease. A unit vector in that direction is $\mathbf{u} = \left(\frac{1}{\sqrt{5}}\right)\langle -1, 2\rangle$.

15. The level curves are $\dfrac{y}{x^2} = k$. For $\mathbf{p} = (1, 2)$, $k = 2$, so the level curve through $(1, 2)$ is $\dfrac{y}{x^2} = 2$ or $y = 2x^2$ $(x \neq 0)$.

$\nabla f(x, y) = \langle -2yx^{-3}, x^{-2}\rangle$

$\nabla f(1, 2) = \langle -4, 1\rangle$, which is perpendicular to the parabola at $(1, 2)$.

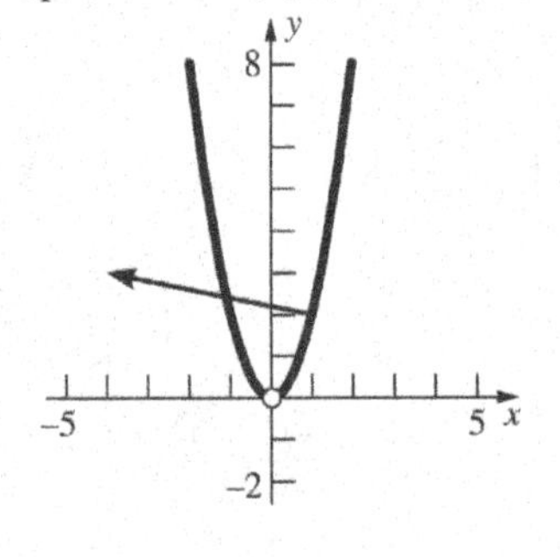

17. $u = \left\langle \frac{2}{3}, -\frac{2}{3}, \frac{1}{3}\right\rangle$

$D_{\mathbf{u}}f(x, y, z) = \langle y, x, 2z\rangle \cdot \left\langle \frac{2}{3}, -\frac{2}{3}, \frac{1}{3}\right\rangle$

$D_{\mathbf{u}}f(1, 1, 1) = \dfrac{2}{3}$

19. a. Hottest if denominator is smallest; i.e., at the origin.

b. $\nabla T(x, y, z) = \dfrac{-200\langle 2x, 2y, 2z\rangle}{(5 + x^2 + y^2 + z^2)^2}$;

$\nabla T(1, -1, 1) = \left(-\frac{25}{4}\right)\langle 1, -1, 1\rangle$

$\langle -1, 1, -1\rangle$ is one vector in the direction of greatest increase.

c. Yes

21. $\nabla f(x, y, z)$

$$= \left\langle x\left(x^2+y^2+z^2\right)^{-\frac{1}{2}} \cos\sqrt{x^2+y^2+z^2},\right.$$

$$y\left(x^2+y^2+z^2\right)^{-\frac{1}{2}} \cos\sqrt{x^2+y^2+z^2},$$

$$\left. z\left(x^2+y^2+z^2\right)^{-\frac{1}{2}} \cos\sqrt{x^2+y^2+z^2}\right\rangle$$

$$= \left(\left(x^2+y^2+z^2\right)^{-\frac{1}{2}} \cos\sqrt{x^2+y^2+z^2}\right)\langle x, y, z\rangle$$

which either points towards or away from the origin.

23. He should move in the direction of

$$-\nabla f(\mathbf{p}) = -\langle f_x(\mathbf{p}), f_y(\mathbf{p})\rangle = -\left\langle -\frac{1}{2}, -\frac{1}{4}\right\rangle$$

$= \left(\frac{1}{4}\right)\langle 2, 1\rangle$. Or use $\langle 2, 1\rangle$. The angle α formed with the East is $\tan^{-1}\left(\frac{1}{2}\right) \approx 26.57°$ (N63.43°E).

N

$\langle 2, 1\rangle$

P α

E

25. The climber is moving in the direction of $\mathbf{u} = \left(\frac{1}{\sqrt{2}}\right)\langle -1, 1\rangle$. Let

$$f(x, y) = 3000e^{-(x^2+2y^2)/100}.$$

$$\nabla f(x, y) = 3000e^{-(x^2+2y^2)/100}\left\langle -\frac{x}{50}, -\frac{y}{25}\right\rangle;$$

$$f(10, 10) = -600e^{-3}\langle 1, 2\rangle$$

She will move at a slope of

$$D_u(10, 10) = -600e^{-3}\langle 1, 2\rangle \cdot \left(\frac{1}{\sqrt{2}}\right)\langle -1, 1\rangle$$

$$= \left(-300\sqrt{2}\right)e^{-3} \approx -21.1229.$$

She will descend. Slope is about –21.

27. $\nabla T(x, y) = \langle -4x, -2y\rangle$

$$\frac{dx}{dt} = -4x, \frac{dy}{dt} = -2y$$

$\dfrac{\frac{dx}{dt}}{-4x} = \dfrac{\frac{dy}{dt}}{-2y}$ has solution $|x| = 2y^2$. Since the particle starts at (–2, 1), this simplifies to $x = -2y^2$.

29. a. $\nabla T(x, y, z) = \left\langle -\dfrac{10(2x)}{(x^2+y^2+z^2)^2}, -\dfrac{10(2y)}{(x^2+y^2+z^2)^2}, -\dfrac{10(2z)}{(x^2+y^2+z^2)^2}\right\rangle$

$$= -\frac{20}{(x^2+y^2+z^2)^2}\langle x, y, z\rangle$$

$r(t) = \langle t\cos\pi t, t\sin\pi t, t\rangle$, so $r(1) = \langle -1, 0, 1\rangle$. Therefore, when $t = 1$, the bee is at (–1, 0, 1), and $\nabla T(-1, 0, 1) = -5\langle -1, 0, 1\rangle$.

$\mathbf{r}'(t) = \langle \cos\pi t - \pi t\sin\pi t, \sin\pi t + \pi t\cos\pi t, 1\rangle$, so $\mathbf{r}'(1) = \langle -1, -\pi, 1\rangle$.

$U = \dfrac{r'(1)}{|r'(1)|} = \dfrac{\langle -1, -\pi, 1\rangle}{2+\pi^2}$ is the unit tangent vector at (–1, 0, 1).

$$D_{\mathbf{u}}T(-1, 0, 1) = \mathbf{u} \cdot \nabla T(-1, 0, 1)$$

$$= \frac{\langle -1, -\pi, 1\rangle \cdot \langle 5, 0, -5\rangle}{\sqrt{2+\pi^2}} = -\frac{10}{\sqrt{2+\pi^2}} \approx -2.9026$$

Therefore, the temperature is decreasing at about 2.9°C per meter traveled when the bee is at (–1, 0, 1); i.e., when $t = 1$ s.

b. Method 1: (First express T in terms of t.)

$$T = \frac{10}{x^2 + y^2 + z^2} = \frac{10}{(t\cos \pi t)^2 + (t\sin \pi t)^2 + (t)^2} = \frac{10}{2t^2} = \frac{5}{t^2}$$

$T(t) = 5t^{-2}; T'(t) = -10t^{-3}; t'(1) = -10$

Method 2: (Use Chain Rule.)

$$D_t T(t) = \frac{dT}{ds}\frac{ds}{dt} = (D_{\mathbf{u}}T)(|\mathbf{r}'(t)|), \text{ so } D_t T(t) = [D_{\mathbf{u}}T(-1, 0, 1)](|\mathbf{r}'(1)|) = -\frac{10}{\sqrt{2+\pi^2}}\left(\sqrt{2+\pi^2}\right) = -10$$

Therefore, the temperature is decreasing at about 10°C per second when the bee is at $(-1, 0, 1)$; i.e., when $t = 1$ s.

31.

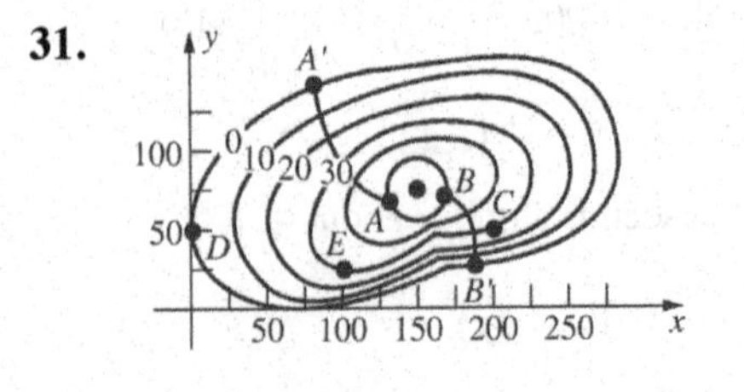

a. $A'(100, 120)$

b. $B'(190, 25)$

c. $f_x(C) \approx \frac{20-30}{230-200} = -\frac{1}{3}; f_y(D) = 0; \; D_{\mathbf{u}}f(E) \approx \frac{40-30}{25} = \frac{2}{5}$

33. Leave: $(-0.1, -5)$

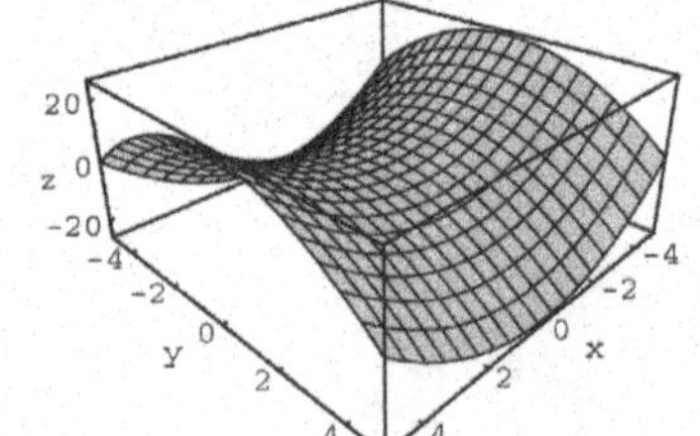

$x^2 - y^2$

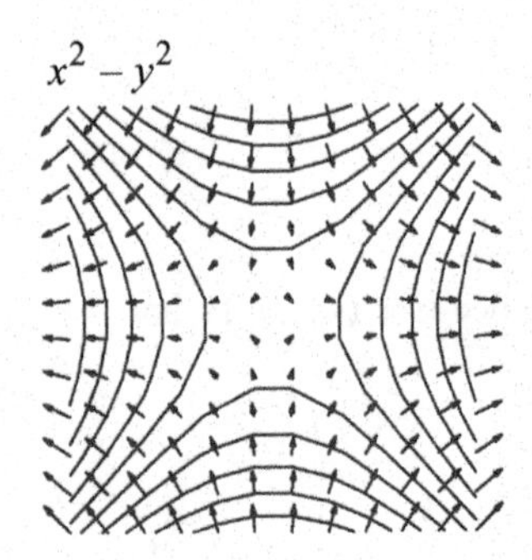

35. Leave: $(3, 5)$

12.6 Concepts Review

1. $\frac{\partial z}{\partial x}\frac{dx}{dt}+\frac{\partial z}{\partial y}\frac{dy}{dt}$

3. $\frac{\partial z}{\partial x}\frac{\partial x}{\partial t}+\frac{\partial z}{\partial y}\frac{\partial y}{\partial t}$

Problem Set 12.6

1. $\frac{dw}{dt}=(2xy^3)(3t^2)+(3x^2y^2)(2t)$
   $=(2t^9)(3t^2)+(3t^{10})(2t)=12t^{11}$

3. $\frac{dw}{dt}=(e^x\sin y+e^y\cos x)(3)+(e^x\cos y+e^y\sin x)(2)=3e^{3t}\sin 2t+3e^{2t}\cos 3t+2e^{3t}\cos 2t+2e^{2t}\sin 3t$

5. $\frac{dw}{dt}=[yz^2(\cos(xyz^2)](3t^2)+[xz^2\cos(xyz^2)](2t)+[2xyz\cos(xyz^2)](1)$
   $=(3yz^2t^2+2xz^2t+2xyz)\cos(xyz^2)=(3t^6+2t^6+2t^6)\cos(t^7)=7t^6\cos(t^7)$

7. $\frac{\partial w}{\partial t}=(2xy)(s)+(x^2)(-1)=2st(s-t)s-s^2t^2=s^2t(2s-3t)$

9. $\frac{\partial w}{\partial t}=e^{x^2+y^2}(2x)(s\cos t)+e^{x^2+y^2}(2y)(\sin s)=2e^{x^2+y^2}(xs\cos t+y\sin s)$
   $=2(s^2\sin t\cos t+t\sin^2 s)\exp(s^2\sin^2 t+t^2\sin^2 s)$

11. $\frac{\partial w}{\partial t}=\frac{x(-s\sin st)}{(x^2+y^2+z^2)^{1/2}}+\frac{y(s\cos st)}{(x^2+y^2+z^2)^{1/2}}+\frac{z(s^2)}{(x^2+y^2+z^2)^{1/2}}=s^4t(1+s^4t^2)^{-1/2}$

13. $\frac{\partial z}{\partial t}=(2xy)(2)+(x^2)(-2st)=4(2t+s)(1-st^2)-2st(2t+s)^2;\left.\left(\frac{\partial z}{\partial t}\right)\right|_{(1,\,-2)}=72$

15. $\frac{dw}{dx}=(2u-\tan v)(1)+(-u\sec^2 v)(\pi)=2x-\tan \pi x-\pi x\sec^2 \pi x$
   $\left.\frac{dw}{dx}\right|_{x=\frac{1}{4}}=\left(\frac{1}{2}\right)-1-\left(\frac{\pi}{2}\right)=-\frac{1+\pi}{2}\approx -2.0708$

17. $V(r,h)=\pi r^2h,\ \frac{dr}{dt}=0.5$ in./yr,
   $\frac{dh}{dt}=8$ in./yr
   $\frac{dV}{dt}=(2\pi rh)\left(\frac{dr}{dt}\right)+(\pi r^2)\left(\frac{dh}{dt}\right);\ \left.\left(\frac{dV}{dt}\right)\right|_{(20,\,400)}=11200\pi\ \text{in.}^3/\text{yr}$
   $=\frac{11200\pi\ \text{in.}^3}{1\ \text{yr}}\times\frac{1\ \text{board ft}}{144\ \text{in.}^3}\approx 244.35$ board ft/yr

19. The stream carries the boat along at 2 ft/s with respect to the boy.

$\frac{dx}{dt} = 2, \frac{dy}{dt} = 4, s^2 = x^2 + y^2$

$2s\left(\frac{ds}{dt}\right) = 2x\left(\frac{dx}{dt}\right) + 2y\left(\frac{dy}{dt}\right)$

$\frac{ds}{dt} = \frac{(2x+4y)}{s}$

When $t = 3, x = 6, y = 12, s = 6\sqrt{5}$. Thus,

$\left(\frac{ds}{dt}\right)\Big|_{t=3} = \sqrt{20} \approx 4.47$ ft/s

21. Let $F(x, y) = x^3 + 2x^2y - y^3 = 0$.

Then $\frac{dy}{dx} = \frac{-\frac{\partial F}{\partial x}}{\frac{\partial F}{\partial y}} = -\frac{(3x^2 + 4xy)}{2x^2 - 3y^2} = \frac{3x^2 + 4xy}{3y^2 - 2x^2}$.

23. Let $F(x, y) = x \sin y + y \cos x = 0$.

$\frac{dy}{dx} = -\frac{(\sin y - y \sin x)}{x \cos y + \cos x} = \frac{y \sin x - \sin y}{x \cos y + \cos x}$

25. Let $F(x, y, z) = 3x^2z + y^3 - xyz^3 = 0$.

$\frac{\partial z}{\partial x} = -\frac{(6xz - yz^3)}{3x^2 - 3xyz^2} = \frac{yz^3 - 6xz}{3x^2 - 3xyz^2}$

27. $\frac{\partial T}{\partial s} = \frac{\partial T}{\partial x}\frac{\partial x}{\partial s} + \frac{\partial T}{\partial y}\frac{\partial y}{\partial s} + \frac{\partial T}{\partial z}\frac{\partial z}{\partial s} + \frac{\partial T}{\partial w}\frac{\partial w}{\partial s}$

29. $y = \left(\frac{1}{2}\right)[f(u) + f(v)]$,

where $u = x - ct, v = x + ct$.

$y_x = \left(\frac{1}{2}\right)[f'(u)(1) + f'(v)(1)] = \left(\frac{1}{2}\right)[f'(u) + f'(v)]$

$y_{xx} = \left(\frac{1}{2}\right)[f''(u)(1) + f''(v)(1)]$

$= \left(\frac{1}{2}\right)[f''(u) + f''(v)]$

$y_t = \left(\frac{1}{2}\right)[f'(u)(-c) + f'(v)(c)]$

$= \left(-\frac{c}{2}\right)[f'(u) - f'(u)]$

$y_{tt} = \left(-\frac{c}{2}\right)[f''(u)(-c) - f''(v)(c)]$

$= \left(\frac{c^2}{2}\right)[f''(u) + f''(v)] = c^2 y_{xx}$

31. Let $w = \int_x^y f(u)du = -\int_y^x f(u)du$, where $x = g(t)$, $y = h(t)$.

Then

$\frac{dw}{dt} = \frac{\partial w}{\partial x}\frac{dx}{dt} + \frac{\partial w}{\partial y}\frac{dy}{dt} = -f(x)g'(t) + f(y)h'(t)$

$= f(h(t))h'(t) - f(g(t))g'(t)$.

Thus, for the particular function given

$F'(t) = \sqrt{9 + (t^2)^4}(2t)$

$- \sqrt{9 + \left(\sin \sqrt{2}\pi t\right)^4}\left(\sqrt{2}\pi \cos \sqrt{2}\pi t\right)$;

$F'\left(\sqrt{2}\right) = (5)\left(2\sqrt{2}\right) - (3)\left(\sqrt{2}\pi\right)$

$= 10\sqrt{2} - 3\sqrt{2}\pi \approx 0.8135$.

33. $c^2 = a^2 + b^2 - 2ab \cos 40°$ (Law of Cosines) where a, b, and c are functions of t.

$2cc' = 2aa' + 2bb' - 2(a'b + ab')\cos 40°$ so $c' = \frac{aa' + bb' - (a'b + ab')\cos 40°}{c}$.

When $a = 200$ and $b = 150$, $c^2 = (200)^2 + (150)^2 - 2(200)(150)\cos 40° = 62{,}500 - 60{,}000 \cos 40°$.

It is given that $a' = 450$ and $b' = 400$, so at that instant,

$c' = \frac{(200)(450) + (150)(400) - [(450)(150) + (200)(400)]\cos 40°}{\sqrt{62{,}500 - 60{,}000 \cos 40°}} \approx 288$.

Thus, the distance between the airplanes is increasing at about 288 mph.

12.7 Concepts Review

1. perpendicular

3. $x - +4(y-1)+6(z-1)=0$

Problem Set 12.7

1. $\nabla F(x, y, z) = 2\langle x, y, z\rangle;$

   $\nabla F\left(2, 3, \sqrt{3}\right) = 2\left\langle 2, 3, \sqrt{3}\right\rangle$

   Tangent Plane:

   $2(x-2)+3(y-3)+\sqrt{3}\left(z-\sqrt{3}\right)=0,$ or

   $2x+3y+\sqrt{3}z=16$

3. Let $F(x, y, z) = x^2 - y^2 + z^2 + 1 = 0.$

   $\nabla F(x, y, z) = \langle 2x, -2y, 2z\rangle = 2\langle x, -y, z\rangle$

   $\nabla F\left(1, 3, \sqrt{7}\right) = 2\left\langle 1, -3, \sqrt{7}\right\rangle$, so $\left\langle 1, -3, \sqrt{7}\right\rangle$ is normal to the surface at the point. Then the tangent plane is

   $1(x-1)-3(y-3)+\sqrt{7}\left(z-\sqrt{7}\right)=0,$ or more simply, $x-3y+\sqrt{7}z=-1.$

5. $\nabla f(x, y) = \left(\frac{1}{2}\right)\langle x, y\rangle; \nabla f(2, 2) = \langle 1, 1\rangle$

   Tangent plane: $z-2=1(x-2)+1(y-2)$, or $x+y-z=2$.

7. $\nabla f(x, y) = \left\langle -4e^{3y}\sin 2x, 6e^{3y}\cos 2x\right\rangle;$

   $\nabla f\left(\frac{\pi}{3}, 0\right) = \left\langle -2\sqrt{3}, -3\right\rangle$

   Tangent plane: $z+1=-2\sqrt{3}\left(x-\frac{\pi}{3}\right)-3(y-0)$,

   or $2\sqrt{3}x+3y+z=\frac{\left(2\sqrt{3}\pi-3\right)}{3}.$

9. Let $z = f(x, y) = 2x^2y^3;$

   $dz = 4xy^3\,dx + 6x^2y^2\,dy$. For the points given, $dx = -0.01,\ dy = 0.02$,

   $dz = 4(-0.01)+6(0.02) = 0.08.$

   $\Delta z = f(0.99, 1.02) - f(1, 1)$

   $= 2(0.99)^2(1.02)^3 - 2(1)^2(1)^3 \approx 0.08017992$

11. $dz = 2x^{-1}dx + y^{-1}dy = (-1)(0.02) + \left(\frac{1}{4}\right)(-0.04)$

    $= -0.03$

    $\Delta z = f(-1.98, 3.96) - f(-2.4)$

    $= \ln[(-1.98)^2(3.96)] - \ln 16 \approx -0.030151$

13. Let

    $F(x, y, z) = x^2 - 2xy - y^2 - 8x + 4y - z = 0;$

    $\nabla F(x, y, z) = \langle 2x-2y-8, -2x-2y+4, -1\rangle$

    Tangent plane is horizontal if $\nabla F = \langle 0, 0, k\rangle$ for any $k \neq 0$.

    $2x - 2y - 8 = 0$ and $-2x - 2y + 4 = 0$ if $x = 3$ and $y = -1$. Then $z = -14$. There is a horizontal tangent plane at $(3, -1, -14)$.

15. For $F(x, y, z) = x^2 + 4y + z^2 = 0,$

    $\nabla F(x, y, z) = \langle 2x, 4, 2z\rangle = 2\langle x, 2, z\rangle.$

    $F(0, -1, 2) = 0$, and

    $\nabla F(0, -1, 2) = 2\langle 0, 2, 2\rangle = 4\langle 0, 1, 1\rangle.$

    For $G(x, y, z) = x^2 + y^2 + z^2 - 6z + 7 = 0,$

    $\nabla G(x, y, z) = \langle 2x, 2y, 2z-6\rangle = 2\langle x, y, z-3\rangle.$

    $G(0, -1, 2) = 0$, and

    $\nabla G(0, -1, 2) = 2\langle 0, -1, -1\rangle = -2\langle 0, 1, 1\rangle.$

    $\langle 0, 1, 1\rangle$ is normal to both surfaces at $(0, -1, 2)$ so the surfaces have the same tangent plane; hence, they are tangent to each other at $(0, -1, 2)$.

17. Let $F(x, y, z) = x^2 + 2y^2 + 3z^2 - 12 = 0;$

    $\nabla F(x, y, z) = 2\langle x, 2y, 3z\rangle$ is normal to the plane.

    A vector in the direction of the line, $\langle 2, 8, -6\rangle = 2\langle 1, 4, -3\rangle$, is normal to the plane.

    $\langle x, 2y, 3z\rangle = k\langle 1, 4, -3\rangle$ and (x, y, z) is on the surface for points $(1, 2, -1)$ [when $k = 1$] and $(-1, -2, 1)$ [when $k = -1$].

19. $\nabla f(x, y, z) = 2\langle 9x, 4y, 4z\rangle;$

    $\nabla f(1, 2, 2) = 2\langle 9, 8, 8\rangle$

    $\nabla g(x, y, z) = 2\langle 2x, -y, 3z\rangle;$

    $\nabla f(1, 2, 2) = 4\langle 1, -1, 3\rangle$

    $\langle 9, 8, 8\rangle \times \langle 1, -1, 3\rangle = \langle 32, -19, -17\rangle$

    Line: $x = 1+32t, y = 2-19t, z = 2-17t$

21. $dS = S_A dA + S_W dW$

$= -\frac{W}{(A-W)^2}dA + \frac{A}{(A-W)^2}dW = \frac{-WdA + AdW}{(A-W)^2}$

At $W = 20, A = 36$:

$dS = \frac{-20dA + 36dW}{256} = \frac{-5dA + 9dW}{64}$.

Thus, $|dS| \le \frac{5|dA| + 9|dW|}{64} \le \frac{5(0.02) + 9(0.02)}{64}$

$= 0.004375$

23. $V = \pi r^2 h, dV = 2\pi rh\,dr + \pi r^2 dh$

$|dV| \le 2\pi rh|dr| + \pi r^2|dh| \le 2\pi rh(0.02r) + \pi r^2(0.03h)$

$= 0.04\pi r^2 h + 0.03\pi r^2 h = 0.07V$

Maximum error in V is 7%.

25. Solving for R, $R = \frac{R_1 R_2}{R_1 + R_2}$, so

$\frac{\partial R}{\partial R_1} = \frac{R_2^2}{(R_1 + R_2)^2}$ and $\frac{\partial R}{\partial R_2} = \frac{R_1^2}{(R_1 + R_2)^2}$.

Therefore, $dR = \frac{R_2^2 dR_1 + R_1^2 dR_2}{(R_1 + R_2)^2}$;

$|dR| \le \frac{R_2^2|dR_1| + R_1^2|dR_2|}{(R_1 + R_2)^2}$. Then at $R_1 = 25$,

$R_2 = 100, dR_1 = dR_2 = 0.5,\ R = \frac{(25)(100)}{25 + 100} = 20$

and $|dR| \le \frac{(100)^2(0.5) + (25)^2(0.5)}{(125)^2} = 0.34$.

27. Let $F(x, y, z) = xyz = k$; let (a, b, c) be any point on the surface of F.

$\nabla F(x, y, z) = \langle yz, xz, xy\rangle = \left\langle \frac{k}{x}, \frac{k}{y}, \frac{k}{z}\right\rangle$

$= k\left\langle \frac{1}{x}, \frac{1}{y}, \frac{1}{z}\right\rangle$

$\nabla F(a, b, c) = k\left\langle \frac{1}{a}, \frac{1}{b}, \frac{1}{c}\right\rangle$

An equation of the tangent plane at the point is

$\left(\frac{1}{a}\right)(x-a) + \left(\frac{1}{b}\right)(x-b) + \left(\frac{1}{c}\right)(x-c) = 0$, or

$\frac{x}{a} + \frac{y}{b} + \frac{z}{c} = 3$.

Points of intersection of the tangent plane on the coordinate axes are $(3a, 0, 0)$, $(0, 3b, 0)$, and $(0, 0, 3c)$.

The volume of the tetrahedron is

$\left(\frac{1}{3}\right)$(area of base)(altitude)$=\frac{1}{3}\left(\frac{1}{2}|3a||3b|\right)(|3c|)$

$= \frac{9|abc|}{2} = \frac{9|k|}{2}$ (a constant).

29. $f(x, y) = (x^2 + y^2)^{1/2}$; $f(3, 4) = 5$

$f_x(x, y) = x(x^2 + y^2)^{-1/2}$; $f_x(3, 4) = \frac{3}{5} = 0.6$; $f_y = (x, y) = y(x^2 + y^2)^{-1/2}$; $f_x(3, 4) = \frac{4}{5} = 0.8$

$f_{xx}(x, y) = y^2(x^2 + y^2)^{-3/2}$; $f_x(3, 4) = \frac{16}{125} = 0.128$; $f_{xy}(x, y) = -xy(x^2 + y^2)^{-3/2}$;

$f_{xy}(3, 4) = -\frac{12}{125} = -0.096$

$f_{yy} = x^2(x^2 + y^2)^{-3/2}$; $f_{xx}(3, 4) = \frac{9}{125} = 0.072$

Therefore, the second order Taylor approximation is

$f(x, y) = 5 + 0.6(x-3) + 0.8(y-4) + 0.5[0.128(x-3)^2 + 2(-0.096)(x-3)(y-4) + 0.072(y-4)^2]$

a. First order Taylor approximation: $f(x, y) = 5 + 0.6(x-3) + 0.8(y-4)$.

Thus, $f(3.1, 3.9) \approx 5 + 0.6(0.1) + 0.8(-0.1) = 4.98$.

b. $f(3.1, 3.9) \approx 5 + 0.6(-0.1) + 0.8(0.1) + 0.5[0.128(0.1)^2 + 2(-0.096)(0.1)(-0.1) + 0.072(-0.1)^2] = 4.98196$

c. $f(3.1,\ 3.9) \approx 4.9819675$

12.8 Concepts Review

1. closed bounded

3. $\nabla f(x_0, y_0) = \mathbf{0}$

Problem Set 12.8

1. $\nabla f(x, y) = \langle 2x - 4, 8y \rangle = \langle 0, 0 \rangle$ at $(2, 0)$, a stationary point.
$D = f_{xx} f_{yy} - f_{xy}^2 = (2)(8) - (0)^2 = 16 > 0$ and $f_{xx} = 2 > 0$. Local minimum at $(2, 0)$.

3. $\nabla f(x, y) = \left\langle 8x^3 - 2x, 6y \right\rangle = \left\langle 2x(4x^2 - 1), 6y \right\rangle$
$= \langle 0, 0 \rangle$, at $(0, 0), (0.5, 0), (-0.5, 0)$ all stationary points.
$f_{xx} = 24x^2 - 2$; $D = f_{xx} f_{yy} - f_{xy}^2 = (24x^2 - 2)(6) - (0)^2 = 12(12x^2 - 1)$.
At $(0, 0)$: $D = -12$, so $(0, 0)$ is a saddle point.
At $(0.5, 0)$ and $(-0.5, 0)$: $D = 24$ and $f_{xx} = 6$, so local minima occur at these points.

5. $\nabla f(x, y) = \langle y, x \rangle = \langle 0, 0 \rangle$ at $(0, 0)$, a stationary point.
$D = f_{xx} f_{yy} - f_{xy}^2 = (0)(0) - (1)^2 = -1$, so $(0, 0)$ is a saddle point.

7. $\nabla f(x, y) = \left\langle \dfrac{x^2 y - 2}{x^2}, \dfrac{xy^2 - 4}{y^2} \right\rangle = \langle 0, 0 \rangle$ at $(1, 2)$.
$D = f_{xx} f_{yy} - f_{xy}^2 = (4x - 3)(8y - 3) - (1)^2 = 32x^{-3} y^{-3} - 1$, $f_{xx} = 4x^{-3}$
At $(1, 2)$: $D = 3 > 0$, and $f_{xx} > 0$, so a local minimum at $(1, 2)$.

9. Let $\nabla f(x, y)$
$= \langle -\sin x - \sin(x + y), -\sin y - \sin(x + y) \rangle = \langle 0, 0 \rangle$
Then $\begin{pmatrix} -\sin x - \sin(x + y) = 0 \\ \sin y + \sin(x + y) = 0 \end{pmatrix}$. Therefore,
$\sin x = \sin y$, so $x = y = \dfrac{\pi}{4}$. However, these values satisfy neither equation. Therefore, the gradient is defined but never zero in its domain, and the boundary of the domain is outside the domain, so there are no critical points.

11. We do not need to use calculus for this one. $3x$ is minimum at 0 and $4y$ is minimum at -1. $(0, -1)$ is in S, so $3x + 4y$ is minimum at $(0, -1)$; the minimum value is -4. Similarly, $3x$ and $4y$ are each maximum at 1. $(1, 1)$ is in S, so $3x + 4y$ is maximum at $(1, 1)$; the maximum value is 7. (Use calculus techniques and compare.)

13. $\nabla f(x, y) = \langle 2x, -2y \rangle = \langle 0, 0 \rangle$ at $(0, 0)$.
$D = f_{xx} f_{yy} - f_{xy}^2 = (2)(-2) - (0)2 < 0$, so $(0, 0)$ is a saddle point. A parametric representation of the boundary of S is $x = \cos t$, $y = \sin t$, t in $[0, 2\pi]$.
$f(x, y) = f(x(t), y(t)) = \cos^2 t - \sin^2 t + 1$
$= \cos 2t - 1$
$\cos 2t - 1$ is maximum if $\cos 2t = 1$, which occurs for $t = 0, \pi, 2\pi$. The points of the curve are $(\pm 1, 0)$. $f(\pm 1, 0) = 2$
$f(x, y) = \cos 2t - 1$ is minimum if $\cos 2t = -1$, which occurs for $t = \dfrac{\pi}{2}, \dfrac{3\pi}{2}$. The points of the curve are $(0, \pm 1)$. $f(0, \pm 1) = 0$. Global minimum of 0 at $(0, \pm 1)$; global maximum of 2 at $(\pm 1, 0)$.

15. Let x, y, z denote the numbers, so $x + y + z = N$. Maximize

$P = xyz = xy(N - x - y) = Nxy - x^2y - xy^2$.

Let $\nabla P(x, y) = \langle Ny - 2xy - y^2, Nx - x^2 - 2xy \rangle$

$= \langle 0, 0 \rangle$.

Then $\begin{pmatrix} Ny - 2xy - y^2 = 0 \\ Nx - x^2 - 2xy = 0 \end{pmatrix}$.

$N(x, -y) = x^2 - y^2 = (x + y)(x - y)$. $x = y$ or $N = x + y$.

Therefore, $x = y$ (since $N = x + y$ would mean that $P = 0$, certainly not a maximum value).

Then, substituting into $Nx - x^2 - 2xy = 0$, we obtain $Nx - x^2 - 2x^2 = 0$, from which we obtain $x(N - 3x) = 0$, so $x = \frac{N}{3}$ (since $x = 0 \Rightarrow P = 0$).

$P_{xx} = -2y$;

$D = P_{xx}P_{yy} - P_{xy}^2$

$= (-2y)(-2x) - (N - 2x - 2y)^2$

$= 4xy - (N - 2x - 2y)^2$

At $x = y = \frac{N}{3}$: $D = \frac{N^2}{3} > 0$, $P_{xx} = -\frac{2N}{3} < 0$ (so local maximum)

If $x = y = \frac{N}{3}$, then $z = \frac{N}{3}$.

Conclusion: Each number is $\frac{N}{3}$. (If the intent is to find three distinct numbers, then there is no maximum value of P that satisfies that condition.)

17. Let S denote the surface area of the box with dimensions x, y, z.

$S = 2xy + 2xz + 2yz$ and $V_0 = xyz$, so

$S = 2(xy + V_0y^{-1} + V_0x^{-1})$.

Minimize $f(x, y) = xy + V_0y^{-1} + V_0x^{-1}$ subject to $x > 0, y > 0$.

$\nabla f(x, y) = \langle y - V_0x^{-2}, x - V_0y^{-2} \rangle = \langle 0, 0 \rangle$ at $(V_0^{1/3}, V_0^{1/3})$.

$D = f_{xx}f_{yy} - f_{xy}^2 = 4V_0^2x^{-3}y^{-3} - 1$,

$f_{xx} = 2V_0x^{-3}$.

At $\left(V_0^{1/3}, V_0^{1/3}\right)$: $D = 3 > 0$, $f_{xx} = 2 > 0$, so local minimum.

Conclusion: The box is a cube with edge $V_0^{1/3}$.

19. Let S denote the area of the sides and bottom of the tank with base l by w and depth h.

$S = lw + 2lh + 2wh$ and $lwh = 256$.

$S(l, w) = lw + 2l\left(\frac{256}{lw}\right) + 2w\left(\frac{256}{lw}\right)$, $w > 0, l > 0$.

$S(l\ w) = \langle w - 5121^{-2}, l - 512w^{-2} \rangle = \langle 0, 0 \rangle$ at $(8, 8)$. $h = 4$ there. At $(8, 8)$ $D > 0$ and $S_{11} > 0$, so local minimum. Dimensions are 8' × 8' × 4'.

21. Let $\langle x, y, z \rangle$ denote the vector; let S be the sum of its components.

$x^2 + y^2 + z^2 = 81$, so $z = (81 - x^2 - y^2)^{1/2}$.

Maximize $S(x, y) = x + y + (81 - x^2 - y^2)^{1/2}$, $0 \le x^2 + y^2 \le 9$.

Let $\nabla S(x, y) = \langle 1 - x(81 - x^2 - y^2)^{-1/2}, 1 - y(81 - x^2 - y^2)^{-1/2} \rangle = \langle 0, 0 \rangle$.

Therefore, $x = (81 - x^2 - y^2)^{1/2}$ and $y = (81 - x^2 - y^2)^{1/2}$. We then obtain $x = y = 3\sqrt{3}$ as the only stationary point. For these values of x and y, $z = 3\sqrt{3}$ and $S = 9\sqrt{3} \approx 15.59$.

The boundary needs to be checked. It is fairly easy to check each edge of the boundary separately. The largest value of S at a boundary point occurs at three places and turns out to be $\frac{18}{\sqrt{2}} \approx 12.73$.

Conclusion: the vector is $3\sqrt{3}\langle 1, 1, 1 \rangle$.

23. Let $P(x, y, z)$ be any point on $z = x^2 + y^2$. The square of the distance between the point $(1, 2, 0)$ and P can be expressed as $d^2 = f(x, y) = (x-1)^2 + (y-2)^2 + z^2$. To find the critical points, set $f_x(x, y)$ $= 4x^3 + 2x + 4xy^2 - 2 = 0$ and $f_y(x, y) = 4y^3 + 2y + 4x^2y - 4 = 0$. Multiplying the first equation by y and the second equation by x and summing the results leads to the equation $-2y + 4x = 0$. Thus, $y = 2x$. Substituting into the first equation yields $10x^3 + x - 1 = 0$, whose solution is $x \approx 0.393$. Consequently, $y \approx 0.786$. $f_{xx}(x, y) = 2 + 12x^2 + 4y^2$, $f_{yy}(x, y) = 2 + 12y^2 + 4x^2$, and $f_{xy}(x, y) = 8xy$. The value of D for the critical point $(0.393, 0.786)$ is approximately 57 and since $f_{xx}(0.393, 0.786) > 0$, $(0.393, 0.786)$ yields a minimum distance. The point on the surface $z = x^2 + y^2$ is $(0.393, 0.786, 0.772)$ and this minimum distance is approximately 1.56.

25. $A = \left(\frac{1}{2}\right)[y + (y + 2x\sin\alpha)](x\cos\alpha)$ and

$2x + y = 12$. Maximize $A(x, \alpha) = 12x\cos\alpha - 2x^2\cos\alpha + \left(\frac{1}{2}\right)x^2\sin 2\alpha$, x in $(0, 6]$, a in $\left(0, \frac{\pi}{2}\right)$.

$A(x, \alpha) = \left\langle 12\cos\alpha - 4x\cos\alpha + 2x\sin\alpha\cos\alpha,\ -12x\sin\alpha + 2x^2\sin\alpha + x^2\cos 2\alpha \right\rangle = \langle 0, 0\rangle$ at $\left(4, \frac{\pi}{6}\right)$.

At $\left(4, \frac{\pi}{6}\right)$, $D > 0$ and $A_{xx} < 0$, so local maximum, and $A = 12\sqrt{3} \approx 20.78$. At the boundary point of $x = 6$, we get $\alpha = \frac{\pi}{4}$, $A = 18$. Thus, the maximum occurs for width of turned-up sides = 4", and base angle $= \frac{\pi}{2} + \frac{\pi}{6} = \frac{2\pi}{3}$.

27. Let M be the maximum value of $f(x, y)$ on the polygonal region, P. Then $ax + by + (c - M) = 0$ is a line that either contains a vertex of P or divides P into two subregions. In the latter case $ax + by + (c - M)$ is positive in one of the regions and negative in the other. $ax + by + (c - M) > 0$ contradicts that M is the maximum value of $ax + by + c$ on P. (Similar argument for minimum.)

a.

x	y	$2x + 3y + 4$
–1	2	8
0	1	7
1	0	6
–3	0	–2
0	–4	–8

Maximum at $(-1, 2)$

b.

x	y	$-3x + 2y + 1$
–3	0	10
0	5	11
2	3	1
4	0	–11
1	–4	–10

Minimum at $(4, 0)$

29. $z(x, y) = y^2 - x^2$

$z(x, y) = \langle -2x, 2y\rangle = \langle 0, 0\rangle$ at $(0, 0)$.

There are no stationary points and no singular points, so consider boundary points.

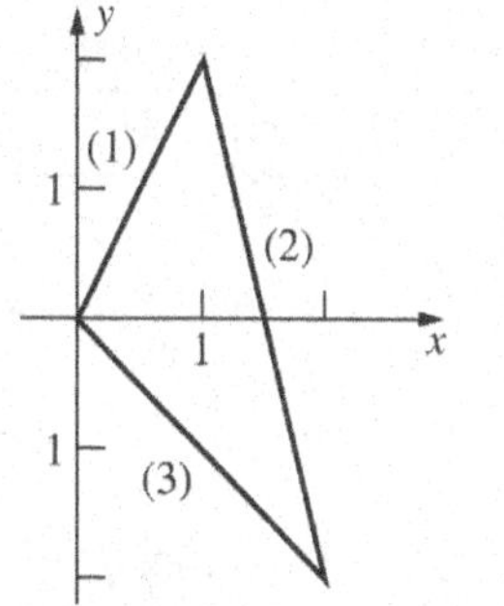

On side 1:

$y = 2x$, so $z = 4x^2 - x^2 = 3x^2$

$z'(x) = 6x = 0$ if $x = 0$.

Therefore, $(0, 0)$ is a candidate.

On side 2:

$y = -4x + 6$, so

$z = (-4x + 6)^2 - x^2 = 15x^2 - 48x + 36$.

$z'(x) = 30x - 48 = 0$ if $x = 1.6$.

Therefore, $(1.6, -0.4)$ is a candidate.

On side 3:

$y = -x$, so $z = (-x)^2 - x^2 = 0$.

Also, all vertices are candidates.

x	y	z
0	0	0
1.6	–0.4	–2.4
2	–2	0
1	2	3

Minimum value of –2.4; maximum value of 3

31.

	x_i	y_i	x_i^2	$x_i y_i$
	3	2	9	6
	4	3	16	12
	5	4	25	20
	6	4	36	24
	7	5	49	35
$\Sigma_{i=1}^{5}$	25	18	135	97

$m(135) + b(25) = (97)$ and $m(25) + (5)b = (18)$. Solve simultaneously and obtain $m = 0.7, b = 0.1$. The least-squares line is $y = 0.7x + 0.1$.

33. Let x and y be defined as shown in Figure 4 from Section 12.8. The total cost is given by

$$C(x, y) = 400\sqrt{x^2 + 50^2} + 200(200 - x - y) + 300\sqrt{y^2 + 100^2}$$

Taking partial derivatives and setting them equal to 0 gives

$$C_x(x, y) = 200(x^2 + 50^2)^{-1/2}(2x) - 200 = 0$$

$$C_y(x, y) = 150(y^2 + 100^2)^{-1/2}(2y) - 200 = 0$$

The solution of these equations is

$$x = \frac{50}{\sqrt{3}} \approx 28.8675 \text{ and } y = \frac{100}{\sqrt{1.25}} \approx 89.4427$$

We now apply the second derivative test:

$$C_{xx}(x, y) = \frac{400\sqrt{x^2 + 50^2} - 400x^2/\sqrt{x^2 + 50^2}}{x^2 + 50^2}$$

$$C_{yy}(x, y) = \frac{300\sqrt{y^2 + 100^2} - 300y^2/\sqrt{y^2 + 100^2}}{y^2 + 100^2}$$

$$C_{xy}(x, y) = 0$$

Evaluated at $x = 50/\sqrt{3}$ and $y = 100/\sqrt{1.25}$,

$$D \approx (5.196)(1.24) - 0^2 > 0$$

Thus, a local minimum occurs with

$$C\left(50/\sqrt{3}, 100/\sqrt{1.25}\right) \approx \$79,681$$

We must also check the boundary. When x = 0,

$$C_1(y) = C(0, y) = 200(200 - y) + 300\sqrt{y^2 + 100^2}$$

and when y = 0,

$$C_2(x) = C(x, 0) = 400\sqrt{x^2 + 50^2} + 200(200 - x)$$

Using the methods from Chapter 3, we find that C_1 reaches a minimum of about \$82,361 when $y = \sqrt{8000}$ and C_2 reaches a minimum of about \$87,321 when $x = \sqrt{2500/3}$. Addressing the boundary x + y = 200, we find that

$$C_3(x) = C(x, 200 - x) = 400\sqrt{x^2 + 50^2} + 300\sqrt{(200 - x)^2 + 100^2}$$

This function reaches a minimum of about \$82,214 when $x \approx 41.08$.

Thus, the minimum cost path is when $x = 50/\sqrt{3} \approx 28.8675$ ft and $y = 100/\sqrt{1.25} \approx 89.4427$ ft, which produces a cost of about \$79,681.

35. $f(x, y) = 10 + x + y$

$\nabla f = \langle 1, 1\rangle \neq \mathbf{0}$; thus no interior critical points exist. Letting $x = 3\cos t$, $y = 3\sin t$, $0 \le t \le 2\pi$, $g(t) = f(3\cos t, 3\sin t)$ and $g'(t) = 3\cos t - 3\sin t$. Setting $g'(t) = 0$ yields $t = \pi/4$ or $5\pi/4$.

The critical points are $\left(3/\sqrt{2}, 3/\sqrt{2}\right)$ and $\left(-3/\sqrt{2}, -3/\sqrt{2}\right)$.

Since $f\left(3/\sqrt{2}, 3/\sqrt{2}\right) = 10 + 6/\sqrt{2}$ and $f\left(-3/\sqrt{2}, -3/\sqrt{2}\right) = 10 - 6/\sqrt{2}$, the minimum value of f on $x^2 + y^2 \le 3$ is $10 - 6/\sqrt{2}$ and the maximum value of f is $10 + 6/\sqrt{2}$.

37. The volume of the box can be expressed as $V(l, w, h) = lwh = 2$ and the surface area as $S(l, w, h) = 2lh + 2wh + lw + lw$. Since $h = \frac{2}{lw}$,

$S(l, w) = \frac{4}{w} + \frac{4}{l} + lw + lw$ When cost is factored,

$$C(l, w) = \frac{1}{w} + \frac{1}{l} + 0.65lw \text{ with } w > 0, l > 0$$

$$C_l(l, w) = -\frac{1}{l^2} + 0.65w = 0$$

$$C_w(l, w) = -\frac{1}{w^2} + 0.65l = 0$$

Solving this system of equations leads to

$$w = \sqrt[3]{\frac{0.65}{0.4225}} \approx 1.1544 \text{ and } l = w \approx 1.1544.$$

Consequently, $h \approx 1.501$. Applying the second derivative test with $C_{ll}(l, w) = \frac{2}{l^3}$,

$C_{ww}(l, w) = \frac{2}{w^3}$ and $C_{lw}(l, w) = 0.65$,

$D \approx 1.268 > 0$. Thus, the minimum cost occurs when the length is approximately 1.1544 feet, the width is approximately 1.1544 feet and the height is approximately 1.501 feet.

39. $T(x, y) = 2x^2 + y^2 - y$

$\nabla T = \langle 4x, 2y - 1 \rangle = \mathbf{0}$

If $x = 0$ and $y = \frac{1}{2}$, so $\left(0, \frac{1}{2}\right)$ is the only interior critical point.

On the boundary $x^2 = 1 - y^2$, so **T** is a function of y there.

$T(y) = 2(1 - y^2) + y^2 - y = 2 - y - y^2$, $y = [-1, 1]$

$T'(y) = -1 - 2y = 0$ if $y = -\frac{1}{2}$, so on the boundary, critical points occur where y is -1, $-\frac{1}{2}, 1$.

Thus, points to consider are $\left(0, \frac{1}{2}\right)$, $(0, -1)$, $\left(\frac{\sqrt{3}}{2}, -\frac{1}{2}\right)$, $\left(-\frac{\sqrt{3}}{2}, -\frac{1}{2}\right)$ and $(0, 1)$. Substituting these into $T(x, y)$ yields that the coldest spot is $\left(0, \frac{1}{2}\right)$ where the temperature is $-\frac{1}{4}$, and there is a tie for the hottest spot at $\left(\pm\frac{\sqrt{3}}{2}, -\frac{1}{2}\right)$ where the temperature is $\frac{9}{4}$.

41. Without loss of generality we will assume that $\alpha \le \beta \le \gamma$. We will consider it intuitively clear that for a triangle of maximum area the center of the circle will be inside or on the boundary of the triangle; i.e., α, β, and γ are in the interval $[0, \pi]$. Along with $\alpha + \beta + \gamma = 2\pi$, this implies that $\alpha + \beta \ge \pi$.

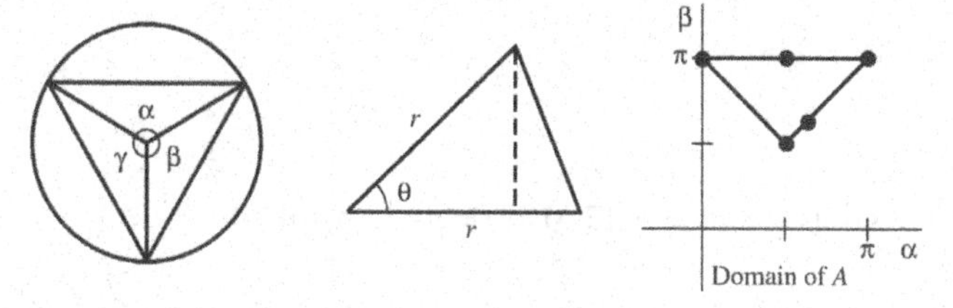

The area of an isosceles triangle with congruent sides of length r and included angle θ is $\frac{1}{2}r^2 \sin\theta$.

$$\begin{aligned}\text{Area}(\Delta ABC) &= \frac{1}{2}r^2 \sin\alpha + \frac{1}{2}r^2 \sin\beta + \frac{1}{2}r^2 \sin\gamma \\ &= \frac{1}{2}r^2(\sin\alpha + \sin\beta + \sin[2\pi - (\alpha + \beta)] \\ &= \frac{1}{2}r^2[\sin\alpha + \sin\beta - \sin(\alpha + \beta)]\end{aligned}$$

Area(ΔABC) will be maximum if (*) $A(\alpha, \beta) = \sin\alpha + \sin\beta - \sin(\alpha + \beta)$ is maximum.

Restrictions are $0 \le \alpha \le \beta \le \pi$, and $\alpha + \beta \ge \pi$.

Three critical points are the vertices of the triangular domain of A: $\left(\frac{\pi}{2}, \frac{\pi}{2}\right)$, $(0, \pi)$, and (π, π). We will now search for others.

$\Delta A(\alpha, \beta) = \langle \cos\alpha - \cos(\alpha + \beta), \cos\beta - \cos(\alpha + \beta) \rangle = 0$ if

$\cos\alpha = \cos(\alpha + \beta) = \cos\beta$.

Therefore, $\cos\alpha = \cos\beta$, so $\alpha = \beta$ [due to the restrictions stated]. Then

$\cos\alpha = \cos(\alpha+\alpha) = \cos 2\alpha = 2\cos^2\alpha - 1$, so $\cos\alpha = 2\cos^2\alpha - 1$.

Solve for α: $2\cos^2\alpha - \cos\alpha - 1 = 0$; $(2\cos\alpha + 1)(\cos\alpha - 1) = 0$;

$\cos\alpha = -\frac{1}{2}$ or $\cos\alpha = 1$; $\alpha = \frac{2\pi}{3}$ or $\alpha = 0$.

(We are still in the case where $\alpha = \beta$.) $\left(\frac{2\pi}{3}, \frac{2\pi}{3}\right)$ is a new critical point, but $(0, 0)$ is out of the domain of A.

There are no critical points in the interior of the domain of A.
On the $\beta = \pi$ edge of the domain of A; $A(\alpha) = \sin\alpha - \sin(\alpha - \pi) = 2\sin\alpha$ so $A'(\alpha) = 2\cos\alpha$.

$A'(\alpha) = 0$ if $\alpha = \frac{\pi}{2}$. $\left(\frac{\pi}{2}, \pi\right)$ is a new critical point.

On the $\beta = \pi - \alpha$ edge of the domain of A:

$A(\alpha) = \sin\alpha + \sin(\pi - \alpha) - \sin(2\alpha - \pi) = 2\sin\alpha + \sin 2\alpha$, so

$A'(\alpha) = 2\cos\alpha + 2\cos 2\alpha = 2[\cos\alpha + (2\cos^2\alpha - 1)] = 2(2\cos\alpha - 1)(\cos\alpha + 1)$.

$A'(\alpha) = 0$ if $\cos\alpha = \frac{1}{2}$ or $\cos\alpha = -1$, so $\alpha = \frac{\pi}{3}$ or $\alpha = \pi$.

$\left(\frac{\pi}{3}, \frac{2\pi}{3}\right)$ and $(\pi, 0)$ are outside the domain of A.

(The critical points are indicated on the graph of the domain of A.)

α	β	A	
$\frac{\pi}{2}$	$\frac{\pi}{2}$	2	
0	π	0	
π	π	0	
$\frac{2\pi}{3}$	$\frac{2\pi}{3}$	$\frac{3\sqrt{3}}{2}$	Maximum value of A. The triangle is equilateral.
$\frac{\pi}{2}$	π	2	

43. Local max: $f(1.75, 0) = 1.15$; Global max: $f(-3.8, 0) = 2.30$

45. Global min: $f(0, 1) = f(0, -1) = -0.12$

47. Global max: $f(1.13, 0.79) = f(1.13, -0.79) = 0.53$; Global min: $f(-1.13, 0.79) = f(-1.13, -0.79) = -0.53$

49. Global max: $f(3,3) = f(-3,3) \approx 74.9225$
Global min: $f(1.5708, 0) = f(-1.5708, 0) = -8$

51. Global max: $f(0.67, 0) = 5.06$
Global min: $f(-0.75, 0) = -3.54$

53. Global max: $f(2.1, 2.1) = 3.5$
Global min: $f(4.2, 4.2) = -3.5$

12.9 Concepts Review

1. free; constrained

3. $g(x, y) = 0$

Problem Set 12.9

1. $\langle 2x, 2y\rangle = \lambda\langle y, x\rangle$
   $2x = \lambda y$, $2y = \lambda x$, $xy = 3$
   Critical points are
   $(\pm\sqrt{3}, \pm\sqrt{3})$, $f(\pm\sqrt{3}, \pm\sqrt{3}) = 6$.
   It is not clear whether 6 is the minimum or maximum, so take any other point on $xy = 3$, for example (1, 3). $f(1, 3) = 10$, so 6 is the minimum value.

3. Let $\nabla f(x, y) = \lambda \nabla g(x, y)$, where
   $g(x, y) = x^2 + y^2 - 1 = 0$.
   $\langle 8x - 4y, -4x + 2y\rangle = \lambda\langle 2x, 2y\rangle$
   1. $4x - 2y = \lambda x$
   2. $-2x + y = \lambda y$
   3. $x^2 + y^2 = 1$
   4. $0 = \lambda x + 2\lambda y$ (From equations 1 and 2)
   5. $\lambda = 0$ or $x + 2y = 0$ (4)

   $\lambda = 0$:
   6. $y = 2x$ (1)
   7. $x = \pm\frac{1}{\sqrt{5}}$ (6, 3)
   8. $y = \pm\frac{2}{\sqrt{5}}$ (7, 6)

   $x + 2y = 0$:
   9. $x = -2y$
   10. $y = \pm\frac{1}{\sqrt{5}}$ (9, 3)
   11. $x = \frac{2}{\sqrt{5}}$ (10, 9)

   Critical points: $\left(\frac{1}{\sqrt{5}}, \frac{2}{\sqrt{5}}\right)$, $\left(-\frac{1}{\sqrt{5}}, -\frac{2}{\sqrt{5}}\right)$, $\left(\frac{2}{\sqrt{5}}, -\frac{1}{\sqrt{5}}\right)$, $\left(-\frac{2}{\sqrt{5}}, \frac{1}{\sqrt{5}}\right)$
   $f(x, y)$ is 0 at the first two critical points and 5 at the last two. Therefore, the maximum value of $f(x, y)$ is 5.

5. $\langle 2x, 2y, 2z\rangle = \lambda\langle 1, 3, -2\rangle$
   $2x = \lambda$, $2y = 3\lambda$, $2z = -2\lambda$, $x + 3y - 2z = 12$
   Critical point is $\left(\frac{6}{7}, \frac{18}{7}, -\frac{12}{7}\right)$.
   $f\left(\frac{6}{7}, \frac{18}{7}, -\frac{12}{7}\right) = \frac{72}{7}$ is the minimum.

7. Let l and w denote the dimensions of the base, h denote the depth. Maximize $V(l, w, h) = lwh$ subject to $g(l, w, h) = lw + 2lh + 2wh = 48$.
   $\langle wh, lh, lw\rangle = \lambda\langle w + 2h, l + 2h, 2l + 2w\rangle$
   $wh = \lambda(w + 2h)$, $lh = \lambda(l + 2h)$, $lw = \lambda(2l + 2w)$, $lw + 2lh + 2wh = 48$
   Critical point is (4, 4, 2).
   $V(4, 4, 2) = 32$ is the maximum.

9. Let l and w denote the dimensions of the base, h the depth. Maximize $V(l, w, h) = lwh$ subject to $0.601w + 0.20(lw + 2lh + 2wh) = 12$, which simplifies to $21w + lh + wh = 30$, or $g(l, w, h) = 2lw + lh + wh - 30$.
   Let $\nabla V(l, w, h) = \lambda \nabla g(l, w, h)$;
   $\langle wh, lh, lw\rangle = \lambda\langle 2w + h, 2l + h, l + w\rangle$.
   1. $wh = \lambda(2w + h)$
   2. $lh = \lambda(2l + h)$
   3. $lw = \lambda(l + w)$
   4. $2lw + lh + wh = 30$
   5. $(w - l)h = 2\lambda(w - l)$ (1, 2)
   6. $w = l$ or $h = 2\lambda$

   $w = l$:
   7. $l = 2\lambda = w$ (3) Note: $w \neq 0$, for then $V = 0$.
   8. $h = 4\lambda$ (7, 2)
   9. $\lambda = \frac{\sqrt{5}}{2}$ (7, 8, 4)
   10. $l = w = \sqrt{5}$, $h = 2\sqrt{5}$ (9, 7, 8)

   $h = 2\lambda$:
   11. $\lambda = 0$ (2)
   12. $l = w = h = 0$ (11, 1 – 3)
   (Not possible since this does not satisfy 4.)
   $(\sqrt{5}, \sqrt{5}, 2\sqrt{5})$ is a critical point and
   $V(\sqrt{5}, \sqrt{5}, 2\sqrt{5}) = 10\sqrt{5} \approx 22.36$ ft³ is the maximum volume (rather than the minimum volume since, for example, $g(1, 1, 14) = 30$ and $V(1, 1, 14) = 14$ which is less than 22.36).

11. Maximize $f(x, y, z) = xyz$, subject to
    $g(x, y, z) = b^2c^2x^2 + a^2c^2y^2 + a^2b^2z^2 - a^2b^2c^2 = 0$
    $\langle yz, xz, xy\rangle = \lambda\langle 2b^2c^2x, 2a^2c^2y, 2a^2b^2z\rangle$
    $yz = 2b^2c^2x$, $xz = 2a^2c^2y$, $xy = 2a^2b^2z$,
    $b^2c^2x^2 + a^2c^2y^2 + a^2b^2z^2 = a^2b^2c^2$
    Critical point is $\left(\frac{a}{\sqrt{3}}, \frac{b}{\sqrt{3}}, \frac{c}{\sqrt{3}}\right)$.
    $V\left(\frac{a}{\sqrt{3}}, \frac{b}{\sqrt{3}}, \frac{c}{\sqrt{3}}\right) = \frac{8abc}{3\sqrt{3}}$, which is the maximum.

13. Maximize $f(x,y,z) = x + y + z$ with the constraint $g(x,y,z) = x^2 + y^2 + z^2 - 81 = 0$. Let $\nabla f(x,y,z) = \lambda \nabla g(x,y,z)$, so $\langle 1,1,1 \rangle = \lambda \langle 2x, 2y, 2z \rangle$; Thus, $x = y = z$ and $3x^2 = 81$ or $x = y = z = \pm 3\sqrt{3}$.

The maximum value of f is $9\sqrt{3}$ when $\langle x,y,z \rangle = \langle 3\sqrt{3}, 3\sqrt{3}, 3\sqrt{3} \rangle$

15. Minimize $d^2 = f(x,y,z)$ $= (x-1)^2 + (y-2)^2 + z^2$ with the constraint $g(x,y,z) = x^2 + y^2 - z = 0$;

$\langle 2x-2, 2y-4, 2z \rangle = \lambda \langle 2x, 2y, -1 \rangle$

Setting up, solving each equation for λ, and substituting into equation $x^2 + y^2 - z = 0$ produces $\lambda \approx -1.5445$; The resulting critical point is approximately (0.393, 0.786, 0.772). The nature of this problem indicates this will give a minimum value rather than a maximum. The minimum distance is approximately 1.5616.

17. (See problem 37, section 12.8). Let the dimensions of the box be l, w, and h. Then the cost of the box is

$.25(2hl + 2hw + lw) + .4(lw)$ or

$C(l,w,h) = .5hl + .5hw + .65lw$.

We want to minimize C subject to the constraint $lhw = 2$; set $V(l,h,w) = lwh - 2$.

Now:

$\nabla C(l,w,h) = (.5h + .65w)\mathbf{i} + (.5h + .65l)\mathbf{j} + .5(l+w)\mathbf{k}$

and

$\nabla V(l,w,h) = wh\mathbf{i} + lh\mathbf{j} + lw\mathbf{k}$

Thus the Lagrange equations are

$.5h + .65w = \lambda wh \quad (1)$

$.5h + .65l = \lambda lh \quad (2)$

$.5(l+w) = \lambda lw \quad (3)$

$lwh = 2 \quad (4)$

Solving (4) for h and putting the result in (1) and (2), we get

$$\frac{1}{lw} + .65w = \frac{2\lambda}{l} \quad (5)$$

$$\frac{1}{lw} + .65l = \frac{2\lambda}{w} \quad (6)$$

Multiply (5) by l and (6) by w to get

$$\frac{1}{w} + .65lw = 2\lambda \quad (7)$$

$$\frac{1}{l} + .65lw = 2\lambda \quad (8)$$

from which we conclude that $l = w$. Putting this result into (3) we have

$$l = \lambda l^2 \quad (9)$$

Since $V \neq 0$, $l \neq 0$ and (9) tells us that $l = \frac{1}{\lambda}$;

thus $l = \frac{1}{\lambda}$, $w = l = \frac{1}{\lambda}$, $h = \frac{2}{lw} = 2\lambda^2$.

Putting these results into equation (1), we conclude

$$.5(2\lambda^2) + .65\left(\frac{1}{\lambda}\right) = \lambda\left(\frac{1}{\lambda}\right)(2\lambda^2) \text{ or}$$

$$.65\left(\frac{1}{\lambda}\right) = \lambda^2.$$

Hence: $\lambda = \sqrt[3]{.65} \approx .866$, so the minimum cost is obtained when:

$$l = w = \frac{1}{\lambda} \approx 1.154 \quad \text{and} \quad h = 2\lambda^2 \approx 1.5$$

19. (See problem 40, section 12.8)

Let

c = circumference of circle

p = perimeter of first square

q = perimeter of second square

Then the sum of the areas is

$$A(c,p,q) = \frac{c^2}{4\pi} + \frac{p^2}{16} + \frac{q^2}{16} = \frac{1}{4}\left[\frac{c^2}{\pi} + \frac{p^2}{4} + \frac{q^2}{4}\right]$$

so we wish to maximize and minimize

$\tilde{A}(c,p,q) = \frac{c^2}{\pi} + \frac{p^2}{4} + \frac{q^2}{4}$ subject to the constraint $L(c,p,q) = c + p + q - k = 0$.

Now

$$\nabla \tilde{A}(c,p,q) = \frac{2c}{\pi}\mathbf{i} + \frac{p}{2}\mathbf{j} + \frac{q}{2}\mathbf{k}$$

$$\nabla L(c,p,q) = \mathbf{i} + \mathbf{j} + \mathbf{k}$$

so the Lagrange equations are

$$\frac{2c}{\pi} = \lambda \quad (1)$$

$$\frac{p}{2} = \lambda \quad (2)$$

$$\frac{q}{2} = \lambda \quad (3)$$

$$c + p + q = k \quad (4)$$

Putting (1), (2) and (3) into (4) we get

$$\left(4 + \frac{\pi}{2}\right)\lambda = k \text{ or } \lambda = \frac{2k}{8+\pi}$$

Therefore:

$$c_0 = \frac{\pi k}{8+\pi} \approx 0.282k$$

$$p_0 = \frac{4k}{8+\pi} \approx 0.359k$$

$$q_0 = \frac{4k}{8+\pi} \approx 0.359k$$

Now $A(c_0, p_0, q_0) \approx 0.0224k^2$ while $A(k,0,0) \approx .079k^2$, so we conclude that $A(c_0, p_0, q_0)$ is a minimum value. There is also a maximum value (see problem 40, section 12.8) but our Lagrange approach does not capture this. The reason is that the maximum exists because *c*, *p*, and *q* must all be ≥ 0. Our constraint, however, does not require this and allows negative values for any or all of the variables. Under these conditions, there is no global maximum.

21. Finding critical points on the interior first:
$\frac{\partial f}{\partial x} = 1 \neq 0$ $\frac{\partial f}{\partial y} = 1 \neq 0$; There are no critical points on the interior. Finding critical points on the boundary: $\nabla f(x,y) = \lambda \nabla g(x,y)$;
$\langle 1,1 \rangle = \lambda \langle 2x, 2y \rangle$; The solution to the system $1 = \lambda \cdot 2x, 1 = \lambda \cdot 2y, x^2 + y^2 = 1$ is $\lambda = \pm\frac{1}{\sqrt{2}}$, $x = \pm\frac{1}{\sqrt{2}}$, $y = \pm\frac{1}{\sqrt{2}}$ The four critical points are $\left(\frac{1}{\sqrt{2}}, \pm\frac{1}{\sqrt{2}}\right)$ and $\left(-\frac{1}{\sqrt{2}}, \pm\frac{1}{\sqrt{2}}\right)$.
$f\left(\frac{1}{\sqrt{2}}, \frac{1}{\sqrt{2}}\right) = 10 + \sqrt{2}$ is the maximum value.
$f\left(-\frac{1}{\sqrt{2}}, -\frac{1}{\sqrt{2}}\right) = 10 - \sqrt{2}$ is the minimum value.

23. Finding critical points on the interior:
$\frac{\partial f}{\partial x} = 2x + 3 - y = 0$; $\frac{\partial f}{\partial y} = 2y - x = 0$
The solution to this system is the only critical point on the interior, $c_1 = (-2,-1)$.
Critical points on the boundary will come from the solutions to the following system of equations:
$2x + 3 - y = \lambda \cdot 2x$, $2y - x = \lambda \cdot 2y$, $x^2 + y^2 = 9$. From the solutions to this system, the critical points are $c_2 = (0,3)$,
$c_3 = \left(\frac{3\sqrt{3}}{2}, -\frac{3}{2}\right)$, $c_4 = \left(-\frac{3\sqrt{3}}{2}, -\frac{3}{2}\right)$
$f(c_1) = -3$, $f(c_2) = 9$, $f(c_3) \approx 20.6913$, $f(c_4) \approx -2.6913$ The max value of *f* is ≈ 20.6913 and the min value is -3.

25. $\frac{\partial f}{\partial x} = \frac{\partial f}{\partial y} = 2(1 + x + y) = 0 \Rightarrow x + y = -1$
There is no minimum or maximum value on the interior since there are an infinite number of critical points. The critical points on the boundary will come from the solutions to the following system of equations:
$$2(1 + x + y) = \lambda \cdot \frac{1}{2} x$$
$$2(1 + x + y) = \lambda \cdot \frac{1}{8} y$$
Solving these two equations for λ leads to $y = -x - 1$ or $y = 4x$. Together with the constraint $\frac{x^2}{4} + \frac{y^2}{16} - 1 = 0$ leads to the critical points on the boundary: $\left(\frac{-1 - 2\sqrt{19}}{5}, \frac{-4 + 2\sqrt{19}}{5}\right)$, $\left(\frac{-1 + 2\sqrt{19}}{5}, \frac{-4 - 2\sqrt{19}}{5}\right)$, $\left(-\frac{2}{\sqrt{5}}, -\frac{8}{\sqrt{5}}\right)$ and $\left(\frac{2}{\sqrt{5}}, \frac{8}{\sqrt{5}}\right)$. Respectively, the maximum value is ≈ 29.9443 and the minimum value is 0.

27. Let $\alpha + \beta + \gamma = 1$, $\alpha > 0$, $\beta > 0$, and $\gamma > 0$.
Maximize $P(x, y, z) = kx^\alpha y^\beta z^\gamma$, subject to $g(x, y, z) = ax + by + cz - d = 0$.
Let $\nabla P(x, y, z) = \lambda \nabla g(x, y, z)$. Then $\left\langle k\alpha x^{\alpha-1} y^\beta z^\gamma, k\beta x^\alpha y^{\beta-1} z^\gamma, k\gamma x^\alpha y^\beta z^{\gamma-1} \right\rangle = \lambda \langle a, b, c \rangle$.
Therefore, $\frac{\lambda ax}{\alpha} = \frac{\lambda by}{\beta} = \frac{\lambda cz}{\gamma}$ (since each equals $kx^\alpha y^\beta z^\gamma$).
$\lambda \neq 0$ since $\lambda = 0$ would imply $x = y = z = 0$ which would imply $P = 0$.
Therefore, $\frac{ax}{\alpha} = \frac{by}{\beta} = \frac{cz}{\gamma}$ (*).

The constraints $ax + by + cz = d$ in the form $\alpha\left(\frac{ax}{\alpha}\right) + \beta\left(\frac{by}{\beta}\right) + \gamma\left(\frac{cz}{\gamma}\right) = d$ becomes

$\alpha\left(\frac{ax}{\alpha}\right) + \beta\left(\frac{ax}{\alpha}\right) + \gamma\left(\frac{ax}{\alpha}\right) = d$, using (*).

Then $(\alpha+\beta+\gamma)\left(\frac{ax}{\alpha}\right)=d$, or $\frac{ax}{\alpha}=d$ (since $\alpha+\beta+\gamma=1$).

$x=\frac{\alpha d}{a}$(**); $y=\frac{\beta d}{b}$ and $z=\frac{\gamma d}{c}$ then following using (*) and (**).

Since there is only one interior critical point, and since P is 0 on the boundary, P is maximum when $x=\frac{\alpha d}{a}$, $y=\frac{\beta d}{b}$, $z=\frac{\gamma d}{c}$.

29. $\langle -1,2,2\rangle=\lambda\langle 2x,2y,0\rangle+\mu\langle 0,1,2\rangle$

$-1=2\lambda x$, $2=2\lambda y+\mu$, $2=2\mu$, $x^2+y^2=2$, $y+2z=1$

Critical points are $(-1, 1, 0)$ and $(1, -1, 1)$.

$f(-1, 1, 0) = 3$, the maximum value;

$f(1, -1, 1) = -1$, the minimum value.

31. Let $\langle a_1, a_2, \ldots a_n\rangle=\lambda\langle 2x_1, 2x_2, \ldots, 2x_n\rangle$.

Therefore, $a_i=2\lambda x_i$, for each $i = 1, 2, \ldots, n$ (since $\lambda=0$ implies $a_i=0$, contrary to the hypothesis).

$\frac{x_i}{a_i}=\frac{x_j}{a_j}$ for all i, j $\left(\text{since each equals } \frac{1}{2\lambda}\right)$.

The constraint equation can be expressed

$$a_1^2\left(\frac{x_1}{a_1}\right)^2+a_2^2\left(\frac{x_2}{a_2}\right)^2+\ldots+a_n^2\left(\frac{x_n}{a_n}\right)^2=1.$$

Therefore, $\left(a_1^2+a_2^2+\ldots+a_n^2\right)\left(\frac{x_1}{a_1}\right)^2=1$.

$x_1^2=\frac{a_1^2}{a_1^2+\ldots+a_n^2}$; similar for each other x_i^2.

The function to be maximized in a hyperplane with positive coefficients and constant (so intercepts on all axes are positive), and the constraint is a hypersphere of radius 1, so the maximum will occur where each x_i is positive. There is only one such critical point, the one obtained from the above by taking the principal square root to solve for x_i.

Then the maximum value of w is

$$a_1\left(\frac{a_1}{\sqrt{A}}\right)+a_2\left(\frac{a_2}{\sqrt{A}}\right)+\ldots+a_n\left(\frac{a_n}{\sqrt{A}}\right)=\frac{A}{\sqrt{A}}=\sqrt{A}$$

where $A=a_1^2+a_2^2+\ldots+a_n^2$.

33. Min: $f(4, 0) = -4$

35. Min: $f(0, 3) = f(0, -3) = -0.99$

12.10 Chapter Review

Concepts Test

1. True: Except for the trivial case of $z = 0$, which gives a point.

3. True: Since $g'(0)=f_x(0, 0)$

5. True: Use "Continuity of a Product" Theorem.

7. False: See Problem 25, Section 12.4.

9. True: Since $\langle 0,0,-1\rangle$ is normal to the tangent plane

11. True: It will point in the direction of greatest increase of heat, and at the origin, $\nabla T(0, 0)=\langle 1, 0\rangle$ is that direction.

13. True: Along the x-axis, $f(x, 0)\to\pm\infty$ as $x\to\pm\infty$.

15. True: $-D_{\mathbf{u}}f(x, y)=-[\nabla f(x, y)\cdot\mathbf{u}]$ $=\nabla f(x, y)\cdot(-\mathbf{u})=D_{-\mathbf{u}}f(x, y)$

17. True: By the Min-Max Existence Theorem

19. False: $f\left(\frac{\pi}{2}, 1\right)=\sin\left(\frac{\pi}{2}\right)=1$, the maximum value of f, and $(\pi/2, 1)$ is in the set.

Sample Test Problems

1. a. $x^2 + 4y^2 - 100 \ge 0$

$\frac{x^2}{100} + \frac{y^2}{25} \ge 1$

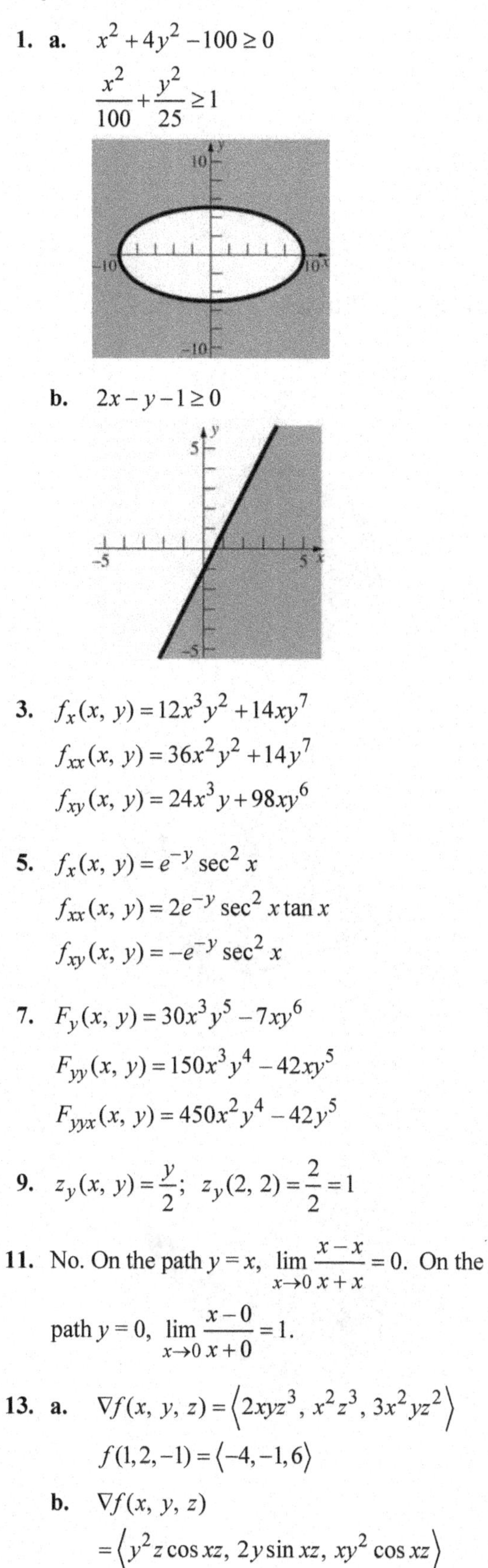

b. $2x - y - 1 \ge 0$

3. $f_x(x, y) = 12x^3y^2 + 14xy^7$

$f_{xx}(x, y) = 36x^2y^2 + 14y^7$

$f_{xy}(x, y) = 24x^3y + 98xy^6$

5. $f_x(x, y) = e^{-y}\sec^2 x$

$f_{xx}(x, y) = 2e^{-y}\sec^2 x \tan x$

$f_{xy}(x, y) = -e^{-y}\sec^2 x$

7. $F_y(x, y) = 30x^3y^5 - 7xy^6$

$F_{yy}(x, y) = 150x^3y^4 - 42xy^5$

$F_{yyx}(x, y) = 450x^2y^4 - 42y^5$

9. $z_y(x, y) = \frac{y}{2};\ z_y(2, 2) = \frac{2}{2} = 1$

11. No. On the path $y = x$, $\lim_{x\to 0} \frac{x - x}{x + x} = 0$. On the path $y = 0$, $\lim_{x\to 0} \frac{x - 0}{x + 0} = 1$.

13. a. $\nabla f(x, y, z) = \langle 2xyz^3, x^2z^3, 3x^2yz^2 \rangle$

$f(1,2,-1) = \langle -4, -1, 6 \rangle$

b. $\nabla f(x, y, z)$

$= \langle y^2 z\cos xz, 2y\sin xz, xy^2\cos xz \rangle$

$\nabla f(1, 2, -1) = -4\langle \cos(1), \sin(1), -\cos(1) \rangle$

$\approx \langle -2.1612, -3.3659, 2.1612 \rangle$

15. $z = f(x, y) = x^2 + y^2$

$\langle 1, -\sqrt{3}, 0 \rangle$ is horizontal and is normal to the vertical plane that is given. By inspection, $\langle \sqrt{3}, 1, 0 \rangle$ is also a horizontal vector and is perpendicular to $\langle 1, -\sqrt{3}, 0 \rangle$ and therefore is parallel to the vertical plane. Then $\mathbf{u} = \left\langle \frac{\sqrt{3}}{2}, \frac{1}{2} \right\rangle$ is the corresponding 2-dimensional unit vector.

$D_{\mathbf{u}} f(x, y) = \nabla f(x, y) \cdot \mathbf{u}$

$= \langle 2x, 2y \rangle \cdot \left\langle \frac{\sqrt{3}}{2}, \frac{1}{2} \right\rangle = \sqrt{3}x + y$

$D_{\mathbf{u}} f(1, 2) = \sqrt{3} + 2 \approx 3.7321$ is the slope of the tangent to the curve.

17. a. $f(4, 1) = 9$, so $\frac{x^2}{2} + y^2 = 9$, or $\frac{x^2}{18} + \frac{y^2}{9} = 1$.

b. $\nabla f(x, y) = \langle x, 2y \rangle$, so $f(4,1) = \langle 4, 2 \rangle$.

c.

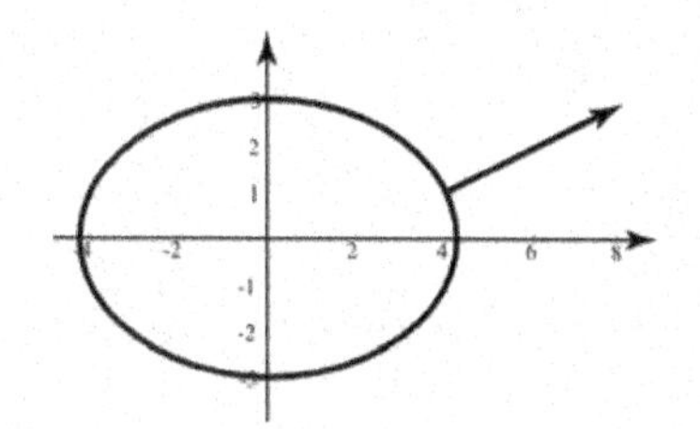

19. $f_x = f_u u_x + f_v u_y = \left(\frac{1}{v}\right)(2x) + \left(-\frac{u}{v^2}\right)(yz)$

$= x^{-2}y^{-1}z^{-1}(x^2 + 3y - 4z)$

$f_y = f_u u_y + f_v v_y = \left(\frac{1}{v}\right)(-3) + \left(-\frac{u}{v^2}\right)(xz)$

$= -x^{-1}y^{-2}z^{-1}(x^2 + 4z)$

$f_z = f_u u_z + f_v v_z = \left(\frac{1}{v}\right)(4) + \left(-\frac{u}{v^2}\right)(xy)$

$= x^{-1}y^{-1}z^{-2}(3y - x^2)$

21. $F_t = F_x x_t + F_y y_t + F_z z_t$

$= \left(\frac{10xy}{z^3}\right)\left(\frac{3t^{1/2}}{2}\right) + \left(\frac{5x^2}{z^3}\right)\left(\frac{1}{t}\right) + \left(-\frac{15x^2y}{z^4}\right)(3e^{3t})$

$= \frac{15xy\sqrt{t}}{z^3} + \frac{5x^2}{z^3 t} - \frac{45x^2ye^{3t}}{z^4}$

23. Let $F(x, y, z) = 9x^2 + 4y^2 + 9z^2 - 34 = 0$
$\nabla F(x, y, z) = \langle 18x, 8y, 18z \rangle$, so $\nabla f(1, 2, -1) = 2\langle 9, 8, -9 \rangle$.
Tangent plane is $9(x - 1) + 8(y - 2) - 9(z + 1) = 0$, or $9x + 8y - 9z = 34$.

25. $df = y^2(1+z^2)^{-1}dx + 2xy(1+z^2)^{-1}dy - 2xy^2z(1+z^2)^{-2}dz$
If $x = 1, y = 2, z = 2, dx = 0.01, dy = -0.02, dz = 0.03$, then $df = -0.0272$.
Therefore, $f(1.01, 1.98, 2.03) \approx f(1, 2, 2) + df = 0.8 - 0.0272 = 0.7728$

27. Let (x, y, z) denote the coordinates of the 1st octant vertex of the box. Maximize $f(x, y, z) = xyz$ subject to
$g(x, y, z) = 36x^2 + 4y^2 + 9z^2 - 36 = 0$
(where $x, y, z > 0$ and the box's volume is $V(x, y, z) = f(x, y, z)$).
Let $\nabla f(x, y, z) = \lambda \nabla g(x, y, z)$.
$\langle yz, xz, xy \rangle 8 = \lambda \langle 72x, 8y, 18z \rangle$
1. $8yz = 72\lambda x$
2. $8xz = 8\lambda y$
3. $8xy = 18\lambda z$
4. $36x^2 + 4y^2 + 9z^2 = 36$
5. $\frac{yz}{xz} = \frac{72\lambda x}{8\lambda y}$, so $y^2 = 9x^2$. (1, 2)
6. $\frac{yz}{xz} = \frac{72\lambda x}{18\lambda y}$, so $z^2 = 4x^2$. (1, 3)
7. $36x^2 + 36x^2 + 36x^2 = 36$, so $x = \frac{1}{\sqrt{3}}$.
(5, 6, 4)
8. $y = \frac{3}{\sqrt{3}}, z = \frac{2}{\sqrt{3}}$ (7, 5, 6)
$V\left(\frac{1}{\sqrt{3}}, \frac{3}{\sqrt{3}}, \frac{2}{\sqrt{3}}\right) = 8\left(\frac{1}{\sqrt{3}}\right)\left(\frac{3}{\sqrt{3}}\right)\left(\frac{2}{\sqrt{3}}\right)$
$= \frac{16}{\sqrt{3}} \approx 9.2376$
The nature of the problem indicates that the critical point yields a maximum value rather than a minimum value.

29. Maximize $V(r, h) = \pi r^2 h$, subject to
$S(r, h) = 2\pi r^2 + 2\pi rh - 24\pi = 0$.
$\langle 2\pi rh, \pi r^2 \rangle = \lambda \langle 4\pi r + 2\pi h, 2\pi r \rangle$
$rh = \lambda(2r + h), r = 2\lambda, r^2 + rh = 12$
Critical point is (2, 4). The nature of the problem indicates that the critical point yields a maximum value rather than a minimum value. Conclusion: The dimensions are radius of 2 and height of 4.

Review and Preview Problems

1.

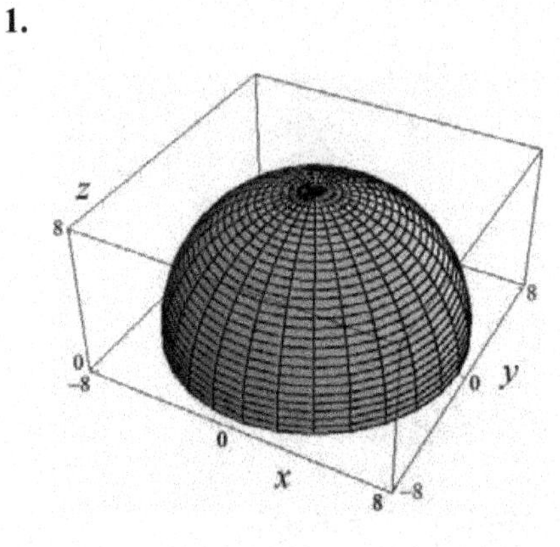

3. 

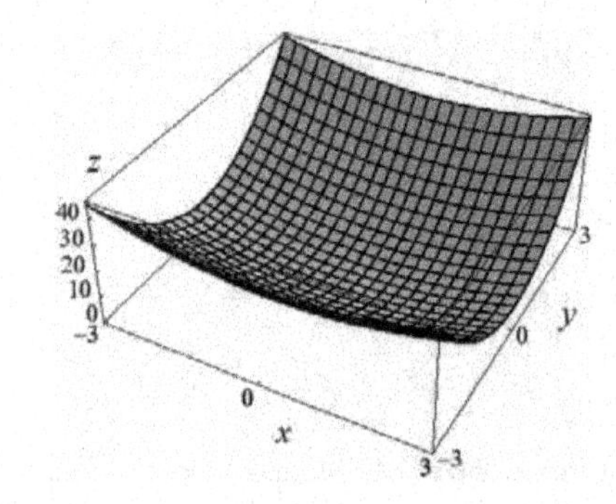

5.

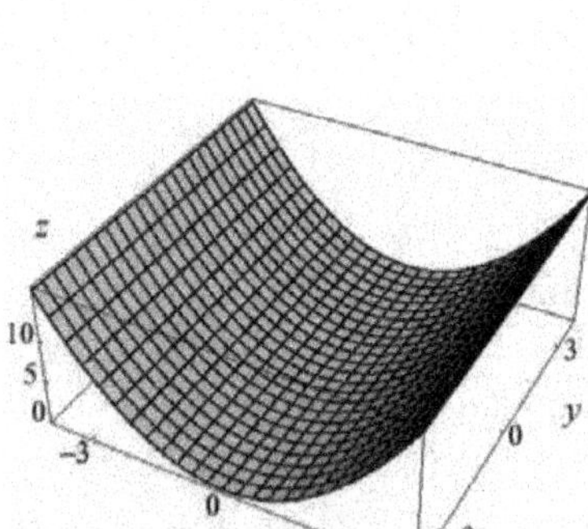

7.

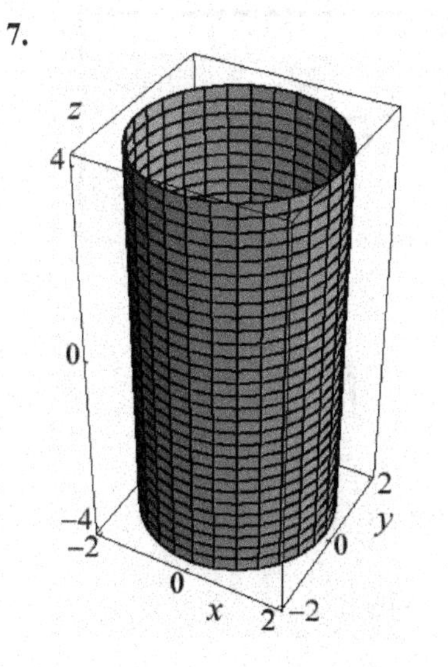

9.

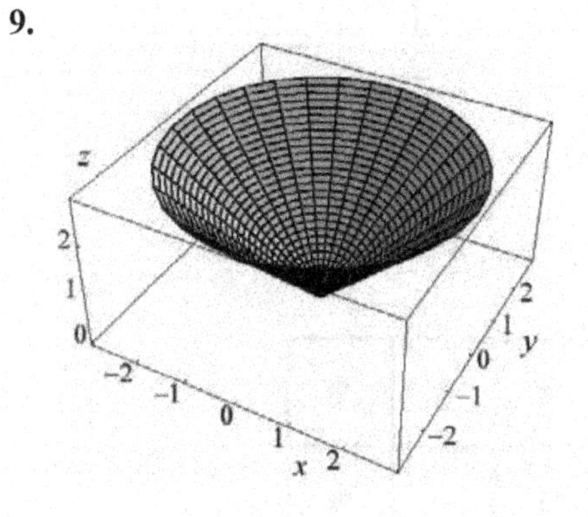

11.

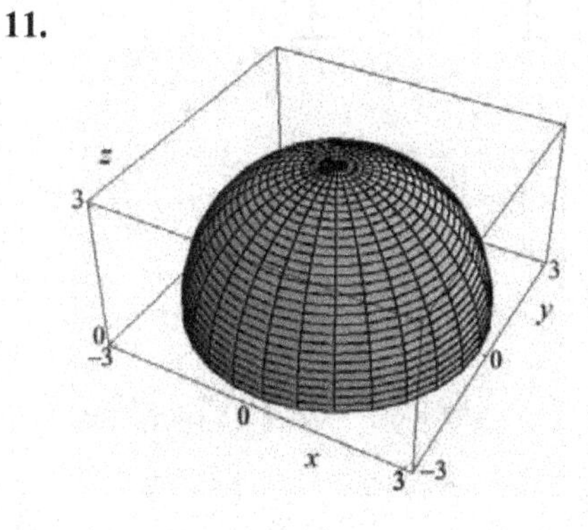

13.

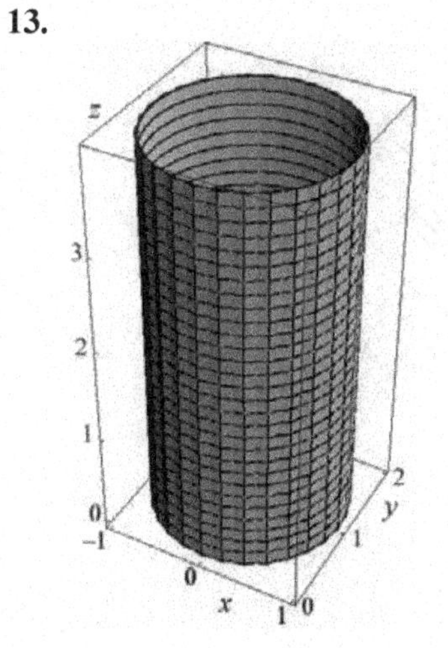

15. $\int e^{-2x}\,dx = -\frac{1}{2}e^{-2x} + C$

17. $\int_{-a/2}^{a/2} \cos\left(\frac{x\pi}{a}\right) dx = \frac{a}{\pi}\sin\left(\frac{x\pi}{a}\right)\Big|_{-a/2}^{a/2} = \frac{2a}{\pi}$

19. $\int_0^{\pi} \sin^2 x\,dx = \left[\frac{1}{2}x - \frac{1}{4}\sin 2x\right]_0^{\pi} = \frac{\pi}{2}$

21. $\int_{x=1/4}^{3/4} \frac{1}{1+u}\,du = \frac{1}{2}\ln\left|1+x^2\right|\Big|_{1/4}^{3/4} = \frac{1}{2}\ln 2$

$u = 4r^2 + 1;\ du = 8r\,dr$

23. $\int_0^3 r\sqrt{4r^2+1}\,dr = \frac{1}{8}\int_{r=0}^{3} \sqrt{u}\,du$

$= \left[\frac{1}{8}\cdot\frac{2}{3}\left(4r^2+1\right)^{3/2}\right]_0^3 = \frac{-1+37^{3/2}}{12}$

25. $\int_0^{\pi/2}\left(\frac{1}{2}+\frac{1}{2}\cos 2\theta\right)d\theta$

$= \left[\frac{1}{2}\theta + \frac{1}{4}\sin 2\theta\right]_0^{\pi/2} = \frac{\pi}{4}$

27. $2\pi\left(\sqrt{a^2-b^2} - \sqrt{a^2-c^2}\right)$

θ is not part of the integrand.

29. The solid is half of a right circular cylinder of radius 3 and height 8.

$V = \frac{1}{2}\pi r^2 h = \frac{\pi}{2}(9)(8) = 36\pi$

31. The solid looks similar to a football.

$V = \pi\int_0^{\pi} \sin^2 x\,dx = \pi\left[\frac{1}{2}x - \frac{1}{4}\sin 2x\right]_0^{\pi} = \frac{\pi^2}{2}$

33. The solid is half an elliptic paraboloid.
In the xz-plane, we can consider rotating the graph of $z = 9 - x^2$ around the z-axis for $0 \le x \le 3$. Using the Shell Method, we would get

$V = 2\pi\int_0^3 x\left(9-x^2\right)dx$

$= 2\pi\left[\frac{9x^2}{2} - \frac{x^4}{4}\right]_0^3 = 2\pi\left[\frac{81}{2} - \frac{81}{4}\right] = \frac{81\pi}{2}$

CHAPTER 13

Multiple Integrals

13.1 Concepts Review

1. $\sum_{k=1}^{n} f(\overline{x}_k, \overline{y}_k)\Delta A_k$

3. continuous

Problem Set 13.1

1. $\iint_{R_1} 2\,dA + \iint_{R_2} 3\,dA = 2A(R_1) + 3A(R_2)$
$= 2(4) + 3(2) = 14$

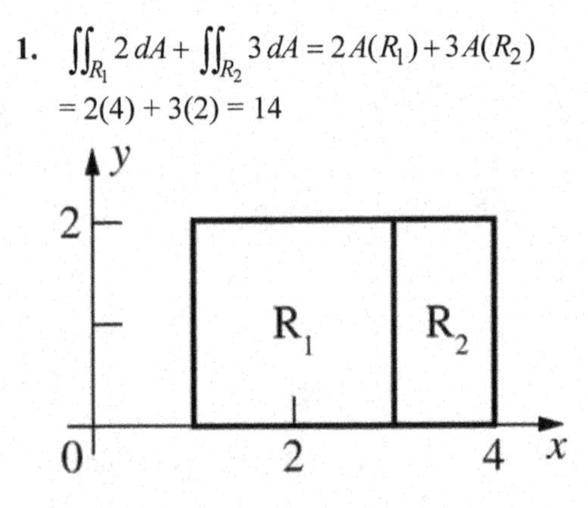

3. $\iint_R f(x, y)dA = \iint_{R_1} 2\,dA + \iint_{R_2} 1\,dA + \iint_{R_3} 3\,dA$
$= 2A(R_1) + 1A(R_2) + 3A(R_3)$
$= 2(2) + 1(2) + 3(2) = 12$

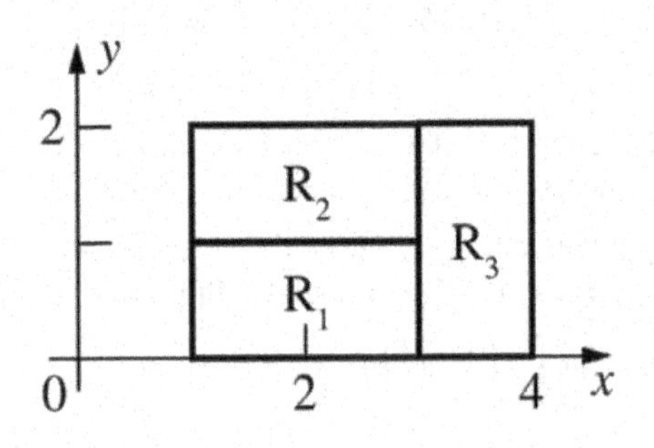

5. $3\iint_R f(x, y)dA - \iint_R g(x, y)dA = 3(3) - (5) = 4$

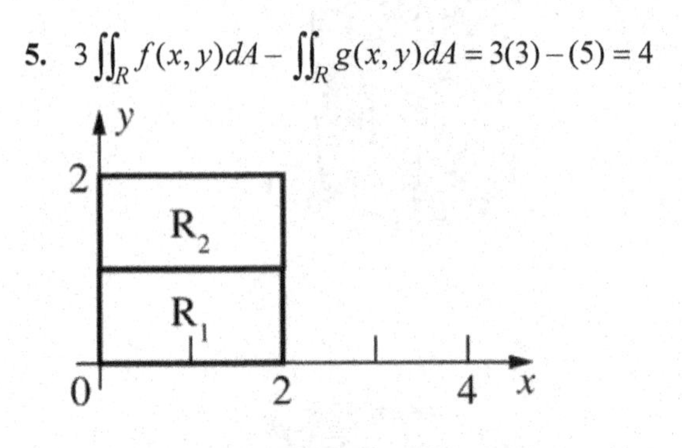

7. $\iint_R g(x, y)dA - \iint_{R_1} g(x, y)dA = (5) - (2) = 3$

9.

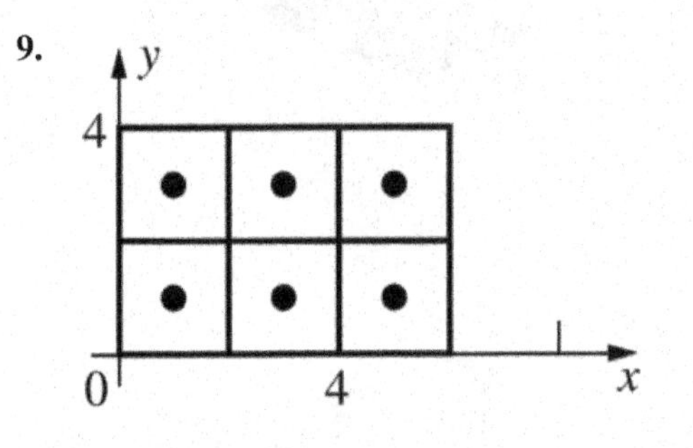

$[f(1, 1) + f(3, 1) + f(5, 1) + f(1, 3) + f(3, 3) + f(5, 3)](4) = [(10) + (8) + (6) + (8) + (6) + (4)](4) = 168$

11. $4(3 + 11 + 27 + 19 + 27 + 43) = 520$

13. $4\left(\sqrt{2} + \sqrt{4} + \sqrt{6} + \sqrt{4} + \sqrt{6} + \sqrt{8}\right) \approx 52.5665$

15.

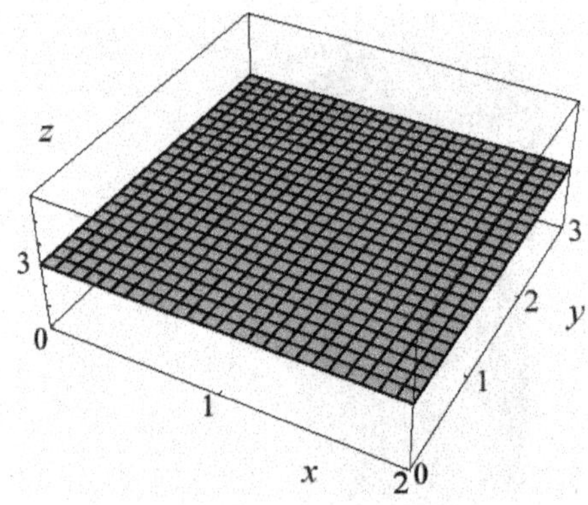

17.

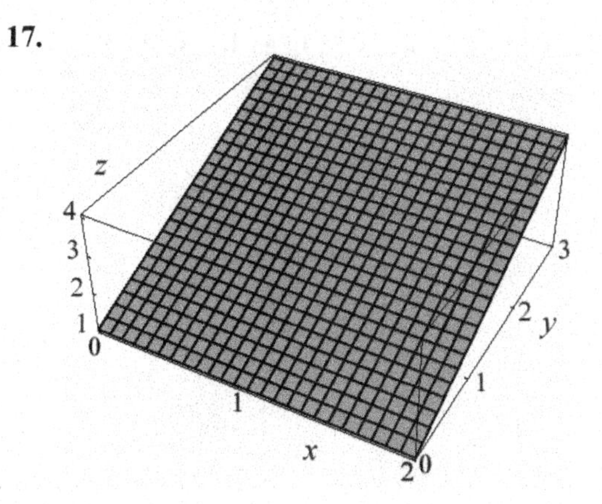

19.

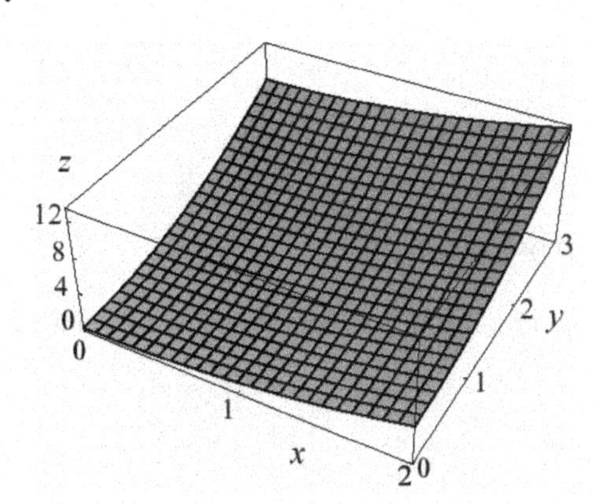

21.

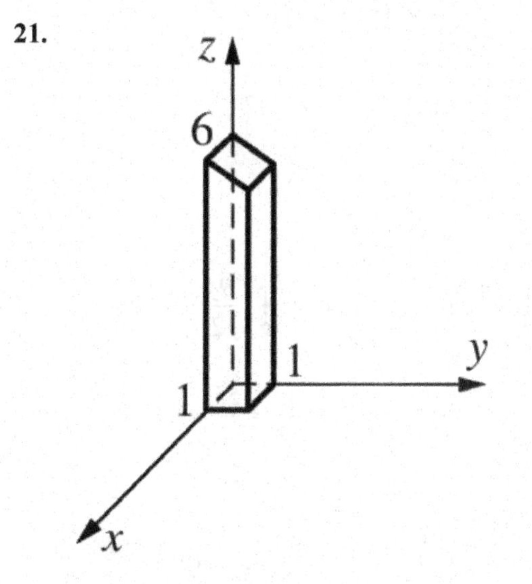

$z = 6 - y$ is a plane parallel to the x-axis. Let T be the area of the front trapezoidal face; let D be the distance between the front and back faces.

$\iint_R (6 - y)dA = \text{volume of solid} = (T)(D)$

$= \left[\left(\frac{1}{2}\right)(6+5)\right](1) = 5.5$

23. $\iint_R 0\,dA = 0A(R) = 0$

The conclusion follows.

25.

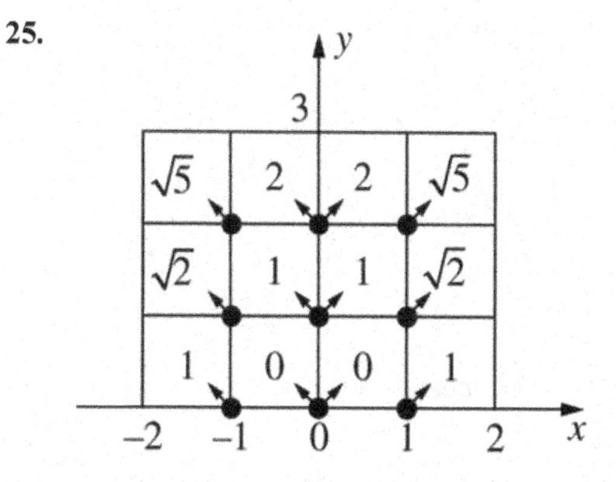

For c, take the sample point in each square to be the point of the square that is closest to the origin. Then $c = 2\sqrt{5} + 2\sqrt{2} + 2(2) + 4(1) \approx 15.3006$

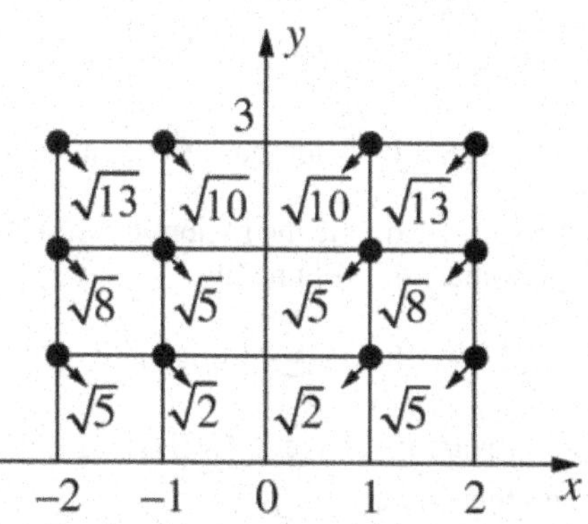

For C, take the sample point in each square to be the point of the square that is farthest from the origin. Then,

$C = 2\sqrt{13} + 2\sqrt{10} + 2\sqrt{8} + 4\sqrt{5} + 2\sqrt{2} \approx 30.9652.$

27. The values of $[\![x]\!][\![y]\!]$ and $[\![x]\!]+[\![y]\!]$ are indicated in the various square subregions of R. In each case the value of the integral on R is the sum of the values in the squares since the area of each square is 1.

a. The integral equals –6.

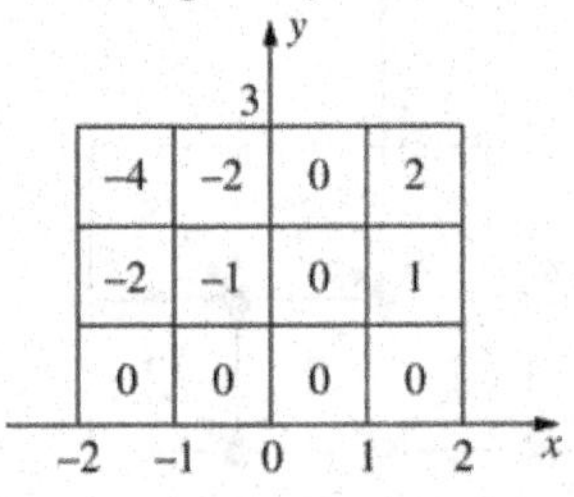

b. The integral equals 6.

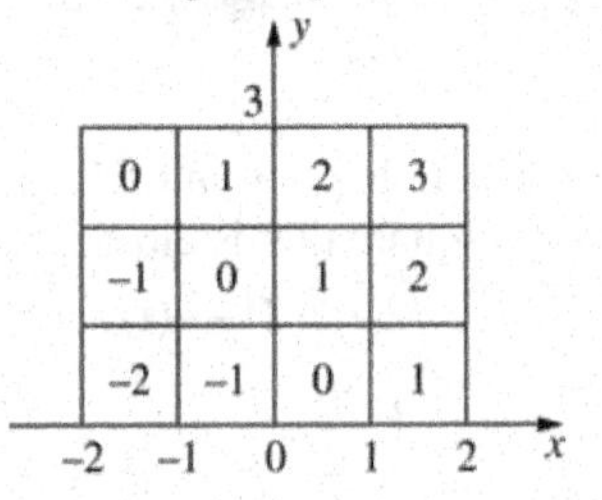

29. Total rainfall in Colorado in 2005; average rainfall in Colorado in 2005.

31. To begin, we divide the region R (we will use the outline of the contour plot) into 16 equal squares. Then we can approximate the volume by

$$V = \iint_R f(x,y)\,dA \approx \sum_{k=1}^{16} f(\overline{x}_k, \overline{y}_k)\Delta A_k .$$

Each square will have $\Delta A = (1\cdot 1) = 1$ and we will use the height at the center of each square as $f(\overline{x}_k, \overline{y}_k)$. Therefore, we get

$$V \approx \sum_{k=1}^{16} f(\overline{x}_k, \overline{y}_k) = 20+21+24+29+22+23+26+32+26+27+30+35+32+33+36+42$$

$$= 458 \text{ cubic units}$$

13.2 Concepts Review

1. iterated

3. signed; plus; minus

Problem Set 13.2

1. $\int_0^2 [9y - xy]_0^3\, dx = \int_0^2 [27 - 3x]\, dx$

$$= \left[27x - \frac{3}{2}x^2\right]_0^2 = 48$$

3. $\int_0^2 \left[\left(\frac{1}{2}\right)x^2y^2\right]_{y=1}^3 dx = \int_0^2 4x^2\, dx = \frac{32}{3}$

5. $\int_1^2 \left[\frac{x^2y}{2} + xy^2\right]_{x=0}^3 dx = \int_1^2 \left(\frac{9y}{2} + 3y^2\right) dy$

$$= \left[\frac{9y^2}{4} + y^3\right]_1^2 = 17 - \frac{13}{4} = \frac{55}{4} = 13.75$$

7. $\int_0^{\pi} \left[\left(\frac{1}{2}\right)x^2 \sin y\right]_{x=0}^1 dy = \int_0^{\pi} \left(\frac{1}{2}\right) \sin y\, dy = 1$

9. $\int_0^{\pi/2} [-\cos xy]_{y=0}^1\, dx = \int_0^{\pi/2} (1 - \cos x)\, dx$

$$= \frac{\pi}{2} - 1 \approx 0.5708$$

11. $\int_0^3 \left[\frac{2(x^2+y)^{3/2}}{3}\right]_{x=0}^1 dy$

$= \int_0^3 \frac{2[(1+y)^{3/2} - y^{3/2}]}{3} dy$

$= \left[\frac{4[(1+y)^{5/2} - y^{5/2}]}{15}\right]_0^3 = \frac{4\left(32-9\sqrt{3}\right)-4}{15}$

$= \frac{4\left(31-9\sqrt{3}\right)}{15} \approx 4.1097$

13. $\int_0^{\ln 3} \left[\left(\frac{1}{2}\right)\exp(xy^2)\right]_{y=0}^1 dx = \int_0^{\ln 3} \left(\frac{1}{2}\right)(e^x - 1)dx$

$= 1 - \left(\frac{1}{2}\right)\ln 3 \approx 0.4507$

15. $\int_0^{\pi} \left[\frac{1}{2}y^2 \cos^2 x\right]_0^3 dx = \int_0^{\pi} \frac{9}{2}\cos^2 x\, dx$

$= \left[\frac{9}{4}x + \frac{9}{8}\cos 2x\right]_0^{\pi} = \frac{9\pi}{4}$

17. $\int_0^1 0\, dx = 0$ (since xy^3 defines an odd function in y).

19. $\int_0^{\pi/2}\int_0^{\pi/2} \sin(x+y)dx\, dy$

$= \int_0^{\pi/2} [-\cos(x+y)]_{x=0}^{\pi/2} dy$

$= \int_0^{\pi/2} \left[-\cos\left(\frac{\pi}{2}+y\right) + \cos y\right] dy$

$= \int_0^{\pi/2} (\sin y + \cos y)dy = [-\cos y + \sin y]_0^{\pi/2}$

$= (0 + 1) - (-1 + 0) = 2$

21. $V = \int_0^2 \int_0^3 (20 - x - y)\, dy\, dx$

$= \int_0^2 \left(20y - xy - \frac{1}{2}y^2\right)_0^3 dx$

$= \int_0^2 \left(60 - 3x - \frac{9}{2}\right) dy = 105$

23. $V = \int_0^3 \int_0^4 \left(1 + x^2 + y^2\right) dx\, dy$

$= \int_0^3 \left(x + \frac{1}{3}x^3 + xy^2\right)_0^3 dy$

$= \int_0^3 \left(4 + \frac{64}{3} + 4y^2\right) dy = 112$

25. $z = \frac{x}{2}$ is a plane.

$x - 2z = 0$

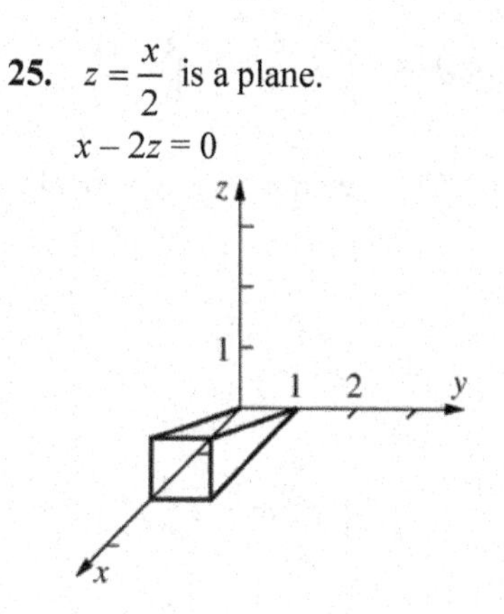

27. $z = x^2 + y^2$ is a paraboloid opening upward with z-axis.

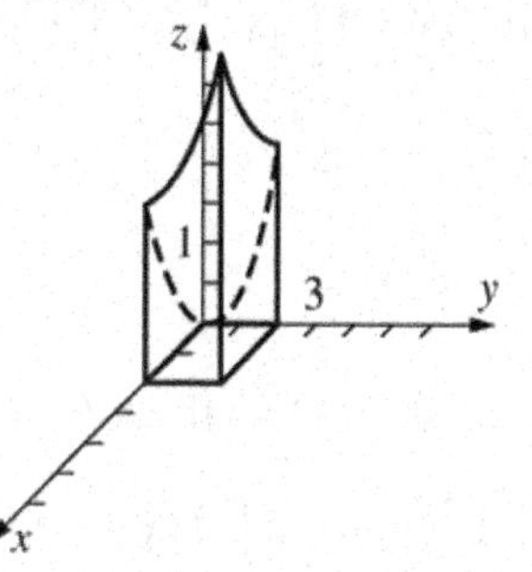

29. $\int_1^3 \int_0^1 (x + y + 1)dx\, dy = \int_1^3 \left[\left(\frac{1}{2}\right)x^2 + yx + x\right]_{x=0}^1 dy$

$= \int_1^3 \left(y + \frac{3}{2}\right) dy = 7$

31. $x^2 + y^2 + 2 > 1$

$\int_{-1}^1 \int_0^1 [(x^2 + y^2 + 2) - 1] dy\, dx$

$= \int_{-1}^1 \left[x^2 y + \left(\frac{1}{3}\right)y^3 + y\right]_{y=0}^1 dx$

$= \int_{-1}^1 \left(x^2 + \frac{4}{3}\right) dx = \frac{10}{3}$

33. $\int_a^b \int_c^d g(x)h(y)dy\, dx = \int_a^b g(x) \int_c^d h(y)dy\, dx$

$= \int_c^d h(y)\, dy \int_a^b g(x)dx$

(First step used linearity of integration with respect to y; second step used linearity of integration with respect to x; now commute.)

35. $\int_0^1\int_0^1 xye^{x^2}e^{y^2}\,dy\,dx = \left(\int_0^1 xe^{x^2}\,dx\right)\left(\int_0^1 ye^{y^2}\,dy\right)$

$= \left(\int_0^1 xe^{x^2}\,dx\right)^2$ (Changed the dummy variable y to the dummy variable x.)

$= \left(\left[\frac{e^{x^2}}{2}\right]_0^1\right)^2 = \left(\frac{e-1}{2}\right)^2 \approx 0.7381$

37. $\int_{-2}^2 x^2 dx \int_{-1}^1 |y^3|\,dy = \left(2\int_0^2 x^2 dx\right)\left(2\int_0^1 y^3 dy\right)$

$= 2\left(\frac{8}{3}\right)2\left(\frac{1}{4}\right) = \frac{8}{3}$

39. $\int_{-2}^2 [\![x^2]\!]\,dx \int_{-1}^1 |y^3|\,dy = 2\int_0^2 [\![x^2]\!]\,dx\, 2\int_0^1 y^3 dy$

$= 2\left[\int_0^1 0\,dx + \int_1^{\sqrt{2}} 1\,dx + \int_{\sqrt{2}}^{\sqrt{3}} 2\,dx + \int_{\sqrt{3}}^2 3\,dx\right]\left[2\left(\frac{1}{4}\right)\right] = 2\left[0 + (\sqrt{2}-1) + 2(\sqrt{3}-\sqrt{2}) + 3(2-\sqrt{3})\right]\left[\frac{1}{2}\right]$

$= 5 - \sqrt{3} - \sqrt{2} \approx 1.8537$

41. $0 \le \int_a^b\int_a^b [f(x)g(y) - f(y)g(x)]^2\,dx\,dy = \int_a^b\int_a^b [f^2(x)g^2(y) - 2f(x)g(x)f(y)g(y) + f^2(y)g^2(x)]\,dx\,dy$

$= \int_a^b f^2(x)dx \int_a^b g^2(y)dy - 2\int_a^b f(x)g(x)dx \int_a^b f(y)g(y)dy + \int_a^b f^2(y)dy \int_a^b g^2(x)dx$

$= 2\int_a^b f^2(x)dx \int_a^b g^2(x)dx - 2\left[\int_a^b f(x)g(x)dx\right]^2$

Therefore, $\left[\int_a^b f(x)g(x)dx\right]^2 \le \int_a^b f^2(x)dx \int_a^b g^2(x)dx.$

13.3 Concepts Review

1. A rectangle containing S; 0

3. $\int_a^b \int_{\phi_1(x)}^{\phi_2(x)} f(x, y)\,dy\,dx$

Problem Set 13.3

1. $\int_0^1 [x^2 y]_{y=0}^{3x}\,dx = \int_0^1 3x^3\,dx = \frac{3}{4}$

3. $\int_{-1}^3 \left[\frac{x^3}{3} + y^2 x\right]_{x=0}^{3y} dy = \int_{-1}^3 (9y^3 + 3y^3)dy$

$= [3y^4]_{-1}^3 = 243 - 3 = 240$

5. $\int_1^3 \left[\left(\frac{1}{2}\right)x^2 \exp(y^3)\right]_{x=-y}^{2y} dy = \int_1^3 \left(\frac{3}{2}\right)y^2 \exp(y^3)dy$

$= \left(\frac{1}{2}\right)(e^{27} - e) \approx 2.660 \times 10^{11}$

7. $\int_{1/2}^1 [y\cos(\pi x^2)]_{y=0}^{2x}\,dx = \int_{1/2}^1 2x\cos(\pi x^2)dx$

$= -\frac{\sqrt{2}}{2\pi} \approx -0.2251$

9. $\int_0^{\pi/9} [\tan\theta]_{\theta=\pi/4}^{3r}\,dr = \int_0^{\pi/9} (\tan 3r - 1)dr$

$= \left[-\frac{\ln|\cos 3r|}{3} - r\right]_0^{\pi/9}$

$= \left(\frac{-\ln\left(\frac{1}{2}\right)}{3} - \frac{\pi}{9}\right) - \left(-\frac{\ln(1)}{3} - 0\right)$

$= \frac{3\ln 2 - \pi}{9} \approx -0.1180$

11. $\int_0^{\pi/2} [e^x \cos y]_{x=0}^{\sin y}\,dy = \int_0^{\pi/2} (e^{\sin y}\cos y - \cos y)dy$

$= e - 2 \approx 0.7183$

13. $\int_0^2 \left[xy + \left(\frac{1}{2}\right)y^2\right]_{y=0}^{\sqrt{4-x^2}} dx$

$= \int_0^2 \left[x(4-x^2)^{1/2} + 2 - \left(\frac{1}{2}\right)x^2\right] dx = \frac{16}{3}$

15. $\int_{-1}^{1} \int_{x^2}^{1} xy \, dy \, dx = 0$

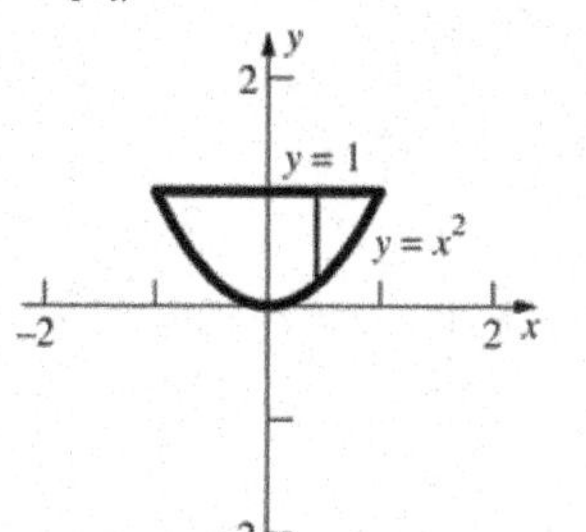

17.

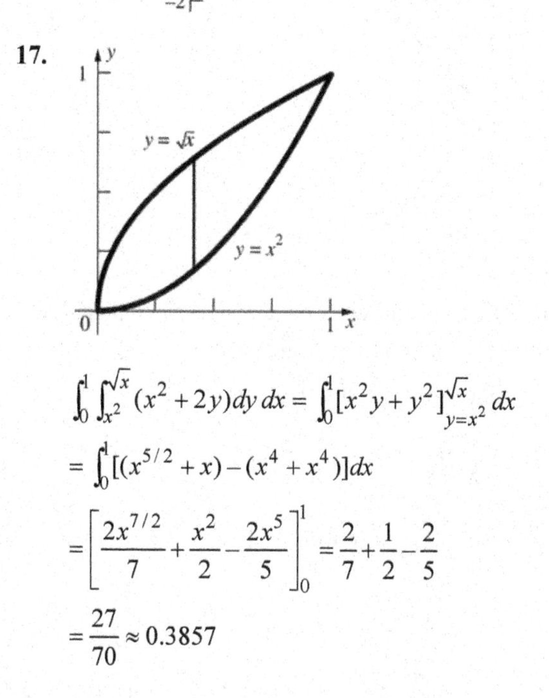

$\int_0^1 \int_{x^2}^{\sqrt{x}} (x^2 + 2y) dy \, dx = \int_0^1 [x^2 y + y^2]_{y=x^2}^{\sqrt{x}} dx$

$= \int_0^1 [(x^{5/2} + x) - (x^4 + x^4)] dx$

$= \left[\frac{2x^{7/2}}{7} + \frac{x^2}{2} - \frac{2x^5}{5}\right]_0^1 = \frac{2}{7} + \frac{1}{2} - \frac{2}{5}$

$= \frac{27}{70} \approx 0.3857$

19. $\int_0^2 \int_x^2 2(1+x^2)^{-1} dy \, dx = 4 \tan^{-1} 2 - \ln 5 \approx 2.8192$

21.

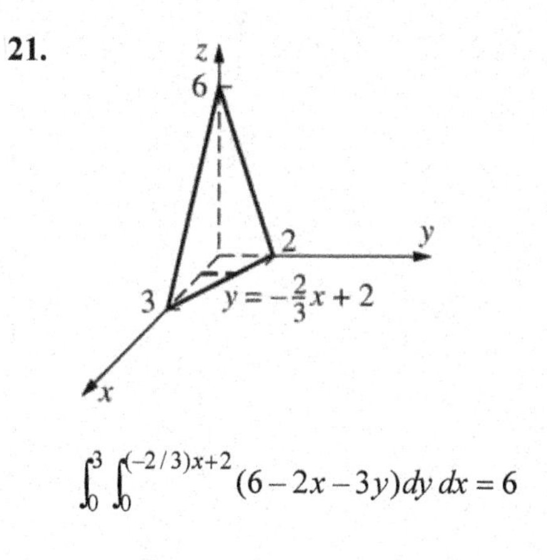

$\int_0^3 \int_0^{(-2/3)x+2} (6 - 2x - 3y) dy \, dx = 6$

23.

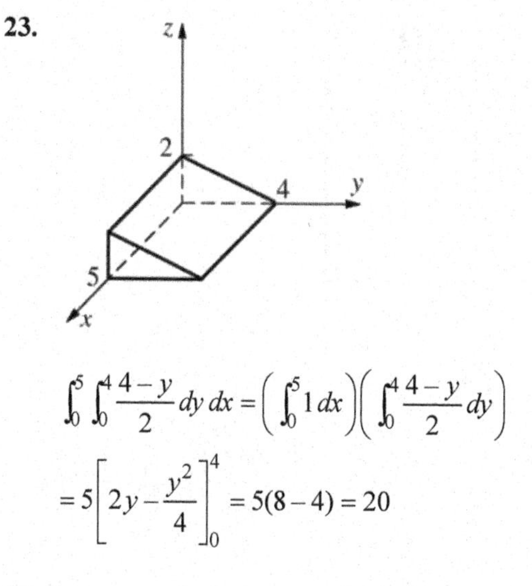

$\int_0^5 \int_0^4 \frac{4-y}{2} dy \, dx = \left(\int_0^5 1 \, dx\right)\left(\int_0^4 \frac{4-y}{2} dy\right)$

$= 5\left[2y - \frac{y^2}{4}\right]_0^4 = 5(8-4) = 20$

25.

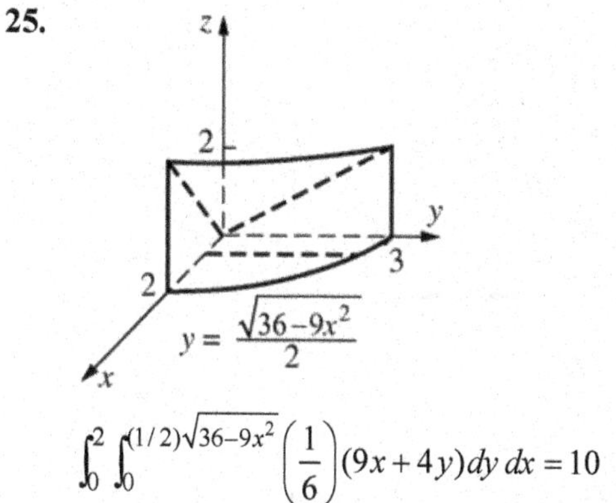

$\int_0^2 \int_0^{(1/2)\sqrt{36-9x^2}} \left(\frac{1}{6}\right)(9x + 4y) dy \, dx = 10$

27.

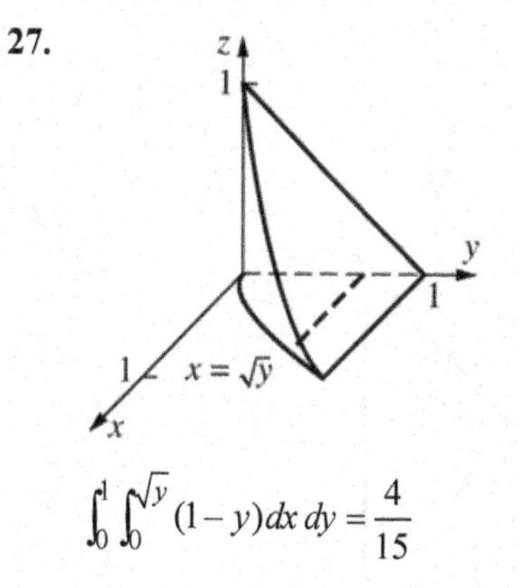

$\int_0^1 \int_0^{\sqrt{y}} (1-y) dx \, dy = \frac{4}{15}$

29.

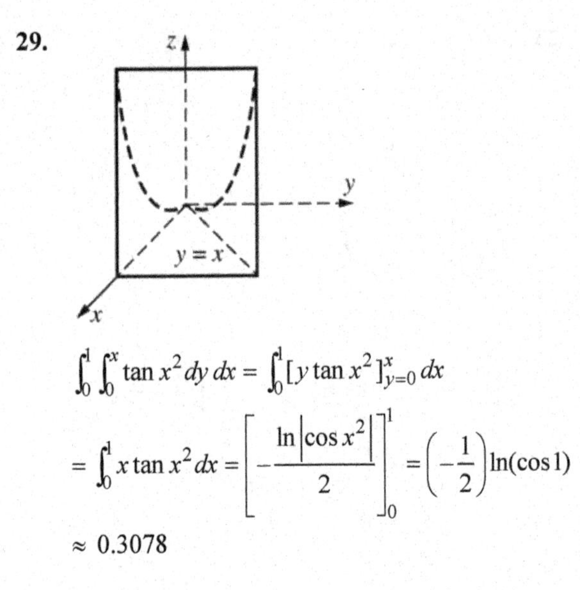

$$\int_0^1 \int_0^x \tan x^2\, dy\, dx = \int_0^1 [y \tan x^2]_{y=0}^{x}\, dx$$

$$= \int_0^1 x \tan x^2\, dx = \left[-\frac{\ln\left|\cos x^2\right|}{2} \right]_0^1 = \left(-\frac{1}{2}\right)\ln(\cos 1)$$

≈ 0.3078

31.

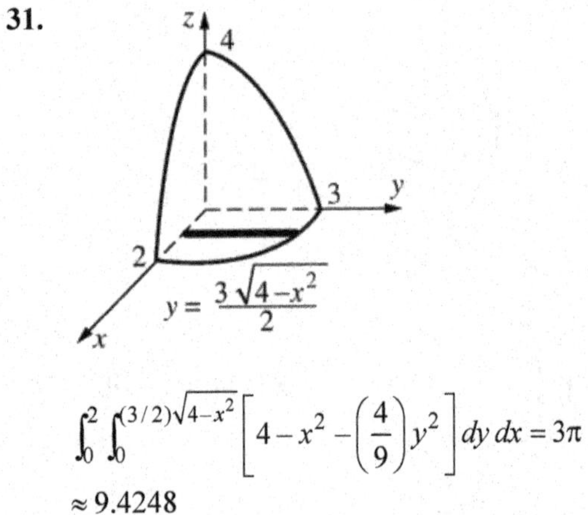

$$\int_0^2 \int_0^{(3/2)\sqrt{4-x^2}} \left[4 - x^2 - \left(\frac{4}{9}\right) y^2 \right] dy\, dx = 3\pi$$

≈ 9.4248

33. $\int_0^1 \int_y^1 f(x, y)\, dx\, dy$

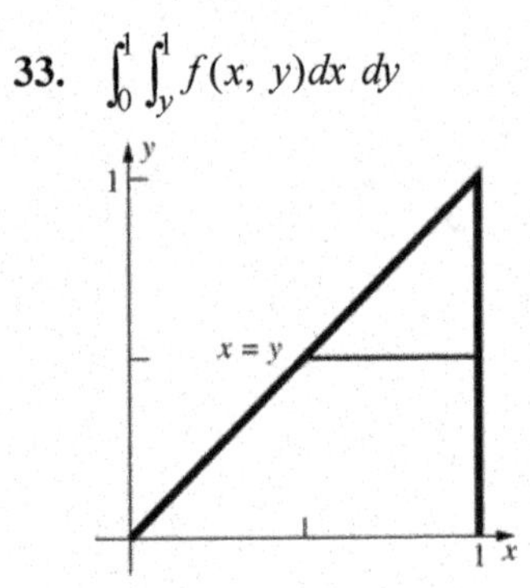

35. $\int_0^1 \int_{y^4}^{\sqrt{y}} f(x, y)\, dx\, dy$

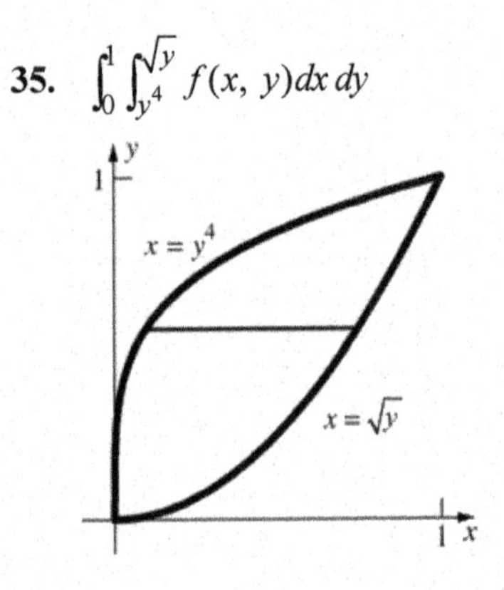

37. $\int_{-1}^0 \int_{-x}^1 f(x, y)\, dy\, dx + \int_0^1 \int_x^1 f(x, y)\, dy\, dx$

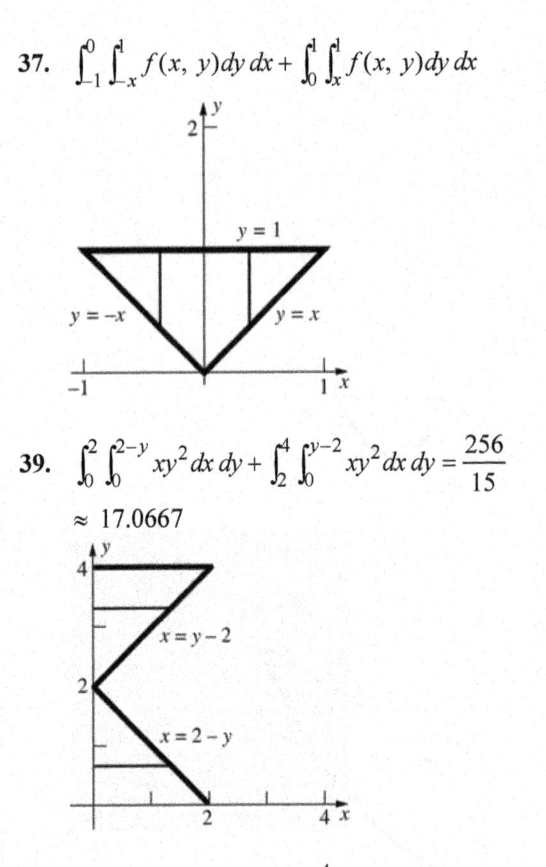

39. $\int_0^2 \int_0^{2-y} xy^2\, dx\, dy + \int_2^4 \int_0^{y-2} xy^2\, dx\, dy = \frac{256}{15}$

≈ 17.0667

41. The integral over S of $x^4 y$ is 0 since this is an odd function of y. Therefore,

$$\iint_S (x^2 + x^4 y)\, dA = \iint_S x^2\, dA$$

$$= 4\left(\iint_{S_1} x^2\, dA + \iint_{S_2} x^2\, dA \right)$$

$$= 4\left(\int_0^1 \int_{\sqrt{1-x^2}}^{\sqrt{4-x^2}} x^2\, dy\, dx + \int_1^2 \int_0^{\sqrt{4-x^2}} x^2\, dy\, dx \right)$$

$$= 4\left(\int_0^2 x^2 \sqrt{4 - x^2}\, dx - \int_0^1 x^2 \sqrt{1 - x^2}\, dx \right)$$

$$= 4\left(16 \int_0^{\pi/2} \sin^2\theta \cos^2\theta\, d\theta - \int_0^{\pi/2} \sin^2\phi \cos^2\phi\, d\phi \right)$$

(using $x = 2\sin\theta$ in 1st integral; $x = \sin\phi$ in 2nd)

$$= 60 \int_0^{\pi/2} \sin^2\theta \cos^2\theta\, d\theta = \frac{15\pi}{4}$$

≈ 11.7810

43.

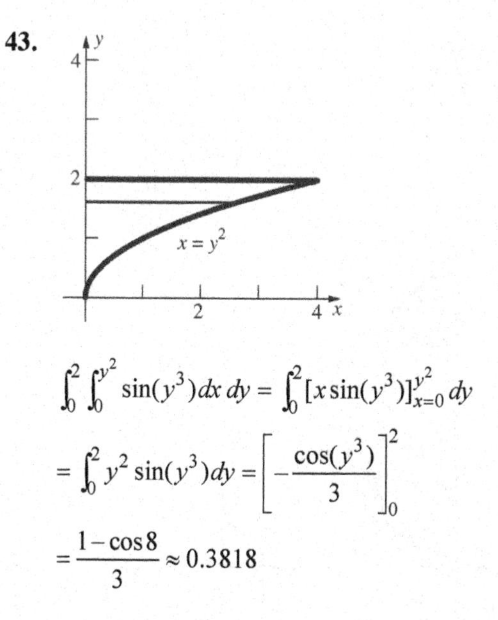

$$\int_0^2 \int_0^{y^2} \sin(y^3)\,dx\,dy = \int_0^2 [x\sin(y^3)]_{x=0}^{y^2} dy$$

$$= \int_0^2 y^2 \sin(y^3)dy = \left[-\frac{\cos(y^3)}{3}\right]_0^2$$

$$= \frac{1-\cos 8}{3} \approx 0.3818$$

45. We first slice the river into eleven 100' sections parallel to the bridge. We will assume that the cross-section of the river is roughly the shape of an isosceles triangle and that the cross-sectional area is uniform across a slice. We can then approximate the volume of the water by

$$V \approx \sum_{k=1}^{11} A_k(y_k)\Delta y = \sum_{k=1}^{11} \frac{1}{2}(w_k)(d_k)100$$

$$= 50\sum_{k=1}^{11}(w_k)(d_k)$$

where w_k is the width across the river at the left side of the kth slice, and d_k is the center depth of the river at the left side of the kth slice. This gives

$$V \approx 50[300\cdot 40 + 300\cdot 39 + 300\cdot 35 + 300\cdot 31$$
$$+290\cdot 28 + 275\cdot 26 + 250\cdot 25 + 225\cdot 24$$
$$+205\cdot 23 + 200\cdot 21 + 175\cdot 19]$$
$$= 4{,}133{,}000 \text{ ft}^3$$

13.4 Concepts Review

1. $a \le r \le b; \alpha \le \theta \le \beta$

3. $\int_0^{\pi}\int_0^2 r^3\,dr\,d\theta$

Problem Set 13.4

1. $$\int_0^{\pi/2}\left[\left(\frac{1}{3}\right)r^3 \sin\theta\right]_{r=0}^{\cos\theta} d\theta$$

$$= \int_0^{\pi/2}\left(\frac{1}{3}\right)\cos^3\theta \sin\theta\,d\theta$$

$$= \frac{1}{12} \approx 0.0833$$

3. $$\int_0^{\pi}\left[\frac{r^3}{3}\right]_{r=0}^{\sin\theta} d\theta = \int_0^{\pi}\frac{\sin^3\theta}{3}d\theta$$

$$= \int_0^{\pi}\frac{(1-\cos^2\theta)\sin\theta}{3}d\theta$$

$$= \left[\frac{-\cos\theta}{3} + \frac{\cos^3\theta}{9}\right]_0^{\pi}$$

$$= \left(\frac{1}{3}-\frac{1}{9}\right) - \left(-\frac{1}{3}+\frac{1}{9}\right) = \frac{4}{9}$$

5. $$\int_0^{\pi}\left[\frac{1}{2}r^2\cos\frac{\theta}{4}\right]_0^2 d\theta = \int_0^{\pi} 2\cos\frac{\theta}{4}d\theta$$

$$= \left[8\sin\frac{\theta}{4}\right]_0^{\pi} = 4\sqrt{2}$$

7.

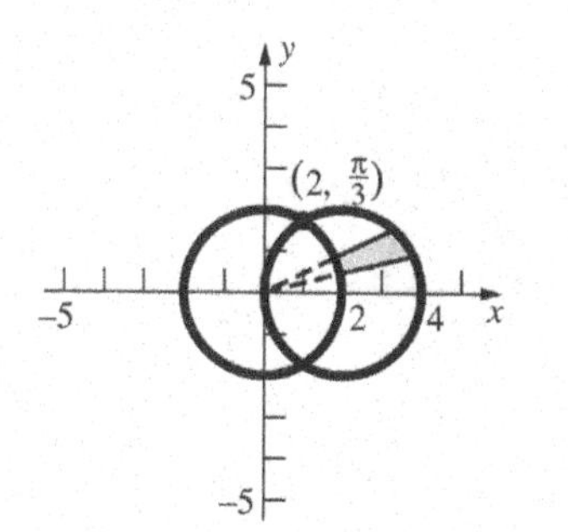

$$2\int_0^{\pi/3}\int_2^{4\cos\theta} r\,dr\,d\theta = 2\left[\frac{2\pi}{3}+\sqrt{3}\right] \approx 7.6529$$

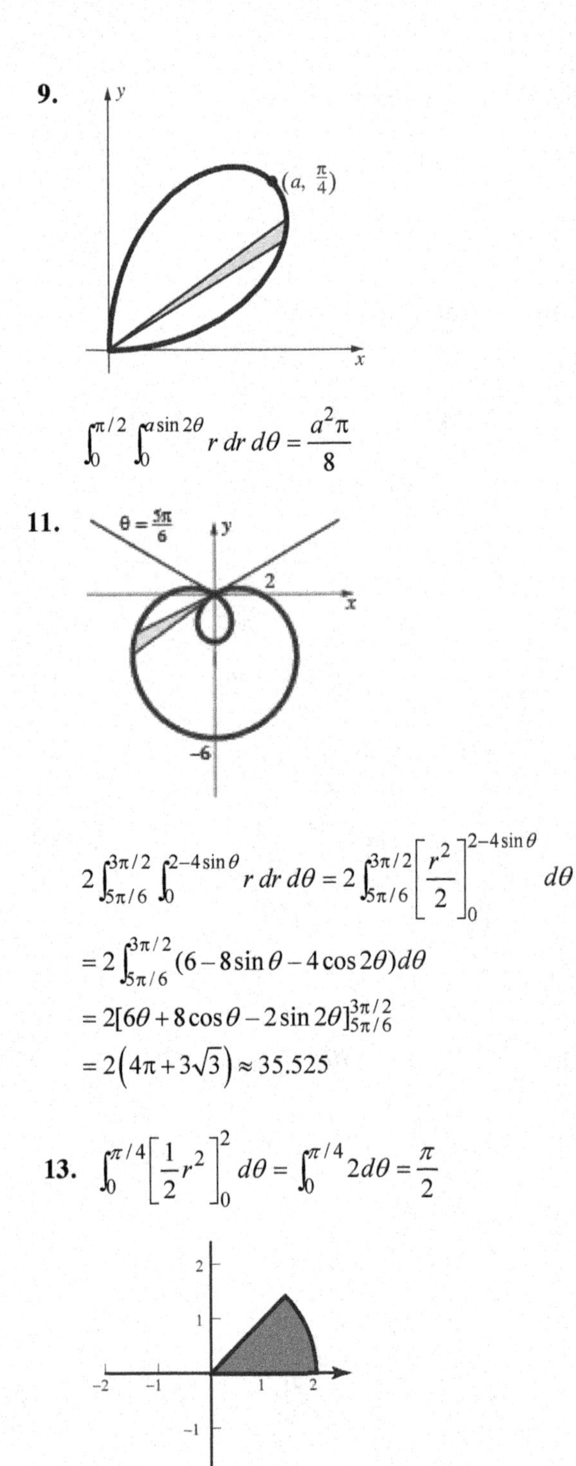

9.

$$\int_0^{\pi/2}\int_0^{a\sin 2\theta} r\,dr\,d\theta = \frac{a^2\pi}{8}$$

11.

$$2\int_{5\pi/6}^{3\pi/2}\int_0^{2-4\sin\theta} r\,dr\,d\theta = 2\int_{5\pi/6}^{3\pi/2}\left[\frac{r^2}{2}\right]_0^{2-4\sin\theta} d\theta$$

$$= 2\int_{5\pi/6}^{3\pi/2}(6-8\sin\theta-4\cos 2\theta)d\theta$$

$$= 2[6\theta+8\cos\theta-2\sin 2\theta]_{5\pi/6}^{3\pi/2}$$

$$= 2\left(4\pi+3\sqrt{3}\right)\approx 35.525$$

13. $\int_0^{\pi/4}\left[\frac{1}{2}r^2\right]_0^2 d\theta = \int_0^{\pi/4} 2d\theta = \frac{\pi}{2}$

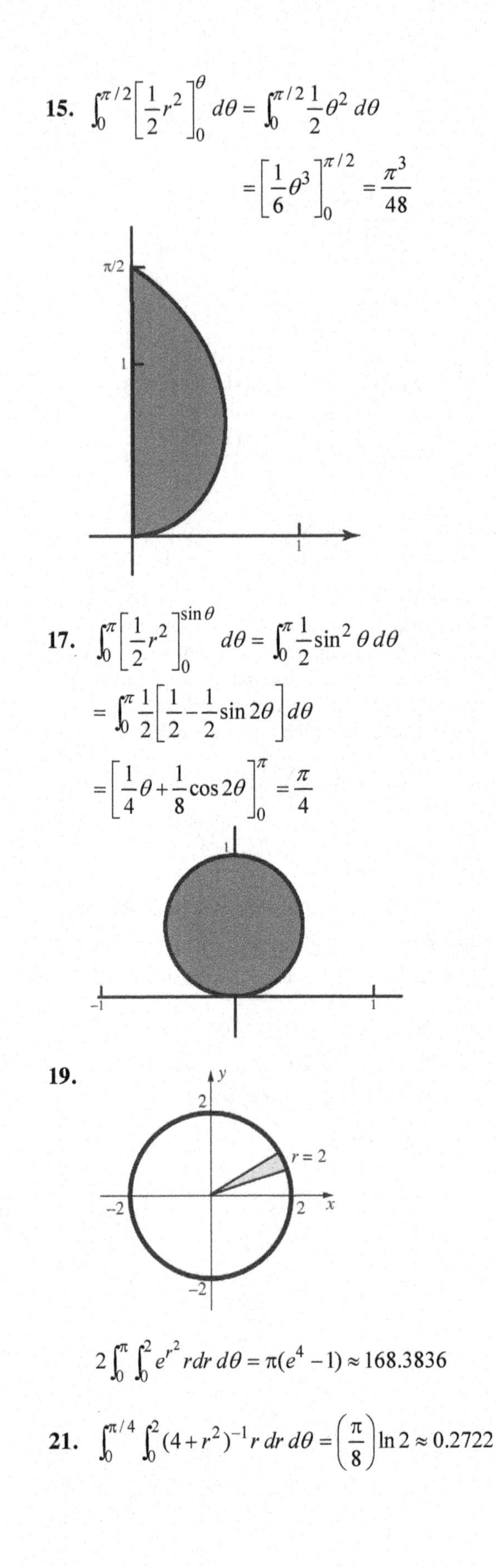

15. $\int_0^{\pi/2}\left[\frac{1}{2}r^2\right]_0^{\theta} d\theta = \int_0^{\pi/2}\frac{1}{2}\theta^2\,d\theta$

$$= \left[\frac{1}{6}\theta^3\right]_0^{\pi/2} = \frac{\pi^3}{48}$$

17. $\int_0^{\pi}\left[\frac{1}{2}r^2\right]_0^{\sin\theta} d\theta = \int_0^{\pi}\frac{1}{2}\sin^2\theta\,d\theta$

$$= \int_0^{\pi}\frac{1}{2}\left[\frac{1}{2}-\frac{1}{2}\sin 2\theta\right]d\theta$$

$$= \left[\frac{1}{4}\theta+\frac{1}{8}\cos 2\theta\right]_0^{\pi} = \frac{\pi}{4}$$

19.

$$2\int_0^{\pi}\int_0^{2} e^{r^2}r\,dr\,d\theta = \pi(e^4-1)\approx 168.3836$$

21. $\int_0^{\pi/4}\int_0^{2}(4+r^2)^{-1}r\,dr\,d\theta = \left(\frac{\pi}{8}\right)\ln 2 \approx 0.2722$

23.

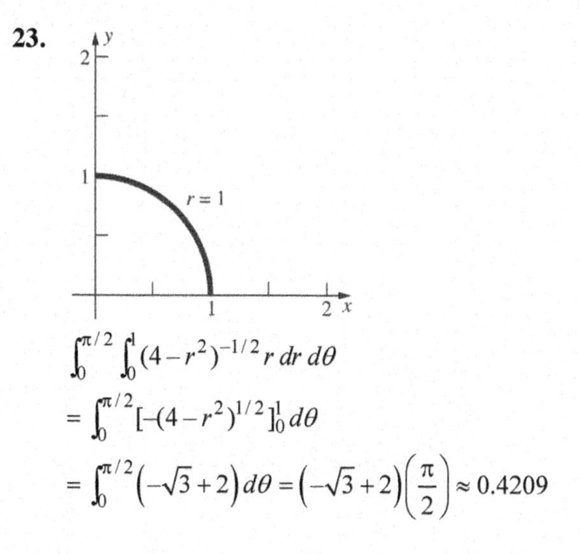

$$\int_0^{\pi/2}\int_0^1 (4-r^2)^{-1/2} r\,dr\,d\theta$$
$$= \int_0^{\pi/2}[-(4-r^2)^{1/2}]_0^1\,d\theta$$
$$= \int_0^{\pi/2}\left(-\sqrt{3}+2\right)d\theta = \left(-\sqrt{3}+2\right)\left(\frac{\pi}{2}\right) \approx 0.4209$$

25. $\int_{\pi/4}^{\pi/2}\int_0^{\csc\theta} r^2\cos^2\theta\, r\,dr\,d\theta = \frac{1}{12} \approx 0.0833$

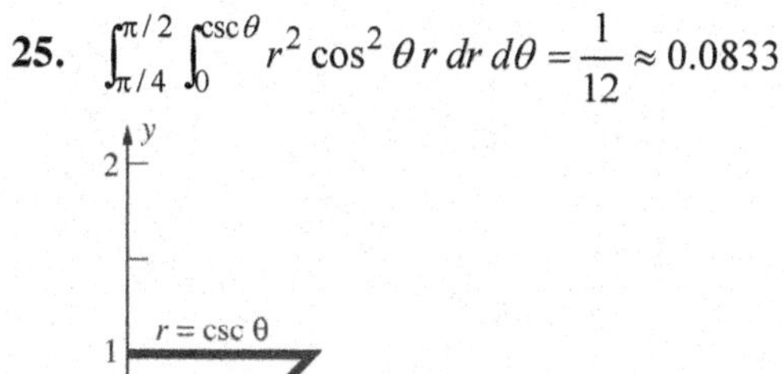

27.

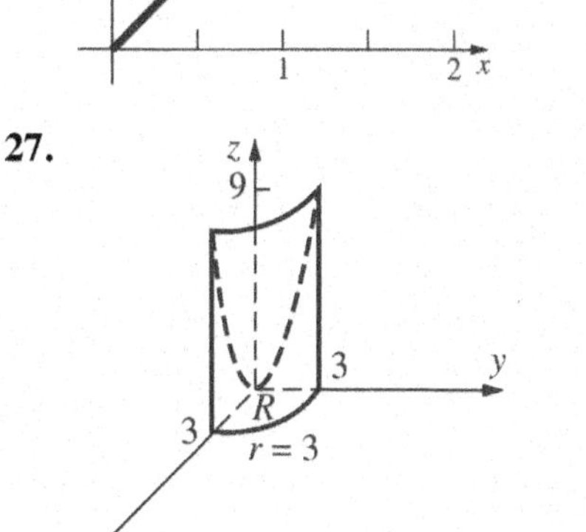

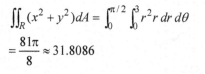

$$\iint_R (x^2+y^2)\,dA = \int_0^{\pi/2}\int_0^3 r^2 r\,dr\,d\theta$$
$$= \frac{81\pi}{8} \approx 31.8086$$

29.

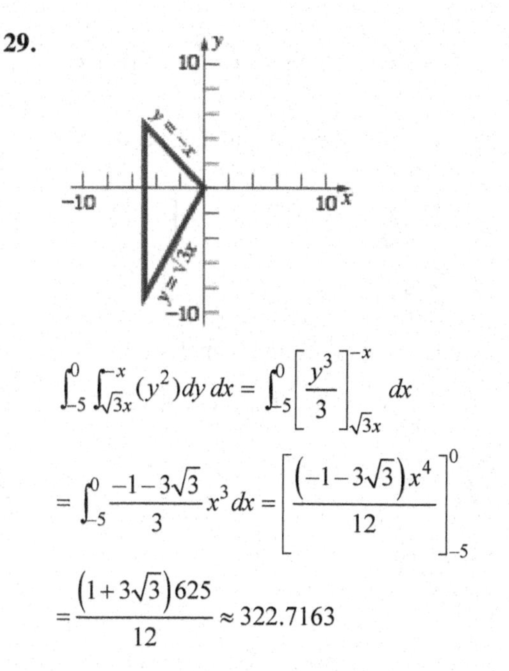

$$\int_{-5}^0\int_{\sqrt{3}x}^{-x}(y^2)\,dy\,dx = \int_{-5}^0\left[\frac{y^3}{3}\right]_{\sqrt{3}x}^{-x}dx$$
$$= \int_{-5}^0 \frac{-1-3\sqrt{3}}{3}x^3\,dx = \left[\frac{\left(-1-3\sqrt{3}\right)x^4}{12}\right]_{-5}^0$$
$$= \frac{\left(1+3\sqrt{3}\right)625}{12} \approx 322.7163$$

31. This can be done by the methods of this section, but an easier way to do it is to realize that the intersection is the union of two congruent segments (of one base) of the spheres, so (see Problem 20, Section 5.2, with $d = h$ and $a = r$) the volume is $2\left[\left(\frac{1}{3}\right)\pi d^2(3a-d)\right] = 2\pi d^2\frac{(3a-d)}{3}$.

33.

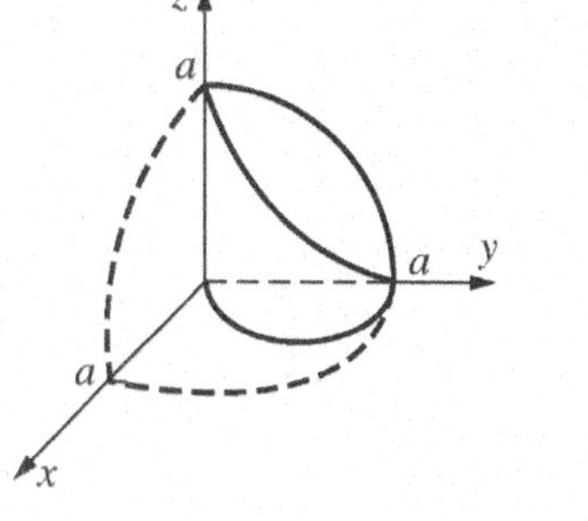

$$\text{Volume} = 4\int_0^{\pi/2}\int_0^{a\sin\theta}\sqrt{a^2-r^2}\,r\,dr\,d\theta$$
$$= \int_0^{\pi/2}\left[\left(-\frac{1}{3}\right)(a^3\cos^3\theta - a^3)\right]d\theta$$
$$= \left(-\frac{4}{3}\right)a^3\left[\frac{2}{3}-\frac{\pi}{2}\right] = \left(\frac{2}{9}\right)a^3(3\pi-4)$$

35. Choose a coordinate system so the center of the sphere is the origin and the axis of the part removed is the z-axis.
Volume (Ring) = Volume (Sphere of radius a) – Volume (Part removed)

$$= \frac{4}{3}\pi a^3 - 2\int_0^{2\pi}\int_0^{\sqrt{a^2-b^2}}\sqrt{a^2-r^2}\,r\,dr\,d\theta = \frac{4}{3}\pi a^3 - 2(2\pi)\int_0^{\sqrt{a^2-b^2}}(a^2-r^2)^{1/2}r\,dr$$

$$= \frac{4}{3}\pi a^3 + 4\pi\left[\frac{1}{3}(a^2-r^2)^{3/2}\right]_0^{\sqrt{a^2-b^2}} = \frac{4}{3}\pi a^3 + 4\pi\frac{1}{3}(b^3-a^3) = \frac{4}{3}\pi b^3$$

37. $\int_0^{\pi/2}\left[\lim_{b\to\infty}\int_0^b(1+r^2)^{-2}r\,dr\right]d\theta = \int_0^{\pi/2}\left(\lim_{b\to\infty}\left[\left(-\frac{1}{2}\right)(1+b^2)^{-1}-\left(-\frac{1}{2}\right)\right]\right)d\theta = \int_0^{\pi/2}\left(\frac{1}{2}\right)d\theta = \frac{\pi}{4} \approx 0.7854$

39. Using the substitution $u = \frac{x-\mu}{\sigma\sqrt{2}}$ we get

$du = \frac{dx}{\sigma\sqrt{2}}$. Our integral then becomes

$$\int_{-\infty}^{\infty}\frac{1}{\sigma\sqrt{2\pi}}e^{-(x-\mu)^2/2\sigma^2}dx = \int_{-\infty}^{\infty}\frac{1}{\sqrt{\pi}}e^{-u^2}du$$

$$= \frac{2}{\sqrt{\pi}}\int_0^{\infty}e^{-u^2}du$$

Using the result from Example 4, we see that

$\int_0^{\infty}e^{-u^2}du = \frac{\sqrt{\pi}}{2}$. Thus we have

$$\int_{-\infty}^{\infty}\frac{1}{\sigma\sqrt{2\pi}}e^{-(x-\mu)^2/2\sigma^2}dx$$

$$= \frac{2}{\sqrt{\pi}}\cdot\frac{\sqrt{\pi}}{2} = 1.$$

13.5 Concepts Review

1. $\iint_S x^2y^4\,dA$

3. $\iint_S x^4y^4\,dA$

Problem Set 13.5

1.

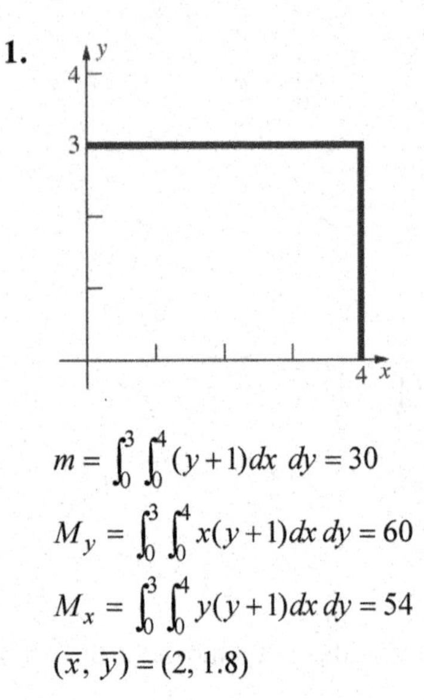

$$m = \int_0^3\int_0^4 (y+1)dx\,dy = 30$$

$$M_y = \int_0^3\int_0^4 x(y+1)dx\,dy = 60$$

$$M_x = \int_0^3\int_0^4 y(y+1)dx\,dy = 54$$

$(\bar{x}, \bar{y}) = (2, 1.8)$

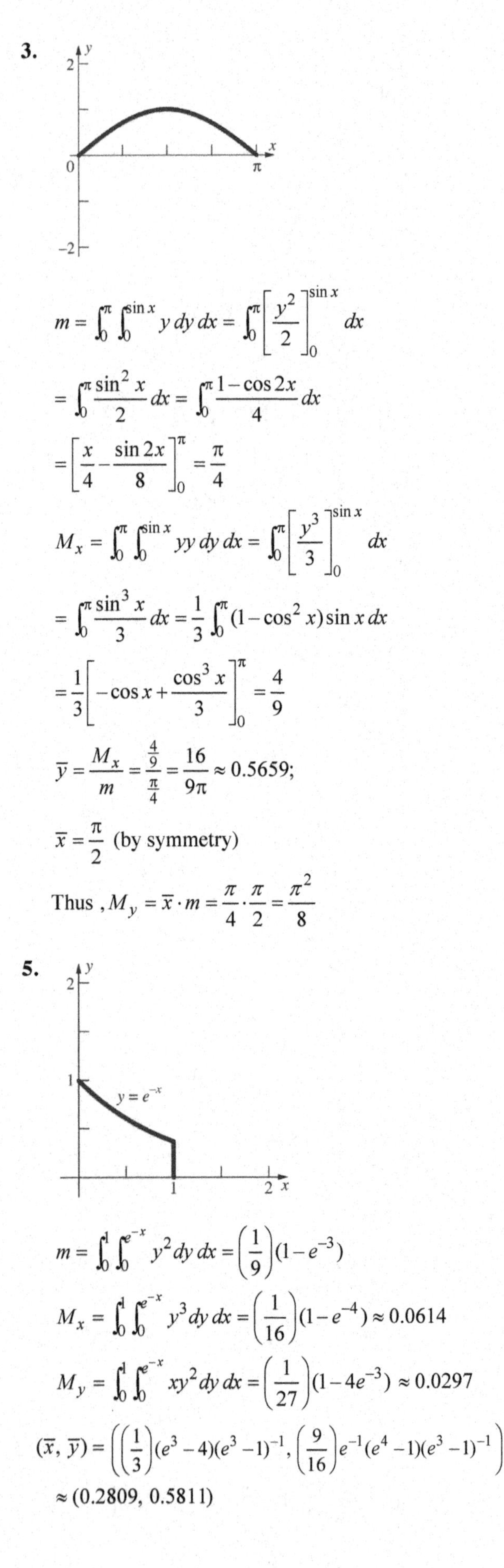

3.

$$m = \int_0^{\pi}\int_0^{\sin x} y\,dy\,dx = \int_0^{\pi}\left[\frac{y^2}{2}\right]_0^{\sin x}dx$$

$$= \int_0^{\pi}\frac{\sin^2 x}{2}dx = \int_0^{\pi}\frac{1-\cos 2x}{4}dx$$

$$= \left[\frac{x}{4} - \frac{\sin 2x}{8}\right]_0^{\pi} = \frac{\pi}{4}$$

$$M_x = \int_0^{\pi}\int_0^{\sin x} yy\,dy\,dx = \int_0^{\pi}\left[\frac{y^3}{3}\right]_0^{\sin x}dx$$

$$= \int_0^{\pi}\frac{\sin^3 x}{3}dx = \frac{1}{3}\int_0^{\pi}(1-\cos^2 x)\sin x\,dx$$

$$= \frac{1}{3}\left[-\cos x + \frac{\cos^3 x}{3}\right]_0^{\pi} = \frac{4}{9}$$

$$\bar{y} = \frac{M_x}{m} = \frac{\frac{4}{9}}{\frac{\pi}{4}} = \frac{16}{9\pi} \approx 0.5659;$$

$\bar{x} = \frac{\pi}{2}$ (by symmetry)

Thus, $M_y = \bar{x}\cdot m = \frac{\pi}{4}\cdot\frac{\pi}{2} = \frac{\pi^2}{8}$

5.

$$m = \int_0^1\int_0^{e^{-x}} y^2\,dy\,dx = \left(\frac{1}{9}\right)(1-e^{-3})$$

$$M_x = \int_0^1\int_0^{e^{-x}} y^3\,dy\,dx = \left(\frac{1}{16}\right)(1-e^{-4}) \approx 0.0614$$

$$M_y = \int_0^1\int_0^{e^{-x}} xy^2\,dy\,dx = \left(\frac{1}{27}\right)(1-4e^{-3}) \approx 0.0297$$

$$(\bar{x}, \bar{y}) = \left(\left(\frac{1}{3}\right)(e^3-4)(e^3-1)^{-1}, \left(\frac{9}{16}\right)e^{-1}(e^4-1)(e^3-1)^{-1}\right)$$

$\approx (0.2809, 0.5811)$

7.

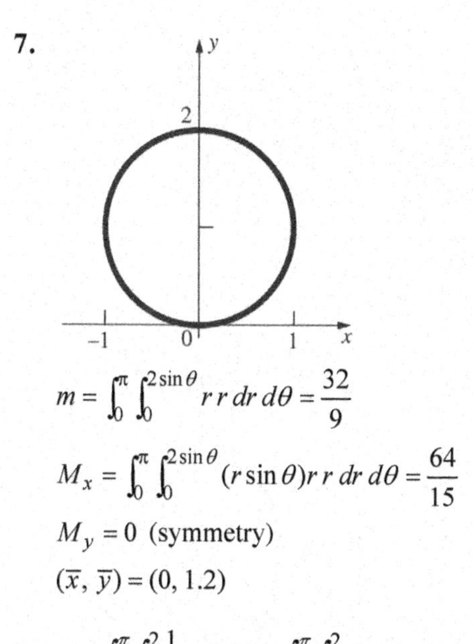

$m = \int_0^{\pi} \int_0^{2\sin\theta} r\, r\, dr\, d\theta = \frac{32}{9}$

$M_x = \int_0^{\pi} \int_0^{2\sin\theta} (r\sin\theta) r\, r\, dr\, d\theta = \frac{64}{15}$

$M_y = 0$ (symmetry)

$(\bar{x}, \bar{y}) = (0, 1.2)$

9. $m = \int_0^{\pi} \int_1^{2} \frac{1}{r} r\, dr d\theta = \int_0^{\pi} \int_1^{2} dr d\theta = \pi$

$M_x = \int_0^{\pi} \int_1^{2} \frac{1}{r} r\cos\theta\, r\, dr d\theta = \int_0^{\pi} \frac{3}{2}\cos\theta d\theta = 3$

$M_y = 0$ by symmetry

$(\bar{x}, \bar{y}) = \left(0, \frac{3}{\pi}\right)$

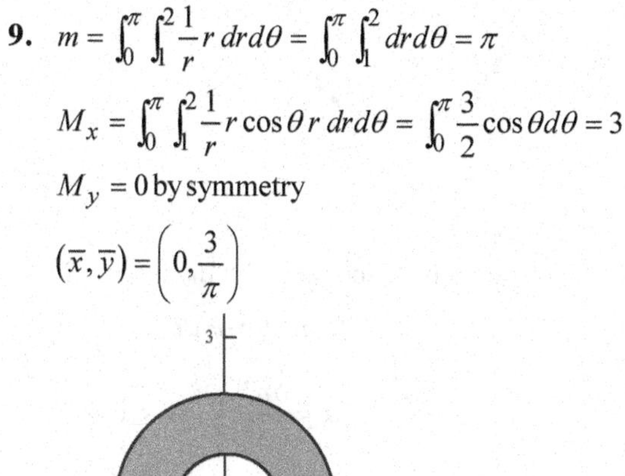

11.

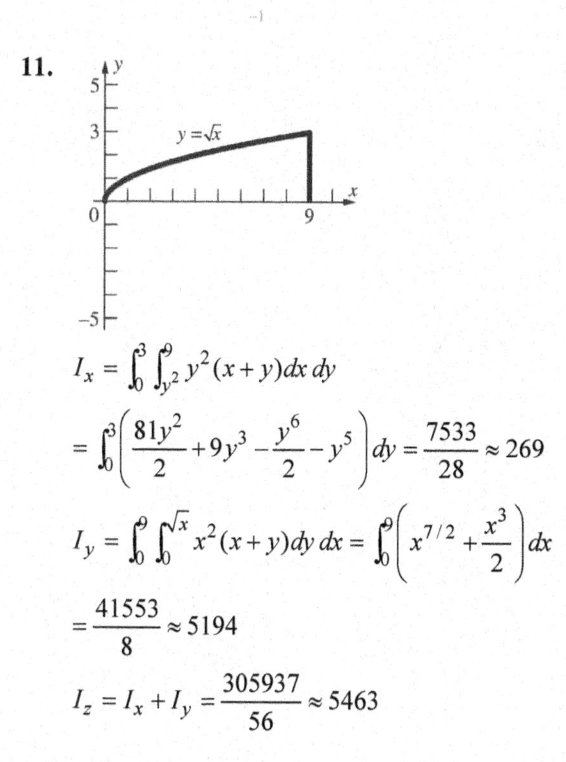

$I_x = \int_0^3 \int_{y^2}^9 y^2(x+y) dx\, dy$

$= \int_0^3 \left(\frac{81y^2}{2} + 9y^3 - \frac{y^6}{2} - y^5 \right) dy = \frac{7533}{28} \approx 269$

$I_y = \int_0^9 \int_0^{\sqrt{x}} x^2(x+y) dy\, dx = \int_0^9 \left(x^{7/2} + \frac{x^3}{2} \right) dx$

$= \frac{41553}{8} \approx 5194$

$I_z = I_x + I_y = \frac{305937}{56} \approx 5463$

13.

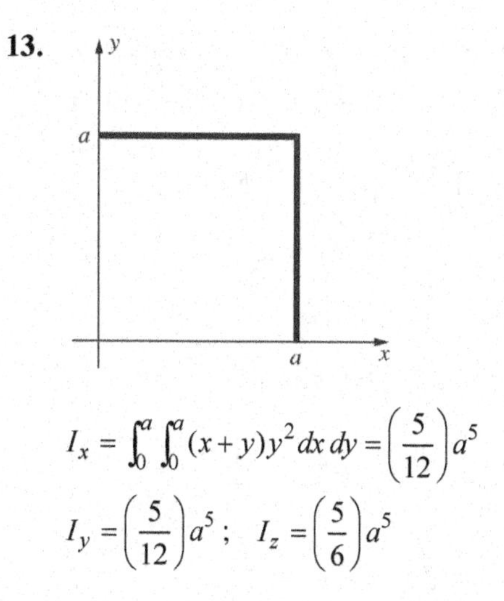

$I_x = \int_0^a \int_0^a (x+y)y^2 dx\, dy = \left(\frac{5}{12}\right) a^5$

$I_y = \left(\frac{5}{12}\right) a^5; \quad I_z = \left(\frac{5}{6}\right) a^5$

15. The density is constant , $\delta(x, y) = k$.

$m = \int_0^2 kx\, dx = \left[\frac{k}{2}x^2\right]_0^2 = 2k$

$M_x = \int_0^2 \int_0^x ky\, dy\, dx = \int_0^2 \left[\frac{k}{2}y^2\right]_0^x dx$

$= \int_0^2 \frac{k}{2}x^2\, dx = \left[\frac{k}{6}x^3\right]_0^2 = \frac{4k}{3}$

$M_y = \int_0^2 \int_0^x kx\, dy\, dx = \int_0^2 [kxy]_0^x\, dx$

$= \int_0^2 kx^2\, dx = \left[\frac{k}{3}x^3\right]_0^2 = \frac{8k}{3}$

$(\bar{x}, \bar{y}) = \left(\frac{8k/3}{2k}, \frac{4k/3}{2k}\right) = \left(\frac{4}{3}, \frac{2}{3}\right)$

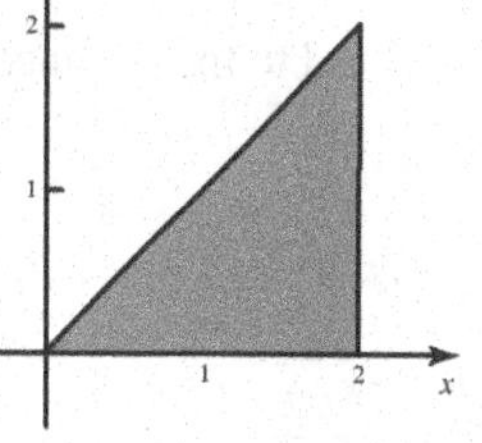

17. The density is proportional to the squared distance from the origin, $\delta(x, y) = k(x^2 + y^2)$.

$$m = \int_{-3}^{3}\int_{0}^{9-x^2} k(x^2 + y^2)\,dy\,dx$$

$$= \int_{-3}^{3} k\left[x^2y + \frac{1}{3}y^3\right]_0^{9-x^2} dx$$

$$= \int_{-3}^{3} k\left[246 - 72x^2 + 8x^4 - \frac{1}{3}x^6\right] dx$$

$$= k\left[246x - 24x^3 + \frac{8}{5}x^5 - \frac{1}{21}x^7\right]_{-3}^{3} = \frac{25596k}{35}$$

$$M_x = \int_{-3}^{3}\int_{0}^{9-x^2} ky(x^2 + y^2)\,dy\,dx$$

$$= \int_{-3}^{3} k\left[\frac{1}{2}x^2y^2 + \frac{1}{4}y^4\right]_0^{9-x^2} dx$$

$$= \int_{-3}^{3} k\left[\frac{6561}{4} - \frac{1377x^2}{2} + \frac{225}{2}x^4 - \frac{17x^6}{2} + \frac{x^8}{4}\right] dx$$

$$= \frac{29160k}{7}$$

$M_y = 0$ by symmetry

$$(\overline{x}, \overline{y}) = \left(0, \frac{29160k/7}{25596k/35}\right) = \left(0, \frac{450}{79}\right)$$

19. The density is proportional to the distance from the origin, $\delta(r, \theta) = k \cdot r$.

$$m = \int_0^{\pi}\int_1^3 kr^2\,dr\,d\theta = \int_0^{\pi}\left[\frac{k}{3}r^3\right]_1^3 d\theta$$

$$= \int_0^{\pi}\frac{26}{3}k\,d\theta = \frac{26k\pi}{3}$$

$$M_x = \int_0^{\pi}\int_1^3 kr^2\,r\sin\theta\,dr\,d\theta$$

$$= \int_0^{\pi}\left[\frac{k}{4}r^4\sin\theta\right]_1^3 d\theta = \int_0^{\pi} 20k\sin\theta\,d\theta$$

$$= [-20k\cos\theta]_0^{\pi} = 40k$$

$M_y = 0$ by symmetry

$$(\overline{x}, \overline{y}) = \left(0, \frac{40k}{26k\pi/3}\right) = \left(0, \frac{60}{13\pi}\right)$$

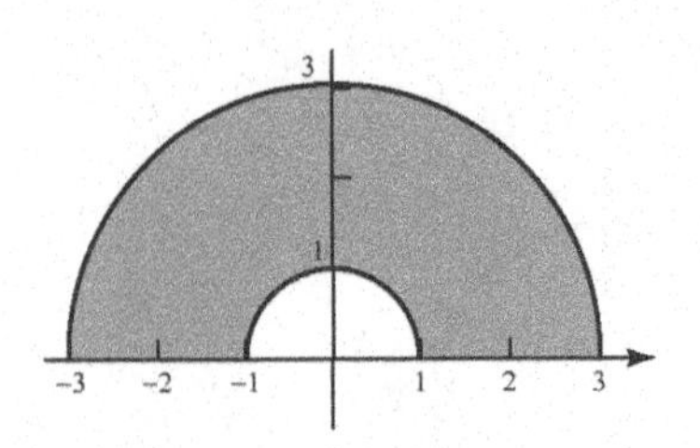

21. $m = \int_0^a\int_0^a (x + y)\,dx\,dy = a^3$

$$\overline{r} = \left(\frac{I_x}{m}\right)^{1/2} = \left(\frac{5}{12}\right)^{1/2} a \approx 0.6455a$$

23.

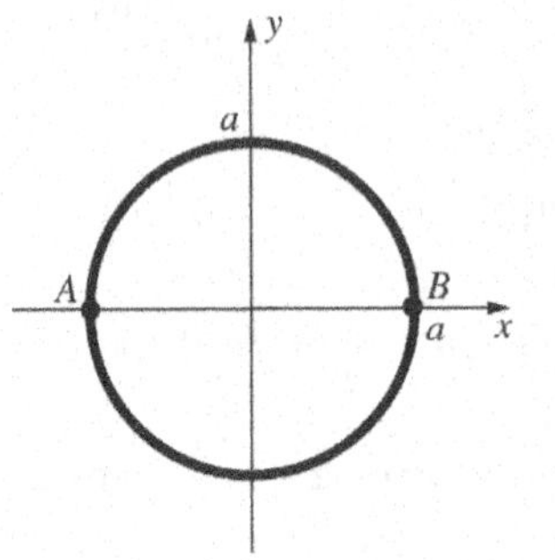

$m = \delta\pi a^2$

The moment of inertia about diameter *AB* is

$$I = I_x = \int_0^{2\pi}\int_0^a \delta r^2\sin^2\theta\,r\,dr\,d\theta$$

$$= \int_0^{2\pi}\frac{\delta a^4\sin^2\theta}{4}d\theta = \frac{\delta a^4}{8}\int_0^{2\pi}(1 - \cos 2\theta)d\theta$$

$$= \frac{\delta a^4}{8}\left[\theta - \frac{\sin 2\theta}{2}\right]_0^{2\pi} = \frac{\delta a^4\pi}{4}$$

$$\overline{r} = \left(\frac{I}{m}\right)^{1/2} = \left(\frac{\frac{\delta a^4\pi}{4}}{\delta\pi a^2}\right)^{1/2} = \frac{a}{2}$$

25.

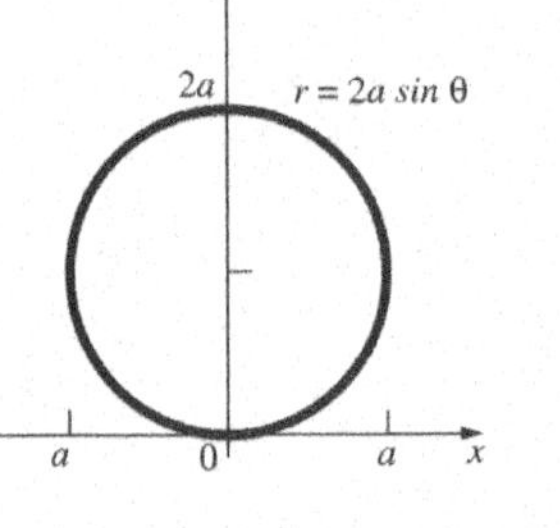

$$I_x = \iint_S \delta y^2\,dA$$

$$= 2\delta\int_0^{\pi/2}\int_0^{2a\sin\theta}(r\sin\theta)^2 r\,dr\,d\theta$$

$$= 2\delta\int_0^{\pi/2} 4a^4\sin^6\theta\,d\theta$$

$$= 8a^4\delta\frac{(1)(3)(5)}{(2)(4)(6)}\frac{\pi}{2} = \frac{5a^4\delta\pi}{4}$$

27.

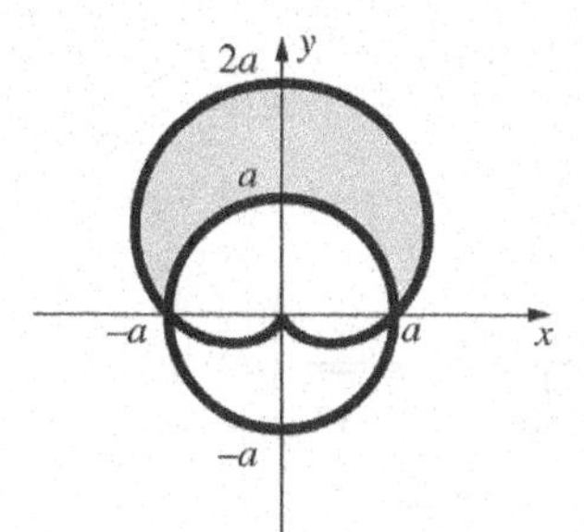

$\bar{x} = 0$ (by symmetry)

$$M_x = \iint_S ky\,dA = 2k\int_{-\pi/2}^{\pi/2}\int_0^{a(1+\sin\theta)}(r\sin\theta)r\,dr\,d\theta = 2k\int_0^{\pi/2}\left(\frac{a^3}{3}\right)(3\sin^2\theta + 3\sin^3\theta + \sin^4\theta)d\theta$$

$$= \left(\frac{2}{3}\right)ka^3\left[\frac{(15\pi+32)}{16}\right] = \left(\frac{1}{24}\right)ka^3(15\pi+32)$$

$$m = \iint_S k\,dA = 2k\int_{-\pi/2}^{\pi/2}\int_0^{a(1+\sin\theta)} r\,dr\,d\theta = 2k\int_0^{\pi/2}\left(\frac{1}{2}\right)a^2(2\sin\theta + \sin^2\theta)d\theta = ka^2\left[\frac{(8+\pi)}{4}\right] = \left(\frac{1}{4}\right)ka^2(\pi+8)$$

Therefore, $\bar{y} = \dfrac{M_x}{m} = \dfrac{\left(\frac{1}{24}\right)ka^3(15\pi+32)}{\left(\frac{1}{4}\right)ka^2(\pi+8)} = \dfrac{a(15\pi+32)}{6(\pi+8)} \approx 1.1836a$

29. a.

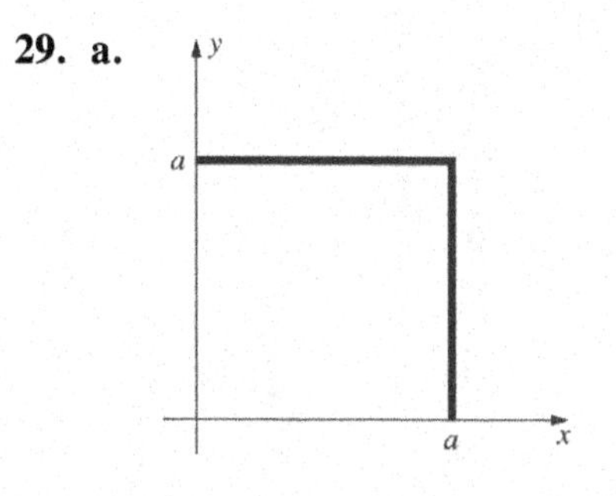

$$m = \iint_S (x+y)dA = \int_0^a\int_0^a (x+y)dx\,dy$$

$$= \int_0^a\left(\left[\frac{x^2}{2} + xy\right]_{x=0}^a\right)dy = \int_0^a\left(\frac{a^2}{2} + ay\right)dy$$

$$= \left[\frac{a^2y}{2} + \frac{ay^2}{2}\right]_0^a = a^3$$

b. $M_y = \iint_S x(x+y)dA = \int_0^a\int_0^a (x^2+xy)dy\,dx$

$$= \int_0^a\left(\left[x^2y + \frac{xy^2}{2}\right]_{y=0}^a\right)dx = \int_0^a\left(ax^2 + \frac{a^2x}{2}\right)dx$$

$$= \left[\frac{ax^3}{3} + \frac{a^2x^2}{4}\right]_0^a = \frac{7a^4}{12}$$

Therefore, $\bar{x} = \dfrac{M_y}{m} = \dfrac{7a}{12}$.

c.

$I_y = I_L + d^2m$, so $\dfrac{5a^5}{12} = I_L + \left(\dfrac{7a}{12}\right)^2(a^3)$;

$I_L = \dfrac{11a^5}{144}$

31. $I_x = 2[I_{23}] = \dfrac{ka^4\pi}{2}$; $I_y = 2[I_{23} + md^2]$

$= 2[0.25a^4\pi + (k\pi a^2)(2a)^2] = 8.5ka^4\pi$

$I_z = I_x + I_y = 9ka^4\pi$

33. $M_y = \iint_{S_1 \cup S_2} x\delta(x, y)dA$

$= \iint_{S_1} x\delta(x, y)dA + \iint_{S_2} x\delta(x, y)dA$

$= \dfrac{m_1 \iint_{S_1} x\delta(x, y)dA}{m_1} + \dfrac{m_2 \iint_{S_2} x\delta(x, y)dA}{m_2}$

$= m_1\bar{x}_1 + m_2\bar{x}_2$

Thus, $\bar{x} = \dfrac{M_y}{m} = \dfrac{m_1 x_1 + m_2 x_2}{m_1 + m_2}$ which is equal to what we are to obtain and which is what we would obtain using the center of mass formula for two point masses. (Similar result can be obtained for $\bar{y}$.)

35.

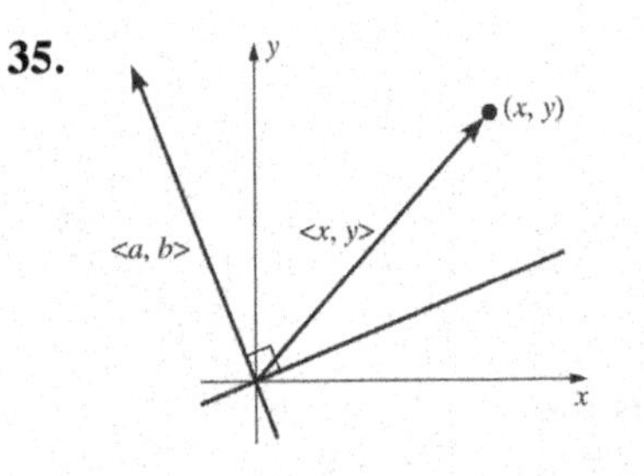

$\langle a,b\rangle$ is perpendicular to the line $ax + by = 0$. Therefore, the (signed) distance of (x, y) to L is the scalar projection of $\langle x,y\rangle$ onto $\langle a,b\rangle$, which is $d(x, y) = \dfrac{\langle x, y\rangle \cdot \langle a, b\rangle}{|\langle a, b\rangle|} = \dfrac{ax + by}{|\langle a, b\rangle|}$.

$M_L = \iint_S d(x, y)\delta(x, y)dA$

$= \iint_S \dfrac{ax + by}{|\langle a, b\rangle|}\delta(x, y)dA$

$= \dfrac{a}{|\langle a, b\rangle|}\iint_S x\delta(x, y)dA + \dfrac{b}{|\langle a, b\rangle|}\iint_S y\delta(x, y)dA$

$= \dfrac{a}{|\langle a, b\rangle|}(0) + \dfrac{b}{|\langle a, b\rangle|}(0) = 0$

[since $(\bar{x}, \bar{y}) = (0, 0)$]

13.6 Concepts Review

1. $\|\mathbf{u} \times \mathbf{v}\|$

3. $\displaystyle\int_{-a}^{a} \int_{-\sqrt{a^2 - x^2}}^{\sqrt{a^2 - x^2}} \left(\frac{a}{\sqrt{a^2 - x^2 - y^2}} \right) dy\, dx$

$= \displaystyle\int_0^{2\pi} \int_0^{a} \left(\frac{ar}{\sqrt{a^2 - r^2}} \right) dr\, d\theta$; $2\pi a^2$

Problem Set 13.6

1.

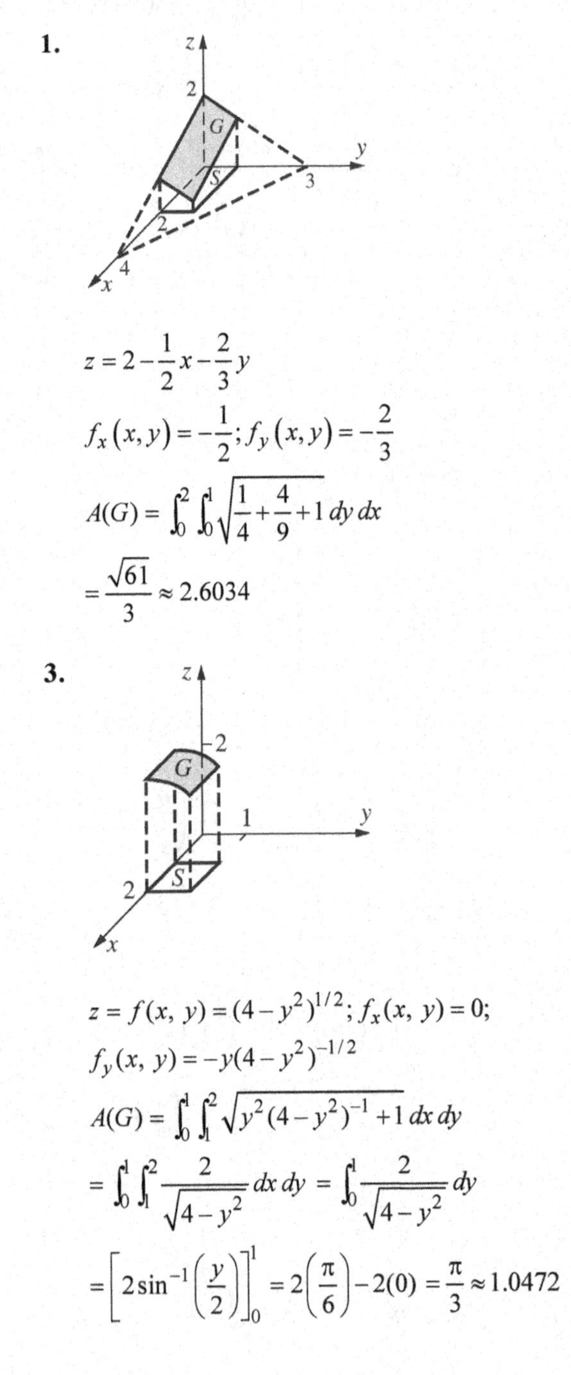

$z = 2 - \dfrac{1}{2}x - \dfrac{2}{3}y$

$f_x(x, y) = -\dfrac{1}{2}$; $f_y(x, y) = -\dfrac{2}{3}$

$A(G) = \displaystyle\int_0^2 \int_0^1 \sqrt{\frac{1}{4} + \frac{4}{9} + 1}\, dy\, dx$

$= \dfrac{\sqrt{61}}{3} \approx 2.6034$

3.

$z = f(x, y) = (4 - y^2)^{1/2}$; $f_x(x, y) = 0$;

$f_y(x, y) = -y(4 - y^2)^{-1/2}$

$A(G) = \displaystyle\int_0^1 \int_1^2 \sqrt{y^2(4 - y^2)^{-1} + 1}\, dx\, dy$

$= \displaystyle\int_0^1 \int_1^2 \frac{2}{\sqrt{4 - y^2}}\, dx\, dy = \int_0^1 \frac{2}{\sqrt{4 - y^2}}\, dy$

$= \left[2\sin^{-1}\left(\dfrac{y}{2}\right) \right]_0^1 = 2\left(\dfrac{\pi}{6}\right) - 2(0) = \dfrac{\pi}{3} \approx 1.0472$

5.

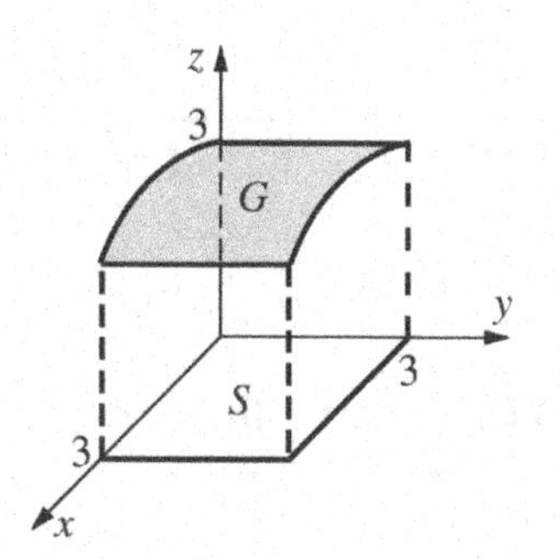

Let $z = f(x, y) = (9 - x^2)^{1/2}$.

$f_x(x, y) = -x(9 - x^2)^{-1/2}$, $f_y(x, y) = 0$

$$A(G) = \int_0^2 \int_0^3 [x^2(9 - x^2)^{-1} + 1] dy\, dx = \int_0^2 \int_0^3 3(9 - x^2)^{-1/2} dy\, dx = 9 \sin^{-1}\left(\frac{2}{3}\right) \approx 6.5675$$

7.

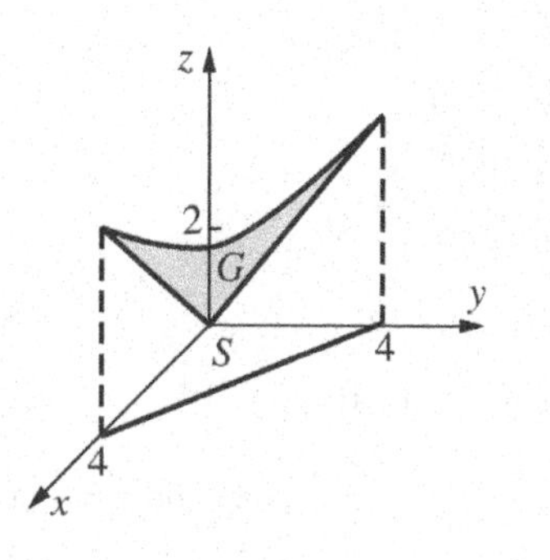

$z = f(x, y) = (x^2 + y^2)^{1/2}$

$f_x(x, y) = x(x^2 + y^2)^{-1/2}$, $f_y(x, y) = y(x^2 + y^2)^{-1/2}$

$$A(G) = \int_0^4 \int_0^{4-x} [x^2(x^2 + y^2)^{-1} + y^2(x^2 + y^2)^{-1} + 1]^{1/2} dy\, dx = \int_0^4 \int_0^{4-x} \sqrt{2} dy\, dx = 8\sqrt{2}$$

9.

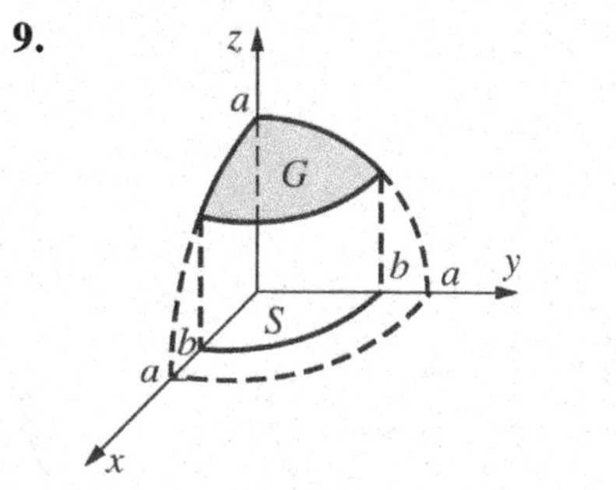

$f_x(x, y) = \frac{x^2}{a^2 - x^2 - y^2}$; $f_y(x, y) = \frac{y^2}{a^2 - x^2 - y^2}$

(See Example 3.)

$$A(G) = 8 \iint_S \frac{a}{\sqrt{a^2 - x^2 - y^2}} dA$$

$$= 8 \int_0^{\pi/2} \int_0^b \frac{a}{\sqrt{a^2 - r^2}} r\, dr\, d\theta$$

$$= 8a\left(\frac{\pi}{2}\right) \int_0^b (a^2 - r^2)^{-1/2} r\, dr$$

$$= -4a\pi\left[(a^2 - r^2)^{1/2}\right]_0^b = 4\pi a\left(a - \sqrt{a^2 - b^2}\right)$$

11.

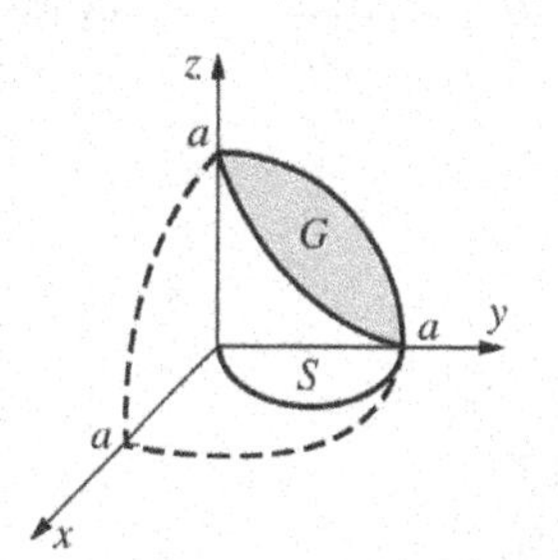

$f_x(x, y) = \frac{x^2}{a^2 - x^2 - y^2}$; $f_y(x, y) = \frac{y^2}{a^2 - x^2 - y^2}$

(See Example 3.)

$$A(G) = 4 \int_0^{\pi/2} \int_0^{a \sin\theta} \frac{a}{\sqrt{a^2 - r^2}} r\, dr\, d\theta$$

$$= 4a^2 \int_0^{\pi/2} (1 - \cos\theta) d\theta = 2a^2(\pi - 2)$$

13.

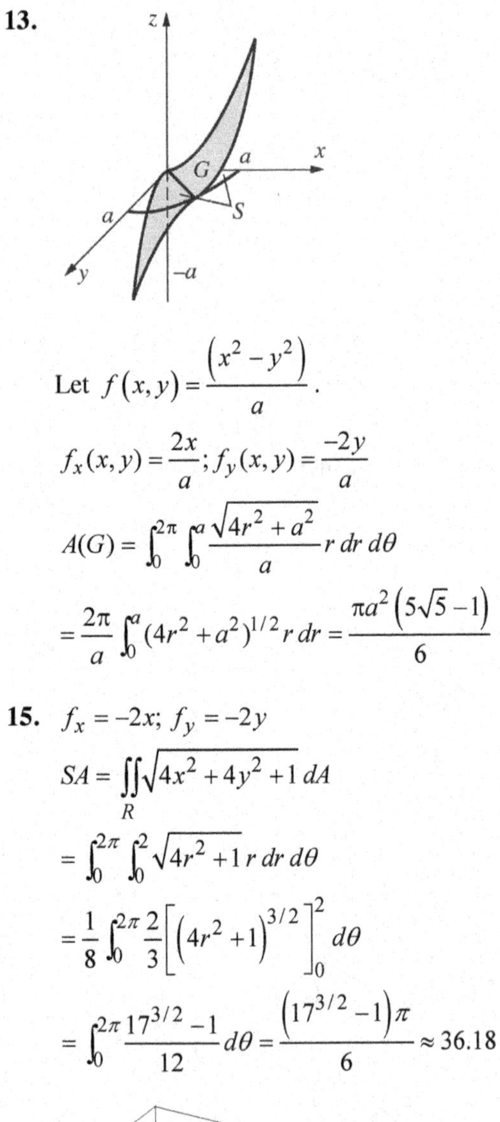

Let $f(x,y) = \frac{(x^2 - y^2)}{a}$.

$$f_x(x,y) = \frac{2x}{a};\ f_y(x,y) = \frac{-2y}{a}$$

$$A(G) = \int_0^{2\pi}\int_0^a \frac{\sqrt{4r^2 + a^2}}{a} r\,dr\,d\theta$$

$$= \frac{2\pi}{a}\int_0^a (4r^2 + a^2)^{1/2} r\,dr = \frac{\pi a^2\left(5\sqrt{5} - 1\right)}{6}$$

15. $f_x = -2x;\ f_y = -2y$

$$SA = \iint_R \sqrt{4x^2 + 4y^2 + 1}\,dA$$

$$= \int_0^{2\pi}\int_0^2 \sqrt{4r^2 + 1}\,r\,dr\,d\theta$$

$$= \frac{1}{8}\int_0^{2\pi} \frac{2}{3}\left[\left(4r^2 + 1\right)^{3/2}\right]_0^2 d\theta$$

$$= \int_0^{2\pi} \frac{17^{3/2} - 1}{12} d\theta = \frac{\left(17^{3/2} - 1\right)\pi}{6} \approx 36.18$$

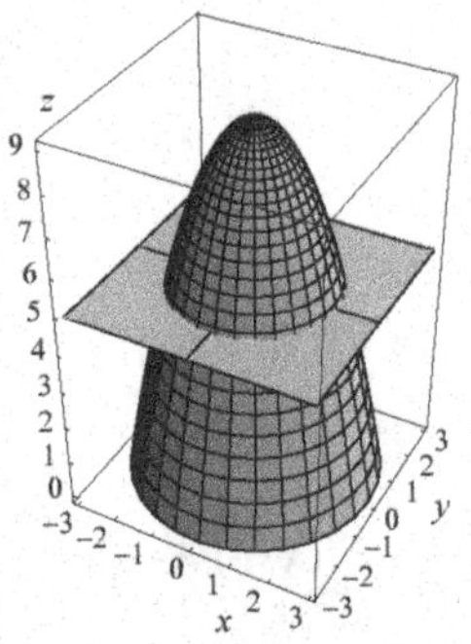

17. $z = f(x,y) = (D - Ax - By)/C$

$$A(G) = \iint_R \sqrt{(A/C)^2 + (B/C)^2 + 1}\,dA$$

$$= \int_0^{D/A}\int_0^{D/B-(A/B)x} \sqrt{(A/C)^2 + (B/C)^2 + 1}\,dy\,dx$$

$$= \sqrt{(A/C)^2 + (B/C)^2 + 1} \times Area(Triangle)$$

$$= \frac{1}{2}\frac{D}{A}\frac{D}{B}\sqrt{(A/C)^2 + (B/C)^2 + 1}$$

$$= \frac{D^2}{2ABC}\sqrt{A^2 + B^2 + C^2}$$

z
D/C
D/B
D/A
y
x

19. $\overline{x} = \overline{y} = 0$ (by symmetry)

Let $h = \frac{h_1 + h_2}{2}$. Planes $z = h_1$ and $z = h$ cut out the same surface area as planes $z = h$ and $z = h_2$. Therefore, $\overline{z} = h$, the arithmetic average of h_1 and h_2.

21. a. $A = \pi b^2$

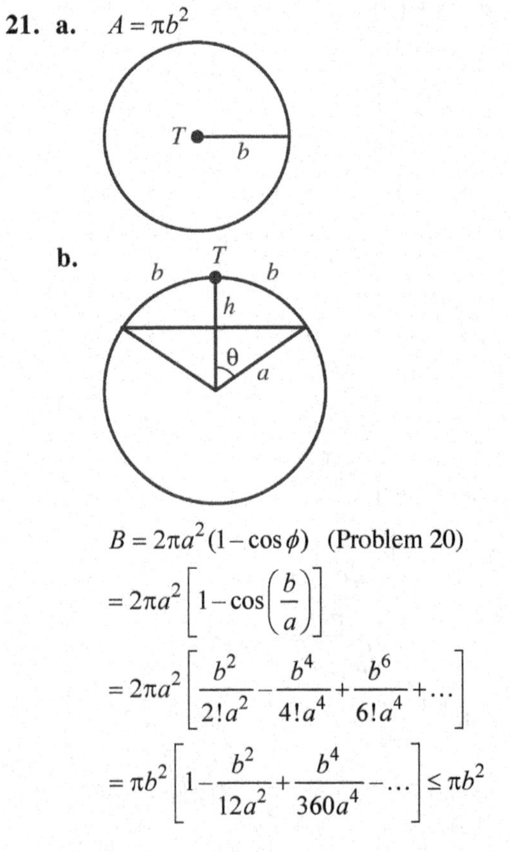

b.

$B = 2\pi a^2(1 - \cos\phi)$ (Problem 20)

$$= 2\pi a^2\left[1 - \cos\left(\frac{b}{a}\right)\right]$$

$$= 2\pi a^2\left[\frac{b^2}{2!a^2} - \frac{b^4}{4!a^4} + \frac{b^6}{6!a^4} + \ldots\right]$$

$$= \pi b^2\left[1 - \frac{b^2}{12a^2} + \frac{b^4}{360a^4} - \ldots\right] \le \pi b^2$$

c.

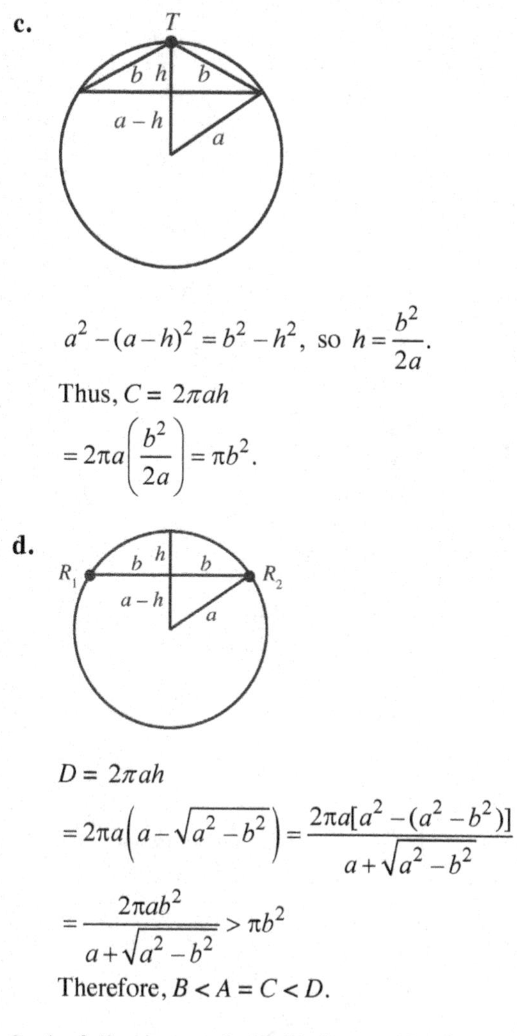

$a^2-(a-h)^2=b^2-h^2$, so $h=\dfrac{b^2}{2a}$.

Thus, $C=2\pi ah$

$=2\pi a\left(\dfrac{b^2}{2a}\right)=\pi b^2$.

d.

$D=2\pi ah$

$=2\pi a\left(a-\sqrt{a^2-b^2}\right)=\dfrac{2\pi a[a^2-(a^2-b^2)]}{a+\sqrt{a^2-b^2}}$

$=\dfrac{2\pi ab^2}{a+\sqrt{a^2-b^2}}>\pi b^2$

Therefore, $B<A=C<D$.

23. In the following, each double integral is over S_{xy}

$A(S_{xy})f(\bar{x},\bar{y})=A(S_{xy})(a\bar{x}+b\bar{y}+c)$

$=\iint dA\left[a\dfrac{\iint x\,dA}{\iint dA}+b\dfrac{\iint y\,dA}{\iint dA}+c\right]$

$=a\iint x\,dA+b\iint y\,dA+c\iint dA$

$=\iint(ax+by+c)dA$

= Volume of solid cylinder under S_{xy}

25. Let G denote the surface of that part of the plane $z=Ax+By+C$ over the region S. First, suppose that S is the rectangle $a\le x\le b,\ c\le y\le d$. Then the vectors **u** and **v** that form the edge of the parallelogram G are $\mathbf{u}=(b-a)\mathbf{i}+0\mathbf{j}+A(b-a)\mathbf{k}$ and $\mathbf{v}=0\mathbf{i}+(d-c)\mathbf{j}+B(d-c)\mathbf{k}$. The surface area of G is thus

$|\mathbf{u}\times\mathbf{v}|=$

$|-A(b-a)(d-c)\mathbf{i}-B(b-a)(d-c)\mathbf{j}+(b-a)(d-c)\mathbf{k}|$

$=(b-a)(d-c)\sqrt{A^2+B^2+1}$

A normal vector to the plane is $\mathbf{n}=-A\mathbf{i}-B\mathbf{j}+\mathbf{k}$. Thus,

$\cos\gamma=\dfrac{\mathbf{n}\cdot\mathbf{k}}{|\mathbf{n}||\mathbf{k}|}=\dfrac{\langle -A,-B,1\rangle\cdot\langle 0,0,1\rangle}{\sqrt{A^2+B^2+1}\cdot 1}$

$=\dfrac{1}{\sqrt{A^2+B^2+1}}$

$\sec\gamma=\dfrac{1}{\cos\gamma}=\sqrt{A^2+B^2+1}=\dfrac{|\mathbf{u}\times\mathbf{v}|}{A(S)}=\dfrac{A(G)}{A(S)}$.

If S is not a rectangle, then make a partition of S with rectangles $R_1,R_2,\ldots,R_n$. The Riemann sum will be $\sum_{m=1}^{n}A(G_m)=\sec\gamma\sum_{m=1}^{n}A(R_m)$.

As we take the limit as $|P|\to 0$ the sum converges to the area of S. Thus the surface area will be

$A(G)=\lim_{|P|\to 0}\sec\gamma\sum_{m=1}^{n}A(R_m)=\sec\gamma A(S)$.

27. a. $f_x=2x, f_y=2y$

$A(G)=\iint_S\sqrt{4x^2+4y^2+1}\,dA$

$=\int_0^{\pi/2}\int_0^3\sqrt{4r^2+1}\,r\,dr\,d\theta$

$=\dfrac{\pi}{2}\left[\dfrac{1}{12}\left(4r^2+1\right)^{3/2}\right]_0^3$

$=\dfrac{\pi}{24}\left(37^{3/2}-1\right)\approx 29.3297$

b. $f_x=2x, f_y=2y$

$A(G)=\int_0^3\int_0^{3-x}\sqrt{4x^2+4y^2+1}\,dy\,dx$

Parabolic rule with $n=10$ gives $SA\approx 15.4233$

29. The surface area of a paraboloid and a hyperbolic paraboloid are the same over identical regions. So, the areas depend on the regions.
$E=F<A=B<C=D$

13.7 Concepts Review

1. volume

3. y; $\sqrt{y}$

Problem Set 13.7

1. $\int_{-3}^{7}\int_{0}^{2x}(x-1-y)\,dy\,dx = \int_{-3}^{7}-2x\,dx = -40$

3. $\int_{1}^{4}\int_{z-1}^{2z}\int_{0}^{y+2z}dx\,dy\,dz = \int_{1}^{4}\int_{z-1}^{2z}(y+2z)\,dy\,dz$

$= \int_{1}^{4}\left[\frac{y^2}{2}+2yz\right]_{y=z-1}^{2z}dz$

$= \int_{1}^{4}\left(\frac{7z^2}{2}+3z-\frac{1}{2}\right)dz$

$= \left[\frac{7z^3}{6}+\frac{3z^2}{2}-\frac{z}{2}\right]_{1}^{4} = \frac{189}{2} = 94.5$

5. $\int_{4}^{24}\int_{0}^{24-x}\left[\frac{1}{x}\left(yz+\frac{1}{2}z^2\right)\right]_{0}^{24-x-y}dy\,dx$

$= \int_{4}^{24}\int_{0}^{24-x}\left[\frac{(x+y-24)(x-y-24)}{2x}\right]dy\,dx$

$= \int_{4}^{24}\left[-\frac{y\left(y^2-3(x-24)^2\right)}{6x}\right]_{0}^{24-x}dx$

$= \int_{4}^{24}-\frac{(x-24)^3}{3x}dx$

$= \left[-576x+12x^2-\frac{x^3}{9}+4608\ln x\right]_{4}^{24} \approx 1927.54$

7. $\int_{0}^{2}\int_{1}^{z}x^2\,dx\,dz = \int_{0}^{2}\left(\frac{1}{3}\right)(z^3-1)\,dz = \frac{2}{3}$

9. $\int_{-2}^{4}\int_{x-1}^{x+1}3y^2\,dy\,dx = \int_{-2}^{4}(6x^2+2)\,dx = 156$

11. $\int_{0}^{1}\int_{0}^{3}\int_{0}^{(12-3x-2y)/6}f(x, y, z)\,dz\,dy\,dx$

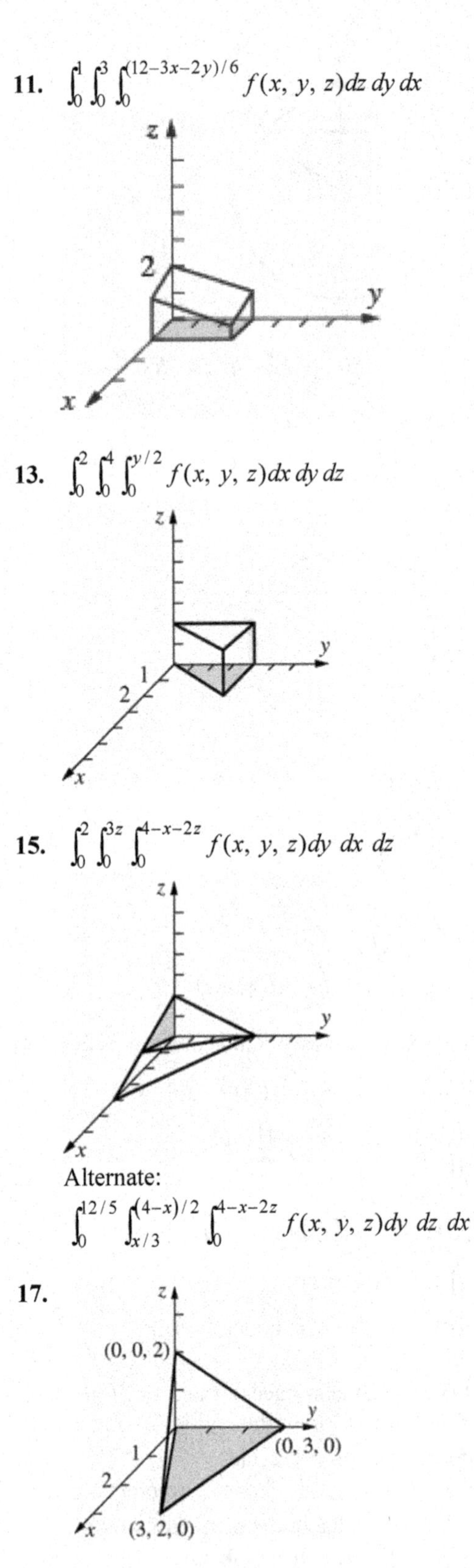

13. $\int_{0}^{2}\int_{0}^{4}\int_{0}^{y/2}f(x, y, z)\,dx\,dy\,dz$

15. $\int_{0}^{2}\int_{0}^{3z}\int_{0}^{4-x-2z}f(x, y, z)\,dy\,dx\,dz$

Alternate:

$\int_{0}^{12/5}\int_{x/3}^{(4-x)/2}\int_{0}^{4-x-2z}f(x, y, z)\,dy\,dz\,dx$

17.

Using the cross product of vectors along edges, it is easy to show that $\langle 2,6,9\rangle$ is normal to the upward face. Then obtain that its equation is $2x + 6y + 9z = 18$.

$\int_{0}^{3}\int_{2x/3}^{(9-x)/3}\int_{0}^{(18-2x-6y)/9}f(x, y, z)\,dz\,dy\,dx$

19. $\int_1^4 \int_0^1 \int_0^{\sqrt{1-y^2}} f(x, y, z)dz\,dy\,dx$

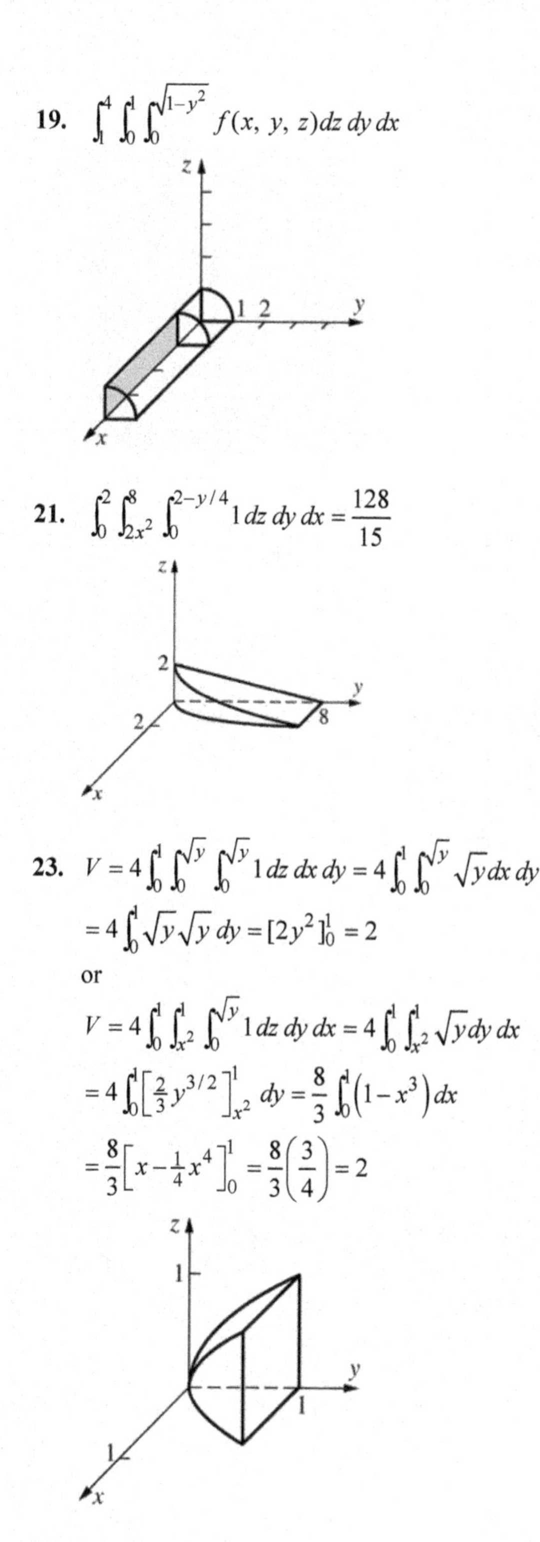

21. $\int_0^2 \int_{2x^2}^8 \int_0^{2-y/4} 1\,dz\,dy\,dx = \frac{128}{15}$

23. $V = 4\int_0^1 \int_0^{\sqrt{y}} \int_0^{\sqrt{y}} 1\,dz\,dx\,dy = 4\int_0^1 \int_0^{\sqrt{y}} \sqrt{y}\,dx\,dy$

$= 4\int_0^1 \sqrt{y}\sqrt{y}\,dy = [2y^2]_0^1 = 2$

or

$V = 4\int_0^1 \int_{x^2}^1 \int_0^{\sqrt{y}} 1\,dz\,dy\,dx = 4\int_0^1 \int_{x^2}^1 \sqrt{y}\,dy\,dx$

$= 4\int_0^1 \left[\frac{2}{3}y^{3/2}\right]_{x^2}^1 dy = \frac{8}{3}\int_0^1 \left(1 - x^3\right)dx$

$= \frac{8}{3}\left[x - \frac{1}{4}x^4\right]_0^1 = \frac{8}{3}\left(\frac{3}{4}\right) = 2$

25. Let $\delta(x, y, z) = x + y + z$. (See note with next problem.)

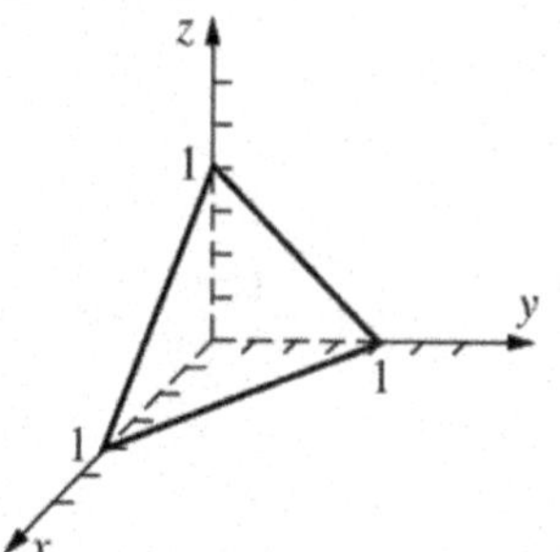

$m = \int_0^1 \int_0^{1-x} \int_0^{1-x-y} (x+y+z)dz\,dy\,dx = \frac{1}{8}$

$M_{yz} = \int_0^1 \int_0^{1-x} \int_0^{1-x-y} x(x+y+z)dz\,dy\,dx = \frac{1}{30}$

$\overline{x} = \frac{4}{15}$; Then $\overline{y} = \overline{z} = \frac{4}{15}$ (symmetry).

27. Let $\delta(x, y, z) = 1$. (See note with previous problem.)

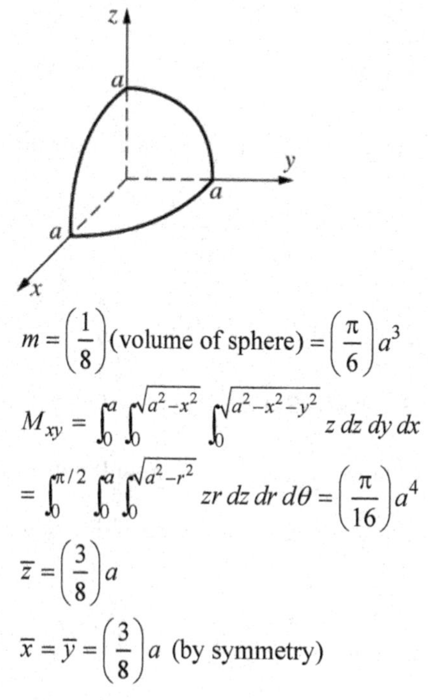

$m = \left(\frac{1}{8}\right)$(volume of sphere) $= \left(\frac{\pi}{6}\right)a^3$

$M_{xy} = \int_0^a \int_0^{\sqrt{a^2-x^2}} \int_0^{\sqrt{a^2-x^2-y^2}} z\,dz\,dy\,dx$

$= \int_0^{\pi/2} \int_0^a \int_0^{\sqrt{a^2-r^2}} zr\,dz\,dr\,d\theta = \left(\frac{\pi}{16}\right)a^4$

$\overline{z} = \left(\frac{3}{8}\right)a$

$\overline{x} = \overline{y} = \left(\frac{3}{8}\right)a$ (by symmetry)

29.

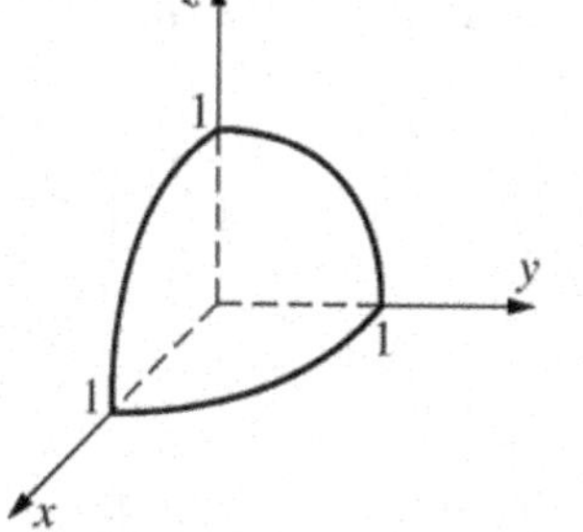

The limits of integration are those for the first octant part of a sphere of radius 1.

$\int_0^1 \int_0^{\sqrt{1-x^2}} \int_0^{\sqrt{1-x^2-y^2}} f(x, y, z)dz\,dy\,dx$

31. $\int_0^2 \int_0^{2-z} \int_0^{9-x^2} f(x, y, z)dy\,dx\,dz$

Figure is same as for Problem 32 except that the solid doesn't need to be divided into two parts.

33.

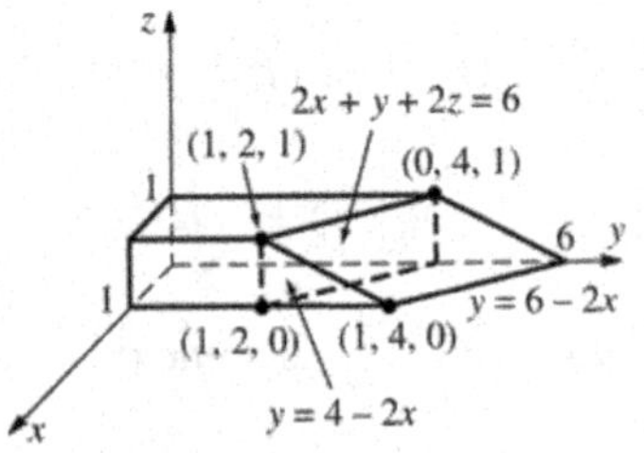

a. $\int_0^1\int_0^{4-2x}\int_0^1 dz\,dy\,dx + \int_0^1\int_{4-2x}^{6-2x}\int_0^{3-x-y/2} dz\,dy\,dx = 3+1 = 4$

b. $\int_0^1\int_0^1\int_0^{6-2x-2z} 1\,dy\,dx\,dz = 4$

c. $A(S_{xz})f(\overline{x},\overline{z})$ (S_{xz} is the unit square in the corner of xz-plane; and $(\overline{x},\overline{z}) = \left(\frac{1}{2},\frac{1}{2}\right)$ is the centroid of S_{xz}.)

$= (1)\left[6-2\left(\frac{1}{2}\right)-2\left(\frac{1}{2}\right)\right] = 4$

35. $\int_0^1\int_0^1\int_0^{6-2x-2z}(30-z)dy\,dx\,dz = \int_0^1\int_0^1(30-z)(6-2x-2z)dx\,dz = \int_0^1([(30-z)(6x-x^2-2xz)]_{x=0}^1)dz$

$= \int_0^1(30-z)(5-2z)dz = \int_0^1(150-65z+2z^2)dz = \left[150z-\frac{65z^2}{2}+\frac{2z^3}{3}\right]_0^1 = \frac{709}{6}$

The volume of the solid is 4 (from Problem 33).

Hence, the average temperature of the solid is $\frac{\frac{709}{6}}{4} = \frac{709}{24} \approx 29.54°$.

37. $M_{yz} = \int_0^1\int_0^1\int_0^{6-2x-2z} x\,dy\,dx\,dz = \int_0^1\int_0^1 x(6-2x-2z)\,dx\,dz = \int_0^1\left[3x^2-\frac{2}{3}x^3-x^2z\right]_{x=0}^1 dz$

$= \int_0^1\left(\frac{7}{3}-z\right)dz = \left[\frac{7}{3}z-\frac{1}{2}z^2\right]_0^1 = \frac{11}{6}$

$M_{xz} = \int_0^1\int_0^1\int_0^{6-2x-2z} y\,dy\,dx\,dz = \int_0^1\int_0^1\left[\frac{1}{2}(6-2x-2z)^2\right]dx\,dz = \int_0^1\left[18x-6x^2+\frac{2}{3}x^3-12xz+2x^2z+2xz^2\right]_{x=0}^1$

$= \int_0^1\left(\frac{38}{3}-10z+2z^2\right)dz = \frac{25}{3}$

$M_{xy} = \int_0^1\int_0^1\int_0^{6-2x-2z} z\,dy\,dx\,dz = \int_0^1\int_0^1 z(6-2x-2z)dx\,dz = \int_0^1([z(6x-x^2-2xz)]_{x=0}^1)dz$

$= \int_0^1(5z-2z^2)dz = \left[\frac{5z^2}{2}-\frac{2z^3}{3}\right]_0^1 = \frac{11}{6}$

Hence, $(\overline{x},\overline{y},\overline{z}) = \left(\frac{11/6}{4},\frac{25/3}{4},\frac{11/6}{4}\right) = \left(\frac{11}{24},\frac{25}{12},\frac{11}{24}\right)$

39. $m = \int_0^1 \int_0^1 \int_0^{4-x-y} k\,dz\,dy\,dx = \int_0^1 \int_0^1 k(4-x-y)\,dy\,dx = k\int_0^1 \left(\frac{7}{2}-x\right)dx = 3k$

$$M_{yz} = \int_0^1 \int_0^1 \int_0^{4-x-y} kx\,dz\,dy\,dx = k\int_0^1 \int_0^1 \left(4x - x^2 - xy\right)dy\,dx = k\int_0^1 \left(\frac{7}{2}x - x^2\right)dx = \frac{17k}{12}$$

$$M_{xz} = \int_0^1 \int_0^1 \int_0^{4-x-y} ky\,dz\,dy\,dx = k\int_0^1 \int_0^1 \left(4y - xy - y^2\right)dy\,dx = k\int_0^1 \left(\frac{5}{3}x - \frac{x}{2}\right)dx = \frac{17k}{12}$$

$$M_{xy} = \int_0^1 \int_0^1 \int_0^{4-x-y} kz\,dz\,dy\,dx = k\int_0^1 \int_0^1 \left(8 - 4x + \frac{x^2}{2} - 4y + xy + \frac{y^2}{2}\right)dy\,dx = k\int_0^1 \left(\frac{37}{6} - \frac{7x}{2} + \frac{x^2}{2}\right)dx = \frac{55k}{12}$$

$\bar{x} = \frac{Myz}{m} = \frac{17k/12}{3k} = \frac{17}{36}$ $\quad \bar{y} = \frac{Mxz}{m} = \frac{17k/12}{3k} = \frac{17}{36}$ $\quad \bar{z} = \frac{M_{xy}}{m} = \frac{55k/12}{3k} = \frac{55}{36}$

41.

z z z

1 2 3

Figure 1: When the center of mass is in this position, it *will* go lower when a little more soda leaks out since mass above the center of mass is being removed.
Figure 2: When the center of mass is in this position, it *was* lower moments before since mass that was below the center of mass was removed, causing the center of mass to rise.
Therefore, the center of mass is lowest when it is at the height of the soda, as in Figure 3. The same argument would hold for a soda bottle.

43. a. $1 = \int_0^{12} \int_0^x ky\,dy\,dx = \int_0^{12} \frac{k}{2}x^2\,dx$

$= \left[\frac{k}{6}x^3\right]_0^{12} = 288k \Rightarrow k = \frac{1}{288}$

b. $P(Y>4) = \int_4^{12} \int_4^x \frac{1}{288}y\,dydx$

$= \int_4^{12} \left[\frac{1}{576}y^2\right]_4^x dx = \int_4^{12} \frac{1}{576}\left(x^2 - 16\right)dx$

$= \left[\frac{1}{576}\left(x^3 - 16x\right)\right]_4^{12} = \frac{20}{27}$

c. $E[X] = \int_0^{12} \int_0^x x\frac{1}{288}y\,dy\,dx$

$= \int_0^{12} \left[\frac{1}{576}xy^2\right]_0^x dx = \int_0^{12} \frac{1}{576}x^3\,dx$

$= \left[\frac{1}{2304}x^4\right]_0^{12} = 9$

45. a. $P(X>2) = \int_2^4 \int_x^4 \frac{3}{256}\left(x^2+y^2\right)dy\,dx$

$= \int_2^4 \left[\frac{3}{256}\left(x^2 y + \frac{1}{3}y^3\right)\right]_x^4 dx$

$= \int_2^4 \frac{1}{64}\left(-x^3 + 3x^2 + 16\right)dx$

$= \left[\frac{1}{64}\left(-\frac{1}{4}x^4 + x^3 + 16x\right)\right]_2^4 = \frac{7}{16}$

b. $P(X+Y<4) = \int_0^2 \int_x^{4-x} \frac{3}{256}\left(x^2+y^2\right)dy\,dx$

$= \int_0^2 \left[\frac{3}{256}\left(x^2 y + \frac{1}{3}y^3\right)\right]_x^{4-x} dx$

$= \int_0^2 \frac{1}{32}\left(-x^3 + 3x^2 - 6x + 8\right)dx$

$= \frac{1}{32}\left[-\frac{1}{4}x^4 + x^3 - 3x^2 + 8x\right]_0^2 = \frac{1}{4}$

c. $E[X+Y]$

$$= \int_0^4 \int_0^y (x+y)\frac{3}{256}(x^2+y^2)\,dx\,dy$$

$$= \int_0^4 \left[\frac{3}{256}\left(\frac{1}{4}x^4+\frac{1}{2}x^2y^2+\frac{1}{3}x^3y+xy^3\right)\right]_0^y dx$$

$$= \int_0^4 \frac{25}{1024}y^4\,dy = \left[\frac{5}{1024}y^5\right]_0^4 = 5$$

47. $f_X(x) = \int_0^x \frac{1}{288}y\,dy$

$$= \left[\frac{1}{576}y^2\right]_0^x = \frac{1}{576}x^2;\ 0 \le x \le 12$$

$$E(X) = \int_0^{12} x \cdot f_X(x)\,dx = \frac{1}{576}\int_0^{12} x^3\,dx$$

$$= \frac{1}{576}\left[\frac{1}{4}x^4\right]_0^{12} = 9$$

13.8 Concepts Review

1. $r\,dz\,dr\,d\theta$, $\rho^2 \sin\phi\,d\rho\,d\theta\,d\phi$

3. $\int_0^{\pi}\int_0^{2\pi}\int_0^1 \rho^4 \cos^2\phi \sin\phi\,d\rho\,d\theta\,d\phi$

Problem Set 13.8

1. The region is a right circular cylinder about the z-axis with radius 3 and height 12.

$$\int_0^{2\pi}\int_0^3 [rz]_0^{12}\,dr\,d\theta = \int_0^{2\pi}\int_0^3 12r\,dr\,d\theta$$

$$= \int_0^{2\pi}\left[6r^2\right]_0^3 d\theta = \int_0^{2\pi} 54\,d\theta = 108\pi$$

3. The region is the region under the paraboloid $z = 9 - r^2$ above the xy-plane in that part of the first quadrant satisfying $0 \le \theta \le \frac{\pi}{4}$.

$$\int_0^{\pi/4}\int_0^3 \left[\frac{1}{2}rz^2\right]_0^{9-r^2} dr\,d\theta$$

$$= \int_0^{\pi/4}\int_0^3 \left[\frac{1}{2}r\left(81-18r+r^2\right)\right] dr\,d\theta$$

$$= \int_0^{\pi/4}\frac{1}{2}\left[\frac{81}{2}r^2-6r^3+\frac{1}{4}r^4\right]_0^3 d\theta$$

$$= \int_0^{\pi/4}\frac{243}{4}d\theta = \frac{243\pi}{16}$$

5. The region is a sphere centered at the origin with radius a.

$$\int_0^{\pi}\int_0^{2\pi}\left[\frac{1}{3}\rho^3 \sin\varphi\right]_0^a d\theta\,d\varphi$$

$$= \int_0^{\pi}\int_0^{2\pi}\frac{1}{3}a^3 \sin\varphi\,d\theta\,d\varphi$$

$$= \int_0^{\pi}\frac{2\pi}{3}a^3 \sin\varphi\,d\varphi$$

$$= \left[-\frac{2\pi}{3}a^3 \cos\varphi\right]_0^{\pi} = \frac{4\pi a^3}{3}$$

7. $\int_0^{2\pi}\int_0^2\int_{r^2}^4 r\,dz\,dr\,d\theta = 8\pi \approx 25.1327$

9. $V = \int_0^{2\pi}\int_0^3\int_4^{\sqrt{25-r^2}} r\,dz\,dr\,d\theta$

$$= \int_0^{2\pi}\int_0^3 [rz]_4^{\sqrt{25-r^2}}\,dr\,d\theta$$

$$= \int_0^{2\pi}\int_0^3\left[r\sqrt{25-r^2}-4r\right]dr\,d\theta$$

$$= \int_0^{2\pi}\left[-\frac{1}{3}\left(25-r^2\right)^{3/2}-2r^2\right]_0^3 d\theta$$

$$= \int_0^{2\pi}\frac{7}{3}d\theta = \frac{14\pi}{3}$$

11. $\int_0^{2\pi}\int_0^2\int_{r^2/4}^{\sqrt{5-r^2}} r\,dz\,dr\,d\theta$

$$= \int_0^{2\pi}\int_0^2\left[r(5-r^2)^{1/2}-\frac{r^3}{4}\right]dr\,d\theta$$

$$= \int_0^{2\pi}\frac{5^{3/2}-4}{3}d\theta = \frac{2\pi(5^{3/2}-4)}{3} \approx 15.0385$$

13. Let $\delta(x, y, z) = 1$.
(See write-up of Problem 26, Section 13.7.)

$$m = \int_0^{2\pi}\int_0^2\int_{r^2}^{12-2r^2} r\,dz\,dr\,d\theta = 24\pi$$

$$M_{xy} = \int_0^{2\pi}\int_0^2\int_{r^2}^{12-2r^2} zr\,dz\,dr\,d\theta = 128\pi$$

$$\bar{z} = \frac{16}{3}$$

$\bar{x} = \bar{y} = 0$ (by symmetry)

15. Let $\delta(x, y, z) = k\rho$

$$m = \int_0^{\pi}\int_0^{2\pi}\int_a^b k\rho\rho^2 \sin\phi\,d\rho\,d\theta\,d\phi = k\pi(b^4 - a^4)$$

17. Assume that the hemisphere lies above the xy-plane.
Let $\delta(x, y, z) = \rho$.
(Letting $k = 1$ - - see comment at the beginning of the write-up of Problem 26 of the previous section.)

$$m = \int_0^{\pi/2}\int_0^{2\pi}\int_0^a \rho^3 \sin\phi \, d\rho \, d\theta \, d\phi$$

$$= \int_0^{\pi/2}\int_0^{2\pi} \frac{a^4 \sin\phi}{4} d\theta \, d\phi$$

$$= \int_0^{\pi/2} \frac{\pi a^4 \sin\phi}{2} d\phi = \frac{\pi a^4}{2}$$

$$M_{xy} = \int_0^{\pi/2}\int_0^{2\pi}\int_0^a \rho^4 \sin\phi\cos\phi \, d\rho \, d\theta \, d\phi$$

$(z = \rho\cos\phi)$

$$= \int_0^{\pi/2}\int_0^{2\pi} \frac{a^5 \sin 2\phi}{10} d\theta \, d\phi$$

$$= \int_0^{\pi/2} \frac{\pi a^5 \sin 2\phi}{5} d\phi = \left(\frac{\pi}{5}\right)a^5$$

$$\bar{z} = \frac{\frac{\pi a^5}{5}}{\frac{\pi a^4}{2}} = \frac{2}{5}a; \bar{x} = \bar{y} = 0 \text{ (by symmetry)}$$

19. $I_z = \iiint_S (x^2 + y^2)k(x^2 + y^2)^{1/2} dV$

$$= \int_0^{\pi/2}\int_0^{2\pi}\int_0^a k\rho^5 \sin^4\phi \, d\rho \, d\theta \, d\phi = \left(\frac{k}{16}\right)\pi^2 a^6$$

21. $\int_0^{\pi}\int_0^{\pi/6}\int_0^1 \rho^2 \sin\phi \, d\rho \, d\theta \, d\phi = \frac{\pi}{9} \approx 0.3491$

23. Volume $= \int_0^{\pi}\int_0^{\sin\theta}\int_{r^2}^{r\sin\theta} r \, dz \, dr \, d\theta$

$$= \int_0^{\pi}\int_0^{\sin\theta} r(r\sin\theta - r^2)dr \, d\theta = \int_0^{\pi} \frac{\sin^4\theta}{12} d\theta$$

$$= \frac{1}{48}\int_0^{\pi}\left[1 - 2\cos 2\theta + \frac{1+\cos 4\theta}{2}\right]d\theta$$

$$= \frac{\pi}{32} \approx 0.0982$$

25. a. Position the ball with its center at the origin. The distance of (x, y, z) from the origin is $(x^2 + y^2 + z^2)^{1/2} = \rho$.

$$\iiint_S (x^2 + y^2 + z^2)^{1/2} dV = 8\int_0^{\pi/2}\int_0^{\pi/2}\int_0^a \rho(\sin\theta\rho^2)d\rho \, d\theta \, d\phi = \pi a^4$$

Then the average distance from the center is $\dfrac{\pi a^4}{\left[\left(\frac{4}{3}\right)\pi a^3\right]} = \dfrac{3a}{4}$.

b. Position the ball with its center at the origin and consider the diameter along the z-axis. The distance of (x, y, z) from the z-axis is $(x^2 + y^2)^{1/2} = \rho\sin\phi$.

$$\iiint_S (x^2 + y^2)^{1/2} dV = 8\int_0^{\pi/2}\int_0^{\pi/2}\int_0^a (\rho\sin\phi)(\rho^2\sin\theta)d\rho \, d\theta \, d\phi = \frac{a^4\pi^2}{4}$$

Then the average distance from a diameter is $\dfrac{\left[\frac{a^4\pi^2}{4}\right]}{\left[\left(\frac{4}{3}\right)\pi a^3\right]} = \dfrac{3\pi a}{16}$.

c. Position the sphere above and tangent to the xy-plane at the origin and consider the point on the boundary to be the origin. The equation of the sphere is $\rho = 2a\cos\phi$, and the distance of (x, y, z) from the origin is ρ.

$$\iiint_S (x^2 + y^2 + z^2)^{1/2} dV = \int_0^{\pi/2}\int_0^{2\pi}\int_0^{2a\cos\phi} \rho(\rho^2\sin\theta)d\rho \, d\theta \, d\phi = \frac{8\pi a^4}{5}$$

Then the average distance from the origin is $\dfrac{\left[\frac{8\pi a^4}{5}\right]}{\left[\left(\frac{4}{3}\right)\pi a^3\right]} = \dfrac{6a}{5}$.

27. a. $M_{yz} = \iiint_S kx\,dV = 4k\int_0^{\pi/2}\int_0^{\alpha}\int_0^{a}(\rho\sin\phi\cos\theta)(\rho^2\sin\phi)d\rho\,d\theta\,d\phi = ka^4\pi\frac{(\sin\alpha)}{4}$

$m = \iiint_S k\,dV = 4k\int_0^{\pi/2}\int_0^{\alpha}\int_0^{a}\rho^2\sin\phi\,d\rho\,d\theta\,d\phi = \frac{4a^3k\alpha}{3}$

Therefore, $\bar{x} = \frac{\left[\frac{ka^4\pi(\sin\alpha)}{4}\right]}{\left[\frac{4a^3k\alpha}{3}\right]} = \frac{3a\pi(\sin\alpha)}{16\alpha}.$

b. $\frac{3\pi a}{16}$ (See Problem 25b.)

29. Let m_1 and m_2 be the masses of the left and right balls, respectively. Then $m_1 = \frac{4}{3}\pi a^3k$ and $m_2 = \frac{4}{3}\pi a^3(ck)$, so $m_2 = cm_1$.

$$\bar{y} = \frac{m_1(-a-b)+m_2(a+b)}{m_1+m_2} = \frac{m_1(-a-b)+cm_1(a+b)}{m_1+cm_1} = \frac{-a-b+c(a+b)}{1+c} = \frac{(a+b)(-1+c)}{1+c} = \frac{c-1}{c+1}(a+b)$$

(Analogue) $\bar{y} = \frac{m_1\bar{y}_1 + m_2\bar{y}_2}{m_1+m_2} = \bar{y}_1\frac{m_1}{m_1+m_2} + \bar{y}_2\frac{m_2}{m_1+m_2}$

13.9 Concepts Review

1. u-curve; v-curve

3. Jacobian

Problem Set 13.9

1.

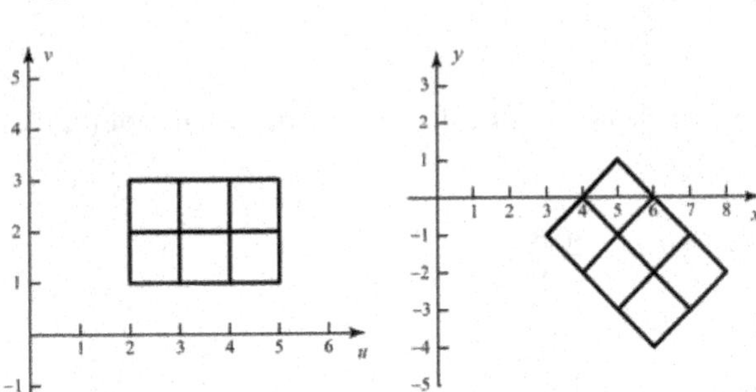

3.

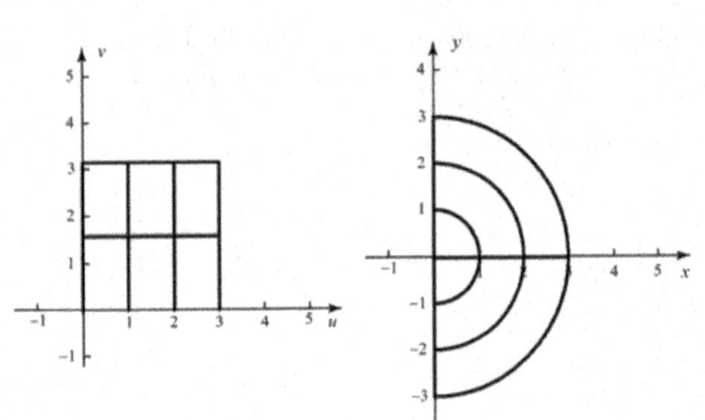

5.

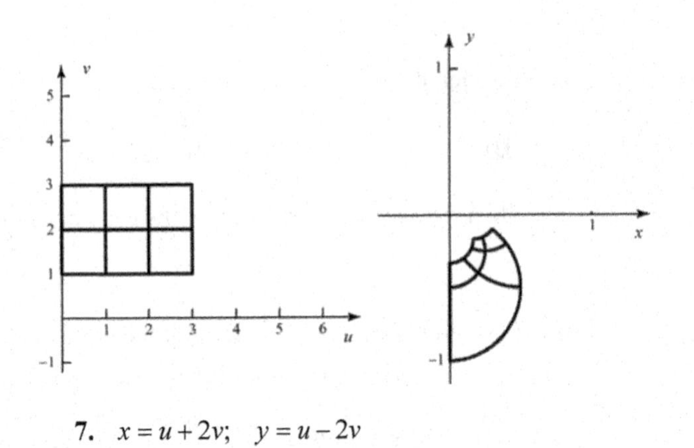

7. $x = u + 2v$; $y = u - 2v$

$G(0,0) = (0,0)$; $G(2,0) = (2,2)$
$G(2,1) = (4,0)$; $G(0,1) = (2,-2)$

The image is the square with corners $(0,0)$, $(2,2)$, $(4,0)$, and $(2,-2)$. The Jacobian is

$J = \begin{vmatrix} 1 & 2 \\ 1 & -2 \end{vmatrix} = -4.$

9. $x = u^2 + v^2; \quad y = v$

$G(0,0) = (0,0); \quad G(1,0) = (1,0)$
$G(1,1) = (2,1); \quad G(0,1) = (1,1)$

Solving for u and v gives
$u = \sqrt{x - y^2}$ and $v = y$
The $u = 0$ curve is $0 = \sqrt{x - y^2} \Rightarrow x = y^2$.
The $u = 1$ curve is $1 = \sqrt{x - y^2} \Rightarrow x = y^2 + 1$.
The $v = 0$ curve is $y = 0$, and the $v = 1$ curve is $y = 1$. The image is the set of (x, y) that satisfy $y^2 \le x \le y^2 + 1, 0 \le y \le 1$. The Jacobian is
$J = \begin{vmatrix} 2u & 2v \\ 0 & 1 \end{vmatrix} = 2u$

11. $u = x + 2y; \quad v = x - 2y$ Solving for x and y gives $x = u/2 + v/2; \quad y = u/4 - v/4$. The Jacobian is $J = \begin{vmatrix} 1/2 & 1/2 \\ 1/4 & -1/4 \end{vmatrix} = -\frac{1}{8} - \frac{1}{8} = -\frac{1}{4}$.

13. $u = x^2 + y^2; \quad v = x$. Solving for x and y gives $x = v; \quad y = \sqrt{u - v^2}$. The Jacobian is

$$J = \begin{vmatrix} 0 & 1 \\ \frac{1}{2\sqrt{u - v^2}} & \frac{-v}{\sqrt{u - v^2}} \end{vmatrix} = -\frac{1}{2\sqrt{u - v^2}}$$

15. $u = xy; \quad v = x$. Solving for x and y gives $x = v; \quad y = u/v$. The Jacobian is

$$J = \begin{vmatrix} 0 & 1 \\ \frac{1}{v} & -\frac{u}{v^2} \end{vmatrix} = -\frac{1}{v}.$$

17. Let $u = x + y, v = x - y$. Solving for x and y gives $x = u/2 + v/2$ and $y = u/2 - v/2$. The Jacobian is $J = \begin{vmatrix} \frac{1}{2} & \frac{1}{2} \\ \frac{1}{2} & -\frac{1}{2} \end{vmatrix} = -\frac{1}{2}$.

The region of integration gets transformed to the triangle in the uv-plane with vertices (1,1), (4,4), and (7,1). The integral in the uv-plane is more easily done by holding v fixed and integrating u. Thus,

$$\iint_R \ln \frac{x+y}{x-y} \, dA = \int_1^4 \int_v^{8-v} \ln \frac{u}{v} \left| \frac{1}{2} \right| du \, dv$$

$$= \frac{1}{2} \int_1^4 \left[-u + u \ln \frac{u}{v} \right]_v^{8-v} dv$$

$$= \frac{1}{2} \int_1^4 \left[-8 + 2v + (8 - v) \ln \frac{8 - v}{v} \right] dv$$

$$= \frac{1}{2} \left[3 - 64 \ln 4 + \frac{49}{2} \ln 7 + 16 \ln 16 \right]$$

$$\approx 3.15669$$

19. Let $u = 2x - y$ and $v = y$. Then $x = u/2 + v/2$ and $y = v$. The Jacobian is

$J = \begin{vmatrix} 1/2 & 1/2 \\ 0 & 1 \end{vmatrix} = \frac{1}{2}$. Thus,

$$\iint_R \sin(\pi(2x - y)) \cos(\pi(y - 2x)) \, dA$$

$$= \int_0^3 \int_{v+2}^{8-v} \sin(\pi u) \cos(\pi(-u)) \left| \frac{1}{2} \right| du \, dv$$

$$= \frac{1}{2} \int_0^3 \int_{v+2}^{8-v} \frac{1}{2} 2 \sin(\pi u) \cos(\pi u) \, du \, dv$$

$$= \frac{1}{4} \int_0^3 \int_{v+2}^{8-v} \sin(2\pi u) \, du \, dv$$

$$= \frac{1}{4} \int_0^3 \left[-\frac{\cos 2\pi u}{2\pi} \right]_{v+2}^{8-v} dv$$

$$= \frac{1}{8\pi} \int_0^3 \left[\cos(2\pi(v + 2)) - \cos(2\pi(8 - v)) \right] dv$$

$$= \frac{1}{8\pi} \int_0^3 \left[\cos(2\pi v + 4\pi) - \cos(16\pi - 2\pi v) \right] dv$$

$$= \frac{1}{8\pi} \int_0^3 \left[\cos(2\pi v) - \cos(2\pi v) \right] dv = 0$$

21. The transformation to spherical coordinates is

$x = \rho \sin\phi \cos\theta$

$y = \rho \sin\phi \sin\theta$

$z = \rho \cos\phi$

The Jacobian is

$$J = \begin{vmatrix} \frac{\partial x}{\partial \rho} & \frac{\partial x}{\partial \theta} & \frac{\partial x}{\partial \phi} \\ \frac{\partial y}{\partial \rho} & \frac{\partial y}{\partial \theta} & \frac{\partial y}{\partial \phi} \\ \frac{\partial z}{\partial \rho} & \frac{\partial z}{\partial \theta} & \frac{\partial z}{\partial \phi} \end{vmatrix} = \begin{vmatrix} \sin\phi\cos\theta & -\rho\sin\phi\sin\theta & \rho\cos\phi\cos\theta \\ \sin\phi\sin\theta & \rho\sin\phi\cos\theta & \rho\cos\phi\sin\theta \\ \cos\phi & 0 & -\rho\sin\phi \end{vmatrix}$$

$$= \cos\phi \begin{vmatrix} -\rho\sin\phi\sin\theta & \rho\cos\phi\cos\theta \\ \rho\sin\phi\cos\theta & \rho\cos\phi\sin\theta \end{vmatrix} - 0 \begin{vmatrix} \sin\phi\cos\theta & \rho\cos\phi\cos\theta \\ \sin\phi\sin\theta & \rho\cos\phi\sin\theta \end{vmatrix} + (-\rho\sin\phi) \begin{vmatrix} \sin\phi\cos\theta & -\rho\sin\phi\sin\theta \\ \sin\phi\sin\theta & \rho\sin\phi\cos\theta \end{vmatrix}$$

$$= \cos\phi\left(-\rho^2 \sin\phi\sin\theta\cos\phi\sin\theta - \rho^2\cos\phi\sin\phi\cos^2\theta\right) - \rho\sin\phi\left(\rho\sin^2\phi\cos^2\theta + \rho\sin^2\phi\sin^2\theta\right)$$

$$= -\rho^2\cos^2\phi\,\sin\phi\left(\sin^2\theta + \cos^2\theta\right) - \rho^2\sin^3\phi$$

$$= -\rho^2\sin\phi\left(\cos^2\phi + \sin^2\phi\right)$$

$$= -\rho^2\sin\phi$$

23. Let $X = x(U,V)$ and $Y = y(U,V)$. If R is a region in the xy-plane with preimage S in the uv-plane, then

$P((X,Y) \in R) = P((U,V) \in S)$

Writing each of these as a double integral over the appropriate PDF and region gives

$$\iint_R f(x,y)\,dy\,dx = \iint_S g(u,v)\,dv\,du$$

Now, make the change of variable $x = x(u,v)$ and $y = y(u,v)$ in the integral on the left. Therefore,

$$\iint_S g(u,v)\,dv\,du = \iint_R f(x,y)\,dy\,dx = \iint_S f\left(x(u,v), y(u,v)\right)\left|J(u,v)\right|\,dv\,du$$

Thus probabilities involving (U,V) can be obtained by integrating $f\left(x(u,v), y(u,v)\right)\left|J(u,v)\right|$. Thus, $f\left(x(u,v), y(u,v)\right)\left|J(u,v)\right|$ is the joint PDF of (U,V).

25. **a.** Let $u = x + y,\ v = x$. Solving for x and y gives $x = v,\ y = u - v$. The Jacobian is

$$J = \begin{vmatrix} 0 & 1 \\ 1 & -1 \end{vmatrix} = -1.$$

Thus,

$$g(u,v) = f(v, u-v)\left|-1\right|$$

$$= \begin{cases} e^{-v-(u-v)}, & \text{if } 0 \le v, 0 \le u - v \\ 0, & \text{otherwise} \end{cases}$$

$$= \begin{cases} e^{-u}, & \text{if } 0 \le v \le u \\ 0, & \text{otherwise} \end{cases}$$

b. The marginal PDF for U is obtained by integrating over all possible v for a fixed u. If $u \ge 0$, then

$$g_U(u) = \int_0^u e^{-u}\,dv = ue^{-u}$$

Thus, $g_U(u) = \begin{cases} ue^{-u}, & \text{if } 0 \le u \\ 0, & \text{otherwise} \end{cases}$

13.10 Chapter Review

Concepts Test

1. True: Use result of Problem 33, Section 13.2, and then change dummy variable y to dummy variable x.

3. True: Inside integral is 0 since $\sin(x^3y^3)$ is an odd function in x.

5. True: It is less than or equal to $\int_1^2 \int_0^2 1\,dx\,dy$ which equals 2.

7. False: Let $f(x, y) = x$, $g(x, y) = x^2$, $R = \{(x, y): x \text{ in } [0, 2], y \text{ in } [0, 1]\}$. The inequality holds for the integrals but $f(0.5, 0) > g(0.5, 0)$.

9. True: See the write-up of Problem 26, Section 13.7.

11. True: The integral is the volume between concentric spheres of radii 4 and 1. That volume is 84π.

13. False: There are 6.

15. True: $\sqrt{f_x^2 + f_y^2 + 1} \le \sqrt{9} = 3$

17. False: $J(u, v) = \begin{vmatrix} 2 & 0 \\ 0 & 2 \end{vmatrix} = 4$

Sample Test Problems

1. $\int_0^1 \left(\frac{1}{2}\right)(x^2 - x^3)dx = \frac{1}{24} \approx 0.0417$

3. $\int_0^{\pi/2} \left[\frac{r^2 \cos\theta}{2}\right]_{r=0}^{2\sin\theta} d\theta = \int_0^{\pi/2} 2\sin^2\theta\cos\theta\,d\theta$

$= \left[\frac{2\sin^3\theta}{3}\right]_0^{\pi/2} = \frac{2}{3}$

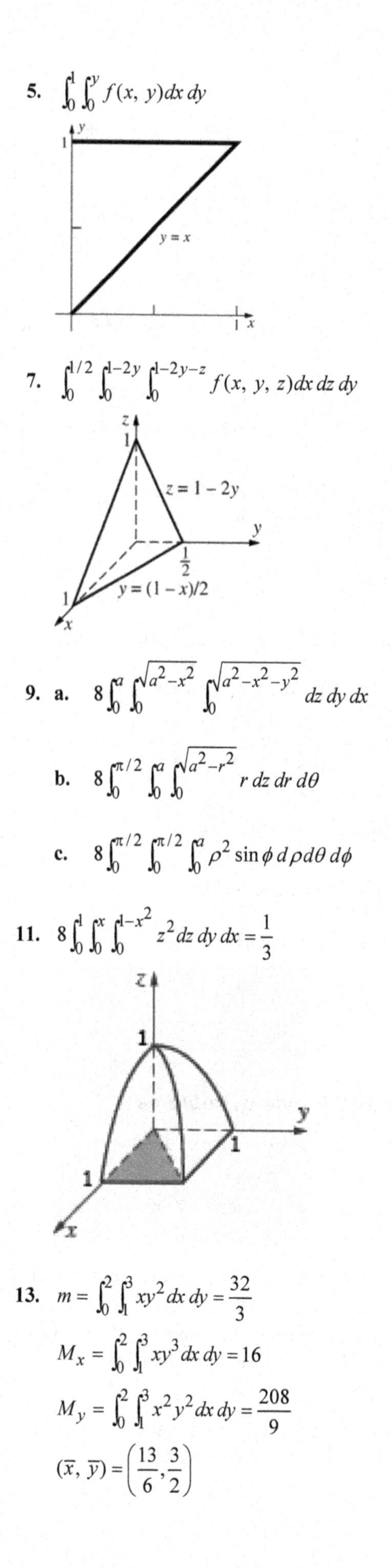

5. $\int_0^1 \int_0^y f(x, y)dx\,dy$

7. $\int_0^{1/2} \int_0^{1-2y} \int_0^{1-2y-z} f(x, y, z)dx\,dz\,dy$

9. a. $8\int_0^a \int_0^{\sqrt{a^2-x^2}} \int_0^{\sqrt{a^2-x^2-y^2}} dz\,dy\,dx$

b. $8\int_0^{\pi/2} \int_0^a \int_0^{\sqrt{a^2-r^2}} r\,dz\,dr\,d\theta$

c. $8\int_0^{\pi/2} \int_0^{\pi/2} \int_0^a \rho^2 \sin\phi\,d\rho\,d\theta\,d\phi$

11. $8\int_0^1 \int_0^x \int_0^{1-x^2} z^2\,dz\,dy\,dx = \frac{1}{3}$

13. $m = \int_0^2 \int_1^3 xy^2\,dx\,dy = \frac{32}{3}$

$M_x = \int_0^2 \int_1^3 xy^3\,dx\,dy = 16$

$M_y = \int_0^2 \int_1^3 x^2y^2\,dx\,dy = \frac{208}{9}$

$(\bar{x}, \bar{y}) = \left(\frac{13}{6}, \frac{3}{2}\right)$

15. $z = f(x, y) = (9 - y^2)^{1/2}$; $f_x(x, y) = 0$;

$f_y(x, y) = -y(9 - y^2)^{-1/2}$

$$\text{Area} = \int_0^3 \int_{y/3}^{y} \sqrt{y^2(9-y^2)^{-1} + 1}\, dx\, dy$$

$$= \int_0^3 \int_{y/3}^{y} 3(9-y^2)^{-1/2}\, dx\, dy$$

$$= \int_0^3 (9-y^2)^{-1/2}(2y)dy = [-2(9-y^2)^{1/2}]_0^3 = 6$$

17. $\delta(x, y, z) = k\rho$

$$m = \int_0^{\pi} \int_0^{2\pi} \int_1^3 k\rho\,\rho^2 \sin\phi\, d\rho\, d\theta\, d\phi = 80\pi k$$

19. $$m = \int_0^a \int_0^{(b/a)(a-x)} \int_0^{(c/ab)(ab-bx-ay)} kx\, dz\, dy\, dx$$

$$= \left(\frac{k}{24}\right) a^2 bc$$

21. Let $x = \frac{u+v}{2}$ and $y = \frac{v-u}{2}$. Then we have,

$\sin(x-y)\cos(x+y) = \sin u \cos v$ and

$$J(u,v) = \begin{vmatrix} \frac{1}{2} & \frac{1}{2} \\ -\frac{1}{2} & \frac{1}{2} \end{vmatrix} = \frac{1}{4} - \left(-\frac{1}{4}\right) = \frac{1}{2}.$$

Thus,

$$\iint_R \sin(x-y)\cos(x+y)\, dx dy$$

$$= \int_0^{\pi} \int_0^{\pi} \frac{1}{2} \sin u \cos v\, du\, dv$$

$$= \frac{1}{2} \int_0^{\pi} \cos v [-\cos u]_0^{\pi}\, dv$$

$$= \int_0^{\pi} \cos v\, dv = 0$$

Review and Preview Problems

1. Answers may vary. One solution is
$x = 3\cos t$, $y = 3\sin t$, $0 \le t < 2\pi$

Then: $x^2 + y^2 = 9\cos^2 t + 9\sin^2 t = 9$ as required.

3. Answers may vary. One solution for the circle is
$x = 2\cos t$, $y = 2\sin t$

To have the semicircle where $y > 0$, we need $\sin t > 0$, so we restrict the domain of t to $0 < t < \pi$.

5. Answers may vary. One solution is
$x = -2 + 5t$, $y = 2$ for $t \in [0,1]$.

7. Note that $x + y = 9$, so a simple parameterization is to let one variable be t and the other be $9 - t$. Since we are restricted to the first quadrant, we must have $t > 0$ and $9 - t > 0$; hence the domain is $t \in (0,9)$. Finally, since orientation is to be up and to the left, we want y to increase and x to decrease as t increases. Thus we use $x = 9 - t$ and $y = t$.

9. Since we are restricting the parabola to the points where $y > 0$, a simple parameterization is

$x = -t$, $y = 9 - t^2$, $t \in [-3,3]$

Note that the orientation is right to left.

11. $\nabla f(x,y) = f_x(x,y)\,\mathbf{i} + f_y(x,y)\,\mathbf{j}$. Now if

$f(x,y) = x\sin x + y\cos y$, then

$f_x(x,y) = x\cos x + \sin x$, $f_y(x,y) = \cos y - y\sin y$

Thus:

$\nabla f(x,y) = (x\cos x + \sin x)\,\mathbf{i} + (\cos y - y\sin y)\,\mathbf{j}$

13. $\nabla f(x,y,z) = f_x\,\mathbf{i} + f_y\,\mathbf{j} + f_z\,\mathbf{k}$. Now if

$f(x,y,z) = x^2 + y^2 + z^2$, then

$f_x(x,y,z) = 2x$, $f_y(x,y,z) = 2y$, $f_z(x,y,z) = 2z$

so

$\nabla f(x,y,z) = 2x\,\mathbf{i} + 2y\,\mathbf{j} + 2z\,\mathbf{k}$

15. $\nabla f(x,y,z) = f_x\,\mathbf{i} + f_y\,\mathbf{j} + f_z\,\mathbf{k}$. Now if

$f(x,y,z) = xy + xz + yz$, then $f_x(x,y,z) = y + z$, $f_y(x,y,z) = x + z$, and $f_z(x,y,z) = x + y$ so

$\nabla f(x,y,z) = (y+z)\mathbf{i} + (x+z)\mathbf{j} + (x+y)\mathbf{k}$

17. $$\int_0^{\pi} \sin^2 t\, dt = \int_0^{\pi} \frac{1-\cos 2t}{2}\, dt$$

$$= \frac{1}{2}\int_0^{\pi} (1 - \cos 2t)\, dt$$

$$= \frac{1}{2}\left[t - \frac{1}{2}\sin 2t\right]_0^{\pi}$$

$$= \frac{1}{2}\left[\left(\pi - \frac{1}{2}\sin(2\pi)\right) - \left(0 - \frac{1}{2}\sin(2\cdot 0)\right)\right]$$

$$= \frac{\pi}{2}$$

19. $$\int_0^1 \int_1^2 xy\, dy\, dx = \int_0^1 \left[\int_1^2 xy\, dy\right] dx = \int_0^1 \left[\frac{xy^2}{2}\right]_{y=1}^{y=2} dx$$

$$= \int_0^1 \left(\frac{3x}{2}\right) dx = \left[\frac{3x^2}{4}\right]_0^1 = \frac{3}{4}$$

21. $\int_0^{2\pi}\int_1^2 r^2\,dr\,d\theta = \int_0^{2\pi}\left[\int_1^2 r^2\,dr\right]d\theta$

$= \int_0^{2\pi}\left[\frac{r^3}{3}\right]_{r=1}^{r=2} d\theta$

$= \int_0^{2\pi}\frac{7}{3}d\theta = \frac{7}{3}\theta\Big|_0^{2\pi}$

$= \frac{14\pi}{3} \approx 14.66$

23. Note that

$\int_0^{2\pi}\int_0^{\pi}\int_r^R \rho^2 \sin\phi\,d\rho\,d\phi\,d\theta =$

$\int_0^{2\pi}\int_0^{\pi}\left[\int_r^R \rho^2 \sin\phi\,d\rho\right]d\phi\,d\theta =$

$\int_0^{2\pi}\int_0^{\pi}\left[\frac{\rho^3}{3}\sin\phi\right]_{\rho=r}^{\rho=R} d\phi\,d\theta =$

$\int_0^{2\pi}\left[\int_0^{\pi}\frac{1}{3}(R^3-r^3)\sin\phi\,d\phi\right]d\theta =$

$\int_0^{2\pi}\left[-\frac{1}{3}(R^3-r^3)\cos\phi\right]_{\phi=0}^{\phi=\pi} d\theta =$

$\int_0^{2\pi}\left[\frac{2}{3}(R^3-r^3)\right]d\theta = \frac{4}{3}\pi R^3 - \frac{4}{3}\pi r^3$

which we recognize as the difference between the volume of a sphere with radius R and the volume of a sphere with radius r. Thus the volume in problem 22 is that of a spherical shell with center at (0,0,0) and, in this case, outer radius = 2 and inner radius = 1 .

25. This will be the unit vector, at the point $(3,4,12)$, in the direction of ∇F, where $F(x,y,z) = x^2+y^2+z^2$.

Now

$\nabla F(x,y,z) = F_x\,\mathbf{i} + F_y\,\mathbf{j} + F_z\,\mathbf{k} = 2x\,\mathbf{i} + 2y\,\mathbf{j} + 2z\,\mathbf{k}$

so that $\nabla F(3,4,12) = \langle 6,8,24\rangle$ and the unit vector in the same direction is

$\frac{\nabla F}{\|\nabla F\|} = \left\langle \frac{6}{26},\frac{8}{26},\frac{24}{26}\right\rangle = \left\langle \frac{3}{13},\frac{4}{13},\frac{12}{13}\right\rangle.$

This agrees with our geometric intuition, since $x^2+y^2+z^2=169$ is the surface of a sphere with center at $O=(0,0,0)$ and radius = 13. Now the plane tangent to a sphere (center at (0,0,0) and radius r) at any point $P=(a,b,c)$ is perpendicular to the radius at that point; so it would follow that the vector $\overrightarrow{OP} = \langle a,b,c\rangle$ is perpendicular to the tangent plane and hence normal to the surface. The unit normal in the direction of $\overrightarrow{OP}$ is simply $\frac{1}{r}\langle a,b,c\rangle$, or in our case $\frac{1}{13}\langle 3,4,12\rangle$.

CHAPTER 14 Vector Calculus

14.1 Concepts Review

1. vector-valued function of three real variables or a vector field

3. gravitational fields; electric fields

Problem Set 14.1

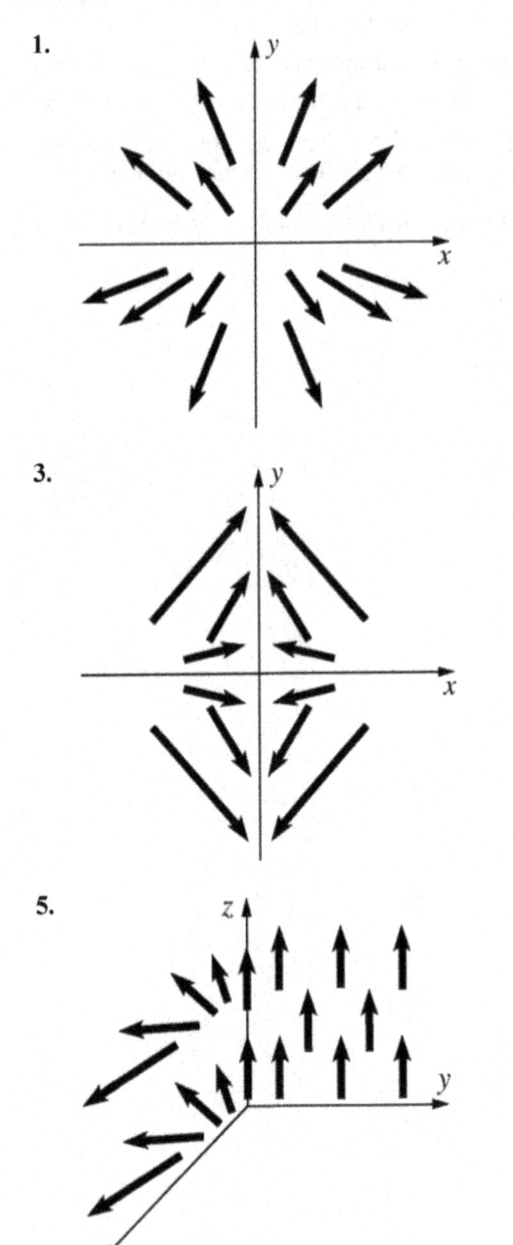

7. $\langle 2x-3y, -3x, 2\rangle$

9. $f(x, y, z) = \ln|x| + \ln|y| + \ln|z|$;
$\nabla f(x, y, z) = \left\langle x^{-1}, y^{-1}, z^{-1}\right\rangle$

11. $e^y \langle \cos z, x\cos z, -x\sin z\rangle$

13. div $\mathbf{F} = 2x - 2x + 2yz = 2yz$
curl $\mathbf{F} = \left\langle z^2, 0, -2y\right\rangle$

15. div $\mathbf{F} = \nabla \cdot \mathbf{F} = 0+0+0=0$
curl $\mathbf{F} = \nabla \times \mathbf{F} = \langle x-x, y-y, z-z\rangle = \mathbf{0}$

17. div $\mathbf{F} = e^x \cos y + e^x \cos y + 1 = 2e^x \cos y + 1$
curl $\mathbf{F} = \left\langle 0, 0, 2e^x \sin y\right\rangle$

19. **a.** meaningless

b. vector field

c. vector field

d. scalar field

e. vector field

f. vector field

g. vector field

h. meaningless

i. meaningless

j. scalar field

k. meaningless

21. Let $f(x, y, z) = -c|\mathbf{r}|^{-3}$, so

$$\text{grad}(f) = 3c|\mathbf{r}|^{-4}\frac{\mathbf{r}}{|\mathbf{r}|}$$
$$= 3c|\mathbf{r}|^{-5}\mathbf{r}$$

Then $\text{curl}\,\mathbf{F} = \text{curl}\left[\left(-c|\mathbf{r}|^{-3}\right)\mathbf{r}\right]$

$$= \left(-c|\mathbf{r}|^{-3}\right)(\text{curl}\,\mathbf{r}) + \left(3c|\mathbf{r}|^{-5}r\right)\times\mathbf{r} \text{ (by 20d)}$$
$$= \left(-c|\mathbf{r}|^{-3}\right)(\mathbf{0}) + \left(3c|\mathbf{r}|^{-5}\right)(\mathbf{r}\times\mathbf{r}) = 0 + 0 = 0$$

$\text{div}\,\mathbf{F} = \text{div}\left[\left(-c|\mathbf{r}|^{-3}\right)\mathbf{r}\right]$

$$= \left(-c|\mathbf{r}|^{-3}\right)(\text{div}\,\mathbf{r}) + \left(3c|\mathbf{r}|^{-5}\mathbf{r}\right)\cdot\mathbf{r} \text{ (by 20c)}$$
$$= \left(-c|\mathbf{r}|^{-3}\right)(1+1+1) + \left(3c|\mathbf{r}|^{-5}\right)|\mathbf{r}|^2$$
$$= \left(-3c|\mathbf{r}|^{-3}\right) + 3c|\mathbf{r}|^3 = 0$$

23. $\text{grad}\,f = \left\langle f'(r)xr^{-1/2}, f'(r)yr^{-1/2}, f'(r)zr^{-1/2}\right\rangle$ (if $r \neq 0$)

$$= f'(r)r^{-1/2}\langle x, y, z\rangle$$
$$= f'(r)r^{-1/2}\mathbf{r}$$

$$\text{curl}\,\mathbf{F} = [f(r)][\text{curl}\,\mathbf{r}] + [f'(r)r^{-1/2}\mathbf{r}]\times\mathbf{r}$$
$$= [f(r)][\text{curl}\,\mathbf{r}] + [f'(r)r^{-1/2}\mathbf{r}]\times\mathbf{r}$$
$$= \mathbf{0} + \mathbf{0} = \mathbf{0}$$

25. a. Let $P = (x_0, y_0)$.
div $\mathbf{F}$ = div $\mathbf{H} = 0$ since there is no tendency toward P except along the line $x = x_0$, and along that line the tendencies toward and away from P are balanced; div $\mathbf{G} < 0$ since there is no tendency toward P except along the line $x = x_0$, and along that line there is more tendency toward than away from P; div $\mathbf{L} > 0$ since the tendency away from P is greater than the tendency toward P.

b. No rotation for **F**, **G**, **L**; clockwise rotation for **H** since the magnitudes of the forces to the right of P are less than those to the left.

c. div $\mathbf{F} = 0$; curl $\mathbf{F} = \mathbf{0}$

div $\mathbf{G} = -2ye^{-y^2} < 0$ since $y > 0$ at P; curl $\mathbf{G} = \mathbf{0}$

div $\mathbf{L} = (x^2 + y^2)^{-1/2}$; curl $\mathbf{L} = \mathbf{0}$

div $\mathbf{H} = 0$; curl $\mathbf{H} = \left\langle 0, 0, -2xe^{-x^2}\right\rangle$ which points downward at P, so the rotation is clockwise in a right-hand system.

27. $\mathbf{F}(x, y, z) = M\,\mathbf{i} + N\,\mathbf{j} + P\,\mathbf{k}$, where

$$M = -\frac{x}{(1+x^2+y^2)^{3/2}},\quad N = -\frac{y}{(1+x^2+y^2)^{3/2}},$$
$P = 0$

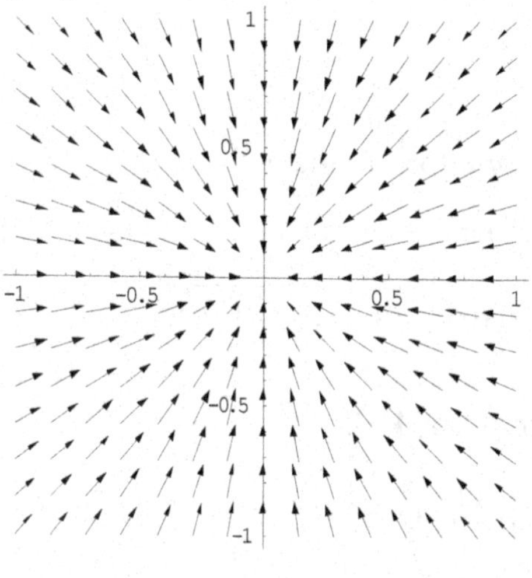

a. Since all the vectors are directed toward the origin, we would expect accumulation at that point; thus $\text{div}\,\mathbf{F}(0,0,0)$ .should be negative. Calculating,

$$\text{div}\,\mathbf{F}(x, y, z) = \frac{3(x^2+y^2)}{(1+x^2+y^2)^{5/2}} - \frac{2}{(1+x^2+y^2)^{3/2}} + 0$$

so that $\text{div}\,\mathbf{F}(0,0,0) = -2$

b. If a paddle wheel is placed at the origin (with its axis perpendicular to the plane), the force vectors all act radially along the wheel and so will have no component acting tangentially along the wheel. Thus the wheel will not turn at all, and we would expect $\text{curl}\,\mathbf{F} = \mathbf{0}$. By calculating

$$\text{curl}\,\mathbf{F} = (0-0)\mathbf{i} + (0-0)\mathbf{j} + \left(\frac{3yx}{(1+x^2+y^2)^{3/2}} - \frac{3xy}{(1+x^2+y^2)^{3/2}}\right)\mathbf{k} = \mathbf{0}$$

29. $\nabla f(x, y, z) = \frac{1}{2}m\omega^2\langle 2x, 2y, 2z\rangle = m\omega^2\langle x, y, z\rangle = \mathbf{F}(x, y, z)$

31. a. $\mathbf{F}\times\mathbf{G} = (f_y g_z - f_z g_y)\mathbf{i} - (f_x g_z - f_z g_x)\mathbf{j} + (f_x g_z - f_z g_x)\mathbf{k}$. Therefore,

$$div(\mathbf{F}\times\mathbf{G}) = \frac{\partial}{\partial x}(f_y g_z - f_z g_y) - \frac{\partial}{\partial y}(f_x g_z - f_z g_x) + \frac{\partial}{\partial z}(f_x g_y - f_y g_x).$$

Using the product rule for partials and some algebra gives

$$div(\mathbf{F}\times\mathbf{G}) = g_x\left[\frac{\partial f_z}{\partial y} - \frac{\partial f_y}{\partial z}\right] + g_y\left[\frac{\partial f_x}{\partial z} - \frac{\partial f_z}{\partial x}\right] + g_z\left[\frac{\partial f_y}{\partial x} - \frac{\partial f_x}{\partial y}\right]$$

$$+ f_x\left[\frac{\partial g_z}{\partial y} - \frac{\partial g_y}{\partial z}\right] + f_y\left[\frac{\partial g_x}{\partial z} - \frac{\partial g_z}{\partial x}\right] + f_z\left[\frac{\partial g_y}{\partial x} - \frac{\partial g_x}{\partial y}\right]$$

$$= \mathbf{G}\cdot curl(\mathbf{F}) - \mathbf{F}\cdot curl(\mathbf{G})$$

b. $\nabla f\times\nabla g = \left(\frac{\partial f}{\partial y}\frac{\partial g}{\partial z} - \frac{\partial f}{\partial z}\frac{\partial g}{\partial y}\right)\mathbf{i} - \left(\frac{\partial f}{\partial x}\frac{\partial g}{\partial z} - \frac{\partial f}{\partial z}\frac{\partial g}{\partial x}\right)\mathbf{j} + \left(\frac{\partial f}{\partial x}\frac{\partial g}{\partial y} - \frac{\partial f}{\partial y}\frac{\partial g}{\partial x}\right)\mathbf{k}$

Therefore, $div(\nabla f\times\nabla g) = \frac{\partial}{\partial x}\left(\frac{\partial f}{\partial y}\frac{\partial g}{\partial z} - \frac{\partial f}{\partial z}\frac{\partial g}{\partial y}\right) - \frac{\partial}{\partial y}\left(\frac{\partial f}{\partial x}\frac{\partial g}{\partial z} - \frac{\partial f}{\partial z}\frac{\partial g}{\partial x}\right) + \frac{\partial}{\partial z}\left(\frac{\partial f}{\partial x}\frac{\partial g}{\partial y} - \frac{\partial f}{\partial y}\frac{\partial g}{\partial x}\right).$

Using the product rule for partials and some algebra will yield the result
$div(\nabla f\times\nabla g) = 0$

14.2 Concepts Review

1. Increasing values of t

3. $f(x(t), y(t))\sqrt{[x'(t)]^2 + [y'(t)]^2}$

Problem Set 14.2

1. $\int_0^1 (27t^3 + t^3)(9 + 9t^4)^{1/2}\,dt = 14\left(2\sqrt{2} - 1\right) \approx 25.5980$

3. Let $x = t$, $y = 2t$, t in $[0, \pi]$.

   Then $\int_C (\sin x + \cos y)ds = \int_0^\pi (\sin t + \cos 2t)\sqrt{1+4}\,dt = 2\sqrt{5} \approx 4.4721$

5. $\int_0^1 (2t + 9t^3)(1 + 4t^2 + 9t^4)^{1/2}\,dt = \left(\frac{1}{6}\right)(14^{3/2} - 1) \approx 8.5639$

7. $\int_0^2 [(t^2 - 1)(2) + (4t^2)(2t)]dt = \frac{100}{3}$

9. $\int_C y^3\,dx + x^3\,dy = \int_{C_1} y^3\,dx + x^3\,dy + \int_{C_2} y^3\,dx + x^3\,dy = \int_1^{-2}(-4)^3\,dy + \int_{-4}^{2}(-2)^3\,dx = 192 + (-48) = 144$

11. $y = -x + 2$

$\int_1^3 ([x + 2(-x+2)](1) + [x - 2(-x+2)](-1))dx = 0$

13. $\langle x, y, z\rangle = \langle 1, 2, 1\rangle + t\langle 1, -1, 0\rangle$

$\int_0^1 [(4 - t)(1) + (1 + t)(-1) - (2 - 3t + t^2)(-1)]dt = \frac{17}{6} \approx 2.8333$

15. On $C_1: y = z = dy = dz = 0$

On $C_2: x = 2, z = dx = dz = 0$

On $C_3: x = 2, y = 3, dx = dy = 0$

$\int_0^2 x\,dx + \int_0^3 (2 - 2y)dy + \int_0^4 (4 + 3 - z)dz = \left[\frac{x^2}{2}\right]_0^2 + [2y - y^2]_0^3 + \left[7z - \frac{z^2}{2}\right]_0^4 = 2 + (-3) + 20 = 19$

17. $m = \int_C k|x|\,ds = \int_{-2}^{2} k|x|(1 + 4x^2)^{1/2}\,dx = \left(\frac{k}{6}\right)(17^{3/2} - 1) \approx 11.6821k$

19 $\int_C (x^3 - y^3)dx + xy^2\,dy = \int_{-1}^{0} [(t^6 - t^9)(2t) + (t^2)(t^6)(3t^2)]dt = -\frac{7}{44} \approx -0.1591$

21. $W = \int_C \mathbf{F}\cdot d\mathbf{r} = \int_C (x + y)dx + (x - y)dy = \int_0^{\pi/2} [(a\cos t + b\sin t)(-a\sin t) + (a\cos t - b\sin t)(b\cos t)]dt$

$= \int_0^{\pi/2} [-(a^2 + b^2)\sin t\cos t + ab(\cos^2 t - \sin^2 t)]dt = \int_0^{\pi/2} \frac{-(a^2 + b^2)\sin 2t}{2} + ab\cos 2t\,dt$

$= \left[\frac{(a^2 + b^2)\cos 2t}{4} + \frac{ab\sin 2t}{2}\right]_0^{\pi/2} = \frac{a^2 + b^2}{-2}$

23. $\int_0^{\pi}\left[\left(\frac{\pi}{2}\right)\sin\left(\frac{\pi t}{2}\right)\cos\left(\frac{\pi t}{2}\right) + \pi t\cos\left(\frac{\pi t}{2}\right) + \sin\left(\frac{\pi t}{2}\right) - t\right]dt = 2 - \frac{2}{\pi} \approx 1.3634$

25. The line integral $\int_{C_i} \mathbf{F} \bullet d\mathbf{r}$ represents the *work* done in moving a particle through the force field $\mathbf{F}$ along the curve C_i, $i = 1, 2, 3,$

a. In the first quadrant, the tangential component to C_1 of each force vector is in the positive y-direction , the same direction as the object moves along C_1. Thus the line integral (work) should be positive.

b. The force vector at each point on C_2 appears to be tangential to the curve, but in the opposite direction as the object moves along C_2. Thus the line integral (work) should be negative.

c. The force vector at each point on C_3 appears to be perpendicular to the curve, and hence has no component in the direction the object is moving. Thus the line integral (work) should be zero

27. $\int_C \left(1+\frac{y}{3}\right) ds = \int_0^{2\pi} (1+10\sin^3 t)[(-90\cos^2 t\sin t)^2 + (90\sin^2 t\cos t)^2]^{1/2}\, dt = 225$

Christy needs $\frac{450}{200} = 2.25$ gal of paint.

29. $C: x + y = a$

Let $x = t$, $y = a - t$, t in $[0, a]$.

Cylinder: $x + y = a$; $(x+y)^2 = a^2$; $x^2 + 2xy + y^2 = a^2$

Sphere: $x^2 + y^2 + z^2 = a^2$

The curve of intersection satisfies: $z^2 = 2xy$; $z = \sqrt{2xy}$.

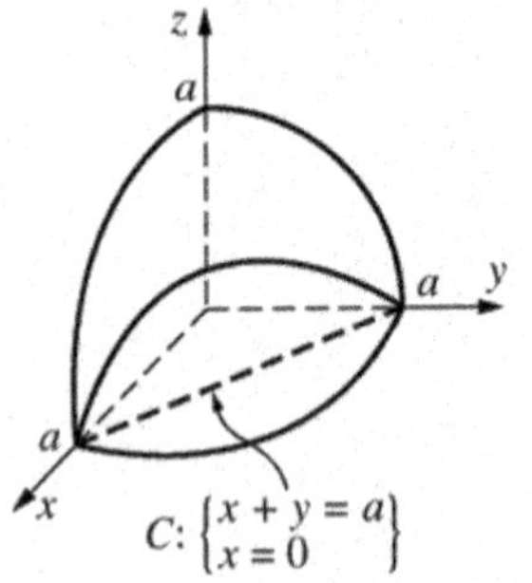

$C: \begin{Bmatrix} x + y = a \\ x = 0 \end{Bmatrix}$

Area $= 8\int_C \sqrt{2xy}\, ds = 8\int_0^a \sqrt{2t(a-t)}\sqrt{(1)^2 + (-1)^2}\, dt = 16\int_0^a \sqrt{at - t^2}\, dt$

$$= 16\left[\frac{t-\frac{a}{2}}{2}\sqrt{at-t^2} + \frac{\left(\frac{a}{2}\right)^2}{2}\sin^{-1}\left(\frac{t-\frac{a}{2}}{\frac{a}{2}}\right)\right]_0^a = 16\left(\left[0+\left(\frac{a^2}{8}\right)\left(\frac{\pi}{2}\right)\right]-\left[0+\left(\frac{a^2}{8}\right)\left(\frac{-\pi}{2}\right)\right]\right)$$

$$= 2a^2\pi$$

Trivial way: Each side of the cylinder is part of a plane that intersects the sphere in a circle. The radius of each circle is the value of z in $z = \sqrt{2xy}$ when $x = y = \frac{a}{2}$. That is, the radius is $\sqrt{2\left(\frac{a}{2}\right)\left(\frac{a}{2}\right)} = \frac{a\sqrt{2}}{2}$. Therefore, the total area of the part cut out is $r\left[\pi\left(\frac{a\sqrt{2}}{2}\right)^2\right] = 2a^2\pi.$

31. $C: x^2 + y^2 = a^2$

Let $x = a\cos\theta, y = a\sin\theta, \theta$ in $\left[0, \frac{\pi}{2}\right]$.

$$\text{Area} = 8\int_C \sqrt{a^2 - x^2}\, ds = 8\int_0^{\pi/2} (a\sin\theta)\sqrt{(-a\sin\theta)^2 + (a\cos\theta)^2}\, d\theta = 8\int_0^{\pi/2} (a\sin\theta)\sqrt{a^2}\, d\theta = 8a^2[-\cos\theta]_0^{\pi/2} = 8a^2$$

33. a. $$\int_C x^2 y\, ds = \int_0^{\pi/2} (3\sin t)^2 (3\cos t)[(3\cos t)^2 + (-3\sin t)^2]^{1/2} dt = 81\int_0^{\pi/2} \sin^2 t\cos t\, dt = 81\left[\left(\frac{1}{3}\right)\sin^3 t\right]_0^{\pi/2} = 27$$

b. $$\int_{C_4} xy^2 dx + xy^2 dy = \int_0^3 (3-t)(5-t)^2(-1)dt + \int_0^3 (3-t)(5-t)^2(-1)dt = 2\int_0^3 (t^3 - 13t^2 + 55t - 75)dt = -148.5$$

14.3 Concepts Review

1. $f(\mathbf{b}) - f(\mathbf{a})$

3. 0; 0

Problem Set 14.3

1. $M_y = -7 = N_x$, so **F** is conservative.

$f(x, y) = 5x^2 - 7xy + y^2 + C$

3. $M_y = 90x^4 y - 36y^5 \neq N_x$ since

$N_x = 90x^4 y - 12y^5$, so **F** is not conservative.

5. $M_y = \left(-\frac{12}{5}\right)x^2 y^{-3} = N_x$, so **F** is conservative.

$f(x, y) = \left(\frac{2}{5}\right)x^3 y^{-2} + C$

7. $M_y = 2e^y - e^x = N_x$ so **F** is conservative.

$f(x, y) = 2xe^y - ye^x + C$

9. $M_y = 0 = N_x, M_z = 0 = P_x$, and $N_z = 0 = P_y$, so **F** is conservative. f satisfies

$f_x(x, y, z) = 3x^2$, $f_y(x, y, z) = 6y^2$, and

$f_z(x, y, z) = 9z^2$.

Therefore, f satisfies

1. $f(x, y, z) = x^3 + C_1(y, z)$,
2. $f(x, y, z) = 2y^3 + C_2(x, z)$, and
3. $f(x, y, z) = 3z^3 + C_3(x, y)$.

A function with an arbitrary constant that satisfies 1, 2, and 3 is

$f(x, y, z) = x^3 + 2y^3 + 3z^3 + C$.

11. Writing **F** in the form

$\mathbf{F}(x, y, z) = M(x, y, z)\mathbf{i} + N(x, y, z)\mathbf{j} + P(x, y, z)\mathbf{k}$

we have $M(x, y, z) = \frac{-2x}{(x^2 + z^2)}$, $N(x, y, z) = 0$,

$$P(x, y, z) = \frac{-2z}{(x^2 + z^2)}$$

so that

$\frac{\partial M}{\partial y} = 0 = \frac{\partial N}{\partial x}, \frac{\partial M}{\partial z} = \frac{4xz}{(x^2 + z^2)^2} = \frac{\partial P}{\partial x}$,

$\frac{\partial N}{\partial z} = 0 = \frac{\partial P}{\partial y}$. Thus **F** is conservative by Thm. D.

We must now find a function $f(x, y, z)$ such that

$\frac{\partial f}{\partial x} = \frac{-2x}{(x^2 + z^2)}, \frac{\partial f}{\partial y} = 0, \frac{\partial f}{\partial z} = \frac{-2z}{(x^2 + z^2)}$.

Note that **F** is a function of x and z alone so f will be a function of x and z alone.

a. Applying the first condition gives

$$f(x,y,z) = \int \frac{-2x}{x^2+z^2}\,dx$$
$$= \ln\left(\frac{1}{x^2+z^2}\right) + C_1(z)$$

b. Applying the second condition,

$$\frac{-2z}{(x^2+z^2)} = \frac{\partial f}{\partial z} = \frac{\partial}{\partial z}\ln\left(\frac{1}{x^2+z^2}\right) + \frac{\partial C_1}{\partial z} =$$
$$\frac{-2z}{(x^2+z^2)} + \frac{\partial C_1}{\partial z} \text{ which requires } \frac{\partial C_1}{\partial z} = 0$$

Hence $f(x,y,z) = \ln\left(\frac{1}{x^2+z^2}\right) + C$

13. $M_y = 2y + 2x = N_x$, so the integral is independent of the path. $f(x, y) = xy^2 + x^2y$

$$\int_{(-1,2)}^{(3,1)} (y^2+2xy)dx + (x^2+2xy)dy$$
$$= [xy^2 + x^2y]_{(-1,2)}^{(3,1)} = 14$$

15. For this problem, we will restrict our consideration to the set
$D = \{(x,y) \mid x > 0, y > 0\}$
(that is, the first quadrant), which is open and simply connected.

a. Now, $\mathbf{F}(x,y) = M\,\mathbf{i} + N\,\mathbf{j}$ where

$$M(x,y) = \frac{x^3}{(x^4+y^4)^2},\quad N(x,y) = \frac{y^3}{(x^4+y^4)^2};$$

thus $\frac{\partial M}{\partial y} = \frac{-8x^3y^3}{(x^4+y^4)^3} = \frac{\partial N}{\partial x}$ so $\mathbf{F}$ is conservative by Thm. D, and hence $\int_C \mathbf{F}(\mathbf{r})\cdot d\mathbf{r}$ is independent of path in D by Theorem C.

b. Since $\mathbf{F}$ is conservative, we can find a function $f(x,y)$ such that

$$\frac{\partial f}{\partial x} = \frac{x^3}{(x^4+y^4)^2},\quad \frac{\partial f}{\partial y} = \frac{y^3}{(x^4+y^4)^2}.$$

Applying the first condition gives

$$f(x,y) = \int \frac{x^3}{(x^4+y^4)^2}\,dx = \frac{-1}{4(x^4+y^4)} + C(y)$$

Appling the second condition gives

$$\frac{y^3}{(x^4+y^4)^2} = \frac{\partial f}{\partial y} = \frac{\partial}{\partial y}\left(\frac{-1}{4(x^4+y^4)}\right) + \frac{\partial C}{\partial y} =$$
$$\frac{y^3}{(x^4+y^4)^2} + \frac{\partial C}{\partial y};$$

hence $\frac{\partial C}{\partial y} = 0$ and $C(y) = constant$. Therefore

$f(x,y) = \frac{-1}{4(x^4+y^4)} + c$, and so, by Thm. A,

$$\int_{(2,1)}^{(6,3)} \frac{x^3}{(x^4+y^4)^2}\,dx + \frac{y^3}{(x^4+y^4)^2}\,dy$$
$$= f(6,3) - f(2,1) = \left(-\frac{1}{5508} + c\right) - \left(-\frac{1}{68} + c\right)$$
$$= \frac{20}{1377}$$

c. Consider the linear path $C: y = \frac{x}{2},\ 2 \le x \le 6$ in D which connects the points $(2, 1)$ and $(6, 3)$; then $dy = \frac{1}{2}dx$. Thus

$$\int_{(2,1)}^{(6,3)} \frac{x^3}{(x^4+y^4)^2}\,dx + \frac{y^3}{(x^4+y^4)^2}\,dy =$$
$$\int_2^6 \frac{x^3}{(x^4+(\frac{x}{2})^4)^2}\,dx + \frac{(\frac{x}{2})^3}{(x^4+(\frac{x}{2})^4)^2}\left(\tfrac{1}{2}dx\right) =$$
$$\int_2^6 \frac{16}{17x^5}\,dx = \left[\frac{-4}{17x^4}\right]_2^6 = -\frac{1}{5508} + \frac{1}{68} =$$
$$\frac{20}{1377}$$

17. $M_y = 18xy^2 = N_x, M_z = 4x = P_x,\ N_z = 0 = P_y$.
By paths (0, 0, 0) to (1, 0, 0); (1, 0, 0) to (1, 1, 0); (1, 1, 0) to (1, 1, 1)

$$\int_0^1 0\,dx + \int_0^1 9y^2\,dy + \int_0^1 (4z+1)dz = 6$$

(Or use $f(x, y) = 3x^2y^3 + 2xz^2 + z$.)

19. $M_y = 1 = N_x, M_z = 1 = P_x,\ N_z = 1 = P_y$ (so path independent). From inspection observe that $f(x, y, z) = xy + xz + yz$ satisfies $f = \langle y+z, x+z, x+y\rangle$, so the integral equals $[xy + xz + yz]_{(0,\,0,\,0)}^{(-1,\,0,\,\pi)} = -\pi$. (Or use line segments $(0,1,\ 0)$ to $(1,1,0)$, then $(1,1,0)$ to $(1,1,1)$.)

21. $f_x = M, f_y = N, f_z = P$
$f_{xy} = M_y$, and $f_{yx} = N_x$, so $M_y = N_x$.
$f_{xz} = M_z$ and $f_{zx} = P_x$, so $M_z = P_x$.
$f_{yz} = N_z$ and $f_{zy} = P_y$, so $N_z = P_y$.

23. $\mathbf{F}(x, y, z) = k|\mathbf{r}|\dfrac{\mathbf{r}}{|\mathbf{r}|} = k\mathbf{r} = k\langle x, y, z\rangle$

$f(x, y, z) = \left(\dfrac{k}{2}\right)(x^2 + y^2 + z^2)$ works.

25. $\int_C \mathbf{F} \cdot d\mathbf{r} = \int_a^b (m\mathbf{r}'' \cdot \mathbf{r}')dt$

$= m\int_a^b (x''x' + y''y' + z''z')dt$

$= m\left[\dfrac{(x')^2}{2} + \dfrac{(y')^2}{2} + \dfrac{(z')^2}{2}\right]_a^b$

$= \dfrac{m}{2}\left[|\mathbf{r}'(t)|^2\right]_a^b = \dfrac{m}{2}\left[|\mathbf{r}'(b)|^2 - |\mathbf{r}'(a)|^2\right]$

27. $f(x,y,z) = -gmz$ satisfies $\nabla f(x, y, z) = \langle 0, 0, -gm\rangle = \mathbf{F}$. Then, assuming the path is piecewise smooth,

$\text{Work} = \int_C \mathbf{F} \cdot d\mathbf{r} = [-gmz]_{(x_1, y_1, z_1)}^{(x_2, y_2, z_2)}$

$= -gm(z_2 - z_1) = gm(z_1 - z_2).$

29. a. $M = \dfrac{y}{(x^2 + y^2)}; M_y = \dfrac{(x^2 - y^2)}{(x^2 + y^2)^2}$

$N = -\dfrac{x}{(x^2 + y^2)}; N_x = \dfrac{(x^2 - y^2)}{(x^2 + y^2)^2}$

b. $M = \dfrac{y}{(x^2 + y^2)} = \dfrac{(\sin t)}{(\cos^2 t + \sin^2 t)} = \sin t$

$N = -\dfrac{x}{(x^2 + y^2)} = \dfrac{(-\cos t)}{(\cos^2 t + \sin^2 t)} = -\cos t$

$\int_C \mathbf{F} \cdot d\mathbf{r} = \int_C M dx + N dy$

$= \int_0^{2\pi} [(\sin t)(-\sin t) + (-\cos t)(\cos t)]dt$

$= -\int_0^{2\pi} 1\, dt = -2\pi \neq 0$

31. Assume the basic hypotheses of Theorem C are satisfied and assume $\int_C \mathbf{F}(\mathbf{r}) \bullet d\mathbf{r} = 0$ for every closed path in D. Choose any two distinct points A and B in D and let C_1, C_2 be arbitrary positively oriented paths *from A to B* in *D.* We must show that

$\int_{C_1} \mathbf{F}(\mathbf{r}) \bullet d\mathbf{r} = \int_{C_2} \mathbf{F}(\mathbf{r}) \bullet d\mathbf{r}$

Let $-C_2$ be the curve C_2 with opposite orientation; then $-C_2$ is a positively oriented path *from B to A* in *D.* Thus the curve $C = C_1 \cup -C_2$ is a closed path (in *D*) between *A* and *B* and so, by our assumption,

$0 = \int_C \mathbf{F}(\mathbf{r}) \bullet d\mathbf{r} = \int_{C_1} \mathbf{F}(\mathbf{r}) \bullet d\mathbf{r} + \int_{-C_2} \mathbf{F}(\mathbf{r}) \bullet d\mathbf{r} =$

$\int_{C_1} \mathbf{F}(\mathbf{r}) \bullet d\mathbf{r} - \int_{C_2} \mathbf{F}(\mathbf{r}) \bullet d\mathbf{r}$

Thus we have $\int_{C_1} \mathbf{F}(\mathbf{r}) \bullet d\mathbf{r} = \int_{C_2} \mathbf{F}(\mathbf{r}) \bullet d\mathbf{r}$ which proves independence of path.

14.4 Concepts Review

1. $\dfrac{\partial N}{\partial x} - \dfrac{\partial M}{\partial y}$

3. source; sink

Problem Set 14.4

1.

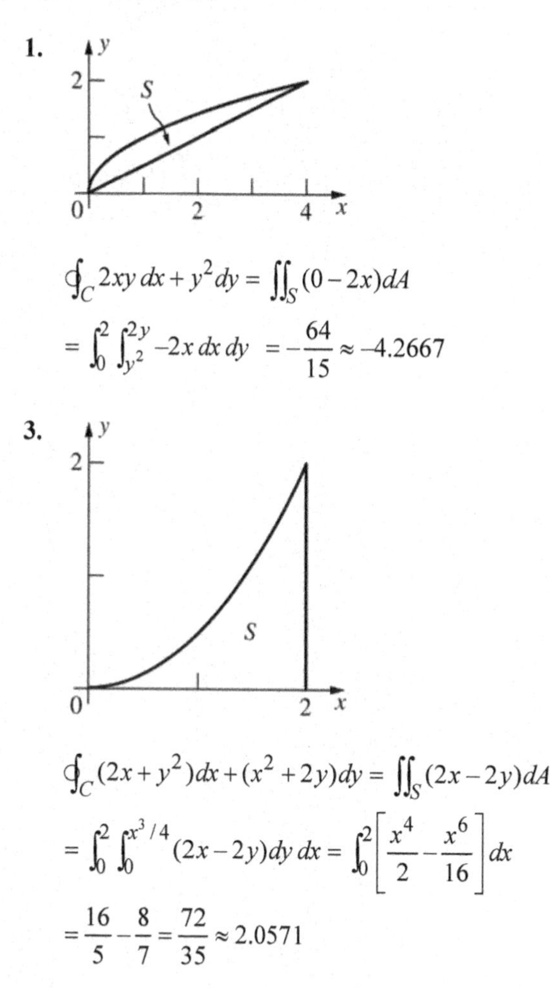

$\oint_C 2xy\, dx + y^2 dy = \iint_S (0 - 2x)dA$

$= \int_0^2 \int_{y^2}^{2y} -2x\, dx\, dy = -\dfrac{64}{15} \approx -4.2667$

3.

$\oint_C (2x + y^2)dx + (x^2 + 2y)dy = \iint_S (2x - 2y)dA$

$= \int_0^2 \int_0^{x^3/4} (2x - 2y)dy\, dx = \int_0^2 \left[\dfrac{x^4}{2} - \dfrac{x^6}{16}\right]dx$

$= \dfrac{16}{5} - \dfrac{8}{7} = \dfrac{72}{35} \approx 2.0571$

5.

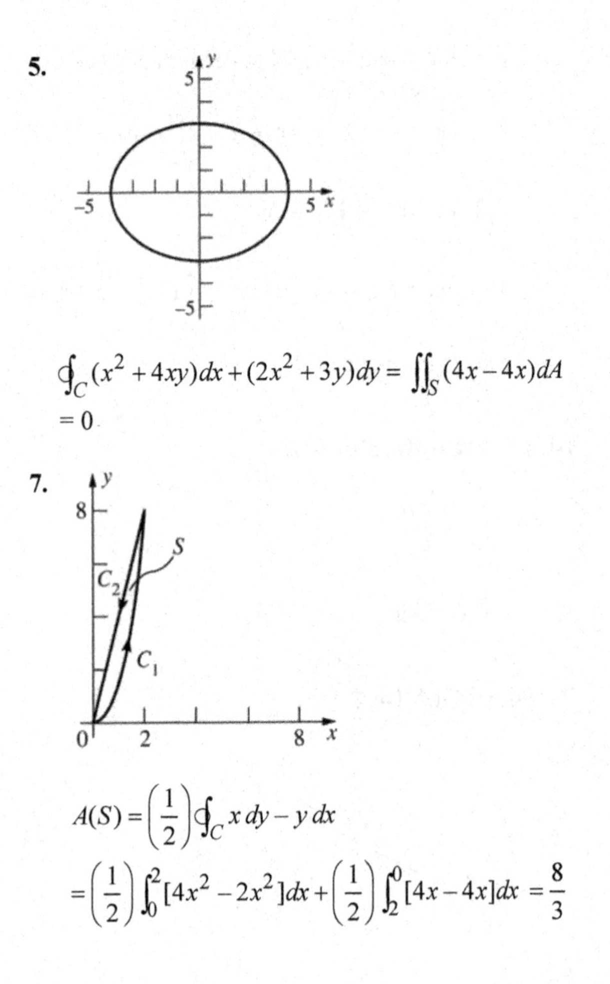

$\oint_C (x^2+4xy)dx+(2x^2+3y)dy = \iint_S (4x-4x)dA$
$= 0$

7.

$A(S) = \left(\frac{1}{2}\right)\oint_C x\,dy - y\,dx$

$= \left(\frac{1}{2}\right)\int_0^2 [4x^2-2x^2]dx + \left(\frac{1}{2}\right)\int_2^0 [4x-4x]dx = \frac{8}{3}$

9.

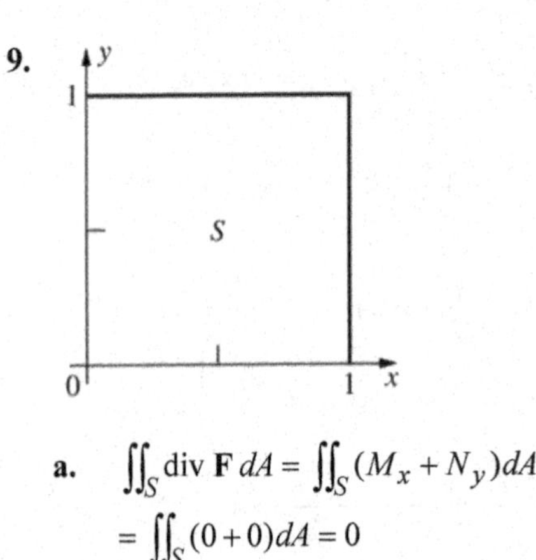

a. $\iint_S \text{div } \mathbf{F}\, dA = \iint_S (M_x+N_y)dA$
$= \iint_S (0+0)dA = 0$

b. $\iint_S (\text{curl } \mathbf{F})\cdot\mathbf{k}\, dA = \iint_S (N_x - M_y)dA$
$= \iint_S (2x-2y)dA = \int_0^1\int_0^1 (2x-2y)dx\,dy$
$= \int_0^1 (1-2y)dy = 0$

11. a. $\iint_S (0+0)dA = 0$

b. $\iint_S (3x^2-3y^2)dA = 0$, since for the integrand, $f(y, x) = -f(x, y)$.

13. $\iint_S (\text{curl } \mathbf{F})\cdot\mathbf{k}dA = \int_{C_1} \mathbf{F}\cdot\mathbf{T}ds - \int_{C_2} \mathbf{F}\cdot\mathbf{T}ds$
$= 30 - (-20) = 50$

15.

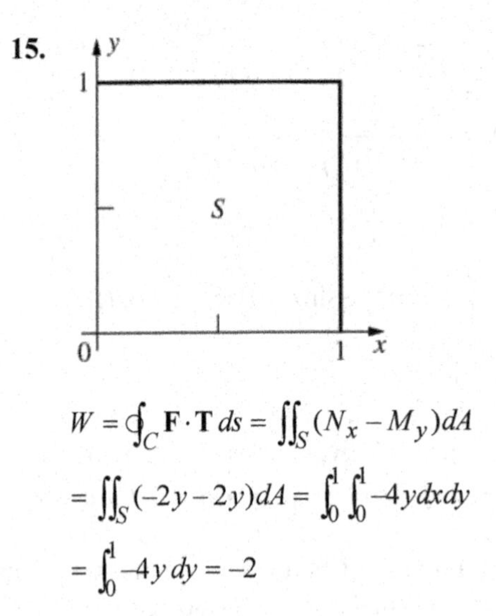

$W = \oint_C \mathbf{F}\cdot\mathbf{T}\,ds = \iint_S (N_x - M_y)dA$
$= \iint_S (-2y-2y)dA = \int_0^1\int_0^1 -4y\,dx\,dy$
$= \int_0^1 -4y\,dy = -2$

17. $\mathbf{F}$ is a constant, so $N_x = M_y = 0$.
$\oint_C \mathbf{F}\cdot\mathbf{T}ds = \iint_S (N_x - M_y)dA = 0$

19. a. Each equals $(x^2 - y^2)(x^2 + y^2)^{-2}$.

b. $\oint_C y(x^2 + y^2)^{-1}dx - x(x^2 + y^2)^{-1}dy = \int_0^{2\pi}(-\sin^2 t - \cos^2 t)dt = \int_0^{2\pi} -1\,dt$

c. M and N are discontinuous at (0, 0).

21. Use Green's Theorem with $M(x, y) = -y$ and $N(x, y) = 0$.
$\oint_C (-y)dx = \iint_S [0 - (-1)]dA = A(S)$
Now use Green's Theorem with $M(x, y) = 0$ and $N(x, y) = x$.
$\oint_C x\,dy = \iint_S (1 - 0)dA = A(S)$

23. $A(S) = \left(\frac{1}{2}\right)\oint_C x\,dy - y\,dx = \left(\frac{1}{2}\right)\int_0^{2\pi}[(a\cos^3 t)(3a\sin^2 t)(\cos t) - (a\sin^3 t)(3a\cos^2 t)(-\sin t)]dt = \left(\frac{3}{8}\right)a^2\pi$

25. a. $\mathbf{F}\cdot\mathbf{n} = \frac{x^2 + y^2}{(x^2 + y^2)^{3/2}} = \frac{1}{(x^2 + y^2)^{1/2}} = \frac{1}{a}$

Therefore, $\int_C \mathbf{F}\cdot\mathbf{n}\,ds = \frac{1}{a}\int_C 1\,ds = \frac{1}{a}(2\pi a) = 2\pi$.

b. $\text{div}\,\mathbf{F} = \frac{(x^2 + y^2)(1) - (x)(2x)}{(x^2 + y^2)^2} + \frac{(x^2 + y^2)(1) - (y)(2y)}{(x^2 + y^2)^2} = 0$

c. $M = \frac{x}{(x^2 + y^2)}$ is not defined at (0, 0) which is inside C.

d. If origin is outside C, then $\oint_C \mathbf{F}\cdot\mathbf{n}\,ds = \iint_S \text{div}\,\mathbf{F}\,dA = \iint_S 0\,dA = 0$.
If origin is inside C, let C' be a circle (centered at the origin) inside C and oriented clockwise. Let S be the region between C and C'. Then $0 = \iint_S \text{div}\,\mathbf{F}\,dA$ (by "origin outside C" case)

$= \int_C \mathbf{F}\cdot\mathbf{n}\,ds - \int_{C'} \mathbf{F}\cdot\mathbf{n}\,ds$ (by Green's Theorem) $= \int_C \mathbf{F}\cdot\mathbf{n}\,ds - 2\pi$ (by part a), so $\int_C \mathbf{F}\cdot\mathbf{n}\,ds = 2\pi$.

27. a. $\text{div}\,\mathbf{F} = 4$

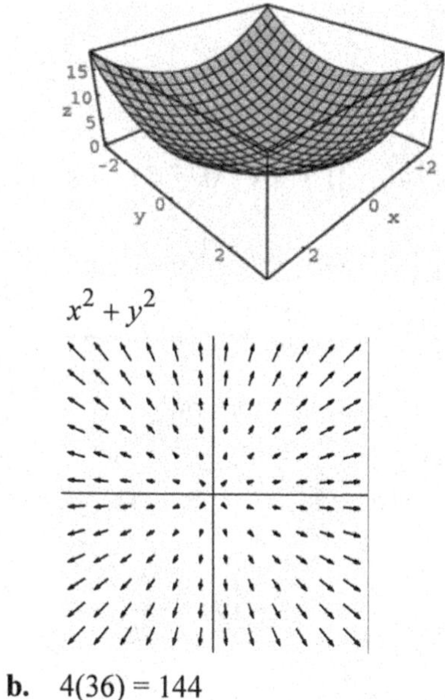

b. $4(36) = 144$

29. a. $\text{div}\,\mathbf{F} = -2\sin x \sin y$
$\text{div}\,\mathbf{F} < 0$ in quadrants I and III
$\text{div}\,\mathbf{F} > 0$ in quadrants II and IV

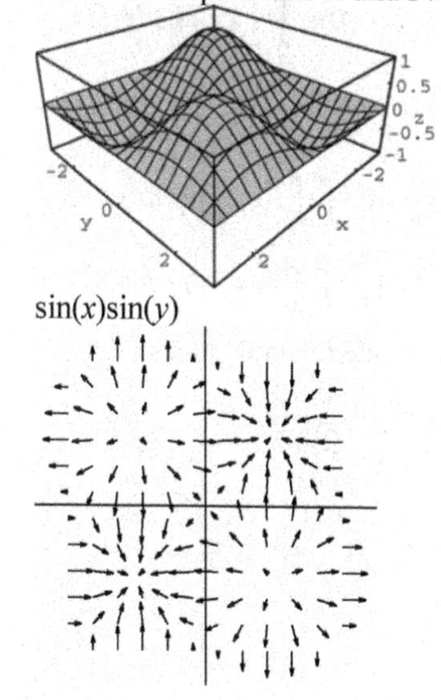

b. Flux across boundary of S is 0.
Flux across boundary T is $-2(1 - \cos 3)^2$.

14.5 Concepts Review

1. surface integral

3. $\sqrt{f_x^2 + f_y^2 + 1}$

Problem Set 14.5

1. $\iint_R [x^2 + y^2 + (x + y + 1)](1 + 1 + 1)^{1/2}\, dA$

$= \int_0^1 \int_0^1 \sqrt{3}(x^2 + y^2 + x + y + 1)\,dx\,dy = \frac{8\sqrt{3}}{3}$

3 $\iint_R (x + y)\sqrt{[-x(4 - x^2)^{-1/2}]^2 + 0 + 1}\, dA$

$= \int_0^{\sqrt{3}} \int_0^1 \frac{2(x + y)}{(4 - x^2)^{1/2}}\,dy\,dx$

$= \int_0^{\sqrt{3}} \frac{2x + 1}{(4 - x^2)^{1/2}}\,dx$

$= \left[-2(4 - x^2)^{1/2} + \sin^{-1}\left(\frac{x}{2}\right)\right]_0^{\sqrt{3}}$

$= \frac{\pi + 6}{3} = 2 + \frac{\pi}{3} \approx 3.0472$

5. $\int_0^{\pi} \int_0^{\sin\theta} (4r^2 + 1)r\,dr\,d\theta = \left(\frac{5}{8}\right)\pi \approx 1.9635$

7. $\iint_R (x + y)(0 + 0 + 1)^{1/2}\, dA$

Bottom ($z = 0$): $\int_0^1 \int_0^1 (x + y)\,dx\,dy = 1$

Top ($z = 1$): Same integral

Left side ($y = 0$): $\int_0^1 \int_0^1 (x + 0)\,dx\,dz = \frac{1}{2}$

Right side ($y = 1$): $\int_0^1 \int_0^1 (x + 1)\,dx\,dz = \frac{3}{2}$

Back ($x = 0$): $\int_0^1 \int_0^1 (0 + y)\,dy\,dz = \frac{1}{2}$

Front ($x = 1$): $\int_0^1 \int_0^1 (1 + y)\,dy\,dz = \frac{3}{2}$

Therefore, the integral equals

$1 + 1 + \frac{1}{2} + \frac{3}{2} + \frac{1}{2} + \frac{3}{2} = 6.$

9.

z
1
y
1
x

$\iint_G \mathbf{F} \cdot \mathbf{n}\, ds = \iint_R (-Mf_x - Nf_y + P)dA$

$= \int_0^1 \int_0^{1-y} (8y + 4x + 0)\,dx\,dy$

$= \int_0^1 [8(1 - y)y + 2(1 - y)^2]dy$

$= \int_0^1 (-6y^2 + 4y + 2)dy = 2$

11. $\int_0^5 \int_{-1}^1 [-xy(1 - y^2)^{-1/2} + 2]dy\,dx = 20$

(In the inside integral, note that the first term is odd in y.)

13. $m = \iint_G kx^2\, ds = \iint_R kx^2\sqrt{3}dA$

$= \sqrt{3}k \int_0^a \int_0^{a-x} x^2\,dy\,dx = \left(\frac{\sqrt{3}k}{12}\right)a^4$

15.

z
a
y
a
a
x

Let $\delta = 1$.

$m = \iint_S 1\, ds = \iint_R (1 + 1 + 1)^{1/2}\, dA$

$= \sqrt{3} \int_0^a \int_0^{a-y} dx\,dy = \sqrt{3} \int_0^a (a - y)dy = \frac{a^2\sqrt{3}}{2}$

$M_{xy} = \iint_S z\, ds = \iint_R (a - x - y)\sqrt{3}\, dA$

$= \sqrt{3} \int_0^a \int_0^{a-y} (a - x - y)\,dx\,dy$

$= \sqrt{3} \int_0^a \left[a(a - y) - \frac{(a - y)^2}{2} - y(a - y)\right]dy$

$= \sqrt{3} \int_0^a \left(\frac{a^2}{2} - ay + \frac{y^2}{2}\right)dy = \frac{a^3\sqrt{3}}{6}$

$\bar{z} = \frac{M_{xy}}{m} = \frac{a}{3}$; then $\bar{x} = \bar{y} = \frac{a}{3}$ (by symmetry).

17. $\mathbf{r}(u,v) = u\,\mathbf{i} + 3v\,\mathbf{j} + \left(4 - u^2 - v^2\right)\mathbf{k}$

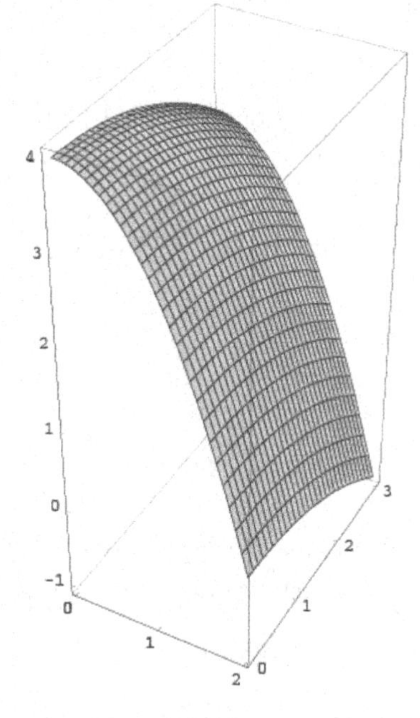

19. $\mathbf{r}(u,v) = 2\cos v\,\mathbf{i} + 3\sin v\,\mathbf{j} + u\,\mathbf{k}$

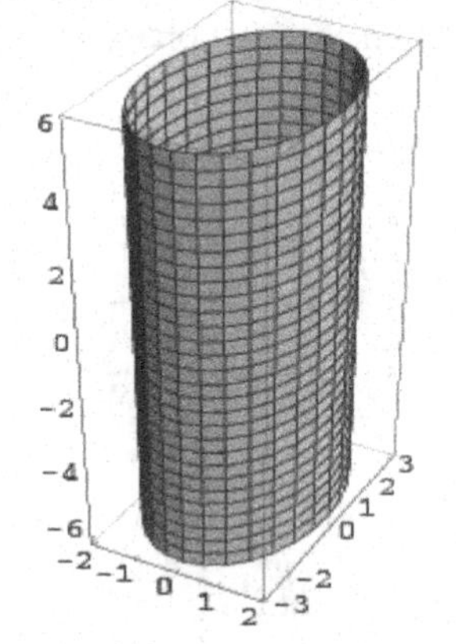

21. $\mathbf{r_u}(u,v) = \sin v\,\mathbf{i} + \cos v\,\mathbf{j} + 0\mathbf{k}$,

$\mathbf{r_v}(u,v) = u\cos v\,\mathbf{i} - u\sin v\,\mathbf{j} + 1\mathbf{k}$

$\mathbf{r_u} \times \mathbf{r_v} = \cos v\,\mathbf{i} - \sin v\,\mathbf{j} - u\,\mathbf{k}$

$\|\mathbf{r_u} \times \mathbf{r_v}\| = \sqrt{\cos^2 v + \sin^2 v + u^2} = \sqrt{1+u^2}$

Using integration formula 44 in the back of the book we get

$$A = \int_{-6}^{6}\int_{0}^{\pi} \sqrt{u^2+1}\,dv\,du = \pi\int_{-6}^{6}\sqrt{u^2+1}\,du =$$

$$\pi\left[\frac{u}{2}\sqrt{u^2+1} + \frac{1}{2}\ln\left|u+\sqrt{u^2+1}\right|\right]_{-6}^{6} =$$

$$\pi\left[6\sqrt{37} + \ln\sqrt{\frac{\sqrt{37}+6}{\sqrt{37}-6}}\right] \approx 122.49$$

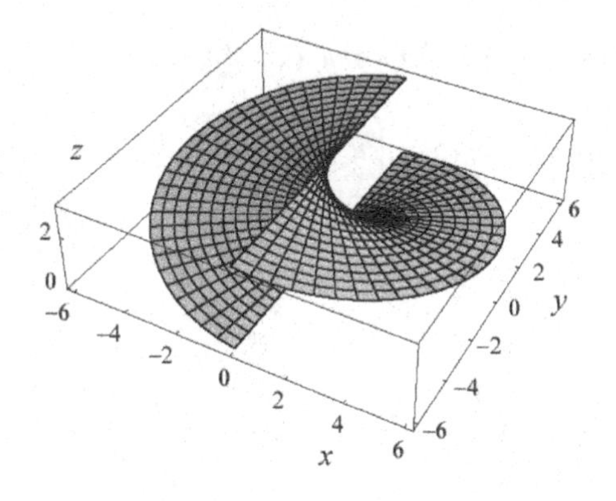

23. $\mathbf{r_u}(u,v) = 2u\cos v\,\mathbf{i} + 2u\sin v\,\mathbf{j} + 5\mathbf{k}$,

$\mathbf{r_v}(u,v) = -u^2\sin v\,\mathbf{i} + u^2\cos v\,\mathbf{j} + 0\mathbf{k}$

$\mathbf{r_u} \times \mathbf{r_v} = -5u^2\cos v\,\mathbf{i} - 5u^2\sin v\,\mathbf{j} + 2u^3\,\mathbf{k}$

$= -u^2\left[5\cos v\,\mathbf{i} + 5\sin v\,\mathbf{j} - 2u\,\mathbf{k}\right]$

$\|\mathbf{r_u} \times \mathbf{r_v}\| = u^2\sqrt{25+4u^2}$

Thus

$$A = \int_{0}^{2\pi}\int_{0}^{2\pi} u^2\sqrt{4u^2+25}\,dv\,du$$

$$= 2\pi\int_{0}^{2\pi} u^2\sqrt{4u^2+25}\,du$$

$$\underset{\substack{w=2u\\dw=2\,du}}{=} 2\pi\int_{0}^{4\pi}\frac{w^2}{4}\sqrt{w^2+(5)^2}\,\left(\tfrac{1}{2}dw\right) =$$

$$\frac{\pi}{4}\int_{0}^{4\pi} w^2\sqrt{w^2+(5)^2}\,dw.$$

Using integration formula 48 in the back of the book we get

$$A = \frac{\pi}{4}\left[\frac{w}{8}\left(2w^2+25\right)\sqrt{w^2+25} - \frac{625}{8}\ln\left|w+\sqrt{w^2+25}\right|\right]_{0}^{4\pi}$$

$$= \frac{\pi}{4}\left[\frac{\pi}{2}\left(32\pi^2+25\right)\sqrt{16\pi^2+25} - \frac{625}{8}\ln\left|4\pi+\sqrt{16\pi^2+25}\right| + \frac{625}{8}\ln 5\right] \approx 5585.42$$

25. $\delta(x,y,z) = k|z| = 5ku \ (k > 0)$. Thus

$$m = \int_0^{2\pi}\int_0^{2\pi}(5ku)\left(u^2\sqrt{4u^2+25}\right)dv\,du =$$

$$5k\int_0^{2\pi} u^3\sqrt{4u^2+25}\,du \underset{\substack{t=4u^2+25 \\ dt=8u\,du}}{=}$$

$$\frac{5k}{8}\int_{25}^{16\pi^2+25}\left(\frac{t-25}{4}\right)\sqrt{t}\,dt =$$

$$\frac{5k}{32}\left[\frac{2}{5}t^{5/2} - \frac{50}{3}t^{3/2}\right]_{25}^{16\pi^2+25} \approx$$

$$\frac{5k}{32}[139760 + 833] \approx 21968k$$

27. $\mathbf{r_u} = -5\sin u \sin v\,\mathbf{i} + 5\cos u \sin v\,\mathbf{j} + 0\mathbf{k}$ and
$\mathbf{r_v} = 5\cos u \cos v\,\mathbf{i} + 5\sin u \cos v\,\mathbf{j} + -5\sin v\,\mathbf{k}$.
Thus,

$$\mathbf{r_u}\times\mathbf{r_v} = \begin{vmatrix} \mathbf{i} & \mathbf{j} & \mathbf{k} \\ -5\sin u \sin v & 5\cos u \sin v & 0 \\ 5\cos u \cos v & 5\sin u \cos v & -5\sin v \end{vmatrix}$$

$$= (-25\cos u \sin^2 v)\mathbf{i} + (-25\sin u \sin^2 v)\mathbf{j} +$$
$$(-25\sin^2 u \sin v \cos v - 25\cos^2 u \sin v \cos v)\mathbf{k}$$
$$= (-25\cos u \sin^2 v)\mathbf{i} + (-25\sin u \sin^2 v)\mathbf{j} +$$
$$+(-25\sin v \cos v)\mathbf{k}$$
$$= (-25\sin v)[\cos u \sin v\,\mathbf{i} + \sin u \sin v\,\mathbf{j} + \cos v\,\mathbf{k}]$$

Thus:

$$\|\mathbf{r_u}\times\mathbf{r_v}\|$$
$$= |-25\sin v|\sqrt{(\cos u \sin v)^2 + (\sin u \sin v)^2 + \cos^2 v}$$
$$= 25|\sin v|\sqrt{[(\cos^2 u + \sin^2 u)\sin^2 v + \cos^2 v}$$
$$= 25|\sin v|\sqrt{\sin^2 v + \cos^2 v} = 25|\sin v|$$

29. a. 0 (By symmetry, since $g(x,y,-z) = -g(x,y,z)$.)

b. 0 (By symmetry, since $g(x,y,-z) = -g(x,y,z)$.)

c. $\iint_G (x^2+y^2+z^2)dS = \iint_G a^2 dS$
$= a^2\text{Area}(G) = a^2(4\pi a^2) = 4\pi a^4$

d. Note:
$\iint_G (x^2+y^2+z^2)dS = \iint_G x^2 dS + \iint_G y^2 dS$
$= \iint_G z^2 dS = 3\iint_G x^2 dS$
(due to symmetry of the sphere with respect to the origin.)
Therefore,
$\iint_G x^2 dS = \left(\frac{1}{3}\right)\iint_G (x^2+y^2+z^2)dS$
$= \left(\frac{1}{3}\right)4\pi a^4 = \frac{4\pi a^4}{3}.$

e. $\iint_G (x^2+y^2)dS = \left(\frac{2}{3}\right)4\pi a^4 = \frac{8\pi a^4}{3}$

31. a. Place center of sphere at the origin.
$F = \iint_G k(a-z)dS = ka\iint_G 1\,dS - k\iint_G z\,dS$
$= ka(4\pi a^2) - 0 = 4\pi a^3 k$

b. Place hemisphere above xy-plane with center at origin and circular base in xy-plane.
F = Force on hemisphere + Force on circular base
$= \iint_G k(a-z)dS + ka(\pi a^2)$
$= ka\iint_G 1\,dS - k\iint_G z\,dS + \pi a^3 k$
$= ka(2\pi a^2) - k\iint_R z\sqrt{\frac{a^2}{a^2-x^2-y^2}}\,dA + \pi a^3 k$
$= 3\pi a^3 k - k\iint_R z\frac{a}{z}\,dA$
$= 3\pi a^3 k - ka(\pi a^2) = 2\pi a^3 k$

c. Place the cylinder above xy-plane with circular base in xy-plane with the center at the origin.
F = Force on top + Force on cylindrical side + Force on base
$= 0 + \iint_G k(h-z)dS + kh(\pi a^2)$
$= kh\iint_G 1\,dS - k\iint_G z\,dS + \pi a^2 hk$
$= kh(2\pi ah) - 4k\iint_R z\sqrt{\frac{a^2}{a^2-y^2} + 0 + 1}\,dA + \pi a^2 hk$
(where R is a region in the yz-plane: $0 \le y = a, 0 \le z \le h$)
$= 2\pi ah^2 k + \pi a^2 hk - 4k\int_0^a\int_0^h \frac{az}{\sqrt{a^2-y^2}}\,dz\,dy$
$= 2\pi ah^2 k + \pi a^2 hk - \pi kah^2$
$= \pi ah^2 k + \pi a^2 hk = \pi ahk(h+a)$

14.6 Concepts Review

1. boundary; ∂S

3. div **F**

Problem Set 14.6

1. $\iiint_S (0+0+0)dV = 0$

3. $\iint_{\partial S} \mathbf{F}\cdot\mathbf{n}\, dS = \iiint_S (M_x + N_y + P_z)dV = \int_{-1}^{1}\int_{-1}^{1}\int_{-1}^{1}(0+1+0)dx\, dy\, dz = 8$

5. $\iint_{\partial S} \mathbf{F}\cdot\mathbf{n}\, dS = \iiint_S (M_x + N_y + P_z)dV = \int_0^c\int_0^b\int_0^a (2xyz+2xyz+2xyz)dx\, dy\, dz = \int_0^c\int_0^b 3a^2yz\, dy\, dz = \int_0^c \frac{3a^2b^2z}{2}dz$
$= \frac{3a^2b^2c^2}{4}$

7. $2\iiint_S (x+y+z)dV = 2\int_0^{2\pi}\int_0^2\int_0^{4-r^2}(r\cos\theta + r\sin\theta + z)r\, dz\, dr\, d\theta = \frac{64\pi}{3} \approx 67.02$

9. $\iiint_S (1+1+0)dV = 2(\text{volume of cylinder}) = 2\pi(1)^2(2) = 4\pi \approx 12.5664$

11. $\iiint_S (M_x + N_y + P_z)dV = \iiint_S (2+3+4)dV$
$= 9(\text{Volume of spherical shell})$
$= 9\left(\frac{4\pi}{3}\right)(5^3 - 3^3) = 1176\pi \approx 3694.51$

13. $\iint_{\partial S} \mathbf{F}\cdot\mathbf{n}\, dS = \iiint_S (M_x + N_y + P_z)\, dS =$
$\iiint_S (0+2y+0)\, dV = 2\iiint_S y\, dx\, dy\, dz$
Using the change of variable (from (x,y,z) to (r,y,θ)) defined by
$x = r\cos\theta,\ y = y,\ z = r\sin\theta$ yields the Jacobian

$$J(r,y,\theta) = \begin{vmatrix} \frac{\partial x}{\partial r} & \frac{\partial y}{\partial r} & \frac{\partial z}{\partial r} \\ \frac{\partial x}{\partial y} & \frac{\partial y}{\partial y} & \frac{\partial z}{\partial y} \\ \frac{\partial x}{\partial \theta} & \frac{\partial y}{\partial \theta} & \frac{\partial z}{\partial \theta} \end{vmatrix} = \begin{vmatrix} \cos\theta & 0 & \sin\theta \\ 0 & 1 & 0 \\ -r\sin\theta & 0 & r\cos\theta \end{vmatrix} =$$

$r\cos^2\theta + r\sin^2\theta = r$. Further, the region S is now defined by $r^2 \le 1,\ 0 \le y \le 10$. Hence, by the change of variable formula in Section 13.9,
$2\iiint_S y\, dx\, dy\, dz = 2\int_0^{2\pi}\int_0^{10}\int_0^1 yr\, dr\, dy\, d\theta =$
$\int_0^{2\pi}\int_0^{10} y\, dy\, d\theta = \int_0^{2\pi} 50\, d\theta = 100\pi \approx 314.16$

15. $\left(\frac{1}{3}\right)\iiint_S (1+1+1)dV = V(S)$

17. Note:
1. $\iint_R (ax+by+cz)dS = \iint_R d\, dS = dD$ (R is the slanted face.)
2. $\mathbf{n} = \frac{\langle a, b, c\rangle}{(a^2+b^2+c^2)^{1/2}}$ (for slanted face)
3. $\mathbf{F}\cdot\mathbf{n} = 0$ on each coordinate-plane face.

Volume $= \left(\frac{1}{3}\right)\iint_S \mathbf{F}\cdot\mathbf{n}\, dS$ (where $\mathbf{F} = \langle x, y, z\rangle$).
$= \left(\frac{1}{3}\right)\iint_R \mathbf{F}\cdot\mathbf{n}\, dS$ (by Note 3)
$= \left(\frac{1}{3}\right)\iint_R \frac{(ax+by+cz)}{\sqrt{a^2+b^2+c^2}}dS$
$= \frac{dD}{3\sqrt{a^2+b^2+c^2}}$

19. a. div $\mathbf{F} = 2 + 3 + 2z = 5 + 2z$

$$\iint_{\partial S} \mathbf{F}\cdot\mathbf{n}\,dS = \iiint_S (5+2z)dV = \iiint_S 5\,dV + 2\iiint_S z\,dV = 5\text{ (Volume of } S) + 2M_{xy}$$

$$= 5\left(\frac{4\pi}{3}\right) + 2\bar{z}\text{ (Volume of } S) = \frac{20\pi}{3} + 2(0)(\text{Volume}) = \frac{20\pi}{3}$$

b. $\mathbf{F}\cdot\mathbf{n} = (x^2+y^2+z^2)^{3/2}\langle x, y, z\rangle \cdot \dfrac{\langle x, y, z\rangle}{\sqrt{x^2+y^2+z^2}} = (x^2+y^2+z^2)^2 = 1$ on ∂S.

$$\iint_{\partial S} \mathbf{F}\cdot\mathbf{n}\,dS = \iiint_S 1\,dV = 4\pi(1)^2 = 4\pi$$

c. div $\mathbf{F} = 2x + 2y + 2z$

$$\iint_{\partial S} \mathbf{F}\cdot\mathbf{n}\,dS = \iiint_S 2(x+y+z)dV$$

$= 2\iiint_S x\,dV$ (Since $\bar{x} = \bar{z} = 0$ as in a.)

$$= 2M_{yz} = 2(\bar{x})(\text{Volume of } S) = 2(2)\left(\frac{4\pi}{3}\right) = \frac{16\pi}{3}$$

d. $\mathbf{F}\cdot\mathbf{n} = 0$ on each face except the face R in the plane $x = 1$.

$$\iint_{\partial S} \mathbf{F}\cdot\mathbf{n}\,dS = \iint_R \mathbf{F}\cdot\mathbf{n}\,dS = \iint_R \langle 1, 0, 0\rangle\cdot\langle 1, 0, 0\rangle dS = \iint_R 1\,dS = (1)^2 = 1$$

e. div $\mathbf{F} = 1 + 1 + 1 = 3$

$$\iint_{\partial S} \mathbf{F}\cdot\mathbf{n}\,dS = \iiint_S 3\,dV = 3(\text{Volume of } S) = 3\left(\frac{1}{3}\left[\frac{1}{2}(4)(3)\right](6)\right) = 36$$

f.

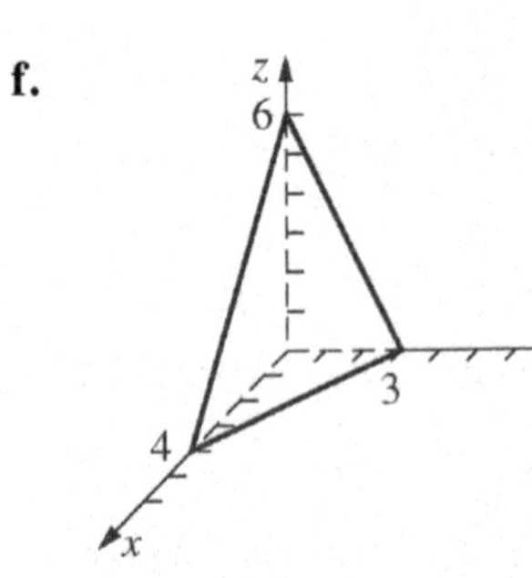

div $\mathbf{F} = 3x^2 + 3y^2 + 3z^2 = 3(x^2+y^2+z^2) = 3$ on ∂S.

$$\iint_{\partial S} \mathbf{F}\cdot\mathbf{n}\,dS = 3\iiint_S (x^2+y^2+z^2)dV = 3\left(\frac{3}{2}\frac{8\pi}{15}\right) = \frac{12\pi}{5}$$

(That answer can be obtained by making use of symmetry and a change to spherical coordinates. Or you could go to the solution for Problem 22, Section 13.9, and realize that the value of the integral in this problem is $\frac{3}{2}$.

g. $\mathbf{F}\cdot\mathbf{n} = [\ln(x^2+y^2)]\langle x, y, 0\rangle\cdot\langle 0, 0, 1\rangle = 0$ on top and bottom.

$\mathbf{F}\cdot\mathbf{n} = (\ln 4)\langle x, y, 0\rangle \cdot \dfrac{\langle x, y, 0\rangle}{\sqrt{x^2+y^2}} = (\ln 4)\sqrt{x^2+y^2} = (\ln 4)\sqrt{4} = 2\ln 4 = 4\ln 2$ on the side.

$$\iint_{\partial S} \mathbf{F}\cdot\mathbf{n}\,dS = \iint_R 4\ln 2\,dS = (4\ln 2)[2\pi(2)(2)] = 32\pi\ln 2$$

21. $\iint_{\partial S} D_{\mathbf{n}} f\,dS = \iint_{\partial S} \nabla f\cdot\mathbf{n}\,dS = \iiint_S \text{div}(\nabla f)dV = \iiint_S \nabla^2 f\,dV$ (See next problem.)

23. $\iint_{\partial S} fD_{\mathbf{n}} g\,dS = \iint_{\partial S} f(\nabla g\cdot\mathbf{n})dS = \iint_{\partial S} (f\nabla g)\cdot\mathbf{n}\,dS = \iiint_S \text{div}(f\nabla g)dV$ (Gauss)

$= \iiint_S [f(\text{div}\,\nabla g) + (\nabla f)\cdot(\nabla g)]dV = \iiint_S (f\nabla^2 g + \nabla f\cdot\nabla g)dV$ (See Problem 20c, Section 14.1.)

14.7 Concepts Review

1. $(\text{curl } \mathbf{F}) \cdot \mathbf{n}$

3. $\iint_S (\text{curl } \mathbf{F}) \cdot \mathbf{n}\, dS$

Problem Set 14.7

1. $\oint_{\partial S} \mathbf{F} \cdot \mathbf{T}\, dS = \iint_R (N_x - M_y)\, dA = \iint_R 0\, dA = 0$

3.

$$\iint_S (\text{curl } \mathbf{F}) \cdot \mathbf{n}\, dS = \oint_{\partial S} \mathbf{F} \cdot \mathbf{T}\, ds = \oint_{\partial S} (y+z)dx + (x^2+z^2)dy + y\, dz$$

$$= \int_0^1 1\, dt + \int_0^{\pi} [(1+\sin t)(-\sin t) + \cos t]dt + \int_0^1 -1\, dt + \int_0^{\pi} \sin^2 t\, dt \quad (*)$$

$$= \int_0^{\pi} (-\sin t + \cos t)dt = -2$$

The result at (*) was obtained by integrating along S by doing so along C_1, C_2, C_3, C_4 in that order.

Along C_1 : $x = 1, y = t, z = 0, dx = dz = 0, dy = dt, t$ in $[0, 1]$

Along C_2 : $x = \cos t, y = 1, z = \sin t, dx = -\sin dt, dy = 0, dz = \cos t\, dt, t$ in $[0, \pi]$

Along C_3 : $x = -1, y = 1 - t, z = 0, dx = dz = 0, dy = dt, t$ in $[0, 1]$

Along C_4 : $x = -\cos t, y = 0, z = \sin t, dx = \sin t\, dt, dy = 0, dz = \cos t\, dt, t$ in $[0, \pi]$

5. ∂S is the circle $x^2 + y^2 = 12,\ z = 2$.

Parameterization of circle:

$x = \sqrt{12} \sin t, y = \sqrt{12} \cos t,\ z = 2, t$ in $[0, 2\pi]$

$$\oint_{\partial S} \mathbf{F} \cdot \mathbf{T}\, ds = \oint_{\partial S} yz\, dx + 3xz\, dy + z^2 dz$$

$$= \int_0^{2\pi} (24 \sin^2 t - 72 \cos^2 t)dt = -48\pi \approx -150.80$$

7. $(\text{curl } \mathbf{F}) \cdot \mathbf{n} = \langle 3, 2, 1 \rangle \cdot \left\langle \left(\frac{1}{\sqrt{2}}\right) 1, 0, -1 \right\rangle = \sqrt{2}$

$$\oint_{\partial S} \mathbf{F} \cdot \mathbf{T}\, ds = \iint_S \sqrt{2} dS = \sqrt{2} A(S)$$

$= \sqrt{2}[\sec(45°)]$(Area of a circle)$= 8\pi \approx 25.1327$

9. $(\text{curl } \mathbf{F}) = \langle -1+1, 0-1, 1-1 \rangle = \langle 0, -1, 0 \rangle$

The unit normal vector that is needed to apply Stokes' Theorem points downward. It is

$$n = \frac{\langle -1, -2, -1 \rangle}{\sqrt{6}}.$$

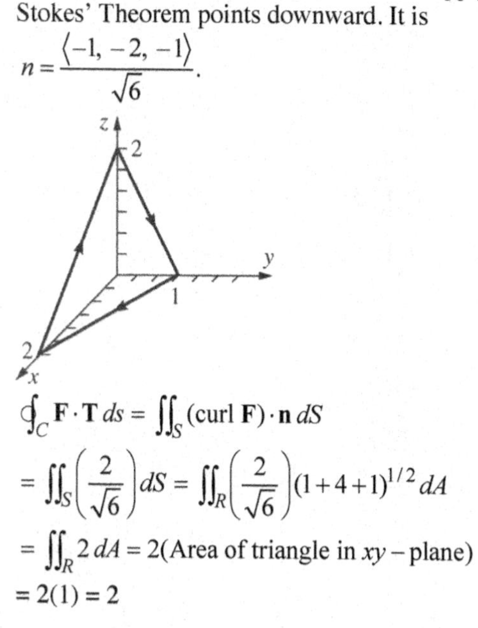

$$\oint_C \mathbf{F} \cdot \mathbf{T}\, ds = \iint_S (\text{curl } \mathbf{F}) \cdot \mathbf{n}\, dS$$

$$= \iint_S \left(\frac{2}{\sqrt{6}}\right) dS = \iint_R \left(\frac{2}{\sqrt{6}}\right) (1+4+1)^{1/2} dA$$

$= \iint_R 2\, dA = 2$(Area of triangle in xy – plane)

$= 2(1) = 2$

11. $(\text{curl } \mathbf{F}) \cdot \mathbf{n} = \langle 0, 0, 1\rangle \cdot \langle x, y, z\rangle = z$

$\iint_S z\, dS = \iint_R 1\, dA = \text{Area of } R$

$\pi\left(\frac{1}{2}\right)^2 = \frac{\pi}{4} \approx 0.7854$

13. Let $H(x, y, z) = z - g(x, y) = 0$.

Then $\mathbf{n} = \frac{\nabla H}{|\nabla H|} = \frac{\langle -g_x, g_y, 1\rangle}{\sqrt{1 + g_x^2 + g_y^2}}$ points upward.

Thus,

$\iint_S (\text{curl } \mathbf{F}) \cdot \mathbf{n}\, dS = \iint_{S_{xy}} (\text{curl } \mathbf{F}) \cdot \mathbf{n} \sec\gamma\, dA$

$= \iint_{S_{xy}} (\text{curl } \mathbf{F}) \frac{\langle -g_x, -g_y, 1\rangle}{\sqrt{g_x^2 + g_y^2 + 1}} \sqrt{g_x^2 + g_y^2 + 1}\, dA$

(Theorem A, Section 14.5)

$= \iint_{S_{xy}} (\text{curl } \mathbf{F}) \cdot \langle -g_x, -g_y, 1\rangle\, dA$

15. $\text{curl } F = \langle 0 - x, 0 - 0, z - 0\rangle = \langle -x, 0, z\rangle$

$\oint_C \mathbf{F} \cdot \mathbf{T}\, ds = \iint_S (\text{curl } \mathbf{F}) \cdot \mathbf{n}\, dS$

$= \iint_{S_{xy}} (\text{curl } \mathbf{F}) \cdot \langle -g_x, -g_y, 1\rangle\, dA$ where

$z = g(x, y) = xy^2$. (Problem 13)

$= \iint_{S_{xy}} \langle -x, 0, z\rangle \cdot \langle -y^2, -2xy, 1\rangle\, dA$

$= \int_0^1 \int_0^1 (xy^2 + 0 + xy^2)\, dx\, dy$

$= \int_0^1 \left([x^2 y^2]_{x=0}^1\right) dy = \int_0^1 y^2\, dy = \frac{1}{3}$

17. $\oint_{\partial S} \mathbf{F} \cdot \mathbf{T}\, ds = \iint_S (\text{curl } \mathbf{F}) \cdot \mathbf{n}\, dS$

$= \iint_{S_{xy}} (\text{curl } \mathbf{F}) \cdot \langle -g_x, -g_y, 1\rangle\, dA$

$= \iint_{S_{xy}} \langle 2, 2, 0\rangle \cdot \left[\frac{\langle x, y, (a^2 - x^2 - y^2)^{-1/2}\rangle}{(a^2 - x^2 - y^2)^{-1/2}}\right] dA$

$= 2\iint_{S_{xy}} (x + y)(a^2 - x^2 - y^2)^{-1/2}\, dA$

$= 2\iint_{S_{xy}} y(a^2 - x^2 - y^2)^{-1/2}\, dA$

$= 4\int_0^{\pi/2} \int_0^{a\sin\theta} (r\sin\theta)(a^2 - r^2)^{-1/2}\, dr\, d\theta$

$= \frac{4a^2}{3}$ joules

19. a. Let C be any piecewise smooth simple closed oriented curve C that separates the "nice" surface into two "nice" surfaces, S_1 and S_2.

$\iint_{\partial S} (\text{curl } \mathbf{F}) \cdot \mathbf{n}\, dS$

$= \iint_{S_1} (\text{curl } \mathbf{F}) \cdot \mathbf{n}\, dS + \iint_{S_2} (\text{curl } \mathbf{F}) \cdot \mathbf{n}\, dS$

$= \oint_C \mathbf{F} \cdot \mathbf{T}\, ds + \oint_{-C} \mathbf{F} \cdot \mathbf{T}\, ds = 0$ ($-C$ is C with opposite orientation.)

b. div(curl **F**) = 0 (See Problem 20, Section 14.1.) Result follows.

14.8 Chapter Review

Concepts Test

1. True: See Example 4, Section 14.1

3. False: grad(curl **F**) is not defined since curl **F** is not a scalar field.

5. True: See the three equivalent conditions in Section 14.3.

7. False: $N_z = 0 \neq z^2 = P_y$

9. True: It is the case in which the surface is in a plane.

11. True: See discussion on text page 752.

Sample Test Problems

1.

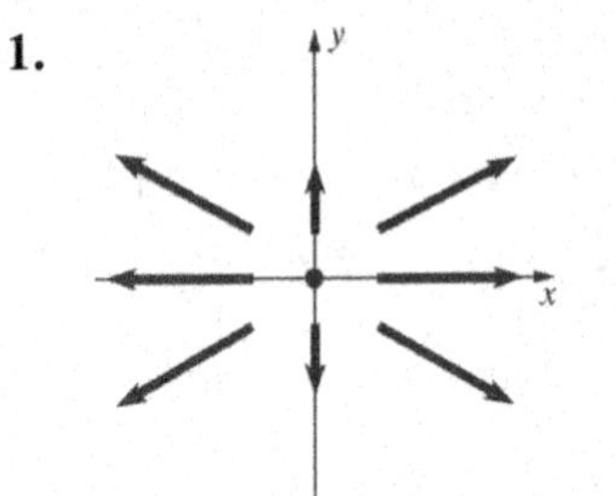

3. $\text{curl}(f\nabla f) = (f)(\text{curl } \nabla f) + (\nabla f \times \nabla f)$

$= (f)(\mathbf{0}) + \mathbf{0} = \mathbf{0}$

5. a. Parameterization is $x = \sin t$, $y = -\cos t$, t in $\left[0, \frac{\pi}{2}\right]$.

$\int_0^{\pi/2} (1 - \cos^2 t)(\sin^2 t + \cos^2 t)^{1/2}\, dt = \frac{\pi}{4} \approx 0.7854$

b $\int_0^{\pi/2} [t\cos t - \sin^2 t\cos t + \sin t\cos t]dt$

$= \frac{(3\pi - 5)}{6} \approx 0.7375$

7. $[xy^2]_{(1,1)}^{(3,4)} = 47$

9. a.

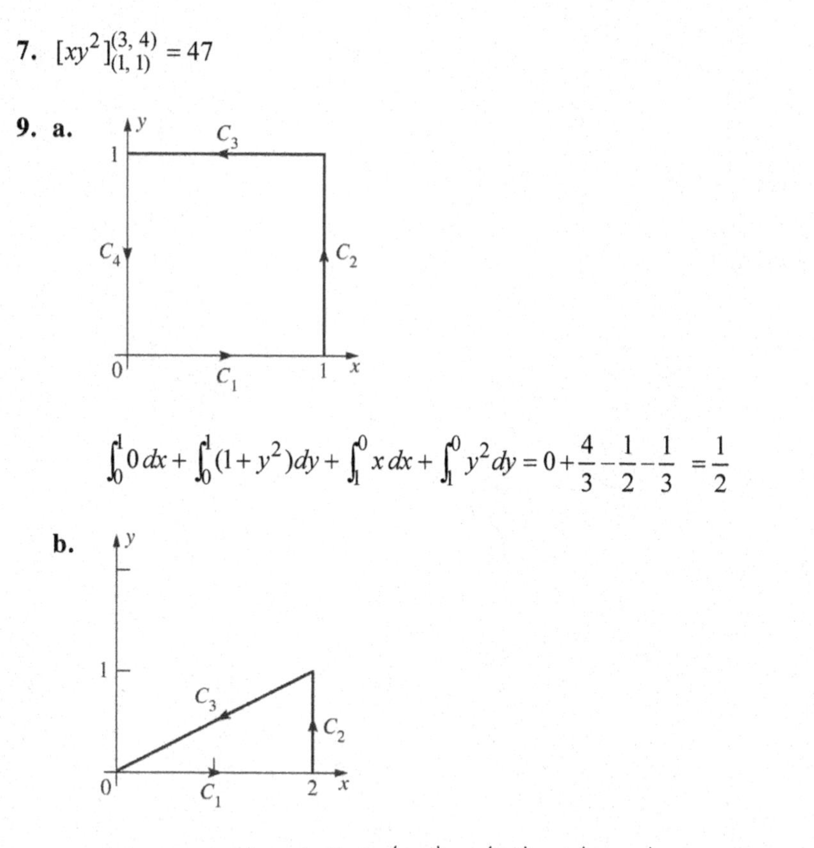

$$\int_0^1 0\,dx + \int_0^1 (1+y^2)dy + \int_1^0 x\,dx + \int_1^0 y^2 dy = 0 + \frac{4}{3} - \frac{1}{2} - \frac{1}{3} = \frac{1}{2}$$

b.

A vector equation of C_3 is $\langle x, y\rangle = \langle 2,1\rangle + t\langle -2,-1\rangle$ for t in $[0, 1]$, so let $x = 2 - 2t$, $y = 1 - t$ for t in $[0, 1]$ be parametric equations of C_3.

$$\int_0^2 0\,dx + \int_0^1 (4+y^2)dy + \int_0^1 [2(1-t)^2(-2) + 5(1-t)^2(-1)]dt = 0 + \frac{13}{3} - 3 = \frac{4}{3}$$

c.

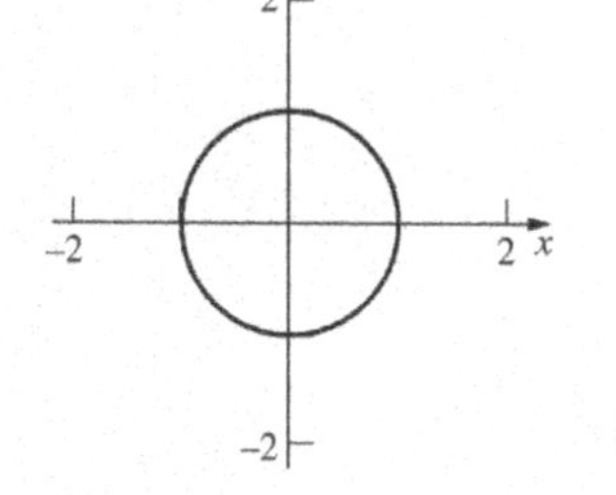

$x = \cos t$
$y = -\sin t$
t in $[0, 2\pi]$

$$\int_0^{2\pi} [(\cos t)(\sin t)(-\sin t) + (\cos^2 t + \sin^2 t)(\cos t)]dt = \int_0^{2\pi} (1 - \sin^2 t)\cos t\,dt = \left[\sin t - \frac{\sin^3 t}{3}\right]_0^{2\pi} = 0$$

11. Let $f(x, y) = (1 - x^2 - y^2)$ and $g(x, y) = -(1 - x^2 - y^2)$, the upper and lower hemispheres.

Then $\text{Flux} = \iint_G \mathbf{F} \cdot \mathbf{n}\,dS$

$= \iint_R [-Mf_x - Nf_y + P]dA + \iint_R [-Mg_x - Ng_y + P]dA = \iint_R 2P\,dA$ (since $f_x = -g_x$ and $f_y = -g_y$)

$= \iint_R 6\,dA = 6$ (Area of R, the circle $x^2 + y^2 = 1$, $z = 0$)

$= 6\pi \approx 18.8496$

13. ∂S is the circle $x^2 + y^2 = 1,\ z = 1$.
A parameterization of the circle is $x = \cos t$, $y = \sin t$, $z = 1$, t in $[0, 2\pi]$.

$$\oint_{\partial S} \mathbf{F} \cdot \mathbf{T}\, ds = \oint_{\partial S} \left[x^3 y\, dx + e^y\, dy + z \tan\left(\frac{xyz}{4}\right) dz \right] = \oint_{\partial S} (x^3 y + e^y\, dy)$$

$$= \int_0^{2\pi} [(\cos t)^3 (\sin t)(-\sin t) + (e^{\sin t})(\cos t)]\, dt = 0$$

15. curl $\mathbf{F} = \langle 3-0, 0-0, -1-1 \rangle = \langle 3, 0, -2 \rangle$

$$\mathbf{n} = \frac{\langle a, b, 1 \rangle}{\sqrt{a^2 + b^2 + 1}}$$

$$\oint_C \mathbf{F} \cdot \mathbf{T}\, ds = \iint_S (\text{curl } \mathbf{F}) \cdot \mathbf{n}\, dS = \iint_S \frac{3a-2}{\sqrt{a^2+b^2+1}}\, dS = \frac{3a-2}{\sqrt{a^2+b^2+1}} [A(S)]$$

$$= \frac{3a-2}{\sqrt{a^2+b^2+1}} (9\pi) \quad (S \text{ is a circle of radius } 3.)$$

$$= \frac{9\pi(3a-2)}{\sqrt{a^2+b^2+1}}$$

CHAPTER 15

Differential Equations

15.1 Concepts Review

1. $r^2 + a_1 r + a_2 = 0$; complex conjugate roots

3. $(C_1 + C_2 x)e^x$

Problem Set 15.1

1. Roots are 2 and 3. General solution is $y = C_1 e^{2x} + C_2 e^{3x}$.

3. Auxiliary equation: $r^2 + 6r - 7 = 0$, $(r + 7)(r - 1) = 0$ has roots $-7, 1$.
General solution: $y = C_1 e^{-7x} + C_2 e^x$
$y' = -7C_1 e^{-7x} + C_2 e^x$
If $x = 0, y = 0, y' = 4$, then $0 = C_1 + C_2$ and $4 = -7C_1 + C_2$, so $C_1 = -\frac{1}{2}$ and $C_2 = \frac{1}{2}$.
Therefore, $y = \frac{e^x - e^{-7x}}{2}$.

5. Repeated root 2. General solution is $y = (C_1 + C_2 x)e^{2x}$.

7. Roots are $2 \pm \sqrt{3}$. General solution is $y = e^{2x}(C_1 e^{\sqrt{3}x} + C_2 e^{-\sqrt{3}x})$.

9. Auxiliary equation: $r^2 + 4 = 0$ has roots $\pm 2i$.
General solution: $y = C_1 \cos 2x + C_2 \sin 2x$
If $x = 0$ and $y = 2$, then $2 = C_1$; if $x = \frac{\pi}{4}$ and $y = 3$, then $3 = C_2$.
Therefore, $y = 2\cos 2x + 3\sin 2x$.

11. Roots are $-1 \pm i$. General solution is $y = e^{-x}(C_1 \cos x + C_2 \sin x)$.

13. Roots are $0, 0, -4, 1$.
General solution is
$y = C_1 + C_2 x + C_3 e^{-4x} + C_4 e^x$.

15. Auxiliary equation: $r^4 + 3r^2 - 4 = 0$, $(r+1)(r-1)(r^2+4) = 0$ has roots $-1, 1, \pm 2i$.
General solution:
$y = C_1 e^{-x} + C_2 e^x + C_3 \cos 2x + C_4 \sin 2x$

17. Roots are $-2, 2$. General solution is $y = C_1 e^{-2x} + C_2 e^{2x}$.
$y = C_1(\cosh 2x - \sinh 2x) + C_2(\sinh 2x + \cosh 2x) = (-C_1 + C_2)\sinh 2x + (C_1 + C_2)\cosh 2x$
$= D_1 \sinh 2x + D_2 \cosh 2x$

19. Repeated roots $\left(-\frac{1}{2}\right) \pm \left(\frac{\sqrt{3}}{2}\right)i$.

General solution is $y = e^{-x/2}\left[(C_1 + C_2 x)\cos\left(\frac{\sqrt{3}}{2}\right)x + (C_3 + C_4 x)\sin\left(\frac{\sqrt{3}}{2}\right)x\right]$.

21. $(*) x^2y'' + 5xy' + 4y = 0$

Let $x = e^z$. Then $z = \ln x$; $y' = \frac{dy}{dx} = \frac{dy}{dz}\frac{dz}{dx} = \frac{dy}{dz}\frac{1}{x}$;

$$y'' = \frac{dy'}{dx} = \frac{d}{dx}\left(\frac{dy}{dz}\frac{1}{x}\right) = \frac{dy}{dz}\frac{-1}{x^2} + \frac{1}{x}\frac{d^2y}{dz^2}\frac{dz}{dx}$$

$$= \frac{dy}{dz}\frac{-1}{x^2} + \frac{1}{x}\frac{d^2y}{dz^2}\frac{1}{x}$$

$$\left(-\frac{dy}{dz} + \frac{d^2y}{dz^2}\right) + \left(5\frac{dy}{dz}\right) + 4y = 0$$

(Substituting y' and y'' into (*))

$$\frac{d^2y}{dz^2} + 4\frac{dy}{dz} + 4y = 0$$

Auxiliary equation: $r^2 + 4r + 4 = 0$, $(r+2)^2 = 0$ has roots $-2, -2$.

General solution: $y = (C_1 + C_2 z)e^{-2z}$, $y = (C_1 + C_2 \ln x)e^{-2\ln x}$

$y = (C_1 + C_2 \ln x)x^{-2}$

23. We need to show that $y'' + a_1y' + a_2y = 0$ if r_1 and r_2 are distinct real roots of the auxiliary equation. We have,

$$y' = C_1r_1e^{r_1x} + C_2r_2e^{r_2x}$$
$$y'' = C_1r_1^2e^{r_1x} + C_2r_2^2e^{r_2x}.$$

When put into the differential equation, we obtain

$$y'' + a_1y' + a_2y = C_1r_1^2e^{r_1x} + C_2r_2^2e^{r_2x} + a_1\left(C_1r_1e^{r_1x} + C_2r_2e^{r_2x}\right) + a_2\left(C_1e^{r_1x} + C_2e^{r_2x}\right) \quad (*)$$

The solutions to the auxiliary equation are given by

$r_1 = \frac{1}{2}\left(-a_1 - \sqrt{a_1^2 - 4a_2}\right)$ and

$r_2 = \frac{1}{2}\left(-a_1 + \sqrt{a_1^2 - 4a_2}\right)$.

Putting these values into (*) and simplifying yields the desired result: $y'' + a_1y' + a_2y = 0$.

25. a. $e^{bi} = 1 + (bi) + \frac{(bi)^2}{2!} + \frac{(bi)^3}{3!} + \frac{(bi)^4}{4!} + \frac{(bi)^5}{5!} + \ldots = \left(1 - \frac{b^2}{2!} + \frac{b^4}{4!} - \frac{b^6}{6!} + \right)\ldots + i\left(b - \frac{b^3}{3!} + \frac{b^5}{5!} - \frac{b^7}{7!} + \ldots\right)$

$= \cos(b) + i\sin(b)$

b. $e^{a+bi} = e^ae^{bi} = e^a[\cos(b) + i\sin(b)]$

c. $D_x\left[e^{(\alpha+\beta i)x}\right] = D_x[e^{\alpha x}(\cos\beta x + i\sin\beta x)] = \alpha e^{\alpha x}(\cos\beta x + i\sin\beta x) + e^{\alpha x}(-i\beta\sin\beta x + i\beta\cos\beta x)$

$= e^{\alpha x}[(\alpha + \beta i)\cos\beta x + (\alpha i - \beta)\sin\beta x]$

$(\alpha + \beta)e^{(\alpha+\beta i)x} = (\alpha + \beta i)[e^{\alpha x}(\cos\beta x + i\sin\beta x)] = e^{\alpha x}[(\alpha + \beta i)\cos\beta x + (\alpha i - \beta)\sin\beta x]$

Therefore, $D_x[e^{(\alpha+\beta i)x}] = (\alpha + i\beta)e^{(\alpha+\beta i)x}$

27. $y = 0.5e^{5.16228x} + 0.5e^{-1.162278x}$

29. $y = 1.29099e^{-0.25x}\sin(0.968246x)$

15.2 Concepts Review

1. particular solution to the nonhomogeneous equation; homogeneous equation

3. $y = Ax^2 + Bx + C$

Problem Set 15.2

1. $y_h = C_1e^{-3x} + C_2e^{3x}$

$y_p = \left(-\frac{1}{9}\right)x + 0$

$y = \left(-\frac{1}{9}\right)x + C_1e^{-3x} + C_2e^{3x}$

3. Auxiliary equation: $r^2 - 2r + 1 = 0$ has roots 1, 1.

$y_h = (C_1 + C_2x)e^x$

Let $y_p = Ax^2 + Bx + C$; $y'_p = 2Ax + B$;

$y''_p = 2A$.

Then $(2A) - 2(2Ax + B) + (Ax^2 + Bx + C) = x^2 + x$.

$Ax^2 + (-4A + B)x + (2A - 2B + C) = x^2 + x$

Thus, $A = 1, -4A + B = 1, 2A - 2B + C = 0$, so $A = 1, B = 5, C = 8$.

General solution: $y = x^2 + 5x + 8 + (C_1 + C_2x)e^x$

5. $y_h = C_1e^{2x} + C_2e^{3x} \cdot y_p = \left(\frac{1}{2}\right)e^x \cdot y$

$= \left(\frac{1}{2}\right)e^x + C_1e^{2x} + C_2e^{3x}$

7. $y_h = C_1e^{-3x} + C_2e^{-x}$

$y_p = \left(-\frac{1}{2}\right)xe^{-3x}$

$y = \left(-\frac{1}{2}\right)xe^{-3x} + C_1e^{-3x} + C_2e^{-x}$

9. Auxiliary equation: $r^2 - r - 2 = 0$,

$(r + 1)(r - 2) = 0$ has roots $-1, 2$.

$y_h = C_1e^{-x} + C_2e^{2x}$

Let $y_p = B\cos x + C\sin x$; $y'_p = -B\sin x + C\cos x$; $y''_p = -B\cos x - C\sin x$.

Then $(-B\cos x - C\sin x) - (-B\sin x + C\cos x)$
$-2(B\cos x + C\sin x) = 2\sin x$.

$(-3B - C)\cos x + (B - 3C)\sin x = 2\sin x$, so $-3B - C = 0$ so $-3B - C = 0$ and $B - 3C = 2$; $B = \frac{1}{5}; C = \frac{-3}{5}$.

General solution: $\left(\frac{1}{5}\right)\cos x - \left(\frac{3}{5}\right)\sin x + C_1e^{2x} + C_2e^{-x}$

11. $y_h = C_1 \cos 2x + C_2 \sin 2x$

$y_p = (0)x\cos 2x + \left(\frac{1}{2}\right)x\sin 2x$

$y = \left(\frac{1}{2}\right)x\sin 2x + C_1 \cos 2x + C_2 \sin 2x$

13. $y_h = C_1 \cos 3x + C_2 \sin 3x$

$y_p = (0)\cos x + \left(\frac{1}{8}\right)\sin x + \left(\frac{1}{13}\right)e^{2x}$

$y = \left(\frac{1}{8}\right)\sin x + \left(\frac{1}{13}\right)e^{2x} + C_1 \cos 3x + C_2 \sin 3x$

15. Auxiliary equation: $r^2 - 5r + 6 = 0$ has roots 2 and 3, so $y_h = C_1e^{2x} + C_2e^{3x}$.

Let $y_p = Be^x; y_p' = Be^x; y_p'' = Be^x$.

Then $(Be^x) - 5(Be^x) + 6(Be^x) = 2e^x;\ 2Be^x = 2e^x; B = 1$.

General solution: $y = e^x + C_1e^{2x} + C_2e^{3x}$

$y' = e^x + 2C_1e^{2x} + 3C_2e^{3x}$

If $x = 0, y = 1,\ y' = 0,$ then $1 = 1 + C_1 + C_2$ and $0 = 1 + 2C_1 + 3C_2;\ C_1 = 1, C_2 = -1.$

Therefore, $y = e^x + e^{2x} - e^{3x}$.

17. $y_h = C_1e^x + C_2e^{2x}$

$y_p = \left(\frac{1}{4}\right)(10x + 19)$

$y = \left(\frac{1}{4}\right)(10x + 19) + C_1e^x + C_2e^{2x}$

19. $y_h = C_1 \cos x + C_2 \sin x$

$y_p = -\cos \ln|\sin x| - \cos x - x\sin x$

$y = -\cos x \ln|\sin x| - x\sin x + C_3 \cos x + C_2 \sin x$ (combined $\cos x$ terms)

21. Auxiliary equation: $r^2 - 3r + 2 = 0$ has roots $1, 2$, so $y_h = C_1e^x + C_2e^{2x}$.

Let $y_p = v_1e^x + v_2e^{2x}$ subject to $v_1'e^x + v_2'e^{2x} = 0$, and $v_1'(e^x) + v_2'(2e^{2x}) = e^x(ex + 1)^{-1}$.

Then $v_1' = \frac{-e^x}{e^x(e^x + 1)}$ so $v_1 = \int \frac{-e^x}{e^x(e^x + 1)}dx = \int \frac{-1}{u(u + 1)}du$

$= \int\left(\frac{-1}{u} + \frac{1}{u + 1}\right)du = -\ln u + \ln(u + 1) = \ln\left(\frac{u + 1}{u}\right) = \ln\frac{e^x + 1}{e^x} = \ln(1 + e^{-x})$

$v_2' = \frac{e^x}{e^{2x}(e^x + 1)}$ so $v_2 = -e^{-x} + \ln(1 + e^{-x})$

(similar to finding v_1)

General solution: $y = e^x \ln(1 + e^{-x}) - e^x + e^{2x}\ln(1 + e^{-x}) + C_1e^x + C_2e^{2x}$

$y = (e^x + e^{2x})\ln(1 + e^{-x}) + D_1e^x + D_2e^{2x}$

23. $L(y_p) = (v_1u_1 + v_2u_2)'' + b(v_1u_1 + v_2u_2)' + c(v_1u_1 + v_2u_2)$
$= (v_1'u_1 + v_1u_1' + v_2'u_2 + v_2u_2') + b(v_1'u_1 + v_1u_1' + v_2'u_2 + v_2u_2') + c(v_1u_1 + v_2u_2)$
$= (v_1''u_1 + v_1'u_1' + v_1'u_1' + v_1u_1'' + v_2''u_2 + v_2'u_2' + v_2'u_2' + v_2u_2'') + b(v_1'u_1 + v_1u_1' + v_2'u_2 + v_2u_2') + c(v_1u_1 + v_2u_2)$
$= v_1(u_1'' + bu_1' + cu_1) + v_2(u_2'' + bu_2' + cu_2) + b(v_1'u_1 + v_2'u_2) + (v_1''u_1 + v_1'u_1' + v_2''u_2 + v_2'u_2) + (v_1'u_1' + v_2'u_2')$
$= v_1(u_1'' + bu_1' + cu_1) + v_2(u_2'' + bu_2' + cu_2) + b(v_1'u_1 + v_2'u_2) + (v_1'u_1 + v_2'u_2)' + (v_1'u_1' + v_2'u_2')$
$= v_1(0) + v_2(0) + b(0) + (0) + k(x) = k(x)$

15.3 Concepts Review

1. 3; π

3. 0

Problem Set 15.3

1. $k = 250, m = 10, B^2 = k/m = 250/10 = 25, B = 5$
(the problem gives the mass as $m = 10\,\text{kg}$)
Thus, $y'' = -25y$. The general solution is $y = C_1 \cos 5t + C_2 \sin 5t$. Apply the initial condition to get $y = 0.1\cos 5t$.
The period is $\frac{2\pi}{5}$ seconds.

3. $y = 0.1\cos 5t = 0$ whenever $5t = \frac{\pi}{2} + \pi k$ or $t = \frac{\pi}{10} + \frac{\pi}{5}k$.

$$\left|y'\left(\frac{\pi}{10} + \frac{\pi}{5}k\right)\right| = 0.5\left|\sin 5\left(\frac{\pi}{10} + \frac{\pi}{5}k\right)\right| = 0.5\left|\sin\left(\frac{\pi}{2} + \pi k\right)\right| = 0.5 \text{ meters per second}$$

5. $k = 20$ lb/ft; $w = 10$ lb; $y_0 = 1$ ft; $q = \frac{1}{10}$ s-lb/ft, $B = 8, E = 0.32$

$E^2 - 4B^2 < 0$, so there is damped motion. Roots of auxiliary equation are approximately $-0.16 \pm 8i$.
General solution is $y \approx e^{-0.16t}(C_1 \cos 8t + C_2 \sin 8t)$. $y \approx e^{-0.16t}(\cos 8t + 0.02\sin 8t)$ satisfies the initial conditions.

7. Original amplitude is 1 ft. Considering the contribution of the sine term to be negligible due to the 0.02 coefficient, the amplitude is approximately $e^{-0.16t}$.
$e^{-0.16t} \approx 0.1$ if $t \approx 14.39$, so amplitude will be about one-tenth of original in about 14.4 s.

9. $LQ'' + RQ' + \frac{Q}{C} = E(t)$; $10^6 Q' + 10^6 Q = 1$; $Q' + Q = 10^{-6}$

Integrating factor: e^t
$D[Qe^t] = 10^{-6}e^t$; $Qe^t = 10^{-6}e^t + C$;
$Q = 10^{-6} + Ce^{-t}$
If $t = 0, Q = 0$, then $C = -10^{-6}$.
Therefore, $Q(t) = 10^{-6} - 10^{-6}e^{-t} = 10^{-6}(1 - e^{-t})$.

11. $\frac{Q}{[2(10^{-6})]} = 120\sin 377t$

a. $Q(t) = 0.00024 \sin 377t$

b. $I(t) = Q'(t) = 0.09048\cos 377t$

13. $3.5Q'' + 1000Q + \frac{Q}{[2(10^{-6})]} = 120\sin 377t$

(Values are approximated to 6 significant figures for the remainder of the problem.)

$Q'' + 285.714Q' + 142857Q = 34.2857\sin 377t$

Roots of the auxiliary equation are

$-142.857 \pm 349.927i$.

$Q_h = e^{-142.857t}(C_1\cos 349.927t + C_2\sin 349.927t)$

$Q_p = -3.18288(10^{-4})\cos 377t + 2.15119(10^{-6})\sin 377t$

Then, $Q = -3.18288(10^{-4})\cos 377t + 2.15119(10^{-6})\sin 377t + Q_h$.

$I = Q' = 0.119995\sin 377t + 0.000810998\cos 377t + Q_h'$

$0.000888\cos 377t$ is small and $Q_h' \to 0$ as $t \to \infty$, so the steady-state current is $I \approx 0.12\sin 377t$.

15. $A\sin(\beta t + \gamma) = A(\sin\beta t\cos\gamma + \cos\beta t\sin\gamma)$

$= (A\cos\gamma)\sin\beta t + (A\sin\gamma)\cos\beta t$

$= C_1\sin\beta t + C_2\cos\beta t$, where $C_1 = A\cos\gamma$ and $C_2 = A\sin\gamma$.

[Note that $C_1^2 + C_2^2 = A^2\cos^2\gamma + A^2\sin^2\gamma = A^2$.)

17. The magnitudes of the tangential components of the forces acting on the pendulum bob must be equal.

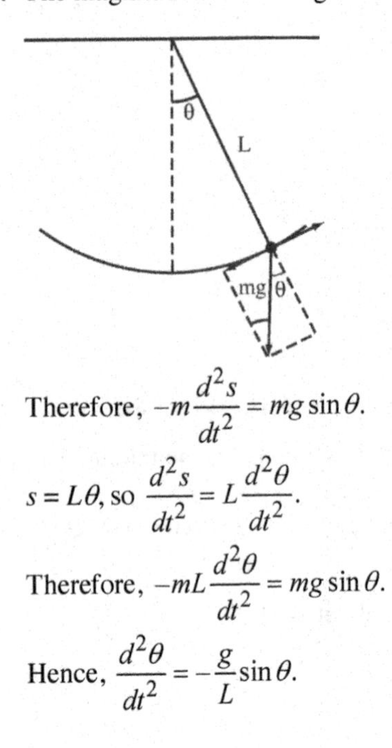

Therefore, $-m\frac{d^2s}{dt^2} = mg\sin\theta$.

$s = L\theta$, so $\frac{d^2s}{dt^2} = L\frac{d^2\theta}{dt^2}$.

Therefore, $-mL\frac{d^2\theta}{dt^2} = mg\sin\theta$.

Hence, $\frac{d^2\theta}{dt^2} = -\frac{g}{L}\sin\theta$.

15.4 Chapter Review

Concepts Test

1. False: y^2 is not linear in y.

3. True: $y' = \sec^2 x + \sec x \tan x$
$$2y' - y^2 = (2\sec^2 x + 2\sec x \tan x) - (\tan^2 x + 2\sec x \tan x + \sec^2 x)$$
$$= \sec^2 x - \tan^2 x = 1$$

5. True: D^2 adheres to the conditions for linear operators.
$$D^2(kf) = kD^2(f)$$
$$D^2(f+g) = D^2 f + D^2 g$$

7. True: –1 is a repeated root, with multiplicity 3, of the auxiliary equation.

9. False: That is the form of y_h. y_p should have the form $Bx \cos 3x + Cx \sin 3x$.

Sample Test Problems

1. $u' + 3u = e^x$. Integrating factor is e^{3x}.
$$D[ue^{3x}] = e^{4x}$$
$$y = \left(\frac{1}{4}\right)e^x + C_1 e^{-3x}$$
$$y' = \left(\frac{1}{4}\right)e^x + C_1 e^{-3x}$$
$$y = \left(\frac{1}{4}\right)e^x + C_3 e^{-3x} + C_2$$

3. (Second order homogeneous)
The auxiliary equation, $r^2 - 3r + 2 = 0$, has roots 1, 2. The general solution is $y = C_1 e^x + C_2 e^{2x}$.
$$y' = C_1 e^x + 2C_2 e^{2x}$$
If $x = 0, y = 0,\ y' = 3$, then $0 = C_1 + C_2$ and $3 = C_1 + 2C_2$, so $C_1 = -3, C_2 = 3$.
Therefore, $y = -3e^x + 3e^{2x}$.

5. $y_h = C_1 e^{-x} + C_2 e^x$ (Problem 2)
$$y_p = -1 + C_1 e^{-x} + C_2 e^x$$

7. $y_h = (C_1 + C_2 x)e^{-2x}$ (Problem 12)
$$y_p = \left(\frac{1}{2}\right)x^2 e^{-2x}$$
$$y = \left[\left(\frac{1}{2}\right)x^2 + C_1 + C_2 x\right]e^{-2x}$$

9. (Second-order homogeneous)
The auxiliary equation, $r^2 + 6r + 25 = 0$, has roots $-3 \pm 4i$. General solution:
$$y = e^{-3x}(C_1 \cos 4x + C_2 \sin 4x)$$

11. Roots are –4, 0, 2. $y = C_1 e^{-4x} + C_2 + C_3 e^{2x}$

13. Repeated roots $\pm\sqrt{2}$
$$y = (C_1 + C_2 x)e^{-\sqrt{2}x} + (C_3 + C_4 x)e^{\sqrt{2}x}$$

15. (Simple harmonic motion)
$k = 5;\ w = 10;\ y_0 = -1$
$$B = \sqrt{\frac{(5)(32)}{10}} = 4$$
Then the equation of motion is $y = -\cos 4t$.
The amplitude is $|-1| = 1$; the period is $\frac{2\pi}{4} = \frac{\pi}{2}$.

17. $Q'' + 2Q' + 2Q = 1$
Roots are $-1 \pm i$.
$$Q_h = e^{-t}(C_1 \cos t + C_2 \sin t) \text{ and } Q_p = \frac{1}{2};$$
$$Q = e^{-t}(C_1 \cos t + C_2 \sin t) + \frac{1}{2}$$
$$I(t) = Q'(t) = -e^{-t}[(C_1 - C_2)\cos t + (C_1 + C_2)\sin t]$$
$I(t) = e^{-t} \sin t$ satisfies the initial conditions.